DIRECTORY OF ON-GOING RESEARCH IN CANCER EPIDEMIOLOGY

1992

DIRECTORY OF ON-GOING RESEARCH IN CANCER EPIDEMIOLOGY

1992

IARC
DKFZ

M. Coleman
J. Wahrendorf
E. Démaret

In collaboration with

H.J. Baur
A.M. Bêh
S. Whelan

This Directory is produced jointly by the International Agency for Research on Cancer (IARC), Lyon, France, and the German Cancer Research Center (DKFZ), Heidelberg, Germany. The Division of Cancer Etiology of the US National Cancer Institute partially supports the Directory (contract NO1-CO-84340).

INTERNATIONAL AGENCY FOR RESEARCH ON CANCER

The International Agency for Research on Cancer (IARC) was established in 1965 by the World Health Assembly, as an independently financed organization of the World Health Organization. The headquarters of the Agency are at Lyon, France.

The Agency conducts a programme of research concentrating particularly on the epidemiology of cancer and the study of potential carcinogens in the human environment. Its field studies are supplemented by biological and chemical research carried out in the Agency's laboratories in Lyon and, through collaborative research agreements, in national research institutions in many countries. The Agency also conducts a programme for the education and training of personnel for cancer research.

The publications of the Agency are intended to contribute to the dissemination of authoritative information on different aspects of cancer research. A complete list is printed at the back of this book.

Published by
the International Agency for Research on Cancer,
150 cours Albert Thomas, 69372 Lyon Cedex 08, France
© International Agency for Research on Cancer, 1992

Distributed by Oxford University Press,
Walton Street, Oxford OX2 6DP, UK
and in the USA by Oxford University Press,
200 Madison Avenue, New York, NY 10016

All rights reserved. No part of this publication may be reproduced, stored in a retrieval system, or transmitted, in any form or by any means, electronic, mechanical, photocopying, recording, or otherwise without the prior permission of the copyright holder.

ISBN 92 832 2117 6

Printed in the United Kingdom

TABLE OF CONTENTS

PREFACE vii
INTRODUCTION ix
  Managing the Directory ix
  Selection of Material ix
  Content and Quality of Abstracts x
  Keeping the Directory Current x
  References x
  Addresses x
  Directories 1976-1992 xi
  Indexing xi
  Using the Indexes on Microcomputer xii
  Cancer Registries xii
  Biological Materials Banks xii
  Future of the Directory xii
  Acknowledgements xii
HOW TO USE THE DIRECTORY xiii
ELECTRONIC SEARCHING OF THE DIRECTORY – PROSE xv
LIST OF ABBREVIATIONS xxi

PROJECTS

Algeria	3	Iceland	161
Argentina	4	India	163
Australia	5	Indonesia	176
Austria	21	Ireland	178
Belgium	23	Israel	179
Brazil	27	Italy	186
Canada	29	Japan	213
Chile	54	Malaysia	236
China	55	Mexico	238
Costa Rica	79	Netherlands	239
Cuba	80	New Caledonia	254
Czechoslovakia	81	New Zealand	255
Denmark	85	Nigeria	260
Egypt	98	Norway	261
Estonia	99	Pakistan	265
Finland	100	Papua-New Guinea	266
France	108	Paraguay	267
Gambia	138	Poland	268
Germany	139	Romania	273
Greece	155	South Africa	274
Hong Kong	156	Spain	278
Hungary	157	Sri Lanka	280

Sweden 281
Switzerland 304
Tanzania 307
Thailand 308
Turkey 309
United Kingdom 310
United States of America 358
Uruguay 473
USSR 476
Yugoslavia 484

INDEXES

Investigators 491
Terms 525
Sites 593
Types of Study 649
Chemicals 661
Occupations 687
Countries 703
Cancer Registries 723

LIST OF CANCER REGISTRIES 729
LIST OF BIOLOGICAL MATERIALS BANKS 749
LIST OF IARC PUBLICATIONS 775

PREFACE

The 1992 Directory of On-Going Research in Cancer Epidemiology has now been published, thanks to the contribution of a large number of epidemiologists throughout the world, several collaborating for many years, but many contributing for the first time.

For the first time also, the production of the 1992 Directory has been managed entirely on microcomputer. It contains 1197 descriptions of research projects being carried out by some 900 investigators in 60 countries. It can be consulted as an ordinary book with indexes, or can be searched electronically, using PROSE (PROject SEarch), a user-friendly program provided on diskette for IBM-compatible microcomputers. This permits more precise and flexible searches than are possible when searching the Directory manually, and thus considerably increases its usefulness.

The Directory also contains a comprehensive list of population-based cancer registries, as well as a list of biological materials collections.

A more powerful electronic version of the Directory is currently being developed in the form of a CDROM, which will contain not only the entire Directory, but a number of other IARC publications: the complete series of IARC Monographs on the Evaluation of Carcinogenic Risks to Humans, the Cross-Index of Synonyms and Trade Names and the Directory of Agents being Tested for Carcinogenicity. This will provide a unique set of information on published human and animal studies, with expert evaluations of carcinogenicity to humans, together with up-to-date listings of current animal and epidemiological research.

The Directory has appeared annually since 1976. Due to the heavy budgetary constraints which the project is now facing, a biennial cycle will be adopted for the time being. The next issue will thus appear in 1994. Modifications to the content of the Directory may also have to be made, but we are confident that everything will be done to maintain the high quality of the Directory and that it will continue to provide useful information to epidemiologists all over the world.

Dr L. Tomatis
Director
International Agency for
Research on Cancer (IARC)

Prof. H. zur Hausen
Scientific Director
Deutsches Krebsforschungszentrum (DKFZ)

INTRODUCTION

The Directory of On-Going Research in Cancer Epidemiology has three aims:

(1) to inform cancer epidemiologists and other interested scientists about current work in the field of cancer epidemiology;

(2) to facilitate direct contacts between research workers;

(3) to enable unnecessary duplication of work to be avoided.

The Directory is produced jointly by the International Agency for Research on Cancer (IARC), Lyon, France and the German Cancer Research Centre (Deutsches Krebsforschungszentrum - DKFZ), Heidelberg, Germany.

Managing the Directory

The Directory is now being managed entirely at IARC. A database management system, EPIBASE, has been developed during the last year and all operations are carried out on a microcomputer in Lyon: mailings, maintenance of address file, maintenance of project and key-word databases, and creation of output to Interleaf to produce camera-ready copies for the printer.

This transfer of operations to IARC will make production procedures more efficient and reduce the error rate; data management will improve, as well as the quality of the Directory.

Selection of Material

The main criterion for inclusion of a project in the Directory is that it deals with cancer epidemiology. Originally, 'epidemiology' was interpreted fairly broadly, but editorial policy has been more restrictive in recent years, partly due to increasing pressure on Directory space. The large number of projects received which are judged to fall outside the scope of the Directory has also prompted us to attempt a more precise working definition of cancer epidemiology.

For the purposes of inclusion in the Directory, 'cancer epidemiology' is taken to refer to studies of cancer which are designed to estimate cancer morbidity or mortality in human populations, or to investigate potential risk factors in cancer aetiology, or to evaluate methods of cancer prevention. Studies in the field of genetic and molecular epidemiology are also accepted.

Purely descriptive studies from developed countries, e.g. of the incidence of cancer in the general population, are not usually included, while such studies are still accepted from developing countries, where analytical studies may be difficult to carry out. Two other types of project are also accepted, namely those which involve:

(1) development of methods to assess human exposures and to test the application of such methods in epidemiological studies. Methods of assessing exposure to potential carcinogens in large numbers of individuals, e.g. by the use of DNA adducts, are undergoing rapid development, and are increasingly important in monitoring of actual rather than indirectly estimated exposure;

(2) biological materials banks (BMB): biological material is increasingly being used, or stored for later use, in epidemiological studies, to look for tumour markers or biochemical markers of premalignant lesions. More detailed descriptions of biological materials banks are accepted as project abstracts, provided the project also comprises a current epidemiological study. A list of biological materials banks has been maintained in the Directory for several years and this section has been expanded.

Studies which primarily deal with diagnosis, treatment, prognosis, survival, health care programmes and, of course, animal studies will not normally be included. Any of these topics may well form part of a study included in the Directory, but the primary purpose of the study should fall within the definition of cancer epidemiology given above.

Content and Quality of Abstracts

The amount of information given for each project abstract is variable, but all abstracts should cover the following key points:

- aim of the study or hypothesis being tested;
- study type or design;
- size of study population;
- nature of any control group; and
- outline of methods of data collection and analysis.

Investigators should limit background information (e.g. the rationale for the study) to a bare minimum. Directory users are mainly interested in what is being done; they will normally know why it is being done.

In a proportion of all new projects reported, essential information is missing. In such cases the principal investigator is contacted. If no reply is received the project is excluded.

A number of projects were deleted from the Directory last year, in order to improve the quality of the information (incomplete abstract, no contact with principal investigator for three years or more, no update of abstract for five years or more etc.). The principal investigators of these projects were contacted and offered to submit a revised abstract, if the study was still current.

We must stress that the Directory editors do not act as referees of the quality of research. We simply try to include clear, concise accounts of all relevant research, in order to make these available to the scientific community, since one of the aims of the Directory is to reflect, as far as possible, what is being done in the field. Many excellent research projects are not included only because they fall outside the scope of the Directory, while inclusion in the Directory implies only an adequate description of relevant research, not an assessment of its quality.

Keeping the Directory Current

To keep the Directory free from studies which have already been completed and published, all contributors to the current Directory are contacted every year and asked to update their contribution.

New projects for which 1992 was given as termination date, or with abstracts which seemed to report completed work, were checked in CANCERLINE to ensure that the study had not been published. Any projects for which apparently final results have been published were removed.

We would like to stress that investigators can report studies at any time. Abstract forms can be obtained from Mrs E. Demaret, International Agency for Research on Cancer, 150 cours Albert Thomas, 69372 Lyon Cedex 08, France.

References

Up to three references to interim publications from a study are accepted in the project abstracts. It is not necessary to reference background information, which should in any case be very brief. Journal articles and book chapters are accepted, but not meeting abstracts, letters or editorials, or articles submitted or in press. The reference will give the journal title, volume number, page number and year of publication in that order. For the sake of brevity, the title of the article will be excluded, and the name of the first author will be given only if other than the principal investigator cited in the Directory abstract.

Addresses

Addresses of potential contributors are extracted from various sources – literature data bases, annual reports from research institutes, lists of participants at meetings, membership lists of scientific associations, etc. Some 2500 research workers are contacted every year.

An important part of keeping the Directory current is constant vigilance over the accuracy and usefulness of addresses. Our address list is constantly updated, in order to avoid, for example, sending an invitation to an erroneous address, or recontacting investigators who have informed us that they do not work in cancer epidemiology, or have retired.

Directories 1976-1992

The table below shows, for each issue of the Directory, the number of projects included, the number and percentage of completely new studies and the number of countries in which data were being collected.

Year	No. of projects	New projects (%)	No. of countries
1976	622	622 (100.0)	65
1977	908	467 (51.4)	69
1978	1025	341 (33.3)	70
1979	1092	295 (27.0)	74
1980	1261	353 (28.0)	78
1981	1313	299 (22.8)	80
1982	1247	275 (22.1)	74
1983	1302	256 (19.7)	80
1984	1213	200 (16.5)	80
1985	1229	261 (21.2)	88
1986	1352	334 (24.7)	89
1987	1320	240 (18.2)	84
1988	1237	173 (14.0)	84
1989/90	1300	278 (21.3)	86
1991	1147	208 (18.1)	86
1992	1197	284 (23.7)	70

Indexing

Each project has a unique serial number, which is used for all references to that project in the indexes. This number appears at the top left of each entry.

Projects are listed by serial number in the main body of the Directory. They are in alphabetic order of the country from which they are reported, usually the same as the country in which they are being carried out. Within a country, the listing is alphabetic by town, and within town, alphabetic by principal investigator. (For previous users: please note that for USA, projects are no longer in alphabetic order of state, town and name, but are listed alphabetically by town and name).

In order to find projects which fit any particular description, eight separate indexes can be used:

Name of investigator

Term (key-word)

Cancer site

Study type

Specific **chemical** exposure(s)

Specific **occupation(s)**

Country in which data are being collected

Cancer registry providing data for a study.

Every project is indexed in at least five ways: by name of each investigator; by key-word(s); by cancer site; by study type; and by country. Where relevant, studies are also indexed to chemical exposure or occupation, and to collaborating cancer registries.

All index headings currently in use are listed at the front of each index, including cross-references to more specific index headings and to other indexes. For example, the list of headings at the front of the TERMS (key-words) index shows that studies indexed to the generic term 'Reproductive Factors' may also be sought under any of the more specific headings 'Abortion', 'Birthweight', 'Menarche', etc., and that studies indexed to the general term 'Metals' may also be searched for in the CHEMICALS index under any of the specific metals 'Aluminium', 'Beryllium', etc. The lists of index headings should simplify the selection of efficient search terms. The main features of the indexes, together with major changes in 1992, are outlined at the front of each index.

Using the Indexes on Microcomputer

The indexes are also provided on diskette with specially written software, enabling simultaneous use of two or more indexes to identify projects which fit a precise description.

Guidelines on use of the diskette to search the indexes with PROSE software can be found after the introduction.

Cancer Registries

The index of Cancer Registries was created in 1989 and identifies each project in the Directory in which a given cancer registry is involved. 'Involvement' implies either that the project is carried out within the cancer registry or that the registry releases individual data for a project carried out by an external investigator. Each registry has been given a short name, by which it appears in the index. The short name of the registry also appears beneath the abstract in the body of the Directory, under the heading REGI.

The address list of population-based cancer registries, which follows the index itself, enables registries to be readily contacted. This list contains 266 population-based general and site-specific registries. The registries appear alphabetically by short name within country.

Biological Materials Banks

Biological materials banks can provide valuable historical records for epidemiological studies. This section of the Directory provides a list of banks storing many different types of biological materials, including e.g. cells in culture, faeces, hair, saliva, serum, toenail clippings, urine, etc.

The information provided for each biological materials collection includes the number of persons sampled (or, rarely, the number of samples stored), the year in which the collection was started (and finished if appropriate), the temperature at which the material is stored, and whether personal identifiers (ID), collateral data (CD) and follow-up information (FU) are available.

'BMB' (biological materials bank) is also a key-word in the TERMS index and is used to index any study using biological materials for long-term follow-up of study subjects. The type of specimen stored is also indexed (e.g. 'Sputum', 'Tissue' etc.).

A list of addresses for the biological materials banks is also provided, so that the holders of the listed collections can be contacted directly.

We invite anyone we have not yet contacted and who is creating or using a collection of human biological materials, and who would be interested in collaborating in epidemiological studies, to contact Mrs E. Demaret, International Agency for Research on Cancer, 150 cours Albert Thomas, 69372 Lyon Cedex 08, France.

Future of the Directory

Due to budgetary constraints, we will not be able to produce a Directory in 1993. We hope that a volume will appear in 1994 and that it will be possible to continue publishing the Directory.

Acknowledgements

Production of this Directory would not have been possible without the contribution of all research workers and we would like to thank them for reporting their studies and thus ensuring that other researchers are aware of their work. We apologize to scientists whose reports arrived too late for inclusion in this edition (they will be considered for inclusion in the next issue), and to those scientists whom we have yet to contact.

Lyon and Heidelberg, January 1992   M.P. Coleman
J. Wahrendorf
E. Démaret

HOW TO USE THE DIRECTORY

Layout of Information about a Study

The layout of a typical project abstract is shown in Fig. 1. The numbers circled in the figure correspond to the numbers in the description below.

1. Serial number (an asterisk denotes a project included for the first time)
2. Principal investigator or study co-ordinator
3. Address of principal investigator or study co-ordinator. Telephone, telex and telefax numbers are given when available.
4. Project collaborator(s)
5. Title of project
6. Abstract of project
7. TYPE: The type of study, e.g. Cohort, Cross-Sectional, etc. Some projects involve more than one type of study.
8. TERM: The key-words summarizing the main features of the study.
9. SITE: The cancer sites or types under study. For many studies, particularly prospective ones, all sites of cancer are of interest, and hence 'All Sites' is a frequent entry in this section.
10. CHEM: Specific chemical(s) under study.
11. OCCU: Specific occupation(s) under study.
12. LOCA: The country (location) where the data are being collected. Only shown if different from country of investigator reporting the project, or if more than one country.
13. TIME: The period over which the investigators consider their study will take place. We have tried to avoid open-ended studies, but this has not always been possible, particularly for long-term prospective studies or registries.
14. REGI: Cancer registry involved in the study.
15. Unique identification number, which remains the same throughout the 'life' of the project. For internal use.

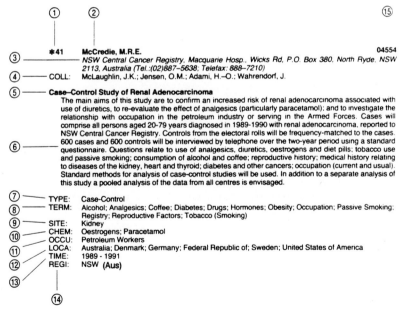

Fig. 1 Layout of a sample project

xiii

Electronic searching of the Directory – PROSE

1.1 Introduction

PROSE (PROject SEarch) provides a new electronic search and display facility for research projects included in the Directory. As when using the Directory indexes, the result of a search comprises a list of project serial numbers which meet the search description, and this can be shown on screen, printed or saved for review. In order to fit all the information on a single diskette, only the indexes and retrieval software are included. Future developments may include provision of the abstracts themselves in electronic form.

1.2 Why electronic searching?

With the Directory published just as a conventional book, only one of the eight indexes can be searched at a time. Although in most indexes the projects are subclassified by cancer site, it is difficult to exploit the indexes to the full in a single search. Electronic access solves this problem.

All the indexes (except the investigator index) can be used, either singly or in combination, to identify projects meeting any description, such as case-control studies of bladder cancer in the USA or Canada involving sweeteners or tobacco smoking. It would be time-consuming to identify such projects using only the printed indexes.

PROSE has been written and developed by Hans-Jurgen Baur at DKFZ, Heidelberg.

2.1 What you need to run PROSE

PROSE will run on any IBM PC, /XT, /AT or 100 per cent IBM-compatible microcomputer. The computer must have at least 512 kilobytes of RAM, and a 5.25 inch floppy diskette drive. PROSE can be run from a high-density floppy diskette drive (see section 2.4 below), but it is more convenient and quicker to operate PROSE after copying it to a hard disk.

2.2 Installation

PROSE is distributed on one 5.25 inch double-density diskette containing the indexes and software, in compressed format, and an installation program (INSTALL.EXE) to transfer the system to a hard disk. The diskette contains two files:

    INSTALL.EXE about 310 kbytes
    READ.ME about 13 kbytes

The second file contains this description of PROSE.

2.3 Installing PROSE on a hard disk

You will need at least 512 kilobytes of RAM, a 5.25 inch disk drive and a fixed disk drive with about 1 megabyte of free space. The CONFIG.SYS file in the root (C:\) directory should contain the lines:

    BUFFERS = 15
    FILES = 20

but PROSE will perform better if the figures are 25 and 40, respectively. To install PROSE, first create a directory on your hard disk on which you want to install PROSE by entering a command such as:

    MD PROSE

at the DOS prompt C:\>. Then change to the new directory by entering

    CD PROSE

and insert the PROSE-diskette in the diskette drive and enter

    COPY A:*.* C:

in order to copy all files to the hard disk, where A: is the diskette drive and C: is the hard disk. For installation of PROSE enter at the DOS prompt C:\PROSE>:

    INSTALL

When all files have been installed, PROSE can be run from your hard disk by entering the command PROSE at the DOS prompt.

2.4 Installing PROSE without a hard disk drive

If your computer does not have a hard disk, you will need at least 512 kilobytes of RAM, a 5.25 inch low-density drive, and a high density drive (either 5.25 inch or 3.5 inch). The CONFIG.SYS file should be as described above. From the DOS prompt A:>, copy PROSE to a high-density disk, by entering:

   COPY A:*.* B:

where A: is the low-density drive and B: is the high-density drive. Then change to the high-density drive and install PROSE by entering successively:

   B:
   INSTALL

PROSE can now be started from drive B: by entering the command PROSE at the DOS prompt.

3.1 PROSE facilities

PROSE is a menu-driven program, and all available options are shown on each screen. Context-sensitive help is available on-line from most screens, and error messages provide instructions if the wrong key is inadvertently used. To start PROSE from the DOS prompt, simply type:

   PROSE

followed by the carriage return key (<RETURN>). The master screen (Fig. 2) displays the six main functions, which can be selected either by pressing the relevant numeric key, or by moving the reverse video bar with the cursor keys to highlight the chosen option, then pressing <RETURN>. Each option, when highlighted, is briefly described below the list of options. Use of the special function keys (<F1>, <F2>, etc) is indicated in the bottom margin of each screen.

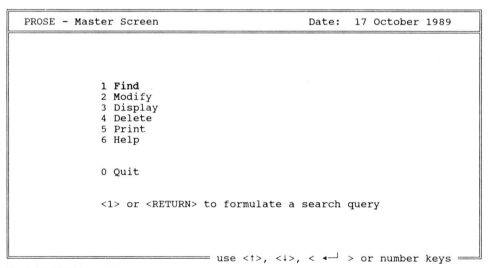

Fig. 2: PROSE Master Screen

3.2 Formulating a search

A search is begun by selecting the option FIND from the master screen. PROSE offers two search modes, EXPERT and ASSIST. The EXPERT mode allows a query to be formulated immediately, provided the user is familiar with the usual rules of syntax (use of AND, OR, * and brackets, etc), while the ASSIST mode - activated with the <F2> key - will prompt the user at each stage of the search. EXPERT mode is provided as the default. Either upper-case or lower-case letters can be used. At any point in formulating a search, each of the indexes can be consulted to check the correct term to be used, or the entire index can be scrolled on the screen to search for suitable terms.

3.3 Search in EXPERT mode

This is the default mode for executing a search. The query screen (Fig. 3) provides four lines for entering a search formulation, which consists of one or more search terms, of the type:

INDEX = 'item'

where the index to be searched is selected from among:

1. SITE cancer site
2. TERM key-word
3. TYPE study type
4. CHEM chemical
5. OCCU occupation
6. LOCA country (location)
7. REGI cancer registry
8. QUERY previous query

and the item to be selected from that index is placed after the 'equals' sign, in quote marks (' '). Further search terms can then be added, using AND, OR and AND NOT (logical operators). 'Word-wrapping' at the end of each line is not supported, but will often work correctly; if the text of the search request is too long for one line, it is safer to begin the next line with a new word. The usual editing keys can be used (cursor keys, Home, End, Del, etc).

Example search formulations would be:

1. SITE = 'LUNG' AND LOCA = 'FRANCE'
2. (SITE = 'STOMACH' OR SITE = 'COLON') AND LOCA = 'FRANCE'
3. TYPE = 'COHORT' AND SITE = 'COLON' AND (LOCA = 'USA' OR LOCA = 'CANADA')
4. SITE = 'LUNG' AND NOT TYPE = 'CASE SERIES'
5. CHEM = 'CHLOR*' AND TYPE = 'COHORT'
6. QUERY = 'FILENAME' AND TYPE = 'COHORT'

Example 4 would identify lung cancer studies of all study types except case series. Example 5 would identify all cohort studies involving exposure to any chemical beginning with 'chlor-'.

If no brackets are specified, PROSE evaluates the search terms from right to left. Omission of the brackets in example 3 would thus not alter the result, whereas in example 2 the brackets are essential.

```
┌──────────────────────────────────────────────────────────────────────────────┐
│   PROSE - Query Screen                          Date:   17 October 1989      │
├──────────────────────────────────────────────────────────────────────────────┤
│                                                                              │
│         Please enter your search query:                                      │
│                                                                              │
│                                                                              │
│                                                                              │
│                                                                              │
│            Examples:                                                         │
│                                                                              │
│            1. Site = 'Lung' and Loca = 'France'                              │
│            2. (Site = 'Breast (F)' or Site = 'Colon') and                    │
│               loca = 'France'                                                │
│            3. Chem = 'Chlor*' and Type = 'cohort'                            │
│            <F2> for ASSIST mode to prompt for search terms                   │
│                                                                              │
│                                                                              │
│   ═══ <F1> Help ═ <F2> Assist ═══════════ <F9> Execute ═══════ <ESC> Quit ═══│
└──────────────────────────────────────────────────────────────────────────────┘
```
Fig. 3: PROSE Query Screen

The eighth option available from the query screen is to incorporate a previously saved query into a search, e.g. to refine a search that has identified too many projects to be useful, by adding further search terms.

Saved searches are stored in a QUERY index under a name supplied by the user. They can be incorporated into a new search in exactly the same way as terms from other indexes (example 6 above).

If the name used to save a query has been forgotten, a list of all saved queries can be examined in ASSIST mode (see below). Select option 8 (Previous Query) and use the function key <F3> (LIST).

3.4 Search in ASSIST mode

This mode is activated from the query screen with the function key <F2>. The user is prompted to select one of the seven indexes or the index of previous searches by highlighting it with the reverse video bar and pressing <RETURN> (see Fig. 4).

The program then offers a choice between include ('equal to') and exclude ('not equal to'), again selected with the reverse video bar. Finally, the entire index is displayed. The item of interest can be selected in two ways:

a) Enter the first (up to 8) letters of the item of interest. A beep indicates that the item does not exist. If the item of interest is found, it can be selected by pressing <RETURN>.

b) The index can be scrolled up and down with the <PgUp> and <PgDn> (Page+ and Page-) keys, and the item of interest located with the arrow keys and selected by pressing <RETURN>.

The complete search term is then displayed in the search field, and another search term can be defined, either in EXPERT mode or in ASSIST mode, or else the search can be executed (see below).

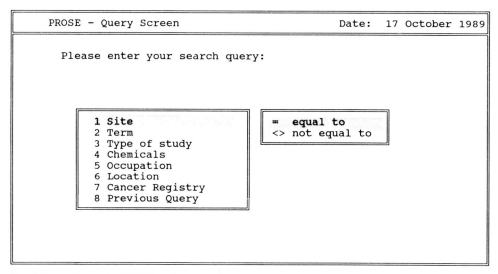

Fig. 4: Query Screen in ASSIST mode

3.5 Executing a search formulation

Once the search has been formulated, the function key <F9> will cause it to be executed. An error in the syntax of the search request will be reported in a highlighted message at the top of the screen; this disappears after several seconds. The error will be pin-pointed by a marker against the line containing the error and another marker beneath the position where it occurs. A message at the bottom of the screen invites the user to continue by pressing any key; the faulty search request can then be edited.

The screen indicates the progress of the search, then displays the result, showing the number of projects which satisfy the search description, and the serial number of each project (Fig. 5).

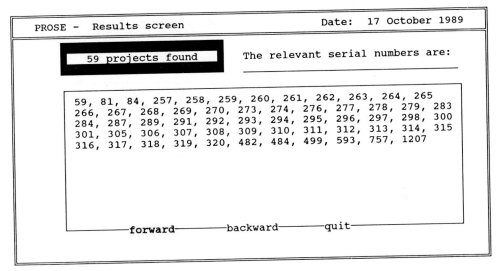

Fig. 5: Result of a PROSE search

3.6 Saving and printing

The user is then offered the options of printing and/or saving the result of the search:

    PRINT THE RESULTS ? (Y/N) N

Type Y (for Yes) and <RETURN> to print the search, or simply <RETURN> to accept the default option (No).

The printed result will show the search request and the project serial numbers which correspond to it. Whichever print option is selected, the option to save the search is then offered:

    SAVE THIS SEARCH ? (Y/N) N

Type Y (for Yes) and <RETURN> to save the search, or simply <RETURN> to accept the default option (No), which returns the user to the main menu. PROSE will prompt for a filename (up to eight letters) under which to save the search:

    SAVE QUERY UNDER WHICH NAME ?

When a name has been entered, the user is returned to the main menu.

4. Help

General hints on using PROSE can be obtained by selecting the HELP option <F1> from the master screen, and this function key will provide context-sensitive help from many other screens.

5. Modify, Display, Delete and Print

These functions are available as options 2–5 from the master screen. They allow a previously saved search formulation to be edited (it can then be modified, executed or deleted) and the results of a previously saved search can be displayed on screen or printed.

In each case, the name of the saved search must be supplied: the function key <F3> can be used to examine an alphabetic list of previously saved searches. This shows the name supplied by the user, the date it was saved and the first few words of the search formulation (Fig. 6). The list can be scrolled with the cursor keys, and selection is made either numerically or with the reverse video bar and <RETURN>.

```
PROSE  -    Print Screen                        Date:    17 October 1989

     Enter name of search to be processed:

        0: EXAMPLE1 (10/09/89) : (site = 'Breast (F)' or site = 'co
        1: EXAMPLE2 (10/09/89) : chem = 'chlor*' and type = 'cohort
        2: EXAMPLE3 (10/09/89) : site = 'LUNG' and LOCA = 'france'
        3: EXAMPLE4 (10/09/89) : query = 'example1' and type = 'cas




        <PgUp>, <PgDn>, <ESC>, <RETURN>    or number
```

Fig. 6: List from PRINT screen

6. Quitting PROSE

Use the ESC (escape) key as many times as needed to return to the main menu, then select the 'Quit' option to return to DOS.

LIST OF ABBREVIATIONS

This list includes the most common acronyms and abbreviations used in the Directory. The meaning of other abbreviations is given in full in the abstract in which they appear.

AFP	Alphafoetoprotein
AHH	Aryl Hydrocarbon Hydroxylase
AIDS	Acquired Immunodeficiency Syndrome
ALL	Acute Lymphocytic Leukaemia
AML	Acute Myeloid Leukaemia
ANLL	Acute Non-Lymphoblastic Leukaemia
ATLL	Acute T-Cell Leukaemia-Lymphoma
BCC	Basal Cell Carcinoma
BCG	Bacille Calmette-Guerin
BCME	Bis(chloromethyl)Ether
BL	Burkitt's Lymphoma
BMB	Biological Materials Banks
CEA	Carcinoembryonic Antigen
CIN	Cervical Intraepithelial Neoplasia
CIS	Carcinoma In-Situ
CLL	Chronic Lymphocytic Leukaemia
CML	Chronic Myeloid Leukaemia
CMME	Chloromethyl Methyl Ether
CMV	Cytomegalovirus
CNRS	Centre National de la Recherche Scientifique (France)
CNS	Central Nervous System
2,4-D	2,4-Dichlorophenoxyacetic Acid
DBCP	Dibromochloropropane
DES	Diethylstilboestrol
DMPA	Depot Medroxyprogesterone Acetate
DNA	Deoxyribonucleic Acid
EBV	Epstein-Barr Virus
ELF	Extremely Low Frequency (0-300 Hz)
ELISA	Enzyme-Linked Immuno-Sorbent Assay
EMF	Electromagnetic Fields
ENT	Ear, Nose and Throat
EORTC	European Organisation for Research and Treatment of Cancer
EPA	Environmental Protection Agency (USA)
FAB	French-American-British Classification
FAP	Familial Adenomatous Polyposis
FDA	Food & Drug Administration (USA)
FSH	Follicle Stimulating Hormone
GC	Gas Chromatography
G6PD	Glucose-6-Phosphate Dehydrogenase
HBsAg	Hepatitis B Surface Antigen
HBV	Hepatitis B Virus
HCG	Human Chorionic Gonadotrophin
HCV	Hepatitis C Virus
HD	Hodgkin's Disease
HDV	Hepatitis D Virus
HIV	Human Immunodeficiency Virus
HLA	Histocompatibility Locus Antigen
HPLC	High Pressure Liquid Chromatography
HPV	Human Papilloma Virus
HSV	Herpes Simplex Virus
HTLV	Human T-Cell Leukaemia Virus
IARC	International Agency for Research on Cancer
ICD	International Classification of Diseases
ICD-O	International Classification of Diseases - Oncology
ILO	International Labour Office

ISIC	International Standard Industrial Classification
KS	Kaposi's Sarcoma
LH	Luteinizing Hormone
MDA	4-4'-Methylene Dianiline
MOCA	4,4-Methylene-bis(2-Chloroaniline)
MRC	Medical Research Council (UK)
MS	Mass Spectrometry
NCI	National Cancer Institute (USA)
NCIC	National Cancer Institute of Canada
NHL	Non-Hodgkin's Lymphoma
NIEHS	National Institute of Environmental Health Sciences (USA)
NIH	National Institutes of Health (USA)
NIOSH	National Institute for Occupational Safety and Health (USA)
NPC	Nasopharyngeal Cancer
OPCS	Office of Population Censuses and Surveys (England & Wales)
OSHA	Occupational Safety and Health Administration (USA)
PAH	Polycyclic Aromatic Hydrocarbons
PBB	Polybrominated Biphenyls
PCB	Polychlorinated Biphenyls
PCR	Polymerase Chain Reaction
PHC	Primary Hepatocellular (Liver) Carcinoma
PMR	Proportional Mortality (or Morbidity) Ratio
PVC	Polyvinyl Chloride
RFLP	Restriction Fragment Length Polymorphism
RIA	Radioimmunoassay
RNA	Ribonucleic Acid
SCE	Sister Chromatid Exchange
SEARCH	Surveillance of Environmental Aspects Related to Cancer in Humans (IARC programme)
SEER	Surveillance, Epidemiology and End Results program (USA)
SIR	Standardized Incidence Ratio
SMR	Standardized Mortality Ratio
2,4,5-T	2,4,5-Trichlorophenoxyacetic Acid
TLV	Threshold Limit Value
TNM	Tumour-Nodes-Metastasis classification (UICC)
TSH	Thyroid Stimulating Hormone
UICC	Union Internationale Contre le Cancer (Switzerland)
WHO	World Health Organization

LIST OF PROJECTS

ALGERIA

SETIF

1 **Hamdi-Chérif, M.** 04702
Centre Hospitalier de Sétif, Serv. de Médecine Prév., & d'Epidémiologie, Sétif, Algeria (Tel.: 903001; Tlx: 86958)
COLL: Coleman, M.P.; Sekfali, N.; Benlatreche, K.; Allouache, A.; Rahal, D.

Retrospective Study of Cancer Morbidity in Sétif, Algeria

Cancer is a major public health problem in Algeria. This project has two major objectives: to provide an accurate description of cancer morbidity in the wilaya (province) of Setif, and to serve as a feasibility study for a population-based cancer registry in Setif. A team of 12 public health interns has been formed to collect a simple data set on each tumour diagnosed in the 3-year period 1 January 1986 to 31 December 1988 in persons normally resident in the wilaya. Data sources include hospital records, pathology laboratories, private physicians and town halls (death certificates), and referral hospitals outside the wilaya. Only 15 data items will be recorded, including sex, birthdate and commune of residence of the person; and date of diagnosis, site, morphology and basis of diagnosis of the tumour. These data will enable incidence rates to be calculated by sex, age and cancer site and by geographical region within the county. The study is expected to identify over 6,000 cancers, and to establish the contacts and data collection mechanisms required for setting up the cancer registry, which will record prospectively, from 1 January 1990, all cancers diagnosed in the wilaya of Setif.

TYPE: Incidence
TERM: Geographic Factors; Registry
SITE: All Sites
TIME: 1989 – 1992

ARGENTINA

BUENOS AIRES

2   **Dominguez, F.V.**   04341
Univ. of Buenos Aires, Fac. of Odontology, Oral Pathology Dept., M.T. de Alvear 2142 , Buenos Aires, Argentina (Tel.: +54 1 838983)
COLL: Keszler, A.; Guglielmotti, M.B.

Odontogenic Cysts and Neoplasms in Children
The purpose of this study is to review and collect cases of dentigerous and other developmental cysts of the jaws in children and teenagers, from laboratories of pathology and oral diagnosis centers in Argentina, to establish the pathogenic association and frequency of these lesions as precursors of odontogenic neoplasms (i.e. ameloblastoma) or other tumour-related jaw diseases. Preliminary studies on this material have shown associations and anatomical correlations not suspected for this age group. Inflammatory and genetic factors, as well as chemical therapy on teeth of the regions of the jaws involved with these cysts are also considered a tumour initiators. More than 250 cases are included in the study and control groups are used when incidence analysis or quantitative methods are applied. The general study methods are clinical, radiographic, histological, and morphometrical and genetic. Papers have been published in J. Oral. Maxillofac. Surg. 44:609, 1986; rev. fac. Odont. (Buenos Aires) 8:34, 1988; J. Oral. Path., 17 39, 1988.

TYPE: Cross-Sectional
TERM: Adolescence; Childhood; Drugs; Genetic Factors; Prevalence
SITE: Odontogenic Neoplasms
TIME: 1985 - 1992

AUSTRALIA

ADELAIDE

3   **Woodward, A.**   01999
Univ. of Adelaide, Dept. of Community Medicine, Box 498 G.P.O., Adelaide SA 5001, Australia (Tel.: (08)2245135)
COLL: McMichael, A.J.; Roder, D.M.; Crouch, P.; Mylvaganam, A.

Lung Cancer and Other Cancers among Uranium Miners, as Related to Radiation Exposure
2,580 individuals were employed in the South Australian Radium Hill uranium mine during its operational period from 1951 to 1961. Death records and cancer registry records are being examined progressively to identify lung cancers and other cancer cases among these individuals, by occupational category at Radium Hill. Electoral rolls, driving licences and health insurance records have been employed to identify the miners still alive and resident in Australia. Accessible miners, or next-of-kin have been interviewed to determine their smoking and occupational histories. Individual workers' exposures to radiation are being estimated from historical records of radon gas levels in the mine, and detailed job histories. Smoking histories obtained from miners who are still alive show a slightly greater smoking prevalence among underground workers but not an association of smoking with estimated exposure to radon daughters at Radium Hill. By the end of 1988, about 25% of the Radium Hill workforce were found to have died. Preliminary analyses show that workers receiving 40 Working Level Months of radiation or more have experienced approximately four times the rate of lung cancer among surface workers at the mine.

TYPE: Case-Control; Cohort
TERM: Metals; Mining; Occupation; Radiation, Ionizing; Registry; Tobacco (Smoking)
SITE: Lung
CHEM: Uranium
OCCU: Miners, Uranium
REGI: S. Australia (Aus)
TIME: 1981 - 1992

BRISBANE

4   **Battistutta, D.**   04340
Queensland Inst. of Medical Research, Dept. of Cancer Epidemiology, Bramston Terrace, Herston Brisbane Qld 4006, Australia (Tel.: +61 7 2536222; Tlx: AA145420)
COLL: Knight, N.; MacLennan, R.; Chenevix-Trench, G.

Population-Based Registry of Colorectal Familial Adenomatous Polyposis
A register has been established of all affected and at-risk members of familial adenomatous polyposis (FAP) families in Queensland. It has identified 28 FAP families, of which 57 are affected and 88 at risk for FAP. In collaboration with the responsible clinician, a screening protocol for each at-risk individual is developed. The register provides reminders to clinicians and at-risk individuals of forthcoming screening tests, traces at-risk individuals lost to follow-up, and acts as a source of information on FAP for family members. Similar registers in other countries (UK, Scandinavia) have been shown to be very effective in preventing large bowel cancer in persons identified to be at risk for FAP. In collaboration with similar registers being set up in other Australian states, it is envisaged that research into the genetic and environmental factors that play a role in the development of FAP, and ultimately large bowel cancer, will be possible.

AUSTRALIA

TYPE: Incidence; Registry
TERM: Environmental Factors; Familial Adenomatous Polyposis; Familial Factors; High-Risk Groups; Premalignant Lesion
SITE: Colon; Rectum
REGI: Brisbane II (Aus)
TIME: 1988 – 1993

5   Khoo, S.K.   02575
Univ. of Queensland, Royal Brisbane Hosp., Dept. of Obstetrics & Gynecology, Clinical Sciences Bldg, Brisbane Qld 4029, Australia (Tel.: (07)2535211)

Registry of Gestational Trophoblastic Disease in Queensland
The rarity of the disease in Australia necessitates centralization of facilities and collection of information. The Registry continues to function with the support of the Queensland Cancer Fund. Since June 1976, 431 patients have been registered. The objectives of the Registry are to: (1) study the incidence of hydatidiform mole and other neoplastic disorders of the trophoblast in Queensland; (2) provide a comprehensive follow-up programme, using HCG-beta measurements in blood; (3) centralize facilities for consultation and treatment of complications of the diseases; (4) determine long-term effects of cytotoxic chemotherapy in successfully treated patients in terms of fertility and pregnancy outcome; (5) evaluate various factors as a guide to prognosis. Patients with gestational trophoblastic disease are registered and the clinical data recorded by the Registry. Follow-up studies are done with the cooperation of the doctors in Queensland. Papers appeared in Aust. NZ J Obstet. Gynaecol. 20:35-42, 1980; 141-150, 1982 and 26:129-135, 1986.

TYPE: Incidence; Registry
TERM: Chemotherapy; Fertility; Pregnancy; Premalignant Lesion; Prognosis
SITE: Choriocarcinoma; Hydatidiform Mole
TIME: 1976 – 1993

6   MacLennan, R.   03493
Queensland Inst. Medical Research, Cancer Epidemiology Unit, Bramston Terrace , Herston, Brisbane Qld 4006, Australia (Tel.: +61 7 2536247; Fax: 2525499)
COLL: McLeod, G.R.C.; Little, J.H.

Fluorescent Light and the Risk of Malignant Melanoma in Brisbane
A case-control study of melanoma of the skin is being conducted in Brisbane, Australia, to estimate the possible risk of prior exposure to fluorescent light. Due to the high incidence rate it has been feasible to recruit over 400 new cases for a study powerful enough to detect risks smaller than those reported in previous studies. All incident cases in the Greater Brisbane Metropolitan area have been included in the sample approached with permission of their doctors. Controls are a sample of the general population matched by sex and age. Data collection includes a detailed history of exposure to fluorescent light. The history will be validated by information on lighting at former work places. The study includes all significant risk factors found in recent studies in Australia and North America, to enable control of potential confounding factors which might falsely produce apparently increased risk for exposure to fluorescent light.

TYPE: Case-Control
TERM: Fluorescent Light; Radiation, Ultraviolet
SITE: Melanoma
TIME: 1985 – 1992

7   MacLennan, R.   03494
Queensland Inst. Medical Research, Cancer Epidemiology Unit, Bramston Terrace , Herston, Brisbane Qld 4006, Australia (Tel.: +61 7 2536247; Fax: 2525499)
COLL: Battistutta, D.; Grattan, H.; Ward, M.; Cowen, A.; Bain, C.J.; Goulston, K.; Bokey, L.; Chapuis, P.; Killingback, M.; Lambert, J.; Korman, M.; Eaves, E.R.; McIntyre, O.R.; McLeish, J.; Macrae, F.A.; Penfold, J.C.B.; St John, D.J.B.; Wahlqvist, M.L.

Multicentre Collaborative Clinical Trial for Prevention of Large Bowel Adenomas
This multicentre clinical trial aims to assess the effects of dietary modification on the incidence and growth of colorectal adenomas, the precursors of one of the most common cancers. Subjects with histologically confirmed adenomatous polyps will be selected. In this high risk and well motivated group of people, the effects of a low fat diet, increased dietary fibre, and beta carotene on the incidence and size of adenomatous polyps after two years follow-up will be measured. This has significant implications for

AUSTRALIA

the frequency of colonoscopy and related health service costs. The design will be 2x2x2 factorial with at least 420 subjects. 400 subjects had been recruited by the end of 1987. The trial should be concluded by 1989 and analysis by 1990.

TYPE: Incidence
TERM: Diet; Fat; Fibre; Premalignant Lesion; Prevention; Vitamins
SITE: Rectum
CHEM: Beta Carotene
TIME: 1987 - 1992

8   Martin, N.G.   04193
Queensland Inst. of Medical Research, Cancer Epidemiology Unit, Bramston Terrace, Herston Brisbane Qld 4006, Australia (Tel.: +61 7 2536278; Fax: 2525499)
COLL: Green, A.C.; MacLennan, R.; McLeod, G.R.C.; Little, J.H.; Ring, I.

Genotype and Environment Interactions in the Aetiology of Malignant Melanoma

This collaborative project with population-based familial registries of melanoma in Queensland and New South Wales will contact all patients with melanoma reported to population-based cancer registries to establish if there is a family history of melanoma and whether the patient is a twin. If either is present, patients will be further investigated together with first degree relatives. The combined populations of Queensland and New South Wales yield some 2,000 new cases of melanoma per year, of which some 300 will be familial, and 50 will be a twin. Cases back to 1982 will be included in the initial survey.

TYPE: Case Series
TERM: Familial Factors; Genetic Factors; Genetic Markers; Registry; Twins
SITE: Melanoma
REGI: NSW (Aus); Queensland (Aus)
TIME: 1987 - 1992

* 9   Muller, J.M.   05152
Queensland Dept. of Health, Epidemiology and Prevention Unit, 147-163 Charlotte St., G.P.O. Box 48, Brisbane Qld 4000, Australia (Tel.: +61 7 2340907; Fax: 2210951)
COLL: Ring, I.; Balanda, K.P.; Cannon, L.; Clutton, S.

Cervical Cancer Screening in Queensland

A range of studies has been initiated to evaluate several aspects of cervical cancer screening, using Pap smears, as a means of reducing morbidity and mortality from cancer of the cervix among Queensland women. The primary aim of these studies is to develop and evaluate strategies to increase screening coverage among the most at-risk women in the community, a secondary aim being to increase overall levels of coverage among all eligible women aged 20-69 years. The various components of this research include: cross-sectional postal surveys of 2,500 randomly selected Queensland women, covering knowledge, attitudes, psychosocial issues and behaviour; a cohort study of four groups of at-risk women, being women who have not had a Pap smear for two years or more, exposed to different strategies to promote Pap smear screening; evaluation of a mass media campaign using television, through analysis of pathology laboratory data and health services data; and the evaluation of specific programmes piloted for women living in rural areas and Aboriginal and Islander women which focus on service provision issues.

TYPE: Cohort; Methodology
TERM: Cytology; Screening
SITE: Uterus (Cervix)
TIME: 1989 - 1993

* 10   Muller, J.M.   05153
Queensland Dept. of Health, Epidemiology and Prevention Unit, 147-163 Charlotte St., G.P.O. Box 48, Brisbane Qld 4000, Australia (Tel.: +61 7 2340907; Fax: 2210951)
COLL: Ring, I.; Balanda, K.P.; Cannon, L.; Clutton, S.

Breast Cancer Screening in Queensland

This project will evaluate various aspects of breast cancer screening, using mammography, as a mean of reducing morbidity and mortality from breast cancer among Queensland women. The primary aim is to assist in the development and implementation of the breast cancer prevention programme. The project includes monitoring and evaluation of breast screening services using routine data collected at the clinics; an assessment of women's satisfaction with the services and experience of mammography from

AUSTRALIA

point of service surveys; an examination of factors that influence and motivate attendance for screening, through point-of-service surveys; and the monitoring and evaluation of non-individualised strategies used to recruit eligible women from the target population, through population-based community surveys which will include an examination of knowledge, attitudes, psychosocial issues and behaviour.

TYPE: Methodology
TERM: Mammography; Screening
SITE: Breast (F)
TIME: 1989 - 1993

11 Ring, I. 04144
Queensland Dept. of Health, Cancer Epidemiology, and Prevention Unit, 147-163 Charlotte St., State Health Bldg, G.P.O. Box 48, Brisbane Qld 4001, Australia (Tel.: (7)2344187/2340929; Fax: 2210951 ; Tlx: AA 42531)
COLL: Williams, M.; MacLennan, R.; Chick, J.

Estimated Incidence of Skin Cancer in Queensland
The aim of this project is to estimate the incidence of diagnosed skin cancer in Queensland both for selected areas of the state and for the state as a whole. Queensland has probably the highest incidence of skin cancer in the world and this project will provide basic information not presently available on the level and pattern of this important disease. This knowledge is a basic prerequisite of programmes designed to have an impact on skin cancer. The Queensland Cancer Registry now collects histologically verified squamous cell carcinomas and basal cell carcinomas and a survey is to be conducted to obtain an estimate of clinically diagnosed non-melanoma skin cancers, to combine with the Registry data to provide an overall estimate of diagnosed skin cancer. Fourty doctors and a hospital outpatient department were selected randomly in each of four parts of the state, so that the four areas combine to form a state estimate.

TYPE: Incidence
TERM: Histology; Registry
SITE: Skin
REGI: Queensland (Aus)
TIME: 1984 - 1992

DARWIN

12 Mathews, J.D. 04067
Menzies School of Health Research, P.O. Box 41096, Casuarina, Darwin NT 5792, Australia
COLL: Parsons, W.J.; Giles, G.G.

Mortality and Cancer Incidence in a Large Cohort of Men Employed as Herbicide Sprayers in Victoria
From the 1950s, phenoxyherbicides have been used in increasing quantities for weed control in Victoria, Australia. Spraying on government land and on some private land was carried out by men employed by the Victorian Department of Crown Lands. A mortality follow-up of over 2,000 men to 1983 showed that there was no significant increase in cancer overall, nor in soft-tissue sarcoma or lymphoma mortality. Data are being linked to the Victoria Cancer Registry to look for possible effects on cancer incidence.

TYPE: Cohort
TERM: Chemical Exposure; Herbicides; Occupation; Pesticides; Registry
SITE: Lymphoma; Sarcoma
CHEM: Phenoxy Acids
OCCU: Herbicide Manufacturers; Herbicide Sprayers
REGI: Victoria (Aus)
TIME: 1982 - 1993

13 Mathews, J.D. 02632
Menzies School of Health Research, P.O. Box 41096, Casuarina, Darwin NT 5792, Australia
COLL: Clifford, C.; Hopper, J.L.; Giles, G.G.

Australian Twin Registry
Previous studies have shown that cancer concordance rates in monozygous (MZ) twins are quite low, and this evidence has sometimes been cited to support the view that genetic factors are unimportant in the origin of cancer. The alternative view is that genetic factors are necessary, but not sufficient, as causes

AUSTRALIA

of cancer; the low concordance rates in MZ twins would then be attributed to the effect of environmental differences and/or to the effects of chance (stochastic) events which may influence the age of onset of cancer. The Australian National Health and Medical Research Council Twin Registry comprises 15,000 pairs of twins (volunteers); the incidence of cancer will be followed by linking Twin Registry information to Cancer Registry data and by follow-up of twins. On a collaborative basis, it is hoped to obtain additional information from other twin registries on the age of onset of cancer in twins. Data on the age of onset of cancer in twins will be used to develop and test statistical models for the origin of cancer which allow for the effect of genetic and environmental factors as well as for the effects of random (stochastic) processes.

TYPE: Incidence
TERM: Environmental Factors; Genetic Factors; Mathematical Models; Record Linkage; Registry; Twins
SITE: All Sites
TIME: 1981 - 1993

HOBART

* 14   **Dwyer, T.**   04992
Univ. of Tasmania, Menzies Center for Population Health Research, Tasmanian Cancer Registry, 43 Collins St., Hobart Tas. 7000, Australia (Tel.: +61 2 354880; Fax: 354816)
COLL: Berwick, M.; Roy, C.R.; Gies, H.P.

Human Exposure to Ultraviolet Radiation at Varying Latitudes
Currently available data linking environmental exposure to ultraviolet radiation and the incidence of malignant melanoma in humans lacks objective validation. This study aims to measure ultraviolet dose in a cohort of people and to validate an exposure questionnaire. Individual environmental exposure to ultraviolet radiation during specific activities will be measured by polysulphone film badges located on seven anatomical sites (hand, shoulder, thigh, calf, cheek, back and chest). An extra badge is placed on the surface where the individual undertook the activity. The activities will cover leisure activities (e.g. boating, swimming, tennis) and work-related exposures (e.g. outdoor work, office work). Current and typical sun exposure patterns, duration of activity, dress and use of sunscreen (blockouts) will be measured by questionnaire and related back to the actual exposure recorded by the badge and by environmental monitoring of ultraviolet radiation. Measurements will take place during different seasons and at different latitudes in Australia. The age range of the subjects will be 5-69 years and the sample size will be 1,800 of which 1,200 will be in Hobart.

TYPE: Methodology
TERM: Radiation, Ultraviolet
SITE: Melanoma
LOCA: Australia; United States of America
TIME: 1992 - 1993

15   **Lowenthal, R.M.**   04244
Royal Hobart Hosp., Liverpool St., Hobart Tas. 7000, Australia (Tel.: +61 02 388157; Fax: 312043)
COLL: Marsden, K.; Bicevskis, M.; Jupe, D.M.L.

Environmental and Occupational Exposures of Patients with Myelodysplastic Syndromes
The myelodysplastic syndromes (MDS) are a group of pre-leukaemic conditions that have been little studied epidemiologically. Some cases have developed as a consequence of treatment with anti-cancer agents and others have followed exposure to benzene, but in the majority of cases no such factor can be identified. The aim of the study is to identify environmental and other substances associated with increased risk of developing MDS. From 1988 to 1990, all patients with MDS in southern Tasmania were identified and categorized by a series of laboratory tests. Detailed interviews using a questionnaire which enables a Siemiatycki exposure index to be calculated were commenced in 1990. Eighty patients were alive on 1 January 1990 and a further 20 new patients are expected annually. Both epidemiological and laboratory aspects will be continued in 1991 and 1992 to obtain details of lifetime exposure to possible aetiological factors. Two age- and sex-matched controls will be interviewed for each patients. Patients visiting hospital outpatient clinics will serve as controls for hospital patients, and patients of private physicians and general practitioners as controls for non-hospital patients.

AUSTRALIA

TYPE: Case-Control
TERM: Chemical Exposure; Environmental Factors; Occupation; Premalignant Lesion
SITE: Myelodysplastic Syndrome
REGI: Tasmania (Aus)
TIME: 1988 - 1993

MELBOURNE

16 Christie, D.G. 02892
Univ. of Melbourne, Dept. of Community Medicine, 159 Barry St., Carlton, Melbourne Vict. 3053, Australia (Tel.: +61 3 3447449; Fax: 3476136 ; Tlx: AA35185)
COLL: Robinson, K.; Gordon, I.; Potter, A.

Health Watch: The Australian Petroleum Industry Health Surveillance Programme
This prospective study within the Australian oil industry is designed to investigate possible associations between cancer mortality/morbidity and occupationally determined exposure to hydrocarbon chemicals. Data are collected on all deaths; cancer deaths are confirmed by reference to biopsy or autopsy reports. Subjects have been classified by industrial hygienists into four ranked exposure groups. Study members have contributed 50,000 person-years at risk to the mortality analyses with 170 deaths. The all-cause SMR for males is 0.7; for all malignant neoplasms in males the SMR is 0.9. 40,000 person-years have accumulated for cancer incidence with 116 incident cancers; the all-cancer SIR in males is 0.9. Twenty incident cancers of the lymphatic and haematopoietic tissue have been reported and the SIR for this group of cancer is 2.0 (95%C.I. 1.2-3.2). A case-control study of lympho-haematopoietic cancer is underway. A paper has been published in Med. J. Aust. 147:222-225, 1987.

TYPE: Case-Control; Cohort
TERM: Chemical Exposure; Occupation; Petroleum Products
SITE: All Sites; Haemopoietic; Lymphoma
CHEM: Hydrocarbons
OCCU: Petroleum Workers
TIME: 1981 - 1993

17 Giles, G.G. 03862
Anti-Cancer Council of Victoria, Cancer Epidemiology Centre, 1 Rathdowne St. , Carlton South, Melbourne Vict. 3053, Australia (Tel.: +61 3 6623300; Fax: 6633412 ; Tlx: AA 34158)
COLL: Marks, R.; Staples, M.

National Skin Cancer Survey
The aims are to estimate the incidence of skin cancers in Australia, principally, basal cell carcinoma, and squamous cell carcinoma and premalignant solar keratosis, and to assess the role of sunlight in the development of skin cancer, including exogenous geographical factors and endogenous individual response to sunlight. This survey was first conducted in 1985. In the 1990 re-survey 50,000 persons representing a population-based sample across Australia were interviewed concerning their history of skin cancer treatment in the last 12 months. Individual reponses to sunlight including burning and tanning ability, are being recorded and correlated with prevalence of skin cancer. Medical details concerning the site and histology of the lesions treated are being confirmed with the treating medical practitioners. Demographic data including lifestyle factors, smoking, occupation, and socio-economic status, were also collected. The prevalence of skin cancers will be stratified according to latitude to assess the aetiological role of sunlight, having controlled for the other variables. To examine trends, the survey will be repeated at five yearly intervals. This is the only national survey of skin cancer of all types in Australia, a country with purportedly the highest incidence of these tumours in the world. A paper has been published in Br.Med.J. 296:13-17, 1988.

AUSTRALIA

TYPE: Correlation; Incidence
TERM: Geographic Factors; Lifestyle; Occupation; Premalignant Lesion; Prevalence; Radiation; Ultraviolet; Registry; Socio-Economic Factors; Tobacco (Smoking)
SITE: Melanoma; Skin
REGI: Victoria (Aus)
TIME: 1985 - 1996

18   Giles, G.G.   04745
Anti-Cancer Council of Victoria, Cancer Epidemiology Centre, 1 Rathdowne St., Carlton South, Melbourne Vict. 3053, Australia (Tel.: +61 3 6623300; Fax: 6633412 ; Tlx: AA 34158)
COLL: Hopper, J.L.; Ireland, P.; Ktenas, D.; Larkins, R.; O'Dea, K.; Powles, J.W.; Proietto, J.; Williams, J.

Melbourne Collaborative Cohort Study

The aim of the programme is to investigate the relationship between dietary intakes and specific disease outcomes, particularly cancer of the colon and rectum, breast and prostate. Emphasis is placed on the role of lipids, fibre and free radical scavengers and their food sources. The study population will be a mix of Italian-born, Greek-born and Australian-born residents of the Melbourne statistical division aged between 40 and 69 years at recruitment. The cohort structure will be 50,000 persons with a mix of approximately equal numbers of males and females and 60% migrant and 40% Australian-born. Optically scannable questionnaires will be administered at the survey center. The questionnaires will include questions on demographic, medical and dietary and other lifestyle factors. A food frequency questionnaire will also be completed. A physical examination will include anthropometry, body impedance and blood pressure measurements. Blood samples will be taken and total cholesterol, HDL cholesterol, and triglycerides will be assayed immediately. Four 1ml aliquots of plasma and three aliquots of resuspended white cells (buffy coat) and 1 aliquot of red cells will be stored in liquid nitrogen. Each participant will be followed up on the third anniversary of recruitment to collect details of health events and dietary status. Follow-up is planned for 20 years. Record linkage to the Victorian Cancer Registry and to all death certificates will be performed annually. A paper was published in Proc. Nutr. Soc. Aust. 15:61-68, 1990.

TYPE: Cohort
TERM: BMB; Diet; Lifestyle; Lipids; Migrants; Plasma; Red Cells; White Cells
SITE: Breast (F); Colon; Prostate; Rectum
REGI: Victoria (Aus)
TIME: 1990 - 2010

* 19   Giles, G.G.   05054
Anti-Cancer Council of Victoria, Cancer Epidemiology Centre, 1 Rathdowne St., Carlton South, Melbourne Vict. 3053, Australia (Tel.: +61 3 6623300; Fax: 6633412 ; Tlx: AA 34158)
COLL: Kaye, A.; Gonzales, M.

Australian Brain Tumour Register

The register includes all intracranial tumours and all tumours of the spinal cord. Its broad aim is to describe the incidence of these tumours in more detail than is usually possible in cancer registries, with a view to increasing their descriptive epidemiology and to monitor changes over time. The register records about 1,600 tumours annually. Notification is usually via neurosurgeons, neurosurgical units and neuropathologists. Follow-up details are obtained by writing to surgeons. Detailed socio-demographic data are collected in addition to review diagnoses and clinical information and treatment details. It is anticipated that the register will serve as a resource for research into the epidemiology and pathology of these tumours. To this end, a database and repository of slides is kept centrally.

TYPE: Incidence; Registry
TERM: Demographic Factors; Registry; Survival; Trends
SITE: Brain; Spinal Cord
REGI: Victoria (Aus)
TIME: 1986 - 1999

* 20   Giles, G.G.   05055
Anti-Cancer Council of Victoria, Cancer Epidemiology Centre, 1 Rathdowne St., Carlton South, Melbourne Vict. 3053, Australia (Tel.: +61 3 6623300; Fax: 6633412 ; Tlx: AA 34158)
COLL: Russell, I.; Reed, R.

AUSTRALIA

Register of In Situ and Small Size Breast Cancers
The register was established prior to the widespread use of mammographic screening in Victoria, primarily to elucidate the classification, treatment and natural history of in situ carcinoma and invasive cancers less than 10mm in diameter. These data will be useful in evaluating the impact of mammographic screening. Eligible cases (currently about 150 annually) are identified via pathology notifications to the Victoria Cancer Registry and demographic, diagnostic, clinical and pathological data are then collected from the treating surgeons and pathologists. Annual follow-up of recurrences and new primary tumours is conducted via the treating surgeon.

TYPE: Incidence
TERM: In Situ Carcinoma; Mammography; Registry; Survival
SITE: Breast (F)
REGI: Victoria (Aus)
TIME: 1988 - 1999

* 21 Giles, G.G.   05056
Anti-Cancer Council of Victoria, Cancer Epidemiology Centre, 1 Rathdowne St. , Carlton South, Melbourne Vict. 3053, Australia (Tel.: +61 3 6623300; Fax: 6633412 ; Tlx: AA 34158)
COLL: St John, D.J.B.; Macrae, F.A.; Carden, A.; Bankier, A.; Watts, C.

Familial Adenomatous Polyposis Register for Victoria
The principal aim of the registry is to prevent death from colorectal cancer in a small but high-risk group of individuals who have a family history of familial adenomatous polyposis (FAP). It is also expected to contribute to research into the environmental and genetic causes of cancer. The registry and data collection protocols are being developed in close consultation with the Leeds Castle group. Cases of FAP are usually recruited via their doctors. The registrar then arranges for family members at risk to be entered on the register; clinicians and persons at risk are then sent reminders of forthcoming screening tests. The register also serves as a source of information and education about FAP. Currently 49 pedigrees, including 53 affected and 45 persons at risk of FAP are registered.

TYPE: Incidence; Registry
TERM: Environmental Factors; Familial Factors; High-Risk Groups
SITE: Colon; Rectum
TIME: 1987 - 1999

* 22 Hopper, J.L.   05111
Univ. of Melbourne, Epidemiology Unit, 151 Barry St., Carlton, Melbourne Vict. 3053, Australia (Tel.: +61 3 3446990; Fax: 3447014)
COLL: Flander, L.; Giles, G.G.; Carlin, J.; Green, M.; Collins, J.; Russell, I.

Breast Cancer in Families
The study aims to measure the risk of breast cancer in genetically related relatives of breast cancer patients, and specifically to measure the variation in risk by age at onset and laterality. It will also determine the extent to which risk is a consequence of other factors including current and past weight, reproductive, contraceptive and menstrual history, and use of alcohol. Patients with a recent diagnosis of breast cancer are selected at random, stratified by early (age < 45) and late onset, from new registrations with the Victoria Cancer Registry. With the consent of both the treating surgeon and the patient, a questionnaire on major risk factors is administered at face-to-face interview or by telephone. Each patient is asked to identify and encourage the cooperation of living relatives. If permission is given by the patient, these relatives will be asked to complete the same questionnaire. A similar process will be undertaken to study the relatives of the spouse of partner of the patient, to obtain 'control' pedigrees. Stored DNA will allow testing of hypotheses relating susceptibility to putative genetic markers.

TYPE: Case Series; Genetic Epidemiology
TERM: Alcohol; BMB; Contraception; DNA; Genetic Factors; Genetic Markers; Menstruation; Physical Factors; Registry; Reproductive Factors
SITE: Breast (F)
REGI: Victoria (Aus)
TIME: 1990 - 1995

23 Rich, A.M.   03990
Univ. of Melbourne, Dept. of Dental Medicine & Surgery, 711 Elizabeth St. , Melbourne Vict., Australia (Tel.: 3410255; Tlx: AA35185)
COLL: Radden, B.G.; Reade, P.C.

AUSTRALIA

Oral Cancer in Victoria, Australia
This is an on-going study collecting data from patients who develop squamous cell carcinoma of the oral mucosa. The patients are all diagnosed in one Oral Pathology Biopsy Service and detailed, standardized data are available. Information collected includes age and sex of patients, extent of lesion at diagnosis, site of involvement in the oral cavity, suspected aetiological factors, in particular tobacco and alcohol habits, occupation and haematological status and survival data. By 1984, 244 patients were included in the study and the results from this group were reported. This information is being extended and updated.

TYPE: Case Series
TERM: Alcohol; Biopsy; Occupation; Survival; Tobacco (Smoking)
SITE: Oral Cavity
TIME: 1975 - 1992

24   **St John, D.J.B.**   02984
Royal Melbourne Hosp., Dept. of Gastroenterology, P.O., Melbourne Vict. 3050, Australia (Tel.: +61 3 3427470; Fax: 3427802)
COLL: Young, G.P.; Macrae, F.A.; Alexeyeff, M.; Deacon, M.C.; Evans, G.; Fenwick, M.; Gange, D.; Hankinson, D.

Familial Colorectal Cancer and Early Diagnosis of Colorectal Cancer
A programme of selective screening for colorectal cancer was established in the Department of Gastroenterology at The Royal Melbourne Hospital in 1979. A register of families with hereditary non-polyposis colon cancer (HNPCC), familial adenomatous polyposis (FAP) and individuals having first-degree relatives with sporadic colorectal cancer includes over 35 separate kindreds with HNPCC, 26 kindreds with FAP and a total of over 1,500 individuals having regular cancer surveillance. As a result of studies of the quality of surveillance of FAP families, a central State-wide register, the ESSO Familial Polyposis Register, was established within the Victorian State Cancer Registry in April 1988 to promote efficient management of these families throughout the community. A special aspect of the research programme is evaluation of methods of surveillance, particularly application of new faecal occult blood tests to screening. Recent papers were published by Macrae et al. in Med. J. Aust. 151:552-557, 1989, by Young et al. in J. Gastroenterol. Hepatol. 5:194-203, 1990 and by St. John in Aust. N.Z.J. Surg. 60:415-7, 1990.

TYPE: Intervention; Registry
TERM: Familial Factors; Genetic Factors; High-Risk Groups; Registry; Screening
SITE: Colon; Rectum
REGI: Brisbane II (Aus)
TIME: 1979 - 1992

NEWCASTLE

* 25   **Christie, D.**   04958
Univ. of Newcastle, Discipline of Environmental, & Occupational Health, 86 Platt St., Newcastle NSW 2298, Australia (Tel.: + 61 49 211234; Fax: 601197)
COLL: Brown, A.; Devey, P.; Nie, V.

Brain Tumours in a Coal-Mine
Three historical cohorts have been established from 1 January 1975 to 31 December 1989 in an index coal-mine reported to have excess brain tumours, and in two control coal-mines. The latter work the same seam, have had similar operational histories and use a workforce recruited from the same region. Incidence of primary brain tumours is being examined in each mine for the last 15 years. Ionising and non-ionising radiation is being assessed in each mine; chemical usage over 20 years is also being compared between mines.

AUSTRALIA

TYPE: Cohort
TERM: Chemical Exposure; Electromagnetic Fields; Mining; Occupation; Radiation, Ionizing; Radiation, Non-Ionizing; Registry
SITE: Brain
OCCU: Miners, Coal
REGI: NSW (Aus)
TIME: 1990 - 1992

PERTH

26 de Klerk, N.H. 04825
Univ. of Western Australia, Dept. of Social and Prev. Med., Queen Elizabeth II Med. Cent., Verdun St., Nedlands, Perth WA 6009, Australia (Tel.: +61 9 3893456)
COLL: Musk, A.W.; Hobbs, M.S.T.; Eccles, J.

Occupational Exposures in Subjects with Mesothelioma but no Known Asbestos Exposure
The aim of the study is to determine if subjects with malignant mesothelioma but no history of asbestos exposure have other exposures which may be responsible for their disease. All subjects reported to the Mesothelioma Registry of Western Australia and in whom there is no evidence of asbestos exposure will be included as cases. An interviewer-administered questionnaire on occupational and environmental exposures will be conducted. Comparison subjects will include subjects with other (non-asbestos-related) cancers who will receive the same questionnaires.

TYPE: Case-Control
TERM: Environmental Factors; Occupation; Registry
SITE: Mesothelioma
REGI: Sydney (Aus); W. Australia (Aus)
TIME: 1990 - 1995

27 English, D.R. 03033
The Queen Elizabeth II Medical Center, NH and MRC Research Unit in, Epidemiology and Preventive Medicine, Nedlands, Perth WA 6009, Australia (Tel.: +61 9 3893134; Fax: 3893648; Tlx: AA93446)
COLL: Armstrong, B.K.

Aetiology of Benign Pigmented Naevi in Children and Adolescents
The aim of this project is to identify the aetiological factors of benign pigmented naevi in children and adolescents. 2,450 white children, aged 5-14 years, were interviewed in 1985 and had their naevi counted. A subset of 1,000 children had their backs photographed. 700 of these children were reexamined in 1988 and will be examined again in 1991. Analysis of the data collected in 1985 shows that constitutional sensitivity to the sun and sun exposure are important risk factors for the development of naevi.

TYPE: Cohort
TERM: Adolescence; Childhood; Lifestyle; Naevi; Pigmentation; Premalignant Lesion; Radiation, Ultraviolet
SITE: Benign Tumours; Melanoma
TIME: 1984 - 1993

28 Jacobs, I. 04326
Curtin Univ. of Technology, Center for Advanced Studies, in Health Sciences, Kent St., Bentley, Perth WA 6102, Australia
COLL: Phillips, M.; Spickett, J.T.

Health Effects of Agricultural Chemicals
This is a prospective study of 8,178 cereal farmers and their families in Western Australia, to determine if there are any long term adverse effects associated with exposure to herbicides or other agricultural pesticides. Initial health status was assessed by postal questionnaire. Follow-up of health assessments will be undertaken by postal questionnaire, together with tracing of national registry information on cancers, birth anomalies and deaths. Approximately 100 incident cases of cancer are expected in the first five years of follow-up. Cancers of all sites are of interest, with special emphasis on those of the soft tissue and leukaemias. Base-line data on pesticide usage and farm practices were collected in 1984, prior to administration of the health questionnaires, by postal questionnaire and validated against governmental

AUSTRALIA

agricultural statistics. The use of pesticides and related farm practices since collection of the base-line data is monitored by annual questionnaires.

TYPE: Cohort
TERM: Chemical Exposure; Herbicides; Occupation; Pesticides; Registry
SITE: All Sites; Leukaemia; Soft Tissue
OCCU: Farmers
TIME: 1984 - 1993

29   Kricker, A.   04837
Univ. of Western Australia, NH & MRC Research Unit in Epidemiology, Nedlands, Perth WA 6009, Australia (Tel.: +61 9 3893219; Fax: 3893648)
COLL: English, D.R.; Heenan, P.T.; Randell, P.L.; Armstrong, B.K.

Prevalence, Incidence, and Aetiology of Non-Melanocytic Skin Cancer in Western Australia
The aims of the study are to estimate the incidence and prevalence of non-melanocytic skin cancer (NMSC) in an Australian population, to ascertain the most efficient, accurate way of monitoring the incidence, and to describe in detail the association between NMSC and cutaneous indicators of sun-sensitivity and exposure to the sun. The prevalence survey was undertaken in Geraldton, W. Australia, in November 1987 in a study population of 5,475 persons aged 40-64 years and listed on the state electoral roll. The final response rate was 82%. Each subject had a whole body examination by a dermatologist. Altogether 808 suspected skin cancers in 443 study subjects were diagnosed, and 74% were excised and examined histopathologically. The most common cancer was basal cell carcinoma (BBC). Approximately half of these lesions were on th trunk (Med. J. Aust. 1990). For the case-control study 1,263 of the 1987 survey attenders were interviewed in 1988. Cases were survey subjects with histopathologically confirmed BCC or squamous cell carcinoma (SCC) diagnosed during the survey or in the previous 12 months. Controls (4 per case within each stratum defined by age and sex) were selected from survey participants who did not have a confirmed skin cancer. A highly structured interview questionnaire was developed. The questionnaire sought data on topics related to factors relevant to the aetiology of NMSC: constitutional factors and history of skin conditions, and sun exposure. It is proposed to repeat the cross-sectional study in 1992 in order to calculate age-specific incidence and prevalence of NMSC. Subjects have been contacted by mailed questionnaire in 1989 and 1991 to ascertain incident cases of treated skin cancer. It will thus be possible to calculate the actual incidence of skin cancer in those subjects who were seen in both 1987 and 1992; the incident cases being those diagnosed and treated between 1987 and 1992 plus the prevalent cases diagnosed at the 1992 survey. A paper was published in Int. J. Cancer 48, 1991.

TYPE: Case-Control; Cross-Sectional; Incidence
TERM: Radiation, Ultraviolet
SITE: Skin
TIME: 1987 - 1992

30   Musk, A.W.   00919
Sir Charles Gairdner Hosp., Dept. of Respiratory Medicine, Verdun St., Nedlands, Perth WA 6009, Australia (Tel.: +61 9 3983251)
COLL: Hobbs, M.S.T.; de Klerk, N.H.; Hansen, J.; Eccles, J.

Occupational and Environmental Exposure to Crocidolite
This study aims to describe the mortality of persons formerly in the workforce of the crocidolite mine and mill at Wittenoom, Western Australia and relate their mortality to measures of crocidolite exposure. The mortality from asbestos-related diseases of persons in the previous resident population of the township of Wittenoom, who had no occupational exposure to asbestos will also be described. Interactions between smoking and exposure to crocidolite in causing disease will be investigated. It is intended to determine whether pleural thickening or pleural plaque formation are antecedents of malignant pleural mesothelioma independently of level of exposure to crocidolite, whether asbestosis is a necessary precursor of crocidolite-caused lung cancer, and to describe other aspects of the causation, natural history and outcome of asbestos-related disease in those exposed to crocidolite at Wittenoom.

AUSTRALIA

TYPE: Cohort
TERM: Dusts; Environmental Factors; Mining; Occupation; Tobacco (Smoking)
SITE: Colon; Lung; Mesothelioma; Peritoneum; Stomach
CHEM: Asbestos, Crocidolite
OCCU: Miners, Asbestos
TIME: 1975 - 1993

31   Musk, A.W.   04467
Sir Charles Gairdner Hosp., Dept. of Respiratory Medicine, Verdun St., Nedlands, Perth WA 6009, Australia (Tel.: +61 9 3983251)
COLL: Hobbs, M.S.T.; de Klerk, N.H.; Hickling, C.

Prevention of Malignant Disease among Workers Exposed to Crocidolite at Wittenoom Gorge
The study aims to determine the effect of dietary supplements of beta carotene or retinol on the incidence of malignancy in subjects who were exposed to crocidolite at Wittenoom Gorge in Western Australia between 1943 and 1966. Over 2,000 subjects have been contacted by mail and will be invited to participate. Following initial assessment with chest radiography, dietary and smoking histories, and measurement of serum beta carotene and retinol, subjects will be randomly assigned to receive supplements of beta carotene or retinol. They will be reviewed annually for 5 years, with periodic measurement of plasma beta carotene and retinol levels. The incidence of malignancies of all sites will be compared between the two groups and with the expected incidence of cancer derived from previous calculations.

TYPE: Intervention
TERM: Diet; Dusts; Occupation; Prevention; Tobacco (Smoking); Vitamins
SITE: Lung; Mesothelioma
CHEM: Asbestos, Crocidolite; Beta Carotene; Retinoids
OCCU: Asbestos Workers; Miners, Asbestos; Transport Workers
TIME: 1988 - 1993

* 32   Musk, A.W.   05120
Sir Charles Gairdner Hosp., Dept. of Respiratory Medicine, Verdun St., Nedlands, Perth WA 6009, Australia (Tel.: +61 9 3983251)
COLL: Hobbs, M.S.T.; de Klerk, N.H.; Eccles, J.

Mortality in Gold Miners
A group of 2,000 Kalgoorlie gold miners were included in a cross-sectional morbidity study of respiratory symptoms and lung function in 1966. Details of their occupational histories and smoking histories were obtained. It is proposed to ascertain the vital status and cause of death of all members of the cohort and, in addition, the incidence of lung cancer, through the Registrar General's Department of Births, Deaths and Marriages to obtain death certificates and cancer registries throughout Australia to ascertain the incidence of lung cancer. This study will test the hypothesis that exposure to silica is associated with an increased rate of lung cancer.

TYPE: Cohort
TERM: Dusts; Mining; Occupation; Registry
SITE: Lung
CHEM: Silica
OCCU: Miners, Gold
REGI: W. Australia (Aus)
TIME: 1991 - 1993

PRAHRAN

* 33   Fairley, C.K.   05021
Monash Univ., Alfred Hosp., Dept. of Social and Preventive Medicine, Commercial Rd, Prahran Vict. 3181, Australia (Tel.: +61 3 2762645; Fax: 5298580)
COLL: McNeil, J.J.; Abramson, M.; Garland, S.; Quinn, M.A.

Human Papillomavirus and Cervical Cancer
This is a case-control study of incident cases of invasive cervical cancer and community-based controls, to identify whether infection with HPV is more prevalent (as determined by PCR) in cervical scrapes of cases than in controls. The study will aim to enrol 100 cases of invasive cervical cancer (squamous) and 200 community controls over an 18-month period up to September 1992. Controls will

AUSTRALIA

be obtained from the local medical officer by whom the case was referred. A questionnaire for confounding variables will be administered.

TYPE: Case-Control
TERM: BMB; HPV; PCR
SITE: Uterus (Cervix)
TIME: 1991 - 1992

* 34   Hurley, S.   05069
       Monash Univ., Alfred Hosp., Dept. of Social and Preventive Medicine, Commercial Rd, Prahran
       Vict. 3181, Australia (Tel.: +61 3 2762651; Fax: 5298580)
COLL: McNeil, J.J.; Donnan, G.A.; Giles, G.G.

Adult Cerebral Glioma and Occupational Exposures

A case-control study of malignant brain tumours is being conducted, testing the following hypotheses about occupational exposure: (1) certain workplace chemicals (formaldehyde, organic solvents, PAHs, phenols, vinyl chloride, acrylonitrile) are associated with malignant brain tumours; (2) persons who have worked in occupations with presumed high exposure to electromagnetic fields, particularly in the extremely low frequency range, are at elevated risk of malignant brain tumours. Approximately 400 cases and 400 controls will be interviewed, using a structured interview, administered by a nurse interviewer. The usefulness of expert panels in performing retrospective assessments of occupational exposures will be assessed. This study includes tests of (a) the feasibility of using such a panel; and (b) validity and reliability.

TYPE: Case-Control
TERM: Electromagnetic Fields; Occupation; Pesticides; Plastics; Solvents
SITE: Brain
CHEM: Acrylonitrile; Formaldehyde; PAH; Phenol; Vinyl Chloride
TIME: 1989 - 1992

SYDNEY

35   Coates, M.S.   04342
      New South Wales Central Cancer Registry, P.O. Box 380, North Ryde, Sydney NSW 2113, Australia
      (Tel.: (02)8875634; Fax: 8887210 ; Tlx: 71675)
COLL: MacLennan, R.

New South Wales Familial Melanoma Registry

The aim is to investigate genetic and environmental factors and their interaction in the aetiology of melanoma by contacting all cases of melanoma incident in New South Wales in the study period (1,000). Postal questionnaires will be used to ask about melanoma in first degree other relatives, and spouses; twin status; morbidity and risk factors, e.g. sun exposure for the index case and his or her spouse and first degree relatives. The validity of postal questionnaires will be assessed by interview.

TYPE: Cross-Sectional
TERM: Environmental Factors; Familial Factors; Radiation, Ultraviolet; Registry; Twins
SITE: Melanoma; Skin
REGI: NSW (Aus)
TIME: 1987 - 1992

36   Corbett, S.J.   04588
      New South Wales Dept. of Health, Epidemiology Branch, Locked Mail Bag 961, Sydney NSW 2059,
      Australia (Tel.: (02)3919207; Fax: 3919232)
COLL: O'Neill, B.J.

Lymphoid Malignancy and Occupation: A Population-Based Case-Referent Study in a Coal and Steel Producing Region

An apparent occupational outbreak of non Hodgkin's lymphoma among the employees of one of the underground coal mines on the southern New South Wales Coalfield prompted this investigation. The primary aim of the study is to explore any relationship between a history of work in an underground coal mine and the risk of developing one of these malignancies. The Illawarra region of Southern New South Wales is a coal-mining and steel-producing area with a population of approximately 250,000. All male cases of Hodgkin's Disease, non-Hodgkin's lymphoma and chronic lymphocytic leukaemia (ICD9

AUSTRALIA

codes 200, 202 and 204.1) occurring in the region between 1977 and 1986 were identified from hospital and general practitioner records and from the New South Wales Central Cancer Registry. 155 male cases have been identified. For each case two population controls from the region were chosen from a stratified sample of the national medical insurance database, which contains the names, addresses and dates of birth of 96% of the Australian population. Personal interviews were conducted with each of the subjects enrolled in the study, using a structured questionnaire to obtain a consecutive occupational history and details of specific occupational exposures, and a medical, family and residential history. Proxy interviews with the nearest family members were conducted for deceased cases and their matched controls. Validation of occupational histories with respect to coal mining will be made by reference to the NSW Miners Superannuation Fund.

TYPE: Case-Control
TERM: Coal; Mining; Occupation; Registry
SITE: Hodgkin's Disease; Leukaemia (CLL); Non-Hodgkin's Lymphoma
OCCU: Miners, Coal
REGI: NSW (Aus)
TIME: 1986 – 1992

37 **Driscoll, T.R.** 04547
National Inst. of Occupational Health, G.P.O. Box 58, Sydney NSW 2001, Australia (Tel.: (02)5659303; Fax: 5659300 ; Tlx: 177243)
COLL: Leigh, J.; Rogers, A.J.; Thompson, R.

Risk Factors for Mesothelioma
All 74 cases of histologically confirmed malignant mesothelioma (pleural and peritoneal) referred to Royal Prince Alfred Hospital, Sydney in the period 1 January 1980 – 31 December 1985 will be compared with 74 age-sex-residence matched controls with non-skin, non-respiratory cancer. Detailed work and environmental exposure histories will be recorded, as well as data on potential confounding factors. Odds ratios for various exposures will be estimated. Sample size calculations suggest that relative risks of 3.0 will be detectable ($P < 0.05$) with a power of 80%. Case ascertainment is through the Australia Mesothelioma Program.

TYPE: Case-Control
TERM: Dusts; Environmental Factors; Occupation; Registry
SITE: Mesothelioma
CHEM: Asbestos
OCCU: Asbestos Workers; Miners, Asbestos
REGI: Sydney (Aus)
TIME: 1989 – 1992

* 38 **Kaldor, J.** 05200
Univ. of New South Wales, Epidemiology and Clinical Research, 150 Albion St., Sydney NSW 2010, Australia (Tel.: +61 2 3324648; Fax: 3321837)
COLL: Cooper, D.

Aetiology of HIV-Related Malignancies
Kaposi's sarcoma (KS), specific forms of lymphoma and perhaps other cancers occur at an increased risk in people with HIV infection. KS is particularly frequent among people whose HIV infection was acquired through homosexual contact. A study of cancer in people with HIV infection is being undertaken in two parts. First, people newly diagnosed with AIDS are being interviewed with regard to history of specific sexual activities and sexually transmissible disease, and other factors. Comparison will then be made between those who develop KS and those who do not. About 75 cases of KS are expected in the study. A similar comparison will be made between those who develop lymphoma and those who do not. In the second part of the study, people who have been treated with the anti-HIV drug zidovudine are being followed up for the occurrence of malignancy, in particular leukaemia.

AUSTRALIA

TYPE: Case-Control; Cohort
TERM: AIDS; BMB; Drugs; HIV; Sexual Activity; Sexually Transmitted Diseases; Treatment
SITE: Kaposi's Sarcoma; Lymphoma
TIME: 1991 - 1993

39   Leigh, J.   02454
National Inst. of Occupational Health, and Safety, GPO Box 58, Camperdown, Sydney NSW 2001, Australia (Tel.: (02)5659303; Fax: 5659303 ; Tlx: 177243)
COLL: Driscoll, T.R.; Baker, G.; Rogers, A.J.; Shilkin, K.B.; Steele, R.H.; Lee, J.; Ferguson, D.A.

Australian Mesothelioma Surveillance Program. Australian Mesothelioma Register
The Mesothelioma Surveillance Program began on 9 July 1980 with the aim of standardizing the diagnosis, classification and further epidemiological knowledge of the disease. Particularly it aims to identify potential causal agents other than asbestos, to determine the relative importance of different fibre types in causation and to establish dose-response relationships. The Program sought notification of all cases of mesothelioma in Australia from 1 January 1980 to 31 December 1985 (903 case). Since that time the Register has collected less detailed information on all incidence cases of mesothelioma in Australia. A recent paper arising from these data appeared in Cancer 67:1912-1920, 1991.

TYPE: Incidence; Registry
TERM: Chemical Exposure; Dose-Response; Dusts; Environmental Factors; Mining; Occupation
SITE: All Sites; Mesothelioma; Peritoneum; Pleura
CHEM: Asbestos; Mineral Fibres
REGI: Sydney (Aus)
TIME: 1980 - 1993

40   Leigh, J.   03416
National Inst. of Occupational Health, and Safety, GPO Box 58, Camperdown, Sydney NSW 2001, Australia (Tel.: (02)5659303; Fax: 5659303 ; Tlx: 177243)
COLL: Rogers, A.J.; Morgan, G.G.

Carcinogenic Effects of Diesel Exhaust in Coal Workers
This project involves cross-sectional and longitudinal studies of cancer incidence in the entire working coal industry (20,000 men) as well as follow-up of ex-workers and deceased workers. Direct analyses of polycyclic aromatic hydrocarbons in diesel-using coal mines plus full clinical examination of the whole group will be performed. Retrospective and prospective studies using a computerized medical data base will be undertaken.

TYPE: Cohort; Correlation
TERM: Chemical Exposure; Coal; Environmental Factors; Mining; Occupation
SITE: Lung
CHEM: Diesel Exhaust; PAH
OCCU: Miners, Coal
TIME: 1984 - 1994

41   McCredie, M.R.E.   04330
NSW Central Cancer Registry, Macquarie Hosp., Wicks Rd, P.O. Box 380, North Ryde, Sydney NSW 2113, Australia (Tel.: (02)8875638; Fax: 8887210)
COLL: Ford, J.M.

Case-Control Study of Childhood Brain Tumours in New South Wales
A case-control study of childhood brain tumours is being conducted in New South Wales (NSW). Potential cases comprise all children aged 0-19 years diagnosed in 1985-1989, reported to the NSW Central Cancer Registry and living in the Sydney Metropolitan area, Newcastle, Wollongong or the Australian Capital Territory. Two controls for each case, matched for age and sex are being obtained from the electoral rolls. Mothers and fathers of 100 cases and 200 controls are to be interviewed with the standard SEARCH questionnaires. Questions relate to parental and childhood exposures to N-nitroso compounds, ionizing radiation, genetic predisposition, head trauma, barbiturate consumption and parental occupational exposures.

AUSTRALIA

TYPE: Case-Control
TERM: Childhood; Drugs; Intra-Uterine Exposure; Radiation, Ionizing; Registry; Trauma
SITE: Brain
CHEM: Barbiturates; N-Nitroso Compounds
REGI: NSW (Aus)
TIME: 1988 - 1992

42 McCredie, M.R.E. 04554
NSW Central Cancer Registry, Macquarie Hosp., Wicks Rd, P.O. Box 380, North Ryde, Sydney NSW 2113, Australia (Tel.: (02)8875638; Fax: 8887210)
COLL: McLaughlin, J.K.; Jensen, O.M.; Adami, H.O.; Wahrendorf, J.

Case-Control Study of Renal Adenocarcinoma

The main aims of this study are to confirm an increased risk of renal adenocarcinoma associated with use of diuretics, to re-evaluate the effect of analgesics (particularly paracetamol); and to investigate the relationship with occupation in the petroleum industry or serving in the Armed Forces. Cases will comprise all persons aged 20-79 years diagnosed in 1989-1990 with renal adenocarcinoma, reported to NSW Central Cancer Registry. Controls from the electoral rolls will be frequency-matched to the cases. 600 cases and 600 controls will be interviewed by telephone over the two-year period using a standard questionnaire. Questions relate to use of analgesics, diuretics, oestrogens and diet pills; tobacco use and passive smoking; consumption of alcohol and coffee; reproductive history; medical history relating to diseases of the kidney, heart and thyroid; diabetes and other cancers; occupation (current and usual). Standard methods for analysis of case-control studies will be used. In addition to a seperate analysis of this study a pooled analysis of the data from all centres is envisaged.

TYPE: Case-Control
TERM: Alcohol; Analgesics; Coffee; Diabetes; Drugs; Hormones; Obesity; Occupation; Passive Smoking; Registry; Reproductive Factors; Tobacco (Smoking)
SITE: Kidney
CHEM: Oestrogens; Paracetamol
OCCU: Petroleum Workers
LOCA: Australia; Denmark; Germany; Sweden; United States of America
REGI: NSW (Aus)
TIME: 1989 - 1992

43 Shaw, H.M. 04713
Univ. of Sydney, Royal Prince Alfred Hosp., Melanoma Unit, Dept. of Surger, Missenden Rd, Camperdown, Sydney NSW 2050, Australia (Tel.: (02)5167156)
COLL: Kefford, R.F.; McCarthy, W.H.

Molecular Genetics of Melanoma

The hypotheses being tested are that (1) hereditary melanoma and the dysplastic naevus syndrome are pleiotropic effects of an autosomally inherited gene, or gene defect, located on the short arm of chromosome 1 (1p), and (2) that sporadic (non-hereditary) melanoma is associated with somatic defects on the same gene. This study aims to test the above hypotheses by (1) documentation of the incidence of melanoma and dysplastic naevus syndrome in 300 kindreds with a family history of melanoma and (2) genetic linkage analysis of a carefully documented group of Australian kindreds affected by hereditary melanoma/dysplastic naevus syndrome using RFLP analysis of the inheritance within those kindreds of alleles detected by a series of probes mapped to the short arm of chromosome 1, and (3) analysis of loss of constitutional heterozygocity in sporadic and familial melanoma tumour samples using this analysis on chromosome 1p.

TYPE: Genetic Epidemiology
TERM: Genetic Markers; Heredity; RFLP
SITE: Melanoma
TIME: 1985 - 1993

AUSTRIA

GRAZ

* 44   Marth, E.   05119
   Hygiene-Inst. der Karl-Franzens Univ., Universitätsplatz 4, 8010 Graz, Austria (Tel.: (0316)3804360; Fax: 382218)
COLL: Pfeiffer, K.P.; Möse, J.R.; Köck, M.; Pichler-Semmelrock, F.

Environmental, Dietary and Social Factors in Regional Cancer Mortality in Styria

This is an analysis of cancer mortality between 1978 and 1987 in small regions of Styria, Austria, with different respiratory cancer mortality rates in regions with substantial environmental pollution (air and/or water pollution), compared with small regions without substantial environmental pollution. In some regions the respiratory cancer mortality rate was twice as high as in regions without substantial environmental pollution (Zbl. Bakt. Hyg. 1991). Starting from these results and considering additional socio-economic data in the next few years, studies about the multifactorial genesis of cancer will be conducted, which also consider the social situation and food habits together with environmental conditions. With multivariate statistical methods the importance of single factors and the interaction between these factors will be studied.

TYPE: Correlation; Mortality
TERM: Air Pollution; Diet; Environmental Factors; Socio-Economic Factors; Water
SITE: All Sites
TIME: 1990 - 1993

VIENNA

45   Karner-Hanusch, J.   04721
   Univ. of Vienna, Dept. of Surgery, Alserstr. 4, 1097 Vienna, Austria (Tel.: 40400/2243)
COLL: Roth, E.; Hufnagl, A.

Genetic Alterations in Patients with Colorectal Tumours

The aim is to monitor genetic alterations in colorectal tumours in order to examine allelic losses in (1) patients with spontaneous tumours; (2) patients with a positive family history of colorectal tumours; (3) patients with adenomas. A second objective is to identify genetic alterations in normal colonic mucosae in the three groups in order to reveal chromosomal defects even in normal mucosae. Peripheral blood leucocytes are used as a source of normal DNA. This should make it possible to detect patients at risk of developing colorectal cancer. Tumour tissue, normal mucosa and blood from every patient operated on for colorectal tumours are analysed for RFLP. 20 samples from which DNA has been successfully collected have been taken to date.

TYPE: Genetic Epidemiology
TERM: Chromosome Effects; DNA; Familial Factors; RFLP
SITE: Colon; Rectum
TIME: 1990 - 1995

46   Neuberger, M.   04755
   Univ. of Vienna, Inst. of Environmental Hygiene, Dept. of Preventive Medicine, Kinderspitalgasse 15, 1090 Vienna, Austria (Tel.: (222)431595)
COLL: Haider, M.; Kundi, M.

Individual Asbestos Exposure, Smoking and Mortality

In the oldest asbestos cement factory of the world, a historical prospective cohort study started in 1973 including all persons employed in production in 1950-1981 for at least three years. From 2,816 persons eligible for the study, factory records, measurements of dust and fibres and interview-based smoking histories were used to estimate exposures over time. Underlying causes of death are compared with regional and national rates. In addition, best available diagnoses from autopsy records, hospital records and other sources are collected. SMR and life-table methods are used to calculate lung cancer risks compared to the general population and between different exposure groups within the cohort. Nested case-control studies are used for analysing details of exposure such as fibre type. A paper was published in Br. J. Indust. Med. 47:615-620, 1990.

AUSTRIA

TYPE: Case-Control; Cohort
TERM: Clinical Records; Dusts; Occupation; Tobacco (Smoking)
SITE: All Sites; Lung; Mesothelioma
CHEM: Asbestos, Chrysotile; Asbestos, Crocidolite; Mineral Fibres
OCCU: Asbestos Workers
TIME: 1973 - 1993

WIENER NEUSTADT

47 Niessner, H. 04851
General Hosp. Wiener Neustadt, Dept. of Internal Medicine, Corvinusing 3-5 , 2700 Wiener Neustadt, Austria (Tel.: (02622)23521)
COLL: Schulte-Hermann, R.

Haemopoietic Malignancies in Wiener Neustadt County
The aims are to demonstrate whether there is an increase in incidence of haemopoietic malignancies in Wiener Neustadt county in comparison with the neighbouring and other counties in Austria, and to conduct a case-control study to examine whether there is any association between haemopoietic malignancies and certain environmental factors typical for this industrialised county (water and soil pollutants). The study group will include some 200 cases with haemopoietic malignancies over five years; the two control groups (one from the hospital population of Wiener Neustadt, the other from distant counties with different types of environmental pollution) will both be of the same size. Data collection of treated haemopoietic malignancy cases will be carried out in haematology departments of Wiener Neustadt and the neighbouring hospitals (patients resident in Wiener Neustadt town and county), with the help of data from the Austrian Cancer Register. It is planned to expand this project into a breast and colorectal cancer study in collaboration with Hungary, CSFR (Slovakia) and Yugoslavia (Slovenian Cancer Study Group).

TYPE: Case-Control; Correlation; Incidence
TERM: Soil; Water
SITE: Haemopoietic
CHEM: Hydrocarbons, Halogenated
TIME: 1989 - 1992

BELGIUM

ANTWERP

48 Engels, H. 04619
Univ. of Antwerp, Dept. of Epidemiology, and Community Medicine, Universiteitsplein 1 , 2610 Antwerp, Belgium (Tel.: (03)8202523; Fax: 8202640 ; Tlx: 33646)
COLL: Eylenbosch, W.J.; Weyler, J.; Van Marck, E.; Ramael, M.; Temmerman; Nyong'o, A.

Genital HPV Infection and Cervical Cancer
Prevalence of genital HPV-infection is being estimated among 1,000 women participating in a cervical screening programme, and among 500 women at risk for sexually transmitted diseases (STDs). Different detection methods are used: physical examination, colposcopy, cytology and HPV-DNA detection techniques (PCR, In Situ Hybridisation). Prevalence will be compared by method of detection. A case-control study will also be conducted in Nairobi, Kenya. The hypothesis that HPV infection is a more important risk factor for precancerous lesions than other STDs (HSV, Chlamydia T., Gonorrhoea, Syphilis) will be tested. 100 cases (women with abnormal PAP smears) and 200 controls will be recruited from women participating in a cervical cancer screening programme in city council clinics and STD clinics in Nairobi. A feasibility study is now being carried out.

TYPE: Case-Control; Cross-Sectional
TERM: Cytology; HPV; HSV; PCR; Prevalence; Screening; Sexually Transmitted Diseases
SITE: Uterus (Cervix)
LOCA: Belgium; Kenya
REGI: Belgium (Bel)
TIME: 1988 - 1993

BRUSSELS

* 49 De Quint, P. 05105
Free Univ. of Brussels, Faculty of Medicine and Pharmacy, Sect. of Public Health, Laarbeeklaan 103, 1090 Brussels, Belgium (Tel.: (02)4774213; Fax: 4774311)
COLL: Depoorter, A.M.

Parental Occupational Exposures and the Development of Cancer in the Offspring
The aim of the study is to investigate the association between parental occupational exposures and the incidence of leukaemia, lymphoma, brain tumours and neuroblastoma in their offspring. The cases and the controls are recruited in two large hospitals for children in Brussels. The case group includes children less than 15 years of age with newly diagnosed tumours. The controls are surgery patients, pair-matched to the cases on sex and age at diagnosis. A face-to-face interview of the parents is used to assess information on the personal and familial antecedents of the index child, the family habits, their living environment and the occupations and occupational exposures of both parents. Information on the occupational exposures is also obtained from occupational medical officers, specialists in industrial techniques and industrial hygienists. At least 50 cases and 50 controls will be included in the study. A conditional logistic regression model for matched sets will be used to analyse the data.

TYPE: Case-Control
TERM: Childhood; Parental Occupation; Registry
SITE: Brain; Leukaemia; Lymphoma; Neuroblastoma
REGI: Belgium (Bel)
TIME: 1988 - 1992

50 Lechat, M.F. 03936
Catholic Univ. of Louvain, Dept. of Epidemiology, School of Public Health, Clos Chapelle-aux-Champs 30, UCL 30.34, 1200 Brussels, Belgium (Tel.: (02)215878)
COLL: De Wals, P.; Weatherall, J.W.; Beckers, R.; Borlee, I.; Goujard, J.; Stoll, C.; Karkut, G.; Lillis, D.F.; Radic, A.; Calabro, A.; Calzolari, E.; Galanti, C.; Hansen-Koenig, D.; Ten Kate, L.P.; Stone, D.; Harris, F.; Nevin, N.; Pexieder, T.; Svel, I.; Laurence, K.; Tenconi, R.; Ayme, S.; Cuschieri, A.; Mastroiacovo, P.

EUROCAT - Registration of Congenital Anomalies
The long term objective of the EUROCAT project is to test the feasibility of carrying out epidemiological surveillance in the countries of the EEC taking congenital anomalies as an example. The specific objectives of the study are: (1) to establish in each country of the EEC one or several regional registers of

BELGIUM

congenital anomalies providing reliable epidemiological information on the occurrence of the registered conditions in the progeny of the population living in a defined geographical area; (2) to harmonize the methods of diagnosis and of data collection in order to ensure the comparability of statistics between participating centres; (3) to monitor the occurrence rate of congenital anomalies in the progeny of different population groups in order to identify any expected frequency; (4) to investigate the risk associated with possible teratogenic factors; (5) to create in each country an area where reporting is reliable, so that base line rates are available for calibrating any information system established at national level for the detection of adverse environmental influences; (6) to evaluate the effectiveness and efficiency of screening programmes, of preventive measures, and of treatment methods; (7) to provide a well documented set of cases recorded in a defined population for clinical research. (EUROCAT Report No. 2, Surveillance of Congenital Anomalies. Years 1980-1984 De Wals, P. and Lechat, M.F. (eds), Dept. of Epidemiology, Chatolic University of Louvain, Brussels, 1987.)

TYPE: Registry
TERM: Chemical Exposure; Congenital Abnormalities; Mutation, Germinal; Mutation, Somatic; Prevention; Registry; Screening
SITE: Inapplicable
LOCA: Belgium; Denmark; France; Germany; Ireland; Italy; Luxembourg; Netherlands; Switzerland; United Kingdom; Yugoslavia
TIME: 1979 - 1993

51 Van den Oever, R. 03141
National Confederation, of Christian Sickness Funds, Medical Direction, Wetstr. 121, 1040 Brussels, Belgium
COLL: Van den Berghe, H.; Lahaye, D.

Detection of Occupational Cancer Based on the Health Insurance Cancer Registry
Of the total of 10 million Belgians, 46% are affiliated with the Confederation of Christian Sickness Funds (CCSF) where a cancer morbidity register has been kept since 1963. Studies of the complete occupational history of each new case of cancer of the nasal cavity and sinuses and of mesothelioma reported each year to the CCSF registry have been carried out since 1978 (in 1982 a total of 155 nasal cancers and 104 mesotheliomas). The relationship between these malignant neoplasms and exposure to occupational and extra-occupational risks has been shown in numerous published reports. Since 1980 a total of 532 new cases of cancer of the larynx have been included in the study to detect occupational influences (e.g., asbestos) as opposed to alcohol and tobacco use. Occupational exposures to arsenic, asbestos, asphalt, chromates, mineral oil, nickel and wood dust are specifically investigated. Detailed clinical information on histology, site, stage and treatment of the cancer is obtained from the hospital and the attending physicians. If occupational exposure to a carcinogen is shown, an application is made to the Belgian Fund of Occupational Diseases for compensation. The results of these continuing epidemiological investigations are published for each type of cancer separately at regular intervals.

TYPE: Incidence
TERM: Alcohol; Dusts; Histology; Metals; Occupation; Registry; Stage; Tobacco (Smoking); Treatment; Wood
SITE: Larynx; Mesothelioma; Nasopharynx
CHEM: Arsenic; Asbestos; Chromium; Mineral Oil; Nickel; PAH; Tars
TIME: 1978 - 1993

GHENT

* 52 Bleyen, L.J. 05044
State Univ. of Ghent, Dept. of Hygiene and Social Medicine, Block A , De Pintelaan 185, 9000 Ghent, Belgium (Tel.: (091)403636; Fax: 404994)
COLL: de Backer, G.; Roels, H.; Vandevelde, E.

Feasibility and Efficiency of Breast Cancer Screening
The objectives of this prospective cohort study in Ghent are: (a) to assess whether screening for breast cancer is as feasible and efficient using the existing infrastructure (private and state), in order to obtain a given reduction in mortality, as with the set-up of a special unit; (b) a study of risk factors in breast cancer aetiology (weight, height, Quetelet-index and body-surface, age at menarche, children and age at each birth, breastfeeding, personal and family history, mammographical breast pattern, etc.); (c) the development of measures and strategies, in order to increase the attendance rate and to maintain

BELGIUM

compliance with the programme; and (d) the contribution of breast examination and breast self-examination towards mammography, corrected for age, and the impact on interval cancer mortality. The target population consists of 26,000 women aged 25-39, and 42,000 women aged 40-69. The control group will either be a historical control, or the female population aged 25-69 in another city with similar characteristics. The data collection will be through specially designed registration forms, surveys, and through the National Mortality Statistics and the National Cancer Registry.

TYPE: Cohort; Methodology
TERM: Lactation; Menarche; Menopause; Parity; Physical Factors; Registry; Screening
SITE: Breast (F)
REGI: Belgium (Bel)
TIME: 1991 - 1998

LEUVEN

* 53   **Kesteloot, H.**   05197
University Hospital St. Rafaël, School of Public Health, Dept. of Epidemiology, Capucijnenvoer 33, 3000 Leuven, Belgium (Tel.: +32 16 216894; Fax: 215500 ; Tlx: 24484 AZRAFL)
COLL: Lesaffre, E.; Joossens, J.V.

Relationship between Nutrients and Cancer
This is a study of the relationship between macro-nutrients and organ-specific and total cancer mortality. Statistical methods are being used to study the relations involved, both univariate and multivariate. The studies are both cross-sectional and longitudinal and use data obtained from a total population of more than 1 billion people. Use is made of the Leuven Mortality Monitoring System and of data provided by FAO and other sources. Each country is compared with other countries (which can be considered as control or reference populations). Changes in nutrient intake are also related to changes in organ-specific cancer mortality. Special attention is given to the role of fat (saturated, mono-unsaturated, poly-unsaturated) for all cancers and to the role of salt in stomach cancer in particular. Papers have been published in Cancer Causes and Control 2:79-83, 1991 and Prev. Med. 20:226-236, 1991.

TYPE: Correlation; Mortality
TERM: Diet; Fat; Nutrition
SITE: All Sites
TIME: 1987 - 1993

54   **Van den Berghe, H.**   03590
Univ. of Louvain, Center for Human Genetics, Gasthuisberg Campus 0 en N , Herestr. 49, 3000 Leuven, Belgium (Tel.: +32 16 215878; Fax: 215992)
COLL: Cassiman, J.J.; Fryns, J.P.; Vlietinck, R.; Oosterlinck, A.; David, G.; Van Leeuwen, F.E.; Marynen, P.; Mecucci, C.; Vercauteren, P.; van de Ven, W.; Dalcin, P.

Cancer Genetics and Cytogenetics
This project involves: (1) A cytogenetic investigation of patients with leukaemia, preleukaemia, lymphoma and allied disorders, as well as solid tumours. Specific or non-randomly occurring chromosome aberrations in the malignant cells are being identified and the relation between the chromosome pattern observed and the professional activities and/or exposure of the individuals with the disease studied. 1,500 patients are being studied per year. A control group has not been set up so far; (2) Assessment of the relation between the observed chromosome aberrations and oncogene expression; (3) Monitoring with cytogenetic techniques of workers exposed to ionizing radiation (chromosome breakage and SCE); (4) A computer program to score SCE and statistical treatment of the data; (5) A semi-automated cytogenetic analysis allowing for large-scale investigations. Recent papers appeared in Cancer Genet. Cytogenet. 25:233-245, 1987, 26:5-13 and 51-58, 1987 and in "Leukaemia", (J.A. Whittaker and I.W. Delamore, (eds)), Blackwell Scientific Publications, London, Boston, Melbourne, 1987, pp. 137-151.

BELGIUM

TYPE: Molecular Epidemiology
TERM: Chromosome Effects; Monitoring; Mutation, Somatic; Occupation; Radiation, Ionizing; SCE
SITE: All Sites; Leukaemia; Lymphoma
TIME: 1970 - 1993

MOL

* 55   **Laleman, G.R.**   05114
S.C.K./C.E.N., Dept. of Medicine, Boeretang 200, 2400 Mol, Belgium (Tel.: (0032)14332808; Fax: 14315021 ; Tlx: 31922 ATOMOL)
COLL: Holmstock, L.; Van Mieghem, E.; Haelterman, M.; Van Damme, K.

Cancer Risk among Workers Employed at a Nuclear Research Centre

The primary objective of this retrospective cohort study with a nested case-control study is to examine whether an excess in cancer morbidity and mortality can be observed among 4,300 workers employed between 1954 and 1990 at the Nuclear Research Centre (S.C.K./C.E.N.) in Mol. The second objective is to examine whether an excess in leukaemia or lymphoma morbidity and mortality can be observed among the children of those workers. The personnel register and the medical record will be used to define the study cohort and to collect data on exposure to low doses of ionising radiation. Additional data (e.g. vital status) will be gathered through questionnaires, home visits and contacting physicians (exact data on morbidity and mortality). The experience of the cohort will be compared with figures from the National Institute of Statistics, the National Cancer Register and eventually morbidity and mortality registers of neighbouring countries.

TYPE: Cohort
TERM: Childhood; Occupation; Radiation, Ionizing; Registry
SITE: All Sites; Leukaemia; Lymphoma
OCCU: Power Plant Workers
REGI: Belgium (Bel)
TIME: 1991 - 1995

BRAZIL

MARINGA

* 56 Sichieri, R. 05207
      Univ. Estadual de Maringa, Av. Colombo, 3690, 87100 Maringa, Brazil (Tel.: (0442)262727; Fax: 222754)
      COLL: Everhart, J.

Anthropometric Measures and Diet as Risk Factors for Chronic Diseases: an Ecological Study
This study comprises an ecological analysis of anthropometric and dietary data from a national survey conducted in 1974 among 55,000 households in the 25 state capitals of Brazil, as predictors for cancer mortality of the following sites: stomach, oesophagus, lung, breast and large intestine. Dietary intake was measured for a period of seven days. Units of analysis will be the 25 capitals. For each state, age-adjusted mortality rates and proportional mortality are being calculated for the age-group 30 years and older. Predictive variables include height, body mass index, and the dietary factors fat composition (animal and vegetable), amount of cereals, type of staple food, consumption of sugar, alcohol, vitamin A, vitamin C, calcium, meat, fish. Measures of association between the predictors of diet or body composition and the outcome variables will be based on Spearman correlation coefficients.

TYPE: Correlation
TERM: Alcohol; Diet; Fat; Physical Factors; Vitamins
SITE: Breast (F); Colon; Lung; Oesophagus; Stomach
CHEM: Calcium
TIME: 1991 - 1993

RIO DE JANEIRO

* 57 Pinto, C.B. 05080
      Oswaldo Cruz Foundation, Health Information Centre, Av. Brasil 4365, Manguinhos, 21040 Rio de Janeiro, Brazil (Tel.: (021)2901696; Fax: 5909741 ; Tlx: 23239)
      COLL: Szwarcwald, C.L.; Castilho, E.A.

Cancer Mortality in Rio de Janeiro
The main purposes of this study are (1) to describe cancer mortality for sites of the digestive tract and (2) to analyse regional differentials. Average annual age- and sex-specific mortality rates per 100,000 were calculated for the period 1979-1981 for each cancer site and for five-year age groups up to 79 years and for 80 years and above, for each of the 64 municipalities of the state. Age-standardised rates, cumulative rates and years of life lost were calculated. Standardised mortality ratios (SMR) were also calculated to compare different geographical areas. Factor and cluster analysis will be used. As soon as population data for 1990 are available from the census, temporal trends will be evaluated.

TYPE: Mortality
TERM: Cluster; Geographic Factors; Time Factors; Trends
SITE: Gastrointestinal
TIME: 1989 - 1992

* 58 Pinto, C.B. 05081
      Oswaldo Cruz Foundation, Health Information Centre, Av. Brasil 4365, Manguinhos, 21040 Rio de Janeiro, Brazil (Tel.: (021)2901696; Fax: 5909741 ; Tlx: 23239)
      COLL: Szwarcwald, C.L.; Castilho, E.A.

Cancer Mortality in Brazilian Metropolitan Areas
Brazil is the fifth largest country in the world, with a wide variety of socio-economic conditions, and national morbidity and mortality data conceal a considerable range of regional variations for many diseases, including cancer. There are nine so-called "metropolitan regions" consisting of the capitals of nine states and their neighbouring municipalities. These regions are supposed to have more reliable mortality data and to have socio-demographic characteristics representative of their regions. The objective of this study is to describe the most important types of cancer for each region. Sex-specific age-standardised (world) mortality rates (ASR) per 100,000 will be calculated for the period 1977-1987, in order to analyse temporal trends. Some other variables from death certification, such as place of birth, may also be analysed, to assess any effect of migration.

BRAZIL

TYPE: Mortality
TERM: Trends
SITE: Breast (F); Colon; Liver; Lung; Oesophagus; Prostate; Rectum; Stomach; Uterus (Cervix)
TIME: 1989 – 1992

SAO PAULO

59 Brasilino de Carvalho, M. 04103
Heliopolis Hosp., Dept. of Head & Neck Surgery, Rua Abilio Soares 639, Apto 93-B, 04005 Sao Paulo, Brazil (Tel.: 8844595)
COLL: Sobrinho, J.A.; Franco, E.L.F.; Rapoport, A.; Fava, A.S.; Gois Filho, J.F.; Chagas, J.F.S.; Kowalski, L.P.; Kanda, J.L.

Relationship between Cancer of the Upper Aero-Digestive Tract and Consumption of Alcohol and Tobacco

Although the relation between the role of alcohol and tobacco consumption in the development of cancer of the upper aero-digestive tract is today well established, the carcinogenic mechanisms of these agents remain obscure. The objective of this study is to follow up, for a period of ten years, asymptomatic subjects belonging to a high risk group due to their alcohol and tobacco habits, and to identify the various factors directly or indirectly related to the development of tumours. 300 chronic alcoholics, smokers or non-smokers with no symptoms of disease of the upper aero-digestive tract, will be examined every six months at the Department of the Head and Neck Surgery of the Hospital Heliopolis in Sao Paulo, Brazil. At the first examination a specific questionnaire will be completed and at each subsequent visit new data will be reported. A physical examination will also be carried out. If a neoplastic lesion is found the patient will be treated. If the examination is normal, a cytological smear of the mouth and oropharynx will be taken. At the end of the ten-year period an evaluation of the influence of alcohol and tobacco on the development of neoplasms at the various sites will be made, as an assessment of the influence of initiation and duration of consumption.

TYPE: Cohort; Incidence
TERM: Alcohol; Cytology; High-Risk Groups; Tobacco (Smoking)
SITE: Oral Cavity; Oropharynx
TIME: 1986 – 1996

CANADA

EDMONTON

60 Fincham, S.F. 03916
Alberta Cancer Board, Div. of Epidemiology, and Preventive Oncology, 9707-110th St., Edmonton Alberta, Canada T5K 2L9 (Tel.: (403)4829375; Fax: 4887809)
COLL: Berkel, J.

Occupation as a Factor in Alberta Cancer Patients

A long-term study programme has been established to monitor cancer incidence by occupation, using the Alberta Cancer Registry as a source of cases. A detailed occupational and exposure history, demographic data and histories of smoking and alcohol consumption are routinely collected, using a self-reporting questionnaire. Subjects are all new cancer registrants, aged 25-74, in Alberta. Data are analysed annually, using descriptive and analytical statistics for effect estimation and control of confounders, for all sites.

TYPE: Incidence
TERM: Alcohol; Monitoring; Occupation; Registry; Tobacco (Smoking)
SITE: All Sites
REGI: Alberta (Can)
TIME: 1983 - 1993

61 Fincham, S.M. 02142
Alberta Cancer Board, Dept. of Epidemiology, & Preventive Oncology, 9707-110th St., Edmonton Alberta, Canada T5K 2L9 (Tel.: (403)4829375)
COLL: Hanson, J.

Occupational Diseases in Alberta Workers Exposed to Chemicals

A data base was established on a cohort of 5,777 Alberta workers occupationally exposed to chemicals and 1,645 workers not exposed. The data file will be matched periodically with the Alberta Cancer Registry using record-linkage techniques. Registry linkage in 1988 permitted cancer incidence risk estimation after a duration of 10 years and linkage with provincial vital statistics files for other causes of death is planned.

TYPE: Cohort
TERM: Chemical Exposure; Occupation; Record Linkage; Registry
SITE: All Sites
REGI: Alberta (Can)
TIME: 1978 - 1992

62 Lees, A.W. 00353
Cross Cancer Inst., Breast Unit, 11560 University Ave., Edmonton Alberta, Canada T6G 1Z2 (Tel.: (403)4328518; Fax: 4920884)
COLL: Jensen, J.; de Metz, C.; Krause, B.E.; Allan, S.; Starreveld, A.A.; Berkel, J.

Breast Unit in Northern Alberta

The Breast Unit has 7000 patients on its data base and is accruing 500 patients a year. Epidemiological, treatment and survival data are recorded. A link has been established with the molecular genetic laboratory to study oncogenes and a repository of frozen tissue started. Studies planned or in progress are MRI spectroscopy of advanced breast cancers and monoclonal antibody imaging for staging of patients. The group will be closely linked with a province wide screening program directed to women aged 50-69, and a multicentre, multiprovince study on the effect of dietary fat reduction on the incidence of mammographic dysplasia and breast cancer. An Alberta study of in situ carcinoma of the breast has been completed and one on the effect of risk factors on survival. Papers were published in Annals of Surgery, (Temple, W.); Breast Cancer Research and Treatment, 210, 653-657, 1989. Patients continue to be entered into NSABP, NCIC adjuvant studies and innovative Phase 1, 2 and 3 studies of treatment of reccurrent cancer. The next international symposium is planned for 1993 in conjunction with the International Association for Breast Cancer Research.

CANADA

TYPE: Case Series
TERM: BMB; Fat; Oncogenes; Prognosis; Survival; Tissue; Treatment
SITE: Breast (F)
TIME: 1982 - 1993

HAMILTON

63   Walter, S.D.   04513
McMaster Univ., Health Science Centre, Room 3H4, 1200 Main St. W., Hamilton Ontario, Canada L8N 3Z5 (Tel.: (416)5259140; Fax: 5770017 ; Tlx: 0618347)
COLL: Taylor, S.M.; Davies, J.A.; Marrett, L.D.; Drake, J.J.; Hayes, M.V.

Geographical Variation in Ontario Cancer Rates
Site-specific and total cancer incidence and mortality data for Ontario in the period 1964-86 will be examined for spatial and temporal variation. The accuracy and completeness of address registration in the population-based provincial registry will be evaluated using capture-recapture and related methodology. Various methods of address imputation and correction will be considered, together with an assessment of their effects on spatial variation in cancer rates across the province. Several statistical indices will be computed to measure the degree of spatial auto-correlation in the data, and the persistence of spatial auto-correlation over time. Finally, selected ecologic comparisons will be made between cancer rates, demographic profiles, and environmental variables derived from the census and other data sources; the comparisons will use weighted maximum likelihood regression, taking into account systematic and statistical variation between small areas of the region.

TYPE: Correlation; Incidence; Methodology; Mortality
TERM: Geographic Factors; Registry
SITE: All Sites
REGI: Ontario (Can)
TIME: 1988 - 1992

LAVAL-DES-RAPIDES

64   Franco, E.L.F.   04232
Univ. of Quebec, Inst. Armand-Frappier, Dept. of Epidemiology, 531 Blvd des Prairies , Laval-des-Rapides Québec, Canada H7V 1B7 (Tel.: (514)6875010; Fax: 6865501)
COLL: Kowalski, L.P.; Curado, M.P.; Oliveira, B.V.; Fava, A.S.; Kanda, J.L.

Risk Factors for Upper Respiratory and Digestive System Cancers in Brazil
A case-control study has been used to analyse risk factors for upper aero-digestive system cancers in three metropolitan areas in Southeastern Brazil (Sao Paulo, Curitiba, Goiania). Between February 1986 and December 1988, complete interviews have been obtained from 784 cases and 1,568 age- and sex-matched hospital controls. The distribution of cases according to anatomical site was follows: 67 lip, 306 oral cavity, 217 pharynx and 194 larynx cancers. Preliminary analyses for the effect to tobacco and alcohol and definitive results for cancers of the oral cavity have been published (Rev. Bras. Cir. Cab. Pesc. 11:23-33, 1987; Int. J. Cancer 43:992-1000, 1989). The substantive analyses for the other sites are currently being performed. A cohort of patients admitted to one of the participating hospitals in Sao Paulo has been actively followed up to monitor for the occurrence of second tumours and their determinants.

TYPE: Case-Control
TERM: Alcohol; Diet; Multiple Primary; Tobacco (Smoking)
SITE: Larynx; Lip; Oral Cavity; Pharynx
TIME: 1987 - 1992

65   Franco, E.L.F.   04233
Univ. of Quebec, Inst. Armand-Frappier, Dept. of Epidemiology, 531 Blvd des Prairies , Laval-des-Rapides Québec, Canada H7V 1B7 (Tel.: (514)6875010; Fax: 6865501)
COLL: Camargo, B.; Lopes, L.F.; Barreto, J.H.S.; Johnsson, R.R.; Mauad, M.A.; Saba, L.M.B.

Risk Factors for Wilms' Tumour in Brazil
The aim was to investigate risk factors for Wilms' tumour in different areas in Brazil as a parallel project to the Brazilian Cooperative Group for the Treatment of Wilms' Tumour (GCBTTW). Hypotheses to be investigated are: parental occupation with particular emphasis on chemical and crude oil refinery work,

maternal exposure to medications from the most common pharmacological groups during pregnancy, food and beverage consumption for selected items (carotene-containing vegetables and fruits, tea and coffee, and alcoholic beverages), smoking history, history of common infectious diseases, herbicide exposure, and general demographic and socioeconomic information. The study uses a case-control design which identifies prospectively all incident cases of histologically confirmed Wilms' tumour in three large metropolitan areas in Brazil during a period of two years. Two hospital controls were identified for each case matched for age (+/- one year), sex, hospital catchment area, and trimester of admission. Controls were children without neoplastic diseases attending both outpatient and inpatient clinics in hospitals in the three areas. Between April 1987 and January 1989 complete interviews were obtained with the parents of 112 cases and 224 controls. Ninety-seven of the Wilms' tumour patients (86.6%) were identified through the GCBTTW trial. In-depth histopathologic typing and staging have been completed and the substantive analysis of risk factor information is being currently performed. The analysis of epidemiological correlates of genetic characteristics of Wilms' tumour patients will be published in 1991.

TYPE: Case-Control
TERM: Age; Alcohol; Chemical Exposure; Coffee; Diet; Drugs; Fruit; Herbicides; Infection; Occupation; Socio-Economic Factors; Tea; Tobacco (Smoking); Vegetables
SITE: Wilms' Tumour
OCCU: Chemical Industry Workers; Dyestuff Workers; Leather Workers; Petroleum Workers
LOCA: Brazil
TIME: 1987 - 1992

66   Siemiatycki, J.   02469
Inst. Armand-Frappier, Univ. du Québec, 531, Blvd des Prairies, Laval-des-Rapides Québec, Canada H7V 1B7 (Tel.: (514)6875010; Fax: 6865501)
COLL: Gerin, M.; Richardson, L.

Exposure-Based Case-Control Approach to Discovering Occupational Carcinogens
In an investigation of associations between cancer and environmental exposures in the workplace, begun in 1979, all male patients aged 35-70 in the Montreal area with newly diagnosed cancers of certain sites, and a population control series, were interviewed by specially trained interviewers who recorded all the subjects' jobs and industries and detailed descriptions of tasks performed and materials manipulated. Information on smoking, environment and diet was also requested. The job descriptions were assessed by specially trained chemists and hygienists, who listed all the chemicals to which the subject may have been exposed using a checklist, intended to include the most common exposures in the Montreal industrial environment, which contains 275 potential occupational exposures. For each substance judged to have been present in a given job, the chemists provided semi-quantitative (low, medium, or high) indications of the level and frequency of exposure, and their confidence that the person was exposed (possible, probable, definite). For analysis, cancers at each site constitute a case group, and can be compared with two control groups: the population controls and the other types of cancer. It is possible to analyse relative risks between any of the 14 sites of cancer and the 275 chemical exposures taking into account a host of potential confounding factors. Over 4,000 subjects were interviewed. In-depth statistical analyses have thus far been carried out in relation to 50 substances, including organic and inorganic dusts, petroleum-derived substances, PAH's and formaldehyde. Recent papers have been published in Int. J. Cancer 44:53-58, 1989 and in Am. J. Ind. Med. 16:547-567, 1989

TYPE: Case-Control; Methodology
TERM: Alcohol; Chemical Exposure; Diet; Dusts; Environmental Factors; Metals; Occupation; Petroleum Products; Plastics; Socio-Economic Factors; Solvents; Tobacco (Smoking); Wood
SITE: Bladder; Colon; Hodgkin's Disease; Kidney; Liver; Lung; Melanoma; Non-Hodgkin's Lymphoma; Oesophagus; Pancreas; Prostate; Stomach; Testis
CHEM: Diesel Exhaust; Gasoline; Silica
TIME: 1978 - 1992

67   Siemiatycki, J.   04785
Inst. Armand-Frappier, Univ. du Québec, 531, Blvd des Prairies, Laval-des-Rapides Québec, Canada H7V 1B7 (Tel.: (514)6875010; Fax: 6865501)

Health Risks due to Chrysotile Asbestos in the Non-Occupational Environment
This ecological study of the effects of non-occupational exposures to chrysotile asbestos, will (1) determine whether women having lived near chrysotile asbestos mines have had excess mortality from lung diseases and asbestos-related cancers; and (2) evaluate the EPA linear risk model of lung cancer, estimating the mean cumulative exposure in women exposed and comparing the risk projected by the model with that observed at the estimated dose. The exposed population consists of 10,000 women

CANADA

> = 30 years of age, in the two asbestos areas in Quebec Province, from 1950 to 1989 (400,000 person-years). 600,000 women from other comparable regions of Quebec will serve as control population. The study will have enough power (alpha = 5%, beta = 80%) to detect SMRs of 108 for all cancers, 114 for gastric and peritoneal cancers, 127 for lung cancer and 117 for non neoplastic respiratory diseases. Data will be obtained from death certificates for 1950 to 1989, from population census figures by municipality and age for 1951 to 1986, and from a survey of residential history of 1,000 women aged +50 years. Past exposures will be estimated applying regression techniques, using information on fibre concentrations in 1984, dust measures (1972-1990), volume and granulometry of fibres retained in chimney filters in asbestos mills, computer simulation of aerosol dispersion, and eye-witnesses. Using these data, five specialists will infer a probabilistic distribution of probable past atmospheric concentrations.

TYPE: Incidence; Mortality
TERM: Dusts; Environmental Factors; Female
SITE: Gastrointestinal; Lung; Mesothelioma
CHEM: Asbestos, Chrysotile
TIME: 1989 - 1992

* 68   **Siemiatycki, J.**   05222
Inst. Armand-Frappier, Univ. du Québec, 531, Blvd des Prairies , Laval-des-Rapides Québec, Canada H7V 1B7 (Tel.: (514)6875010; Fax: 6865501)

Multivariate Regression Analyses of Occupational Risk Factors for Cancer
Among recognized human carcinogenic agents and circumstances, about half are substances which were first found to be carcinogenic in the occupational milieu. Although it it important to discover occupational carcinogens for the sake of preventing occupational cancer, the potential benefit of such discoveries goes beyond the factory walls since most occupational exposures find their way into the general environment, sometimes at higher concentrations than in the workplace. In 1979, a large case-control study was undertaken in Montreal, designed to provide evidence on the associations between hundreds of relatively common occupational exposures and many sites of cancer. Between 1979 and 1986, interviews were carried out with 3,730 cancer cases and 533 population controls. Each subject's job history was scrutinized by a team of specially trained chemists and hygienists, who inferred a list of chemical exposures. The first several years of the study were devoted to development and evaluation of methods, and to data collection. Most recently a sweeping analysis of the possible interactions between all types of cancer in the data set and 294 occupational exposures has been completed. Further analyses will be carried out by logistic regression methods and should entail a tailored site-by-site approach rather than the totally mechanised analysis that was possible at the first stage. The objective of this stage of the study is to exploit the existing data base from the case-control study to detect and characterize occupational risk factors for the 14 main types of cancer in the data base, namely: oesophagus, stomach, colon, rectum, pancreas, all lung, lung (squamous cell, oat cell, and adenocarcinoma), prostate, bladder, kidney, melanoma of skin and non-Hodgkin's lymphoma.

TYPE: Case-Control
TERM: Chemical Exposure; Occupation
SITE: All Sites
TIME: 1979 - 1993

MONTRÉAL

69   **Ghadirian, P.**   04346
Hôtel-Dieu de Montréal, Unité de Recherche en Epidémiologie, 3840, rue St-Urbain, Montréal Québec, Canada H2W 1T8 (Tel.: (514)8432742; Fax: 8496880 ; Tlx: 05562117 hdm)
COLL: Lacroix, A.; Bhat, B.; Gavino, V.

Diet and Cancers of the Breast, Colon and Prostate: Biological Markers
Breast cancer is the first common cancer among women in Canada; colorectal cancer is the second most common cancer in females and the third in males, after prostate cancer. Therefore these three nutritional-fat-related cancers make up a very large proportion of total cancer incidence and mortality rates. Although several causative factors have been suggested, the aetiology of these diseases, in particular the role of diet and nutrition are poorly understood. A total of 1,200 cases (400 cases for each cancer site) and 1,200 population-based controls chosen by random digit dialling will be studied. In addition to a retrospective study on nutrition and other risk factors, biological specimens such as blood, adipose tissue, cheek cells and toenails will be collected from both case and control groups for

CANADA

determination of: (1) fatty acid patterns in fat biopsies, red blood cells and cheek cells; (2) carotenoids in plasma, fat biopsies and cheek cells; (3) tocopherols in plasma and fat biopsies; (4) selenium in toenail clippings and glutathione peroxidase activity in red blood cells. The results of biological specimens will be compared with the information in the food frequency questionnaire. Biological specimens will be collected prior to surgery or initiation of therapy. The breast and colo-rectal surveys are part of the IARC SEARCH project and the case-control study of cancer of the prostate is part of the North American Study Group.

TYPE: Case-Control
TERM: Diet
SITE: Breast (F); Colon; Prostate
TIME: 1987 - 1992

70   **Ghys, R.**   04055
Laboratoire Radio-Médical Inc., 750 Blvd Henri-Bourassa E., Montréal Québec, Canada H2C 1E6 (Tel.: (514)3842232)

Risk Factors Involved in Breast Cancer
A complex relationship exists between chronic mammary dysplasia (fibrocystic disease) and breast cancer. A number of high-risk factors are known to increase the risk of malignant breast tumours. It is proposed to quantify these factors and to establish whether they favour carcinogenesis rather than "premalignant" dysplastic changes. It is felt that women seen in "Breast Clinics" lend themselves better to epidemiological studies comparing benign and malignant breast lesions. All patients seen at the study clinics since 1971 have complete breast examinations, including mammography and thermography. Diaphanoscopy is also routinely performed since 1980. Cytopunctures are done since 1986, using mammographic or echographic control, in all patients with severe dysplasia (15% of the cases). Presently, 80 patients are seen per week: all of them have breast "problems"; 3% of the new cases have a cancer. Patients with verified malignant or benign conditions will be compared for the following factors: family history of breast and other malignancies (close to 90% of the patients are French-Canadian, a uniquely homogeneous ethnic group for a developed country), body weight (body mass index), age at onset of menstruation and menopause, use of oral contraceptives, parity and lactation, cigarette smoking and caffeine consumption.

TYPE: Cross-Sectional
TERM: Coffee; Familial Factors; Lactation; Menopause; Menstruation; Oral Contraceptives; Parity; Physical Factors; Screening; Tobacco (Smoking)
SITE: Breast (F)
TIME: 1982 - 1992

71   **Guttmann, R.**   03638
Royal Victoria Hosp., Dept. of Transplantation, & Immunogenetics, 687 Pine Ave. W., Montréal Québec, Canada H3A 1A1 (Tel.: (514)8421231)
COLL: Hanley, J.

Relationship of Delayed Type Hypersensitivity to Cancer Risk in Immunosuppressed Renal Allograft Recipients
1,000 patients have been entered into a study that will observe over a ten year period the incidence of tumours, and viral and other infections and relate these to annual examinations and delayed hypersensitivity skin testing. The hypothesis to be tested is that anergy is a risk factor and predictor for tumour development.

TYPE: Cohort
TERM: High-Risk Groups; Immunosuppression; Infection; Transplantation; Virus
SITE: Kidney
LOCA: France; Germany; Italy; Spain; Switzerland; United States of America
TIME: 1983 - 1992

* 72   **Infante-Rivard, C.**   04951
McGill Univ., School of Occupational Health, 1130 Pine Ave. W., Montréal Québec, Canada H3A 1A3 (Tel.: (514)3984231; Fax: 3987435)
COLL: Siemiatycki, J.

CANADA

Parental Occupation and Incidence of Acute Lymphoblastic Leukaemia in Children
It has been postulated that parental exposure to solvents during the course of their work, and in particular during the months before conception for both parents, and during pregnancy for the mother, is a risk factor for the development of acute lymphoblastic leukaemia (ALL) in their offspring. To study this hypothesis, a case-control study will be carried out. Approximately 500 cases of ALL, up to nine years of age, diagnosed between 1980 and 1993 will be included. Cases from six regions of the province of Quebec, accounting for 85% to 90% of the total Quebec population, will be ascertained. Two controls per case, matched for age, will be chosen: one from all other cancers and serious haematological disorders diagnosed at the same hospital as the case, and one population control from the same geographical area as the case at time of diagnosis. Data will be collected using a telephone interview. Exposures will be coded by chemists according to the method developed by Siemiatycki. The children's exposures to pesticides and passive smoking are also of interest.

TYPE: Case-Control
TERM: Chemical Exposure; Intra-Uterine Exposure; Parental Occupation; Passive Smoking; Pesticides; Solvents
SITE: Leukaemia (ALL)
TIME: 1990 - 1994

* 73 **McDonald, J.C.** 05142
McGill Univ., School of Occupational Health, 1130 Pine Ave. West, Montréal Québec, Canada H3A 1A3 (Tel.: (514)3984238; Fax: 3988981)
COLL: Liddell, F.D.K.; McDonald, A.D.

Cancer Mortality in Quebec Chrysotile Miners and Millers
This study will extend follow-up to the end of 1989 of a cohort of over 11,000 chrysotile miners and millers, previously followed through 1975. The specific objectives are (1) to estimate exposure-response relationships between fibre and dust measurements of workers and mortality from malignant diseases at several sites including lung, pleura and peritoneum, gastrointestinal tract, kidney, and larynx, accounting for duration and intensity of exposure particularly as regards industrial process, time-related variables, and low exposure levels; (2) to examine the interaction between smoking and asbestos exposure in lung cancer; (3) to explore the relationship between laryngeal cancer mortality, alcohol consumption, smoking, and asbestos. These questions will be answered by a series of case-control studies nested within the cohort, where cases will comprise those deceased from the cause of interest and controls for each case will be selected from those who survived the case.

TYPE: Case-Control; Cohort
TERM: Alcohol; Dose-Response; Dusts; Occupation; Time Factors; Tobacco (Smoking)
SITE: Gastrointestinal; Kidney; Larynx; Lung; Peritoneum; Pleura
CHEM: Asbestos, Chrysotile; Mineral Fibres
OCCU: Miners, Asbestos
TIME: 1989 - 1992

74 **Messing, K.** 04403
Univ. du Québec à Montréal, Groupe de Recherche-Action, en Biologie du Travail, C.P. 8888, Succ. "A", Montréal Québec, Canada (Tel.: (514)2823334)
COLL: Bradley, W.E.C.; Dubeau, H.; Arsenault, A.; Dupras, G.

Frequencies and Types of Lymphocyte Mutants Induced in Humans Exposed to Ionizing Radiation
In Project 1, 60 workers in a nuclear power plant exposed to 100 mrem-5 rem annual doses of radiation, are being compared with 20 office workers at the same plant and 20 laboratory controls. HPRT-mutant T-lymphocytes are cloned and frequency determined by the Albertini method. DNA of mutant lymphocytes is examined by Southern transfer analysis to determine whether damage is typical of exposure to radiation. In Project 2, 25 nuclear medicine patients exposed to 5 mCi of Th. 201 are studied before and after exposure using the same techniques as above. A paper was published in Mutat. Res. 262:1-6, 1991.

CANADA

TYPE: Cross-Sectional; Molecular Epidemiology
TERM: DNA; HPRT-Mutants; Lymphocytes; Metals; Occupation; Radiation, Ionizing
SITE: Inapplicable
CHEM: Technetium
OCCU: Radiation Workers
TIME: 1987 - 1992

75   **Thériault, G.P.**   04502
     *McGill Univ., School of Occupational Health, 1130 Pine Ave. W., Montréal Québec, Canada H3A
     1A3 (Tel.: (514)3984227; Fax: 3987435)*
COLL: Goldberg, M.; Miller, A.B.

Long-Term Effects of to 50/60 Hz Electric and Magnetic Fields
A case/control study will be carried out to determine whether there exists an aetiological relationship between exposure to 50 and 60 hertz electric and magnetic fields and cancer among exposed workers at Electricite de France (EDF), Hydro-Quebec (HQ), and Ontario Hydro (OH). the cohort will consist of 150,000 employees from the three utilities. Cases will comprise all cancers identified in the relevant period (1978-1988 for EDF, 1970-1988 for HQ and OH). they will be ascertained through the existing records at the company (and from tumour registry data or death certificates for HQ and OH). One to four controls will be selected from the files of the corresponding elective utility, according to cancer site (or electric utility). Analysis will be performed using methods appropriate to the matched design of the study. the validity of combining the three data sets will be tested by standard epidemiological and biostatistical procedures.

TYPE: Case-Control; Cohort
TERM: Electromagnetic Fields; Occupation
SITE: All Sites
LOCA: Canada; France
TIME: 1989 - 1992

OTTAWA

76   **Ashmore, J.P.**   04520
     *Bureau of Radiation & Medical Devices, Dept. of National Health & Welfare, 775 Brookfield Rd,
     Ottawa Ontario, Canada K1A 1C1 (Tel.: (613)9546660; Fax: 9571089 ; Tlx: 0533679)*
COLL: Grogan, D.; Krewski, D.; Wigle, D.T.; Mao, Y.S.; Semenciw, R.; Fair, M.E.; Werner, M.

Mortality Study of Radiation Workers
The study involves the computer linkage of the Department of National Health and Welfare's National Dose Registry for Radiation Workers with the Mortality Data Base at Statistics Canada. The National Dose Registry is a centralised occupational radiation dose record system which includes annual dose summaries of all monitored radiation workers in Canada from 1951 to the present. Dose records originate from the Department's National Dosimetry Services and from a number of Nuclear Power facilities which perform their own dosimetry. The Registry covers 80 different occupations at over 17,500 different organisations. The Mortality Data Base at Statistics Canada includes a record of all deaths in Canada from 1951 to the present. The cohort consists of an estimated 325,000 individuals representing a collective dose of 1,750 person-sieverts and 1,500,000 person-years of radiation monitoring. Computer linkages are complete and analysis is in progress.

TYPE: Cohort
TERM: Occupation; Radiation, Ionizing; Registry
SITE: All Sites
OCCU: Radiation Workers
REGI: Canada (Can)
TIME: 1984 - 1992

77   **Gaudette, L.A.**   04522
     Statistics Canada, Tunney's Pasture, Ottawa Ontario, Canada K1A 0L2 (Tel.: (613)9571765)
COLL: Miller, A.B.; Freitag, S.; Ball, D.; Dufour, R.

CANADA

Cancer Among the Inuit of Canada
The first phase of the study is to produce a valid account of the descriptive epidemiology of cancer in the Inuit of Canada for comparison with other circumpolar Inuit populations. The registry so created will then be available for analytical epidemiological studies and (potentially) intervention studies to reduce the prevalence of current cancer public health problems in the Inuit and to prevent lifestyle changes which could lead to further increases. Data on about 500 Inuit cases are being identified for the time period 1969 to present from cancer registries in the Northwest Territories, Quebec and Newfoundland for inclusion in an international monograph. The distribution by site of observed cases will be compared to that expected based on Canadian rates. The registry will be maintained by the Vital Statistics and Disease Registries Section, Health Division, Statistics Canada.

TYPE: Incidence
TERM: Eskimos; Ethnic Group; Registry
SITE: All Sites
REGI: N.W. Territory (Can); Newfoundland (Can); Quebec (Can)
TIME: 1987 – 1993

78 Létourneau, E.G. 03957
Dept. of National Health & Welfare, Health Protection Branch, 775 Brookfield Rd , Ottawa Ontario, Canada K1A 1C1
COLL: Choi, N.W.; Eaton, R.S.; McGregor, R.G.; Walker, W.B.

Lung Cancer and Domestic Exposure to Radon and Radon Daughters
The objective is to determine whether domestic radiation exposure due to inhalation of air containing radon and radon daughters results in a detectable increase in lung cancer. The elevated levels of radon daughters found in some Canadian homes may be approaching those found in uranium mines and may give a radiation dose to the bronchi which is equal or superior to that received by uranium miners. Uranium miners have been found in other studies to experience an increase rate of lung cancer. The study will evaluate approximately 750 lung cancer cases in Winnipeg, Manitoba, Canada, and a similar number of controls. Winnipeg is known to be the town with the highest known levels of natural radon in Canada. An attempt will be made to measure and evaluate lifetime exposure to radon daughters using radon integrating dosimeters. Dosimeters are being placed in houses for two consecutive six month periods. Dosimeters will also be placed at the locations of previously occupied houses. Sex- and age-(± 4 years) matched controls are randomly selected from the Winnipeg City telephone directory.

TYPE: Case-Control
TERM: Dose-Response; Environmental Factors; Metals; Occupation; Radiation, Ionizing
SITE: Lung
CHEM: Radon; Uranium
OCCU: Miners, Uranium
TIME: 1985 – 1992

79 Mao, Y.S. 02296
Health and Welfare Canada, Lab. Centre for Disease Control, Bureau of Chronic Disease Epidemiology, Tunney's Pasture, Ottawa Ontario, Canada K1A 0L2 (Tel.: (613)9571765; Fax: 9527009)
COLL: Semenciw, R.; Wigle, D.T.

Canadian Farm Operators Mortality Study
The objective of this record-linkage study is to examine the mortality patterns in a cohort of about 360,000 men identified as farm operators in 1971, relative to the referent population(s). Record linkage of census records to the Canadian Mortality Data Base is underway at Statistics Canada, using the Gneralized Iterative Record Linkage System. rates of death from particular causes which exceed the corresponding rates of death in the referent population(s) will be examined carefully for possible associations with certain variables, including age of the farm operators, the type and size of farming operations, use of agricultural chemicals (herbicides and insecticides or fertilizers) the amount of fuel used for farm machinery, geographical location, and some non-farming aspects such as off-farm wor or off-farm residence. Poisson regression analysis will be conducted using both internal and external control groups. While the zero exposure group within the cohort will be used as an internal control, the general population will serve as an external control group.

CANADA

TYPE: Case-Control; Cohort
TERM: Fertilizers; Geographic Factors; Herbicides; Insecticides; Record Linkage
SITE: All Sites; Sarcoma
TIME: 1985 - 1992

* 80 **Mao, Y.S.** 05236
Health and Welfare Canada, Lab. Centre for Disease Control, Bureau of Chronic Disease Epidemiology, Tunney's Pasture, Ottawa Ontario, Canada K1A 0L2 (Tel.: (613)9571765; Fax: 9527009)
COLL: Semenciw, R.; Mills, C.; Eliasziw, M.; Marrett, L.D.

Great Lakes Cancer Risk Assessment Study
The first part of this study is an ecological study of water quality and cancer among Great Lakes communities. Descriptive statistics regarding water quality, socio-demographic information and cancer mortality for roughly 70 communities which border on or obtain drinking water from the Great Lakes will be examined. Water quality information is available on 35 of these communities. Secondly, a pilot case-control study is under way which will examine the following cancers: colorectal, stomach, bladder, kidney, brain, leukaemia, prostate, and lymphoma in the light of contaminants in drinking water.

TYPE: Case-Control; Correlation
TERM: Chemical Exposure; Chlorination; Geographic Factors; Water
SITE: All Sites; Bladder; Brain; Colon; Kidney; Leukaemia; Lymphoma; Prostate; Rectum; Stomach
TIME: 1989 - 1993

* 81 **Mao, Y.S.** 05237
Health and Welfare Canada, Lab. Centre for Disease Control, Bureau of Chronic Disease Epidemiology, Tunney's Pasture, Ottawa Ontario, Canada K1A 0L2 (Tel.: (613)9571765; Fax: 9527009)
COLL: Desmeules, M.; Robson, D.; Mikkelsen, T.; Gaudette, L.A.; Hill, G.B.; Thomas, J.; Montpetit, V.; Lach, B.; Downey, R.

Brain Cancer Study
This is a four-part study of brain cancer in Canada. The first part will consist of a descriptive study of temporal trends in brain cancer. The second part will examine the role of changes in diagnostic practices as a possible explanation for upward trends in brain cancer mortality. This will involve a chart review of roughly 350 cases of individuals with neurological disorders (about 200 brain cancer cases) in which diagnostic information from computerized tomopgraphy and magnetic resonance imaging has been removed. A case-control study of brain cancer in which prescription drug use is determined by way of record linkage to the Saskatchewan drug plan dataset is under way. A slide review of roughly 800 brain cancer cases from two time periods (1960's versus 1980's) has also been undertaken in order to investigate changes in the relative frequency of brain cancer histological types.

TYPE: Case-Control; Incidence
TERM: Clinical Records; Drugs; Histology; Trends
SITE: Brain
TIME: 1991 - 1994

* 82 **Morrison, H.** 05239
Health and Welfare Canada, Lab. Centre for Disease Control, Bureau of Chronic Disease Epidemiology, LCDC Bldg, Tunney's Pasture, Ottawa Ontario, Canada K1A 0L2 (Tel.: (613)9411286; Fax: 9527009)
COLL: Semenciw, R.; Wigle, D.T.; Morison, D.; Chaffey, C.

Canadian Farm Operators Study
The objective of this record-linkage study is to examine mortality patterns in a cohort of about 360,000 men identified as farm operators in 1971, relative to referent populations. Record linkage of census records to the Canadian Mortality database has been completed by Statistics Canada using the Generalized Iterative Record Linkage System. Linkage to cancer incidence records is in progress. Selected cancers which have been reported as being increased among farm populations will be examined for possible associations with farm and socio-demographic variables, including age of the farm operators, the type and size of farming operations, use of agricultural chemicals (herbicides, insecticides or fertilizers), fuel/oil use, numbers of farm animals and location.

CANADA

TYPE: Case-Control; Cohort
TERM: Fertilizers; Herbicides; Insecticides; Occupation; Pesticides; Petroleum Products; Record Linkage
SITE: All Sites; Brain; Leukaemia; Multiple Myeloma; Non-Hodgkin's Lymphoma; Prostate
CHEM: Phenoxy Acids
OCCU: Farmers
TIME: 1985 – 1993

* 83   Morrison, H.   05240
Health and Welfare Canada, Lab. Centre for Disease Control, Bureau of Chronic Disease Epidemiology, LCDC Bldg, Tunney's Pasture, Ottawa Ontario, Canada K1A 0L2 (Tel.: (613)9411286; Fax: 9527009)
COLL: Semenciw, R.; Stocker, H.; Mao, Y.S.

Fluorspar Miner Mortality Update Study

A cohort of approximately 2,000 fluorspar miners occupationally exposed to radon has been identified. A previous analysis of this cohort was based on mortality follow-up to the end of 1984. An update is under way to extend the follow-up period to the end of 1989. A stratified analysis will be performed.

TYPE: Cohort
TERM: Occupation; Radiation, Ionizing
SITE: All Sites; Lung
CHEM: Radon
OCCU: Miners, Fluorspar
TIME: 1991 – 1993

84   Semenciw, R.   03346
Health and Welfare Canada, Lab. Centre for Disease Control, Bureau of Chronic Disease Epidemiology, Tunney's Pasture, Ottawa Ontario, Canada K1A 0L2 (Tel.: (613)9571768; Fax: 9527009)
COLL: Mao, Y.S.; Wigle, D.T.; Mark, E.

Cancer Surveillance in the Laboratory Centre for Disease Control

The Cancer Section of the Laboratory Centre for Disease Control was established in 1977 to: (1) monitor the risk of cancer in the Canadian population to identify high-risk groups and to determine if cancer risks change significantly over time; (2) evaluate factors (lifestyle, environment, human biology) which contribute to the risk of cancer; (3) provide epidemiological and biostatistical expertise to other programmes in the Health Protection Branch and to other government agencies and institutions; (4) disseminate cancer surveillance information to health planners, administrators and research workers (5) establish national priorities for cancer research and control programmes; (6) assess progress in disease control in Canada and identify countries with apparently successful programmes to learn from their experience; (7) develop improved methods for the national surveillance of cancer. Current work is concentrated on major cancer risk factors. The main data source is Statistics Canada, which operates mortality, cancer incidence and hospital morbidity data bases.

TYPE: Cross-Sectional
TERM: Diet; Environmental Factors; High-Risk Groups; Lifestyle; Occupation
SITE: All Sites
TIME: 1977 – 1993

* 85   Semenciw, R.   05242
Health and Welfare Canada, Lab. Centre for Disease Control, Bureau of Chronic Disease Epidemiology, Tunney's Pasture, Ottawa Ontario, Canada K1A 0L2 (Tel.: (613)9571768; Fax: 9527009)
COLL: Morrison, H.; Fair, M.E.; Verdier, P.; Mao, Y.S.

Cancer Incidence Follow-Up of the Nutrition Canada Survey Cohort

A nutrition survey of about 12,000 Canadians was conducted during the period 1970–1972. Dietary history, medical history and anthropometric measurements are available for participants. The cancer incidence experience of this cohort is being determined by way of record linkage using the Generalized Iterative Record Linkage System. The objective of this study is to investigate the relationship between potential dietary risk factors and major cancers.

CANADA

TYPE: Cohort
TERM: Alcohol; Diet; Nutrition; Obesity; Registry; Tobacco (Smoking)
SITE: All Sites; Breast (F); Colon
REGI: Canada (Can)
TIME: 1991 - 1993

86   Silins, J.   03063
Health Division, Vital Statistics, & Disease Registries Section, 18-A R.H. Coats Bldg , Tunney's Pasture, Ottawa Ontario, Canada K1A 0T6 (Tel.: (613)9518553; Tlx: 0533585)
COLL: Fair, M.E.; Coppock, E.A.; Carpenter, M.; Jordan-Simpson, D.; Brancker, A.; Gaudette, L.A.

Vital Statistics and Disease Registries Section - Statistics Canada

Statistics Canada has developed the files and facilities for (and is currently collaborating with a number of agencies in carrying out) a variety of long term medical follow-up studies, including cancer incidence and mortality related to occupational and other defined population groups. In particular, three inter-related computer systems have been developed to permit optional use of a number of different records for the entire country for health-related research. These include: (1) a computerized generalized record linkage system which uses probabilistic matching techniques to link occupational or other records with mortality or cancer incidence records; (2) the Mortality Data Base which contains all Canadian death records, including coded cause of death, on magnetic tape from 1950 onwards; and (3) the National Cancer Incidence Reporting System which includes all incident cases reported by provincial cancer registries since 1969. Quality assessment programmes and trends in cancer incidence and mortality are also carried out. Mortality atlases have been prepared to illustrate the spatial variation of cancer mortality rates in Canada in order to facilitate the detection of high risk regions and any general patterns of disease distribution. A number of epidemiological studies, primarily concerned with excess cancer risk are currently being conducted in five broad areas including: (1) follow-up of occupational groups such as uranium and other miners, nickel workers, farmers, radiation workers and asbestos workers; (2) long-term follow-up of patients with specific medical diagnoses and/or receiving various treatments; (3) reproductive outcomes; (4) follow-up of persons categorized by various lifestyle or environmental exposures; and (5) death clearance of cancer registry files. Several of these projects are carried out on a cost-recovery basis for a number of different agencies (e.g., NCIC, Health and Welfare, universities, provincial health departments, private industries, etc.). Further work is underway to investigate the usefulness of a number of starting point files for follow-up of certain population groups, to evaluate the accuracy and improve the strategies of the record linkage procedures, and to further develop cancer and reproductive outcome files.

TYPE: Incidence; Mortality
TERM: Chemical Exposure; Dusts; Environmental Factors; Geographic Factors; Lifestyle; Mapping; Metals; Mining; Occupation; Pesticides; Petroleum Products; Plastics; Radiation, Ionizing; Record Linkage; Registry; Reproductive Factors
SITE: All Sites
CHEM: Asbestos; Formaldehyde; Glass Fibres; Gold; Nickel; Uranium; Vinyl Chloride
OCCU: Asbestos Workers; Chemical Industry Workers; Farmers; Firemen; Mineral Fibre Workers; Miners; Miners, Uranium; Morticians; Nickel Workers; Radiation Workers
LOCA: Canada; United States of America
REGI: Canada (Can)
TIME: 1974 - 1993

87   Wigle, D.T.   04319
Dept. of National Health & Welfare, Health Protection Branch, Lab. Center for Disease Control, Tunney's Pasture, Ottawa Ontario, Canada K1A 0L2 (Tel.: (613)9571852; Tlx: 0533679)
COLL: Mao, Y.S.

Mortality Follow-Up of Canada Health Survey Cohort

A health survey of 33,000 Canadians was conducted in 1978-1979. The survey involved a detailed interview consisting of medical history, physical examination and biological tests. The mortality experience of the survey cohort to the end of 1985 has been determined by linkage to the National Death Index of Statistics Canada. The objectives of this study are to investigate the associations between potential risk factors (alcohol, smoking, obesity, socio-economic status) and major causes of death, and to determine the number of deaths, attributable to smoking and passive smoking. Multivariate analysis will be used to control for confounding factors. Mortality follow-up to the end of 1988 was completed during 1990.

CANADA

TYPE: Cohort
TERM: Alcohol; Passive Smoking; Record Linkage; Socio-Economic Factors; Tobacco (Smoking)
SITE: All Sites
TIME: 1987 - 1992

QUÉBEC

88   Brisson, J.   04358
  Univ. Laval, Fac. de Médecine, Dépt. de Médecine Sociale, et Préventive, Québec Québec,
  Canada G1K 7P4 (Tel.: (418)6567448; Fax: 6562627)
COLL: Meisels, A.; Morin, C.

Risk Factors for HPV 16 Lesions of the Uterine Cervix

HPV 16 has been linked to cervical cancer and intraepithelial lesions of the uterine cervix, including cervical intraepithelial neoplasia (CIN) and cervical condyloma. The main objectives of the proposed case-control study are: (1) to evaluate the relation of dietary intake of vitamin A, serum levels of retinol and beta carotene, cigarette smoking, oral contraceptive (OC) use, and age at first intercourse to the occurrence of HPV 16 positive CIN, and (2) to compare the association of vitamin A, cigarette smoking, OC use and age at first intercourse to occurrence of HPV 16 positive CIN with the association of these factors to occurrence of HPV 16 negative CIN and cervical condyloma. Cases and controls (about 2,000) will be women seen for the first time at the Colposcopy Clinic of St-Sacrement Hospital in Quebec City. All will undergo a personal interview, sampling of blood for measurement of serum retinol and betacarotene, gynaecological examination, Pap smear, and colposcopically-directed biopsy of visible lesions. A histological review of biopsy material will be done. Southern blot hybridization will be performed under stringent conditions to identify women with HPV positive cervical lesions. Relative risks will be estimated taking confounding factors into account.

TYPE: Case-Control
TERM: Age; Condyloma; Diet; HPV; Histology; Nutrition; Oral Contraceptives; Premalignant Lesion; Sexual Activity; Tobacco (Smoking); Vitamins
SITE: Uterus (Cervix)
CHEM: Beta Carotene; Retinoids
TIME: 1988 - 1992

89   Meyer, F.   04635
  Univ. Laval, Fac. de Médecine, Equipe de Recherche, en Epidémiologie, Québec Québec,
  Canada G1K 7P4 (Tel.: (418)6565186; Fax: 6562627)

Dietary Determinants of Menarche: A Prospective Study

Diet in childhood and adolescence is considered to be an important determinant of the risk of breast cancer later in life. Early menarche is associated with an increased risk of breast cancer. It is possible that the same dietary factors influence the timing of menarche and the risk of breast cancer. A cohort study was initiated in 1986 in Quebec City, Canada, to investigate the dietary determinants of menarche. About 3,000 premenarcheal girls aged 10 were enrolled in the study population, and they provided a three-day dietary record and a seven-day physical activity diary on three occasions during the school year. Health information and anthropometric measurements were also obtained at the time of enrolment. This cohort is followed up every year to ascertain when each girl's menarche occured. The relationships between diet and age at menarche will be assessed, mainly using the Cox model, and the influence of other determinants of menarche will be taken into account in the analyses. This cohort study will document the role of specific elements of the diet on the timing of menarche. It will also provide baseline information with which to evaluate the influence of diet on the later development of breast cancer.

TYPE: Cohort
TERM: Adolescence; Diet; Menarche; Physical Activity; Physical Factors
SITE: Breast (F)
TIME: 1986 - 1992

90   Meyer, F.   04881
  Univ. Laval, Fac. de Médecine, Equipe de Recherche, en Epidémiologie, Québec Québec,
  Canada G1K 7P4 (Tel.: (418)6565186; Fax: 6562627)
COLL: Bairati, I.; Fradet, Y.

CANADA

Influence of Diet on the Occurrence and Progression of Prostate Cancer
This case-control study will assess the relationship between diet and prostate cancer. Cases (about 410 men) will be all newly diagnosed prostate cancer patients, aged 55 to 75, treated in seven Quebec City hospitals from September 1990 to December 1992. Controls (about 460 men) will be randomly selected among patients of similar ages treated in the same hospitals for a first transurethral resection for benign prostatic hyperplasia. Nutritionists, blinded to the disease status, will interview all subjects on their dietary intakes during the past year using a detailed food frequency questionnaire and food models. All pathological samples will be reviewed to assess histological grade of cancer and presence and severity of precancerous lesions. Diet, especially fat and carotene intakes, will be compared using polytomous logistic regression across six groups of subjects. These will include patients with cancer at various stages (latent, localised, regional and metastatic), those with precancerous lesions and men free of both cancer and precancerous lesions.

TYPE: Case-Control
TERM: Diet; Fat; Premalignant Lesion; Stage; Vitamins
SITE: Prostate
CHEM: Beta Carotene
TIME: 1990 - 1993

SASKATOON

91 Dosman, J.   04741
    Univ. of Saskatchewan, Dept. of Medicine, Centre for Agricultural Medicine, Room 539, Ellis Hall, Saskatoon Saskatchewan, Canada S7N OX0 (Tel.: (306)9668286; Tlx: 9668799)
COLL: McDuffie, H.H.; Hill, G.B.; Thériault, G.P.; McLaughlin, J.K.; Choi, N.W.; Robson, D.; Fincham, S.F.; Skinnider, L.; Pahwa, P.; White, D.; To, T.; West, R.; Spinelli, J.J.

National Study of Pesticide Exposure and Health
Recent studies suggest that use of agricultural pesticides is associated with an increased risk of developing soft tissue sarcoma. A group of investigators from Alberta, Saskatchewan, Manitoba, Quebec, Ontario, Newfoundland and British Columbia proposes to conduct a case-control study involving about 1,600 male patients, newly diagnosed as having one of the above types of cancer, together with an equal number of male controls selected from the general population. Patients and controls will be sent questionnaires to ascertain exposure to pesticides and the presence of other relevant factors. Subjects who indicate exposure to pesticides will be interviewed to determine the type and level of exposure. The risk of each of the above types of cancer in relation to pesticide exposure will be derived and the attributable proportion of such cancers among males will be estimated.

TYPE: Case-Control
TERM: Occupation; Pesticides
SITE: Hodgkin's Disease; Multiple Myeloma; Non-Hodgkin's Lymphoma; Sarcoma; Soft Tissue
OCCU: Farmers; Forest Workers; Horticulturists
TIME: 1989 - 1993

92 McDuffie, H.H.   03239
    Univ. of Saskatchewan, Centre for Agricultural Medicine, Dept. of Medicine, Royal Univ. Hosp., Saskatoon Saskatchewan, Canada S7N OXO (Tel.: (306)9666154; Fax: 9668799)
COLL: Dosman, J.; Klaassen, D.J.

Collaborative Studies of Primary Lung Cancer
A series of case-control studies is being conducted to investigate the relationship between various risk factors (positive family history, occupational exposures, atopy, gender, radon levels in homes) and cigarette smoking habits and the aetiology of primary lung cancer. Data recorded by the cancer registry have been compiled on over 2,900 cases and postal questionnaires or personal interviews conducted with more than 1,500 cases and similar number of controls. Manuscripts concerning histology (Chest 98:1187-1193, 1990 and 99:404-407, 1991) and family history (J. Clin. Epidemiol. 44:69-76, 1991) have recently been published.

CANADA

TYPE: Case-Control
TERM: Chemical Exposure; Dusts; Enzymes; Familial Factors; Fungicides; Genetic Factors; Herbicides; Insecticides; Occupation; Tobacco (Smoking)
SITE: Lung
CHEM: Radon
OCCU: Agricultural Workers; Farmers
REGI: Saskatchewan (Can)
TIME: 1982 - 1992

93 McDuffie, H.H. 04752
Univ. of Saskatchewan, Centre for Agricultural Medicine, Dept. of Medicine, Royal Univ. Hosp., Saskatoon Saskatchewan, Canada S7N OXO (Tel.: (306)9666154; Fax: 9668799)

Radon Exposure in Homes and Primary Lung Cancer in Women

Various mechanisms have been proposed to explain sex differences in the occurrence of primary lung cancer. Women patients have lower exposure to cigarette smoke and putative occupational carcinogens, while developing this disease at younger ages than men. There is limited evidence that decreases in pulmonary function related to cigarette smoking may be one biological mechanism which helps to explain these results. Exposure in the home to alpha radiation derived from radon and radon progeny is a viable alternative hypothesis. In Saskatchewan, exposure to this naturally occurring radiation in homes is possible over a wide geographical area. Women in the age group most at risk of developing lung cancer have typically spent more time in the home than men. Women have two to three times the levels of body contamination than men exposed to the same ambient living-area radon concentrations. Personal interviews are being conducted and alpha radiation levels being measured in the homes of female patients with primary lung cancer and age-matched control women.

TYPE: Case-Control
TERM: Female; Radiation, Ionizing
SITE: Lung
CHEM: Radon
REGI: Saskatchewan (Can)
TIME: 1989 - 1993

94 McDuffie, H.H. 04843
Univ. of Saskatchewan, Centre for Agricultural Medicine, Dept. of Medicine, Royal Univ. Hosp., Saskatoon Saskatchewan, Canada S7N OXO (Tel.: (306)9666154; Fax: 9668799)
COLL: White, M.D

Pesticide Exposure and Cancer in Females

Several cancer sites have been epidemiologically associated with rural residence, farming, specific farm practices and pesticide exposure in men. Women are often arbitrarily excluded from studies of putative carcinogens when occupational exposure is the primary focus. A population-based case-control study (45 soft tissue sarcoma, 144 non-Hodgkin's lymphoma, 16 gliomas, 255 community controls, 128 sisters of patients) is being conducted, using a postal questionnaire to ascertain medical history, reproductive factors, pesticide exposure and smoking patterns. Cancer Registry records are used to access cases and verify the diagnosis, controls are obtained through the Saskatchewan Hospital Services Plan. Age - adjusted odds ratios will be used to assess the hypothesis regarding increased risk of site specific tumours and pesticide exposure.

TYPE: Case-Control
TERM: Female; Occupation; Pesticides; Reproductive Factors; Tobacco (Smoking)
SITE: Brain; Multiple Myeloma; Non-Hodgkin's Lymphoma; Sarcoma; Soft Tissue
OCCU: Farmers; Health Care Workers; Horticulturists
TIME: 1989 - 1992

SHERBROOKE

95 Leduc, C.P. 04187
Univ. de Sherbrooke, Faculté de Médecine, Dept. Sciences de Sante Communautaire, 3001-12eme Ave. N, Sherbrooke Quebec, Canada J1H 5N4 (Tel.: (819)5645367)
COLL: Charlin, B.; Tahan, T.; Pageau, R.; Rochon, M.; Vobecky, J.S.; Vobecky, J.; Madarnas, P.

CANADA

Nutritional Determinants in the Aetiology of Head and Neck Cancer
Smoking and heavy alcohol consumption are the most important risk factors for head and neck cancer. Because of this association, other relationships have been difficult to identify. One important aspect in need of analysis is the importance of previous food habits and intake in the aetiology of these cancers. All new cases of head and neck cancer in the regional treatment centre will be studied and information recorded on previous nutritional status. An estimated 150 cases will be compared with two age- and sex-matched controls. Known risk factors will also be controlled. An important by-product of this study will be evaluation of the role of antecedent nutritional status in treatment tolerance and survival.

TYPE: Case-Control
TERM: Alcohol; Diet; Tobacco (Smoking)
SITE: Head and Neck
TIME: 1987 - 1992

96 Lepine, J. 04372
Center Hospitalier Univ. de Sherbrooke, Unite de Recherche en Nutrition Humaine, & Departement de Medecine, Sherbrooke Quebec, Canada J1H 5N4
COLL: Vobecky, J.; Vobecky, J.S.

Body Mass Regulation and Nutrition in Breast Cancer Patients
It is intended to evaluate the nutritional status of newly diagnosed breast cancer patients, as compared to known population standards from the same geographical area, in order to assess the significance of nutritional status (in particular obesity) in the aetiology of breast cancer and as a predictive factor in treatment outcome. Obesity and increased body mass index represent a risk factor for breast cancer. Newly diagnosed patients will be evaluated for their eating habits and nutritional status, using dietetic, anthropometric and biochemical measurements. These will be correlated with the disease status, including extent of disease, age, menopausal status and oestrogen receptor status of the tumour. Patients who receive chemotherapy will be followed prospectively for changes in their nutritional status, eating habits and clinical outcome.

TYPE: Case Series
TERM: Chemotherapy; Nutrition; Obesity; Physical Factors
SITE: Breast (F)
TIME: 1987 - 1992

97 Vobecky, J. 04337
Univ. of Sherbrooke, Fac. Med., Human Nutrition Research Unit, Dept. Comm. Hlth Scien., 3001 12e Ave. Nord, Sherbrooke Quebec, Canada J1H 5N4
COLL: Echavé, L.V.; Langlois, S.P.; Abdulnour, E.; Vobecky, J.S.

Role of Vitamin A and Cholesterol in the Aetiology of Lung Cancer
To study the possible protective effect of vitamin A and the increased risk associated with a high cholesterol content of the diet in relation to lung cancer, a case-control study on 180 incident cases of primary lung carcinoma is being conducted. Two controls will be selected for each case, one among patients not suffering with lung carcinoma, the other from the neighbourhood of the cases. The hospital controls will have a lung X-ray to confirm the absence of cancer. Cases and hospital controls come from a university hospital. For the neighbourhood controls, the presence or absence of symptoms associated with lung cancer will be established by a questionnaire. If symptoms are present, the subject will be referred to a physician for confirmation of the diagnosis. Data on food intake of vitamin A, beta carotene, cholesterol and other relevant nutrients, including amino and fatty acids, will be gathered for the period prior to diagnosis of lung cancer as well as for the assessment of present nutritional status. A complete history of environmental and tobacco smoke exposure will be collected in order to control for confounding. Special attention will be paid to passive smoking. Stratified and multivariate analysis will be used for the identification of potential associations.

CANADA

TYPE: Case-Control
TERM: Cholesterol; Diet; Nutrition; Passive Smoking; Vitamins
SITE: Lung
CHEM: Beta Carotene
TIME: 1987 - 1992

ST JOHN'S

98 **Farid, N.R.** 04016
Memorial Univ. of Newfoundland, Faculty of Medicine, Div. Endocrinology, & Metabolism/Thyroid Research, 300 Prince Phillip Dr. , St John's Newfoundland, Canada A1B 3V6 (Tel.: (709)376549; Tlx: 0164101)
COLL: Stenszky, V.

Immunogenetics of Differentiated Thyroid Carcinoma
The aim is to compare genetic liability to thyroid cancer in an iodide sufficient area (Newfoundland) with that in an iodide deficient area (Eastern Hungary). 60 cases in Newfoundland, and 100 in Debrecen, Hungary, have been recruited retrospectively in order to permit study of the immunogenetic phenotype in relation to tumour behaviour. 140 controls will be selected in Newfoundland, and 160 in Debrecen. The tumour histology will be reviewed blind and charts perused for basic demographic data, evidence of metastases at surgery, therapy, recurrence or late metastases. Preliminary results suggest that major histocompatibility complex linked gene(s) determine susceptibility to and the biological behaviour of thyroid cancer.

TYPE: Case-Control
TERM: Genetic Factors; Genetic Markers; HLA; Iodine Deficiency
SITE: Thyroid
LOCA: Canada; Hungary
TIME: 1985 - 1992

TORONTO

* 99 **Chen, J.K.** 05018
Univ. of Toronto, Fac. of Dentistry, 4384 Medical Sciences Bldg, Toronto Ontario, Canada M5S 1A8 (Tel.: (416)9786624)
COLL: Katz, R.; Krutchkoff, D.; Eisenberg, E.

Trends in Lip Cancer in Connecticut, 1935-1985
The primary purpose of this study is to investigate secular trends in lip cancer incidence in Connecticut from 1935-1985. A secondary purpose is to determine the degree to which lip cancer alone influenced dramatic differences in trends previously reported for all oral cancer as compared to intra-oral squamous cell carcinoma in the same population (Cancer 65:2796-2802, 1990 and 66:1288-1296, 1990). The data source for this study is the Connecticut Tumor Registry (CTR). Established in 1935, it is both the oldest population-based cancer registry in the world and the largest in the United States. Three types of incidence rates (crude, age-specific and age-adjusted) of lip cancer (n=2,219) are calculated for the 51-year study period. The study also examines age, sex, geographical distribution and lesion subsite analyses in addition to histopathological patterns of differentiation in lip cancer. According to the primary results, two major possible causal factors of lip cancer, solar radiation and smoking, among the local residents will be extensively analysed.

TYPE: Incidence
TERM: Geographic Factors; Radiation, Ultraviolet; Registry; Tobacco (Smoking); Trends
SITE: Lip
REGI: Connecticut (USA)
TIME: 1990 - 1992

100 **Choi, B.C.K.** 04325
Univ. of Toronto, Unit of Occupational, and Environmental Health, 150 College St. , Toronto Ontario, Canada M5S 1A8 (Tel.: (416)9786236)
COLL: Connolly, J.

CANADA

Factors Affecting Urinary Mutagenic Level of Workers with Bladder Cancer
The Ames/Salmonella Microsome Test (Ames Assay) is widely used to assay mutagenicity. The fact that cigarette smokers generally have mutagenic urine is now documented, but it has also been shown that mutagens can be detected in non-smoking workers in the rubber industry. In addition, there have been reports that bladder cancer patients have higher levels of excretion of tryptophan metabolites than do comparison subjects, and bladder cancer patients and normal donors differ in their urinary nitrosamines. It has been suggested that bladder cancer cases and normal controls may differ in their urinary mutagenicity in terms of potency of the mutagens and their type (i.e. base-pair or frame-shift). This provides a basis for the use of urinary Ames Test as a possible screening technique for bladder cancer cases in the workplace. Forty bladder cancer patients and 40 hospital controls will be selected for the study. Subjects will be non-smokers and will not have been under any radiation treatment or chemotherapy within the previous 60 days. An epidemiological questionnaire will be administered concerning the individual's exposure to industrial chemicals, artificial sweeteners, beverages and selected food items. Two 24-hour urine specimens, one taken while the individual is in the working environment and the other after he/she has been out of the working environment for at least 48 hours, will be extracted, frozen and tested for mutagenicity. For each specimen, Ames Test involving 2 bacterial strains TA98 and TA100, with and without S-9 liver homogenates, will be carried out. The following information will result from this study: (1) the bladder cancer status of the patients, (2) urinary mutagenicity as assayed by the Ames Test, and (3) epidemiological data obtained from the questionnaire. Multivariate analysis will be used to relate the type and potency of urinary mutagens to the bladder cancer status, simultaneously adjusted for epidemiological variables obtained in the questionnaires.

TYPE: Case-Control
TERM: Chemical Exposure; Mutagen; Occupation; Sweeteners
SITE: Bladder
TIME: 1987 - 1992

* 101   Feuer, G.M.   05035
Univ. of Toronto, Depts of Clinical Biochemistry, and Pharmacology, 100 College St., Toronto Ontario, Canada M5G 1L5 (Tel.: (416)9782262/9782263; Fax: 9785650)
COLL: Kerenyi, N.A.; Szalai, J.P.

Light Pollution and Increased Cancer Incidence
The objective of this study is to investigate cancer incidence among factory workers who work mostly in darkness compared with workers in another factory with normal lighting exposure, and with the general population. At present, cancer is responsible for almost half of all deaths among men in the same age group. The study will investigate the hypothesis that one of the most important aetiological factors in the rapid increase in malignant melanoma is the change in exposure to light in the last 100 years. Further studies are planned to examine by laboratory methods whether a higher level of serum melatonin, the oncostatic pineal hormone, is associated with the lower cancer rate which may occur in the group working in a dark environment.

TYPE: Incidence
TERM: Fluorescent Light; Occupation; Pigmentation
SITE: All Sites; Melanoma
TIME: 1991 - 1994

102   Finkelstein, M.M.   02819
Ontario Ministry of Labour, 400 University Ave., 8th Floor, Toronto Ontario, Canada M7A 1T7 (Tel.: (416)3267879; Fax: 3267889)

Mortality in Asbestos Cement Factory Workers
Mortality occurring among employees of an asbestos cement factory is being studied in relation to estimates of past dust exposure. Both chrysotile and crocidolite asbestos had been used at this factory and mortality rates from lung cancer and mesothelioma are known to be markedly elevated among exposed employees. The control groups consist of workers from non-asbestos areas of the factory as well as the general population of the province. Papers have been published in Br. J.Ind.Med. 40:138-144, 1983; Am.Rev.Resp.Dis. 129:754-761, 1984 and in Toxicol. Ind. Hlth 6:623-627, 1990.

CANADA

TYPE: Case-Control; Cohort
TERM: Dusts; Occupation
SITE: Gastrointestinal; Mesothelioma; Respiratory
CHEM: Asbestos, Chrysotile; Asbestos, Crocidolite
OCCU: Cement Workers
TIME: 1978 - 1992

103 Finkelstein, M.M. 04300
Ontario Ministry of Labour, 400 University Ave., 8th Floor, Toronto Ontario, Canada M7A 1T7 (Tel.: (416)3267879; Fax: 3267889)
COLL: Beck, P.

Association between Radium in Drinking Water Supplies and the Risk of Bone Cancer Death in Ontario Youths

The aim of this case-control study is to measure the association between exposure to radium in domestic water supplies and the risk of death from bone sarcoma. The study subjects are the 327 Ontario born individuals, 25 years of age or less, who died of bone sarcoma between 1950 and 1983. Each has been matched with a control of the same age who died of any other disease cause. Water samples are collected at the residences of the subjects and radium-226 is measured by the radon emanation method. Statistical analysis is by Mantel - Haenszel and conditional logistic regression methods.

TYPE: Case-Control
TERM: Metals; Water
SITE: Bone
CHEM: Radium
TIME: 1985 - 1992

104 Finkelstein, M.M. 04742
Ontario Ministry of Labour, 400 University Ave., 8th Floor, Toronto Ontario, Canada M7A 1T7 (Tel.: (416)3267879; Fax: 3267889)

Lung Cancer in the Ontario Steel Industry

This is a population-based case-control study in two Ontario steelmaking cities, to test the hypothesis that steelworkers are at increased risk of lung cancer. 576 cases, each with two matched controls, were identified from the Ontario death registry. The occupations of the subjects were identified from the death certificates and the major steel companies provided work histories for their employees in the sample. Specific job activities in the steel industry are being tested for association with lung cancer.

TYPE: Case-Control
TERM: Occupation
SITE: Lung
CHEM: Steel
OCCU: Steel Workers
TIME: 1988 - 1992

105 Howe, G.R. 02379
Univ. of Toronto, National Cancer Inst. of Canada, Dept. of Prev. Med. and Biostatistics, 12 Queen's Park Crescent West, McMurrich Bldg, 3rd Floor, Toronto Ontario, Canada M5S 1A8 (Tel.: (416)9785097)
COLL: Baines, C.J.; Miller, A.B.

Cancer Incidence and Mortality in a Cohort of Diabetics

The primary objective of the present study is to monitor cancer incidence and mortality among diabetics in order to examine possible associations between (1) artificial sweetener use and bladder cancer; (2) factors relating to diabetes and cancer of the pancreas; (3) smoking patterns among diabetics and lung cancer. The secondary objective is to relate mortality from non-neoplastic diseases, such as cardiovascular diseases, to factors relating to diabetic status, such as control of disease. A self-administered questionnaire has been distributed containing questions on identifying information, details relating to onset and control of diabetes, and a number of lifestyle habits such as smoking and artificial sweetener use. 13,000 questionnaires have now been received. The subsequent cancer incidence and general mortality experience of the cohort will be monitored using computerized record linkage techniques to link the data to the national mortality data base and national cancer incidence reporting system maintained by Statistics Canada. In 1986 a supplementary questionnaire was distributed to the cohort to get data on current aspartame use and more specific data on past artificial sweetener use. A 55% response rate was achieved. A paper was published in J. Can. Dietetic Assoc. 46(4):288-291, 1985.

TYPE: Cohort; Incidence; Mortality
TERM: Diabetes; Lifestyle; Record Linkage; Registry; Sweeteners; Tobacco (Smoking)
SITE: Bladder; Lung; Pancreas
REGI: Canada (Can)
TIME: 1980 – 1992

106 Howe, G.R. 02908
Univ. of Toronto, National Cancer Inst. of Canada, Dept. of Prev. Med. and Biostatistics, 12 Queen's Park Crescent West, McMurrich Bldg, 3rd Floor, Toronto Ontario, Canada M5S 1A8 (Tel.: (416)9785097)
COLL: Baines, C.J.; Howe, G.R.; Jain, M.

Diet and Cancer in a Cohort Study of Women
Participants in the National Study of Breast Cancer Screening have been asked to complete a self-administered dietary questionnaire previously validated on a sample of 200 participants in the Breast Screening Study in Toronto in 1981. Approximately 60% of the total sample in the Breast Screening Study of 89,000 women age 40-59 completed the questionnaire. Using the follow-up proposed for the National Breast Screening Study and extending this if necessary by computerized record linkage with National Cancer Incidence and Mortality files, will enable the relationship between cancers in women and diet to be further evaluated. This dietary information will complement the information on risk factors for breast and other cancers already being collected in the National Breast Screening Study.

TYPE: Cohort
TERM: Diet; Female; Record Linkage; Screening
SITE: Breast (F); Colon; Kidney; Ovary; Pancreas; Rectum; Uterus (Corpus)
TIME: 1982 – 1996

107 L'Abbé, K.A. 04878
Ontario Ministry of Labour, Industrial Diseases Standards Panel, 10 King St. E., 7th Floor, Toronto M5C 1C3, Canada M5C 1C3 (Tel.: (416)9655056; Fax: 3244536)

Metropolitan Toronto Fire-Fighters Study
The aim of this study is to determine whether or not there is excess mortality among fire-fighters, in terms of cancer or any specific cause of death. A historical cohort study will examine the mortality experience of approximately 5,000 professional fire-fighters from six Metropolitan Toronto fire departments. Mortality follow-up from 1955 to 1988 will be compared to the general population experience in the province of Ontario. Relationships between occupational exposure variables and patterns of mortality will also be investigated.

TYPE: Cohort
TERM: Occupation
SITE: All Sites
OCCU: Firemen
TIME: 1989 – 1992

108 Marrett, L.D. 04551
Ontario Cancer Treatment, & Research Foundation, Div. of Epidemiology & Statistics, 7 Overlea Blvd, Toronto Ontario, Canada M4H 1A8 (Tel.: (416)4234240; Fax: 4232017)
COLL: Weir, H.K.

Aetiology of Malignant Germ Cell Tumours
Malignant germ cell tumours (MGCT) occur mainly in the testis but can also occur in the ovary and at extra-gonadal sites. The majority of testicular tumours and ovarian tumours in females under age 15 are of germ cell origin. Incidence rates for testicular cancer in Ontario, as well as in other parts of the world, have been rising in recent years. The childhood incidence of ovarian cancer and MGCT (all sites) also appears to be increasing. The reasons for the increases are unknown. This case-control study is designed to elucidate the aetiology of malignant germ cell tumours (MGCT) in Ontario, Canada. Cases (approximately 555 expected), are being identified through the Ontario Cancer Registry and will constitute all residents of Ontario, aged 59 or less, diagnosed with MGCT between 1986 and 1990. Controls (n = 1430) have been selected from the general population of Ontario, frequency-matched to the cases for age and sex. Relevant exposure information (medical, occupational, and hormonal)

CANADA

covering the period from pre-conception to diagnosis, will be collected through mailed questionnaires to subjects and their mothers (for living subjects aged 16 and over) or to the mothers of deceased cases and those subjects under the age of 16 at time of diagnosis. In addition, this study is including pathology review of all identified cases for the purpose of classifying tumours according to histological subgroups (pure seminoma, pure nonseminoma, and mixed seminoma and nonseminoma). Classifying disease state according to these histological subgroups may elucidate aetiological risk factors specific to these subgroups.

TYPE: Case-Control
TERM: Classification; Histology; Hormones; Occupation; Registry
SITE: Ovary; Testis
REGI: Ontario (Can)
TIME: 1988 - 1992

109   Miller, A.B.   00251
      Univ. of Toronto, Dept. Prev. Medicine & Biostatistics, McMurrich Bldg, 4th Floor, 12 Queen's Park Crescent West, Toronto Ontario, Canada M5S 1A8 (Tel.: (416)9785097)
COLL: Howe, G.R.

Cancer Following Multiple Fluoroscopies

This cohort study is designed to evaluate the carcinogenic risk from ionizing radiation in the dose range 0 to 100 rad and, for the major cancer sites, to establish if there is a possible dose-response relationship. The study population consists of patients first admitted to Canadian tuberculosis sanatoria between the years 1930 and 1952. A substantial proportion (40 to 50%) were fluoroscoped regularly in conjunction with artificial pneumothorax treatment and consequently accumulated a substantial radiation dose consisting of many exposures over a period of time. Information on the admission and treatment records of these patients was extracted from the original sanitoria files. The basic patient data consist of approximately 130,000 patient records. Three computerized record linkage steps using the identifying information present on the records have been undertaken, the first internally linking records to bring together admissions referring to the same individual, the second linking these composite records to the mortality data base maintained by Statistics Canada for the years 1950 through 1975, in order to ascertain the mortality experience of the cohort, and the third to the mortality data prior to 1950 and for 1976 through 1980. A dosimetry study has been completed and published (Health Phys. 35:259-70, 1978) which determined the dose experienced by various body organs exposed to radiation from fluoroscopy. The dosimetry information has been applied to the individual patient records in order to estimate the radiation dose received by each patient. The analysis will consist of estimation of relative risks with respect to mortality for different cancer sites as a function of dose calculated relative to those members of the study population who were not fluoroscoped. A preliminary report by Howe on breast cancer mortality has been published; (Radiation Carcinogenesis: Epidemiology and Biological Significance, (eds) J.D. Boice, Jr and J. F. Fraumeni, Jr; Raven Press, New York 1984, pp. 119-129). A definitive report on this experience has been published in N. Eng. J. Med. 321:1285-1289, 1989. A fourth linkage with national cancer incidence files is now planned.

TYPE: Cohort
TERM: Clinical Records; Radiation, Ionizing; Record Linkage; Treatment
SITE: All Sites
TIME: 1976 - 1992

110   Miller, A.B.   01835
      Univ. of Toronto, Dept. Prev. Medicine & Biostatistics, McMurrich Bldg, 4th Floor, 12 Queen's Park Crescent West, Toronto Ontario, Canada M5S 1A8 (Tel.: (416)9785097)
COLL: Thompson, D.W.; Clarke, E.A.; Wall, C.

Natural History of Dysplasia of the Uterine Cervix

In most models of the natural history of squamous carcinoma of the cervix, dysplasia is regarded as the first recognizable pathological process in an orderly progression from normal to malignant epithelium. Since not all dysplasia become malignant, there is need to clarify the extent to which regression forms part of the natural history of dysplasia. The present study aims at seeking information on the natural history of cytologically identified but untreated dysplasia in comparison with a sample of normal women with smears examined in the same laboratory. There were 22,753 women with minimal dysplasia, 23,321 with mild, 11,264 with moderate and 2,697 with severe dysplasia. The control series included 20,976 women. Information was also sought on the possible association of oral contraceptive use and the risk of progression of dysplasia. Data extraction is complete. Analyses to date show an increasing risk of invasive cancer with increasing degrees of dysplasia. The relative risks for the manifestation of a malignancy in a subsequent cervical smear are 1.48, 3.42, 20.9 and 71.5 for women with minimal, mild, moderate and severe dysplasia, respectively. The cohort will now be linked to the Ontario Cancer Registry to provide further data on risk of progression.

TYPE: Case-Control
TERM: Oral Contraceptives; Premalignant Lesion
SITE: Uterus (Cervix)
TIME: 1978 - 1992

111   **Miller, A.B.**   02001
Univ. of Toronto, Dept. Prev. Medicine & Biostatistics, McMurrich Bldg, 4th Floor, 12 Queen's Park Crescent West, Toronto Ontario, Canada M5S 1A8 (Tel.: (416)9785097)
COLL: Baines, C.J.; Howe, G.R.; Wall, C.; Boyes, D.A.; Hill, G.B.; Bowman, D.M.; Simard, A.; Fabia, J.; Bassett, A.A.; Simor, I.; Deschenes, L.; Bethune, G.; Catton, G.; Cantin, J.; Bush, H.

Breast Cancer Screening
A long-term randomized trial of screening for breast cancer has been initiated in Canada to find whether screening will reduce the mortality rate from breast cancer. Several projects have indicated that screening is of benefit in women over 50, but that more needs to be known about its benefit in women in their forties. For women over 50, the question remains whether screening should include mammography or whether physical examination alone will suffice. Fifteen centres in Canada recruited 1,000-15,000 women between the ages of 40 and 59. Women aged 40-49 either underment mammography and physical examination annually, or were followed by mail annually. Women aged 50-59 had annual screening, half by physical examination alone and half by both physical examination and mammography. The screening continued for three or four years, and all women are being followed for at least a further five years. Every participant had an initial physical examination and instruction in breast self-examination. Recruitment into the study ceased on 31 March 1985. 89,968 women were enrolled. Annual rescreening continued to 1988. Follow-up continues.

TYPE: Cohort
TERM: Mammography; Screening
SITE: Breast (F)
TIME: 1980 - 1996

112   **Miller, A.B.**   02628
Univ. of Toronto, Dept. Prev. Medicine & Biostatistics, McMurrich Bldg, 4th Floor, 12 Queen's Park Crescent West, Toronto Ontario, Canada M5S 1A8 (Tel.: (416)9785097)
COLL: Coldman, A.J.; Howe, G.R.

Extension of the British Columbia Cohort Study
In a previous study, two cohorts of women born in 1914-18 and 1929-33 were identified and studied from the records of the British Columbia Screening programme 1949-1969. The analysis showed that regression is an important part of the natural history of carcinoma of the cervix. In the present study, the period of observation for the two cohorts has been updated to 1980 and a third cohort born in 1944-1948 indentified. This will permit the answering of questions relating to the efficacy of screening and evaluate the incidence of the precursors of invasive cancer of the cervix in a presumed high risk young cohort. This will enable predictions from mathematical models of screening based on the previous analysis to be refined and will provide the opportunity to include a large body of data in that currently being collected from a number of countries by the IARC. The three cohorts comprise over 300,000 longitudinal records of women. The basic records of the cohorts have been extracted and internal record linkage, will be conducted to eliminate duplication.

TYPE: Cohort
TERM: High-Risk Groups; Mathematical Models; Screening
SITE: Uterus (Cervix)
TIME: 1981 - 1992

VANCOUVER

113   **Band, P.R.**   03605
BC Cancer Control Agency, Div. of Epidemiology, Biometry, and Occupational Oncology, 600 W 10th Ave., Vancouver BC, Canada V5Z 4E6 (Tel.: (604)8776000)
COLL: Spinelli, J.J.; Gallágher, R.P.; Threlfall, W.J.; Raynor, D.; Wong, M.

CANADA

Identification of Occupational Cancer Risk Factors
The objective of this study is to identify occupational cancer risk factors. Life-time occupational histories of male cancer patients aged 20 years and over ascertained by the British Columbia Cancer Registry, and of controls from the general population are being obtained to determine whether employment in individual occupations or industries is associated with an increased risk for specific cancers. A self-administered questionnaire requesting detailed occupational histories as well as information on ethnic origin, education, smoking history and alcohol consumption will be mailed to each case and control. An estimated number of 5,000 cancer cases and 1,500 controls per year will be studied. For each individual cancer site, occupations and industry titles as well as combinations of similar occupations and industries will be examined. In addition, exposure-response and latency effects will be analysed.

TYPE: Case-Control
TERM: Alcohol; Dose-Response; Education; Environmental Factors; Ethnic Group; Latency; Occupation; Registry; Tobacco (Smoking)
SITE: All Sites
REGI: Br. Columbia (Can)
TIME: 1983 - 1992

114 Band, P.R. 03606
BC Cancer Control Agency, Div. of Epidemiology, Biometry, and Occupational Oncology, 600 W 10th Ave., Vancouver BC, Canada V5Z 4E6 (Tel.: (604)8776000)
COLL: Spinelli, J.J.; Gallágher, R.P.; Bert, J.L.; Grace, J.R.; Moody, J.; Svirchev, L.M.

Detection of Carcinogens in the Workplace
The long-term objective of this project is to create a job-exposure matrix for the Province of British Columbia which can be used to establish relationships between specific types of cancer and specific chemical substances. The job exposure matrix will be used in conjunction with the identification of occupational cancer risk factors by allowing exposures to be inferred from the work history of cancer cases. To initiate the job exposure matrix, the study of a specific industry, the Trail Operations of Cominco, Limited, has been undertaken. The Trail Operations, involving zinc recovery and refining, lead smelting, sulphuric acid production and fertilizer production, is sufficiently complex to provide an excellent case study upon which the methodology of developing the exposure matrix can be tested and improved.

TYPE: Methodology
TERM: Chemical Exposure; Fertilizers; Metals; Occupation
SITE: All Sites
CHEM: Lead; Sulphuric Acid; Zinc
OCCU: Chemical Industry Workers; Fertilizer Workers; Miners, Zinc-Lead; Smelters, Lead
TIME: 1984 - 1992

115 Horsman, D.E. 04875
BCCA, Cytogenetic Lab., 600 W. 10th Ave., Vancouver BC, Canada V5Z 4E6 (Tel.: (604)8776000)
COLL: Sparling, T.

Effects of Smoking on the Development of Myelodysplasia
This case-control study will assess the risk of myelodysplasia (MDS) associated with smoking. A smoking history will be obtained on all newly diagnosed patients with MDS, as diagnosed by haematological, marrow morphologic and cytogenetic investigations. About 200 cases will be recruited, with 200 population controls. Smoking data will be obtained by a questionnaire mailed to all identified patients' physicians.

TYPE: Case-Control
TERM: Tobacco (Smoking)
SITE: Leukaemia; Myelodysplastic Syndrome
TIME: 1989 - 1992

116 McBride, M.L. 04880
Cancer Control Agency of BC, Div. of Epidemiology, Biometry and, Occupational Oncology, 600 W 10th Ave., Vancouver BC, Canada V5Z 4E6 (Tel.: (604)8776000; Fax: 8724596)
COLL: Thériault, G.P.; Fincham, S.F.; Choi, N.W.; Robson, D.; Gallágher, R.P.

CANADA

Childhood Leukaemia and Extremely Low Frequency Electromagnetic Field Exposure
Several studies have suggested that children exposed to intense electric or magnetic fields at extremely low frequency may be at higher risk of leukaemia. A population-based case-control study will be conducted of the association between exposure to electric and magnetic fields and risk of leukaemia in children aged up to 14, with particular emphasis on leukaemia subtypes. All children newly diagnosed with leukaemia from 1 January 1989, to 31 December 1993, and resident in the principal cities and surrounding areas of each of the participating provinces will be eligible. For each case, one control child will be matched on birthdate, sex and city of residence from the most appropriate available source. Data on magnetic field exposure will be collected using questionnaires, activity diaries, wiring configuration assessment, and a personal field dosimeter. Average exposure, duration of high exposure, and various components of electromagnetic field exposure can be analysed separately. The risk of all childhood leukaemia and each subtype of leukaemia in relation to 60 Hz field exposure will be examined. Associations will also be assessed with family history of cancer and congenital anomalies; parental occupational and recreational exposures; gestational factors and other exposures surrounding pregnancy; post-natal factors, including radiation, and exposure to viral infections, chemicals, and medications; and socio-economic status.

TYPE: Case-Control
TERM: Chemical Exposure; Childhood; Congenital Abnormalities; Drugs; Electromagnetic Fields; Infection; Intra-Uterine Exposure; Occupation; Radiation, Ionizing; Socio-Economic Factors
SITE: Leukaemia
TIME: 1990 - 1994

117   Stich, H.F.   03691
BC Cancer Research Centre, Environmental Carcinogenesis Unit, 601 W. 10th Ave., Vancouver BC, Canada V5Z 1L3 (Tel.: (604)8776010; Tlx: 04507648)
COLL: Krishnan Nair, M.; Mathew, B.; Sankaranarayanan, R.

Intervention Trials with Carotenoids and Retinoids on Tobacco/Areca Nut Chewers with Emphasis on Maintenance Doses
Tobacco/areca nut chewers (Kerala, India) with well-developed oral leukoplakias were chosen for several short-term intervention trials. The participants were randomly distributed into four groups: (I) receiving 200,000 IU vitamin A per week (0.14 mg/kg body weight/day), (II) receiving vitamin A (100,000 IU/week) and beta carotene (180 mg/week), (III) receiving beta carotene (180 mg/week), and (IV) receiving placebo capsules, for up to six months. Endpoints are the frequency of micronucleated cells in the oral mucosa, the complete remission of oral leukoplakias, and the development of new leukoplakias. Preliminary results indicate a remission of oral leukoplakias of 57.1% in group I, 27.5% in group II, 14.8% in group III, and 3.0% in group IV, and the development of new leukoplakias in 0% of group I, 7.8% in group II, 14.8% in group III, and 21.2% in group IV. The search for maintenance doses was recently initiated by placing chewers with remission of leukoplakias on various weekly doses of beta carotene and low doses of vitamin A. Papers are in press in Int. J. Cancer and Cancer Letter.

TYPE: Intervention
TERM: Betel (Chewing); Premalignant Lesion; Prevention; Tobacco (Chewing); Vitamins
SITE: Oral Cavity
CHEM: Beta Carotene; Retinoids
LOCA: India
TIME: 1984 - 1992

118   Stich, H.F.   04402
BC Cancer Research Centre, Environmental Carcinogenesis Unit, 601 W. 10th Ave., Vancouver BC, Canada V5Z 1L3 (Tel.: (604)8776010; Tlx: 04507648)
COLL: Brunnemann, K.D.; Parida, B.B.

Carcinogens in the Saliva of Chewers Using Various Types of Smokeless Tobacco
A link between levels of tobacco-specific nitrosamines (TSNA) in the saliva, micronucleated cells in the oral mucosa, oral leukoplakia, and carcinomas is being sought by examining population groups with various oral habits, including snuff dippers (Inuit, Northwest Territories, Canada), users of Khaini tobacco (Bihar, India), users of tobacco-containing toothpaste (Orissa, India), and reverse smokers (Orissa, India). In addition, TSNA will be determined in the nasal mucosa of individuals using nasal snuff daily (Germany, and Kerala, India). The following compounds will be quantitated: N-nitrosonornicotine, N-nitrosoanatabine, N-nitrosoanabasine, and 4-(methyl-N-nitrosamino)-1-(3-pyridyl)-1-butanone. For counting the frequency of micronuclei, exfoliated cells will be sampled from the buccal mucosa,

CANADA

palate, lower gingival groove and tongue. The question of whether intraoral areas with a high frequency of micronucleated cells coincide with the distribution of oral carcinomas will be investigated.

TYPE: Correlation
TERM: Chemical Exposure; Micronuclei; Premalignant Lesion; Tobacco (Chewing); Tobacco (Smoking)
SITE: Oral Cavity
CHEM: N-Nitroso Compounds
LOCA: Canada; Germany; India
TIME: 1986 - 1992

* 119   Yang, C.   05097
BC Cancer Control Agency, Div. of Epidemiology, Biometry, & Occupational Oncology, 600 W. 10th Ave., Vancouver BC, Canada V5Z 4E6 (Tel.: (604)8776000; Fax: 8724596)
COLL: Band, P.R.; Gallágher, R.P.; Spinelli, J.J.; Daling, J.R.; Weiss, N.S.

Influence of Reproductive and Occupational Exposures and Passive Smoking on the Incidence of Breast Cancer

The specific aim of this population-based case-control study is to examine the association of breast cancer with the following: prior abortion, postpartum use of lactation suppressing agents, postmenopausal use of progestin, occupation and history of passive smoking. All women in British Columbia (BC) aged 75 years or less diagnosed with breast cancer between June 1988 and November 1989 were identified through the BC Cancer Registry. The 1989 Provincial Voters' List was used to select randomly a group of controls with an age distribution similar to that of the cases. Information regarding personal characteristics, pregnancy, and medical and occupational histories was ascertained by postal questionnaire. Telephone calls were used to follow-up non-responders and to obtain missing information. At the close of recruitment, 1109 cases and 1074 controls had returned the questionnaire. A logistic regression model will be used as the primary analytic tool.

TYPE: Case-Control
TERM: Abortion; Hormones; Lactation; Occupation; Passive Smoking; Registry
SITE: Breast (F)
REGI: Br. Columbia (Can)
TIME: 1988 - 1992

WINNIPEG

120   Choi, N.W.   04870
Manitoba Cancer Treatment & Res. Found., Dept. of Epidemiology & Biostatistics, 100 Olivia St., Winnipeg Manitoba, Canada R3E 0V9 (Tel.: (204)7872178; Fax: 7836875)
COLL: Nelson, N.A.; Moodie, P.F.; Oh, H.C.; West, P.; Musto, R.

Cancer among Registered Indians in Manitoba

The Manitoba Cancer Registry will be linked to the Medical Services Bureau's data base, which includes some 60,000 registered Indians who have resided in Manitoba at any time during a nine-year period, 1978-1986, in order to identify Indians who have been diagnosed with cancer during this period. Expected numbers of various cancers will be calculated on the basis of cancer incidence in the non-Indian population of Manitoba. Certain cancers in excess or deficit will be further investigated by means of case-control studies. A standardized questionnaire will be used with the assistance of nursing station personnel.

TYPE: Incidence
TERM: Indians; Record Linkage; Registry
SITE: All Sites
REGI: Manitoba (Can)
TIME: 1991 - 1993

* 121   Oh, H.C.   05139
Manitoba Cancer Treatm. & Res. Found., Dept. of Epidemiology & Biostatistics, 100 Olivia St., Winnipeg Manitoba, Canada R3E 0V9 (Tel.: (204)7872178; Fax: 7836875)
COLL: Choi, N.W.

Passive Smoking and Lung Cancer

In this study, the effect of involuntary smoking on non-smokers and mild smokers is of interest. Lung cancer risk will be assessed in relation to both involuntary smoking volume and that of voluntary

CANADA

smoking. Collection of data has already started on known risk factors of lung cancer cases who come from the Manitoba Cancer Registry, including past smoking and involuntary smoking in conjuction with the study on lung cancer and home environment. This is a community based case-control study using systematic interview questionnaires. The possibility of expanding this study to other centres in Canada is being explored.

TYPE: Case-Control
TERM: Passive Smoking; Registry
SITE: Lung
REGI: Manitoba (Can)
TIME: 1988 - 1992

CHILE

SANTIAGO

122 Nervi, F. 04728
Pontificia Univ. Catolica de Chile, Depto. de Gastroenterologia, Casilla 114-D , Santiago, Chile (Tel.: +56 2 397780; Fax: 2225515 ; Tlx: 240395 pucva cl)
COLL: Chianale, J.

Gallbladder Cancer: Prevalence, Early Detection, and Aetiology in a High-Risk Area.
The aim of this study is to detect subjects with gallbladder cancer in the pre-symptomatic stage in a Chilean population. The prevalence of cholesterol gallstones (the major risk factor for gallbladder cancer) is approximately 45% in Chileans older than 20 years. Clinical survey and echography will be used in a selected population of 10,000 subjects and also a small Indian village in Southern Chile (1,000). This first stage of the study will last from 1990-1995. Cases and subjects with potential risk factors (e.g. chronic carriers of S. Typhi) will be followed to estimate incidence of both gallstone disease and gallbladder cancer. Analysis of data will include univariate and multivariate logistic regression methods. A paper was published in Int. J. Cancer 46:1131-1133, 1990 (J. Chianale et al.).

TYPE: Cross-Sectional
TERM: Gallstones
SITE: Gallbladder
TIME: 1990 - 1995

* 123 Serra, I. 05221
Univ. of Chile, Faculty of Medicine, School of Public Health, Independencia 939, Santiago Correo 7, Chile (Tel.: 776560/5352; Fax: 776560/5363)
COLL: Báez, S.; Calvo, A.; Decinti, E.; Maturana, M.

Socio-Economic Status and Urban-Rural Residence as Risk Factors for Biliary Tract Cancer in Chile
The aim of this retrospective study is to investigate whether particular subgroups of the Chilean population, stratified by socio-economic levels and by residential districts (comunas, categorised as urban, rural or intermediate), have a greater risk of dying from biliary tract cancer than the general population. The hypothesis is that the poor are more prone to this cancer. A random sample of approximately 130 of the national total of 1,178 deaths from biliary tract cancer in 1986 are being analysed to establish the risk of dying associated to the socio-economic status, place of residence and occupation. Socio-economic level will be measured through years of school. The 1982 census will provide the national figures.

TYPE: Case Series
TERM: Occupation; Rural; Socio-Economic Factors; Urban
SITE: Bile Duct; Gallbladder
TIME: 1991 - 1992

CHINA

BEIJING

* 124 Cai, S.X. 05011
Chinese Academy of Preventive Medicine, Inst. of Occupational Medicine, 29 Nan Wei Rd, Beijing 100050, China, People's Republic of (Tel.: 338761/260; Fax: 3011875)
COLL: Huang, M.Y.; Wang, A.R.; He, Y.P.; Su, C.Y.; Zhang, Z.Q.; Zhang, Z.; Zhou, Q.L.

Cancer Morbidity and Mortality Among Workers in the Chromate Production Industry
A cooperative group is undertaking a national survey to strengthen the prevention and treatment of occupational cancer in China, to study the morbidity of some common occupational tumours, to develope guidelines for the control of occupational tumours, and to provide scientific bases for the formulation of prevention and treatment measures. A retrospective survey is being carried out among 2,548 workers in chromate producing plants in Chongging, Shanghai, Tinghou, Qingdao, Suzhou and Tiangjin, and 5,458 control workers in the same cities (a total of 8,006 from start of operations to 31 December 1989, more than 30 years). Deaths from all causes and morbidity of all malignant neoplasms are being investigated.

TYPE: Cohort
TERM: Occupation
SITE: All Sites; Lung
CHEM: Chromium
TIME: 1982 - 1992

125 Chen, J.S. 03616
Chinese Academy of Preventive Medicine, Inst. of Nutrition and Food Hygiene, 29 Nan Wei Rd, Beijing 100050, China, People's Republic of (Tel.: 3011878)
COLL: Campbell, T.C.; Li, J.; Peto, R.

Diet and Nutritional Status and Cancer Mortality in China
The first ecological study on diet, lifestyle and cancer mortality (14 sites) was carried out in 65 counties in China in 1983. Information on food and nutrient intake, blood and urine assays for nutritional and other indices, and lifestyle/medical history questionnaire with a total of 367 variables were compiled in a monograph published in 1990 (Diet, Lifestyle and Mortality in China: A Study of the Characteristics of 65 Chinese Counties, Oxford University Press, Cornell University Press and the People's Medical Publishing House). About 20 papers have been published using this database and more papers are under preparation. In order to collect more information on diet and lifestyle, a second ecological study covered 69 counties (65 old ones plus four new ones) was carried out in 1989. In addition to the above data items, more information on blood pressure, lung function and socio-economic status (seven questionnaires) was collected. Blood and urine assays will be started in 1991 and the entry of questionnaire data is in progress. In order to update the mortality data for correlation analysis, a retrospective mortality survey including all causes of death in 1986-1988 was carried out in the 69 survey counties. Disease classification (ICD-9) and data entry are now being carried out.

TYPE: Correlation
TERM: Diet; Lifestyle; Nutrition; Socio-Economic Factors; Trace Elements; Vitamins
SITE: Breast (F); Colon; Leukaemia; Liver; Lung; Nasopharynx; Oesophagus; Rectum; Stomach
TIME: 1983 - 1995

* 126 Huang, M.Y. 05012
Chinese Academy of Preventive Medicine, Inst. of Occupational Medicine, 29 Nan Wei Rd, Beijing 100050, China, People's Republic of (Tel.: 338761/260; Fax: 3011875)
COLL: Cai, S.X.; Zhang, X.M.; Hung, K.L.; Wu, H.C.; Hu, X.L.; Zhang, M.Q.; Liang, X.Z.; Zhou, D.N.; Wang, Z.Y.; Lian, X.X.; Ding, C.Y.; Cheng, S.M.; Yang, C.; Chen, N.H.; Xie, C.L.; Cai, X.C.

Occupational Tumours in Workers Exposed to Vinyl Chloride
In order to improve the prevention and treatment of occupational malignant tumours and to understand their prevalence among workers exposed to vinyl chloride, a cooperative group was organised and 13 polyvinyl chloride production plants in China investigated. This survey covered about 6,000 workers exposed to vinyl chloride and about 7,000 unexposed workers who served as controls. A retrospective investigation has been completed, and a prospective study will examine all causes of death, all causes of malignant tumours and liver cancer on the basis of the previous survey.

CHINA

TYPE: Cohort
TERM: Occupation; Plastics
SITE: All Sites; Liver
CHEM: Vinyl Chloride
OCCU: Vinyl Chloride Workers
TIME: 1982 - 1992

* 127 Liang, S. 05013
Chinese Academy of Preventive Medicine, Inst. of Occupational Medicine, 29 Nan Wei Rd, Beijing 100050, China, People's Republic of (Tel.: 338761)
COLL: Duan, B.R.; Chen, S.Y.

Retrospective Cohort Mortality Study of Lung Cancer in Workers Exposed to Dust
The objective of this study is to investigate whether workers in the mining industry have an increased risk of dying from lung cancer compared with members of the general population. Possible associations between (1) mining and lung cancer; (2) factors related to lung cancer such as pneumoconiosis, dust levels, chemicals; and (3) smoking patterns in relation to dust exposure and lung cancer will be examined. A paper was published in the Chinese J. Ind. Hyg. and Occup. Dis. 5:26-28, 1987.

TYPE: Cohort
TERM: Dusts; Mining; Occupation; Pneumoconiosis; Tobacco (Smoking)
SITE: Lung
OCCU: Miners
TIME: 1990 - 1992

128 Lu, S.H. 04681
Chinese Academy of Medical Sciences, Cancer Institute, Zuoanmen Wai , Panjiayao, Beijing 100021, China, People's Republic of (Tel.: 781331/461)
COLL: Chui, S.X.; Luo, F.J.; Li, F.M.; Guo, L.P.; Yang, C.S.; Montesano, R.; Yang, W.X.

Relevance of N-Nitroso Compounds to Oesophageal Cancer in China
N-Nitroso compounds occur widely in the environment and most of them are carcinogenic to many species of animals. Previous work suggests that human exposure to nitroso compounds and their precursors is high in Lin-xian, a high oesophageal cancer incidence area in northern China. This project is intended to examine the role of these compounds in the aetiology of oesophageal cancer in Lin-xian with specific aims outlined below: (1) to assess the human exposure to N-nitroso compounds by studying the identities and quantities of nitroso compounds in the foods that are suspected risk factors, such as "pickled vegetables" and "mouldy grain", and determining the level of N-nitroso compounds in representative food samples before and after cooking; (2) to analyse the N-nitroso compounds in urine and gastric juice collected from three groups of individuals: early oesophageal cancer patients, those with severe oesophageal dysplasia, and normal individuals (controls); (3) to study N-nitroso compound-activated oncogenes in oesophageal epithelia of fetus and monkey; 4) to analyse DNA adducts in adjacent tissues and cancerous tissues of oesophagus collected from Lin-xian; 5) to determine whether supplementation with vitamins C and E decreases N-nitroso compounds in urine.

TYPE: Cross-Sectional
TERM: DNA; Prevention; Vitamins
SITE: Oesophagus
CHEM: N-Nitroso Compounds
TIME: 1985 - 1992

129 Tao, S.C. 04280
Beijing Inst. for Cancer Research, Dept. of Epidemiology, Da-Hong-Luo-Chang St. , Western District, Beijing 100034, China, People's Republic of (Tel.: 662974)
COLL: Guo, J.

Dietary Factors in the Aetiology of Breast Cancer among Postmenopausal Women
Certain epidemiological and experimental data indicate that there is a relation between diet and breast cancer incidence; relevant hypotheses concern the effects of fat, protein and total calories. The aim of this case-control study is to assess dietary factors and other characteristics of potential aetiological significance in the occurrence of breast cancer among postmenopausal women. Dietary and nutritional factors will be studied among 250-300 histologically diagnosed female breast cancer cases resident in

the Beijing urban area. Nearest neighbours will be used as controls, matched for age, sex and marital status.

TYPE: Case-Control
TERM: Diet; Menopause
SITE: Breast (F)
TIME: 1988 - 1992

130 Wei, L.   03940
Ministry of Public Health, Laboratory of Industrial Hygiene, Dept. of Radiation Health, 2 Xinkang St., Deshengmenwai, Beijing 100088, China, People's Republic of
COLL: Zha, Y.; Tao, Z.; Chen, D.; Yuan, R.; He, W.

Cancer Mortality and Mutational Diseases in a High-Background Radiation Area of Yangjiang, China
Measurements revealed that individual external exposure to the environmental radiation is 330mRmR/yr (average) in the high background radiation area (HBRA), and 114mRmR/yr (average) in a neighbouring control area (CA). About 80,000 inhabitants in each area whose families lived there for two or more generations are being observed to study whether there is any increase in cancer mortality or frequency of mutational diseases in the HBRA. Studies of environmental carcinogens and mutagens other than natural radiation by means of case - control study and commune survey showed that the factors believed to be related to mutational diseases were not statistically different between these two areas, except that the frequency of medical X-ray exposure of people in CA was a little higher than that in HBRA. About one million person-years in HBRA and as many in CA were observed for cancer mortality and sex-, age- and site-specific cancer mortality analysed. The total numbers of 31 kinds of hereditary diseases and congenital deformities in children below 12 years old between two areas (13,425 examinees in HBRA, 13,087 in CA) were almost identical. However, the frequency of Down's Syndrome in HBRA was statistically higher than that in CA, but the incidence in HBRA was within the range of spontaneous morbidity and dependent of material age. 25,258 and 21, 837 children were examined in HBRA and CA respectively.

TYPE: Cohort; Correlation; Mortality
TERM: Congenital Abnormalities; Down's Syndrome; Environmental Factors; Mutation Rate; Mutation, Germinal; Radiation, Ionizing
SITE: All Sites
TIME: 1972 - 1992

131 Wu, M.   03948
Chinese Academy of Medical Sciences, Cancer Inst., Department of Cellular Biology, Chaoyang Qu, Beijing 100021, China, People's Republic of (Tel.: 7715058; Fax: 5058)

Genetic Epidemiology of Susceptibility to Oesophageal Cancer in Yangcheng County
A previous retrospective survey revealed a strong tendency to familial aggregation of oesophageal cancer patients in Yangcheng, a county in Shanxi Province in the Taihang mountain range with a very high oesophageal cancer mortality rate (150/100,000). Follow-up of the 622 families positive for oesophageal cancer found in the 1979 survey was carried out in 1989. Results confirmed the existence of familial aggregation of oesophageal cancer in this county. Genetics and the cellular and molecular bases of susceptibility to oesophageal cancer are now being studied using laboratory methods, such as DNA repair capacity and linkage studies with various DNA probes. Papers have been published in Nat. Med. J. China 68:659, 1988 and 70:679-681, 1990.

TYPE: Cohort; Genetic Epidemiology
TERM: DNA Repair; Genetic Factors; High-Risk Groups; Linkage Analysis
SITE: Oesophagus
TIME: 1985 - 1995

* 132 Xie, D.   05040
Chinese Academy of Preventive Medicine, Inst. of Occupational Medicine, 29 Nan Wei Rd, Beijing 100050, China, People's Republic of (Tel.: 338761/312)
COLL: Fang, F.

Genetic Epidemiology in the Population of a High Incidence Area for Lung Cancer
This project aims to explore the relationship between pre-malignant lesions and exposure level to benzo(a)pyrene in the population of a high incidence area for lung cancer, Xuanwei county. Levels of UDS (unscheduled DNA synthesis) and micronucleus frequency in human lymphocytes will be

determined in individuals exposed to high, moderate and low levels of benzo(a)pyrene for 20 years and in a control group. The study population will be confined to women.

TYPE: Molecular Epidemiology
TERM: BMB; DNA; Environmental Factors; Female; Lymphocytes; Micronuclei; Premalignant Lesion
SITE: Lung
CHEM: Benzo(a)pyrene
TIME: 1991 - 1993

133　Xie, H.J.　04811
Beijing Tuberculosis & Thoracic, Tumor Research Inst., Bldg 11, 3rd Floor, West Yan-Jing St., Zhao Wei, Beijing 100025, China, People's Republic of (Tel.: 582864)
COLL: Fan, R.L.; Wu, Z.R.; Chen, S.H

Beijing Lung Cancer Study

There is a special problem of lung cancer in Beijing, China. It is known that the Chinese female risk of lung cancer is quite high, but the smoking prevalence is extremely low, around 10%. The risk factors in female lung cancer cannot only be smoking and other factors will be explored. A surveillance network including 60 points and a population of 700,000-1,000,000 has been established, and this population will be observed for five years. The indoor air pollution will be measured as well as carcinogenic substances of coal smoke from excised lung tissue and 24 hour urine specimens. It is planned to compare these measurements between female lung cancer cases and appropriate controls.

TYPE: Case-Control; Cohort
TERM: Coal; Cooking Methods; Female
SITE: Lung
CHEM: Benzo(a)pyrene
TIME: 1990 - 1994

134　Yin, S.　04809
Chinese Academy of Preventive Medicine, Inst. of Occupational Medicine, 29 Nan Wei Rd, Beijing, China, People's Republic of (Tel.: 3014187; Fax: 3011875)
COLL: Blot, W.J.; Li, G.

Study of Benzene Exposure in China

The aim of the study is to examine the dose-response relationship between benzene and leukaemia and other blood disorders. It is planned to carry out a retrospective cohort study. About 70,000 workers exposed to benzene and 30,000 control workers employed during 1972-1987 in 12 cities are being studied. Information on industrial hygiene including benzene concentration in the factories is collected. The benzene level for each job title is estimated. The personal information includes the details of occupational history, vital status, cause of death, and diagnostic material for leukaemia and blood disorders. The main sources of data are abstracted from the salary lists, job registry form, medical and physical exam records, and records of benzene measurement in the factories.

TYPE: Cohort
TERM: Occupation; Solvents
SITE: Leukaemia
CHEM: Benzene
TIME: 1986 - 1992

135　You, W.C.　04810
Beijing Inst. for Cancer Research, Da-Hong-Luo-Chang St., Beijing 100034, China, People's Republic of (Tel.: 6023658)
COLL: Li, J.Y.; Jin, M.L.; Zhang, L.; Chang, Y.S.; Zhao, L.; Yang, Z.T.; Feng, H.M.; Liu, W.D.; Yang, B.Q.; Shen, H.; Hen, Z.X.; Zhao, R.X.; Liu, X.Q.; Xu, G.W.; Blot, W.J.; Kneller, R.; Keefer, L.

Precancerous Gastric Lesions in Relation to Stomach Cancer in China

The gastric cancer precursor study in Shandong province is a seven year study designed to investigate the aetiology of precancerous lesions of the stomach. A variety of possible aetiological factors, including diet and smoking are being investigated. In addition, the study attempts to quantify rates of transition in gastric mucosal status over time, including transition from chronic atrophic gastritis and dysplasia to cancer. Subjects are identified from population rosters in 14 randomly selected villages in five towns located in Linqu county, a rural area in Shandong, People's Republic of China. 3,434 subjects were interviewed and given a physical and endoscopic examination. Biopsies, blood, urine and gastric juice were collected from individuals in the study. The characteristics of prevalence rates of precancerous gastric lesions, their relationships with N-nitroso compounds in urine and gastric juice, micronutrients in serum and environmental factors are being investigated. This is a collaborative study between China and the United States.

TYPE: Cohort
TERM: Atrophic Gastritis; BMB; Diet; Nutrition; Premalignant Lesion; Serum; Tissue; Tobacco (Smoking); Urine
SITE: Stomach
CHEM: N-Nitroso Compounds
TIME: 1989 - 1995

136 Zhang, S.Z. 04765
Lab. of Industrial Hygiene, Ministry of Public Health, Div. of Radiation Health, 2 Xinkang St., Deshengmenwai, Beijing 100088, China, People's Republic of (Tel.: 1166-238)
COLL: Sun, Q.; Tao, Z.

Cancer Mortality in a Steel Company

Cases comprise male workers who died of liver cancer (51 cases) and stomach cancer (42 cases) in Baotou Steel-Iron Company during the period 1972-1987. Two controls were individually matched to each case on the basis of nationality, sex, age and year of death. All cases and controls were listed from the death registry in the company, and their occupational files were then reviewed; wives, parents or children were also interviewed to collect detailed information on factors such as occupational exposure to X-ray and inhaled thorium-containing dusts, as well as smoking, alcohol history, medical X-ray, diet, and socio-economic status. Odds ratios will be used as the measure of association between the occupational risk factors and cancers studied. The purpose of the study is to explore whether the occupational factors, especially radiation exposure, lead to an increased risk of liver cancer and stomach cancer.

TYPE: Case-Control
TERM: Alcohol; Diet; Dusts; Occupation; Radiation, Ionizing; Socio-Economic Factors; Tobacco (Smoking)
SITE: Liver; Stomach
CHEM: Thorium
TIME: 1989 - 1992

* 137 Zhang, S.Z. 05033
Lab. of Industrial Hygiene, Ministry of Public Health, Div. of Radiation Health, 2 Xinkang St., Deshengmenwai, Beijing 100088, China, People's Republic of (Tel.: 1166-238)
COLL: Wang, Q.L.; Tin, D.Y.

Lung Cancer Risk from Indoor Radon and Its Daughters

A case-control study is being carried out to evaluate the lung cancer risk from indoor radon and its daughters. Approximately 400 cases of lung cancer diagnosed during the period 1981-1990 will be included in the study, and an equal number of controls living in the same areas selected at random will be matched by age, sex and calendar year of death. A personal interview will obtain information on socio-demographic and environmental factors. The level of radon, radon daughters and gamma radiation in parts of dwellings will be measured. The results of primary measurements indicated that the concentration of radon reaches 50-300 Bq/M3. Logistic regression analysis will be used to examine the relationship between lung cancer risk and radon exposure in the home.

TYPE: Case-Control
TERM: Radiation, Ionizing
SITE: Lung
CHEM: Radon
TIME: 1990 - 1995

138 Zhang, Z.W. 04006
Beijing Inst. for Cancer Research, Da-Hong-Luo-Chang St., Beijing 100034, China, People's Republic of (Tel.: 6023658)
COLL: Gou, Y.R.; Zhao, L.; Song, W.Z.; Zhang, Y.

CHINA

Psychosocial Factors and Cancer: A Case-Control Study
In this case-control study 200 cases with stomach cancer and 200 with breast cancer diagnosed after July 1986 in the Beijing area will be selected. 400 healthy individuals and 400 other cancer patients will be selected as controls. Each subject will be investigated using a questionnaire on personality and life events. The study aims to investigate: (1) the relationship between personality type and cancer incidence; (2) the personality and characters of cancer patients and (3) the relationship between life events and cancer incidence.

TYPE: Case-Control
TERM: Psychological Factors
SITE: All Sites; Breast (F); Stomach
TIME: 1986 - 1992

139 Zhang, Z.W. 04585
Beijing Inst. for Cancer Research, Da-Hong-Luo-Chang St., Beijing 100034, China, People's Republic of (Tel.: 6023658)
COLL: Guo, Y.R.; Zhao, L.; Song, W.Z.; Zhang, Y.

Psychosocial Factors and Cancer: A Cohort Study
In this prospective study, 15,000 people living in the Beijing area will be observed for at least five years for cancer incidence. Their age is from 30 to 60 years, and they have middle or higher education degrees. Personality type and life events of every subject will be investigated and scored in the first year. Cancer incidence of each personality type (including depression, anxiety, social control etc.) will be compared to estimate the combinative action of life events. This is the first prospective study on psychosocial factors and cancer in China. This project is in cooperation with the National Institute of Psychology and Beijing Public Security Bureau. One characteristic is to use local Security Policemen as investigators. The aim is to investigate the relationship between cancer incidence and personality type, and to develop new methods of cancer prevention and treatment.

TYPE: Cohort
TERM: Psychosocial Factors
SITE: All Sites
TIME: 1989 - 1993

GUANGZHOU

140 Hu, M.X. 04078
Sun Yat Sen Univ. Medical Sciences, Department of Medical Statistics, & Community Medicine, Zhongshan Rd II, Guangzhou, China, People's Republic of (Tel.: 778223; Tlx: 0074)
COLL: Liu, Q.

Nasopharyngeal Carcinoma in Sihui County
In order to study further the relationship between dietary habits, especially the consumption of salted fish, and nasopharyngeal carcinoma incidence in a high-risk population, another case-control study on nasopharyngeal carcinoma is being carried out in Sihui County. All the patients with nasopharyngeal carcinoma diagnosed with histological verification during the study period in this area are selected as cases. The controls are matched to the cases on the basis of age, sex and residence. Conditional logistic regression techniques will be applied to analyse the data.

TYPE: Case-Control
TERM: Diet
SITE: Nasopharynx
TIME: 1986 - 1992

141 Li, C.C. 03657
Tumor Hosp., Cancer Inst., Sun Yat Sen Univ. Medical Sciences, Head & Neck Dept., Dongfeng 4th Rd East, Guangzhou, China, People's Republic of (Tel.: 778223/2792)
COLL: Liu, M.Z.

Serum VCA-IgA and MA-IgA Investigation for Nasopharyngeal Cancer (NPC) in a High Risk Area
In a prospective study in a high risk area of Guangdong, sera from normal individuals and from persons with nasopharyngeal mucosal hyperplastic lesions (NPHL) will be taken. VAC-IgA and MA-IgA antibody investigations will be carried out by IE or IF tests using genetically engineered antigen as a diagnostic

CHINA

antigen. 100 persons with positive VCA-IgA and MA-IgA antibody results and 100 persons with negative results, as well as the individuals presenting with NPHL will be followed-up for a five-year period, to determine the occurrence of NPC in different individuals.

TYPE: Cohort
TERM: Antibodies; Premalignant Lesion; Sero-Epidemiology
SITE: Nasopharynx
TIME: 1986 - 1992

142   Li, C.C.   04328
Tumor Hosp., Cancer Inst., Sun Yat Sen Univ. Medical Sciences, Head & Neck Dept., Dongfeng 4th Rd East, Guangzhou, China, People's Republic of (Tel.: 778223/2792)
COLL: Soo, R.

Nasopharyngeal Mucosal Hyperplastic Lesions and their Relationship with Early Malignant Change
Nasopharyngeal Mucosal Hyperplastic Lesions (NMHL) are considered to be precancerous lesions of NPC and the aim is to study the relationship between NMHL and NPC. Serological analysis (VCA-IgA, EA-IgA and MA-IgA) and biopsies will be carried out during cancer surveys in high risk areas. In addition the ultrastructural characteristics of NMHL under transmission electron microscope (TEM) and the expression of epidermal growth factor (EGF) receptor on nasopharyngeal epithelium will be studied. Detection of serum VCA-IgA, EA-IgA, and MA-IgA will be carried out every year during cancer surveys. Incidence among people with NMHL will be compared with that of a control group. Biopsy tissues are examined under TEM. Monoclonal antibody against EGF receptor is used to study the expression of EGF receptor on epithelium, which may have some relationship with cell transformation. The research is being carried out in high risk areas such as Zongshan County and Guangzhou city.

TYPE: Cross-Sectional
TERM: Antibodies; Biopsy; Premalignant Lesion
SITE: Nasopharynx
TIME: 1986 - 1992

HANGZHOU

* 143   Jiao, D.A.   05226
Zhejiang Medical Univ., Cancer Inst., 68# Jiefang Rd, Hangzhou 310009, China, People's Republic of (Tel.: (0571)727427; Fax: 771571 ; Tlx: 35036 ZMU CN)
COLL: Chen, K.

Assessment of Validity and Reliability of Retrospective Dietary Survey Method
Dietary factors are very important risk factors or protective factors for cancers and cardiovascular diseases. A valid and reliable survey method is an urgent requirement. A quantitative food survey method has been designed. From the quantity of food and size of portions taken by the subjects, the nutritional elements (e.g. protein, fat, crude fibre, vitamins) are calculated. Assessment of the validity and reliability of this method is being carried out. A total of 233 research subjects are included (105 males, 128 females).

TYPE: Methodology
TERM: Diet; Nutrition
SITE: All Sites
TIME: 1988 - 1992

* 144   Yang, G.   05225
Zhejiang Medical Univ., Cancer Inst., 68# Jiefang Rd, Hangzhou 310009, China, People's Republic of (Tel.: (0571)727427; Fax: 771571 ; Tlx: 35036 ZMU CN)
COLL: Zheng, S.; Yu, H.; Ma, X.Y.; Jiao, D.A.; Yuan, Y.C.; Yao, K.Y.; Zhou, L.; Chen, K.

Family Aggregation and Genetic Factors in Common Cancers
The aim of this population-based case-control study is to assess the relationship between several common cancers in China (including stomach, colorectum, liver, breast, lung) and their genetic background, as well as the interrelation between environmental and genetic factors. 3,200 cancer cases diagnosed in the period 1987-1991 will be compared with 3,200 non-consanguineous controls matched for sex, age and residence. Information collected includes pedigree, sex, age, history of residence, residence of family, diet, smoking habits and family disease history (especially history of

cancer). Information is collected from cases and their spouses, their first-degree relatives and their second-degree relatives. The analysis indicators are the cumulative incidence rate, familial aggregation, relative risk, heritability and genetic model.

TYPE: Case-Control; Genetic Epidemiology
TERM: Diet; Environmental Factors; Familial Factors; Genetic Factors; Genetic Markers; Geographic Factors; Multiple Primary; Pedigree; Segregation Analysis; Tobacco (Smoking)
SITE: Breast (F); Colon; Liver; Lung; Rectum; Stomach
TIME: 1991 - 1995

* 145   Yu, H.   05230
Zhejiang Medical Univ., Cancer Inst., 68# Jiefang Rd, Hangzhou 310009, China, People's Republic of (Tel.: (0571)727427; Fax: 771571 ; Tlx: 35036 ZMU CN)
COLL: Zheng, S.; Jiao, D.A.; Chen, K.; Yang, G.; Zhou, L.; Yuan, Y.C.; Ma, X.Y.; Shao, Y.W.

Intervention Trial of Colorectal Cancer in a High Incidence Area

Jiashan County of Zhejiang Province has had the highest mortality rate of colorectal cancer in China (22.65/100,000 in 1974-1976, 21.30/100,000 in 1986-1989). Epidemiological studies previously conducted in the county showed that dietary habits, contaminated drinking water, parasite infestation and low selenium content in soil and grains were closely associated with colorectal cancer. The county is randomly divided into trial and control areas with attention given to avoid contamination of control areas from trial areas. The intervention measures include health education through community and mass media, safe drinking water supply, prevention and treatment of parasite diseases (schistosomiasis and hookworm infestation), provision of selenium-salt in trial area. The receptive rate of health education, prevalence rate of parasitic diseases, the consumption rate of selenium salt and the compliance with advice on healthy lifestyles are used as intermediate indicators for evaluation. The final indicator for evaluation will be the reduction in incidence rate and mortality from colorectal cancer as well as cancers of all sites. An economic evaluation (cost-benefit and cost-effectiveness analysis) will also be made.

TYPE: Intervention
TERM: Cost-Benefit Analysis; High-Risk Groups; Lifestyle; Parasitic Disease; Water
SITE: Colon; Rectum
CHEM: Selenium
TIME: 1991 - 2000

146   Zhang, S.Q.   03053
People's Hospital of Zhejiang, Department of Colorectal Cancer, Hangzhou Zhejiang, China, People's Republic of

Relationship between Schistosomiasis and Colorectal Cancer

As schistosomiasis ova produce proliferation and anaplasia of intestinal epithelium, it is proposed to follow individuals with and without evidence of schistosomal infection as determined during mass screening of a county population.

TYPE: Cohort; Correlation
TERM: Parasitic Disease; Premalignant Lesion; Screening
SITE: Colon; Rectum
TIME: 1977 - 1992

* 147   Zheng, S.   05227
Zhejiang Medical Univ., Cancer Inst., 68# Jiefang Rd, Hangzhou 310009, China, People's Republic of (Tel.: (0571)727427; Fax: 771571 ; Tlx: 35036 ZMU CN)
COLL: Wu, J.M.; Yu, H.; Jiao, D.A.; Chen, K.; Zhou, L.; Ma, X.Y.; Shoa, Y.W.; Yuan, Y.C.; Sun, Q.R.; Shu, W.X.; Tang, H.X.

Screening for Colorectal Cancer in China

The aim of this study is to establish a practical and cost-effective screening model for colorectal cancer in China. Based on results of previous laboratory and epidemiological studies, a screening programme is being implemented in a high incidence area. More than 70,000 residents (30+ years old) are recruited for screening. In this programme, the combination of immunological faecal occult blood test and computerized risk assessment for screenees are being used as a primary screening procedure, fiberoptic sigmoidoscopy as the secondary screening and the full colonoscopy and/or air-barium enema as the diagnostic procedure. Evaluation of the efficacy of the programme will be based on

CHINA

whether there is a change of mortality from cancer of all sites as well as in mortality from colorectal cancer; a cost-benefit and cost-effectiveness analysis will also be done.

TYPE: Cohort
TERM: Cost-Benefit Analysis; Screening
SITE: Colon; Rectum
TIME: 1986 - 1995

* 148   Zheng, S.   05229
Zhejiang Medical Univ., Cancer Inst., 68# Jiefang Rd, Hangzhou 310009, China, People's Republic of (Tel.: (0571)727427; Fax: 771571 ; Tlx: 35036 ZMU CN)
COLL: Jiao, D.A.; Yu, H.; Chen, K.; Zhou, L.; Yang, G.; Lai, K.D.; Cheng, Z.Y.; Gao, P.; Xia, L.F.; Yang, W.X.

Colorectal Cancer in Six Regions of China

The incidence and mortality of colorectal cancer varies greatly in different areas of China. The hypothesis is that this variation is due to differences in lifestyle and dietary habits. To verify the hypothesis, a population-based case-control study of colorectal cancer is conducted in six regions of China: Dalian, Shenyang (Northeast), Yingchuan (Northwest), Guiyang (Southwest), Zhengzhou (Middle) and Hangzhou (East). The study includes 200 pairs of colorectal adenocarcinoma cases and controls (1:1 match) from each region (1,200 pairs in total). Past lifestyle, including diet, food energy intake, physical activity, family history and reproductive history (females) are assessed. A comparison of risk factors will be made between different regions.

TYPE: Case-Control
TERM: Alcohol; Diet; Family History; Lifestyle; Physical Activity; Reproductive Factors; Tobacco (Smoking)
SITE: Colon; Rectum
TIME: 1988 - 1993

* 149   Zhou, L.   05228
Zhejiang Medical Univ., Cancer Inst., 68# Jiefang Rd, Hangzhou 310009, China, People's Republic of (Tel.: (0571)727427; Fax: 771571 ; Tlx: 35036 ZMU CN)
COLL: Yu, H.; Chen, K.; Yang, G.; Zheng, S.

Risk Factors for Colorectal Cancer

A case-control study is being conducted to study risk factors in populations at high risk for developing colorectal cancer. All patients detected in a screening programme carried out in Jia Shan County are included in the study. Information on diet, medical history, family history, etc. is being collected. Cases will be matched with controls for age, sex and residence. Conditional regression analysis will be performed.

TYPE: Case-Control
TERM: Diet; Family History; High-Risk Groups
SITE: Colon; Rectum
TIME: 1991 - 1993

HARBIN

150   Dai, X.D.   03787
Harbin Medical Univ., Cancer Research Inst., Dept. of Cancer Epidemiology, Ha-Ping Rd, Harbin 150040, China, People's Republic of (Tel.: 65003)
COLL: Wan, X.R.; Lin, C.Y.; Zhang, C.Y.; Gao, Y.J.; Jiang, J.S.

Nasopharyngeal Carcinoma in North China

This is a comparison study of epidemiological characteristics and aetiological factors related to NPC in endemic and non-endemic regions. The study relates to the low-incidence areas. (1) In a case-control study 100 patients with NPC will be compared with 200 matched controls, 100 patients with other tumours and 100 normal individuals. The EBV antibodies of serum will be determined by immunoperoxidase and immunofluorescence assay. (2) Chromosomal analysis of tumours of different histological subtypes will be undertaken. (3) Differences in patterns of smoking, exposure to potential carcinogens, etc. will be explored in patients with the more differentiated NPCs and those with undifferentiated NPCs. (4) The role of genetics in NPC will be investigated by collecting pedigrees from multiple-case NPC families. In a case-control study, 100 patients with NPC will be matched with 100

CHINA

patients with tumours of the head and neck. All these patients must belong to families of which the last three generations were born in Heilongjiang Province.

TYPE: Case-Control
TERM: Antibodies; Chemical Exposure; Genetic Factors; Infection; Pedigree; Tobacco (Smoking); Virus
SITE: Nasopharynx
TIME: 1985 - 1992

* 151   Dai, X.D.   05147
Harbin Medical Univ., Cancer Research Inst., Dept. of Cancer Epidemiology, Ha-Ping Rd, Harbin 150040, China, People's Republic of (Tel.: 65003)
COLL: Sun, X.W.; Liu, C.B.; Shi, Y.B.; Len, C.Y.

Risk Factors for Lung Cancer among Women in North-East China

A case-control study among women in Harbin is being carried out. The study aims to investigate the role of smoking and other risk factors in the aetiology of lung cancer in women. The cases include all newly diagnosed primary lung cancer in females aged less than 70 in the study areas, between 1991 and 1994. A population control will be selected for each case, matched on sex and five-year age group. A pre-coded questionnaire will be used. Information about demographic factors, active and passive smoking exposure, lifetime residential and occupational histories, diet and cooking practices, personal history of non-malignant lung diseases, history of tuberculosis and cancer of lung in first-degree relatives will be collected. Logistic regression will be used for the analysis.

TYPE: Case-Control
TERM: Coal; Cooking Methods; Diet; Female; Occupation; Passive Smoking; Tobacco (Smoking)
SITE: Lung
TIME: 1991 - 1995

152   Hu, J.   04770
Harbin Medical Univ., Dept. of Epidemiology, 41 Da Zhi St., Harbin 150001, China, People's Republic of
COLL: Yunyuan, L.

Estimate of Latency in Stomach Cancer

This study will estimate the latency of stomach cancer caused by smoking and drinking and provide a theoretical basis for prevention. Data on smoking, drinking and diet for different time periods will be obtained by interview in about 200 cases with histologically confirmed cancer of the stomach and 400 controls. Analysis by logistic regression for risk factors will be carried out together with the estimation of latency of stomach cancer.

TYPE: Case-Control
TERM: Alcohol; Diet; Latency; Tobacco (Smoking)
SITE: Stomach
TIME: 1987 - 1992

* 153   Hu, J.   05068
Harbin Medical Univ., Dept. of Epidemiology, 41 Da Zhi St., Harbin 150001, China, People's Republic of
COLL: Liu, Y.Y.

Estimation of Latency of Lung Cancer Caused by Smoking

This study will estimate the latency of lung cancer caused by different types of cigarette. Data concerning smoking were obtained by interview and questionnaire in about 170 cases with histologically confirmed cancers of the lung and about 340 controls. By means of calculating accumulated effective exposure score and excess exposure fraction, the latency of lung cancer will be estimated using a multivariate model.

CHINA

TYPE: Case-Control
TERM: Latency; Tobacco (Smoking)
SITE: Lung
TIME: 1987 - 1992

154   Yu, Z.F.   04118
Harbin Medical Univ., Dept. of Epidemiology, 41 Dong Da Zhi St., Harbin 150001, China, People's Republic of (Tel.: 341157)
COLL: Li, L.; Dong, J.

Aetiological Study of Cervical Cancer in North China
Cervical cancer, despite its declining mortality in China, is still frequent. In order to study the aetiology of cervical cancer, a case-control study is being carried out at nine hospitals in North China. All new cases diagnosed over a one year period will be identified (200 expected) and controls (400) matched with cases according to age at the time of diagnosis are selected from hospitals in the same region for each case. A detailed questionnaire will elicit information on genital hygiene, sexual activity, socio-economic status, familial cancer history, cigarette smoking habits, chronic cervicitis and other risk factors. Logistic regression methods will be used to analyse the data. In addition, trends in mortality from cervical cancer over some 20 years in the Daowai district of Harbin city will be analysed by birth cohort.

TYPE: Case-Control; Mortality
TERM: Birth Cohort; Reproductive Factors; Sexual Activity; Socio-Economic Factors; Tobacco (Smoking); Trends
SITE: Uterus (Cervix)
TIME: 1982 - 1992

* 155   Yu, Z.F.   05180
Harbin Medical Univ., Dept. of Epidemiology, 41 Dong Da Zhi St., Harbin 150001, China, People's Republic of (Tel.: 341157)
COLL: Dong, J.; Pu, L.M.; Love, E.J.

Risk Factors for Breast Cancer with Different Histological Types
A case-control study will be carried out to investigate the aetiology of breast cancer of different histological types. The cases will be in-patients with newly diagnosed and histologically confirmed breast cancer, selected from the seven major hospitals in Harbin and the major hospitals in the other large cities of the Heilongjiang. Two controls will be matched to each case on age (within 2.5 years) and usual place of residence. The interview questionnaire will include environmental, genetic, menstrual, marital, dietary, and reproductive factors, history of breast feeding, family history etc. Conditional logistic regression will be used in the analysis.

TYPE: Case-Control
TERM: Diet; Environmental Factors; Genetic Factors; Histology; Reproductive Factors
SITE: Breast (F)
TIME: 1991 - 1994

HEFEI

* 156   Chen, R.   04957
Anhui Medical Univ., School of Public Health, Hefei 230032, Anhui, China, People's Republic of
COLL: Hu, M.X.; Coggon, D.

Lung Cancer Mortality in Copper Miners
The objective of the study is to evaluate lung cancer mortality in copper miners. A historical cohort has been established at Tongling copper mines in Anhui, China from 1970 to 1990. Approximately 7,000 men who worked at the copper mines for five years or more have been identified from the occupational records of the miners. Their lung cancer mortality will be compared with that of the general population in Tongling city. Preliminary analyses of lung cancer mortality for the period 1970 to 1985 show that there is a high risk of lung cancer in copper miners, especially in underground miners and drilling miners.

CHINA

TYPE: Cohort
TERM: Dusts; Mining; Occupation
SITE: Lung
CHEM: Copper; Iron; Silica
OCCU: Miners, Copper
TIME: 1990 - 1992

HENYANG

157 Fan, J.X.   03488
Ministry of Nuclear Industry, Hunan No 415 Hospital, Huan Henyang, China, People's Republic of (Tel.: 22851)
COLL: Wang, L.H.; Zhou, C.T.; Nie, G.H.; Huang, Y.H.

Primary Lung Cancer in Uranium Miners
A number of retrospective epidemiological investigations aim to identify the relationship between lung cancer in uranium miners and exposure to radon decay products, to provide a basis for protection of workers in uranium mines. However several questions remain unanswered, in particular, the relationship between the risk of cancer and the cumulative radiation exposure in terms of working level-months. Uranium miners in the Middle-South of China will be grouped into cohorts. Information required includes: date of birth, date of entry into mine, work history, date of leaving mine, health and cumulative exposure doses (WLM). Those suffering from lung cancer will be ascertained with date of diagnosis, tumour histology type, X-ray appearance, presence of metastases and date of death being recorded. About 15,000 people exposed to radon and its daughters are available for study. The control group is to be made up of general regional populations. A paper has been published in Huazhong J. Ind. Hyg. 1:23, 1985.

TYPE: Cohort
TERM: Histology; Metals; Occupation; Radiation, Ionizing; Time Factors
SITE: Lung
CHEM: Radon; Uranium
OCCU: Miners, Uranium
TIME: 1985 - 1992

NANJING

158 Ding, J.H.   04716
Jangsu Cancer Research Inst., Dept. of Epidemiology, No 42, Bei Zi Ting Rd , Nanjing 210009, China, People's Republic of (Tel.: 713125)
COLL: Lin, Y.T.

Risk Factors for Colorectal Cancer
This case-control study aims to identify risk factors associated with large bowel cancer and to compare differences in the risk factors between colon and rectum cancer. 100 colon and 100 rectum cancer cases, newly diagnosed, will be interviewed using a specially designed questionnaire to obtain information on items such as dietary habits, lifestyle, medical history and familial cancer history. Individuals with diseases other than cancers and digestive system diseases, matched individually with cases for age and sex, will be selected from local hospitals to serve as controls. Methods of analysis will include Mantel-Haenszel analysis and conditional logistic regression analysis.

TYPE: Case-Control
TERM: Diet; Familial Factors; Lifestyle
SITE: Colon; Rectum
TIME: 1989 - 1992

* 159 Jin, Z.G.   05112
Inst. of Occupational Medicine, of Jiang-Su Province, 212 He-Ban-Chun , Mia-Gao-Qiao, Nanjing 210038, China, People's Republic of (Tel.: 651176)
COLL: Xue, S.Z.

Relationship between Bladder Cancer and Chlordimeform
The carcinogenic potential of chlordimeform (CDM) and its metabolites has been suspected since the 1970's. An increased incidence of carcinomas of the urinary bladder among workers exposed to

CHINA

p-chloro-o-toluidine, the main metabolite of CDM in humans, has been reported. A preliminary case-control study (55 bladder cancer cases and 55 controls) has been carried out to assess the relationship between bladder cancer and the use of CDM by pesticide operators. Preliminary results show that there is no relationship between CDM and bladder cancer after CDM has been used in the area for 10 years, but there is a significant relationship between smoking and bladder cancer. Recruitement will be continued for a further five-year period.

TYPE: Case-Control
TERM: Occupation; Pesticides
SITE: Bladder
CHEM: Chlordimeform
OCCU: Pesticide Workers
TIME: 1988 - 1995

* 160   Jin, Z.G.   05113
Inst. of Occupational Medicine, of Jiang-Su Province, 212 He-Ban-Chun, Mia-Gao-Qiao, Nanjing 210038, China, People's Republic of (Tel.: 651176)
COLL: Tao, B.; Xue, S.Z.; Lin, M.

Cancer and Polymorphism of N-Acetyltransferases in Chinese

N-acetyltransferase polymorphism is related to various types of cancer, for example, urinary bladder cancer to slow acetylator phenotype, colon cancer to rapid phenotype, etc. There is a significant difference in the gene frequency of N-acetyltransferases between Chinese and other races. This study will examine the difference between N-acetylation ratio of patients with cancer and the general population in areas of high cancer mortality in China, for example, hepatic cancer in Qin Dong county, oesophageal cancer in Huai-An county. A local, genetically balanced group of 300 students, and 50-100 patients with cancer will be assessed by chemical methods for their acetylation type. The study will test whether there is an association between specific cancer risks and the Chinese genetic make-up.

TYPE: Cross-Sectional; Molecular Epidemiology
TERM: Chinese
SITE: Liver; Oesophagus; Stomach
TIME: 1991 - 1993

NANNING

161   Liu, Q.F.   04548
Cancer Institute of Guangxi, 6 Bin-hu Rd, Nanning Guangxi, 530027, China, People's Republic of (Tel.: 21477344)
COLL: Mo, C.C.; Zhong, S.C.; Ran, R.Q.; Yeh, F.S.

Randomised Trial of Green Tea and Herbs in Prevention of Primary Hepatocellular Carcinoma in High Risk Area

PHC is closely related to aflatoxin B1 intake from food and to hepatitis B infection in Guangxi. Green tea, as well as certain Chinese traditional herbs, seems to have an inhibitory effect on hepatocarcinogenesis induced by aflatoxin B1 in rats. The aim of this follow-up study is to confirm the role of the factors suspected of involvement in PHC and the potential inhibitory effect of green tea and herbs. Since 1986, more than 15,000 serum specimens have been collected from the inhabitants of a high incidence area for PHC in Guangxi and tested for HBsAG, anti-HBs and anti-HBc. Meanwhile, urine specimens from children as well as adults have been quantitatively assayed by HPLC for aflatoxin M1. According to the results of the examinations, 3,000 individuals, considered to be at high risk for PHC were randomly divided into three subcohorts. All the members of each subcohort were routinely given tablets of green tea, radix salvia miltiorrhizae (a Chinese traditional herb) and a placebo, respectively. The risk of developing PHC between the three subcohorts will be compared in follow-up.

CHINA

TYPE: Intervention
TERM: HBV; Mycotoxins; Tea
SITE: Liver
CHEM: Aflatoxin
TIME: 1986 - 1992

* 162 Shang, E.R. 05127
Guangxi Medical College, Guangxi Cancer Inst., Dept. of Gynecologic Oncology, Bin-hu Rd, Nanning Guangxi, China, People's Republic of (Tel.: 21477924)
COLL: Wu, J.N.; Chen, X.Q.; Xie, X.L.; Li, L.

Cytogenetic Study of Patients with Malignant Neoplasms in GuangXi
Cells with similar abnormal karyotypes have been found with a higher frequency in patients with gynaecological malignancies (other than trophoblastic tumours) in a previous investigation. This project aims to confirm that observation by studying G-banded chromosomes (at least 50 cells per case, using peripheral lymphocytes) using more cases. The expected number of cases is 100, including at least 30 malignant ovarian tumours and 30 liver cancer cases.

TYPE: Molecular Epidemiology
TERM: Chromosome Effects; G-Banding; Lymphocytes
SITE: Liver; Ovary; Uterus (Cervix); Uterus (Corpus)
TIME: 1991 - 1992

163 Gao, R.N. 02754
Shanghai Cancer Inst., Dept. of Epidemiology, 2200 Xie Tu Rd, Shanghai 200032, China, People's Republic of (Tel.: 376550)
COLL: Gao, Y.T.; Guo, B.C.; Wu, X.N.; Yu, E.X.

Primary Liver Cancer in Shanghai
The study aims to determine the association of primary liver cancer with HBV infection and other known or suspected factors. 150 in-patients with primary liver cancer in the urban area of Shanghai will be interviewed. Two patient controls (other cancer and other diseases with the exception of liver diseases) will be matched for each case in the same hospital by sex, age (+/- five years) and place of residence. Sera of cases and controls will be tested for various HBV markers (HBsAG, Anti-HBs, Anti-HBc, HBeAG, Anti-HBe). Information on occupation, smoking habits, alcohol consumption, previous liver diseases, and family cancer history will also be collected by trained interviewers. The results will be expressed in the relative risk of these factors to primary liver cancer.

TYPE: Case-Control
TERM: Alcohol; Familial Factors; Infection; Sero-Epidemiology; Tobacco (Smoking); Virus
SITE: Liver
TIME: 1982 - 1992

164 Gao, R.N. 04436
Shanghai Cancer Inst., Dept. of Epidemiology, 2200 Xie Tu Rd, Shanghai 200032, China, People's Republic of (Tel.: 376550)
COLL: Tu, J.T.; Yang, G.; Ji, B.T.; Gao, Y.T.; Xu, G.X.; Mo, S.Q.

Colorectal Cancer in Shanghai
This population-based case-control study of colorectal cancer aims to examine the association of colorectal cancer of different anatomical locations and histological types with specific nutrients, food consumption, physical activity, reproductive history, hormone use, family cancer history, bowel disease history, psychological factors, exposure to tobacco, alcohol, occupation, radiation, drugs, etc. About 1,600 newly diagnosed colorectal cancer cases aged 15-69 years among the permanent residents of Shanghai urban area will be identified through the Shanghai Cancer Registry during the period from September 1988 to August 1990. Controls will be randomly selected from the general population by age-sex frequency matching. All cases and controls will be interviewed by trained interviewers with a structured questionnaire. The data will be analysed by univariate and multivariate methods.

TYPE: Case-Control
TERM: Alcohol; Diet; Drugs; Familial Factors; Histology; Hormones; Nutrition; Occupation; Physical Activity; Psychosocial Factors; Radiation, Ionizing; Registry; Reproductive Factors; Tobacco (Smoking)
SITE: Colon; Rectum
REGI: Shanghai (Chi)
TIME: 1988 - 1992

* 165   Gao, Y.T.   05109
Shanghai Cancer Inst., Dept. of Epidemiology, 2200 Xie Tu Rd, Shanghai 200032, China, People's Republic of (Tel.: 4315510; Fax: 4331428)
COLL: Ji, B.T.; Jin, F.; Blot, W.J.; McLaughlin, J.K.

Cancers of the Oesophagus, Colon, Rectum and Pancreas in Shanghai
The aim of the study is to identify risk factors for cancers of the oesophagus, colon, rectum and pancreas. Newly diagnosed cases of these cancers between 30-74 years of age among permanent residents of Shanghai urban area for the period 1 October 1990 to 30 September 1992 will be included in the study. The numbers of cases for each site to be interviewed are approximately 900 for oesophageal cancer, 900 each for colon and rectum cancers, and 460 for pancreatic cancer. Frequency matching will be used in the study according to the age and sex distribution of the cases of four cancer sites registered by the Shanghai Cancer Registry during 1984. 1500 controls will be randomly selected from the general population of Shanghai urban area. This control group will be used for each cancer site investigated in the study. A common questionnaire will be used for all study subjects, covering demographic factors, residential history, tobacco use, alcohol consumption, tea drinking, occupational history, dietary intake, medical history, family history of cancer, menstruation and reproductive history. Analysis of each cancer site and its subgroups (by histological type and anatomic site) will be conducted by logistic regression.

TYPE: Case-Control
TERM: Alcohol; Diet; Familial Factors; Menstruation; Occupation; Physical Activity; Registry; Reproductive Factors; Tea; Tobacco (Smoking)
SITE: Colon; Oesophagus; Pancreas; Rectum
REGI: Shanghai (Chi)
TIME: 1990 - 1993

166   Jin, F.   04632
Shanghai Cancer Inst., 2200 Xie Tu Rd, Shanghai 200032, China, People's Republic of (Tel.: 315360)
COLL: Qin, D.; Gao, Y.T.

Shanghai Cancer Registration System with Chinese Characters
The Shanghai cancer registration system, with Chinese characters, is a package established on micro-computer (IBM-compatible) for managing and analysing cancer incidence and mortality data in Chinese script. The Registry was established in 1963 and receives more than 15,000 new cancer notifications per year. The basic functions of the system mainly consist of four parts: data management, deleting duplications, statistics and graphs. The system is written in DBASE-III and BASIC, and uses Chinese characters for entering name, address and other data.

TYPE: Methodology
TERM: Registry
SITE: All Sites
REGI: Shanghai (Chi)
TIME: 1988 - 1993

167   Liu, P.L.   04308
Shanghai Medical Univ., School of Public Health, Dept. of Epidemiology, 138 Yi Xue Yuan Rd, Shanghai 200032, China, People's Republic of (Tel.: 431190043; Fax: 4330543 ; Tlx: 33325 SFMC CN)
COLL: Jin, T.H.; Chen, G.C.

Sero-Epidemiological Study of HBV in a Population with High Incidence of PLC
PLC is one of the most common malignant neoplasms in China, approximately 100,000 people dying from the disease annually. Qidong, Nanhuei and Haimen counties are endemic areas with the highest mortality. This project is designed to investigate some 5,000 persons from these areas and to examine their HBV markers. The aim is to determine whether the highest mortality from PLC is accompanied by a

CHINA

high frequency of HBV markers, and to describe the distribution of HBV markers and immunity to HBV in the three counties. It is hoped that a relationship between PLC and hepatitis B infection will be established.

TYPE: Correlation
TERM: Infection; Virus
SITE: Liver
TIME: 1988 - 1992

168 Liu, P.L. 04879
Shanghai Medical Univ., School of Public Health, Dept. of Epidemiology, 138 Yi Xue Yuan Rd, Shanghai 200032, China, People's Republic of (Tel.: 431190043; Fax: 4330543 ; Tlx: 33325 SFMC CN)
COLL: Wong, Y.F.

The Role of Viruses in Cervical Cancer

The aim of the study is to evaluate a possible association between viral infections and cervical cancer. Viruses known and suspected to be involved in the aetiology of cervical cancer, including HPV, HSV-2 and CMV will be examined together with their interaction with other possible factors, such as oral contraceptive use and sexual behaviour. Blood and fresh tissue samples from biopsies or surgical specimens will be collected from about 100 untreated cases and 100 controls. Protocols will be developed to assess the presence of markers of exposure to potential carcinogens in the specimens. DNA hybridization methods will be used to assess the proportion of subjects who are positive for specific types of HPV. Serological markers will be used for HSV-2, CMV and other viruses.

TYPE: Case-Control
TERM: Biopsy; CMV; DNA; HPV; HSV; Oral Contraceptives; Sero-Epidemiology; Sexual Activity
SITE: Uterus (Cervix)
TIME: 1991 - 1993

169 Shen, F.M. 04314
Shanghai Medical Univ., School of Public Health, Dept. of Epidemiology, Yi Xue Yuan Rd 138, Shanghai 200032, China, People's Republic of (Tel.: 4311900/216)
COLL: Haile, R.W.C.

Genetic Epidemiology of Breast Cancer in Shanghai

Breast cancer (BC) is a heterogeneous disease. It is hypothesized that the susceptibility to BC is inheritable, and its mode of inheritance is autosomal dominant. The major objectives of this study are: (1) to describe and quantify the observed family aggregation of BC using hospital-based and municipal cancer registry data on laterality, age at menarche or at menopause, age at first delivery, parity, family history and by different clinical and pathological types; (2) to carry out simple segregation analysis followed by complex segregation analysis by fitting genetic models to data collected from pedigrees with probands and showing excess kinship in order to examine the mode of inheritance and (3) using genetic markers and the computer program LIPED to perform genetic linkage analysis for those families in which cancer susceptibility is inherited as a single major gene locus. As bilateral disease probands with a definite family history and complicated with other gynaecological malignancies are very rare in the population, sample size in this study will be limited to about 100 families, and this will include five years' registered data.

TYPE: Genetic Epidemiology
TERM: Age; Genetic Markers; Linkage Analysis; Menarche; Menopause; Parity; Registry; Segregation Analysis
SITE: Breast (F)
REGI: Shanghai (Chi)
TIME: 1987 - 1992

170 Shu, X.O. 04492
Shanghai Cancer Inst., 2200 Xie Tu Rd, Shanghai 200032, China, People's Republic of (Tel.: 315360)
COLL: Gao, Y.T.; Zheng, W.; Yin, D.M.; Gao, R.N.; Jin, F.

Childhood Cancer in Shanghai

The general purpose of this case-control study is to determine the importance of various prenatal and childhood environmental exposures in the aetiology of childhood neoplasms such as leukaemia, brain tumour, lymphomas and cancer of liver and bone. Special emphasis will be placed on prenatal and early childhood exposures to chemicals, radiation, drugs, tobacco and alcohol, and also on the influence of high risk occupations and industries of the parents in the development of childhood cancer in their offspring. About 900 neoplasms diagnosed in children aged 0-14 between January 1980 and December 1989 will be identified through the Shanghai Cancer Registry. There will be about 300 leukaemia cases, 150 brain tumour cases, 100 lymphoma cases and 350 other cancer cases. An age-sex-frequency matched control group with 500 healthy children will be randomly selected from the general population. The mother and father of the study subjects will be interviewed in person. The information collected from the cases and controls will be compared, in order to identify factors associated with the development of different sites of childhood cancer. After 1990, the data collection will be continued for some rare childhood cancers so that the number of cases and controls will be large enough to evaluate the risk factors of these rare diseases as well. Clinical information for all cases will be abstracted from medical records to examine the prognostic factors of childhood cancer.

TYPE: Case-Control
TERM: Alcohol; Chemical Exposure; Drugs; Intra-Uterine Exposure; Occupation; Prognosis; Radiation, Ionizing; Registry; Tobacco (Smoking)
SITE: Bone; Brain; Leukaemia; Liver; Lymphoma
REGI: Shanghai (Chi)
TIME: 1988 - 1992

* 171   Tao, X.G.   04974
Shanghai Medical Univ., School of Public Health, Dept. of Environmental Health, 138 Yi Xue Yuan Rd, Shanghai 200032, China, People's Republic of (Tel.: (021)4311900; Fax: 4330543; Tlx: 33325 sfmc cn)
COLL: Hong, C.J.; Yu, S.Z.

Risk of Male Lung Cancer Attributed to Indoor Coal Combustion
Lung cancer is now the leading cause of death from cancer among males in Shanghai. Indoor air pollution from coal combustion may make some contribution. The purpose of this study is to explore the risk of lung cancer death in male residents who live in coal-using families. Stratified by two extreme levels of ambient sulphur dioxide and inhalable particulate concentrations, four areas were chosen in the city. In each of the areas, two neighbouring communities were chosen, one using coal and the other using coal gas as fuel. The percentage of smokers and the ambient environment of the two chosen communities in each area are similar. A total of 117,039 person-years was observed from 1 January 1978 to 31 December 1987 for all males in the eight communities. Standardized mortality rates for lung cancer among males in the eight communities are calculated. Relative risk and population attributable risk are being used to evaluate the risks.

TYPE: Cohort
TERM: Air Pollution; Coal; Male
SITE: Lung
CHEM: Sulphur Dioxide
TIME: 1987 - 1992

172   Tu, J.T.   04352
Shanghai Cancer Inst., Dept. of Epidemiology, No. 25, Lane 2200, 2200 Xie Tu Rd, Shanghai 200032, China, People's Republic of (Tel.: 375590/139)
COLL: Chen, H.Q.; Chen, X.L.

Cancer Screening Among Factory Workers
The benefit of mass screening for several cancer sites (breast, cervix, stomach, etc.) is largely accepted. The aim of this study is to evaluate the efficacy of a cancer screening programme conducted among about 100,000 factory workers. The target cancer sites are stomach, rectum, liver and lung in both sexes and breast and cervix in females. A standard questionnaire, including important risk factors for the target cancers, will be developed and used in interviewing. High risk individuals for different cancer sites will then be drawn out of the general population. Specified screening procedures will be used for early detection of different cancers. After the first mass screening, health monitoring for high risk individuals will subsequently be conducted by the trained doctors of the clinic in the factory. The length of the interval between mass screenings will be decided according to the staging of the new cases detected each year. All cancer cases at the target sites will be included in the analysis whether discovered by screening or not. The staging and survival time of cancer cases will be used as an intermediate index and changing cancer incidence and mortality as a final index for evaluating the efficacy of this cancer screening programme.

CHINA

TYPE: Intervention
TERM: High-Risk Groups; Monitoring; Screening; Stage; Survival
SITE: Breast (F); Liver; Lung; Rectum; Stomach; Uterus (Cervix)
TIME: 1988 - 2003

* 173   Wang, Y.X.   05095
Chinese Academy of Sciences, Shanghai Inst. of Nuclear Research, P.O. Box 800204 , Shanghai 201800, China, People's Republic of (Tel.: (8621)9528476; Fax: 950021 ; Tlx: 30910 SINRS CN)

Trace Elements and Liver Cancer
This project on Chongming Island, a high incidence area for liver cancer with an uneven geographical distribution, investigates the relation between trace elements and liver cancer. We selected two communities with high liver cancer mortality (> 40 per 100,000) and two with lower mortality (< 20 per 100,000) for the study. Samples of scalp hair and blood (5 ml) were collected three times, in March, July and November. Each time, 100 healthy adults aged between 20 and 40 (75 females, 25 males) were randomly selected from each community (600 samples in total). A case-control study of the association of liver cancer among the permanent farmers of Chongmin Island will be identified during the study period. Controls of the same age, sex and similar economic background will be randomly selected from the general population on the island. All subjects are peasants from locations with no industrial contamination. The hair trace element content is analysed by X-ray fluorescence, with ICAP for serum analysis, and differential pulse polarography catalytic wave analysis for selenium. The data will be analysed by univariate and multivariate methods.

TYPE: Case-Control; Correlation
TERM: Trace Elements
SITE: Liver
CHEM: Cadmium; Copper; Iron; Lead; Selenium; Zinc
TIME: 1984 - 1992

174   Xiang, Y.B.   04577
Shanghai Cancer Inst., Shanghai Cancer Registry, 2200 Xie Tu Rd, Shanghai 200032, China, People's Republic of (Tel.: 315360)
COLL: Gao, Y.T.; Jin, F.

Cancer in Shanghai: Relative Survival Analysis
Shanghai Cancer Registry has been collecting data on cancer patients since 1972. There are about 14,000 new cancer cases every year. Incidence, mortality and survival rates can be calculated. Survival analysis of lung cancer has been reported in China, but population-based survival analysis of all cancers has not been reported in detail, and relative survival rates have not been used. About 150,000 cancer cases were known to have died by 30 December 1987, the closing date for this analysis, but for about 40,000 cancer cases the vital status at this date is unknown, and they are being followed up to determine whether they are dead, alive or lost to follow-up. Using these data and the general population mortality data the observed and relative survival rates for cancer will be obtained to evaluate the results of activities for prevention and medical care of cancer in Shanghai. Comparisons by age and sex, and with other countries will be considered. Time trends in survival will also be analysed. Main methods of analysis are provided by a specialised statistics package of relative survival analysis.

TYPE: Incidence; Mortality
TERM: Registry; Survival; Trends
SITE: All Sites
REGI: Shanghai (Chi)
TIME: 1988 - 1992

175   Xue, S.Z.   04674
Shanghai Medical Univ., School of Public Health, Dept. of Occupational Health, Box 206, Shanghai 200032, China, People's Republic of (Tel.: (21)4311900/214; Fax: 4330543 ; Tlx: 33325 SFMC CN)
COLL: Wang, Z.Q.; Zhou, D.H.

Lung Cancer Incidence Among Employees Exposed to Low Levels of Technical CMME
A previous study on lung cancer mortality among employees exposed to relatively high concentrations of technical CMME demonstrated an excess of lung cancer. To study whether lung cancer risk will be similarly raised under improved working conditions with much lowered CMME concentrations, a new cohort of 500 employees has been followed since the early 1970's. The results will be helpful in verifying

the dose-response relationship and checking the effectiveness of preventive measures and the adequacy of the current exposure limit. Local mortality data are used as the basis for calculation of indirect SMRs. Data are collected through field survey and interview.

TYPE: Cohort
TERM: Dose-Response; Occupation; Prevention
SITE: Lung
OCCU: Chemical Industry Workers
TIME: 1988 - 1996

176 Xue, S.Z. 04807
Shanghai Medical Univ., School of Public Health, Dept. of Occupational Health, Box 206, Shanghai 200032, China, People's Republic of (Tel.: (21)4311900/214; Fax: 4330543 ; Tlx: 33325 SFMC CN)
COLL: Jin, Z.G.; Tao, B.; Ling, M.

Susceptibility to Naphthylamine Carcinogenesis, and N-Acetyl-Transferase Polymorphism.
Thousands of tons of antioxidants A & D derived from naphthylamines (technical grade alpha- and beta-isomers) have been produced and used in rubber processing in China since the mid 1950's. Cases of urinary bladder cancer have rarely been reported. The phenotype of rapid N-acetyltransferase activity might be responsible for this. A cohort study of the incidence and mortality of urinary bladder cancer among the exposed workers and a survey on the polymorphism of N-Acetyltransferase in those people is designed. The size of the cohort is estimated at over 500 and the period of follow-up would be around 20-25 years. Data will be collected through interviewing and urine analysis. Preliminary results reveal that most of the general population subjects are rapid acetylators and the few cases of bladder cancer are all slow acetylators.

TYPE: Cohort
TERM: Analgesics; Genetic Factors; Occupation
SITE: Bladder
CHEM: 1-Naphthylamine; 2-Naphthylamine; Amines, Aromatic
TIME: 1991 - 1995

177 Yu, S.Z. 04864
Shanghai Medical Univ., School of Public Health, Inst. of Preventive Medicine, 138 Yi Xue Yuan Rd, Shanghai 200032, China, People's Republic of (Tel.: 4311900/30; Tlx: 33325 CN)
COLL: Yang, X.Z.; Chen, Y.X.; Liao, C.S.; Xu, J.L.; Ren, U.F.

Evaluation of a Screening Programme to Identify High-Risk Populations
Using information on three major risk factors for cervical cancer, probabilities of developing the disease for individual women were calculated through multiple regression analysis in 24,633 married women of Jingan County, Jiangxi Province. Sensitivity and specificity analyses were carried out to determine a high-risk population for cervical cancer. Separate cervical cancer screening programmes for the high-risk and non-high-risk population are now being performed to evaluate the practical effects of risk factor screening for determining high-risk populations.

TYPE: Cohort; Methodology
TERM: High-Risk Groups; Screening
SITE: Uterus (Cervix)
TIME: 1989 - 1992

178 Yuan, J.M. 04581
Shanghai Cancer Inst., 2200 Xie Tu Rd, Shanghai 200032, China, People's Republic of (Tel.: 315360)
COLL: Yu, M.C.; Gao, Y.T.; Henderson, B.E.

Nasopharyngeal Carcinoma in Shanghai
This population-based case-control study is designed to investigate the association of NPC with (1) intake of preserved food commonly used in Shanghai; (2)occupational exposure to dust, smoke and fumes; (3) active and passive exposure to tobacco smoke; (4) exposure to incense; and (5) history of chronic ear and nose diseases. Cases will be histologically confirmed incident cases of NPC among permanent residents of Shanghai Urban aged 15-74, diagnosed from 1 January 1987 to 31 December 1990 and identified through the Shanghai Cancer Registry. It is expected to interview about 800 cases. The mothers of all cases and controls under age 50 years who reside in Shanghai will be interviewed

CHINA

(mothers of about 150 cases and 150 controls). 800 controls will be randomly selected from the general population of Shanghai Urban by age-sex frequency-matching. All interviews will be conducted in person by four trained nurses using a structured questionnaire. All information will be analysed using univariate and multivariate methods.

TYPE: Case-Control
TERM: Diet; Dusts; Occupation; Passive Smoking; Registry; Tobacco (Smoking)
SITE: Nasopharynx
REGI: Shanghai (Chi)
TIME: 1988 - 1992

179　Zheng, W.　04354
Shanghai Cancer Inst., Dept. of Epidemiology, 2200 Xie Tu Rd 25, Shanghai 200032, China, People's Republic of
COLL: Shu, X.O.; Dan, R.P.

Leukaemia in Urban Shanghai
A population-based case-control study is underway to investigate the risk factors for leukaemia in urban Shanghai. Approximately 500 new cases classified by FAB are expected to be identified through the Shanghai Cancer Registry during the period 1987-1989. A control will be selected for each case by multistage random sampling, matched on age and sex. Trained interviewers will use a standard questionnaire to obtain information on factors such as residential history, occupation and occupational exposure, drug usage, previous diseases, different sources of radiation exposure, dietary history and personal habits. The clinical information for all new cases diagnosed during 1985 to 1989 will be abstracted from medical records to examine the prognostic factors for leukaemia and to estimate the incidence rates for leukaemia of different histological cell types.

TYPE: Case-Control
TERM: Diet; Drugs; Histology; Lifestyle; Occupation; Radiation, Ionizing; Registry
SITE: Leukaemia
REGI: Shanghai (Chi)
TIME: 1987 - 1992

180　Zheng, W.　04586
Shanghai Cancer Inst., Dept. of Epidemiology, 2200 Xie Tu Rd 25, Shanghai 200032, China, People's Republic of
COLL: Gao, R.N.; Shu, X.O.; Ji, B.T.; Gao, Y.T.

Case-Control Study of Cancers of the Oral Cavity, Nose, Nasal Sinus and Larynx
The objectives of this study are to (1) investigate the association of cancers of the oral cavity, pharynx (nasopharynx excluded), nose, nasal sinus and larynx with the consumption of tobacco and alcohol; (2) investigate the relationship of these cancers to dietary related variables, radiation exposure, dental hygiene habits and previous diseases; (3) identify high risk occupations, industries and occupational carcinogenic exposure. Incident cases under 75 years old for carcinoma of oral cavity, pharynx (200 cases), nose, nasal sinus (150 cases) and larynx (300 cases) will be identified through Shanghai Cancer Registry during the period from Jan. 1988 to Dec. 1989 in Urban Shanghai. An age-sex frequency-matched control group with 400 subjects will be randomly selected from the general population. Interviews will be conducted in person with a structured questionnaire.

TYPE: Case-Control
TERM: Alcohol; Diet; Hygiene; Occupation; Radiation, Ionizing; Registry; Tobacco (Smoking)
SITE: Larynx; Nasal Cavity; Oral Cavity; Pharynx
REGI: Shanghai (Chi)
TIME: 1988 - 1992

* 181　Zhu, H.G.　04977
Shanghai Medical Univ., School of Public Health, Dept. of Environmental Health, 138 Yi Xue Yuan Rd, Shanghai 200032, China, People's Republic of (Tel.: (021)4311900; Fax: 4330543 ; Tlx: 3325 sfmc cn)
COLL: Tao, X.G.; Zhao, Q.Y.

Mutagenicity of Drinking Water and Risk of Stomach Cancer
It has been reported that the mutagenicity of drinking water increases progressively along the Huangpu River from the upper to the lower reaches. The positive samples (1983-1985) in both tap and raw water

CHINA

from the upper reaches are 0%, while those from the lower reaches are 100% and 96.9% respectively according to the Ames test. The purpose of this study is to assess the relationship between the mutagenicity of drinking water and the risk of male stomach and liver cancer, which now comprise about 45% of male cancer deaths among residents of Shanghai. Two communities have been chosen, one drinking tap water and another drinking river water directly from the upper reach area, and two further communities with the same practices from the lower reach area. 178,442 person-years were observed from 1 January 1984 to 31 December 1988, including all males born after 31 December 1959 in the four communities. Standardized mortality rates for stomach and liver cancer are calculated. Odds ratios from unconditional logistic regression with 14 independent variables are used to evaluate the risks of stomach and liver cancers when drinking lower reach water.

TYPE: Correlation
TERM: Mutagen; Water
SITE: Liver; Stomach
TIME: 1990 - 1992

TAIYUAN

182   Sun, S.   04803
China Inst. for Radiation Protection, Dept. of Radiation Medicine, P.O. Box 120, Taiyuan 030006, China, People's Republic of (Tel.: +86 351 774005; Fax: 740707)
COLL: Wang, Y.; Li, W.; Yuan, L.; Li, S.

Epidemiological Studies of Workers in the Nuclear Industry in China
The nuclear industry has been operated for about 30 years in China. It is important to know the possible late effects among workers with a history of occupational exposure. Historical cohort studies were conducted in the following four sub-cohorts: uranium prospecting teams; miners in uranium mines; workers in uranium plants and nuclear power plants. The period of follow-up is 1971-1986; the total number of subjects about 40,000. SMR and RR were analysed in different dose groups. Preliminary results show that there is no significant difference of incidence of cancer between exposure and control groups in nuclear plants. The only finding was an increase of lung cancer in uranium miners.

TYPE: Cohort
TERM: Occupation; Radiation, Ionizing
SITE: All Sites
CHEM: Radon; Uranium
OCCU: Miners; Power Plant Workers
TIME: 1986 - 1992

TIANJIN

183   Wang, Q.S.   04285
Tianjin Cancer Inst., Dept. of Epidemiology, Ti-Yuan-Bei, Tianjin, China, People's Republic of

Occupational Cancer Risks
Occupational hazards might be risk factors for several cancers. This study aims to assess the risk for different professions or industries by cancer site. Data on patients will be based on the Tianjin Cancer Registry. Data on the population by profession or industry are based on the database of the Third Chinese Census in 1982. All cancer patients diagnosed since 1981 will be included. Cancer sites will be coded to ICD-9 plus grouped sites, for example, cancers of digestive system. Professions or industries will be coded to 16 groups. Some industrial groups will be subdivided. Cancer mortality rate, incidence rate and relative risk by profession or industry for every site of cancer will be assessed.

CHINA

TYPE: Incidence
TERM: Chemical Exposure; Occupation; Registry
SITE: All Sites
REGI: Tianjin (Chi)
TIME: 1987 - 1992

XILINHAOTE

184 Latan, A. 04192
Hospital of Xilinguole League, Dept. of Surgery, Xilinhaote, China, People's Republic of

Gastric Cancer in Mongolian Herdsmen
The results of a mortality study carried out in 1977 showed an adjusted death rate for stomach cancer of 61.93 for Mongolian herdsmen compared with 15.93 for the general population. In the county of Xianghuang the adjusted death rate was as high as 107.25. The aim of the present study is to assess the relationship between dietary habits of Mongolian herdsmen and their high gastric cancer mortality. Their main diet consists of large amounts of Mongolian tea with milk, wheat porridge, dried meat, mouldy cheese and sour milk. Most of them are smokers and also heavy drinkers. Their annual consumption of green vegetables and fruit is very low. A preliminary study included 39 cases from Xianghuang and two control groups of similar size, one consisting of people who died from cardiovascular diseases, one of apparently healthy subjects. Controls were matched with cases for sex, age, socio-economic status and residence. Relatives of the deceased persons were interviewed, as well as the healthy people and medical records and life style investigated. Results of the preliminary study show an association between gastric cancer and intake of tea, dried meat, mouldy cheese, sour milk and alcohol, but not with smoked and salted food.

TYPE: Case-Control
TERM: Alcohol; Diet; Lifestyle; Tobacco (Smoking)
SITE: Stomach
TIME: 1977 - 1993

ZHENGZHOU

185 Lu, J.B. 02826
Henan Cancer Research Inst., Dept. of Epidemiology, Zhengzhou Henan Province, China, People's Republic of (Tel.: 552149/461)
COLL: Muñoz, N.; Yang, C.S.

Epidemiological Studies of Oesophageal Cancer
The aim of this work is (1) registration of cancer and of death from all causes to enable incidence and mortality studies to be carried out in the population of Lin Xian and Yu Xian counties (1,600,000) and in 10 counties in Henan (population 7,000,000). (2) The natural history of oesophageal pre-cancerous lesions is being studied by observation of 400 cases of oesophagitis; in addition, intervention studies are carried out on 80 cases of dysplasia under treatment, 80 cases of dysplasia pending treatment, and 80 normal subjects acting as controls. Every year blood samples are collected from these subjects for determination of vitamin A, vitamin B, carotene, vitamin C, etc. (3) Studies on prevention of oesophageal cancer will be carried out in Lin Xian county. A paper has been published in Int. J. Cancer 36:645-645, 1985.

TYPE: Incidence; Mortality
TERM: High-Risk Groups; Premalignant Lesion; Prevention; Vitamins
SITE: Oesophagus
CHEM: Beta Carotene
TIME: 1975 - 1992

186 Lu, J.B. 04095
Henan Cancer Research Inst., Dept. of Epidemiology, Zhengzhou Henan Province, China, People's Republic of (Tel.: 552149/461)
COLL: Sang, J.Y.; Dueng, W.Z.

CHINA

Precursor Lesions of Oesophageal Cancer: A Prospective Study of Oesophageal Epithelial Dysplasia in Linxian County

In 1975, 19,250 subjects aged over 30 were examined cytologically in a screening programme for oesophageal cancer in the high-risk population of Linxian county. 958 subjects were selected from these for a prospective study. They were divided into three groups: group A, composed of 294 cases which were diagnosed through cytology as having severe dysplasia; group B, made up of 328 cases with mild dysplasia, and group C consisting of 336 subjects with a normal oesophagus. Age, sex and residence in all the three groups were matched. They will be followed-up for 15 years (1975-1990), and some of the suspected risk factors and co-factors for cancer such as smoking habits, type of water used and socio-economic status, will be investigated by interview on a continuing basis. Papers have been published in Clin. Oncol. 12:195, 1985; Chinese J. Cancer 6(4):250-253, 1987 and in Int. J. Cancer 41:805-808, 1988.

TYPE: Cohort
TERM: Cytology; Premalignant Lesion; Screening; Socio-Economic Factors; Tobacco (Smoking); Water
SITE: Oesophagus
TIME: 1975 - 1992

187   Qiu, S.L.   04729
Henan Medical Univ., Inst. of Medical Sciences, 40 University Rd, Zhengzhou 450052, China, People's Republic of (Tel.: (0371)23817; Fax: 3109)
COLL: Yang, G.R.; Yang, C.S.

Dietary Calcium Supplementation and Oesophageal Epithelial Cell Proliferation in a Population at High Risk for Oesophageal Cancer

After a pilot study on the effect of supplementary calcium on oesophageal epithelial cell proliferation in animals and humans in a high-risk area for oesophageal cancer, a randomised double-blind intervention trial has been under way in Huxian county, Henan Province since late 1989 to determine whether oral calcium supplementation could reduce the proliferation of oesophageal epithelial cells. 604 people living in a high-risk area for oesophageal carcinoma were examined by endoscopy, and four or more biopsy specimens were taken from the middle- and lower-third of the oesophagus, or from macroscopic lesions for routine histopathological diagnosis and incubation with tritiated thymidine. From them, 200 cases suffering from oesophagitis with basal cell hyperplasia and/or dysplasia were selected, and then divided into two groups with equal numbers of males and females of the same age (+/-five years) and with similar severity of lesions. The participants were randomly allocated to treatment with 1200 mg of calcium or to a placebo daily for one year. At the end of the study, the effect will be evaluated by histopathology and micro-auto-radiography of the endoscopic biopsy mucosa.

TYPE: Intervention
TERM: High-Risk Groups
SITE: Oesophagus
CHEM: Calcium
TIME: 1989 - 1992

188   Yang, G.R.   04808
Henan Medical Univ., Henan Inst. of Medical Sciences, 40 University Rd , Zhengzhou Henan 450052, China, People's Republic of (Tel.: (0371)663817; Tlx: 3109)
COLL: Qiu, S.L.

Prevention and Treatment of Oesophageal Carcinoma in a High-Risk Population in China.

To reduce the morbidity and mortality of oesophageal cancer in a high-risk population, endoscopic surveys and intervention studies have been carried out in Huxian county, Henan Province during the last seven years in 1,793 subjects submitted to endoscopic examinations and biopsy. 36 cases have been diagnosed as oesophageal or cardial carcinoma, including 29 early and 7 advanced cancer. All of the early cancers were treated by surgery or endoscopic Nd-YAG laser therapy with good curative results. 791 cases suffering from chronic oesophagitis with basal cell hyperplasia and/or dysplasia have been assigned to a double-blind intervention study. Of these, 176 were treated with 10,000 IU retinol, 200mg riboflavine and 200mg vitamin C once a week; 60 cases received 200mg new retinoic analogue RII once a week in the period 1984-1985; and 126 cases were treated with 600mg of calcium per day between 1986-1987. Another 200 cases have been randomly allocated to treatment with 1200mg of calcium or to placebo per day since Dec. 1989. A regular cancer registry and follow-up study have been carried out in order to evaluate whether the secondary prevention mentioned above could reduce the morbidity and mortality of oesophageal cancer in these populations.

CHINA

TYPE: Intervention
TERM: Vitamins
SITE: Oesophagus
CHEM: Calcium
TIME: 1983 – 1992

COSTA RICA

SAN JOSÉ

189　Sittenfeld, A.N.A.　　　　　　　　　　　　　　　　　　　　　　　　04252
　　　　Univ. de Costa Rica, Centro de Investigacion en Biologia, Celular y Molecular, San José, Costa Rica (Tel.: 246749)
COLL:　Arce, V.; Brenes, F.; Visona, K.; Marten, A.

Correlation between Hepatocellular Carcinoma and HBV Markers in the General Population of Costa Rica

　　Hepatocellular carcinoma (HCC) has an incidence of 1.21/100,000 (approximately 40 cases a year) in Costa Rica, although there are geographical variations and areas of relatively high risk, which are endemic for HBV. It has been proposed that areas of low HBV incidence have a relatively small risk for HCC. The main objective of the present study is to determine the relationship between HBV (serum and tissue markers) and aflatoxin in patients with HCC. It is also hoped to establish the usefulness of serial alpha-foetoprotein molecular variants as an early indicator of HCC in a population of 200 HBsAg carriers from one of the high HBV incidence areas. Retrospective studies will be performed on a bank of serum samples from the HBsAg carriers, collected during a period of 12 years (1974-1986). Papers have been published in "Viral Hepatitis and AIDS". International Symposium of Viral Hepatitis and AIDS-LSU-ICMRT. San José, Costa Rica, Dec. 1-5, 1986. Villarejos, V.M. (ed.). Trejos Hnos, 1987, pp. 379-382 and 383-386, and J. Immunol. Methods 106:19-26, 1988.

TYPE:　　Correlation
TERM:　　HBV; Mycotoxins
SITE:　　 Liver
CHEM:　　Aflatoxin
TIME:　　 1974 - 1992

CUBA

HABANA

190   Fernández, L.M.   04621
Inst. Nacional de Oncologia, 29 y F. Vedado, Habana 4, Cuba (Tel.: 327531)
COLL: Caraballoso, M.; Lence, J.; Diaz, A.

Cancer Mortality in Cuba
   The main objective of this study is the analysis of cancer mortality data by factors such as person-years of life lost, geographical distribution, time trends, and the effects of age, calendar period and birth cohort. Data are provided by the statistical office of the Ministry of Public Health: these are tabulated by sex, residence and site (all sites, lung, breast), to produce age-specific and adjusted rates; these will be used for life-tables, regression models and generalised linear models. These studies will provide a description of geographic mortality patterns, for use in further analytical studies and in formulation of health policies.

TYPE: Mortality
TERM: Age; Birth Cohort; Geographic Factors; Time Factors; Trends
SITE: All Sites; Breast (F); Lung
TIME: 1987 - 1992

CZECHOSLOVAKIA

BRATISLAVA

191   **Dimitrova, E.**   03714
Inst. of Experimental Oncology, Slovak Academy of Sciences, Dept. of Epidemiology, Ul. Csl. armady 21., CZ 812 32 Bratislava, Czechoslovakia (Tel.: (47)57541)
COLL: Plesko, I.; Somogyi, J.; Kiss, J.; Kramárová, E.

Projections of Cancer Incidence and Mortality in Czechoslovakia
The main objective of this study is to determine projections of cancer incidence and mortality until the year 2000 and 2020 in Czechoslovakia. The projections are based on the evolution of incidence and mortality of individual cancer sites in the past 10-15 years and the population age-sex projections for the given years. For selected cancer sites the possible changes in risk factors and the effects of preventive measures will be taken into account in forecasting future incidence rates. Recent result were published in Cs. Zdrav. 34:281, 1986 and Neoplasma 35(6):635, 1988 and 36(4), 1989.

TYPE: Incidence; Mortality
TERM: Anatomical Distribution; Prevention; Prognosis; Projection; Time Factors
SITE: All Sites
TIME: 1986 - 1992

192   **Plesko, I.**   04781
Slovak Academy of Sciences, Inst. of Cancer Research, Ul. Csl. armady 21, 812 32 Bratislava, Czechoslovakia (Tel.: (7)57541)
COLL: Vlasák, V.; Obsiniková, A.; Kramárová, E.

Skin Cancer in Slovakia
Incidence and mortality data for non-melanoma skin cancer in the National Cancer Registry for the territory of Slovakia since 1968 are highly complete and detailed, including those on basal cell carcinoma. The main aim of this descriptive study is the evaluation of time-trends of individual histological types of skin cancer and of malignant skin melanoma as well as their different and changing distribution in the various anatomical parts of the skin. The possible causes of geographical aggregation and the risk of developing either another basal cell skin cancer or cancer of another organ in persons with basal cell cancer will also be investigated.

TYPE: Incidence
TERM: Anatomical Distribution; Geographic Factors; Histology; Multiple Primary; Registry; Trends
SITE: All Sites; Melanoma; Skin
REGI: Slovakia (Cze)
TIME: 1990 - 1992

KUTNA HORA

193   **Jílek, V.**   01174
Inst. of Mineral Raw Materials, Dept. of Geochemistry, Vitezna 425, 28403 Kutna Hora, Czechoslovakia (Tel.: 61234; Tlx: 123452 nskh)
COLL: Ujházy, V.; Plesko, I.; Krejcirík, L.; Kafka, S.; Charvátová, S.; Koutová, E.

Geographical Distribution of Malignant Tumours in Districts of Czechoslovakia in Relation to the Environment
This study aims to evaluate the effect of factors in the geochemical environment. The chemical composition of selected components of human environment in some districts of Czechoslovakia will be assessed in relation to the geographical distribution of malignant tumours. Maps of the data obtained will be prepared, and the level of concentration of the chemical elements received by the population will be calculated. Relationships between external factors in the geochemical environment and prevalence of malignant tumours will be evaluated statistically. At the same time, levels of some trace elements in the environment, and in certain organs of the body using necropsy material, will be studied using neutron activation analysis (NAA), atomic absorption spectrophotometry (AAS) and emission plasma spectrophotometry (DCP). Results will be used as a basis for improving the human environment.

CZECHOSLOVAKIA

TYPE: Correlation
TERM: Air Pollution; Chemical Exposure; Environmental Factors; Geographic Factors; Radiation, Ionizing; Soil; Water
SITE: All Sites
TIME: 1986 - 1992

* 194 Kubik, A. 05201
Research Inst. of Tuberculosis, & Respiratory Diseases, 67, Budinova, 180 71 Prague 8 - Bulovka, Czechoslovakia (Tel.: 8182060)
COLL: Walter, S.D.; Parkin, D.M.; Reissigova, J.; Adamec, M.

Implications of the Results of a Screening Programme for the Natural History of Lung Cancer
The objective of this study is to estimate some important characteristics of the natural history of lung cancer in a population of heavily smoking middle-aged males in Czechoslovakia. The baseline data, such as the observed prevalence of lung cancer at screening and the incidence during intervals between screens, have been obtained in a randomized controlled trial of over 6,000 high-risk individuals (Int. J. Cancer 45:26-33, 1990). To estimate (1) the duration of the pre-clinical state and of the lead time; (2) the sensitivity of the screening method; and (3) the underlying incidence rate in the screened group, the statistical model developed by Walter and Day (Am. J. Epidemiol. 118, 868-886, 1983) will be used. This model has proved useful in analysing data from breast cancer screening, but lung cancer has not so far been approached in this way.

TYPE: Cohort
TERM: High-Risk Groups; Screening; Tobacco (Smoking)
SITE: Lung
TIME: 1990 - 1992

195 Sram, R.J. 04222
Psychiatric Research Inst., Lab. of Pharmacogenetics, 181 03 Prague 8 - Bohnice, Czechoslovakia (Tel.: 8552247, 8552005)
COLL: Holá, N.

Cytogenetic Analysis in Groups Occupationally Exposed to Mutagens and Carcinogens
The effect of occupational exposure to mutagens and carcinogens on the frequency of chromosome aberrations in peripheral lymphocytes is studied in groups of workers exposed to the following agents: (1) halogenated ethers, bis(chloromethyl) ether and CMME, (2) substances released during soft-coal pressure gasification, (3) substances released during soft-coal open-cast mining, (4) mineral oils, (5) substances in the petrochemical industry. Approximately 300 workers and 50 matching controls are analysed each year. The aim of the study is to evaluate the effect of preventive hygiene measures in decreasing the frequency of chromosomal aberrations, and the possible relationship between cytogenetic data and types of biological markers.

TYPE: Cohort; Molecular Epidemiology
TERM: Chromosome Effects; Lymphocytes; Mining; Occupation; Petroleum Products; Prevalence; Prevention
SITE: Inapplicable
CHEM: BCME; Mineral Oil
OCCU: Chemical Industry Workers; Glass Workers; Miners, Coal; Petrochemical Workers
TIME: 1985 - 1992

196 Sram, R.J. 04223
Psychiatric Research Inst., Lab. of Pharmacogenetics, 181 03 Prague 8 - Bohnice, Czechoslovakia (Tel.: 8552247, 8552005)
COLL: Holá, N.

Occupational Exposure to Cytostatics in Hospitals
The effect of cytostatics in various clinical departments is followed using the cytogenetic analysis of peripheral lymphocytes in nurses and medical doctors. The aim of this study is to check working conditions and by new hygiene measures to decrease the level of cytostatics so as not to induce an increase in chromosomal aberrations. Approximately 60 nurses and medical doctors and 25 matching controls are analysed each year.

CZECHOSLOVAKIA

TYPE: Cohort; Molecular Epidemiology
TERM: Chemotherapy; Chromosome Effects; Lymphocytes; Occupation; Prevention
SITE: Inapplicable
OCCU: Health Care Workers
TIME: 1986 - 1992

197 Vonka, V. 03310
Inst. of Sera and Vaccines, Dept. of Experimental Virology, W. Pieck 108 , 101 03 Prague - 10, Czechoslovakia (Tel.: 732015)
COLL: Kanka, J.; Roth, Z.; Hirsch, I.; Suchánková, A.

Role of a Virus in the Aetiology of Cervical Cancer
The prospective study on the cervical neoplasia-HSV2 relationship has been completed. In the course of the study more than 10,000 women were enrolled and subsequently followed. A total of 150 cases of moderate to severe dysplasia, 83 cases of carcinoma in situ and 21 cases of invasive carcinoma were detected. More than 60% of patients were found at enrolment, the other cases developing in originally healthy subjects. The epidemiological characteristics of the patients were determined and on the basis of these findings control subjects were selected for the serological studies. For determining HSV2 antibody presence in sera taken at enrolment, the microneutralization test and type 2-specific solid phase radioimmunoassay were used. No difference in the prevalence of HSV2 antibody between the patients and the matched control subjects was revealed by either test. Moreover, no significant differences in the prevalence of HSV2 antibody were observed between patients suffering from the various pathological conditions or between those diagnosed at the enrolment and in the later course of the study. Present results do not provide any support for the hypothesis that HSV2 is aetiologically involved in cervical neoplasia. The material collected in the course of the study will be utilised for estimating the risk of cervical dysplasia, carcinoma in situ and invasive carcinoma, associated with past infections with other viral agents. This information will be obtained by determining the antibody presence in sera collected at enrolment and in the subsequent course of the study in patients and matched control subjects. Papers have been published in Int. J. Cancer 33:49-60 and 61-66, 1984; 38:161-165, 1986 and in Adv. Cancer Res. 48:141-191, 1987.

TYPE: Case-Control; Cohort
TERM: Antibodies; BMB; HSV; In Situ Carcinoma; Infection; Premalignant Lesion; Sero-Epidemiology; Serum; Virus
SITE: Uterus (Cervix)
TIME: 1975 - 1992

ROZNAVA

198 Icsó, J. 03865
District Institution, of National Health, Dept. of Occupational Medicine, Nemocnicna 18, 04801 Roznava, Czechoslovakia (Tel.: 244449, 568088; Tlx: 77306)
COLL: Szollosová, M.

Lung Cancer in Iron-Ore Miners in Slovakia
A significantly higher incidence of lung cancer in iron-ore miners has been observed in some parts of Slovakia. Besides alpha-radiation it is clear that other factors play a role in the aetiology: the chemical composition of iron-ore, underground dust from diesel-mechanisms with its inorganic and organic fraction, the mould in the water or air and other possible contributing aetiological factors. The aim of the study is to follow the occurrence of lung cancer in a group of approximately 1,000 iron-ore miners in Slovakia and to study their occupational and environmental hazards. The incidence of lung cancer in the miners will be compared with that in approximately 25,000 non-miners, over 20 years of age, living in the same district area.

CZECHOSLOVAKIA

TYPE: Cohort
TERM: Chemical Exposure; Dusts; Environmental Factors; Metals; Mining; Occupation; Radiation, Ionizing
SITE: Lung
CHEM: Iron; Radon
OCCU: Miners, Iron
TIME: 1980 - 1995

VYSNE HAGY

199 Jurikovic, M. 03801
Inst. for Tuberculosis, & Respiratory Diseases, 05984 Vysne Hagy, Czechoslovakia
COLL: Rovenský, E.; Komada, J.; Bereza, D.; Dolezel, L.; Oruzinsky, J.

Influence of Chrysotile on Development of Lung Malignancies in Asbestos Mining and Milling in Slovakia

This is a prospective clinical and epidemiological study, which explores the occurrence of lung cancer and malignant mesothelioma in employees of the mining and milling of asbestos (chrysotile). 150 employees are being regularly examined: personal and professional history, duration of asbestos exposure, estimating the load of asbestos by semi-quantitative investigation of asbestos fibre in sputum by microscopy, cytological investigation of the sputum, smoking, chest X-ray in P-A and lateral projection, and lung function. The relationships between these data are monitored. At the mid-point of the study - 1985 - one case of lung carcinoma (with asbestosis) has been found after 30 years of continuous exposure. The mean length of exposure of the whole cohort is 18 years. No malignant mesothelioma has been found, nor malignancies of other sites. Partial results suggest a probable low carcinogenic effect of the chrysotile being mined in this small area of Slovakia. Definitive conclusions will be stated after ten years of research in 1990.

TYPE: Cohort
TERM: Chemical Exposure; Cytology; Dusts; Minerals; Mining; Occupation; Sputum; Tobacco (Smoking)
SITE: Lung
CHEM: Asbestos, Chrysotile; Mineral Fibres
OCCU: Miners, Asbestos
TIME: 1981 - 1992

DENMARK

AARHUS

200   **Bundgaard, T.**   04769
Aarhus Univ. Hosp., ENT Dept., 8000 Aarhus C, Denmark (Tel.: +45 86 125555)
COLL: Wildt, J.; Elbrond, O.

Cancer of the Oral Cavity
The purpose of this matched case-control study is to estimate the relative risks of the use of tobacco, chewing tobacco, snuff, and alcohol, of dental status and the pursuit of certain professions in the development of oral cancer. The cases are consecutively referred patients with oral cancer in a geographical area with 1.4 million inhabitants. Controls (3 per case) will be matched to cancer cases for age and sex using the Danish Central Registry of Persons. About 180 cases and 480 referents will take part in the study. Methods of analysis will include Mantel-Haenszel analysis and logistic regression.

TYPE: Case-Control
TERM: Alcohol; Hygiene; Registry; Tobacco (Chewing); Tobacco (Smoking)
SITE: Oral Cavity
REGI: Denmark (Den)
TIME: 1986 - 1992

201   **Olsen, J.**   00923
Aarhus Univ., Inst. of Social Medicine, Høegh Guldbergsgade 8, 8000 Aarhus C, Denmark (Tel.: +45 86 138822; Fax: 139919)
COLL: Sabroe, S.

Cancer among 10,000 New Members of the Carpenters' and Cabinet Makers' Trade Union
Most cohorts in occupational epidemiology are based on cross-sectional sampled active workers, thus making the studies vulnerable to the so called "healthy worker selection". In this study cohort members enter at the beginning of their work career and are followed over time to estimate cancer incidence according to occupational exposure. 10,000 carpenters and cabinet makers were recruited in 1977-1980 and data on cancer will be obtained by linkage with the Danish Cancer Registry during the period 1990 to 2000. Detailed data on the occupational exposures are based on questionnaires sent out in 1980 and updated information from samples of the entire study population (both cases and controls - "case-base" study).

TYPE: Case-Control; Cohort
TERM: Chemical Exposure; Dusts; Occupation; Registry; Wood
SITE: Leukaemia; Nasal Cavity; Respiratory
OCCU: Cabinet Makers; Carpenters, Joiners; Lacquerers
REGI: Denmark (Den)
TIME: 1977 - 2000

202   **Sabroe, S.**   04334
Aarhus Univ., Inst. of Social Medicine, Høegh-Guldbergsgade 8, 8000 Aarhus C, Denmark
COLL: Olsen, J.; Søgaard, J.; Sell, A.; Schultz, H.

Testis Cancer and Birthweight
In Denmark more than 200 new testis-cancer cases are diagnosed every year. Apparently the incidence rate in Denmark is among the highest in the world. This is probably due to both genetic and environmental causes, and the epidemiology of the disease indicates that the first stages in the cancer process take place very early in life, maybe even in utero. The aim of the study is, in a case-control design, to compare birthweight and age of gestation in a group of unselected testis cancer patients with men of the same age without testis cancer. Thus birthweight is an indirect measure of damaging influences on the foetus. Cases are the 300 youngest patients from the radiotherapy unit in Aarhus. Age-matched males born in the same municipality have been selected as controls. Birth information has been obtained from the midwifery case records. An analysis of the seasonal variation in birth month of the testis cancer patients will also be carried out.

DENMARK

TYPE: Case-Control
TERM: Birthweight; Seasonality
SITE: Testis
TIME: 1987 - 1992

COPENHAGEN

203 Anderson, L. 04517
Danish Cancer Society, Danish Cancer Registry, Rosenvængets Hovedvej 35 , Box 839, 2100 Copenhagen, Denmark (Tel.: +45 31 268866; Fax: 260090)
COLL: Storm, H.H.; Jensen, O.M.; Johansen, H.

Late Effects of Thorotrast among Danish Patients 1932-1947
Information on approximately 1,100 patients injected with Thorotrast by cerebral angiography during the period 1935-1946 was collected at the Finsen Institute, and these patients have been followed for cancer occurrence. (Faber, Environ, Res. 18:37-43, 1979). The aim of this study is to re-establish the cohort with enough detail to permit record linkage with the Danish Cancer Registry, the Danish death registry and the National patient discharge registry. Cancer occurrence and mortality (all diseases) will be studied in relation to latency and radiation dose from Thorotrast, as well as other background variables. Data collection is based on hospital records, the old Thorotrast file and routine data collection from registries.

TYPE: Cohort
TERM: Dose-Response; Drugs; Late Effects; Latency; Radiation, Ionizing; Registry
SITE: All Sites
CHEM: Thorotrast
REGI: Denmark (Den)
TIME: 1989 - 1992

204 Autrup, H. 04677
Danish National Inst. of, Occupational Health, Dept. of Chemistry and Biochemistry, Lersopark Allé 105, 2100 Copenhagen, Denmark (Tel.: +45 31 299711; Fax: 270107)
COLL: Christensen, J.M.; Sorsa, M.I.; Farmer, P.B.

Biomonitoring of Styrene Exposed Individuals
The aim of the study is to develop and validate new methods to assess exposures to styrene, and to apply these methods in studies of exposed populations. Styrene has been classified by IARC as a possible carcinogen. There is, however, an urgent need for more research to establish the actual cancer risk of the more than 10,000 styrene workers in Denmark and for the more than 100,00 workers in the rest of Europe. Exposure occurs in situations where manufacturing of glassfibre-reinforced materials such as boats and windmill whips. 50 individuals with a well defined occupational exposure to styrene and a non-exposed control group (25 individuals) will be investigated. The methods to develop and validate are: (1) Identification and quantification of styrene-DNA adducts by &S'32.P-post-labelling technique (2) Quantification of styrene hemoglobin adducts by GC-MS. (3) Detection and quantification of urinary styrene-DNA repair products by combination of HPLC and GC-MS methods. The proposed study also includes: (4) Air-sampling, and sampling of biological materials, and anlysis for styrene in air, serum and urine, and analysis of micronuclei in lymphocytes. Detection of DNA adducts in urine may be an indirect measure of the biological activity within hours prior to sampling. It is expected that the present study will provide new information on the relation between external dose, biological active dose and cytogenetic parameters. The methods developed and used in the project may also be valuable for future investigations of exposure to other potential carcinogens.

TYPE: Methodology; Molecular Epidemiology
TERM: BMB; DNA Adducts; DNA Repair; Lymphocytes; Micronuclei; Occupation; Plastics; Serum; Urine
SITE: Inapplicable
CHEM: Styrene
OCCU: Mineral Fibre Workers
TIME: 1989 - 1992

205 Ewertz, M. 04591
Danish Cancer Society, Danish Cancer Registry, Rosenvængets Hovedvej 35 , Box 839, 2100 Copenhagen Ø, Denmark (Tel.: +45 31 268866)
COLL: Jensen, O.M.; Tjønneland, A.; Overvad, K.

DENMARK

Prospective Cohort Study of Diet and Cancer
It has been estimated that 10-70% of cancer mortality in the USA may be related to diet. It can therefore be assumed that a similar proportion of all cancer is attributable to diet in Denmark, where up to 80% of cancers occur at sites for which dietary components, nutrients or nutritional status have been suggested as possible risk factors. A large prospective cohort study is planned, to examine the association between specific dietary components, foods or nutrients -alone or in combination - and the risks for cancer at specific sites. A further objective is to create a data bank of information on dietary factors, personal habits and biological samples, which can be used for testing future hypotheses on cancer risk. The study population will consist of 60,000 persons aged 50-64 years, drawn from the Central Population Register. Information on usual adult diet and known important risk factors for cancer, such as smoking and reproductive history, will be obtained by a self-administered and interviewer-checked questionnaire. Anthropometric measurements (e.g., height, weight, body impedance) will be made, and biological samples (e.g., blood, urine, nails) stored. The biological sampling will be repeated, and follow-up questionnaires mailed to participants every five years to asses dietary changes. The study population will be followed to record cases of cancer, death and emigration by linkage to the Cancer Registry and the Central Population Register. Follow-up for other diseases will be feasible by record linkage with other disease registries. With these combined approaches, an almost complete follow-up is envisaged.

TYPE: Cohort
TERM: BMB; Diet; Nutrition; Physical Factors; Registry; Reproductive Factors; Tobacco (Smoking); Toenails; Urine
SITE: All Sites
REGI: Denmark (Den)
TIME: 1989 - 1993

206 Ewertz, M. 04592
Danish Cancer Society, Danish Cancer Registry, Rosenvængets Hovedvej 35 , Box 839, 2100 Copenhagen Ø, Denmark (Tel.: +45 31 268866)
COLL: Adami, H.O.; Holmberg, L.; Karjalainen, S.; Tretli, S.

Breast Cancer among Males
While breast cancer is common among women, it is an extremly rare disease in men: about 1% of all malignant breast neoplasms arise in males. Probably due to the rarity of the disease, little is known about the aetiology. In order to investigate the epidemiology of male breast cancer, a collaboration was agreed upon between the cancer registries of Denmark, Finland, Norway and Sweden. A first study describes the trend in male breast cancer incidence in the Nordic countries, confirming an approximately two-fold variation in the rates, while a second study includes a detailed analysis of the survival of 1,429 male patients with breast cancer. A case-control study is now underway to study aetiological factors. It includes all men diagnosed with primary carcinoma of the breast between 1 August 1987 and 30 July 1991 in the four Nordic countries. Data will be collected by self-administered (mailed) questionnaires, supplemented by telephone interviews when necessary. Papers have been published in Int. J. Cancer 43:27-31, 1989 and in Cancer 64:1177-1182, 1989 (Adami, H.O. et al.).

TYPE: Case-Control; Incidence
TERM: Registry; Survival; Trends
SITE: Breast (M)
LOCA: Denmark; Finland; Norway; Sweden
REGI: Denmark (Den); Finland (Fin); Norway (Nor); Sweden (Swe)
TIME: 1987 - 1992

207 Giwercman, A. 04439
Rigshospitalet, Univ. Dept. of Growth and Reproduction, Blegdamsvej 9 , 2100 Copenhagen Ø, Denmark (Tel.: 35455068)
COLL: Skakkebæk, N.E.

Prevalence of Carcinoma-In-Situ of the Testis among Infertile Men
Infertile men are known to have an increased risk of testicular cancer. Invasive cancer of the testis can be prevented if the malignancy is detected at the stage of carcinoma-in-situ(CIS). It is the aim of this study to assess the prevalence of CIS of the testis in the above-mentioned group of men in order to decide whether routine screening for early testicular neoplasia should be offered to these men. Testicular biopsies will be performed in 500 infertile men. Biopsy specimens will be evaluated by means of light microscopic examination. The prevalence of CIS of the testis in this group of men will be compared to the frequency of CIS in an autopsy series of gonads from 400 men who suffered sudden, unexpected death.

DENMARK

TYPE: Cross-Sectional
TERM: Autopsy; Biopsy; In Situ Carcinoma; Infertility
SITE: Testis
TIME: 1984 - 1992

208 Hansen, E.S. 03066
Univ. of Copenhagen, Inst. of Social Medicine, Blegdamsvej 3, 2200 Copenhagen N, Denmark (Tel.: +45 31 357900)

Role of Inhalation of Combustion Products in the Aetiology of Cancer and Atherosclerosis
Inhalation exposure to particulate and gaseous compounds from fossil fuel combustion is expected to increase cancer risk as well as the risk of developing atherosclerosis. The study will be carried out as a comparison of the follow-up experience of nine exposed occupational cohorts and three non-exposed cohorts. Comparisons will concern mortality from diseases related to atherosclerosis and incidence of carcinomas of the respiratory and gastrointestinal tracts, the urinary system, the reproductive organs, the CNS, the skin, the blood- and lymph-forming tissues and sarcomas. In this study the effects of occupational exposure will be evaluated in the light of the relative significance of particular characteristics of the exposure. Effect modification by other risk factors will be considered. Papers have been published in Scand. J. Work Environm. Hlth 15:45-46, 1989, Br. J. Ind. Med. 46:582-585, 1989 and 47:805-809, 1990.

TYPE: Cohort
TERM: Air Pollution; Chemical Exposure; Environmental Factors; Metals; Occupation; Petroleum Products; Registry
SITE: Bladder; Brain; Gastrointestinal; Haemopoietic; Male Genital; Respiratory; Sarcoma; Skin
CHEM: Arsenic; Chromium; Nickel; PAH
REGI: Denmark (Den)
TIME: 1983 - 1992

209 Johansen, H. 01932
Rigshospitalet, Blegdamsvej 9, 2100 Copenhagen, Denmark (Tel.: (01)31386633/5839)

The Danish Thorotrast Register Project
Continuous follow-up of some 1,000 patients injected with Thorotrast in Denmark prior to 1949, primarily for cerebral angiography, commenced in 1949. The cohort is followed systematically for cancer occurrence and death. The Thorotrast register is used for a linkage study, evaluating cancer incidence and mortality in collaboration with the Danish Cancer Registry and the Danish Institute for Clinical Epidemiology. The Register has been used in linkage studies with defined cohorts, such as epileptics, to control for the influence of X-ray media on cancer risk in the study of other risk factors. It is further planned that the Thorotrast Register form the basis for a histo-pathological study on cancers and other conditions among Thorotrast injected patients.

TYPE: Cohort
TERM: Drugs; Radiation, Ionizing; Registry
SITE: All Sites
CHEM: Thorotrast
TIME: 1949 - 1989

210 Kruger Kjær, S. 04593
Danish Cancer Society, Danish Cancer Registry, Rosenvængets Hovedvej 35, Box 839, 2100 Copenhagen Ø, Denmark (Tel.: +45 31 268866)
COLL: Bock, J.; Lynge, E.; Dahl, C.; Faber Vestergaard, B.; de Villiers, E.M.; Jensen, O.M.

Case-Control Study Focussing on "The Male Factor" for Cervical Cancer
Cervical cancer is the sixth most frequent type of cancer among women in Denmark, and the number of precancerous lesions diagnosed increased up to 1982, especially in younger women. The sexual habits of a woman are known to be associated with her risk for developing cervical cancer, but recent studies have indicated that the number of female partners that a man has had may also be of importance. The aim of the study is to test the following specific hypotheses: (1) Men with early onset of sexual activity and multiple sexual partners (including prostitutes) increase the risk for cervical cancer in their female partners; (2) Women who have first inercourse at an early age and multiple partners are at a high risk of developing cervical cancer; (3) Use of oral contraceptives increases the risk for cervical cancer; (4) Smoking is an independent risk factor for cervical cancer; (5) Early age of onset of smoking increases the risk for cervical cancer. The investigation is a population-based case-control study with a two-phase

DENMARK

data collection. In the first phase, women living in Greater Copenhagen and diagnosed with carcinoma in situ of cervical cancer in 1985-86 at the age of 20-49 years have been identified in the Cancer Registry. In the second phase, a subgroup who report having had only one sexual partner will be investigated. Controls have been selected as a random sample of women from the Central Population Register, stratified for age and area of residence to match the age-residence distribution of the cases. The controls for the second phase will be a subgroup with the same eligibility as the cases. Examination of case and control women together with their male partners is currently under way.

TYPE: Case-Control
TERM: Age; Oral Contraceptives; Registry; Sexual Activity; Time Factors; Tobacco (Smoking)
SITE: Uterus (Cervix)
REGI: Denmark (Den)
TIME: 1985 - 1992

* 211   Lynge, E.   05183
Danish Cancer Society, Danish Cancer Registry, Rosenvængets Hovedvej 35 , Box 839, 2100 Copenhagen Ø, Denmark (Tel.: +45 31 268866)
COLL: Johanning, H.

Heredity of Colorectal Cancer
Spouses of patients with colorectal cancer in Denmark did not have an increased risk of this disease. Parents of patients with colorectal cancer, however, did have a 1.6-1.9 fold risk of colorectal cancer. The two studies in combination point to a possible genetic factor in the aetiology of colorectal cancer of importance for the general population. To pursue this hypothesis further, a study is planned on the risk of colorectal cancer among other identifiable relatives of the index patients in the two previous studies. A statistical analysis will be made assuming various modes of inheritance.

TYPE: Genetic Epidemiology
TERM: Genetic Factors; Heredity; High-Risk Groups; Registry
SITE: Colon; Rectum
REGI: Denmark (Den)
TIME: 1991 - 1993

* 212   Lynge, E.   05184
Danish Cancer Society, Danish Cancer Registry, Rosenvængets Hovedvej 35 , Box 839, 2100 Copenhagen Ø, Denmark (Tel.: +45 31 268866)
COLL: Engholm, G.

Lung Cancer and the Environment
The purpose is to estimate the independent contribution to lung cancer occurrence in Denmark from occupational and environmental exposures. Both the overall and histology-specific lung cancer incidence will be analysed, based on data from the Occupational Cancer Register 1970-1985. In order to control for smoking, multivariate methods for standardization of lung cancer risk in occupational groups, geographical region, etc. will be made. The standardization will use data from about 100,000 personal interviews on smoking performed by Gallup during 1969-1975.

TYPE: Incidence
TERM: Environmental Factors; Histology; Occupation; Registry; Tobacco (Smoking)
SITE: Lung
REGI: Denmark (Den)
TIME: 1990 - 1992

213   Melbye, M.   03040
Danish Cancer Society, Danish Cancer Registry, Rosenvængets Hovedvej 35 , P.O. Box 839, 2100 Copenhagen Ø, Denmark (Tel.: +45 31 268866; Fax: 260090)
COLL: Biggar, R.J.; Ebbesen, P.

Natural History of Human Immunodeficiency Virus (HIV) Infection among Homosexual Men
A random selection of 260 homosexual men, all members of a Danish Gay organisation, has been studied approximately every year since 1981. Questionnaire information covering demographic data and sexual behaviour history has been obtained, and serum, lymphocytes, urine, saliva, and semen have been collected. The objective is to study the long-term consequences of HIV infection with respect to immunological changes, and clinical criteria (including AIDS and related cancers). Much of the

DENMARK

information already gathered from the first year's study has been covered in a review in Br. Med. J. 292:5-12, 1986.

TYPE: Cohort
TERM: AIDS; Homosexuality; Infection; Sero-Epidemiology; Sexual Activity; Virus
SITE: All Sites; Hodgkin's Disease
CHEM: Nitrites
TIME: 1981 – 2001

214 Mellemgaard, A. 04594
Danish Cancer Society, Danish Cancer Registry, Rosenvængets Hovedvej 35 , Box 839, 2100 Copenhagen Ø, Denmark (Tel.: +45 31 268866)
COLL: Lynge, E.; Jensen, O.M.

Colorectal Cancer: Incidence of Cancer in Spouses of Patients

Colorectal cancer is one of the most frequent cancers in Denmark, occurring in aproximately 4% of the population. A diet rich in fat and meat, but low in fibre, vegetables and perhaps calcium, is believed to increase the risk for colorectal cancer. Married couples are likely to eat the same diet throughout a substantial part of their lives, and we are investigating whether spouses of patients with colorectal cancer have a higher risk than the general Danish population. The incidence of colorectal cancer is therefore being investigated among the spouses of 9,554 patients with colorectal cancer notified to the Danish Cancer Registry in 1968-72. Of these, 1,010 patients were single, and the marital status was uncertain for 15 patients. Of the remaining 8,529, 8,345 eligible spouses have been traced, and 8,095 were alive on 1 January 1943. The incidence of cancers of the right and left sides of the colon and of the rectum are being investigated for these spouses through 1984.

TYPE: Cohort
TERM: Diet; Marital Status; Registry
SITE: Colon; Rectum
REGI: Denmark (Den)
TIME: 1984 – 1992

215 Møller, H. 04558
Danish Cancer Society, Danish Cancer Registry, Rosenvængets Hovedvej 35 , Box 839, 2100 Copenhagen Ø, Denmark (Tel.: +45 31 268866; Fax: 260090)
COLL: Jensen, O.M.; Ewertz, M.; Skakkebæk, N.E.; Giwercman, A.; Rørth, M.; Maase, H.

Risk Factors for Testicular Cancer and Cryptorchidism

Denmark has the highest registered incidence of testicular cancer (TC) in the world. The disease is most frequent in the young; 70% of cases are under 40 years of age at the time of diagnosis. The only known strong risk factor for TC is cryptorchidism, but several observations indicate that this association may be indirect, and caused by underlying aetiologic factors common to both TC and cryptorchidism. This population-based case-control study will comprise 700 TC cases, 550 men who have been treated for cryptorchidism, and 1,400 population controls. The men are interviewed by telephone and their mothers are approached with a self-administered questionnaire. Factors to be evaluated are: the mother's use of hormones, her height and weight, details about the pregnancy, her use of tobacco and alcohol, her occupational exposures before and during pregnancy, weight of the son at birth, occurrence of congenital malformations, infectious diseases during childhood, history of testicular trauma, the man's use of tobacco and alcohol, and occupational exposures.

TYPE: Case-Control
TERM: Alcohol; Congenital Abnormalities; Cryptorchidism; Hormones; Infection; Intra-Uterine Exposure; Occupation; Physical Factors; Pregnancy; Registry; Tobacco (Smoking); Trauma
SITE: Testis
REGI: Denmark (Den)
TIME: 1988 – 1992

* 216 Nielsen, N.H. 05076
Danish Cancer Society, Danish Cancer Registry, Inst. of Cancer Epidemiology, Rosenvængets Hovedvej 35, Box 839, 2100 Copenhagen Ø, Denmark (Tel.: +45 31 268866; Fax: 269000)
COLL: Lanier, A.K.; Gaudette, L.A.; Nikitin, Y.P.; Storm, H.H.

DENMARK

Cancer in Circumpolar Inuit Eskimos
Systematically collected cancer incidence data 1969-1988 from circumpolar areas is being compiled with specific reference to Inuit (Eskimo) population groups. The data originate from population-based Inuit cancer registries in Alaska (Center for Disease Control, Anchorage), Canada (Statistics Canada, Ottawa), Greenland (Danish Cancer Registry) and USSR (Institute of Internal Medicine, Novosibirsk). In most circumpolar areas the patterns are changing rapidly, parallel to changes in lifestyle. A comprehensive description of the findings in the world Inuit population will be presented including trends and comparisons between Inuit groups and with cancer incidence in other parts of the world. The results will be published in a monograph which will also include related background information. A summary is planned in the Inuit language.

TYPE: Incidence
TERM: Eskimos; Ethnic Group; Lifestyle; Registry; Trends
SITE: All Sites
LOCA: Canada; Greenland; USSR; United States of America
REGI: Canada (Can); Denmark (Den)
TIME: 1990 - 1992

217   Olsen, J.H.   04595
Danish Cancer Society, Danish Cancer Registry, Rosenvængets Hovedvej 35 , Box 839, 2100 Copenhagen Ø, Denmark (Tel.: +45 31 268866)
COLL: Boice, J.D.; Fraumeni, J.F.; Jensen, J.

Cancer Risk in Epileptics and their Children Following Treatment with Anticonvulsive Drugs
The aims of the study are to evaluate the risk for cancer among children of epileptic mothers who were treated intensively with phenobarbital and other anticonvulsive drugs to prevent seizures. The risk for cancer among epileptics, especially liver cancer and leukaemia, will also be evaluated. Account will be taken of the fact that some patients received injections of Thorotrast for cerebral angiography. Approximately 8,000 patients admitted for epilepsy to the Danish epilepsy treatment centre in Filadelfia, Dianalund, during 1933-62 have been linked to the files of the Danish Cancer Registry. In order to identify children born after the date of admission of a patient to the centre; the patients are being traced in population registries all over the country. Similarly, information on their children will be linked to the files of the Cancer Registry in order to detect cancers among the children.

TYPE: Cohort
TERM: Drugs; Epilepsy; Radiation, Ionizing; Registry
SITE: All Sites; Childhood Neoplasms; Leukaemia; Liver
CHEM: Barbiturates; Thorotrast
REGI: Denmark (Den)
TIME: 1989 - 1993

218   Olsen, J.H.   04596
Danish Cancer Society, Danish Cancer Registry, Rosenvængets Hovedvej 35 , Box 839, 2100 Copenhagen Ø, Denmark (Tel.: +45 31 268866)
COLL: de Nully Brown, P.; Holländer, H.C.

Parental Occupation and Risk for Childhood Cancer
Parental occupation is one suspected risk factor for the occurrence of cancer in childhood. Reports point to occupations that involve exposure to hydrocarbons, lead or chemicals, and to the occupations of higher social classes. The aim of this case-control study is to detect possible associations between the industrial employment of the parents at the time of conception and during pregnancy and the risk of the child subsequently developing a malignancy. Approximately 2,000 cases of childhood cancer and 10,000 controls are included in the study. The parents of both cases and controls are being identified in the files of the Central Population Register, and occupational histories are established through the Supplementary Pension Fund Danish Cancer Registry linkage study file, with special attention to the period around conception of the index child and the pregnancy of the mother. By applying the evaluation of chemical exposures used in other case-control studies (on brain cancer and multiple myeloma) being carried out in the Registry to the combinations of industrial activity and job-title held by the parents of children included in the present study, it may be possible to reanalyse the material according to type of chemical exposure at the work place of the parents. Included among the exposures are chemicals such as hydrocarbons, organic solvents, paints, rubber, plastics and metals.

DENMARK

TYPE: Case-Control
TERM: Intra-Uterine Exposure; Metals; Occupation; Paints; Plastics; Registry; Rubber; Solvents
SITE: Childhood Neoplasms
CHEM: Hydrocarbons
REGI: Denmark (Den)
TIME: 1989 - 1993

219 Prener, A. 04597
Danish Cancer Socieety, Danish Cancer Registry, Rosenvængets Hovedvej 35 , Box 839, 2100 Copenhagen Ø, Denmark (Tel.: +45 31 268866)
COLL: Høgaard Nielsen, N.; Storm, H.H.; Hart Hansen, J.P.; Jensen, O.M.

Cancer in Greenland

Since the late 1960s, cancer registration has covered the population of Greenland, which has the largest population of Inuit descent in the world. Registration was deficient until the early 1980s; incidence data have now been established for the period 1950-74 retrospectively. assembled on an ad-hoc basis, These data reveal a characteristic pattern of cancer similar to that observed in Inuit populations in Alaska and Canada. The pattern is changing rapidly, in parallel with changes in Greenlandic life style. Data from 1975 to 1985 have been revised and supplemented with information from the Department of Pathology of Rigshospitalet, Copenhagen, Denmark which serves all of Greenland. A comprehensive description of the findings, including trends and comparisons with the incidence of cancer in Denmark, is being undertaken. The results will be published in a monograph, which will also include a description of the characteristics of the Inuit population and of the country itself. A summary publication is planned in the inuit language.

TYPE: Incidence
TERM: Eskimos; Ethnic Group; Lifestyle; Registry; Trends
SITE: All Sites
LOCA: Greenland
REGI: Denmark (Den)
TIME: 1989 - 1992

220 Storm, H.H. 03887
Danish Cancer Society, Danish Cancer Registry, Rosenvængets Hovedvej 35 , Box 839, 2100 Copenhagen Ø, Denmark (Tel.: +45 31 268866; Fax: 269000)
COLL: Boice, J.D.; Jensen, O.M.; Andersson, M.A.; Rose, C.; Dombernowsky, P.; Pedersen, M.; Johansen, H.; Jacobsen, A.; Stovall, M.

Contralateral Breast Cancer in Denmark. A study of the Risk Associated with Adjuvant Radiotherapy

The risk of cancer in the opposite breast eight years or more following a primary breast cancer is studied. The association between radiation dose received by the contralateral breast during the course of primary breast cancer treatment and the subsequent risk of contralateral breast cancer is evaluated. The study is designed as a matched case-control study within the cohort of breast cancer patients diagnosed in Denmark 1943-1985. Approximately 800 women with contralateral breast cancer and 800 controls will be abstracted emphasizing radiation dose, and including other known risk factors for breast cancer. The radiation dose to the contralateral breast will be assessed for each patient and the data analysed with respect to dose response, age at treatment of first breast cancer and latency time between first breast cancer and the contralateral. A cohort study in Denmark (Brit. J. Cancer 54:483-492, 1986) revealed a relative risk for contralateral breast cancer of 2.8, inversely related to age at diagnosis. A significant increased risk was observed among women with primary localized irradiated breast cancer compared to women with primary localized breast cancer ten or more years after primary treatment.

DENMARK

TYPE: Case-Control
TERM: Age; Dose-Response; Latency; Radiation, Ionizing; Registry
SITE: Breast (F)
REGI: Denmark (Den)
TIME: 1985 - 1992

221   Storm, H.H.   04598
Danish Cancer Society, Danish Cancer Registry, Rosenvængets Hovedvej 35 , Box 839, 2100 Copenhagen Ø, Denmark (Tel.: +45 31 268866; Fax: 269000)

Risk for Leukaemia Following Treatment for Uterine Cancer

An international collaborative study is being carried out between cancer registries in Denmark, Norway, Finland, Canada and the USA to investigate the risk for leukaemia following radiotherapy for cancer of the uterine corpus. Dosimetry and patient characteristics were obtained from hospital records for each patient. The design is a matched case-control study within each registry catchment area. A total of 160 cases of leukaemia, of which 120 are non-chronic lymphocytic leukaemia, is expected to be abstracted. From Denmark, 21 cases of non-chronic lymphocytic leukaemia and 13 of chronic lymphocytic leukaemia will be abstracted, with corresponding controls.

TYPE: Case-Control
TERM: Dose-Response; Multiple Primary; Radiotherapy; Registry
SITE: Leukaemia; Leukaemia (ALL); Leukaemia (CLL); Uterus (Corpus)
LOCA: Canada; Denmark; Finland; Norway; United States of America
REGI: Denmark (Den); Finland (Fin); Norway (Nor)
TIME: 1987 - 1992

222   Storm, H.H.   04599
Danish Cancer Society, Danish Cancer Registry, Rosenvængets Hovedvej 35 , Box 839, 2100 Copenhagen Ø, Denmark (Tel.: +45 31 268866; Fax: 269000)
COLL: Boice, J.D.; Kleinerman, R.A.; Jensen, O.M.; Stovall, M.

Lung Cancer Risk Following Radiation for Cancer of the Uterine Cervix

An international cohort study followed by several case-control studies addressing the risk of new cancers after treatment for cancer of the uterine cervix is the basis of this project. Information on exposure to radiation and potential confounding factors was extracted from hospital records. An important observation from the studies is that cancer risk seems to be elevated for more than 30 years after primary treatment, and the risk seems to increase with time. In collaboration with the Radiation Epidemiology Branch, NCI, USA, data are being collected on cervical cancer patients who developed lung cancer ten or more years following treatment for first cancer. The cases will be matched to suitable controls. A separate analysis is being carried out on the Danish data. The first risk that will be examined is for second primary breast cancer. Additional potential risk factors have been taken into consideration, and the registration and abstraction of cases may be considered more uniform than in the international material. The approach taken is a case-control study, in which the matching ratio is as large as possible in order to narrow the confidence intervals of observed risks. The dose to the breast is approximately 30 rad, and no elevation of risk for breast cancer as a function of dose has been observed. Surgical and radiogenic ablation of the ovaries had an overall protective effect against the development of breast cancer.

TYPE: Case-Control; Cohort
TERM: Multiple Primary; Radiotherapy; Registry
SITE: All Sites; Breast (F); Lung; Uterus (Cervix)
LOCA: Denmark; United States of America
REGI: Denmark (Den)
TIME: 1987 - 1992

223   Storm, H.H.   04600
Danish Cancer Society, Danish Cancer Registry, Rosenvængets Hovedvej 35 , Box 839, 2100 Copenhagen Ø, Denmark (Tel.: +45 31 268866; Fax: 269000)
COLL: Stovall, M.; Engholm, G.; Anderson, M; Jensen, O.M.; Curtis, R.E.

Risk for Leukaemia Following Treatment for Breast Cancer

In a study of multiple primary cancers in Denmark occurring in 1943-82, a relative risk of 2.3 was observed for acute nonlymphocytic leukaemia following treatment for breast cancer during the entire observation period. Some 70% of the breast cancer patients had received combination treatment with

DENMARK

surgery and radiotherapy. In order to evaluate the leukaemogenic effect of radiation treatment for breast cancer, a matched case-control study nested in the cohort of breast cancer patients was begun in 1985. The study involved 21 cases of chronic lymphocytic leukaemia, matched to 39 controls, and 72 cases of other leukaemias, matched to 167 controls. The relative risks for non-chronic lymphocytic leukaemia were calculated by radiation dose to the bone marrow: a small, non-significant increase in risk was found, but no meaningful trend with received dose of radiation to the bone marrow was observed. A parallel study has been carried out in Connecticut, USA, by the Radiation Epidemiology Branch of NCI. The two studies will be evaluated together.

TYPE: Case-Control
TERM: Dose-Response; Multiple Primary; Radiotherapy; Registry
SITE: Breast (F); Leukaemia; Leukaemia (CLL)
REGI: Denmark (Den)
TIME: 1985 - 1992

* 224   Storm, H.H.   05091
Danish Cancer Society, Danish Cancer Registry, Rosenvængets Hovedvej 35, Box 839, 2100 Copenhagen Ø, Denmark (Tel.: +45 31 268866; Fax: 269000)
COLL: Blettner, M.; Engholm, G.; Curtis, R.E.

Radiation-Induced Leukaemia

With the aim of studying the effect of radiation to the bone marrow on the risk of leukaemia and to clarify whether a linear-exponential function with cell killing, a quadratic exponential, linear-exponential or linear function of radiation dose to the bone marrow exists, it is proposed to combine and analyse three Danish case-control studies on leukaemia following radiation exposure. The three studies to be combined are leukaemia following treatment of cancer of the cervix, the breast, and the corpus uteri. Data abstraction and dosimetry was done in a similar way in all three studies. The wide dose range from 0-750 cGy in the studies enables the study of dose-models, and is expected to add to the knowledge to be used for extrapolation and estimation of the effects on humans of low-dose radiation exposure. The analysis will include use of the background incidence rates from a non-exposed population.

TYPE: Case-Control
TERM: Dose-Response; Radiotherapy; Registry
SITE: Leukaemia
REGI: Denmark (Den)
TIME: 1990 - 1992

* 225   Storm, H.H.   05092
Danish Cancer Society, Danish Cancer Registry, Rosenvængets Hovedvej 35, Box 839, 2100 Copenhagen Ø, Denmark (Tel.: +45 31 268866; Fax: 269000)
COLL: Brinton, L.A.; Persson, I.R.; Clemmesen, I.H.; Manders, T.

Cancer Risk in Patients Having Gynaecological Operations

In order to study the risk of cancer among women subsequent to a gynaecological operation such as oophorectomy and/or hysterectomy, a cohort of Danish patients undergoing such surgery between 1976 and 1987 has been identified in the Danish National Patient Discharge Registry. Based on the frequency counts and the occurrence of relevant disease entities in the master file, sub-cohorts will be constructed. These cohorts will be linked to the Cancer Registry and observed cancers obtained and expected numbers calculated. Standard procedures in operation at the Cancer Registry will be used under the assumption of a proportionate hazards model. The relative risk taken as the O/E ratio will be calculated by time since surgery, after stratifying the cohort according to surgical procedure, index disease and age. All cancers will be studied, but specific emphasis will be on cancers of breast, lung and ovaries.

DENMARK

TYPE: Cohort
TERM: Hysterectomy; Oophorectomy; Record Linkage; Registry
SITE: All Sites
LOCA: Denmark; Sweden
REGI: Denmark (Den)
TIME: 1989 – 1992

HERLEV

* 226 Binder, V. 05134
Herlev Hosp., Medical Gastrointestinal Dept., Herlev Ringvej 75, 2730 Herlev, Denmark (Tel.: +45 44 535300; Fax: 535332)

Survival and Bowel Cancer Occurrence in a Cohort of Patients with Ulcerative Colitis and Crohn's Disease

All patients diagnosed with ulcerative colitis (N = 1,161) or Crohn's disease (N = 373) in 1962-1987 and residents of Copenhagen County have been followed up from diagnosis of inflammatory bowel disease to the end of 1987 (mean 11.7 years for ulcerative colitis and 8.5 years for Crohn's disease). Eight and three bowel cancers, respectively, have been reported to date, which is not more than expected in the background population.

TYPE: Cohort
TERM: Crohn's Disease; Ulcerative Colitis
SITE: Colon; Small Intestine
TIME: 1962 – 1992

HVIDOVRE

227 Bülow, S. 02636
Hvidovre Hosp., The Polyposis Register, Dept. of Surgical Gastroenterology, 2650 Hvidovre, Denmark (Tel.: +45 31 471411/2236; Fax: 473311)

The Danish Polyposis Register

A central Danish register of all cases of familial adenomatous polyposis was established in 1971. The aim of the register is (1) to collect all Danish cases of polyposis; (2) to offer a coordinating and consultative service for present and future prophylactic examination, detection and treatment of the disease in hospitals all over the country. At present 401 cases in 130 families have been registered. Papers appeared in Dan. Med. Bull. 34:1-15, 1987, in Ann. Med. 21:299-307 and in World J. Surg. 15:41-46, 1991.

TYPE: Genetic Epidemiology
TERM: Clinical Effects; Familial Adenomatous Polyposis; Familial Factors; Mutation, Germinal; Premalignant Lesion; Registry
SITE: Colon; Rectum
TIME: 1976 – 1993

ODENSE

228 Birkeland, S.A. 03399
Odense University Hospital, Dept. of Nephrology, 5000 Odense C, Denmark

Scandinavian Project on renal transplantation and cancer

One complication in renal transplantation with lifelong immunosuppression is development of patients with tumours. The aim of this study, which includes all patients with renal transplantation performed in the Nordic countries, is to measure cancer risk relative to that of the general population, and to evaluate factors such as living related donor vs. cadaver kidney donor, type and site of tumours, HLA-donor-recipient mismatch, donor-recipient differences in age, sex and ABO blood group, presence of lymphocytotoxic antibodies in the recipients or not, time for tumour occurrence after transplantation, recipient kidney disease, number of transplantations performed per recipient, patient- and graft-survival, number of blood transfusions given to the recipients, type and amount of immunosuppressive drugs, period of uraemia before transplantation and occurrence or not of any viral disease. The project at present includes the period up to 1982 with about 6,000 transplanted patients

DENMARK

observed for about 30,000 person years, in whom 400-500 tumours have been observed. The original data from the transplantation centres are registered and updated with the civil registries, the Scandiatransplant registry and the cancer registries, and risk is evaluated using observed/expected ratios.

TYPE: Cohort
TERM: Age; Blood Group; Drugs; HLA; Immunosuppression; Registry; Sex Ratio; Time Factors; Transplantation; Virus
SITE: All Sites
CHEM: Azathioprine
LOCA: Denmark; Finland; Iceland; Norway; Sweden
REGI: Denmark (Den); Finland (Fin); Iceland (Ice); Norway (Nor); Sweden (Swe)
TIME: 1971 - 1992

229 Brok, K.E. 04798
Odense University Hospital, Dept. of Urology, Sdr Blvd 29, 5000 Odense C, Denmark (Tel.: +45 66 143333)
COLL: Mommsen, S.

Case-Control Study of Prostatic Cancer

The aim of this study is to examine risk factors in the aetiology of prostatic cancer. Cases comprise all patients with newly diagnosed prostatic cancer in the period 1990-93 living in the island of Fuen, Denmark. About 300 cases are expected in the period. Controls are matched for race, sex and age and are randomly selected from the national population register. All cases and controls are interviewed for their lifestyle habits, dietary habits, occupational exposure, sexual history, history of general and urological diseases and urological symptoms. Biochemical epidemiology of prostatic cancer is examined by blood samples from cases and controls (androgens, oestrogens, lipids).

TYPE: Case-Control
TERM: Biochemical Markers; Diet; Hormones; Lifestyle; Occupation; Sexual Activity
SITE: Prostate
TIME: 1989 - 1993

230 Grandjean, P. 03723
Odense Univ., Inst. of Community Health, Dept. of Environmental Medicine, J.B. Winslows Vej 17, 5000 Odense C, Denmark (Tel.: +45 66 158600; Fax: 65918296 ; Tlx: 59964 os dk)
COLL: Jensen, O.M.; Juel, K.; Lynge, E.

Mortality and Cancer Morbidity of Cryolite Workers

Heavy occupational fluoride exposure occurred in cryolite workers in the past. In 1931, skeletal fluorosis was discovered as a disease new to medicine when cryolite workers were examined. At least 80 cases have occurred at the factory in Copenhagen where cryolite ore from Greenland is processed. Detailed personnel records are available from 1924. An excess occurrence of respiratory cancer in male workers employed for at least six months between 1924 and 1961 was seen during 1943-1978. Follow-up to the end of 1989 will include 430 male and 100 female cryolite workers. A paper appeared in Am. J. Epidemiol. 121:57-64, 1985.

TYPE: Cohort
TERM: Chemical Exposure; Environmental Factors; Occupation
SITE: All Sites
CHEM: Fluorides
OCCU: Cryolite Workers
TIME: 1981 - 1992

231 Holm, N.V. 02306
Univ. of Odense, Inst. of Clinical Genetics, Winslowsvej 17, 5000 Odense, Denmark (Tel.: +45 66 158600)
COLL: Jensen, O.M.

Cancer Aetiology Studies in the Danish Twin Population

The Danish Twin Registry consists of the total population of twin pairs born in Denmark during a certain period. The present study includes all same-sexed twin pairs born during the period 1881-1930 in which both partners survived the age of 14 years, in all about 11,000 pairs. The study comprises an evaluation of the importance of genetic factors in the development of breast cancer, gastro-intestinal cancer, uterus

DENMARK

cancer, testicular cancer, malignant melanoma and leukaemia. In twin pairs with one of these neoplasms, cancer of all sites will be studied. More intensive studies are carried out in breast cancer twins. A matched case-control study of monozygotic twin pairs discordant for breast cancer will evaluate various risk factors (number of births, age at first birth, menopause and height). To evaluate genetic heterogeneity a twin family study will be performed. A report (J. Nat. Cancer Inst., 65:285-298, 1980) dealt with concordance rates and various risk factors of breast cancer and another one (Cancer Surv. 1:17-32, 1982) with concordance rates of breast cancer, colorectal cancer and leukaemia.

TYPE: Case-Control; Cohort
TERM: Age; Genetic Factors; Menopause; Parity; Physical Factors; Twins
SITE: Breast (F); Gastrointestinal; Leukaemia; Melanoma; Testis; Uterus (Cervix); Uterus (Corpus)
TIME: 1979 - 1993

232   Jørgensen, K.E.   04241
Univ. Hospital of Odense, Inst. of Oto-Rhino-Laryngology, ENT-Dept., Sødra Blvd, 5000 Odense C, Denmark (Tel.: +45 66 113333)

Occurrence of Secondary Cancers in a Population of Patients Treated for Laryngeal Cancer
From 1965 to 1987 close to 600 patients were treated for laryngeal cancer at the Oncological Centre and ENT Department, University Hospital of Odense. Since 1979 all patients have been prospectively registered in a computer-system, partly in order to be able to analyse the results of treatment, and partly as patients have been included in a countrywide randomized project concerning a radiosensitizer. During the whole period all cases of secondary cancer have been registered and an abnormally high frequency of pulmonary cancer has already been observed. The termination date for the project has not been decided. The total population of Denmark is being used as a control group in cooperation with the Danish Cancer Registry.

TYPE: Cohort
TERM: Multiple Primary; Registry; Treatment
SITE: Larynx; Lung
REGI: Denmark (Den)
TIME: 1979 - 2000

RISSKOV

233   Mortensen, P.B.   04777
Aarhus Psychiatric Hosp., Inst. of Psychiatric Demography, Skovagervej 2, 8240 Risskov, Denmark (Tel.: +45 86 177777)

Neuroleptics and Cancer Risk in Schizophrenics and Other Psychotic Patients
The aim of these studies is to clarify the association between use of neuroleptics and other psychotropic drugs and cancer risk in psychotic patients. The study consists of a longitudinal case-register-based study of: (1) a cohort of 6,178 schizophrenic in-patients identified in 1957 and followed-up through 1984 and (2) a cohort consisting of all patients (approx. 70,000) admitted for the first time to psychiatric hospitals during the period 1970-1986 with a diagnosis of functional psychosis (ICD 295-299). These two cohorts are checked against the Danish Cancer Registry and site-specific cancer incidence rates determined. A series of case-control within cohort studies are performed in order to determine the impact of neuroleptics and other psychotropic drugs (including reserpine) on cancer risk. So far, studies of male cancer of the bladder and prostate in males and cancer of the breast and uterine cervix in females have been performed in the cohort of schizophrenic patients.

TYPE: Case-Control; Cohort
TERM: Drugs; Record Linkage; Registry
SITE: Bladder; Breast (F); Lung; Prostate; Uterus (Cervix)
CHEM: Phenothiazines; Reserpine
REGI: Denmark (Den)
TIME: 1984 - 1993

EGYPT

CAIRO

234 **Ebeid, N.I.** 04297
Military Medical Academy, Depts of Occupational Medicine, Gynaecology & Biochemistry, 99, 26 July St., Bonlac, Cairo, Egypt (Tel.: 3488928)
COLL: Wahba, R.; Hafez, A.H.

Schistosomiasis and Exposure to Industrial Pollutants in the Aetiology of Liver Cancer

Both Schistosomiasis and, to a lesser extent, HBV infection are common in Egypt. The aim of this study is to examine the role of concomitant exposure to welding fumes (as an industrial pollutant) as an added carcinogenic factor. Few positive HBV cases have been detected in this study, so viral hepatitis has been excluded. The study has been extended to comprise 98 individuals of whom 48 are exposed to welding fumes, while 50 are not exposed (control group). Ultrasonography of the liver is being carried out for the majority of cases. Alpha phytolic protein testing will be done shortly afterwards. Smoking habits are also being investigated.

TYPE: Cross-Sectional
TERM: Metals; Occupation; Parasitic Disease; Tobacco (Smoking); Trace Elements; Welding
SITE: Liver
CHEM: Cadmium; Copper; Lead
OCCU: Welders
TIME: 1987 - 1992

235 **Ebeid, N.I.** 04452
Military Medical Academy, Depts of Occupational Medicine, Gynaecology & Biochemistry, 99, 26 July St., Bonlac, Cairo, Egypt (Tel.: 3488928)
COLL: Rashad, S.; Daoud, S.; Milner, M.

Manufacture of Contraceptive Pills: Possible Carcinogenic Hazard and Hormonal Disturbance

The endogenous hormones, oestradiol, oestrone, oestriol, progesterone and testosterone have all been considered as potentially carcinogenic by OSHA. Female workers in the contraceptive pill manufacturing industry are susceptible to hormone inhalation, and may thus incur carcinogenic risk. Estimation of hormonal levels has been done in 30 of these workers attending gynaecology, outpatient clinics, together with controls. Clinical and pathological studies of both uterus and breasts of these patients have revealed a diversity of disturbances and precancerous lesions. The results are still inconclusive, and wider study of more workers and controls is planned.

TYPE: Cross-Sectional
TERM: Hormones; Occupation; Oral Contraceptives
SITE: Breast (F); Uterus (Corpus)
CHEM: Oestradiol; Oestriol; Oestrone; Progesterone; Testosterone
OCCU: Pharmaceutical Industry Workers
TIME: 1986 - 1992

236 **Ebeid, N.I.** 04532
Military Medical Academy, Depts of Occupational Medicine, Gynaecology & Biochemistry, 99, 26 July St., Bonlac, Cairo, Egypt (Tel.: 3488928)
COLL: Omar, S.; El Asser, A.K.; Nasser, A.; El Din Shebl, S.

Impact of Smoking on Workers Engaged in Cement Industry

The health hazards of smoking may be compounded by some occupational exposures: smoke particles may act as vectors of work-place chemicals. The cement industry is a major employer in Egypt, and this cross-sectional study will investigate possible cancer hazards and their interaction with smoking. Interviews, physical examinations, lung function tests and assay of various biological markers will be carried out, and the results stratified by cement dust exposure and smoking habit. About 200 workers will be examined initially. Marker studies will include haptoglobin and other acute phase proteins, retinol and its binding protein, and carcinoembryonic antigen.

TYPE: Cross-Sectional
TERM: Biochemical Markers; Dusts; Occupation; Tobacco (Smoking)
SITE: All Sites; Liver; Respiratory
OCCU: Cement Workers
TIME: 1988 - 1992

ESTONIA

TALLINN

237   Hint, E.   04831
      Inst. of Exp. & Clinical Medicine, Ministry of Health, Dept. of Clinical Oncology, 42 Hiiu St., Tallinn 200107, Estonia (Tel.: (014)514333)
COLL: Purde, M.; Eomois, M.; Lilleorg, A.

Registry for Population at High Risk of Familial Breast Cancer

Epidemiological investigations in Estonia since 1968 have revealed some peculiarities in familial breast cancer. Breast cancer in relatives increases the risk for both malignant and benign breast disease. Familial breast cancer is detected earlier and survival rates are better than non-familial breast cancer. These findings have prompted creation of a registry of women including healthy relatives of breast cancer patients. Information will be collected by questionnaires and the verification of breast cancer will be checked through the Estonian Cancer Registry. Relatives will be invited to take part in mammographic screening. Of specific interest will be the study of hormonal profile (oestradiol and prolactin) and of immune status by MCA and Ca-125 markers in breast cancer patients' daughters and sisters. To date approximately 3,000 relatives have been registered.

TYPE: Cohort; Registry
TERM: BMB; Biochemical Markers; Familial Factors; High-Risk Groups; Hormones; Mammography; Registry; Serum
SITE: Breast (F)
REGI: Estonia (Est)
TIME: 1990 - 1993

* 238   Hint, E.   05140
      Inst. of Exp. & Clinical Medicine, Ministry of Health, Dept. of Clinical Oncology, 42 Hiiu St., Tallinn 200107, Estonia (Tel.: (014)514333)
COLL: Tekkel, M.

Identification of Precancerous Breast Disease

In Estonia a prospective study of breast cancer was started in 1974. To date information about risk factors has been collected from more than 30,000 women by self-administered questionnaire, checked by the physician and completed if necessary. The risk factors under study include: family history of breast cancer, early menarche, age at first intercourse, age at first delivery, nulliparity, prolonged breast feeding, and late menopause. Four study groups of women aged 25 to 64 have been created: (1) 6,500 women from the general female population of 13 regions, undergoing medical examination within a public health programme and considered to be representative of the adult female population of Estonia as a whole; (2) 7,800 healthy women; (3) 15,600 women with benign breast diseases; and (4) 700 women with breast cancer. Groups 2-4 are a result of regular consultations with specialists on breast disease. The list of all women under study is checked through the Estonian Cancer Registry. It is hypothesized that if the distribution (%) of one or more of the risk factors examined in the group of patients with benign breast diseases is higher than in the groups of "healthy women" or "general female population", but lower than in the group of "breast cancer patients", then the variable would be of importance for identification of precancerous conditions. On the basis of the results obtained, measures for prevention of breast cancer will be taken.

TYPE: Cohort
TERM: High-Risk Groups; Prevention; Registry
SITE: Breast (F)
REGI: Estonia (Est)
TIME: 1991 - 1993

FINLAND

ESPOO

239 Sipponen, P.I. 02195
Jorvi Hosp., Dept. of Pathology, 002740 Espoo 74, Finland
COLL: Seppälä, K.; Varis, K.S.; Kekki, M.; Siurala, M.

Frequency of Intestinal Metaplasia with Colonic Type Characteristics in the Stomach of Patients with Gastric Carcinoma and Benign Lesions
The investigation is designed to clarify the frequency of intestinal metaplasia (IM) with histochemical colonic-type characteristics in the stomach of asymptomatic subjects and of patients with gastric lesions of different types. The aim is to find out whether the occurrence of IM with colonic-like characteristics is associated with gastric carcinoma or with lesions considered to be pre-malignant. The study is performed by using histochemical staining methods, known to be characteristic to the colonic mucosa and mucosubstances, on mucosa specimens obtained by gastroscopy. 125 patients with verified gastric carcinoma and 301 first-degree relatives, 62 patients with pernicious anaemia, and 183 first-degree relatives, 406 consecutive outpatients and 358 controls matched from a large Finnish population study are being examined. The investigation will continue as a follow-up study in patient groups selected according to the presence or absence of IM with colonic type characteristics in the tissue samples. Results were published in Acta path. microbiol. scand. Sect. A, 88:217-224, 1980 and Annals of Clinical Res. 13:139-143, 1981.

TYPE: Case-Control
TERM: Histology; Premalignant Lesion
SITE: Stomach
TIME: 1979 - 1992

240 Siurala, M. 04494
Louhentie 1 A 2, Espoo 13, Finland (Tel.: (90)466355)
COLL: Sipponen, P.I.; Ihamäki, T.; Kekki, M.

Benign and Malignant Gastric Tumours in Gastric Carcinoma Families and Controls
In 1972 and 1976, 301 first-degree relatives of 73 consecutive probands with gastric carcinoma were examined by endoscopy with direct vision biopsy (average 11.2 specimens per subject), penta-gastrin test, serum pepsinogen I and II etc, and in a similar way 358 first-degree relatives of 73 control probands, matched to the cancer probands for age, sex, place of birth, residence and occupation (Scand. J. Gastroent. 14:801-812, 1979 and APMIS 87:457-462, 1979). In the first part of the study the occurrence of benign and malignant stomach tumours in gastric carcinoma and control families will be evaluated using data obtained from Finnish death statistics and the Finnish Cancer Registry. The follow-up time will be on average 14.5 years. The occurrence of stomach tumours will also be referred to the morphological and functional status of the subjects at the first examination.

TYPE: Cohort
TERM: Familial Factors; Registry
SITE: Stomach
REGI: Finland (Fin)
TIME: 1988 - 1992

HELSINKI

241 Husgafvel-Pursiainen, K. 04833
Inst. of Occupational Health, Topeliuksenkatu 41 a A, 00250 Helsinki, Finland (Tel.: +358 0 47471; Fax: 4747208)
COLL: Anttila, S.; Hackman, P.; Heikkilä, L.; Hietanen, E.; Hirvonen, A.; Karjalainen, A.; Ridanpää, M.; Taikina-aho, O.; Vainio, H.

Lung Cancer, Asbestos and Smoking in Finland
The aims of this study are (I) to establish practical criteria based on occupational history, pulmonary fibre content and histopathology for the identification of lung cancer related to occupational asbestos exposure; (2) to study host-related factors with biochemical and molecular biology methods; and (3) to study biological aspects of lung cancer, such as activation of cellular proto-oncogenes using the same material, and DNA adduct formation in lung vs. white blood cell DNA. Lung tissue will be collected from randomly selected patients with lung cancer. On the day before surgery, patients will be interviewed on

FINLAND

their occupational and smoking histories. Resected lung material will be used to study fibre content, macroscopical changes, histological type and extent of the tumour, microscopical fibrosis, asbestos bodies, several enzyme parameters, oncogene activation (tumour, peripheral lung tissue, white blood cells), and DNA adduct formation (tumour, peripheral lung tissue, white blood cells). So far about 65 cases have been included; the project aims to double the number of subjects. No control group is included. Various electron-microscopical, biochemical, immunohistochemical, histological and molecular methods will be used.

TYPE: Case Series; Molecular Epidemiology
TERM: BMB; Biochemical Markers; DNA Adducts; Fibrosis; Histology; Occupation; Oncogenes; Tissue; Tobacco (Smoking)
SITE: Lung
CHEM: Asbestos; Mineral Fibres
TIME: 1988 - 1993

242 Kaprio, J. 01228
Univ. of Helsinki, Dept. of Public Health, Haartmaninkatu 3, 0029 Helsinki, Finland (Tel.: (0)43461)
COLL: Koskenvuo, M.J.; Pukkala, E.I.; Teppo, L.

Cancer in Twins

The study of twins may elucidate the relative role of different environmental factors in cancer aetiology. The Finnish Twin Cohort Study contains data on some 34,500 adult twins of the same sex. Zygosity was determined by questionnaire in 1975. Record linkage with the Finnish Cancer Registry for cancer cases incident to 1989 has been carried out. Site-specific subsamples are being identified to examine how discordant identical pairs differ. The twin cohort has been extended to include all twin births between 1958 and 1986. A total of 23,500 twin pairs (of same or different sex), were found. This data base can be used for various genetic epidemiological studies.

TYPE: Case-Control; Cohort
TERM: Environmental Factors; Record Linkage; Registry; Twins
SITE: All Sites
REGI: Finland (Fin)
TIME: 1975 - 1992

243 Knekt, P.B. 04461
Social Insurance Institution, P.O. Box 78, 00381 Helsinki, Finland (Tel.: (0)4343591; Fax: 4343333; Tlx: 122375 KELA-SF)
COLL: Aromaa, A.; Maatela, J.; Hakulinen, T.R.; Saxén, E.A.; Teppo, L.; Aaran, R.K.; Hakama, M.K.; Nikkari, T.; Alfthan, G.; Peto, R.

Serum Antioxidants and Risk of Cancer

The aim of the present epidemiological study is to examine the role of vitamin A, vitamin E and selenium in the aetiology of cancer by studying the association between serum beta carotene, retinol, retinol-binding protein, alpha-tocopherol, and selenium levels and the incidence of cancer. During 1966-1972 the Social Insurance Institution's Mobile Clinic Health Examination Survey carried out multiphasic screening examinations in various parts of Finland. As a part of that survey, serum samples were taken and stored at -20° C. Ten years later the present longitudinal study was started to investigate the association between serum micronutrient levels and the incidence of cancer. The study population comprised about 40,000 men and women, initially aged 15-99 years. Cancer incidence data from the nationwide Finnish Cancer Registry were linked to the health examination data set. During a mean follow-up of about ten years, cancer was diagnosed in 1,100 persons. The serum micronutrient levels were determined from the stored serum samples collected from these cancer patients and from 2,000 controls matched for sex, municipality and age. The beta carotene, retinol, and alpha-tocopherol measurements were performed using HPLC and the selenium determinations by electrothermal atomic absorption spectrometry. The data analyses are mainly based on the conditional logistic model.

FINLAND

TYPE: Case-Control; Cohort
TERM: BMB; Metals; Registry; Sero-Epidemiology; Serum; Vitamins
SITE: All Sites; Breast (F); Colon; Lung; Prostate; Rectum; Stomach; Uterus (Cervix)
CHEM: Beta Carotene; Retinoids; Selenium; Tocopherol
REGI: Finland (Fin)
TIME: 1980 - 1992

244   Knekt, P.B.   04539
Social Insurance Institution, P.O. Box 78, 00381 Helsinki, Finland (Tel.: (0)4343591; Fax: 4343333; Tlx: 122375 KELA-SF)
COLL: Seppänen, R.; Aromaa, A.; Teppo, L.

Diet and Cancer

The aim of the present epidemiological study is to examine the predictive role of different nutrients on the risk of cancer. During 1966-1972 the Social Insurance Institution's Mobile Clinic Health Examination Survey carried out multiphasic screening examinations in various parts of Finland. As a part of that study, a dietary survey of about 10,000 men and women, initially aged 15-99 years, was carried out. the data on the average daily food consumption during the previous year were collected by a dietary history interview. Cancer incidence data from the nationwide Finnish Cancer Registry were linked to the health examination data set. During a mean follow-up of 15 years, cancer was diagnosed in about 500 persons. The associations between the nutrients and the incidence of cancer are being estimated, mainly based on Cox's life-table regression model.

TYPE: Cohort
TERM: Diet; Registry
SITE: Breast (F); Colon; Lung; Prostate; Rectum; Stomach
REGI: Finland (Fin)
TIME: 1986 - 1992

245   Koskinen, H.   04542
Inst. of Occupational Health, Topeliuksenkatu 41 a A, 00250 Helsinki, Finland (Tel.: (90)47471)
COLL: Hernberg, S.G.; Heikkilä, P.; Kauppinen, T.P.; Kurppa, K.; Lahdensuo, A.; Liippo, K.; Nurminen, M.; Partanen, T.J.; Tala, E.; Tossavainen, A.

Lung Cancer and Occupational Exposure

There are about 2,000 cases of lung cancer annually in a population of 4.9 million in Finland. In this pilot study it is proposed to evaluate a large scale multi-centre epidemiological study on occupational causes of lung cancer in Finland. The feasibility study was conducted in the catchment area of two university lung clinics where incident, living lung cancer patients are interviewed. The exposure experience of the study cases will be supplemented with an interview of referents from the base population of the clinics (random sample of the base population as well as referents with some other cancers than lung cancer).

TYPE: Case-Control
TERM: Occupation
SITE: Lung
TIME: 1988 - 1992

246   Partanen, T.J.   04780
Inst. of Occupational Health, Topeliuksenkatu 41a A, 00290 Helsinki, Finland (Tel.: +358 047471; Tlx: 121394)
COLL: Hernberg, S.G.; Koskinen, H.; Kauppinen, T.P.; Degerth, R.; Kaipainen, P.; Teppo, L.; Pukkala, E.I.

Primary Pancreatic Cancer and Chemical Exposures at Work

This national case-referent study aims at revealing connections between pancreatic cancer and industries, occupations, and occupational chemical exposures previously prevalent in Finland. The case series will comprise all histologically confirmed primary exocrine pancreatic cancer cases aged 40-74 years at diagnosis notified to the Finnish Cancer Registry during the period 1984-87 and deceased during the same period. The referent series will consist of the deceased cases of stomach, colon, and rectal cancers, diagnosed in Finland during the same period. The next-of-kin of both the cases and the referents will receive a postal questionnaire on the job history of the case or the referent as well as on smoking; alcohol, coffee and sugar consumption; and diabetes mellitus. The work histories will be coded into industry, job titles and chemical exposures. The translation of work histories into exposures will be done by two experienced industrial hygienists, using all the available sources of information. The data

FINLAND

analysis will concentrate on the estimation of the risk ratios (point estimates and confidence intervals) for the different industries, job titles, and exposures.

TYPE: Case-Control
TERM: Alcohol; Chemical Exposure; Coffee; Diabetes; Occupation; Registry; Tobacco (Smoking)
SITE: Pancreas
REGI: Finland (Fin)
TIME: 1989 - 1992

247 Pukkala, E.I. 03242
Finnish Cancer Registry, Liisankatu 21 B, 00170 Helsinki 17, Finland (Tel.: 176290)

Occupation, Socio-Economic Status and Education as Risk Determinants of Cancer

All cancer cases diagnosed in Finland after the year 1970 are linked with the national census file from 31 December 1970. The data on occupation, socio-economic status and education are combined with data on cancer, and differences in cancer incidence are analysed by taking into account relevant information from various sources in Finland about possible risk or protective factors for the population groups in question.

TYPE: Incidence
TERM: Alcohol; Diet; Education; High-Risk Groups; Occupation; Record Linkage; Registry; Socio-Economic Factors; Tobacco (Smoking)
SITE: All Sites
REGI: Finland (Fin)
TIME: 1988 - 1993

248 Pukkala, E.I. 04143
Finnish Cancer Registry, Liisankatu 21 B, 00170 Helsinki 17, Finland (Tel.: 176290)
COLL: Lahermo, P.; Gustavsson, N.; Hakulinen, T.R.

Comparison of Cancer Incidence and Geochemical Data

All cancer incidence data of the Finnish Cancer Registry from the year 1953 (approximately 400,000 cases) are connected with the trace elements data obtained by the Geochemical Survey of Finland covering the whole of the area of Finland. After adjustment for known confounding factors asssociations between trace elements and cancer incidence will be sought.

TYPE: Correlation; Incidence
TERM: Registry
SITE: All Sites
REGI: Finland (Fin)
TIME: 1990 - 1993

249 Riihimaki, V. 03991
Inst. of Occupational Health, Dept. of Industrial Hygiene, and Toxicology, Topeliuksenkatu 41a A, 00250 Helsinki, Finland (Tel.: 47471; Tlx: 121394)
COLL: Asp, S.; Hernberg, S.G.; Pukkala, E.I.

Mortality and Cancer Morbidity among Chlorinated Phenoxyacid Applicators in Finland

Chlorinated phenoxyacids have been associated with increased risk of soft tissue sarcomas and lymphomas. This ongoing cohort study aims to assess mortality and cancer morbidity patterns in nearly 2,000 men involved from 1955 through 1972 with the spraying of phenoxyacids (2, 4-D, 2,4,5-T) to control brushwood in forest plantations, by roadsides and railways and along electric power lines. The cohort was formed in 1972 from the personnel records of four authorities/companies. All workers who, by that time, had been assigned to spraying for at least two weeks were eligible. Subsequent exposure histories and smoking habits of the workers have recently been explored. The cohort has been followed-up to date of death or to appearance in the cancer registry as a collaborative effort of the Institute of Occupational Health and the Finnish Cancer Registry, and the mortality as well as cancer morbidity experience will be compared with that of the general population.

FINLAND

TYPE: Cohort
TERM: Chemical Exposure; Herbicides; Occupation; Pesticides; Registry
SITE: All Sites; Lymphoma; Sarcoma
CHEM: 2,4,5-T; 2,4-D; Phenoxy Acids
OCCU: Herbicide Manufacturers; Herbicide Sprayers
REGI: Finland (Fin)
TIME: 1972 - 1992

250 Sankila, R. 03246
Finnish Cancer Registry, Liisankatu 21 B, 00170 Helsinki 17, Finland (Tel.: (0)176290)
COLL: Hakulinen, T.R.; Pukkala, E.I.; Teppo, L.

Multiple Cancer
The data base of the Finnish Cancer Registry on multiple cancer (about 8,000 cases in 1953 to 1985) is analysed in order to detect positive and negative associations between different types of cancer. The effects of several factors (including age, histology, treatment, etc.) will be analysed in depth.

TYPE: Cohort
TERM: Age; Histology; Multiple Primary; Prevention; Registry; Treatment
SITE: All Sites
REGI: Finland (Fin)
TIME: 1988 - 1992

251 Sorsa, M.I. 04608
Inst. of Occupational Health, Topeliuksenkatu 41 a A, 00250 Helsinki, Finland (Tel.: +358 0 47471; Fax: 413644 ; Tlx: 12139)
COLL: Peltonen, K.; Osterman-Golkar, S.

Development of Biological and Chemical Monitoring Methods for Risk Evaluation in 1,3-Butadiene Exposure
1,3-Butadiene is an unusually potent indirectly genotoxic chemical which is among the top 30 chemicals in the world production list. The aims of the present study are to develop biological (cytogenetics, Hb-alkylation) and chemical (different metabolites and conjugation products) methods applicable for human exposure assessment. The study includes both experimental (rat exposures) and worker studies in butadiene manufacturing (30 workers) and in the use of butadiene in paper lamination (30 workers). All studies include both on-site and off-site controls.

TYPE: Cross-Sectional; Methodology; Molecular Epidemiology
TERM: Animal; Chromosome Effects; Monitoring; Occupation; Plastics
SITE: Inapplicable
CHEM: Butadiene
OCCU: Laminators; Plastics Workers
TIME: 1988 - 1992

IMATRA

252 Jäppinen, P.T. 04720
Enso Gutzeit Oy, Center for Occupational Health, 55800 Imatra, Finland (Tel.: +358 54 291331; Fax: 291330)
COLL: Tola, S.; Karjalainen, S.

Lung Cancer and Occupation in the Province of Kymi, Finland.
The aim is to assess the risk of lung cancer in relation to occupational exposures in the province of Kymi, Finland. In a case-referent study, 898 lung cancer cases (registered in the Finnish Cancer Registry from the province during 1982-86) and at least one control per case matched by age (+/-5 years), sex, and vital status will be examined. Occupational history and smoking habits are examined by questionnaire.

FINLAND

TYPE: Case-Control
TERM: Occupation
SITE: Lung
OCCU: Paper (and Pulp) Workers
REGI: Finland (Fin)
TIME: 1989 - 1992

JOENSUU

253   Herva, M.A.   04399
Univ. of Joensu, Dept. of Economics & Statistics, Yliopistokatu 7, 80101 Joensuu, Finland (Tel.: (385)9731511; Tlx: 46813 JOY SF)

Projection of Cancer Mortality in North Karelia
The purpose of the study is to analyse some common types of cancer using previous morbidity and mortality data in North Karelia Province of Finland (population: 177,000). Chiang's program will be used to apply the lifetable technique to five-year age- and sex-specific proportions of cancer deaths in current statistics for North Karelia. This will be done for lung, stomach, colon and rectum, and prostate and breast (F) cancers, to project cancer mortality up to the year.

TYPE: Methodology; Mortality
TERM: Mathematical Models; Projection
SITE: Breast (F); Colon; Larynx; Leukaemia; Lip; Lung; Oesophagus; Prostate; Rectum; Stomach; Uterus (Cervix); Uterus (Corpus)
TIME: 1987 - 1992

KUOPIO

254   Notkola, V.J.   04248
Kuopio Regional Inst., of Occupational Health, P.O. Box 93, 70701 Kuopio, Finland

Occupational Cancer Mortality among Farmers in Finland 1979-1985
The aim of this study is to compare the cause-specific cancer mortality among farmers in Finland during 1979-85 to the mortality of the total economically active population. The study is based on the Finnish farm register at 31.12.1978. Male and female farmers working in the farms at registered were defined as the population at risk. Data on all deaths during 1979-85 are obtained from death certificate data in the Finnish register on causes of death. The total study population includes about 150,000 farmers. Estimation of relative cancer risks is made by use of a log-linear model and indirect age standardization has also been used.

TYPE: Cohort
TERM: Occupation; Registry
SITE: Brain; Hodgkin's Disease; Leukaemia; Lung; Prostate
OCCU: Farmers
TIME: 1983 - 1992

255   Salonen, J.T.   04154
Univ. of Kuopio, Faculty of Medicine, Dept. of Community Health, Box 6, 702 11 Kuopio 1, Finland (Tel.: +358 71 162302)
COLL: Hakama, M.K.

Health Habits and Cancers
The purpose of this study is to investigate the role of fats and antioxidants, smoking, and alcohol consumption in the aetiology of cancer. The study hypotheses concern breast, lung, colon-rectum and basal cell cancers. The study sample is the total Finnish population aged 45 to 64 (approximately one million). A participation rate of at least 70% is expected. Cancer incidence in the cohort is followed by the Finnish Cancer Registry. Those participating in the study will fill in a questionnaire about e.g. diet, smoking, physical exercise, and other aspects of health behaviour, personal and social background, socio-economic background, disease history, personality traits and social interaction. In addition, nutrient intake is assessed with a 48-hour food-recording. Toe-nail clippings are obtained by mail.

FINLAND

TYPE: Case-Control; Cohort
TERM: Alcohol; Diet; Physical Activity; Psychological Factors; Registry; Socio-Economic Factors; Tobacco (Smoking)
SITE: Breast (F); Colon; Lung; Rectum; Skin
REGI: Finland (Fin)
TIME: 1989 - 2000

256 Syrjänen, K.J. 03693
Univ. of Kuopio, Dept. of Pathology, Kuopio Cancer Research Centre, P.O.B. 1627, 70211 Kuopio, Finland (Tel.: +358 71 162740; Fax: 162753)
COLL: Parkkinen, S.; Castrén, O.; Saarikoski, S.; Väyrynen, M.; Syrjänen, S.; Yliskoski, M.; Kellokoski, J.; Kataja, V.; Hippeläinen, M.; Chang, F.; Kallio, P.

Natural History of Genital HPV Infections and Cervical Cancer

A long-term prospective follow-up study was started in October 1981 to assess the natural history of human papillomavirus (HPV) infections in the uterine cervix and their associations with CIN and cervical cancer. To date a total of 532 women have been followed-up for a mean of 60 (SD: 22) months, by cervical punch biopsy or Pap smears, and colposcopy repeated at 6-monthly intervals. The clinical course of the lesions is analysed using the life-table technique, as well as the Cox-model. Cervical swabs (for C.trachomatis and HSV), and HPV, HSV and CMV serology are done. PAP smears and punch biopsies are analysed for the cytopathic changes of HPV and for concomitant CIN. Expression of HPV-encoded proteins is demonstrated by the indirect immunoperoxidase (IP-PAP) technique on frozen sections. In fresh biopsy samples (or frozen sections), the immunocompetent cells within the in-situ inflammatory infiltrate are phenotypically characterised using monoclonal antibodies to define T cell subsets, NK (natural killer) cells and Langerhans cells. In-situ hybridisation, Southern blot and PCR techniques are used to detect HPV DNA with cloned HPV types 6,11,16,18,31,33,42 DNA probes. Biopsy specimens will also be tested for amplification of the major cellular oncogenes (c-onc) and growth factor receptors. All patients are interviewed for their sexual and smoking habits by a detailed questionnaire. HLA-typing is now being completed. The incidence and prevalence of cervical HPV infections in an unselected Finnish female population are being assessed in the mass-screening programme of the Finnish Cancer Society. Since 1985, females have been also invited to a third (treatment) group, now comprising 530 women, and randomised to four treatments: ionization, laser, cryotherapy and interferon. Since 1986, male partners of the treated women have been examined for HPV, and treated by either laser or interferon. A paper has been published in Acta Pathol. Microbiol. Scand. 97:957-970, 1989.

TYPE: Cohort; Intervention
TERM: Antibodies; Biopsy; Condyloma; DNA; HPV; Infection; Prevention; Screening; Sero-Epidemiology; Sexual Activity; Tobacco (Smoking); Treatment
SITE: Female Genital; Male Genital
TIME: 1981 - 1992

* 257 Syrjänen, K.J. 05029
Univ. of Kuopio, Dept. of Pathology, Kuopio Cancer Research Centre, P.O.B. 1627, 70211 Kuopio, Finland (Tel.: +358 71 162740; Fax: 162753)
COLL: Eskelinen, M.; Johansson, R.; Tuomisto, J.; Jägeroos, H.; Jänne, J.; Alhonen, L.; Kettunen, K.; Länsimies, E.; Soimakallio, S.; Partanen, K.; Airaksinen, O.; Vehviläinen-Julkunen, K.; Meriläinen, P.; Naukkarinen, A.; Kosma, V.M.; Syrjänen, S.; Penttilä, I.; Hämäläinen, E.; Mahlamäki, E.; Heinonen, K.; Mononen, I.; Lehtonen, J.; Ahonen, R.; Mäntyjärvi, R.; Hippeläinen, M.

Female Breast Cancer

A long-term prospective follow-up study was started in March 1990 to assess the epidemiology and risk factors as well as the biology of female breast cancer in Kuopio province, Finland. Three series of women are included: (1) all women with primary breast cancer in Kuopio province; (2) all women with a benign breast lesion diagnosed in Kuopio University Hospital, and a series of age-matched healthy controls from the population registry. The study includes: (1) evaluation and assessment of risk factors; (2) the possibilities of prevention; (3) early diagnosis and (4) prognostic evaluation as well as treatment and follow-up. In addition, samples of tumours and of blood will be taken from all patients enrolled to be analysed using modern biomedical technology, including DNA-techniques, chromosomal analyses and biochemical assays. This is a multi-institutional project with workers from 15 different institutions in Kuopio University and Kuopio University Hospital. It is hoped that this multi-disciplinary approach, will permit assessment of the risk factors for breast cancer, as well as the mechanisms regulating its biological behaviour which could be used as prognostic indicators of this disease.

FINLAND

TYPE: Case-Control
TERM: BMB; Blood; Prevention; Prognosis; Registry; Treatment
SITE: Benign Tumours; Breast (F)
REGI: Finland (Fin)
TIME: 1990 - 2005

TAMPERE

258 Hakama, M.K. 00198
Univ. of Tampere, Dept. of Public Health, PB 607, 33101 Tampere, Finland (Tel.: (31)156111; Fax: 156057 ; Tlx: 22415)
COLL: Pukkala, E.I.; Louhivuori, K.

Effect of a Mass Screening Programme on the Risk of Cervical Cancer

The aim of the project is to evaluate the effect of an organized mass screening programme for cervical cancer on the incidence and mortality of cervical cancer and to monitor changes in efficacy by time and by population groups. The programme in nationwide and population based and consists of essential elements such as identification of target population, quality control and information on attendance and risks. Data collection is based on the national population registry, cancer registry and registration of deaths. The Finnish female population is about 2.5 million and the organized programmes cover the ages 30 to 60 years.

TYPE: Cohort
TERM: Registry; Screening
SITE: Uterus (Cervix)
REGI: Finland (Fin)
TIME: 1970 - 2000

FRANCE

ALBI

* 259   Grosclaude, P. 05110
Registre des Cancers du Tarn, Recherche en Epidémiologie, et Prévention, Chemin des 3 Tarn, 81000 Albi, France (Tel.: +33 63475951)
COLL: Roumagnac, M.; Duchene, Y.

Aetiology of Prostatic Cancer
This is a case-control study of prostatic cancer and potential risk factors such as sexual activity, various occupational risk factors, and dietary habits. Patients with prostatic cancer (histologically diagnosed from 1 October 1991) and controls living in the 'departement' of Tarn will be studied. Two controls will be obtained for each case, one from the general population and one hospital patient. The information will be collected by investigating doctors with the use of a questionnaire. The survey will be carried out on 200 patients and 400 controls. Recruitment will last one year. This study is part of a multi-centre inquiry in collaboration with two other cancer registries (Martinique and Geneva).

TYPE: Case-Control
TERM: Diet; Occupation; Registry; Sexual Activity
SITE: Prostate
REGI: Geneva (Swi); Martinique (Mar); Tarn (Fra)
TIME: 1991 - 1993

BESANÇON

260   Schraub, S. 04568
Registre des Tumeurs du Doubs, CHR Jean Minjoz, 1, blvd Fleming, 25030 Besançon Cédex, France (Tel.: +33 81668309; Fax: 81668859)
COLL: Belon-Leneutre, M.; Mercier, M.

Occupational Cancer: Cerebral Tumours and Vinyl Chloride Exposure
This is a primarily retrospective study among patients (or their proxies) registered with a cerebral tumour by the cancer registry, to see if there is a cluster of workers exposed to vinyl chloride. A case-control study will then be carried out of all patients with cerebral tumours, in which vinyl chloride exposure will be the principal risk factor addressed.

TYPE: Case-Control
TERM: Occupation; Plastics; Registry
SITE: Brain
CHEM: Vinyl Chloride
OCCU: Vinyl Chloride Workers
REGI: Doubs (Fra)
TIME: 1989 - 1992

CAEN

261   Gignoux, M. 03721
CHU Côte de Nacre, Reg. des Tumeurs Digestives du Calvados, Serv. de Chirurgie Digestive, 14040 Caen Cédex, France (Tel.: +33 31448112)
COLL: Launoy, G.; Pottier, D.

Cancers of the Digestive Tract in Calvados
The aim of the study (covering a population of over 500,000) is to evaluate the annual incidence by localization, age and sex, of carcinomas of the digestive tract. Data are collected from public and private hospitals, laboratories and physicians. Special attention is accorded to environmental and geographical patterns, profession, pre-cancerous diseases, methods of treatment and survival, particularly for carcinoma of the oesophagus for which Calvados is a high risk area. Specific studies include: (1) Identification of high-risk groups for oesophageal cancer. Areas having higher incidence of oesophagel cancer than others in the 'departement', mainly rural, have been identified and important differences between socio-professional categories within these areas noted. Populations at risk from the three French digestive tract cancer registries (Cote d'Or, Haut Garonne, Calvados) will be compared. (2) Case-control study on the role of alcohol in oesophageal cancer. This study will enable the assessment

FRANCE

of possible differences between the effect of apple-derived versus grape-derived alcohol on the oesophagus, as well as between strong and light alcohol drinks. It will also look into the possible protective role of vitamines. (3) Epidemiological characteristics of sub-types of colorectal cancer. The role of residence on tumour localisation and survival outcome will be studied. (4) Hepatocarcinoma and liver cirrhosis. A cohort of 418 cirrhotics, replying to a dietary questionnaire between 1975 and 1978 will be followed up. The aims are to assess the incidence of hepatosarcoma in this group, and to determine the role of alcohol on the risk of developing cancer of the liver and on survival.

TYPE: Case-Control; Cohort; Incidence
TERM: Alcohol; Cirrhosis; Environmental Factors; Geographic Factors; High-Risk Groups; Occupation; Premalignant Lesion; Registry; Survival; Vitamins
SITE: Colon; Liver; Oesophagus; Rectum
REGI: Caen (Fra)
TIME: 1978 - 1993

DIJON

262   Faivre, J.   03860
Registre des Cancers Digestifs, Fac. de Médecine, Registre des Tumeurs, 7, blvd Jeanne d'Arc, 21033 Dijon Cédex, France (Tel.: +33 80652323)
COLL: Backley, D.; Biasco, G.; Boutron, M.C.; De Oliveira, H.; Estève, J.; Giacosa, A.; Hill, M.J.; Kasper, H.; Maskens, A.; Thomson, M.H.; Wiebecke, B.; Wilpart, M.

Case-Control Study of Patients with Adenomatous Polyps or Cancer of the Large Bowel

Two separate studies have been prepared. In an epidemiological study three groups of cases will be included in each centre: small adenomas (n=150), large adenomas (n=150), adenocarcinomas (n=150), and two groups of controls: general population controls (n=300), and polyp free controls (n=300). Most information will be obtained by means of questionnaires administered by specially trained dieticians. The following information will be collected: personal history (demographic data, medical history, anthropometric data), description of adenomas or cancers, dietary history (using a diet history questionnaire collecting dietary information by meal). The objective of the second study, a clinical study, is to obtain in each centre 15 cases with small adenomas, 15 cases with large adenomas, 15 with adenocarcinomas, and 15 polyp free controls, for whom diet history, faeces, serum and cell kinetic studies will be available.

TYPE: Case-Control
TERM: Alcohol; BMB; Diet; Faeces; Lifestyle; Premalignant Lesion; Serum
SITE: Colon
LOCA: Belgium; France; Germany; Italy; Portugal; United Kingdom
TIME: 1986 - 1992

263   Faivre, J.   04428
Registre des Cancers Digestifs, Fac. de Médecine, Registre des Tumeurs, 7, blvd Jeanne d'Arc, 21033 Dijon Cédex, France (Tel.: +33 80652323)
COLL: Bedenne, L.; Durand, G.; Milan, C.; Arveux, P.; Boutron, M.C.

Evaluation of Mass Screening for Colorectal Cancer

This controlled trial compares the effect of systematic screening for colorectal adenomas in a well defined population of 45,000 subjects aged 45-74 with a comparable population in which there is no screening. In the test group the subjects will be offered faecal occult blood testing. Initially the whole population will be sent an explanatory letter and a document on colorectal cancer and the possibilities for screening. Information will also be given in local newspapers, municipal newsletters, local and regional radio and television programmes. The test will be prescribed by the general practitioners (GP) during the first three months of the study, then mailed to the subjects who have not received it from their GP. Tests will be sent to a reference laboratory. A total colonoscopy will be offered to individuals with a positive test. The first analyses will examine the acceptance rate and the positive predictive value of the tests for cancers and polyps. Refusal and control subjects will be followed through the records of the digestive tract cancer registry for the area. Long term evaluation of efficacy of screening wil be judged on reduction of mortality from colorectal cancer in the screened population.

FRANCE

TYPE: Cohort; Intervention
TERM: Faeces; Registry; Screening
SITE: Colon; Rectum
REGI: Dijon I (Fra)
TIME: 1987 - 1993

FONTENAY-AUX-ROSES

264   Tirmarche, M.   03473
Commissariat à l'Energie Atomique, IPSN, Lab. d'Etudes Biomédicales, Ave. General Leclerc, B.P.
no 6, 92260 Fontenay-aux-Roses, France (Tel.: +33 16 46547194; Tlx: 204841)
COLL: Chameaud, J.; Flamant, R.

Uranium Miners in France
Lung cancer has been identified as being linked to exposure to radon daughters in uranium miners in the USA, Czechoslovakia and Canada. The aim of this study is to verify this relation for French uranium miners and to study, if possible, the risk of death due to other cancers in this group of workers. The mortality of this cohort of uranium underground miners will be compared with that expected from the national statistics for the French male population of the same age groups and the same calendar periods. The vital status of these miners has been studied up to December 1985. The analysis of cancer mortality is limited to those miners having experienced more than two years of underground mining. The mean period of survey is 26 years, and the mean duration of exposure to radon and its decay products is 14 years. As mean annual exposure and working conditions were modified during the year 1956, the analysis of cancer mortality is based on two groups, separated by year of first exposure to radon and its decay products (before or since 1956). In both cohorts, a statistically significant excess of lung cancer can be observed. The study of the dose-response relationship will be published during the next year. Initial results have been published in the Proceedings of the International Congress of IRPA VII, Sydney, 10-17 April 1988, vol. 1, p. 171-175, Pergamon Press.

TYPE: Cohort
TERM: Dose-Response; Metals; Mining; Occupation; Radiation, Ionizing
SITE: All Sites; Lung
CHEM: Radon; Uranium
OCCU: Miners, Uranium
TIME: 1981 - 1993

GIF-SUR-YVETTE

265   Dousset, M.   04296
Conseiller Scientifique, du Service Central de Protection, contre les Rayonnements Ionisants, 7,
Rue de la Gruerie, 91190 Gif-sur-Yvette, France (Tel.: +33 69075174)

Lung Cancer Mortality and Radon 222 Concentration in Dwellings
The age-specific lung cancer mortality rate is to be followed from 1968 to 1982 in two populations living respectively in the Limousin (750,000) and in the Poitou-Charentes (1,500,000) regions. There is a factor 2 between the average radon concentration in dwellings of these two French regions. Taking into account known confounding factors such as tobacco, the ratio of the age-specific lung cancer mortality rate of these populations will be correlated with the differences (which increase with age) in the accumulated dose equivalent delivered to bronchial epithelium.

TYPE: Correlation; Mortality
TERM: Radiation, Ionizing
SITE: All Sites
TIME: 1986 - 1992

LA TRONCHE

266   Faure, J.R.   02612
Fac. de Médecine, Lab. de Médecine Légale, et de Toxicologie, Chemin Duhamel, 38700 La
Tronche, France (Tel.: +33 76424888/377)
COLL: Barrett, L.; De Gaudemaris, R.; Thony, C.

FRANCE

Cancer Death Rates in a Population of Screw-Cutters
The role of solvents and cutting oils in the occurrence of cancers other than skin cancers has been suspected, as a result of experiments and of some epidemiological surveys. Death statistics for the population of a district where the screw-cutting industry employs one quarter of the employed population (12,000 workers, of a total population of 115,000 with 53,000 gainfully employed), are being analysed to find out whether there is an excess of cancer mortality among screw-cutting workers. This study will be complemented by subsequent establishment of a cancer registry in the district. The following information is available in death certificates: age, sex, residence, occupation at time of death or before retirement, cause of death, anatomical site of cancer. The control group is made up of the general population, for which the same information is available. A prior study established the levels of trichloroethylene in the workshop atmosphere, and in the blood, urine and exhaled air of 188 exposed workers.

TYPE: Mortality
TERM: Chemical Exposure; Environmental Factors; Occupation; Solvents
SITE: All Sites
CHEM: Hydrocarbons, Halogenated; Mineral Oil; Trichloroethylene
OCCU: Screw Cutters
TIME: 1979 - 1992

LILLE

267 Adenis, L. 04515
Centre Oscar Lambret, Rue F. Combemale, BP 307, 59020 Lille Cédex, France (Tel.: +33 20493434)
COLL: Lefebvre, J.L.; Cambier, L.

Head and Neck Cancer Registry of Northern France
Prospective registration of all new head and neck cancers occurring in Northern France (Nord et Pas-de-Calais district) is now being carried out. Data are collected from all ENT specialists, stomatologists, head and neck surgeons, and radiotherapy specialists (public or private activity) in both districts. The data items collected are name, age, address, primary site, and histological types. About 1,400 new cases are recorded yearly in a population of 4 million inhabitants. Histological types are known for 99.2 per cent of cases.

TYPE: Incidence; Registry
TERM: Histology
SITE: Head and Neck
TIME: 1984 - 1993

* 268 Fenaux, P. 04959
Centre Hospitalier Universitaire, Serv. des Maladies du Sang, 1, Place de Verdun, 59700 Lille, France (Tel.: +33 20444348; Fax: 20444094)
COLL: Haguenoer, J.M.Y.; Nisse, C.; Pamart, B.; Quiquandon, I.; Preudhomme, C.

Environmental Factors in the Pathogenesis of Myelodysplastic Syndromes
Myelodysplastic syndromes (MDS) are the most frequent preleukaemic states. About 65 new cases of MDS are referred to this institution every year. A case-control study on the role of environmental factors in the pathogenesis of MDS is being conducted, using the method of Siemiatycki (J. Nat. Cancer Inst. 66:217-225, 1981). The study includes one control for each patient, sex- and age-matched with the patient, and living in the same area. Interviewers, who are not informed of the diagnosis, carry out an in-depth occupational, environmental and health study at the homes of cases and controls, using a checklist of specific chemicals and substances. Results of the study will be pooled with those from similar studies, using the same methods, performed in Cardiff, Bournemouth and Leeds in the United Kingdom.

FRANCE

TYPE: Case-Control
TERM: Chemical Exposure; Environmental Factors; Occupation; Premalignant Lesion
SITE: Myelodysplastic Syndrome
TIME: 1991 - 1993

269 Lefebvre, J.L. 04838
Centre Oscar Lambret, Head and Neck Oncology Dept., Rue F. Combemale, 59020 Lille Cédex, France
COLL: Adenis, L.; Joveniaux, A.; Cambier, L.

Prospective Epidemiological Study of Head and Neck Cancers
This prospective study concerns new patients presenting with at least one cancer originating in the upper aerodigestive tract, particularly oral cavity, oropharynx, larynx and hypopharynx. The following variables are recorded and compared with for the overall population and between these four sites; sex, age, marital status, smoking (duration of exposure, type of tobacco smoked, daily consumption, total consumption before diagnosis, pack years, etc.), drinking habits and occupational exposures (heat, cold, humidity, dryness, coal dust, plaster, cement or sand dust, wood dust, metal dusts or fumes, vegetal dust, chemical dusts or fumes). Special attention is given to the comparison of smoking and drinking habits and occupational exposure.

TYPE: Case Series
TERM: Alcohol; Hygiene; Occupation; Tobacco (Smoking)
SITE: Hypopharynx; Oral Cavity; Oropharynx
TIME: 1986 - 1993

LYON

* 270 Boffetta, P. 05193
Int. Agency for Research on Cancer, Unit of Analytical Epidemiology, 150, cours Albert Thomas, 69372 Lyon Cédex 08, France (Tel.: +33 72738485; Fax: 72738575 ; Tlx: 380023 circ f)
COLL: Saracci, R.; Kogevinas, M.; Peto, J.; Easton, D.; Wong, O.; Boreiko, C.; Steenland, K.N.; Nordberg, G.F.; Fanning, D.; Kazantzis, G.; Davies, J.

International Collaborative Study on Lead-Exposed Workers
Cohort studies of workers exposed to lead have suggested increases in lung and stomach cancers. The first phase of this project is a combination of existing cohorts, which will be reanalysed in an identical manner with respect to duration of exposure and time since first exposure. On the basis of the results of the first phase and recruitment of new cohorts, further steps of the project may include a combined update of existing and new cohorts, a comprehensive industrial hygiene survey and the conduct of nested case-control studies.

TYPE: Cohort
TERM: Metals; Occupation; Registry; Time Factors
SITE: All Sites; Lung; Stomach
CHEM: Lead
OCCU: Battery Plant Workers; Smelters, Lead
REGI: Sweden (Swe)
TIME: 1991 - 1992

271 Bosch, F.X. 03857
Int. Agency for Research on Cancer, Unit of Field and Intervention Studies, 150, cours Albert Thomas, 69372 Lyon Cédex 08, France (Tel.: +33 72738485; Fax: 72738575 ; Tlx: 380023 circ f)
COLL: Muñoz, N.; Rodriguez, M.C.; Hernandez, J.M.; Castillo, R.; Grifols, R.; Lluch, A.; Plasencia, J.

Follow-up of a Cohort of HBsAg-positive Blood Donors in Catalonia
Mortality data from the region of Catalonia in Spain suggest that it is a high-risk area for liver cancer within the context of the Western industrialized world: death rates in 1983 were 12.8 among males and 8.3 among females. A study has been initiated to assemble a cohort of about 3,000 HbsAg-positive blood donors and to link their names with local death certificate files to determine liver cancer risk. Data are being obtained from five major blood banks operating in the area. An analysis was conducted in 1987 linking the names of the HbsAg carriers to the 1981 census. No excess of liver cancer could be identified. Current work is in progress using the 1986 census data.

FRANCE

TYPE: Cohort
TERM: Antigens; Infection; Virus
SITE: Liver
LOCA: Spain
TIME: 1985 – 1992

272 Bosch, F.X. 04710
Int. Agency for Research on Cancer, Unit of Field and Intervention Studies, 150, cours Albert Thomas, 69372 Lyon Cédex 08, France (Tel.: +33 72738485; Fax: 72738575 ; Tlx: 380023 circ f)
COLL: Muñoz, N.; Peto, J.

International Biological Study on Cervical Cancer
The International Biological Study on Cervical Cancer (IBSCC) is an international project aimed at creating a repository of cervical cancer tissue at the IARC for use in studying markers of exposure to known or suspected risk factors. In the first phase, samples of invasive cervical cancer are being collected in about 20 countries with varying incidence rates of cervical cancer. Protocols will then be developed to assess the presence of markers of exposure to potential carcinogens in the specimens. In the first protocol, DNA/RNA hybridization methods will be used to assess the prevalence of specific types of HPV. All the assays will be performed in the same reference laboratory. A brief questionnaire will also be used to assess exposure to other known risk factors for cervical cancer. In the future, a collection of exfoliated cervical cells from appropriate samples of healthy women from the same countries will then be organized.

TYPE: Cross-Sectional
TERM: BMB; Biochemical Markers; Biopsy; Cytology; HPV; Tissue
SITE: Uterus (Cervix)
TIME: 1989 – 1993

273 Bosch, F.X. 04868
Int. Agency for Research on Cancer, Unit of Field and Intervention Studies, 150, cours Albert Thomas, 69372 Lyon Cédex 08, France (Tel.: +33 72738485; Fax: 72738575 ; Tlx: 380023 circ f)
COLL: Benito, E.; Esteva, M.; Mulet, M.; Obrador, A.; Estève, J.; Muñoz, N.

Familial Study of Diet and Colorectal Cancer
The role of diet in colorectal cancer has been examined mainly in case-control studies. Recent studies suggest inherited susceptibility to adenomatous polyps and subsequent adenocarcinoma of the colon and rectum in a segment of the population. The use of controls with similar genetical susceptibility to the cases would presumably provide a more precise estimation of the role of diet in the development of colorectal cancer. This study will evaluate the association of colorectal cancer and dietary factors among cases and controls using siblings of cases as controls. Cases will be all newly diagnosed colorectal cancers in the population of Majorca from January 1990; approximately 300 cases are expected. Controls will be the traceable sibling of cases during the same period. To estimate food consumption, a semi-quantitative food frequency questionnaire will be administered by trained interviewers.

TYPE: Case-Control
TERM: Diet; Genetic Factors
SITE: Colon; Rectum
LOCA: Spain
TIME: 1990 – 1993

274 Brémond, A.G. 04709
INSERM U 265, 151, Cours Albert Thomas, 69424 Lyon Cédex 03, France (Tel.: +33 72330123; Fax: 72348784)
COLL: Coste, I.; Victoria, J.; Courtial, I.

Evaluation of a Breast Cancer Mass Screening Programme
A breast cancer screening programme began in April 1987 in the Rhone 'departement'. All women aged 50 to 69 living in this 'departement' receive a personal invitation to participate. The screening test used is a single oblique view mammogram every two years. For each positive test (abnormal mammogram) the final diagnosis and, if cancerous, stage of cancer are recorded. The objectives of the study are (1) to assess the cost/benefit of screening, and (2) to measure the effectiveness and efficiency of this type of screening programme in the context of the French health system. A breast cancer registry was established in 1988 to evaluate the screening programme. The quality of the screening test has been

FRANCE

evaluated and cancers compared in the following three groups: new cancers, cancers appearing in the interval between two screening rounds, cancers in women not responding to the screening programme.

TYPE: Cross-Sectional; Incidence; Registry
TERM: Cost-Benefit Analysis; Mammography; Registry; Screening
SITE: Breast (F)
REGI: Rhône (Fra)
TIME: 1987 – 1992

275 Duclos, J.C. 03054
Inst. Univ. de Médecine du Travail, 8, Ave. Rockefeller, Domaine Rockefeller, 69373 Lyon Cedex 08, France (Tel.: 78777000/4363)
COLL: Pignat, J.C.

Ethmoid Tumours and Harmful Dusts

In this study analyses of wood dust (composition and particle size) are carried out in plants where ethmoid adenocarcinomas have been reported among the workforce. The existence of specific histological types of ethmoid cancer in such plants is being investigated. Four to six cases are anticipated annually. A systematic study of histological changes in the nasal mucosa of wood workers, classified by 5-year age-groups, according to the type of wood used and the granulometry of the work places where they are exposed, is also being carried out. The aim of this histological study is to assess the architectural and cytological changes in relation to length of exposure, granulometric concentrations and type of wood used. The approximate number of exposed workers is 250. Other patients visiting the ENT department will serve as controls.

TYPE: Case Series; Cross-Sectional
TERM: Dusts; Environmental Factors; Histology; Occupation; Wood
SITE: Nasal Cavity
OCCU: Wood Workers
TIME: 1984 – 1992

276 Estève, J. 04422
Int. Agency for Research on Cancer, Unit of Biostatistics and Informatics, 150, cours Albert Thomas, 69372 Lyon Cédex 08, France (Tel.: +33 72738485; Fax: 72738575 ; Tlx: 380023 circ f)
COLL: Roy, P.; Kaldor, J.; English, D.R.; Fichtinger-Schepman, A.; Natarajan, N.; Kyrtopoulos, S.; Wild, C.; Hall, J.; Somers, R.; Van Leeuwen, F.E.; Hagenbeek, A.; Bron, D.; Hayat, M.; Henry-Amar, M.; ten Bokkel Huinink, G.; Stoter, G.; Jones, W.R.; Kaye, S.; Sleijfer, D; Pangalis, G.A.

DNA Damage Following Chemotherapy for Cancer

Cytostatic agents which act by interfering with cellular DNA are also likely to be carcinogenic and mutagenic. Recent developments in biochemistry and cytogenetics permit the measurement of DNA adducts and other forms of DNA damage. In collaboration with the Lymphoma and Genito-urinary Groups of the European Organisation for Research and Treatment of Cancer, studies are being carried out of the relationship between DNA adducts formed by chemotherapeutic agents and long-term end points, particularly survival (as a measure of therapeutic efficacy) and second cancer risk. Chemotherapeutic agents for which DNA adducts are being measured by immuno-assay are cis-platinum, used in the treatment of non-seminomatous testicular cancer, and procarbazine, which is used in combination chemotherapy for lymphoma. Studies of oncogene mutation and cytogenetic endpoints are also being carried out.

TYPE: Molecular Epidemiology
TERM: Chemotherapy; DNA Adducts; Drugs; Multiple Primary; Survival
SITE: All Sites
CHEM: Cis-Platinum; Procarbazine
LOCA: Belgium; France; Greece; Netherlands; United Kingdom
TIME: 1988 – 1992

277 Estève, J. 04703
Int. Agency for Research on Cancer, Unit of Biostatistics and Informatics, 150, cours Albert Thomas, 69372 Lyon Cédex 08, France (Tel.: +33 72738485; Fax: 72738575 ; Tlx: 380023 circ f)
COLL: Coleman, M.P.; Parkin, D.M.; Damiecki, P.; Schifflers, E.; Renard, H.; Arslan, A.

FRANCE

Global Time Trends in Cancer Incidence and Mortality
A comprehensive analysis of time trends in incidence and mortality of cancer at 28 different sites, using data from 39 regional and national cancer registries covering a period of up to 25 years, is in progress. Incidence data are derived from the five volumes of 'Cancer Incidence in Five Continents, and mortality data from WHO. The irregular structure of incidence data has required the development of new statistical methods and software; this work is largely complete, and preliminary analyses will begin in the second half of 1988. Results will be presented in the form of time trends by broad age-group, in the form of cancer trends by calendar period, and by year of birth (cohort analyses), showing the patterns by global region and by sex for each type of cancer.

TYPE: Incidence; Methodology; Mortality
TERM: Geographic Factors; Registry; Trends
SITE: All Sites
TIME: 1987 - 1992

278 Hours, M. 04445
Univ. Claude Bernard, Inst. d'Epidémiologie, 8, Ave. Rockfeller, Domaine Rockfeller, 69373 Lyon Cédex 08, France (Tel.: +33 78777000/4691)
COLL: Ayzac, L.; Bergeret, A.; Dananche, B.; Fabry, J.; Févotte, J.

Occupational Risks for Urinary Bladder and Lung Neoplasms
The main purpose of this study is to identify relative risks of bladder or lung cancer which are significantly high for some occupational activities or exposures, whether previously suspected or not. This case-control study includes 500-600 cases detected over five years in a highly industrialized area; matched controls are chosen from among other in-patients. The study involves a detailed occupational interview and a review of this interview by industrial experts in order to identify probable occupational exposures.

TYPE: Case-Control
TERM: Chemical Exposure; Occupation
SITE: Bladder; Lung
TIME: 1984 - 1992

* 279 Hours, M. 05164
Univ. Claude Bernard, Inst. d'Epidémiologie, 8, Ave. Rockfeller, Domaine Rockfeller, 69373 Lyon Cédex 08, France (Tel.: +33 78777000/4691)
COLL: Févotte, J.; Dananche, B.; Ayzac, L.; Bergeret, A.; Pollen, C.; Fière, D.; Hollard, D.; Broustet, A.; Cicollela, A.; Philippe, J.

Acute Myeloid Leukaemia and Exposure to Pesticides or Glycol Ethers
The aim of the study is to identify risk of acute myeloid leukaemia associated with two groups of chemicals used by many people during occupational activities: pesticides and glycol ethers. This case-control study is conducted in hopitals. 210 cases are expected over three years. The study involves a detailed occupational interview and a review of data from this interview by industrial experts in order to identify probable exposures to pesticides and glycol ethers.

TYPE: Case-Control
TERM: Occupation; Pesticides; Solvents
SITE: Leukaemia (ALL)
CHEM: Glycol Ethers
TIME: 1990 - 1994

* 280 Kogevinas, M. 05195
Int. Agency for Research on Cancer, Unit of Analytical Epidemiology, 150, cours Albert-Thomas, 69372 Lyon Cédex 08, France (Tel.: +33 72738485; Fax: 72738575 ; Tlx: 380023 circ f)
COLL: Saracci, R.; Winkelmann, R.; Boffetta, P.; Andersen, A.; Benn, T.; Biocca, M.; Coggon, D.; Kurppa, K.; Lundberg, I.S.; Lynge, E.; Astrup-Jensen, A.; Bellander, T.; Bjerk, J.E.; Pannett, B.; Pfäffli, P.; Tolomei, S.

IARC International Cohort Study on Workers Exposed to Styrene
The objective of the study is to examine cancer risk among workers exposed to styrene. Previous epidemiological studies have indicated increased risk of leukaemia and lymphoma. A collaborative group of researchers from six European countries was formed in order to collect and evaluate data on exposure to styrene, mainly in the glass-reinforced plastics industry. The total population suitable for

FRANCE

study is approximately 20,000 workers. All cohorts will be followed for mortality, while cancer incidence will be available for some countries. A group of industrial hygienists has evaluated levels of present and past exposures to styrene in the industry. Exposure data available for early periods of production indicate that exposure levels were very similar throughout Europe.

TYPE: Cohort
TERM: Occupation; Plastics; Registry
SITE: All Sites
CHEM: Styrene
OCCU: Laminators; Plastics Workers
LOCA: Denmark; Finland; Italy; Norway; Sweden; United Kingdom
REGI: Denmark (Den); Finland (Fin); Norway (Nor); Sweden (Swe)
TIME: 1988 - 1992

* 281   **Kogevinas, M.**   05196
Int. Agency for Research on Cancer, Unit of Analytical Epidemiology, 150, cours Albert-Thomas, 69372 Lyon Cédex 08, France (Tel.: +33 72738485; Fax: 72738575 ; Tlx: 380023 circ f)
COLL: Boffetta, P.; Saracci, R.; Vainio, H.; Winkelmann, R.; Ferro, G.; Andersen, A.; Facchini, L.; Hours, M.; Henneberger, P.; Jäppinen, P.T.; Lynge, E.; Persson, B.; Pearce, N.E.; Rodrigues, V.; Soskolne, C.L.; Boal, W.; Coggon, D.; Merletti, F.; Sunyer, J.; Raymond, L.; Lutz, J.-M.; Katsouyiannopoulos, V.; Heederik, D.; Wild, P.; Miyake, H.; Szadkowska-Stanczyk, I.

IARC International Cohort Study on Cancer Risk among Workers in the Pulp and Paper Industry
The paper and pulp industry is spread world-wide, and employs hundreds of thousands of workers. The few prospective epidemiological studies in this industry indicate that cancer risk, particularly lung cancer, gastrointestinal cancer and neoplasms of the lymphatic tissue, may be elevated, but evidence is still not convincing. The objective of this study is to evaluate cancer risk in relation to specific processes and exposures in this industry. Personnel employed in plants producing pulp, paper and paper products and in mills involved in recycling will be included. Cohorts are currently being assembled, and it is expected that the international study will include data for more than 100,000 workers. Two distinct phases are planned. In a first phase, a retrospective cohort study will be conducted. Depending on the results, nested case-control studies on specific neoplasms will be considered. A parallel industrial hygiene study is being conducted.

TYPE: Cohort
TERM: Chemical Exposure; Dusts; Dyes; Occupation; Pesticides; Plastics; Registry
SITE: All Sites
CHEM: Chlorine; Chlorophenols; Dioxins; Formaldehyde
OCCU: Paper (and Pulp) Workers
LOCA: Algeria; Brazil; Canada; Denmark; Finland; France; Greece; Italy; Japan; Netherlands; New Zealand; Norway; Poland; Portugal; Spain; Sweden; Switzerland; United Kingdom; United States of America
REGI: Canada (Can); Denmark (Den); Finland (Fin); Geneva (Swi); Isère (Fra); New Zealand (NZ); Norway (Nor); Sweden (Swe)
TIME: 1991 - 1996

282   **Muñoz, N.**   02769
Int. Agency for Research on Cancer, Unit of Field and Intervention Studies, 150, cours Albert Thomas, 69372 Lyon Cédex 08, France (Tel.: +33 72738485; Fax: 72738575 ; Tlx: 380023 circ f)
COLL: Wahrendorf, J.; Bang, L.J.; Crespi, M.; Grassi, A.; Thurnham, D.

Precancerous Lesions of the Oesophagus
Follow-up of over 1,000 subjects in whom the prevalence of precancerous lesions was determined in 1980 and 1984 in Linxian and Huixian continues. The additional data pertaining to the intervention study are being analysed, including: (1) the dietary surveys on the 610 study subjects who were interviewed three times during the study period, i.e. at the beginning of the study, six months later and at the end of the study. These surveys include general dietary information and a 24 hour recall; (2) correlation of the dietary data with the blood levels of retinol, riboflavin and zinc, before and after intervention.

FRANCE

TYPE: Intervention
TERM: Biochemical Markers; Biopsy; Diet; Metals; Micronuclei; Premalignant Lesion; Prevention; Vitamins
SITE: Oesophagus
CHEM: Beta Carotene; Retinoids; Zinc
LOCA: China
TIME: 1982 – 1993

283 Muñoz, N. 03879
Int. Agency for Research on Cancer, Unit of Field and Intervention Studies, 150, cours Albert Thomas, 69372 Lyon Cédex 08, France (Tel.: +33 72738485; Fax: 72738575 ; Tlx: 380023 circ f)
COLL: Bosch, F.X.; Aristizabal, N.; Ascunce, N.; Gili, M.; Gonzalez, L.C.; Izarzugaza, I.; Moreo, P.; Navarro, C.; Tafur, L.; Viladiu, P.; Shah, K.V.; de Sanjosé, S.; Guerrero, E.; Santamaria, M.; Alonso de Ruiz, P.

Cervical Cancer, Male Sexual Behaviour and Papilloma Virus in High- and Low-Risk Areas for Cervical Cancer

Colombia and Spain are the two countries in which one of the highest and one of the lowest incidence rates for cervical cancer have been reported, respectively. A collaborative case-control study has been conducted to determine: (1) how much of the 10-fold differential in risk between these two countries is due to female sexual behaviour; (2) how much is due to male sexual behaviour; and (3) the role of HPV in the development of this tumour in both populations. In Spain, the study included 250 in situ and 226 invasive cases, and in Colombia 276 in situ and 180 invasive cases. 636 male partners in Spain and 472 in Colombia have also participated. A one-to-one control was selected from the files of the laboratory where the in situ cases are diagnosed and from a representative sample of the female population of the area (controls for the invasive cases). Case accrual was terminated in Spain in June 1988 and in April 1989 in Colombia. The number of study subjects interviewed is 2,968. 89% of these agreed to provide a cytological sample and 93% a serum sample. HPV markers are being looked for in cervical cells obtained from both cases and controls, in fresh tissue samples from biopsies or surgical specimens collected from untreated cases, and in cell specimens taken from uterine cervix and male urethras. The Virapap, Southern blot and PCR techniques for HPV-DNA hybridization were used. Serological markers for HSV-2, CMV, chlamydia, gonorrhoea and syphilis and serum levels of vitamin E, carotenes and retinol have also been measured. The study has been conducted in nine provinces in Spain (Gerona health district, Zaragoza, Sevilla, Murcia, Salamanca, Navarra, Alava, Guipuzcoa and Vizcaya) and in Cali, Colombia. Analysis of the questionnaires revealed that early age at first sexual intercourse, a high number of sexual partners, low level of education and practice of prostitution were the main risk factors among women. A high number of sexual partners of the husbands increased the risk of cervical cancer in their wives, but only in Spain. A strong association with HPV (mainly HPV16) was observed in both countries. Residual moderate associations were also found with chlamydia and HSV2. Statistical analysis to assess the combined effect of the various risk factors is in progress.

TYPE: Case-Control
TERM: Biochemical Markers; HPV; HSV; Infection; Sero-Epidemiology; Sexual Activity; Sexually Transmitted Diseases; Vitamins
SITE: Uterus (Cervix)
CHEM: Beta Carotene; Retinoids
LOCA: Colombia; Spain
TIME: 1985 – 1992

* 284 Muñoz, N. 05185
Int. Agency for Research on Cancer, Unit of Field and Intervention Studies, 150, cours Albert Thomas, 69372 Lyon Cédex 08, France (Tel.: +33 72738485; Fax: 72738575 ; Tlx: 380023 circ f)
COLL: Oliver, W.; Vivas, J.; Peraza, S.; Cano, E.; Alvarez, N.; Castro, D.; Sanchez, V.; de Contreras, O.; de Sanjosé, S.; Correa, P.; Sobala, G.

Chemoprevention Trial on Precancerous Lesions of the Stomach

A double-blind placebo-controlled randomized trial will be carried out to assess the ability of treatment for H. pylori and treatment with antioxidants (beta carotene and vitamins C and E) to induce regression of chronic gastritis (with or without atrophy or intestinal metaplasia) or to block its progression to dysplasia, as compared with a group receiving a placebo. It will include 3,000 subjects aged 35-64 years. Histological, histochemical and biochemical end-points will be used to assess the efficacy of the two treatments. Pilot studies to assess the prevalence of H. pylori infection and to compare the bioavailability of two antioxidant preparations have been carried out.

FRANCE

TYPE: Intervention
TERM: Antioxidants; BMB; Blood; H. pylori; Premalignant Lesion; Prevention; Vitamins
SITE: Stomach
CHEM: Beta Carotene
LOCA: Venezuela
TIME: 1991 - 1996

* 285   Muñoz, N.   05186
Int. Agency for Research on Cancer, Unit of Field and Intervention Studies, 150, cours Albert Thomas, 69372 Lyon Cédex 08, France (Tel.: +33 72738485; Fax: 72738575 ; Tlx: 380023 circ f)
COLL: Oliver, W.; Vivas, J.; Peraza, S.; Alvarez, N.; Parkin, D.M.; de Sanjosé, S.

Case-Control Study of Stomach Cancer

The aims of this case-control study are: (a) to identify the main risk factors for stomach cancer; and (b) to evaluate the efficacy of a screening programme for stomach cancer in Tachira state. It will include 300 cases of histologically confirmed stomach cancer and two groups of 300 controls each (hospital and neighbourhood controls), matched by sex and age. Exposure to the risk factors and intensity of screening will be investigated by means of a questionnaire and review of screening records. Antibodies to H. pylori will be measured in serum from cases and controls and genetic alterations in cancerous and normal gastric mucosa will be assessed.

TYPE: Case-Control
TERM: BMB; H. pylori; Screening; Serum
SITE: Stomach
LOCA: Venezuela
TIME: 1991 - 1995

286   Parkin, D.M.   04202
Int. Agency for Research on Cancer, Unit of Descriptive Epidemiology, 150, cours Albert-Thomas, 69372 Lyon Cédex 08, France (Tel.: +33 72738485; Fax: 72738575 ; Tlx: 380023 circ f)
COLL: Balzi, D.; Biggeri, A.; Buiatti, E.; Geddes, M.; Kaldor, J.; Khlat, M.

Cancer in Italian Migrant Populations

Populations of Italian origin have migrated to several environmentally distinct parts of the world. The objective of the study is to compare rates for the major cancer sites in these migrant populations (1) with each other, (2) with the locally-born, and (3) with those remaining in Italy. The countries studied are Canada, USA, Brazil, Uruguay, Argentina, Australia, France, Switzerland and UK. Both mortality and incidence data have been used. The results are presented as relative risks (Italy: host country, and Migrants: host country) derived from poisson regression models when population denominators are available, logistic models, comparing one cancer with all others, when they are not. For Australia and the USA, data are available on time of migration and the effect of duration of residence in the new host country has been studied. A paper was published in Cancer Causes and Control 2:133-140, 1991 (Geddes et al.).

TYPE: Incidence; Mortality
TERM: Geographic Factors; Migrants; Registry
SITE: All Sites
LOCA: Argentina; Australia; Brazil; Canada; France; Switzerland; United Kingdom; United States of America; Uruguay
TIME: 1987 - 1992

287   Parkin, D.M.   04552
Int. Agency for Research on Cancer, Unit of Descriptive Epidemiology, 150, cours Albert-Thomas, 69372 Lyon Cédex 08, France (Tel.: +33 72738485; Fax: 72738575 ; Tlx: 380023 circ f)
COLL: Cardis, E.; Kaldor, J.; Masuyer, E.; Hansluwka, H.E.; Augustin, J.; Plesko, I.; Storm, H.H.; Karjalainen, S.; Lutz, J.-M.; Staneczeck, W.; Michaelis, J.H.; Vargha, M.; Terracini, B.; Coebergh, J.W.W.; Langmark, F.; Zatonski, W.A.; Barlow, L.; Raymond, L.; Stiller, C.A.; Black, R.; Rahu, M.A.; Kriauciunas, R.; Merabishivili, V.; Pompe-Kirn, V.; Bobev, D.; Ivanov, E.

European Childhood Leukaemia-Lymphoma Incidence Survey

The nuclear power plant accident at Chernobyl in 1986 resulted in radiation exposure to much of the European population. Even though the best available estimates suggest that any consequent increase in adverse health effects will not be detectable epidemiologically, there has been substantial public concern about the risk of leukaemia and other cancers. A collaborative group of population-based

FRANCE

cancer registries has been established, to study childhood leukaemia and lymphoma incidence in Europe, by calendar year, age, sex and geographical region. Of all malignancies, leukaemia is the most appropriate to study, because it is probably the most radiosensitive, appears relatively early following exposure, varies little in incidence over time and space and has few known causes. Estimates of radiation exposure are available through collaboration with UNSCEAR, and geographic and temporal relationship with leukaemia incidence will be studied. Incidence data are being collected from 1980, to allow the evaluation of any preexisting trends, until at least 1996, 10 years after the accident. An abstract was published in Radiation Research 124:370-371, 1990.

TYPE: Incidence
TERM: Childhood; Radiation, Ionizing; Registry; Trends
SITE: Leukaemia
LOCA: Austria; Czechoslovakia; Denmark; Finland; France; Germany; Hungary; Italy; Netherlands; Norway; Poland; Sweden; Switzerland; USSR; United Kingdom; Yugoslavia
REGI: Berlin (FRG); Denmark (Den); Estonia (Est); Finland (Fin); Geneva (Swi); Isère (Fra); Leningrad (USSR); Lithuania (Lit); Mainz (FRG); Norway (Nor); Oxford I (UK); Piedmont (Ita); Poland (Pol); Scotland (UK); Slovakia (Cze); Slovenia (Yug); Sweden (Swe); The Hague (Net)
TIME: 1988 - 1996

288   Parkin, D.M.   04659
Int. Agency for Research on Cancer, Unit of Descriptive Epidemiology, 150, cours Albert-Thomas, 69372 Lyon Cédex 08, France (Tel.: +33 72738485; Fax: 72738575 ; Tlx: 380023 circ f)
COLL: Bouchardy, C.; Zatonski, W.A.; Tyczynski, J.; Iscovich, J.M.; Bernstein, L.; Mirra, A.P.; Marmot, M.G.; Swerdlow, A.J.

Cancer Mortality in Migrant Populations around the World

This is a study of major migrant groups in different parts of the world, for which mortality or incidence rates for specific cancer sites are compared with the locally-born, and with the population in the country of origin of the migrants. Particular emphasis is placed on extension of risk of different cancers in relation to ag at migration, or duration of residence in the new host country (migrants to the USA and Australia). In the case of Israel, the risk of cancers in second-generation migrants (born in Israel) is now under study. Regression methods are used to calculate odds ratios, taking into account possible confounding factors such as time, place of residence within the host country, and social class.

TYPE: Mortality
TERM: Geographic Factors; Migrants; Time Factors
SITE: Bladder; Breast (F); Colon; Lung; Melanoma; Oesophagus; Prostate; Rectum; Stomach; Uterus (Cervix); Uterus (Corpus)
LOCA: Australia; France; Israel; Poland; United Kingdom; United States of America
TIME: 1988 - 1993

* 289   Parkin, D.M.   05216
Int. Agency for Research on Cancer, Unit of Descriptive Epidemiology, 150, cours Albert-Thomas, 69372 Lyon Cédex 08, France (Tel.: +33 72738485; Fax: 72738575 ; Tlx: 380023 circ f)
COLL: Ngelangel, C.A.; Laudico, A.V.; Reyes, M.G.

Screening for Breast Cancer in the Population of Metropolitan Manila

Screening for breast cancer by mammography, with or without physical examination of the breast, has been shown to be effective in reducing mortality from breast cancer in women over 50 years. However, since the technological requirements are expensive, such programmes are inappropriate for those developing countries where breast cancer incidence is moderately elevated. The objectives of this study are: (1) to determine whether physical examination of the breast combined with the teaching of breast self-examination, performed once a year by trained midwives/nurses, reduces mortality from breast cancer by 20 to 30% over a period of five to 10 years, and (2) to determine whether this intervention will lead to a reduction in the cumulative prevalence of advanced (stage 2) breast cancer, and to a change in the nature of the treatment and follow-up of breast cancer cases in the community. A protocol has been developed for a randomized controlled trial of screening for breast cancer in 330, 000 women aged 35-64, using physical examination by trained nurses as the sole screening modality. A pilot study will commence in 1991 in six health centres with a total female population of 12,000 to examine various aspects of feasibility and compliance, and to estimate predictive value of physical examination in this population.

FRANCE

TYPE: Methodology
TERM: Screening
SITE: Breast (F)
LOCA: Philippines
TIME: 1991 - 1994

* 290   Parkin, D.M.   05218
Int. Agency for Research on Cancer, Unit of Descriptive Epidemiology, 150, cours Albert-Thomas, 69372 Lyon Cédex 08, France (Tel.: +33 72738485; Fax: 72738575 ; Tlx: 380023 circ f)
COLL: Rolón, P.A.; Khlat, M.

Plantar Melanoma in Paraguay
Data from Paraguay indicate that malignant melanoma of the sole of the foot is relatively common, as it is in African countries. Aetiological hypotheses, such as occurrence of naevi, trauma and thermal injury, have never been tested. The situation in Paraguay, where the entire population of approximately 4 million is well covered by hospital services, seems particularly favourable to a case-control study of aetiological factors of plantar melanoma. This case-control study will examine not only possible relevant factors, such as wearing shoes, presence of plantar naevi, trauma, thermal injury, but also major risk factors for melanoma in European populations, such as history of exposure to ultraviolet light. 50 cases and 200 controls will be interviewed. The questionnaire has also been translated into local languages.

TYPE: Case-Control
TERM: Naevi; Radiation, Ultraviolet; Trauma
SITE: Melanoma
LOCA: Paraguay
TIME: 1988 - 1993

* 291   Parkin, D.M.   05219
Int. Agency for Research on Cancer, Unit of Descriptive Epidemiology, 150, cours Albert-Thomas, 69372 Lyon Cédex 08, France (Tel.: +33 72738485; Fax: 72738575 ; Tlx: 380023 circ f)
COLL: Zatonski, W.A.; Tyczynski, J.; Tarkowski, W.; Swerdlow, A.J.; Brancker, A.; Bernstein, L.; Iskovich, J.; Matos, E.

Cancer in Polish Migrant Populations
People of Polish origin have migrated to several countries which differ in their environment and lifestyle. The aim of the study is to compare incidence and/or mortality for major cancer sites in migrant populations with the local-born and with the population of the country of origin. Data from Argentina, Australia, Brazil, Canada, England and Wales, France, Israel and United States will be used. For countries like Australia, Israel and the US the risk of cancer according to duration of stay (and/or age at arrival) will be studied, and for Canada the evaluation of cancer risk in Canada-born individuals with Polish parents will be done. If appropriate census denominators are available, Poisson regression will be employed to evaluate cancer risk. If not, logistic regression will be used.

TYPE: Incidence; Mortality
TERM: Geographic Factors; Migrants
SITE: All Sites
LOCA: Argentina; Australia; Brazil; Canada; France; Israel; United Kingdom; United States of America
TIME: 1991 - 1993

292   Riboli, E.   04874
Int. Agency for Research on Cancer, Unit of Analytical Epidemiology, 150, cours Albert Thomas, 69372 Lyon Cédex 08, France (Tel.: +33 72738485; Fax: 72738575 ; Tlx: 380023 circ f)
COLL: Ahrens, W.; Benhamou, S.; Boffetta, P.; Calheiros, J.M.; González, C.A.; Haley, N.J.; Howe, G.R.; Jindal, S.K.; Kalandidi, A.; Levi, F.G.; Medes, A.; Merletti, F.; Pershagen, G.; Rodriguez, F.F.; Saracci, R.; Simonato, L.; Winck, J.C.

Lung Cancer in Nonsmokers and Environmental Tobacco Smoke
A case-control study has been started in 11 collaborating centres in Europe, North America and Asia to investigate the relationship between exposure to environmental tobacco smoke and to other environmental risk factors (occupational exposures, air pollution, diet) and the risk of lung cancer in subjects who have never smoked tobacco. Data are collected by personal interview. Self-reported smoking or nonsmoking status will be cross-checked by interview of spouses in a subsample of subjects. Biological samples (urine and blood) will also be collected in a subsample to obtain biochemical validation of current smoking status. To date about 175 cases and 328 controls have been interviewed. Data collection will be extended to the end of 1992.

FRANCE

TYPE: Case-Control
TERM: Air Pollution; Biochemical Markers; Diet; Occupation; Passive Smoking
SITE: Lung
LOCA: Canada; France; Germany; India; Italy; Portugal; Spain; Sweden; Switzerland
TIME: 1989 - 1992

293   Riboli, E.   04895
Int. Agency for Research on Cancer, Unit of Analytical Epidemiology, 150, cours Albert Thomas, 69372 Lyon Cédex 08, France (Tel.: +33 72738485; Fax: 72738575 ; Tlx: 380023 circ f)
COLL: Saracci, R.; Kaaks, R.; Slimani, N.; Clavel, F.; Villelminot, S.; Berrino, F.; Pisani, P.; Vineis, P.; Gafà, L.; Tumino, R.; Day, N.E.; Khaw, K.T.; Bingham, S.; Key, T.; Forman, D.; Trichopoulou, A.; Katsouyanni, K.; Collette, H.J.A.; Kromhout, D.; Wahrendorf, J.; Boeing, H.; González, C.A.; Torrent, M.; Quiros Garcia, J.R.; Navarro, C.; Del Moral, A.; Martinez, C.

Prospective Studies on the Role of Nutrition in the Aetiology of Cancer
The objective of the project is to investigate the relationship between diet, nutritional status and the risk of cancer at several sites. The plan is to set up large prospective studies in European populations characterized by different dietary habits and different patterns of cancer incidence. In each collaborating country, cohorts of 70,000 middle-aged subjects will be recruited using appropriate sampling techniques suitable for each specific population. Blood and other biological specimens will be collected from study participants and stored at a low temperature. Laboratory analyses will be carried out at a later stage on samples from subjects who do or do not develop cancer. Cohorts will be followed up through population cancer registries or other suitable sources until a sizeable number of cancer cases has been collected. Cancer incidence at the most common sites (stomach, colon, rectum, breast, lung, urinary tract, uterus) will be analyzed in relation to information on diet and nutritional status. A pilot phase started in 1989 in Italy, France, Spain and the UK and was completed in 1991. Extension of the study to Greece, Germany and the Netherlands was started in 1990. Data collection for the main study was started in 1991 in France, and will start in 1992 in Italy, Greece and the UK and is planned for 1993 in Germany and the Netherlands. Data collection should be completed within four years and follow-up will start in 1995-1996.

TYPE: Cohort
TERM: BMB; Blood; Diet; Environmental Factors; Nutrition; Urine
SITE: All Sites
LOCA: France; Germany; Greece; Italy; Netherlands; Spain; Sweden; United Kingdom
TIME: 1989 - 2000

294   Saracci, R.   02700
Int. Agency for Research on Cancer, Unit of Analytical Epidemiology, 150, cours Albert Thomas, 69372 Lyon Cédex 08, France (Tel.: +33 72738485; Fax: 72738575 ; Tlx: 380023 circ f)
COLL: Kogevinas, M.; Winkelmann, R.; Boffetta, P.; Ferro, G.; Johnson, E.; Becher, H.; Bertazzi, P.A.; Coggon, D.; Green, L.M.; Kauppinen, T.P.; Littorin, L.; Lynge, E.; Mathews, J.D.; Neuberger, M.; Osman, J.; Pearce, N.E.; Bueno de Mesquita, H.B.

International Registry of Persons Exposed to Phenoxy Acid Herbicides and Contaminants
The objective of the project conducted through joint IARC and NIEHS collaboration is to maintain an International Register of workers exposed to chlorophenoxy herbicides, chlorophenols and contaminants (principally chlorinated dibenzo dioxins), and to serve as a basis for follow-up of possible long term health effects. Previous epidemiological studies have reported an increased risk of cancer, particularly soft-tissue sarcoma and non-Hodgkin's lymphoma in subjects occupationally exposed to these chemicals. The register currently includes information on 18,910 production workers or sprayers forming 20 cohorts from 10 countries. Exposure of workers was reconstructed through questionnaires and other documents at factory or spraying cohort level combined with job histories. The first mortality follow-up has been completed. The Register is being enlarged with incoporation of cohorts from Germany, and nested case-control studies on soft tissue sarcoma and non-Hogkin's lymphoma are planned. The next follow-up for mortality and cancer incidence is underway.

FRANCE

TYPE: Cohort; Incidence
TERM: Herbicides; Occupation; Pesticides; Registry
SITE: All Sites
CHEM: Chlorophenols; Dioxins; Phenoxy Acids
OCCU: Agricultural Workers; Chemical Industry Workers; Forest Workers; Herbicide Manufacturers; Herbicide Sprayers; Railroad Workers
LOCA: Australia; Austria; Canada; Denmark; Finland; Germany; Italy; Netherlands; New Zealand; Sweden; United Kingdom; United States of America
REGI: Denmark (Den); Finland (Fin); New Zealand (NZ); Sweden (Swe)
TIME: 1984 – 1992

295   Saracci, R.   04893
Int. Agency for Research on Cancer, Unit of Analytical Epidemiology, 150, cours Albert Thomas, 69372 Lyon Cédex 08, France (Tel.: +33 72738485; Fax: 72738575 ; Tlx: 380023 circ f)
COLL: Boffetta, P.; Kogevinas, M.; Andersen, A.; Bertazzi, P.A.; Frentzel-Beyme, R.R.; Gardner, M.J.; Olsen, J.; Simonato, L.; Teppo, L.; Westerholm, P.; Ferro, G.; Cherrie, J.

Health Effects of Man-Made Mineral Fibres in the Producer Industry

A prospective study of workers employed in 13 man-made mineral fibre (MMMF) plants from seven European countries was started in 1976. Environmental measurements of MMMF concentration have been carried out by the Institute of Occupational Medicine in Edinburgh, UK. Follow-up through 1982 (Scand. J. Work Environ. Health 12 (Suppl. 1):34-47, 1986) showed an increase in lung cancer mortality (189 observed, 151.2 expected), which was correlated with time since first exposure, but not duration of employment. The increase was concentrated among workers employed in the early technological phase, when no dust-suppressing agent was used. In order to investigate this finding, an extension of the follow-up until the end of 1987 is now in progress. A nested case-control study of lung cancer has been planned. Its aims are: (a) the investigation of the confounding effect of tobacco smoking and other occupational and non-occupational exposures to known or suspected carcinogens, (b) a detailed assessment of MMMF exposure. Feasibility of the case-control study is now under way.

TYPE: Case-Control; Cohort
TERM: Occupation; Registry; Tobacco (Smoking)
SITE: All Sites; Lung
CHEM: Mineral Fibres
OCCU: Mineral Fibre Workers
LOCA: Denmark; Finland; Germany; Italy; Norway; Sweden; United Kingdom
REGI: Denmark (Den); Finland (Fin); Norway (Nor); Sweden (Swe)
TIME: 1990 – 1992

296   Sasco, A.J.   04155
Int. Agency for Research on Cancer, Unit of Analytical Epidemiology, 150, cours Albert Thomas, 69372 Lyon Cédex 08, France (Tel.: +33 72738485; Fax: 72738575 ; Tlx: 380023 circ f)
COLL: Saracci, R.; Berrino, F.; Delendi, M.; Lowenfels, A.B.; Muti, P.

International Multicentre Case-Control Study of Male Breast Cancer

About one per cent of breast cancer cases occur in males. The present study will evaluate the role of reproductive life, personal history of disease and drug habits, body build, hepatic function and hormonal profiles. Of specific interest will be the evaluation of the role of hormones (androgens, oestrogens) in the aetiology of male breast cancer. The understanding of the occurrence of breast cancer in males may shed a useful light on the physiopathology of breast cancer in females. Cytogenetic investigations will also be carried out to investigate the genetic component of this disease. Several case-control groups will be assembled in participating centres. It is planned to enrol centres which could provide 15 cases, at least 10 of which are incident cases over a 12- to 18-month period. Three controls will be age matched to each case. Blood samples will be collected from each subject.

FRANCE

TYPE: Case-Control
TERM: Alcohol; Diet; Drugs; Hormones; Physical Factors; Reproductive Factors; Tobacco (Smoking)
SITE: Breast (M)
LOCA: Czechoslovakia; France; Greece; Italy; Netherlands; Switzerland; United Kingdom; United States of America; Yugoslavia
TIME: 1987 - 1995

297 Sasco, A.J. 04487
Int. Agency for Research on Cancer, Unit of Analytical Epidemiology, 150, cours Albert Thomas, 69372 Lyon Cédex 08, France (Tel.: +33 72738485; Fax: 72738575 ; Tlx: 380023 circ f)
COLL: Riboli, E.; Saracci, R.; Hu, M.X.; Ging, L.

Breast Cancer and Reproductive and Endocrine Factors
The aim of the study is to evaluate the relationship between hormonal profiles and breast cancer incidence in premenopausal women. The study uses a case-control approach in a population with a low incidence of the disease. Study group was selected from the population of Guangdong province in China. Incident cases, all premenopausal, were pair-matched to control women on the basis of age (within two years), and residence. Women taking contraceptive pills or any other hormonal treatment, reserpine or tranquilliser were excluded, as were women having or having had a pregnancy in the preceding 12 months, whether carried to full term or ending in a spontaneous or induced abortion, women having lactated in the preceding six months, and women having documented hormonal diseases, gynaecological conditions or chronic debilitating conditions. A detailed questionnaire is administered to cases and controls, including the following items: personal identification data, details of diagnosis, repproductive life history, age at menarche, personal history of disease, family history of cancer, diet history and other factors. Saliva and blood specimens were collected between day 20 and day 24 of the menstrual cycle. Laboratory determinations of testosterone, oestradiol, progesterone, prolactin and growth hormone will be carried out. The Chinese component of the study has now been completed and analyses of the questionnaire data were done. Preliminary results show a positive association of breast cancer with late age at first full term birth, early age at menarche, university education and some aspects of diet. Analysis of the biological samples and further statistical analysis will be carried out in 1992.

TYPE: Case-Control
TERM: Alcohol; Chinese; Diet; Drugs; Familial Factors; Hormones; Menarche; Race; Reproductive Factors; Tobacco (Smoking)
SITE: Breast (F)
CHEM: Oestradiol; Progesterone; Prolactin; Testosterone
LOCA: China
TIME: 1988 - 1993

298 Sasco, A.J. 04488
Int. Agency for Research on Cancer, Unit of Analytical Epidemiology, 150, cours Albert Thomas, 69372 Lyon Cédex 08, France (Tel.: +33 72738485; Fax: 72738575 ; Tlx: 380023 circ f)
COLL: Saracci, R.; Ahlbom, A.N.; Becker, N.; Belli, S.; Benhamou, S.; Bourke, G.J.; Chilvers, C.E.D.; Hatton, F.; Iversen, O.H.; Kauppinen, T.P.; Laporte, J.; Maximilien, R.; Moulin, J.J.; Tessier, C.; Van Leeuwen, F.E.

International Study of Cancer Risk in Biology Research Laboratory Workers
A review of the literature on health risks linked to work in laboratories has shown the paucity of studies on cancer risk for populations of research workers, with the exception of chemists, who have an increased risk of leukaemia/lymphoma. This underlines the need for an evaluation of cancer risk among biology laboratory workers, with an emphasis on risk linked to genetic engineering and molecular biology. A three-step approach has been decided: (1) retrospective cohort study (started in 1991), with about 70,000 subjects; (2) prospective cohort study, with periodical reassessment of exposure and outcome (to be started in 1994 and pursued until 2016); (3) case-control studies nested within the cohort. Death registries and cancer registries will be used to ascertain cases for the retrospective component, and registries (where available) and ad hoc surveillance for the prospective component. A feasibility study, aiming at a detailed assessment of exposures, present and historical, has been conducted in 1988-1989 and results will be published shortly. A paper has been published in Medecine/Sciences 5:489-498, 1989. The retrospective cohort study is presently being conducted in 14 public research institutions of nine European countries (Finland, France, Germany, Ireland, Italy, the Netherlands, Norway, Sweden and United Kingdom), covering biomedical and agronomic research.

FRANCE

TYPE: Case-Control; Cohort
TERM: Chemical Exposure; Occupation; Registry
SITE: All Sites
OCCU: Laboratory Workers
LOCA: Finland; France; Germany; Ireland; Italy; Netherlands; Sweden; United Kingdom
TIME: 1988 - 2016

299   Sasco, A.J.   05223
Int. Agency for Research on Cancer, Unit of Analytical Epidemiology, 150, cours Albert Thomas, 69372 Lyon Cédex 08, France (Tel.: +33 72738485; Fax: 72738575 ; Tlx: 380023 circ f)
COLL: Parkin, D.M.; Gupta, P.C.; Peto, R.

Tobacco Use and All Causes of Mortality in India
Little information is available on precise estimates of mortality and morbidity attributable to tobacco in developing countries. Few studies have been conducted on the use of tobacco other than in the form of cigarettes. It is therefore proposed to explore the feasibility of a prospective cohort study of 100,000 men, aged 35 and over, chosen from the lists of voters in Bombay. Provided the results of the feasibility study are acceptable, a five- to ten-year follow-up study will be started to assess the cause-specific mortality of the cohort, with particular attention being given to cancer and other causes of death in relation to the various forms of tobacco use (smoking, chewing, snuff, etc.).

TYPE: Cohort
TERM: Tobacco (Chewing); Tobacco (Smoking); Tobacco (Snuff)
SITE: All Sites
LOCA: India
TIME: 1990 - 2000

300   SEARCH Programme,   03299
Int. Agency for Research on Cancer, Unit of Analytical Epidemiology, 150, cours Albert Thomas, 69372 Lyon Cédex 08, France (Tel.: +33 72738485; Fax: 72738575 ; Tlx: 380023 circ f)
COLL: Boyle, P.; Walker, A.M.; Baghurst, P.A.; Choi, W.C.; Ghadirian, P.; Howe, G.R.; Jain, M.; McMichael, A.J.; Miller, A.B.; de Waard, F.; Velema, J.; Zatonski, W.A.; Bueno de Mesquita, H.B.

SEARCH Case-Control Study of Cancers of the Pancreas, Bile Duct and Gallbladder
The study comprises a series of population-based case-control studies conducted in a number of international centres, with the SEARCH programme of IARC acting as the central co-ordinating centre, with the aim of examining the hypotheses that (1) dietary fats, in particular those of animal origin, affect pancreatic cancer risk, (2) stimulators of cholecystokinin release elevate pancreatic cancer risk, (3) some aetiological factors for pancreatic cancer are also toxic to islet cells in a way which can be identified by altered patterns of onset of diabetes prior to pancreatic cancer onset. Cases are incident cases, controls are population-based and the questionnaire has been standardized to the extent feasible in a multinational study.

TYPE: Case-Control
TERM: Diet; Hormones
SITE: Bile Duct; Gallbladder; Pancreas
LOCA: Australia; Canada; Netherlands; Poland
TIME: 1983 - 1992

301   SEARCH Programme,   03498
Int. Agency for Research on Cancer, Unit of Analytical Epidemiology, 150, cours Albert Thomas, 69372 Lyon Cédex 08, France (Tel.: +33 72738485; Fax: 72738575 ; Tlx: 380023 circ f)
COLL: Preston-Martin, S.; Boyle, P.; Choi, N.W.; Filippini, G.; McCredie, M.R.E.; Peris-Bonet, R.; Shu, X.O.; Holly, E.A.; Daling, J.R.; Cordier, S.E.

Brain Tumours in Children
The objectives are to evaluate the role of parental and childhood exposure to N-nitroso compounds, to quantify the importance of known risk factors (ionizing radiation and genetic predisposition), and to explore the relationship with other suggested risk factors including head trauma, barbiturates and parental occupational exposures. In utero and perinatal experiences are of particular interest. The biological mothers of 1,000 children aged 0-19 with a primary tumour of the brain or cranial meninges and of an equal or greater number of population controls will be interviewed, as will fathers of cases and controls.

FRANCE

TYPE: Case-Control
TERM: Childhood; Drugs; Environmental Factors; Intra-Uterine Exposure; Radiation, Ionizing; Trauma
SITE: Brain
CHEM: Barbiturates; N-Nitroso Compounds
LOCA: Australia; Canada; China; France; Italy; Spain
TIME: 1985 - 1992

302   **SEARCH Programme,**   03914
       Int. Agency for Research on Cancer, Unit of Analytical Epidemiology, 150, cours Albert Thomas, 69372 Lyon Cédex 08, France (Tel.: +33 72738485; Fax: 72738575 ; Tlx: 380023 circ f)
COLL: Boyle, P.; Howe, G.R.; Ahlbom, A.N.; Becker, N.; Choi, N.W.; McMichael, A.J.; Salzburg, M.; Wahrendorf, J.; Ménégoz, F.; Griciute, L.; Trichopoulos, D.; Gurevicius, R; Modan, B.; Cordier, S.E.

SEARCH Study of Adult Brain Tumours
This study will investigate the aetiology of tumours of the brain and cranial meninges in adults (aged over 25). The study has been designed specifically to investigate (1) the role of exposure to dietary N-Nitroso compounds, associated chemicals and known inhibitors of nitrosation; (2) occupational exposures; and (3) other putative risk factors in the aetiology of brain cancers. The study comprises a series of population-based case-control studies conducted in a number of international centres, with the SEARCH programme of IARC acting as the central co-ordinating centre. It is anticipated that data collection will continue until the end of 1989 and that over 1,000 cases will be assembled. Data analysis will start during 1990.

TYPE: Case-Control
TERM: Chemical Exposure; Diet; Occupation
SITE: Brain
CHEM: N-Nitroso Compounds
LOCA: Australia; Canada; France; Germany; Greece; Israel; Sweden; USSR
TIME: 1985 - 1992

303   **SEARCH Programme,**   04401
       Int. Agency for Research on Cancer, Unit of Analytical Epidemiology, 150, cours Albert Thomas, 69372 Lyon Cédex 08, France (Tel.: +33 72738485; Fax: 72738575 ; Tlx: 380023 circ f)
COLL: Boyle, P.; MacMahon, B.; Trichopoulos, D.; Ghadirian, P.; Zaridze, D.; LaVecchia, C.; James, W.P.T.; Piettanen, P.; Bartsch, H.; Hietanen, E.; O'Higgins, N.; Martin-Moreno, J.; Leake, R.E.; Katsouyanni, K.; George, W.O.; Mahon, G; Dicato, M; Zatonski, W.A.; Levi, F.G.; Salonen, J.T.; Franceschi, S.; Bruzzi, P.A.; Plesko, I.

SEARCH Study of Breast and Colorectal Cancer
The aims of this international, collaborative series of population-based case - control studies of breast cancer and colorectal cancer are to examine the relationships between intake of alcohol, specific nutrients, foods and food groups, energy intake, output and balance, reproductive and hormone use history, family history and risk of breast cancer and colorectal cancer. Among the principal hypotheses to be tested are (1) that cases of breast cancer will have a higher intake of alcoholic beverages than controls; and (2) that cases of colon and breast cancer will have higher intakes of total fat, saturated fat and protein, higher total energy intake and lower total energy output, lower vegetable intake and more first-degree relatives with a history of the same cancer than age- and sex-matched controls. Controls will be selected from the general population.

TYPE: Case-Control
TERM: Alcohol; Diet; Familial Factors; Hormones; Reproductive Factors
SITE: Breast (F); Colon; Rectum
LOCA: Canada; China; Czechoslovakia; Finland; France; Greece; Ireland; Italy; Poland; Spain; Switzerland; United Kingdom
TIME: 1988 - 1993

304   **SEARCH Programme,**   04844
       Int. Agency for Research on Cancer, Unit of Analytical Epidemiology, 150, cours Albert Thomas, 69372 Lyon Cédex 08, France (Tel.: +33 72738485; Fax: 72738575 ; Tlx: 380023 circ f)
COLL: Boyle, P.; Elwood, J.M.; Jensen, O.M.; Østerlind, A.; Walter, S.D.; Gallágher, R.P.; Lejeune, F.; Little, J.H.; Dore, J.F.; Cesarini, J.P.

SEARCH Study of Malignant Melanoma
A series of multicentric, collaborative, population-based case-control studies of malignant melanoma are being undertaken to investigate the influence on the risk of malignant melanoma of a number of risk

FRANCE

factors including host factors (skin complexion, naevi with emphasis on dysplastic naevi and number of raised naevi on the arms, eye colour, hair colour), aspects of sun exposure history, exposure to sources of artificial ultraviolet light, sunlight awareness, dietary factors, occupational exposures, hormonal factors, family history of melanoma and dysplastic naevi, photosensitising agents, history of eczema, psoriasis and use of certain pharmacological compounds and several other factors. In parallel with this study, a multicentric cohort study of patients with dysplastic naevi is being undertaken in collaboration with the Malignant Melanoma Group of EORTC and other clinical groups. The aim is to form a group of patients with dysplastic naevi large enough to assess the incidence of subsequent melanoma and to relate this risk to various histological aspects of the naevi.

TYPE: Case-Control; Cohort
TERM: Diet; Histology; Hormones; Naevi; Occupation; Pigmentation; Radiation, Ultraviolet
SITE: Melanoma
LOCA: Belgium; Canada; Denmark; France; Germany; Netherlands; New Zealand; USSR
TIME: 1989 - 1995

305 SEARCH Programme, 04845
Int. Agency for Research on Cancer, Unit of Analytical Epidemiology, 150, cours Albert Thomas, 69372 Lyon Cédex 08, France (Tel.: +33 72738485; Fax: 72738575 ; Tlx: 380023 circ f)
COLL: Boyle, P.; Linet, M.S.; Cartwright, R.A.; O'Conor, G.T.; Mitelman, F.; Urquhart, J.; Collet, J.; Elwood, J.M.; McBride, M.L.; Zaridze, D.; Simonato, L.

SEARCH Study of Childhood Leukaemia and Related Haematological Malignancies

The aims of this international, collaborative series of population-based case-control studies of childhood leukaemia and related haematologic malignancies are to focus on: perturbations in the postnatal development of the child's immune system, certain parental occupational exposures and various childhood residential exposures. In addition, the study will attempt to confirm a number of hypotheses generated from recent studies. Centres should start data collection before the end of 1990. A study protocol, outlining hypotheses, methods and requirements for participating centres, is available upon request from IARC.

TYPE: Case-Control
TERM: Childhood; Environmental Factors; Immunodeficiency; Occupation
SITE: Haemopoietic; Leukaemia
LOCA: Canada; France; Italy; New Zealand; USSR; United Kingdom
TIME: 1989 - 1995

* 306 Vainio, H. 05205
Int. Agency for Research on Cancer, Unit of Carcinogen Identification, & Evaluation, 150, cours Albert Thomas, 69372 Lyon Cédex 08, France (Tel.: +33 72738485; Fax: 72738575 ; Tlx: 380023 circ f)
COLL: Boffetta, P.; Kogevinas, M.; Matos, E.; Saracci, R.

Re-Analysis of the Carcinogenic Risk of Exposures in the Wood and Leather Industries

Work in the furniture and cabinet making is considered to entail a carcinogenic risk to humans (IARC group 1), and carpentry and joinery to be possibly carcinogenic to humans (IARC group 2B), whereas the remaining two wood industries (lumber and sawmill, pulp and paper) were not classifiable as to their carcinogenicity to humans (IARC group 3). When leather industries were evaluated concerning their carcinogenic risk to humans, work in boot and shoe manufacture and repair was considered also as IARC group 1, whereas leather tanning and processing and other leather goods manufacture industries were included in group 3. Exposures to specific chemicals or mixtures were not evaluated on these occasions, but a number of chemicals occurring in the wood and leather industries have also been tested for carcinogenicity (e.g. formaldehyde, chlorophenols, tannins, benzene, chromium and chromium compounds). The aim of the study is to conduct a combined re-analysis of the available epidemiological studies of carcinogenic risk from exposure to wood or leather dust and specific chemicals in the wood and leather industries. The project will consist of: (1) the combination of relevant cohort studies of wood and leather workers and case-control studies of sinonasal cancer into separate data sets; (2) a systematic assessment of available information on exposures, including data not used or not presented in previous reports; (3) a combined analysis according to specific exposures and job titles.

FRANCE

TYPE: Case-Control; Cohort
TERM: Dusts; Leather; Occupation; Plastics; Solvents; Wood
SITE: All Sites; Bladder; Nasal Cavity
CHEM: Benzene; Formaldehyde
OCCU: Leather Workers; Wood Workers
LOCA: Canada; China; Finland; France; Germany; Italy; Japan; United Kingdom; United States of America
TIME: 1991 - 1993

MARSEILLE

307 Bernard, J.L. 03962
Fac. de Médecine, Registre des Cancers de l'Enfant, 27, Blvd Jean Moulin, 13385 Marseille Cédex 5, France (Tel.: +3391921418)
COLL: Bernard, E.; Scheiner, C.; Perrimond, H.; Gouvernet, J.; Hassoun, J.; Raybaud, C.; Mariani, R.; Thyss, A.

Paediatric Tumour Registry
A population-based paediatric tumour registry was set up in 1984 in two South-Eastern regions of France, Provence-Alpes-Cote d'Azur and Corsica. This study surveys a population of 809,000 children under age 15. All incident cases of malignant tumours and leukaemias are registered. In addition, a case-control study is being conducted with a special focus on genetic risk factors. On a nationwide basis, information is being collected on cancer in sibs and twins and on unusual pathological associations in children with cancer and their families.

TYPE: Case-Control; Incidence; Registry
TERM: Familial Factors; Genetic Factors; Registry
SITE: Childhood Neoplasms
REGI: Marseille (Fra)
TIME: 1984 - 1993

308 Erny, R. 03756
Hôp. de la Conception, Serv. d'Obstétrique, et de Gynécologie, 144, Rue St. Pierre, 13385 Marseille Cédex 5, France

Early Detection of Endometrial Cancer and Hyperplasia
The aim is to detect endometrial cancer and hyperplasia by applying a progestin test to all women who, two years after menopause, are not taking cyclic hormonal treatment, do not have metrorraghia, and whose genital system is normal. If bleeding occurs after the test, hysterography is carried out, followed by pelvic echography and biopsy currettage when pathology is demonstrated. At this stage 1,500 cases have been identified. This study has been criticized for its lack of a control population. It is intended to carry out a similar study on women who did not bleed after the test. Publications have appeared in Rev. Fr. Gynecol. Obstet. 81(4):195-198, 1986 and in "Mises a Jour en Gynecologie et Obstetrique", D.P. Vigo, Paris, 1987, pp. 5-28.

TYPE: Intervention
TERM: High-Risk Groups; Screening
SITE: Uterus (Corpus)
TIME: 1981 - 1993

METZ

309 Viel, J.F. 04733
Hop. Régional de Metz, Dépt. de Santé Publique, 1 Place Philippe de Vigneulles, 57038 Metz Cédex 1, France (Tel.: (87)553131)
COLL: Richardson, S.

Cancer Mortality and Pesticides in Farming
The aim of the study is to examine possible associations between cancer mortality among farmers and pesticide exposure. The cases comprise all French male farmers, between 35 and 75 years old, who died between 1984 and 1986 from cancers already shown to appear more frequently in this occupation. Failure to document pesticide exposure at an individual level has made it necessary to use ecological exposure data. A comprehensive quantification of pesticide exposure for geographical units is being

FRANCE

determined. The associations between cancer mortality rates and pesticide exposure will be tested, using adequate statistical techniques (modified test on the correlation coefficient, regression analysis with spatially parameterised error variance-covariance matrix) taking account of their spatial auto-correlation and of potential confounding variables.

TYPE: Correlation
TERM: Occupation; Pesticides
SITE: Bladder; Brain; Kidney; Leukaemia; Lymphoma; Multiple Myeloma; Pancreas; Prostate; Sarcoma
OCCU: Farmers
TIME: 1989 - 1992

MEYLAN

* 310   Lutz, J.-M.   04966
Registre du Cancer de l'Isère, 21 Chemin des Sources, 38240 Meylan, France (Tel.: +33 76907610; Fax: 76518667)
COLL: Romazzini, S.; Colonna, M.; Laydevant, G.

Study of Chloroprene Industry
This retrospective follow-up study investigates the possible association between exposure to chloroprene (or to its manufacturing process) and cancer mortality and morbidity, using cancer registry data. Workers will be followed from 1965 up to 1990. A continuation of follow-up is planned in case of a positive association of mortality with the exposure of interest.

TYPE: Cohort
TERM: Chemical Exposure; Occupation; Registry
SITE: All Sites
CHEM: Chloroprene
REGI: Isère (Fra)
TIME: 1990 - 1992

MONTPELLIER

311   Daurès, J.   04531
D.I.M. Hôp. Lapeyronie, 555 Rte de Ganges, 34000 Montpellier, France (Tel.: (67)338146)
COLL: Momas, I.

Bladder Cancer in the Département of Hérault
This case-control investigation on bladder cancer has four aims: indentification of risk factors in bladder cancer, of groups of dangerous products, of interactions between factors, and of dose-response relations between exposure and bladder cancer. 250 incident cases with transitional-cell carcinoma and without any other cancer and resident in Herault, who came to public and private urology departments in Herault between 1 January 1987 and 30 December 1988 are matched to two randomly selected controls, one in the Herault population, the other at Lapeyronie Hospital, free of any cancer. The inquiry deals with: (1) way of life (smoking, nutrition, water, alcohol, coffee, artificial sweeteners); (2) drugs (phenacetin); (3) occupational exposures; (4) agricultural activities. The analysis of anonymous data will use statistical methods for matched data: estimates of odds-ratios (Mantel-Haenszel), heterogeneity test, interaction test, logistic regression etc.

TYPE: Case-Control
TERM: Alcohol; Chemical Exposure; Coffee; Drugs; Nutrition; Occupation; Sweeteners; Tobacco (Smoking); Water
SITE: Bladder
CHEM: Phenacetin
OCCU: Agricultural Workers; Dyestuff Workers; Rubber Workers
TIME: 1987 - 1992

312   Domergue, J.   04204
Centre Paul Lamarque, Inst. du Cancer, Dépt. de Chirurgie, 2, Ave. Bertin Sans , B.P. 5054, 34000 Montpellier, France (Tel.: (67)613100)
COLL: Pujol, H.; Corbin, A.

FRANCE

Role of Hormonal Therapy in Preventing Breast Cancer in Patients with Ductal or Lobular Hyperplasia
It has been proved that women with lobular and/or ductal hyperplasia are at high risk of breast cancer. While several studies are being conducted to evaluate the role of tamoxifen in reducing the incidence of breast cancer in women at familial risk, to date a method of following this high risk-group has not been evolved. The aim is to compare modifications in breast samples of patients randomly assigned to receive hormonal therapy with either progesterone or tamoxifen and patients not treated. Patients included are those with ductal or lobular "atypical hyperplasia", diagnosed by a pathologist, with or without a familial history of breast cancer. Blocks are kept for further cancer marker studies. About 200 patients recruited over 2 years will be allocated by blocked randomisation into two groups: (1) no treatment (2) progesterone for days 5-25 of cycle (women aged 30-39) plus tamoxifen (women aged 40-50). Every six months patients are seen in clinics for clinical examination and mammography. After two years of follow-up, patients will be offered new biopsies to evaluate histological changes.

TYPE: Intervention
TERM: Biopsy; Drugs; Familial Factors; High-Risk Groups; Hormones; Mammography; Premalignant Lesion; Prevention
SITE: Breast (F)
CHEM: Progesterone; Tamoxifen
TIME: 1985 - 1992

PARIS

313   Blanc, C.   04925
Electricite de France - Gaz de France, Serv. Général, de Médecine de Contrôle, 22-30 Ave. de Wagram, 75008 Paris, France (Tel.: (1)47647209; Fax: 47648201)
COLL: Puppinck, C.; Chevalier, A.; Godard, C.; Callet, B.; Goldberg, M.; Guenel, P.

Establishment of a Cancer Registry among Workers in the French National Electricity and Gas Company
Sick leave, invalidity and mortality among the 150,000 workers of EDF-GDF are recorded, with diagnosis, by the Social Security Department of the company. For every case of cancer detected the Department physicians complete a questionnaire, stating the date of diagnosis, the site of the tumour and its morphology. From 1978 to 1989 about 3,000 cases have been detected. New cases are recorded directly. The registry will permit descriptive morbidity and mortality studies and aetiological studies of occupational hazards. Presently a survey of exposure to electromagnetic fields is being carried out.

TYPE: Incidence; Mortality; Registry
TERM: Electromagnetic Fields; Occupation
SITE: All Sites
OCCU: Electrical Workers; Gas Workers
TIME: 1978 - 1993

314   Calmettes, C.   04903
INSERM U113, 27, Rue de Chaligny, 75571 Paris Cédex 12, France (Tel.: +33 1 40011313; Fax: 2011499)
COLL: Body, J.J.; Brandi, M.L.; Fischer, J.; Garcia, A.; Hansen, H.S.; Lenoir, G.; Limbert, E.; Moller, P.; Ponder, B.A.J.; Raue, F.; Telenius Berg, M.; Vasen, H.F.A.; Weissel, M.

Genetic Study of Medullary Thyroid Carcinoma
Medullary thyroid cancer (MTC) is a well-known hereditary syndrome which occurs either isolated, or in association with pheochromocytoma and hyperparathyroidism as multiple endocrine neoplasia (MEN) type II; it is characterised by an autosomal dominant inheritence. Cure relies on rapid surgical removal and diagnosis is supported by assay of calcitonin, its biological marker. Recently, minisatellite DNA probes have been applied to linkage (chromosome 10) for MEN diagnosis. A collaborative study is necessary to localise the gene defect and allow a reliable diagnosis of the hereditary disease; for this purpose, three objectives have to be realised: (1) to set up a central register of MCT cases to identify informative families to be offered DNA diagnosis; (2) to collect and exchange blood DNA and tumours from patients and family members; and (3) to provide a framework for similar collaboration in DNA screening methods for inherited cancers. The participating countries are establishing groups for clinical and laboratory studies of MEN. Results will be pooled to achieve optimum patient care.

FRANCE

TYPE: Genetic Epidemiology
TERM: BMB; Blood; DNA; Genetic Factors; Screening; Tissue
SITE: Thyroid
LOCA: Austria; Belgium; Denmark; France; Germany; Italy; Netherlands; Norway; Portugal; Spain; Sweden; Switzerland; United Kingdom
TIME: 1990 - 1993

* 315 **Hubert, A.** 05165
CNRS, Lab. d'Epidémiologie et, Immunovirologie des Tumeurs, ISD 15, 28, rue du Dr Roux, 75724 Paris Cedex 15, France (Tel.: +33 1 45688930; Fax: 45688931)
COLL: Jeannel, D.; Crognier, E.; de Thé, G.; Sancho-Garnier, H.; Eschwege, E.; Slimane, B.

Dietary and Environmental Risk Factors for Nasopharyngeal Carcinoma in Morocco and among Moroccan Migrants

Nasopharyngeal carcinoma will be studied in an endemic country, Morocco, where no epidemiological research on this subject has, as yet, been performed; in addition, NPC will also be studied in a migrant population from Morocco in southern France. The programme includes three parts: (1) estimation of the incidence of NPC in Morocco and in Moroccans of the Provence-Alpes-Cote d'Azur region; (2) anthropological study of dietary changes among Moroccan migrants; (3) study of the environmental risk factors in Morocco through a case-control study in Morocco, together with genetic (HLA profiles) and virological (EBV) studies.

TYPE: Case-Control; Incidence
TERM: BMB; Diet; EBV; Environmental Factors; HLA; Maghrebians; Migrants; Serum
SITE: Nasopharynx
LOCA: France; Morocco
TIME: 1991 - 1993

316 **Leclerc, A.** 03765
INSERM, Unité 88, 91, Blvd de l'Hopital, 75634 Paris Cédex 13, France
COLL: Luce, D.; Gerin, M.; Brugère, J.

Nasal Cancer and Formaldehyde Exposure

The objective of the study is to determine whether formaldehyde exposure (professional and domestic) can increase the risk of nasal cancer, and to study interactions with other occupational and non-occupational factors. In a case-control design 200 cases collected in specialized departments of French hospitals are compared with 400 controls, selected among hospital cancer patients (first control group) and aquaintances of the cases (second control group). Each case and control is interviewed, with a detailed questionnaire on past occupations; for each occupation, questions are asked about the task performed, and materials manipulated. Information concerning other factors (alcohol and tobacco consumption, passive smoking, past history of nasal non-cancer problems) is also requested. Coding of selected occupational exposures is performed by specialists in occupational hygiene (M. GERIN, Canada). The analysis will use classical methods for case-control studies (unmatched logistic models). A separate analysis will be performed for adenocarcinoma (half of the cases). The effects of wood exposure, including kind of wood and occupation, and histology of cases, will be studied in detail.

TYPE: Case-Control
TERM: Alcohol; Dusts; Leather; Metals; Occupation; Passive Smoking; Plastics; Tobacco (Smoking); Wood
SITE: Nasal Cavity
CHEM: Chromium; Formaldehyde; Nickel
TIME: 1986 - 1992

317 **Leclerc, A.** 04546
INSERM, Unité 88, 91, Blvd de l'Hopital, 75634 Paris Cédex 13, France
COLL: Brugère, J.

Laryngeal Cancer and Occupation

In this case-control study, patients with cancer of the glottis (n=150), supraglottis (n=150) and hypopharynx (n=150) diagnosed in 12 specialised departments of French hospitals are compared with unmatched hospital patients (n=300). Every case and control is interviewed, with a detailed questionnaire on alcohol (amount and type), tobacco (amount and type), passive smoking, and questions about past occupations. For each occupation, questions are asked about the tasks performed, and materials manipulated. Interviews are taking place in 1989, 1990 and 1991. The analysis

FRANCE

will be performed in two stages, based on past occupations and based on selected occupational exposures, according to occupational hygiene experts, or job-exposure matrices. Occupational exposures to asbestos, sulphuric acid, chromium, nickel, silica, and vehicle exhaust will be studied. In the two stages of the analysis, age, alcohol and tobacco consumption will be controlled for. The role of other risk factors such as type of alcohol and tobacco, passive smoking will be studied. Differences between the three sites of cancer will be stressed.

TYPE: Case-Control
TERM: Alcohol; Chemical Exposure; Dusts; Occupation; Passive Smoking; Tobacco (Smoking)
SITE: Hypopharynx; Larynx
CHEM: Asbestos; Chromium; Nickel; Silica; Sulphuric Acid; Vehicle Exhaust
TIME: 1989 - 1992

REIMS

318   Delisle, M.J.   04425
Inst. Jean Godinot, Serv. de Médecine Nucléaire, Registre des Cancers de la Thyroïde, 1, Rue du General Koenig, 51100 Reims, France (Tel.: +33 26360504; Fax: 26060045 ; Tlx: 830048)
COLL: Maès, B.; Schvartz, C.; Vaudrey, C.; Pochart, J.M.

Thyroid Carcinoma in Champagne-Ardenne
Since 1967, 1,080 clinically diagnosed cases of thyroid cancer have been treated and followed up in an area of about 1.5 million inhabitants. Inclusion of all patients in a well-defined region has been ensured since 1979 by creating a regional multidisciplinary group. By the end of 1990, 637 patients had been registered in this local organ-specific registry. Data on all patients include sex, age at diagnosis, histology, clinical extension using the TNM system, therapeutic method and survival. Incidence rates have been analysed with respect to age, sex, geographical distribution, associated thyroid disease and second cancer. The main purpose is to evaluate the programme by comparing evolution of all the parameters over 25 years, e.g. mean age at diagnosis, percentage of aggressive lesions. In addition, prognostic factors have been identified by a multifactorial analysis which enables a scoring system to assess the risk of relapse or death for each individual patient at the time of diagnosis and to provide a better therapeutic approach. A paper has been published in Bull. Cancer 73(2):165-170, 1986.

TYPE: Case Series; Incidence; Registry
TERM: Clinical Records; Histology; Prognosis; Survival; Treatment
SITE: Thyroid
TIME: 1967 - 1993

VANDOEUVRE

* 319   L'Huillier, M.C.   05235
Hôpital d'Enfants, Serv. de Médecine Infantile II, Rue du Morvan, 54511 Vandoeuvre Cédex, France (Tel.: +33 83558120; Fax: 83596069; Tlx: 960561)
COLL: Sommelet, D.

Childhood Cancer Registry and Risk Factors
Following the results of a case-control study based on cases collected by the Childhood Cancer Registry of Lorraine (population-based registry created in 1983, registering solid tumours, leukaemias and all malignant and benign brain tumours; population base: 535,236 children aged 0 to 15), a data abstract form is completed by the parents of each child entered into the registry as from 1 January 1991. The form takes into consideration many potential risk factors (genetic, personal, environmental), focussing on history of infections and familial cancers.

FRANCE

TYPE: Registry
TERM: Environmental Factors; Family History; Genetic Factors; Infection; Registry
SITE: Childhood Neoplasms
REGI: Lorraine (Fra)
TIME: 1991 - 1993

320 Moulin, J.J. 03982
Inst. National de Recherche, et de Sécurité, Dépt. d'Epidémiologie, Ave. de Bourgogne, B.P. 27, 54501 Vandoeuvre-les-Nancy Cédex, France (Tel.: (83)510775; Tlx: 850778 F)
COLL: Haguenoer, J.M.Y.

Mortality Study of Welders Exposed to Chronium and Nickel
Following the publication of a protocol by the WHO a mortality study was set up in France, to assess the risk of cancer of stainless steel welders. Cohorts under study will include welders (stainless steel and mild steel) and controls matched on year of birth, sex, and socio-economic group. In order to obtain a justifiable cohort size, several plants will be included in the study. Causes of death will be considered throughout the period 1975-1984, since stainless steel welding was introduced at the beginning of the 1960s. Environmental measurements of airborne chromium and nickel concentrations will be performed. Data on smoking habits will be available for each man (welder or control). For the group of welders, job histories and welding processes will be ascertained. Relative risks will be used for the internal comparisons (stainless steel welders, mild steel welders, and control groups). In addition, those groups will be compared with the general population mortality (SMRs).

TYPE: Cohort
TERM: Chemical Exposure; Environmental Factors; Metals; Occupation; Tobacco (Smoking)
SITE: All Sites
CHEM: Chromium; Nickel
OCCU: Welders
TIME: 1985 - 1992

321 Moulin, J.J. 03984
Inst. National de Recherche, et de Sécurité, Dépt. d'Epidémiologie, Ave. de Bourgogne, B.P. 27, 54501 Vandoeuvre-les-Nancy Cédex, France (Tel.: (83)510775; Tlx: 850778 F)
COLL: Mur, J.M.; Mantout, B.; Portefaix, P.; Smagghe, G.

Mortality Study of Stainless Steel Foundry Workers
The aim of the study is to determine whether mortality from respiratory cancers (lung and larynx) and from buccal cavity and pharyngeal cancers is higher among chromium and nickel exposed workers than in the general population. In order to investigate this problem, a historical prospective study is being carried out in two plants, for the years 1952 to 1984. The cohorts under study include 8,000 workers. Information required includes date of birth, date of first employment in the foundry, job history and date of leaving. Deaths are indentified by investigations in administrative registries, and causes of death are ascertained by hospital or medical records. Expected numbers of deaths are calculated from the general French population, adjusted for sex, age, and year of death. SMRs will be calculated according to duration of exposure and time since first exposure.

TYPE: Cohort
TERM: Chemical Exposure; Metals; Occupation
SITE: All Sites; Larynx; Lung; Oral Cavity; Pharynx
CHEM: Chromium; Nickel
OCCU: Foundry Workers
TIME: 1984 - 1992

322 Pham, Q.T. 03989
INSERM U115, Santé au Travail, et Santé Publique, Méthodes et Applications, Rte de la Foret de Haye, 54505 Vandoeuvre-les-Nancy, France (Tel.: (83)565656)
COLL: Gabiano, P.

Mortality Among Iron Miners: A Prospective Study
Previous studies have shown an increase of lung cancer in iron miners. This study will investigate for this occupation the risk of developing cancers of other sites. A mortality study has been initiated in a cohort of 13,921 miners, active and retired, alive on 1 January 1982, in which the distribution of underground and surface workers and their length of employment are globally known. A record is established for each deceased person, containing information on name, date of birth, date of entering the mine, length of

work in the mine or elsewhere, previous and subsequent main occupations and duration. Information on height, weight, tobacco and alcohol consumption, hospitalization, accidents at work, place of death, sources of death information, cause(s) of death (immediate, consecutive and contributing) and autopsy results, is also recorded. The relative frequency of causes of death will be calculated for the whole cohort and possible relationship with occupational exposures established. After five years 1,806 were deceased. Causes of death are known for 1,395. Follow-up continues.

TYPE: Cohort
TERM: Alcohol; Chemical Exposure; Metals; Mining; Occupation; Physical Factors; Time Factors; Tobacco (Smoking)
SITE: Lung; Stomach
CHEM: Iron
OCCU: Miners, Iron
TIME: 1982 - 1992

VERNEUIL EN HALATTE

323   Auburtin, G.   04714
CERCHAR, Boite Postale no. 2, 60550 Verneuil en Halatte, France (Tel.: +33 44 556596; Fax: 556699 ; Tlx: 140094 F)

Lung Cancer and Pneumoconiosis in Coal Miners and Ex-Miners
The objective of this study is to determine the relationship between lung cancer and coalworker's pneumoconiosis (CWP) while controlling for smoking. It is a population-based case-control study among underground miners and ex-miners in the Nord-Pas de Calais collieries. Incident cases of primary bronchiolar and lung cancer over a period of two years among all men living in the Nord-Pas de Calais Region will be registered through a network of pneumologists and pathologists. Only cases from the defined population will be included. Controls, selected from the same population, will be individually matched to cases on age and years of experience underground. The main risk factor to be studied is the presence of CWP and its degree of severity. Definitions of CWP according to medico-legal criteria and to radiological criteria alone, will both be used in the analyses. This study should provide a better understanding of the role of CWP in the development of broncho-pulmonary lung cancer. Positive findings could provide a scientific basis for compensation for lung cancer in a person suffering from CWP, and encourage prevention strategies before diagnosis of pneumoconiosis and during follow-up of pneumoconiotic patients.

TYPE: Case-Control
TERM: Occupation; Pneumoconiosis
SITE: Lung
OCCU: Miners, Coal
TIME: 1990 - 1993

VILLEJUIF

324   Benhamou, S.   04815
Inst. Gustave Roussy, INSERM U287, Rue Camille Desmoulins, 94805 Villejuif Cédex, France (Tel.: (1)45594139; Fax: 46787430)
COLL: Benhamou, E.; Le Mab, G.; Grosclaude, P.; Raymond, L.; Paoletti, C.; Sancho-Garnier, H.

Prostate Cancer, Sexual Behaviour and Diet
It is proposed to carry out a collaborative population-based case-control study in Geneva, Switzerland and in two counties (departements) in France for which prostate cancer represents the highest incidence rate in males. The study will also be performed in Paris in different hospitals. The aim of this study is to determine the influence of sexual activity, sexually transmitted diseases and diet (in particular fats) in prostate cancer. Cases will be taken from the Tarn, Martinique and Geneva cancer registries. Sex- and age-matched controls will be chosen from the population census list. Risk factors will be assessed by interview. In each area, 150 cases and 300 controls are expected.

FRANCE

TYPE: Case-Control
TERM: Diet; Fat; Registry; Sexual Activity; Sexually Transmitted Diseases
SITE: Prostate
LOCA: France; Switzerland
REGI: Geneva (Swi); Martinique (Mar); Tarn (Fra)
TIME: 1990 - 1993

325 Benhamou, S. 04816
Inst. Gustave Roussy, INSERM U287, Rue Camille Desmoulins, 94805 Villejuif Cédex, France (Tel.: (1)45594139; Fax: 46787430)

Lifestyle and Medical Drug Consumption in Renal Cell Carcinoma and Renal Pelvis Cancer.
A hospital-based case-control study has been undertaken since 1987 to study possible causes of renal cell carcinoma and renal pelvis cancer. The main environmental risk factors investigated are consumption of tobacco, alcohol, coffee and the use of analgesics and diuretics. All cases are histologically confirmed. Two controls per case, hospitalised for a non-tobacco-related disease (one for a malignant disease and one for a non-malignant disease) are matched for sex, age at interview (± 5 yrs) and hospital. 300 cases and 600 controls are expected.

TYPE: Case-Control
TERM: Alcohol; Analgesics; Coffee; Diuretics; Occupation; Tobacco (Smoking)
SITE: Kidney
TIME: 1987 - 1992

326 Benhamou, S. 04817
Inst. Gustave Roussy, INSERM U287, Rue Camille Desmoulins, 94805 Villejuif Cédex, France (Tel.: (1)45594139; Fax: 46787430)
COLL: Benhamou, E.; Schrameck, C.

Smoking and Female Lung Cancer
As part of an international case-control study on lung cancer and environmental tobacco smoke exposure, a hospital-based case-control study on lung cancer among women is being carried out. The aim is to evaluate the influence of different characteristics of smoking habits (in particular type of cigarettes and tar levels). Lung cancer will be histologically confirmed and 2 female controls hospitalised for a non-tobacco-related disease and matched for age (± 5 yrs), and hospital will be included.

TYPE: Case-Control
TERM: Female; Occupation; Passive Smoking; Tobacco (Smoking)
SITE: Lung
TIME: 1990 - 1993

327 Clavel, F. 02301
INSERM, U 21, Dépt. de Statistique Médicale, Inst. G. Roussy-Hautes-Bruyères, Rue Camille Desmoulins, 94805 Villejuif Cédex, France
COLL: Flamant, R.; Laplanche, A.

Vinyl Chloride in Relation to Malignant Tumours
The purpose of this prospective study is to examine the relationship between vinyl chloride (VC) exposure and pathology (in particular malignant tumours) and causes of death. It is being conducted in factories producing VC (monomer or polymer). Only male workers, aged between 40 and 55 years, are considered. For each worker exposed, a worker never exposed to VC, of the same age (+/-two years), working in the same plant, is used as a control. Both exposed and non-exposed workers are interrogated by the same physician. The form contains some identification items (age, civil status, place of residence), information on exposure, including the whole occupational history, on tobacco and alcohol habits and health status. 2,282 workers were entered into the study and will be seen yearly for several years in order to follow their health status and note possible changes in relation to VC exposure. Preliminary analysis will now start.

FRANCE

TYPE: Case-Control; Cohort
TERM: Alcohol; Chemical Exposure; Male; Occupation; Plastics; Tobacco (Smoking)
SITE: All Sites
CHEM: Hydrocarbons, Halogenated; Vinyl Chloride
TIME: 1980 – 1992

* 328 Clavel, F. 04980
INSERM, U 21, Dépt. de Statistique Médicale, Inst. G. Roussy-Hautes-Bruyères, Rue Camille Desmoulins, 94805 Villejuif Cédex, France
COLL: Cordier, S.E.; Flandrin, G.; Conso, F.; Limasset, J.G.

Occupational Risk Factors in Hairy Cell Leukaemias

A high proportion of mechanics, farmers, and physicians has been reported in a case series of hairy cell leukaemias (HCL). This observation has given rise to the hypothesis that there is a relationship between HCL and exposure to benzene, other solvents, pesticides and radiations, already known or suspected to be involved in other types of leukaemia. A case-control study including all HCL cases diagnosed in France between 1 January 1980 and 1 January 1989 has been set up in 27 departments of haematology. Two hospital controls are matched to each case on sex, age, year and place of hospitalization, and area of residence. The study will include about 400 cases and 800 controls. A detailed interview of patients investigates the exposures of interest and results are reviewed for coding by a panel of experts in industrial hygiene and occupational medicine. For pesticides, exposure assessment is performed by specialized physicians on the farms concerned.

TYPE: Case-Control
TERM: Electromagnetic Fields; Occupation; Pesticides; Radiation, Ionizing; Solvents
SITE: Leukaemia
CHEM: Benzene
TIME: 1988 – 1992

* 329 Cordier, S.E. 05049
INSERM U170, 16 bis Ave. Paul Vaillant-Couturier, 94807 Villejuif, France (Tel.: + 33 1 45595034; Fax: 45595151)

Cancer Risk among Laboratory Workers

A cohort study was set up in order to investigate a cluster of rare cancers among young scientists at the Institut Pasteur in Paris. 3,765 persons having worked at the Institute for at least six months between 1971 and 1986 were identified. Their mortality was studied up to the end of 1987. Presently, the mortality study is being completed by the search for missing causes of death (15%). The study of occupational exposures within the Institute will be completed by a nested case-control study on incident and fatal cancers of pancreas, bone, brain and the haematopoietic system.

TYPE: Case-Control; Cohort
TERM: Chemical Exposure; Occupation
SITE: All Sites
OCCU: Laboratory Workers
TIME: 1986 – 1992

330 Flamant, R. 03793
Inst. Gustave Roussy, Dépt. de Statistiques Medicales, Rue Camille Desmoulins , 94805 Villejuif Cédex, France (Tel.: (1)45594144)
COLL: Dutreix, A.; Sarrazin, D.; Oberlin, O.; de Vathaire, F.; Rodary, C.; Hill, C.; François, P.

Second Cancer in Children in Relation to Radiotherapy Dose

The aim of this cohort study, which includes 626 patients, is to examine the relation between the radiation dose received by children for a first cancer at an anatomical site, and the appearance of another cancer on this same site. Only five-year survivors treated at the Institut Gustave-Roussy from 1950 to 1969 have been included. The reason for this limitation in dates is to have patients with a long period of follow-up and no massive polychemotherapy. 28 second cancers have been observed. Radiotherapy doses received by all 628 children on the site of these 28 second cancers will be re-calculated.

FRANCE

TYPE: Cohort
TERM: Multiple Primary; Radiation, Ionizing; Radiotherapy; Survival; Time Factors
SITE: All Sites
TIME: 1985 - 1992

331 Henry-Amar, M. 03726
Inst. Gustave-Roussy, Dépt. de Biostatistique, et d'Epidémiologie, 39-53, Rue Camille Desmoulins, 94805 Villejuif Cédex, France (Tel.: (1)45594142; Fax: 46787430)

Second Cancers after Radiotherapy and/or Chemotherapy for Hodgkin's Disease
The aim of this study is to determine whether radiotherapy or chemotherapy or their combination used in the treatment of Hodgkin's disease is a factor of increased risk for a second cancer. If an increased risk is observed, it will be related to exposure, time relationships and type of second primary. This is a retrospective study on a cohort of approximately 1,600 patients, based on medical records dating from 1960. Particular attention will be paid to the following second primaries: acute leukaemia, non-Hodgkin's lymphoma and bronchus carcinoma. 19 secondary acute leukaemias were observed among 871 adult patients followed at least for one year post initial therapy. The sole factor related to leukaemia risk was the total dose of nitrogen mustard given. Subsequent analyses will be performed taking into account other second primary cancers than leukaemia. A paper has been published in Rec. Res. Cancer Res. 17:270-283, 1989.

TYPE: Cohort
TERM: Chemotherapy; Multiple Primary; Radiation, Ionizing; Time Factors; Treatment
SITE: All Sites; Leukaemia; Lung; Non-Hodgkin's Lymphoma
TIME: 1985 - 1992

332 Henry-Amar, M. 04270
Inst. Gustave-Roussy, Dépt. de Biostatistique, et d'Epidémiologie, 39-53, Rue Camille Desmoulins, 94805 Villejuif Cédex, France (Tel.: (1)45594142; Fax: 46787430)

Second Malignancies Following Radiotherapy and Chemotherapy for Hodgkin's Disease
The aim of this multi-centre case-control study is to ascertain the risk of second acute leukaemia, non-Hodgkin's lymphoma and various solid tumours in relation to the type of treatment given, radiotherapy and/or chemotherapy. The extent of radiotherapy and doses of each drug administered have been recorded. Cases were patients who developed a second malignancy at least one year after having been treated according to one of the six clinical trials conducted from 1963 to 1987 by the EORTC Lymphoma Cooperative Group or the Groupe Pierre et Marie Curie. Overall, 74 second primary cancers have been observed, 10 acute leukaemias, 11 non-Hodgkin's lymphomas, and 53 solid tumours. Three controls per case have been chosen at random using the following criteria: sex, age within 1 year of case's age, date of diagnosis of Hodgkin's disease within 3 years of case's diagnosis, duration of survival at least as long as the interval between the diagnosis of the first and second primary cancers in the case. The histological material has been reviewed, both that of the second primary for cases, and that of the initial Hodgkin's disease material for all cases and controls.

TYPE: Case-Control
TERM: Chemotherapy; Multiple Primary; Radiotherapy
SITE: Gastrointestinal; Leukaemia; Lung; Non-Hodgkin's Lymphoma
LOCA: Belgium; France; Germany; Italy; Netherlands
TIME: 1987 - 1992

* 333 Stücker, I. 05157
INSERM U170, 16 Av. Paul Vaillant-Couturier, 94807 Villejuif Cédex, France (Tel.: +33 1 45595033; Fax: 595151)
COLL: Laurent, P.; Bechtel, P.; Brochard, P.

Cytochrome P450: Genetic Polymorphism and Environmental Risk Factors in Lung Cancer
This study aims to investigate the statistical interaction between exposure to PAH and capacity to metabolise them (i.e. inducibility of aryl hydrocarbon hydroxylase (AHH) by the regulator gene Ah). The case-control design includes 300 incident lung cancer cases and 300 hospital controls. All subjects are males aged 75 or less, born in metropolitan France from parents born in metropolitan France. One control per case is matched on age and area of residence. Investigations include a detailed questionnaire on tobacco smoking and occupational PAH exposure; Western blot analysis for the measurement of AHH inducibility; measure of glutathione transferase; activity of AHH estimated by caffeine test on urine of smokers.

FRANCE

TYPE: Case-Control; Molecular Epidemiology
TERM: BMB; Dusts; Occupation; Tobacco (Smoking)
SITE: Lung
CHEM: Asbestos; Chromium; Iron; Nickel; PAH
TIME: 1989 – 1992

GAMBIA

BANJUL

* 334   Bah, E.   05135
          Medical Research Council Labs, Gambia Hepatitis Intervention Study, P.O. Box 273, Banjul,
          Gambia (Tel.: (220)95229)
COLL: Jack, A.; Inskip, H.M.

The Gambia Cancer Registry
This national population-based cancer registry was started in July 1986 as part of the Gambia Hepatitis Intervention Study. The primary objective of the registry is to monitor cancer incidence in the country. Data collected by the registry for the next 30 years will be used to evaluate the protective effectiveness of infant hepatitis B immunization in presenting subsequent primary hepatocellular carcinoma in the adult. Information on all malignant tumours occurring within the country is obtained from hospitals and clinics both in the public and private sector and analysed annually by age, sex and site. For the past four years (July 1986 to June 1990), a total of 1,002 cancers have been registered. A paper describing the first two years of operation was published in Br. J. Cancer 62:647-650, 1990.

TYPE: Incidence; Intervention; Registry
TERM: Data Resource; HBV; Registry; Vaccination
SITE: All Sites
REGI: Gambia (Gam)
TIME: 1986 - 2021

FAJARA

335   Inskip, H.M.   03900
          International Agency, for Research on Cancer, c/o MRC Laboratories, Fajara near Banjul, Gambia
          (Tel.: (220)95229; Fax: 96117)
COLL: Jack, A.; Fortvin, M.; Chotard, J.; Tomatis, L.T.; Bosch, F.X.; Greenwood, B.; Whittle, H.C.; Mendy,
          M.; Bah, E.; George, M.; Cham, K.; Hall, A.J.; Smith, P.G.; Kane, M.; Parkin, D.M.; Day, N.E.

Gambia Hepatitis Intervention Study
The long-term aim of this study is to assess the effectiveness of hepatitis B vaccine in the prevention of PHC. This cancer is common in The Gambia, a country in which approximately 15% of adults are carriers of the hepatitis B virus. The first phase of this project was completed at the end of 1990, by which time information on 125,000 children had been collected. Almost half of these children received hepatitis B vaccine, which had been progressively incorporated into The Gambia's Expanded Programme of Immunization during a five-year period. Subsets of this cohort have been studied in detail and have shown that the vaccine has been very effective in preventing infection and the carrier state. The effectiveness of the vaccine in the prevention of liver cancer will be determined by linkage of the database with The Gambia's cancer registry, but results are not expected for many years. Many ancillary studies have been conducted to examine the natural history of hepatitis B virus infection and response to the vaccine and its cost-effectiveness. New areas of work are concentrating on the effects of aflatoxin exposure and the examination of mutant hepatitis B viruses against which the vaccine may fail to protect.

TYPE: Cohort; Incidence; Intervention
TERM: BMB; HBV; Prevention; Registry; Vaccination
SITE: Liver
CHEM: Aflatoxin
REGI: Gambia (Gam)
TIME: 1986 - 2021

GERMANY

BERLIN

336 Garbe, C. 04625
Freie Univ. Berlin, Klinikum Steiglitz, Hautklinik & Poliklinik, Hindenburgdamm 30, W 1000 Berlin 45, Germany (Tel.: (030)7983923)
COLL: Breibart, E.W.; Roser, M.; Burg, G.; Jung, G.; Weiss, J.; Kresbach, H.; Soyer, H.P.; Kreysel, H.W.; Weckbecker, J.; Schnyder, U.W.; Panizzon, R.; Wolff, H.H.; Zaun, H.; Bahmer, F.A.; Tilgen, W.; Orfanos, C.E.

Early Diagnosis and Risk Factors of Malignant Melanoma

A multi-centre case-control study including over 600 malignant melanoma (MM) patients and the same number of age- and sex-matched controls is underway. Persons under investigation answer a questionnaire about their sun exposure habits and sun reaction, about their awareness of pigmented moles and, for MM patients, about the reasons which underlie their decision to consult a doctor. In a whole body examination, pigmentational characteristics and all pigmented moles are carefully documented. Histological examination is performed for clinically atypical naevi. Relative risks for developing MM will be calculated in relation to potential risk factors such as sunburn, different recreational activities, occupation, hormonal factors and immunosuppression, as well as in relation to risk indicators like numbers of common and dysplastic naevi, further pigmented moles and pigmentational characteristics. Factors influencing early diagnosis will be analysed.

TYPE: Case-Control
TERM: Hormones; Immunosuppression; Lifestyle; Naevi; Occupation; Radiation, Ultraviolet
SITE: Melanoma
LOCA: Austria; Germany; Switzerland
TIME: 1989 - 1993

337 Kohlmeier, L. 04942
Inst. für Sozialmedizin, und Epidemiologie, Bundesgesundheitsamt, General-Pape-Str. 64, W 1000 Berlin 42, Germany (Tel.: +49 30 78007197; Fax: 78007109 ; Tlx: 184016)
COLL: Neusetzer,; Krumhaar,; Loddenkemper,; Kaiser, D.; Lohde, D.; Broll, D.; Mainz, D.

Pet Birds and Lung Cancer

To test whether contact with pet birds within or outside the home is related to lung cancer, a population based case-control study is being conducted with 200 cases obtained from the Berlin population and 400 controls under 65 years of age. Lap-top computers are being used to record information at interview in the clinic. All cases and controls are drawn from the Berlin population.

TYPE: Case-Control
TERM: Animal
SITE: Lung
TIME: 1989 - 1992

* 338 Kohlmeier, L. 05166
Inst. für Sozialmedizin, und Epidemiologie, Bundesgesundheitsamt, General-Pape-Str. 64, W 1000 Berlin 42, Germany (Tel.: +49 30 78007197; Fax: 78007109 ; Tlx: 184016)
COLL: Thamm, M.; Mund-Hoym; Randow

Antioxidant Status and Breast Cancer in East and West Berlin

The study is designed to test the hypothesis that a dose-response relationship exists between antioxidant status (levels of beta carotene, tocopherols and selenium) and risk of breast cancer. A comparison of the risk profiles in both eastern and western German populations is also planned. In each clinic a minimum of 100 cases are to be recruited and matched 1:2 with population controls. The exposure assessment will be via analysis of toenails and adipose tissue as well as computerized interviews.

GERMANY

TYPE: Case-Control
TERM: Antioxidants; Dose-Response
SITE: Breast (F)
CHEM: Beta Carotene; Selenium; Tocopherol
TIME: 1990 - 1992

339 **Geissler, E.** 04744
Akademie der Wissenschaften, Zentralinst. für Molekularbiologie, Robert Rössle Str. 10, O 1115 Berlin-Buch, Germany (Tel.: +37 346 2328)
COLL: Staneczeck, W.

Simian Virus (SV40) and Human Intracranial Tumours
Millions of people were vaccinated with polio vaccine contaminated with SV40 during 1955-1962. This study will try to assess whether this viral contaminant might be involved in malignant late effects. In 30 out of 110 human cerebral tumours, footprints of SV40 or of SV40-like viruses could be found by indirect immuno-fluorescence and by DNA hybridization. About 750,000 persons treated with SV40-contaminated polio vaccine between 1960-1962 are now being followed up for incidence of malignancies, compared with a similar number of persons who were treated with SV40-free polio vaccine (Progr. Med. Virol. 37:211-22, 1990).

TYPE: Cohort
TERM: Vaccination; Virus
SITE: Brain
REGI: Berlin (FRG)
TIME: 1965 - 1993

BREMEN

* 340 **Greiser, E.** 04986
Bremen Inst. for Prevention, Research and Social Medicine, Grünenstr. 120, W 2800 Bremen, Germany (Tel.: (0421)595960; Fax: 5959665)
COLL: Lotz, I.; Brand, H.

Leukaemia and Malignant Lymphoma in the Vicinity of an Industrial Waste Dump
A previous study revealed an increased incidence of leukaemias and malignant lymphomas in a rural district near a disused waste-dump. The current study begins with the registration of all cases of leukaemia and malignant lymphoma (ICD-9 200-208), diagnosed between 1984 and 1990 in all municipalities neighbouring the waste dump. Data are collected from all local public and private hospitals, physicians, public health offices and laboratories. A case-control study, including all registered leukaemia patients with age- and sex-matched population controls, has been set up to identify and examine the role of risk factors for leukaemias and malignant lymphomas. 700 cases and 1400 age- and sex-matched population controls are expected. Risk factors covered will include exposure to solvents, PAH, ionizing radiation, herbicides and pesticides, certain therapeutic drugs, occupations known to be associated with an elevated risk for this group of diseases and vicinity to the dump site. Cases and controls will be interviewed by trained personnel, using a standardized questionnaire.

TYPE: Case-Control
TERM: Drugs; Environmental Factors; Herbicides; Occupation; Pesticides; Radiation, Ionizing; Solvents; Waste Dumps
SITE: Leukaemia; Lymphoma
CHEM: PAH
TIME: 1990 - 1993

341 **Jöckel, K.H.** 04306
Bremen Inst. for Prevention, Research, and Social Medicine, Div. of Biometry and Data Processing, Grünenstr. 120, W 2800 Bremen 1, Germany (Tel.: (0421)5959650; Fax: 5959665)
COLL: Ahrens, W.; Bolm-Audorff, U.; Jahn, I.

Occupational Risk Factors for Lung Cancer
A case-control study with 1,000 cases and the same number of population controls will be carried out, investigating the association between risk of developing a bronchial carcinoma and serveral

GERMANY

occupational factors. A detailed job history including description of the work place and exposure-information about known occupational carcinogens will be obtained by means of a standardized interview. Detailed information about smoking habits will also be obtained from each subject. It is intended to establish a more detailed description of occupational hazards for lung cancer from the identification of occupational risk-groups and possible associations with certain substances.

TYPE: Case-Control
TERM: Chemical Exposure; Occupation; Tobacco (Smoking)
SITE: Lung
TIME: 1987 - 1993

ERFURT

* 342 Schunk, W.W.   05126
Medizinische Akademie Erfurt, Inst. für Arbeitsmedizin, Gustav-Freytag-Str. 1 , O 5082 Erfurt, Germany (Tel.: 387311)

Cancer in Rubber Workers in Thuringia
This study of workers in the rubber industry has been initiated with the aim of measuring exposure to asbestos in talc in the last 20 years. Over 200 workers were exposed. They will be compared with 100 non-exposed workers. Many exposed workers are now retired. Medical histories will be collected and clinical, functional and x-ray examinations carried out. The study is mainly aimed at assessing the relation between exposure to asbestos (quantity and type) in talc and cancer, especially cancer of the lung and mesothelioma.

TYPE: Cohort
TERM: Dusts; Occupation; Rubber
SITE: Lung; Mesothelioma
CHEM: Asbestos; Talc
OCCU: Rubber Workers
TIME: 1991 - 1993

ESSEN

* 343 Popp, W.   04999
Univ. of Essen (GHS), Inst. for Hygiene and Industrial Med., Hufelandstr. 55 , W 4300 Essen 1, Germany (Tel.: (0201)7234574)
COLL: Norpoth, K.; Vahrenholz, C.; Schürfeld, C.; Müller, C.

Validity of Molecular Epidemiological Methods in Occupational Medicine
The aim of the study is to test the validity and usefulness of certain methods of molecular epidemiology (SCE, alkaline elution) in occupationally exposed groups. These methods, partly combined with methods of biological monitoring, are being applied to groups of workers who are exposed to chromium and nickel (welders), benzene and toluene (shoemakers) and ethylene oxide (hospital personnel). For every group, consisting of 20-40 persons each, control groups, matched for age, sex and smoking habits, are formed. An additional positive control group consists of multiple myeloma patients receiving therapy with the alkylating agent melphalan.

GERMANY

TYPE: Methodology; Molecular Epidemiology
TERM: Drugs; Occupation; SCE; Solvents
SITE: Inapplicable
CHEM: Benzene; Chromium; Ethylene Oxide; Melphalan; Nickel; Toluene
OCCU: Health Care Workers; Shoemakers-Repairers; Welders
TIME: 1990 - 1992

FRIEDRICHSHAFEN

344 Schäfer, T. 03994
Dornier System GmbH, Abt. PBIC, Planungsberatung im, Gesundheitswesen, Postfach 1360, W 7990 Friedrichshafen, Germany (Tel.: (07541)209218)

Model Project for the Regional Analysis of Health and Environmental Data in Saarland
In this project, data on cancer morbidity and mortality in the cancer registry of Saarland, Federal Republic of Germany, are analysed with respect to ecological and environmental data collected for the pilot application of the "Manual on Ecological planning". The purpose of the first phase was to adjust the usual methodology of cancer mortality cartography to an extremely high level of regional disaggregation. On this basis an unbiased description of the health status of the population and its regional variation should be provided with special reference to random fluctuations and their statistical significance. In the second phase changes in regional morbidity and mortality are related to regional variations in environmental, ecological and occupational exposures with adjustment for smoking. Recent publications appeared in Saarlaendisches Aerzteblatt 9:558-573, 1978 (Grueger, J.) and in Sozialmed. Schwerpunkte Rheuma und Krebs (Ed. E.O. Kraseman et al), Berlin, 1987, pp. 240-253.

TYPE: Correlation
TERM: Environmental Factors; Geographic Factors; Mapping; Occupation; Registry; Tobacco (Smoking)
SITE: All Sites; Gastrointestinal; Respiratory
REGI: Saarland (FRG)
TIME: 1983 - 1992

GIEßEN

345 Paul, E. 04388
Zentrum für Dermatologie, und Andrologie, Gaffkystr. 14, W 6300 Gießen, Germany (Tel.: (0641)7023560)
COLL: Rauh, M.

Incidence of Malignant Melanoma in Central Hesse
A geographically limited catchment area around a clinic specialising in melanoma therapy has been defined, on the assumption that all melanoma patients from this area, which comprises one million inhabitants, are seen in the clinic. Only patients coming from the defined area are included. Basic epidemiological data are analysed statistically and incidence and distribution of melanomas as well as the different types of melanoma, stage and thickness calculated. Results to date show a statistically significant trend of thinner melanomas and earlier stage tumours. This early detection may be the result of increasing awareness concerning melanoma during the last decade, resulting from education programmes. A paper has been published in Anticancer Res. 7:447-448, 1987 (Rauh et al.).

TYPE: Incidence
TERM: Stage
SITE: Melanoma
TIME: 1982 - 1992

346 Woitowitz, H.J. 04322
Justus-Liebig Univ. Gießen, Inst. für Arbeits-, und Sozialmedizin, Aulweg 129/III, W 6300 Gießen, Germany (Tel.: (0641)7024240; Fax: 7024495)
COLL: Römer, W.; Rödelsperger, K.; Schulz, A.; Calavrezos, A.

Risk Factors for Diffuse Malignant Mesothelioma
Diffuse malignant mesothelioma is a signal tumour of occupational asbestos exposure. Among risk factors other than asbestos, natural and artificial mineral fibres have been implicated. Concerning erionite, an epidemiological proof for a causal relation already exists. Age, sex, state of health and type of tissue may also modify risk. In a multi-centre case-control study, several hundred mesothelioma cases

GERMANY

from eight hospitals are being compared with controls from these hospitals. Mesotheliomas are diagnosed by a panel of pathologists. By an extensive interview, occupational and environmental exposure to mineral fibres as well as health history is documented. For mesothelioma cases, the influence of effect modifiers and therapy is analysed. Finally, lung tissue fibre analysis is performed, mainly for cases and controls without occupational asbestos exposure. It is intended to examine whether environmental asbestos exposure or exposure to certain other mineral fibres dominates the fibre burden of lung tissue for mesothelioma cases as compared with controls.

TYPE: Case-Control
TERM: Chemical Exposure; Environmental Factors; Occupation
SITE: Mesothelioma; Pleura
CHEM: Mineral Fibres
TIME: 1987 - 1992

GUMMERSBACH

347 Schmauz, R. 04385
Postfach 10 08 08, W 5270 Gummersbach, Germany
COLL: Okong, P.; Owor, R.

Case-Control Study of Squamous Cell Genital Cancer

Current aetiological hypotheses for genital cancers will be examined in a case-control study in Uganda, an area of high incidence. Cancer will be squamous tumours of cervix, vulva and penis. Controls will be non-cancer patients seen in the same hospital for other surgical procedures. About 300 cases are expected over 3 years. Interviews will focus on sexual habits, socio-economic factors and tribal origin. Tissue will be examined for HPV by Southern blot.

TYPE: Case-Control
TERM: Ethnic Group; HPV; Sexual Activity; Socio-Economic Factors
SITE: Penis; Uterus (Cervix); Vulva
LOCA: Uganda
TIME: 1987 - 1992

348 Schmauz, R. 04386
Postfach 10 08 08, W 5270 Gummersbach, Germany
COLL: Müller-Hermelink, K.H.; Ngendahayo, P.; Wabinga, H.

Lymphadenopathies, Tropical Malaria, and/or AIDS

Tissue from malignant (Non-Hodgkin's lymphoma) and inflammatory lymphadenopathies will be examined histopathologically in a survey in Uganda (AIDS and malaria both prevalent), and Rwanda (mainly AIDS). About 100 Non-Hodgkin's lymphoma cases are expected. The frequency of various effects of immunosuppression by these two diseases on the lymphoreticular system will be compared.

TYPE: Cross-Sectional
TERM: AIDS; Immunosuppression
SITE: Non-Hodgkin's Lymphoma
LOCA: Rwanda; Uganda
TIME: 1987 - 1992

HAMBURG

349 Doerken, H. 03529
Universitätsklin. Eppendorf, Martinistr. 52, W 2000 Hamburg 20, Germany
COLL: Kuse, R.; Hossfeld, D.K.

Leukaemia in Man and Dog

According to the Chinese experience there is at least one tumour (oesophageal cancer) occurring in man and at the same time in his domestic animals, namely, the chicken, well known in experimental cancer research since 1908-1911. Other correlations between tumours in humans and domestic animals have not been proved to date. In particular there is no detectable direct connection between humans and cattle or between humans and leukaemia in cats. According to some casual observations in Northern Germany cases of human and dog leukaemia have been found in the same family or house. To

GERMANY

investigate this, all 170 practising veterinarians in the State of Hamburg are being contacted to collect all cases of leukaemia in dogs for the last five years. Following this, the owners will be contacted to ask about corresponding diseases in the families. A paper was published in Med. Klin. 82(17):551-557, 1987.

TYPE: Case Series
TERM: Animal; Cluster
SITE: Leukaemia
TIME: 1984 – 1992

HANNOVER

350   **Baltrusch, H.J.F.**   03317
Medizinische Hochschule, Immunohämatologie, Transfusionsmedizin – Blutbank 8350, Konstanty-Gutschow-Str. 8, W 3000 Hannover 61, Germany (Tel.: (0511)5322083; Fax: 5325550)
COLL: Schedel, I

Psychosocial Setting of Plasmocytoma

Accidental observations suggest that the onset of plasmocytoma is precipitated by psychosocial stress. A serial investigation has been initiated in order to assess the role of adverse life events in the clinical manifestation and outcome of plasmocytoma, and how this is possibly correlated with haematological and immunological parameters. Consecutive or randomly selected plasmocytoma patients are studied by a structured clinical interview technique and a series of psychological tests, including Family Attitudes Questionnaire, Habits of Nervous Tension, Courtauld Emotional Control Scale, Defense Scale, Rorschach and Thematic Apperception Test. Assessment of the patients' coping behaviour is also included in the study, as well as aspects of guidance and treatment. So far 15 patients have been studied; at least 50 patients will be included. The results of this investigation will be compared with those of a similar investigation in patients with polycythaemia vera.

TYPE: Case-Control
TERM: Psychosocial Factors; Stress
SITE: Haemopoietic
TIME: 1984 – 1992

* 351   **Baltrusch, H.J.F.**   05146
Medizinische Hochschule, Immunohämatologie, Transfusionsmedizin – Blutbank 8350, Konstanty-Gutschow-Str. 8, W 3000 Hannover 61, Germany (Tel.: (0511)5322083; Fax: 5325550)
COLL: Stangel, W.

Biopsychosocial Setting of Polycythaemia Vera

Polycythaemia vera (PV) is a rare myeloproliferative disease of largely unknown aetiology. Clinical observations suggest that the disease might be precipitated by psychosocial stress. An interdisciplinary controlled study has been initiated in order to ascertain the role of such stress in the aetiology and clinical course of PV. Consecutive PV patients and their matched pair controls are studied by a structured interview technique and a series of psychological tests, including a questionnaire measuring Type C (cancer-prone) behaviour. Family Attitudes Questionnaire, State-Trait Anxiety Inventory, Habits of Nervous Tension, Courtauld Emotional Control Scale, Rorschach and Thematic Apperception Test. Assessment of patients' coping behaviour is also included in the study as well as aspects of guidance and treatment. So far, 40 patients have been studied. Partial results were published in Psychiat. Fenn. Suppl.:133-142, 1981 and Ann. NY Acad. Sci. 521:1-15, 1988.

TYPE: Case-Control
TERM: Psychosocial Factors; Stress
SITE: Haemopoietic
TIME: 1979 – 1993

HEIDELBERG

352   **Abel, U.R.**   04149
German Cancer Research Centre, Inst. for Epidemiology and Biometry, Im Neuenheimer Feld 280, W 6900 Heidelberg, Germany (Tel.: (06221)484391; Fax: 401271 ; Tlx: 461562 dkfz d)
COLL: Becker, N.; Frentzel-Beyme, R.R.; Kaufmann, M; Schlag, P.; Schulz, G.; Wahrendorf, J.

GERMANY

History of Infections and Cancer Risk
There have been some communications in the literature that cancer patients report in their medical history less frequent and less severe episodes of feverish infections than healthy controls. Together with clinical observations of a beneficial effect of post-operative bacterial infections on long-term prognosis in various malignancies as well as the fact of an increased cancer risk in patients treated with immunosuppressive agents, this leads to hypothesize a protective role of an active immune system in the aetiology of cancer. To confirm this a case-control study is being carried out, with cancer patients from two major clinics of Heidelberg University. Five cancer sites will be considered, recruiting 50 patients for each site (colon, rectum, stomach, breast, and ovary). These patients and equal numbers of hospital controls and population controls will be interviewed with a standardized questionnaire inquiring predominantly into history of infections. The interview-derived information will be partly validated by contacting the respective general practitioners.

TYPE: Case-Control
TERM: Immunosuppression; Infection
SITE: Breast (F); Colon; Ovary; Rectum; Stomach
TIME: 1987 - 1993

353   Becker, N.   03574
German Cancer Research Centre, Inst. for Epidemiology and Biometry, Im Neuenheimer Feld 280, W 6900 Heidelberg, Germany (Tel.: (06221)484385; Fax: 45516 ; Tlx: 461562 dkfz d)
COLL: Abel, U.R.; Frentzel-Beyme, R.R.

Allergy and Cancer
Participation of the immune system in cancer defense mechanisms has been discussed for a long time. It is known that there is an increased cancer incidence in persons with immune suppression. Whether the reverse also applies, namely, that persons with an enhanced reaction of the immune system, such as those with various allergies, experience a reduced cancer incidence is however disputed. Studies already carried out on this subject showed a wide variety of results. This study aims to investigate again the question of the part played by the immune system in the pathogenesis of cancer in persons with a previous diagnosis of allergy. A retrospective follow-up study is being carried out of 5,000 patients attending a university dermatological hospital for the years 1973-1974. Persons with a negative allergy-test admitted to the same hospital for the purpose of excluding an allergic disease serve as controls.

TYPE: Cohort
TERM: Allergy; Immunology
SITE: All Sites
TIME: 1985 - 1992

354   Blettner, M.   04943
German Cancer Research Centre, Inst. for Epidemiology and Biometry, Im Neuenheimer Feld 280, W 6900 Heidelberg, Germany (Tel.: +49 6221 484383; Fax: 45516)
COLL: Renz, K.; Seitz, G.; Wahrendorf, J.

Historical Cohort Study among Workers in Nuclear Power Stations
The study investigates the relationship between occupational exposure to low-level ionising radiation and causes of death. A retrospective cohort study will include approximately 4000 current and past employees from several nuclear power stations in Germany. Exposure information will be obtained from personal records for all workers who had been monitored for occupational exposure to radiation. Internal comparisons will be made between groups with different duration and degree of exposure. The study will be part of a large international cohort study, in preparation at IARC.

TYPE: Cohort
TERM: Occupation; Radiation, Ionizing
SITE: All Sites
OCCU: Radiation Workers
TIME: 1989 - 1995

355   Boeing, H.   04944
German Cancer Research Centre, Inst. for Epidemiology and Biometry, Im Neuenheimer Feld 280, W 6900 Heidelberg, Germany (Tel.: +49 6221 484898; Fax: 45516)
COLL: Weisgerber, U.; Raedsch, R.; Waldherr, R.; Rozen, P.

GERMANY

Randomized Double-Blind Intervention Trial with Calcium in Polypectomized Patients
In a double-blind randomized trial with 100 polypectomized patients it will be tested whether 2 g of calcium daily, given as lactate gluconate in effervescent tablets, reduces cell proliferation of the colon crypts compared to placebo treated patients. All participants will undergo a run-in phase for 3 months with placebo and are then assigned randomly to calcium or placebo treatment for 9 months. Biopsies will be taken before and at the end of the study from the recto-sigmoidal region. Two different methods are considered for measuring cell proliferation: bromodeoxyuridine (BrdU) and proliferating cell nuclear antigen (PCNA). Each participant will be visited every 3 months at home for compliance assessment and the distribution of new tablets. In connection with these visits urine will be collected, as well as a 48-h stool on dry ice. The biological material will be utilized to study the metabolic effect of calcium on the lumen of the colon and to control for compliance. Parameters of interest are bile acids and fatty acid metabolism and their binding to calcium. Aliquots of stool will be stored for additional analyses.

TYPE: Intervention
TERM: BMB; Diet; Faeces; High-Risk Groups; Minerals; Premalignant Lesion; Prevention; Urine
SITE: Colon; Rectum
CHEM: Calcium
TIME: 1989 - 1992

356 Boeing, H. 04928
German Cancer Research Centre, Inst. for Epidemiology and Biometry, Im Neuenheimer Feld 280, W 6900 Heidelberg, Germany (Tel.: +49 6221 484898; Fax: 45516)
COLL: Wahrendorf, J.; Pfeifer, H.

Cohort Study on Diet and Cancer
As part of a large European cohort study, instruments and procedures for a study on diet and cancer will be developed in Germany. At least 50,000 participants of both sexes between 30 and 65 years of age will be recruited, in cooperation with institutions of the medical insurance system. In the first phase, a national nutrition survey conducted in 1985 - 1988 with 22,000 participants of all ages will be utilized for developing instruments and procedures. On a small sample it will be tested whether an active follow-up for major disease ocurrences is possible, by repeatedly contacting survey participants. The survey will also be used to validate a short food frequency questionnaire to measure general dietary intake. The reference method will be 7-day records filled in at the survey. The dietary questionnaire will then be further tested for reliability and validity using nitrogen output in urine and 24-h-dietary recall.

TYPE: Cohort; Methodology
TERM: Diet; Nutrition
SITE: All Sites
TIME: 1990 - 1992

357 Chang-Claude, J. 04704
German Cancer Research Centre, Inst. for Epidemiology and Biometry, Im Neuenheimer Feld 280, W 6900 Heidelberg, Germany (Tel.: +49 6221 484898; Fax: 45516 ; Tlx: 461562 dkfz d)
COLL: Wahrendorf, J.; Qiu, S.L.; Yang, G.R.; Muñoz, N.; Crespi, M.; Raedsch, R.; Thurnham, D.; Correa, P.

Precancerous Lesions of the Oesophagus among Young Persons in China
This investigation is conducted among persons between 15 and 26 years of age in Huixian, Henan Province, a high risk area for oesophageal cancer in the People's Republic of China. The objective is to collect information on the prevalence of precancerous lesions of the oesophagus at early ages and to identify risk factors associated with the development of such lesions. Eligible subjects are selected from households in which a case of oesophageal cancer has been diagnosed in the past six years and from twice the number of control households. The survey includes an oesophagoscopy with guided biopsies, a physical examination, an interview with a standardised questionnaire including known and suspected risk factors for oesophageal cancer, a 10ml blood sample and overnight urine. A case-control analysis of subjects with chronic oesophagitis (considered a precursor lesion of oesophageal cancer) versus controls will be performed to identify risk factors with regard to diet, lifestyle and environmental exposures. The measurement of vitamin levels and trace elements in plasma, cell proliferation studies of oesophageal epithelium, the micronuclei test on oesophageal smears and the analysis of urinary excretion of some N-nitrosamino acids will provide additional data for the elucidation of parameters that may have aetiological importance.

GERMANY

TYPE: Case-Control
TERM: Age; Alcohol; BMB; Biopsy; Cytology; Diet; Lifestyle; Micronuclei; Plasma; Premalignant Lesion; Prevalence; Tobacco (Smoking); Urine; Vitamins
SITE: Oesophagus
CHEM: N-Nitroso Compounds
LOCA: China
TIME: 1987 - 1992

358 Frentzel-Beyme, R.R. 04828
German Cancer Research Centre, Inst. for Epidemiology and Biometry, Im Neuenheimer Feld 280, W 6900 Heidelberg, Germany (Tel.: +49 6221 484378; Fax: 45516 ; Tlx: 461562 dkfz d)
COLL: Winter, J.; Buhr, P.

Risk Factors for Thyroid Gland Neoplasms
This case-control study is based on the clinical regional registry of thyroid cancer patients, which was established in 1986 at the oncological centre Heidelberg-Mannheim. The number of patients registered annually (about 60) will permit inclusion of a minimum of 200 cases in the interview study. One control group is drawn from the extended pool of patients with benign thyroid tumours diagnosed at the same treatment centres, the second comparison group is a population sample obtained from the offices of population registration in the study area. The interview covers aetiologically relevant information, concentrating on reproductive history and central nervous stimulation.

TYPE: Case-Control
TERM: Occupation; Reproductive Factors
SITE: Thyroid
TIME: 1988 - 1992

* 359 Kahn, T.M. 05070
Deutsches Krebsforschungszentrum, Inst. fü Angewandte Tumorvirologie, Im Neuenheimer Feld 506, W 6900 Heidelberg, Germany (Tel.: (06221)424650; Fax: 470333)
COLL: Bercovich, J.; Chang-Claude, J.; Turazza, E.; Claudiani, J.; Gurucharri, C.; Sprovieri, O.; Valacco, A.; Pires Torres, C.; Ojeda, R.; Amestoy, G.; Bayo, J.; Collalto, A.; Haimovici, L.; Nishijama, S.; Gissmann, L.; Grinstein, S.; zur Hausen, H.

HPV Infection and Cancer of the Cervix and the Upper Respiratory Tract in Argentina
Mortality rates for cervical cancers in Argentina vary widely by geographical region, and mortality rates for oropharyngeal and laryngeal cancers are high. This study aims to determine the prevalence of HPV infection in women with a normal cervix and patients presenting with premalignant and malignant lesions of the genital and upper respiratory tracts from different geographical, ethnic and socio-economic origins. Patients were recruited from 10 hospitals in the Buenos Aires area, which includes persons from all over the country. Cervical smears obtained from colposcopically normal patients will be tested for HPV using filter in situ hybridization, and biopsy material from cancers and lesions will be tested by Southern blot hybridization. A case-control study for cervical cancers is being carried out to recruit 250 cases and 500 age-matched controls using a questionnaire to gain information on risk factors for cervical cancer. Analysis will be carried out to elucidate the risk factors explanatory of the large regional differences in cervical cancer mortality rates.

TYPE: Case Series; Case-Control
TERM: Alcohol; HPV; Premalignant Lesion; Reproductive Factors; Socio-Economic Factors; Tobacco (Smoking)
SITE: Oral Cavity; Oropharynx; Respiratory; Uterus (Cervix)
LOCA: Argentina
TIME: 1989 - 1993

* 360 Maier, H. 05169
Univ. Heidelberg, Hals-, Nasen- & Ohrenklinik, Im Neuenheimer Feld 400 , W 6900 Heidelberg, Germany (Tel.: (06221)566703; Fax: 565913; Tlx: 461745 unikl hd)
COLL: Lessner, K.; Dietz, A.; Gewelke, U.; Heller, W.D.

Multiple Primary Tumours in Laryngeal Cancer
The primary objective of this prospective study is to investigate the risk for multiple primary tumours associated with tumour stage, tobacco consumption, alcohol consumption, diet, occupation and social status in patients with laryngeal cancer. In the first part of the study (1988/1989), 164 patients with squamous cell carcinoma of the larynx were interviewed by a trained interviewer using a structured

GERMANY

questionnaire. In the second part of the study, cases are followed up at monthly intervals for five years in order to detect local recurrences, metastases and secondary or multiple primary tumours. By May 1991, a total of 36 patients had died, 25 due to the occurrence of secondary or multiple tumours. A first analysis of the data is planned in 1992.

TYPE: Cohort
TERM: Alcohol; Diet; Multiple Primary; Occupation; Stage; Tobacco (Smoking)
SITE: Larynx
TIME: 1988 - 1994

361 Merzenich, H. 04945
German Cancer Research Centre, Inst. for Epidemiology and Biometry, Im Neuenheimer Feld 280, W 6900 Heidelberg, Germany (Tel.: +49 6221 484382; Fax: 45516)
COLL: Boeing, H.

Relationship of Nutrition and Physical Activity in Childhood with Age at Menarche

Since early age of menarche is known as a risk factor for pre-menopausal breast cancer, it is important to know whether factors which mofify the age at menarche also modify breast cancer risk. This study will document the role of specific life-style factors (nutrition, physical activity etc.) on the timing of menarche. The study is prospective. The study population consists of about 280 9- to 14-year-old girls who participated in a nation-wide representative nutrition survey (1986-1989) which included a food record and a 7-day activity protocol. In a baseline examination the parents of the participants were asked to fill in a mailed questionnaire about reproductive history of the mother, maturation and anthropometric data of the daughter, physical activity of the daughter, health information and the actual nutritional habits of both, the household and the daughter, measured by a short food-frequency questionnaire. Participants were then followed for a maximum of two years until the event of menarche. For this purpose each girl was contacted by mail and asked about the occurrence of menarche and current body weight and height. These baseline data are important in the development of strategies for primary breast cancer prevention.

TYPE: Cohort
TERM: Lifestyle; Menarche; Nutrition; Physical Activity; Physical Factors; Prevention; Reproductive Factors
SITE: Breast (F)
TIME: 1988 - 1992

* 362 Raue, F. 05000
Medizinische Universitätsklinik, Bergheimerstr. 58, W 6900 Heidelberg, Germany (Tel.: (06221)568605; Fax: 565999)

Register of Medullary Thyroid Carcinoma

The objective of this retrospective and prospective register is to identify patients with hereditary medullary thyroid carcinoma (MTC). Approximately 25% of this rare type of thyroid carcinoma is genetically determined (multiple endocrine neoplasia Type II) and can be detected at an early stage by family screening. A questionnaire has been distributed containing questions on personal information, details relating to the age at diagnosis and histological findings, the present status of the patients, and about family screening and surgical therapy. Lymphocytes of members of MTC-families were stored for further genetic studies. 650 patients with MTC (25% familial cases) have been reported by 27 cooperative centers (Med. Klin. 85:113-117, 1990); about 40 new cases have been registered annually since 1988. Survival data and prognostic factors have been calculated (Acta Endocrinol. 124, Suppl. 1:Abstr. 17, 1991). This register will provide a basis for further collaborative diagnostic and therapeutic studies.

TYPE: Genetic Epidemiology
TERM: BMB; Genetic Factors; Lymphocytes; Registry
SITE: Thyroid
TIME: 1988 - 1993

* 363 Schlaefer, K.O.H. 05175
Deutsches Krebsforschungszentrum, Inst. für Epidemiologie, und Biometrie, Im Neuenheimer Feld 280, W 6900 Heidelberg, Germany (Tel.: +49 6221 422367; Fax: 45516)
COLL: Wahrendorf, J.; Möslein, G.; Herfarth, C.; Friedl, W.; Jäger, K.

GERMANY

Aetiology of Familial Adenomatous Polyposis
The aim of this study is to detect co-factors of familial adenomatous polyposis (FAP). FAP has an autosomal dominant inheritance, the gene mapping to the long arm of chromosome 5 (5q21). Factors responsible for the age of onset are unclear, and gene-environment-interactions may be important. The aim is to investigate the following factors: (1) nutrition, especially fat, vitamins, and fibres, and nutrients enhancing the production of bile acid; (2) earlier viral and bacterial infections, especially polio, arbo and cytomegalo viral infections, and streptoccocal and clostridial infections; and (3) the proliferation rate of single pre-FAP polyps, to determine whether this might be a prognostic factor for the proximity of the onset. A collaboration with two major centers for surgical treatment of FAP has been established. This will be a cohort study with a nested case-control study. The control will be the next elder sib, if not diseased, or the sib closest in age. Families with at least one person with FAP will be selected and followed up for the next 10 years. At the regular clinical check-up, an interviewer will administer a questionnaire, and blood, stool and tissue samples (removed polyps) will be collected. It is expected to identify a cohort of 200-300 families (600-700 people).

TYPE: Case-Control; Cohort; Genetic Epidemiology
TERM: BMB; Bile Acid; Blood; Diet; Faeces; Familial Adenomatous Polyposis; Genetic Factors; Heredity; Infection; Linkage Analysis; Nutrition; Pedigree; Polyps; Premalignant Lesion; Tissue; Virus
SITE: Colon; Rectum
TIME: 1991 - 2001

364 Schlehofer, B. 04700
German Cancer Research Center, Inst. of Epidemiology & Biometry, Im Neuenheimer Feld 280, W 6900 Heidelberg, Germany (Tel.: +49 6221 484351; Fax: 45516; Tlx: 461562 dkfz d)
COLL: Michaelis, J.H.; Kaatsch, P.; Schlehofer, J.; Wahrendorf, J.

Case-control study of Parvovirus Infections and Childhood Leukaemia
In this study protective effects of parvovirus-infections in humans will be investigated. Animal experiments have shown that incidence and growth rate of tumours of different localizations could be reduced by Parvoviruses (H1- or adeno-associated parvoviruses). For adeno-associated virus (AAV) serological findings also indicate suppressive effects in human cancer. Parvovirus B-19, a virus which causes diseases (exanthema subitum) in humans, has not yet been studied for a protective effect. The target tissue for B-19 replication is the bone marrow. In a nationwide case-control study performed jointly by the German Cancer Research Center (Dept. of Epidemiology and Dept. of Tumorvirology) and the National Registry of Childhood Malignancies at the University of Mainz, about 400 incident cases of childhood leukaemia and the same number of age-sex-matched hospital controls will be investigated. Serological analysis will assess the presence and titres of antibodies against H1-, adeno-associated- and B-19-parvoviruses. In addition to the serological tests, the parents will be asked to respond to a self-administered questionnaire about possible risk factors in the aetiology of leukaemia, including life-style, occupational and medical history 3 months before and during pregnancy as well as environmental and health history. Infectious diseases of the children prior to diagnosis will be taken into consideration. In a parallel study sera of patients with different solid tumours and healthy controls will also be tested for the presence of antibodies of H1, adeno-associated and B-19 viruses.

TYPE: Case-Control
TERM: Antibodies; Childhood; Lifestyle; Occupation; Registry; Virus
SITE: Leukaemia
REGI: Mainz (FRG)
TIME: 1989 - 1992

365 Schlehofer, B. 04701
German Cancer Research Center, Inst. of Epidemiology & Biometry, Im Neuenheimer Feld 280, W 6900 Heidelberg, Germany (Tel.: +49 6221 484351; Fax: 45516; Tlx: 461562 dkfz d)
COLL: Boeing, H.; Wahrendorf, J.; Waldherr, R.

Case-control Study on Renal Cell Carcinoma
The rates of renal cell cancer have been increasing in several countries. To identify aetiological risk factors for this tumour, a population-based case-control study in the Rhein-Neckar-Odenwald area is being carried out. In 10 hospitals all incident cases age 20 to 75 years will be interviewed during the two years 1989-90. 350 cases are expected, and the same number of population controls drawn randomly from the residential list will be recruited. The main aspects of the face-to-face interview are questions on drug consumption (diuretics and analgesics), medical and smoking history, but there is also a focus on exposures to occupational risk factors. Several questions are related to the risk of obesity and physical activities and a self-administered questionnaire for dietary habits is also given to cases and controls.

GERMANY

Special attention will be paid to the possible association with exposure to pesticides and herbicides: in a substudy for a 30% sample of the cases, fat tissue of the renal capsule will be analysed for persistent chlorinated hydrocarbons and compared with results from the renal fat tissue of age- and sex-matched autopsy controls. This study is part of an international study conducted with similar protocols in Australia, Austria, China, Denmark Sweden and the USA.

TYPE: Case-Control
TERM: Diet; Drugs; Herbicides; Obesity; Occupation; Pesticides; Physical Activity; Tobacco (Smoking)
SITE: Kidney
TIME: 1989 - 1992

366 Schroeder-Kurth, T.M. 03509
Inst. für Anthropologie, und Humangenetik, Abt. für Cytogenetik, Im Neuenheimer Feld 328, W 6900 Heidelberg, Germany (Tel.: (06221)563877)
COLL: Auerbach, A.D.

International Fanconi Anaemia Registry

Fanconi's anaemia is an autosomal recessive trait of which more than 300 cases have been reported. Many patients die from bleeding in early childhood and some die from cancer. The underlying genetic cause may be a defect in the repair capacity of the cells. Clinical heterogeneity is evident, as well as a broad variation in the amount of increase in chromosomal instability, which is useful for differential diagnosis, particularly when combined with sensitivity tests. The International Fanconi Anaemia Registry tries to get large-scale information about the course of the disease, haemological data, frequency of malformations connected with the disease, cytogenetic findings and incidence of cancer. Today, about 200 cases have been listed in the registry. Cases from Europe and China are collected in Heidelberg, and from America and Japan in New York. The cases are merged into one registry in New York. Directory readers are invited to contribute to the registry.

TYPE: Case Series
TERM: Familial Factors; Genetic Factors; Genetic Markers; Registry
SITE: All Sites
TIME: 1982 - 1992

367 ter Meulen, J. 04862
German Cancer Research Centre, Inst. for Virological Research, Im Neuenheimer Feld 280, W 6900 Heidelberg, Germany (Tel.: (06221)484629; Fax: 401271)
COLL: Pawlita, M.; Chang-Claude, J.; Mgaya, H.N.; Luande, J.

HPV Infection and Other Sexually Transmitted Diseases in Tanzanian Women

The present study aims to identify risk groups for HPV infection which is associated with a high risk of CIN and cervical cancer. The study subjects will be some 600 female inpatients, admitted for a variety of gynaecological and obstetrical disorders, excluding those associated with cervical cancer. Investigations include: application of a questionnaire on recognized risk factors for cervical cancer, blood sampling for HBV, syphilis and HIV serology, and collection of cervical swabs for Papstaining and of cervical scrapings for detection of HPV-DNA with PCR-technology. The occurrence of specific HPV-types in Tanzanian cervical carcinomas, which is important for the determination of the appropriate primers for the PCR-reaction, is currently being investigated. The data will be analysed by log-linear models.

TYPE: Cross-Sectional
TERM: Cytology; HBV; HIV; HPV; PCR; Premalignant Lesion; Sexually Transmitted Diseases
SITE: Uterus (Cervix)
LOCA: Tanzania
TIME: 1989 - 1992

368 van Kaick, G. 02529
German Cancer Research Center, Inst. of Nuclear Medicine, Im Neuenheimer Feld 280, W 6900 Heidelberg, Germany (Tel.: (6221)484563; Fax: 401271; Tlx: 461562 DKFZ D)
COLL: Muth, H.; Wagner, G.; Kaul, A.

"Thorotrast" Research Project

Thorotrast, a 25% colloidal solution of Thorium dioxide was used as a contrast medium between 1930-1950. After intravenous injection the $ThOy_2$ particles were stored in the reticuloendothelial system (RES). The project aims at discovering the late effects of chronic radiation exposure by studying a large

GERMANY

number of patients injected with Thorotrast, comparing these results with a corresponding control group and examining the relationship between late effects and radiation dose. The study includes: biophysical investigations to calculate the total effective radiation dose; clinical, biochemical and radiological examinations to determine the state of health; classification of the causes of death for patients who died; statistical analysis of the results obtained. To date the names of 5,159 Thorotrast patients and 5,160 control patients have been obtained from records in different hospitals of West Germany: 901 Thorotrast patients and 669 control patients have been examined clinically. Living patients were re-examined every two years. Apart from this epidemiological study the following investigations are being carried out: animal experiments to (1) investigate the distribution of Thorium dioxide particles within the body and calculate the resulting tissue doses and (2) estimate the non-radiation effects of deposited Thorium dioxide. Papers appeared in Radiation Carcinogenesis: Epidemiology and Biological Significance, (eds) J.D. Boice, Jr and J.F. Fraumeni, Jr, Raven Press, New York, 1984, pp. 253-262 and in BIR Report 21:Risks from Radium and Thorotrast, (eds) D.M. Taylor et al., British Institute of Radiology, London, 1989, pp. 97-104.

TYPE: Cohort
TERM: Dose-Response; Drugs; Late Effects; Radiation, Ionizing
SITE: Bone; Leukaemia; Liver; Lung; Lymphoma; Spleen
CHEM: Thorotrast
TIME: 1968 – 1992

369   Wahrendorf, J. 04158
German Cancer Research Center, Inst. of Epidemiology & Biometry, Im Neuenheimer Feld 280, W 6900 Heidelberg 1, Germany (Tel.: +49 6221 422200; Fax: 45516 ; Tlx: 461562 dkfz d)
COLL: Dhom, G.; Robra, B.P.; Wiebelt, H.; Oberhausen, R.; Weiland, M.

Evaluation of Colorectal Cancer Screening Programmes in the Federal Republic of Germany
Screening for early signs of colorectal cancer by occult blood testing is offered to men and women aged 45 and above within the statutory health insurance system in the Federal Republic of Germany since 1977. Annual participation rates are low (10 to 20%), however cumulative participation rates (ever participated) are estimated to be between 50 and 60%. Centrally kept records of this screening programme cannot be linked on an individual basis in a longitudinal fashion due to regulations on data privacy. The feasibility of having access to deaths of colorectal cancer, healthy controls and the complete screening histories of these persons has been explored in the Saarland and has lead to a population-based case-control evaluation of the colorectal cancer screening programme comparing screening histories of persons who died of colorectal cancer with screening histories of persons who did not. Starting with about 3,000 bioptic cases of colorectal cancer about 700 deaths were identified. Te exact cause of death and screening history is currently retrieved together with information on controls.

TYPE: Case-Control
TERM: Clinical Records; Screening
SITE: Colon; Rectum
TIME: 1986 – 1992

370   Wahrendorf, J. 04367
German Cancer Research Center, Inst. of Epidemiology & Biometry, Im Neuenheimer Feld 280, W 6900 Heidelberg 1, Germany (Tel.: +49 6221 422200; Fax: 45516 ; Tlx: 461562 dkfz d)
COLL: Becher, H.; Saracci, R.

Cohort Studies of Persons Exposed to Phenoxy Acid Herbicides and their Contaminants
Phenoxy acid herbicides have been widely used throughout the world since the early 1940's. A group of related compounds are the chlorophenols which are used in the manufacture of some of the above herbicides and as a preservative or a pesticide. These compounds which are of great commercial importance have been known to be contaminated with dioxins and furans which are formed during their manufacture and for which well-documented acute toxic effects in man are known. There is also concern as to whether human exposure does result in long-term effects, such as cancer. Contributing to a European IARC-coordinated study workers, employed in three chemical companies from the early 1950's are enrolled in this study. Establishment of the cohort and provisions for a mortality follow-up are underway.

GERMANY

TYPE: Cohort
TERM: Chemical Exposure; Herbicides; Occupation; Pesticides
SITE: All Sites
CHEM: Chlorophenols; Dioxins; Phenoxy Acids
OCCU: Chemical Industry Workers
TIME: 1987 - 1992

KÖLN

371　　Flatten, G.　　03260
Zentralinst. für die Kassenärztliche Versorgung in der Bundesrepublik Deutschland, Herbert-Lewinstr. 5, W 5000 Köln 41, Germany (Tel.: +494 221 4005130; Fax: 8883242)

Continuous Evaluation of the German National Screening Programme
In the German statutory health insurance system (93% of the population) an annual cancer check-up is offered to women (from age 20, pap smear; from age 30, pap smear and breast physical examination with instruction in breast self examination; from age 45, faecal occult blood testing (FOBT) is added) and for men (from age 45, digital rectal examination + FOBT). Screening contacts are documented in a standard way, data are collected and analysed centrally by the Zentralinstitut. Attendance rates and detection rates are computed (by age and sex) and a number of special analyses conducted. Information obtained will be used for evaluation and further development of screening programmes (screening interval, inclusion of mammography, follow-up of positive FOBT).

TYPE: Cross-Sectional
TERM: Screening
SITE: Breast (F); Colon; Prostate; Rectum; Skin; Uterus (Cervix)
TIME: 1972 - 1993

MÜNCHEN

372　　Spiess, H.　　03997
Univ. München, Pettenkoferstr. 8A, W 8000 München 2, Germany (Tel.: (89)51603677)
COLL: Mays, C.W.

Effects of Ra-224 on Humans
The hypothesis is that radiation can induce specific diseases in humans. 899 patients who received repeated injections of Ra-224 after World War II, mostly for the treatment of tuberculosis or ankylosing spondylitis, are being followed. The diseases inducible by Ra-224 include malignant bone sarcomas, benign exostoses, growth retardation, tooth breakage, kidney diseases, liver diseases, and cataracts. At three-year intervals the patients are contacted by questionnaire. The information is used to predict the risk from other types of radiation exposure.

TYPE: Cohort
TERM: Radiation, Ionizing
SITE: Bone; Kidney; Liver
TIME: 1952 - 2000

MAINZ

* 373　　Michaelis, J.H.　　05170
Univ. Mainz, Inst. für Medizin, Statistik, & Dokumentation, Langenbeckstr. 1 , W 6500 Mainz, Germany (Tel.: (06131)173252; Fax: 172968)
COLL: Kaatsch, P.

Nationwide Registry of Childhood Malignancies
A nationwide registry of childhood malignancies was established in 1980 by the two German societies of paediatric oncology and haematology as a combination of a population-based and a clinical registry. Documentation of German controlled clinical trials is integrated in the system. More than 95% of all childhood malignancies diagnosed in the FRG are registered. Notification of cases by over 100 hospitals and departments of paediatric oncology and haematology is voluntary. After admission of a newly diagnosed child, a basic form is sent to the registry. In response the cooperating physician gets a tumour-specific questionnaire which contains items about epidemiological aspects, the patient's history

GERMANY

and diagnostic procedures. Long-term follow-up data are collected periodically to calculate survival and to get information about late effects or secondary malignancies. Incidence by sex, age, site and geographical location are estimated annually, based on more than 1,000 registered children per year. Based on subgroups of the more than 12,000 registered patients, some specific epidemiological studies are being performed, e.g. on viral aetiology of leukaemia, regional clusters, low-dose ionizing radiation effects. Further information about the registry is published in annual technical reports in German, available on request.

TYPE: Incidence; Registry
TERM: Cluster; Late Effects; Multiple Primary; Radiation, Ionizing; Registry; Survival; Virus
SITE: Childhood Neoplasms
REGI: Mainz (FRG)
TIME: 1980 - 1992

* 374 **Michaelis, J.H.** 05171
Univ. Mainz, Inst. für Medizin, Statistik, & Dokumentation, Langenbeckstr. 1 , W 6500 Mainz, Germany (Tel.: (06131)173252; Fax: 172968)
COLL: Kaatsch, P.

Potential Influence of the Proximity to Nuclear Plants on the Incidence of Childhood Malignancies
Following public discussion on a possible increase in incidence of childhood malignancy in the proximity of nuclear plants, a study was started in 1989. The aim of the study is to estimate cancer incidence around all 20 nuclear plants in Western Germany in a defined region up to 15 km from each plant (divided into three zones) and to compare this with the incidence in comparable regions, matched for each plant, by distance, density of population, and other parameters of demographic structure. In order to evaluate further potential differences in incidence, a special questionnaire is used for obtaining information on lifestyle variables and occupation of the parents, possibly elevated genetic risk, course of pregnancy and childhood development, and exposure to environmental factors.

TYPE: Incidence
TERM: Environmental Factors; Genetic Factors; Lifestyle; Parental Occupation; Radiation, Ionizing; Registry
SITE: Childhood Neoplasms; Leukaemia; Lymphoma; Neuroblastoma; Wilms' Tumour
REGI: Mainz (FRG)
TIME: 1989 - 1992

OLDENBURG

375 **Baltrusch, H.J.F.** 02866
International Psycho-Oncology Project, Bergstr. 10, W 2900 Oldenburg, Germany (Tel.: (541)44112147)
COLL: Bastecky, J.; Ebigbo, P.O.; Forsén, A.; Gehde, E.; Grassi, L.; Illiger, H.J.; Németh, G.; Stangel, W.; Zhang, Z.W.

Biobehavioural Precursors of Cancer
Clinical studies in cancer patients have revealed a number of personality traits, such as lack of closeness to parents, suppression of unacceptable feelings, in particular anger, avoidance of emotional conflicts, harmonizing behaviour, exaggerated altruism and negligence of own health. The use of these variables for possible screening of persons being at a high cancer risk (biobehavioural cancer risk profile or Type C behaviour pattern) is studied. To date, 1,300 European and 120 Nigerian cancer patients (both sexes and at all sites) have been investigated using the Thomas Family Attitude Questionnaire (FAQ). 1600 cancer-free individuals served as controls. Evaluation of the results showed that male cancer patients (testis, large bowel) and female cancer patients (breast, uterus, ovary, vulva) showed a statistically higher significant lack of relation to parents in the FAQ than their age-matched controls. As a second step, the Courtauld Emotional Contral Scale (CECS), measuring the suppression of anger, anxiety and depressive mood, was administered to 130 male and 80 female cancer patients and to 150 controls. As a third step, a study, measuring "Type C" behaviour, assessing anger-in, anger-out, emotional control, rationality, depression, anxiety, social support and optimism has been set up and is in progress. It is hoped to assess the possible relationships between personality factors, coping style, neuro-endocrine and psychoneuroimmunological parameters and cancer progression. Partial resulta have been published in Ann. NY Acad. Sci. 521:1-15, 1988 and in Int. J. Neurosci. 51:257-260, 1990.

GERMANY

TYPE: Case-Control
TERM: Psychological Factors; Stress
SITE: All Sites
LOCA: Austria; China; Czechoslovakia; Finland; Germany; Greece; Hungary; Italy; Nigeria; Yugoslavia
TIME: 1981 – 1993

GREECE

ATHENS

376 Linos, D. 03736
Athens Medical School, Aretaiion Hosp., Dept. of Surgery, 76 Vasilissis Sofias Ave., Athens, Greece
COLL: Koutras, D.

Aetiology of Thyroid Cancer
The study aims to investigate the role of several risk factors in the development of thyroid cancer in the Greek population. Among the factors being investigated are radiation, socio-economic and personal characteristics, dietary habits and prior medical conditions. The study design is that of the matched case-control study (group matching) with two groups of controls. The first group of controls will include persons with benign thyroid disease; the second will be hospital controls, seen in the hospital for emergency medical care. Matching factors will include age (+/- 5 years) and sex. The diagnosis of thyroid cancer or benign thyroid disease will be confirmed by histological examination, and absence of thyroid disease in the general control group will be verified by clinical and laboratory examinations.

TYPE: Case-Control
TERM: Diet; Histology; Physical Factors; Radiation, Ionizing; Socio-Economic Factors
SITE: Thyroid
TIME: 1985 - 1992

377 Papaevangelou, G.J. 04274
Athens School of Hygiene, National Center for Viral Hepatitis, National Center for AIDS, 196 Alexandras Ave., 11521 Athens, Greece (Tel.: 6467473)
COLL: Roumeliotou, A.; Kallinikos, G.; Economidou, J.

Epidemiology of AIDS in Greece
This is a study of factors responsible for Kaposi's Sarcoma in AIDS cases as well as in HTLV III negative patients. Anti-HTLV-III positive individuals are followed up to study the natural history of the infection. The sample includes 200 homosexuals, 120 haemophiliacs, 20 drug addicts and 20 other asymptomatic HTLV-III positive subjects. Demographic, socio-economic and environmental factors as well as detailed biochemical, serological and other laboratory and personal characteristics are studied and correlated to the development of Kaposi's Sarcoma and other malignancies. The epidemiological characteristics of classical Kaposi's Sarcoma are also studied in an effort to discern the difference between these two entities.

TYPE: Cohort
TERM: AIDS; Drugs; Environmental Factors; HTLV; Haemophilia; Homosexuality; Infection; Sero-Epidemiology; Socio-Economic Factors; Virus
SITE: Kaposi's Sarcoma
TIME: 1980 - 1992

378 Papaevangelou, G.J. 04275
Athens School of Hygiene, National Center for Viral Hepatitis, National Center for AIDS, 196 Alexandras Ave., 11521 Athens, Greece (Tel.: 6467473)
COLL: Roumeliotou, A.; Kallinikos, G.; Economidou, J.

Natural History of Viral Hepatitis B
This is a prospective study to understand the factors responsible for establishment of chronicity and development of primary liver cancer. It includes several aspects of the natural history of viral hepatitis B. More than 2,000 patients hospitalized in the Infectious Diseases Hospital of Athens have been included in this study. Patients are followed up to understand the factors responsible for the establishment of chronic hepatitis B. HBV, HDV markers, biochemical profiles, immunological, histological and immunochemical studies are performed. The relation of demographic, socio-economic, environmental and genetic factors to the development of the chronic disease and its further evolution are examined. In parallel, chronic HBsAg carriers detected through existing screening programmes, are studied in detail and followed up.

TYPE: Cohort
TERM: Environmental Factors; Familial Factors; HBV; Immunology; Infection; Socio-Economic Factors
SITE: Liver
TIME: 1980 - 1992

HONG KONG

HONG KONG

379   Koo, L.C. 02246
Univ. of Hong Kong, Medical Faculty, Dept. of Community Medicine, 5 Sassoon Rd , Li Shu Fan Bldg, Hong Kong, Hong Kong (Tel.: (5)8199289)
COLL: Ho, J.H.C.

Cultural, Environmental, and Familial Backgrounds of Female Lung Cancer Patients in Hong Kong
This retrospective study intends to identify the environmental and/or ethnic risk factors which contribute to the unusually high rates of lung cancer among Chinese females in Hong Kong. Chinese females have among the highest world age-adjusted incidence rates and a very low proportion of ever-smokers. 200 female lung cancer patients and 200 neighbourhood female controls matched for age, socio-economic status, and residence were interviewed between 1981-1983 on their life histories. The semi-structured interviews focused on: (1) basic demographic background; (2) residential history; (3) smoking history (including passive smoking at home and/or workplace); (4) occupational history; (5) personal habits and hygiene; (6) medical history; and (7) dietary history. Emphasis was placed on recall data concerning environmental exposure to suspected initiators, promoters, or protectors 20 or more years before diagnosis. Analysis of the possible interrelationships of these factors in increasing or reducing risk among ever- or never-smokers is continuing. Papers have been published in Nutr. Cancer 11:155-172, 1988, Soc. Sci. Med. 26:751-760, 1988 and in Environm. Res. 52-23-33, 1990.

TYPE: Case-Control
TERM: Chemical Exposure; Chinese; Diet; Environmental Factors; Female; Hygiene; Lifestyle; Occupation; Passive Smoking; Socio-Economic Factors; Tobacco (Smoking)
SITE: Lung
TIME: 1980 - 1993

380   Lam, T.H. 03287
Univ. of Hong Kong, Medical Faculty, Dept. of Community Medicine, Unit for Behavioural Sciences, Li Shu Fan Bldg, 5 Sassoon Rd, Hong Kong, Hong Kong (Tel.: (5)8199280)
COLL: Kleevens, J.W.L.; Kung, I.T.M.; Saw, D.; Lam, W.K.; Lo, K.K.; Hsu, C.; Seneviratne, S.; Lam, S.Y.; Chan, W.C.; Wong, C.M.; Cheang, J.; Chan, J.

Lung Cancer in Chinese Women in Hong Kong
This study aims (1) to examine whether passive smoking is a likely causal factor in lung cancer in non-smoking Chinese women and, if so, the relationship between passive smoking and histological type; and (2) to study the role of other factors such as cooking habits, residential, occupational and dietary factors. About 400-500 female patients diagnosed clinically with primary lung cancer in the major hospitals in Hong Kong are interviewed and followed up until a definitive pathological diagnosis is made. Patients are accepted as cases only when the specimens are reviewed and confirmed independently by at least three of a panel of six pathologists. Difficult or controversial cases are reviewed by the whole panel. There are two series of controls with the same number in each series as the cases. The first series consists of women matched by age and residence to the cases and they are interviewed by telephone. The second series consists of women in the neighbourhood of the cases, matched by age and interviewed in person. A structured questionnaire is used in the interview. Particular attention is paid to the smoking history of the respondent, husbands, parents and relatives living together as well as workmates. Questions are also directed to obtaining information on residential history, occupational history and exposure to dust and chemicals, past history of respiratory diseases, family history of cancer and lung diseases, fuel and oil used in cooking and consumption of certain food items. Eight hospitals have participated in the study. A paper has been published in Br. J. Cancer 56:673-678, 1987.

TYPE: Case-Control
TERM: Chemical Exposure; Chinese; Cooking Methods; Diet; Dusts; Familial Factors; Female; Histology; Occupation; Passive Smoking
SITE: Lung
TIME: 1983 - 1992

HUNGARY

BUDAPEST

381 Ábrahám, E. 02058
Fövárosi Bajcsy-Zsilinszky Korhaz, Tüdögondozó Intézet, X. Köbányai ut 45, Budapest, Hungary
COLL: Czanik, P.; Dinya, E.; Karácsonyi, L; Sali, A.

Prospective Study to Determine High-Risk Groups for Lung Cancer
Screening of the population between the ages of 40-74, both sexes, in an industrial area of Budapest with 87,000 inhabitants, was started in 1968 with the aim of early detection of pulmonary diseases. Between 1975-1978, at the time of screening, supposed risk factors for lung cancer: smoking, occupational exposure, chronic respiratory complaints, fibrotic lung lesions, were determined in about 30,000 persons, 90-97% of the population group concerned. Until 31 December 1985, 343 people were diagnosed with lung cancer. Mathematical-statistical analysis (with adaptation of the log-linear model) made possible the ranking of the single risk factors for lung cancer both separately and cumulatively. The number of the possible risk factor variants is 180. On the basis of this ranking the population was classified into four groups in respect of lung cancer: risk free, moderate risk, high risk and super-high risk groups. The risk of lung cancer for people belonging to the super-high risk groups is about 20 times higher on average than for those in the risk-free groups. Immunological and bronchopathological examinations aiming to clear up the carcinogenic effects of the single risk-factors are continuing.

TYPE: Cohort
TERM: High-Risk Groups; Premalignant Lesion; Screening; Tobacco (Smoking)
SITE: Lung
TIME: 1968 - 1992

382 Bánóczy, J.E. 04009
Semmelweis Medical School, Dept. of Conservative Dentistry, Mikszáth Kálmán Tér 5, 1088 Budapest VIII, Hungary (Tel.: (1)131854)
COLL: Bosnyák, M.; Rigo, O.

Epidemiological Investigations and Screening of Oral Carcinomas and Precancerous Lesions
Due to the high and still increasing incidence of oral carcinomas and precancerous lesions, a need for new methods of early diagnosis arose, as well as the aim to establish the prevalence and incidence rates of oral tumours and precancerous lesions, and to find the appropriate methods by which the widest range of a given population might be screened regularly. In Hungary, complex multiphasic screening in connection with compulsory lung screening has been practiced for several years. On this basis regular screening examinations were started at the lung screening station of the 6th district, Budapest, including clinical, cytological and histological examinations, and introducing early treatment of detected cases. The sample size will comprise about 20,000 individuals, and it is planned to carry out a longitudinal survey, for a period of 7 to 8 years. Cost-effectiveness will be evaluated.

TYPE: Cohort
TERM: Cost-Benefit Analysis; Cytology; Histology; Premalignant Lesion; Screening; Treatment
SITE: Head and Neck; Oral Cavity
TIME: 1986 - 1996

383 Czeizel, A. 03932
National Inst. of Hygiene, Dept. of Human Genetics and Teratology, Gyáli ut 2-6, 1966 Budapest IX, Hungary (Tel.: 335773; Tlx: 225349 oki h)

Mutagenic Effects of Chemical Poisoning
Foetal death, congenital anomalies, birthweight, etc. in the offspring and chromosome aberrations, SCEs and other mutagenic endpoints in the peripheral lymphocyte cultures of persons surviving from suicide attempt using high doses of chemicals for self-poisoning are studied. The aim is to detect the somatic and germinal mutagenic consequences of high doses of chemicals, e.g. drugs and pesticides. In addition the teratogenic consequences of suicide attempts during pregnancy are examined in the offspring (including childhood tumours, and behavioural development). From 1990 the effects of some antioxidants (natural Vitamin E, and beta carotene) are studied on the chromosome aberrations of peripheral lymphocytes in persons who attempt suicide with high doses of chemicals. Papers have been published in Mutat. Res. 198:255, 1988 (Huang et al.); Arch. Toxicol. 62:1-7, 1988 and in Am. J. Obstet. Gynecol. 161:(2) 497, 1989.

HUNGARY

TYPE: Molecular Epidemiology; Registry
TERM: Chemical Mutagenesis; Chromosome Effects; Congenital Abnormalities; Drugs; Lymphocytes; Mutation, Germinal; Mutation, Somatic; SCE
SITE: Inapplicable
TIME: 1984 - 1992

384 Czeizel, A. 03933
National Inst. of Hygiene, Dept. of Human Genetics and Teratology, Gyáli ut 2-6, 1966 Budapest IX, Hungary (Tel.: 335773; Tlx: 225349 oki h)

Surveillance of Indicator Conditions Caused by Germinal Mutations

New mutations are determined in the first generation of a large population-based database (10.6 million). There is a high ascertainment rate with a total registered number of congenital anomalies exceeding 4.7%. The aim is to improve clinical services for patients with indicator conditions (identification of syndromes, genetic counselling, etc.). Indicator conditions under surveillance include sentinel anomalies (13 sentinel abnormalities and two sentinel childhood tumours: retinoblastoma and Wilm's tumour), Down's syndrome and unidentified multiple congenital abnormalities. The environmental history in families having babies with new mutations i.e. in sporadic cases, is also being obtained. Papers have been published in Mut. Res. 212:3-9, 1989 and 238:87-97, 1990, and in Hum. Genet. 82:359-366, 1989.

TYPE: Molecular Epidemiology; Registry
TERM: Chemical Mutagenesis; Congenital Abnormalities; Down's Syndrome; Environmental Factors; Mutation, Germinal; Radiation, Ionizing
SITE: Inapplicable
TIME: 1980 - 1992

385 Gundy, S. 04404
National Inst. of Hygiene, Dept. of Human Genetics & Teratology, Lab. of Human Mutagenesis, Gyáli ut 2-6, Budapest IX. 1966, Hungary
COLL: Bodrogi, I.; Baki, M.

Chromosomal Changes in Peripheral Blood Lymphocytes of Young Testicular Cancer Patients Treated with Chemotherapy

Lymphocytes of young males aged 20-30 years with primary neoplasms of the testis are examined for chromosomal aberrations and SCEs. Possible relationships will be established between chromosomal changes and the dose (for body weight), type and duration of chemotherapy treatment, including the individual sensitivity of patients, and the repair time of somatic cells and blood transfusions. Extrapolation will be made in the future from somatic cells (peripheral blood lymphocytes) to germinal cell injury. The duration of the study will be at least five years. Peripheral blood lymphocytes are used for microculture methods. Chromosomal preparations are strained with FPG technique. Chromosome and chromatid-type aberrations' distribution of I-II-III cell cycles and the SCE frequency are recorded. The chromosomal changes will be examined before, during and following the different chemotherapy programme. A paper has been published in Neoplasma 36 (4):457-464, 1989.

TYPE: Molecular Epidemiology
TERM: Chemotherapy; Lymphocytes; SCE
SITE: Testis
TIME: 1987 - 1992

MISKOLC

386 Takács, S. 02862
Inst. of Public Health & Epidemiology, Lab. of Community Hygiene, Bacsó Béla u. 12, P.O.Box 186, 3501 Miskolc, Hungary (Tel.: (46)54612; Fax: 58060)
COLL: Ujszászy, L.; Bokros, F.; Tatár, A.; Ferencz, T.

Environmental Exposures of Cases of Digestive and Bladder Cancer

Geographical differences between tumour incidence may be influenced by environmental effects. According to animal experiments and human observations, nitrate/nitrite, N-nitrosamines may play a role. The purpose of the present study is to examine the geographical distribution of the incidence of cancers of the digestive tracts and bladder and determine the level of these environmental exposures. Information on the lifestyle and environment of cancer patients is being obtained by personal interview

HUNGARY

and/or questionnaire and the nitrate, nitrite, and ammonium content of the drinking water usually consumed by these persons, estimated. The cases will be analysed according to histology and localisation of cancer. The control group will be chosen from the lowest incidence areas. In investigations completed in 1982-1983, levels of four trace elements were determined in foodstuffs, drinking water, sediments and human organs (liver, lung, kidney, adrenal glands). Data on 172 persons who died from cancer of various sites were analysed. Trace element concentration was measured in the tumours and in healthy tissue from the same organ as the tumour. Tissue from stomach and colon tumours and normal tissue from regions close to the tumours, taken from 49 patients during operation, was analysed. It was found that copper content was significantly higher in the tumour tissue. Further elements investigated are Al, B, Ba, Cr, Hg, Li, Mo, Ni, Pb, Sr, Se and Ti. The studied subjects are: blood, cerebrospinal- and amniotic fluid.

TYPE: Case-Control
TERM: Environmental Factors; Geographic Factors; Lifestyle; Water
SITE: Bladder; Gastrointestinal
CHEM: Chromium; N-Nitroso Compounds; Nickel; Nitrates; Nitrites; Selenium
TIME: 1979 - 1995

387 Takács, S.   04670
Inst. of Public Health & Epidemiology, Lab. of Community Hygiene, Bacsó Béla u. 12, P.O.Box 186, 3501 Miskolc, Hungary (Tel.: (46)54612; Fax: 58060)
COLL: Radóczy, M.; Déri, Z.

Indoor Radon Concentrations and Lung Cancer
The goal of this investigation is to study the correlation between respiratory cancer occurrence and exposure to radon. The concentration of radon gas is measured in the flats of the diseased and of a reference group. The diseased group is selected from the data base of the oncological screening centre. The first reference group is selected at random from people living in the same area and not suffering from respiratory cancer. Members of the second reference group are selected from an area where the occurrence of respiratory diseases is very low. It is planned to carry out measurements in 200 flats with two detectors in a room and one detector in the soil. In this way the correlation between the radon emanation of the soil and the radon concentration of the living room can be investigated. The detector equipment consists of two alpha trace-detectors, which can detect the total alpha-exposure (radon and thoron) and the alpha exposure solely from radon, so that thoron exposure can be evaluated from the difference between the two. The latest results do not show any direct evidence for a causative relation between radon daughter exposure and respiratory cancer. The study of the relation between haemopoietic tumours (leukaemia) and concentration of indoor radon is now being planned.

TYPE: Case-Control
TERM: Metals; Radiation, Ionizing
SITE: Haemopoietic; Leukaemia; Lung
CHEM: Radon; Thoron
TIME: 1983 - 1995

NYIREGYHÁZA

388 Juhász, L.   00573
County Hosp., Dept. of Oncology, Cancer Registry of the County, Szabolcz-Szatmár-Bereg, Szent István u. 68, 4401 Nyiregyháza, Hungary (Tel.: +36 42 12222/600; Fax: 14392)
COLL: Dauda, G.

Gastric Cancers in the County of Szabolcs-Szatmár-Bereg
This is a study of gastric cancer in a defined county in Hungary. For each newly diagnosed case of gastric cancer two hospital controls are chosen, matched by age and residence, with a view to better characterising high-risk groups. In the second phase of the study 284 cases and 429 controls were interviewed. Questions are asked on food habits and family history. Biopsy and autopsy material from patients is typed according to the Järvi-Laurén system. In 1972-1987, 423 cases were so classified. The distribution was: intestinal 31.2%, diffuse 55.3%, mixed 13.5%, figures which differ from those obtained by others. This information is compared with data on epidemiological factors. Data show that not only stomach cancer but also ulcers occur significantly more often in families of cases than in families of controls. Papers have been published in: DAB: Development of County Cancer Registry and the Use of Data, 1989, pp 96-136.

HUNGARY

TYPE: Case-Control
TERM: Diet; Familial Factors; High-Risk Groups; Histology; Registry
SITE: Stomach
REGI: Szabolcs-Szatmár (Hun)
TIME: 1976 - 1993

389 Juhász, L. 00691
County Hosp., Dept. of Oncology, Cancer Registry of the County, Szabolcz-Szatmár-Bereg, Szent István u. 68, 4401 Nyiregyháza, Hungary (Tel.: +36 42 12222/600; Fax: 14392)

High-Risk Groups in Breast Cancer
For each newly diagnosed breast cancer case, two age- and residence-matched hospital controls are chosen. Questions are asked on demographic variables, reproduction-associated variables and family history. A series of articles relating to this study has been published in Hungarian. In the second phase of the study 531 cases and 390 controls were interviewed. The data confirm the previous finding, namely that breast cancer occurs more often in the families of patients than in controls. No differences were found in the occurrence of other cancers between families of patients and controls, or by degrees of relationship. Prostate cancer showed a clustering in families of patients with breast cancer but not in families of controls. (Familial Cancer. Ist Int. Res. Conf. Basel, 1985. (Eds: Mueller & Weber) Karger, Basel pp. 63-65). A paper has been published in Development of County Cancer Registry and the Use of Data, 1989, pp 137-152.

TYPE: Case-Control
TERM: Familial Factors; High-Risk Groups; Histology; Registry; Reproductive Factors
SITE: Breast (F)
REGI: Szabolcs-Szatmár (Hun)
TIME: 1976 - 1993

ICELAND

REYKJAVIK

390   Hallgrímsson, J.   01818
Univ. of Iceland, Dept. of Pathology, P.O. Box 1465, 121 Reykjavik, Iceland
COLL: Tulinius, H.; Bjarnason, Ó.; Magnússon, B.; Thorhallsson, P.; Geirsson, G.B.; Blondal, H.; Arnorsson, J.V.; Benediktsdottir, K.; Agnarsson, B.A.; Jónasson, J.G.; Isaksson, H.J.; Bjornsson, J.

Histological Classification of Tumours in Iceland according to the WHO Classification

This is a comprehensive study of tumours and tumour-like lesions in Iceland according to the WHO's International Histological Classifications of Tumours. All tumours occurring in the country in the periods 1955-1974 and 1955-1984 will be classified and, when desirable, tumour-like lesions at some sites will be included. This will represent a standard approach that is hoped to yield valuable data on tumours, their incidence and comparisons with other geographical areas. At the same time information will become available on the implementation of the WHO classifications. The WHO has given both scientific and financial support to this work. Results for skin, bone, kidney, upper respiratory tract, cervix, urinary bladder, lung, lymphomas, ovary, breast thyroid, endometrium and testis have been published. Studies of stomach, are near completion. CNS, soft tissue, oral, oesophageal, intestinal, appendiceal, pancreatic, biliary and prostatic tumours are now being studied.

TYPE: Incidence; Methodology
TERM: Autopsy; Biopsy; Classification; Histology
SITE: Appendix; Bile Duct; Colon; Nervous System; Oesophagus; Oral Cavity; Pancreas; Prostate; Small Intestine; Soft Tissue
TIME: 1970 - 1993

* 391   Tryggvadottir, L.   05177
Icelandic Cancer Registry, Icelandic Cancer Society, Skogarhlid 8, Box 5420, 125 Reykjavik, Iceland (Tel.: +354 1 621414; Fax: 621417)
COLL: Tulinius, H.; Olafsdottir, G.

Age at Diagnosis of Familial and Sporadic Breast Cancer

Two related hypotheses are tested: (1) that familial breast cancer cases are, on average, younger at diagnosis than sporadic cases; and (2) that this relationship is obscured for recently diagnosed cases, because in this group the probability of having a first degree relative with diagnosed breast cancer will be higher with increasing age of the woman. This effect should diminish with time since diagnosis. A familial case is defined as a woman who has a first degree relative with breast cancer. Information on families of breast cancer patients comes from the Icelandic Cancer Registry, which is population-based, and contains extensive data on families of about half of all breast cancer cases diagnosed in the country since 1910, around 1000 cases. It is planned to use multiple regression analysis with age at diagnosis as the dependent variable and year of diagnosis and familiality as independent variables.

TYPE: Incidence
TERM: Age; Familial Factors; Registry
SITE: Breast (F)
REGI: Iceland (Ice)
TIME: 1983 - 1992

* 392   Tryggvadottir, L.   05178
Icelandic Cancer Registry, Icelandic Cancer Society, Skogarhlid 8, Box 5420, 125 Reykjavik, Iceland (Tel.: +354 1 621414; Fax: 621417)
COLL: Tulinius, H.; Sigurdsson, K.; Larusdottir, M.K.; Johannesson, B.

Total Number of Menstrual Cycles and the Risk of Breast Cancer

The hypothesis of this study is that: a woman's probability of being diagnosed with breast cancer increases with the total number of ovulatory menstrual cycles. Information on length of menstrual cycles, age at menarche and menopause, number of completed pregnancies and abortions, total duration of lactation and use of oral contraceptives comes from an Icelandic databank. The data has been gathered as part of population-based cervical cancer screening in Iceland, via interviewer-administered questionnaires. Over 45,000 women have answered these questions since 1979. Around 900 of these women have been diagnosed with breast cancer. The design is a nested case-control study.

ICELAND

TYPE: Case-Control; Cohort
TERM: Abortion; Lactation; Menarche; Menopause; Menstruation; Oral Contraceptives; Pregnancy; Registry
SITE: Breast (F)
REGI: Iceland (Ice)
TIME: 1990 - 1992

* 393   Tryggvadottir, L.   05179
Icelandic Cancer Registry, Icelandic Cancer Society, Skogarhlid 8, Box 5420 , 125 Reykjavik, Iceland (Tel.: +354 1 621414; Fax: 621417)
COLL: Tulinius, H.; Olafsdottir, G.

Linkage of a Breast Cancer Susceptibility Locus to the ABO Locus

The aim of this study is to test the hypothesis that a breast cancer susceptibility gene is linked to the ABO locus. Information on families of breast cancer patients comes from the population-based Icelandic Cancer Registry, which contains extensive data on families of about half of all breast cancer cases diagnosed in the country since 1910 around 1000 cases. ABO blood group information is obtained by record linkage with data from the Icelandic Blood Bank and from hospitals. Linkage analysis will be carried out to evaluate the association of breast cancer risk with blood group.

TYPE: Genetic Epidemiology
TERM: Blood Group; Record Linkage; Registry
SITE: Breast (F)
REGI: Iceland (Ice)
TIME: 1988 - 1992

INDIA

AHMEDABAD

394 Adhvaryu, S.G. 03595
Gujarat Cancer and Research Inst., Dept. of Cancer Biology, New Civil Hospital Campus, Ahmedabad 380016, India (Tel.: 378454/378459)
COLL: Dave, B.J.; Trivedi, A.H.

Cytogenetic Studies in Individuals Chewing Tobacco and Areca Nut

The habit of chewing tobacco in combination with areca nut and/or lime is a very common practice in South East Asia. The habit has been shown to be strongly associated with cancer of the oral cavity, which is the predominant type of cancer among males in India. Since not all tobacco chewers develop oral cancer, the individual's genome might have an important role in susceptibility. The aim is to develop parameters which may prove useful in identifying individuals prone to tobacco induced genomic damage. Cytogenetic parameters such as C-band heteromorphism in chromosomes 1, 9 and 16, spontaneous and mutagen induced SCE frequencies and chromosome aberrations in lymphocytes as well as frequency of micronucleated cells in exfoliated buccal mucosa cells are being studied in the following three groups: (1) normal healthy individuals, not consuming tobacco in any form, (2) individuals chewing tobacco for the last two years at least, but without morphological alterations in the mucosa and (3) individuals with sub-mucous fibrosis, leukoplakia or oral cancer.

TYPE: Molecular Epidemiology
TERM: Areca Nut; Chromosome Effects; Genetic Factors; Mutation, Somatic; SCE; Tobacco (Chewing)
SITE: Oral Cavity
TIME: 1984 - 1993

395 Balar, D.B. 04394
Gujarat Cancer & Research Inst., Dept. of Pathology, New Civil Hospital Campus, Asarwa Ahmedabad 380016, India (Tel.: 377463; Tlx: 121-680 gcri in)
COLL: Patel, T.B.; Patel, R.D.

Role of HPV Infection in Cervical Cancer

HPVs have been suggested as aetiological agents in cancer of the cervix. Delineation of disease patterns has been hindered by the long disease latency: based on earlier studies of known aetiological factors, i.e. age at marriage, duration of sexual activity and multiplicity of partners, an incubation period of 20-25 years has been suggested. In 1986, 711 cases of cervical cancer, all in married women of low socio-economic status, were seen at the M.P. Shah Cancer Hospital. 14 cases were under age 30, and for the rest age at marriage was between 20 and 30 years, giving a latent period of 20 to 30 years between beginning sexual activity and onset of the disease. All cases of cervical cancer attending the hospital for a period of three years from 1988 to 1990 will be studied to test the hypothesis of a possible role of HPV. Factors to be examined include age at marriage, details of sexual habits, age at disease onset, and personal hygiene. Immunohistochemical techniques will be used to study the strain of papilloma virus present in the cervical biopsies of patients to assess the role of HPV infection.

TYPE: Case Series
TERM: HPV; Infection; Sexual Activity
SITE: Uterus (Cervix)
TIME: 1985 - 1992

396 Bhatavdekar, J.M. 04262
Gujarat Cancer & Research Inst., Div. of Endocrinology, Dept. of Cancer Biology, New Civil Hosp. Campus, Asarwa Ahmedabad 380016, India (Tel.: 37845159)
COLL: Balar, D.B.; Patel, T.B.

Risk Factors for Epithelial Ovarian Carcinoma and their Relationship to Oestrogen- and Progesterone-Receptor Status

Epithelial ovarian cancer is the third most frequently observed malignancy in the state of Gujarat. The risk factors for this malignancy (age at menarche, menstruation, parity, abortions, age at first full term pregnancy, age at menopause, family history of ovarian cancer) will be compared with hormonal and oestrogen- and progesterone-receptor status in approximately 150 histologically confirmed epithelial ovarian cancer patients and in 60 healthy controls matched for age and parity. The aim is to determine if any association exists between the aetiological risk factors, hormones and oestrogen- and

INDIA

progesterone-receptor status. Preliminary data suggest a higher occurrence of epithelial ovarian cancer in women with more than four children, abnormal levels of pituitary gonadotrophins, and lower socio-economic group (income less than Rs. 500/-per month).

TYPE: Case-Control
TERM: Abortion; Hormones; Menarche; Menopause; Menstruation; Parity; Socio-Economic Factors
SITE: Ovary
TIME: 1987 - 1992

397 Bhatavdekar, J.M. 04369
Gujarat Cancer & Research Inst., Div. of Endocrinology, Dept. of Cancer Biology, New Civil Hosp. Campus, Asarwa Ahmedabad 380016, India (Tel.: 37845159)
COLL: Balar, D.B.; Patel, T.B.

Risk Factors in Oropharyngeal Cancer and Sialoglycoprotein Levels

Oropharyngeal cancer is the number one malignancy in males in Gujarat. The risk factors for oropharyngeal carcinoma are e.g. ulcers in the oral cavity, tobacco chewing, snuffing, betel quid consumption, cigarette or bidi (indigenous cigarette) smoking and alcohol consumption. In women, habits of tobacco snuff inhalation and use of tobacco powder for cleaning teeth have also been found to be associated in some cases with oropharyngeal cancer. The aim of this study is to determine if any association exists between the risk factors and abnormal levels of protein-bound-sialic acid, lipid-soluble-sialic acid and free sialic acid in oropharyngeal carcinoma patients. Sialoglycoproteins will be measured in 100 oropharyngeal cancer patients and 100 age-matched healthy controls who will be followed for two years.

TYPE: Case-Control
TERM: Alcohol; Betel (Chewing); Biochemical Markers; Tobacco (Chewing); Tobacco (Smoking)
SITE: Oropharynx
TIME: 1987 - 1992

* 398 Bhatavdekar, J.M. 04955
Gujarat Cancer & Research Inst., Div. of Endocrinology, Dept. of Cancer Biology, New Civil Hosp. Campus, Asarwa Ahmedabad 380016, India (Tel.: 37845159)
COLL: Shah, N.G.; Kapadia, A.; Giri, D.D.; Balar, D.B.; Patel, N.L.

Epidemiological and Endocrinological Study of Ovarian Cancer

Only a few studies of hormone concentrations in ovarian cancer have been carried out. Results for breast cancer indicate that hormonal abnormalities may also precede and presumably favour the onset of epithelial ovarian cancer or they may be related to the evolution of the disease. Various endocrinological aspects of epithelial ovarian cancer will be examined in 100 Indian females. At the same time, known epidemiological risk factors such as age at menarche, parity, age at first full term birth, dietary habits, socio-economic status, and lifestyle, will be assessed to permit examination of the possible impact of the various risk factors on endocrinological and biological parameters. The endocrine parameters luteinizing hormone, follicle stimulating hormone, prolactin, oestradiol, progesterone, testosterone, dehydroepiandrosterone sulphate, androstenedione, etc. will be studied by RIA.

TYPE: Case Series
TERM: Diet; Hormones; Lifestyle; Menarche; Parity; Socio-Economic Factors
SITE: Ovary
TIME: 1991 - 1993

* 399 Giri, D.D. 04960
The Gujarat Cancer & Research Inst., New Civil Hosp. Campus, Asarwa Ahmedabad 380016, India (Tel.: 37845459; Tlx: 121 6680 gcri in)
COLL: Bhatavdekar, J.M.; Balar, D.B.

Biochemical Characteristics in Female Breast Cancer

Breast cancer ranks second among Indian female cancers (age standardised rate approximately 20/100,000), with indications that incidence is increasing. Hyperprolactinaemia has been found in breast cancer patients at this Institute, a finding not always reflected in studies published in the West. The aim of this case-control study is to examine various risk factors, i.e., age at menarche, parity, age at first birth, dietary habits, height, weight, socio-economic status, etc., among 200 breast cancer patients and controls belonging to the various socio-religious communities which make up the Gujarat population. The study will also investigate the steroid and peptide hormone profile, as well as the steroid receptor expression patterns in patients belonging to the various communities in an attempt to better understand the epidemiology of breast cancer in this region.

TYPE: Case-Control
TERM: Biochemical Markers; Diet; Hormones; Menarche; Parity; Physical Factors; Socio-Economic Factors
SITE: Breast (F)
TIME: 1991 - 1994

400   Patel, T.B. 02856
Gujarat Cancer & Research Inst., Div. Epidemiology & Biostatistics, M.P. Shah Hosp., New Civil Hosp. Campus, Asarwa Ahmedabad 380016, India (Tel.: 37845459; Tlx: 121 6680 gcri in)
COLL: Bhaduri, A.; Patel, D.; Balar, D.B.; Patel, N.L.; Patel, R.D.; Kapadia, A.

Epidemiology of Breast and Cervix Cancer in Gujarat

Both breast and cervical cancer are frequent in Gujarat State. The main objective of the project is to study the effect of income, age and parity on the occurrence of breast and cervical cancer. Analysis of preliminary data for 1980-1984 revealed the maximum number of cervical cancer cases in women in the 40-49 years age group (35.46%) in the income group < 500 rupees per month and with four or more children. This observation has been corroborated by a further study of 284 cases in 1985-1986, in which 60.91% cases were in the group with low income and four or more children. 77% of the stage III and IV tumours were in women with four or more children. Change of sex partners is extremely uncommon in this part of the country. Breast cancer was most frequent in the 40-49 years age group (35.46%). As compared to cancer of the cervix, twice as many cases of breast cancer were seen in nulliparous women. Only 7.37% cases were observed in women with more than four children and in income group > 500 rupees per month while 44.9% of cases were in females of income group < 500 rupees per month and having four or fewer children. 54.5% women with four or fewer children were seen to have stage III and IV disease at presentation. The study is on-going to correlate dietary history and age at first marriage with the occurrence of the tumour.

TYPE: Case Series
TERM: Age; High-Risk Groups; Marital Status; Parity; Sexual Activity; Socio-Economic Factors
SITE: Breast (F); Uterus (Cervix)
TIME: 1980 - 1992

401   Patel, T.B. 03815
Gujarat Cancer & Research Inst., Div. Epidemiology & Biostatistics, M.P. Shah Hosp., New Civil Hosp. Campus, Asarwa Ahmedabad 380016, India (Tel.: 37845459; Tlx: 121 6680 gcri in)
COLL: Balar, D.B.; Bhaduri, A.; Patel, D.; Patel, R.D.

Histology Correlated to Occupational Exposure and Smoking

Cancer of the lung is observed in both sexes in Gujarat State, although it is more frequent in males. Tobacco smoking has a close association with lung cancer in males, while alcohol is assuming a very limited role. The most frequently observed histological subtype was epidermoid carcinoma, forming 76% of cases of which all were active smokers of more than 10 bidis (indigenous cigarettes) per day. The age group with the maximum number of cases was 35-64 years (76%) with a peak at 50-59 years. The male: female ratio was 11:1 in this age group. A larger number of cases were from rural areas. The role of passive smoking and occupational risk in relation to specific histological types will be investigated.

TYPE: Case Series
TERM: Chemical Exposure; Occupation; Passive Smoking; Tobacco (Smoking)
SITE: Lung
TIME: 1983 - 1992

* 402   Patel, T.B. 04968
Gujarat Cancer & Research Inst., Div. Epidemiology & Biostatistics, M.P. Shah Hosp., New Civil Hosp. Campus, Asarwa Ahmedabad 380016, India (Tel.: 37845459; Tlx: 121 6680 gcri in)
COLL: Patel, N.L.; Patel, D.; Balar, D.B.; Giri, D.D.

Nasopharyngeal Cancer (Lympho-epithelioma) in West India

During 1986 to 1990, 37 cases of nasopharyngeal carcinoma were treated at the Institute. 67% were children or adolescents (under 20 years of age). 33 cases came from various districts of Gujarat and four from the province of Rajasthan. This study aims to detect the presence, if any, of regional or familial clustering of the disease among cases already registered in the Institute and normally residing in Gujarat

INDIA

State. A second objective is to examine the association of EBV and lympho-epithelioma in this population, and to look for expression of epithelial/lymphoid and other transfers by the tumour cells to understand the cell biology. Immunohistochemical methods will be used to detect epithelial and lymphoid cell markers. In situ hybridization techniques will be used to localise EBV DNA on archival material.

TYPE: Case Series
TERM: Cluster; EBV
SITE: Nasopharynx
TIME: 1991 – 1994

BOMBAY

* 403   Desai, P.B.   05106
Tata Memorial Hosp., Dr E. Borges Marg, Bombay 400012, India (Tel.: (022)4129750; Fax: 4129937 Tlx: 01173649 TMC IN)
COLL: Rao, D.N.

Epidemiology of Childhood Cancer

This is a continuing study on the epidemiology of childhood cancer as seen in India at a large referral centre like the Tata Memorial Centre, Bombay, over a 20-year period (1970-1990), with a hospital-based cancer registry operating alongside a population-based cancer registry. The predominance of retinoblastoma and the differences in the pattern of Hodgkin's disease in the material (histological type) may offer a clue to variations in risk and related aetiological agents. Paediatric cancer cases will be analysed for factors like cell type, hereditary and genetic factors, past history of parental exposure to radiation and identification of high-risk groups in the community. Availability of incidence data will provide an opportunity for international comparison in the distribution of paediatric cancers. Over 9,000 cases are expected to be included in this study.

TYPE: Case Series
TERM: Genetic Factors; Heredity; High-Risk Groups; Radiation, Ionizing
SITE: Childhood Neoplasms
TIME: 1980 – 1992

404   Gupta, P.C.   03327
Tata Inst. of Fundamental Research, Basic Dental Research Unit, Homi Bhabha Rd , Bombay 400005, India (Tel.: 219111; Tlx: (011)3009)
COLL: Mehta, F.S.; Pindborg, J.J.

Mortality Experience in Relation to Tobacco Chewing and Smoking Habits

Chewing of tobacco with lime, betel leaf, areca nut and smoking of bidis, chuttas etc. is practised widely in India. Recent studies have shown that the age-adjusted mortality among bidi smokers is greater than among tobacco chewers and mortality among betel tobacco chewers, bidi smokers and chutta smokers is significantly higher compared to non-users of tobacco. In the present study attempts are being made to determine the possible causes for this excess mortality. In a house-to-house survey, a sample of 36,000 tobacco users (age 15 years and over) in three districts of India is being interviewed annually and examined for the presence of oral cancer. Since there is no death certification system in rural India and often expert medical help is neither sought nor available, it is difficult in general to find out the precise cause of death. Therefore a list of various possible symptoms is prepared and information on symptoms before death is being collected from the next-of-kin. Similar information is also being collected from a control group of non-users of tobacco.

TYPE: Cohort
TERM: Betel (Chewing); Tobacco (Chewing); Tobacco (Smoking)
SITE: Oral Cavity
TIME: 1977 – 1992

405   Gupta, P.C.   04535
Tata Inst. of Fundamental Research, Basic Dental Research Unit, Homi Bhabha Rd , Bombay 400005, India (Tel.: 219111; Tlx: (011)3009)
COLL: Mehta, F.S.; Pindborg, J.J.

INDIA

Cancer Risk Following Premalignant Oral Lesions
Certain oral mucosal lesions such as leukoplakia, the palatal changes associated with reverse smoking and conditions such as submucous fibrosis are known to increase the risk of development of oral cancer. Precise relative risk estimates are however not yet available. One complicating factor is a strong association of tobacco chewing and smoking with oral cancer, as well as with precancerous lesions and conditions. In this study, cohorts consisting of 36,000 tobacco chewers and smokers are being followed annually over a period of ten years in three areas of India, by house-to-house visits. These individuals are examined every year by dentists and the clinical diagnosis is carefully documented. The diagnosis of oral cancer is histologically confirmed. This study will provide incidence rates of oral cancer among those with specific types of oral precancerous lesions and conditions and also among those without any preceding oral precancerous lesions or conditions. The incidence rates will be calculated by the method of person-years and will be adjusted for age, sex and the type of tobacco usage. Possible effects of changes in tobacco habits would also be studied.

TYPE: Cohort
TERM: Betel (Chewing); Premalignant Lesion; Tobacco (Chewing); Tobacco (Smoking)
SITE: Oral Cavity
TIME: 1977 - 1992

* 406   Jayant, K.   05024
Tata Memorial Center, Cancer Research Int., Epidemiology Unit, Dr Ernest Borges Marg, Bombay 400012, India
COLL: Gulati, S.S.; Notani, P.N.; Kamat, M.R.

Role of Penile Hygiene in the Aetiology of Cancer of the Cervix Uteri
A study of differential cervical cancer rates in the various religious groups in Bombay showed that the combined effect of two risk factors, early age at first coitus and poor penile hygiene, was the sum of their separate independent effects (Br. J. Cancer 56:685, 1987). To confirm this finding, a case-control study has been initiated. Cervical cancer patients from Bombay attending the Tata Memorial Hospital, as well as age-, community- and social class-matched controls from the general population are interviewed by a trained medical social worker. The questionnaire includes questions on tobacco habits, besides those on sexual activity and hygiene, for patients, controls and partners. To date, data on 60 cases and an equal number of controls have been collected. When about 200 cases have been accumulated, the data will be analysed using univariate and multivariate methods.

TYPE: Case-Control
TERM: Hygiene; Sexual Activity
SITE: Uterus (Cervix)
TIME: 1988 - 1992

407   Jussawalla, D.J.   04458
Bombay Cancer Registry, Indian Cancer Society, Lady Ratan Tata Medical & Res. Center, M. Karve Rd, Bombay 400021, India (Tel.: 2047436)
COLL: Yeole, B.B.; Jayant, K.

Survival of Breast Cancer Patients in Bombay
Data collected by the Bombay Cancer Registry on breast cancer patients for the years 1973 to 1982 and death records maintained by the Department of Vital Statistics of Bombay Municipal Corporation for the years 1973-1987 will be reviewed to study the survival rates. Firstly incident cases will be matched with the death certificates maintained by the Municipal Corporation. The vital status of the remaining unmatched patients will be ascertained either by post or by home visits, and five and 10-year relative survival rates will be calculated.

TYPE: Incidence
TERM: Registry; Survival
SITE: Breast (F)
REGI: Bombay (Ind)
TIME: 1989 - 1992

408   Mehta, F.S.   01032
Tata Inst. of Fundamental Research, Basic Dental Research Unit, Homi Bhabha Rd, Bombay 400005, India (Tel.: 219111; Tlx: 011-3009)
COLL: Pindborg, J.J.; Gupta, P.C.; Aghi, M.B.; Daftary, D.K.

INDIA

Oral Cancer and Precancerous Lesions in Relation to Tobacco Use in Rural Indian Populations
Tobacco usage as practised in different parts of India is associated with oral cancer and precancerous lesions. This study aims to demonstrate a reduction in the incidence of precancerous lesions when tobacco habits are discontinued. A sample of 12,000 individuals with tobacco habits has been examined and interviewed in each of three rural areas. Villagers are being exposed to a concentrated programme of education, motivation and guidance. These individuals are re-examined annually to assess changes in tobacco habits, development of new oral precancerous lesions and changes in previously diagnosed lesions. Preliminary results showed that among the individuals who stopped or substantially reduced their tobacco habits, the regression of leukoplakia was significantly higher and the incidence of leukoplakia significantly lower (Lancet.i: 1235-1238, 1986). The health education was of significant help in stopping the tobacco habits of the individuals (IARC Sci. Publ. 74) 307-17, 1986) and of special help to difficult subgroups like tobacco chewers (Am. Publ. Health 76:709, 1986). The tenth follow-up examination is complete and the data are being analysed.

TYPE: Cross-Sectional
TERM: Premalignant Lesion; Prevention; Tobacco (Chewing); Tobacco (Smoking)
SITE: Oral Cavity
TIME: 1977 - 1992

* 409 Notani, P.N. 05077
Tata Memorial Centre, Cancer Research Inst., Epidemiology Unit., Dr E. Borges Marg, Bombay 400012, India
COLL: Jayant, K.

Role of Diet in Cancer of the Female Breast
Nutritional factors, in particular fat intake, appear to play an important role in endocrine-dependent cancers. A case-control study has been initiated to investigate the possible relationship between dietary factors and cancer of the female breast, taking into account the other risk factors related with reproductive history, obesity, personal habits (tobacco, alcohol) and family history of cancer. A questionnaire is administered to patients attending the Tata Memorial Hospital, which is the largest cancer treatment facility in the region. A first control group is obtained from patients who are diagnosed as not having cancerous or precancerous lesions. A second group is composed of individuals from the general population. To date, a little over 200 cases and around 100 each of hospital and population controls have been interviewed. Recruitment of cases and controls is in progress.

TYPE: Case-Control
TERM: Alcohol; Diet; Obesity; Reproductive Factors; Tobacco (Smoking)
SITE: Breast (F)
TIME: 1988 - 1993

* 410 Rao, N.D. 05124
Tata Memorial Hosp., Dr E. Borges Marg, Bombay 400012, India (Tel.: 4129761; Fax: 4129937; Tlx: 01173649 TMC IN)
COLL: Desai, P.B.

Digestive Tract Cancers in Western India
The aim of this hospital-based case-control study is to identify high-risk groups for gastrointestinal cancer. Available data indicate that cancer of the oesophagus has a high incidence in northern India, whereas stomach cancer incidence is high in the south. The incidence of colon cancer is low but it is likely to increase due to changes in lifestyle. Patients will be interviewed by social investigators for habits and customs, dietary practices and other related information before being examined by the clinician. Four major gastrointestinal gastrointestinal cancers - oesophagus, stomach, colon and rectum will be included. Over 2,000 cases and 4,000 control patients are expected. Stratified analysis will be carried out mainly to test the hypothesis of a protective effect of vegetarian diet, and the additional influence of alcohol and tobacco. It is proposed to stratify for age, sex and residence (Bombay, rest of Maharashtra state and others). Relative risk for each of the factors will be calculated by using test-based estimation procedure. Patient accrual will be completed by 1991.

INDIA

TYPE: Case-Control
TERM: Alcohol; Diet; Tobacco (Smoking)
SITE: Colon; Oesophagus; Rectum; Stomach
TIME: 1987 - 1992

* 411   Yeole, B.B.   05098
Indian Cancer Society, Bombay Cancer Registry, 74 Jerbai Wadia Rd, Parel, Bombay 400012, India (Tel.: 4121519)
COLL: Jayant, K.; Jussawalla, D.J.; Natekar, M.V.

Trends in Tobacco-Related Cancers in Females of Greater Bombay

Upper alimentary and respiratory cancers are known to be aetiologically associated with the habit of chewing or smoking tobacco. Data collected by the Bombay Cancer Registry for the period 1964 to 1987 will be used to study incidence trends for cancers of the tongue, buccal mucosa, oropharynx, hypopharynx, larynx and lung. Differences in cancer incidence for these sites will be examined, taking into consideration the pattern of tobacco habits in the different population groups. Log-linear regression models will be used to examine trends in the crude, age-adjusted and age-specific incidence rates.

TYPE: Incidence
TERM: Female; Registry; Tobacco (Chewing); Tobacco (Smoking); Trends
SITE: Hypopharynx; Larynx; Lung; Oral Cavity; Oropharynx; Tongue
REGI: Bombay (Ind)
TIME: 1991 - 1992

* 412   Yeole, B.B.   05100
Indian Cancer Society, Bombay Cancer Registry, 74 Jerbai Wadia Rd, Parel, Bombay 400012, India (Tel.: 4121519)
COLL: Jayant, K.; Jussawalla, D.J.; Kadam, V.T.

Cancer in the Gujarati Population in Bombay

The Gujarati population in Bombay comprises Hindu, Muslim, Jain and Parsi religious groups. Amongst these different groups, there is wide variation in habits, customs and socio-economic status. Striking differences in the relative frequency of cancer at various sites are also observed. The site pattern of cancer in this community will be assessed by religion and the magnitude and nature of differences observed will be examined to assess the extent to which these could be ascribed to variations noted in the different lifestyles. The basic data for the study will be obtained from the Bombay Cancer Registry for the period 1986-1988. As Indian Census figures do not provide information on population by mother tongue and age, age-standardised cancer ratios (ASCAR) will be calculated.

TYPE: Incidence
TERM: Lifestyle; Registry; Religion; Socio-Economic Factors
SITE: All Sites
REGI: Bombay (Ind)
TIME: 1991 - 1992

GOA

413   Vaidya, S.G.   04572
Goa Cancer Society, Dr E. Borges Rd, Dona Paula Goa, India (Tel.: 3426/5884)
COLL: Vaidya, N.S.; Kamat, V.; Shetye, S.B.

Assessment of the Efficacy of an Anti-Tobacco Community Education Programme

A project is under way in Goa to assess the efficacy of an anti-smoking and anti-tobacco community education programme. The aim of the project is to test the effectiveness of educating the community through school children. The study population is the rural population of about 685,000 (1981 census), divided into three zones, two experimental and one control. About 10,000 individuals aged 15 and above are selected at random in each zone by a two-stage sampling procedure (villages, individuals). The selected individuals are interviewed and examined by dental surgeons and researchers for tobacco habits and oral precancerous lesions. Since the first survey was carried out (1986-1988), education programmes have been introduced in the experimental zones. In one zone, education is carried out through schools and school children, and in the other through schools, primary health centres and anganwadis (centres for health care of pre-school children and their mothers, run by the Department of Social Welfare of the Government). Baseline results among the 14,363 males interviewed were 30.2%

INDIA

smokers and 3.5% using smokeless tobacco, and among the 15,350 females, 7.4% smokers and 13.1% using smokeless tobacco. 21.8% of the 4,829 male tobacco habituees, have oral precancerous lesions and 12.3% of the 3,151 female habituees. Efficacy of the intervention programmes will be tested after three surveys.

TYPE: Intervention
TERM: Premalignant Lesion; Prevention; Tobacco (Chewing); Tobacco (Smoking)
SITE: Oral Cavity
TIME: 1986 - 1992

LUCKNOW

414   Kushwaha, M.S.   03340
King Georges Medical College, Postgrad. Dept. Pathol. & Bacteriology, Lymphoma & Leukaemia Registry, Lucknow 226003, India (Tel.: 82554)
COLL: Misra, N.C.; Nath, P.; Misra, P.K.; Jaiswal, M.S.D.; Kumar, A.

Lymphoma-Leukaemia Registry

A Lymphoma-Leukaemia Registry was established in 1971. More than 1,000 cases of lymphomas including Hodgkin's disease, and more than 1,000 cases of all types of leukaemias have been recorded. The centre records diagnosis, clinical, haematological and histopathological data, therapy and follow-up of the cases encountered in the city of Lucknow and province of Uttar Pradesh, with a population of nearly 110 million. An attempt is being made to identify factor(s) responsible for the genesis and promotion of these diseases through family and environmental histories. Morphological findings are also being correlated with the biological behaviour of the disease. For the last four years multiple myeloma has been regularly registered and studied in a similar manner. So far 70 cases are on record. Cytochemical and immunological marker studies are done in cases of lymphoproliferative disorders. The centre also provides teaching and research facilities for undergraduates, postgraduates and young scientists, acts as a referral service for expert opinion in the field, and investigates various problems associated with these diseases. A paper has been published in Leukem. Res. 9(6):799-802, 1985.

TYPE: Registry
TERM: Biochemical Markers; Environmental Factors; Histology
SITE: Hodgkin's Disease; Leukaemia; Lymphoma; Multiple Myeloma
TIME: 1985 - 1993

MADRAS

415   Bharati Arumugam, S.   03319
Madras Medical College, Inst. of Pathology, & Electron Microscopy, Park Town, Madras 600 003, India (Tel.: 30001, 39181)
COLL: Narendran, P.; Kanaka, T.S.; Balakrishnan, T.S.; Logamuthukrishnan, T.; Kalyanaraman, S.; Ramamurthy, B.; Reginald, S.; Arumugam, S.; Govindan, S.

Brain Tumours in India

The main object of the study is to assess the frequency of occurrence and type of brain tumours in India. The project includes the preparation of statistics on the occurrence of these tumours in studies reported by neuropathologists in various parts of India and the establishment of a Tamil Nadu State neuropathology registry, including histopathology. Since the inception of the registry in 1972, a total of nearly 4,300 intracranial solid tumours have been diagnosed by one neuropathologist. The slides, paraffin blocks and tissue material are available for more extensive studies.

TYPE: Registry
TERM: Classification; Histology
SITE: Brain
TIME: 1972 - 1993

416   Chittukadu, G.K.   04611
Cancer Inst., Epidemiology Unit, Canal Bank Rd, Madras 600020, India (Tel.: 412714, 412185)
COLL: Shanta, V.

INDIA

Breast Cancer in Madras, India
Breast cancer is the second most important cancer site in Madras. This study will examine the hypothesis that (1) premenopausal and postmenopausal women have different risk factors; (2) that age at first child birth greater than 25 years confers a higher risk in postmenopausal women; and (3) that premenopausal women with less than four children or nulliparous women a have higher risk. A case-control study will be done, using all breast cancer cases (about 1,000) registered in the population-based cancer registry from 1989 to 1992, and two groups of controls for each case. The first group will consist of cancer patients who have no disease in the breast, gynaecological organs or endocrine glands (all sites except ICD-O 174, 180-184, 193 & 194); the second group will be patients who attend the general out-patient department, without cancer. Information will be obtained on age, sex, religion, education, income, hypertension, diabetes, age at menarche, age at first child birth, number of children, age at last child birth, age at menopause, history of hysterectomy and oophorectomy and use of contraceptives.

TYPE: Case-Control
TERM: Contraception; Diabetes; Menarche; Menopause; Parity; Registry; Religion; Socio-Economic Factors
SITE: Breast (F)
REGI: Madras (Ind)
TIME: 1989 - 1992

* 417   Shanta, V.   05155
Madras Metropolitan Tumor Registry, Cancer Inst. (WIA), Canal Bank Rd , Adyar, Tamil Nadu, Madras 600020, India (Tel.: 412714; Fax: 412185)
COLL: Vasanthi, L.; Swaminathan, R.; Chacko, P.

Childhood Leukaemia and Lymphoma
This retrospective survey will establish the epidemiological profile of childhood leukaemias and lymphoma. Case records of children admitted between 1984-1990 with lymphoma (n = 141) or leukaemia (n = 174) will serve as the source of data. Demographic, environmental, socio-economic and cultural particulars of the study subjects and their families will be obtained. Medical history of the study subjects from the time of conception to presentation with clinical features will be abstracted, with particular reference to parental consanguinity and age at the time of pregnancy, birth order of the patients, growth and developmental mile-stones and childhood infections. Relevant details about the siblings of the study subjects will also be obtained. Clinical profiles of the study subjects will be collected to find out the usual modes of presentation to the physician.

TYPE: Case Series
TERM: Birth Order; Childhood; Clinical Records; Environmental Factors; Infection; Sib; Socio-Economic Factors
SITE: Leukaemia; Lymphoma
TIME: 1991 - 1992

NEW DELHI

418   Luthra, U.K.   03737
Indian Council of Medical Research, Cytology Research Centre, Maulana Azad Medical College Campus, Bahadur Shah Zafar Marg, New Delhi 110002, India (Tel.: 3311889, 3319271)
COLL: Das, D.K.; Gupta, M.M.; Seghal, A.; Murthy, N.S.; Singh, V.; Bhatnagar, P.; Sharma, B.K.; Das, B.C.; Agarwal, S.S.; Bhambhani, S.; Sodhani, P.; Menon, R.; Kumar, D.; Dutta, S.; Sharma, J.K.; Kashyap, V.; Gupta, S.; Attam, K.; Singh, K.; Singh, M.; Chadha, B.; Chadha, P.

Uterine Cervical Dysplasia
The biological behaviour of precancerous and early cancerous lesions of the cervix is not clearly understood. Identification of relevant risk factors and detection and management of such lesions are important in the prevention and management of invasive cancer of the cervix. A prospective multi-disciplinary study involving various parameters such as clinical, epidemiological, cytopathological, colposcopic, virological, cytogenetic, immunological and ultrastructural studies on precancerous and early cancerous lesions of the cervix has been underway at the Cytology Research Centre, New Delhi, India since 1976 with the aim of understanding the biological behaviour of these lesions of cervix, and of discovering relevant risk factors, particularly the delineation of possible high risk dysplasia from relatively low risk dysplasia. This prospective cohort study proposes to register 1,000 dysplasias with twice the number of controls, one group of controls matched for age and parity and the other for age only and follow them for 5-15 years. The virological parameter includes sero-epidemiology of HSV and

INDIA

detection of HPV in cervical smears as well as in the tissues. The cytogenetic component includes study of chromatid exchange, silver stained nucleolus organizing regions and chromosomal aberrations. At present a cohort of 950 dysplasias with matched controls is being followed with a view to determining the importance of the above factors in progression or regression of the dysplastic lesions.

TYPE: Cohort
TERM: Chromosome Effects; Cytology; HPV; HSV; Premalignant Lesion; Screening; Sero-Epidemiology
SITE: Uterus (Cervix)
TIME: 1976 - 1992

* 419 Luthra, U.K. 05168
Indian Council of Medical Research, Cytology Research Centre, Maulana Azad Medical College Campus, Bahadur Shah Zafar Marg, New Delhi 110002, India (Tel.: 3311889, 3319271)
COLL: Kumar, D.; Dutta, S.; Sharma, J.K.; Kashyap, V.; Chadha, P.; Chadha, B.; Gupta, S.; Attam, K.; Sunderwa, J.; Singh, K.; Singh, M.; Padubidri

Uterine Cervical Dysplasia

A study carried out here during 1976-1988 highlighted some socio-demographic and biological parameters as important risk factors for the development of cervical dysplasia and its progression to cervical cancer. Nutritional studies and molecular techniques for analysis of HPV genotypes were introduced towards the end of the study. This study was initiated from 1988 to elucidate the role of HPV along with other possibly preventable co-factors in the process of cervical carcinogenesis and management of precancerous and early cancerous lesions of the cervix. It is proposed to register a cohort of 400 mild-to-moderate dysplasia subjects and 800 age-matched controls, along with their male spouses, for follow-up of three to six years. The end-point of the study is progression to severe dysplasia or carcinoma in situ. So far 80 women with confirmed dysplasia, along with 160 controls, have been recruited. The subjects are recruited from women attending gynaecology outpatient departments of eight collaborating hospitals from the metropolitan city of Delhi.

TYPE: Cohort
TERM: Cytology; Diet; HPV; HSV; Premalignant Lesion; Sexual Activity
SITE: Uterus (Cervix)
TIME: 1988 - 1995

PUNE

* 420 Vaidya, R. 04975
Armed Forces Medical College, Dept. of Preventive, and Social Medicine, Sholapur Rd, Pune 411040, India
COLL: Ghosh, M.K.

Cancer of the Upper Aerodigestive Tract

The aim of this hospital-based study is to improve understanding of the epidemiology of carcinomas of the upper aerodigestive tract among Indian armed forces personnel and their families undergoing treatment at the Malignant Diseases Treatment Centre in Pune. The role of aetiological factors such as smoking, alcohol consumption, use of snuff, dietary habits and state of oral hygiene in upper aerodigestive cancers will be examined. An attempt will be made to analyse the factors which result in delay in reporting to medical centres after onset. Approximately 100 cases and an equal number of controls will be studied. The controls will be patients admitted to the same hospital, matched for age, sex and socio-economic status, who have never suffered from any malignancy. All cases and controls will be personally interviewed, using a specially designed and tested questionnaire.

INDIA

TYPE: Case-Control
TERM: Alcohol; Diet; Hygiene; Tobacco (Smoking); Tobacco (Snuff)
SITE: Hypopharynx; Larynx; Oesophagus; Oral Cavity; Oropharynx
TIME: 1991 - 1992

TRIVANDRUM

* 421   **Krishnan Nair, M.**   05075
Regional Cancer Centre, Medical College Campus, Trivandrum 695011 Kerala, India (Tel.: (0471)71904; Fax: 65347 ; Tlx: 435413 td in)
COLL: Padmanabhan, T.K.; Gangadharan, P.; Sreedevi Amma, N.; Ramachandran, T.P.; Sankaranarayanan, R.

Cancer Registry in an Area of High Natural Radiation

Radiation-emitting sands in some coastal areas in Kerala, South India, have caused much concern, especially about carcinogenesis. A population-based cancer registry has been set up in order to estimate cancer incidence in the area. The rural population covered is 500,000. Agriculture and traditional jobs like coil making or fishing are the major occupations. The area of high natural background radiation (annual average 700 millirems) covers a population of 100,000. The cancer registry will cover the entire taluk (administrative unit composed of more than one village). Radiation measurements will be undertaken to estimate the individual doses received by the residents. Cancer incidence in the population of the radiation-exposed area will be compared with that of the population in the rest of the taluk. The radiation measurements are being undertaken with assistance from the Bhabha Atomic Research Centre, Bombay.

TYPE: Incidence; Registry
TERM: Radiation, Ionizing
SITE: All Sites
TIME: 1990 - 2000

* 422   **Parukutty Amma, K.**   05079
Regional Cancer Center, Medical College Campus, P.O. Box 2417, Trivandrum 695011 Kerala, India (Tel.: 74541/71541; Fax:91047165347 ; Tlx: 435413 TD IN)
COLL: Sankaranarayanan, R.; Cherian, V.; Padmakumary, G.; Rajeevkumar, S.; Krishnan Nair, M.; Gangadharan, P.

Childhood Cancer in Kerala

Every year, about 200 cases of childhood cancer are registered by the hospital registry in the Regional Cancer Centre, Trivandrum. This project aims to study the relative frequency of various cancers, sex ratio, age distribution, yearly trends and survival. All malignancies have been coded to ICD-O from 1982 onwards. Efforts are under way to organise a statewide Paediatric Cancer Regsitry. The aim is to establish a population-based register for paediatric cancer in Kerala and to record clinical, pathological, therapeutic and survival data on all children newly diagnosed with cancer.

TYPE: Relative Frequency
TERM: Sex Ratio; Survival; Trends
SITE: Childhood Neoplasms
TIME: 1982 - 1994

423   **Sankaranarayanan, R.**   04890
Regional Cancer Centre, Medical College Campus, Trivandrum 695011, Kerala, India (Tel.: (0471)71904; Fax: 65347 ; Tlx: 435413 TD IN)
COLL: Krishnan Nair, M.; Gangadharan, P.; Nair, M.; Padmakumary, G.; Mayadevi, S.; Duffy, S.W.; Day, N.E.

Diet and Oesophageal, Oral, Larynx and Lung Cancer in Kerala

To test the hypothesis that a diet rich in milk, vegetables and fruit is protective against cancers of certain sites, the association between dietary factors and cancer of the oral cavity, oesophagus, larynx and lung is being investigated using a case-control design. Information on diet and lifestyle will be collected from cancer patients seen at the Regional Cancer Centre in Trivandrum. Controls will be apparently healthy visitors of patients admitted to the Medical College Hospital with medical or surgical conditions unrelated to malignancy. Data will be collected through personal interview from all oral cancer patients registered between 1990-1992 and their controls; 800 oesophageal cancer cases and 2,000 controls over three

INDIA

years; 1,500 lung and larynx cancer cases and 4,000 controls over four years. The data will be analysed by logistic regression to produce odds ratio estimates of relative risk.

TYPE: Case-Control
TERM: Diet; Fruit; Tobacco (Smoking); Vegetables
SITE: Larynx; Lung; Oesophagus; Oral Cavity
TIME: 1990 - 1993

* 424   **Sankaranarayanan, R.**   05084
Regional Cancer Centre, Medical College Campus, Trivandrum 695011, Kerala, India (Tel.: (0471)71904; Fax: 65347 ; Tlx: 435413 TD IN)
COLL: Krishnan Nair, M.; Padmakumary, G.; Cherian, V.; Duffy, S.W.; Day, N.E.; Gangadharan, P.

Risk Factors for Breast Cancer in Kerala

Breast cancer, which accounts for 21% of cancers in women in Kerala, is a major malignancy, revealing an increase in occurrence. The aim of this unmatched case-control study is to evaluate the relationship between socio-economic status, reproductive factors, diet history and menstrual status. A detailed questionnaire is administered to cases and controls concerning reproductive history, age at menarche and marriage, parity and diet history. Cases (n = 1,000) will be women diagnosed with breast cancer at the Regional Cancer Centre, Trivandrum from 1990-1992. Controls (n = 2,000) will be selected among apparently healthy visitors of patients with non-malignant conditions at the Medical College, Trivandrum.

TYPE: Case-Control
TERM: Diet; Menstruation; Parity; Reproductive Factors; Socio-Economic Factors
SITE: Breast (F)
TIME: 1990 - 1993

* 425   **Sankaranarayanan, R.**   05085
Regional Cancer Centre, Medical College Campus, Trivandrum 695011, Kerala, India (Tel.: (0471)71904; Fax: 65347 ; Tlx: 435413 TD IN)
COLL: Krishnan Nair, M.; Padmakumary, G.; Cherian, V.; Duffy, S.W.; Day, N.E.; Gangadharan, P.

Cervical Cancer in Kerala

Cervical cancer is the most common cancer diagnosed among women at the Regional Cancer Centre, Trivandrum (23%). In this unmatched case-control study, a detailed questionnaire will be administered to 1,500 cervical cancer cases to elicit information on education, socio-economic status, familial cancer history, marital history, number of pregnancies, age at first coitus and pregnancy, number of sexual partners, genital hygiene, diet and promiscuity of the partner. 2,000 controls will be selected at the time of diagnosis of the case from visitors to patients with non-malignant conditions at the Medical School, Trivandrum. Unconditional logistic regression will be used to analyse the data.

TYPE: Case-Control
TERM: Diet; Education; Hygiene; Parity; Sexual Activity; Socio-Economic Factors
SITE: Uterus (Cervix)
TIME: 1990 - 1992

* 426   **Sankaranarayanan, R.**   05086
Regional Cancer Centre, Medical College Campus, Trivandrum 695011, Kerala, India (Tel.: (0471)71904; Fax: 65347 ; Tlx: 435413 TD IN)
COLL: Krishnan Nair, M.; Sreedevi Amma, N.; Mathew, B.; Cherian, V.; Sureshchandra Dutt, G.; Padmavathy Amma, B.

Randomised Clinical Trial to Evaluate the Chemopreventive Potential of Spirulina Algae in Oral Leukoplakia and Submucous Firbosis

Oral leukoplakia and submucous fibrosis (SMF) are well-established oral precancerous lesions among Indians. Systemic administration of vitamin A and beta-carotene has resulted in regression of oral leukoplakias, even in the presence of continuing tobacco use. Spirulina, a blue-green alga, is rich in micronutrients like beta-carotene, lutein and several vitamins. The carotene available in these algae are completely bioavailable and they could thus be a low-cost alternative to ensure adequate intake of vitamin A. Spirulina, at a dose of 1 gm daily, will be administered to 75 individuals and placebo to 75 individuals with leukoplakia and SMF for one year. The end-point will be clinical regression of the oral leukoplakia and objective improvements in the symptoms from oral SMF.

INDIA

TYPE: Intervention
TERM: Premalignant Lesion; Vitamins
SITE: Oral Cavity
CHEM: Beta Carotene
TIME: 1991 - 1993

* 427 **Sebastian, P.** 05028
Regional Cancer Centre, Medical College Campus, Trivandrum 695011 Kerala, India (Tel.: 72963, 76799; Fax: 65347 ; Tlx: 435413 td in)
COLL: Bhattathiri, V.N.; Sreelekha, T.T.; Ramani, P.; Vijayakumar, T.; Cherian, T.; Krishnan Nair, M.

Plasma Glutathione - A Tumour Marker in Oral Cancer
Plasma glutathione levels have been reported to be low in malignancy, returning to normal levels after successful treatment. Serial estimations of plasma glutathione may therefore show alterations before the recurrence becomes manifest clinically. The low values of plasma glutathione in cancer are probably due to high levels of gamma-glutamyl transpeptidase enzyme associated with several tumours. This study aims to evaluate the role of plasma glutathione as a tumour marker in oral cancer, the commonest cancer in India. The plasma glutathione levels will be estimated before treatment in 50 oral cancer patients undergoing radical radiotherapy, 50 patients undergoing surgery and in 50 age- and sex-matched healthy controls. The post-treatment glutathione values will be estimated two months after treatment and then every three months for two years. Plasma glutathione values before and after treatment will be correlated to clinical status of the disease and will be compared to the levels in the controls. The study is to be done over a period of three years.

TYPE: Case-Control
TERM: Tumour Markers
SITE: Oral Cavity
TIME: 1989 - 1992

VARANASI

* 428 **Mohapatra, S.C.** 05214
Banaras Hindu Univ., Inst. of Medical Sciences, Dept. of Preventive and Social Medicine, BHU Campus, Varanasi 221005, India (Tel.: 42668)
COLL: Meenakshi, M.; Shukla, H.S.

Nutritional Factors in Breast Cancer
This hospital-based case-control study will examine the role of nutritional factors in breast cancer. Approximately 200 cases of breast cancer will be registered between March 1989 and February 1992. Controls will be selected from other surgical units, matched by age, sex, geographical area, parity and socio-economic status. Each case will be matched with one normal control and with one woman with benign breast disease. Nutritional data will be obtained by use of standardized utensils which contain known weights of raw and cooked foods prevalent in the study area. Local nutrient conversion tables will be used. Serum levels of certain amino-acids will also be recorded.

TYPE: Case-Control
TERM: BMB; DNA; Diet; Nutrition; Serum
SITE: Breast (F)
TIME: 1989 - 1992

INDONESIA

SEMARANG

429   Sarjadi,   04034
Diponegoro Univ., Medical Faculty, Kariadi Teaching Hosp., Dept. of Pathology, Jalan dr. Soetomo no. 16, Semarang 50231, Indonesia (Tel.: (24)311476/311523)
COLL: Indrawijaya,; Tjahjono; Sugondo, T.

Minimum Incidence Rates of Cancer in the Semarang City Population, Based on Microscopically Diagnosed Cancers
Population-based cancer data are still difficult to obtain in Indonesia. Starting in 1970, minimum incidence rates of cancer have been calculated in the Semarang City population, based on cases microscopically diagnosed in the Department of Pathology, Diponegoro University. Despite being based only on histopathologically confirmed cases, the age-standardized rates for nasopharynx and cervix cancer were high, both exceeding those observed in Singapore Malays. Rates for cervix cancer were similar to those observed in Hongkong and Rangoon. An increase in the incidence of lung cancer from 1970 to 1981 was also observed. Data for 1981-1985 were published in 1987. This is a cooperative study between the Indonesian Cancer Society and the Dept. of Pathology, Diponegro University.

TYPE: Incidence
TERM: Geographic Factors
SITE: Lung; Nasopharynx; Uterus (Cervix)
TIME: 1970 - 1992

430   Sarjadi,   04758
Diponegoro Univ., Medical Faculty, Kariadi Teaching Hosp., Dept. of Pathology, Jalan dr. Soetomo no. 16, Semarang 50231, Indonesia (Tel.: (24)311476/311523)
COLL: Indrawijaya,; Tirtosugondo; Kasno

Cancer Incidence 1985-1994 in Semarang
This project involves the creation of the first population-based cancer registry in Indonesia, for the city of Semarang. Data will be collected from all health care facilities in the city, and incidence rates will be derived. The aim is to generate a database for epidemiological research, to identify the most frequent malignancies and to provide the information required for health care planning and cancer control activities.

TYPE: Incidence
TERM: Registry
SITE: All Sites
TIME: 1985 - 1994

YOGYAKARTA

431   Soeripto,   04641
Gadjah Mada Univ., Faculty of Medicine, Dept. of Pathology, Jl. Kesehatan Sekip , Yogyakarta, Indonesia (Tel.: 888688/564)
COLL: Prijono Tirtoprodjo; Zuchairi Dahlan; Achmad Ghozali

Cancer Incidence in Yogyakarta: Population
Population-based cancer registration at Srandakan district, Bantul county, Yogyakarta is being developed. The aim of this study is to determine cancer incidence in Bantul county and Yogyakarta town during the period 1989 to 1993, through population-based cancer registration. The form containing identifying information, clinical diagnosis, and histopathological diagnosis has been distributed to the health centres, private practice physicians, and hospitals in both areas. Methods of analysis will be age-adjusted standardization, using as the denominator the combined population in both areas.

INDONESIA

TYPE: Incidence; Registry
TERM: Registry
SITE: All Sites
TIME: 1989 – 1993

432   Soeripto,   04642
Gadjah Mada Univ., Faculty of Medicine, Dept. of Pathology, Jl. Kesehatan Sekip , Yogyakarta, Indonesia (Tel.: 888688/564)
COLL: Endang Soetristi; Irianiwati

Incidence of Childhood Cancer in Yogyakarta
The objective of the study is to determine the incidence of childhood cancer in Yogyakarta with hospital-based cancer registration. All childhood cancer patients at the hospitals in this area were collected from medical records. Patients from outside Yogyakarta were excluded. Age-adjusted standardisation will be used for analysis. The childhood population of Yogyakarta will be used for denominator. Data have been published in "International Incidence of Childhood Cancer", Parkin D.M., Stiller C.A., Draper G.J., Bieber C.A., Terracini B. and Young J.L. (Eds). IARC Sci. Publ. No 87, Lyon, France, 1988.

TYPE: Incidence
TERM: Childhood; Clinical Records; Registry
SITE: All Sites
TIME: 1989 – 1998

IRELAND

DUBLIN

* 433　　Bourke, G.J.　　05016
　　　　　Univ. College of Dublin, Dept. of Community Medicine, and Epidemiology, Earlsfort Terrace, Dublin 2, Ireland (Tel.: (353)12693244/6345; Fax: 1754568)
COLL:　　Daly, L.; Herity, B.; Barry, W.; Brogan, J.

Total and Cancer Mortality in Research Workers in Agriculture
　　　The aim of this historical cohort study is to ascertain if employees in an agricultural research institute have an excess all-cause or cancer mortality compared to the general population and, if so, to identify groups of workers at increased risk. The study cohort consists of approximately 1,500 administrative, research and technical staff who were employed for at least two consecutive years in the institute from 1960 to 1986. Mortality follow-up to the end of 1988 is done with personnel records, and causes of death will be determined from the Registrar General's records. Statistical analysis of the data will be based on person-years of exposure, with the use of regression techniques where appropriate. The study cohort forms part of the IARC International Study of Cancer Risk in Biology Research Laboratory Workers.

TYPE:　　Cohort
TERM:　　Occupation
SITE:　　All Sites
OCCU:　　Administrative Workers; Laboratory Workers
TIME:　　1989 - 1992

ISRAEL

BEER SHEVA

434 Goldsmith, J.R. 04746
Ben Gurion Univ. of the Negev, Faculty of Health Sciences, P.O. Box 653, 84120 Beer Sheva, Israel
(Tel.: (057)660813; Fax: 37342 ; Tlx: 5253 unast il)
COLL: Sofer, T.

Childhood Cancer in the Negev

Using cancer incidence data for 1960-85 from the Israel Cancer Registry for the Negev (Southern) Region of the country, geographical and ethnic (Jewish/Beduin) distributions are being looked at in order to see if clusters have occurred. At the same time the possible associations of childhood cancer incidence with birth defects and genetic abnormalities are being investigated. Two kibbutzim (co-operative settlements) with increased incidence of cancer have been detected and a case-referent study is now being conducted, using non-cancer residents of the kibbutzim as one referent population and residents of adjacent kibbutzim of the same age group, sex, and country of origin as a second referent population. Of special concern is the possible exposures to agricultural chemicals, such as herbicides and pesticides. The area includes a large chemical industry complex and waste dump as well as a nuclear research facility.

TYPE: Case-Control
TERM: Cluster; Congenital Abnormalities; Ethnic Group; Herbicides; Pesticides
SITE: Childhood Neoplasms
REGI: Israel (Isr)
TIME: 1989 - 1992

435 Odes, S. 04946
Soroka Medical Center, Gastroenterology Unit, P.O. Box 151, 84101 Beer Sheva, Israel (Tel.: (057)660242; Fax: 74696)
COLL: Vardi, H.

Oesophageal Carcinoma in Indian Jews in Israel

A prospective cohort study of oesophageal cancer is being carried out in the Indian Jewish (n = 8,461) population of Southern Israel. The study district has a total population of 450,700, and the cohort includes 70% of all Jewish migrants from India. Cases are obtained from clinical, radiological, endoscopic and pathology records at the hospitals and clinics and from the Central Cancer Registry in Jerusalem. Interviews with patients and/or relatives provide data about aetiological factors, including diet. Incidence data are age-adjusted to the population of Israel. Data are processed utilizing SPSS-X including Student's test and the chi-square method. A recent report (J.Clin.Gastroenterol. 12(2):222-7) 1990 for the period 1961 to 1985 shows a significantly higher mean age-adjusted incidence rate in immigrant Indian males, 6.5 (per 100,000), and Indian females, 17.2, than non-Indian males, 2.7 and non-Indian females, 2.1. The mean female to male ratio was: Indian 1.75, non-Indians 0.59. The age at diagnosis was lower in Indian females (54.6 years) than Indian males (66.2 years) Risk factors appear to be poverty in all Indians, smoking in men and spicy diets in women. The study is expected to continue for the next 20 years.

TYPE: Cohort
TERM: Diet; Indians; Lifestyle; Tobacco (Smoking)
SITE: Oesophagus
REGI: Israel (Isr)
TIME: 1977 - 1997

* 436 Odes, S. 05121
Soroka Medical Center, Gastroenterology Unit, P.O. Box 151, 84101 Beer Sheva, Israel (Tel.: (057)660242; Fax: 74696)
COLL: Krugliak, P.; Fraser, G.

Frequency of Colonic Adenomas and Cancer in High, Medium and Average Risk Populations in Southern Israel

A screening programme for colorectal adenomas and cancer has operated from 1983-1987 and again from 1990 in the population (age-group 40-70) in Beer Sheba in southern Israel in order to determine the frequency of positive findings and the efficacy of faecal occult blood testing, sigmoidoscopy and colonoscopy. Screenees are interviewed to determine the level of their risk according to the criteria of Rozen et al. (Front. Gastrointest. Res. 10:164, 1986). The categories are: high risk - previous colorectal

ISRAEL

tumours (I) or inflammatory bowel disease (II); moderate risk - female genital tumour (III) or family history of colorectal tumour (IV); average risk - asymptomatic persons over the age of 40 (V). In 1983-1987 the number of screenees in these categories was 22, 11, 11, 77 and 980 respectively. In the total group there were 41 cases wih polyps and 1 case with colon cancer. Faecal occult blood tests are done on the whole screened population, while total colonoscopy is offered to people in groups I and II and flexible sigmoidoscopy carried out in groups III, IV and V. The faecal occult blood test had poor sensitivity and positive predictive value compared with sigmoidoscopy and colonoscopy. The study was reactivated in 1990, including 750 new screenees.

TYPE: Case Series
TERM: High-Risk Groups; Polyps; Screening
SITE: Colon; Rectum
TIME: 1983 - 1994

HAIFA

437   Rennert, G.   04194
Carmel Hosp., Kupat Holim Klalit, & Technion Faculty of Medicine, Dept. of Family and Community Health, 7 Michal St., 34362 Haifa, Israel (Tel.: (04)254473; Fax: 242414)
COLL: Epstein, L.M.

Time Trends in Lung Cancer Incidence among Jewish Immigrants to Israel
This study is designed to evaluate the trends in incidence rates of lung cancer among Jews who have immigrated to Israel from different countries in the world. The incidence rates will be compared to rates in the country of origin (country of birth) and to the total rates in Israel according to the length of time which has elapsed since immigration. The cases will include all 9,894 Jews born outside of Israel with lung cancer diagnosed in Israel after immigration and reported to the Israel Cancer Registry between the years 1962-1982. This study will be limited by the availability of rates of lung cancer in the corresponding countries from which the cases emigrated.

TYPE: Incidence
TERM: Ethnic Group; Migrants; Registry; Time Factors
SITE: Lung
REGI: Israel (Isr)
TIME: 1986 - 1992

438   Rennert, G.   04911
Carmel Hosp., Kupat Holim Klalit, & Technion Faculty of Medicine, Dept. of Family and Community Health, 7 Michal St., 34362 Haifa, Israel (Tel.: (04)254473; Fax: 242414)
COLL: Ben Harush, M.; Perez, J.

Childhood Malignancies in Northern Israel
The study is designed to evaluate possible risk-factors for childhood malignancies among historical cases treated in the only paediatric oncology facility in Northern Israel. Age- and sex-matched hospitalized controls will be assigned to the cases. Paternal and maternal (including pregnancy-related) factors will be studied as well as various childhood exposures. The estimated number of cases is about 400. Hospital controls will be recruited prospectively and matched on same current age as cases to ensure similar number of years of recall.

TYPE: Case-Control
TERM: Intra-Uterine Exposure; Occupation
SITE: Childhood Neoplasms
TIME: 1990 - 1993

439   Rennert, G.   04913
Carmel Hosp., Kupat Holim Klalit, & Technion Faculty of Medicine, Dept. of Family and Community Health, 7 Michal St., 34362 Haifa, Israel (Tel.: (04)254473; Fax: 242414)

Risk Factors and Protective Factors for Cancer - A Prevalence Study
A sample of 3,000 Israeli adults of both sexes will be studied in Haifa. Information will be obtained through home interviews on smoking, diet, sun-exposure and reproductive factors, occupational history and history of radiation exposure, as well as on family history of cancer. The study population will be identified through stratified sampling (by sex and age groups 30-44 and 45-65) of the list of all people insured by

the major insurance company in Israel (covering more than 80% of the population) and will be followed up for changes in habits and occurrence of cancer. Analysis will include estimating prevalence of certain health habits with comparisons made between the studied groups in the different follow-up periods.

TYPE: Cross-Sectional
TERM: Diet; Familial Factors; Occupation; Radiation, Ionizing; Radiation, Ultraviolet; Reproductive Factors; Tobacco (Smoking)
SITE: All Sites
TIME: 1990 - 1995

440 Rennert, G. 04914
Carmel Hosp., Kupat Holim Klalit, & Technion Faculty of Medicine, Dept. of Family and Community Health, 7 Michal St., 34362 Haifa, Israel (Tel.: (04)254473; Fax: 242414)

Immunization History in Childhood Malignancies

This case-control study is designed to evaluate the role of lack of compliance with routine immunization schedules in infants as a risk factor for future development of childhood malignancies. All records of all children diagnosed with malignancies between the years 1985-1989 in Israel will be studied for their immunization history as specified in the regional health departments. The case group currently includes 87 subjects. Controls will be age-matched children from the general population. The data collected will include history of diphtheria, tetanus and pertussis (DTP) and polio immunization, number of immunization courses and adherence to immunization schedule. Analysis will be multivariate employing conditional logistic regression.

TYPE: Case-Control
TERM: Immunology; Vaccination
SITE: Childhood Neoplasms
REGI: Israel (Isr)
TIME: 1990 - 1992

441 Robinson, E. 04312
RAMBAM Medical Center, TECHNION-Israel Inst. of Technology, Northern Israel Oncology Center, Haifa 35 254, Israel (Tel.: (04)514481)
COLL: Rennert, G.; Nasrallah, S.; Adler, Z.; Neugut, A.

Clinical Characteristics of Patients with Multiple Primary Neoplasms

Multiple primary neoplasms (MPNs) constitute almost 10% of all cancer seen in the USA. Very little is known about how they differ from first primary neoplasms. It is the purpose of the proposed investigation to explore the stage and grade distribution, and survival of various MPNs and to compare these characteristics to those of first primary neoplasms. Three population-based cancer registries will be studied. At Columbia, the Connecticut Tumor Registry and other SEER Registry data will be analysed, while the Israel Cancer Registry data will be analysed at the Technion in Israel. For each registry, second neoplasms of interest will be compared to first neoplasms at the same site for the characteristics listed above. For example, the survival of secondary acute leukaemia will be compared to the survival of de novo acute leukaemia. The use of both an American and an Israeli registry will allow generalizability of results and confirmation of important findings. Second neoplasms to be studied will be those that occur at a rate greater than expected by chance. A monograph detailing these results will be prepared. Knowledge about differences in clinical behavior may lead to changes in therapeutic approach; to more aggressive surveillance for second neoplasms; and to further clues about the biology of carcinogenesis. Papers have been published in J. Nat. Cancer Inst. 80:233-240, 1988; Cancer Detect. Prev. 13 (5/6):287-293, 1989 and in Radiat. Oncol. 17:109-113, 1990.

ISRAEL

TYPE: Cohort
TERM: Multiple Primary; Registry; Stage; Survival
SITE: Breast (F); Colon; Head and Neck; Leukaemia; Lymphoma; Thyroid
LOCA: Israel; United States of America
REGI: Connecticut (USA); Israel (Isr)
TIME: 1987 - 1992

JERUSALEM

442   Iscovich, J.M.   04834
Ministry of Health, Dept. of Epidemiology, Israel Cancer Registry, 20 King David St. , Jerusalem, Israel (Tel.: +972 2 247172; Fax: 381772)
COLL: Steinitz, R.; Robinson, R.; Rennert, G.; Katz, L.

Contralateral Breast Cancer: Definition and Implication for Breast Cancer and Incidence Rates.
The question of how to distinguish contralateral (bilateral) breast cancer from spread or multicentric site of the disease and its implications for incidence rates is being investigated in the Israel Cancer Registry. From a cohort of 22,653 female Jews with first breast cancer diagnosed in Israel during the period 1960-1986 and followed until August 1988, 1,395 women have subsequently developed contralateral disease. A review of the pathological and medical reports available in the Israel Cancer Registry is being undertaken, using the following criteria (separately or in combination): stage at first diagnosis, time elapsed between diagnoses, histology and appearance of metastasis in relation to the date of the second breast cancer diagnosis.

TYPE: Cohort
TERM: Histology; Multiple Primary; Registry; Stage; Time Factors
SITE: Breast (F)
REGI: Israel (Isr)
TIME: 1988 - 1992

443   Kark, J.D.   03915
Hebrew Univ., Faculty of Medicine, Hadassah SPH & Community Medicine, Dept. of Social Medicine, P.O. Box 1172, Ein Karem, 91000 Jerusalem, Israel (Tel.: (02)447113)
COLL: Goldbourt, U.; Levine, C.; Wahrendorf, J.; Martinsohn, C.

Ethnic, Dietary, Clinical, Socio-Demographic and Biochemical Risk Factors for Cancer
Cancer incidence is being studied in a cohort of 10,000 civil servants and municipal employees first examined in 1963. 1,200 cases of cancer have been identified though the Israel National Cancer Registry, matched by name and identity card number. Histology is available for over 80%. The goals of the project are to identify ethnic, dietary, clinical, socio-demographic and biochemical risk factors for cancer. The large number of cases should permit analysis of incidence by site as well as by histological sub-type. A peculiar aspect of the analysis is the examination of religious orthodoxy (assessed through type of religious education as well as through self-definition) in subjects eventually contracting cancer compared with those without cancer. A number of case-control studies are envisaged to investigate: coffee and cigarette smoking synergism in lung cancer; fruit and vegetable consumption and lung cancer; dietary fat and colon cancer. In a later stage, reexamination of some 20,000 frozen sera (from the study examination of 1963, 1965 and 1968) is envisaged, in order to correlate cancer incidence to various parameters not evaluated earlier in the study period. Given the number of cases and the availability of dates of diagnosis, factors associated with survival of cancer patients might also be elucidated. A paper on the association of blood pressure and cancer mortality has been published in J. Nat. Cancer Inst. 77:63-70, 1986.

TYPE: Case-Control; Cohort
TERM: Coffee; Diet; Ethnic Group; Fat; Fruit; Lipids; Registry; Religion; Socio-Economic Factors; Tobacco (Smoking); Vegetables
SITE: All Sites; Colon; Lung
REGI: Israel (Isr)
TIME: 1985 - 1992

444   Katz, L.   04750
Ministry of Health, Israel Cancer Registry, 20 King David St., 91000 Jerusalem, Israel (Tel.: (02)247172; Fax: 381772)
COLL: Avni, A.; Iscovich, J.M.; Steinitz, R.

ISRAEL

Geographical Mapping of Cancer Incidence in Israel
The aim of the study is to examine the geographical distribution of Cancer Incidence in Israel. The results are expected to point to occupational/environmental exposure to carcinogens. The study will be based on the detailed medical and demographic information recorded in the Israel Cancer Registry and on population data by sex, age and locality from the 1972 and 1983 censuses of population and housing conducted by the Central Bureau of Statistics. The study will be of approximately 50,000 cases diagnosed in 1971-73 and 1982-84. SIRs will be calculated for each district, municipality, and regions within cities.

TYPE: Incidence
TERM: Environmental Factors; Geographic Factors; Mapping; Occupation
SITE: All Sites
REGI: Israel (Isr)
TIME: 1990 - 1992

445   Steinitz, R.   03852
Israel Cancer Registry, Ministry of Health, Dept. of Epidemiology, 20 King David St. , 91000 Jerusalem, Israel (Tel.: (02)247172; Fax: 381772)
COLL: Katz, L.; Iscovich, J.M.

Multiple Primary Malignancies in the Israel Cancer Registry
The Israel Cancer Registry's system of registering each tumour separately and linking tumours in the same patient by a common accession number in the index file carrying the demographic information, offers the possibility of studying the incidence of cancer at multiple sites. Approximately 4,000 patients with multiple primary malignancies will be available for the study, not including basal cell and squamous cell carcinoma of the skin. Expected number of cases will be established by observing person-years of exposure to the risk of developing additional primary cancers, and then applying the appropriate incidence rates for each site, sex, age group and continent of origin. From this, the relative risk of developing additional primary cancers will be calculated.

TYPE: Cohort
TERM: Multiple Primary; Registry
SITE: All Sites
REGI: Israel (Isr)
TIME: 1985 - 1992

TEL AVIV

446   Chaitchik, S.   04190
Tel Aviv Medical Center, Int. of Oncology, Ichilov Hosp., Dept of Oncology, 6 Weizman St., Tel Aviv 64239, Israel (Tel.: (03)210453; Tlx: 342298 tlvmc)
COLL: Shenberg, C.; Biran, T.; Mantel, M.; Weininger, J.

Evaluation of the Levels of Selenium in a Population with Breast Cancer and a Normal Population at Risk
The purpose of the study is to evaluate levels of selenium in patients with breast cancer. The patients are referred to the Tel-Aviv Medical Center for diagnosis and surgery and will be compared with a matched normal population. 60 patients with breast cancer will be included in the study and about 120 women at risk will be evaluated using a highly sensitive XRF method. A preliminary study was carried out in order to compare the selenium concentration in breast cancer patients and healthy subjects (controls). The plans are to continue until at least 120 humans have been compared.

TYPE: Case-Control
TERM: Biochemical Markers
SITE: Breast (F)
CHEM: Selenium
TIME: 1987 - 1992

447   Chaitchik, S.   04686
Tel Aviv Medical Center, Int. of Oncology, Ichilov Hosp., Dept of Oncology, 6 Weizman St., Tel Aviv 64239, Israel (Tel.: (03)210453; Tlx: 342298 tlvmc)
COLL: Shenberg, C.; Biran, T.; Mantel, M.; Weininger, J.

ISRAEL

Serum Selenium and Colo-Rectal Cancer
The aim of this study is to evaluate serum selenium levels in patients with colo-rectal cancer. Analyses will be performed on surgical specimens, both on tumour tissue and adjacent normal tissue. In a preliminary study of breast cancer cases and healthy controls, serum selenium was significantly lower in cases (weighted mean 0.076 ppm, SE 0.014) than in controls (0.119, SE 0.023), and selenium level was inversely related to clinical stage. In the pilot study, 20 patients with colo-rectal cancer will be studied.

TYPE: Case-Control
TERM: Biochemical Markers; Trace Elements
SITE: Breast (F); Colon; Rectum
CHEM: Selenium
TIME: 1991 - 1992

448   Kahan, E.   03763
Tel Aviv Univ., Sackler Sch. Med., Inst. of Occupational Health, Dept. of Epidemiology, University St., Ramat Aviv, 69978 Tel Aviv 69978, Israel (Tel.: (03)425827)
COLL: Hare, C.

Survey of Latent Carcinoma of Prostate in Israel
Carcinoma of prostate (CP) is defined as latent (LCP) when detected in surgical specimens from patients clinically not suspected, or at autopsy. In post mortems and biopsies carried out during the last decade, a very high frequency of LCP was found. The Jewish population in Israel comprises people originating from over 100 countries. A previous study showed that the annual incidence of CP in Israel's population over 50 years old is 72 per 100,000, but in two groups there is a significant difference. Jews originating from Greece and Bulgaria showed a high annual rate: 103.9; Jews originating from Yemen had an extremely low rate: 31.4. The present survey aims to collect demographic information about LCP in order to compare prevalence rates of LCP by country of origin with the existing incidence rate for CP in the population. If the incidence rate differences between Jews from various origins is the expression of different exposures to risk, similar proportional differences in prevalence of LCP in Jews, by country origin, would be expected.

TYPE: Cross-Sectional
TERM: Autopsy; Ethnic Group; Prevalence
SITE: Prostate
TIME: 1986 - 1992

449   Kahan, E.   04459
Tel Aviv Univ., Sackler Sch. Med., Inst. of Occupational Health, Dept. of Epidemiology, University St., Ramat Aviv, 69978 Tel Aviv 69978, Israel (Tel.: (03)425827)
COLL: Luria,; Derazne, E.; Peretz, H.

Physical Activities and Colon Cancer
Colon cancer is one of the most frequent malignancies in Western countries and in Israel. Studies suggest a decreased risk among high-activity occupations and among members of kibbutzim (cooperative settlements) compared to urban dwellers. This may be due to the high-activity life-style in kibbutzim. The purpose of this study is to evaluate whether there is a negative correlation between physical activity (recreation, home, work, etc.) and incidence of colon cancer. This will be a hospital-based case-control study of 200 cases and 200 matched controls. Cases will be randomly chosen among living colon cancer patients aged 25-69 years, diagnosed in the Institute of Cancer of Beilinson Medical Center, Israel, from 1986-1990. Matched controls will be chosen among living patients with cancers other than colon cancer. A questionnaire about physical activity will be performed and confounding factors, such as socio-demographic, diet components, vitamin A, family history of cancer and anthropometric characteristics will also be assessed. Data analysis will be done using logistic analysis of non-parametric data on physical activity, (moderate, intermediate and high activity), against incidence of colon cancer.

ISRAEL

TYPE: Case-Control
TERM: Diet; Familial Factors; Physical Activity; Vitamins
SITE: Colon
TIME: 1989 - 1992

TEL HASHOMER

450 Modan, B. 04556
Tel Aviv Univ., Chaim Sheba Medical Center, Bitan 27, 52621 Tel Hashomer Ramat-Gan, Israel
COLL: Boice, J.D.; Kellerman, R.

Potential Radiocarcinogenicity of Heart Catheterisation in Childhood
This is a pilot study to determine the availability of data for a long-term follow-up study of possible carcinogenic effects among about 15,000 children who underwent cardiac catheterisation for congenital heart disease, under X-ray control, during the period 1955-1969. Dosimetry will be obtained by reconstruction of old measurements and treatment plans, and cancer cases and deaths will be ascertained through registries. The objective will be to re-assess the carcinogenic effect of low-dose irradiation. Thus far +/- 12,000 records have been accrued from hospitals in the USA, United Kingdom, Netherlands and Israel.

TYPE: Cohort
TERM: Childhood; Radiation, Ionizing
SITE: All Sites
LOCA: Israel; United Kingdom; United States of America
TIME: 1986 - 1992

451 Modan, B. 05172
Tel Aviv Univ., Chaim Sheba Medical Center, Bitan 27, 52621 Tel Hashomer Ramat-Gan, Israel

Cancer and Benign Tumours Following Scalp Irradiation
This is a cohort study of 10,834 subjects irradiated in childhood and approximately 16,000 sibling and neighbourhood controls. The cohort is linked with the cancer registry and a central population register to identify cases and the vital status of all subjects. The aim is to estimate cancer rates and the duration of any excess risk. Recent emphasis is on low dose radiation, in view of the increased risk of thyroid and breast cancer observed thus far following low-dose exposures (Lancet 1:629-631, 1989).

TYPE: Cohort
TERM: Childhood; Latency; Radiation, Ionizing; Registry
SITE: All Sites; Benign Tumours
REGI: Israel (Isr)
TIME: 1965 - 2060

452 Pines, A. 04692
Chaim Sheba Medical Center, Dept. of Gastroenterology, Tel Hashomer 52621, Israel (Tel.: +972 35 310660)
COLL: Bat, L.; Shemesh, E.

Epidemiological Aspects of Malignant and Benign Colorectal Polyps
It was well established that adenomatous colon polyps and cancer are more common in Ashkenazi than in non-Ashkenazi Jews. The aim of this study is to compare three types of adenomatous polyps with increasing level of neoplastic proliferation (benign polyps, carcinoma in-situ within a polyp, and invasive-malignant polyps). The study population includes all patients undergoing colonoscopy in the gastroenterology institute since 1978. Data are collected from computerised files held at the institute and from clinical records. It is expected to evaluate a total of 60 patients with invasive adenomas, 90 with in-situ carcinoma and 400 with benign polyps. The parameters to be examined comprise: origin, sex, age, family history of colorectal cancer and morphological and histological characteristics of the adenomas.

TYPE: Case Series
TERM: Biopsy; Clinical Records; Familial Factors; Histology; Polyps; Premalignant Lesion
SITE: Colon; Rectum
TIME: 1988 - 1992

ITALY

AVIANO

453　Franceschi, S.　　　　　　　　　　　　　　　　　　　　　　　　　　　　03718
　　　　Aviano Cancer Centre, Epidemiology Unit, Via Pedemontana Occ., 33081 Aviano PN, Italy (Tel.:
　　　　(0434)659232/9354; Fax: 652182)
COLL:　Talamini, R.; Serraino, D.; Bidoli, E.; La Vecchia, C.; Barra, S.; Baron, J.A.; Cristofolini, M.

Epidemiological Surveillance of Neoplastic Disease

A Cancer Institute opened in Aviano (Province of Pordenone) in 1984, serving a large area in the North Eastern part of Italy. Although no Cancer Registry has been activated, the concentration of patients in this institution and in a few others in nearby towns (chiefly Pordenone, Trieste, Udine, Padua and Trento) provides an opportunity to investigate with a case-control approach a number of neoplastic diseases, particularly those presenting mortality rates 30-100% higher in this area than in the rest of Italy (cancer of the oesophagus, oral cavity and urinary tract). Controls are persons admitted at the same network of hospitals where cancer patients are interviewed, with acute non-neoplastic diseases. Trained interviewers identify and question cases and controls using a standard questionnaire to obtain information on personal characteristics and habits such as cigarette smoking, alcohol and methylxanthine consumption, occupational exposures, selected dietary variables, problem-oriented medical history and history of use of selected drugs (including, for females, contraceptive practices and hormonal therapy). At present information has been collected on the following sites (approximate number of interviews in brackets):urinary tract (750), upper digestive tract (900), lympho-haematopoietic organs (750), thyroid gland (190), melanoma skin cancer (120). New investigations have started on non-melanoma skin cancer, cancers of the endometrium, breast and colon-rectum. A number of papers were published in 1990.

TYPE:　Case-Control
TERM:　Alcohol; Diet; Drugs; Hormones; Occupation; Oral Contraceptives; Pesticides; Premalignant
　　　　Lesion; Tobacco (Smoking)
SITE:　Female Genital; Gastrointestinal; Lymphoma; Skin; Thyroid; Urinary Tract
CHEM:　Methylxanthines
TIME:　1985 – 1992

* 454　Franceschi, S.　　　　　　　　　　　　　　　　　　　　　　　　　　　05051
　　　　Aviano Cancer Centre, Epidemiology Unit, Via Pedemontana Occ., 33081 Aviano PN, Italy (Tel.:
　　　　(0434)659232/9354; Fax: 652182)
COLL:　Amadori, D.; Nanni, O.; Filiberti, R.; Giacosa, A.; Merlo, F.; D'Avanzo, B.; La Vecchia, C.; Negri, E.;
　　　　Parazzini, F.; Decarli, A.; Ferraroni, M.; Levi, F.G.; Palli, D.; Salvini, S.; Bruzzi, P.A.; Montella, M.

Validation of a Dietary Questionnaire for Case-Control Studies

The aim of the present study is to evaluate the reproducibility and validity of a new food frequency questionnaire to be used in case-control studies in different parts of Italy (provinces of Pordenone, Milan, Genoa, Forli, and Naples) and the Canton of Vaud (Switzerland). The questionnaire includes 93 foods and beverages in addition to 13 questions concerning general eating patterns. This form will be administered by trained interviewers twice (once in winter and once in summer) to approximately 600 volunteers, aged 35-74, recruited in various ways (e.g. local newspapers and magazines, health organizations, open university programmes, etc.), and will be compared to the results of two one-week diet records, including detailed descriptions of each food (i.e. weight, brand, method of preparation and recipes). Comparison of nutrient intake score computed from this relatively simple questionnaire with absolute intake obtained from extensive diet record collection will be performed, both before and after adjustment for total caloric intake. Additional analysis will also be carried out in order to define further the influence of season on data collection (reproducibility), the specific foods that contribute nutrients of interest and the importance of recording portion size.

ITALY

TYPE: Methodology
TERM: Diet; Nutrition
SITE: All Sites
LOCA: Italy; Switzerland
TIME: 1991 - 1994

455   Tirelli, U.   04503
Centro di Riferimento Oncologico, Div. di Oncologia Medicale, Grupo Italiano Cooperat. AIDS e Tumori, Via Pedemontana Occidentale, 33081 Aviano, Italy (Tel.: (434)652997; Fax: 652182 ; Tlx: 461277 cerion i)
COLL: Monfardini, S.; Carbone, A.; Franceschi, S.; Gavosto, F.

Malignant Tumours in Association with HIV Infection
Since 1981 AIDS has been associated with an increased number of malignancies, in particular Kaposi's Sarcoma (KS) and non-Hodgkin's lymphoma (NHL). In Italy, a cooperative study group on HIV-related tumours was established in 1986 - GICAT (Gruppo Italiano Cooperativo AIDS e Tumori). Through questionnaires sent to infectious disease specialists, haematologists, oncologists, etc. and in cooperation with the Central Office of the Ministry of Health, where it is mandatory to report all new cases of AIDS seen in Italy, GICAT is monitoring HIV-related tumours. As of September 1988, 435 HIV-related tumours have been recorded (198 KS, 116 high-grade NHL, 20 other NHL, 52 Hodgkin's disease, 49 other rare tumours), mainly in intravenous drug abusers, in accordance with the overall epidemiology of the epidemic in Italy. In Italy, at least 15% of patients with AIDS are also affected by tumours, in particular high-grade NHL and KS. Some HIV-related tumours, (NHL, KS, Hodgkin's disease have peculiar clinico-pathological features while others (testicular and cervical cancer) are similar to those observed in the general population. Recently a European working party on HIV-associated tumours has been established with coordinators Dr Silvio Monfardini from the Centro di Riferimento Oncologico, Aviano, Italy and Prof. Jean Marie Andrieu from the Laennec Hospital, Paris, France.

TYPE: Incidence
TERM: AIDS; HIV; High-Risk Groups
SITE: Hodgkin's Disease; Kaposi's Sarcoma; Non-Hodgkin's Lymphoma
TIME: 1986 - 1992

BARI

* 456   Assennato, G.   04952
Univ. of Bari, Inst. of Occupational Health, Piazza Giulio Cesare, 77124 Bari, Italy (Tel.: +39 80 278216; Fax: 5575451)
COLL: Ferri, G.M.; Porro, A.; Misciagna, G.; Colucci, G.; Centonze, S.; Boeing, H.

Diet and Lung Cancer
The main objective of this study is to evaluate the association between specific nutrients (Vitamin A, E, D, retinol binding proteins) and incidence of lung cancer in a cohort of 20,000 subjects aged 55 years or older, living in five health districts of the Province of Bari in southern Italy. The cohort will be selected from general practitioners' offices. Baseline data will be measured by food frequency questionnaires validated on a cohort sub-group. Blood samples will be stored. Outcome (lung cancer) will be identified by a new integrated notification system based on general practitioner and pathology services. All cases will be histologically confirmed. Periodic linkages between the cohort database and administrative registries will be performed to keep track of migrations and deaths. A case-referent study will be nested using stored blood samples to compare levels of nutrients and other variables between lung cancer cases and controls matched on sex, age, and date of blood collection. Statistical analysis will be performed using multivariate methods.

TYPE: Case-Control; Cohort
TERM: BMB; Blood; Diet; Nutrition; Vitamins
SITE: Lung
TIME: 1990 - 1994

* 457   Assennato, G.   04953
Univ. of Bari, Inst. of Occupational Health, Piazza Giulio Cesare, 77124 Bari, Italy (Tel.: +39 80 278216; Fax: 5575451)
COLL: Ferri, G.M.; Porro, A.; Strickland, P.; Tockman, M.S.; Poirier, M.C.

ITALY

DNA Adducts of Polycyclic Aromatic Hydrocarbons and Membrane Antigens in Occupational Groups at High Risk of Lung Cancer

The objective of the study is to evaluate the association between indicators of biological effective dose indicators (DNA adducts of PAH) and indicators of early biological effects (membrane antigens stained with immuno-assay procedures based on monoclonal antibodies against small cells lung cancer and non-small cell lung cancer antigens) among 350 coke oven workers at a steel plant in southern Italy, at high risk of lung cancer. A pilot investigation (to test feasibility) was carried out on a sub-group of 23 workers in 1988. Data collection began in May 1990. Every worker gives a blood sample, and undergoes sputum induction, spirometry and a questionnaire interview (including respiratory symptoms, smoking and occupational history, and potential sources of confounding hydrocarbon exposure). PAH-DNA adducts will be measured in peripheral lymphocytes by immunoassay (ELISA). Levels of adducts will be compared with those of an unexposed group. Sputum slides prepared and fixed in the field centre will be studied by the Papanicolaou method. Monoclonal antibodies (against small cell lung cancer and non-small cell lung cancer antigens) immuno-stained with Avidin-Biotin complex with a double bridge amplification system will be applied on sputum specimens preserved in Saccomanno fixative. Follow-up of the workers will be carried out to detect incidence and mortality for lung cancer. The data will be analysed by univariate and multivariate methods.

TYPE: Cohort; Molecular Epidemiology
TERM: Antigens; BMB; Blood; DNA Adducts; Occupation; Sputum; Tobacco (Smoking)
SITE: Lung
CHEM: PAH
OCCU: Coke-Oven Workers
TIME: 1988 - 1995

CAGLIARI

458   Carta, P.   04736
Univ. di Cagliari, Ist. di Medicina del Lavoro, Via S. Giorgio 12, 09124 Cagliari, Italy (Tel.: +39 70 670481; Fax: 654350)
COLL: Cocco, P.L.

Mortality Study of Lead and Zinc Miners in Sardinia

The cohort study of Sardinian miners affected by silicosis found an excess lung cancer risk in those silicotics who had airway obstruction at entry, after adjusting for smoking (Br. J. Ind. Med. 48:122-129, 1991). Follow-up of two cohorts of lead and zinc miners has now been completed. Exposure to radon daughters is currently being assessed by recent measurements and estimates for the past made using available information on ventilation devices. Exposure to silica and nitrogen oxides was assessed by the Institute during the period 1973-1983. The first cohort includes 5,000 miners with at least one year's employment in 1960-1971, followed-up through 31 December 1987. The second consists of 1,700 miners, surveyed in 1973, for whom respiratory function data and chest X-ray at entry are also available. Standardised mortality ratios for lung cancer and other groups of diseases and single causes of interest will be calculated using the Sardinian general male population age- and year-specific mortality as reference.

TYPE: Cohort
TERM: Mining; Occupation; Silicosis
SITE: Lung; Stomach
CHEM: Lead; Nitrogen Oxide; Radon; Silica; Zinc
OCCU: Miners, Zinc-Lead
TIME: 1984 - 1992

* 459   Carta, P.   05160
Univ. di Cagliari, Ist. di Medicina del Lavoro, Via S. Giorgio 12, 09124 Cagliari, Italy (Tel.: +39 70 670481; Fax: 654350)
COLL: Cocco, P.L.; Flore, C.; Cherchi, P.

Cohort Study of Aluminium Smelters

The appearance of cancer cases at various sites among the workers of an aluminium smelting plant of south-western Sardinia suggested the opportunity of following up their mortality by a cohort study. A list of 1,984 employees who had ever worked at the plant was provided by the company. A preliminary study has been conducted on 1,148 male blue collar workers with at least one year's employment in 1971-1980, followed up through December 1990. Expected deaths were derived from age- and year-specific mortality rates for the Sardinian male population. A total of 48 deaths occurred in the cohort. An excess risk for pancreatic cancer (3 observed vs 0.8 expected) was the only significant finding. Other excesses were observed for cancer of the oral cavity (2 observed vs 0.8 expected) and lymphomas and multiple myeloma considered jointly (3 observed vs 1.2 expected). An attempt to relate the mortality experience to workplace exposure is currently under way.

TYPE: Cohort
TERM: Occupation
SITE: Bladder; Lung; Lymphoma; Pancreas
CHEM: Aluminium; Fluorides; PAH; Tars
OCCU: Smelters, Aluminium
TIME: 1990 - 1992

460   Cocco, P.L.   04737
      Univ. di Cagliari, Ist. di Medicina del Lavoro, Via S. Giorgio 12, 09124 Cagliari, Italy (Tel.: +39 70 670481; Fax: 654350)
COLL: Todde, P.F.; Manca, P.; Dessi, S.

Mortality of G6PD-Deficient Subjects
The importance of investigating the consequences of the genetic deficiency of the enzyme glucose-6-phosphate dehydrogenase (G6PD) has been stressed, particularly concerning cancer and cardiovascular diseases. To test the hypothesis of health consequences of G6PD deficiency, a cohort about 2,000 male subjects (all ages), found to be G6PD deficient during a regional screening campaign conducted in Sardinia in 1981 was collected. The mortality of the cohort will be ascertained from 1982 through 1991, evaluating the causes of death reported on the death certificates. Standardized mortality ratios will be calculated for cardiovascular diseases and all cancers, as well as for other selected causes of death, using the Sardinian male population age- and year-specific mortality rates as reference.

TYPE: Cohort
TERM: Enzymes; G6PD; Genetic Factors
SITE: All Sites
TIME: 1990 - 1992

* 461   Cocco, P.L.   05161
      Univ. di Cagliari, Ist. di Medicina del Lavoro, Via S. Giorgio 12, 09124 Cagliari, Italy (Tel.: +39 70 670481; Fax: 654350)
COLL: Carta, P.; Flore, C.; Cherchi, P.

Mortality in Lead and Zinc Foundries: Follow-Up 1932-1990
The mortality experience of workers in two lead and zinc foundries in southern Sardinia for the periods 1932-1971 and 1970-1989 respectively, is currently under investigation. Members of the older cohort were 1,950 male employees with a minimum of one year of employment in the same foundry. Vital status has been ascertained for about 95% of this cohort; death certificates have been obtained for 505 out of 548 deceased subjects and coded to ICD-9. Limited exposure data are available for lead, cadmium, arsenic and silica. Standardized mortality ratios will be calculated for groups of diseases and single causes of interest deriving expected numbers from age- and year-specific mortality rates for the Sardinian male population. The ascertainment of the vital status of the members of the younger cohort is still in progress.

TYPE: Cohort
TERM: Dusts; Metals; Occupation
SITE: Bladder; Kidney; Leukaemia; Liver; Lung
CHEM: Arsenic; Cadmium; Lead; PAH; Silica; Zinc
OCCU: Smelters, Lead
TIME: 1984 - 1992

CASALE MONFERRATO

462   Betta, P.G.   03754
      Santo Spirito Hosp., Lab. of Pathological Anatomy, Viale Giolitti 2 , 15033 Casale Monferrato, Italy (Tel.: (0142)334368)
COLL: Robutti, F.; Crosignani, P.; Donna, A.F.; Malfatto, F.

ITALY

Exposure to Agricultural Chemicals and Occurrence of Malignant Lymphomas
This hospital-based case-control study is designed to investigate the relationship between exposure to chemicals in rural areas and the incidence of Hodgkin's and non-Hodgkin's lymphomas (NHL). Results of previous experimental studies in mice suggest a positive association between exposure to some herbicides and NHL. 60 cases have been identified from the tumour registry at the City Hospital. All cases have been histologically confirmed. 120 controls have been randomly selected from patients admitted with clinical picture similar to that of the cases, but with benign pathology. Interviews of both cases and controls are underway and particular attention is paid to residential and occupational history.

TYPE: Case-Control
TERM: Chemical Exposure; Environmental Factors; Fungicides; Herbicides; Occupation; Pesticides; Registry; Rural
SITE: Hodgkin's Disease; Non-Hodgkin's Lymphoma
OCCU: Farmers
TIME: 1986 - 1992

463 Botta, M. 04084
Osp. Santo Spirito, Reparto Medicina Donne, Via Giolitti 2, 15033 Casale Monferrato, Italy (Tel.: +39 142 344928)
COLL: Castagneto, B.; Cocito, V.; Magnani, C.; Terracini, B.; Brusa, M.; Degiovanni, D.

Cohort Study of Workers in Asbestos-Cement Production in Casale Monferrato, Italy
A cohort of 3,370 workers in a factory producing asbestos-cement products in 1950 or after is being followed-up. The plant started the activity in 1907 and employed up to 1,600 workers (both sexes). Environmental conditions are described as rather poor in the past but no measures of asbestos concentration are available before 1978. Mortality in the cohort has been compared with Italian mortality rates and with local rates as well. A paper has been published in Med. Lav., 78(6):451-453, 1987 (Magnani, C. et al.). The study is being extended to clerks working in the same plant. Work history within the plant is being reconstructed for all workers in the cohort from administrative and medical records. The feasibility of a nested case-control study of respiratory cancers is under evaluation.

TYPE: Cohort
TERM: Chemical Exposure; Dusts; Occupation
SITE: Respiratory
CHEM: Asbestos
OCCU: Cement Workers; Office Workers
TIME: 1985 - 1992

464 Botta, M. 04357
Osp. Santo Spirito, Reparto Medicina Donne, Via Giolitti 2, 15033 Casale Monferrato, Italy (Tel.: +39 142 344928)
COLL: Cocito, V.; Magnani, C.; Terracini, B.; Brusa, M.; Degiovanni, D.; Castagneto, B.

Relatives of Workers in Asbestos Cement Production in Casale Monferrato, Italy
Several reports have suggested a carcinogenic risk associated with non-occupational exposure to asbestos but only a few formal epidemiological studies have been carried out. The present project is an expansion of an occupational cohort mortality study of workers in asbestos-cement production in Casale Monferrato at work in 1950 or afterwards. A large excess in mortality from respiratory cancers and asbestosis has been found (Magnani C., Med. Lav. 78(6):451-453, 1987). Wives of asbestos cement workers included in the cohort are being identified through data available from the registrar's offices of the municipalities where their husbands lived. They will be followed through 1964-1988 and the causes of death will be collected. It is estimated that number of person-years at risk will be around 37,000, which corresponds to a minimum detectable relative risk of about 3 for lung cancer ($p = 0.05$, power = 80%). A pilot study has been conducted with good results. Wives of clerks working in the same plant will be included as referents. The feasibility of the inclusion of the offspring in the cohort will be evaluated.

ITALY

TYPE: Cohort
TERM: Chemical Exposure; Dusts; Environmental Factors; Minerals
SITE: All Sites; Lung
CHEM: Asbestos
TIME: 1988 - 1992

CASTELLANA GROTTE

465   Leandro, G.   04464
Osp. Spec. Gastroenterologia, "S. De Bellis", Ist. di Ricovero, e Cura a Carattere Scientifico, Viale Valente, 70013 Castellana Grotte (BA), Italy (Tel.: +39 80 735122)

Hepatocellular Carcinoma in Cirrhotic Patients. A Follow-up Study

Hepatocellular carcinoma (HCC) is one of the most common cancers in the world. Liver cirrhosis could itself be a premalignant condition acting as a promoter, since about 75% of HCC are associated with cirrhosis. This study, therefore, aims at following cirrhotic patients in order to determine which possible risk factors are associated with HCC. The clinical and serological data of cirrhotic patients admitted to our hospital will be analysed through probabilistic models (linear discriminant fuction analysis, multiple logistic regression, etc.). These models provide a powerful method of computer-assisted diagnosis. This should permit (1) determination of the risk factors for HCC in cirrhotic patients; (2) evaluation of their possible epidemiological relevance (e.g. the relation of the HBsAG carrier state to HCC); (3) making diagnostic laboratory tests more effective. In this way, sub-populations at highest risk of HCC can be identified: intensive clinical screening and periodic follow-up would make earlier diagnosis possible and consequently improve the prognosis. To date 645 cirrhotic patients have been included in the study. About 1,000 are expected.

TYPE: Case Series; Cohort
TERM: Cirrhosis; HBV; High-Risk Groups; Premalignant Lesion
SITE: Liver
TIME: 1985 - 1992

CATANIA

466   Belfiore, A.   04924
Univ. of Catania, Cattedra di Endocrinologia, Osp. Garibaldi, Piazza S. Maria di Gesu , 95124 Catania, Italy (Tel.: +39 95 2544856; Fax: 7158072)
COLL: Garofalo, R.; Giuffrida, D.; La Rosa, G.L.; Ippolito, O.; Fiumara, A.

Thyroid Cancer Associated with Graves' Disease

The aim of the study is to test the hypothesis that hyperthyroidism, in particular Graves' Disease, is a risk factor for thyroid cancer. In a previous study (J. Clin. Endocrinol. Metab. 70:830-835, 1990) 359 hyperthyroid patients and 582 euthyroid patients who underwent surgery during a 6-year period were examined retrospectively. It was found that thyroid cancer was most aggressive when associated with Graves' disease, least aggressive when associated with toxic adenoma, and with intermediate aggressiveness in euthyroid patients. It was therefore suggested that thyroid stimulating anti-bodies in Graves' disease may stimulate thyroid cancer growth and aggressiveness by acting through the TSH receptor. All patients seen in this clinic (which represent the great majority of all thyroid patients seen in an area of about one million inhabitants), with one or more thyroid nodules, will be selected for surgery and followed up for occurrence of thyroid cancer.

TYPE: Cohort
TERM: Hyperthyroidism
SITE: Thyroid
TIME: 1988 - 1992

FLORENCE

467   Bechi, P.   03479
Clinica Chirurgica III, Viale Morgagni 85, 50134 Florence, Italy (Tel.: +39 55 412029; Fax: 4240133)
COLL: Becciolini, A.; Balzi, M.; Amorosi, A.; Bogani, S.; Bonanomi, A.; Dei, R.

ITALY

Partial Gastrectomy and Cancer of the Gastric Stump
Partial gastrectomy has been considered as predisposing to gastric cancer. The study aims to identify phenotypic changes in the pre-neoplastic state. Studying a series of 850 patients partially gastrectomized 15-25 years previously for peptic ulcer it was possible to demonstrate a correlation between foveolar hyperplasia (the most typical post-gastrectomy histological feature) and bile reflux. Since bile acids have been experimentally shown to stimulate ileal and colonic cell proliferation and cell proliferation is central in the initation of carcinogenesis, the relationships between bile reflux, cell proliferation kinetics and post-gastrectomy gastric histological changes seem worth of careful investigation. For this reason concomittantly with the main study on the total number of subjects (with mulitple biopsies after an initial evaluation of gstric nitrites and enterogstric reflux), a second study has started in 40 subjects randomly chosen from the total. In these subjects gastric proliferative activity will also be assessed by means of 3H Thymidine uptake in biopsy specimens. Cell kinetics findings will be correlated with histological and reflux findings, in order to find out whether foveolar hyperplasia is associated with significant qualitative and quantitative cell kinetics modification. Moreover, these results will be correlated with those obtained in the progression of the endoscopic-histological follow-up and with possible cancer-developement in the different subjects. A paper was published in Am. J. Gastroenterol., Oct. 1991.

TYPE: Cohort; Incidence
TERM: Biopsy; Gastrectomy; High-Risk Groups; Premalignant Lesion; Prevalence
SITE: Stomach
CHEM: N-Nitroso Compounds; Nitrites
TIME: 1982 - 1993

468 Carli, P. 04904
Univ. di Firenze, 2nda Clinica di Dermatologia, Via della Pergola 58 , 50121 Florence, Italy (Tel.: (055)2478356; Fax: 2478356 (2 pm-8 am))
COLL: Biggeri, A.; Bondi, R.; Urso, C.; Giannotti, B.

Dysplastic and Atypical Naevi as Risk Factors for Malignant Melanoma
The aim of this study is to investigate, in a mediterranean population, the relationship between cutaneous melanoma (CM) and the presence of atypical and dysplastic naevi. The number of common acquired naevi (and their topographic distribution) and the presence of small congenital naevi will also be evaluated. Other known risk factors will be investigated as potential confounders (acute and chronic sunlight exposure, phenotype, phototype). Physical assessment of skin colour and measurement of minimal erythemal dose will be performed. Cases are patients admitted in the period 1990-1992 to the 2nd Dermatology Clinic of the University of Florence, with a histologically confirmed diagnosis of CM, residents in the Florence area and aged 20-69 years. 150 or more cases are expected. A random sample of the resident population drawn from computerized demographic files will be utilized as the source of the control group. Two controls per case will be matched for sex, age and place of residence.

TYPE: Case-Control
TERM: Naevi; Pigmentation; Premalignant Lesion; Radiation, Ultraviolet
SITE: Melanoma
TIME: 1990 - 1992

469 Chellini, E. 04610
Centre for Study & Prevention of Cancer, (CSPO), Viale A. Volta 171 , 50131 Florence, Italy (Tel.: (055)578062; Fax: 578955)
COLL: Buiatti, E.; Merler, E.; Paci, E.; Seniori Costantini, A.R.; Zappa, M.; Biancalani, M.; Dini, S.; Negri, G.; Pingitore, R.; Megha, T.; Tosi, P.; Paoletti, L.

Malignant Mesothelioma in Tuscany
All incident cases of malignant mesothelioma, histologically diagnosed in the Pathology Institutes involved in the project, are collected for the region of Tuscany. In order to obtain a reliable histological diagnosis, the histological sections of all cases are reviewed by the pathologists of the regional panel for malignant mesothelioma, and a consensus diagnosis is made, also using immunohistochemical techniques. Cases or, if deceased, their nearest relatives are interviewed by a trained interviewer in order to collect data on occupational and environmental exposure to asbestos or other risk factors. Detection of mineral fibres in pleura and lung tissue of those cases of pleural malignant mesothelioma who undergo surgical treatment is done by transmission electron microscopy and associated methods. A retrospective study was carried out on cases diagnosed in three Pathology Institutes in the period 1970-1987. This mapping project permits the evaluation of past occupational asbestos exposure, which

may also be present today, and subsequently the implementation of preventive measures. The relevance of non occupational exposures will also be evaluated.

TYPE: Case Series; Cross-Sectional
TERM: Dusts; Histology; Immunologic Markers
SITE: Lung; Mesothelioma; Pleura
CHEM: Asbestos; Mineral Fibres
TIME: 1988 - 1993

470 Geddes, M. 04627
Center for Study and Prevention, of Cancer, Epidemiology Sect., Tuscany Cancer Registry, 171 Viale Volta, 50131 Florence, Italy (Tel.: (055)576226; Fax: 578955)
COLL: Paci, E.; Masala, G.; Vannucchi, G.; Zappa, M.

Adequacy of Information Collected from Substitutes in Case-control Studies
The study compares information on smoking habits and occupational history collected from two sources. The first source is represented by personal interviews with 255 males with lung cancer, resident in the province of Florence, who were interviewed in the period 1980-84 during a case-control study. All the cases died before 1988. The second source is a self-administered questionnaire completed by a close relative, generally the wife, to whom a questionnaire similar to the one used in the first interview was sent by mail. Non-responders are interviewed by telephone. The aims of the study are (1) to evaluate the concordance between the original information and the information obtained from the proxy; and (2) to verify if the substitutes' compliance with the questionnaire depends on variables such as degree of relationship with the deceased, age, place of residence and level of education.

TYPE: Methodology
TERM: Occupation; Tobacco (Smoking)
SITE: Lung
TIME: 1987 - 1992

471 Geddes, M. 04201
Center for Study and Prevention, of Cancer, Epidemiology Sect., Tuscany Cancer Registry, 171 Viale Volta, 50131 Florence, Italy (Tel.: (055)576226; Fax: 578955)
COLL: Balzi, D.; Biggeri, A.; Buiatti, E.; Chellini, E.; Cecconi, R.; Gaspari, R.

Cancer Risk in Italian Migrants within Italy
The aim of the study is to verify the risk for all cancers and for some specific cancer sites in the Italian population migrated from the South and North-East to some northern areas of Italy, compared with the local population. The hypothesised protective effect of being born in the South or in the North-East of Italy will be estimated in relation to the duration of the period of migration, to the age at migration and to the specific area of birth. The study is a multicentric case-control study. Cases are all cancer decendents aged 35-84 years in 1985-89 who were resident in the municipalities of Bologna, Firenze, Genova, and Torino. Two controls are matched to each case by sex, year of birth and place of residence. For each person information regarding demographic data, place of birth and year of immigration will be collected. Relative risks and mortality rates will be calculated for several cancer sites, evaluating the effect of sex, age at migration, period of immigration and duration of residence in the immigration area.

TYPE: Case-Control
TERM: Age; Geographic Factors; Migrants; Registry; Time Factors
SITE: All Sites
TIME: 1987 - 1992

472 Paci, E. 04640
Center for Cancer Study & Prevention, Epidemiology Sect., Viale Volta 171 , 50131 Florence, Italy (Tel.: +39 55 578062; Fax: 578955)
COLL: Zappa, M.

Lung Tumours in Textile Workers in Prato (Italy)
Following a previous report where an excess of risk for lung tumours among textile workers in Prato was found, a case-control study is being carried out to re-evaluate this evidence. Cases are incident cases which occurred over the period 1984-1987, collected by the Cancer Registry. Controls are deaths for selected causes, not related to asbestos exposure and matched (1:3) to cases by age and sex. A postal questionnaire is sent to the next of kin of both the cases and controls. Analysis will produce estimates of relative risks using stratified and logistic analysis. 224 cases have been enrolled in the study so far.

ITALY

TYPE: Case-Control
TERM: Dusts; Occupation; Registry
SITE: Lung
CHEM: Asbestos
OCCU: Textile Workers
REGI: Tuscany (Ita)
TIME: 1986 – 1992

473 Palli, D. 04273
Center for Study and Prev. of Cancer, Epidemiology Sect., Viale Volta 171 , 50131 Florence, Italy (Tel.: +39 55 578062; Fax: 578955)
COLL: Buiatti, E.; Carli, S.; Ciatto, S.; Paci, E.; Rosselli del Turco, M.

Screening for Breast Cancer: Evaluation by a Case-Control Study

A population based screening programme for breast cancer has been started in 1970 in a rural area near the city of Florence. The resident female population in the age-group 40-70 years (34,000 women at the 1981 census), is invited, every two years, to have a mammographic examination (double-view) in a mobile unit. Recently a case-control study has been carried out (Palli et al. Int. J. Cancer 38:501-4, 1986) showing a significant protective effect for participating women in the older age-group (50+ years). Monitoring of the results by means of an extension of the case-control study has been planned. Additional cases (women dead from breast cancer with a diagnosis after the start of the programme) will be identified every year on the basis of the mortality data available for the area; a matched group of controls will be periodically sampled from the general population. An average of 20 cases is expected every year.

TYPE: Case-Control
TERM: Screening
SITE: Breast (F)
TIME: 1987 – 1992

474 Seniori Costantini, A.R. 03510
Center for Study and Prev. of Cancer, Epidemiology Sect., Viale Volta 171 , 50131 Florence, Italy (Tel.: +39 55 578062; Fax: 578955)
COLL: Chellini, E.; Miligi, L.R.; Merler, E.; Scarpelli, A.R.

Occupational Cancer Risk Assessment in Tuscany

The aim of the project is to define "maps" of exposure in several major industrial sectors in Tuscany. In the period 1984-1988 data about leather and textile industries were collected with the cooperation of the Occupational Local Health Services of the Tuscan Region. Information concerns number and type of departments, number of workers in different jobs, chemicals used. A computer software has been implemented with the aim of defining job/exposure matrices to carcinogens. Extension of the project will concern wood and furniture, glass, pottery and clothing industry. Mapping of exposure to asbestos in all industrial archives of the region is now under way. Updating of data is planned annually.

TYPE: Methodology
TERM: Dusts; Glass; Leather; Occupation; Solvents; Textiles; Wood
SITE: All Sites
CHEM: Asbestos
OCCU: Asbestos Workers; Cabinet Makers; Glass Workers; Leather Workers; Potters; Textile Workers; Wood Workers
TIME: 1989 – 1993

GENOA

* 475 Gennaro, V. 05211
Cancer Research Inst., Dept. of Epidemiology & Biostatistics, Viale Benedetto XV, 10 , 16132 Genoa, Italy (Tel.: +39 10 357787; Fax: 352999 ; Tlx: 286356 istex i)
COLL: Fontana, V.; Ceppi, M.; Puntoni, R.; Doria, M.; Beggi, A.; Verganelli, E.; Manti, A.

Cohort Study Among Oil Refinery Workers in Genoa

The cause-specific mortality in about 1,500 blue- and white-collar workers employed between 1949 and 1988 at an oil refinery in Genoa is being analysed, adjusting for sex, age and calendar year. Mortality is compared with the general population. Blue-collar workers are considered as potentially exposed to hydrocarbons, asbestos, nickel, chromium and other agents. 101 persons (6.7%) have been lost to follow-up. 196 deaths have been observed and evaluated by Cox regression, SMR and Kaplan-Meier analyses. Preliminary analysis show, using internal comparisons, a statistically significant excess ($p < 0.05$) in the exposed groups for all causes of death, accidents, ill-defined diseases, non-neoplastic diseases of the digestive and respiratory tract, cancer of the oesophagus, respiratory tract, mesothelioma and leukaemias.

TYPE: Cohort
TERM: Dusts; Occupation; Solvents
SITE: All Sites
CHEM: Asbestos; Benzene; Chromium; Hydrocarbons; Nickel
OCCU: Petroleum Workers
TIME: 1988 - 1993

476 Puntoni, R. 03243
Inst. of Oncology, Dept. Epidemiology & Biostatistics, Viale Benedetto XV, 10, 16132 Genoa, Italy (Tel.: +39 10 300767)
COLL: Manniello, M.A.; Patrone, M.G.

Italian Dietary Habits in Relation to Cancer

The aim of this study is to collect detailed information on the diet of the Italian population by sex, age and geographical area. Dietary interviews of a random sample of people with a telephone are being organized. Questionnaires based on 24-hour recall and weekly frequency of food consumption will be used. It is intended to interview by telephone at least 1,000 persons in different parts of Italy selected on the basis of differing incidence of cancer of the gastro-intestinal tract. It will then be possible to correlate geographically cancer mortality rates (particularly cancer of the gastro-intestinal tract) with dietary habits.

TYPE: Correlation
TERM: Diet
SITE: All Sites; Gastrointestinal
TIME: 1988 - 1992

477 Puntoni, R. 03818
Inst. of Oncology, Dept. Epidemiology & Biostatistics, Viale Benedetto XV, 10, 16132 Genoa, Italy (Tel.: +39 10 300767)
COLL: Merlo, F.; Ceppi, M.; Stagnaro, E.

Cancer Mortality among Genova Dockyard Workers

The objective of the present study is to evaluate the mortality experience in longshoremen employed in the shipyard of Genova. Since several substances regularly handled in this branch of the shipyard are likely to represent a hazard for the workers, the aims of the study are: (1) to identify sub-groups of the worker population at risk; (2) to identify the substances reponsible for the occurrence of higher than expected cancer specific mortality figures; (3) to define cancer as well as non-cancer mortality rates by age, length of exposure, age at entry and job title. Smoking habits will be accounted for using indirect estimates as reported by some authors. The study refers to the period 1974-1985. Mortality data will be collected using standard techniques in order to guarantee comparability between study and referent groups. 4,500 workers have been include in the study, accounting for about 50,000 person-years of observation.

TYPE: Cohort
TERM: Age; Chemical Exposure; Occupation; Time Factors; Tobacco (Smoking)
SITE: All Sites
OCCU: Shipyard Workers
TIME: 1985 - 1992

478 Vercelli, M. 04511
Inst. of Clinical, & Experimental Oncology, Dept. of Descriptive Epidemiology, Viale Benedetto XV, 10, 16132 Genoa, Italy (Tel.: +39 10 3534961; Fax: 352999 ; Tlx: 216353 istex i)
COLL: Puntoni, R.; Ceppi, M.; Orengo, A.; Reggiardo, G.; Orlandini, C.; Casella, C.; De Lucia, G.

Cancer Incidence in Genoa by Place of Birth and Length of Residence

Previous mortality study for breast, stomach and lung tumours in Genoa's residents coming from other areas of Italy and treated in Genoa, has pointed out interesting results which confirm the importance of migration as a risk factor for these tumours. This study proposes to analyse with the same method,

ITALY

1986-1987 incidence data deviving from Genoa Cancer Registry (25,000 inhabitants, 3,500 cancer cases/year) and to verify the concordance of incidence and mortality risks. In particular, the aim is to understand the role of incidence data in giving more indications on cell morphology.

TYPE: Incidence
TERM: Environmental Factors; Geographic Factors; Latency; Migrants; Registry; Time Factors
SITE: All Sites
REGI: Genoa (Ita)
TIME: 1991 - 1993

L'AQUILA

479 Corrao, G. 03967
Univ. of L'Aquila, School of Medicine, Dept. of Int. Medicine & Publ. Health, Via Verdi 28, 67100 L'Aquila, Italy (Tel.: +39 862 433485; Fax: 433433)
COLL: Russo, R.; Carle, F.; Lepore, A.R.; di Orio, F.; Vineis, P.; Ciccone, G.

Evaluation of Cancer Risk in Farmers with Professional Exposure to Pesticides
A follow-up study is being conducted on 40,000 men living in different regions of Italy (South Piedmont, Abruzzo, Molise) who were licensed to use potentially toxic pesticides during the period 1970-74. In comparison with the total experience of the population in the study area, the Piedmont sub-cohort showed an apparent excess of skin malignancies and lymphomas. A follow-up for mortality is now underway in order to confirm the preliminary findings. Mortality rates in the cohort will be compared with the experience of the population in the study area. A limitation of the follow-up study is the non-availability of information about exposure to specific pesticides. Therefore, a case-control study within the cohort is planned. In this study, cases with malignancies for which an excess has been suggested in the follow-up study will be compared with a sample of the total cohort. The purpose is to interpret any cancer excess in the cohort in terms of specific exposures. Information will be collected through interviews with cases and controls or their relatives.

TYPE: Case-Control; Cohort
TERM: Chemical Exposure; Occupation; Pesticides
SITE: All Sites; Brain; Haemopoietic; Skin
OCCU: Farmers
TIME: 1985 - 1993

480 Corrao, G. 04822
Univ. of L'Aquila, School of Medicine, Dept. of Int. Medicine & Publ. Health, Via Verdi 28, 67100 L'Aquila, Italy (Tel.: +39 862 433485; Fax: 433433)
COLL: Busellu, G.P.; Di Placido, R.; Recchia, C.; di Orio, F.

Diet and Other Risk Factors for Gastric Cancer
Excess mortality from gastric cancer has been observed in L'Aquila district. A case-control study has been designed in order to evaluate risk factors for this cancer. Cases will be 120 new histological diagnoses of gastric cancer admitted to endoscopic services of L'Aquila Province in the course of two years. In the same period 240 control patients admitted to provincial hospitals for acute, non-neoplastic and non-digestive disease will be recruited. Trained interviewers will identify and question cases and controls, using standard questionnaires to obtain information on personal characteristics, alcohol and smoking habits, dietary history and use of certain drugs. The role of these factors and their interactions will be quantitatively estimated.

TYPE: Case-Control
TERM: Alcohol; Diet; Drugs; Tobacco (Smoking)
SITE: Stomach
TIME: 1990 - 1992

MILAN

481 Berrino, F. 03480
Ist. Nazionale per lo Studio, e la Cura dei Tumori, Servizio di Epidemiologia, Via G. Venezian 1, 20133 Milan, Italy (Tel.: +39 2 2366342; Fax: 2362692 ; Tlx: 333290 tumist i)
COLL: Ferrario, F.; Macaluso, M.; Pisani, P.; Baldasseroni, A.; Axerio, M.

ITALY

Occupational Exposures and the Aetiolology of Kidney Cancer and Malignant Lymphomas
Population-based case-control studies are increasingly used to test associations between chemicals and neoplasms through the use of job-histories as proxies for the history of exposure to chemicals. This study aims to apply this methodology in evaluating the role of a series of substances and occupations in the aetiology of kidney cancer and malignant lymphomas. All incident cases in the Varese Province (where the Lombardy Cancer Registry is situated) during the period 1986-1989 will be included. Expected numbers are 200 and 400 respectively for kidney cancer and lymphoma cases. Two controls per case are extracted from the electoral rolls of the province. Occupational histories are collected by means of a structured questionnaire, developed in detail for any occupational setting where exposures are likely to occur. A handbook for easy consultation by interviewers was developed. The list of exposures/occupations to be tested is the following: asbestos, polycyclic aromatic hydrocarbons, arsenic, beryllium, chromium, cadmium, lead, nickel and their compounds, chlorophenols and phenoxy-acids, BCME and CMME, chloropyrene, vinyl chloride, styrene, benzol chloride, dimethylsulphate, epichlorohydrin, formaldehyde, trichloroethylene, tetrachloroethylene; leather and allied industries, wood and allied industries, pulp and paper manufacturing, and the rubber industry. Information about known potential confounders is also collected through interview.

TYPE: Case-Control
TERM: Chemical Exposure; Dusts; Leather; Metals; Occupation; Pesticides; Plastics; Registry; Rubber; Solvents; Wood
SITE: Kidney; Lymphoma
CHEM: Arsenic; Asbestos; Benzoyl Chloride; Beryllium; Cadmium; Chlorophenols; Chloropyrenes; Chromium; Dimethylsulphate; Epichlorohydrin; Formaldehyde; Hydrocarbons; Hydrocarbons, Halogenated; Lead; Nickel; PAH; Phenoxy Acids; Styrene; Tetrachloroethylene; Trichloroethylene; Vinyl Chloride
REGI: Varese (Ita)
TIME: 1984 - 1992

482 **Berrino, F.** 04421
Ist. Nazionale per lo Studio, e la Cura dei Tumori, Servizio di Epidemiologia, Via G. Venezian 1, 20133 Milan, Italy (Tel.: +39 2 2366342; Fax: 2362692 ; Tlx: 333290 tumist i)
COLL: Pisani, P.; Muti, P.; Crosignani, P.; Micheli, A.; Totis, A.; Fissi, R.; Mazzoleni, C.; Pierotti, M.; Secreto, G.

Hormones and Diet in the Aetiology of Breast Cancer
The ORDET (Ormoni e Dieta nella Eziologia dei Tumori) is a prospective study of 10,000 healthy women to be recruited in Varese Province and to be followed up through the local Cancer Registry. The major aims are to clarify the role of endogenous hormones and dietary habits in the aetiology of breast cancer, and to study the relation between these two factors. Recruitment began in 1987 and is expected to be completed by 1991. Each woman is requested to provide both standardised information and biological specimens to be stored at -80° C. Information is available on: social and cultural factors, menstrual and reproductive history, diet and alcohol consumption, history of nutritional status and physical activity, various anthropometric measurements, clinical breast examination or mammography, blood pressure, pulse rate, cutaneous sebum production, hirsutism score and psychological tests. The biological bank includes aliquots of serum, plasma, red blood cell membranes, red blood cell cytoplasm, leucocytes, prepared from blood samples taken in the early morning on an empty stomach, aliquots of 12hr urine collections and a toenail sample. 135 incident cases are expected within the first ten years of follow-up.

TYPE: Cohort
TERM: Alcohol; BMB; Diet; Hormones; Mammography; Physical Activity; Physical Factors; Registry; Reproductive Factors; Socio-Economic Factors; Toenails; Urine
SITE: Breast (F)
REGI: Varese (Ita)
TIME: 1987 - 2000

483 **Bertazzi, P.A.** 03709
Univ. Milano, Inst. Occupational Health, Labour Clinic "Luigi Devoto", Dept. of Epidemiology, Via San Barnaba 8, 20122 Milan, Italy (Tel.: +39 2 5450812; Fax: 55187172)
COLL: Pesatori, A.C.; Zocchetti, E.; Guercilena, S.; Consonni, D.; Landi, M.T.; Tironi, A.

Long-Term Investigation of the Cancer Risk of Persons Potentially Exposed to TCDD after the Seveso Accident in 1976
The aim of the project is to investigate the possible long term effects of the accident on the health experience of the population living in the affected area. Nearly 300,000 subjects were included in the

ITALY

study; ten percent of them had been living in zones definitely or probably contaminated by TCDD, and the remaining were enrolled and concurrently examined as reference group. The types of studies feasible turned out to be mortality and cancer incidence. Mortality results for the period 1976-1986 have been published (Am. J. Epidemiol. 129, 1187, 1989; Scand. J. Work Environ. Health 15, 85, 1989; Med. Lav. 80, 316, 1989); they showed an increase of cardiovascular deaths early after the accident, and suggestive increases of certain types of cancers without, however, definite and consistent patterns. The collection of cancer incidence data, 1976-1986, is nearly completed. All hospital records in the Region of Lombardy containing any kind of indication of cancer problems have been linked with the vital statistics records of people residing in the Seveso area since the accident date. After the linkage, each record is checked against the original medical documentation in the relevant hospitals. Hypotheses of specific associations between particular types of cancer and the exposure of interest will be tested by means of case-control studies with both cases and referents sampled from the same source population. Papers were published in Am. J. Epidemiol. 129:1187-1200, 1989 and Scand. J. Work Environ. Hlth 15:85-100, 1989.

TYPE: Case-Control; Cohort
TERM: Chemical Exposure; Environmental Factors; Record Linkage
SITE: All Sites; Leukaemia; Lymphoma; Soft Tissue
CHEM: Dioxins
TIME: 1985 - 1993

* 484 Bertazzi, P.A. 05220
Univ. Milano, Inst. Occupational Health, Labour Clinic "Luigi Devoto", Dept. of Epidemiology, Via San Barnaba 8, 20122 Milan, Italy (Tel.: +39 2 5450812; Fax: 55187172)
COLL: Della Foglia, M.; Castiglione, G.; Donelli, S.; Zocchetti, C.; Pesatori, A.C.; Consonni, D.

Cancer Risk in Workers Exposed to Dyes and Formaldehyde in Textile Production

After identification of all textile dyeing and finishing shops located within the territories of three Local Health Units in northern Italy, enumeration of all persons employed during the period 1950-1979 has been completed. The production process in each shop was examined in detail and each job characterised in terms of the type of chemicals used and the time period in use. Individual work histories are available. It is anticipated that the minimum exposure categorisation will be: dyes (and possibly specific types of dyestuff); dyes and formaldehyde; formaldehyde; others. Additional exposure information is also being retrieved. Around 5,000 workers will meet the admission criteria. Their mortality will be examined for the period 1950-1990. Local and regional mortality rates are available to compute age-, sex-, year-, and cause-specific expected figures. For workers alive as of 1977, cancer morbidity will also be examined with the cooperation of the regional hospital discharge registration system of the Lombardy Tumour Registry. The focus of the study is on bladder cancer risk (18-22 cases expected) and exposure to suspected carcinogenic dyes, and formaldehyde exposure and nasal and pharyngeal cancer (6-8 cases expected).

TYPE: Cohort
TERM: Dyes; Occupation
SITE: All Sites; Bladder; Nasal Cavity; Pharynx; Respiratory
CHEM: Benzidine; Chromium; Formaldehyde
OCCU: Textile Dyers
REGI: Varese (Ita)
TIME: 1991 - 1994

485 Crosignani, P. 03452
Ist. Nazionale per lo Studio, e la Cura dei Tumori, Serv. di Epidemiologia, Via Venezian 1, 20133 Milan, Italy (Tel.: +39 2 2366342; Tlx: 333290 tumist i)
COLL: Viganò, C.; Berrino, F.; Gatta, G.; Sant, M.

Incidence of Malignant Tumours among Native and Immigrant Groups in Varese Province

The study aims to (1) evaluate birthplace and eventual subsequent migration in relation to incidence of malignant neoplasms among residents in the Varese Province; and (2) estimate the proportion of observed differences which can be explained in terms of different distributions of known causal (or preventive) agents in diet, work environments, and tobacco smoking. Information about the birthplace of incident cases registered by the Lombardy Cancer Registry during 1976-1987 will be integrated with a complete migration history collected through municipalities. Comparable information will be obtained for a random sample of the resident population, extracted from electoral rolls. The distribution of age-specific proportions of natives and immigrants will be used to estimate the denominators of incidence rates. A case-control analysis will be applied to evaluate age at immigration as a risk indicator.

A subset consisting of 1,000 out of the 3,500 controls have already been interviewed about current diet, smoking habits, and occupational history. The distribution of these variables according to birthplace and migration history will be analysed. Associations found will help to evaluate the plausibility of explaining risk differences observed between natives and immigrant groups, in terms of confounding by known environmental risk factors.

TYPE: Correlation; Incidence
TERM: Age; Diet; Environmental Factors; Migrants; Occupation; Prevalence; Registry; Tobacco (Smoking)
SITE: All Sites
REGI: Varese (Ita)
TIME: 1984 - 1992

486 Decarli, A. 03790
Univ. of Milano, Inst. of Biometry & Medical Statistics, Via Venezian 1 , 20133 Milan, Italy (Tel.: +39 2 2361302; Fax: 2362930)
COLL: La Vecchia, C.; Mazzoleni, C.; Salvini, S.; Negri, E.

Risk Factors for Multiple Myeloma in Italy

The present research project is aimed at analysing risk factors for multiple myeloma through a standard case-control design. At the moment, the descriptive epidemiology of this neoplasm in various geographical areas has been considered in detail, chiefly on account of increasing incidence and mortality rates in several countries. However the causes of myelomatosis (with the possible exception of ionising radiation) are still poorly understood. Data collection for this study started in January 1985, on the basis of information obtained in a previous pilot phase and of a network of cooperating hospitals in which a case-control surveillance of several neoplastic diseases has been conducted since 1979. A total of 400 cases and 800 age-sex matched hospital controls will be enrolled over a five year period. At the moment about 180 cases and 320 controls have been interviewed. The present research project will give the opportunity for analysing in detail trends in death certification rates from multiple myeloma in Italy from 1955 onwards as well as the geographical distribution of mortality from this neoplasm within various Italian provinces, thus producing a comprehensive analysis of the epidemiology of myelomatosis in Italy.

TYPE: Case-Control
TERM: Geographic Factors
SITE: Multiple Myeloma
TIME: 1985 - 1992

487 Duca, P. 04905
Univ. of Milan, Inst. of Biometry & Medical Statistics, Via Venezian 1 , 20133 Milan, Italy (Tel.: (02)2361302; Fax: 2362930)
COLL: Bisanti, L.; Braga, M.; Bellini, A.; Zucchi, A.; Musch, G.; Marubini, E.

Mortality Study of Workers Occupationally Exposed to Motor Vehicle Exhaust

This study is part of a larger collaborative project aimed to evaluate the cancer risk in workers (taxi-drivers, petrol station workers, etc.) occupationally exposed to environmental pollutants (motor-vehicle exhausts, heating exhausts, dusts) in the Italian metropolitan areas. In particular, this study will deal with a cohort of 5,000 traffic wardens active in Milan at any time from 1945-1950 to date. They are expected to contribute almost 75,000 person-years and 320 deaths. Exposure will be estimated based on length of employment and on direct (air sample analyses) and indirect (annual consumption of motor-vehicle and heating fuel) indicators of environmental pollution. Information on date and cause of death will be obtained from the Census Office at the place of residence. ICD-code of the cause of death will be provided by the National Bureau of Statistics.

ITALY

TYPE: Cohort
TERM: Air Pollution; Dusts; Occupation
SITE: All Sites; Larynx; Lung
CHEM: Vehicle Exhaust
OCCU: Drivers; Petrol Station Attendants
TIME: 1990 - 1992

488 La Vecchia, C. 03370
"M. Negri" Inst. Pharmacol. Research, Dept. of Clinical Pharmacology, Unit of Epidemiology, Via Eritrea 62, 20157 Milan, Italy (Tel.: (02)35794427; Tlx: 331268)
COLL: D'Avanzo, B.; Fasoli, M.; Franceschi, S.; Gentile, A.; Liberati, C.; Negri, E.; Parazzini, F.; Tognoni, G.; Decarli, A.

Case-Control Surveillance of Neoplastic Disease

Case-control studies on histologically confirmed cancers of several sites have been underway since 1979. Cases are patients admitted to University and General Hospitals in the Greater Milan area, and controls are persons admitted to the same network of hospitals with acute, non-neoplastic disease. Trained interviewers identify and question cases and controls, using standard questionnaires to obtain information on personal characteristics and habits including cigarette smoking, alcohol and methylxanthine consumption, a few selected dietary variables, a problem-oriented medical history, and use of selected drugs (including, for females, contraceptive practices and hormonal therapy). At present the following sites are included (approximate numbers per year given in brackets):oral cavity (20), oesophagus (40), stomach (100), intestines (colon and rectum) (150), liver (30), gallbladder and bile ducts (20), pancreas (30), larynx (20), breast (400), cervix (50), corpus uteri (50), ovary (50), prostate (80), bladder (50), kidney (30), thyroid (50), Hodgkin's disease (30), other lymphomas (30), multiple myelomas (50). All cases are histologically confirmed and diagnosed within the year prior to interview. Some 1,000 controls per year are interviewed. Recent papers appeared in Br. J. Cancer 60:385-388, 1989; Eur. J. Cancer 25:1613-1618, 1989; and in Int. J. Cancer 45:275-279, 1990.

TYPE: Case-Control
TERM: Age; Alcohol; Diet; Drugs; Hormones; Oral Contraceptives; Premalignant Lesion; Tobacco (Smoking)
SITE: All Sites
CHEM: Methylxanthines
TIME: 1979 - 1993

489 La Vecchia, C. 04062
"M. Negri" Inst. Pharmacol. Research, Dept. of Clinical Pharmacology, Unit of Epidemiology, Via Eritrea 62, 20157 Milan, Italy (Tel.: (02)35794427; Tlx: 331268)
COLL: Decarli, A.; Parazzini, F.; Negri, E.

Smoking and Tobacco Related Neoplasms in Italy

Of all cancers, tobacco-related sites showed the most pronounced rises in mortality over the last decades. The present project aims to: (1) Analyse trends in total tobacco and cigarette sales (from 1900 onwards) and smoking (from 1950 onwards) in Italy as a whole and in various geographical areas (provinces or regions) for each sex and age group; (2) measure tar, nicotine and carbon monoxide yields and market shares of various cigarette brands; (3) study trends in mortality from lung and other neoplasms, in the whole of Italy from 1955 onwards, and in various regions or provinces from 1970 onwards using cross-sectional and cohort approaches; analysis of tobacco-related risks from the case-control surveillance of neoplastic diseases contracted by this study group; (4) offer wide coverage to data on smoking and tobacco-related neoplasms to ultimately reduce smoking prevalence in the population. Trends in tobacco consumption are estimated from published and unpublished data of the State Monopoly, the Central Institute of Statistics and the Departments of Health and Finance, the major problem being to give a reasonable estimate of smuggling. Death certification rates are derived from publications and copies of the original computer tapes obtained from the Central Institute of Statistics with extracts of all primary death records. Analyses of mortality rates are based both on standard cross sectional approach (age-specific and age-standardized rates, with major attention given to the younger groups) and cohort methods including age period and cohort models. Recent papers appeared in Cancer Res. 48:7285-7293, 1988; Int. J. Epidemiol. 18:556-562, 1989; Int. J. Cancer 43:784-785, 1989.

ITALY

TYPE: Correlation; Mortality
TERM: Birth Cohort; Drugs; Mathematical Models; Time Factors; Tobacco (Smoking); Trends
SITE: Bladder; Kidney; Larynx; Lung; Oesophagus; Oral Cavity; Pancreas; Pharynx
CHEM: Carbon Monoxide; Nicotine; Tars
TIME: 1984 - 1993

490   Parazzini, F.   03183
Ist. di Ricerche Farmacologiche, "Mario Negri", Dipto. di Farmacologia Clinicale, Via Eritrea, 62,
20157 Milan, Italy (Tel.: (02)357941; Fax: 3546277 ; Tlx: 331268 NEGRI I)
COLL: La Vecchia, C.; Fasoli, M.; Decarli, A.

Risk Factors for Gestational Trophoblastic Diseases in Italy

In order to evaluate the relation of genetic and environmental factors to the aetiology of trophoblastic diseases, a case-control study was started in June 1981. Todate about 330 cases of complete and partial histologically confirmed hydatidiform mole and 60 cases of choriocarcinoma, have been collected. Controls are women admitted for normal delivery matched with cases within five year age groups. Cases and controls are identified and interviewed in a network of collaborating teaching and general hospitals in the greater Milan area. A standard questionnaire is used to obtain information on socio-demographic factors, smoking habits, alcohol and methylxanthine consumption, a few selected dietary habits, gynaecological and obstetric data, history of lifetime contraceptive practices or other female hormone use, history of diseases or other factors thought to be important in the aetiology of trophoblastic diseases, and family history of repeated spontaneous abortions or trophoblastic diseases. Papers appeared in Br. J. Obstet. Gynaecol. 93:582-585, 1986, Am. J. Obstet. Gynecol. 158:93-100, 1988 and Gynecol. Oncol. 31:310-314, 1988.

TYPE: Case-Control
TERM: Abortion; Alcohol; Contraception; Diet; Familial Factors; Hormones; Parity; Tobacco (Smoking)
SITE: Choriocarcinoma; Hydatidiform Mole
CHEM: Methylxanthines
TIME: 1981 - 1993

491   Parazzini, F.   04477
Ist. di Ricerche Farmacologiche, "Mario Negri", Dipto. di Farmacologia Clinicale, Via Eritrea, 62,
20157 Milan, Italy (Tel.: (02)357941; Fax: 3546277 ; Tlx: 331268 NEGRI I)
COLL: La Vecchia, C.; Decarli, A.; Franceschi, S.; Gramenzi, A.

Case-Control Surveillance of Breast and Female Genital Tract Cancers

Although the role of hormonal, reproductive and general lifestyle habits in breast and female genital tract carcinogenesis is well understood, several questions remain unsettled, as for instance, oral contraceptive use or oestrogen replacement therapy and the risk of breast cancer, or the interaction between parity and sexual factors in cervical cancer. To identify and better quantify relative and attributable risks in female cancers, a large case-control surveillance of histologically confirmed benign and malignant breast, ovarian, endometrial, cervical and vulvar tumours has been conducted since 1982. Controls (age-matched in five-year groups with cases) are women hospitalized for conditions other than hormonal, gynaecological or neoplastic acute disorders. Cases and controls are identified in a large network of teaching and general hospitals in the Greater Milan area. A total of about 400 cases of breast cancer, 50 ovarian, endometrial, and cervical cancer, 20 vulvar cancer and 800 controls per year will be collected. Recent papers appeared in Int. J. Epidemiol. 19:259-263, 1990, Int. J. Cancer 45:428-430, 1990 and in Am. J. Obstet. Gyn. 164:522-527, 1991.

TYPE: Case-Control
TERM: Diet; Hormones; Oral Contraceptives; Parity; Sexual Activity
SITE: Benign Tumours; Breast (F); Ovary; Uterus (Cervix); Uterus (Corpus); Vulva
TIME: 1982 - 1993

* 492   Veronesi, U.   05231
Ist. Nazionale dei Tumori, Via G. Venezian 1, 20133 Milan, Italy (Tel.: +39 2 2390; Fax: 70602991)
COLL: Costa, A.; De Palo, G.; Formelli, F.; Maltoni, C.; Marubini, E.; Coopmans de Yoldi, G.

Breast Cancer Chemoprevention with Fenretinide

The main aim of this study is to evaluate the effectiveness of the synthetic retinoid fenretinide in preventing breast cancer. Fenretinide inhibits chemically induced mammary carcinomas in different experimental animals. Due to its peculiar concentration in the mammary gland and fat (also demonstrated in human) fenretinide is here proposed as the agent of choice for a breast cancer

ITALY

prevention study. The design of the study is to orally administer for five years 200 mg daily of fenretinide (with a three-day drug holiday at the end of each month) to stage I-II breast cancer patients between 33-68 years of age. This intervention schedule is to be randomly assigned versus no treatment in a population of 3,500-5,000 subjects, depending on the biological activity of the retinoid. The principal endpoint of the study is the possible decrease of incidence of new primaries in the contralateral breast. Local, regional and distant recurrences of the disease as well as new primaries in organs other than the breast will also be recorded and analyzed. No placebo is foreseen for the control arm as extensive data are already available on both acute and chronic toxicity of fenretinide: moreover, ethical considerations, too, do not recommend it due to the length of the trial. However, the protocol foresees a blind, unbiased review of all mammograms; cytological and histological examinations of all biopsied lumps are also performed blindly. This study started on 1 March 1987, and by 15 April 1991, 2,308 patients had already been randomized: 1,160 in the fenretinide arm and 1,148 in the control arm. The major analysis will be a comparison between the curves for the cumulative incidence of new contralateral primaries over time, when all patients have been followed for seven years. The log-rank two-sample statistics will be used to compare the time for the first tumour to occur among control patients with that of the fenretinide treated patients.

TYPE: Intervention
TERM: BMB; Prevention; Recurrence
SITE: Breast (F)
CHEM: Retinoids
TIME: 1985 - 1992

MODENA

* 493   Aggazzotti, G.   05041
Univ. di Modena, Ist. di Igiene, Via Campi 287, 41100 Modena, Italy (Tel.: (059)360084; Fax: 363057)
COLL: Fantuzzi, G.; Predieri, G.; Righi, E.

Exposure to Trihalomethanes in Indoor Swimming Pools

The study involves biological monitoring of human exposure to trihalomethanes (THM) in indoor swimming pools, particularly to chloroform, a potential carcinogen (IARC group 2B, U.S. National Toxicology Program group 2). Blood levels of chloroform have been evaluated in exposed (127 swimmers and visitors in indoor swimming pools) and 40 non exposed subjects. Samples of end-expired air (alveolar air) collected from 160 swimmers and 26 controls are also being analysed in order to assess the validity of this index when evaluating environmental exposure. Data about subjects are collected by questionnaire and also include information on possible exposure to solvents outside the swimming pool (e.g. in hobby activities); the type of activity practised in the swimming pool (visitor, swimmer, style of swimming, etc.) is recorded, as well as time spent at the pool. Samples are analysed by gas chromatography with an electron capture detector, and confirmed by mass-spectrometry.

TYPE: Methodology
TERM: Monitoring
SITE: All Sites
CHEM: Chloroform; Trihalomethanes
TIME: 1985 - 1995

494   Ponz de Leon, M.   04480
Ist. di Patologia Medica, Policlinico, Via del Pozzo 71, 41100 Modena, Italy (Tel.: + 39 59 379269)
COLL: Roncucci, L.; Sassatelli, R.; Scalmati, A.; Pedroni, M.; Benatti, P.; Zanghieri, G.

Registry of Digestive Tract Cancers

Digestive tract cancers, in particular colorectal tumours, have been registered since 1984 in a predominantly urban population in Northern Italy. Of special interest is familial occurrence of these and other neoplasms. For this purpose a detailed genealogical tree is drawn for the registered patients, limited to first-degree relatives. Occurrence of cancer in relatives is recorded and compared with that in a suitable control group drawn from the general population, matched for age (+/- 5 years) and sex. The specific objective of this study is to determine the frequency of cancer among relatives, to ascertain the total burden of cancer in the general population attributable to hereditary tumours (Lynch syndromes I and II), and to explore the possible implication of different types of genetic transmission in colorectal cancer.

ITALY

TYPE: Cohort
TERM: Familial Factors; Heredity; Registry
SITE: Colon; Rectum
REGI: Modena (Ita)
TIME: 1984 - 1993

MONFALCONE

495   Bianchi, C.   01963
Hosp. of Monfalcone, Lab. of Pathological Anatomy, Via Galvani, 34074 Monfalcone, Italy (Tel.: (0481)487500)
COLL: Brollo, A.; Ramani, L.; Zuch, C.

Asbestos Exposure and Malignancies in the Monfalcone Area
The objectives are: (1) to obtain data on asbestos exposure in the Monfalcone area, north-eastern Italy (about 60,000 inhabitants); (2) to identify asbestos related tumours; and (3) to prevent adverse effects in subjects exposed to asbestos in the past. Three parameters are investigated in a necropsy series (about 1,700 cases):hyaline pleural plaques, lung asbestos bodies and work history. The relationships between the above parameters and smoking and alcohol habits and malignancies are analysed. Available data indicate that severe exposure to asbestos occurred in the Monfalcone area until the 1970s. Working in shipyards was the main source of exposure. A large number of pleural and pulmonary asbestos-related tumours were observed and a possible relationship between asbestos and liver cancer emerged. A publication appeared in IARC Scientific Publications No. 112, E. Riboli and M. Delendi (eds), Lyon, 1991, pp 127-140.

TYPE: Correlation
TERM: Air Pollution; Alcohol; Autopsy; Chemical Exposure; Dusts; Environmental Factors; Occupation; Tobacco (Smoking)
SITE: All Sites
CHEM: Asbestos
TIME: 1980 - 1992

PADOVA

496   Chieco-Bianchi, L.   04376
Ist. di Oncologia, Univ. degli Studi di Padova, Via Gattamelata 64, 35128 Padova, Italy
COLL: De Rossi, A.; Amadori, A.; Del Mistro, A.; Calabro, L.; Gallegaro, L.; Miazzo, G.; Whittle, H.C.; Hall, A.J.; Bosch, F.X.

Natural History of Human Retrovirus Infections in The Gambia
The principal objective of this study is to evaluate the prevalence and the pattern of spread of retrovirus infections in Gambian children. This will be done by evaluating virus-specific serum antibodies as well as by monitoring other markers for viral infection (antigenaemia, Ig in vitro synthesis, in situ hybridization virus isolation). This study will provide information on virus transmission in paediatric populations, types of retrovirus involved and, possibly, their pathogenic spectrum. Children with signs of AIDS or related pathological conditions will be evaluated for virological and immunological parameters. Tumour samples, either frozen or paraffin embedded, will be processed for in situ hybridization using virus probes. If available, viable biological material will be studied for virus detection.

TYPE: Cross-Sectional
TERM: AIDS; Childhood; Infection; Prevalence; Virus
SITE: All Sites
LOCA: Gambia
TIME: 1988 - 1992

PALERMO

* 497   Pinzone, F.   05082
Sicilian Regional Health Administration, Epidemiology Unit, Via Vaccaro 5, 90145 Palermo, Italy (Tel.: (091)6969220)
COLL: Dardanoni, L.; Pagliaro, L.; Brignone, G.; Palazzotto, G.; Traina, A.; Romano, F.M.; Cusimano, R.; Pisa, R.

ITALY

Palermo Cancer Registry
A population-based cancer registry has been created in Palermo, registering all incident cases of neoplastic disease (ICD-9 140-208) among residents in Palermo urban area (approximately 750,000 inhabitants).

TYPE: Incidence; Registry
TERM: Data Resource
SITE: All Sites
TIME: 1991 - 1999

PERUGIA

498   La Rosa, F.   04723
Univ. of Perugia, Dept. of Hygiene, Via del Giochetto, 06100 Perugia, Italy (Tel.: + 39 75 4693331; Fax: 4693317)
COLL: Alberti Fidanza, A.; Blasi, B.; Fidanza, F.; Greco, M.; Mastrandrea, V.; Minelli, L.; Petrinelli, A.M.; Rivosecchi, P.; Vitali, R.

Diet and Other Risk Factors for Colorectal Cancer
The aim of the project is to examine a large random sample (more than 10,000 people) of the Umbrian resident population (about 800,000 inhabitants) to clarify the role of some individual characteristics and lifestyles in the occurrence of colorectal cancer. A questionnaire will be administered to obtain base information about dietary and smoking habits, occupation and family history; body mass index, blood pressure and ECG will be registered; blood group, haematocrit and cholesterol, lipids, selenium and concentration of some vitamins will be determined from venous blood samples; a sample of the blood will be stored for future use. The cohort will be followed-up until the year 2000, and colorectal cancer occurrence compared with that of the general Umbrian population.

TYPE: Cohort
TERM: BMB; Blood; Blood Group; Cholesterol; Diet; Familial Factors; Lipids; Occupation; Physical Factors; Tobacco (Smoking); Trace Elements; Vitamins
SITE: Colon; Rectum
CHEM: Selenium
TIME: 1990 - 2000

499   Mastrandrea, V.   03943
Univ. of Perugia, Dept. of Hygiene, Via del Giochetto, 06100 Perugia, Italy
COLL: Cicioni, C.; Siracusa, A.; Vitali, R.; Fiordi, T.; Mollichella, E.; Saba, G.; Petrinelli, A.M.; Ragni, G.

Lung Cancer in the Umbria Region (Italy)
Umbria is a small region of Central Italy (approximately 800,000 inhabitants) which shows some variation in certain characteristics pertaining to social and economic development, types of industry, pollution, density, etc. The objectives of the study are: (1) to analyse the incidence of lung cancer in the total population; (2) to assess the distribution of incidence in workers exposed to different occupational risk factors; and (3) to set up a case-control study to quantify the role of smoking, occupation and other putative risk factors associated with this problem. Cases are identified from periodic surveys of the pathology services, the bronchoscopic clinics and the oncology services of the region. Information on occupational history and smoking habits is collected by interview of the subjects or their next of kin. In 1985 about 360 newly diagnosed cases were identified; for approximately half of them it was possible to obtain a personal interview. It is expected that this percentage will increase in the next years. For subjects under 65 years old, the current job title and industry of employment will be used to compute occupation specific incidence rates using the census data as denominators. All the new cases undergoing bronchoscopy as out- or in-patients in the Perugia hospital oncology service (about one third or the new cases in the whole region) are interviewed immediately after the diagnostic bronchoscopy and compared with a hospital-based control group, matched by age, sex, and place of residence.

ITALY

TYPE: Case-Control
TERM: Chemical Exposure; Occupation; Tobacco (Smoking)
SITE: Lung
TIME: 1985 - 1992

ROME

500 Cerimele, D.M. 01443
Catholic Univ., A. Gemelli Polyclinic, Dept. of Dermatology, Via Pineta Sacchetti, 00168 Rome, Italy (Tel.: +39 6 33054227)
COLL: Tulli, A.; Rusciani, L.; Capizzi, R.

Environmental and Genetic Factors in Multiple Skin Cancer
The main aim of this research is to study the interrelationships between environmental and genetic factors in the development of multiple skin cancers. A questionnaire developed for the study, is being used to collect information from skin cancer patients on colour of the skin, eyes and hair, capacity to tan, duration of sun exposure to the skin, and presence now or in the past of other skin cancers. HLA typing of patients affected by two or more skin cancers will be undertaken, to confirm the possible negative correlation between HLA-B17 and the presence of multiple skin cancers. A paper has been published (Dermatologica 176, 176-181; 1988).

TYPE: Cross-Sectional
TERM: Environmental Factors; Genetic Factors; HLA; Immunology; Multiple Primary; Pigmentation; Radiation, Ultraviolet
SITE: Skin
TIME: 1988 - 1992

501 Cerimele, D.M. 04601
Catholic Univ., A. Gemelli Polyclinic, Dept. of Dermatology, Via Pineta Sacchetti, 00168 Rome, Italy (Tel.: +39 6 33054227)
COLL: Rotoli, M.; Celleno, L.; Cottoni, F.; Borroni, G.; Giannetti, A.; Amerio, P.L.; Lomuto, M.

Epidemiology of Kaposi's Sarcoma in Italy
The main objective of the present study is to collect information on the incidence of classic Kaposi's Sarcoma in Italy. Southern Italy is thought to be an area of high prevalence of Kaposi's Sarcoma, but no data are available, except for a very small area, north Sardinia. A preliminary investigation addressed to all dermatologists working in southern Italy has given unsatisfactory results. A sample study has been planned, to investigate all cases of Kaposi's Sarcoma observed in four areas comparable for size and population. Two areas are thought to be areas of high prevalence (Sassari, Chieti), the other two to be areas of low prevalence (Pavia, Modena). The collaboration of the dermatologists working in these four areas has been obtained. A questionnaire, similar to that used to gather the data in the Sassari area, has been distributed to collect the main clinical, pathological and genetic features of patients affected by Kaposi's Sarcoma. HLA typing has been planned to check if the immunogenetic features previously observed in the Sassari area (Acta Med. Rom. 25, 270-278; 1987) are common also to other Italian areas. A cytogenetic investigation is in progress to expand the data previously reported (Human Genet. 72, 311-317; 1986).

TYPE: Incidence
TERM: Antigens; HLA; Immunology
SITE: Kaposi's Sarcoma
TIME: 1988 - 1992

502 Comba, P. 04614
Ist. Superiore di Sanita, Lab. di Igiene Ambientali, Viale Regina Elena 299, 00161 Rome, Italy (Tel.: +39 6 4990; Fax: 4957621; Tlx: 610071 istisan)
COLL: Vetrugno, T.; Bruno, C.; Savelli, D.; Grignoli, M.

Epidemiology of Asbestos-Related Disease
This project encompasses mortality and morbidity studies concerning asbestos-related disease in Italy, including both descriptive and aetiological studies. Mortality from pleural malignant neoplasms has been studied in the Italian provinces, with reference to asbestos exposure. The same geographical analysis has been applied to peritoneal malignant neoplasms. Information about cases of pleural mesothelioma diagnosed and/or treated in over 500 Italian hospitals in the years 1984-88 is being collected with

reference to occupational and environmental exposure. Occupational cohort studies are being conducted among asbestos cement workers, subjects who worked in the construction and maintenance of trains, and subjects compensated for asbestosis. The mortality of subjects living in the neighbourhood of an asbestos mine is being studied. Papers were published in Med. Lav. 81:407-413, 1990 (Blasetti, F. et al) and in Epidemiologia e Prevenzione 45:39-47, 1990 (Bruno, C. et al).

TYPE: Cohort; Correlation
TERM: Asbestosis; Dusts; Occupation
SITE: Mesothelioma; Peritoneum; Pleura
CHEM: Asbestos
OCCU: Asbestos Workers
TIME: 1988 - 1992

503 Crespi, M. 04166
National Cancer Inst. "Regina Elena", Dept. Environmental Carcinogenesis, Epidemiology & Prevention, Viale Regina Elena 291, 00161 Rome, Italy (Tel.: +39 6 4457086; Fax: 4457086)
COLL: Giacosa, A.; Tortorelli, A.; Castellaneta, A.; Saragoni, A.; Bonaguri, C.; Falcini, F.; Ridolfi, R.; Morettini, A.; Zampi, G.C.; Guarnieri, C.; Gabrielli, M.; D'Albasio, G.; Casale, V.; Tarquini, M.; Grassi, A.; Caperle, M.; Sacripanti, P.; Scala, S.; Franze, A.; Bordi, C.; Missale, G.

Diet and Chronic Atrophic Gastritis

This is the Italian section of the European Collaborative Study on the role of diet and other factors on the aetiology of atrophic gastritis. Using a frequency dietary questionnaire, the study will aim to define different dietary profiles for cases (those with atrophic gastritis type B) and matched controls. Every participating centre (Rome, Florence, Forli, Genoa, Parma) will collect data on 120 subjects: 40 cases, 40 endoscopic controls and 40 population or hospital non-endoscopic controls (for evaluation of the possible bias inbuilt in the endoscopic population). The endoscopic protocol, the histopathological characterisation and the modalities of dietary interviews, have been standardised among the participating centres. For the Italian section, data on about one third of the whole population should be available in mid 1989.

TYPE: Case-Control
TERM: Atrophic Gastritis; Diet; Premalignant Lesion
SITE: Stomach
TIME: 1986 - 1992

* 504 Crespi, M. 05162
National Cancer Inst. "Regina Elena", Dept. Environmental Carcinogenesis, Epidemiology & Prevention, Viale Regina Elena 291, 00161 Rome, Italy (Tel.: +39 6 4457086; Fax: 4457086)
COLL: Conti, E.M.S.; Ramazzotti, V.; Genovese, O.

Cancer in Farmers Employed in Greenhouses

The study is conducted in the Province of Latina where many greenhouses for flower and vegetable farming exist since the 1960's. All diagnosed cases (1983-1987) of soft-tissue sarcoma, non-Hodgkin's lymphoma and malignant brain tumour have been identified by the Latina Cancer Registry. The controls are represented by two randomly selected groups, one in the general population, the other in the Latina hospital from subjects free of cancer. All cases (or their relatives) and controls are interviewed by questionnaire on their occupational histories, habits, etc. The occupational histories will be analysed using a pre-planned job-exposure matrix, specific for type of agricultural production, calendar period, geographical location and pesticide use. In the same area a cohort study of approximately 9,000 men licensed to buy and use pesticides during the period 1971-1984 is planned. Their mortality for all cancers and specific sites will be compared with the general population. Information on the pesticides actually handled and occupational history and exposure in greenhouses will be obtained from a 10% random sample of the cohort. Finally, approximately 40 farmers employed in greenhouses will be asked to participate in biological monitoring for the assessment of exposure to the pesticides.

ITALY

TYPE: Case-Control; Cohort
TERM: Occupation; Pesticides; Registry
SITE: All Sites; Brain; Non-Hodgkin's Lymphoma; Soft Tissue
OCCU: Agricultural Workers; Farmers; Horticulturists
REGI: Latina (Ita)
TIME: 1990 - 1993

* 505 Crespi, M. 05163
National Cancer Inst. "Regina Elena", Dept. Environmental Carcinogenesis, Epidemiology & Prevention, Viale Regina Elena 291, 00161 Rome, Italy (Tel.: +39 6 4457086; Fax: 4457086)
COLL: Conti, E.M.S.; Caperle, M.; Ramazzotti, V.; Ferro-Luzzi, A.; Maiani, G.; Spaziani, E.

Ecological Study on Diet, Vitamins and Cancer Frequency
The study is conducted within two areas of Latina Province: Campodimele and Roccagorga. The first (800 inhabitants) seems to be characterized by prolonged life-expectancy due also to a low frequency of diet-related cancers and cardiovascular diseases; migration may be a possible bias. The second (4,500 inhabitants) represents a comparison area. Both are covered by the Latina Cancer Registry. The population under study is composed randomly by two comparable groups (400 from Campodimele and 800 from Roccagorga). Each subject will be interviewed by a semi-quantitative food consumption questionnaire. Anthropometric measurements, blood pressure and lifestyle will also be assessed. Vitamins (carotenoids and vitamin C), cholesterol and its fractions and triglycerides will be analysed in blood. Data analysis will be performed to evaluate a possible protective effect of dietary habits and vitamin levels.

TYPE: Cross-Sectional
TERM: BMB; Cholesterol; Diet; Registry; Vitamins
SITE: All Sites; Breast (F); Colon; Stomach
CHEM: Beta Carotene
REGI: Latina (Ita)
TIME: 1991 - 1993

506 Greggi, S. 04269
Catholic Univ. of San Cuore, Dept. Gynaecology & Obstetrics, EORTC, Gynaecological Oncology Group, Largo A. Gemelli 8, 00168 Rome, Italy (Tel.: (06)33504496-4979; Fax: 3051343 ; Tlx: 611330 ucatro i)
COLL: Genuardi, M.; Neri, G.; Vermorken, J.B.; Pecorelli, S.; Mancuso, S.

European Registry of Familial Ovarian Cancer
After breast, bowel, and lung, ovarian cancer is the leading cause of death from cancer in women. The discovery of any aspect of the aetiology of ovarian neoplasms would be helpful, especially in ovarian cancer kindreds where genetic counselling may be beneficial. To date 140 kindreds with more than one member affected by ovarian cancer have been reported in the literature. The increased incidence of genetically determined ovarian cancer during the last decade may be related not only to improved data collection but also to new environmental factors. An EORTC cooperative study has been planned, to collect data on familial ovarian cancer, and to create a European Registry in Rome. Data will be elicited on year of diagnosis, age at diagnosis, race, religion, occupation (unusual environmental exposure), gravidity, parity, use of oral contraceptives, oestrogens, use and duration of talc, history of endometrial/breast cancer, pedigree, FIGO Stage, histology, grade of tumour differentiation, disease status, treatment and survival for all affected subjects. Registry plans are to evaluate the following aspects: (1) identification and registration of ovarian cancer-prone families, (2) knowledge of detailed epidemiological characteristics of affected subjects to see how they resemble or differ from other patients, (3) establishment of a bank of biological specimens from family members of the registry, (4) identification, registration and surveillance of high-risk subjects, (5) exchange of data with the US registry.

ITALY

TYPE: Case Series
TERM: BMB; Biochemical Markers; Dusts; Familial Factors; High-Risk Groups; Hormones; Occupation; Oral Contraceptives; Pedigree; Race; Religion; Reproductive Factors; Stage; Survival; Treatment
SITE: Ovary
CHEM: Oestrogens; Talc
LOCA: Austria; Belgium; Bulgaria; Denmark; France; Germany; Greece; Hungary; Israel; Italy; Netherlands; Sweden; Switzerland; Turkey; USSR; United Kingdom; Yugoslavia
TIME: 1989 – 1994

507 Lagorio, S. 04634
Ist. Superiore di Sanita, Lab. Igiene degli Ambienti Confinati, Viale Regina Elena 299, 00161 Rome, Italy (Tel.: +39 6 4990; Tlx: 610071 istisan)
COLL: Comba, P.; Belli, S.; Grignoli, M.; Pirastu, R.; Seniori Costantini, A.R.; Chellini, E.; Merler, E.; Masina, A.; Lodi, P.; Carmentano, P.; Brisanti, L.A.; Duca, G.; Braga, M.; Costa, G.; Borgia, P.; Forastiere, R.; Rizzelli, R.

Pooled Analysis of Occupational Cohorts Exposed to Motor Vehicle Exhaust
This study aims to assess possible excess cancer risks linked to exposure to engine exhausts. The study design is a pooled analysis of eight retrospective cohort mortality studies in progress at different research centers in Italy. The three taxi-driver cohorts will enrol about 6,000 subjects through the taxi driver associations in Rome, Florence and Bologna. The police officer cohort will recruit about 3,000 subjects in service at the beginning of the 1960's in the municipalities of Florence, Bologna and Milan. The bus driver cohort will be enrolled through the plant personnel records of the Transport Authorities of Florence and Bologna (about 3,000 workers in service in the 1950's. All the cohorts will be followed up from 1965 to 1984. SMRs, overall and cause-specific, will be estimated taking the mortality experience of the Italian and regional populations as standards. The pooled analysis of each professional group will reduce the power constraints, while the stratified analysis by town will address the exposure issue in terms of historical levels of air pollution. Information on employment duration and jobs held will be retrieved for all cohort members. Information on smoking habits is presently available for a sub-cohort of police officers from Florence and Bologna and for a sub-cohort of bus drivers from Bologna who underwent medical surveillance.

TYPE: Cohort
TERM: Air Pollution; Occupation; Tobacco (Smoking)
SITE: All Sites
CHEM: Vehicle Exhaust
OCCU: Bus Drivers; Policemen; Taxi Drivers
TIME: 1989 – 1992

508 Settimi, L. 04894
Ist. Superiore di Sanita, Lab. di Igiene Ambientale, Viale Regina Elena 299, 00161 Rome, Italy (Tel.: +39 6 4990; Fax: 4040064)
COLL: Boffetta, P.; Comba, P.; Iannarilli, R.; Magnani, C.; Terracini, B.; Grignoli, M.; Martuzzi, M.

Cancer Risk and Pesticide Exposure
The study has been planned to investigate on possible associations between several cancer sites and occupational exposure to pesticides and to test previously reported associations. It is a multicentric hospital-based case-control study specifically designed to investigate the subjects' occupational chemical exposures. During a two-year period, histologically confirmed incident cases of major cancers in adults of both sexes are identified in hospitals in five Italian rural areas (Asti, Imola, Pescia, Pistoia, Grosseto). Trained personnel interview patients about their life-time work experience, with special emphasis on activities in agriculture. Other information includes pathological data, smoking history and diet. A team of coders analyse the occupational histories using a pre-planned job-exposure matrix, specific with respect to type of agricultural production, calendar years, geographical location and pest. Each cancer site will be compared with a selected pool of the other sites. During the first year of the study approximately 1,000 cases have been identified.

ITALY

TYPE: Case-Control
TERM: Occupation; Pesticides
SITE: All Sites
OCCU: Agricultural Workers
TIME: 1990 - 1994

TORINO

509 Ciccone, G. 03695
Univ. of Torino, Dept. Biomed. Science & Human Oncology, Serv. Epidemiology of Tumours, Via Santena 7, 10126 Torino, Italy (Tel.: +39 11 678872)
COLL: Comba, P.; Vineis, P.

Phenoxy Herbicide Exposure and Soft-Tissue Sarcoma

The study is conducted in a rural area where large amounts of 2,4,5-T and 2,4-D have been used in rice-growing. All newly diagnosed cases of soft-tissue sarcomas which occurred in 1984-1988 in both sexes are identified (the final number will be approximately 150). Histological specimens have been collected and are read independently by two pathologists. The control group is represented by a sex-stratified random sample of the general population (approximately 300 individuals). A second control group of randomly selected deceased individuals will be used for comparison with the deceased cases. All cases and controls (or their relatives) are interviewed by mail questionnaire on their occupational histories (among other items). The questionnaires are evaluated blindly by a working group (including agronomists), in order to assess the probability of exposure to phenoxy herbicides for each individual. The relative risk of soft-tissue sarcoma following exposure to phenoxy herbicides for different jobs in agriculture will be estimated. A paper has been published in Scand. J. Work Env. Hlth 13:9-17, 1986, describing the first phase of the study (1981-83).

TYPE: Case-Control
TERM: Chemical Exposure; Environmental Factors; Herbicides; Occupation; Pesticides; Rural
SITE: Soft Tissue
CHEM: 2,4,5-T; 2,4-D; Phenoxy Acids
TIME: 1984 - 1992

510 Ciccone, G. 04612
Univ. of Torino, Dept. Biomed. Science & Human Oncology, Serv. Epidemiology of Tumours, Via Santena 7, 10126 Torino, Italy (Tel.: +39 11 678872)
COLL: Vineis, P.; Avanzi, G.C.; Rege Cambrin, G.; Ponzio, G.

Solvent Exposure and Cytogenetics in Leukaemias and MDS

A few previous investigations have suggested that environmental exposure to chemicals, particularly in the occupational setting, could be associated with cytogenetic abnormalities in patients affected by leukaemia. Turin is a large industrialised town, where approximately 10% of the adult population is estimated to be occupationally exposed to organic solvents. A pilot study on cytogenetic abnormalities in leukaemias and MDS has therefore been undertaken. The study consists in the enrolment of all cases of these pathologies in the Main Hospital of the town (approximately 100 leukaemias and 70 MDS in one year). Such patients are interviewed with a detailed questionnaire, including the collection of specific information on solvent exposure at work. Other items which are investigated are exposure to welding fumes and cigarette smoking. All cases (except those with chronic lymphocytic leukaemia) will have a cytogenetic characterisation, performed with standardised banding techniques. The purposes of the study are: (1) to investigate possible associations between cytogenetic abnormalities (e.g. deletions or translocations) and environmental exposure; (2) to assess the possibility of including cytogenetic tests within a larger population-based investigation on leukaemia, which is now in the early planning phase.

TYPE: Case Series; Molecular Epidemiology
TERM: Chromosome Effects; Occupation; Solvents; Tobacco (Smoking); Welding
SITE: Leukaemia; Myelodysplastic Syndrome
TIME: 1988 - 1992

511 Terracini, B. 02703
Univ. di Torino, Depto di Scienze Biol. & Oncol. Umana, Serv. di Epidemiologia dei Tumori, Via Santena 7, 10126 Torino, Italy (Tel.: (011)678872/6966383)
COLL: Ceci, A.; Fossati-Bellani, F.; Lo Curto, M.; Magnani, C.; Mancini, M.; Masera, G.; Paolucci, G.; Rossi, M.; Rosso, P.

ITALY

Second Primary Neoplasms among Long-Term Survivors from Childhood Cancer
The occurrence of second primary cancers is being monitored among long-term survivors after some types of childhood cancer reported to the Italian Registry of off-therapy children. Risks for second cancers have been estimated in a cohort including subjects who (1) were off therapy before 1 January 1981 (onset of the Registry) and were alive on that date, or (2) were off therapy during 1981-1988. This corresponded to a total of 3,141 subjects, 505 with Hodgkin's Disease, 223 with NHL, 299 with neuroblastoma, 347 with Wilms' Tumour, 1,630 with ALL and 137 with other acute leukaemias. 1,025 subjects were born before 1970 and 2,116 after. 1,745 were males and 1,396 females. Deaths and second primary incident cancers were identified respectively through the Registrars of the towns of residence and through the Institutions where patients are being followed up. The study includes 19,259 person-years at risk, 9,571 following ALL. A total of 28 second primary cancers were diagnosed histologically during the follow-up period vs. 3.12 expected (from the sex- and age-specific rates from the Varese Cancer Registry in 1982-1987). Among long-term survivors after ALL, a primary cancer of the CNS occurred in 9 vs 0.20 expected. This observation led to the design of a case-control study nested in the ALL sub-cohort, which is presently under way. Risk factors being investigated relate to the therapy originally used for treatment of ALL.

TYPE: Cohort
TERM: Childhood; Multiple Primary; Survival; Treatment
SITE: Hodgkin's Disease; Leukaemia (ALL); Leukaemia (ANLL); Neuroblastoma; Non-Hodgkin's Lymphoma; Wilms' Tumour
TIME: 1991 - 1992

512 Vineis, P. 04353
Univ. di Torino, Dipto. di Scienze Biomediche, e Oncologia Umana, Via Santena 7, 10126 Torino, Italy (Tel.: (011)678872/690327; Fax: 635267)
COLL: Bartsch, H.; Caporaso, N.E.; Harris, C.C.; Kadlubar, F.; Tannenbaum, S.R.; Wogan, G.N.; Estève, J.; Talaska, G.; Terracini, B.

Biochemical Epidemiology of Smoking and Bladder Cancer
A pilot study has suggested that aromatic amines contained in tobacco smoke, particularly 4-aminobiphenyl, are associated with the excess of bladder cancer observed in smokers. 4-aminobiphenyl and other aromatic amines contained in tobacco smoke bind covalently to macromolecules, including haemoglobin. The concentration of such adducts may be influenced by the variable ability to metabolize aromatic amines shown by different individuals. In this collaborative project, biological samples are being collected from healthy subjects and bladder cancer patients, and haemoglobin and DNA adducts are being measured in exfoliated cells and bladder biopsies. The subjects are also phenotyped for the activity of N-acetyltransferase and other enzymes relevant to aromatic amine metabolism, using a method based on caffeine administration. Haemoglobin adducts of 4-aminobiphenyl and other aromatic amines are measured in different metabolic subgroups (i.e. slow vs. fast metabolizers). In addition, urinary mutagenicity and the concentration of nitrosamines and aromatic amines in the urines are measured. A paper by Bartsch, H. et al. has been published in J. Nat. Cancer Inst. 82:1826, 1991.

TYPE: Case-Control
TERM: BMB; Blood; DNA Adducts; Tobacco (Smoking); Urine
SITE: Bladder
CHEM: Amines, Aromatic; N-Nitroso Compounds
TIME: 1988 - 1993

513 Vineis, P. 04788
Univ. di Torino, Dipto. di Scienze Biomediche, e Oncologia Umana, Via Santena 7, 10126 Torino, Italy (Tel.: (011)678872/690327; Fax: 635267)
COLL: Seniori Costantini, A.R.; Ciccone, G.; Crosignani, P.; Miligi, L.R.; Pisani, P.; Ricci, M.; Rodella, S.; Stagnaro, E.; Tumino, R.

Haematolymphopoietic Malignancies in Italy
High prevalence of exposure to solvents in industrial areas (estimated 10-16% in adult males in the provinces of Torino, Varese and Firenze), and to herbicides in agricultural areas is found in some parts of Italy. A multicenter population-based case-control study on haemato-lymphopoietic malignancies is planned in 12 Italian areas, chosen on the basis of high exposure prevalence. In three years, it is intended to interview approximately 1,100 cases of leukaemia, 1,900 cases of lymphoma (Hodgkin and non-Hodgkin) and 400 cases of myeloma; 2,000 randomly selected individuals, residents of the same areas, will serve as a population control group. Particular care will be devoted to diagnostic aspects,

ITALY

including B- and T-cell characterization and, in four of the areas, cytogenetics. The occupational histories will be blindly evaluated by industrial hygienists and agronomists, in order to assess the probability of exposure to the chemicals of interest. The study will have a statistical power of 90% to detect a relative risk of 2.0 at the 5% significance level for associations between organic solvent exposure and leukaemias and between herbicide exposure and non-Hodgkin's lymphomas.

TYPE: Case-Control
TERM: Herbicides; Occupation; Pesticides; Solvents; Welding
SITE: Hodgkin's Disease; Leukaemia; Multiple Myeloma; Non-Hodgkin's Lymphoma
CHEM: Phenoxy Acids
OCCU: Electrical Workers; Farmers; Leather Workers; Mechanics
TIME: 1990 - 1993

514   Zanetti, R.   04185
Registro Tumori Piemonte, Via San Francesco da Paola 31, 10123 Torino, Italy (Tel.: (011)547718/548732)
COLL: Vineis, P.; Rosso, S.; Martina, G.; Colonna, S.; Roffino, R.

Case-Control Study on Skin Carcinoma
The aim of the study is the identification of risk factors for skin carcinoma. Subjects include all incident cases of skin carcinoma diagnosed in the population of Turin. Controls are randomly sampled from the population file provided by the National Health Service. Variables under consideration are: skin characteristics; tanning and cosmetics habits; UV radiation from solar light and artificial sources; X radiation; cosmetic and therapeutic use of Psoralens; tobacco consumption; previous skin diseases; immunodeficiencies; professional exposure to tar, polycyclic aromatic hydrocarbons, mineral oils. For each factor the study will estimate the relative risk and the population attributable risk. Total interviews will include data from 550 cases and as many controls.

TYPE: Case-Control
TERM: Cosmetics; Drugs; Immunodeficiency; Radiation, Ionizing; Radiation, Ultraviolet; Registry; Tobacco (Smoking)
SITE: Skin
CHEM: Mineral Oil; PAH; Tars
REGI: Piedmont (Ita)
TIME: 1987 - 1992

TRENTO

* 515   Amichetti, M.   05224
Osp. Santa Chiara, Centro per Oncologia, USL C5, Largo Medaglie d'Oro, 38100 Trento, Italy (Tel.: (461)903203; Fax: 920030)
COLL: Piffer, S.; Valentini, A.; Graiff, C.; Perani, B.

Breast Self-Examination in Patients with Breast Cancer Diagnosed in the Province of Trento Between 1982-1986
The primary objective of this retrospective study is to evaluate the use of breast self-examination (BSE) among patients with breast cancer in the Province of Trento (population 500,000). The hypothesis is that women with breast cancer rarely perform BSE, a simple method of early detection of breast cancer. All cases diganosed between 1982 and 1986 have been identified by searching pathological and clinical records in local hospitals: this was necessary because there is no tumour registry. Specific questionnaires have been mailed or direct interviews performed. About 650 cases have been identified so far.

ITALY

TYPE: Case Series
TERM: Prevention
SITE: Breast (F)
TIME: 1988 - 1992

VERONA

516   Ricci, P.   04188
Univ. di Verona, Ist. di Anatomia Patologica, Strada delle Grazie, 37134 Verona, Italy (Tel.: +39 045 933323)
COLL: Merler, E.; Darra, F.; Piva, C.; Carretta, D.; Lestani, M.; Domenici, R.; Magnani, C.; Terracini, B.

Carcinogenic Risks among Railway Coach Repair Workers
The cancer mortality of an historical cohort of workers employed in coach repair at the Bologna Repair Department of the Italian Railways will be studied. Included are all workers employed between 1957 and 1977 (around 3,000 subjects), whose working histories will be reconstructed using records which include the periods during which each task was performed. Vital status will be determined for each subject, and cause of death will be assessed through death certificates. The aim of the study is to evaluate the mortality risk for exposure to asbestos (exposure to a mixture of chrysotile, amosite and crocidolite since 1957) and welding fumes. The opportunity to perform some nested case-control studies will be evaluated. Cases and controls will be identified from this and other cohorts of Italian coach repair factories already enumerated.

TYPE: Cohort
TERM: Chemical Exposure; Dusts; Occupation; Welding
SITE: All Sites
CHEM: Asbestos
OCCU: Railroad Workers
TIME: 1987 - 1992

JAPAN

CHIBA

517 Murata, M. 02680
Chiba Cancer Center, Research Inst., Div. of Epidemiology, 666-2, Nitona-cho, Chiba 280, Japan (Tel.: (0472)645431; Fax: 62868)
COLL: Shimamura, K.; Fukuma, S.; Takayama, K.; Nagashima, Y.; Ishikawa, A.; Hayashi, M.

Mass Surveys for Cancer

Two cohort studies are being conducted on participants in cancer screening programmes using a population-based cancer registry. One is for about 40,000 normal women who attended a gynaecological mass survey in 1977. They have already been followed up for 10 years. Among them 638 cancer cases were detected, including 143 stomach, 64 colorectal, 22 lung, 133 breast and 88 uterine. By matching city and birth year, a double number of non-affected controls were chosen from the original cohorts. Family history of cancer and reproductive histories were compared between them, using information obtained from a simple questionnaire at the mass survey. Positive effect of family history was found for breast and colorectal cancer, but not for stomach and unterine cervical cancers. Cases with breast cancer showed an older average age at marriage and lower parity than controls. The other study was for about 77,000 participants in a gastric mass survey in 1984. Their computer file is now being constructed, primarily to evaluate effectiveness of the mass screening. In those screened for the second stage examination, detection rate of stomach cancer was examined in relation to prior attendance at a gastric mass survey. It was two-fold higher in those who did not attend in the previous year as compared to those who did.

TYPE: Cohort
TERM: Familial Factors; Reproductive Factors; Screening
SITE: All Sites; Stomach; Uterus (Cervix)
TIME: 1978 - 1995

FUKUI CITY

518 Yamazaki, S. 04675
Fukui Prefectural Hospital, Fukui Cancer Registry, 2-8-1, Yotsui, Fukui City 910, Japan (Tel.: (0776)545151; Fax: 545151)
COLL: Fujita, M.; Hattori, M.

Incidence of Gastric Cancer in Fukui Prefecture

The Fukui Prefectural Government has run a cancer registry in Fukui Prefecture since 1984, in cooperation with the Fukui Medical Association, in order to estimate cancer incidence. Gastric cancer is the leading cause of cancer death in Fukui Prefecture. In 1984, Fukui Prefecture had a population of about 800,000. There were 330 deaths from gastric cancer and 792 new cases were diagnosed. About 27% of new cases were diagnosed at an early stage. Epidemiological and survival studies by depth of invasion, type of treatment and other factors have been conducted for gastric cancer and the data are now being analysed.

TYPE: Registry
TERM: Screening; Treatment
SITE: Stomach
TIME: 1984 - 1992

GIFU

519 Shimizu, H. 04491
Gifu Univ., School of Medicine, Dept. of Public Health, 40 Tsukasamachi, Gifu 500, Japan (Tel.: (0582)651241; Fax: 659020)
COLL: Higashiiwai, H.; Sugawara, N.; Morita, N.; Hisamichi, S.; Komatsu, S.

Vitamin A and Dysplasia of Uterine Cervix

To examine the relationship between vitamin A intake or serum vitamin A (retinol) level and the occurrence of dysplasia of uterine cervix, a case-control study is being conducted. About 300 cases of uterine dysplasia are diagnosed annually in Miyagi prefecture. Study cases comprise almost half of the patients randomly selected through a mass screening programme in 1987-1988. Control series are

JAPAN

women who visit the Miyagi Cancer Society for general health checks, matched for age and residential area. A health-nurse is interviewing both cases and controls. All of the cases and controls are being asked to provide blood samples for the analysis of serum retinol. Conditional logistic regression analysis will be used to estimate the relative risks for several risk factors for cervical dysplasia, such as number of children and frequency of sexual intercourse, as well as of vitamin A intake.

TYPE: Case-Control
TERM: Parity; Premalignant Lesion; Sexual Activity; Vitamins
SITE: Uterus (Cervix)
CHEM: Retinoids
TIME: 1987 - 1992

HIROSHIMA

520   Kurihara, M.   01097
Research Inst. for Nuclear Medicine, and Biology, Minami-ku, 2-3 Kasumi 1-chome, Minami-ku Hiroshima 734, Japan (Tel.: (082)2511111; Fax: 2558339)
COLL: Munaka, M.; Hayakawa, N.; Ohtaki, M.; Matsuura, M.

Cancer Mortality among Atomic Bomb Survivors in Hiroshima Prefecture
The atomic bomb survivors in Hiroshima Prefecture numbered about 173,000 in 1982. In order to observe the late effects of low dose radiation, a survey on cancer mortality among the survivors from 1968 onwards is being conducted, using the data bank of the survivors in this institute. This includes comparison with mortality among the general population of all Japan and among the non-exposed population of Hiroshima Prefecture. The mortality statistics for 1968 to 1972 was published in J. Radiat. Res. 22:456-471, 1981, and those for 1973 to 1977 in the Proceedings of the Institute No.26, 134-147, 1985 (in Japanese). The mortality for 1978 to 1982 has been investigated and is now being analysed.

TYPE: Cohort; Mortality
TERM: Atomic Bomb; Radiation, Ionizing; Survival
SITE: All Sites
TIME: 1975 - 1993

521   Shigematsu, I.   00665
Radiation Effects Research Found., 5-2 Hijiyama Park, Minami-ku, Hiroshima 732, Japan (Tel.: (082)2613131; Fax: 2637279)
COLL: Thiessen, J.W.; Trosko, J.E.; Schull, W.J.; Hasegawa, Y.; Kono, T.

Late Radiation Effects of Atomic Bomb Exposure in the Population of Hiroshima and Nagasaki, Japan
The object of the study is to determine the late medical consequences of previous atomic bomb exposure in the populations of Hiroshima and Nagasaki. The project was initiated in 1947 as the Atomic Bomb Casualty Commission in collaboration with the Japanese National Institute of Health. In 1975 the name of the project was changed to the Radiation Effects Research Foundation with equal funding from Japan and the United States. Examination facilities, computer centres and research laboratories are maintained in both Hiroshima and Nagasaki. Mortality surveillance of a cohort of about 120,000 persons in Hiroshima and Nagasaki and about 75,000 F0I offspring including controls is being conducted. Currently, approximately 9,000 persons exposed as children or adults and about 1,000 exposed in utero, with their controls, receive regular biennial clinical examinations. The most important radiation-related effects are increased incidence rates of leukaemia, multiple myeloma, and cancers of the thyroid, female breast, lung, stomach, colon and ovary. Case-control studies are now in progress to determine possible interactions between radiation exposure and a number of environmental factors. Particular attention is being focused on early cancer detection, precancerous changes, cancer education and possible radiation-induced acceleration of aging. Major laboratory emphasis is placed on various cytogenetic, histological, immunological, hormonal and biochemical changes which may represent evidence of somatic or germinal mutation. Chromosomal aberrations, impairment of cell-mediated immunity, oncogene activation, and certain tissue mutations are being correlated with radiation dose, and the development of cancer. The frequency of biochemical or DNA mutagenesis in the children of the exposed is being actively investigated. Recent refinements of total body and specific organ radiation dose estimates provide the basis for new reports regarding dose-response relationships for all radiation-induced effects.

JAPAN

TYPE: Case-Control; Cohort; Molecular Epidemiology
TERM: Atomic Bomb; Chromosome Effects; DNA; Dose-Response; Histology; Late Effects; Oncogenes; Radiation, Ionizing
SITE: All Sites
TIME: 1947 - 1993

* 522 Tahara, E. 04973
Hiroshima Univ. School of Medicine, Dept. of Pathology I, 1-2-3 Kasumi, Minami-ku, Hiroshima 734, Japan (Tel.: (082)2511111; Fax: 2515250)
COLL: Ito, H.; Yasui, W.; Yamakido, M.

Gastric Cancer in Former Workers of a Mustard Gas Factory

The primary objective of the study is to examine the association of gastric cancer in former workers at a mustard gas factory with occupation at the factory, the type of gas exposure, and period of exposure. This study also aims to obtain macroscopic and histological characteristics of gastric cancer in these populations in comparison with control cases which will be randomly selected from gastric cancers diagnosed in the general population. 64 cases of gastric cancer aged 39-84 years among about 5,000 former workers of the factory have been collected during the period from 1974 to 1990.

TYPE: Cohort
TERM: Histology; Occupation
SITE: Stomach
CHEM: Mustard Gas
TIME: 1974 - 1993

HYOGO

523 Hashimoto, T. 04212
Hyogo College of Medicine, Dept. of Genetics, Mukogawa-cho, 1-1, Nishinomiya, Hyogo 663, Japan (Tel.: (798)456587)
COLL: Furuyama, J.

Establishment of Cell Lines from Fanconi Anaemia Patients and Heterozygote, and Assay of the Sensitivity to Carcinogens of their Cell Lines

Fanconi anaemia (FA) is an autosomal recessive disease and FA cells that show hypersensitivity to cross-linking agents. But it has been unclear that the cells of FA heterozygotes showed hypersensitivity to these agents, though FA heterozygotes were estimated to have a higher risk of developing various cancers. Lymphoblastoid cell lines from FA patients and FA family members were established, and their hypersensitivity to these agents, using cytotoxic analysis and cytogenetical studies checked. 25 cell lines of 6 Japanese FA families were established and contacts with the Department of Pediatrics are being continued in order to find new FA families.

TYPE: Case Series
TERM: Genetic Factors; High-Risk Groups; Lymphocytes; Mutagen; Premalignant Lesion
SITE: Inapplicable
LOCA: United States of America
TIME: 1984 - 1992

KAGOSHIMA

524 Sato, E. 02659
Kagoshima Univ., Faculty of Medicine, IInd Dept. of Pathology, Usuki-cho 1208-1, Kagoshima, Japan (Tel.: (0992)642211/2118)
COLL: Tokunaga, M.

Study of Endemic Adult T-Cell Lymphoma-Leukaemia

Viroepidemiological studies of Adult T-cell lymphoma-leukaemia have shown a close relationship to C type retrovirus. This lymphoma is characerized by the geographical clustering; peripheral T-cell nature of the tumour cells with pleomorphic nuclear morphology, adult onset with no thymic mass, cutaneous manifestation and a rapid clinical course with no response to anticancer agents. The objectives of this project are: (1) to determine the incidence in the population of Kagoshima Prefecture using the cases detected by the surface markers study; (2) to identify the cases from the retrospective paraffin blocks; (3) to study the relationship of antibody titre to the specific virus of the patients and histopathology of the

JAPAN

lymphoma; (4) to study the histopathology of the pre-malignant lesion of adult T-cell lymphoma; (5) to study the relationship between proviral DNA interaction of HTLV-1 and histopathology of lymphoma. Papers have been published in Rec. Adv. Cancer Res. 25:145-155, 1985, Acta Patol. Jap. 37:179-192, 1987 and in J. Int. Soc. Prev. Oncol. 14:1989.

TYPE: Case Series
TERM: HTLV; Histology; Sero-Epidemiology; Virus
SITE: Lymphoma; Skin
TIME: 1980 - 1992

KITA-KYUSHU

525 Tsuchiya, K. 00703
Univ. of Occupational, and Environmental Health, Iseigaoka, Yahatanishi-ku, Kita-Kyushu 807, Japan (Tel.: (093)603-1611; Fax: 6017324)
COLL: Okubo, T.; Takahashi, K.

Lung Cancer among Chromium Plating Workers
To examine frequency of lung cancer among chromium platers, 626 chromium platers and, as a reference, 567 platers with history of plating other than chromium have been followed up from 1976. The majority of the chromium handling occurred before 1970. Follow up is carried out through the Japanese Family Registration System and information on cause of death is obtained from death certificates. A preliminary report was made, based on results until 1987, in Arch. Environm. Hlth 45:107-111, 1990. The result showed that lung cancer was the only cause of death that was significantly higher than expected for all platers (16 observed, 8.9 expected; SMR 179; 95% CI 102-290). This elevated SMR, however, was not statistically significant in either of the two plater subgroups. Follow-up will continue.

TYPE: Cohort
TERM: Metals; Occupation
SITE: Liver; Lung
CHEM: Chromium
OCCU: Chromium Plating Workers
TIME: 1977 - 1997

526 Yoshimura, T. 04199
Univ. of Occupational, and Environmental Health, Dept. of Clinical Epidemiology, 1-1, Iseigaoka, Yahatanishi-ku, Kita-Kyushu 807, Japan (Tel.: (093)6031611)
COLL: Ogimoto, I.; Ikeda, M.

Gastric Cancer in the Young Adult and Reproductive History
In Japan the incidence of gastric cancer among young adults is found to be even lower in males than in females. It has been suggested by clinicians that the occurrence of female gastric cancer in young adults might be related to reproductive history. In this study, it is intended to clarify the relationship between gastric cancer in the young adult and reproductive history. 100 gastric cancer cases in young adults (less than 40 years old), 200 hospital controls and 200 population controls will be collected for a matched case-control study. Information on general risk factors, such as dietary habits drinking and smoking history and reproductive history will be obtained by self-administered questionnaire.

TYPE: Case-Control
TERM: Adolescence; Alcohol; Diet; Reproductive Factors; Tobacco (Smoking)
SITE: Stomach
TIME: 1986 - 1992

KORIYAMA

527 Nomizu, T. 04654
Hoshi General Hosp., Dept. of Surgery, 2-1-16, Ohmachi, Koriyama 963, Japan (Tel.: (0249)233711; Fax: 393141)
COLL: Watanabe, F.; Sato, H.; Takita, K.; Tsuchiya, A.; Yamaki, Y.; Abe, R.

Cancer-Prone Families
It is suggested that genetic as well as environmental factors are implicated when cancers are clustered in one family. The aims of the study are to define the clinical features of familial breast, gastric or colorectal

JAPAN

cancer and the clinical and genetic biological markers for familial cancer in order to apply the results to cancer prevention. More than 250 cases of familial cancer are being investigated. Clinical features of familial cancer are compared with those of non-familial cancer, such as age of onset, side or site of cancer, pathology, multicentricity or frequency of multiple primary cancers. Oncogenes, chromosome analysis and flowcytometric DNA analysis of carcinoma cells are also being investigated. Papers have appeared in J. Jap. Soc. Clin. Surg. 48:1597-1599, 1987 and Jap. J. Cancer Clin. 32:485-493, 1986.

TYPE: Case Series
TERM: Biochemical Markers; DNA; Familial Factors; Genetic Factors; Multiple Primary; Oncogenes; Prevention; Tumour Markers
SITE: Breast (F); Colon; Rectum; Stomach
TIME: 1984 - 1993

KURUME CITY

528 Fukuda, K. 04624
Kurume Univ., School of Medicine, Dept. of Public Health, 67 Asahi-machi, Kurume City 830, Japan (Tel.: (0942)353311)
COLL: Shibata, A.; Hirohata, I.; Hirohata, T.

Primary Liver Cancer and Lifestyle
This study aims to identify any risk factors for primary liver cancer including HBV infection, blood transfusion, alcohol drinking and several aspects of daily lifestyle. 200 cases and an equal number of age and sex-matched hospital controls will be chosen. Information is collected by an interviewer and will be analysed by a multivariate analysis.

TYPE: Case-Control
TERM: Alcohol; HBV; Lifestyle
SITE: Liver
TIME: 1986 - 1993

KYOTO

529 Watanabe, H. 04885
Kyoto Prefectural Univ. of Medicine, Dept. of Urology, Kawaramachi-Hirokoji, Kamigyo-ku, Kyoto 602, Japan (Tel.: (075)2515593; Fax: 2117093)
COLL: Hirayama, T.; Hirohata, T.; Aoki, K.; Watanabe, S.; Minowa, M.; Tsuchihashi, Y.

Case-Control Study for High Risk Group of Renal Cell Carcinoma
Mortality from renal cell carcinoma is increasing in Japan. To define the risk factors for this disease, we designed a matched case-control study on 200 cases of renal cell carcinoma and 200 controls. Cases will be patients aged 40-69 with renal cell carcinoma, selected from patients visiting 15 clinics in Kyoto, Osaka, Shiga and Shizuoka in the next five years. Controls will be matched for sex, age (within three years) and residence in the same prefecture. In this study, we will use two questionnaires, with 146 questions covering past and family history, physical condition, physical constitution, eating habits, drinking, smoking, drug use, occupation, income, social environment, residence, lifestyle, marriage and sexual activity. By March 1991, 100 case-control pairs had been recruited.

TYPE: Case-Control
TERM: Alcohol; Diet; Drugs; Lifestyle; Socio-Economic Factors; Tobacco (Smoking)
SITE: Kidney
TIME: 1990 - 1993

NAGASAKI

* 530 Amemiya, T. 05212
Nagasaki University, School of Medicine, Dept. of Ophthalmology, 7-1 Sakamoto-machi, Nagasaki 852, Japan
COLL: Takano, J.; Kusuki, Y.

JAPAN

Retinoblastoma in Nagasaki Prefecture
Between 1965 and 1986, the incidence of retinoblastoma in Nagasaki Prefecture was about 1 in 16,053 births, which is the average for Japan. In Shimabara District (Nagasaki Prefecture), the incidence was however 1 in 10,331, which is higher than in any other district in Nagasaki Prefecture. The population of Shimabara has mainly immigrated from Hyogo, Aichi and Shizuoka Prefectures. The incidence of retinoblastoma in Tamba and Tajima Districts (Hyogo Prefecture) was very close to that of Shimabara, 1/10,570 and 1/10,411 respectively. The aim of this study is to assess the role of the atomic bomb in retinoblastoma. A questionnaire will be sent to retinoblastoma patients in Nagasaki and Hiroshima if possible. Survivors will be contacted for personal details and radiation dose will be calculated on the basis of data accumulated by the Atomic Bomb Effect Research Centers in Nagasaki and Hiroshima.

TYPE: Correlation; Incidence
TERM: Atomic Bomb; Radiation, Ionizing
SITE: Retinoblastoma
TIME: 1990 – 1993

531 Ikeda, T. 04690
Nagasaki Univ. School of Medicine, Dept. of Pathology, 12-4 Sakamotomachi, Nagasaki 852, Japan (Tel.: (0958)472111; Fax: 478514)
COLL: Shimokawa, I.; Matsuo, T.; Soda, M.; Mine, M.; Mori, H.

District Specificity of Cancer Incidence in Nagasaki, Japan
Mortality and morbidity rates in Nagasaki prefecture were the highest in Japan in 1986, and some 35% of the population of Nagasaki City were A-bomb survivors. Moreover, the high frequency of T-cell lymphoma and ATLL in Southwestern Japan including Nagasaki Prefecture is well known. This study aims to clarify the causes of increased cancer incidence, and correlation between cancer incidence and related factors, such as radiation exposure, HBV, and HTLV. Nagasaki City Cancer Registry was established in 1957, Nagasaki Tumor Tissue Registry in 1974, and Nagasaki Prefectural Cancer Registry in 1985, covering populations of 450,000, 1,000,000, and 1,600,000 respectively. This population-based study (1973–1982) revealed that the cancer risk was definitely increased for leukaemia, multiple myeloma, cancers of the breast, thyroid, skin, liver, gall bladder and pancreas. Recent publications include GANN Monograph on Cancer Research 32:41-52, 1986, Jap. J. Cancer Clin. 33:807-814, 1987, and Nagasaki Med. J. 63:306-311, 1988.

TYPE: Correlation; Incidence
TERM: Atomic Bomb; HBV; HTLV; Radiation, Ionizing; Registry
SITE: All Sites
REGI: Nagasaki (Jap)
TIME: 1978 – 1995

* 532 Mine, M. 05238
Nagasaki Univ. School of Medicine, Scientific Data Center for the, Atomic Bomb Disaster, 12-4 Sakamoto-machi, Nagasaki, Japan (Tel.: (0958)472111/2453)
COLL: Okumura, Y.; Mori, H.; Kondo, H.; Ikeda, T.; Kondo, S.

Cancer Mortality among Atomic Bomb Survivors in Nagasaki Prefecture
A cohort of 110,000 survivors of the Nagasaki atomic bomb is being followed up for mortality. The Scientific Data Center for the Atomic Bomb Disaster at Nagasaki University has collected medical data on A-bomb survivors since 1970. This study aims to investigate the correlation between cancer mortality and related factors such as radiation exposure and health examination. The mortality statistics for 1970 to 1984 were published in Radiat. Res. 103:419-431, 1985 and those for 1970 to 1988 in Int. J. Radiat. Biol. 58:1035-1043, 1990

JAPAN

TYPE: Cohort; Mortality
TERM: Atomic Bomb; Radiation, Ionizing; Survival
SITE: All Sites
TIME: 1970 - 1993

NAGOYA

533 Aoki, K. 04697
Aichi Cancer Center, 1-1 Kanokoden, Chikusa-ku, Nagoya 464, Japan (Tel.: (052)7626111; Fax: 7635233)
COLL: Miyake, K.; Kamiyama, S.; Takizawa, Y.; Hisamichi, S.; Yanagawa, H.; Murata, M.; Inaba, Y.; Nakamura, K.; Anzai, S.; Toyoshima, H.; Ishibashi, H.; Ohno, Y.; Kawai, K.; Suzuki, R.; Tanaka, H.; Watanabe, H.; Hashimoto, T.; Nose, T.; Hayakawa, N.; Yoshimura, T.; Fukuda, K.; Tokudome, S.

Evaluation of Risk Factors for Cancer - A Large Scale Cohort Study

An inquiry on lifestyles (smoking, drinking, dietary habits), occupational, personal and family disease history, health status, attitudes to health and health behaviours was carried out in 1988-1990 on about 120,000 inhabitants of 50 cities and towns throughout Japan and about 30,000 industrial workers in four cities. A questionnaire was first completed by each subject, then health nurses or health services persons in the district confirmed the answers by interview or by telephone. About 40% of the inhabitants in this study were screened by a standard health check-up including blood pressure, urine, ECG, serum protein and cholesterol, liver function tests etc. A serum bank was established of 50,000 volunteers among them, (-80 degrees, five aliquots of 0.3-0.5 ml per person). Cancer incidence is being investigated using cancer registry or other local system using records from hospital, health insurance, health check-ups, health nurse visits, etc. Deaths are also monitored by death certificate with Government permission. Follow-up will be continued until 1988-2000. Interim analysis will be done in 1993-1995.

TYPE: Cohort
TERM: Alcohol; BMB; Diet; Nutrition; Occupation; Registry; Serum; Tobacco (Smoking)
SITE: All Sites
CHEM: Beta Carotene
REGI: Aichi (Jap)
TIME: 1988 - 2000

534 Ohno, Y. 04638
Nagoya Univ., School of Medicine, Dept. of Preventive Medicine, Tsurumai-cho , Showa-ku, Nagoya 466, Japan (Tel.: (052)7412111; Fax: 7336729)
COLL: Kubo, N.; Sakamoto, G.; Watanabe, S.; Cornain, S.; Tjindarbumi, D.; Ramli, M.; Darwis, I.; Tjahjadi, G.; Sutrisno, E.; Sri Roostini, E.; Prihartono, J.; Budiningsih, S.

Japan-Indonesia Study on Aetiology and Clinicopathology of Breast Cancer

This is a hospital-based case-control study. Elegible cases are female breast cancer patients aged 25-69 years, histologically confirmed and newly diagnosed at the Department of Surgery, Centre Hospital, University of Indonesia, Jakarta. Controls (two per case) are female patients with diseases other than cancer admitted to or visiting the same department. Cases and controls are matched, by pair matching, for sex, age (+/- 3 years) and date of admittance/visit (+/-3 months). Controls are examined by palpation to confirm absence of lumps in the breast. Routine demographic and epidemiological information is collected by direct interview of the study subject herself, at hospital, by two academic nurses. A pilot study was already done in 1988, and the study started officially from 1 April 1989. All previously reported risk factors are to be tested. A case-control study with the same study protocols will be considered in 1990 in Tokyo, Japan. In parallel with this case-control study, clinical and histopathological studies are to be performed.

JAPAN

TYPE: Case-Control
TERM: Alcohol; Diet; Oral Contraceptives; Physical Factors; Reproductive Factors; Tobacco (Smoking)
SITE: Breast (F)
LOCA: Indonesia; Japan
TIME: 1989 - 1993

535 Ohno, Y. 04639
Nagoya Univ., School of Medicine, Dept. of Preventive Medicine, Tsurumai-cho , Showa-ku, Nagoya 466, Japan (Tel.: (052)7412111; Fax: 7336729)
COLL: Kubo, N.; Hayashi, Y.; Nagao, K.; Genka, K.; Ohmine, K.; Aoki, K.; Fukuma, S.

Lung Cancer and its Risk Factors in Okinawa, Japan

This case-control study aims to explore the risk factors for lung cancer in Okinawa, particularly among males, since they have experienced the highest lung cancer mortality in Japan over a recent decade. Eligible cases, identified at National Okinawa Hospital and two major hospitals, include all patients aged 40-89 years with histologically confirmed primary lung cancer, diagnosed since 1 January 1988. Controls are randomly selected from the general population, using the electoral registers; matching individually for residence, sex and age to within two years with a 2:1 allocation ratio. Routine demographic and epidemiological information, including smoking and drinking habits, occupation, and dietary practices, are collected principally at the subject's home by 7-10 trained public health nurses, using a 14-page questionnaire. Interviews started in October, 1988. A main hypothesis concerns smoking habits, but the study is intended to be hypothesis-generating, since smoking in Okinawa is lower than the national average, approximately 20% in males and 10% in females. In parallel with this case-control study, comparative clinical and pathological studies are performed between Okinawa and Chiba prefecture.

TYPE: Case-Control
TERM: Alcohol; Diet; Male; Occupation; Tobacco (Smoking)
SITE: Lung
TIME: 1988 - 1993

536 Tominaga, S. 04504
Aichi Cancer Centre Research Inst., 1-1 Kanokoden, Chikusa-ku, Nagoya 464, Japan (Tel.: (052)7626111; Fax: 635233)
COLL: Kato, I.; Kobayashi, S.

Risk Factors and Natural History of Atrophic Gastritis and Stomach Cancer

To elucidate risk factors and to study the natural history of stomach cancer and chronic atrophic gastritis (CAG), a questionnaire survey was conducted on about 5,000 patients who received a gastric endoscopic examination at the Aichi Cancer Center Hospital. Common and specific risk factors will be studied from analyses of baseline data. The study subjects free from stomach cancer at baseline will be followed for at least five years to study risk factors and natural history of stomach cancer and CAG.

TYPE: Cohort; Cross-Sectional
TERM: Atrophic Gastritis; Premalignant Lesion
SITE: Stomach
TIME: 1986 - 1995

NATORI

537 Yamamoto, R. 03440
Shokei Women's Jr College, 10-1 Yurigaoka 4 chome, Natori 981-12, Japan (Tel.: (022)3830111; Fax: 3830130)
COLL: Shimizu, H.; Hisamichi, S.; Fukao, A.

Digestive Tract Cancers and Food Consumption

To elucidate nutritional factors for digestive cancers, a time lag analysis between trends in cancer incidence and food intake in Miyagi Prefecture (Japan) has been performed, comparing population-based nutritional status (intake per capita) with the age-adjusted cancer mortality or incidence. Daily food intake estimated from a nutritional survey in Miyagi (1947-1977) was adjusted using the tables of the sex-and age-specific food constitution (1965, 1970 and 1975) to determine daily food intake by sex and age. The results of time lag (0, 3, 6, 9, 12 years) analysis by correlation matrix indicated that gastric cancer was influenced by recent food intake, while colorectal cancer was most effected by foods consumed more than ten years ago. Multiple regression analysis and geographical analysis between cancer mortality or incidence and age-adjusted food consumption continue.

TYPE: Correlation; Incidence
TERM: Diet; Fat; Trends; Vitamins
SITE: Colon; Oesophagus; Rectum; Stomach
CHEM: Sodium Chloride
TIME: 1983 - 1989

NIIGATA

* 538   Toyoshima, H.   05008
Niigata Univ. School of Medicine, Dept. of Public Health, 1-757 Asahimashidori, Niigata 951, Japan (Tel.: (025)2236161; Fax: 2255885)

Factors Related to Carcinogenesis in an Agricultural Village in Niigata Prefecture
This is part of a large scale cohort project which aims to evaluate factors related to carcinogenesis in Japan. The total cohort population is 100,000 and members will be followed up for 10 years. Information concerning lifestyle, e.g. food intake, occupation, habits, family history, etc. have been obtained by questionnaire and serum samples are stored. In the present study, DNA samples from 3,000 individuals, in addition to the data mentioned above, are being stored to be analysed for DNA abnormalities in the future.

TYPE: Cohort; Molecular Epidemiology
TERM: BMB; DNA; Diet; Lifestyle; Occupation; Serum
SITE: All Sites
TIME: 1990 - 2000

NISHIYAMA

539   Hayakawa, K.   03197
Kinki University, School of Medicine, Department of Public Health, Sayama-cho, Osaka-fu 589, Nishiyama 380, Japan
COLL: Shimizu, T.

Ageing Twins and Cancer
This is a longitudinal prospective study of 2,000 pairs of twins born before 1935 and living all over Japan. About one third of the subjects are identical twins and 70 pairs are identical twins reared apart. At the present time, a questionnaire survey has been conducted including questions about personal, family, and medical histories, present symptoms, food, alcohol and cigarette consumption and signs of ageing (such as presbyopia or grey hair). The subject pairs will be surveyed by home visits as long as one of the twins is alive. A paper has been published in Jap. J. Hyg. 45(1):121-124, 1990.

TYPE: Cohort; Incidence
TERM: Lifestyle; Twins
SITE: All Sites
TIME: 1981 - 1993

OHTSU

540   Watanabe, S.   04643
Shiga Univ. of Medical Science, Dept. of Preventive Medicine, Seta-tsukinova-cho, Ohtsu 520-21, Japan (Tel.: (0775)482186)

Cancer Among Workers in Chemical Industries and a Carbon Electrode Industry
Mortality from cancer and all causes is being determined in a small cohort of 259 current and retired chromate workers who were exposed to chromates for at least 5 years before 1972, and who are being followed up from 1960. A similar study is in progress for 341 workers from a carbon electrode industry and 1,000 coke-oven workers from a chemical industry, who have been followed up since 1980. Their cancer mortality by site is being compared to that of the general population of Japan.

JAPAN

TYPE: Cohort
TERM: Metals; Occupation
SITE: All Sites
CHEM: Chromium; Tars
OCCU: Chromate Producing Workers; Coke-Oven Workers; Electrode Manufacturers
TIME: 1989 - 1993

OKINAWA

541   Sakai, R.   03908
Ryukyu Univ., School of Health Science, Dept. of Epidemiology, Nishihara-cho , 207 Uehara, Okinawa 903-01, Japan (Tel.: (9889)53331)
COLL: Mori, W.; Machinami, M.

Smoking, Alcohol Drinking and Cancer Risk

Cigarette smoking and alcohol consumption are increasing in the general population in Japan. A case-control study is being conducted to investigate the influences of cigarette-smoking and alcohol-drinking in the population. About 30,000 autopsy cases (6,000 cancer cases and 18,000 controls) in the Department of Pathology, University of Tokyo from 1930 to the present day are being analysed. Control cases are the autopsy cases which did not have malignant neoplasms, or diseases associated with malignant neoplasms, cigarette-smoking and alcohol-consumption. Cancer cases are divided by site of cancer. Information on smoking and alcohol consumption is obtained from the clinical protocol. Descriptive epidemiological and matched pair analyses are performed.

TYPE: Case-Control
TERM: Alcohol; Tobacco (Smoking)
SITE: Liver; Lung
TIME: 1978 - 1992

542   Sakai, R.   04486
Ryukyu Univ., School of Health Science, Dept. of Epidemiology, Nishihara-cho , 207 Uehara, Okinawa 903-01, Japan (Tel.: (9889)53331)

Dietary Epidemiology

Okinawa has very low cancer mortality and morbidity, especially for the digestive system (such as stomach, liver), in both sexes. There are many kinds of dietary plants, including medical herbs. This study aims to examine the carcinogenic and anti-carcinogenic properties of these dietary plants. A case-control study has been conducted among several general hospitals in Okinawa, using a questionnaire-interview. The present sample size is about 1,000 in-patients, including the cancer cases. In-patients with benign diseases among the same hospitals are used as controls. There are 2-3 controls per case, matched by sex, age, residence and smoking- and drinking habits. Matched and unmatched techniques will be used in the analysis. Over 5,000 in-patients will be collected during the study period.

TYPE: Case-Control
TERM: Diet; Plants; Projection
SITE: Gastrointestinal
TIME: 1982 - 1995

* 543   Sakai, R.   05125
Ryukyu Univ., School of Health Science, Dept. of Epidemiology, Nishihara-cho , 207 Uehara, Okinawa 903-01, Japan (Tel.: (9889)53331)

Dietary Epidemiology in Okinawa

This hospital-based case-control study is being carried out to assess the carcinogenicity and anti-tumour activity of dietary plants and herbs in Okinawa, which has the lowest cancer incidence in Japan, especially of cancers of the digestive tract. Daily ingestion of these plants is evaluated by interview of all inpatients in five major general hospitals. The controls are selected among the respondents in the interview survey, after matching for age, sex, etc. The total number of subjects will be over 10,000.

JAPAN

TYPE: Case-Control
TERM: Diet; Plants
SITE: All Sites
TIME: 1982 - 1995

OSAKA

544 Hanai, A. 04689
Center for Adult Diseases, 1-3-3 Nakamichi, Higashinari-ku, Osaka 537, Japan (Tel.: (06)9721181; Fax: 9727749)
COLL: Fujimoto, I.; Taniguchi, H.; Berg, J.; Ries, L.; Young, J.; Percy, C.; Van Holten, V.; Hankey, B.

Stomach Cancer Incidence Trends among Japanese in Japan and US
The relative proportions of Lauren's two main histological types of stomach carcinoma, intestinal and diffuse, have shown that (1) the intestinal type of carcinoma is seen more in older age groups, in males and in high-risk areas; (2) the diffuse type is seen more in younger age-groups, in females, and in low-risk areas; (3) the ratio of the intestinal type to the diffuse type decreases when the total incidence decreases. The histological diagnoses reported to the Osaka Cancer Registry are being analysed to assess whether or not patterns (1) and (2) are also observed in Osaka, and to ascertain if the ratio (intestinal/diffuse) will fall with an anticipated decrease of incidence in Osaka in the future. The 60,779 stomach cancers registered at the Osaka Cancer Registry from 1973 to 1985 have already been analysed by sex and age-group. In the next two years, stomach cancer incidence by histological type in Osaka from 1977 to 1987 will be compared with the same data on whites, blacks and Japanese in the USA registered by the SEER program in the USA.

TYPE: Incidence
TERM: Age; Ethnic Group; Histology; Registry; Trends
SITE: Stomach
LOCA: Japan; United States of America
REGI: Osaka (Jap); SEER (USA)
TIME: 1989 - 1992

545 Morimoto, K. 04691
Osaka Univ., School of Medicine, Dept. of Hygiene, and Preventive Medicine, Nakanoshima 4-3-57, Kita-ku, Osaka 530, Japan (Tel.: (6)4435531/263; Fax: 4419116)
COLL: Takeshita, T.; Takeuchi, T.; Shirakawa, T.; Mure, K.; Cui, Y.S.; Zu, Z.H.; Ogura, H.

Epidemiology of Chromosomal Damage in Human Peripheral Lymphocytes
It has been suggested that an increase in the frequencies of chromosomal alterations such as SCEs, chromosomal aberrations and micronuclei in somatic cells may have a close relationship with cancer development. In this study, the effects of lifestyle factors such as cigarette smoking, alcohol consumption, hours of work, exposure to hazardous environmental factors such as ionizing radiation, and/or various chemical muta/carcinogens on the chromosomal alterations in peripheral blood lymphocytes will be assessed. The cells of about 200 healthy persons, several hundred cancer patients and controls, and persons with cancer-prone diseases such as Fanconi anaemia, xeroderma pigmentosum, and ataxia telangiectasia will be tested for sensitivity to these agents. The ability of the immune response to prevent cancer development, especially natural killer cell activity, is being investigated. Data are analysed to elucidate the interaction between lifestyles, hereditary predisposition and exposure to environmental hazardous agents, to obtain basic data for the primary prevention of cancer.

TYPE: Case-Control; Molecular Epidemiology
TERM: Alcohol; Chemical Exposure; Chromosome Effects; Heredity; Lifestyle; Lymphocytes; Micronuclei; SCE; Tobacco (Smoking)
SITE: All Sites
TIME: 1983 - 1992

546 Morinaga, K. 04025
Center for Adult Diseases, Dept. of Field Research, 3 Nakamichi 1-chome, Higashinari-ku, Osaka 537, Japan (Tel.: (06)9721181; Fax: 9727749)
COLL: Kohyama, N.

JAPAN

Mesotheliomas and Mineral Fibres
Specimens of lung tissues are collected for the cases with malignant mesotheliomas reviewed by the Osaka Mesothelioma Panel. Two controls are matched for age and hospital. Mineral fibres are detected and counted by analytical electron microscopy. Preliminary results suggest that only chrysotile can induce peritoneal mesotheliomas. 25 cases and 50 controls will be analysed.

TYPE: Case-Control
TERM: Dusts
SITE: Mesothelioma; Peritoneum; Pleura
CHEM: Asbestos; Mineral Fibres
TIME: 1983 - 1992

547　　Morinaga, K.　　04883
Center for Adult Diseases, Dept. of Field Research, 3 Nakamichi 1-chome, Higashinari-ku, Osaka 537, Japan (Tel.: (06)9721181; Fax: 9727749)
COLL: Sakato, J.; Tukuma, H.; Sobue, T.

Case-Control Study on Lung Cancer in Sennan District, Osaka
There have been more than 70 asbestos factories in Sennan district Osaka. Between 1985 and 1988, 79 deaths of lung cancer (age less than 80) were registered at Ozaki Health Center. Trained nurses interviewed the relatives of 63 cases and collected the information on occupational history, smoking habit, residence history, and the dietary habit on the frequency of taking green vegetables. The same questionnaire cards were sent to about 2,300 inhabitants who were randomly chosen as one to 20 from the resident file of City Offices. Nearly 1,200 were collected. Four respondents were selected as a control for each case, matched for sex and age.

TYPE: Case-Control
TERM: Dusts; Occupation; Tobacco (Smoking); Vegetables
SITE: Lung
CHEM: Asbestos
TIME: 1988 - 1992

* 548　　Morinaga, K.　　04967
Center for Adult Diseases, Dept. of Field Research, 3 Nakamichi 1-chome, Higashinari-ku, Osaka 537, Japan (Tel.: (06)9721181; Fax: 9727749)
COLL: Kohyama, N.; Satoh, Y.

Mortality in Workers Exposed to Quartz, Kaolinite, Clinoptilolite and Mordenite
A retrospective cohort study of workers exposed to quartz, kaolinite, and zeolite is being undertaken to ascertain whether there is an excess of lung cancer and high mortality from mesothelioma. The plant has two mines: quartz and kaolinite operating since 1941, and zeolite since 1966. Analytical electron microscopic findings show that the zeolite is of two types, clinoptilolite and mordenite and that the mordenite is fibrous in form. The cohort consists of nearly 950 male and 250 female workers retired since 1965.

TYPE: Cohort
TERM: Mining; Occupation
SITE: Lung; Mesothelioma
CHEM: Mineral Fibres
OCCU: Miners
TIME: 1990 - 1995

* 549　　Saji, F.　　05187
Osaka Univ. Medical School, Dept. of Obstetrics & Gynecology, Nakanoshima 4-3-57, Kita-Ku, Osaka 530, Japan
COLL: Tanizawa, O.; Ohashi, K.; Okada, S.

Perinatal Infection of HTLV-I: A Prospective Study
HTLV-I is the aetiological agent of some adult T-cell leukaemia/lymphoma (ATL) in humans. Epidemiological studies of ATL in Japan have revealed a tendency to familial clustering, suggesting vertical transmission of the virus. The object of this study is to determine the mode of perinatal infection of HTLV-I and the prophylactic effect on vertical transmission by prohibiting the HTLV-I carrier mothers from breast-feeding their neonates. A serological survey was undertaken among pregnant women attending University-affiliated hospitals in Osaka (9,000 deliveries a year). Blood samples of neonates

born to seropositive parturients are collected at delivery and every third month after birth, and assayed for HTLV-I infection by DNA analysis. Preliminary data (J. Clin. Immunol. 9:409-414, 1989 and Cancer 66:1933-1937, 1990) suggest that postpartum infection via breast milk is a likely major perinatal transmission route.

TYPE: Case Series
TERM: HTLV; Sero-Epidemiology
SITE: Leukaemia
TIME: 1987 - 1992

550 Ubukata, T. 04373
Center for Adult Diseases, Dept. of Epidemiology, and Gastroenterological Mass Survey, Higashinari-ku, 1-3, Nakamichi, Osaka 537, Japan (Tel.: (06)9721181)
COLL: Morinaga, K.; Fujimoto, I.; Oshima, A.

Smoking, Drinking and Stomach Cancer

Cancer of the stomach is still the first leading cause of cancer death in Japan, although the age-adjusted death rate for stomach cancer has shown a marked declining trend. Alcohol and tobacco are of particular interest because they are known to be risk factors for cancers of upper digestive and respiratory tracts. In order to determine if drinking and smoking are risk factors for stomach cancer in Japan, a case-control study is being carried out in a cohort of about 14,000 persons screened for stomach cancer at the Center for Adult Diseases, Osaka during the period of 1961-1967, followed up through 1985 by means of record linkage to the Osaka Cancer Registry. The number of gastric cancers diagnosed during the follow-up period was 486. For each case, 3 controls will be selected randomly from the cohort, matched for sex, age (within 5 years), occupation and year of initial screening. All screenees were interviewed routinely regarding demographic and medical variables. The data will be analysed by multiple logistic regression methods.

TYPE: Case-Control
TERM: Alcohol; Record Linkage; Registry; Tobacco (Smoking)
SITE: Stomach
REGI: Nagoya (Jap)
TIME: 1987 - 1992

OTSU

* 551 Aoyama, T. 03004
Shiga University of Medical Science, School of Medicine, Department of Experimental Radiology, Tsukinowa-cho, Seta, Otsu 520-21, Japan
COLL: Sugahara, T.; Yamamoto, Y.; Kato, H.; Shimizu, Y.

Mortality and Health Study of Radiological Technologists to Evaluate the Risks Involved in Exposure to Low-Dose Radiation

Japanese radiological technologists are a suitable population for the study of the risk from low-level exposure to radiation. This study aims to survey the mortality and cause of death, health status (cell-mediated immunity test, 27 serum biochemical examinations, CRP test, 4 urine examinations, pulse, blood pressure, ECG, peripheral blood cell examinations, visual acuity, chromosome aberrations and physical strength), and cumulative radiation doses of this population. The total registered number of the technologists was about 22,000 at the end of March, 1986. 9,179 members born before 1950 were traced for the mortality study from 1969 to 1982 as the first follow-up. The second follow-up of this population between 1983 to 1986 was started in 1986. The health study has so far been carried out on 1,000 members of 20 branches of the Japan Association of Radiological Technologists. Comparisons will be made with the general population. Papers appeared in "Biological Effects of Low-Level Radiation", International Atomic Energy Agency, Vienna, 1983, pp. 319-328 (IAEA-SM-266/43) and in J. UOEH II (Suppl.):432-442, 1989.

JAPAN

TYPE: Cohort
TERM: Occupation; Radiation, Ionizing
SITE: All Sites; Brain
OCCU: Health Care Workers
TIME: 1981 - 1992

SAGA

552 Tokudome, S. 03109
Saga Cancer Registry, Saga Medical School, Dept. of Community Health Science, Saga 849, Japan (Tel.: (0952)316511)
COLL: Kono, S.; Ikeda, M.; Ogata, M.; Shimono, M.; Makimoto, K.; Uchimura, H.

Cancer and Other Causes of Death among Psychiatric Patients

This is a follow-up study of the mortality of about 3,000 psychiatric patients institutionalized from 1945 to 1982 in order to investigate whether they are at high risk of mortality from cancer (and other causes). The patients are divided by the psychiatric diagnosis to examine possible carcinogenic effect of drugs used in psychiatric practice. Smoking and drinking habits are also studied to evaluate their possible influence on cancer mortality of the patients.

TYPE: Cohort; Mortality
TERM: Alcohol; Drugs; Tobacco (Smoking); Treatment
SITE: All Sites
TIME: 1982 - 1992

553 Tokudome, S. 03113
Saga Cancer Registry, Saga Medical School, Dept. of Community Health Science, Saga 849, Japan (Tel.: (0952)316511)
COLL: Ikeda, M.; Kuratsune, M.

Lung Cancer among Copper Smelters

Following a study which showed a high mortality from lung cancer among copper smelters at a metal refinery (Int. J. Cancer: 13, 552-558, 1974; Int. J. Cancer: 17, 310-317, 1976) the data are being used for two studies: (1) a case-control study, using mortality cards kept at the health centre to evaluate the interaction of smoking and occupational exposure on occupational lung cancer; (2) a prospective study, extending the study period and study subjects to include employees whose Honseki (permanent address) is outside Oita Prefecture, a group excluded from the last analysis.

TYPE: Case-Control; Cohort
TERM: Chemical Exposure; Metals; Occupation; Tobacco (Smoking)
SITE: All Sites; Lung
CHEM: Arsenic; Copper
OCCU: Smelters, Copper
TIME: 1983 - 1992

554 Tokudome, S. 04112
Saga Cancer Registry, Saga Medical School, Dept. of Community Health Science, Saga 849, Japan (Tel.: (0952)316511)
COLL: Ikeda, M.

Follow-Up Study of Blood-Transfused Patients

About 4,000 blood-transfused patients undergoing partial gastrectomy at some institutions from about 1950 to 1970 are being followed. This study aims to examine whether patients who have received blood transfusion and have incurred non-A, non-B hepatitis (C hepatitis) are a high risk of cirrhosis and hepatocellular carcinoma.

JAPAN

TYPE: Cohort
TERM: Cirrhosis; Infection; Virus
SITE: Liver
TIME: 1986 - 1992

555   Tokudome, S.   04113
Saga Cancer Registry, Saga Medical School, Dept. of Community Health Science, Saga 849, Japan (Tel.: (0952)316511)
COLL: Ikeda, M.

Long-Term Follow-up Study of HTLV-I Carriers

A prospective study of about 8,000 HTLV-I carriers among blood donors at 7 Red Cross Blood Centres in Kyushu, where HTLV-I carriers are highly prevalent, is proposed. The aim of this study is to investigate the incidence rate of ATL (Adult T-cell Leukaemia/lymphoma) among HTLV-I carriers and the latency period from the infection of HTLV-I to the onset of ATL. It is also proposed to study aetiological (prognostic) factors of ATL in a case-control study.

TYPE: Cohort
TERM: HTLV; Infection; Latency; Virus
SITE: Leukaemia; Lymphoma
TIME: 1986 - 1992

SAITAMA

* 556   Hoshiyama, Y.   05067
Saitama Cancer Center Research Inst., Komuro 818, Ina-machi, Saitama 362, Japan (Tel.: (048)7221111; Fax: 7221739)
COLL: Tagashira, Y.; Sasaba, T.; Nakachi, K.; Imai, K.

High-Risk Groups and Primary Prevention of Stomach Cancer

A hospital-based case-control study is being conducted on 294 matched pairs. Cases are histologically confirmed in Saitama Cancer Center Hospital. Controls are selected from the general population in the same cities (or towns) as the cases, and matched for sex and age (within two years). Subjects are interviewed with a questionnaire, mainly concerned with dietary habit, smoking and drinking histories. On the basis of the frequency of consumption of the selected factors, an individual risk score will be calculated for all 588 subjects.

TYPE: Case-Control
TERM: Alcohol; Diet; Lifestyle; Prevention; Tobacco (Smoking)
SITE: Stomach
TIME: 1991 - 1994

557   Nakachi, K.   04366
Saitama Cancer Center Research Inst., Dept. of Epidemiology, 818 Komuro, Ina-Machi , Saitama 362, Japan (Tel.: (0487)221111/327; Fax: 221739)
COLL: Imai, K.; Kawajiri, K.

Cancer Incidence in Relation to Epidemiological, Immunological and Genetic Factors

The purpose of this study is to examine host-environmental interactions in carcinogenesis in terms of epidemiological, immunological and genetic factors. Data are being collected annually on residents over 40 years old in one town. All participants answer a self-administered questionnaire on dietary habits/history, mental stress, personality occupation, family history of cancer, smoking and drinking habits, etc. Blood tests include cell-mediated immunity selenium, vitamin A, and lipo-proteins. Data on 4,000 individuals have been collected since 1986. DNA has also been isolated from peripheral lymphocytes and stored since 1989 for a study of DNA polymorphisms. The screening of genetically high risk people for lung cancer is carried out by RFLPs of the cytochrome P450IA1 gene.

JAPAN

TYPE: Cohort; Molecular Epidemiology
TERM: Alcohol; BMB; DNA; Diet; Familial Factors; Genetic Factors; Immunology; Lymphocytes; Occupation; RFLP; Stress; Tobacco (Smoking); Trace Elements; Vitamins
SITE: All Sites
CHEM: Selenium
TIME: 1986 - 1993

558 Sasaba, T. 02128
Saitama Cancer Center Research Inst., 818 Komuro, Ina-Machi, Saitama 362, Japan (Tel.: (048)7221111; Fax: 7221739)
COLL: Tagashira, Y.; Nakachi, K.; Imai, K.; Hoshiyama, Y.

Liver Cancer in Relation to Clonorchiasis and HBV

Long-term residents of the eastern part of Saitama have been shown to experience a liver cancer risk nearly three times that of the prefectural average. The striking similarity in the geographical clustering between high death rates from liver cancer and an endemic parasitic disease was described in Jap. J. Publ. Hlth. 34:121-127, 1987. On the contrary, there was no clustering of high or low rates for HBsAg positivity, suggesting HBV does not play an important role in clustering of liver cancer deaths in Saitama. A follow-up study on 700 individuals (300 cases of parasitic diseases and 400 HBV carriers) is now underway.

TYPE: Cohort
TERM: Geographic Factors; HBV; Parasitic Disease
SITE: Bile Duct; Liver
TIME: 1987 - 1993

559 Tagashira, Y. 03594
Saitama Cancer Center Research Inst., Komuro 818, Ina-machi, Saitama 362, Japan (Tel.: (048)7221111; Fax: 7221739)
COLL: Sasaba, T.; Nakachi, K.; Imai, K.; Hoshiyama, Y.

Case-Control Study in Saitama Cancer Center Hospital

This is a hospital-based case-control study. Cases are all first admission in-patients to the hospital (1,000/year), and are grouped according to cancer sites and histological types. Controls are selected from the general public living in the same cities or towns as cases and matched for sex and age. Department staff interview cases and controls with a questionnaire, which is mainly concerned with dietary habits, smoking and drinking histories, and other environmental factors. So far, more than 4,500 cases and about 1,000 controls have been interviewed. Analysis will include single factor risk estimations and interactions between more than one factor.

TYPE: Case-Control
TERM: Alcohol; Diet; Lifestyle; Tobacco (Smoking)
SITE: All Sites
TIME: 1985 - 1993

SAPPORO

560 Miyake, H. 03979
Sapporo Medical College, Dept. of Public Health, South 1, West 17, Chuo-ku, Sapporo 060, Japan (Tel.: (011)6112111; Fax: 612586)
COLL: Mori, M.; Goto, R.; Masuoka, H.; Yoshida, K.; Ohba, S.; Mitsuhashi, T.

Prospective Cohort Study of the Elderly in Hokkaido

This prospective cohort study was designed in 1983 to clarify the association between nutritional factors and adult diseases including various cancers among the elderly. The baseline surveys were performed in 1984 or in 1985 by health nurses for persons over 40 years of age who lived in Hokkaido Prefecture. In total, 3,185 persons (1,531 males and 1,654 females) have been involved in the study. The questionnaire used for the baseline survey included items concerning demographic factors, health status, lifestyle, and consumption frequency of 38 food items. A follow-up survey with regard to health status is carried out every year. As a result of the recent survey, a more frequent intake of instant noodles was associated with increased mortality. A paper has been published in Jpn. J. Cancer Clin. 37:255-260, 1991.

JAPAN

TYPE: Cohort
TERM: Age; Diet; Lifestyle
SITE: All Sites
TIME: 1984 - 1993

561   Miyake, H.   04727
Sapporo Medical College, Dept. of Public Health, South 1, West 17, Chuo-ku, Sapporo 060, Japan (Tel.: (011)6112111; Fax: 612586)
COLL: Takeda, T.; Nishi, M.

Incidence of Malignant Disease in Children in Hokkaido
A Registry of Childhood Malignancies has been operating in Hokkaido Prefecture, Japan since 1969. Since application for public assistance to pay medical expenses depends on registration, the Registry covers almost 100% of the incidence of childhood cancer. There are about 70,000 live births annually. There has been no report on the exact incidence of malignant diseases of children in Japan, and the aim is to describe the incidence of childhood cancer (leukaemia, malignant lymphoma, Wilms' tumour, etc.) during the 20 years of registration. A paper has been published in J. Pediatr. Surg. 25:545-546, 1990.

TYPE: Incidence
TERM: Registry
SITE: Childhood Neoplasms
TIME: 1969 - 1993

562   Takeda, T.   04732
Sapporo National Hosp., Dept. of Pediatrics, Kikusui 4-2, Shiroishi-ku, Sapporo 003, Japan (Tel.: (011)8119111; Fax: 8320652)
COLL: Takasugi, N.; Hanai, J.; Miyake, H.; Nishi, M.

Mass Screening for Neuroblastoma in Sapporo City
Mass screening for neuroblastoma, aimed at six-month-old infants, has been performed in Sapporo City since 1981. There are about 20,000 live births annually. Some epidemiological problems are being studied: (1) influence of screening on the incidence by birth cohort, (2) difference in incidence between groups screened and not screened, (3) cost-benefit relations of the mass screening, (4) necessity of re-screening, (5) final survival rate of the patients screened (including both true positive cases and false negative cases). Several papers were published in Cancer 60:433-436, 1987; Eur. J. Pediatr. 147:308-311, 1988 and in Pediatr. Res. 26:603-607, 1989.

TYPE: Incidence
TERM: Birth Cohort; Cost-Benefit Analysis; Screening
SITE: Neuroblastoma
TIME: 1981 - 1993

SENDAI

563   Fukao, A.   04267
Tohoku Univ., School of Medicine, Dept. of Public Health, 2-1 Seiryo-machi, Sendai 980, Japan (Tel.: (022)2741111; Fax: 2754877)
COLL: Minami, Y.; Hisamichi, S.; Sugawara, N.

Case-Control Study of Scirrhous Cancer of the Stomach
From many epidemiological studies, it has been clarified that the major risk factor of gastric cancer is environmental and that mortality and incidence are higher in males than in females. However, mortality and incidence under 40 years of age are higher in females than in males, and the prognosis of female patients with scirrhous gastric cancer is very poor. Hypotheses are: (1) there are different risk factors between the scirrhous (undifferentiated) and the other histological type of gastric cancer (differentiated), (2) hormonal or genetic environment are important factors. Approximately 150 cases and 300 controls (females from a mass screening programme for gastric cancer) will be identified from the member hospitals and interviewed by doctors and nurses. Of particular interest are environmental, hormonal and genetic factors. Data collection is continuing and analysis will be done by the matched-pair method.

JAPAN

TYPE: Case-Control
TERM: Diet; Environmental Factors; Genetic Factors; Hormones; Lifestyle; Menstruation; Occupation; Prognosis; Sexual Activity
SITE: Stomach
TIME: 1987 - 1993

* 564 Fukao, A. 05052
Tohoku Univ., School of Medicine, Dept. of Public Health, 2-1 Seiryo-machi, Sendai 980, Japan (Tel.: (022)2741111; Fax: 2754877)
COLL: Komatsu, S.; Hisamichi, S.; Sugawara, N.; Ida, Y.; Takano, A.

Lifestyle, Personality and Cancer Incidence

The aim of this cohort study is to investigate the influence of both lifestyle and personality on cancer incidence. The cohort, comprising about 50,000 people aged 40-64 and living in Miyagi Prefecture, Japan, was questioned about lifestyle and personality (Eysenck Personality Questionnaire) in 1990. About 10,000 serum samples were collected and beta-carotene and some substances related to cancer will be measured. The state of attendance at periodic health examinations and cancer screening programmes will be followed annually. Incidence of cancer will be measured using the Miyagi Cancer Registry.

TYPE: Cohort
TERM: BMB; Lifestyle; Psychological Factors; Registry; Serum
SITE: All Sites
CHEM: Beta Carotene
REGI: Miyagi (Jap)
TIME: 1990 - 2000

TOKOROZAWA

565 Kono, S. 04271
National Defense Medical College, Dept. of Public Health, 3-2 Namiki, Tokorozawa 359, Japan (Tel.: (0429)951211/2283)
COLL: Ikeda, N.; Yanai, F.; Shinchi, K.; Imanishi, K.; Nishikawa, H.

Risk Factors for Colorectal Polyps

Since colorectal polyps are regarded as precursor lesions of colorectal cancer, aetiological factors for colorectal cancer should be more easily found in studies of colorectal polyps. The study is cross-sectional. Colorectal polyps detected by fibre-optic endoscopy will be related to diet and other lifestyle factors. Study subjects are middle-aged men of the Self-Defense Forces (SDF) in nothern Kyushu who are examined before retirement under the health examination programme at the SDF Fukuoka Hospital. Data for the period October 1986 to December 1990 on 2,760 men in their early 50's have been computerised. Analyses of earlier data showed a strong association between non-distilled alcoholic beverages and colon polyps. To establish a cohort of 5,000 men with comprehensive medical information, the study has been extended to the SDF Kumamoto Hospital.

TYPE: Cross-Sectional
TERM: Alcohol; Diet; Physical Activity; Premalignant Lesion; Prevalence
SITE: Colon; Rectum
TIME: 1987 - 1993

TOKYO

566 Anzai, S. 04227
Showa Univ., School of Medicine, Dept. of Public Health, 1-5-8, Hatanodai, Shinagawa-ku, Tokyo 142, Japan (Tel.: (03)7848133; Fax: 7860072)
COLL: Aoki, K.; Tominaga, S.; Kurihara, N.; Inaba, Y.; Fujiki, H.; Hirohata, T.; Kuroki, T.

Evaluation of Factors Affecting Cancer Incidence

Nationwide re-evaluation of the positive or negative risk for cancer, of factors such as smoking, alcohol, animal fat, green and yellow vegetables and vitamins is being undertaken, using cohort groups. Cancer sites of interest are stomach, lung, liver, large intestine, uterus, and some other sites with relatively high incidence rates. Approximately 1,000 residents and employees aged over 40 will be examined; their biological and epidemiological characteristics will be described for the first few years, and subsequently their cancer incidence will be assessed at least for the next five to ten years. The sample size, protocols, and follow-up methods including ethical problems are now being discussed.

TYPE: Cohort
TERM: Alcohol; Diet; Fat; Tobacco (Smoking); Vitamins
SITE: Colon; Liver; Lung; Stomach; Uterus (Corpus)
TIME: 1987 - 2002

567　Hayata, Y.　01486
Tokyo Medical College Hosp., Dep. of Surgery, 6-7-1 Nishishinjuku, Shinjuku-ku, Tokyo 160, Japan (Tel.: (3)3426111)
COLL: Oho, K.; Saito, Y.; Funatsu, H.; Kato, H.

Mass Surveys of Tokyo Metropolitan Government Employees by Chest X-Ray and in High Risk Cases also by Sputum Cytology

The aim is to detect lung cancer, hopefully at an early stage. Chest X-rays have been performed since 1953 on employees of the Tokyo Metropolitan Government, and 235 lung cancer cases have been detected among a total of 2,337,621 examinations. The overall detection rate per 100,000 examinations was 10.1, 24.9 in males aged 40 years and over and 10.4 in females aged 40 years and over. Among these 41.7% were stage I, 13.2% stage II and 45.1% stage III and IV. Resectability rates were 87.8%, 64.5% and 29.2% respectively. There are currently three groups of examinees: a high-risk group examined semi-annually, a group examined annually, and the group of cases not examined the previous year. The rate of early stage lung cancer detection was 24.3% in the semi-annual group, 12.3% in the annual group and 9.5% in the third group. There were also significant differences between the three groups in resectability and five-year survival rates. Since 1974 sputum cytology surveys have been conducted on a high-risk group of 4,898 examinees to detect lung cancer and atypical squamous metaplasia cells. Cases with squamous metaplasia cells are followed up at four month intervals. Metaplastic cells are examined for DNA content. Evaluation of squamous metaplasia cells and/or DNA analysis results were based on serial findings in several canine experimental series, in which lung cancer was induced in the major bronchi by intramucosal injections of 20-methylcholanthrene. The high-risk group is defined as those aged 45 or more with a cigarette index of over 400, those involved in occupations involving exposure to pollution, and those with cough or bloody sputum. Papers appeared in Cancer Res. 82:163-173, 1982 and in Lung Cancer Diagnosis (Hayata, Y. ed., Barron, J.P. trans., Igaku-Shoin, Tokyo, 1982, p. 59-67).

TYPE: Cohort
TERM: Cytology; DNA; High-Risk Groups; Occupation; Screening; Sputum; Tobacco (Smoking)
SITE: Lung
TIME: 1953 - 1992

568　Inaba, Y.　04305
Jutendo Univ. School of Medicine, Dept. of Hygiene, 1-1 Hongo, 2-chome , Bunkyo-ku, Tokyo 113, Japan (Tel.: (03)8133111)
COLL: Narnihisa, T.; Ichikawa, S.; Sato, N.; Kikuchi, S.

Chronic Hepatitis and Liver Cirrhosis

In order to elucidate the relationship between type of chronic hepatitis or liver cirrhosis and liver cancer, a retrospective cohort study is to be carried out. The cohort will consist of about 1,000 cases of chronic hepatitis or liver cirrhosis whose liver tissue was biopsied at Juntendo University Hospital from 1973 to 1982. Each case is to be studied for type of disease, age, sex, HBV markers, history of blood transfusion, alcohol intake and history of smoking. Information is taken from their personal hospital record. The vital status of each case as of the end of 1990 will be checked by out-patient or in-patient hospital records and the 'Koseki' (permanent residence record). When the case is dead, the causes of death will be confirmed by the autopsy record or death certificate. Analysis of the cohort will be done by the life table method with inner cohort comparison.

JAPAN

TYPE: Cohort
TERM: Alcohol; Cirrhosis; HBV; Tobacco (Smoking); Virus
SITE: Liver
TIME: 1986 - 1992

569　Kobayashi, N.　02244
National Children's Hosp., 3-35-31 Taishido, Setagaya-ku, Tokyo 154, Japan (Tel.: (3)4148121; Fax: 4190381)
COLL: Matsui, I.; Tanimura, M.

Japan Children's Cancer Registry

This is a nationwide and physician-based registry of children with cancers. The registry is controlled by the Committee of the Japan Children' Cancer Registry, part of the Children's Cancer Association of Japan. Cancer morbidity in Japanese children is being assessed. During the 17 years from 1969 to 1987, 23,746 cases (13,223 males and 10,523 females) were registered including acute lymphatic leukaemias accountinng for 23%, acute myeloid leukaemias 12%, neuroblastomas 10%, retinoblastomas 8%, Wilms' tumours 4%, primary liver cell carcinomas 2.2%, lymphosarcomas 2.0%, rhabdomyosarcomas 1.8%, acute monocytic leukaemias 1.3%, medulloblastomas 1.2% of total cases. The fourth report of the registry, covering 1984-1988, will be published in December 1990.

TYPE: Incidence; Registry
TERM: Childhood
SITE: Childhood Neoplasms
TIME: 1969 - 1993

570　Kobayashi, N.　03892
National Children's Hosp., 3-35-31 Taishido, Setagaya-ku, Tokyo 154, Japan (Tel.: (3)4148121; Fax: 4190381)
COLL: Hayakawa, N.; Matsui, I.

Children's Cancer-Immunodeficiency Registry

The registry is a nationwide and physician-based registry of cases with primary immunodeficiency syndromes. From 1975 to 1989, 812 cases of immunodeficiency syndromes have been registered. The most frequent syndromes recorded are common variable immunodeficiency (15.9%), chronic granulomatous disease (12.5%), selective IgA deficiency (12.0%), infantil X-linked agammaglobulinemia (9.3%), severe combined immunodeficiency (9.2%), and ataxia-telangiectasia (7.5%). In 812 registered cases, 27 cases with malignant neoplasms were reported. The incidence of malignancies in infant cases was 3.2% (21/663). This is more than 300 times that in the general population in Japan. The rate is especially high in the Chediak-Higashi syndrome (33.3%; including possible cases of blastic crisis) and ataxia-telangiectasia (15.5%). These relationships are being studied.

TYPE: Registry
TERM: Immunodeficiency
SITE: Childhood Neoplasms
TIME: 1975 - 1993

571　Kobayashi, N.　04836
National Children's Hosp., 3-35-31 Taishido, Setagaya-ku, Tokyo 154, Japan (Tel.: (3)4148121; Fax: 4190381)
COLL: Matsui, I.

Risk Factors for Childhood Malignancy in Japan

This study will examine the hypothesis that childhood malignancy is caused by genetic factors, by intra-uterine exposure to environmental factors, and by infection. Registered cases from the Japan Children's Cancer Registry (a few hundred cases of each type of malignancy, including leukaemia) will be subjected to analysis in respect of twins, associated abnormalities including chromosomal aberration and translocation, and history of exposure to environmental factors. Controls will be healthy children, if possible, and also exposure comparisons will be made between solid tumour and leukaemia cancer.

TYPE: Case-Control
TERM: Congenital Abnormalities; Drugs; Environmental Factors; Intra-Uterine Exposure; Radiation, Ionizing; Registry; Twins
SITE: Childhood Neoplasms; Leukaemia
TIME: 1985 - 1993

572   Nakamura, K.   04247
Showa Univ., School of Medicine, Dept. of Hygiene, and Preventive Medicine, 1-5-8, Hatanodai, Shinagawa-ku, Tokyo 142, Japan
COLL: Tadera, M.; Masaki, M.

Cancer in a Working Population in the Tokyo Area

The aim is to elucidate the effects of life style (diet, alcohol drinking, smoking, marital and socio-economic status, etc.) and several biological factors such as body build, blood pressure, serum cholesterol, etc., in the occurrence of cancer. Subjects are approximately 7,000 employees of stockbroking firms, aged 40 or more, and living in the Tokyo Area. A baseline survey will be carried out in 1988. As long as the subjects are employed in a company using the Health Insurance Society, occurrence of cancer is easily detected. After resignation, information on death can be collected by means of inquiring into their family or their 'Koseki' (family register in local government).

TYPE: Cohort
TERM: Alcohol; Cholesterol; Diet; Lifestyle; Marital Status; Physical Factors; Socio-Economic Factors; Tobacco (Smoking)
SITE: All Sites; Colon; Liver; Lung; Rectum; Stomach
TIME: 1987 - 1993

573   Tsugane, S.   04786
National Cancer Centre Research Inst., Epidemiology Div., 5-1-1 Tsukiji , Chuo-ku, Tokyo 104, Japan (Tel.: (3)35422511; Fax: 35460630)
COLL: Hamada, G.S.; Watanabe, S.; Laurenti, R.

Lifestyle Factors among Japanese Residents in Sao Paulo

The aim of this study is to collect data on the prevalence of lifestyle factors, including biological and anthropometric characteristics, from a random sample of Japanese residents in the city of Sao Paulo and to correlate these data with known disease mortality rates which differ between Japanese living in Japan and those living in the United States (Int. J. Epidemiol. 18(3):647-651, 1989; Int. J. Cancer 45(3):436-439, 1990). The subjects are 410 Japanese residents aged 40 to 69, 146 Japanese immigrants and 264 second-generation descendants, selected by a random sampling method according to geographic location. Blood and urine samples, and questionnaires concerning lifestyle, immigration history will be collected. The same study will be conducted in 1990 for other subjects listed with the cultural associations of various Japanese prefectures. Some nutrients (vitamins, trace minerals, lipids) and hormones will be measured in blood and urine.

TYPE: Cross-Sectional
TERM: BMB; Biochemical Markers; Blood; Diet; Hormones; Lifestyle; Lipids; Migrants; Trace Elements; Urine; Vitamins
SITE: All Sites; Breast (F); Liver; Prostate; Stomach
LOCA: Brazil
TIME: 1989 - 1992

574   Tsugane, S.   04609
National Cancer Centre Research Inst., Epidemiology Div., 5-1-1 Tsukiji , Chuo-ku, Tokyo 104, Japan (Tel.: (3)35422511; Fax: 35460630)
COLL: Watanabe, S.; Kabuto, M.; Yamaguchi, M.; Takashima, Y.; Gey, F.

Ecological Study on Cancer Mortality and Lifestyle in Japan

Cancer mortality varies even in the uni-racial population of Japan. The age-adjusted mortality rate of stomach cancer varies almost three-fold between different prefectures. A cross-sectional study in five populations in Japan has been planned in order to clarify the relationship between cancer mortality and lifestyle risk factors using some biochemical markers, i.e. nutrients (vitamins, trace mineral, lipids), hormones, etc. Each area has approximately 100,000 population and a total of 170 randomly selected males aged 40 to 49 are included. Wives of selected men are also included. Blood and urine sampling will also be done, with an interview and medical examination. Surveys were completed in 1989 in Iwate and Okinawa prefectures, and Nagano and Akita prefectures in 1990, and metropolitan Tokyo in 1991. The

JAPAN

laboratory measurements and data analysis will be conducted in 1992. A paper has been published in Cancer Causes and Control 2:165-168, 1991.

TYPE: Correlation
TERM: Hormones; Lifestyle; Lipids; Urine; Vitamins
SITE: All Sites; Lung; Stomach
TIME: 1989 - 1992

* 575   Tsugane, S.   05093
National Cancer Centre Research Inst., Epidemiology Div., 5-1-1 Tsukiji , Chuo-ku, Tokyo 104, Japan (Tel.: (3)35422511; Fax: 35460630)
COLL: Hamada, G.S.; Iriya, K.; Kowalski, L.P.; Torloni, H.; Watanabe, S.

Stomach Cancer among Japanese and Non-Japanese Brazilians

The aim of this case-control study is to identify epidemiological and molecular aspects of stomach cancer among Japanese and non-Japanese Brazilians that could explain the difference in mortality and incidence rates between these populations. Cases will be all those patients diagnosed with stomach cancer at one of several Sao Paulo hospitals from April 1991: 150 first- or second-generation Japanese Brazilians and 150 non-Asian Brazilians of either sex, aged 40-69 and living in Sao Paulo State. Controls will be non-cancer patients matched on hospital, sex, ethnicity and age within three years. All subjects will be interviewed and blood samples taken; fresh tumour tissue will be obtained from cases. Plasma levels of H. pylori antibody and pepsinogen will be measured, and lymphocytes analysed for polymorphism and alkyl-DNA adducts.

TYPE: Case-Control
TERM: Biochemical Markers; DNA Adducts; Diet; H. pylori; Japanese; Lifestyle; Pepsinogen
SITE: Stomach
LOCA: Brazil
TIME: 1991 - 1993

576   Watanabe, S.   03300
National Cancer Center Research Inst., 5-1-1, Tsukiji, Chuo-ku, Tokyo 104, Japan (Tel.: (03)5422511; Fax: 5460630)
COLL: Fujimoto, I.; Hanai, A.; Tsunematsu, Y.; Kobayashi, Y.

Multiple Primary Malignant Neoplasms

The incidence of multiple primary cancers is being analysed by person-years methods, using hospital-based registration data in the National Cancer Center, National Children's Hospital and Center for Adult Diseases. Linkage analysis with the regional cancer registry is also being done in Osaka. Cohort analysis is carried out in cases with childhood cancer. Data on second cancers in breast and lung cancer patients have been published and a publication on second cancers in childhood cancer patients is in press in Cancer Chemotherapy.

TYPE: Cohort
TERM: Childhood; Clinical Records; Multiple Primary; Registry
SITE: All Sites
TIME: 1987 - 1992

YOKOHAMA

577   Okamoto, N.   04655
Kanagawa Cancer Centre, Clinical Research Inst., Dept. of Epidemiology, 54-2, Nakao-cho, Asahi-ku, Yokohama 241, Japan (Tel.: (045)3915761)
COLL: Morio, S.

Lung Cancer and Industrial Dusts

The aim of this study is to examine the correlation between lung cancer and industrial dusts, especially asbestos. About 60,000 people participated a in mass-screening programme for lung cancer from 1988 to 1992. These have been divided into two groups, exposed and never exposed to the dusts, and the two groups will be followed until 1993, by means of record linkage with the population-based cancer registry in Kanagawa Prefecture. Analysis will be undertaken by the person-years method.

JAPAN

TYPE: Cohort
TERM: Dusts; Registry
SITE: Lung
CHEM: Asbestos
REGI: Kanagawa (Jap)
TIME: 1988 – 1993

MALAYSIA

KUALA LUMPUR

578 Prasad, U. 03465
Univ. of Malaya, Dept. of Otorhinolaryngology, Lembah Pantai, Kuala Lumpur 59100, Malaysia
(Tel.: (03)7556554/7574422; Fax: 7556554)
COLL: Pathmanathan, R.; Paramsothy, M; Nuruddin, R.; Singaram, S.; Lim, A.

EBV Serology for Detection of Early Recurrence in Nasopharyngeal Carcinoma
From studies carried out in the context of a project on EBV serology and nasopharynx carcinoma, it has been possible to indicate the significance of EBV serology, particularly IgA/VCA as a marker in the early diagnosis of NPC. The aims is now to establish its role in detecting early recurrence, which would further suggest that activation of EBV plays a part in the malignant process and that a common environmental factor may be responsible. 200 patients with confirmed NPC for whom full clinical, radiological (including CT-scan). immunological and histopathological data are available, will be followed-up in the special NPC clinic in this hospital. As well as detailed clinical examination, EBV serology will be done to detect any rising titres. Those with rising titres will be subjected to tests to detect any recurrence or metastasis. Patients without rising titres and in remission will form a control group.

TYPE: Cohort
TERM: Antibodies; Biochemical Markers; EBV; Infection; Prognosis; Sero-Epidemiology
SITE: Nasopharynx
TIME: 1984 - 1992

579 Prasad, U. 03674
Univ. of Malaya, Dept. of Otorhinolaryngology, Lembah Pantai, Kuala Lumpur 59100, Malaysia
(Tel.: (03)7556554/7574422; Fax: 7556554)
COLL: Kulkarni, M.G.; Rampal, L.; Bachi, R.; Chew, L.N.

Sero-Epidemiological Studies in Sabah, East Malaysia
Sabah is one of the states of Malaysia separated from Peninsular (West) Malaysia by the South China Sea. While 28.1% of the population is comprised of indigenous Kadazan people there is a significant Chinese population living in the state. Both these races have a high incidence of NPC although their culture, life style, eating habits, occupations, religion, etc. are all widely different. EBV serology has been found to be of significance in the early diagnosis of NPC in both races. The aim of the study is to screen the population using IgA/VCA titres as a marker for the diagnosis of NPC and to investigate the epidemiological factor(s) which could have contributed to the causation of this cancer. It is estimated that 100 individuals with high IgA/VCA titres will be picked up out of every 9,000 screened. These will be examined for the presence of NPC, and those without NPC will be followed every six months for five years. Control groups will be identified out of this population for comparison.

TYPE: Case Series
TERM: Chinese; EBV; Race; Screening; Sero-Epidemiology
SITE: Nasopharynx
TIME: 1985 - 1992

580 Prasad, U. 04481
Univ. of Malaya, Dept. of Otorhinolaryngology, Lembah Pantai, Kuala Lumpur 59100, Malaysia
(Tel.: (03)7556554/7574422; Fax: 7556554)
COLL: Rampal, L.; Lim, A.

Traditional Herbal Medicine and Nasopharyngeal Carcinoma
Incidence of NPC in Malaya is highest among the Chinese who are known to keep to their tradition of using herbal medicine for various ailments. A large body of evidence now suggests close association of EBV with the aetiology of NPC. Positivity for EBV antibodies has been consistently noted on studying the serological profile of these patients. The purpose of this epidemiological study is to find out whether a significant relationship exists between traditional herbal medicine use and NPC. A total of 200 histologically confirmed NPC cases and the same number of controls matched for age (within 5 years), sex and ethnic group will be studied. The controls are patients' relatives (excluding blood relatives) or neighbours. Data will be obtained through interview.

MALAYSIA

TYPE: Case-Control
TERM: Chinese; EBV; Ethnic Group; Plants; Race
SITE: Nasopharynx
TIME: 1986 - 1992

* 581 Yadav, M. 05189
Univ. of Malaya, Inst. of Advanced Studies, Jalan Lembah Pantai, 59100 Kuala Lumpur, Malaysia (Tel.: (03)7552744/242; Fax: 7473661 ; Tlx: 39845 UNIMAL MA)
COLL: Norhanom, W.; Nurhayati, Z.A.

Role of HPV in Cervical Neoplasia
Prevalence of genital HPV infection in Malaysia is unknown. The primary objective of this study is to elucidate the distribution of HPV types commonly associated with cervical carcinoma. Investigations include collection of data on potential risk factors through analysis of questionnaires administered at a major public hospital by the researchers. About 50-100 patients will be screened annually with data on diet, reproductive history, medical history, and exposure to tobacco and alcohol. Parallel laboratory studies will include blood antibody titres for HPV, HSV, CMV and HHV-6, determination of HPV-DNA with PCR-technology on cervical scrapes and determination of type-specific HPV by nucleic acid hybridization of cervical biopsies of histologically proven carcinomas.

TYPE: Case Series; Molecular Epidemiology
TERM: Alcohol; BMB; CMV; Diet; EBV; HPV; HSV; PCR; Reproductive Factors; Tobacco (Smoking)
SITE: Uterus (Cervix)
TIME: 1991 - 1995

MEXICO

MEXICO

582 Hernandez, M. 04829
Direccion General de Epidemiologia, Aniceto Ortega 1321, Col. del Valle, Mexico D.F., Mexico (Tel.: +52 5 343263; Fax: 245600)
COLL: Romieu, I.; Meneses, F.; Rojas, R.

Determinants of Uterine Cervix Cancer in Mexico City

This is a population-based case-control study to investigate the determinants of uterine cervix cancer (UCC) among a Mexican population of women to address the following hypotheses: 1) that infection of the uterine cervix by papilloma virus and/or herpes virus type II increases the risk of UCC; 2) that deficiency of some nutrients: vitamin A (both beta carotene and preformed vitamin A), vitamin E, vitamin C, folic acid, and selenium, increases the risk of UCC, with a possible interaction with virus infection. The role of factors such as socio-economic status, use of oral contraceptives, smoking status, and obstetric and gynaecologic history, as well as sexual behaviour will also be investigated. Cases will be selected through the list of positive cytologic exam (pap smears) reported to the General Directorate of Epidemiology and will include all women diagnosed with positive cytology within one year period in Mexico D.F. Two controls for each case will be selected at random from the list of negative cytology; they will be matched by health centres and age (+/-2.5 years). Enrolment of 251 cases and 441 controls is anticipated over a one year period. Questionnaires including information on sexual activity and food frequency will be applied at home by trained interviewers. During the same visit, a blood sample will be obtained from each participant. All cytology will be reexamined by an expert pathologist and viral analysis for papilloma virus and herpes virus type II will be carried out on uterine cervix epithelial cells. After descriptive and stratified analysis, a multivariate conditional logistic regression will be performed to determine the main determinants of UCC as well as to study interactions.

TYPE: Case-Control
TERM: Blood; Diet; HPV; HSV; Nutrition; Oral Contraceptives; Sexual Activity; Socio-Economic Factors; Vitamins
SITE: Uterus (Cervix)
CHEM: Beta Carotene; Folic Acid; Selenium
TIME: 1990 - 1992

583 Romieu, I. 04856
Direccion General de Epidemiologia, Aniceto Ortega 1321, Col. del Valle, Mexico D.F., Mexico (Tel.: +52 5 343263; Fax: 245600)
COLL: Hernandez, M.; Meneses, F.; Rojas, R.

Nutritional Determinants of Breast Cancer

The wide variation in dietary intake makes Mexico a suitable setting for a study of the role of specific nutrients and foods in the genesis of breast cancer. This study is a population-based case-control study to investigate nutritional determinants of breast cancer in a Mexico population. The following specific hypotheses will be addressed: high intake of alcohol and fat increases the risk of breast cancer; intake of vitamin A (both carotenoid and preformed vitamin A), vitamin C and E reduces the risk of breast cancer. During a one year period 500 cases of newly diagnosed localized breast cancer reported to the National Mexican Cancer Registry will be enrolled and matched to neighbourhood controls. Dietary intake will be measured using a validated food frequency questionnaire designed especially for a Mexican population. Blood samples will be obtained for subsequent analysis (retinol, beta carotene, alphatocopherol, and genetic markers), as well as toenail clippings to determine selenium levels. After descriptive and stratified analysis, conditional logistic regression will be performed to determine the major determinants of breast cancer and particularly the role of specific foods and nutrients.

TYPE: Case-Control
TERM: Alcohol; BMB; Diet; Nutrition; Toenails; Trace Elements; Vitamins
SITE: Breast (F)
CHEM: Beta Carotene; Retinoids
TIME: 1990 - 1992

NETHERLANDS

AMSTERDAM

584   Kriek, E.   04679
The Netherlands Cancer Inst., Antoni van Leeuwenhoek Huis, Div. of Chemical Carcinogenesis, Plesmanlaan 121, 1066 CX Amsterdam, Netherlands (Tel.: (020)5122476; Fax: 172625 ; Tlx: 11273 NKI NL)
COLL: Den Engelse, L.; Van Leeuwen, F.E.; Van Schooten, F.J.; Bos, R.P.; Jongeneelen, F.J.

DNA Adducts in Rodent and Human Tissues Exposed to Polycyclic Aromatic Hydrocarbons (PAH)
A number of sensitive methods now permit detection of carcinogen-DNA adducts in human tissues and cells. The purpose of the present study is to develop the application of immunochemical methods and ^{32}P-postlabelling analysis for the determination of PAH-DNA adducts in human and animal tissues. Investigations in operable lung cancer patients are being conducted to measure PAH-DNA adducts in DNA isolated from peripheral blood WBC, and in DNA isolated from lung tissue of selected groups of patients with a smoking history. Occupational exposure to PAH will be investigated in workers in the aluminium industry. Particular attention will be given to persistence of adducts and for this purpose selected groups of workers will be examined over a longer period of time. The metabolite 1-hydroxypyrene will be determined in the urine and will serve as an indicator of more recent exposure. All subjects will be interviewed about possible exposure to other sources of PAH (smoking). The procedures developed and the results obtained are expected to be valuable for future epidemiological studies in the following ways: (1) identification of carcinogen-DNA adducts can provide evidence of exposure of individuals to carcinogens and the amount of a particular DNA-adduct will provide a measure of the degree of exposure; (2) identification of carcinogen-DNA adducts could be used as a marker of biological effects which might be related to the appearance of tumours at a later stage, and thus would provide a means of identifying groups at high risk. Recent publications were published by Van Schooten et al. in Carcinogenesis 11:1677-1681, 1990 and 12:427-433, 1991 and in J. Natl Cancer Inst. 82:927-933, 1990.

TYPE: Methodology
TERM: Biochemical Markers; DNA Adducts; Immunologic Markers; Lymphocytes; Occupation; Tobacco (Smoking)
SITE: Lung
CHEM: PAH
OCCU: Aluminium Workers
TIME: 1988 - 1992

585   Ter Schegget, J.   04804
Univ. of Amsterdam, Academic Medical Centre, Dept. of Virology, Meibergdreef 15 , 1105 AZ Amsterdam, Netherlands (Tel.: (20)5664855; Fax:916531)
COLL: Cornelissen, M.A.; Struyk, A.P.H.B.; Briet, H.A.; Van den Tweel, J.; Lammes, F.; Jebbink, M.; Van der Hoordaa, J.

Human Papilloma Virus (HPV) in Lesions of the Cervix Uteri
The aim of the study is to determine the relative risk of progression to CIN or carcinoma in women who have HPV types 6/11 and 16/18 in the cervix uteri. Prevalence of HPV types 16/18 compared to types 6/11 cervical infection will be assessed in a retrospective longitudinal study, using archival biopsies exhibiting CIN 0, CIN I or CIN II. Evaluation will start with 100 women with CIN 0 lesions, who have already been followed up using colposcopy and histology, and PCR techniques on paraffin sections.

TYPE: Cohort
TERM: Biopsy; HPV; PCR; Premalignant Lesion
SITE: Uterus (Cervix)
TIME: 1989 - 1993

586   Van Leeuwen, F.E.   03163
Netherlands Cancer Inst., Div. of Clinical Oncology, Dept. of Epidemiology, Plesmanlaan 121, 1066 CX Amsterdam, Netherlands (Tel.: (20)5122453; Fax: 6172625 ; Tlx: 11273)
COLL: Kroon, B.B.R.; Peterse, J.L.; Van Dongen, J.A.; Bruning, P.F.; Hart, A.A.M.

Benign Proliferative Breast Lesions and Breast Cancer
The aim is to establish, in a prospective study, the risk of developing a subsequent mammary carcinoma in various groups of patients with different types of histologically defined proliferative breast lesions. A very elaborate form has been designed for this purpose. Additional information is obtained from cytology

NETHERLANDS

and cytophotometry. Included in the study are variables such as cancer in the family, use of contraceptives, alcohol and drugs. Another goal is to detect by continous sampling whether hormonal profiles in women with benign lesions differ from those with mammary cancer and from healthy women. For the main part of the study, 1,250 women with a histologically examined proliferative breast lesion will be followed for at least ten years. Up to now 1,100 patients have entered the study.

TYPE: Cohort
TERM: Alcohol; Drugs; Familial Factors; Genetic Factors; Hormones; Oral Contraceptives
SITE: Benign Tumours; Breast (F)
TIME: 1979 - 1995

587 Van Leeuwen, F.E. 03775
Netherlands Cancer Inst., Div. of Clinical Oncology, Dept. of Epidemiology, Plesmanlaan 121, 1066 CX Amsterdam, Netherlands (Tel.: (20)5122453; Fax: 6172625 ; Tlx: 11273)
COLL: Misdorp, W.; Peterse, J.L.; Kroon, B.B.R.

Breast Cancer and Oral Contraceptives in Women with Benign Breast Disease

Although most studies up to now have failed to find any significant overall relationship between use of oral contraceptives (OC) and breast cancer, some studies have indicated particular sub-categories of OC users to be at increased risk. One controversy in the literature concerns the possibility of an increased risk of breast cancer in women who have used OC for prolonged periods of time and who also have a history of benign breast disease (BBD). Studies published so far have been hindered by small numbers and most of these studies lack data on histology of BBD and information as to whether OC use preceded BBD or not. In 1985 a case-control study nested in a prospective cohort study of women with biopsy-proven benign breast disease was started. The cohort presently consists of approximately 1,100 women; entry of patients will close at N = 1,250 (a follow-up time of 10 years is required). The cohort study aims at assessing the relationship between histological features of the initial (benign) lesion and subsequent breast cancer risk. All breast cancer cases in the cohort (50 expected) will be matched to four "controls" also from the cohort, that did not (yet) develop breast cancer. Matching variables include age, duration of follow-up and histology of the benign lesion. Cases and controls will be interviewed at home as to lifetime use of OC and other hormones and known breast cancer risk factors. An album has been composed, containing colour photographs of all OCs ever marketed in the Netherlands, in order to aid women in recalling the specific OC brands they used.

TYPE: Case-Control; Cohort
TERM: Hormones; Oral Contraceptives
SITE: Benign Tumours; Breast (F)
TIME: 1985 - 1993

588 Van Leeuwen, F.E. 04076
Netherlands Cancer Inst., Div. of Clinical Oncology, Dept. of Epidemiology, Plesmanlaan 121, 1066 CX Amsterdam, Netherlands (Tel.: (20)5122453; Fax: 6172625 ; Tlx: 11273)
COLL: Rookus, M.A.; Braas, P.A.M.; De Rijke, M.E.; Woordes-van Baalen, M.; De Groot-Vlasveld, M.

Breast Cancer and Oral Contraceptives

The study aims are: (1) to evaluate the hypothesis that breast cancer risk is related to oral contraceptive (OC) use in certain sub-categories of users, e.g. women who used OCs before their first full-term pregnancy, women who used OCs before age 25, women with a family history of breast cancer, women with a history of benign breast disease, women aged 45-54 years; (2) to re-evaluate the hypothesis that breast cancer risk is related to overall use of OC. In the Netherlands OC use has been higher than in any other Western country. Also, extensive OC use at young ages started earlier than in other countries. A population-based case-control study is conducted in four regions of the Netherlands. The study will include 1,000 incident cases under 54 years of age (500 of which will be under 45 years of age). Region- and age-matched population controls are selected from the municipal registers in a 1:1 matching design. By personal interview in the participants' homes, lifetime histories will be collected on OC use, use of other hormones, reproductive factors, family history of breast cancer and a possible history of benign breast disease. In order to facilitate recall of all OC brands ever used, subjects will be shown photographs of all OCs (pills and packages) that have been on the market in the Netherlands since 1962. OC use will be checked with the woman's physician(s). This study also enables us to examine the relationship between breast cancer risk and possible confounders, such as diet, alcohol intake, weight history and waist/hip ratio.

NETHERLANDS

TYPE: Case-Control
TERM: Alcohol; Diet; Drugs; Familial Factors; Hormones; Oral Contraceptives; Reproductive Factors
SITE: Breast (F)
CHEM: Oestrogens; Progesterone
TIME: 1986 - 1992

589 Van Leeuwen, F.E. 04508
Netherlands Cancer Inst., Div. of Clinical Oncology, Dept. of Epidemiology, Plesmanlaan 121, 1066 CX Amsterdam, Netherlands (Tel.: (20)5122453; Fax: 6172625 ; Tlx: 11273)
COLL: Stiggelbout, A.M.; Noyon, R.; Schmitz, P.I.M.; Somers, R.

Second Cancer Risk Following Therapy for Cancers of Breast, Testis, Hodgkin's Disease and Non-Hodgkin's Lymphoma
With the advent of improved survival in many patients with advanced cancer and the increasing use of adjuvant therapy in cancer patients with minimal tumour load, the need for quantitative data on second cancer risk following such therapy has increased. The aims of this study are: (1) to assess and quantify the separate and combined effects of radiotherapy and chemotherapy in determining the risk of second lung cancer following breast cancer, Hodgkin's disease, non-Hodgkin's lymphoma (NHL) and testis cancer; (2) to evaluate aspects of treatment (including specific cytostatic drugs) that determine the risk of second leukaemia and the MDS after testis cancer and adjuvant CT for breast cancer; (3) to assess which other factors affect lung cancer and leukaemia risk (e.g. age, sex, treatment toxicity, smoking habits). These aims will be achieved by first conducting four cohort studies of patients with breast cancer, testis cancer, Hodgkin's disease and NHL. Subsequently nested case-control studies within the cohorts will be carried out: one on second lung cancer after all three above-mentioned first primaries and another on second leukaemia after breast and testis cancer. The cohorts will include patients from both the Netherlands Cancer Institute and the Dr. Daniël den Hoed Clinic, Rotterdam. In the cohort studies a limited data set will be collected from the hospital tumour registries; this part of the study serves as a tool to identify cases for the nested case-control studies. In the latter part of the study detailed treatment data (and other second cancer risk factors) will be collected from the medical records for only a small sample of the cohorts, thus making a very efficient study design. Controls will be matched to the cases for survival time, age, sex and year of diagnosis.

TYPE: Case-Control; Cohort
TERM: Age; Chemotherapy; Multiple Primary; Radiotherapy; Sex Ratio; Tobacco (Smoking)
SITE: All Sites; Leukaemia; Lung; Myelodysplastic Syndrome
TIME: 1988 - 1993

* 590 Van Leeuwen, F.E. 05130
Netherlands Cancer Inst., Div. of Clinical Oncology, Dept. of Epidemiology, Plesmanlaan 121, 1066 CX Amsterdam, Netherlands (Tel.: (20)5122453; Fax: 6172625 ; Tlx: 11273)
COLL: Rookus, M.A.; Benraadt, J.; Peterse, J.L.; Lebesque, J.; Van Dongen, J.A.

Risk Factors for Contralateral Breast Cancer
Of all women with unilateral breast cancer surviving 20 years, 15-20% will develop contralateral breast cancer (CLBC). In this study of CLBC, aetiological and genetic factors, lifestyle factors and therapy for the first tumour will be taken into account simultaneously, in a design that also makes it possible to examine controversial risk factors for breast cancer. Specific aims are: (1) to assess the separate and combined effects of genetic factors, known and suspected lifestyle factors and therapy for the first tumour on CLBC risk; (2) to examine controversial risk factors (such as characteristics of the natural menstrual cycle, lactation, oestrogen use, oral contraceptive use, fat consumption and alcohol consumption) for breast cancer; (3) to gain more insight into the biological mechanism of CLBC development by studying genetic and lifestyle factors as determinants of neu-oncogene overexpression in breast tumours; (4) to estimate CLBC incidence in the Netherlands according to various standard definitions and based on reviewed histological slides. A population-based case-control study will be carried out with two matched control groups and one matched second case group. All incident CLBC cases (N = 350) diagnosed in the regions covered by five comprehensive cancer centres will be eligible for the study during an intake period of three years. By comparing CLBC cases with patients with unilateral breast cancer at the age of the first tumour of the CLBC cases, genetic factors, lifestyle factors and therapy for the first tumour can be examined simultaneously. By comparing CLBC cases with a second control group of healthy women that are at least as old as the case was at CLBC diagnosis, an enhanced expression of risk factors can be expected. Therefore, this design facilitates the study of controversial breast cancer risk factors. To be able to interpret the effects thus estimated, relative risks based on a traditional design comparing cases with unilateral breast cancer newly diagnosed at the age of the second tumour of the CLBC cases with healthy controls is necessary as well. Trained interviewers will

NETHERLANDS

interview all women personally at home on lifetime histories of reproductive factors, family history of breast cancer, hormone use and a possible history of benign breast disease. The treatment of the first tumour will be abstracted from medical records.

TYPE: Case-Control
TERM: Alcohol; Chemotherapy; Fat; Genetic Factors; Hormones; Lactation; Lifestyle; Menstruation; Multiple Primary; Oncogenes; Oral Contraceptives; Radiotherapy; Registry
SITE: Breast (F)
CHEM: Oestrogens
REGI: Amsterdam (Net); Eindhoven (Net); Leiden (Net); Nijmegen (Net); Rotterdam (Net)
TIME: 1991 - 1996

BILTHOVEN

* 591   Bueno de Mesquita, H.B.   05208
National Inst. of Public Health, and Environmental Protection, Dept. of Cancer Epidemiology, Antonie van Leeuwenhoeklaan 9, 3720 BA Bilthoven, Netherlands (Tel.: (31)30742971; Fax: 30367370 ; Tlx: 47215 rivm nl)
COLL: Dols, P.W.; Röntgen-Pieper, E.; Van der Kuip, A.M.; Doornbos, G.

Retrospective Mortality Study of Phenoxy Herbicide Manufacturers

During the period 1983-1987 a cohort of 2,310 persons employed in two chemical industries involved in the manufacture of phenoxy herbicides and chlorophenols was assembled in order to evaluate the subsequent cancer mortality, in particular from soft tissue sarcoma and non-Hodgkin's lymphoma. In one of the factories an industrial accident occurred, with concomitant release of dioxin into the environment. The cohort forms part of the International Register of persons occupationally exposed to chlorophenoxy herbicides, chlorophenols and contaminants which was set up in 1984 by IARC to assess the possible carcinogenicity of these compounds to humans. Ascertainment of exposure is based on personnel records with the assistance of company personnel. Vital status of cohort members is assessed through municipal population registers. The Netherlands Central Bureau of Statistics will provide the individual causes of death. Cancer mortality experience will be compared with national mortality rates and mortality of non-exposed workers.

TYPE: Cohort
TERM: Herbicides; Occupation; Pesticides; Registry
SITE: All Sites; Non-Hodgkin's Lymphoma; Sarcoma; Soft Tissue
CHEM: Dioxins; Phenoxy Acids
OCCU: Chemical Industry Workers
TIME: 1983 - 1992

* 592   Bueno de Mesquita, H.B.   05209
National Inst. of Public Health, and Environmental Protection, Dept. of Cancer Epidemiology, Antonie van Leeuwenhoeklaan 9, 3720 BA Bilthoven, Netherlands (Tel.: (31)30742971; Fax: 30367370 ; Tlx: 47215 rivm nl)
COLL: Smeets, F.W.M.; Runia, S.; Hulshof, K.F.A.M.

Reproducibility of a Food Frequency Questionnaire

This study is aimed at investigating the reproducibility of a semi-quantitative food frequency questionnaire developed for a population-based case-control study on the relation between the intake of foods, energy and nutrients and carcinoma of the pancreas and biliary tract. The food frequency questionnaire is aimed at comprehensively assessing diet by estimating usual individual intake of 116 commonly eaten food items and food groups in order to rank subjects according to level of intake approximately one year earlier. Photographs of standard meals and standard portions were used to estimate usual amounts and standard portions. During 1984-1987 the first random sample of 63 population controls, stratified by age, sex and season, was reinterviewed using the same time frame of reference, approximately two years prior to the repeated interview. The other sample of 54 subjects was re-interviewed using the same recall interval. Pearson's correlation coefficient is used to evaluate the relative agreement of food and nutrient intake. Nutrient and food intake scores are divided into tertiles to examine misclassification.

NETHERLANDS

TYPE: Methodology
TERM: Diet; Nutrition
SITE: Bile Duct; Pancreas
TIME: 1984 - 1992

MAASTRICHT

* 593   de Vet, H.C.W.   04981
Univ. of Limburg, Dept. of Epidemiology, & Health Care Research, Postbox 616, 6200 MD Maastricht, Netherlands (Tel.: (043)887382; Fax: 252195)
COLL: Knipschild, P.G.; Schouten, H.J.A.; Sturmans, F.

Relationship between Smoking Behaviour, Sexual Factors and Cervical Dysplasia
The relationship between cervical dysplasia and dietary factors, smoking behaviour and sexual activities is being assessed in a case-control study. The cases (n = 283) were participants in a randomized placebo-controlled trial to assess the effect of beta- carotene on the regression and progression of cervical dysplasia (J. Clin. Epidemiol. 44:273-283, 1991). The controls (n = 744) were recruited from the population registers of six municipalities in areas where most of the cases lived. The cases and controls received a questionnaire which enquired about dietary factors, smoking behaviour and sexual activities. The effect of smoking factors and sexual activities is in the stage of analysis.

TYPE: Case-Control
TERM: Diet; Sexual Activity; Tobacco (Smoking)
SITE: Uterus (Cervix)
TIME: 1989 - 1992

594   Swaen, G.M.H.   04253
Univ. of Limburg, Dept. of Occupational Medicine, Postbox 616, 6200 MD Maastricht, Netherlands (Tel.: (043)888617/888618)
COLL: Sturmans, F.; de Boorder, T.

Epidemiological Study of Workers Exposed to Acrylonitrile in the Netherlands
In a retrospective cohort study cause-specific mortality patterns between 4,000 workers exposed to acrylonitrile in the past and a comparison group comprising 4,000 non-exposed workers will be compared to test the hypothesis that exposed workers experience a higher mortality from lung cancer or cancer of the prostate than those not exposed. Special emphasis will be put on estimation of the quantity of past exposure to acrylonitrile, based on measurement if available, interviews with plant industrial hygienists and other key persons. The study will encompass nine chemical companies located in The Netherlands, where acrylonitrile has been manufactured, stored, handled or used in the production process. Follow-up of the 8,000 workers enrolled in the study will be conducted by means of the Dutch system of population-based cancer registries.

TYPE: Cohort
TERM: Chemical Exposure; Occupation; Plastics; Registry
SITE: All Sites; Lung; Prostate
CHEM: Acrylonitrile
OCCU: Acrylonitrile Workers
REGI: Leiderdorp (Net)
TIME: 1986 - 1992

595   Van den Brandt, P.A.   04506
Univ. of Limburg, Dept. of Epidemiology, P.O.Box 616, 6200 MD Maastricht, Netherlands (Tel.: (043)887313; Fax: 219552 ; Tlx: 56726 NL)
COLL: Bausch-Goldbohm, R.; Van't Veer, P.; Sturmans, F.; Hermus, R.J.J.; Bode, P.; de Bruin, M.

Toenail Selenium and the Risk of Cancer
The hypothesis that selenium reduces the risk of various forms of cancer is being tested within the framework of a prospective cohort study on diet, lifestyle and cancer started in 1986 (Int. J. Epidemiol. 17:472-, 1988 and J. Clin. Epidemiol. 43:285-295, 1990). The cohort originates from the general population and consists of over 120,000 men and women, aged 55-69 years at baseline. Toenail clippings, providing a measure of selenium status, were obtained from 80,000 participants at baseline. Repeated toenail sampling will be performed annually for random samples (n = 250) of the cohort to assess intra-individual variability. The selenium content of the nails of incident cancer cases and of a

NETHERLANDS

random subcohort will be determined by neutron activation analysis. Cancer follow-up consists of record linkage to cancer registries and a pathology registry, for which a linkage has been developed (Int. J. Epidemiol. 19:553-558, 1990). Statistical analysis will involve a case-cohort approach. Special attention will be given to evaluating the suggested gender-specific effect of selenium.

TYPE: Case-Control; Cohort
TERM: BMB; Diet; Lifestyle; Registry; Sex Ratio; Toenails
SITE: Breast (F); Colon; Lung; Prostate; Rectum; Stomach
CHEM: Selenium
REGI: Amsterdam (Net); Eindhoven (Net); Groningen (Net); Leeuwarden (Net); Leiden (Net); Leiderdorp (Net); Maastricht (Net); Nijmegen (Net); Rotterdam (Net); Tilburg (Net); Utrecht (Net)
TIME: 1986 - 1994

596   Van den Brandt, P.A.   04507
Univ. of Limburg, Dept. of Epidemiology, P.O.Box 616, 6200 MD Maastricht, Netherlands (Tel.: (043)887313; Fax: 219552 ; Tlx: 56726 NL)
COLL: Bausch-Goldbohm, R.; Dorant, E.; Sturmans, F.; Hermus, R.J.J.

Dietary Supplements and Drugs in the Aetiology of Cancer

In 1986, a prospective cohort study on diet, other lifestyle habits and the incidence of cancer was started in the Netherlands (Int. J. Epidemiol 17:472-, 1988; J. Clin. Epidemiol. 43:285-295, 1990). Baseline data were obtained by self-administered questionnaire on 120,000 men and women, aged 55-69 years. Questions referred to diet (including supplements), smoking, long-term drug use, medical history, family history of cancer and reproductive history. The present study evaluates the various hypotheses regarding long-term drug use, associated with particular (precursor) conditions (e.g. cimetidine use and gastric ulcer) and the subsequent development of cancer. The interaction with nutritional supplements and alcohol and the effect of supplements per se will also be tested. Follow-up for cancer will be performed by record linkage to a pathology register and the cancer registries in the Netherlands (Int. J. Epidemiol. 19:553-558, 1990).

TYPE: Cohort
TERM: Alcohol; Drugs; Familial Factors; Gastric Ulcer; Registry; Reproductive Factors; Tobacco (Smoking); Vitamins
SITE: All Sites
CHEM: Ascorbic Acid; Beta Carotene; Cimetidine
REGI: Amsterdam (Net); Eindhoven (Net); Groningen (Net); Leeuwarden (Net); Leiden (Net); Leiderdorp (Net); Maastricht (Net); Nijmegen (Net); Rotterdam (Net); Tilburg (Net); Utrecht (Net)
TIME: 1988 - 1994

NIJMEGEN

597   Bos, R.P.   03596
Univ. of Nijmegen, Faculty of Medicine, Dept. of Toxicology, Kapittelweg 54 , 6525 EP Nijmegen, Netherlands (Tel.: (080)514203; Fax: 514090; Tlx: 484211 kunm nl)
COLL: Jongeneelen, D.J.; Sessink, P.J.M.; Scheepers, P.T.J.

Development and Validation of Methods for the Detection of Occupational Exposure to Mutagenic and Carcinogenic Agents

The objective of the project is the development and validation of biological or chemical test methods for the detection of exposure to genotoxic substances (including mutagens and carcinogens). These methods can be used in biological monitoring. After initial chemical analytical research and investigations on laboratory animals, methods can be validated in exposed workers. With respect to PAH-exposure several groups of workers exposed to coal tar and coal tar products are compared with controls (1-hydroxypyrene in urine). For diesel emissions in the first period sensitive chemical analytical methods will be developed. The occupational exposure of workers will be studied. Validations will be made with respect to several parameters, such as environmental measurements and e.g. DNA-adducts (Recent papers: IARC Scientific Publ. 89:389-395, 1988 and Ann. Occup. Hyg. 32:35-43, 1988).

NETHERLANDS

TYPE: Methodology; Molecular Epidemiology
TERM: Chemotherapy; Coal; DNA; Drugs; Monitoring; Mutagen Tests; Occupation
SITE: Inapplicable
CHEM: Cyclophosphamide; Diesel Exhaust; PAH
OCCU: Chemical Industry Workers; Drivers; Health Care Workers
REGI: Amsterdam (Net); Eindhoven (Net); Groningen (Net); Leeuwarden (Net); Leiden (Net); Leiderdorp (Net); Maastricht (Net); Nijmegen (Net); Rotterdam (Net); Tilburg (Net); Utrecht (Net)
TIME: 1980 - 1993

598   Hendriks, J.H.C.L.   01886
St Radboud Hosp., Dept. of Diagnostic Radiology, University Clinics, Geert Grooteplein Zuid 18, 6525 GA Nijmegen, Netherlands (Tel.: (080)514545)
COLL: Holland, R.; Mravunac, M.; Verbeek, A.L.M.

Impact of Mass Screening on Morbidity and Mortality of Breast Cancer
This breast cancer screening project was started in 1975. All women born before 1 January 1940 are invited to participate. The total number of women in the study is 32,500. The attendance rate is 60-70%. The sixth screening round will be completed in December 1986. As a screening-test the single view film mammography is used, with a mean dose to the glandular tissue of 0.07 Rad per breast. Although no randomized controlled trial was performed, the neighbouring cities, where no screening programmes exist, will be used as control groups. Data on incidence and cause of death from 1973 until 1986 in Nijmegen and the control groups will soon be available. Aetiological analyses will be made and the case-control approach used to assess whether screening by mammography prevents death from breast cancer. Using tumour shadows on the mammogram, growth rates of screen-detected cases and interval cases will be calculated. One way to adjust for lead time and length biased sampling with respect to survival curves is to use the distribution of growth rates. Histological and morphological characteristics of invasive and non-invasive cancerous lesions detected at and between screening examinations, are studied. Histo-morphological evaluation of interval cancer and radiologically occult carcinomas has been performed routinely since 1979. Papers were published in Cancer 52:1810-1819, 1983; Lancet I: 591-593, 1222-1224, 1984 and I: 865-866, 1985.

TYPE: Case-Control; Cohort; Incidence; Mortality
TERM: Histology; Mammography; Prevalence; Screening; Survival
SITE: Breast (F)
TIME: 1975 - 1992

599   Nelemans, P.J.   04886
Nijmegen Univ., Dept. of Social Medicine, Epidemiology Unit, Verlengde Groenestr. 75, 6525 EJ Nijmegen, Netherlands (Tel.: (080)513125)
COLL: Verbeek, A.L.M.; Ruiter, D.J.; Rampen, F.H.J.

Malignant Melanoma and Aetiological Indicators in the Netherlands
The aim of the study is to study the role of several risk factors, including sun exposure and chemical exposures, for the development of cutaneous malignant melanoma. A case-control study is being conducted in which patients with cutaneous malignant melanoma (n=200) will be compared with a sample of patients with various other types of cancer, which are not known to be associated with the exposure variables under study (n=400). An interview will provide information about exposure variables and, in a physical examination by a dermatologist, constitutional variables will be defined and the number of naevi will be counted. The source of patients is the registry of the Comprehensive Cancer Centre which registers the incidence of cancer in the region around Nijmegen (IKO). The independent effects of the variables under study will be examined with the use of multiple logistic regression analysis.

TYPE: Case-Control
TERM: Chemical Exposure; Naevi; Occupation; Radiation, Ultraviolet
SITE: Melanoma
REGI: Nijmegen (Net)
TIME: 1989 - 1993

600   Straatman, H.   04075
Nijmegen Univ., Dept. of Social Medicine, Epidemiology Unit, Verlengde Groenestr. 75, 6525 EJ Nijmegen, Netherlands (Tel.: (080)513102)
COLL: Verbeek, A.L.M.; Hendriks, J.H.C.L.; Peeters, P.H.M.; Holland, R.; Mravunac, M.

NETHERLANDS

Estimation of Lead-Time and Incidence Using Nijmegen Breast Cancer Screening Data

One of the important issues in the evaluation of the impact of a screening programme is the estimation of the length of time by which diagnosis is advanced. This is being done for the breast cancer screening programme in Nijmegen, which started in 1975 (age-group 35-64; N = 23,000; attendance rate 80%; single view mammography as a screening test every two years). Assuming an exponential distribution for the pre-clinical detectable disease phase and progressiveness of disease development, preliminary results have been obtained. In the group aged 35-49 at entry the estimates were 1.12 years and in the group 50-64 at entry 2.17 years. The mathematical model is being validated with epidemiological and clinical data. After linkage with mortality data (already available) and validation the resulting model will be used for prediction of the theoretical effects (mortality, morbidity and costs) of different screening scenarios.

TYPE: Methodology
TERM: Cost-Benefit Analysis; Mathematical Models; Screening; Time Factors
SITE: Breast (F)
TIME: 1984 - 1992

* 601 van der Gulden, J.W.J. 05031
Nijmegen Univ., Dept. of Occupational Medicine, Verlengde Groenestr. 75, 6525 EJ Nijmegen, Netherlands (Tel.: (080)613119)
COLL: Kolk, J.; Kiemeney, L.A.L.; Verbeek, A.L.M.

Prostate Cancer and Occupational Exposure

In 1990, a case-referent study was started on the relation between prostate cancer and occupation (job title and function) and occupational exposures. In this study cases are defined as men in whom prostate cancer was histologically confirmed between 1 January and 1 April, 1990. Referents are patients who were treated for benign prostate hyperplasia in this period, and in whom no signs of malignancy were found on histological examination. About 600 cases were selected from the registry of the Comprehensive Cancer Centre IKO. Controls (about 1800) were selected from the National Archive of Pathology Reports (PALGA). Mail questionnaires were sent to all participants to collect information on work history and occupational exposures and on confounding factors such as age, smoking, drinking and socio-economic status. The questionnaire is a slightly modified version of a validated questionnaire on occupational history, developed recently by an EEC Working Party. For farmers, metal workers, repairmen and mechanics, several specific questions were added, since for these occupational groups an increase of risk was found in a feasibility study, executed in 1989. Aetiological analyses commenced early 1991 using the case-referent approach. A paper was published in Int. J. Epidemiol. 19:762-763, 1990.

TYPE: Case-Control
TERM: Fertilizers; Metals; Occupation; Paints; Pesticides; Registry; Solvents; Welding
SITE: Prostate
CHEM: Vehicle Exhaust
OCCU: Farmers; Mechanics; Metal Workers
REGI: Nijmegen (Net)
TIME: 1989 - 1992

602 Verbeek, A.L.M. 03828
Univ. of Nijmegen, Dept. of Epidemiology, Verlengde Groenestr. 75, 6525 EJ Nijmegen, Netherlands (Tel.: (080)513102)
COLL: Hendriks, J.H.C.L.; Holland, R.; Mravunac, M.; Straatman, H.; Peer, P.G.M.

Impact of Mass Screening on Breast Cancer Mortality in Nijmegen

The Nijmegen population screening for breast cancer started in 1975. All Nijmegen women born before 1946 are invited to a biennial mammographical screening (n = 36,000). The attendance rate is 60-70%. The eigth screening is currently under way. The efficacy of screening on breast cancer mortality is analysed by means of a case-control study. The first results up to 1982 (62 breast cancer deaths) suggest a 50% mortality reduction. No effect was observed in the youngest group, aged 35-49 at entry. Using more recent mortality (1983-88) and morbidity data (distant metastasis) further analysis is being undertaken on age-specific effects and the impact of the number and frequency of examinations on mortality. The breast cancer mortality in the total population of Nijmegen will be compared to that of Arnhem, a neighbouring city without a screening programme. The results will be incorporated into a mathematical model presented by Eddy (1980) to determine how frequently women of different age-groups should be screened taking adverse effects such as false-positive and false-negative

NETHERLANDS

screening results into account. Publications appeared in The Lancet I (8433):865–866, 1985 and in the Int. J. Cancer 43:226–230, 1989 and 1055–1060, 1989.

TYPE: Cohort
TERM: Mammography; Mathematical Models; Screening
SITE: Breast (F)
TIME: 1975 – 1993

603 Vooijs, P.G. 04672
Univ. of Nijmegen, Dept. of Pathology, Lab. of Cytopathology, Geert Grooteplein Zuid 24, 6536 HK Nijmegen, Netherlands (Tel.: (31)80614389; Fax: 80540520)
COLL: Habbema, J.D.F.; Lubbe, J.T.N.; Pal, R.; Otto, L.P.

Information System for Cervical Cancer Screening
A nationwide screening programme for cervical cancer is being organized in the Netherlands, based on the results of three pilot projects. All pathology laboratories involved use the same protocol for coding and reporting of cytological findings and adopt the same recommendations for follow-up of abnormalities. They will be linked to a central pathology diagnoses database (PALGA). Linkage of screening results to previous and follow-up cytological and histological findings permits epidemiological studies on a regional and national level, and measurement of the impact of various screening protocols. This will provide an insight into the natural history of cervical cancer. It will offer the opportunity to evaluate quality control protocols. An information system will be developed for a uniform evaluation of registered data. The MISCAN computer program can be used for a detailed evaluation of the screening programme and for comparison of the cost effectiveness of future screening programmes. A paper appeared in Acta Cytol. 33:825–830, 1989.

TYPE: Cohort
TERM: Cost-Benefit Analysis; Cytology; Histology; Mathematical Models; Screening
SITE: Uterus (Cervix)
TIME: 1989 – 1992

SITTARD

* 604 Verduijn, P.G. 05094
Hospital of Sittard, Dept. of Otolaryngology, Walramstr. 23, 6131 BK Sittard, Netherlands (Tel.: (046)597792)
COLL: Hayes, R.B.; Looman, C.; Habbema, J.D.F.; van der Maas, P.J.

Late Effects of Irradiation for Eustachian Tube Dysfunction
This non-concurrent prospective study aims to examine the association of nasopharyngeal radium irradiation and the origin of tumours in the head and neck region. The second question concerns the possible influence of this type of irradiation on hormone balance, as a result of the pituitary gland being in the treatment field. Further, it is investigated whether information can be obtained regarding dose-effect relationships in a dose range (0–50 cGy) for which little is known in man. In exposed (n = 2,542) and control groups, selected from the same clinical records, cause of death will be established for the deceased and the subjects who are still alive will be surveyed with questions concerning their health. First results show that some risk of tumour induction is involved with nasopharyngeal radium irradiation. No increased mortality due to cancer was observed. However, a statistically significant increase in the cumulative incidence for all cancers combined was observed. Separation of the individual tumour sites did not produce a significant excess. Preliminary results of follow-up until February 1, 1985 (mean follow-up 25.3 years) were published in Ann. Otol. Rhinol. Laryngol. 98:839–844, 1989.

NETHERLANDS

TYPE: Cohort
TERM: Dose-Response; Radiotherapy
SITE: Head and Neck
LOCA: Belgium; Germany; Netherlands
TIME: 1982 – 2002

THE HAGUE

605   Coebergh, J.W.W.   04164
Dutch Childhood Leukemia Study Group, Dr Van Welylaan 2, P.O. Box 60604, 2566 ER (2506 LP) The Hague, Netherlands (Tel.: (070)657930; Fax: 617427)
COLL: Van der Does-van den Berg, A.; van Wering, E.R.; Rammeloo, J.A.; Kamps, W.A.

Trends in Incidence, Mortality and Geographical Distribution of Childhood Leukaemia in The Netherlands (1973-1988)
Incidence, mortality and geographical distribution of childhood leukaemia in The Netherlands (1973-1987) are being examined by continuing an epidemiological register-based study of childhood leukaemia in The Netherlands (van Steensel-Moll et al., Br.J.Cancer 47:471-475, 1983). Uniform examination and review of bone marrow slides are performed in the central DCLSG laboratory. Type, age and sex specific leukaemia incidence rates in The Netherlands are compared with those in other countries, analysing rates according to year of diagnosis and year of birth. Time trends in incidence and mortality by age, sex and type of leukaemia, are also studied. Special attention is given to acute lymphocytic leukaemia through analysis of white blood cell count at diagnosis and immunophenotype since 1979.

TYPE: Incidence; Mortality
TERM: Age; Childhood; Geographic Factors; Registry; Trends
SITE: Leukaemia; Leukaemia (ALL)
REGI: The Hague (Net)
TIME: 1988 – 1993

606   Coebergh, J.W.W.   04525
Dutch Childhood Leukemia Study Group, Dr Van Welylaan 2, P.O. Box 60604, 2566 ER (2506 LP) The Hague, Netherlands (Tel.: (070)657930; Fax: 617427)
COLL: Verhagen-Teulings, M.T.; Crommelin, M.A.; Kluck, H.M.; van der Heijden L., H.M.; Friden-Kill, L.M.

Trends in Cancer Incidence and Survival in Southeast Holland
In Southeast Holland the SOOZ-cancer registry has served a population of one million since 1955. It appears to be complete since 1975, but for some tumours such as breast cancer, completeness exists for a longer period in a restricted part of the area, the city of Eindhoven and surrounding communities. This study will explore trends in incidence and survival. As the registry has acquired its data from the hospitals directly, analyses are possible by stage and histology at diagnosis. In the population registers of all municipalities served, the vital status of all patients can be checked as of 1 January 1988. Less than 1% of survivors could not be traced accurately. Relative survival rates will be computed based on the methods and programs of Hakulinen. Results will be evaluated against trends in incidence monitored by the SOOZ-cancer registry as well as in mortality, monitored by the Central Bureau of Statistics. An overview will be made of changes in the supply of relevant consultants and diagnostic and therapeutic facilities, such as mammography, cytology, endoscopy and facilities for megavoltage. Various comparisons will be made between age groups, stratified by age and histological type.

TYPE: Incidence
TERM: Histology; Registry; Survival; Trends
SITE: All Sites; Breast (F)
REGI: Eindhoven (Net)
TIME: 1987 – 1992

UTRECHT

607   Baanders-van Halewijn, E.A.   03316
Preventicon, Inst. Public Health & Epidemiology, Radboudkwartier 261-263 , 3511 CK Utrecht, Netherlands (Tel.: (030)313884)
COLL: de Waard, F.; Collette, H.J.A.; Slotboom, B.J.

NETHERLANDS

Historical Cohort Study of Ovarian Cancer
The purpose of the study is to try and find correlations between ovarian cancer and a number of suspected risk factors, by means of a historical cohort study in the female population of Utrecht and suburbs, aged 50-73 years. The data collected during the DOM- project (a cohort study for early detection of breast cancer covering 38,000 women screened periodically from 1974 to 1980) will be used to discover new hypotheses about the aetiology and natural history of ovarian cancer which could give clues for prevention. These hypotheses will be tested in a subsequent study on a population of women born in the years 1932-1941 (the Lutine project, carried out in 1982-1983). Hypotheses will also be able to be tested using the endocrinological examinations of urinary specimens collected at intake into the study (first screening for breast cancer).

TYPE: Cohort
TERM: Hormones; Prevention; Screening
SITE: Ovary
TIME: 1984 - 1992

608 Baanders-van Halewijn, E.A. 03856
Preventicon, Inst. Public Health & Epidemiology, Radboudkwartier 261-263, 3511 CK Utrecht, Netherlands (Tel.: (030)313884)
COLL: de Waard, F.; Thijssen, J.H.H.; Blankenstein, M.A.; Collette, H.J.A.; van Noord, P.A.H.; Slotboom, B.J.

Relation Between Urinary Hormone Excretion and Ovarian Carcinoma
This prospective study is a further development of a cohort study of correlations between a number of risk factors and ovarian carcinoma. It is now hoped to find clues for prevention by obtaining more insight into endocrinological factors possibly related to these risk factors. Hypotheses are derived from (1) earlier studies of breast and endometrial cancer, given the aetiological relationships between these and ovarian cancer, and (2) from the publications of Cuzick et al. (1983) on ovarian cancer and of Fishman et al. (1984) on breast and endometrial cancer. Correlations between some of the risk factors and the results of biochemical measurement of the concentrations in urine of the three oestrogens (oestrone, oestradiol and oestriol), of the androgenmetabolites (androsterone, aetiocholanolone and dehydroepiandrosterone (DHEA) and of the 16 alpha-hydroxylase product of oestradiol: (16 alpha-hydroxyoesterone) will be analysed by comparing the results of patients and controls. The hypothesis is that clinical evidence of ovarian cancer is preceded by increased concentrations of 16 alpha-hydroxyoestrone and lower levels of androgen metabolites, in particular DHEA in the urine. Urine samples have been collected from the participants of the DOM-project, (a cohort study for early detection of breast cancer) at time of intake. About 19,000 women born between 1911 and 1931 were screened periodically from December 1974 to March 1985 and delivered urine specimens one, two or three times over the period. This provides a unique possibility to carry out chemical analyses of samples taken long before diagnosis. In the cohort about 100 cases of ovarian cancer are expected to be diagnosed in the period December 1974 to the middle of 1987.

TYPE: Cohort
TERM: Drugs; Hormones; Urine
SITE: Ovary
CHEM: Aetiocholanolone; Androsterone; Dehydroepiandrosterone Sulphate; Oestradio.; Oestriol; Oestrone
TIME: 1986 - 1992

609 Collette, H.J.A. 04104
Preventicon, Inst. of Public Health and Epidemiology, Radboudkwartier 261-263, 3511 CK Utrecht, Netherlands (Tel.: (030)313884)
COLL: de Waard, F.; Collette, C.; Slotboom, B.J.; Fracheboud, J.; Beijerinck, D.; Deurenberg, J.J.M.

DOM-Project for Early Diagnosis of Breast Cancer
The DOM-project (Diagnostisch Onderzoek Mammacarcinoom) is a population-based screening project in which use is made of mammography (together with inspection and palpation) to detect breast cancer at an early stage. It was started in the city of Utrecht towards the end of 1974, and a few years later it was introduced in several suburbs as well. At first women born in 1911-1925 were invited to participate, but later on women born 1926-1945 received an invitation as well. To date nearly 60,000 women have participated at least once. There are three steps of control in this study: (1) process evaluation, consisting of response percentage, pick-up rate, predictive value of the test, etc.; (2) early outcomes, such as stage distribution of detected cases, number of false negative tests, etc.; (3) late outcomes, i.e. survival and mortality rates. The last mentioned is most important: the main aim of such a screening programme is

NETHERLANDS

that it should result in a decrease in mortality of the disease the programme is focussed on. As the programme has been going on for more than ten years an enormous amount of data has been collected and computerised. These data are useful both for evaluation of this screening programme, for describing the natural history of breast cancer and for performing cohort nested case-control studies in other cancers.

TYPE: Mortality
TERM: BMB; Screening; Survival
SITE: Breast (F)
TIME: 1974 - 1993

610   de Waard, F.   04826
Preventicon, Radboudkwartier 261-263, 3511 CK Utrecht, Netherlands (Tel.: (030)313884)
COLL: Baanders, A.N.

European Survey of Risk Factors for Breast Cancer in Middle-Aged Women.
The objective of the study is to investigate to what extent the following risk factors: age at menarche, body height, age at first birth, parity and body weight in middle-aged women, correlate with variations in breast cancer incidence in Europe. Present-day incidence rates of breast cancer will be correlated with national or regional data on the prevalence of the risk factors in various European populations. Data will be collected by searching for publications and reports from international and national sources, viz. demographic reports as well as control groups from case-control studies. This is to be achieved by a network of European collaborators, not only in EEC countries, but also in Eastern and Northern Europe.

TYPE: Correlation
TERM: Age; Menarche; Parity; Physical Factors
SITE: Breast (F)
TIME: 1989 - 1992

* 611   Den Tonkelaar, I.   05050
Univ. of Utrecht, Dept. of Public Health and Epidemiology, Radboudkwartier 261-263, 3511 CK Utrecht, Netherlands (Tel.: (030)331529; Fax: 340634)
COLL: Collette, H.J.A.; Seidell, J.C.; van Noord, P.A.H.; Baanders-van Halewijn, E.A.; de Waard, F.; Wynne, H.J.A.

Fat Distribution and Breast Cancer Incidence and Survival in Postmenopausal Women
The purpose of this study is an investigation of the relation between fat distribution (as measured by central and peripheral skin folds) and breast cancer incidence and survival. The study will be performed in about 20,000 post-menopausal women who were aged 49-68 years in 1975-1979 and who are participating in a breast cancer screening project (the DOM-I project). The relationship between fat distribution and known risk factors for malignant breast disease (such as overweight, age, family, history of breast cancer, parity and age at first delivery) will be evaluated.

TYPE: Cohort
TERM: Age; Familial Factors; Fat; Obesity; Parity; Registry; Survival
SITE: Breast (F)
REGI: Utrecht (Net)
TIME: 1991 - 1993

612   van Noord, P.A.H.   04318
Preventicon, Inst. Public Health & Epidemiology, Division of Epidemiology, Radboudkwartier 261-263, 3511 CK Utrecht, Netherlands (Tel.: (030)313884)
COLL: Maas, M.J.; de Bruin, M.; Noordhoek, J.

Prediagnostic Nail Selenium Levels in Colorectal Tumours
It has been suggested that selenium in humans may have a preventive effect in cancer occurrence. Given the observations made in animal studies, breast and colorectal cancers are the tumours best suited for studying the hypothesis in humans. Classical case-control studies in humans have not been able to exclude the possibility that decreased selenium levels found in cancer patients are not caused by the tumour. It is known that selenium can concentrate in or around cancer tissue. Several biological mechanisms have been advanced for selenium: (1) protection against free radical induced damage through the selenium containing enzyme glutathione peroxidase; (2) antioxidative properties of selenium, and (3) antagonism of heavy metals (Hg, As, Cd). The latter are relevant for their local effects in relation to colorectal cancers. This study is of the case-cohort type. Biological samples from

NETHERLANDS

approximately 19,000 women participating in the DOM project for breast cancer screening, have been entered in a biological bank. Selenium behaves biologically as sulphur and is accumulated in keratine, a protein polymer with high sulphur and selenium contents. The collection of toenail clippings to be used in this study was started in mid 1982. First the relation of selenium to breast cancer was addressed: no decreased nail selenium levels were found in a breast cancer case-cohort study (Int. J. Epidemiol. 16(2):318-322, 1987). Given the lower incidence of colorectal tumours a longer follow-up was needed. Selenium levels in the specimens will be determined at IRI Delft by Neutron Activation of the Se 77m isotope.

TYPE: Cohort
TERM: BMB; Toenails; Trace Elements
SITE: Colon; Rectum
CHEM: Selenium
TIME: 1987 - 1992

ZEIST

613 Hermus, R.J.J. 03237
Toxicology and Nutrition Inst. TNO, P.O. Box 360, 3700 AJ Zeist, Netherlands (Tel.: (03404)44144; Fax: 57224)
COLL: Sturmans, F.; Bausch-Goldbohm, R.; Van den Brandt, P.A.; Van't Veer, P.; Dorant, E.

Diet and Cancer of the Breast, Colon, Rectum, Stomach and Lung
Results of epidemiological and animal studies suggest that there is a relation between diet and cancer incidence; relevant hypotheses concern the effects of fat, type of fat, meat, fibre (from different sources), selenium, alcohol and vitamins A, C, E and beta carotene. In the Netherlands, a prospective cohort study has been started to estimate the association between dietary habits as well as nutrient composition of the diet and onset of cancer of the breast, colon, rectum, stomach and lung. In a pilot study (1984-1985) a self-administered semi-quantitative food frequency questionnaire, which includes approximately 170 food items, was developed. In September 1986, a cohort was recruited by inviting a general population sample aged 55-69, randomly selected from 204 municipalities all over the country. Baseline questionnaire data have been collected from over 120,000 men and women. The questionnaire also asked for information on lifestyle factors, occupational, medical and family history. Furthermore, toenail clippings of the majority of participants were collected. Baseline measurements are repeated every year in a subsample. Five year follow-up of the cohort will yield about 300, 400, 200, 700 and 1,200 cases of stomach, colon, rectum, breast and lung cancer respectively. Follow-up information on total and site-specific cancer incidence, is being obtained in collaboration with the cancer registries and the National Data Base of Pathology Records (PALGA). The cohort study will be analysed using the data from the cases and a large subcohort (case-cohort design). Results will be available from 1991 onwards. Papers have been published by Van den Brandt et al. in IARC Techn. Rep. No. 4:79, 1988, J. Clin. Epidemiol. 43 (3):285, 1990 and in Int. J. Epidemiol. 19:553, 1990.

TYPE: Cohort
TERM: Alcohol; Diet; Familial Factors; Fibre; Lifestyle; Metals; Registry; Vitamins
SITE: Breast (F); Colon; Lung; Rectum; Stomach
CHEM: Beta Carotene; Selenium
REGI: Amsterdam (Net); Groningen (Net); Leiden (Net); Maastricht (Net); Nijmegen (Net); Rotterdam (Net); Tilburg (Net); Utrecht (Net)
TIME: 1984 - 1993

614 Kok, F.J. 04633
TNO Toxicology and Nutrition Inst., Epidemiology Sect., P.O.Box 360, 3700 AJ Zeist, Netherlands (Tel.: +31 340444767; Fax: 340457224 ; Tlx: 40022 CIVO NL)
COLL: van Poppel, G.; Wilmer, J.W.G.M.; Schrijver, J.; Hermus, R.J.J.; Atjak, M.T.J.; Vooijs, P.G.

Effect of Beta Carotene Supplementation on DNA Damage in Heavy Smokers
Beta carotene has been associated with decreased risks of lung and bladder cancer in smokers. As an antioxidant, beta carotene may trap free radicals and thus protect against DNA damage and cancer. Smokers have increased levels of micronuclei in urothelial and bronchial mucosal cells, and of SCE in lymphocytes. These intermediate endpoints (in urine, sputum and blood) as well as antioxidants and cotinine in blood are assessed before and after 3 months' beta carotene supplementation (20 mg/day) in a double-blind placebo-controlled intervention trial. Heavy smokers are matched by age, duration and

NETHERLANDS

intensity of smoking and randomly assigned to treatment (n = 75) or placebo group (n = 75). In addition, non-smokers (n = 50) and passive smokers (n = 50) are studied without intervention.

TYPE: Intervention; Molecular Epidemiology
TERM: DNA; Lymphocytes; Micronuclei; Passive Smoking; SCE; Sputum; Tobacco (Smoking); Urine
SITE: Bladder; Lung
CHEM: Beta Carotene
TIME: 1989 - 1992

* 615   Kok, F.J.   05025
TNO Toxicology and Nutrition Inst., Epidemiology Sect., P.O.Box 360 , 3700 AJ Zeist, Netherlands (Tel.: +31 340444767; Fax: 340457224 ; Tlx: 40022 CIVO NL)
COLL: Huttunen, J.K.; Kohlmeier-Arab, L.; Martin-Moreno, J.; Gutzwiller, F.; Manso, C.S.; Riemersma, R.; Strain, J.J.; Trichopoulou, A.; Van't Veer, P.; Zatonski, W.A.

EURAMIC Breast Cancer Study

The EURAMIC breast cancer study is part of the EC-funded European Study on Antioxidants, Myocardial Infarction and Breast Cancer. The breast-cancer arm of this multi-centre case-control study is designed to establish an association between early stage breast cancer and the diet-related antioxidants alpha-tocopherol, beta-carotene and selenium, both separately and in combination. 100 cases and 100 controls are being enrolled from each of 4-8 study centres. Postmenopausal women, 50-69 years of age, with a stable dietary pattern during the past year, are eligible. Breast cancer cases (ICD-O 174, ductal carcinoma, tumour size < 5cm, without fixed axillary nodes or indication of distant metastases at diagnosis) are recruited from collaborating hospitals or screening centres. Controls are recruited from the same source population as the cases , e.g. , via local population registres, health centres, general practitioners, etc. Long-term exposure to the antioxidants alpha-tocopherol and beta-carotene, and selenium is assessed from aspiration and toenail collection is conducted within one to four weeks after diagnosis. In order to guarantee standardixed analysis and to minimize sources of laboratory error, biomarkers are analysed in a central laboratory. Background data from cases and controls include relevant dietary factors, breast cancer risk factors, and general lifestyle indicators. The use of biomarker data, the broad rrange of exposure to antioxidnts in different countries and the simultaneous measurement of several antioxidants offers opportunities for a concise test of the antioxidant hypothesis for breast cancer, as one of the major age-related diseases among women in the EC.

TYPE: Case-Control
TERM: Antioxidants; BMB; Biochemical Markers; Diet; Fat; Nutrition; Toenails
SITE: Breast (F)
CHEM: Beta Carotene; Selenium; Tocopherol
LOCA: Germany; Greece; Ireland; Israel; Netherlands; Poland; Portugal; Spain; Switzerland; United Kingdom
TIME: 1990 - 1992

616   Van't Veer, P.   04787
TNO Toxicology & Nutrition Inst., Sect. of Epidemiology, P.O. Box 360 , 3700 AJ Zeist, Netherlands (Tel.: +31 340444144; Fax: 57224 ; Tlx: 40022 CIVO NL)
COLL: Kampman, E.; Kok, F.J.; Schneijder, P.; Westenbrink, S.; Hermus, R.J.J.

Fermented Milk Products, Calcium and Colon Cancer

Dietary habits characterised by high intake of fat, energy and protein have been linked to a high incidence of colon cancer. Protective effects have been suggested for dietary fibre, vegetables, fruit, calcium and dairy products, including fermented dairy products. To investigate the association of fermented milk products and/or calcium with colon cancer risk, a case-control study will be conducted in the Netherlands from 1990 to 1993. Since the impact of diet on the aetiology of colon cancer may vary for different parts of the colon, these will be studied separately. Newly diagnosed colon cancer patients (n = 250; ICD-0 153; adenocarcinoma) will be compared with 'healthy' population controls (n = 250), frequency-matched for sex and age (5-year interval). Average daily intake of calcium, (fermented) milk products and other dietary factors will be assessed using a dietary history technique (for cases, referring to the year before clinical symptoms). Furthermore, data on major risk factors of colon cancer, life-style habits and medical history will be obtained. Both cases and controls will be interviewed at their homes. Mean intakes among cases and controls will be compared using a t-test of paired samples. Stratified analysis will be used to indentify potential confounders. Logistic regression will be used to evaluate confounding and adjust the odds ratio estimates.

NETHERLANDS

TYPE: Case-Control
TERM: Anatomical Distribution; Dairy Products; Diet; Fat; Fibre; Registry
SITE: Colon
CHEM: Calcium
REGI: Utrecht (Net)
TIME: 1989 - 1993

NEW CALEDONIA

NOUMÉA

617   Bach, F.H.   04355
   South Pacific Commission, Pacific Islands Cancer Registry, Box D5, Nouméa Cédex, New Caledonia (Tel.: (687)262000/363; Fax: 263818 ; Tlx: SOPACOM 3139NM)
COLL: Henderson, B.E.; Kolonel, L.N.; Le Marchand, L.

Pacific Islands Cancer Registry
   The aim of the Pacific Islands Cancer Registry is to provide information on cancer from all the Pacific Islands and compare site specific incidence rates by country and ethnic groups. Registrars visit the islands to collect information from hospitals, laboratories and medical statistics sections or from well established registries. Valuable data are also obtained from tumour registries located near the major referral centres in the United States, New Zealand and Australia. Cross-sectional surveys on lifestyle risk factors are being conducted in several Island countries participating in the registry in order to explain intercountry and ethnic variations in cancer incidence.

TYPE: Incidence; Registry
TERM: Ethnic Group; Geographic Factors; Lifestyle; Registry
SITE: All Sites
REGI: Pacific Islands (Fra)
TIME: 1977 – 1993

NEW ZEALAND

AUCKLAND

618   Kay, R.G.   01417
Univ. of Auckland, Medical School, Dept. of Surgery, Park Rd, Auckland 3, New Zealand (Tel.: (649)795780; Fax: 770956)
COLL: Holdaway, I.M.; Stewart, A.W.; Harvey, V.J.; Carter, J.F.; Gillman, J.C.; Mason, B.H.; Newman, P.; Bierre, A.

Auckland Breast Cancer Study Group - Breast Cancer Survey
This study is concerned with the epidemiology of breast cancer in a multi-racial society in New Zealand. It comprises the 2,706 new cases of breast cancer diagnosed in Auckland in the nine years between 1976 and 1985. Importance is attached to season of tumour detection, height-weight ratios, reproductive history, history of contraceptive practices, previous breast disease, blood group, family history and bilateral disease. Clinical data on oestrogen and progesterone receptor values, adjuvant therapy, grading, nodal status, recurrence and death dates, will be evaluated in relation to these factors. Recent papers have been published in Br. J. Cancer 61:137-141, 1990, Cancer Res. 50:5883-5886, 1990 and Breast Cancer Res. Treat. 15:103-108, 1990.

TYPE: Case Series
TERM: Familial Factors; High-Risk Groups; Hormones; Obesity; Race; Reproductive Factors
SITE: Breast (F)
CHEM: Androgens
TIME: 1976 - 1993

DUNEDIN

619   Chang, A.R.   04324
Univ. of Otago, Medical School, Dept. of Pathology, P.O. Box 913, Dunedin, New Zealand (Tel.: (024)740062)

Characteristics of Women Attending for Colposcopic Examination
Studies carried out overseas have confirmed that cervical malignancies are more prevalent in women of low socio-economic status, smokers and often women from minority ethnic groups. There have been little or no New Zealand studies on the characteristics of women with cervical malignant disease. This study is aiming to see if those women of low socio-economic status, Maoris and Polynesians and smokers have a higher prevalence of the precursor lesions of cervical cancer and also HPV infection. The latter is being assessed by DNA in situ hybridization methods. The study population is approximately 350 per annum and is being assessed at the University Teaching Hospital. The women all reside in the Otago Province, which has a stable population with a mixture of various racial groups including Maoris and Polynesians, as well as a dominant European population. All subjects in the study are interviewed prior to colposcopic assessment and relevant medical records are also scrutinised. Cytological and histological diagnoses are also being evaluated for each patient in the study. Information on methods of contraception and parity is also being obtained. The women in the study group will be compared with those in the general population. Papers appeared in Austr. NZ. J. Obstet. Gynaecol. 29:200-203, and 329-331, 1989.

TYPE: Cross-Sectional
TERM: Contraception; Ethnic Group; HPV; Infection; Reproductive Factors; Socio-Economic Factors; Tobacco (Smoking); Virus
SITE: Uterus (Cervix)
TIME: 1982 - 1993

* 620   Dockerty, J.D.   04982
Univ. of Otago Medical School, Dept. of Preventive & Social Medicine, Hugh Adam Cancer Epidemiology Unit, P.O.Box 913 , Dunedin, New Zealand (Tel.: +64 3 4797203; Fax: 4790529)
COLL: Sharples, K.J.; Skegg, D.C.G.; Elwood, J.M.; Heenan, L.D.B.; Borman, B.

Clustering of Childhood Cancer in New Zealand
Two geographical studies using New Zealand Cancer Registry data are planned. The first is a test of a hypothesis put forward by Kinlen in 1988 on the basis of clustering observed in relatively isolated areas in the United Kingdom which had large population influxes. Kinlen's hypothesis relates to a suggested viral cause of childhood leukaemia. The study will involve defining, on specific criteria, areas of New Zealand

NEW ZEALAND

which have received significant population influxes. The rates of certain childhood cancers in these areas will then be compared with national rates to see whether they are increased. The second study will involve assessing whether or not there is spatial clustering of certain childhood cancers. Data collected by the cancer registry during 1976-1985 will be used. The degree of spatial clustering will be assessed using a method developed by Cuzick and Edwards (1990), which avoids several problems of previously used methods. It will, therefore, be possible to say whether childhood cancers cluster in space to a greater extent than could be expected by chance. The location of any clusters found will be defined, and possible reasons for them considered.

TYPE: Cross-Sectional; Incidence
TERM: Cluster; Geographic Factors; Registry
SITE: Childhood Neoplasms
REGI: New Zealand (NZ)
TIME: 1991 - 1993

* 621 Dockerty, J.D. 04983
Univ. of Otago Medical School, Dept. of Preventive & Social Medicine, Hugh Adam Cancer Epidemiology Unit, P.O.Box 913, Dunedin, New Zealand (Tel.: +64 3 4797203; Fax: 4790529)
COLL: Elwood, J.M.; Skegg, D.C.G.; Sharples, K.J.; Lewis, M.E.

Childhood Cancer in New Zealand

A national case-control study of childhood cancers has begun this year; it aims to identify aetiological factors for specific malignancies (particularly ALL), and includes all childhood cancers except brain cancers. There are about 300 cases and 300 controls in the study, which involves carrying out home interviews of the parents. The cases have been identified through the New Zealand Children's Cancer Registry. The controls have been randomly selected from birth registration records, with matching to cases on age and sex. The exposures of interest include certain chemicals and drugs, diagnostic and therapeutic radiation, parental occupation, immune responsiveness, infectious agents, magnetic fields, and genetic factors. Standard analyses for case-control studies are being conducted. The study contributes to the IARC SEARCH study of childhood cancers.

TYPE: Case-Control
TERM: Chemical Exposure; Drugs; Electromagnetic Fields; Genetic Factors; Marijuana; Parental Occupation; Passive Smoking; Radiation, Ionizing; Registry
SITE: Childhood Neoplasms; Leukaemia (ALL)
CHEM: Chloramphenicol; Phenytoin
TIME: 1992 - 1995

622 Elwood, J.M. 04087
Univ. Otago, Faculty of Medicine, Dept. Preventive & Social Medicine, Hugh Adam Cancer Epidemiology Unit, P.O.Box 913, Dunedin, New Zealand (Tel.: (024)740062; Fax: 741607)
COLL: Whitehead, S.

Malignant Melanoma and Exposure to Natural and Artificial Light

This case-control study assesses the relationship of cutaneous melanoma with acute and chronic sunlight exposure, exposure to fluorescent light and other industrial lighting sources, and host factors such as pigmentation and moles. Cases newly diagnosed from January 1984 to December 1986 in the southern half of the Trent Region in England will be identified, and home interviews carried out on patients aged from 20 to 79. A hospital-based control series will be supplemented by clinical and pathological review for cases. 200 case-control pairs are expected.

TYPE: Case-Control
TERM: Dose-Response; Fluorescent Light; Occupation; Pigmentation; Radiation, Ultraviolet
SITE: Melanoma
LOCA: United Kingdom
TIME: 1986 - 1992

623 Elwood, J.M. 04088
Univ. Otago, Faculty of Medicine, Dept. Preventive & Social Medicine, Hugh Adam Cancer Epidemiology Unit, P.O.Box 913, Dunedin, New Zealand (Tel.: (024)740062; Fax: 741607)
COLL: Jenkinson, C.; Galt, M.

NEW ZEALAND

Female Breast Cancer
This case-control study will assess the relationship of female breast cancer to possible aetiological factors including reproductive history, oral contraceptive and supplemental hormone use, obesity, and a limited assessment of certain aspects of diet. Patients with newly diagnosed breast cancers resident in the Nottingham area, diagnosed between January 1984 and December 1985, a total number of approximately 430, are included along with a hospital-based control group. Home interviews were carried out. The study is linked to one assessing delay behaviour and rapidity of diagnosis, and ultimately survival.

TYPE: Case-Control
TERM: Diet; Hormones; Obesity; Oral Contraceptives; Physical Factors; Reproductive Factors
SITE: Breast (F)
LOCA: United Kingdom
TIME: 1984 – 1992

624 Firth, H.M. 04623
Univ. of Otago Medical School, Dept. of Preventive & Social Medicine, P.O.Box 913, Dunedin, New Zealand (Tel.: +64 3 4797223; Fax: 4790529)
COLL: Herbison, G.P.; Elwood, J.M.; Cooke, K.R.

Cancer by Occupation in New Zealand 1972-1986
The aim of the study is to describe systematically cancer incidence rates and mortality by occupational group in New Zealand for the period 1972-1986. Cancer incidence and mortality data have been obtained from the New Zealand Cancer Registry. The report will discuss any apparent cancer excesses both in terms of the epidemiological information on which the excess is based and in terms of other knowledge of the aetiology of the condition.

TYPE: Incidence; Mortality
TERM: Occupation; Registry
SITE: All Sites
REGI: New Zealand (NZ)
TIME: 1989 – 1992

625 Paul, C.E. 03223
Univ. of Otago Medical School, Dept. of Preventive & Social Medicine, P.O.Box 913, Dunedin, New Zealand (Tel.: (03)4797207; Fax: 4790529)
COLL: Skegg, D.C.G.; Spears, G.F.S.

Steroid Contraception and Breast Cancer
This is a national, population-based case-control study designed to determine whether use of the injectable contraceptive depot medroxyprogesterone acetate (DMPA, Depo Provera) affects the risk of breast cancer in women. In the study 891 women with newly diagnosed breast cancer, aged 25 to 54, were compared with 1,864 control subjects selected at random from the electoral rolls. Cases and controls were interviewed by telephone and information collected about known risk factors as well as about DMPA, oral contraceptives, therapeutic oestrogens, tonsillectomy, abortion, and alcohol use. Data collection is complete. Further analyses of the possible associations between contraceptive steriod use, reproductive factors, alcohol use, tonsillectomy, and the risk of breast cancer are being undertaken. Papers were published in Br. Med. J. 293:723-726, 1986, 299:759-762, 1989 and in Int. J. Cancer 46:366-373, 1990.

TYPE: Case-Control
TERM: Abortion; Alcohol; Contraception; Hormones; Oral Contraceptives; Tonsillectomy
SITE: Breast (F)
CHEM: Oestrogens; Progesterone; Progestogens
TIME: 1983 – 1992

* 626 Richardson, A.K. 05001
Univ. of Otago, Dept. of Preventive and Social Med., Hugh Adam Cancer Epidemiology Unit, P.O.Box 913, Dunedin, New Zealand (Tel.: +64 3 4797203; Fax: 4790529)
COLL: Elwood, J.M.; Williams, S.M.

Evaluation of a Pilot Breast Cancer Screening Programme
This study aims at evaluating two pilot breast cancer screening programmes, in which two-view mammography is offered every two years to women aged 50-64. 40,000 women are eligible to

NEW ZEALAND

participate in the pilot programmes. The evaluation will take place over the first three years of screening (one and a half screening rounds). A series of targets have been agreed, based on the early operating characteristics of programmes that have been successful in the longer term (i.e programmes that have shown decreased mortality from breast cancer among women who were offered a screening). Targets have been set for identification of eligible women, participation, film-reading validity and reliability, sensitivity and specificity, referral rate, biopsy rate, the benign to malignant biopsy ratio, and the stage distribution of cancers detected by screening. Data will be collected directly from the pilot programmes and also from the New Zealand cancer registry. The aim of this evaluation is to ensure the maximum benefit for women offered screening. The results will be used to help decide whether a national screening programme should be started in New Zealand.

TYPE: Cohort
TERM: Registry; Screening
SITE: Breast (F)
REGI: New Zealand (NZ)
TIME: 1990 - 1995

WELLINGTON

* 627　Bethwaite, P.　05043
Wellington School of Medicine, P.O. Box 7343, Wellington South, New Zealand (Tel.: +64 4 855999; Fax: 891661)
COLL: Pearce, N.E.; Carter, J.; Beard, M.; Varcoe, R.; May, S.

Environmental and Occupational Epidemiology of Adult Leukaemia

This is a case-control study contrasting the exposure histories of cases of adult-onset acute leukaemia with those of population controls. Cases are males and females aged 18 to 75 years presenting to the five main treatment centres between 1 January 1988 and 31 December 1991. Estimated enrolment is 130 cases with 4 controls per case. The principal research questions are: (1) is there an association between exposure to tobacco products and the development of adult-onset acute leukaemia, or its major subtypes, in persons aged 18 to 75 years? (2) is there an association between occupational exposure to extremely low frequency electromagnetic fields and the development of adult-onset acute leukaemia, or its major subtypes, in persons aged 18 to 75 years? (3) is there an association between occupational exposures involved in work in farming, chemical spraying, abattoirs, sawmilling, forestry or painting and the development of adult-onset acute leukaemia? (4) are there any differences in the estimates of risk for the various environmental factors between acute leukaemia patients with and without identified clonal chromosomal abnormalities?

TYPE: Case-Control
TERM: Chemical Exposure; Dusts; Electromagnetic Fields; Herbicides; Occupation; Paints; Pesticides; Solvents; Tobacco (Smoking); Wood
SITE: Leukaemia
OCCU: Electrical Workers; Farmers; Forest Workers; Herbicide Sprayers; Meat Workers; Painters; Sawmill Workers
TIME: 1990 - 1992

628　Pearce, N.E.　04249
Wellington Clinical School of Medicine, Wellington Hosp., Dept. of Community Health, Wellington, New Zealand (Tel.: 855999)

Mortality and Cancer Incidence among Phenoxy Herbicide Production Workers and Applicators in New Zealand

This ongoing cohort study will ascertain mortality and cancer incidence in 1,053 workers involved in the manufacture of phenoxy herbicides during the period 1969-1984 and 794 workers involved in the spraying of these chemicals during the period 1973-1984. The production workers cohort comprises all blue-collar workers employed at a single factory for at least one month. The sprayer cohort is based on the national registry of professional chemical applicators. Vital status is being ascertained through the national death index and the national Cancer Registry.

NEW ZEALAND

TYPE: Cohort
TERM: Chemical Exposure; Herbicides; Occupation; Pesticides; Registry
SITE: All Sites; Lymphoma; Soft Tissue
CHEM: Phenoxy Acids
OCCU: Herbicide Manufacturers; Herbicide Sprayers
REGI: New Zealand (NZ)
TIME: 1983 - 1995

629   Pearce, N.E.   04250
Wellington Clinical School of Medicine, Wellington Hosp., Dept. of Community Health, Wellington, New Zealand (Tel.: 855999)
COLL: Bowman, J.D.; Peters, J.M.; Garabrant, D.H.; Thomas, D.C.

Leukaemia in Electrical Workers
A preliminary New Zealand case-control study of 546 male cases of leukaemia registered with the New Zealand Cancer Registry during the period 1979-1983 found an excess risk of leukaemia in electrical workers. Measurements will be obtained of typical exposures to electromagnetic fields (as well as other relevant exposures) in electrical worker occupations. Further cases and controls will be collected and the expanded study will be reanalysed using the occupational hygiene data.

TYPE: Case-Control
TERM: Electromagnetic Fields; Occupation; Registry
SITE: Leukaemia
OCCU: Electrical Workers
REGI: New Zealand (NZ)
TIME: 1987 - 1992

NIGERIA

CALABAR

630 Otu, A.A. 04932
Univ. of Calabar, College of Medical Sciences, Dept. of Surgery, P.M.B. 1115, Calabar, Nigeria (Tel.: (087)222855; Tlx: 65103 UNICAL)
COLL: Tabor, E.

Hepatocellular Carcinoma, Cirrhosis of the Liver and Hepatitis B Virus.
The association of PHC with liver cirrhosis and seropositivity to HBV is investigated by detailed examination of PHC patients and by testing their sera for HBV markers (HBsAg, anti-HBc, anti-HBs) by RIA tests. Samples which are positive for HBsAg are also tested for HBeAg and Anti-HBe. Patients with metastatic hepatic carcinoma and symptom-free, non cancer subjects serve as controls. There are two sex- and age-matched (within five years) controls per case. A communication was published in Cancer 60:2581-2585, 1987.

TYPE: Case-Control
TERM: Antibodies; Antigens; Cirrhosis; HBV; Sero-Epidemiology
SITE: Liver
LOCA: Nigeria; United States of America
TIME: 1978 - 1995

631 Otu, A.A. 04933
Univ. of Calabar, College of Medical Sciences, Dept. of Surgery, P.M.B. 1115, Calabar, Nigeria (Tel.: (087)222855; Tlx: 65103 UNICAL)

HIV Infection and Kaposi's Sarcoma
The present study explores whether Kaposi's sarcoma in Nigeria is associated with HIV infection. Serum samples from KS patients and patients with malignant melanoma of the foot (contemporaneous controls), and age- and sex-matched non-cancer controls are tested for anti-HIV antibody by enzyme-linked immunosorbent assay (ELISA). A communication was published in J. Surg. Oncol. 37:152-155, 1988.

TYPE: Case-Control
TERM: Antibodies; HIV; Sero-Epidemiology
SITE: Kaposi's Sarcoma
TIME: 1980 - 1995

IRRUA

632 Okojie, C.G. 00066
Zuma Memorial Hosp., Irrua Bendel State, Nigeria (Tel.: (055)94300)
COLL: Okojie, P.I.; Wammanda, D.R.

Aetiology of Breast Cancer
Every woman aged between 15 and 70 seen at any clinic in this hospital is examined, breasts carefully palpated, adenopathy tested for after taking history of parity, children alive, nursing period and lactation amenorrhoea. A first report covering 10,670 women over a period of 15 years showed that breast cancer is uncommon among Ishan women of Nigeria, with their early marriage (15 years), first baby by age of 16, grand parity (6-8), prolonged breast feeding leading to high serum prolactin levels and a consequent decrease in circulating gonadotrophins, delay of ovulation and longer lactation amenorrhea. The aims for the study are: (1) to confirm that breast cancer is uncommon in this area in a less developed country where the women are all poorly nourished, and (2) to investigate the important aetiological and/or protective factors in Ishan women. A paper has been published in Curr. Anthropol. 26(4), 1985.

TYPE: Relative Frequency
TERM: Age; Hormones; Lactation; Parity
SITE: Breast (F)
TIME: 1961 - 1992

NORWAY

BERGEN

633   Kvåle, G.   03162
Univ. of Bergen, Inst. of Hygiene & Social Medicine, 5016 Haukeland Sykehus, Bergen, Norway (Tel.: (05)298060)
COLL: Heuch, I.; Nilssen, S.

Reproductive Experience and Cancer Risk
The aim is to examine in detail the effects of reproductive experience (parity, age at first and later births, age at menarche and age at menopause) on cancer occurrence and on mortality from non-cancer causes in a prospective study of women in Norway. In 1956-1959 74% of all women aged 20-69 in three counties in Norway attended a screening examination for breast cancer and responded to a questionnaire. In this cohort of 63,000 some 7,000 cases of cancer had been diagnosed by the end of 1980, more than 1,500 of these being breast cancer. Data on height, weight and ABO blood group are being collected for a sub-group of the cohort to evaluate possible confounding by these variables and to examine possible interactions with the reproductive variables in relation to cancer occurrence. Recent papers appeared in Br. J. Cancer 58:820-824, 1988, J. Epidemiol. Comm. Hlth. 42:30-37, 1988, and Cancer 62:1625-1631, 1988.

TYPE: Cohort
TERM: Blood Group; Menarche; Menopause; Parity; Physical Factors; Screening
SITE: All Sites; Breast (F); Gastrointestinal; Ovary; Uterus (Cervix); Uterus (Corpus)
TIME: 1955 - 1995

DRAGVOLL

634   Flaten, T.P.   04794
Univ. of Trondheim, College of Arts and Science, Dept. of Chemistry, 7055 Dragvoll, Norway (Tel.: +47 7 596118; Fax: 509393)
COLL: Glattre, E.

Chlorination of Drinking Water and Cancer Incidence in Norway
The primary objective of the present study is to investigate whether chlorination of drinking water is associated with the incidence of colorectal or other cancers in Norway. Age-adjusted cancer incidence rates are calculated for different time periods at the county, municipality and municipality aggregate levels. The methods of analysis include (1) comparison of rates between municipalities chlorinating and not chlorinating the water; (2) multiple regression analysis at the county level (there are 19 counties in Norway) between colorectal cancer rates and (a) the mean percentage of the population in the counties receiving chlorinated drinking water over a long period of time and (b) data on dietary intake of fats and fiber and certain socio-economic variables; (3) time series analysis for groups of municipalities that started chlorinating the water in 1951-1964 vs other groups of municipalities.

TYPE: Correlation
TERM: Chlorination; Registry; Water
SITE: Colon; Rectum
REGI: Norway (Nor)
TIME: 1985 - 1992

OSLO

635   Glattre, E.   03505
Cancer Registry of Norway, Montebello, 0310 Oslo 3, Norway (Tel.: +47 2 506050)
COLL: Thoresen, S.Ø.

Aetiology of Thyroid Cancer in Norway
The Thyroid Cancer Project was established in 1985 as a multidisciplinary and multi-institutional study with its secretariat located in the Cancer Registry of Norway and the aim of investigating aetiological aspects of thyroid cancer. On-going studies are: prevalence of occult thyroid cancer in different age-groups and regions of Norway; determination of the weight of the thyroid gland by age, sex and region; case-control study of diet and thyroid cancer; case-control study of prediagnostic s-vitamin D

NORWAY

and thyroid cancer. Papers have been published in Cancer Detect. Prev. 14:625-631, 1990 and Cancer Res. 51:1234-1241, 1991 (Akslen et al.).

TYPE: Case-Control; Cohort; Cross-Sectional
TERM: Diet; Prevalence; Registry; Survival; Trace Elements; Trends; Vitamins
SITE: Thyroid
CHEM: Iodine
REGI: Norway (Nor)
TIME: 1985 - 1995

636 Jellum, E. 04678
The Norwegian Cancer Society, Inst. of Clinical Biochemistry, Pilestredet 32, 0027 Oslo 1, Norway (Tel.: +47 2 867035)
COLL: Andersen, A.; Lund-Larsen, P.; Orjaseter, H.; Theodorsen, L.

The Janus Serum Bank and Early Detection of Cancer

The Janus-project was initiated by the Norwegian Cancer Society in 1973. The Janus serum bank comprises about 400,000 serum samples consolidated from 102,000 donors. The collection includes sera from cohort studies on cardiovascular disease and specimens from 26,000 Red Cross blood donors. From 2-14 (average 4) consecutive samples are available from each donor. The sera are stored at -25° C. At regular intervals the Janus collection is matched against the files in the Norwegian Cancer Registry. From 1973 to 1988, 3500 of the donors have developed some form of cancer. Frozen serum samples collected from a few months to 15 years prior to clinical recognition of their disease are consequently available for research purposes. The aim of the Janus-project is to search in these premorbid sera for chemical, biochemical, immunological or other changes that might be indicative of cancer development at early stages. The following recent results may be cited: the tumour-associated antigen CA-125 is elevated several months prior to diagnosis of ovarian cancer; serum thyroglobulin may be a preclinical tumour marker in subgroups of thyroid cancer; low level of selenium in serum reflects increased risk of thyroid cancer; and raised antibodies in serum against Epstein-Barr virus is a risk factor for development of Hodgkin's disease.

TYPE: Incidence
TERM: BMB; Biochemical Markers; Immunologic Markers; Registry; Serum; Tumour Markers
SITE: All Sites
REGI: Norway (Nor)
TIME: 1973 - 1993

* 637 Kjaerheim, K. 04991
Cancer Registry of Norway, Montebello, 0310 Oslo 3, Norway (Tel.: (02)506050; Fax: 509376)

Cancer Risk in the Restaurant Business

The aim of the study is to analyse the risk of alcohol-associated cancers in male and female waiters in Norway, and the association between cultural and structural work characteristics and alcohol consumption. The hypothesis is that a tolerant alcohol culture and certain job characteristics increase the risk of high alcohol consumption, which in turn increases the risk of cances of the mouth, pharynx, oesophagus, larynx and liver. The study will consist of three consecutive sub-studies. (1) In a historical cohort study (n = 7,000), based on union membership from 1938-1977, observed and expected rates of cancer will be calculated. (2) Data on cultural and structural work characteristics and alcohol consumption will be obtained by mail questionnaire from 7-8,000 persons, including waiters who are union members. A prospective cohort study of this cohort will be undertaken. (3) Nested case-control studies of chosen cancer sites will also be set up.

NORWAY

TYPE: Case-Control; Cohort
TERM: Alcohol; BMB; Occupation; Registry
SITE: Larynx; Liver; Oesophagus; Oral Cavity; Pharynx
OCCU: Waiters
REGI: Norway (Nor)
TIME: 1990 - 1993

638   Larsen, T.E.   04907
Ullevaal Hosp., Dept. of Pathology, 0407 Oslo 7, Norway (Tel.: +47 2 118080)

The Norwegian Melanoma Project

A rapid increase in the incidence of malignant melanoma (MM) has been seen in Norway in the last 20 years. These observations prompted the medical community to initiate a screening programme and during the next 2-3 years the population will be invited to screening for melanoma, dyplastic naevi and clinical risk factors at University Hospital skin departments. Criteria for participating in the screening programme will include mainly (I) having growing or changing naevi; (2) having had melanoma previously or having a family history of melanoma; (3) being sensitive to sunshine or having had sunburn episodes in childhood. Approximately 40 clinical and 10-15 "blind" histological features will be stored at a central data registry at Ulleval Hospital for later statistical analysis. It is also intended (1) to assess if early diagnosis of MM reduces incidence and mortality; (2) to carry out analyses of the prognostic relevance of histological diagnosis, as well as of the criteria on which the clinical and histological diagnosis are based, and to observe the behaviour of non-removed dysplastic naevi in relation to different risk factors; (3) to carry out a case-control study to assess if MM risk is lower in a group of individuals who replied favourably to the invitation to screening, compared with a group of individuals who did not reply.

TYPE: Intervention
TERM: Familial Factors; Naevi; Screening
SITE: Melanoma
REGI: Norway (Nor)
TIME: 1989 - 1995

SKIEN

639   Hoff, G.   04719
Telemark Sentralsjukehus, 3700 Skien, Norway (Tel.: (03)525000)
COLL: Vatn, M.H.; Rosef, O.; Sauar, J.; Larsen, S.; Torgrimsen, T.

Drinking Water and Colorectal Polyps

Chlorination of humus-containing water favours formation of trihalomethanes, particularly chloroform. Epidemiological studies have indicated an association between colorectal cancer and chlorination of drinking water. An endoscopic population screening programme for colorectal polyps in Telemark in 1983 revealed an association between prevalence of colorectal polyps and colour index of drinking water (indicator of humus content). Drinking water variables (including chloroform) are now being measured monthly for two years (1989 and 1990) before 400 randomly chosen individuals aged 50-59 years (200 from each of two residential areas with separate water supplies) will be examined with flexible sigmoidoscopy for detection of polyps. Data will be collected on diet, potential risk factors for colorectal cancer and length of stay in the area.

TYPE: Cross-Sectional
TERM: Chlorination; Diet; Polyps; Time Factors; Water
SITE: Colon; Rectum
TIME: 1989 - 1993

640   Sauar, J.   04489
Telemark Sentralsjukehus, Dept. of Medicine, 3700 Skien, Norway (Tel.: (035)25000)
COLL: Hausken, T.; Hoff, G.; Bjørkheim, A.

Colorectal Polyps in First-Degree Relatives of Patients with Colorectal Carcinoma

For first-degree relatives of patients with colorectal cancer, the risk of developing carcinoma of the large bowel is 2-3 times that of the normal population. Most cases of colorectal cancer develop from adenomas. The prevalence of colorectal adenomas in a normal population has so far, to our knowledge, only been investigated in Telemark, Norway. The aims of the project are: (1) to investigate the prevalence, distribution and degree of dysplasia in adenomas of first-degree relatives as compared to the normal

NORWAY

population, and (2) to evaluate the acceptance and suitability of colonoscopy as a screening method in asymptomatic first-degree relatives. Patients admitted for first diagnosis and treatment of colorectal cancer are interviewed. With the patient's permission, their first-degree relatives aged 45-70 years of age are offered a colonoscopic examination, with polypectomy if polyps are found. The results will be compared with a similar study of a sample of the normal population. Preliminary results suggest a higher adenoma prevalence rate and higher degrees of dysplasia in adenomas from first-degree relatives than in the normal population. Recruitment to the study will continue throughout 1989 to reach an estimated total of 200-300 first degree relatives.

TYPE: Case Series
TERM: Familial Factors; High-Risk Groups; Premalignant Lesion; Prevalence; Screening
SITE: Colon; Rectum
TIME: 1987 - 1992

TRONDHEIM

641 Moe, P.J. 00899
Regionsykehuset, Barneklinikken, Kyrres gate 17, 7000 Trondheim, Norway (Tel.: (07)598171)
COLL: Seip, M; Kolmannskog, S.; Wesenberg, F.

Childhood Leukaemia and Malignant Lymphoma in Norway 1963-1987

All cases of childhood leukaemia diagnosed in Norway during the period 1963-1986 were collected by contacting all paediatric departments, and all larger departments of internal medicine in Norway. Information was also obtained from the National Cancer Registry as cancer registration is compulsory in Norway. Information is obtained as to the type of leukaemia, year of diagnosis and survival time. The first part of this study (1963-1974) has been published. Therapy studies using intermediate doses of methotrexate in ALL in the period 1975-1980 have also been published (Wissenschaft. Information 10:57-64, 1984, Europ. Ped. Hemat. Oncol. 1:113-118, 1984 and Tidsskr. Nor. Laegeforen. 106:2621-2623, 1986). The aim of the second part of this study is to register all current cases of childhood leukaemia (and malignant lymphoma) with emphasis on cell-type and results of therapy.

TYPE: Cohort
TERM: Chemotherapy; Childhood; Classification; Clinical Records; Drugs; Prognosis; Registry; Survival; Time Factors; Treatment
SITE: Leukaemia; Leukaemia (ALL); Lymphoma
CHEM: Methotrexate
REGI: Norway (Nor)
TIME: 1974 - 1992

642 Nygaard, R. 04852
Univ. Hosp., Dept. of Pediatrics, Kyrres gt. 17, 7006 Trondheim, Norway (Tel.: (07)998168; Fax: 997656)
COLL: Clausen, N.; Garvicz, S.; Jonmundsson, G.K.; Kristinsson, J.R.; Lanning, M.; Moe, P.J.; Siimes, M.A.

Relapses and Late Effects following Treatment for Childhood Leukaemia

A cohort of all patients who had completed treatment for childhood leukaemias before 1985 is being followed up with regard to late relapses, fertility, offspring and second malignant neoplasms. Risk and possible risk factors are evaluated by means of Cox analysis and SIR analysis. The first reports from this population-based study are already published, but due to the long-term aspect of late effects, the investigation continues, in cooperation with the Nordic Society for Paediatric Haematology and Oncology. Papers have appeared in Med. Pediatr. Oncol. 17:45-47, 1989, Acta Paediatr. Scand. 79 Suppl. 354:5-17 and 18-242, 1989.

TYPE: Cohort
TERM: Chemotherapy; Childhood; Fertility; Late Effects; Multiple Primary; Radiotherapy; Recurrence
SITE: Leukaemia
LOCA: Denmark; Finland; Iceland; Norway; Sweden
TIME: 1980 - 2000

PAKISTAN

KARACHI

643 Hassan, T.J. 04240
Pakistan Medical Research Council, Research Centre, Jinnah Postgraduate Medical Centre, Karachi 35, Pakistan
COLL: Zuberi, S.J.; Ashraf, S.; Zaidi, S.M.H.

Risk Factors Involved in Gallbladder Cancer in Karachi

A case-control study is being conducted in Karachi to evaluate the risk factors for gallbladder cancer. One hundred biopsied cases will be studied. Two age- and sex-matched controls (one with normal gallbladder, one with gallbladder stones) are selected for each case. All cases and controls are screened for stones by ultra-sound. Risk factors to be evaluated are: gallstones, early pregnancy, high parity, oral contraceptives, diet, cholecystitis, liver disease, smoking, tobacco use, fasting and consumption of ghee.

TYPE: Case-Control
TERM: Diet; Gallstones; Oral Contraceptives; Reproductive Factors; Tobacco (Smoking)
SITE: Gallbladder
TIME: 1987 - 1993

644 Zaidi, S.M.H. 02197
Jinnah Postgraduate Medical Center, Dept. of Radiotherapy, Karachi 35, Pakistan
COLL: Jafarey, N.A.; Ahmed, M.; Askari, A.; Masood, M.A.; Rehman, G.; Siddiqui, M.; Siddiqi, A.M.

Pakistan Medical Research Council Multicentre Tumour Study

Frequency data of tumours of various sites are collected from hospitals in six districts of Pakistan with the aim of (1) determining factors influencing cancer trends, particularly in males; (2) investigating possible environmental risk factors for cancer of the breast, cervix uteri and bone, and evaluating methods of treatment; (3) initiating incidence studies. Interim results have been published in internal Monographs in 1982 and 1991.

TYPE: Incidence; Relative Frequency
TERM: Environmental Factors; Treatment
SITE: All Sites; Bone; Breast (F); Uterus (Cervix)
TIME: 1977 - 1993

PAPUA-NEW GUINEA

GOROKA

645 Alpers, M.P. 03941
Papua-New Guinea Inst., of Medical Research, P.O. Box 60, Goroka, Papua-New Guinea (Tel.: 712200; Tlx: 72654)
COLL: Brabin, B.; Winnett, A.; Moss, D.

Epidemiology and Pathogenesis of Burkitt's Lymphoma in Papua-New Guinea
Patients presenting at Madang and Wewak hospitals with Burkitt's lymphoma will be investigated by surgical biopsy; tumour cell growth in vitro; EBV studies by DNA probe, immunological and virological techniques; and malariometric studies. Over the three year period of the initial phase of this study it is hoped to investigate 15 cases. The epidemiological characteristics of Burkitt's lymphoma and malaria will be closely studied in these communities. Areas of relatively high incidence of Burkitt lymphoma will be identified. Careful studies of malaria in pregnancy (approximately 300 women), EBV infection and genetics will be carried out in these areas. Initial field work has been completed. It is proposed to continue the study once the first stage has been analysed. Preliminary findings indicate a surprisingly poor response to treatment in PNG patients with Burkitt's lymphoma.

TYPE: Case Series
TERM: Biopsy; DNA; Immunologic Markers; Infection; Parasitic Disease; Virus
SITE: Burkitt's Lymphoma
TIME: 1985 – 1993

PARAGUAY

ASUNCIÔN

646   Rolôn, P.A.   04696
Univ. Nacional de Asunciôn, Fac. de Ciencias Médicas, Cancer Registry of Paraguay, Yagatimí 883, Asunciôn, Paraguay (Tel.: 448080, 449683)
COLL: Muñoz, N.; Lopez, B.H.; Bosch, F.X.

Case-Control Study on Oesophageal Cancer in Paraguay
The aim of the study is to determine if there is an association between oesophageal cancer and the habit of drinking mate and terere. In addition, the role of alcohol, tobacco and certain dietary habits will be studied. All incident and resident cases under 75 years of age, diagnosed from January 1988 to December 1990, in 15 hospitals of Paraguay will be included. Three hospital controls per case, matched by sex and age will be included. All cases and controls are being interviewed at the hospitals by trained interviewers. The statistical analysis will be done using Mantel-Haenszel test and conditional logistic regression.

TYPE: Case-Control
TERM: Alcohol; Diet; Mate; Tobacco (Smoking)
SITE: Oesophagus
TIME: 1988 - 1992

POLAND

BRZESKO

647　Moszczynski, P.　03981
Regional Hosp., Provincial Immunology Lab., 68 Kosciuszki, 32-800 Brzesko, Poland (Tel.: 30006; Tlx: 066633)

Haematological and Immunological Disturbances Induced by Occupational Exposure to Organic Solvents

Haematological and immunological examinations have been continuously performed in 109 workers occupationally exposed to benzene, toluene and xylene. Examinations involved the basic haematological indices, histochemical examinations of neutrophils and lymphocytes, the NBT test, the immunoglobulin concentrations in the blood serum, the total T and B cell count, and skin tests of delayed hypersensitivity. A control group consisted of 50 non-exposed healthy subjects in the same industrial factory. As yet no benzene-induced haemopathies have been noted and the main finding was a lowered T lymphocyte count in the peripheral blood and the enzymatic alterations both in neutrophils and lymphocytes. A synergistic depressing effect of organic solvents was observed on certain immunological indexes. This phenomenon reflected in the changes of concentrations of IgA, IgD, IgG, IgM and lysozyme, and the subpopulations of T-cells. Papers appeared in Rev. Roum. Med. Int. 27:137-141, 1989 and in Med. Prac. 60:337-341, 1989.

TYPE:　Molecular Epidemiology
TERM:　Chemical Exposure; Dusts; Occupation; Solvents
SITE:　Haemopoietic
CHEM:　Benzene; Toluene; Xylenes
OCCU:　Lacquerers; Varnishers
TIME:　1980 - 1993

GLIWICE

648　Zemla, B.　03442
Inst. of Oncology, Armii Czerwonej 15, 44-101 Gliwice, Poland (Tel.: 311061; Tlx: 036606 inonk)
COLL:　Banasik, R.; Zielonka, I.; Kolosza, Z.; Skalska-Vorbrodt, J.

Lung Cancer in Relation to Environmental Risk Factors

This is a case-control study of lung cancer (about 300 cases and 600 population controls matched for age and sex) in the upper Silesian region, Poland. The objective is to determine risk in relation to tobacco use, alcohol consumption, and diet. Statistical analysis will take into account length of employment and work conditions, history of migration, and pre- and co-existing diseases. Information will be collected through questionnaire by interviewer. Papers have been published in Wiad. Lek XXXIX(14):946-956, 1986 and in Neoplasma 35(2):135-143, 1988.

TYPE:　Case-Control
TERM:　Alcohol; Diet; Migrants; Occupation; Time Factors; Tobacco (Smoking)
SITE:　Lung
TIME:　1984 - 1992

649　Zemla, B.　04584
Inst. of Oncology, Armii Czerwonej 15, 44-101 Gliwice, Poland (Tel.: 311061; Tlx: 036606 inonk)
COLL:　Kolosza, Z.; Banasik, R.

Cervix Uteri and Breast Cancer Incidence in Upper Silesia

The aim of the study is to describe patterns in the geography and dynamics of cancer in Upper Silesia Region (Katowice voivodship, southern Poland) as part of the atlas of cancer incidence for this region. This is necessary for studies of suspected aetiological factors, as well as for monitoring changes in cancer frequency. Cancer incidence data (cervix, breast) by age and residence are being collected from 1989. Age-specific and age-adjusted rates have been computed and analysed for the whole region and for 43 urban towns and 46 rural county areas, for the period 1975-1985. A paper has been published in Gin. Pol. 3:143-148, 1989.

POLAND

TYPE: Incidence
TERM: Geographic Factors; Rural; Trends; Urban
SITE: Breast (F); Uterus (Cervix); Uterus (Corpus)
TIME: 1989 - 1992

650   Pawlega, J.   03745
Maria Sklodowska-Curie Memorial Inst., of Oncology, Cracow Branch,, Dept. of Cancer Epidemiology, Garncarska 11, 31-115 Krakow, Poland (Tel.: 229900; Tlx: 0325437)

Role of Smoking, Drinking and Diet in Cancer Risk

In 1985, 1,895 people living in the Cracow Region and born in 1938 and later, were covered by a questionnaire survey concerning habits of smoking, drinking, diet and data on weight and height. During the next 15 years these citizens will be followed using the Cracow Cancer Registry in order to find associations between the above factors and particular sites of cancer.

TYPE: Cohort
TERM: Alcohol; Diet; Occupation; Physical Factors; Registry; Tobacco (Smoking)
SITE: All Sites; Breast (F); Lung; Stomach; Uterus (Cervix)
REGI: Cracow (Pol)
TIME: 1986 - 2000

651   Popiela, T.   03848
Medical Academy of Nicolaus Copernicus, 1st Dept. of General, and Gastrointestinal Surgery, Kopernika 40, 31-501 Krakow, Poland (Tel.: 110975)
COLL: Jedrychowski, W.; Zembala, M.

Detection, Treatment and Epidemiology of Gastric Cancer in Poland

The study concentrates on the problems of early gastric cancer detection, and is carried out among hospitalized and out-patients initially in the regions of South and East Poland and in eight main regions of Poland. Recently it has been extended to cover the whole country. The study is based on questionnaire and fibroscopic examinations. In these clinics a study on the effectiveness of operative procedures, taking into consideration long-term survival and the influence of supplementary chemo- and chemo-immunotherapeutic treatment (5FU + BCG, 5FU, FAM, FAM + BCG) in various stages of gastric cancer is being carried out. Independently, an analysis of the influence of epidemiological factors such as cigarette smoking, professional exposure, social status, diet, genetic factors, etc. upon the incidence of the disease is being undertaken.

TYPE: Case Series
TERM: BCG; Chemical Exposure; Diet; Occupation; Socio-Economic Factors; Tobacco (Smoking)
SITE: Stomach
TIME: 1978 - 1992

LÓDZ

652   Stankiewicz, A.K.   04860
Nofer's Inst. of Occupational Medicine, ul. Teresy 8, 90-950 Lódz, Poland (Tel.: (42)552250; Fax: 348331)

G6PD Activity in Cancer Patients and Occupationally Exposed Persons

The aim of the investigation is to determine G6PD activity in erythrocytes in patients with cancer. Data studied previously show that G6PD activity was markedly higher in patients before surgery, as compared to a control group, than after. Investigations will continue to determine whether G6PD activity depends on the progress of the neoplasms. To this effect, G6PD activity in erythrocytes in persons occupationally exposed to carcinogenic factors will also be examined. The study group consists of some 500 persons exposed to aromatic amines. Controls are 53 healthy people, 16-60 years of age, selected randomly from a group of 4,000 not exposed occupationally to carcinogens. G6PD activity is determined by a spectrophotometric method according to Kornberg and Horecker.

POLAND

TYPE: Cross-Sectional
TERM: G6PD; Occupation; Tumour Markers
SITE: All Sites
CHEM: Amines, Aromatic
TIME: 1991 – 1993

653 Stankiewicz, A.K. 04180
Nofer's Inst. of Occupational Medicine, ul. Teresy 8, 90-950 Lódz, Poland (Tel.: (42)552250; Fax: 348331)
COLL: Hanke, J.; Lutz, W.; Krajewska, B.; Pilacik, B.

Genetic Susceptibility to Toxic Substances and its Relationship to Carcinogenesis

This study is aimed at selection of biochemical indicators of predisposition to neoplastic diseases. The following markers of genetically conditioned susceptibility to neoplastic disease were suggested: phenotypes PiZ and PiMZ, alfa 1-antytrypsin, phenotype of free acetylation (to be studied prospectively), high activity G6PD in erythrocytes, and rapid antipyrine metabolism (to be studied retrospectively). Based on examination of families of children afflicted with liver cirrhosis a cohort has been constructed, covering about 200 individuals with PiZ and PiMZ. Phenotypization is carried on with the help of cross immunoelectrophoresis. The cohort of free acetylation includes 50 workers representing this phenotype, occupationally exposed to aromatic amines. Acetylation phenotypes are being determined with the method recommended by WHO. In the retrospective studies, frequency of occurrence of G6PD deficiency, detected with methaemoglobin reduction test in a group of healthy subjects (about 10,000 persons), is compared with frequency of deficiency in patients with neoplastic disease (about 8,000 persons). Similar studies are being carried out based on the antipyrine test (Kellermann) in a group of healthy persons and in a group with lung cancer.

TYPE: Case-Control; Cohort; Molecular Epidemiology
TERM: Chemical Exposure; Cirrhosis; Enzymes; Genetic Factors; Genetic Markers
SITE: Lung
CHEM: Amines, Aromatic
LOCA: Malta; Poland
TIME: 1986 – 1992

* 654 Stankiewicz, A.K. 05156
Nofer's Inst. of Occupational Medicine, ul. Teresy 8, 90-950 Lódz, Poland (Tel.: (42)552250; Fax: 348331)

G6PD Activity in Erythrocytes and the Development of Neoplastic Disease

This study aims to measure the activity of G6PD in erythrocytes (a) in subjects with malignant tumours of various organs and tissues, before and after surgery; (b) in about 100 premenopausal women with benign breast tumours. Dehydroepiandrosterone sulphate will also be measured in these women; (c) in workers occupationally exposed to asbestos dust to evaluate the correlation between development of neoplastic disease and asbestosis.

TYPE: Molecular Epidemiology
TERM: Alcohol; Asbestosis; Dusts; G6PD; Occupation; Tobacco (Smoking); Tumour Markers
SITE: All Sites
CHEM: Asbestos; Dehydroepiandrosterone Sulphate
TIME: 1991 – 1993

655 Sulkowski, W.J. 04498
Nofer's Inst. Occupational Medicine, ENT Division, Teresy 8, 90-950 Lódz, Poland
COLL: Szeszenia-Dabrowska, N.; Kowalska, S.; Laciak, J.; Szymczak, W.; Wilczynska, U.

Evaluation of Occupational Risk of Laryngeal Cancer

The objective of the study is to estimate the contribution of occupational exposures to laryngeal cancer incidence, for which an increasing rate in recent years sets Poland among the countries of highest risk. Although excessive tobacco smoking and alcohol drinking are considered to be the main factors, epidemiological evidence suggests that contact with some industrial chemicals e.g. chromium compounds, polycyclic aromatic hydrocarbons, isopropyl alcohol, mustard gas, asbestos, and others may induce carcinogenesis with the larynx as target organ. A series of about 400 patients with clinically and histopathologically recognised cancer of the larynx, treated in the Laryngological Clinic in Lódz were covered by a questionnaire case-control study with particular emphasis in interviews on occupational exposure, tobacco smoking, alcohol and diet. Two controls per case, matched by age, sex, education

POLAND

and permanent residence, will be examined to establish whether occupational exposure of the group of cases differs significantly from that of the controls. Preliminary reports have been published in Otolaryng. Pol., Suppl. 1984, and Med. Pracy 37, 6:353-361, 1986.

TYPE: Case-Control
TERM: Alcohol; Diet; Dusts; Metals; Occupation; Pesticides; Solvents; Tobacco (Smoking)
SITE: Larynx
CHEM: Asbestos; Chromium; Isopropyl Alcohol; PAH
TIME: 1986 - 1992

656 Górski, T. 04603
Sanitary-Epidemiological Station, Ul. Wodna 40, 90-046 Lódz, Poland (Tel.: +48 42 740846; Tlx: 886747)
COLL: Górecka, D.

Influence of Cigarette Smoking on SCE Frequencies among Nurses Handling Cytostatic Drugs
Genotoxic compounds and various cancer chemotherapeutic agents can interact with tobacco smoke synergistically. The aim of the study is to examine the hypothesis that tobacco smoking has a greater influence on SCE frequencies among nurses handling cytostatic drugs than the cytostatics themselves. The frequencies of SCE in lymphocytes will be investigated among hospital staff who handle anti-cancer drugs and in a control group which does not handle them (smokers and non-smokers).

TYPE: Cross-Sectional; Molecular Epidemiology
TERM: Chemotherapy; Chromosome Effects; Occupation; SCE; Tobacco (Smoking)
SITE: Inapplicable
OCCU: Health Care Workers
TIME: 1988 - 1992

* 657 Starzynski, Z. 05090
Inst. of Occupational Medicine, Dept. of Epidemiology and Statistics, Unit of Epidemiology, Teresy 8 199, 90-950 Lódz, Poland (Tel.: (4842)552250; Fax: 348331 ; Tlx: 885360 IMP PL.)
COLL: Marek, K.; Kujawska, A.; Szeszenia-Dabrowska, N.; Wilczynska, U.; Szymczak, W.

Cancer Mortality in Silicotics
The main objective of the study is to follow-up individuals who developed silicosis and coalworkers' pneumoconiosis and to analyse their causes of death. The immediate objectives are: (1) to evaluate the natural history of the pneumoconioses, i.e. their course and complications as well as concomitant diseases; (2) to determine survival time and cause of death in pneumoconiotics, with particular regard to cancer; (3) to evaluate health and social effects of the pneumoconioses. Data on some 13,000 individuals with pneumoconiosis diagnosed in the years 1970-1985, provided by the Central Register of Occupational Diseases, will serve as the study material, and the general population of Poland will be used as the reference population.

TYPE: Cohort
TERM: Dusts; High-Risk Groups; Pneumoconiosis; Silicosis; Survival
SITE: All Sites; Lung
CHEM: Radon; Silica
TIME: 1991 - 1994

SZCZECIN

* 658 Pilawska, H. 05204
Pomeranian Medical Faculty, Dept. of Epidemiology, Chair of Social Medicine, al. Powstyncow Wlk. 72, 71-164 Szczecin, Poland (Tel.: 824062)
COLL: Walczak, A.; Cipkowska, B.

POLAND

Relationship between Magnesium in the Soil and Cancer
The aim of this study is to correlate soil magnesium levels and the frequency of cancer in three provinces near the Baltic Sea. Soil magnesium levels are obtained from the Agriculture Faculty. These will be correlated with regional cancer rates from cancer registries in the region over a 10-year period.

TYPE: Correlation
TERM: Registry; Soil; Trace Elements
SITE: All Sites
CHEM: Magnesium
TIME: 1991 - 1992

WARSAW

659   Zatonski, W.A.   04764
Oncological Center, Dept. Cancer Control & Epidemiology, Ul. Wawelska 15, 00-973 Warsaw, Poland (Tel.: (022) 233179; Fax: 222429; Tlx: 812704 inonk)
COLL: Estève, J.; Smans, M.; Tyczynski, J.

Atlas of Cancer Mortality in Central Europe
The aim of the study is to present the geographical distribution of cancer mortality in Austria, Czechoslovakia, Bulgaria, Romania, Hungary, Yugoslavia, Germany and Poland. Cancer mortality data for 1982-1986 will be collected from participating countries, broken down by year of death, cause of death, sex, geographical units, age at death etc. Results will be presented in the form of maps.

TYPE: Mortality
TERM: Geographic Factors; Mapping
SITE: All Sites
LOCA: Austria; Bulgaria; Czechoslovakia; Germany; Hungary; Poland; Romania; Yugoslavia
TIME: 1990 - 1992

660   Zatonski, W.A.   04812
Oncological Center, Dept. Cancer Control & Epidemiology, Ul. Wawelska 15, 00-973 Warsaw, Poland (Tel.: (022) 233179; Fax: 222429; Tlx: 812704 inonk)
COLL: Tyczynski, J.

Changes in Geographical Patterns of Cancer Mortality in Poland, 1975-1988
This study is designed to present changes in geographical patterns of cancer mortality in Poland for the time period 1975-1988. The SMR's for 49 districts (voivodships) will be calculated for three time periods: 1975-1979, 1980-1984 and 1985-1988. Direct age-adjustment will be used (world population). Analysis of changes in particular districts and in the whole country will be done, and results will be presented in the form of maps and tables.

TYPE: Mortality
TERM: Geographic Factors; Mapping
SITE: All Sites
TIME: 1990 - 1992

ROMANIA

TULCEA

661   **Georgescu-Tulcea, N.**   04235
Tulcea Departmental Hosp., Unit of Cancerology, 148, 23 August St. , 8800 Tulcea, Romania (Tel.:91513854)

Cancer Incidence by Site, Age and Sex in Tulcea Department
The Tulcea Department Cancer Registry has been in operation since 1981. More than 4,500 cancer cases have been registred. Using data from 1981 - 1987, this study will examine incidence trends by site, age and geographical distribution. Data are compared with data for other Romanian Departments where possible.

TYPE: Incidence
TERM: Age; Anatomical Distribution; Registry
SITE: All Sites
TIME: 1987 - 1992

SOUTH AFRICA

BLOEMFONTEIN

662   **Cronjé, H.S.**   04530
Univ. of the Orange Free State, Dept. of Obstetrics and Gynecology, P.O.Box 339, Bloemfontein 9300, South Africa (Tel.: (051)470741, 4053444; Fax: 473222)

Cervical Intraepithelial Neoplasia in the Orange Free State
The frequency of cervical intraepithelial neoplasia in the Department of Obstetrics and Gynaecology is unexpectedly low in view of the high frequency of infiltrating cervical cancer. A study has just been completed which showed that cervical cytological services are not adequately distributed throughout the Orange Free State. The aim of this study is to determine the prevalence of CIN (and certain sexually transmitted diseases) in black females in Orange Free State. Randomised cluster sampling will be carried out to obtain a sample of 800 subjects evenly distributed by census district, magisterial district, and farms or street blocks. A cervical smear will be taken as well as an endocervical swab and blood to test sexually transmitted disease from each woman (18-65 yrs).

TYPE: Cross-Sectional
TERM: Blacks; Cytology; Premalignant Lesion; Prevalence; Sexually Transmitted Diseases
SITE: Uterus (Cervix)
TIME: 1989 - 1992

CAPE TOWN

663   **Jaskiewicz, K.**   04537
Univ. of Cape Town Medical School, Dept. of Pathology, Anzio Rd, Observatory 7925, Cape Town, South Africa (Tel.: +27 21 471250; Fax: 4171789)
COLL: Stenkop, E.; Louwrens, H.

Diet at a Risk Factor for Atrophic Gastritis
The hypothesis is that endoscopic screening of pre-selected population groups for gastric cancer is an efficient way of establishing diagnosis and prevalence of early cancer and precursor lesions of the stomach. The study of dietary and other factors will give new information concerning aetiology and could be employed in prevention of this disease. Endoscopic and histological examination of gastric mucosa will be undertaken in 600 patients with dyspepsia from a population at increased risk for gastric carcinoma. The pilot study consists of 50 patients with chronic atrophic gastritis, 50 controls with normal gastric mucosa and 50 healthy controls without endoscopic examination. Subjects will be sex- and age-matched, from the same population group. A detailed questionnaire will be employed for each of the 150 subjects, and blood will be analysed for vitamins: C, A, E, B, B6, folates, nicotinic acid and for minerals: Na, K, Ca, Mg, Cu, Zn, Fe, Se. Gastric juices will be analysed for nitrates and nitrites and PH will be estimated.

TYPE: Case-Control
TERM: Atrophic Gastritis; Diet; High-Risk Groups; Nutrition; Premalignant Lesion; Vitamins
SITE: Stomach
CHEM: Copper; Iron; Nitrates; Nitrites; Selenium
TIME: 1987 - 1992

* 664   **Jaskiewicz, K.**   05141
Univ. of Cape Town Medical School, Dept. of Pathology, Anzio Rd, Observatory 7925, Cape Town, South Africa (Tel.: +27 21 471250; Fax: 4171789)
COLL: Robson, S.C.; Williamson, A.L.

Evaluation of HBV and HCV Markers in Biopsy Material from Patients with Hepatocellular Carcinoma
The prevalence of hepatocellular carcinoma (HCC) is extremely high in Southern Africa. The disease is thought to be linked to HBV and HCV infection, to exposure to mycotoxins, alcohol and iron. The expression of viral antigens and anti-HCV antibodies has been studied recently in groups of patients with alcoholic cirrhosis and portal hypertension, presumed non-A non-B chronic active hepatitis, post-transfusional hepatitis, hepatoma and respective control groups. The patients with hepatomas elected for the seroprevalence study of HCV antibodies were predominantly urban, of mixed racial extraction and tended to be older than the patients reported by Kew et al. Only three of the 33 patients with hepatoma had anti-HCV antibodies in contrast to 11 with serum HBsAg positivity. Recent histological studies by immunohistochemistry have shown 52% positivity for liver tissue for HBsAg in

patients with HCC. Over 300 HCC biopsies from selected rural patients, 15% younger than 25 years, are available for study. Liver tissue from sporadic urban cases is also available. Normal liver histology, benign, malignant, epithelial and mesenchymal primary liver tumours; primary biliary cirrhosis; sclerosing cholangitis; presumed non-A non-B hepatitis; hepatoportal sclerosis; haemochromatosis; haemosiderosis and cryptogenic cirrhosis are also available.

TYPE: Case Series
TERM: Biopsy; Cirrhosis; HBV; HCV; Mycotoxins
SITE: Liver
CHEM: Aflatoxin
TIME: 1991 - 1994

665 Warner-Learmonth, G. 04575
Univ. of Cape Town, Medical School, Cytopathology Lab., Room 212, Observatory 7925, Cape Town 7925, South Africa (Tel.: 471250/424; Fax: 478955 ; Tlx: 522208)
COLL: Beck, J.

Abnormal Cervical Smears in Teenagers
This is a prospective study of cervical smears in females aged 19 years and younger. Several authors have documented CIN in sexually active teenagers. However, little attention has been paid to the problem of cervical smears reported as "inflammatory atypia". A preliminary search of the records from January 1987 to August 1988 reveals that 7% of the total number of cervical smears are from patients of 19 years and under. Of these, 78% are pregnant and 21% are using some form of contraception. 1,051 of these 8,349 teenage patients have cytological abnormalities ranging from inflammatory atypia to squamous carcinoma. Every gynaecological smear submitted for review from patients of 19 years and under will be "flagged" on registration. All atypical smears from these patients will be reviewed together with all their previous smears. In this way, it is hoped to follow the progress of inflammatory atypia and record (1) its association with sexually transmitted infections, e.g. chlamydia, herpes, human papilloma virus, trichomonas and candida; and (2) whether it is a forerunner of dysplasia.

TYPE: Case Series
TERM: Adolescence; Clinical Records; Cytology; HPV; HSV; Premalignant Lesion; Sexually Transmitted Diseases
SITE: Uterus (Cervix)
TIME: 1987 - 1997

666 Dijkstra, B.K.S. 02203
Univ. of the Witwatersrand, Johannesburg General Hosp., Dept. of Ear, Nose & Throat, York Rd, Parktown, Johannesburg 2193, South Africa (Tel.: (011)4884911)
COLL: McIntosh, W.A.; Richards, W.

Head, Neck and Larynx Cancers
Data are being collected on all cases of larynx, head and neck cancers to provide a basis for future research. Data on some 6,500 patients from the General Hospital in Johannesburg are on record since 1970 and will be included. Items of data include age, sex, place and date of birth, use of tobacco and alcohol, type and site of tumour (including histological or cytological diagnosis), date of onset of disease and of first visit, treatment (radiotherapy, surgery, laser, chemotherapy) and follow-up after treatment.

TYPE: Case Series
TERM: Alcohol; Cytology; Histology; Tobacco (Smoking); Treatment
SITE: Head and Neck; Larynx
TIME: 1979 - 1993

667 Goldstein, B. 04236
National Centre for Occupational Health, Department of Pathology, 106 Joubert St. Extension , Johannesburg 2000, South Africa (Tel.: (011)7241844; Tlx: 422251 sa)
COLL: Simson, I.W.; Uys, C.J.; Middlecote, B.D.; Webster, I.

Mesothelioma Register and the Asbestos Tumour Reference Panel
Since 1965 histological material on cases of mesothelioma or suspected mesothelioma in South Africa has been submitted to a panel of pathologists for confirmation of the diagnosis. These cases have been recorded in a register which to date contains approximately 1,600 histologically confirmed mesotheliomas, probably the largest collection in the world. Where possible follow-up details on cases

SOUTH AFRICA

are also obtained. The data will be computerised and analysed with respect to: age and sex distribution; site of tumour; occupational and/or environmental exposure to asbestos, and histological type of tumour.

TYPE: Registry
TERM: Chemical Exposure; Dusts; Environmental Factors; Histology; Occupation
SITE: Mesothelioma; Peritoneum; Pleura
CHEM: Asbestos
TIME: 1965 – 1993

668 Kew, M.C. 04460
Univ. of the Witwatersrand, Dept. of Medicine, 7 York Rd, Parktown, Johannesburg 2193, South Africa (Tel.: (011)4883626; Fax: 6434318)
COLL: Song, E.

Aetiology and Diagnosis of Hepatocellular Carcinoma in Southern African Blacks

The purpose of the study is to determine the cause or causes of hepatocellular carcinoma in southern African blacks, who have a very high frequency of this tumour and in whom it frequently occurs at a very young age. Information on the relation between the hepatitis B virus and the tumour has been assembled; aspects that continue to be assessed are relation to age, sex, place of birth and where childhood was spent, subsequent geographical movements, alcohol consumption, cigarette smoking, taking of oral contraceptive steroids. A surveillance programme designed to detect small early heptocellular carcinomas is also being carried out. This involves following patients known to be chronic carriers of the hepatitis B virus (or to have chronic parenchymal hepatic disease) with serial ultrasonographic examination and serum alpha-foetoprotein estimations.

TYPE: Cross-Sectional
TERM: Age; Alcohol; Blacks; Geographic Factors; HBV; Oral Contraceptives; Screening; Sex Ratio; Tobacco (Smoking)
SITE: Liver
CHEM: Steroids
TIME: 1980 – 1993

669 MacDougall, L.G. 04549
Univ. of the Witwatersrand, Dept. of Paediatrics (Haematology), Baragwanath Hosp., P.O. Berthsam, Johannesburg 2013, South Africa (Tel.: (11)9331530)
COLL: Greaves, M.F.; Bernstein, R.; Cohn, R.; Pool, J.E.

Acute Leukaemia in Black and White Children and Factors Influencing Response to Treatment and Survival

The aim is to determine if significant differences exist in the pattern of acute leukaemia between black and white children aged 0-15 years living in the Johannesburg area of South Africa, with special reference to: (1) population based incidence; (2) classification of leukaemia cell type based on morphology (FAB), immunology (cell surface markers), and cytogenetics; (3) analysis of risk factors, response to treatment and survival; (4) pharmacological evaluation of chemotherapeutic drug metabolism, and (5) assessment of drug compliance. All newly diagnosed children with acute leukaemia referred to the Paediatric Haematology/Oncology Clinics at Johannesburg and Baragwanath Hosptials are enrolled in the study categories (1), (2), and (3) above. This comprises approximately 24 children annually (8 blacks, 16 white). Papers appeared in Leuk. Res. 9:765-767, 1985; Am. J. Ped. Hem./ Oncol. 8:43-51, 1985:and in South Afr. Med. J. 75:481-484, 1989.

TYPE: Case Series; Incidence
TERM: Blacks; Caucasians; Chemotherapy; Childhood; Classification; Survival; Treatment
SITE: Leukaemia
TIME: 1975 – 1992

670 Rees, D. 04566
National Center for Occupational Health, P.O.Box 4788, Johannesburg 2000, South Africa (Tel.: (011)7241844; Tlx: 422251 sa)
COLL: Simson, I.W.; Goodman, K.

Occupational and Environmental Exposures Associated with Mesothelioma in South Africa

The two major issues to be investigated in this case-control study are (1) occupational and environmental exposures associated with mesothelioma, including diet and (2) the relatively low reported incidence rate of mesothelioma in black South Africans. Cases will be notified by pathologists,

SOUTH AFRICA

oncologists, cardio-thoracists etc. There will be two controls per case: one cancer control – a subject matched for hospital, age, race, sex with cancer of any organ except the pleura or the lung, and one non-cancer control – a subject matched as above and an in-patient of a medical ward for five days or longer. 150 cases and 300 controls are expected. Data will be collected through detailed exposure questionnaire administered to the cases and controls by trained interviewers.

TYPE: Case-Control
TERM: Blacks; Diet; Environmental Factors; Occupation
SITE: Mesothelioma
TIME: 1988 – 1992

TYGERBERG

671 Hesseling, P.B. 04536
Univ. of Stellenbosch, Tygerberg Hosp., Dept. of Paediatrics, P.O.Box 63, Tygerberg 7505, South Africa (Tel.: (021)9313131; Fax: 9317810 ; Tlx: 526226)
COLL: Wessels, G.

The Tygerberg Hospital Paediatric Tumour Registry

The aim is to record accurately clinical, pathological and therapeutic data on all children newly diagnosed with cancer at this institution. At the same time as standardizing and improving therapy, and effecting a maximum follow up of 5 years, it is intended to establish a population-based register of paediatric cancer for the period 1983-1988 for Namibia. Data will be contributed to a national paediatric cancer register, and incidence and treatment results compared with national and international groups. The data are computerized. 285 new cases have been entered since 1 January 1983. The present rate of follow-up in all patients is 98%.

TYPE: Incidence; Registry
TERM: Treatment
SITE: Childhood Neoplasms
LOCA: Namibia; South Africa
TIME: 1983 – 1993

SPAIN

ALICANTE

672   Vioque, J.   04863
Univ. de Alicante, Dpto. de Salud Comunitaria, Div. de Medicina Preventiva, Campus de San Juan, Apdo 374, 03080 Alicante, Spain (Tel.: (96)5659811; Fax: 5658513)

Diet and Cancers of the Oesophagus, Stomach and Pancreas
This case-control study will evaluate the role of diet in relation to cancers of oesophagus, stomach and pancreas. The major hypotheses to be tested are: (1) a diet with a high fat content is associated with increased risk of pancreatic cancer; (2) a diet rich in fresh fruit and vegetables is associated with a lower risk of oesophagus, stomach and pancreatic cancer; (3) a low intake of vitamins C and A (and specific carotenoid fractions) is associated with a higher risk; (4) a high intake of foods with a high content of salt and carbohydrates is associated with an increased risk of stomach cancer. During a period of one year all newly diagnosed cancers in five Spanish provinces will be collected: 300-400 pancreatic, 600 stomach and 300 oesophageal cancers are expected. Controls will be randomly selected from the census, and matched by sex, age and residence. Data will be collected through personal interview.

TYPE: Case-Control
TERM: Diet; Fat; Fruit; Nutrition; Registry; Vegetables; Vitamins
SITE: Oesophagus; Pancreas; Stomach
REGI: Basque (Spa); Navarra (Spa); Zaragoza (Spa)
TIME: 1991 - 1993

BARCELONA

673   González, C.A.   04302
Hosp. Sant Jaume i Santa Magdalena, Depto Epidemiologia y Biostatistica, Mataro, Barcelona, Spain
COLL: Badia, A.; Cardona, T.; Verge, J.; Viver, J.; Saigi, E.; Batiste, E.; Marcos, G.; Martos, M.C.; Solanilla, P.; Brullet, E.; Badosa, E.; Vida, F.; Riboli, E.

Stomach Cancer in Catalunya and Zaragoza
A multicentre case-control study both in high risk and low risk areas has been initiated to provide information on the role of diet and other factors in gastric cancer. Cases will be all patients with a newly diagnosed cancer of the stomach, on the basis of histological examination in the participating centres. Controls will be chosen from the same hospitals. About 350 cases and 350 controls are expected.

TYPE: Case-Control
TERM: Cooking Methods; Diet; Histology; Occupation; Tobacco (Smoking)
SITE: Stomach
TIME: 1987 - 1992

PALMA DE MALLORCA

674   Benito, E.   04867
Unitat d'Epidemiologia, y Registre de Cancer de Mallorca, Misericordia 2 , 07012 Palma de Mallorca, Spain (Tel.: (071)721056)
COLL: Bosch, F.X.; Obrador, A.; Cabeza, E.; Esteva, M.; Moreno, V.; Roca, P.; Muñoz, N.

Colorectal Polyps and Diet
The aim of this study is to determine possible risk factors in the aetiology of colorectal polyps, such as diet, obesity, parity, exercise and cigarette smoking. The design is a case-control study. 101 diagnosed cases of histologically confirmed adenomas of the large intestine and 292 controls randomly selected from the population census tracts have been inerviewed by specially trained interviewers, focussing on diet, using a semi-quantitative food frequency questionnaire, and lifestyle factors.

TYPE: Case-Control
TERM: Diet; Obesity; Parity; Physical Activity; Polyps; Tobacco (Smoking)
SITE: Colon; Rectum
TIME: 1988 - 1992

675   Garau, I.   04873
Registre de Cancer de Mallorca, Unitat d'Epidemiologia, Misericordia 2 , 07012 Palma de Mallorca, Spain (Tel.: (071)721056)
COLL: Cabeza, E.; Benito, E.; Avella, A.; Roca, P.; Rifa, J.; Moreno, V.; Franch, P.

SPAIN

Majorca Cancer Registry
Majorca, is the largest Balearic Island, with an estimated population of 620,515 in 1989. 25% moved to the Island during the 1960's, mainly from the southern regions of Spain, 72% were born in Majorca, and 3% come from other European countries. Its income is in the highest average in Spain. Since 1982 a population-based registry specific for colorectal cancer with an active follow-up of cases has been in operation in the island. A cancer registry for all sites started in January 1989 to obtain cancer incidence and mortality data in this well-defined population. The registry will be a useful tool for further aetiological cancer research.

TYPE: Incidence; Registry
TERM: Registry
SITE: All Sites
REGI: Mallorca (Spa)
TIME: 1989 - 1993

VALENCIA

* 676   **Cortes Vizcaino, C.**   05019
Univ. de Valencia, Fac. de Medicina, Depto. de Medicina Preventiva, y Salud Publica, Aven. Blasco Ibanez 17, 46010 Valencia, Spain (Tel.: (6)3864166; Fax: 3864173)
COLL: Saiz Sanchez, C.; Gimenez Fernandez, F.J.; Talamante Serrulla, S.; Sabater Pons, A.; Calatayud Sarthou, A.

Digestive Cancers in Spain: Temporal and Spatial Correlations
Mortality from digestive cancers in Spain is being studied to investigate the relationship with risk factors such as consumption of alcohol, coffee and tobacco, and dietary factors. Mortality and demographic data are obtained from the National Institute of Statistics. Data on risk factors are obtained from other surveys. The geographical distribution of several digestive cancers is being correlated with the distribution of the above factors. The temporal evolution of mortality from these cancers will also be correlated with that for consumption of the risk factors in Spain over a 30-year period.

TYPE: Correlation
TERM: Alcohol; Coffee; Diet; Geographic Factors; Time Factors; Tobacco (Smoking)
SITE: Colon; Gallbladder; Liver; Oesophagus; Pancreas; Rectum; Small Intestine; Stomach
TIME: 1991 - 1992

ZARAGOZA

677   **Sinues, B.**   04668
Univ. de Zaragoza, Fac. Med. y Hosp. Clin. Univ., Dpto de Farmacologia, Domingo Miral, 50009 Zaragoza, Spain (Tel.: 3357854)
COLL: Izquierdo, M.; Tres, S.; Bartolome, M.; Perez Viguera, J.

Biological indicators of Occupational Cancer Risk
The aims of the study are: (1) to establish the individual susceptibility to occupationally induced carcinogenesis, and (2), to find some biological indicators for cancer risk. Important genetically determined metabolic differences increase or reduce the occupational cancer risk. The possible relation between the following factors will be investigated: (1) biological effects (measured in blood peripheral lymphocytes, by testing SCEs, chromosome aberrations, PRI, micronuclei), (2) Pharmacogenetics (by phenotyping acetylator and hydroxilator status), (3) liver microsomal activity environmentally-induced (-GT, glucaric and acid excretion) and (4) internal exposure level (by measuring thioether, mutagen and premutagen urinary excretion). A study of smokers, workers exposed to vinyl chloride and workers in the dye industries exposed to arylamine has been initiated.

TYPE: Molecular Epidemiology
TERM: BMB; Biochemical Markers; Chromosome Effects; Dyes; Lymphocytes; Metabolism; Micronuclei; Occupation; Plastics; SCE; Tobacco (Smoking)
SITE: Angiosarcoma; Bladder; Liver; Lung
CHEM: Vinyl Chloride
OCCU: Dyestuff Workers; Plastics Workers; Textile Workers; Vinyl Chloride Workers
TIME: 1989 - 1993

SRI LANKA

PERADENIYA

678   Warnakulasuriya, K.A.   04338
Univ. of Peradeniya, Dept. of Oral Medicine, Augusta Rd, Peradeniya, Sri Lanka (Tel.: (08)88045)

Oral Precancerous Lesions and Conditions
Case-control studies on the aetiology of oral cancer and pre-cancer conducted in India have given clues to aetiology. The current study's objective is to re-evaluate the risk factors in a different geographical setting, Sri Lanka, where chewing habits are similar in many respects. The role of betel quid chewing, particularly with and without tobacco, and smoking will be studied among pre-cancer cases (aged over 20 years). The study is population-based with 456 cases detected following a screening programme. An equal number of age- and sex-matched controls from the same villages were enlisted. All cases were interviewed at the Oral Medicine Clinic and the controls at extended clinics in the field. Interviewing and coding is complete, and analysis is underway.

TYPE: Case-Control
TERM: Betel (Chewing); Premalignant Lesion; Tobacco (Chewing); Tobacco (Smoking)
SITE: Oral Cavity
TIME: 1982 - 1992

SWEDEN

ÖRNSKÖLDSVIK

679 Rutegård, J.N. 04731
Örnsköldsvik Hosp., Dept. of Surgery, 891 89 Örnsköldsvik, Sweden (Tel.: +46 660 89000)
COLL: Stenling, R.; Roos, G.; Jonsson, B.

Ulcerative Colitis, DNA Abnormalities and Colorectal Cancer Risk
In an unselected population of patients with ulcerative colitis from a defined catchment area colorectal cancer risk and the reliability of a colonoscopic surveillance programme will be evaluated, and the frequency of DNA aneuploidy and its association with dysplasia and cancer studied. All patients (n = 127) from the catchment area for the period 1961-1983 are included. Surviving patients who have not been operated have been enrolled into a colonoscopic surveillance programme since 1979. DNA analysis has been performed since 1984. The programme now comprises about 110 patients. Patients are examined regularly and biopsy specimens are collected from the entire colon. Clinical data are recorded and the specimens examined histologically and by flow cytometric DNA analysis. Extra specimens for other cancer markers are also obtained.

TYPE: Cohort
TERM: BMB; DNA; Screening; Tissue; Ulcerative Colitis
SITE: Colon; Rectum
REGI: Sweden (Swe)
TIME: 1979 - 1999

GOTHENBURG

680 Bengtsson, C.B. 04557
Dept. of Primary Health Care, Redbergsv. 6, 416 65 Gothenburg, Sweden (Tel.: +46 31 840170)

Cohort Study of Women in Gothenburg, Sweden
About 1,500 women were studied in a cross-sectional study in 1968-69. Due to the methods of data collection and a high participation rate, over 90%, the participants in the study are representative of women in Gothenburg of the ages studied (initially 38-60 years). The women were studied again in 1974-75 and 1980-81. Analyses of data with respect to cancer mainly refer to the 12-year follow-up study between 1968-69 and 1980-81. A 19-year follwo-up with respect to cancer morbidity and mortality has been carried out. Premorbid characteristics (observations in 1968-69) are related to incidence of cancer during the following 12 and 19 year periods. It is planned to carry out a new cross-sectional study and thus follow-up to 1992-93 (giving a 24-year follow-up). Publications in relation to this study have appeared in Acta Med. Scand. 193:311-318, 1973 and Scan. J. Soc. Med. 6:49-54, 1978 and 17:141-145, 1989.

TYPE: Cohort
TERM: BMB
SITE: All Sites
TIME: 1968 - 1993

681 Granberg, S.B.O. 04604
Univ. of Gothenburg, Sahlgrens Hosp., Dept. of Oncology & Gynaecology, 41345 Gothenburg, Sweden (Tel.: +46 31 601000)
COLL: Köpf, I.

Heteromorphism of Heterochromatic Segment of Chromosome 1 in 25 Families with Several Cases of Ovarian Cancer
During the last decade, evidence has been forthcoming in support of the correlation between heteromorphism of human chromosome 1 q^h. and the incidence of various malignancies, including ovary. The aim of this study is to investigate the degree of variability of the C- band regions of chromosome 1 in human karyotypes and to determine the incidence of heteromorphism in patients with ovarian cancer and their close relatives. 25 families with two or more cases of ovarian cancer are investigated. 192 close family members are included. Peripheral blood samples will be taken for cytogenetic investigations.

SWEDEN

TYPE: Cross-Sectional; Genetic Epidemiology
TERM: Chromosome Effects; Genetic Factors; Heredity
SITE: Ovary
TIME: 1986 - 1992

682 Larsson, S.   03341
Sahlgren Hosp., Dept. of Lung Medicine, Box 17301, 402 64 Gothenburg, Sweden (Tel.: +46 31 840040)
COLL: Järvholm, B.; Sörensen, S.; Flodin, U.; Edling, C.

Attributable Risk of Environmental Factors for Lung Cancer in a Swedish City
The scope of this case-control study is to estimate the attributable risk of some known lung carcinogens, such as smoking and asbestos exposure. 147 cases of lung cancer have been interviewed by questionnaire and compared with 230 controls (109 population controls and 121 hospital controls). Preliminary results showed that the aetiological fraction for smoking/ex-smoking was 95% for men and 78% for women. Male cases were more often asbestos-exposed than controls and there was a tendency towards a dose-response relationship. Risk estimates for asbestos exposure were, however, not significantly increased. The aetiological fraction for asbestos exposure among men was 16%. Few cases and controls stated exposure to other known lung carcinogens.

TYPE: Case-Control
TERM: Dusts; Environmental Factors; Tobacco (Smoking)
SITE: Lung
CHEM: Asbestos
TIME: 1983 - 1992

683 Torén, K.Ö.   04100
Univ. of Gothenburg, Sahlgren Hosp., Dept. of Occupational Medicine, St Sigfridsgatan 85, 412 66 Gothenburg, Sweden (Tel.: +46 31 830615)
COLL: Järvholm, B.; Sällsten, G.

Mortality and Morbidity in Paper Mill Workers in Sweden
Studies from Sweden and the US have indicated that paper and pulp mill workers have an increased mortality of cancer, mainly stomach cancer and cancers of the blood and lymphatic system, but also an increased mortality from chronic obstructive pulmonary diseases and asthma. All these studies have been done with case referent or PMR techniques. With the aim of evaluating mortality among the workers from two Swedish paper mills a case referent study, including about 1,000 subjects has been carried out. The diseases under study were stomach cancers and respiratory diseases, including lung and pleural cancers. Data about the diseases were collected from the local registers of death and burial, the exposures are estimated according to the personal files in the paper mills. (Am. J. Ind. Med. 1991). The health risk among workers in a soft paper mill, mainly with regard to exposure of paper dust, is also being evaluated in an exposed cohort with an unexposed referent cohort. The diseases under study are all forms of cancer and diseases of the respiratory tract. The study includes about 1,000 paper mill workers and 1,500 unexposed controls. These data are collected from the Cancer Registry, from the National Mortality Registry and from a mailed questionnaire. A paper has been published in Am. J. Ind. Med. 19:729-737, 1991.

TYPE: Cohort
TERM: Chemical Exposure; Dusts; Occupation; Registry
SITE: Respiratory; Stomach
OCCU: Paper (and Pulp) Workers
REGI: Sweden (Swe)
TIME: 1986 - 1992

* 684 Torén, K.Ö.   05176
Univ. of Gothenburg, Sahlgren Hosp., Dept. of Occupational Medicine, St Sigfridsgatan 85, 412 66 Gothenburg, Sweden (Tel.: +46 31 830615)
COLL: Hagberg, S.; Nilsson, T.; Persson, B.; Westberg, H.; Wingren, G.B.

Cancer Incidence and Mortality among Workers in the Swedish Pulp Industry
The aim of the study is to investigate if pulp mill workers have an increased risk of cancer, and lymphatic systems. The study is designed as a cohort study conducted in the south of Sweden on sulphate mills and as a case-referent study conducted in the north of Sweden on sulphate and mechanical pulp mills. Cohort members will be identified from the personnel files of the pulp mills. Mortality and cancer

SWEDEN

morbidity will be collected from the National Cancer Registry and the National Mortality Registry. The subjects in the case-referent study will be collected from the Local Registers of Death and Burial. The exposures in both studies will be estimated according to the personnel files in the pulp mills. The size of the cohort study will be about 50,000 person years. The case-referent study will include about 100 cases of lung and stomach cancers and about 30 cases of tumours from the blood and lymphatic system.

TYPE: Case-Control; Cohort
TERM: Dusts; Occupation; Registry; Wood
SITE: Haemopoietic; Lung; Lymphoma; Stomach
CHEM: Sulphur Dioxide; Terpenes
OCCU: Paper (and Pulp) Workers
REGI: Sweden (Swe)
TIME: 1990 - 1993

HUDDINGE

685 Bistoletti, P. 04898
Huddinge Univ. Hosp., Karolinska Inst., Dept. of Obstetrics and Gynaecology, 141 86 Huddinge, Sweden (Tel.: +46 8 7461000)
COLL: Dillner, L.; Dillner, J.; Elfgren, K.

Serum Antibodies to Synthetic E2, E4, E7 Peptides from HPV 16
The primary objective of this investigation is to study HPV antibody responses in patients with cervical neoplasia, in pregnant women and in normal controls. The serum antibody levels after treatment will be measured. We will also investigate if any HPV antibody of the IgG group can serve as a tumour marker in cervical cancer. This is a case-control study, with several samples taken from the same patient. It is expected to study 100 cases and 100 controls. Information on sexually transmitted diseases will be obtained.

TYPE: Case-Control
TERM: HPV; Sero-Epidemiology; Sexually Transmitted Diseases; Tumour Markers
SITE: Uterus (Cervix)
REGI: Sweden (Swe)
TIME: 1989 - 1995

LINKÖPING

686 Arbman, G. 04161
University Hospital, Dept. of Surgery, 581 85 Linköping, Sweden (Tel.: +46 13 191000)
COLL: Axelson, O.; Sjödahl, R.; Nilsson, E.; Fredriksson, M.; Eriksson, A.B.

Occupational Risk Factors for Colon Cancer
In view of findings in earlier studies of colonic cancer in Sweden, this study has been set up to study the risk of asbestos exposure, various chemicals and anti-hypertensive drugs. Cases of colonic cancer will be collected from patient files of the hospitals in the county of Östergöland. One set of referents will be drawn from surgical patients excluding patients with malignant diseases. Another set of referents will be drawn from the population register of the area. About 200 cases and two referent groups of individuals will be collected and included in the material. Information about occupational exposure to a number of agents, anti-hypertensive drugs, diet, smoking and heridity will be obtained through questionnaires.

TYPE: Case-Control
TERM: Chemical Exposure; Diet; Drugs; Heredity; Occupation; Tobacco (Smoking)
SITE: Colon; Rectum
TIME: 1986 - 1992

687 Axelson, O. 04419
University Hospital, Dept. of Occupational Medicine, 581 85 Linköping, Sweden (Tel.: (13)191441/2)
COLL: Noorlind Brage, H.

Health Risks associated with Expoxy-Handling
Although exposure to expoxy compounds is very common, little information is available about the potentially adverse health effects to humans. A cohort study is being undertaken, involving about 2,400

SWEDEN

individuals employed for some time between 1964-1988 at an industry in southern Sweden, where electrical components are founded in epoxy plastics. The intention is to study mortality and cancer incidence among the workers in comparison with national and regional death rates as well as incidence rates. Information on occupation, exposure, etc. will be based on company records.

TYPE: Cohort
TERM: Occupation; Plastics
SITE: All Sites
CHEM: Epoxy Resins
OCCU: Plastics Workers
TIME: 1989 - 1992

688   Wingren, G.B.   04805
University Hospital, Dept. of Occupational Medicine, 581 85 Linköping, Sweden (Tel.: +46 13 221459)
COLL: Noorlind Brage, H.; Hatschek, T.; Carstensen, J.M.; Axelson, O.

Aetiology of Thyroid Cancer

The aim of this study is to find possible aetiological factors of relevance for the development of thyroid cancer. The study is a case-referent study based on questionnaires. Cases are living thyroid cancer patients diagnosed with thyroid cancer some time between 1977-1987 and living in the catchment area of the University Hospital in Linköping. In total, the number of cases are 248 all in the age range 20-60 at time of diagnosis. The cases were selected from the local tumour register at the Oncology Department. 496 referents were selected in a random manner from the same region and period and being of the same age group. The questionnaire, containing questions about occupational exposures, medical history, smoking habits and some dietary habits, was sent to cases and referents and the responses will be coded and analysed using conventional case-referent techniques.

TYPE: Case-Control
TERM: Diet; Occupation; Tobacco (Smoking)
SITE: Thyroid
TIME: 1989 - 1992

LUND

689   Albin, M.P.   04814
Lund University, Dept. of Occupational, & Environmental Medicine, 221 85 Lund, Sweden (Tel.: +46 46 173185; Fax: 143702 ; Tlx: 32764 scanlu)
COLL: Jakobsson, K.; Attewell, R.; Welinder, H.; Johansson, L.G.

Mortality and Cancer Morbidity in Cohorts of Asbestos-Cement Workers

Total and cause-specific mortality and cancer morbidity are being assessed in a cohort of 1,929 male asbestos cement workers with an estimated median exposure of 1.2 fibres per ml. A local cohort of 1,233 industrial workers not exposed to asbesos is used for comparison. A nested case-control study will be carried out using referents from the exposed cohort. Dose-response relationships for lung cancer, mesothelioma, stomach cancer and colo-rectal cancer will be analysed. Data from the follow-up until 1986 are being analysed.

TYPE: Case-Control; Cohort
TERM: Dose-Response; Dusts; Occupation; Registry
SITE: Colon; Lung; Mesothelioma; Rectum; Stomach
CHEM: Asbestos
OCCU: Asbestos Workers
REGI: Lund (Swe)
TIME: 1982 - 1992

* 690   Hagmar, L.E.   05057
Univ. of Lund Hosp., Dept. of Occupational, and Environmental Medicine, 221 85 Lund, Sweden (Tel.: +46 46173173; Fax: 46143702; Tlx: 32764 SCANLU)
COLL: Svensson, B.G.; Möller, T.R.

SWEDEN

Cancer Incidence in Cohorts of Swedish Fishermen
In Sweden, the dominant route for human exposure to persistent organic chlorinated substances such as PCBs and polychlorinated dibenzo-p-dioxins and furans, is consumption of fat fish. Fishermen have a high consumption of fish. Cancer incidence among Swedish fishermen, their wives and children, will be compared to that of regional reference populations. A cohort of about 15,000 Swedish fishermen has been established. Interviews will be performed on dietary habits, smoking habits and alcohol consumption, in random samples from the cohorts and from the regional reference populations. Furthermore, blood samples will be drawn for analysis of n-3 polyunsaturated fatty acids, polychlorinated dibenzo-p-dioxins and furans and PCBs. Nested case-referent studies, considering dietary habits, will be performed for certain cancers within the cohorts.

TYPE: Case-Control; Cohort
TERM: Alcohol; BMB; Blood; Diet; Occupation; Registry; Tobacco (Smoking)
SITE: All Sites
CHEM: Dioxins; Furans; PCB
OCCU: Fishermen
REGI: Sweden (Swe)
TIME: 1988 - 1994

* 691 Hagmar, L.E. 05058
Univ. of Lund Hosp., Dept. of Occupational, and Environmental Medicine, 221 85 Lund, Sweden (Tel.: +46 46173173; Fax: 46143702; Tlx: 32764 SCANLU)
COLL: Gerhardsson, L.; Skerfving, S.; Schütz, A.

Cancer Incidence among Lead-Exposed Smelter Workers
This study aims to test the hypothesis of the carcinogenic potential of lead. A cohort of 848 lead-exposed workers from a secondary lead smelter, employed from 1942 and onwards is established. Since 1968, blood lead is analysed several times a year for each employee. These data will be used in dose-response calculations.

TYPE: Cohort
TERM: BMB; Blood; Dose-Response; Metals; Occupation; Registry
SITE: All Sites
CHEM: Lead
OCCU: Smelters, Lead
REGI: Lund (Swe)
TIME: 1991 - 1993

* 692 Hagmar, L.E. 05059
Univ. of Lund Hosp., Dept. of Occupational, and Environmental Medicine, 221 85 Lund, Sweden (Tel.: +46 46173173; Fax: 46143702; Tlx: 32764 SCANLU)
COLL: Welinder, H.

Cancer Incidence in a Cohort of Workers from the Polyurethane Foam Manufacturing Industry
The aim of the study is to determine whether workers exposed to toluene diisocyanate (TDI) or 4,4'-diphenylmethane diisocyanate (MDI) have increased cancer incidence. TDI is carcinogenic in rodents and both MDI and TDI are mutagenic. These substances are widely used in industry, but no data are available on their carcinogenicity to humans. A cohort of about 8,000 workers from 11 Swedish polyurethane foam manufacturing plants, employed from 1960 onwards, has been identified. Using at least 12 months' employment period for inclusion in the cohort, about 3,500 will be excluded. The final cohort size is expected to be about 4,500. Case-referent studies within the cohort, using individual cumulated exposure estimates, will be performed.

SWEDEN

TYPE: Case-Control; Cohort
TERM: Occupation; Plastics; Registry
SITE: All Sites
CHEM: Diisocyanates
OCCU: Plastics Workers
REGI: Sweden (Swe)
TIME: 1990 - 1993

* 693 **Hagmar, L.E.** 05060
Univ. of Lund Hosp., Dept. of Occupational, and Environmental Medicine, 221 85 Lund, Sweden
(Tel.: +46 46173173; Fax: 46143702; Tlx: 32764 SCANLU)
COLL: Welinder, H.

Cancer Incidence in a Cohort of Workers Exposed to Ethylene Oxide

This is a cohort study of cancer incidence among 2,170 ethylene oxide-exposed workers, employed for at least one year since 1972, in two plants producing disposable medical equipment. The aim of the study is to obtain further epidemiological data on the postulated relationship of ethylene oxide exposure to the development of leukaemia among humans. Individual cumulative exposure to ethylene oxide is calculated and used for analysing exposure-related associations.

TYPE: Cohort
TERM: Dose-Response; Occupation; Registry
SITE: All Sites; Leukaemia; Lymphoma
CHEM: Ethylene Oxide
REGI: Sweden (Swe)
TIME: 1988 - 1995

* 694 **Hagmar, L.E.** 05061
Univ. of Lund Hosp., Dept. of Occupational, and Environmental Medicine, 221 85 Lund, Sweden
(Tel.: +46 46173173; Fax: 46143702; Tlx: 32764 SCANLU)
COLL: Brögger, A.; Hansteen, I.L.; Heim, S.; Högstedt, B.; Knudsen, L.; Lambert, B.; Linnainmaa, K.; Mitelman, F.; Nordenson, I.; Reuterwall, C.; Salomaa, S.; Skerfving, S.; Sorsa, M.I.

Inter-Nordic Prospective Study on Cytogenetic Endpoints and Cancer Risk

To investigate whether high rates of chromosomal aberrations, SCE or micronuclei in peripheral lymphocytes indicate an increased risk for subsequent cancer, a prospective cohort study of 2,969 subjects cytogenetically examined between 1970 and 1988 in any of eight Nordic laboratories, was initiated. To reduce the effects of inter-laboratory variation, the results of the three cytogenetic endpoints were trichotomized for each laboratory (low, medium, high). In the first follow-up there was a positive association between level of chromosome aberrations and subsequent cancer risk, which almost reached statistical significance. However, the follow-up period used is still too short to allow firm conclusions. Further follow-ups will be performed every five years.

TYPE: Cohort; Molecular Epidemiology
TERM: Chromosome Effects; Lymphocytes; Micronuclei; SCE
SITE: All Sites
LOCA: Finland; Norway; Sweden
TIME: 1988 - 2005

* 695 **Hagmar, L.E.** 05062
Univ. of Lund Hosp., Dept. of Occupational, and Environmental Medicine, 221 85 Lund, Sweden
(Tel.: +46 46173173; Fax: 46143702; Tlx: 32764 SCANLU)
COLL: Åkesson, B.; Möller, T.R.

Cancer Incidence in PVC-Processing Workers Exposed to Low Levels of Vinyl Chloride Monomer, Asbestos, and Plasticizers

A significant increase in total cancer incidence, and especially in respiratory cancers, was found in a cohort of 2,031 male workers at a PVC processing plant, employed for at least three months during the period 1945-1980 (Am. J. Ind. Med. 17:553-565, 1990). However, no significant exposure-response associations between exposure estimates for PVC, asbestos, and plasticizers and cancer incidence was found. In 1993, when five more years of observation have been added, a follow-up of the cohort will be performed.

SWEDEN

TYPE: Cohort
TERM: Dusts; Occupation; Plastics; Registry
SITE: All Sites
CHEM: Asbestos; PVC; Vinyl Chloride
OCCU: Plastics Workers
REGI: Sweden (Swe)
TIME: 1986 - 1993

* 696 Hagmar, L.E. 05063
Univ. of Lund Hosp., Dept. of Occupational, and Environmental Medicine, 221 85 Lund, Sweden
(Tel.: +46 46173173; Fax: 46143702; Tlx: 32764 SCANLU)
COLL: Schütz, A.

Cancer Incidence among Leather Tanners in Three Swedish Factories

The general aim of this study is to investigate the association between cancer mortality and exposure to chemical substances in tannery factories. The study includes a cohort of about 3,500 workers from three Swedish tanneries. The cohort analysis will be completed with a nested case-referent study within the cohort, evaluating specific chemical exposures for cases and referents.

TYPE: Case-Control; Cohort
TERM: Chemical Exposure; Leather; Occupation; Pesticides; Plastics
SITE: All Sites
CHEM: Chlorophenols; Chromium; Formaldehyde
OCCU: Leather Workers
TIME: 1990 - 1993

* 697 Hagmar, L.E. 05064
Univ. of Lund Hosp., Dept. of Occupational, and Environmental Medicine, 221 85 Lund, Sweden
(Tel.: +46 46173173; Fax: 46143702; Tlx: 32764 SCANLU)
COLL: Bellander, T.; Möller, T.R.

Cancer Morbidity in Nitrate Fertilizer Workers

In a cohort of 2,131 workers employed in a fertilizer manufacturing factory for at least three months from 1963 (when nitrate fertilizers were first manufactured) to 1985, no increased incidence of stomach cancer or lung cancer was found. The cohort will now be augmented with about 3,200 workers from the other fertilizer manufacturing plant in Sweden. In this plant, nitrate fertilizers have been manufactured since the 1930's.

TYPE: Cohort
TERM: Fertilizers; Occupation; Registry
SITE: Lung; Stomach
CHEM: Nitrates
OCCU: Fertilizer Workers
REGI: Sweden (Swe)
TIME: 1986 - 1992

* 698 Hagmar, L.E. 05065
Univ. of Lund Hosp., Dept. of Occupational, and Environmental Medicine, 221 85 Lund, Sweden
(Tel.: +46 46173173; Fax: 46143702; Tlx: 32764 SCANLU)

Cancer Incidence in a Cohort of Piperazine-Exposed Workers in a Chemical Plant

It has previously been shown that occupational exposure to piperazine dust causes endogenous formation of N-mononitrosopiperazine which is a possible carcinogen (Int. Arch. Occup. Environ. Health 60:25-29, 1988). A previous follow-up of a cohort of 664 workers, mainly exposed to piperazine, and employed during the period 1942-1979, showed a significant increase of malignant lymphomas and myelomas (Scand. J. Work Environ. Health 12:545-551, 1986). A nested case-referent study within the cohort indicated a possible association with piperazine, although the observed numbers were too few to allow any firm conclusions. A further follow-up, including six more years of observation, will now be performed.

SWEDEN

TYPE: Case-Control; Cohort
TERM: Chemical Exposure; Drugs; Dusts; Occupation; Plastics; Registry
SITE: All Sites; Lymphoma; Multiple Myeloma
CHEM: Ethylene Oxide; Piperazine; Urethane
OCCU: Chemical Industry Workers
REGI: Sweden (Swe)
TIME: 1982 - 1992

699 Johansson, L.G. 04835
University Hosp., Dept. of Pathology, 221 85 Lund, Sweden (Tel.: +46 46 173410; Fax: 143307)
COLL: Albin, M.P.

Histological Type of Lung Carcinoma in Asbestos Cement Workers
Several studies have indicated that individuals with lung carcinoma who have been exposed to asbestos have a higher frequency of adenocarcinoma than among lung cancers arising in the general population. In a cohort of 3,144 asbestos-cement workers there were 29 cases of lung carcinoma with preserved histopathological material which were provided with three controls each, matched for type of histopathological sample, age (+/- 5 years), year of death and sex. Smoking habits were recorded. The tumours were classified according to the 1982 revised WHO classification of lung tumours, and peripheral or central location, side and lobe of origin were also recorded. The aim of the present study is to explore whether asbestos-exposed subjects have an increased incidence of adenocarcinoma compared to controls and if there is a dose-response relationship between asbestos dose and adenocarcinoma. Preliminary results indicate that asbestos-exposed individuals do have a higher frequency of adenocarcinoma.

TYPE: Case-Control
TERM: Dose-Response; Histology; Tobacco (Smoking)
SITE: Lung
TIME: 1986 - 1992

700 Mitelman, F. 02282
University Hospital, Dept. of Clinical Genetics, 221 85 Lund, Sweden (Tel.: +46 46 173360; Tlx: +46 46 173226)

Registry of Chromosome Aberrations in Cancer
Chromosome aberrations in human cancer and leukaemia are non-random and tend to cluster to a limited number of specific chromosomes. Experimental evidence indicates that the chromosomal pattern of malignant cells may be related to aetiological factors. The objective of this study is to gain information on this aspect of human cancer by relating the chromosomal pattern to: (1) exposure to known environmental carcinogens, (2) exposure to potential carcinogenic agents, and (3) geographical regions. The material is collected from three main sources: (1) published cases obtained by computer-based literature scans, (2) unpublished cases from this laboratory, and (3) unpublished cases kindly communicated by numerous colleagues all over the world. By the end of 1990, well above 14,000 neoplasms with chromosome aberrations identified by banding techniques studied in Africa, Asia, Europe, the USA and Japan were included in the registry. The data, including information on histopathological diagnosis, sex, age, previous neoplasms or other significant disorders, hereditary disorders, exposure to potential mutagenic and/or carcinogenic agents, chromosome preparation technique, and clinical findings, have been computerized. The total organized material has been published in "Catalog of Chromosome Aberrations in Cancer" (4th edition), Wiley-Liss, Inc., New York, 1991.

TYPE: Molecular Epidemiology
TERM: Chromosome Effects; Environmental Factors; Genetic Markers; Geographic Factors; Mutagen; Mutation, Somatic
SITE: All Sites
TIME: 1975 - 1993

SOLNA

701 Hogstedt, C. 03522
National Inst. Occupational Health, Div. of Occupational Medicine, Ekelundsvagen 16 , 171 84 Solna, Sweden (Tel.: +46 8 7309342; Tlx: 7309000)
COLL: Westerlund, B.

SWEDEN

Cancer Incidence and Mortality among Lumberjacks with and without Exposure to Chlorinated Phenoxy Acid Mixtures
Payrolls for a forestry company in the middle of Sweden have been examined to identify foremen and lumberjacks who had been exposed for at least six days to mixtures of chlorinated phenoxy acids (2,4-D; 2, 4,5-T) and DDT. 141 exposed men with an average exposure of 30 days and 16 foremen with an average exposure of 176 days were identified, as well as 235 unexposed lumberjacks. The national incidence rates are being used to calculate gender, calendar year and five-year age-class specific expected numbers. The first report was published in 1979 and showed a significant increase of cancer among the foremen. The cohorts are now being followed for the period 1979-1985.

TYPE: Cohort
TERM: Chemical Exposure; Herbicides; Occupation; Pesticides
SITE: All Sites
CHEM: 2,4,5-T; 2,4-D; Phenoxy Acids; Silvex
OCCU: Forest Workers
TIME: 1984 - 1992

702 Reuterwall, C. 04730
National Inst. of Occupational Health, Dept. of Occupational Medicine, 171 84 Solna, Sweden (Tel.: +46 8 7309649; Fax: 7309860; Tlx: 15816 ARBSKY S)
COLL: Gustavsson, P.; Bellander, T.

Cancer Incidence and Mortality among Swedish Laboratory Technicians
Laboratory work within medical research may be associated with exposure to several physical and chemical agents with documented mutagenic or carcinogenic effects. Cancer incidence and mortality are being studied in a cohort of laboratory technicians employed at medical research laboratories (approx. 1,500 subjects) and clinical laboratories (approx. 1,000 subjects). Basic exposure information is collected from personnel records and vital status is determined from computerized population registries. Data on underlying causes of death are obtained from Statistics Sweden, data on cancer incidence from the Swedish Cancer Registry. Expected numbers of deaths and cancers are standardized for geographical region, sex, age group, and calendar year. This professional group is still rather young, and the first follow-up can only reveal relatively large increases of risk; later follow-ups, however, will have better statistical power. Depending on the results from the follow-up, an attempt to identify single risk factors will be made in a case-referent study, nested within the cohort. Technicians working at other laboratories might also be included in future studies.

TYPE: Cohort
TERM: Chemical Exposure; Drugs; Occupation; Radiation, Ionizing; Solvents
SITE: All Sites
OCCU: Laboratory Workers
REGI: Sweden (Swe)
TIME: 1989 - 1992

STOCKHOLM

703 Ahlbom, A.N. 03705
Nat. Inst. of Environmental Medicine, Dept. of Epidemiology, Doktorsringen 18, Box 60208, 104 01 Stockholm, Sweden (Tel.: +46 8 7287470/6400; Fax: 313961)
COLL: Rodvall, Y.; Spännare, B.

Brain Tumours in Adults
The aetiology of different histopathological forms of brain tumours is basically unknown. Case reports, animal experiments and epidemiological studies, however, do support the theory that external factors are of importance. The objective of the present study is, within the framework of an international collaborative project (SEARCH), coordinated by the IARC, for each of the major histopathological types of brain tumours (1) to evaluate the importance of N-nitroso compounds; (2) to quantify the effects of certain well known risk-factors, and (3) to investigate the association with several other personal characteristics and exposures. Cases are be all patients in the area covered by the Department of Neurosurgery at the Uppsala University Hospital with malignant primary brain tumour or meningioma during the period 1987 to 1989 between the ages of 25 and 74. 350 cases are be included in the study. One control is selected from the general population of the catchment area for each case. The data collection was performed by use of a questionnaire and subsequent telephone interviews. The analysis of data started in March 1991.

SWEDEN

TYPE: Case-Control
TERM: Chemical Exposure; Environmental Factors; Lifestyle
SITE: Brain
CHEM: N-Nitroso Compounds
TIME: 1985 - 1992

704   Ahlbom, A.N.   04226
Nat. Inst. of Environmental Medicine, Dept. of Epidemiology, Doktorsringen 18, Box 60208, 104 01 Stockholm, Sweden (Tel.: +46 8 7287470/6400; Fax: 313961)
COLL: Feychting, M.

Extremely Low Frequency Electromagnetic Fields and Cancer Risk
Several independent groups have reviewed the epidemiological evidence of an association between exposure to 50 Hz electromagnetic fields and cancer and have concluded that while existing data do not warrant firm conclusions, the hypothesis of possible cancer risk should be followed up. The present study takes advantage of the Swedish census system to define a study base of those people who have lived within 300 metres of any of the 250 and 400 kV power lines in Sweden during 1960-1985. Within this study population a case-control study is being done on all cases of childhood cancer, leukaemia and brain tumours in adults, identified by record linkage to the Swedish Cancer Registry. Exposure is assessed in three different ways: (1) by direct measurements, (2) by estimation of the fields generated by the power lines, and (3) by use of distance between the power line and the homes in the study as an indirect exposure measure. Controls will be matched for sex, age, place and period of residence, and selected at random from the same study population.

TYPE: Case-Control
TERM: Electromagnetic Fields; Environmental Factors; Registry
SITE: Brain; Childhood Neoplasms; Leukaemia
REGI: Sweden (Swe)
TIME: 1987 - 1992

* 705   Ahlbom, A.N.   05042
Nat. Inst. of Environmental Medicine, Dept. of Epidemiology, Doktorsringen 18, Box 60208, 104 01 Stockholm, Sweden (Tel.: +46 8 7287470/6400; Fax: 313961)
COLL: Sasco, A.J.; Wennborg, H.

Cancer Risk in Biological Research Laboratories in Sweden
The aim of the study is to describe cancer incidence and cancer mortality among biological research laboratory workers in Sweden during the last 20 years. Another aim is to make comparisons between different laboratory occupations and, if possible, to connect any excess risk with specific kinds of laboratory work or exposure. The study will be conducted as a retrospective cohort study and will be included in an international multi-center study coordinated by IARC. Biomedical and biological research groups at seven Swedish research institutions will be included in the study. Comparisons will be made with the general population and with research institutions without biological activity. Personal data will be collected by questionnaire. Exposure assessment will be done at the research group level by interviews and questionnaires. The National Swedish Cancer Register will be used to identify the cases.

TYPE: Cohort
TERM: Occupation; Registry
SITE: All Sites
OCCU: Laboratory Workers
REGI: Sweden (Swe)
TIME: 1991 - 1994

706   Bolund, C.   04293
Karolinska Hospital, Radiumhemmet, Department of Psychosomatic Oncology, 104 01 Stockholm, Sweden
COLL: Lindefors, B.M.

Cancer Incidence after Hospitalization Due to Psychiatric Disease
The aim is to investigate the incidence of malignant disease in a population treated for psychiatric disorders. The current debate on the role of psychic factors in cancer aetiology does not take into account the mostly negative correlation between mental illness and cancer found in studies from the first part of the century, based on populations continually cared for in mental hospitals. No data have appeared on cancer morbidity among persons affected by mental/psychic disturbances of different

SWEDEN

types. The national Swedish registries of malignant diseases and in-patient psychiatric care offer a unique opportunity to assess the incidence of malignant disease among psychiatric patients (total; acute or chronic; and divided into groups according to psychiatric diagnosis). Authorization has been acquired to get entry into the computer based registries. The registry of patients diagnosed 1969-1979 and released from hospital and those remaining in hospital care during 1979 will comprise a population of about 175,000 individuals. The study will cover more than 1,000,000 person-years and is expected to yield about 6,000 cancer cases.

TYPE: Cohort; Correlation
TERM: Psychological Factors; Registry
SITE: All Sites
REGI: Sweden (Swe)
TIME: 1986 - 1992

707 Eklund, G. 04086
Karolinska Hosp., Radiumhemmet, Dept. of Cancer Epidemiology, 104 01 Stockholm, Sweden (Tel.: +46 8 7363036)
COLL: Tomenius, L.

Power Lines, 50-Hz Electromagnetic Fields and Childhood Tumours in Sweden

Wertheimer and Leeper published in the Am. J. Epidemiol. 109:273-284, 1979 an epidemiological study from the greater Denver area in Colorado, USA, that showed an excess of electrical wiring configurations suggestive of high current flow near the homes of children who had developed cancer. They proposed that the magnetic field from current in the power lines may directly or indirectly affect cancer incidence. In order to see if the relationship could be found in Sweden a similar study was undertaken in the county of Stockholm. The result showed (Tomenius, Bioelectromagnetics 7:191, 1986) that 200 kV wires and increased 50-Hz magnetic fields, respectively, were more frequent among dwellings of cases than of controls. In order to confirm the results from the statistical point of view and to be able to determine if there is a dose/response relationship between electromagnetic environment and the incidence of childhood tumours a more expanded study, including cases from more counties in Sweden, has begun. Visible electrical constructions will be registered and the 50-Hz magnetic field measured at the birth dwelling of tumour cases reported 1958-1980 to the Swedish Cancer Registry for individuals 0-18 years of age. For each tumour case two individuals matched according to age, sex and church district of birth will be used as controls. To begin with this will be accomplished in seven counties with 1,476 malignant tumours. If necessary more counties in Sweden will later be included. The municipalities where the dwellings are located will be asked to carry out the observations and measurements of the magnetic fields.

TYPE: Case-Control
TERM: Electromagnetic Fields; Environmental Factors; Registry
SITE: Childhood Neoplasms
REGI: Sweden (Swe)
TIME: 1982 - 1992

708 Eklund, G. 04299
Karolinska Hosp., Radiumhemmet, Dept. of Cancer Epidemiology, 104 01 Stockholm, Sweden (Tel.: +46 8 7363036)
COLL: Lindefors, B.M.; Meirik, O.; Rutqvist, L.E.

Relationship between Legal Abortions and Breast Cancer

Several retrospective epidemiological studies have claimed that early abortions in young women are followed by an increased frequency of breast cancer. This observed increased risk is contrasted to the decreased risk observed in association with a full term pregnancy at the same ages. The intention is to use the unique registrations of every legal abortion performed in Sweden for many years. By linkage of an appropriate part of the abortion register and the Swedish cancer register of women with breast cancer, it is expected that a number of cases great enough to indicate an increased frequency of at least 25% will be obtained. The applications for legal abortion from 1966 to 1974, which amount to 165,000, will be used. Women with breast cancer in the abortion cohort will be compared with women belonging to three different cohorts.

SWEDEN

TYPE: Cohort
TERM: Abortion; Registry
SITE: Breast (F)
REGI: Sweden (Swe)
TIME: 1985 – 1992

709 Gerhardsson, M.R. 04172
National Inst., of Environmental Medicine, Dept. of Epidemiology, Doktorsringen 18, Box 60208, 104 01 Stockholm, Sweden (Tel.: +46 8 236900; Tlx: 19530 KIENVS)
COLL: Norell, S.E.; Bylin, G.

Allergy-Related Diseases and Risk of Cancer

Several studies have suggested that allergy is negatively associated with cancer, some have reported a positive association and others have shown no association. The material has often been too small to make a division into cancer subsites. In this 19-year follow-up study of several 100,000 Swedish subjects, the cancer incidence for subjects with allergy-related diseases will be compared with a reference group. The information will be obtained from the in-patient care register, the cause-of-death register and the cancer register. The observation period will be from 1964-83.

TYPE: Cohort
TERM: Allergy; Registry
SITE: All Sites
REGI: Sweden (Swe)
TIME: 1987 – 1992

710 Gustavsson, P. 04747
Karolinska Hosp., Dept. of Occupational Medicine, 104 01 Stockholm, Sweden (Tel.: (8)7292405; Fax: 334333)
COLL: Bellander, T.; Salmonsson, S.

Cancer Risk among Graphite Electrode Workers

Mortality and cancer incidence is being investigated among 900 employees at a Swedish company manufacturing graphite electrodes. Exposure to PAH and respirable dust is under investigation. The cumulate exposures for the workers are estimated by occupational hygienists. The mortality of the cohort is compared to that of the general population, standardising for age, sex, calendar year and geographical region (county). Dose-response relationships are being investigated by internal comparisons within the cohort.

TYPE: Cohort
TERM: Dose-Response; Dusts; Occupation
SITE: All Sites; Lung
CHEM: PAH
OCCU: Electrode Manufacturers
TIME: 1989 – 1992

711 Holm, L.E. 03332
Karolinska Hosp., Radiumhemmet, Dept. of Cancer Prevention, 104 01 Stockholm, Sweden (Tel.: (08)7362266/7363174)
COLL: Fuerst, C.J.; Lundell, M.

Malignant Tumours and Leukaemia Following Radiotherapy for Haemangiomas in Childhood

The incidence of malignant tumours and leukaemia is being analysed in approximately 20.000 patients following radiotherapy for cavernous haemangiomas. The treatment was given between 1920 and 1959 and consisted mainly of application of radium needles on the skin or of X-rays. Approximately 17% of the patients did not receive any radiotherapy. More than 90% of the patients were under 15 years of age at the time of treatment and 88% were less than five years. The great majority were six months old. Dosimetry is being performed on a phantom corresponding to a six months old child. The population of patients will be matched with the Swedish Cause of Death Registry and the Swedish Cancer Registry.

SWEDEN

TYPE: Cohort
TERM: Childhood; Radiation, Ionizing; Radiotherapy; Registry
SITE: All Sites; Leukaemia
REGI: Sweden (Swe)
TIME: 1983 - 1992

712 Lambert, B. 04544
Karolinska Hosp., Dpt. of Clinical Genetics, 104 01 Stockholm, Sweden (Tel.: (08)7295380; Fax: 327734 ; Tlx: 16660 KARO-S)
COLL: Einhorn, N.

Chromosomal Aberrations and Second Malignancy in Melphalan Treated Ovarian Carcinoma Patients
The aim of the study is to evaluate the effect of cytostatic treatment with regard to the persistence of chromosomal aberrations in peripheral T-lymphocytes and the development of second malignancy. A cohort of 50 melphalan-treated ovarian carcinoma patients has been studied for more than ten years after cessation of chemotherapy. Repeated samples for cytogenetic investigations have been taken, and the cohort is regularly followed up with regard to the occurrence of second malignancy. A paper has been published in Mutat. Res. 210:353-358, 1989.

TYPE: Cohort
TERM: Chemotherapy; Chromosome Effects; Drugs; Lymphocytes
SITE: All Sites
CHEM: Melphalan
TIME: 1977 - 1992

713 Lundberg, I.S. 04329
Karolinska Hosp., Dept. of Occupational Medicine, 104 01 Stockholm, Sweden
COLL: Bellander, T.

Mortality and Cancer Incidence among Styrene Exposed Workers
Styrene is a solvent with suggested carcinogenic properties in animal experiments. Some epidemiological studies of styrene exposed workers have shown excess deaths from leukaemia and lymphoma. Workers in the manufacture of glass reinforced polyester plastics have been heavily exposed to styrene. Increased frequencies of chromosomal aberrations, sister chromatid exchanges and micronuclei have been found in peripheral lymphocytes of such workers. This study examines the cancer incidence and mortality pattern of 1,500 to 2,000 workers employed in glass reinforced plastics manufacturing from the 1950's until today. The cohort contains all workers employed in around 30 factories scattered over Sweden. The observed figures from different causes of death and cancer sites will be compared with expected figures generated from the Swedish general population.

TYPE: Cohort
TERM: Chemical Exposure; Occupation; Plastics
SITE: All Sites; Leukaemia; Lymphoma
CHEM: Styrene
OCCU: Plastics Workers
TIME: 1986 - 1993

714 Pershagen, G. 04478
Karolinska Inst., Inst. of Environmental Medicine, Box 60208, 104 01 Stockholm, Sweden (Tel.: +46 8 7287460; Fax: 313961)
COLL: Axelson, O.; Damber, L.

Domestic Radon Exposure and Lung Cancer in Sweden
Radon in dwellings is the major source of exposure in ionising radiation for the Swedish population and a few studies suggest that it is of aetiological importance for lung cancer. The objective of this study is to make a quantitative assessment of dose-response relationship. It is based on 1,500 cases lung cancer and 3,000 controls aged 35-74 from the Swedish cancer registry 1980-1984. There will be radon measurements and inspections of all houses in which study subjects lived in for two or more years between 1947 and 3 years prior to the end of follow-up. A questionnaire will also be submitted to study subjects or next-of-kin on smoking, passive smoking, occupations etc. The analysis will focus on assessment of dose-response relationship as well as on interactions between different exposures.

SWEDEN

TYPE: Case-Control
TERM: Dose-Response; Occupation; Passive Smoking; Radiation, Ionizing; Registry; Tobacco (Smoking)
SITE: Lung
CHEM: Radon
REGI: Sweden (Swe)
TIME: 1987 - 1992

715 Reizenstein, P. 04661
Karolinska Hosp., Dept. of Haematology, Solnavaegen, 104 01 Stockholm, Sweden (Tel.: (08)7294112)
COLL: Ramgren, O.; Beckman, M.

Radiation Exposure During Pregnancy and Childhood Malignancy

The purpose is to see whether the offspring of the 10% of Swedish mothers subjected to X-ray pelvimetry or the higher percentage subjected to other X-rays have an increased risk for malignancy. This study will exploit a computer register for the period 1971-1978 comprising a complete record of some 4 million X-ray examinations in a population base of approximately one million. About 13,000 pelvimetries and an equal number of routine chest X-rays in pregnant women will be matched against the birth register and the cancer register to examine the incidence of childhood malignancy in relation to pre-natal radiation exposure. A data set of more than twice this size on babies born to mothers not subjected to X-rays will serve as an unexposed comparison group. Simultaneously it is proposed to study children to fathers subjected to diagnostic X-ray.

TYPE: Cohort
TERM: Intra-Uterine Exposure; Pregnancy; Radiation, Ionizing; Registry
SITE: Childhood Neoplasms
REGI: Sweden (Swe)
TIME: 1988 - 1992

* 716 Rutqvist, L.E. 04970
Karolinska Hosp., Radiumhemmet, Dept. of Oncology, 104 01 Stockholm 60, Sweden (Tel.: +46 8 7292980; Fax: 348640)
COLL: Lewin, F.

Squamous Cancer of the Mouth, Pharynx, Larynx and Oesophagus

This is a population-based case-control study in two defined geographical regions with a population of 3 million. All cases of squamous cell cancer among males aged 40-79 years in the populations are included. A random sample of males in the general population serve as a control group. Data are collected through personal interviews (smoking and drinking habits, non-smoking tobacco use, dietary and occupational history, etc.). A total of 600 cases and 600 controls will be included.

TYPE: Case-Control
TERM: Alcohol; Diet; Occupation; Registry; Tobacco (Chewing); Tobacco (Smoking); Tobacco (Snuff)
SITE: Larynx; Oesophagus; Oral Cavity; Pharynx
REGI: Lund (Swe); Stockholm (Swe)
TIME: 1988 - 1992

717 Törnberg, S.A. 04861
Karolinska Hosp., Radiumhemmet, Dept. of Oncology, and Cancer Epidemiology, 104 01 Stockholm 60, Sweden (Tel.: +46 8 7293194; Fax: 348640)
COLL: Carstensen, J.M.; Holm, L.E.

Gynaecological Cancer and Serum Cholesterol

A cohort of 46,000 women participated in a general health screening in 1963-1965. Examinations included serum cholesterol and serum beta-lipoprotein levels as well as height and weight. The data were matched with the Swedish Cancer Register and Cause of Death Register until 1987. The purposes of the present studies are to analyse the relationships between the above mentioned parameters and subsequent risk of cancers of the ovary, endometrium and cervix uteri as well as of breast cancer survival. Earlier studies of the same cohort on breast cancer, colorectal cancer, stomach cancer and total cancer incidence have been published, e.g. in JNCI 81:1917-1922, 1989.

SWEDEN

TYPE: Cohort
TERM: Cholesterol; Physical Factors; Registry; Survival
SITE: Breast (F); Ovary; Uterus (Cervix); Uterus (Corpus)
REGI: Sweden (Swe)
TIME: 1990 - 1992

718 Wiklund, K. 04375
Karolinska Hosp., Radiumhemmet, Dept. of Cancer Epidemiology, 104 01 Stockholm, Sweden (Tel.: +46 8 7292398; Fax: 326113)
COLL: Holm, L.E.; Dich, J.; Eklund, G.

Drinking Water and Cancer

The aim is to investigate whether there is an increased risk of lung, colon, rectum and urinary bladder cancers with chlorine in the drinking water, and of stomach, urinary bladder, colon and rectum cancers with nitrate in the drinking water. Associations between potential cancer risk for all sites and different qualities of the drinking water (method of cleaning, number of coliform bacteria, conductivity, permanganate, ammonium, manganese, alkalinity, nitrates, nitrites, phosphorus, carbonic acid and siliconic acid) will also be examined. Cancer risks for different municipalities in Sweden will be compared with the drinking water quality. Information on water quality for the major supplies will be derived for the period since 1961. Cancer morbidity data will be obtained from the nationwide Swedish Cancer Registry, which covers almost all newly diagnosed cancer cases. In the analyses the municipalities will be units: migration between municipalities is known for individuals.

TYPE: Correlation
TERM: Registry; Water
SITE: All Sites; Bladder; Colon; Lung; Rectum; Stomach
CHEM: Nitrates; Nitrites
REGI: Sweden (Swe)
TIME: 1987 - 1993

UMEÅ

719 Emmelin, A. 04470
Univ. of Umeå, Dept. of Epidemiology, and Health Care Research, 901 85 Umeå, Sweden (Tel.: +46 90 101213; Fax: 138977; Tlx: 54005)
COLL: Nyström, L.; Wall, S.

Swedish Dock Workers and Lung Cancer

For several years dock workers have suspected that occupational exposures arising from loading and unloading work by truck may be carcinogenic. In 1979 The National Swedish Board of Occupational Safety initiated a retrospective cohort study of all permanently employed dockers in Sweden who had been first employed before 1974 and for a continuous period of at least six months. The present analysis of the cohort, followed from 1 January 1961 until 1 January 1981 (6,076 workers; 97,076 person-years) showed an increasing excess of lung cancer incidence and mortality. To further analyse whether these trends could be related to the exposure from diesel exhaust fumes a case-referent study within the cohort has been performed of all cases dying of primary lung cancer between 1961 and 1982. Four referents matched by age and port were selected for each case. A questionnaire about residence, work history, special exposure in the port and smoking habits was sent to alive controls and relatives of deceased cases and controls. Work history within the port and smoking habits were validated by interviews with work superintendents and/or safety delegates. Information about the use of trucks and fuel consumption was also collected to generate an exposure index.

TYPE: Case-Control
TERM: Occupation; Tobacco (Smoking)
SITE: Lung
CHEM: Diesel Exhaust
OCCU: Dockers
TIME: 1986 - 1992

720 Hardell, L.O. 03540
University Hospital, Dept. of Oncology, 901 85 Umeå, Sweden (Tel.: +46 90 101000)
COLL: Bengtsson, N.O.; Axelson, O.

SWEDEN

Risk Factors for Colon Cancer
The study includes 400 subjects with colon cancer and two matched population-based controls for each case. Only living cases are included. The cases were diagnosed in the years 1981-1983 and were at that time residents in the three most northern counties of Sweden. Exposure data are collected via self-administered questionnaires and supplemented over the phone if necessary. Occupational histories are obtained for all cases and controls. Previous diseases, intake of drugs, smoking habits, alcohol consumption and food habits are also investigated.

TYPE: Case-Control
TERM: Alcohol; Chemical Exposure; Diet; Drugs; Dusts; Occupation; Pesticides; Tobacco (Smoking)
SITE: Colon
CHEM: Asbestos; Chlorophenols; Phenoxy Acids
TIME: 1984 - 1992

721 Larsson, L.G. 01834
Regional Hosp., Dept. of Oncology, 901 85 Umeå, Sweden
COLL: Damber, L.

Male Lung Cancer in Northern Sweden
In a study of the geographical distribution of all new male lung cancer cases in the three northernmost counties in Sweden from 1959-1977 very large differences were found between different regions. Some communities with mining districts or other heavy industry had several times higher age-adjusted incidence rates than more rural communities. The occupational profile of the cancer cases is being compared with that of a control series - randomly selected through the general population register - and matched by age and place of residence. Information about occupation is obtained for both cases and controls. For about 600 deceased male lung cancer cases registered in the same region from 1972-1977, a more detailed case-control study is being performed based on questionnaires to relatives about occupation, employment and smoking habits. Collection of data was finished in June 1980. The material is now being analysed statistically. Recent reports appeared in Br. J. Cancer 53:673-681, 1986 and Br. J. Ind. Med. 44:446-453, 1987 and Acta Oncol. 26:211-215, 1987. On-going supplementary studies involve iron ore miners and pulp workers.

TYPE: Case-Control
TERM: Geographic Factors; Male; Mining; Occupation; Tobacco (Smoking)
SITE: Lung
OCCU: Miners, Iron; Paper (and Pulp) Workers
TIME: 1978 - 1992

* 722 Lundström, N.G. 04993
Univ. of Umeå, Dept. of Environmental Medicine, 901 87 Umeå, Sweden (Tel.: + 46 90 101700; Fax: 901196)
COLL: Nordberg, G.F.; Malker, H.; Weiner, J.A.; Gerhardsson, L.; Nordberg, M.

Lead and Cancer
Blood lead measurements among occupationally exposed people have been reported to the Swedish National Board of Occupational Health and Safety (NBOSH) since 1971. In this register 10,245 persons were registered until July 1990, when registration ended. The majority (9,000 persons) provided up to 16 samples during the period 1977-1990. Some 87,257 samples are registered. By using data on lead-exposed workers at the Roennskaer smelter and data from the Swedish lead register at the NBOHS, an investigation will be made of a possible association between lead exposure in Sweden and the frequency of cancer and other causes of death. The study concerning cancer is coordinated with a larger international multi-centre study at IARC. The study can be divided into two parts. The first is a follow-up of an earlier study concerning lead-exposed workers at the Roennskaer smelter, so including a larger number of lead-exposed persons, which improves the study capacity to detect possible correlations, for instance with the lead level in blood. The second part concerns the register at the NBOHS. Both cohorts will be compared with the Swedish Cancer Registry. Results will be standardized and compared with the causes of death for Sweden and also with a reference population of industrial workers without any specific risk factors.

SWEDEN

TYPE: Cohort
TERM: Metals; Occupation; Registry
SITE: Kidney; Lung
CHEM: Lead
OCCU: Smelters, Lead
REGI: Sweden (Swe)
TIME: 1991 - 1993

723 Nyström, L. 04637
Univ. of Umeå, Dept. of Epidemiology, and Health Care Research, 901 85 Umeå, Sweden (Tel.: +46 90 101213; Fax: 138977; Tlx: 54005)
COLL: Wall, S.; Larsson, L.G.; Andersson, I.; Bjurstam, N.; Fagerberg, G.; Frisell, J.; Tabar, L.; Rutqvist, L.E.

Pooling of Five Swedish Randomised Controlled Trials of Screening for Breast Cancer with Mammography

The value of mammographic screening has been questioned, especially in the age-group 40-49 years. The three published Swedish randomised controlled trials have failed to prove any statistically significant decrease in the mortality in breast cancer for this age group, probably due to their small size. With this background, the Swedish Cancer Society has initiated a pooling of the five randomised controlled trials that have been performed in Sweden, namely WE-study (Kopparberg and Ostergoetland); Stockholm (south); Goeteborg; and Malmoe, in order to increase the study base, especially in the age-group 40-49 years. The pooled database (n = 318,269) will be linked to the Swedish Cancer Registry and those reported with breast cancer (ICD-7 = 174) and deceased before 1st January 1990 according to the Swedish Cause-of-death Registry will be traced. Information concerning the cause of death i.e. from autopsy protocols, medical records, cause-of-death certificates, will be collected and each case will be blindly evaluated by an independent review group consisting of a surgeon, a pathologist, a radiologist and an oncologist. The effect of mammography screening will then be analysed.

TYPE: Cohort; Intervention
TERM: Mammography; Registry; Screening
SITE: Breast (F)
REGI: Sweden (Swe)
TIME: 1989 - 1992

724 Sandström, A. 02719
Univ. of Umeå, Dept. of Epidemiology, and Health Care Research, 901 85 Umeå, Sweden (Tel.: +46 90 102731; Fax: 138977)
COLL: Dahlgren, L.; Lönnberg, G.; Wall, S.

Occurrence of Cancer at a Smelter

Earlier studies of 3,915 workers first employed at a smelter during 1928-1966 showed an excess mortality up to 1977 of 11% compared to Swedish males, due to cancer (mostly lung) and circulatory diseases. Follow-up of this cohort for a further six years showed a decreasing trend in lung cancer incidence. The cohort has now been extended with workers employed later (up to 1980) and also with white-collar workers to further study mortality of different groups employed at the smelter. A questionnaire study comprising about 7,000 persons employed at the smelter 1928-1979 will add information on smoking habits to the entire cohort and will permit a deeper analysis of the interaction of smelter work and smoking habits. In combination with more in depth interviews with a small number of persons, strategically chosen from the cohort, it is aimed to elucidate the antagonism between the demand for a safe work environment and the demand for a well-paid employment. Papers have been published in Excerpt. Med. Int. Congr. Series 829:155-158, 1988 and in Br. J. Ind. Med. 46:82-89, 1989.

SWEDEN

TYPE: Cohort
TERM: Occupation; Tobacco (Smoking)
SITE: All Sites; Lung
CHEM: Arsenic
OCCU: Smelters, Copper
TIME: 1987 - 1992

UPPSALA

725 Adami, H.O. 04038
University Hosp., Cancer Epidemiology Unit, 751 85 Uppsala, Sweden (Tel.: +46 18 665045; Fax: 503431)
COLL: Pettersson, B.; Johansson, H.; Coleman, M.P.

Descriptive Epidemiology and Survival in Thyroid Cancer in Sweden

A nationwide cancer registry was started in Sweden in 1958. All thyroid cancers are notified to the Cancer Registry. By the end of 1981 a total of 6,935 thyroid cancers had been reported. In all these cases the hospital records and reports from the pathologists and cytologists have been scrutinized and the histopathological specimens reviewed when necessary. By these means all thyroid cancers have been subgrouped by histopathological type: mainly follicular, papillary, mixed, medullary, and anaplastic cancer. There are two major aims of this study. Firstly, the incidence rates will be analysed with respect to age, sex, time trends during the 24 year period of study, geographical differences in Sweden and possibly with respect to birth cohort effects. Secondly, the observed and relative survival rates will be calculated by histopathological subgroup and in relation to the same covariates as the incidence analyses.

TYPE: Incidence
TERM: Age; Birth Cohort; Geographic Factors; Histology; Registry; Survival; Time Factors
SITE: Thyroid
REGI: Sweden (Swe)
TIME: 1986 - 1992

726 Adami, H.O. 03752
University Hosp., Cancer Epidemiology Unit, 751 85 Uppsala, Sweden (Tel.: +46 18 665045; Fax: 503431)
COLL: Lundegårdh, G.; Helmick, C.; Zack, M.M.

Gastric Cancer and Other Malignant Disease after Partial Gastrectomy

The objective is to analyse whether gastric resection for chronic ulcer is a risk factor for the subsequent development of gastric cancer and other malignant diseases. Special emphasis is put on the possible impact of age at operation, duration of follow-up, localization of ulcer, sex and surgical procedure. Through hospital records, operation ledgers and other sources it was possible to identify virtually all patients subjected to gastric resection in a defined geographical area during the period 1950-1958. A total of 6, 500 (90% of all) could be completely followed up with respect to survival and cancer incidence. Gastric cancer risk and large bowel cancer risk has been assessed; on-going analyses aim to clarify the risk for other malignant diseases. Papers have been published in New Engl. J. Med. 319:195-200, 1988 and in Ann. Surg. 212:714-719, 1990.

TYPE: Cohort
TERM: Age; Gastrectomy; Survival; Time Factors
SITE: All Sites
TIME: 1985 - 1992

727 Adami, H.O. 04414
University Hosp., Cancer Epidemiology Unit, 751 85 Uppsala, Sweden (Tel.: +46 18 665045; Fax: 503431)
COLL: Andersson, S.O.; Johansson, J.E.

Risk Factors for Prostate Cancer in Örebro County

The aim is to analyse risk factors for prostate cancer with special emphasis on dietary factors. During a three year period, all patients under 80 years diagnosed as having prostate cancer in a strictly defined geographical area (Örebro county in central Sweden), and an equal number of age-matched controls

SWEDEN

from the general population will be eligible for inclusion in the study. Personal interviews will be carried out with an estimated number of 300 patients and 300 controls. Blood samples will be obtained from all study subjects and stored.

TYPE: Case-Control
TERM: BMB; Diet; Hormones; Occupation; Serum
SITE: Prostate
TIME: 1989 - 1993

728 Adami, H.O. 04790
University Hosp., Cancer Epidemiology Unit, 751 85 Uppsala, Sweden (Tel.: +46 18 665045; Fax: 503431)
COLL: Hansson, L.E.; Nyrén, O.; Hardell, L.O.; Bruce, Å.; Bergström, R.

Risk Factors for Stomach Cancer
A population-based case-control study is being carried out in two counties in northern Sweden and three counties in southern Sweden. The study population consists of all individuals younger than 80 years, born in Sweden and living in any of these counties. The aim is to carry out face-to-face interviews with a total of 300 newly diagnosed cases and with 600 controls who comprise a stratified sample from the general population. Dietary habits are thoroughly characterised during various periods of life from adolescence and later. Detailed information is also gathered concerning socio-economic factors, smoking, use of snuff and alcohol intake and on occupational history.

TYPE: Case-Control
TERM: Alcohol; Diet; Occupation; Socio-Economic Factors; Tobacco (Smoking)
SITE: Stomach
TIME: 1989 - 1992

729 Adami, H.O. 04792
University Hosp., Cancer Epidemiology Unit, 751 85 Uppsala, Sweden (Tel.: +46 18 665045; Fax: 503431)
COLL: Lindblad, P.; Jensen, O.M.; McCredie, M.R.E.; McLaughlin, J.K.; Wahrendorf, J.

Case-Control Study of Renal Adenocarcinoma
The main aim of this international collaborative study is to analyse the risk of renal adenocarcinoma associated with the use of diuretics and analgesics; and to investigate the possible relationship with occupation, notably in the petroleum industry. Cases will comprise all persons aged 20-79 years diagnosed as having renal adenocarcinoma in 1989-1991 in seven counties in central Sweden. Population-based controls will be selected by stratified sampling from the general population. Face to face interviews will be carried out with an estimated 450 cases and the same number of controls using a standardized questionnaire. Questions relate to use of analgesics, diuretics, oestrogens and diet pills; tobacco use and passive smoking; consumption of alcohol and coffee; reproductive history; dietary habits (using a food frequency questionnaire); medical history relating to diseases of the kidney, heart and thyroid; diabetes and other cancers; occupation (current and usual). In addition to a separate analysis of the Swedish component of the study, a pooled analysis of the data from all centers is envisaged.

TYPE: Case-Control
TERM: Alcohol; Analgesics; Coffee; Diabetes; Diet; Diuretics; Drugs; Hormones; Obesity; Occupation; Passive Smoking; Registry; Reproductive Factors; Tobacco (Smoking)
SITE: Kidney
CHEM: Oestrogens; Paracetamol
OCCU: Petroleum Workers
LOCA: Australia; Denmark; Germany; Sweden; United States of America
REGI: Uppsala (Swe)
TIME: 1989 - 1992

* 730 Adami, H.O. 05015
University Hosp., Cancer Epidemiology Unit, 751 85 Uppsala, Sweden (Tel.: +46 18 665045; Fax: 503431)
COLL: Ekbom, A.; Inskip, P.D.

SWEDEN

Diagnostic X-Rays and Risk of Thyroid Cancer
The aim of this study is to clarify whether diagnostic x-rays, including dental x-rays, during various periods of an individual's life, are a risk factor for the development of differentiated (papillary and follicular) thyroid cancer. A population-based case-control study will be carried out in the Uppsala health care region. Eligible patients are the 500 most recently diagnosed cases notified to the regional cancer registry. For each case, one control, matched by age and gender, will be selected from the population register. For each individual, the place of residence during various periods of life will be clarified from parish authorities. From this information follows unequivocally, due to the structure of the Swedish medical system, the hospital to which they belonged at any given time. Information concerning diagnostic x-ray procedures will then be sought in the rosters of the departments of radiology. It is believed that by this method nearly complete lifetime exposure information will be retrieved for all individuals in the study. A sample of surviving cases and controls will be contacted for a short telephone interview to clarify the importance of differential and non-differential misclassification when information on exposure to diagnostic x-rays is obtained by interview rather than by a review of radiology records.

TYPE: Case-Control
TERM: Radiation, Ionizing; Registry
SITE: Thyroid
REGI: Uppsala (Swe)
TIME: 1990 - 1993

731 Bruce, Å. 04424
Swedish National Food Administration, Box 622, 751 26 Uppsala, Sweden (Tel.: +46 18 175500; Fax: 105848 ; Tlx: 76121 SLVUPS S)
COLL: Adami, H.O.; Holmberg, L.; Bergkvist, L.; Bergström, R.; Becker, W.; Ohlander, E.-M.

Diet and Breast Cancer
The aim is to study the importance of diet for the risk of developing breast cancer in middle-aged Swedish women. A case-control study is being coordinated within a general mammographic screening programme in two counties in Sweden. Some 45,000 women aged 40-70 were screened in 1986-1988 in the county of Västmanland. The same number of women will be screened in 1988-1990 in the county of Uppsala. A questionnaire is distributed with the invitation to the mammography. It includes questions on social conditions (marital status, education, children, familial cancer), a few quantitative questions on food items and a food frequency section inquiring about 60 food items. The completed form is brought by the women to the screening and filed at the clinic. In a second phase women with diagnosed breast cancer (500 expected) and an equal number of age-stratified controls are interviewed on the telephone by a nutritionist. The interview starts by identifying the woman's "dietary periods" during which her diet has been virtually constant. These periods are separated e.g. by life events causing changes in her diet, such as moving, marriage, childbirth. The last period is checked against the questionnaire previously sent in. For each period structured questions are asked on the consumption of foods listed in the questionnaire. Portion sizes are estimated. The diets of the cases and the controls will be compared as to the intake of individual foods and nutrients such as fat, fibre, retinol, carotenes and selenium.

TYPE: Case-Control
TERM: Diet; Familial Factors; Fat; Fibre; Mammography; Metals; Nutrition; Socio-Economic Factors; Vitamins
SITE: Breast (F)
CHEM: Beta Carotene; Retinoids; Selenium
TIME: 1986 - 1992

732 Edling, C. 04426
University Hosp., Dept. of Occupational Medicine, 751 85 Uppsala, Sweden (Tel.: (018)663644)

Mining, Radon and Lung Cancer
Lung cancer due to radon exposure is a well-known health hazard among miners. However, there is still debate about the "acceptable" exposure level and the interaction with smoking. In this cohort study 600 miners, who had their first underground exprience in 1970 or later, i.e. when the exposure levels had been reduced to about 200-400 Bq/m&S'3., are followed. Data on health status and smoking habits are collected every third year throughout the study. The lung cancer incidence is compared to that expected computed from local statistics.

SWEDEN

TYPE: Cohort
TERM: Mining; Occupation; Tobacco (Smoking)
SITE: Lung
CHEM: Radon
OCCU: Miners
TIME: 1986 - 1993

733 Edling, C. 04427
University Hosp., Dept. of Occupational Medicine, 751 85 Uppsala, Sweden (Tel.: (018)663644)
COLL: Friis, L.

Cancer Incidence among Sewage Workers

Sewage workers are exposed to several possible mutagens and carcinogens. There is one study reporting urinary mutagens in municipal sewage workers and case reports on cancer in the lung, stomach and kidneys. The aim of this cohort study is to assess whether there is an increased risk of cancer among sewage workers. The cohort consists of about 800 men employed for at least one year during 1965-1986. The incidence is compared to expected numbers computed from national and local statistics.

TYPE: Cohort
TERM: Occupation
SITE: All Sites
OCCU: Sewage Workers
TIME: 1989 - 1993

734 Ekbom, A. 04039
Univ. Hospital of Uppsala, Unit of Cancer Epidemiology, 751 85 Uppsala, Sweden (Tel.: +46 18 665045; Fax: 503431)
COLL: Adami, H.O.; Pinczowski, D.; Helmick, C.; Zack, M.M.

Cancer Risk in Patients with Inflammatory Bowel Disease

The aim of this historical cohort study is to estimate the risk of malignant disease, particularly of the large bowel, in patients with ulcerative colitis (UC) and Crohn's disease (CD). Two sources of information are being used in order to identify all patients who fell ill with an inflammatory bowel disease during the period 1965-1983 in a defined geographical area - the Uppsala health care region with about 1.2 million inhabitants. A computerized in-patient register covering all discharge diagnoses in the study area during the pertinent period and the diagnostic ledgers at all departments of clinical pathology in the area have been scrutinized. In a second phase, which is now ongoing, additional information about clinical characteristics, extent of bowel involvement, disease activity, pharmacological treatment and histopathological features will be retrieved from hospital records and from histopathological reports for a "nested" case-control study. By these means, a cohort of about 3,000 patients with UC and 1,500 patients with CD has now been developed. After computerized record linkages to the death registry and the cancer registry, the observed and expected number of malignant diseases in the cohort will be calculated.

TYPE: Cohort
TERM: Crohn's Disease; Record Linkage; Registry; Ulcerative Colitis
SITE: Colon; Rectum
REGI: Sweden (Swe)
TIME: 1986 - 1993

735 Hillerdal, G.N. 03642
University Hosp., Dept. of Lung Medicine, 750 14 Uppsala, Sweden (Tel.: +46 18 166000)

Swedish Mesothelioma Study

As part of the European study of malignant mesothelioma, a Swedish study is now in progress. All cases of malignant pleural, and as many as possible of peritoneal mesothelioma will be reported and reviewed. An occupational and clinical history is taken. Biopsies will be reviewed by the specially formed Swedish mesothelioma panel. This panel will cooperate closely with the European panel, making comparisons possible. Basically the same investigation protocol will be used all over Western Europe. Patients will be followed and the natural history recorded. As the next step, multicentre prospective therapy studies, also in cooperation with the International Group, will be made. It is expected that about 40 patients a year will be included.

SWEDEN

TYPE: Case Series
TERM: Biopsy; Dusts; Occupation
SITE: Mesothelioma; Peritoneum; Pleura
CHEM: Asbestos
TIME: 1985 - 1992

736 Holmberg, L. 04443
University Hosp., Dept. of Surgery, 751 85 Uppsala, Sweden (Tel.: (18)663000)
COLL: Adami, H.O.; Sparén, P.

Breast Cancer in Women Treated for Cancer in Situ of the Breast

The aim is to analyse the importance of cancer in situ of the female breast as a risk factor for development of invasive breast cancer and death from breast cancer. The natural history of cancer in situ of the breast is largely unknown and available evidence has emerged mainly from small retrospective series. Further knowledge about the possible importance of detecting and treating preinvasive cancers has become a major concern when a large number of such lesions are being detected within mammographic screening programmes. All cases of cancer in situ notified to the Swedish Cancer Registry from 1958 through 1985 have therefore been identified. This cohort, total 1,454, will be followed up by means of computerised linkages - based on the individually unique national registration number - to the National Cancer and Death Registries. By these means it will be possible to calculate person-years at risk and to identify all incident cases and all women deceased with breast cancer as an underlying cause of death. Based on these data, SMRs will be analysed using age of cancer in situ, years of follow-up and calender year of diagnosis (a proxy variable for diagnosiis within or outside screening programmes) as determinants of risk. This approach will be followed by nested case-control studies in order to adjust for possible confounding factors and to enable more detailed studies of histopathological characteristics as determinants of risk.

TYPE: Case-Control; Cohort
TERM: High-Risk Groups; In Situ Carcinoma; Registry
SITE: Breast (F)
REGI: Sweden (Swe)
TIME: 1988 - 1992

737 Persson, I.R. 02084
University Hospital, Unit of Cancer Epidemiology, 751 85 Uppsala 14, Sweden (Tel.: +46 18 554045; Fax: 503431)
COLL: Adami, H.O.; Bergkvist, L.; Kressner, U.; Hoover, R.N.; Shairer, C.

Climacteric Treatment with Hormones: The Risk of Cancer

This prospective cohort study includes all women within a defined geographical area who during the three year period 1977-1980 had oestrogens and oestrogen-progestogen combinations prescribed for menopausal symptoms (about 23,233). A comprehensive mailed questionnaire permitting detailed assessment of epidemiological characteristics, of compliance and of lifetime exposure to oestrogen and progestogen has been sent to a random sample of 735 women among the cohort and to all those who develop breast and endometrial cancer. A similar questionnaire study has been carried out among all women in the cohort who are still alive and who have been prescribed potent oestrogens at any time during the period 1977-1980 (about 16,000). Breast and endometrial cancer incidence in the cohort is continuously registered and related to expected incidence derived from official statistics and from the reported total incidence in the area. Calculation of relative risks will be possible for the whole cohort and for subgroups according to duration of exposure, compounds, dose, presence of added progestogens etc. During 1990, the analyses of breast cancer risk will be extended to include the follow-up period through 1987. In addition, the standardised mortality rates from breast cancer will be analysed in the cohort. Papers by Bergkvist, L. et al have been published in Int. J. Epidemiol. 17:732-737, 1988; Am. J. Epidemiol. 130:503-510, 1989; and in N. Engl. J. Med. 321:293-297, 1989.

SWEDEN

TYPE: Cohort
TERM: Hormones; Menopause; Time Factors; Treatment
SITE: All Sites; Breast (F); Ovary; Uterus (Corpus)
CHEM: Oestrogens; Progestogens
TIME: 1977 - 1992

738   Persson, I.R.   04789
University Hospital, Unit of Cancer Epidemiology, 751 85 Uppsala 14, Sweden (Tel.: +46 18 554045; Fax: 503431)
COLL: Adami, H.O.; Kressner, U.; Naéssen, T.

Oophorectomy and Hysterectomy and the Risk of Breast or Ovarian Cancer
Aim: To investigate whether bilateral oophorectomy or hysterectomy affects the risk of subsequent development of breast or ovarian cancer. Design: A cohort of women who have undergone a gynaecological operation with a hysterectomy and/or bilateral oophorectomy will be defined from the Inpatient Registry. This registry holds data on all hospital admissions during the period 1965 through 1983 in the Uppsala Health Care Region. About 15,000 women will be enrolled. The individual women will be followed up through linkage to the cancer registry, to ascertain the observed number of cases of breast and ovarian cancer. The relative risk estimates will be formed as the ratio of the observed and expected number of cases, the latter calculated from the person-years of observation in the cohort and the age-specific incidence rates of these two cancers in the background population. The risk will be analysed in the cohort as a whole and among subgroups of it according to ovarian and uterine status. To this end, a case-cohort design will be applied. Data in the Inpatient Registry will be supplemented with information from the patients' medical records regarding the exact type of operation, the indication for the operation and other relevant factors. It is estimated that the power to detect a relative risk of breast cancer of 0.7 or smaller among women with a bilateral oophorectomy is 97%.

TYPE: Cohort
TERM: Hysterectomy; Oophorectomy; Registry
SITE: Breast (F); Ovary
REGI: Uppsala (Swe)
TIME: 1990 - 1992

SWITZERLAND

BASEL

739 Müller, H.J. 03985
Kantonsspital Basel, Dept. of Research, Lab. of Human Genetics, Hebelstr. 20, 4031 Basel, Switzerland (Tel.: (061)252525; Fax: 251500)
COLL: Weber, W.; Stalder, G.A.; Torhorst, J.K.H.; Obrecht, J.P.; Speck, B.; Signer, E.

Familial Cancer in Switzerland
This register contains detailed pedigrees of over 3,500 families of cancer patients. Special emphasis is laid on families with (1) breast cancer aggregation in families site-specific or in association with the neoplasms of the Li-Fraumeni/SBLA syndrome; (2) colorectal cancer including FAP and HNPCC, Crohn's disease and ulcerative colitis; (3) Fanconi's anaemia and related disorders; and (4) cancer of the skin. Cancer diagnoses of patients and relatives are verified through standard pathology or cytology reports. The incidence and spectrum of neoplasms in relatives are compared with those in the cancer registry of Basel. DNA from a subset of breast cancer patients and their relatives is being analysed for mutations in the tumour suppressor gene p53 for identifying cancer-prone individuals. Similar DNA studies are performed for colorectal cancer families concentrating on polymorphisms on chromosome 5 (FAP) and 17/18 (HNPCC). Cytogenetic studies and estimations of DNA repair capacities are carried out in families with an assumed increased genomic instability.

TYPE: Genetic Epidemiology; Registry
TERM: DNA Repair; Familial Factors; Genetic Factors; Genetic Markers; Heredity; Multiple Primary; Pedigree; Radiation, Ultraviolet; Registry; Screening
SITE: All Sites; Bone; Breast (F); Childhood Neoplasms; Gastrointestinal; Leukaemia; Lymphoma; Ovary; Testis
REGI: Basle (Swi)
TIME: 1982 - 1992

740 Stähelin, H.B. 04179
Kantonsspital, Medizinisch-geriatrische Klinik, Hebelstr. 10, 4031 Basel, Switzerland (Tel.: (061)252525)
COLL: Ulrich, J.; Torhorst, J.K.H.

Basle Study: Mortality Follow-Up.
This is an on-going prospective study with the aim of analysing risk factors for cardiovascular disease (CVD), cancer and senile dementia. The influence of life style, cardiovascular risk factors, and plasma vitamins on cancer and CVD is investigated. 1960 Basle Study I: Baseline data collection, plasma lipids and clinical data, 6, 000 participants. 1965 Basle Study II: Follow-up. Clinical data, plamsa lipids, 5,000 participants. 1971 Basle Study III: Follow-up. Plasma lipids, plasma vitamins, life style, approximately 4,500 participants. 1980 Mortality follow-up 1965-1980, 600 cases. 1986 Mortality follow-up: 450 additional cases. 1990 Mortality follow-up: 360 additional cases. A mailed survey was sent to approximately 3,600 survivors, asking for information on life style, nutrition, activities of daily life, and medical history. A paper has been published in Am. J. Clin. Nutr. 53:265S-269S, 1991.

TYPE: Cohort
TERM: Diet; Lifestyle; Lipids; Nutrition; Physical Activity; Vitamins
SITE: All Sites; Colon; Lung; Stomach
TIME: 1985 - 1992

GENEVA

741 Bouchardy, C. 04931
Geneva Tumour Registry, La Cluse 55, Geneva, Switzerland (Tel.: 202291)
COLL: Benhamou, S.; Paoletti, C.; Sancho-Garnier, H.; Dayer, P.; Galteau, M.M.

Cancer Risk Related to Tobacco according to Individual Capacity to Metabolize Probe Drugs
The activation of many of the carcinogens identified in cigarette smoke is catalysed by the cytochrome P450-dependent mixed function oxidase system. Some of these enzymes are genetically controlled and there is great variation in activity between individuals. The aim of the study is to identify smokers at very high risk of cancer by testing some of their enzymatic capacity possibly involved in the metabolism of carcinogens in cigarette smoke. This study, involving smokers (or ex-smokers), is a multicentric hospital-based case-control investigation. Specific enzymatic capacity will be investigated in 500

SWITZERLAND

histologically confirmed cancers of lung, larynx or buccal cavity and in 500 non-cancer controls admitted to a network of public hospitals in the Paris area and in Caen. In particular, phenotypic tests using dextromethorphane and mephenytoin will be done in all patients and other probe drugs will be added in a sample of the studied population.

TYPE: Case-Control
TERM: Drugs; Enzymes; Registry; Tobacco (Smoking)
SITE: Larynx; Lung; Oral Cavity; Pharynx
LOCA: France
REGI: Geneva (Swi)
TIME: 1988 - 1992

742 Lopez, A.D. 01083
World Health Organization, GSP Unit, 5 Ave. Appia, 1211 Geneva 27, Switzerland (Tel.: 912111; Tlx: 27821 OMS)

Mortality Statistics
Statistical data concerning mortality following the "A" (ICD-8) and the Basic Tabl. List (ICD-9) list of the International Classification of Diseases (ICD) are collected on a world-wide basis, processed, stored and published annually (in the World Health Statistics Annual). Data on cancer mortality according to the four-digit list of ICD are collected and stored in WHO's data bank. Epidemiological and statistical analyses of cancer mortality are periodically published as a special subject in the World Health Statistics Quarterly (WHSQ) and Technical Report Series. A paper on female lung cancer was published in WHO Wkly Epidem. Rec. 39:297-299, 1986.

TYPE: Mortality
TERM: Classification; Geographic Factors; Time Factors
SITE: All Sites
TIME: 1950 - 1992

LAUSANNE

743 Levi, F.G. 03921
Registre Vaudois des Tumeurs, Inst. Univ. de Médecine Sociale, et Préventive, CHUV Falaises 1, 1011 Lausanne, Switzerland (Tel.: (21)314908, 3143906)
COLL: La Vecchia, C.; Decarli, A.; Cislaghi, C.; Schüler, G.; Bopp, M.; Randriamiharisoa, A.; Negri, E.; Boyle, P.

Analyses of Cancer Death Certification Data in Switzerland
Swiss cancer death certifications from 1950 will be utilized (1) to carry out in a comprehensive analysis of the number of certified deaths; age-specific and age-standardized rates; proportional mortality by site; in graphical and numerical form, and (2) to develop statistical modelling techniques for disentangling the effects of cohort of birth, period of death and age on cancer mortality rates. Further, the distribution of cancer mortality in various Swiss geographical areas will be up-dated and re-considered by means of appropriate modelling techniques (trend surface models) for the analysis and interpretation of geographical patterns. Papers or reports from this study have been published in Oncology 45:313-317, 1988, Rev. Epidemiol. Sante Publ. 38:237-243, 1990 and J. Cancer Res. Clin. Oncol. 116:207-214, 1990.

TYPE: Methodology; Mortality
TERM: Age; Birth Cohort; Geographic Factors; Mathematical Models; Time Factors
SITE: All Sites
TIME: 1987 - 1993

744 Levi, F.G. 04243
Registre Vaudois des Tumeurs, Inst. Univ. de Médecine Sociale, et Préventive, CHUV Falaises 1, 1011 Lausanne, Switzerland (Tel.: (21)314908, 3143906)
COLL: La Vecchia, C.; Scazziga, B.; Portman, L.; Duruz, G.; Gulie, C.; Negri, E.; Torhorst, J.K.H.; Enderlin, F.

Aetiology of Thyroid Cancer
Thyroid diseases and specifically thyroid cancer are an important public health problem in Switzerland. Thyroid cancer mortality rates were the highest in Europe at the beginning of the century, and still remain relatively high, although considerable downward trends have been registered over the last few decades.

SWITZERLAND

Noticeable differences in thyroid cancer incidence are observed between various Swiss Cancer Registries. These temporal and geographical variations indicate that Switzerland constitutes a privileged situation for aetiological research on thyroid neoplasms. This case-control study will investigate: 1) the role of iodine deficiency, suggested by studies of descriptive epidemiology; 2) other dietary indicators and factors (vitamin A, C or specific foods, such as cabbage, which may interfere with the metabolism of thyroid hormones); 3) history of selected diseases (e.g., benign thyroid conditions, other endocrine diseases, immunological diseases); 4) hormonal and reproductive factors for females (parity, use of oral contraceptives). Cases will be subjects with thyroid cancer notified to the Cancer Registry of the Canton of Vaud and to the other cooperating Swiss Cancer Registries. Controls will be subjects admitted for acute, non-neoplastic or hormone-related conditions to the same hospitals where cases are identified, matched for sex, age and region of residence. It is planned to obtain data on 150 cases and 450 controls over a five-year period. Papers have been published in J. Cancer Res. Clin. Oncol. 116:639-647, 1990 and Eur. J. Cancer 27:85-88, 1991.

TYPE: Case-Control
TERM: Diet; Hormones; Iodine Deficiency; Oral Contraceptives; Parity; Radiation, Ionizing; Registry; Vitamins
SITE: Thyroid
REGI: Vaud (Swi)
TIME: 1987 - 1993

745 Levi, F.G. 04465
Registre Vaudois des Tumeurs, Inst. Univ. de Médecine Sociale, et Préventive, CHUV Falaises 1, 1011 Lausanne, Switzerland (Tel.: (21)314908, 3143906)
COLL: La Vecchia, C.; Delaloye, J.F.; de Grandi, P.; Gulie, C.; Monnier, V.; Negri, E.

Risk Factors for Endometrial Cancer in the Canton of Vaud, Switzerland
Endometrial cancer is the most frequent invasive malignancy of the female genital tract in Switzerland and the third most common neoplasm with an estimated number of 800 incident cases per year. It is well known that the risk is higher among women with elevated levels or availability of serum oestrogens, such as obese women or women on menopausal replacement treatment. Most studies, however, have concentrated upon the isolated effect of each single factor, in the absence of an integrated approach to the relative importance of various risk factors and their interactions. Further, there are factors, such as diet or exercise, whose relation with endometrial carcinogenesis is plausible, but far from being proven or quantified. The present proposal is based on a multi-national case-control study of endometrial cancer, originally designed at the Environmental Epidemiology Branch of the American National Cancer Institute to re-assess the role of major identified risk factors for endometrial cancer and their interactions and to explore newer hypotheses, chiefly in the field of diet. The research is based on a case-control design. Cases will be patients with histologically confirmed endometrial cancer registered by the Cancer Registry of the Canton of Vaud. Controls will be subjects admitted for conditions not relating to neoplastic, gynecologic or hormone-related diseases to the same hospitals where cases had been identified, matched by sex, age and region of residence. It is planned to obtain data on 200 cases and 600 controls over a five-year period. The population base provided by the Cancer Registry will allow reliable estimation not only of relative, but also of absolute excess risks, which is a further original aspect of the study, at least in relation to European populations.

TYPE: Case-Control
TERM: Diet; Hormones; Obesity; Physical Activity; Registry; Reproductive Factors
SITE: Uterus (Corpus)
REGI: Vaud (Swi)
TIME: 1988 - 1993

TANZANIA

DAR ES SALAAM

746 Mwakyusa, D. 04636
Muhimbili Medical Centre, Faculty of Medicine, Muhimbili, Dar es Salaam, Tanzania (Tel.: (051)26211; Tlx: 41505 muhmed tz)
COLL: Kitinya, J.N.; Mmuni, K.

Prevalence of Gastritis, Intestinal Metaplasia and Gastric Cancer in Tanzania and their Associated Carcinogenic Factors

The study will include 200 subjects with dyspeptic symptoms and 100 control subjects in each of three areas, two of which, Moshi and Mbeya, have high prevalence rates for stomach cancer, while the third, Dar es Salaam, has a low prevalence rate. These subjects will be gastroscoped and the tissue biopsies will be studied histologically and for mucin histochemistry. In addition, carcinogenic substances such as benzo-pyrenes and aflatoxin will be estimated in soil, food and urine. Prevalence rates of gastritis in dyspeptic and control subjects will be correlated with prevalence rates of gastric cancer.

TYPE: Correlation; Cross-Sectional
TERM: Mycotoxins; Prevalence; Soil; Urine
SITE: Stomach
CHEM: Aflatoxin; Benzo(a)pyrene
TIME: 1989 - 1992

THAILAND

BANGKOK

747 **Srivatanakul, P.** 04497
National Cancer Inst., Research Div., Rama VI Rd, Bangkok 10400, Thailand (Tel.: 2461294/302; Tlx: INSNACA)
COLL: Muñoz, N.; Bosch, F.X.; Estève, J.; Khuhaprema, T.; Rinsurongkawong, S.; Cheirsilpa, A.; Tanprasert, S.

Cohort Study of HBsAg Carriers in Bangkok
The objectives of the present study are (1) to determine the role of other risk factors such as aflatoxin, tobacco, alcohol, steroid hormones and N-nitroso compounds in the development of PHC and other chronic liver diseases; (2) to determine the risk of developing PHC and other chronic liver diseases among a cohort of HBsAg carriers as compared to that of the general population; (3) to evaluate the value of AFP levels and ultrasonography in the detection of early PHC. (These data will provide the necessary information for the formulation of cost-effective screening programmes of high-risk groups for PHC in Thailand and other developing countries); (4) to assess the benefits of early treatment of PHC. A Group of 2,000 HBsAg positive male blood donors over 30 years of age will be selected from among blood donors, Army personnel and other people who come for check up at the Out-Patient Clinic. At the end of the five-year follow-up period a case-control study within the cohort will be carried out to compare both PHC cases and cases of chronic hepatitis and cirrhosis with appropriate controls in relation to the risk factors.

TYPE: Case-Control; Cohort
TERM: Alcohol; Antigens; Cost-Benefit Analysis; HBV; High-Risk Groups; Hormones; Liver Disease; Mycotoxins; Screening; Tobacco (Smoking)
SITE: Liver
CHEM: Aflatoxin; N-Nitroso Compounds; Steroids
TIME: 1987 – 1993

HAT YAI

* 748 **Chongsuvivatwong, V.** 05181
Prince Songkla Univ., Fac. of Medicine, Epidemiology Unit, Hat Yai 90112, Thailand (Tel.: (074)245677; Fax: 235174 ; Tlx: 62168 UNISONE TH)
COLL: Muñoz, N.; Bosch, F.X.; Chichareon, S.

Case-Control Study on Cervical Carcinoma
The hypotheses to be tested are whether squamous cell carcinoma is related to sexually transmitted agents (HPV, HSV-2, chlamydia and CMV), sexual behaviours of the patient and of her partner, use of oral contraceptives, tobacco smoking and the effect of screening for cervical carcinoma. Cases are incident cases in the University Hospital, with histological confirmation. Controls are female patients admitted to the surgical ward of the same hospital, who are in the same age range as the cases and whose admission diagnoses are not related to the above exposure variables. Current husbands of the case or control women are ascertained. 300 cases, 300 controls and their husbands are needed. Information on past sexual behaviour of the subject is obtained at interview in the hospital. Serum is obtained from all subjects. Exfoliated cervical cells from the cases and the controls and exfoliated cells of the glans penis and from the urethra are obtained from the husbands. Analysis will be done using unconditional logistic regression techniques.

TYPE: Case-Control
TERM: BMB; CMV; HPV; HSV; Infection; Oral Contraceptives; Screening; Serum; Sexual Activity; Sexually Transmitted Diseases
SITE: Uterus (Cervix)
TIME: 1991 – 1993

TURKEY

ISTANBUL

749   Akdas, A.   04260
Marmara Univ., School of Medicine, Dept. of Urology, Tophanelioglu cad., No. 13-15 Altunizade, Istanbul 81190, Turkey (Tel.: (1)3255865; Fax: 3250323)
COLL: Simsek, F.; Ersev, D.; Türkeri, L.; Bilir, N.

Aetiology of Testicular Tumours

In this study the possible role of some factors (exposure to X-ray, heat, smoke, dust and chemical substances, testicular trauma, history of cryptorchidism or mumps, wearing tight underwear and pants, sexual activity) in the aetiology of testicular tumours will be investigated. The cases will be all patients with a histologically proven testicular tumour seen in the department of urology. The approximate number is estimated as 15 patients per year. Age-matched male controls will be taken from orthopaedics and general surgery wards. Patients with a diagnosis of malignant disease will be excluded. Necessary information for the cases and controls will be collected by questionnaire.

TYPE: Case-Control
TERM: Chemical Exposure; Radiation, Ionizing; Sexual Activity; Tobacco (Smoking); Trauma
SITE: Testis
TIME: 1985 - 1992

* 750   Akdas, A.   05158
Marmara Univ., School of Medicine, Dept. of Urology, Tophanelioglu cad., No. 13-15 Altunizade, Istanbul 81190, Turkey (Tel.: (1)3255865; Fax: 3250323)
COLL: Simsek, F.; Ilker, Y.; Alican, Y.; Türkeri, L.

Prostate Cancer: Genetic and Epidemiological Factors

In this study the HLA complexes of patients with proven adenocarcinoma of the prostate will be determined in order to find out any possible relationship between genotype and the disease. The possible role of occupation, exposure to drugs and x-ray, heat, chemical substances, medical history of venereal disease, sexual activity and its spectrum in the aetiology of prostate cancer will also be investigated in these patients. The approximate number of cases is estimated at 25 per year. Age-matched controls will be recruited from patients with histologically proven benign prostatic hypertrophy and the information will be obtained by questionnaire.

TYPE: Case-Control
TERM: Chemical Exposure; Drugs; Genetic Factors; HLA; Occupation; Sexual Activity; Sexually Transmitted Diseases
SITE: Prostate
TIME: 1992 - 1996

UNITED KINGDOM

BELFAST

* 751   MacKenzie, G.   05202
Queens Univ. of Belfast, Inst. of Clinical Science, Dept. of Social & Preventive Medicine, Grosvenor Rd, Belfast BT12 6BJ N. Ireland, United Kingdom

Atlas of Cancer Mortality in Northern Ireland

The purpose of this project is to construct a colour atlas showing the temporo-spatial distribution of the major cancers in Northern Ireland during the period 1981-1989. The atlas will also be analytical and use 3-D imaging techniques to illustrate statistical modelling and to identify outliers. An interpretative commentary based on the detailed statistical analyses undertaken will be included.

TYPE: Mortality
TERM: Mapping; Mathematical Models
SITE: All Sites
TIME: 1990 - 1992

BIRMINGHAM

752   Allan, R.N.   04516
General Hosp., Gastroenterology Unit, Steelhouse Lane, Birmingham B4 6NH, United Kingdom (Tel.: (021)2368611/5024; Fax: 2367626)
COLL: Jewell, D.P.; Hellers, G.

Incidence of Cancer in Crohn's Disease

The aim of this study is to establish the pattern and incidence of malignant disease in patients with Crohn's Disease. Patients will be recruited from three centres (Birmingham, Oxford and Stockholm). Only primary referrals within five years of onset of disease between 1954 and 1965 and followed for a minimum of 20 years will be included, which will eliminate any secondary or tertiary referral bias in relation to specialist hospital practice. Some 250 patients will be collected from each centre. The morbidity rates for cancer in patients with Crohn's Disease will be compared with the morbidity rates for cancer in each geographical area in the study. The cancer risk will be calculated for all sites within the digestive system and for extra-intestinal cancer.

TYPE: Cohort
TERM: Crohn's Disease; Ulcerative Colitis
SITE: All Sites; Colon; Rectum
LOCA: Sweden; United Kingdom
TIME: 1989 - 1993

753   Knox, E.G.   04540
Univ. of Birmingham, Dept. of Social Medicine, Edgbaston, Birmingham B15 2TJ, United Kingdom
COLL: Stewart, A.M.; Kneale, G.W.; Gilman, E.A.

Spatial and Temporal Distributions of Childhood Cancers in Great Britain

The study aims to characterise and interpret the spatial, temporal and spatio-temporal distributions of childhood cancers in Britain. Cases include all children who have died of cancer or leukaemia, in England, Scotland and Wales, under the age of 16 years, in the period 1953 to 1981 (who form part of the Oxford Survey of Childhood Cancers, an on-going national case-control study), together with all childhood registrations of leukaemia and non-Hodgkin lymphoma diagnosed in the years 1966 to 1983. Spatial distributions will be mapped by local authority area and area rates will be examined for spatial heterogeneity. Proximity analyses will also be used to examine case distributions with respect to local environmental features which may represent a source of risk. Analyses of the temporal distribution will include examination of cohort childhood cancer rates, specifically looking for age, period and cohort effects. Data will be examined for spatio-temporal clustering, i.e. excesses of cases occurring close together in space and in time, using the method of Knox and Mantel. Identified "clusters" will be examined for associations with local geographical circumstances and temporal events, to test hypotheses on changes in case ascertainment, survival, or migration; on local toxic or radiation hazards; or on transmission of a viral agent.

UNITED KINGDOM

TYPE: Case Series; Incidence; Mortality
TERM: Birth Cohort; Childhood; Cluster; Geographic Factors; Mapping; Radiation, Ionizing; Time Factors; Virus
SITE: Childhood Neoplasms; Leukaemia; Non-Hodgkin's Lymphoma
TIME: 1986 - 1989

754 Knox, E.G. 00858
Queen Elizabeth Medical Centre, Cancer Epidemiology Research Unit, Survey of Childhood Cancers, Birmingham B15 2TH, United Kingdom (Tel.: (021)4727752)
COLL: Stewart, A.M.

National Survey of Childhood Cancers
This long-standing retrospective survey was originally started in Oxford by Dr A. Stewart in 1953 to study the aetiology of childhood cancers. It consists of interviews with the parents of children under 16 dying of cancer or leukaemia in the United Kingdom and with parents of control children matched for age, sex and district; antenatal information is confirmed by application to clinics and doctors, and the case child's medical history is followed up through general practitioner and hospital notes. Surviving children have similarly been interviewed and followed up since 1962 by means of the regional cancer registries. Present projects include: (1) A study of the effects of drugs in pregnancy and of ultrasound during pregnancy (2) A study of parental occupations and their possible relationship to the aetiology of childhood cancers; (3) A study of congenital defects found in children who later develop cancer or leukaemia and congenital defects in their siblings, with a similar study in matched controls and their siblings; (4) A study of second primary tumours occurring in children after cure of their initial cancer with special attention to the effect of radiation and genetics; (5) A study of cancers among siblings of children with cancer.

TYPE: Case-Control; Cohort
TERM: Clinical Effects; Congenital Abnormalities; Drugs; Intra-Uterine Exposure; Multiple Primary; Mutation, Somatic; Occupation; Pregnancy; Radiation, Ionizing; Radiotherapy; Registry; Sib; Ultrasound; Virus
SITE: Childhood Neoplasms; Leukaemia
TIME: 1953 - 1992

755 Mann, J.R. 04245
Birmingham Children's Hosp., Dept. of Oncology, Ladywood Middleway, Birmingham B16 8ET, United Kingdom (Tel.: (021)4544851)
COLL: Cameron, A.H.; Powell, J.E.; Parkes, S.E.; Raafat, F.; Stevens, M.C.G.; Waterhouse, J.A.H.

Childhood Cancer: Epidemiological and Aetiological Investigations
Clinical, epidemiological, treatment and survival data on all children diagnosed with malignancies in the West Midlands region since 1957 (5,000 cases), and prospectively since 1984, have been assembled in the West Midlands Regional Children's Tumour Research Group. The purpose of the project is to use these data for epidemiological studies such as determining changes in incidence in relation to time, variations in lifestyle, geographical and other environmental factors. In addition, retrospective studies on selected diagnostic groups of childhood cancers in relation to clinical features, results of treatment and other variables are being undertaken.

TYPE: Incidence; Registry
TERM: Environmental Factors; Geographic Factors; Lifestyle; Time Factors
SITE: Childhood Neoplasms
REGI: Birmingham (UK)
TIME: 1984 - 1993

756 Sorahan, T.M. 04493
Univ. of Birmingham, Dept. of Social Medicine, Cancer Epidemiology Research Unit, Edgbaston, Birmingham B15 9TJ, United Kingdom (Tel.: (021)4143985; Fax: 4144036)
COLL: Davies, P.

Mortality and Cancer Morbidity of Production Workers in the Polyurethane Foam Industry
Study subjects are all 8,500 past and present workforce employees, from 11 participating factories in England and Wales, with a minimum period of employment of six months and with some employment in the period 1958-1979. The nature of the study is an historical prospective cohort study. Detailed job histories have been collected. Cancer rates among those employees with "higher" levels of exposure to toluene diisocyanate will be compared with cancer rates of all other employees.

UNITED KINGDOM

TYPE: Cohort
TERM: Occupation; Plastics
SITE: All Sites
CHEM: Diisocyanates
OCCU: Plastics Workers
TIME: 1985 – 2000

757 Sorahan, T.M. 04802
Univ. of Birmingham, Dept. of Social Medicine, Cancer Epidemiology Research Unit, Edgbaston, Birmingham B15 9TJ, United Kingdom (Tel.: (021)4143985; Fax: 4144036)
COLL: Basu, M.K.; Robin, P.E.

Aetiology of Salivary Gland Tumours

The aim is to define any risks of salivary gland tumours associated with dental radiography. Cases will be all those men and women diagnosed with salivary gland tumours in the years 1988-1990, aged 20-74 years, and resident in the West Midlands Region (about 150 cases). Names and addresses of 1,000 potential controls will be selected at random from electoral registers of the region. Cases and controls will be interviewed in their own homes in order to obtain information on the following: history of dental X-rays and medical X-rays to the head and neck, occupational exposures, smoking history and use of alcohol. Analysis will be carried out with unconditional logistic regression.

TYPE: Case-Control
TERM: Alcohol; Occupation; Radiation, Ionizing; Tobacco (Smoking)
SITE: Salivary Gland
TIME: 1989 – 1992

758 Stewart, A.M. 04929
Univ. of Birmingham, Faculty of Medicine, Dept. of Social Medicine, Birmingham B15 2TJ, United Kingdom (Tel.: (021)4721311; Fax: 4144036)
COLL: Cummins, C.W.; Nussbaum, R.H.

Late Effects of Low-Level Radiation

The aim is to determine whether military personnel who were stationed during 1943-1962 at Hanford Nuclear Reservation, USA, where there were many radioactive releases, are at greater risk of dying of cancer than a control group at Fort Lewis, also in Washington State. It is expected that 25,000 veterans in each camp will be traced from army records. Those deceased will be identified, as well as their cause of death, and a retrospective cohort study carried out. Exposure data will be available from a Department of Energy study which should be completed by 1994. Some preliminary exposure data and dose reconstructions are available from the State of Washington. Cooperation has been obtained from the US Veterans Administration.

TYPE: Cohort
TERM: Radiation, Ionizing
SITE: All Sites
LOCA: United States of America
TIME: 1988 – 1992

* 759 Wallace, D.M.A. 04976
Queen Elizabeth Hosp., Department of Urology, Clinical Trials Unit, Edgbaston, Birmingham B15 2TH, United Kingdom (Tel.: (021)6272288; Fax: 6272289)
COLL: Harrington, J.M.; Bathers, S.

Feasibility Study for Surveillance of Occupational Urothelial Tumours

The primary objective of this study is to assess the feasibility of establishing a surveillance procedure for occupational urothelial tumours. Further aims will be to identify the areas and occupations where more detailed epidemiological investigations should be carried out, to establish a database of the occupations of workers exposed to the recognised urothelial carcinogens, to establish a register of cases with urothelial cancer exposed at work to urothelial carcinogens, to advise the clinicians, to co-ordinate the further investigation of cases where occupational exposure is suspected and to provide an information service to urologists and other clinicians managing urothelial cancer cases. The cases will be all newly diagnosed patients with urothelial cancer in the West Midlands area over a two-year period. All patients will be screened using an occupational history questionnaire. Suspected exposure will be fully investigated by a trained interviewer. Controls will be selected randomly from patients in the same hospital in the same period of time. To test the reproducibility of interview information a proportion of the

UNITED KINGDOM

interviews will be conducted again by a second interviewer. The assessment of occupational exposure will be retrospective, based on information collected by hygienists in the Institute of Occupational Health, producing a job-exposure matrix for estimation of occupational exposure to certain and suspect carcinogens. The final analysis of the data will concentrate on identifying and investigating unusual clusters of cases that appear to have a common aetiology. The feasibility of further epidemiological studies on occupational urothelial cancer will be examined where the data suggest aetiological hypotheses.

TYPE: Case-Control; Methodology
TERM: Chemical Exposure; Data Resource; Occupation; Registry
SITE: Urinary Tract
REGI: W. Midlands (UK)
TIME: 1990 - 1993

760 Woodman, C.B.J. 04374
Univ. of Birmingham, Dept. of Social Medicine, Vincent Drive, Birmingham, United Kingdom (Tel.: 4721301)
COLL: Knox, E.G.; Young, L.; Gallimore, P.; Luesley, D.; Maitland, N.; Meanwell, C.; Jordan, J.

Aetiology and Natural History of early Cervical Neoplasia
A longitudinal study of a group of about 2,000 young women who have recently embarked on sexual activity will be used to generate sequential observations on the natural history of cervical intraepithelial neoplasia (CIN). The group will be selected from a population where the initial prevalence and subsequent incidence of epithelial abnormality are known to be high. They will be followed for five years. A subset of the study population, those initially found to have CIN or who develop it under observation, will provide data for a case-control investigation of the role of HPV, EBV and HSV in the aetiology of cervical neoplasia. Cases will be compared with controls matched for sexual history. All subjects will be asked for a detailed sexual history. Cytological and serological specimens will be obtained at six-monthly intervals. Stored sequential sera from cases and controls will later be examined for antibodies to EBV and HSV. Stored cytological preparations will be examined subsequently for HPV, EBV, HSV and DNA using dot-blot membrane and in-situ hybridisation techniques. The relative frequency with which specific viral DNA is found in biopsy from patients with and without CIN will be determined and the time of onset of infection related to the development of epithelial abnormality. Cervical punch biopsies specimens will be taken from all patients with cytological abnormality. In-situ hybridisation will be used to detect viral DNA in the biopsy specimens. An estimate will be made of the rate of progression of the earlier forms of dysplasia (CIN I and CIN II) to CIN III and the extent to which this progression is related to specific viral infection.

TYPE: Case-Control; Cohort
TERM: Antibodies; Cytology; DNA; EBV; HPV; HSV; Sexual Activity
SITE: Uterus (Cervix)
TIME: 1987 - 1992

BRADFORD

* 761 Kernohan, E. 05071
Bradford Health Authority, Dept. of Public Health, 109 Duckworth Lane, Bradford BD9 6RL, United Kingdom (Tel.: (0274)366020; Fax: 547509)
COLL: Hantman, J.M.; Rider, P.L.

Incidence of Cancer in Asians
The aims of this study are (1) to improve understanding of the present and likely future patterns of cancer among Asians in the city of Bradford; and (2) to identify any cancers of high incidence for which prevention may be possible. Cancer registrations relating to persons of Asian background will be extracted from the Yorkshire Regional Cancer Registry for the six-year period 1985-1990, using forenames and surnames. Data will be analysed by disease category and age- and sex- specific incidence rates calculated. These will be compared with expected incidence rates for the non-Asian population, and with rates for the Indian sub-continent, derived from the Bombay Cancer Registry. Data collection will be confined to Bradford and Airedale Metropolitan Districts (total population 518,555; OPCS 1989 mid-year estimate). 200 cancer registrations are expected in the Asian population.

UNITED KINGDOM

TYPE: Incidence
TERM: Asians; Registry
SITE: All Sites
LOCA: India; United Kingdom
REGI: Yorkshire (UK)
TIME: 1991 - 1992

BRIDGEND

762 Powell, D.E. 03906
Princess of Wales Hosp., Dept. of Pathology, Coity Rd, Bridgend Wales, United Kingdom

Epidemiology of Leukaemia and Lymphoma
In this longitudinal study of the space and time distribution of leukaemia and lymphoma, a continuous 25-year period is being monitored in a very stable population of 140,000 served by one diagnostic centre. All cases are referred to the one laboratory. Environmental factors studied include housing, schools and water. Banked sera are studied for viruses. Contacts of cases are traced.

TYPE: Correlation
TERM: Cluster; Environmental Factors; Time Factors; Virus
SITE: Leukaemia; Lymphoma
TIME: 1962 - 1993

BRIGHTON

763 Bridges, B.A. 04900
Univ. of Sussex, MRC Cell Mutation Unit, Falmer, Brighton BN1 9RR, United Kingdom (Tel.: (0273)678123; Fax: 678121)
COLL: Arlett, C.F.; Cole, J.

Mutation Frequency of Lymphocytes in Individuals Exposed to Radon
The aim is to see whether individuals living in dwellings with high levels of radon gas have a higher frequency of mutations in their peripheral lymphocytes than those living in dwellings with low levels of radon. The frequency of T-lymphocytes resistant to 6-thioguanine will be measured using a cloning assay. Results of a pilot study are now available as a result of which a further, more extensive study is in progress.

TYPE: Molecular Epidemiology
TERM: Lymphocytes; Radiation, Ionizing
SITE: Inapplicable
CHEM: Radon
TIME: 1990 - 1992

BRISTOL

* 764 Golding, J. 05148
Univ. of Bristol, Hosp. for Sick Children, Inst. of Child Health, St. Michael's Hill , Bristol BS2 8BJ, United Kingdom (Tel.: (0272)225967; Fax: 255051)
COLL: Adam, H.; Parks, S.; Verd, S.; Stejskal, J.; Ignatyeva, R.

Childhood Cancer in Six European Centres
This cohort study will involve prospective collection of information on a geographically-based population from pregnancy through until the children are seven years of age. The study cohort comprises 38,000 children. As the data collection progresses, identification of all children with cancer will be made. Prospective information concerning features of the environment, including type of housing, maternal smoking and alcohol consumption, use of illicit drugs and features of the diet, together with maternal illnesses and drugs taken, will be available for analysis. Features of childhood including exposure to passive smoke and other possible mutagens will also be available. The overall wealth of data should enable features of the background of children who develop cancer to be identified.

UNITED KINGDOM

TYPE: Cohort
TERM: Alcohol; BMB; Blood; Diet; Drugs; Environmental Factors; Mutagen; Passive Smoking; Tissue; Tobacco (Smoking); Urine
SITE: Childhood Neoplasms
CHEM: Nicotine
LOCA: Czechoslovakia; Greece; Spain; USSR; United Kingdom
TIME: 1991 - 2000

* 765 Golding, J. 05149
Univ. of Bristol, Hosp. for Sick Children, Inst. of Child Health, St. Michael's Hill , Bristol BS2 8BJ, United Kingdom (Tel.: (0272)225967; Fax: 255051)
COLL: Mott, M.; von Kries, R.

Vitamin K and Pethidine in the Aetiology of Childhood Cancer
An earlier unexpected finding implicated pethidine and vitamin K in the aetiology of childhood cancer. This study is specifically designed to test the hypothesis that these two substances are implicated in childhood cancer. Information is being abstracted from all case-notes of children born in Bristol since 1965 who have developed cancer. The control group concerns the maternal records of a random sample of all other births from 1965. Abstraction of data is done by clerks who are blind as to whether it is a case or control that is under review. Extensive information on potential confounders is also being abstracted.

TYPE: Case-Control
TERM: Clinical Records; Drugs; Vitamins
SITE: Childhood Neoplasms
CHEM: Pethidine
TIME: 1990 - 1992

* 766 Harte, G.A. 04987
Nuclear Electric plc, Health and Safety Dept., Bedminster Down, Bridgwater Rd , Bristol BS13 8AN, United Kingdom (Tel.: (0272)648403; Fax: 648565; Tlx: 44182)
COLL: Taylor, R.H.; Davies, N.F.; Brodie, P.; Hayward, R.J.; Rodliffe, R.S.

Mortality Study of Employees of Nuclear Electric PLC
Occupational histories and radiation exposure data are being collected for approximately 20,000 employees of Nuclear Electric plc (formerly the Central Electricity Generating Board) who have been classified as radiation workers at any time since the company started operating nuclear power stations in 1962. Mortality in this cohort will be studied in relation to occupational exposure to ionizing radiation and, possibly, other physical and chemical agents.

TYPE: Cohort
TERM: Chemical Exposure; Occupation; Radiation, Ionizing
SITE: All Sites
OCCU: Radiation Workers
TIME: 1990 - 1994

CAMBRIDGE

* 767 Davies, T.W. 05138
Addenbrookes Hosp., Dept. of Community Medicine, Level 5, Hills Rd , Cambridge CB2 2QQ, United Kingdom (Tel.: (0223)336812; Fax: 212920)

East Anglian Diet and Health Study 1991
This case-control study will examine the hypothesis that consumption of dairy products in adolescence is positively correlated with the risk of testicular cancer. Cases will consist of living men who have had testicular cancer. Two cancer controls matched for age (within 2.5 years) are taken from the same cancer registry, as well as two population controls selected by general practitioners and matched for age. Data on current diet will be collected using a self-completed questionnaire and diary, and estimates of past diet will be obtained from the mothers. A pilot study is being conducted first using 20 cases and 80 controls. The main study will include some 200 cases and 800 controls. Response to date to the mailed questionnaire on diet indicates that it may be necessary to recruit a trained interviewer.

UNITED KINGDOM

TYPE: Case-Control
TERM: Adolescence; Dairy Products; Diet; Registry
SITE: Testis
REGI: E. Anglia (UK)
TIME: 1990 – 1992

CARDIFF

768 Clark, R.E. 04687
Univ. Hosp. of Wales, Dept. of Haematology, Heath Park, Cardiff CF4 1LY, United Kingdom (Tel.: (0222)755944)
COLL: Day, N.E.

Haematological Malignancy Following Therapy for Hodgkin's Disease
Among the 241 newly diagnosed cases of Hodgkin's disease (HD) seen at this centre since 1969, no cases of second acute non-lymphocytic leukaemia (s-ANLL) or myelodysplastic syndrome (MDS) have been observed. In contrast, virtually all published series report a high relative risk of s-ANLL in previously treated HD. The hypothesis that this finding is due to a different treatment strategy, especially for cases with chronic relapsing disseminated disease, is being investigated. This will be done by a detailed retrospective study in which precise detail will be obtained on all chemotherapy and radiotherapy treatment received. Comparisons will be made with similar available data from other centres, to determine if the observed difference in the risk of acute leukaemia following HD can be attributed to any difference in HD treatment.

TYPE: Case Series
TERM: Chemotherapy; Multiple Primary; Radiotherapy
SITE: Haemopoietic; Leukaemia (ALL); Myelodysplastic Syndrome
TIME: 1987 – 1992

* 769 Harvey, I.M. 05066
Univ. of Wales College of Medicine, Centre for Applied Public Health Med., Temple of Peace and Health, Cathays Park, Cardiff CF1 3NW, United Kingdom (Tel.: (0222)231021; Fax: 238606)
COLL: Marks, R.; Frankel, S.

Solar Keratosis and Non-Melanoma Skin Cancer in the Elderly Population of South Wales
This descriptive epidemiological study aims to describe the prevalence and incidence of solar keratoses and non-melanoma skin cancer amongst persons over 60 years of age in South Wales. A random sample (n = 1,000) of individuals has been selected and visited by a research dermatologist. Sun exposed areas of the skin have been examined and lesions photographed for validation purposes. Data have been collected on skin type, ancestry and estimated lifetime sun exposure in order to investigate the influence of these factors on prevalence of the lesions.

TYPE: Cross-Sectional
TERM: Pigmentation; Premalignant Lesion; Radiation, Ultraviolet
SITE: Skin
TIME: 1988 – 1992

* 770 Jacobs, A. 04989
Univ. of Wales College of Medicine, Dept. of Epidemiology & Community Med., Heath Park, Cardiff CF4 4XN, Wales, United Kingdom (Tel.: (0222)747747/2318)
COLL: West, R.R.; Fenaux, P.; Lowenthal, R.M.

Chemical and Environmental Exposures in Myelodysplasia
The main aim of this case-control study is to investigate the lifetime occupational and personal history of patients with myelodysplasia (MDS) with respect to exposure to chemicals and radiation. The study seeks 420 case-control pairs. A secondary aim will be to estimate the incidence of MDS in defined populations. All cases of MDS aged 15+ diagnosed since 1 October 1988, resident in defined areas, will be included. The controls will be matched for sex, age, district of residence and year of presentation. Each control will be randomly selected from patients attending out-patient clinics of the hospital which ascertained the case. Cases and controls will be interviewed by trained interviewers using a structured questionnaire and analysis will be carried out using the methods of Breslow and Day for conditional logistic regression and multivariate analysis of matched data.

UNITED KINGDOM

TYPE: Case-Control
TERM: Chemical Exposure; Occupation; Radiation, Ionizing
SITE: Myelodysplastic Syndrome
TIME: 1989 - 1992

* 771 **Mansel, R.E.** 04994
Univ. of Wales College of Medicine, Dept. of Epidemiology & Community Med., Heath Park, Cardiff CF4 4XN, Wales, United Kingdom (Tel.: (0222)747747)
COLL: Gravelle, I.H.; West, R.R.; Miers, M.E.

Wolfe Mammographic Patterns as Predictors of Breast Cancer Risk
Mammographic parenchymal patterns have been proposed as a method of identifying women at high risk of developing breast cancer. In 1976, Wolfe described four breast patterns with reported relative risks as great as 37:1 from the highest (DY) to the lowest risk class (N1). Subsequent studies have produced much smaller quantitative estimates of risk, and several have failed to support Wolfe's hypothesis. This study aims to determine if Wolfe patterns are predictive of breast cancer risk. It is a prospective study of subsequent development of breast cancer among a cohort of 5,000 women who attended the symptomatic breast clinic at the University Hospital of Wales from 1980-1985, and underwent mammography. Each woman was allocated a Wolfe code of N1, P1, P2 or DY. Those with no known breast cancer have been mailed annually and asked to report any breast symptoms. Those known to have a breast cancer have been followed through breast clinic attendance and general practitioner contacts. Data concerning other risk factors have been collected through questionnaires and at clinic attendance. The study will determine the rate of subsequent breast cancer for each of the Wolfe codes allocated at the time of the initial mammogram and will review the role of breast parenchymal patterns in breast cancer risk.

TYPE: Cohort
TERM: High-Risk Groups; Mammography
SITE: Breast (F)
TIME: 1982 - 1992

CARLISLE

* 772 **Cox, N.H.** 05137
Cumberland Infirmary, Dept. of Dermatology, Carlisle Cumbria CA2 7HY, United Kingdom (Tel.: (0228)23444)

Basal Cell Carcinoma in Young Adults
The aim is to determine incidence and trends of BCC in young adults over the period 1979-1989 using Northern Region Cancer Registry data (population about 3 million). The frequency of second tumours in this young age group will also be determined, in order to assess whether young patients with BCC have a high risk of developing further tumours. This will be effected by search of hospital records and contact with general practitioners. At the same time, the frequency of recurrences and metastases will be recorded and histological types of BCC in this age group analysed.

TYPE: Incidence
TERM: Adolescence; Histology; Multiple Primary; Recurrence; Registry; Trends
SITE: Skin
REGI: Northern UK (UK)
TIME: 1990 - 1992

DIDCOT

773 **MacGibbon, B.H.** 01116
National Radiological Protection Board, Chilton, Didcot Oxon OX11 ORQ, United Kingdom (Tel.: (0235)831600; Fax: 833891 ; Tlx: 837124 RADPRO G)
COLL: Kendall, G.M.; Muirhead, C.R.

National Registry for Radiation Workers
A long-term follow-up study is being conducted to investigate any relationship between occupational exposure to radiation and mortality, in particular from cancer. Any worker in the U.K. who is monitored regularly for exposure to radiation, and for whom dose records are kept and can (with consent) be

UNITED KINGDOM

transferred, is entered in the registry. Currently about 100,000 workers are included, having a collective dose in excess of 3000 man Sv. Follow-up information is being obtained principally from the National Health Service Central Registers, but also from the Department of Social Security. Data validation includes checks of health physics records and sample checks of the follow-up mechanisms. The main analysis will consist of a test for trend in mortality rates across different dose groups. The estimate and confidence limits for the trend in cancer rates with dose will be compared with results from studies of populations exposed to high radiation doses, such as the Japanese atomic bomb survivors.

TYPE: Cohort; Incidence; Registry
TERM: Dose-Response; Monitoring; Occupation; Radiation, Ionizing; Trends
SITE: All Sites
OCCU: Radiation Workers
TIME: 1976 - 1993

774   MacGibbon, B.H.   03502
National Radiological Protection Board, Chilton, Didcot Oxon OX11 ORQ, United Kingdom (Tel.: (0235)831600; Fax: 833891 ; Tlx: 837124 RADPRO G)
COLL: Darby, S.C.; Kendall, G.M.; Muirhead, C.R.; Doll, R.

Health of Participants in the UK Atmospheric Nuclear Weapon Tests
At the request of the Ministry of Defence (MOD), a study is being carried out of mortality and cancer incidence among participants in the UK programme of atmospheric nuclear weapons tests, held in Australia and the Pacific. A group of 22,347 participants and a control group of 22,326 servicemen and civilians who did not participate were identified from MOD archives. The National Health Service Central Registers provide the principal mechanism for tracing the study subjects and acquiring death certificates and cancer registration details. However, additional follow-up via the Department of Social Security increased the number of deaths fully identified by 6.5% for the period to the beginning of 1984 (J. Epidemiol. Comm. Hlth 45(1):65-70, 1991). Based on the follow-up over this period, it was concluded that test participation does not have a detectable effect on life expectancy or on the risk of cancer, apart possibly from small hazards of leukaemia and multiple myeloma (Br. Med. J. 296:332-338, 1988). A further five years' follow-up is planned. Analyses based on the combined participant and control cohorts have also been used to examine mortality among UK servicemen who served abroad in the 1950s and 1960s (Br. J. Ind. Med. 47:793-794); mortality from non-violent causes was less than that expected from rates for all men in England and Wales.

TYPE: Cohort
TERM: Radiation, Ionizing
SITE: All Sites
TIME: 1983 - 1993

DUNDEE

775   Lloyd, O.L.L.   04023
Univ. of Dundee, Dept. of Community Medicine, Ninewells Medical School, Dundee DD1 9SY, United Kingdom (Tel.: (0382)60111)
COLL: Williams, F.L.R.

Distribution of Selected Cancers within Scottish Towns
The study analyses the patterns of mortality within subdivisions of towns to identify areas where the incidence (or mortality) of specified cancers is abnormally high. Data on socio-economic factors will also be analysed. Where the data indicate specific environmental or occupational aetiologies, case-control studies will be organized to elucidate the hypotheses generated. A paper by F.L.R. Williams appeared in Scottish Medical Journal 35:136-139, 1990.

TYPE: Incidence; Mortality
TERM: Environmental Factors; Occupation; Socio-Economic Factors
SITE: All Sites
TIME: 1986 - 1993

* 776   Williams, F.L.R.   05009
Ninewells Hosp. and Medical School, Dept. of Epidemiology and Public Health, Dundee DD1 9SY, Scotland, United Kingdom (Tel.: (0382)60111/3104; Fax: 644197 ; Tlx: 76293 uldund g)
COLL: Lloyd, O.L.L.

UNITED KINGDOM

Lung Cancer and Sources of Air-Borne Pollution
It has been suggested that abnormal sex ratios of births may signal the presence of environmental stresses. If so, the detection of such abnormalities would constitute a simple screening procedure to alert medical and environmental health authorities to hazards to health. Previous work has demonstrated a link between abnormal sex ratios of births, a high incidence of lung cancer, and air-borne pollution from foundries. The purpose of this (pilot) study is to evaluate the hypotheses that the sex ratios of births are abnormal (either low or high) in areas where the parents have been residentially exposed to air pollution from foundries or incinerators; and that mortality from all causes and/or from lung cancer in these areas is abnormally high. The sex ratios and mortality will be investigated around 12 sources of air pollution (six incinerators; six foundries); 12 comparison areas will also be investigated.

TYPE: Correlation; Mortality
TERM: Air Pollution; Sex Ratio
SITE: Lung
TIME: 1990 – 1992

EDINBURGH

777   MacLaren, W.M.   04550
Inst. of Occupational Medicine, 8 Roxburgh Place, Edinburgh EH11 1JZ, United Kingdom (Tel.: (031)6675131)
COLL: Agius, R.; Crawford, N.

Exposure to Low Levels of Radon and Thoron Daughters
The aim of the project is to study relationships between exposure to low levels of radon and thoron daughters and mortality, particularly that due to lung cancer, at 10 British coal mines. The study population consists of approximately 15,000 British coal miners employed at 10 mines in Scotland, the North and Midlands of England, and South Wales. All of these men attended either the first round of medical surveys (1953-1958), or the second (1958-1963), of the British National Coal Board's Pneumoconiosis Field Research. Miners involved in this project were medically examined at five-yearly intervals over a period of approximately 27 years. A systematical attendance record system has ensured that information on where the men worked within the mines and for how long, is stored on computer file. The intention is to combine these data with underground measurements of radon and thoron daughter activities, gathered during the 1970's, in order to produce estimates of cumulative exposure. The facilities of the National Health Service Central Register are being used to collect mortality data, and the statistical analysis will compare exposure histories of miners who have died with those of survivors.

TYPE: Cohort
TERM: Metals; Occupation; Radiation, Ionizing; Registry
SITE: Lung
CHEM: Radon; Thoron
OCCU: Miners, Coal
REGI: OPCS (UK)
TIME: 1988 – 1992

HARTSHILL

* 778   Knight, T.   04964
Univ. of Keele, School of Postgrad. Medicine, and Biological Sciences, Thornburrow Drive, Hartshill Stoke-on-Trent ST4 7QB, United Kingdom (Tel.: (0782)49144; Fax: 747319)
COLL: Forman, D.; Newell, D.; Buchanan, D.

Biochemical Markers for Detection of Gastric Precancerous Disease
The aim of this study is to assess the feasibility of using serum levels of pepsinogen I and II to detect atrophic gastritis (AG) and intestinal metaplasia (IM) in a high risk population of 1,000 men aged 25-60 years, employed in local industries. Subjects with Pepsinogen I level < 20 ng/ml or Pepsinogen I;II < 1.5 will be designated cases of AG with/without IM (estimated prevalence 2-4%, i.e. 20-40 cases). All cases of AG and IM and age-matched controls per case, will be endoscoped and biopsied (2 prepyloric, 2 at the incisura, 2 fundic) to confirm the diagnosis. Serum gastrin levels, the prevalence of intragastric infection with H. pylori and current and dietary patterns will be assessed using a three-day semi-quantitative dietary record and a food frequency questionnaire. Serum pepsinogen and gastrin levels will be determined by RIA. Infection with H. pylori will be determined by measurement of antibodies to the

UNITED KINGDOM

organism in serum. Dietary intake data will be analysed by computer software developed at the MRC Dunn Nutrition Unit, Cambridge, UK.

TYPE: Case-Control; Cohort
TERM: Atrophic Gastritis; BMB; Biochemical Markers; Diet; H. pylori; Premalignant Lesion; Serum
SITE: Stomach
TIME: 1990 - 1992

LANCASTER

* 779 Gorst, D.W. 04985
Royal Lancaster Infirmary, Ashton Rd, Lancaster LA1 4RP, United Kingdom (Tel.: (0524)65944)

Northwest Adult Leukaemia Register
This study is a prospective collection of demographic, medical and follow-up data on all cases of acute myeloid, acute lymphoblastic and chronic granulocytic leukaemia occurring in adults (aged > 15) in a geographically predefined area. Data are used to test hypotheses of incidence (e.g. coastal excess) and to establish true outcome and survival figures in unselected patients. At present, there are over 1,000 patients on file. Data are obtained from the haematologist on diagnosis using a proforma, stored and analysed using microcomputer and standard statistical techniques.

TYPE: Incidence; Registry
TERM: Geographic Factors; Survival
SITE: Leukaemia (ALL); Leukaemia (AML); Leukaemia (CML)
TIME: 1987 - 1992

LEEDS

* 780 Cade, J.E. 04956
Dept. of Public Health Medicine, 20 Hyde Terrace, Leeds LS2 9LN, United Kingdom (Tel.: (0532)334862; Fax: 334852)
COLL: Taylor, I.; Waters, W.E.; Perry, M.; Jackson, A.; Campbell, M.

Diet in Women with Asymptomatic Breast Cancer
The aim of the study is to compare nutrient intake (especially fat intake) between women with asymptomatic breast cancer, benign breast disease and normal, control women. Over a two-year period, women attending the assessment clinics of the breast screening programmes in Portsmouth and Southampton will be included in the study. All women are interviewed prior to diagnosis. The interview includes questions on diet and other possible risk factors and women are given a more detailed dietary questionnaire for completion unaided. Height, weight, waist and hip girth are also measured. It is anticipated that about 300 women with breast cancer, 400 women with benign breast disease and 1,400 control women will be included in the study. The control group will be randomly selected from women who are found at the assessment clinic to have no breast disease. Due to the advent of breast screening in the UK for women aged 50-64 years, this is the first oppportunity to study women with asymptomatic breast cancer. The opportunity for bias is reduced since at the time of interview the diagnosis is not known to the women or to the interviewer.

TYPE: Case-Control
TERM: Diet; Fat; Nutrition; Physical Factors
SITE: Breast (F)
TIME: 1989 - 1992

781 Cartwright, R.A. 04046
Univ. of Leeds, Leukaemia Research Fund, Centre for Clinical Epidemiology, 17 Springfield Mount, Leeds LS2 9NG, United Kingdom (Tel.: (0532)443517; Fax: 426065)
COLL: McKinney, P.A.; Ricketts, T.J.

Follow-Up in a Cellophane Plant
Following reports of an excess of leukaemia in a locality in the South West of England (and also the hazards of exposure to carbon disulphide) a follow-up survey has been started in the largest cellophane plant in the UK. The workforce records are available since the 1940s, but the person years of observation have yet to be calculated. The workforce has been approximately 2,000/year and it is anticipated that there will be approximately 50,000 person-years of observation.

UNITED KINGDOM

TYPE: Cohort
TERM: Chemical Exposure; Occupation; Solvents
SITE: Leukaemia
CHEM: Carbon Disulphide
OCCU: Cellophane Manufacturers
TIME: 1986 – 1992

782 Cartwright, R.A. 04049
Univ. of Leeds, Leukaemia Research Fund, Centre for Clinical Epidemiology, 17 Springfield Mount, Leeds LS2 9NG, United Kingdom (Tel.: (0532)443517; Fax: 426065)
COLL: Bodmer, W.; Harnden, D.G.

Epidemiology of Cancer Families
As part of a national effort to identify and study families with multiple similar cases an attempt has been made to identify all such families occurring in the North and Midlands of England. Descriptive epidemiology techniques as well as a molecular biological approach to the material acquired are being employed. So far over 30 information pedigrees with over 250 members have been identified and studied. Cases are identified from the specialist leukaemia/lymphoma registry, testicular tumour registry and case-control studies of leukaemias, breast cancer and testicular tumours. In addition interested clinicians notify the group of case families.

TYPE: Case Series
TERM: Genetic Factors; Genetic Markers; Heredity; Pedigree
SITE: Breast (F); Colon; Leukaemia; Lymphoma; Testis
TIME: 1985 – 1993

* 783 Cartwright, R.A. 05215
Univ. of Leeds, Leukaemia Research Fund, Centre for Clinical Epidemiology, 17 Springfield Mount, Leeds LS2 9NG, United Kingdom (Tel.: (0532)443517; Fax: 426065)
COLL: Alexander, F.E.; Staines, A.

Case-Control Study of Adult Acute Leukaemia
The hypotheses to be tested are: (1) household exposure to radon gas is greater in past case-houses than in control-houses; (2) cases have more occupations related to 'electrical' work; (3) aplasia-forming drugs have been used more frequently by cases; and (4) family history of similar conditions is more frequent in case families. 800 or more cases, and double the number of controls obtained from general practitioners' records, will be interviewed. Data on exposures will be obtained from medical records and from measurements of radon gas.

TYPE: Case-Control
TERM: Chemical Exposure; Drugs; Family History; Occupation
SITE: Leukaemia
CHEM: Radon
OCCU: Electrical Workers
TIME: 1991 – 1996

* 784 Howel, D. 04988
Univ. of Leeds, Dept. of Public Health Medicine, 30 Hyde Terrace, Leeds LS2 9LN, United Kingdom (Tel.: (0532)334853; Fax: 334852)
COLL: Arblaster, L.; Swinburne, L.; Schweiger, M.; Gibbs, A.R.

Asbestos Exposure and Mesothelioma: A Community Study
This is a case-control study aiming to compare asbestos exposure in two groups, and in particular the relative contributions of occupational, household, neighbourhood and incidental contact. Information will be sought from the surviving relatives of recent cases (approximately 200) and matched controls in Yorkshire. The data collected will be related to the results of a quantitative mineral fibre analysis by electron microscopy of lung samples of both cases and controls.

UNITED KINGDOM

TYPE: Case-Control
TERM: Air Pollution; Chemical Exposure; Dusts; Environmental Factors; Occupation; Registry
SITE: Mesothelioma
CHEM: Mineral Fibres
REGI: Yorkshire (UK)
TIME: 1990 - 1992

LEICESTER

785   Jones, D.R.   03837
Leicester Univ. Medical School, Dept. of Community Health, Leicester Royal Infirmary, Clinical Sciences Bldg, Leicester LE2 7LX, United Kingdom (Tel.: (0533)523196; Fax: 523107)
COLL: Goldblatt, P.O.

Cancer Mortality Following Widowhood in the OPCS Longitudinal Study
This study makes use of the OPCS Longitudinal study, in which routinely collected data on deaths, and deaths of a spouse: occurring in a 1% sample of the population of England and Wales in the period 1971-1985, are linked together, and with 1971 and 1981 census records of sample members. The timing and causes of death following the potentially very stressful event of conjugal bereavement may thus be analysed. The effects of several measures of socio-economic status, including social class are being investigated, as are potential effects of social or familial support, measured by household structure and numbers of children. Papers appeared in Stress Med. 2:129-140, 1986 and J. Biosociol. Sci. 19:107-121, 1987.

TYPE: Cohort
TERM: Psychosocial Factors; Socio-Economic Factors; Stress
SITE: All Sites; Breast (F); Lung; Stomach
TIME: 1982 - 1992

786   Jones, D.R.   04448
Leicester Univ. Medical School, Dept. of Community Health, Leicester Royal Infirmary, Clinical Sciences Bldg, Leicester LE2 7LX, United Kingdom (Tel.: (0533)523196; Fax: 523107)
COLL: Ellman, R.; Thomas, B.A.

Bereavement, Coping Strategies, Suppression of Emotion and Breast Cancer Risk
This study aims to test the hypotheses that (1) death of a spouse and (2) a tendency to suppress emotions (particularly anger) are associated with a raised risk of breast cancer. The roles played by coping strategies adopted in response to stressors, and by the availability of social support will be investigated. More than 20,000 women screened at Guildford over the next three years will be asked to provide brief details of any widowhood suffered, and to complete a questionnaire about their emotional and coping responses. Subsequent follow-up will detect new cases of or deaths from breast cancer and hence allow the risk associated with bereavement, etc. to be estimated. Other risk factors which are already measured in evaluating the screening programme will be used as covariates in the analyses.

TYPE: Cohort
TERM: Psychosocial Factors; Registry; Screening; Stress
SITE: Breast (F)
REGI: OPCS (UK)
TIME: 1989 - 1994

LIVERPOOL

787   Osman, J.   02170
Health and Safety Executive, Epidemiology & Medical Statistics Unit, Stanley Precinct, Magdalen House, Bootle Liverpool Merseyside L20 3QZ, United Kingdom (Tel.: (051)9514535; Fax: 951331)
COLL: Hodgson, J.T.

Asbestos Survey
All asbestos workers in the UK are seen at 2-yearly intervals and a morbidity questionnaire completed with clinical examination of lung fields. The mortality experience of the same population is also recorded. Hygiene measurements from some factories are available from 1972. Smoking histories are recorded. A mortality report on male asbestos workers in England and Wales between 1971-1981 was published in

UNITED KINGDOM

Br. J. Ind. Med. 43:158-164, 1986. This study provides continuous surveillance of illness and death in employees known to be exposed to asbestos. An analysis of mortality up to 1990 is planned for 1993.

TYPE: Incidence; Mortality
TERM: Dusts; Minerals; Occupation; Tobacco (Smoking)
SITE: All Sites; Respiratory
CHEM: Asbestos
OCCU: Asbestos Workers
TIME: 1971 - 1993

788 Osman, J. 04473
Health and Safety Executive, Epidemiology & Medical Statistics Unit, Stanley Precinct, Magdalen House, Bootle Liverpool Merseyside L20 3QZ, United Kingdom (Tel.: (051)9514535; Fax: 951331)
COLL: Agius, R.; Hodgson, J.T.

Mortality Study of Scottish Hard Rock Quarry Workers

Current and past workers of about 30 hard rock quarries in Scotland has been enrolled in a prospective mortality study as part of a combined morbidity and mortality study. The total numbers are about 1,500 current and 1,500 past workers. Dust levels and mineralogical composition has been measured at all quarries, and for different quarrying processes. The principal research question for the study was the exposure/response relationship for silicosis, and for any related excesses in respiratory disease mortality. Lung cancer and mortality will also be examined in relation to silica exposure, and compared with national mortality rates.

TYPE: Cohort
TERM: Dusts; Minerals; Occupation
SITE: Lung
CHEM: Silica
OCCU: Quarry Workers
TIME: 1988 - 1993

789 Osman, J. 04475
Health and Safety Executive, Epidemiology & Medical Statistics Unit, Stanley Precinct, Magdalen House, Bootle Liverpool Merseyside L20 3QZ, United Kingdom (Tel.: (051)9514535; Fax: 951331)
COLL: Leon, D.A.

Print Workers Lung Cancer Case-Control Study

A mortality study of 10,791 males in the printing industry found a significant excess of lung cancer amongst machine assistants. In order to make a more definite assessment of the possible lung cancer risk associated with the printing industry, a case-control study was set up to investigate whether there were any relationships with duration of exposure, time period of exposure and place of employment. The study design is case-control, nested within the main cohort study. There are three controls for each of the 121 cases. The controls are selected at random from the population of men of the same age as the case at death and who were at risk in the period that the death occured.

TYPE: Case-Control
TERM: Dose-Response; Latency; Occupation
SITE: Lung
OCCU: Machinists; Printers
TIME: 1987 - 1992

* 790 Osman, J. 05122
Health and Safety Executive, Epidemiology & Medical Statistics Unit, Stanley Precinct, Magdalen House, Bootle Liverpool Merseyside L20 3QZ, United Kingdom (Tel.: (051)9514535; Fax: 951331)
COLL: Hodgson, J.T.; McCaig, R.H.

Childhood Leukaemias in Offspring of Radiation Workers

This is a case-control study based on a population drawn from the past and present workforce of the nuclear fuel reprocessing plant at Sellafield. Information will be gained from official records and other sources. The aim of the study is to evaluate the effects on risk of the occupational histories of the fathers of children with leukaemia and non-Hodgkin's lymphoma. Factors taken into account will include internal and external exposure to ionising radiation, exposure to known carcinogenic chemicals and involvement in contamination incidents.

UNITED KINGDOM

TYPE: Case-Control
TERM: Chemical Exposure; Childhood; Parental Occupation; Radiation, Ionizing; Registry
SITE: Leukaemia; Non-Hodgkin's Lymphoma
OCCU: Power Plant Workers
REGI: OPCS (UK)
TIME: 1990 - 1992

791 Pharoah, P. 04479
Univ. of Liverpool, Dept. of Public Health, P.O. Box 147, Liverpool L69 3BX, United Kingdom (Tel.: (051)7945593; Fax: 7945588 ; Tlx: 627095 unilpl g)
COLL: Ashby, D.; Blettner, M.; Greaves, J.; Roberts, R.J.

Cancer Risk after Exposure to Cutting Oil Mist and Cutting Fluids

A historical prospective mortality study is being undertaken to assess if long term exposure to cutting fluids, during the course of machining, grinding or other cutting operations, is associated with excess mortality, mainly from digestive and respiratory cancers. Data from 25,000 current and past employees, from two automobile factories on Merseyside, form the cohort for the study. Data on the workers are being obtained from personnel records and followed up via OPCS. Site-specific mortality rates will be computed for each exposure category and compared to the rates expected based on the general UK population. Internal comparisons will be made between groups with and without exposure to particular types of cutting fluids. Dose-response models for length and level of exposure will be investigated.

TYPE: Cohort
TERM: Chemical Exposure; Dose-Response; Latency; Occupation; Petroleum Products; Registry
SITE: Colon; Lung; Rectum; Stomach
CHEM: Mineral Oil; PAH
OCCU: Automobile Workers; Machinists
REGI: OPCS (UK)
TIME: 1988 - 1992

* 792 Pharoah, P. 04998
Univ. of Liverpool, Dept. of Public Health, P.O. Box 147, Liverpool L69 3BX, United Kingdom (Tel.: (051)7945593; Fax: 7945588 ; Tlx: 627095 unilpl g)
COLL: Ashby, D.; Blettner, M.

Effects of Exposure to Oil Mist

The aim of this cohort study is to determine whether there is an excess of cancer mortality following exposure to oil mist. A group of workers exposed to oil mist in two car manufacturing industries and a control group of non-exposed persons employed in the same industries are being compared. The cohort, which will involve 25,000 men, is being traced through the National Health Service Central Register, which provides information on cause of death.

TYPE: Cohort
TERM: Occupation; Petroleum Products; Registry
SITE: Lung; Stomach
CHEM: Oil Mist
OCCU: Automobile Workers
REGI: OPCS (UK)
TIME: 1988 - 1992

LONDON

793 Atkin, W.S. 04416
Imperial Cancer Research Fund, Lincoln's Inn Fields, London WC2A 3PX, United Kingdom (Tel.: (01)2420200)
COLL: Morson, B.; Cuzick, J.M.; Bussey, R.; Jass, J.

Cancer Risk in Patients with Adenomas

The aims of this study are to quantify cancer risk in patients found to have adenomas of the large bowel and to determine how this risk is influenced by the number, size and histopathology of the adenomas as well as the age and sex of the patient. The data comprise 2,148 patients who presented at St Mark's Hospital between 1957 and 1979 in whom the only neoplastic lesions were adenomas. These patients have been followed for up to 30 years and the risk of subsequent large bowel cancer determined. In a

UNITED KINGDOM

proportion of cases, repeated examination of the bowel and removal of adenomas occurred. The effect of these examinations on future risk of cancer has been taken into account in the calculation.

TYPE: Cohort
TERM: High-Risk Groups; Histology; Premalignant Lesion; Sex Ratio
SITE: Colon
TIME: 1985 - 1992

* 794   **Barreto, S.M.**   05210
London School of Hygiene, and Tropical Medicine, Dept. of Epidemiology, Keppel St., London WC1E 7HT, United Kingdom (Tel.: (071) 6368636; Fax: 4365389)
COLL: Swerdlow, A.J.; Smith, P.G.; Boffetta, P.; Kogevinas, M.; Andrade, A.

Historial Cohort Study of Brazilian Steelworkers

A retrospective cohort study is being conducted to assess the all-cause mortality of 12,000 Brazilian steelworkers between 1 January 1979 and 31 December 1990. National and regional age-specific death rates will be used to generate expected numbers of deaths for this cohort. Work and health histories are being obtained from company records and cause of death will be taken from death certificates.

TYPE: Cohort
TERM: Metals; Occupation
SITE: All Sites
CHEM: Steel
OCCU: Steel Workers
LOCA: Brazil
TIME: 1989 - 1993

795   **Beral, V.**   04291
London Sch. Hygiene & Tropical Medicine, Dept. Epidemiol. & Population Sciences, Epidemiological Monitoring Unit, Keppel St. (Gower St.) , London WC1E 7HT, United Kingdom (Tel.: (01)63686336)
COLL: Fraser, P.M.; Rooney, C.; MacOnochie, N.

Prostatic Cancer in Employees of the United Kingdom Atomic Energy Authority

In the United Kingdom Atomic Energy Authority Mortality Study a statistically significant increase in mortality from prostatic cancer was found in certain groups of radiation workers: men with the highest recorded cumulative exposures to external radiation, and men who had been monitored for exposure to radionuclides, especially tritium. The availability and validity of the information required for a case-control study based on UKAEA records has been assessed and found satisfactory, and record forms have been developed and piloted. A case - control study involving 137 cases of prostatic cancer and three controls per case is now underway to determine if prostatic cancer risk can be linked to any specific aspect or aspects of the employment history of the individuals concerned.

TYPE: Case-Control
TERM: Chemical Exposure; Occupation; Radiation, Ionizing
SITE: Prostate
TIME: 1986 - 1992

796   **Beral, V.**   04420
London Sch. Hygiene & Tropical Medicine, Dept. Epidemiol. & Population Sciences, Epidemiological Monitoring Unit, Keppel St. (Gower St.) , London WC1E 7HT, United Kingdom (Tel.: (01)63686336)
COLL: Fraser, P.M.; Smith, P.G.; Douglas, A.; Carpenter, L.M.; Booth, M.

Nuclear Industry Combined Epidemiological Analysis

The data from three separate studies of employees in the nuclear industry are being combined so that the relationship between exposure to low-level ionising radiation and mortality can be described with greater precision than is possible in any of the studies individually. The combined study population will be of the order of 75,000, including employees of the United Kingdom Atomic Energy Authority (Brit. Med. J. 291:440-447, 1985) of the Atomic Weapons Establishment (Brit. Med. J. 297:757-770, 1988), and of the Sellafield plant of British Nuclear Fuels (Smith, P., Douglas, A.: Br. Med. J. 293:845-854, 1986). All workers will be followed-up to the end of 1986, and approximately 12,000 deaths are expected of which about 3,000 will be attributed to cancer. Analyses will include comparisons within the workforce of the mortality of employees exposed to external and internal sources of radiation in relation to other employees, and of

UNITED KINGDOM

mortality in relation to the recorded level of exposure to external radiation. Comparisons of mortality rates in the combined workforce with national rates will also be made.

TYPE: Cohort
TERM: Dose-Response; Occupation; Radiation, Ionizing
SITE: All Sites
OCCU: Radiation Workers
TIME: 1987 - 1992

797   Beral, V.   02340
London Sch. Hygiene & Tropical Medicine, Dept. Epidemiol. & Population Sciences, Epidemiological Monitoring Unit, Keppel St. (Gower St.), London WC1E 7HT, United Kingdom (Tel.: (01)63686336)
COLL: Fraser, P.M.; Carpenter, L.M.; Booth, M.; MacOnochie, N.

United Kingdom Atomic Energy Authority Mortality Study
In 1985 and 1987 the mortality in 40,000 employees of the United Kingdom Atomic Energy Authority during 1946-1976 was reported (Brit. Med. J. 291:435-439 and 440-447,1985; Inskip et al, Brit. J. Ind. Med. 44:149-160, 1987). The average duration of follow-up was 16 years, the numbers of deaths from cancers at many sites were small, and wide confidence limits surrounded the estimated risks associated with exposure to specific doses of external radiation. Since that report, the follow-up of the entire study population has been placed on a continuous basis and a new analysis covering the years up to 1986 is planned giving an average duration of follow-up of 23 years with numbers of deaths about two thirds greater than in the previous analysis. Site-specific mortality and latency will thus be examined in greater detail than before, and the risks associated with specific doses of external radiation estimated with greater precision. Ascertainment of registered cancers will permit the examination of non-fatal as well as fatal cancers.

TYPE: Cohort
TERM: Dose-Response; Latency; Occupation; Radiation, Ionizing; Registry; Time Factors
SITE: All Sites
OCCU: Radiation Workers
REGI: OPCS (UK)
TIME: 1980 - 1992

798   Cuzick, J.M.   04589
Imperial Cancer Research Fund, Dept. of Mathematics, Statistics, and Epidemiology, Lincoln's Inn Fields, P.O.Box 123, London WC2A 3PX, United Kingdom (Tel.: (01)2420200; Tlx: 265107 ICRF G)
COLL: DeStavola, B.; Cartwright, R.A.; Glashan, P.

Genetic Basis of Palmar Keratoses in Bladder Cancer
A very high prevalence of palmar keratoses in patients with bladder cancer has been reported. A relationship was found with smoking, but the relation was not strong enough to explain the findings. The hypothesis is that palmar keratoses might be a marker for the way carcinogens are metabolised. The present case-control study examines the genetic component of palmar keratoses in bladder cancer. Cases are first-degree blood relatives of bladder cancer patients. Two control groups are used; the spouse and his/her first-degree relatives and general hospital controls. Cases and controls are interviewed and their palms inspected for keratoses. 200 will be studied, along with their spouses, and family members. 400 hospital controls will be sought.

TYPE: Case-Control
TERM: Genetic Factors; Premalignant Lesion; Tobacco (Smoking)
SITE: Bladder
TIME: 1986 - 1992

799   Cuzick, J.M.   04053
Imperial Cancer Research Fund, Dept. of Mathematics, Statistics, and Epidemiology, Lincoln's Inn Fields, P.O.Box 123, London WC2A 3PX, United Kingdom (Tel.: (01)2420200; Tlx: 265107 ICRF G)
COLL: Singer, A.; Tessy, G.; Hollingworth, T.

Cervix Cancer and Papilloma Virus
The aim is to determine the incidence and progressive potential of papilloma viruses in the causation of cervix cancer, and to examine interaction with other possible factors including smoking, oral contraceptive use, and sexual behaviour. 100 patients with invasive cervix cancer under the age of 40 will

be examined in a case-control study, with controls matched by age and family doctor. A prospective study of 2,000 women attending Family Planning Clinics, who are HPV16 positive but with negative cytologies, will also be carried out. Controls will also be selected from this cohort. A quantitative PCR assay for HPV is being used to relate disease state to the level of HPV16.

TYPE: Case-Control; Cohort
TERM: HPV; Oral Contraceptives; PCR; Sexual Activity; Tobacco (Smoking)
SITE: Uterus (Cervix)
TIME: 1985 - 1992

800   Davies, A.P.   04823
The London Hosp., Dept. of Obstetrics and Gynaecology, Ovarian Cancer Screening Unit, Whitechapel, London E1 1BB, United Kingdom (Tel.: (071)3777674)

Multimodal Screening Programme for Familial Ovarian Cancer
It is planned to create a register of 5,000 women aged 35 years and over who are at increased risk of developing ovarian cancer in view of their family history. An at-risk woman will be defined as having: (1) one first-degree relative with ovarian cancer; or (2) two second-degree relatives with ovarian cancer; or (3) two or more relatives who have developed other cancers before the age of 50 (Lynch II families). Details of their family history and exposure to other suspected risk factors will be collected. All volunteers will undergo annual screening for 3 years using a combination of serum CA-125 and real-time pelvic ultrasonography. It is hoped to: a) assess the specificities and sensitivities of the screening methods individually and in combination for pre-clinical ovarian cancer; b) assess whether screening has had an effect upon the stage distribution of the disease at presentation and upon mortality from ovarian cancer; c) identify the genetic basis for the increased cancer risk.

TYPE: Cohort
TERM: Familial Factors; Genetic Factors; Screening; Stage
SITE: Ovary
TIME: 1990 - 1993

* 801   Elliott, P.   05194
London School of Hygiene, and Tropical Medicine, Keppel St. (Gower St.) , London WC1E 7HT, United Kingdom (Tel.: +44 71 6368636; Fax: 4363611)

Small Area Health Statistics Unit (SAHSU)
This is a new government-funded project set up in reponse to a recommendation of a state enquiry into the raised incidence of childhood leukaemia near the nuclear reprocessing plant at Sellafield. The main objective is to study the risk of disease, especially cancer, near industrial sources of environmental pollution. The project is exploiting British mortality data from 1981 and cancer incidence data from 1974 (1975 in Scotland) as well as small-area population data from the decennial census. Geographical data retrieval is by postcode of residence, there being some 1.6 million postcodes in the UK with an average of 15 households per postcode. Health and population data can be rapidly assembled for circles of arbitrary size around any point in Britain, and the observed number of cases related to those expected from national or regional rates, after standardisation for age, sex and socio-economic classification of area. Current projects include a study of larynx cancer around incinerators of oils and solvents, haemangiosarcoma of the liver around vinyl chloride works, leukaemia around benzene works and respiratory cancers around coke works. A study of the distribution of childhood leukaemia in Britain has been undertaken in collaboration with the Childhood Cancer Research Group in Oxford.

TYPE: Incidence; Mortality
TERM: Air Pollution; Cluster; Data Resource; Environmental Factors; Geographic Factors; Petroleum Products; Plastics; Registry; Solvents
SITE: All Sites; Larynx; Leukaemia; Liver; Lung; Lymphoma
CHEM: Benzene; Vinyl Chloride
REGI: OPCS (UK)
TIME: 1987 - 1993

802   Filakti, H.H.   02178
        OPCS, LS Medical Analysis Section, LS Unit, 10 Kingsway, London WC2B 6JP, United Kingdom
COLL: Macdonald-Davies, I.; Bethune, A.; Harding, S.

UNITED KINGDOM

Cancer Incidence and Survival in a 1% Census Sample
Data on cancer incidence and survival from the OPCS Longitudinal Study are being analysed to separate mortality differentials into those associated with differential incidence and those associated with differential survival. Major innovations include the systematic analysis of national data by survival periods within one year and by cause of death. Factors considered in the analysis of differentials include area of residence, social class, housing, household and family circumstances and marriage and fertility history.

TYPE: Incidence
TERM: Lifestyle; Marital Status; Reproductive Factors; Socio-Economic Factors; Survival
SITE: All Sites
TIME: 1980 - 1994

803 Fletcher, C.D.M. 04231
St. Thomas's Hosp., Dept. of Histopathology, Lambeth Palace Rd, London SE1 7EH, United Kingdom (Tel.: (01)9289292)

Soft Tissue Tumours in Malawi
This department provided a diagnostic surgical pathology service for the East African State of Malawi from 1968-1983, during which time approximately 4,000 soft tissue tumours (including over 700 sarcomas) were received. The distribution of these tumours appears different to that seen in the UK, not only in the light of the very large number of Kaposi's sarcomas. It is hoped to be able to compare this unique collection of material with that seen in Europe. Age, sex and histological type will be compared with those in the UK; DNA hybridization techniques may allow identification of cytomegalovirus in HIV in cases of Kaposi's sarcoma prior to the AIDS epidemic.

TYPE: Case Series
TERM: AIDS; Biopsy; HIV; Histology
SITE: Kaposi's Sarcoma; Soft Tissue
LOCA: Malawi
TIME: 1985 - 1993

804 Fraser, P.M. 04435
London Sch. Hygiene & Tropical Medicine, Dept. Epidemiology & Population Science, Epidemiological Monitoring Unit, Keppel St. (Gower St.), London WC1E 7HT, United Kingdom (Tel.: (071)6368638)
COLL: Day, N.E.

Leukaemogenicity of Cytostatic Drugs Used in Cancer Therapy
The aim of this study is to estimate the relative leukaemogenicity of different cytostatic regimes and to establish dose-response curves for agents shown to induce leukaemia. The relevance is two-fold. First, leukaemia is probably the most serious long-term side effect of cytostatic therapy and quantitative information on the degree of effect should be of value to clinicans in choosing treatment regimes. Second, exposure to cytostatic agents represents one of the few occasions on which humans are deliberately exposed to known carcinogens in a situation where both the dose is closely monitored and long-term follow-up ensured. It is thus the most favourable circumstance in which to estimate in quantitative terms the relationship between dose and duration of treatment, and risk of leukaemia development. Case-control studies will be carried out within cohorts of individuals with one of the following malignancies -Hodgkin's disease, non-Hodgkin's lymphoma, ovarian cancer, testicular cancer and breast cancer. The cohorts will be identified from the records of six cancer registries, covering a population of 22 million. Cases of acute and non-lymphocytic leukaemia occurring in these cohorts during 1960-1986 subsequent to the diagnosis of the initial cancer will be identified. The treatment given for the index cancer of these leukaemia cases and a corresponding set of matched controls will be abstracted from the patients' hospital records at the treating centre. Over 250 leukaemia cases will be available for study. The methods of statistical analysis will be based on conditional logistic regression with general relative risk models.

UNITED KINGDOM

TYPE: Case-Control; Cohort
TERM: Chemotherapy; Dose-Response; Multiple Primary; Radiotherapy; Registry
SITE: Leukaemia
REGI: E. Anglia (UK); Mersey (UK); Oxford I (UK); Thames (UK); W. Midlands (UK); Yorkshire (UK)
TIME: 1988 – 1992

805 Johnson, N.W. 04771
Royal College of Surgeons of England, Dept. of Dental Sciences, 35-43 Lincoln's Inn Fields, London WC2A 3PN, United Kingdom (Tel.: (01)4053474/3171; Fax: 8319438)
COLL: Chang, S.E.; Warnakulasuriya, K.A.

Diagnostic and Prognostic Markers for Oral Cancer and Precancer

The study aims to describe the presence and level of expression of a variety of cell and molecular markers in normal, benign hyperplastic, dysplastic and overtly malignant human oral mucosae. Associations of these markers with clinical and histological features are sought. Etiopathogenic hypotheses are generated as a result of these observations. Follow-up information is sought on all patients and it is hoped in the future to relate behaviour of the lesions to expressions of markers at the start of follow-up. A bank of over 200 tissue samples from UK patients, and over 100 cases from the Indian subcontinent is held. Cytokeratin expression and lectin binding are studied by immunocytochemical methods. Oncogene amplifications and mutations are being studied by Southern blot and PCR techniques. The results to date show a differentiation in the in the keratinocytes of lesions which on clinical and histological grounds would be regarded as at high risk of malignant transformation. Carcinomas associated with tobacco chewing habits show a high (35%) prevalence of ras mutations which are not demonstrable in non-tobacco-associated lesions.

TYPE: Case Series
TERM: BMB; Biochemical Markers; PCR; Premalignant Lesion; Tissue
SITE: Oral Cavity
LOCA: India; Sri Lanka; United Kingdom
TIME: 1988 – 1993

806 Johnson, P.J. 04449
King's College Hosp., Liver Unit, Denmark Hill, London SE5 8RX, United Kingdom (Tel.: (01)3263252)
COLL: Zaman, S.

Risk Factors Involved in the Development of Hepatocellular Carcinoma in Patients with Cirrhosis

In this study, the hypothesis being tested, among others, is that HBV infection is associated with the development of PHC. 1,000 patients with histologically confirmed cirrhosis have been entered since April 1978. The risk factors being studied are age, sex, type of cirrhosis, duration of cirrhosis and chronic liver disease, nationality, smoking history and serum markers of hepatitis B infection. The statistical analysis involves Logistic Regression Analysis and the use of Cox's proportional hazardous method.

TYPE: Cohort
TERM: Cirrhosis; HBV; Liver Disease; Tobacco (Smoking)
SITE: Liver
TIME: 1978 – 1993

807 Kazantzis, G. 01490
Imperial College of Science, Technology & Medicine Sect., Centre for Environmental Technology, 48 Prince's Gardens, London SW7 2PE, United Kingdom (Tel.: (071)5895111/5546; Fax: 5847596 ; Tlx: 929484)
COLL: Sullivan, K.R.; Ades, A.E.; Armstrong, B.G.; Blanks, R.G.

Case-Control Studies of Cadmium Exposed Workers

Case-control sets have been taken from a cohort of 6,995 cadmium exposed workers in England (Lancet i: 1424-1427, 1983) and from two other cohorts of approximately 3,000 men, for more detailed investigation of past cadmium and other environmental exposures. With regard to prostatic cancer, marginally increased risks were observed after "high" or "medium " exposure, but these were not statistically significant (Br. J. Ind. Med. 42:540-545, 1985). With regard to lung cancer, mortality was examined in a cohort of 4,393 men employed in a zinc-lead-cadmium-smelter. An increased lung cancer risk was found, particularly evident in those employed for more than 20 years, with a statistically significant trend in SMRs with increasing duration of employment. Matched logistic regression was used to compare the cumulative exposures of cases of lung cancer to those of controls. The increasing lung

UNITED KINGDOM

cancer risk could not be accounted for by cadmium, but was associated with estimates of cumulative exposure to arsenic and to lead, although it was not possible to determine whether this increased risk was due to arsenic, lead or to other contaminants in the smelter (Br. J. Ind. Med. 45:435-442, 1988). Further case-control sets from the cohorts are now being examined. (Heavy Metals in the Environment, J.P. Vernet (Ed.) Vol. 1, pp 304-307, Geneva). An international case-control study on lung and prostatic cancer is currently being initiated.

TYPE: Case-Control
TERM: Metals; Occupation
SITE: All Sites; Lung; Prostate
CHEM: Arsenic; Cadmium; Lead; PAH
TIME: 1983 - 1994

808   **Kazantzis, G.**   03492
Imperial College of Science, Technology & Medicine Sect., Centre for Environmental Technology, 48 Prince's Gardens, London SW7 2PE, United Kingdom (Tel.: (071)5895111/5546; Fax: 5847596; Tlx: 929484)
COLL: Lam, T.H.; Sullivan, K.R.; Blanks, R.G.

Cohort Mortality Study of Cadmium Workers

A cohort of 6,995 men born before 1940 and exposed to cadmium for more than one year between 1942 and 1970 has been followed up until the end of 1979 (Lancet i: 1424-1427, 1983). No excess of deaths due to prostatic cancer, cerebrovascular disease or renal disease was observed. A significant excess of lung cancer deaths was found in men employed for more than 10 years but this was not related to exposure levels. By contrast, a significant excess of deaths due to bronchitis showed a strong relation to both duration and intensity of exposure, predominantly in men with heavy past exposure to cadmium. This cohort study is being continued prospectively. Further environmental and biological monitoring data which are being collected will allow more precise estimates of exposure in subsequent analysis. A five year update to include mortality experience in the cohort to the end of 1984 has been completed. The findings of the original study have been confirmed, but there is now a stronger indication of an excess lung cancer risk related to intensity of exposure, significant for both the five-year and the total period. Papers have been published in Scand. J. Work Environ. Health 14:220-223, 1988 and Environ. Chem. 27:113-122, 1990. A second five-year update has been completed and is currently being prepared for publication.

TYPE: Cohort
TERM: Chemical Exposure; Environmental Factors; Metals; Occupation
SITE: Kidney; Lung; Prostate
CHEM: Cadmium
TIME: 1979 - 1994

* 809   **Lund, V.J.**   05073
Inst. of Laryngology and Otology, 330 Gray's Inn Rd, London WC1X 8EE, United Kingdom (Tel.: (071)8378855; Fax: 8339480)

Malignant Melanoma of Nose and Sinuses

The aim of the study is (1) to determine the incidence of this rare condition and so to assess geographical and environmental factors in its aetiology; and (2) to correlate prognosis with histopathology. A questionnaire is being sent to all UK otolaryngologists to identify incident cases, and patients willing to participate in the study will be interviewed regarding possible aetiological factors. Histological patterns in a retrospective group of 60 patients are being examined using immunocytochemistry and DNA flow cytometry, to determine possible prognostic criteria in a condition where conventional classifications of melanoma (Clark, Breslow) cannot be applied.

TYPE: Case Series
TERM: BMB; Environmental Factors; Geographic Factors; Prognosis
SITE: Melanoma; Nasal Cavity
CHEM: Formaldehyde
TIME: 1990 - 1992

810   **Marmot, M.G.**   04774
London School of Hygiene and, Tropical Medicine, Keppel St., London WC1E 7HT, United Kingdom (Tel.: +44 71 6368636; Fax: 4363611)
COLL: Davey-Smith, G.; Stansfeld, S.

UNITED KINGDOM

Stress and Health Study
Social class differences in mortality and morbidity from a wide range of diseases persist in the UK, USA and other industrialized countries. In a previous study of the British Civil Service, an unexplained threefold higher mortality from cardiovascular and other disease was found in the lowest compared to the highest employment grade. In a new study set up to investigate this, 10,314 British Civil Servants have been enrolled in a longitudinal study of health and disease. The overall aim is to study the effect on health and disease both of the work environment - psychological workload, control over work (both its pacing and content), opportunity for use of skills and social support at work - and of social supports; and the interaction between these psychosocial factors and other established risk factors in the aetiology of chronic disease. To date, a cross-sectional questionnaire survey and medical examination have been completed and a questionnaire re-survey is in progress. Longitudinal data on incidence of cardiovascular and other diseases will be collected by: (1) repeat medical screening and questionnaire administration to the 10,314 participants; (2) collection of sickness absence data from the Civil Service Central Monitoring System; (3) obtaining medical diagnostic data from General Practitioners on prolonged sickness absences; (4) cancer registration and mortality from the National Health Service Central Register.

TYPE: Incidence
TERM: Occupation; Psychological Factors; Psychosocial Factors; Stress
SITE: All Sites
REGI: OPCS (UK)
TIME: 1985 - 1993

* 811   **Marmot, M.G.**   04995
London School of Hygiene and, Tropical Medicine, Keppel St., London WC1E 7HT, United Kingdom (Tel.: +44 71 6368636; Fax: 4363611)
COLL: Swerdlow, A.J.; Grulich, A.; Head, J.

Cancer Mortality in Migrants to England and Wales
Migrant studies have traditionally been of use in separating the effects of the environment and genetics in cancer aetiology. By linking routinely collected country of birth information from death certificates and the 1971 and 1981 censuses, it is possible to calculate site specific cancer mortality rates, as well as mortality rates from other fatal diseases, for all enumerated immigrant populations. A previous study of migrant mortality based on routinely collected statistics around the 1971 census has been published by OPCS. This study will cover a longer time period and contain additional information. In particular, time trends of disease will be analysed, and possible interacting effects of social class and region of residence examined. Varied projects are planned from the data set. Initially these will include a description of cancer mortality in Italian immigrants, cancer and cardiovascular mortality in South Asian immigrants and cancer mortality in African and Caribbean immigrants.

TYPE: Incidence
TERM: Linkage Analysis; Migrants; Socio-Economic Factors; Trends
SITE: All Sites
TIME: 1990 - 1992

812   **McCartney, A.C.E.**   04847
Inst. of Ophthalmology, Dept. of Pathology, 17-25 Cayton St., London EC1V 9AT, United Kingdom (Tel.: (071)3879621; Fax: 2503207)
COLL: Foster, A.; Thomson, J.

Incidence and Spread of Retinoblastoma in Rural Ghana
An apparently abnormally high frequency of retinoblastoma is being observed in the hospital of Agogo, with a marked preponderance of spread into the orbit and optic nerve. The frequency, tribal patterns and methods of prevention are being investigated and correlated with histopathological data, especially the type of spread from the globe of the eye. The possibility of gene amplification by extensive intermarriage is being investigated. Histopathological data include morphometric analysis and immunohistochemical analysis.

UNITED KINGDOM

TYPE: Case Series
TERM: Ethnic Group; Genetic Factors; Prevention
SITE: Retinoblastoma
LOCA: Ghana; United Kingdom
TIME: 1989 - 1993

813 McPherson, C.K. 04555
London School of Hygiene, and Tropical Medicine, Dept. of Public Health and Policy, Keppel St. (Gower St.), London WC1E 7HT, United Kingdom (Tel.: (071)6368636/9272308; Fax: 4363611 ; Tlx: 8953474)

Early Use of Oral Contraceptives and Breast Cancer in Women Aged 35-44
The proportion of women who are at risk of developing breast cancer and who have had prolonged exposure to oral contraceptives (OC) has only recently become large enough to test for possible delayed effects of prolonged OC use before first pregnancy on breast cancer risk. Since the literature on this association is so conflicting a case-control study of women aged 35-44 in three regions is proposed. Around 15% of these women will have been exposed to OCs for more than four years before first pregnancy. 400 cases and 400 controls matched by age and general practitioner (GP) will be recruited and interviewed to obtain accurate OC use data with specially-designed calendars and a detailed schedule. OC histories will be checked by reference to GP records.

TYPE: Case-Control
TERM: Late Effects; Latency; Oral Contraceptives; Registry
SITE: Breast (F)
REGI: Scotland N. (UK)
TIME: 1988 - 1992

* 814 Phillips, R. 05174
St. Marks' Hosp. for Diseases, of the Colon and Rectum, City Rd, London EC1V 2PS, United Kingdom (Tel.: (071)2531050)
COLL: Spigelman, A.D.; Farmer, K.C.R.; Williams, C.B.; Talbot, I.

Gastrointestinal Polyps and Cancer in Familial Adenomatous Polyposis
The aim of this study is to examine the natural history of upper gastrointestinal polyps in familial adenomatous polyposis and to assess the effect of the non-steroidal anti-inflammatory drug sulindac in the management of these polyps. Approximately 150 patients attend the endoscopy unit for surveillance of upper gastrointestinal polyps and 40 patients have been selected for a placebo-controlled trial of sulindac. Oesophago-gastro-duodenoscopy is performed by a single endoscopist using a side-viewing duodenoscope. A videotape record is kept for future comparison in addition to a diagram. Tissue biopsies are taken for histology and cell proliferation studies (bromodeoxyuridine DNA labelling). The interval between surveillance is either one or three years depending on the severity of the polyps. The role of bile in the aetiology of these polyps is being examined.

TYPE: Case Series; Intervention
TERM: BMB; Bile Acid; DNA; Drugs; Familial Adenomatous Polyposis; Histology; Polyps; Premalignant Lesion; Prevention; Tissue
SITE: Gastrointestinal
TIME: 1988 - 1998

815 Smith, P.G. 04128
London School of Hygiene, and Tropical Medicine, Dept. of Epidemiol. and Popul. Sciences, Keppel St., London WC1E 7HT, United Kingdom (Tel.: ((01)6368636; Fax: 4365389 ; Tlx: 8953474)
COLL: Douglas, A.

Mortality of Workers in a Nuclear Reprocessing Plant
A retrospective-prospective cohort study of the 14,000 workers who worked at the Sellafield nuclear reprocessing plant at any time prior to 1976 is being carried out. The objective is to identify any hazards associated with this form of occupational exposure to ionising radiation. Analyses will include comparison of mortality and cancer incidence rates with the general population and between workers exposed to different doses of internal and external radiation. Cancer incidence data will also be examined. A paper appeared in Br. Med. J. 293:845-854, 1986.

UNITED KINGDOM

TYPE: Cohort
TERM: Occupation; Radiation, Ionizing
SITE: All Sites
TIME: 1976 - 1993

* 816 Szarewski, A.M. 05005
Imperial Cancer Research Fund, P.O.Box 123, Lincoln's Inn Fields, London WC2A 3PX, United Kingdom (Tel.: (071)2693006)
COLL: Cuzick, J.M.; Jenkins, D.; Singer, A.; Turk, J.

Effect of Vitamin Supplementation on CIN-I in Cervical Epithelium

Epidemiological studies have suggested that deficiencies in the consumption of beta carotene, vitamin C and vitamin E may increase the risk of cervical cancer and cervical intraepithelial neoplasia. This appears to be an independent risk factor after controlling for variables such as age at first intercourse and number of sexual partners. The effects of vitamin deficiency or supplementation on Langerhans cells in the cervix have not been previously studied. It is thought that vitamin supplementation is likely to be most effective at the earliest stages of disease and CIN1 offers a unique opportunity for studying the effects of dietary supplementation over a relatively short period of time. 400 women volunteers whose most recent smear shows mild dyskaryosis will be randomised (double blind) to take either a vitamin supplement (containing 18 mg beta carotene, 150 mg vitamin C and 75 mg vitamin E) or placebo for 12 months. Blood samples will be taken at each visit to check levels of these vitamins. The women will be colposcoped at the beginning and end of the study. At the last visit, biopsies will be taken from both the lesion and a normal area for histology and immunocytochemistry, principally regarding Langerhans cells. A vaginal infection screen will be carried out at the same time. Ethical approval has been obtained from both the Middlesex Hospital Ethics Committee and Islington Ethics Committee.

TYPE: Intervention
TERM: BMB; Diet; Vitamins
SITE: Uterus (Cervix)
TIME: 1990 - 1993

* 817 Szarewski, A.M. 05006
Imperial Cancer Research Fund, P.O.Box 123, Lincoln's Inn Fields, London WC2A 3PX, United Kingdom (Tel.: (071)2693006)
COLL: Cuzick, J.M.; Jenkins, D.; Singer, A.; Turk, J.

Assessment of the Effect of Smoking Cessation on Langerhans Cells in Cervical Epithelium and CIN-I

Epidemiological studies have established that cigarette smoking is a risk factor for CIN. No study has yet looked at the effects of smoking cessation on Langerhans (antigen presenting) cells in cervical epithelium. 300 women volunteers who have mild dyskaryosis on their most recent cervical smear will attempt to stop smoking for six months. It is estimated that about 10% will succeed. These women will have a colposcopy at the beginning of the study to confirm that they do not have more severe disease than CIN-I. Only a biopsy of a normal area will be taken so as not to affect the natural history of the lesion. At the six month visit, both another normal biopsy and one from the abnormal area will be taken. All biopsies will be analysed histologically and for a number of immunological parameters, of which the most important will be Langerhans cells. The women's smoking histories will be checked using salivary cotinine assays at each visit. There will also be about 30 women who smoke but have a normal smear, and 30 non-smokers, with or without an abnormal smear to act as controls. Whenever a biopsy for immunological testing is taken, an infection screen will be performed, since vaginal/cervical infection could affect the local immune response. Ethical approval has been obtained from both the Middlesex Hospital Ethics Committee and Islington Ethics Committee.

TYPE: Cohort
TERM: BMB; Cytology; Histology; Tobacco (Smoking)
SITE: Uterus (Cervix)
TIME: 1990 - 1993

818 Wald, N.J. 02088
St Bartholomew's Hosp., Medical College, Dept. of Environmental, and Preventive Medicine, Charterhouse Square, London EC1M 6BQ, United Kingdom (Tel.: (01)2530661/8373)
COLL: Bailey, A.

UNITED KINGDOM

The Oxford Study of Men in a Private Health Insurance Scheme
In this prospective study about 22,000 men, members of the British United Provident Association (BUPA), have been recruited. Follow-up is by notification of death (1,136) or cancer (285) from the National Health Service Central Register. The main aim is to investigate factors which may predispose to cancer, in particular tobacco smoking and blood micronutrient levels such as vitamin A, beta carotene and vitamin E. Carboxyhaemoglobin levels and serum or urinary cotinine levels have been used as an index of tobacco smoke absorption. Both vitamin A (retinol) and vitamin E were significantly associated with the risk of cancer and the most likely explanation is that both these low levels were metabolic consequences, rather than precursors, of the cancer. In contrast, the significantly lower levels of beta carotene found in the subjects who subsequently developed cancer, affecting the risk directly or indirectly. Since the last report in this Directory further work has been carried out on passive smoking, serum cholesterol and cancer, vitamin E and colorectal cancer, Helicobacter pylori and stomach cancer and blood pressure and stroke.

TYPE: Cohort
TERM: Cholesterol; H. pylori; Passive Smoking; Registry; Tobacco (Smoking); Vitamins
SITE: All Sites
CHEM: Beta Carotene; Carbon Monoxide; PAH; Retinoids; Tars
REGI: OPCS (UK)
TIME: 1975 – 1992

819   **Wang, D.Y.**   04574
Imperial Cancer Research Fund, P.O.Box 123, Lincoln's Inn Fields, London WC2A 3PX, United Kingdom (Tel.: (01)2420200)
COLL: Kwa, H.G.; Bulbrook, R.D.; Hayward, J.L.

Prolactin and Aetiology of Breast Cancer
There are a variety of epidemiological reasons for believing that hormones are involved in the aetiology of breast cancer. These and animal studies have suggested that the main hormones are oestrogen and prolactin. Prolactin secretion is also enhanced by oestradiol. To test the hypothesis that endocrine factors (especially oestrogen and prolactin) are involved in the aetiology of breast cancer a series of prospective studies has been conducted on the Island of Guernsey since 1961. In two studies blood specimens were collected from two cohorts of over 5,000 apparently healthy women. The dates of these collections were 1967-1976 and 1977-1984. In the first of these two studies it was reported that postmenopausal women who subsequently developed breast cancer had prolactin levels which were in the top 30 percent of the distribution of those who did not develop the disease. Furthermore, multivariate analysis of the two cohorts established that childbearing was associated with a permanent reduction in blood prolactin levels. Follow-up on these women is being continued to establish and confirm whether subsequent breast cancer development is related to high prolactin secretion. Papers have been published in Int. J. Cancer 28:673-676, 1981 (Kwa, H.G. et al), Eur. J. Cancer Clin. Oncol. 23:1541-1548, 1987 and 24:1225-1231, 1988.

TYPE: Cohort; Incidence
TERM: BMB; Hormones; Reproductive Factors
SITE: Breast (F)
CHEM: Prolactin
TIME: 1976 – 1993

820   **Williams, C.B.**   02966
St. Mark's Hosp., City Rd, London EC1 V2PS, United Kingdom (Tel.: (01)2531050)
COLL: Nicholls, R.J.; Macrae, F.A.; Silman, A.

Colorectal Adenoma Follow-Up Study
This is a prospective, randomised, long-term follow-up study of 1,000 colorectal adenoma patients based on colonoscopy, supple-mented where necessary by double contrast barium enema. Patients accruing to the study are allocated to high or low-risk status (high-risk including all patients over 60 years at entry, those 55 years and over with 2 or more adenomas, or those with 5 or more adenomas and malignant polyps at any age). High-risk patients are then randomised to be examined at either 1 or 3-year intervals, and low-risk patients at 3 or 5-year intervals (all groups are stratified by other risk factors), with surveillance to stop at age 75 years. The object will be to determine whether patients randomised to less frequent observation have a different (worse) incidence of colorectal cancer. A 5-year "control group" is being sought. A preliminary analysis will be attempted in 1989.

UNITED KINGDOM

TYPE: Cohort
TERM: Polyps
SITE: Colon; Rectum
TIME: 1982 - 1993

MACCLESFIELD

821 Paddle, G.M. 01145
Imperial Chemical Industries (ICI), PLC, Epidemiology Unit, Alderley Park, Macclesfield SK10 4TJ Cheshire, United Kingdom (Tel.: (0625)514539; Fax: 582897 ; Tlx: 669095/669388)
COLL: Lawrence-Jones, C.; Paterson, J.C.

Applications of a Computerised Occupational Health Records System
For several years, ICI has had computerised records of the mortality and serious morbidity experience of all its employees in the United Kingdom. These data are analysed annually for geographical or product group correlations by calculating of proportional ratios. The data are also used as an invaluable source for ad hoc studies of specific compounds or occupational groups, particularly those described in the literature as presenting an excess risk. If an in-depth investigation is justified, the health and industrial hygiene records held at the production sites on a distributed computer system can be examined. Sufficient demographic and job histories have now been collected for historical prospective cohort studies to be conducted electronically, and for SMR to be calculated. A paper has been published in Banbury Report 9:177-189, 1981.

TYPE: Cohort
TERM: Chemical Exposure; Occupation
SITE: All Sites
TIME: 1975 - 1993

MANCHESTER

822 Kay, C.R. 01090
Royal College of General Practitioners, Manchester Research Unit, 8 Barlow Moor Rd , Manchester M20 OTR, United Kingdom (Tel.: (061)4457771; Fax: 4459650)
COLL: Hannaford, P.C.; Beral, V.

Cancer in Oral Contraceptive Users
This is a cohort study of 23,000 oral contraceptive users and 23,000 non-users, recruited over 14 months starting in May 1968 by 1,400 general practitioners in the UK. The cohort now consists of approximately 15,000 women. Full details of newly reported morbidity, operations, pregnancy and, when appropriate, cause of death are collected at regular intervals. Patients no longer under observation in the main study are 'flagged' at the National Health Service Central Registry, for notification of death or cancer registration. Comprehensive details of all oral contraceptives used are collected, as well as hormonal replacement treatments. The aim is to investigate all aspects of health with regard to oral contraceptive effects; special interests include cardiovascular effects and genital and breast cancers. Particular emphasis is now placed on past use of oral contraceptives.

TYPE: Cohort
TERM: Oral Contraceptives
SITE: All Sites; Breast (F); Female Genital
TIME: 1968 - 1993

823 Youngson, J.H. 02090
Univ. Hosp. of South Mancheste, Dept. of Epidemiology, Kinnaird Rd , Manchester M20 9QL, United Kingdom (Tel.: (061)4458123)
COLL: Smith, A.

Lymphoid Malignancy in North West England
The purpose of the proposed project is to establish incidence rates by age, sex, occupation and socio-economic group for the various sub-types of lymphoid leukaemia and lymphoma, according to a current histological classification, in the North West Region. Geographical variations in incidence and survival within the region will be examined. All cases of Non-Hodgkin's Lymphoma (NHL), ALL and CLL presenting in patients aged over 14 years in the North West Regional Health Authority area 1983-1987 will

UNITED KINGDOM

be included. This study has been extended to include follow-up of all cases to investigate patterns of referal and survival in the North West region.

TYPE: Incidence
TERM: Geographic Factors; Occupation; Sex Ratio; Socio-Economic Factors; Survival
SITE: Leukaemia; Leukaemia (ALL); Leukaemia (CLL); Non-Hodgkin's Lymphoma
TIME: 1985 - 1992

NEWCASTLE-UPON-TYNE

824 Craft, A.W. 04397
The Royal Victoria Infirmary, Dept. of Child Health, Queen Victoria Rd , Sir James Spence Bldg, Newcastle-upon-Tyne NE1 4LP, United Kingdom (Tel.: (091)2325131/24101)
COLL: Dale, G.

Population Screening for Neuroblastoma

Neuroblastoma has an estimated incidence of about 1 in 10,000. Prognosis is stage-dependent and has improved little over the past decade. A pilot screening programme has been in existence in the Northern region of England since 1968 and has screened 30,000 births. Two cases of neuroblastoma have been identifed by screening in this cohort. There has been 97% coverage at six months of age using a filter paper urine specimen extracted from wet nappies. Levels of homovanillic acid (HVA) and vanillylmandelic acid (VMA) have been estimated in the specimens by gas chromatography-mass spectrometry. This is highly sensitive and specific. A large evaluative programme is now being proposed which will screen half a million births over a period of two to three years with one million unscreened births acting as controls. Papers have been published in Ann. Clin. Biochem. 25 (Suppl.):132-133, 1988 (McGill et al.), 25:233-236, 1988 (Dale et al.) and in Med. Ped. Oncol. 17:373-378, 1989.

TYPE: Incidence
TERM: Biochemical Markers; Cost-Benefit Analysis; Prevention; Registry; Screening; Survival
SITE: Neuroblastoma
REGI: Newcastle (UK); Oxford II (UK)
TIME: 1986 - 1993

825 Parker, L. 04564
Newcastle Univ., Medical School, Children's Cancer Unit, Dept. of Child Health, Framlington Place, Newcastle-upon-Tyne NE2 4HH, United Kingdom (Tel.: (091)2226000/6958; Fax: 2226222 Tlx: 53654 uninew g)
COLL: Craft, A.W.

Maternal Malignancies and Soft Tissue Sarcoma in Children

The Northern Children's Malignant Disease Registry holds information on all children and young adults diagnosed with cancer in the Northern Region since 1968. This study aims to investigate further the relationship between soft tissue sarcoma and maternal malignancies, particularly of breast (Birch et al., Br. J. Cancer 49:325-331, 1984). The families of 400 patients diagnosed with soft tissue sarcoma before age 25, since 1954 will be traced. Questionnaires will be sent to general practitioners and any cases of maternal cancers will be investigated using medical records and pathological specimens whenever possible. Incidence of malignancies in the case mothers will be compared with that of a control population.

TYPE: Incidence
TERM: Childhood; Familial Factors; High-Risk Groups; Registry
SITE: All Sites; Breast (F); Sarcoma; Soft Tissue
REGI: Newcastle (UK)
TIME: 1988 - 1992

* 826 Parker, L. 05038
Newcastle Univ., Medical School, Children's Cancer Unit, Dept. of Child Health, Framlington Place, Newcastle-upon-Tyne NE2 4HH, United Kingdom (Tel.: (091)2226000/6958; Fax: 2226222 Tlx: 53654 uninew g)
COLL: Craft, A.W.; Smith, J.

UNITED KINGDOM

Radiation Exposure in the Parents of Children with Cancer, Congenital Malformation and Down's Syndrome
The aim of the study is to determine whether workers at the Sellafield nuclear fuel reprocessing plant who are parents of children with cancer or congenital malformation or whose children died in infancy have been exposed to an unusually high dose of internal or external radiation when compared with their colleagues whose children do not have these conditions. The children born in Cumbria to the 27,000 men and women employed at the Nuclear Installation, Sellafield (NIS), since 1950 and to around 6,000 contract radiation workers will be identified by linking the workforce file with that of the Cumbrian birth register. It is anticipated that up to 20,000 children will be so identified, 20-30 of whom will have developed a malignant disease and up to 100 will have died. Case children will be identified from cancer registration, congenital malformation registration and death certificates. Radiation records held by the plant will be obtained. Matched control parents will be selected from the workforce file, and radiation dose in the control group will be compared with that in the case-parent group. Matching of parents will be based on age, sex, year of first employment and year of birth of child.

TYPE: Case-Control
TERM: Congenital Abnormalities; Down's Syndrome; Intra-Uterine Exposure; Occupation; Radiation, Ionizing; Registry
SITE: Childhood Neoplasms
OCCU: Radiation Workers
REGI: Newcastle (UK)
TIME: 1990 - 1992

* 827   Parker, L.   05039
Newcastle Univ., Medical School, Children's Cancer Unit, Dept. of Child Health, Framlington Place, Newcastle-upon-Tyne NE2 4HH, United Kingdom (Tel.: (091)2226000/6958; Fax: 2226222 Tlx: 53654 uninew g)
COLL: Craft, A.W.; Smith, J.

Incidence of Cancer, Congenital Malformation (Including Stillbirth) and Down's Syndrome in Cumbria
The purpose of the proposed research is to determine whether (1) children born to workers employed at the nuclear fuel reprocessing plant at Sellafield are at increased risk of cancer, congenital malformation (including stillbirth) or Down's Syndrome, in comparison with the incidence of these conditions in the rest of Cumbria; and (2) there is any increased risk for these conditions with increasing radiation exposure in parents. The incidence of cancer, congenital malformation and infant mortality rate will be determined for Cumbrian children of the plant workforce and for all Cumbrian children born over a similar time period. Any increased risk in the children of the Sellafield workers will be determined. There have been 27,000 employees since 1950, as well as 6,000 contract radiation workers. It is anticipated that up to 20,000 offspring of these workers will be identified and the incidence of ill-health in these children will be compared with the incidence in the remaining 240,000 Cumbrian children born since 1950.

TYPE: Cohort
TERM: Congenital Abnormalities; Dose-Response; Down's Syndrome; Radiation, Ionizing; Registry
SITE: Childhood Neoplasms
REGI: Newcastle (UK)
TIME: 1990 - 1992

NOTTINGHAM

828   Hardcastle, J.D.   03556
Univ. of Nottingham, Univ. Hospital, Queens Medical Centre, Dept. of Surgery, Floor E, West Block, Nottingham NG7 2UH, United Kingdom (Tel.: (0602)709245)
COLL: Chamberlain, J.

Randomised Controlled Trial of Faecal Occult Blood Screening for Colorectal Cancer
The purpose of this trial is to measure the effectiveness of faecal occult blood testing in reducing the mortality from colorectal cancer and to measure the economic effects on health service resources of such a screening service. Data will also be obtained that could be used in the long-term to measure the effectiveness of faecal blood occult screening in reducing the incidence of colorectal cancer by removing adenomas. The effect of different methods of education will also be evaluated in order to achieve the highest compliance. 156,000 persons aged 50-74 years will be identified from General Practitioners' lists and randomly allocated by household to test or control group. The test group will be offered faecal occult blood screening at 2-yearly intervals and each individual will be followed for a

UNITED KINGDOM

minimum of 7 years. The control group are monitored for development of colorectal neoplasia. Persons leaving the trial area have their records flagged at the National Health Service Central Register to permit follow-up for development of colorectal neoplasia and date/cause of death.

TYPE: Cohort; Intervention
TERM: Screening
SITE: Colon; Rectum
TIME: 1981 – 1994

* 829 **Logan, R.F.** 04965
Univ. of Nottingham Medical School, Dept. Publ. Health Medicine & Epidem., Queen's Medical Centre, Clifton Blvd, Nottingham NG7 2UH, United Kingdom (Tel.: (0602)709308; Fax: 709316; Tlx: 37346 uninot g)
COLL: Little, J.H.; Hardcastle, J.D.

Diet and Colorectal Cancer

This case-control study takes advantage of a current colorectal cancer screening trial, which allows the study of asymptomatic as well as symptomatic colorectal cancer patients. Cases and controls are being interviewed in their homes about their past dietary habits using a food frequency approach. Enquiry is also being made about past occupations, leisure activities, past illnesses and drug intakes and family history. Because of difficulties interviewing patients with advanced cancers only patients with Duke's stage A or B cancers are being approached. Two series of control subjects are being assembled: one series who have had recent surgery and a second from family practitioner lists and the records of the screening trial. By February 1991, 80 patients with asymptomatic cancers and 145 with symptomatic cancers had been interviewed. The reponse rate has been 88% in the cancer patients and 78% in the controls. From comparisons with an earlier study of diet in patients with colorectal polyps, it is hoped to identify what dietary constituents or combinations thereof are associated with the development of colorectal cancer. Constituents of particular interest are animal fat and protein intakes and cereal and total fibre intakes. Data collection is scheduled to be complete by early 1992 with a case series of 240.

TYPE: Case-Control
TERM: Diet; Drugs; Family History; Fat; Fibre; Nutrition; Occupation; Physical Activity; Protein
SITE: Colon; Rectum
TIME: 1988 – 1992

* 830 **Logan, R.F.** 05182
Univ. of Nottingham Medical School, Dept. Publ. Health Medicine & Epidem., Queen's Medical Centre, Clifton Blvd, Nottingham NG7 2UH, United Kingdom (Tel.: (0602)709308; Fax: 709316; Tlx: 37346 uninot g)
COLL: Little, J.H.; Hardcastle, J.D.

Colorectal Adenomatous Polyps, Vitamins and Cholesterol

The aim of the study is to determine whether the risk of asymptomatic colorectal adenomas increases with decreasing serum levels of vitamin A and carotene, vitamin E and cholesterol. Blood samples are being obtained at outpatient clinics for the investigation of patients who are found to be faecal occult blood (FOB) positive in the Department of Surgery screening trial and for the follow-up of patients in whom colorectal adenomas or carcinoma were detected as a result. It is anticipated that serum concentrations of these vitamins will be available for about 200 subjects with adenomas, 50 with carcinomas, and 160 FOB-positive subjects in whom no adenoma or carcinoma was found.

TYPE: Case-Control
TERM: BMB; Biochemical Markers; Blood; Cholesterol; Polyps; Premalignant Lesion; Vitamins
SITE: Colon; Rectum
CHEM: Beta Carotene; Retinoids
TIME: 1989 – 1992

OXFORD

831 **Cook-Mozaffari, P.J.** 04527
Imperial Cancer Research Fund, Cancer Epidemiology Unit, Gibson Labs, Radcliffe Infirmary, Woodstock Rd, Oxford OX2 6HE, United Kingdom (Tel.: (0865)56337)
COLL: Draper, G.J.

UNITED KINGDOM

Migration of Young Cancer Patients
Elevated registration ratios, but not mortality ratios, have been observed for leukaemia and for other malignancies among young persons aged 0-24 who were resident in the vicinity of nuclear installations in England and Wales, especially those established before 1955. Much of the discrepancy seems to reflect variations in the efficiency of cancer registration. However, some of the difference may reflect differential migration away from nuclear installations once a child has been diagnosed as having cancer. Records of the Childhood Cancer Research Group in which registration and death information have been linked, will be used to study migration patterns between registration and death.

TYPE: Incidence; Mortality
TERM: Childhood; Geographic Factors; Registry
SITE: Childhood Neoplasms; Leukaemia
REGI: Oxford II (UK)
TIME: 1988 - 1992

832 Cook-Mozaffari, P.J. 04528
Imperial Cancer Research Fund, Cancer Epidemiology Unit, Gibson Labs, Radcliffe Infirmary, Woodstock Rd, Oxford OX2 6HE, United Kingdom (Tel.: (0865)56337)
COLL: Watts, P.; Cochran, S.

Detection of Local Clusters of Cases of Cancer Using Postcodes
Computer files are held which give registration details for cancer patients resident in the South West Region of England and Wales for the years 1961-1980. The postcoding of the address of each patient is almost complete. The data will be used to compare the efficiency of different approaches to the detection of fine-scale clusters of cases of cancer. The methods envisaged are: (1) the calculation of incidence rates at the level of wards and grouped wards; (2) the calculation of ratios of the site of cancer of interest relative to the number of malignancies at other sites within postcode sectors; and (3) cluster analyses in which the distance between cases at the site of interest are compared with the distances between all possible pairs from within the cases and a group of matched controls (choosen from amongst the other cancer cases).

TYPE: Incidence; Methodology
TERM: Cluster; Registry
SITE: All Sites
REGI: S. Western (UK)
TIME: 1988 - 1992

833 Cook-Mozaffari, P.J. 04529
Imperial Cancer Research Fund, Cancer Epidemiology Unit, Gibson Labs, Radcliffe Infirmary, Woodstock Rd, Oxford OX2 6HE, United Kingdom (Tel.: (0865)56337)
COLL: Forman, D.; Darby, S.C.; Doll, R.

Data Bank for Local and Temporal Variations in Cancer Mortality
Tabulations of cancer deaths in England and Wales by pre-1974 Local Authority Areas have been obtained from OPCS for the years 1959-1980 and for the sites of cancer listed in the IARC "Cancer Incidence in Five Continents" series. Aggregations of areas into post-1974 Country Districts or approximations of Country Districts have been made to facilitate the comparison of death rates across the 1974 reorganization of boundaries and to overcome the lack of census information for the year 1981 by pre-1974 Local Authority Areas. Data have so far been analysed for the period 1969-1978 as a basis for studies of geographical variation in the occurrence of leukaemia and cancer of the bladder. It is planned to mobilise the data for the remaining years and to extend investigations to other sites of cancer.

TYPE: Methodology; Mortality
TERM: Geographic Factors; Registry
SITE: All Sites; Bladder; Leukaemia
REGI: OPCS (UK)
TIME: 1988 - 1992

834 Cook-Mozaffari, P.J. 04706
Imperial Cancer Research Fund, Cancer Epidemiology Unit, Gibson Labs, Radcliffe Infirmary, Woodstock Rd, Oxford OX2 6HE, United Kingdom (Tel.: (0865)56337)
COLL: Peto, J.; Davies, J.M.

UNITED KINGDOM

Occurrence of Cancer of the Bladder in Relation to Suspected High-Risk Occupations in England and Wales

Mapping of the death for cancer of the bladder for men aged 25-64 at the level of County District has shown a concentration of elevated rates in certain port and estuarine areas. Multiple regression analyses show a strong geographical association with the chemical industry and with employment in "sea transport and port services". A death-certificate case-control study is being undertaken in areas of elevated risk to explore further the pattern of occupational risk.

TYPE: Case-Control
TERM: High-Risk Groups; Occupation
SITE: Bladder
OCCU: Chemical Industry Workers
TIME: 1988 - 1992

835   Darby, S.C.   02036
       Imperial Cancer Research Fund, Cancer Epidemiology Unit, Radcliffe Infirmary, Gibson Labs.,
       Oxford OX2 6HE, United Kingdom (Tel.: (0865)53762)
COLL: Doll, R.; Smith, P.G.

Mortality of Patients Treated with X-irradiation for Ankylosing Spondylitis

The follow-up of approximately 14,000 patients treated with X-rays for ankylosing spondylitis between 1935 and 1950 has shown, in conjunction with many other studies, that ionising radiation can cause cancer in nearly every organ in the body. The study, which has shown that the cancer risk is increased for at least 20 years after exposure, has been continued, to find out how long the effect lasts and what total risk is associated with a given level of exposure. Results from the extended follow-up of patients who received only a single course of treatment indicate that the increased risk has substantially disappeared by approximately 30 years after exposure. Beyond this the cancer risk in this population returns approximately to normal levels. Detailed dosimetry calculations for a sample of one in every 15 patients in the study have been carried out. Dose-response relationships are now being examined. Follow-up of patients who received more than one course of treatment is also under way. Papers have been published in Br. J. Cancer 55:179-190, 1987 and in Br. J. Radiol. 61:212-220, 1988.

TYPE: Cohort
TERM: Dose-Response; Radiation, Ionizing; Time Factors
SITE: All Sites; Leukaemia
TIME: 1955 - 2000

836   Darby, S.C.   03621
       Imperial Cancer Research Fund, Cancer Epidemiology Unit, Radcliffe Infirmary, Gibson Labs.,
       Oxford OX2 6HE, United Kingdom (Tel.: (0865)53762)
COLL: Doll, R.; Smith, P.G.

Mortality of Women Treated with X-Radiation for Metropathia Haemorrhagica

Previous follow-up of approximately 2,100 patients treated with X-irradiation for metropathia haemorrhagica in Scotland between 1940 and 1960 has shown an excess of deaths from leukaemia and from cancers of the heavily irradiated sites between five and 20 years after treatment. Over the same period the number of deaths from cancer of the breast was below expectation. Preliminary results from an extended follow-up of this group confirm previous findings and indicate that the excess rate of leukaemia is about 1.0 per million women per year per rad in the first 20 years after treatment. This figure is in accord with estimates derived from the survivors of the atomic bomb explosions in Hiroshima and Nagasaki if a simple linear dose-response relationship is assumed. However, studies of women treated with radiation for cancer of the cervix and patients treated with X-rays for ankylosing spondylitis have suggested that this assumption is incorrect when high doses of radiation are delivered to a small volume of marrow. In these studies a lower risk of leukaemia has been observed which may be explained by cell-sterilization at high doses.

UNITED KINGDOM

TYPE: Cohort
TERM: Dose-Response; Radiation, Ionizing
SITE: All Sites; Leukaemia
TIME: 1965 - 2000

837   Darby, S.C.   04370
Imperial Cancer Research Fund, Cancer Epidemiology Unit, Radcliffe Infirmary, Gibson Labs., Oxford OX2 6HE, United Kingdom (Tel.: (0865)53762)
COLL: Silcocks, P.B.; Doll, R.

Lung Cancer in South West England

Radon in houses is one of the most important sources of radiation exposure to which the general public is subject. The principal parts of the body liable to be affected are the bronchi, where radionuclides emitting alpha particles tend to be deposited. The experience of men working in mines where the radon content of the air is unusually high provides good evidence that this type of exposure is a cause of lung cancer. Several studies of the possible effect of radon in houses have already been carried out in other countries. These have confirmed that a small effect is likely to be produced, but further information is needed about the size of the risk. The present study is being carried out in South West England where the highest radon concentrations in houses in Britain are found, and is designed to provide more precise estimates of the size of the risk. It is hoped to collect data for about 600 patients with lung cancer in the course of three years, and at least twice that number of control subjects. Lung cancer patient are interviewed as soon as possible after a presumptive diagnosis of lung cancer has been made. The control group is made up of two parts: first an age-sex-matched sample of patients admitted to the same hospitals with a wide variety of conditions unrelated to smoking, and second an age-sex-matched sample of community controls chosen from the general population of the region. Radon levels in the past houses of both cases and controls are being measured. A paper has been published in Nature 344:824, 1990.

TYPE: Case-Control
TERM: Environmental Factors; Radiation, Ionizing
SITE: Lung
CHEM: Radon
TIME: 1987 - 1992

838   Draper, G.J.   02452
Univ. of Oxford, Radcliff Infirmary, Childhood Cancer Research Group, 57 Woodstock Rd, Oxford OX2 6HJ, United Kingdom (Tel.: (0865)310030; Fax: 514254 ; Tlx: 83147)
COLL: Sanders, B.M.; Stiller, C.A.; Brownbill, P.A.; Bunch, K.J.

Aetiological Studies of Childhood Cancer

A number of related studies are being carried out to investigate possible environmental and genetic factors in the aetiology of childhood cancer. These include (1) analyses of geographical variations internationally; (2) investigation of the occurrence of cancer in the relatives (parents, sibs, twins, children) of children with cancer, particularly children with retinoblastoma; (3) record linkage studies of cohorts of children who may have an increased risk of cancer or for whom possible aetiological can be studied; (4) the effects of parity and parental age. Data from the national registry of childhood tumours maintained by the Group (about 1,200 cases per year) are used for these and other investigations. Recent papers were published in Br. J. Cancer 60:358-365, 1989, and 62:1026-1030 1990 and in Paediatric and Perinatal Epidemiology 4:303-324, 1990.

TYPE: Genetic Epidemiology; Incidence
TERM: Familial Factors; Geographic Factors; Intra-Uterine Exposure; Mutation, Germinal; Record Linkage; Registry; Sib; Twins
SITE: Childhood Neoplasms; Retinoblastoma
REGI: Oxford II (UK)
TIME: 1975 - 1992

839   Draper, G.J.   04688
Univ. of Oxford, Radcliff Infirmary, Childhood Cancer Research Group, 57 Woodstock Rd, Oxford OX2 6HJ, United Kingdom (Tel.: (0865)310030; Fax: 514254 ; Tlx: 83147)
COLL: O'Connor, C.M.; Stiller, C.A.; Vincent, T.J.; Bithell, J.F.; Cook-Mozaffari, P.J.; Muirhead, C.R.

Geographical Studies of Childhood Leukaemia and Cancer

Studies are carried out based on the National Registry of Childhood Tumours which covers virtually all cases of childhood cancer and leukaemia (i.e. those diagnosed at ages 0-14) in England, Scotland and

UNITED KINGDOM

Wales from 1962 onwards, i.e. about 30,000 cases, with addresses, postcodes, wards, (approximate) grid reference and other geographical variables. For children born in 1962 onwards cancer registrations are being linked to birth records and the district of birth added. The objectives of the analyses are to investigate variations in incidence according to both place of diagnosis and place of birth. Plans for the study include analyses of the effect on incidence rates of various geographical and socio-economic factors and of environmental factors such as radon and gamma radiation. Rates around nuclear installations (or other suspected high-incidence areas) will be compared with those for other areas. A particularly appropriate method for investigating the type of hypothesis of most interest for the present study is the 'Poisson maximum' method proposed by Stone (1988). (Reference: 'The Geographical Epidemiology of Childhood Leukaemia and non-Hodgkin Lymphomas in Great Britain 1966-83'. G.J. Draper (ed). HMSO, London, 1991).

TYPE: Correlation; Incidence
TERM: Cluster; Geographic Factors; Radiation, Ionizing; Record Linkage; Registry; Socio-Economic Factors
SITE: Childhood Neoplasms; Leukaemia
REGI: Oxford II (UK)
TIME: 1986 - 1994

840   Forman, D.   04430
Imperial Cancer Research Fund, Cancer Epidemiol. & Clin. Trials Unit, Radcliffe Infirmary, Gibson Lab., Gibson Bldg , Oxford OX2 6HE, United Kingdom (Tel.: (0865)53951; Fax: 310545)
COLL: Pike, M.C.; Chilvers, C.E.D.; Oliver, R.T.D.

Case-Control Study of Testicular Cancer

800 men, aged 15-49 years, with histologically confirmed germ-cell tumours of testis diagnosed between January 1984 and September 1986 have been interviewed in eight different regions of England and Wales. They were asked about their exposure to a number of potential risk factors suggested as being involved in the disease, and their responses are being compared with those from matched population controls. As particular emphasis is being placed on possible in utero hormonal exposures, the mothers of cases and controls have been sent a postal questionnaire concerned with exposures during pregnancy.

TYPE: Case-Control
TERM: Dusts; Hormones; Intra-Uterine Exposure; Occupation; Radiation, Ionizing; Sexual Activity; Solvents
SITE: Testis
CHEM: Asbestos; Mineral Oil
TIME: 1984 - 1992

841   Forman, D.   04431
Imperial Cancer Research Fund, Cancer Epidemiol. & Clin. Trials Unit, Radcliffe Infirmary, Gibson Lab., Gibson Bldg , Oxford OX2 6HE, United Kingdom (Tel.: (0865)53951; Fax: 310545)
COLL: Key, T.; Beral, V.; Hannaford, P.C.; Kay, L.

Prospective Study of Men and Women Aged 45-59 Years

It is proposed to establish a prospective cohort of 50,000 men and women recruited through general practitioners and through breast cancer screening. Subjects will be asked to provide a blood sample and to complete a questionnaire mainly concerned with diet and exercise. As cancers become identified, relevant blood variables and questionnaire information for case groups can be compared with those for disease free subjects. In this way, many fundamental questions concerned with the aetiology of cancer can be addressed. Pilot studies are being undertaken with the Aylesbury District Breast Screening Unit and with some general practitioners. Assessments will be made of compliance, ease of sample collection and validity of the questionnaire. This study will be part of the European Prospective Study of Diet and Cancer, coordinated by IARC.

UNITED KINGDOM

TYPE: Cohort
TERM: BMB; Blood; Diet; Physical Activity
SITE: Breast (F); Colon; Prostate; Stomach
TIME: 1988 – 1993

842 Forman, D. 04432
Imperial Cancer Research Fund, Cancer Epidemiol. & Clin. Trials Unit, Radcliffe Infirmary, Gibson Lab., Gibson Bldg , Oxford OX2 6HE, United Kingdom (Tel.: (0865)53951; Fax: 310545)
COLL: Oliver, R.T.D.; Bodmer, J.

Register of Familial Testicular Tumours

A register has been established to systematically record familial cases of testicular cancer. Such a register should serve firstly to generate more reliable statistics about the familial prevalence of the disease and, secondly, as a resource for biological material for genetic studies. Over 40 confirmed cases of testicular tumours in first degree relatives had been reported to the registry, mainly from oncologists around the country. Worldwide, only about 100 such families have been reported in the literature, and it is clear that the register will add considerably to the number of known cases. In addition, families are also recorded where the index case has a testicular tumour and the relative an undescended testis. To date, 38 such families are registered. At present, blood samples are collected from families reported to the registry in order to examine in more detail the genetic associations with this disease. A sib-pair analysis has been conducted to look at the relationship between tumour and HLA haptotype.

TYPE: Case Series
TERM: BMB; Genetic Factors; HLA
SITE: Testis
TIME: 1985 – 1993

843 Forman, D. 04433
Imperial Cancer Research Fund, Cancer Epidemiol. & Clin. Trials Unit, Radcliffe Infirmary, Gibson Lab., Gibson Bldg , Oxford OX2 6HE, United Kingdom (Tel.: (0865)53951; Fax: 310545)
COLL: Doll, R.; Ferguson-Smith, J.; Bryson, D.; Leech, S.; Packer, P.

Study of Workers Exposed to Nitrate Fertilizers

This study was designed to establish whether employees of a factory producing inorganic fertilizers experience excess mortality from cancer as a result of their exposure to high concentrations of nitrate dust. The cohort, comprising some 1,400 workers, employed for one year or more since 1946, is particularly important in assessing the role of nitrates in cancer aetiology, as many workers have been heavily exposed for several years. The original analysis of these data, covering the period 1946-1981 (Al-Dabbagh et al., Br. J. Ind. Med. 43:507-515, 1986) showed no significant excess for any major form of cancer. The study is currently being updated to include deaths up until 1989. This will allow increased statistical power in analysing the results and a more detailed consideration of rare types of cancer. Also urine samples are being obtained from a representative sample of the employees in order to monitor current exposure levels.

TYPE: Cohort
TERM: BMB; Dusts; Fertilizers; Occupation; Urine
SITE: Bladder; Lung; Oesophagus; Stomach
CHEM: Nitrates
OCCU: Fertilizer Workers
TIME: 1982 – 1993

844 Forman, D. 04453
Imperial Cancer Research Fund, Cancer Epidemiol. & Clin. Trials Unit, Radcliffe Infirmary, Gibson Lab., Gibson Bldg , Oxford OX2 6HE, United Kingdom (Tel.: (0865)53951; Fax: 310545)
COLL: Goss, B.; Swerdlow, A.J.

Epidemiology of Naevi in Childhood

The presence of naevi (moles) is a major risk factor for malignant melanoma. Very little is known about the natural history of naevi, yet there is evidence that congenital naevi and those that appear early in childhood are of special aetiological significance. The aims of this study are to obtain reliable figures for the prevalence of naevi in a sample of Oxford children at birth and 4 years old. A proportion of these children will then be followed up in order to quantify the appearance of naevi during childhood. As well as making naevus counts by anatomical site, the distribution of naevi by size and type will be assessed. The relationship between the appearance of naevi and various risk factors, genetic and environmental, will be investigated. 1,000 new-born babies and 250 4-year old children had been examined and 250 of the babies have been re-examined at the age of one year.

TYPE: Cohort
TERM: Childhood; Environmental Factors; Genetic Factors; Naevi
SITE: Melanoma
TIME: 1988 – 1993

845   Forman, D.   04454
Imperial Cancer Research Fund, Cancer Epidemiol. & Clin. Trials Unit, Radcliffe Infirmary, Gibson Lab., Gibson Bldg , Oxford OX2 6HE, United Kingdom (Tel.: (0865)53951; Fax: 310545)
COLL: Sitas, F.; Newell, F.; Lachlan, G.; Jass, J.; Wild, C.; Wyatt, J.

Epidemiology of Chronic Atrophic Gastritis in Kenya

Parts of rural Kenya are experiencing an epidemic of acute gastritis in young people. Possibly related to this are anecdotal reports of increases in gastric cancer also affecting young people. Two systematic dyspepsia investigations of patients and asymptomatic volunteers in the age range 18-25 years have been carried out. These have shown a high prevalence of atrophic gastritis and a high rate of gastric infection with Heliobacter pylori. 57% of the volunteer group had atrophic gastritis, 70% had evidence of current H. pylori infection and all of them had antibodies of past infection. The occurrence of these findings in such young people is extremely unusual and contrasts strongly with the pattern in developed countries. A survey of about 100 young children has also been carried out to determine the age at which H. pylori infections became established and to examine other factors that may be involved in the aetiology. The role of mycotoxins, particularly aflatoxin, in the aetiology of gastritis has been investigated by examining blood samples from the same groups. Aflatoxin exposure was significantly higher in the dyspeptic patients.

TYPE: Cross-Sectional
TERM: Antibodies; Childhood; Infection; Mycotoxins; Premalignant Lesion; Urine
SITE: Stomach
CHEM: Aflatoxin
LOCA: Kenya
TIME: 1987 – 1992

846   Forman, D.   04455
Imperial Cancer Research Fund, Cancer Epidemiol. & Clin. Trials Unit, Radcliffe Infirmary, Gibson Lab., Gibson Bldg , Oxford OX2 6HE, United Kingdom (Tel.: (0865)53951; Fax: 310545)
COLL: Sitas, F.; Newell, D.; Jewell, D.P.; Buiatti, E.; Palli, D.; Elwood, P.C.; Webb, P.

Epidemiology of Chronic Atrophic Gastritis

Most gastric cancers are preceded by a long period of chronic atrophic gastritis, which may be considered as a premalignant lesion. The role of two serological markers is being evaluated as predictors of gastritis, for eventual use in identifying populations at high risk of gastric cancer. The markers under examination are the serum level of the enzyme pepsinogen I and the presence of antibodies to the bacterium Heliobacter pylori. Parallel serum and gastric biopsy samples have been obtained from 100 patients attending endoscopy clinics in Oxford. The biopsies exhibited a range of gastric histopathology from normal to severely gastritic. Using the markers one can discriminate between those with and without gastritis with a high degree of sensitivity. It is now planned to investigate the serologically defined prevalence of gastritis in a cross-section of populations in Italy (n = 800) and Wales (n = 4,000). Case-control studies will then be conducted, comparing dietary and other factors in persons with and without gastritis.

UNITED KINGDOM

TYPE: Case-Control; Cross-Sectional
TERM: Antibodies; Biochemical Markers; Biopsy; Diet; Enzymes; H. pylori; Pepsinogen; Premalignant Lesion
SITE: Stomach
LOCA: Italy; United Kingdom
TIME: 1987 - 1992

847 Forman, D. 04926
Imperial Cancer Research Fund, Cancer Epidemiol. & Clin. Trials Unit, Radcliffe Infirmary, Gibson Lab., Gibson Bldg , Oxford OX2 6HE, United Kingdom (Tel.: (0865)53951; Fax: 310545)
COLL: Møller, H.; Tjønneland, A.; Palli, D.; Zakelj, M.P.; Zatonski, W.A.; de Backer, G.; Boeing, H.; Tulinius, H.; Calheiros, J.M.; Abid, L.; Manousos, O.; Kyrtopoulos, S.; Coleman, M.P.; Roy, P.; Wild, C.; Newall, D.; Elder, J.; Knight, T.; Webb, P.; Haubrich, T.; Hengels, K.J.; Fukao, A.; Tsugane, S.; Kaye, S.; Potter, J.

European Correlation Study of Biological Markers for Gastritis and Gastric Cancer (EUROGAST)
The primary aim of this international collaborative study is to investigate putative biological correlates of gastric cancer. Centres have been chosen on the basis of contrasting gastric cancer rates and within each centre IOO men and women from each of the two age groups 25-34 and 55-64 years will be randomly selected from the general population. Each participant will provide a blood sample and complete a brief questionnaire. Sera will be assayed for H. pylori antibodies, pepsinogen, and a number of antioxidant micronutrients. All of these parameters, which have been related to the development of gastritis, will be correlated with gastric cancer incidence. Lymphocyte DNA preparations will be assayed for the presence of alkylating agent-DNA adducts, formed by the activity of N-nitroso compounds.

TYPE: Correlation
TERM: BMB; DNA Adducts; H. pylori; Pepsinogen; Serum
SITE: Stomach
CHEM: N-Nitroso Compounds
LOCA: Algeria; Belgium; Denmark; Germany; Greece; Iceland; Italy; Japan; Poland; Portugal; United Kingdom; United States of America; Yugoslavia
TIME: 1989 - 1992

848 Hawkins, M.M. 04799
Univ. of Oxford, Childhood Cancer Research Group, 57 Woodstock Rd, Oxford OX2 6HJ, United Kingdom (Tel.: (0865)310030; Fax: 514254)
COLL: Kinnier Wilson, L.M.; Kingston, J.E.; Draper, G.J.

Second Primary Neoplasms after Childhood Cancer
Survivors of childhood cancer are at a greater risk of developing a subsequent cancer than the corresponding general populations. The purpose of our work is to monitor the incidence of cancers occurring subsequent to childhood cancer diagnosed in Britain, and to investigate the aetiology of these second primary cancers. The study is based on the national population-based register for childhood neoplasms and utilizes information from a national register for multiple primary tumours which has been established. The incidence of secondary leukaemia is being investigated in a cohort of approximately 16,500 one-year survivors of childhood cancer diagnosed between 1962 and 1983. The incidence of solid tumours as second primaries is being studied among about 13,400 three-year survivors of childhood cancer diagnosed between 1940 and 1983. A case-control study of secondary leukaemia is being carried out to investigate the relationship between the relative risk of subsequent leukaemia and the types and doses of chemotherapy and radiotherapy received by patients. Twenty seven secondary leukaemias have so far been included. It is planned to study the aetiology of other types of second primary tumours using a case-control design. Papers have been published in Br. J. Cancer 56:331-338 and 339-347, 1987 and in Int. J. Radiat. Oncol. Biol. Phys.

TYPE: Case-Control; Cohort
TERM: Chemotherapy; Childhood; High-Risk Groups; Multiple Primary; Radiotherapy; Survival
SITE: Leukaemia
REGI: Oxford II (UK)
TIME: 1981 - 1993

849 Hawkins, M.M. 04906
Univ. of Oxford, Childhood Cancer Research Group, 57 Woodstock Rd, Oxford OX2 6HJ, United Kingdom (Tel.: (0865)310030; Fax: 514254)
COLL: Draper, G.J.

UNITED KINGDOM

Pregnancies and Offspring of Childhood Cancer Survivors

As a result of the substantial increases in survival following childhood cancer, an increasing number of survivors reach reproductive age and consider having a family. The outcome of such pregnancies and the health of all offspring are of considerable interest because the therapy which many survivors received is potentially mutagenic to germ cells, and because of the heritable component of childhood cancer and the association between some childhood cancers and congenital abnormalities, and in order to provide reliable estimates of various outcomes for those who counsel survivors and their families. Survivors of reproductive age are identified from the National Registry of Childhood Tumours and information is obtained from general practitioners' records and hospital consultants. Questions include those relating to infertility, spontaneous abortion, termination or stillbirths experienced by survivors, and also any congenital abnormalities, cancer or other genetic disease among offspring born to survivors. There are 2,900 survivors born in 1962 or before, and 2,650 survivors born between 1963 and 1968. A cohort of about 1,500 offspring of survivors has been established and will be followed up to compare mortality and cancer observed with that expected from general population rates. Recent papers include Int. J. Cancer 43:399-402 and 975-978, 1989 and Br. J. Obstet. Gynaecol. 96:378-380, 1989.

TYPE: Cohort
TERM: Abortion; Congenital Abnormalities; Familial Factors; Genetic Factors; Infertility; Mutation, Germinal; Pregnancy; Survival
SITE: Childhood Neoplasms
REGI: Oxford II (UK)
TIME: 1981 - 1993

850 Hawkins, M.M. 04927
Univ. of Oxford, Childhood Cancer Research Group, 57 Woodstock Rd, Oxford OX2 6HJ, United Kingdom (Tel.: (0865)310030; Fax: 514254)
COLL: Kingston, J.E.; Kinnier Wilson, L.M.

Long-Term Survival, Causes of Late Deaths and Cure after Childhood Cancer

There have been great improvements in survival following childhood cancer in recent decades. The purpose of our work is to monitor the mortality occurring after patients have survived several years from diagnosis of childhood cancer in Britain, and to examine in detail the causes of such deaths. The study is based on the National Registry for Childhood Tumours. There are about 11,500 patients who were diagnosed before 1982 and are known to have survived at least three years. Comparison of the numbers of deaths observed with those expected from mortality occurring among the general population identifies departures from expected mortality and provides information relating to cure and late adverse effects of childhood neoplastic disease and its treatment. This work has been of benefit to survivors obtaining life insurance. The number of deaths occurring among individuals who have survived at least five years are sufficiently small (750 among 4,100 five-year survivors diagnosed before 1971, in Britain) to investigate in each case the course of events leading to death from hospital notes, post-mortem reports and death certificates. This detailed look at the sequence of events leading to death has identified some deaths that might have been preventable. Papers were published in Arch. Dis. Child. 64:798-897, 1989 and 65:1356-1363, 1990.

TYPE: Cohort
TERM: Late Effects; Survival; Treatment
SITE: Childhood Neoplasms
REGI: Oxford II (UK)
TIME: 1981 - 1993

851 Key, T. 04876
Imperial Cancer Research Fund, Cancer Epidemiology Unit, Radcliffe Infirmary, Gibson Bldg, Oxford OX2 6HE, United Kingdom (Tel.: (0865)53751; Fax: 310545)
COLL: Bishop, D.T.; Silcocks, P.B.

Case-Control Study of Prostate Cancer

This is a case-control study of men with prostate cancer, matched with healthy controls of the same age selected from the same general practice as the case. Cases and controls are interviewed at home to collect information on diet, exposure to metals, and sexual behaviour, and to collect samples of hair and nails. Preliminary analysis will be done on 100 case-control pairs, to test the hypothesis that risk is directly related to dietary fat, to exposure to cadmium or nickel, and to a venereally transmitted agent.

UNITED KINGDOM

TYPE: Case-Control
TERM: BMB; Diet; Fat; Hair; Metals; Sexual Activity; Sexually Transmitted Diseases; Toenails
SITE: Prostate
CHEM: Cadmium; Nickel
TIME: 1990 - 1992

852 Mant, D. 03839
University of Oxford, Department of Community Medicine, Gibson Labs., Radcliffe Infirmary, Woodstock Rd, Oxford, United Kingdom (Tel.: (0865)511293)
COLL: Pike, M.C.; Vessey, M.P.; Chilvers, C.E.D.

Adenocarcinoma of the Cervix
The objective of the study is to determine the contribution of oral contraception, among other factors, to the aetiology of adenocarcinoma of the cervix in women under 45 years. It is intended to study 150 cases (diagnosed between 1 January 1984 and 31 December 1988) and 450 age-matched controls selected from general practice lists. A comparison group of 300 cases of squamous carcinoma of the cervix in women under 40 years of age and 900 age-matched controls will also be sought. A formal structured questionnaire will be administered to both cases and controls by a trained interviewer. Pathological specimens will be reviewed independently to confirm the diagnosis and evidence of HPV infection will also be sought by DNA-DNA hybridisation.

TYPE: Case-Control
TERM: DNA; HPV; Histology; Infection; Oral Contraceptives; Virus
SITE: Uterus (Cervix)
TIME: 1985 - 1992

853 Neil, H.A.W. 04849
Univ. of Oxford, Dept. of Public Health & Primary Care, Radcliffe Infirmary, Woodstock Rd, Oxford OX2 6HE, United Kingdom (Tel.: (0865)511293)
COLL: Thorogood, M.; Cohen, D.L.; Mann, T.I.

Cholesterol, Cancer and Diabetes
294 diabetic patients identified in a population-based survey in 1982 in Oxford agreed to participate in a clinical study. Various clinical and biochemical variables were measured, including lipoprotein profile. The records of patients were flagged by the National Health Service Central Registry so that deaths could be identified. By 31 December 1990, 93 deaths had occurred. The aim of the study is to examine the predictors of mortality, particularly in type II diabetes, and to examine the relationship between cholesterol concentration and cancer mortality in diabetes. Survival analyses will be undertaken using Cox's proportional hazard model.

TYPE: Cohort
TERM: Biochemical Markers; Cholesterol; Diabetes; Survival
SITE: All Sites
TIME: 1982 - 1992

854 Stiller, C.A. 04930
Univ. of Oxford, Childhood Cancer Research Group, 57 Woodstock Rd, Oxford OX2 6HJ, United Kingdom (Tel.: (0865)310030; Fax: 514254 ; Tlx: 83147 ccrg)
COLL: Bunch, K.J.; Draper, G.J.; Lennox, E.L.; Loach, M.J.; Robertson, C.M.; Sanders, B.M.

Incidence of Childhood Cancer
The National Registry of Childhood Tumours includes virtually all cases of cancer and leukaemia in children aged 0-14 in England, Scotland and Wales from 1962 onwards. Incidence rates are analysed for all childhood cancers and for particular diagnostic groups. Analyses in progress or planned include (i) trends in leukaemia incidence rates; (ii) completeness of ascertainment of cases by national record systems; (iii) neuroblastoma incidence and mortality as background rates for trials of neuroblastoma screening; (iv) incidence rates in relation to residence in new towns and other indicators of population mixing which might be related to infectious factors in the aetiology of leukaemia. Relative frequencies of the major childhood cancers among the principal ethnic groups in Britain are being studied using data of the UK Children's Cancer Study Group, with particular reference to whether children of Asian (Indian sub-continent) ethnic origin have patterns of incidence similar to those occurring in Asia or among White Caucasians in Britain. International variations in incidence rates are studied using data from the IARC study of international childhood cancer incidence.

UNITED KINGDOM

TYPE: Incidence
TERM: Childhood; Ethnic Group; Geographic Factors; Infection; Migrants; Registry; Trends
SITE: Childhood Neoplasms; Leukaemia; Neuroblastoma
REGI: Oxford II (UK)
TIME: 1975 - 1994

855　Thorogood, M.　04671
Univ. of Oxford, Dept. Commun. Medicine & Gen. Practice, Radcliffe Infirmary, Woodstock Rd, Oxford OX2 6HE, United Kingdom (Tel.: (0865)511293; Fax: 310545)
COLL: Vessey, M.P.; Mann, J.R.; McPherson, C.K.

Cancer and Other Diseases in Vegetarians

This is a prospective study of people eating different diets, including some 300 vegans, 800 fish eaters who do not eat meat, 5,000 lacto-ovo-vegetarians and 5,000 omnivorous controls (the friends and relatives of the non-meat-eaters). Subjects have been flagged with the National Health Service Central Register, and information on both deaths and cancer registrations is being provided. Some 3,000 subjects have provided blood samples, and 5,000 have completed 4-day estimated-weight dietary records. The aim of the study is to examine the different mortality experience of non-meat-eaters and meat-eaters, especially with reference to digestive tract cancers, breast cancer, and cardiovascular disease.

TYPE: Cohort
TERM: BMB; Blood; Diet; Registry; Vegetarian
SITE: Breast (F); Gastrointestinal
REGI: OPCS (UK)
TIME: 1980 - 1992

856　Vessey, M.P.　00762
Univ. of Oxford, Dept. of Public Health & Primary Care, Gibson Labs., Radcliffe Infirmary, Gibson Bldg, Oxford OX2 6HE, United Kingdom (Tel.: (0865)511293/4)
COLL: Painter, R.; Mackintosh, L.

Long-Term Follow-Up Study of Women Using Different Methods of Contraception

Seventeen family planning clinics are participating in a prospective study of the beneficial and adverse effects of different methods of contraception. Over 17,000 women are under observation of whom 57% were using oral contraceptives on admission to the study; 25% were using a diaphragm and 18% were using an intrauterine device. During follow-up, each subject is questioned at return visits to the clinic and a record of pregnancies and their outcome, hospital visits, changes in contraceptive methods and the results of cervical smears, is accumulated. Women who default are sent a postal questionnaire and if this is not returned, are telephoned or visited in their homes to collect the necessary information. So far, data representing 275,000 woman-years of follow-up are available for analysis. Losses to follow-up have been fewer than 3 per 1,000 per annum. Analyses completed to date concern the entry characteristics of the subjects, patterns of morbidity and mortality, outcome of pregnancy, fertility after discontinuation of contraception, and efficacy of different methods. Most of the findings fit in well with those of retrospective studies and with those of the Royal College of General Practitioners prospective study. A detailed interim report of the study was published in the Journal of Biosocial Science in October 1976. About 65 publications have now emerged from this study.

TYPE: Cohort
TERM: Contraception; Cytology; Fertility; Oral Contraceptives; Parity; Pregnancy
SITE: All Sites; Breast (F); Uterus (Cervix)
TIME: 1968 - 1994

857　Vessey, M.P.　01289
Univ. of Oxford, Dept. of Public Health & Primary Care, Gibson Labs., Radcliffe Infirmary, Gibson Bldg, Oxford OX2 6HE, United Kingdom (Tel.: (0865)511293/4)
COLL: Hunt, K.

Mortality and Cancer Incidence in Women Receiving Hormone Replacement Therapy

In this investigation women seen at 20 menopause clinics in different parts of the country who were currently using hormone replacemen treatement and had used it continuously for at least one year, were recruited and are being followed up indefinitely using the facilities of the National Health Service Central Registries. Mortality rates and cancer incidence rates are computed and comparisons are made with national mortality and cancer incidence data. Cancer of the endometrium and cancer of the breast are

UNITED KINGDOM

sites of particular interest, as well as cardiovascular disease. About 5,000 women are taking part in the study. Reports were published by Hunt et al. in Br. J. Obstet. Gynaecol. 94:620-635, 1987 and 97:1080-1086, 1990.

TYPE: Cohort
TERM: Hormones; Menopause; Registry
SITE: Breast (F); Uterus (Corpus)
REGI: OPCS (UK)
TIME: 1978 - 1992

858 Vessey, M.P. 01430
Univ. of Oxford, Dept. of Public Health & Primary Care, Gibson Labs., Radcliffe Infirmary, Gibson Bldg, Oxford OX2 6HE, United Kingdom (Tel.: (0865)511293/4)
COLL: Fairweather, D.; Meara, J.

Follow-Up Study of Women Exposed to DES in Pregnancy and their Offspring

In the early 1950s, a randomised double-blind controlled trial of the prophylactic use of stilboestrol in pregnancy was conducted at a London teaching hospital. The study involved over 1,000 primigravidae and the treatment group received stilboestrol according to the Smith and Smith regimen. The participants in the study and their offspring have been identified and traced and death certificates obtained for those who have died. Information about the health of survivors has also been sought from general practitioners and the analysis of the results has been published in Brit. J. Obstet. and Gynaecol. 90:1007-1017, 1983. Long-term follow-up will continue indefinitely. A recent publication appeared in Brit. J. Obstet. and Gynaecol. 90:620-622, 1989.

TYPE: Cohort
TERM: Drugs; Hormones; Intra-Uterine Exposure
SITE: Breast (F); Female Genital
CHEM: DES
TIME: 1978 - 1992

859 Vessey, M.P. 03000
Univ. of Oxford, Dept. of Public Health & Primary Care, Gibson Labs., Radcliffe Infirmary, Gibson Bldg, Oxford OX2 6HE, United Kingdom (Tel.: (0865)511293/4)
COLL: Colin Jones, D.G.; Langman, M.J.S.; Lawson, D.H.

Post-Marketing Surveillance of the Safety of Cimetidine

A total of 10,000 takers of cimetidine and a like number of controls were recruited at Oxford, Portsmouth, Nottingham and Glasgow and followed up for one year. A large excess of gastric ancer was seen in the takers group, but the great majority of the cancers were the result of cimetidine being given to patients with symptoms of undiagnosed malignancy. The takers have been 'labelled' in the National Health Service Central Registries and are being followed up. Papers have been published in Br. J. Med. 291:1084-1088, 1985, Alimentary Pharmacology & Therapeutics 1:167-177, 1987 and in The Lancet i:1453, 1989.

TYPE: Cohort
TERM: Drugs; Registry
SITE: Gastrointestinal
CHEM: Cimetidine
REGI: OPCS (UK)
TIME: 1978 - 1994

PENARTH

860 Wagner, J.C. 00380
MRC, External Staff Team, on Occupational Lung Diseases, Llandough Hosp., Penarth South Glamorgan CF6 1XW, United Kingdom (Tel.: (0222)708761)
COLL: Pooley, F.D.; Gibbs, A.R.

Asbestos Lung Survey in the United Kingdom

The aims of this study are to compare the types and quantities of mineral fibres present in lungs obtained at post-mortem from different sources and to relate the mineral fibre content with the histological findings. Initially this study depended upon the material collected in the United Kingdom in 1977, the

UNITED KINGDOM

preliminary results of which were published in Ann. Occup. Hyg. 26(1-4):423-431, 1982. The findings from the investigation demonstrated the importance of the selective retention of mineral fibres in the lungs. The investigation has now been extended to cover factories in the United Kingdom and the United States and asbestos mines in Canada, Turkey and Cyprus. The problem of mineral fibre contamination of rural communities elsewhere has also been studied. At present there are indications that where there has been major exposure to chrysotile asbestos, the significant fibre found in the lung is usually tremolite, a fibre of amphibole type. Further studies have confirmed that the amphiboles are far more capable of causing disease than chrysotile. The investigation has been extended to include cases of mesothelioma in which there is no evidence of occupational exposure to mineral fibres.

TYPE: Correlation
TERM: Autopsy; Dusts; Histology; Occupation
SITE: Lung; Mesothelioma; Pleura
CHEM: Asbestos; Mineral Fibres
OCCU: Asbestos Workers; Miners, Asbestos; Shipyard Workers
LOCA: Canada; Turkey; United Kingdom; United States of America
TIME: 1977 - 1992

SALISBURY

861   Thomas, H.F.   01915
Wessex Regional Health Authority, Odstock Hosp., Salisbury SP2 8BJ, United Kingdom (Tel.: (0722)336262)
COLL: Winter, P.D.; Donaldson, L.

Mortality Study of Pest Control Officers
A study is being conducted to examine the mortality of pest control officers employed by local authorities in England and Wales. At first an interest was taken in the use of 'antu' (alpha-naphthylthiourea), a rodenticide which chemical tests suggested could be associated with bladder cancer, and which was generally withdrawn from use in the UK in 1967. A report on sub-groups exposed to 'antu' has been published (Br. Med. J. 285:927-931, 1982) suggesting that there was an increased incidence of bladder tumours. A population of 1,500 men presently employed is now being followed-up to examine their mortality experience over the next 10 to 20 years. Information on pesticides handled and personal habits, e.g., smoking and alcohol consumption, has been obtained from a 10% random sample of men (J.R.S.H. 6:204-206, 1986). 115 deaths have been notified to date. An analysis of causes is being undertaken.

TYPE: Cohort
TERM: Alcohol; Chemical Exposure; Drugs; Insecticides; Occupation; Pesticides; Tobacco (Smoking)
SITE: All Sites
CHEM: Lindane; Naphthylthiourea, Alpha; Warfarin
OCCU: Pesticide Workers
TIME: 1979 - 2002

SOUTHAMPTON

862   Coggon, D.   04795
Univ. of Southampton, Southampton General Hosp., MRC Environmental Epidemiology Unit, Tremona Rd, Southampton SO9 4XY, United Kingdom (Tel.: (0703)777624; Fax: 704021)
COLL: Barker, D.J.P.; Winter, P.D.

Food Storage and Domestic Crowding in Childhood and Stomach Cancer
This study tests the hypothesis that domestic crowding and poor food storage facilities in childhood are risk factors for stomach cancer. Information about food storage facilities and the number and size of rooms in 6,000 houses have been abstracted from the records of a housing survey carried out in Chesterfield in 1936. Occupants of these houses will be identified from the 1939 census and followed up for mortality. Risk of stomach cancer will be related to type of food storage facility and measures of crowding in 1939.

UNITED KINGDOM

TYPE: Cohort
TERM: Childhood; Diet; Socio-Economic Factors
SITE: Stomach
TIME: 1989 - 1992

863   Coggon, D. 04796
Univ. of Southampton, Southampton General Hosp., MRC Environmental Epidemiology Unit, Tremona Rd, Southampton SO9 4XY, United Kingdom (Tel.: (0703)777624; Fax: 704021)
COLL: Pannett, B.; Winter, P.D.; Wright, H.

Study of Workers Exposed to Mineral Acid Mists

This study tests the hypothesis that occupational exposure to mineral acid mists is a cause of laryngeal cancer. A retrospective cohort is being assembled to include exposed and unexposed workers from the steel and battery manufacturing industries. Subjects will be followed for mortality and cancer incidence through the National Health Service Central Register. The association between laryngeal cancer and acid mists will be examined principally by a nested case-control approach.

TYPE: Case-Control; Cohort
TERM: Occupation
SITE: Larynx
CHEM: Sulphuric Acid
OCCU: Battery Plant Workers
TIME: 1989 - 1993

* 864   Coggon, D. 05048
Univ. of Southampton, Southampton General Hosp., MRC Environmental Epidemiology Unit, Tremona Rd, Southampton SO9 4XY, United Kingdom (Tel.: (0703)777624; Fax: 704021)
COLL: Maitland, N.

Lung Cancer in Butchers

This study tests the hypothesis that an apparent excess of lung cancer among meat workers in routine statistics of occupational mortality and cancer incidence is explained by a carcinogenic papilloma virus. Lung cancer deaths in butchers in England and Wales during 1987 and 1988 have been identified from death certificates, and histological material is being sought from the tumours where it is available. It is hoped to obtain approximately 50 specimens together with material from a similar number of lung tumours in controls of the same age and sex who were not butchers. Evidence of papillomavirus DNA in the tumours will be sought by a PCR method.

TYPE: Case-Control
TERM: Autopsy; HPV; Occupation; PCR
SITE: Lung
OCCU: Meat Workers
TIME: 1990 - 1992

865   Gardner, M.J. 02533
Univ. of Southampton, Southampton General Hosp., MRC Environmental Epidemiology Unit, Tremona Rd, Southampton SO9 4XY, United Kingdom (Tel.: (0703)777624; Tlx: 47661 sotonu g)
COLL: Winter, P.D.; Powell, C.A.

Effect of Exposure to Amosite Asbestos Dust on Mortality of Insulation Board Manufacturers

This is a historical cohort study on populations of 6,000 workers in two factories in the United Kingdom in which insulation board was manufactured from amosite asbestos for approximately 30 years to 1978. Comparisons made between mortality from cancer and other diseases and the experience of appropriate populations of men of similar age show excess of mesothelioma and lung cancer. A paper was published in Int. J. Epidemiol. 13:3-10, 1984. An individual worker notification study of passing on information on the results of the investigation and anti-smoking advice is underway, including updating of the mortality results, showing a continued doubling of lung cancer risk. A paper was published in J. Soc. Occup. Med. 38:69-72, 1988.

UNITED KINGDOM

TYPE: Cohort
TERM: Chemical Exposure; Dusts; Occupation
SITE: All Sites; Gastrointestinal; Lung; Mesothelioma; Respiratory
CHEM: Asbestos, Amosite
OCCU: Insulation Board Manufacturers
TIME: 1979 - 1992

866 Gardner, M.J. 03632
Univ. of Southampton, Southampton General Hosp., MRC Environmental Epidemiology Unit, Tremona Rd, Southampton S09 4XY, United Kingdom (Tel.: (0703)777624; Tlx: 47661 sotonu g)
COLL: Winter, P.D.; Jones, R.D.

Mortality Follow-Up of Pottery Workers

This is a historical cohort mortality follow-up of 6,000 pottery workers who were previously surveyed for respiratory morbidity in 1970 during a cross-sectional survey. The National Health Service Central Register and the Department of Health and Social Security records have been used to identify the vital status and cause of death of these workers during the ensuing years. The primary objective is to examine the consequences of low exposure to silica dust in a large occupational cohort, particularly levels of lung cancer, which has been shown to be raised among silicotics. Exposure will be assessed by occupational histories, job titles and dust measurements. Cigarette smoking habits were collected in 1970 and can be adjusted for in the analysis. A publication by Winter et al. appeared in IARC Scientific Publication No. 97, 1990.

TYPE: Cohort
TERM: Dusts; Occupation; Registry
SITE: All Sites; Lung
CHEM: Silica
OCCU: Potters
REGI: OPCS (UK)
TIME: 1985 - 1992

867 Gardner, M.J. 03633
Univ. of Southampton, Southampton General Hosp., MRC Environmental Epidemiology Unit, Tremona Rd, Southampton S09 4XY, United Kingdom (Tel.: (0703)777624; Tlx: 47661 sotonu g)
COLL: Coggon, D.; Pannett, B.; Pippard, C.

Mortality Experience of Workers Exposed to Ethylene Oxide

This is a standard historical cohort mortality follow-up of about 1,500 workers employed in manufacture and use of ethylene oxide and 1,500 workers employed in hospital sterilizing units where ethylene oxide has been used. The source of information on workers is factory and hospital personnel, medical and other records. The primary objective is to obtain further epidemiological data on the postulated relationship of ethlyene oxide exposure to the development of leukaemia, in particular, among humans. The majority of exposed British workers are included in the study and preliminary results do not suggest any excess of leukaemia among them. A paper has been published in Br. Ind. Med. 46:860-865, 1989.

TYPE: Cohort
TERM: Occupation
SITE: All Sites; Leukaemia; Lymphoma
CHEM: Ethylene Oxide
TIME: 1985 - 1992

868 Gardner, M.J. 03634
Univ. of Southampton, Southampton General Hosp., MRC Environmental Epidemiology Unit, Tremona Rd, Southampton S09 4XY, United Kingdom (Tel.: (0703)777624; Tlx: 47661 sotonu g)
COLL: Osmond, C.; Pannett, B.; Powell, C.A.

Study of Workers Exposed to Formaldehyde in the British Chemical Industry

This is a historical cohort study on a population of about 15,000 workers in six factories in the United Kingdom which have produced or used formaldehyde for some 30 or more years. The study was set up to address the possibility that formaldehyde may be a human carcinogen following the demonstration that at high doses it caused nasal tumours in rats. The initial analyses have not supported the hypothesis at the levels of exposure experienced in the study groups. Further updates of the mortality findings are in hand and an extended analysis will be finalised during 1991.

UNITED KINGDOM

TYPE: Cohort
TERM: Occupation; Plastics
SITE: All Sites; Nasopharynx; Respiratory
CHEM: Formaldehyde
OCCU: Chemical Industry Workers
TIME: 1981 - 1992

869 Gardner, M.J. 03635
Univ. of Southampton, Southampton General Hosp., MRC Environmental Epidemiology Unit, Tremona Rd, Southampton S09 4XY, United Kingdom (Tel.: (0703)777624; Tlx: 47661 sotonu g)
COLL: Pannett, B.; Winter, P.D.; Powell, C.A.

Mortality of Workers in a Factory Manufacturing Asbestos Cement Products Using Only Chrysotile (White Asbestos)

A standard cohort mortality follow-up has been carried out on 2,167 workers in one factory in England. The National Health Service Central Register has been used to determine vital status and cause of death of these employees. Individuals have been assigned to qualitative exposure levels based upon their re-coded job histories at the factory and extensive discussions about the characters of jobs with long-term employees of the firm. Comparisons have been made between mortality from various causes, including cancers, and the experience of appropriate populations of men of similar ages. No excess of lung cancer has been observed, nor of other asbestos-related diseases. A paper has been published in Br. J. Ind. Med. 43:726-732, 1986 as well as a review article in J. Soc. Occup. Med. 36:124-126, 1986.

TYPE: Cohort
TERM: Chemical Exposure; Dusts; Occupation; Registry
SITE: All Sites
CHEM: Asbestos, Chrysotile
OCCU: Cement Workers
REGI: OPCS (UK)
TIME: 1984 - 1992

870 Gardner, M.J. 04698
Univ. of Southampton, Southampton General Hosp., MRC Environmental Epidemiology Unit, Tremona Rd, Southampton S09 4XY, United Kingdom (Tel.: (0703)777624; Tlx: 47661 sotonu g)
COLL: Winter, P.D.; Pannett, B.; Powell, C.A.

Mortality Study of Workers in the Man-Made Mineral Fibre Production Industry in the United Kingdom

A cohort of some 4,500 workers in England is being followed up to examine possible effects of exposure to glass fibre on long term mortality, in particular in relation to lung cancer and mesothelioma. The study is part of the European collaborative investigation coordinated by IARC. To date results have been published up to the end of 1984 and a further analysis will be carried out to the end of 1989. A recent nested case-control analysis of lung cancer did not reveal any striking associations with fibre dimensions - and overall in the cohort there was no excess of lung cancer compared to local rates; no case of mesothelioma was observed. Papers have been published in Scand. J. Work. Environ. Health 12:85-93, 1986, suppl. 1 and Br. J. Ind. Med. 45:613-618, 1988.

TYPE: Case-Control; Cohort
TERM: Dose-Response; Dusts; Latency; Occupation
SITE: All Sites; Lung; Mesothelioma
CHEM: Glass Fibres
OCCU: Mineral Fibre Workers
TIME: 1978 - 1993

871 Gardner, M.J. 03189
Univ. of Southampton, Southampton General Hosp., MRC Environmental Epidemiology Unit, Tremona Rd, Southampton S09 4XY, United Kingdom (Tel.: (0703)777624; Tlx: 47661 sotonu g)
COLL: Wagner, J.C.; Winter, P.D.

Lung Fibre Burden of Amosite Asbestos in Persons Employed in Insulation Board Manufacture

This is a case-control study within a cohort of ex-workers at an insulation board and panel factory which used amosite asbestos for approximately 30 years. The object is to compare the lung fibre burdens in persons exposed to amosite who died from lung cancer or mesothelioma (approximately 50 cases) with lung fibre burdens in persons who died from other causes (approximately 50 controls).

UNITED KINGDOM

TYPE: Case-Control
TERM: Autopsy; Chemical Exposure; Dusts; Occupation
SITE: Lung; Mesothelioma
CHEM: Asbestos, Amosite; Mineral Fibres
OCCU: Insulation Board Manufacturers
TIME: 1983 - 1992

872 Hall, A.J. 03724
Univ. of Southampton, Dept. of Med. II, Southampton General Hosp., Professorial Medical Unit, Level D, South Lab. & Pathology Block, Tremona Rd, Southampton S09 4XY, United Kingdom (Tel.: (0703)777222)

Mortality of Hepatitis B Positive Blood Donors in England and Wales

Of 2,880 men and 1,054 women who were found to be hepatitis B surface antigen positive by the Blood Transfusion Service in England and Wales between 1971 and 1981, more than 92% were traced to the end of 1983. Five deaths from hepatocellular carcinoma had been reported in men, giving a relative risk compared with the male population of England and Wales as a whole of 42. There had been no deaths from liver cancer in women. In both men and women there was a greater than ten-fold increase in the risk of death from chronic liver disease. Secondary preventive action may be indicated. This is an on-going study which will be up-dated every five years. A paper was published in Lancet I: 91-93, 1985.

TYPE: Cohort
TERM: Antigens; Infection; Virus
SITE: Liver
TIME: 1983 - 1993

873 Hall, A.J. 04379
Univ. of Southampton, Dept. of Med. II, Southampton General Hosp., Professorial Medical Unit, Level D, South Lab. & Pathology Block, Tremona Rd, Southampton S09 4XY, United Kingdom (Tel.: (0703)777222)
COLL: Inskip, H.M.; Corrah, T.; Whittle, H.C.; Rizetto, M.; Bosch, F.X.

Aetiology of Chronic Liver Disease in The Gambia

The aetiology of chronic liver disease in The Gambia is unknown. Alcohol is not consumed by the majority of the population and yet there is a considerable frequency of liver disease. This case-control study proposes to examine the risks of liver disease associated with Hepatitis B infection, delta virus infection, aflatoxin, and bush tea consumption. All cases of hepatocellular carcinoma registered by The Gambian Cancer Registry will be included in the study. The chronic liver disease patients and controls will be recruited from the clinics of Dr Hall and Dr Corrah. The cases will each be matched in the analysis to two controls for age within five years, sex and district of normal residence. It is expected that the relative risk of both chronic liver disease and hepatoma would exceed 2.0, so that a sample size of 100 cases of each disease seems reasonable. The hepatomas should be accumulated at the rate of two per week, so taking a year to collect. It may take two years to recruit 100 incident cases of chronic liver disease.

TYPE: Case-Control
TERM: HBV; Mycotoxins; Registry; Tea
SITE: Liver
CHEM: Aflatoxin
REGI: Gambia (Gam)
TIME: 1988 - 1992

874 Morton, N.E. 04680
Univ. of Southampton, Southampton General Hosp., Dept. of Community Medicine, South Block, Southampton S09 4XY, United Kingdom (Tel.: (0703)777222; Tlx: 47661)
COLL: Iselius, E.L.

Research Programme in Genetic Epidemiology

Two epidemiology projects are conducted. The leukaemia project tests the hypothesis that susceptibility in Down's syndrome depends on mode of origin and that development of acute leukaemia following transient leukaemia of the newborn requires a mutation at an unlinked locus. Mode of origin is being determined from DNA markers. The control group consists of Down's cases without leukaemia. The neurofibromatosis project tests the hypothesis that sporadic cases are mostly germinal mutations, with a high frequency determined by a large gene size, consistently estimated directly from segregation analysis of pedigrees, and indirectly from reproductive performance assuming mutation-selection

UNITED KINGDOM

equilibrum. The linkage project is creating recombination maps of chromosomes 10 and X by multiple pairwise analysis, using DNA markers in families and XY trisomies.

TYPE: Genetic Epidemiology
TERM: DNA; Down's Syndrome; Mutation, Germinal; Pedigree; Segregation Analysis
SITE: Leukaemia
TIME: 1988 - 1992

875   Pippard, C.   04909
Univ. of Southampton, MRC Environmental Epidemiology Unit, Tremona Rd, Southampton, United Kingdom (Tel.: (0703)777624)
COLL: Barker, D.J.P.; Bridges, B.A.

Cohort Study of Ataxia-Telangiectasia Patients
The study aims to determine whether the incidence of any cancer is elevated in homozygotes and heterozygotes. Heterozygotes comprise a significant proportion of the population and may be an at-risk group. The patients and their parents, grandparents, siblings and children are being registered. They will be traced at the National Health Service Central Register and copies of cancer registration and death certificates obtained. Observed numbers of cancers in the cohort will be compared with expected numbers derived from national statistics allowing for age, sex, time period and region of residence. Allowance will be made for the probability of heterozygosity in grandparents.

TYPE: Cohort
TERM: Ataxia Telangiectasia; High-Risk Groups; Registry
SITE: All Sites
REGI: OPCS (UK)
TIME: 1982 - 1993

SUTTON

* 876   Bell, C.M.J.   04954
Thames Cancer Registry, Block E, 15 Cotswold Rd, Sutton SM2 5PY, United Kingdom (Tel.: (081)6427692)
COLL: Horwich, A.

Treatment for Early Stage Seminoma of the Testis
This is a population-based study of a cohort of patients with testicular cancer (N = 831) diagnosed at an early stage between 1961-1985 in the two South Thames Health Regions of England. The patients are being followed up to 1989 to determine the excess risk of second cancers and any excess of causes of death. The aim is to test findings in other studies of an excess of leukaemia and urinary tract cancer following radiotherapy to the pelvis.

TYPE: Cohort
TERM: Multiple Primary; Radiotherapy; Registry
SITE: All Sites; Testis
REGI: Thames (UK)
TIME: 1988 - 1992

877   Moss, S.M.   04882
Inst. of Cancer Research, Dept. of Health, Cancer Screening Evaluation Unit, Block D, 15 Cotswold Rd, Sutton Surrey SM2 5NG, United Kingdom (Tel.: (081)6438901; Fax: 7707876)
COLL: Chamberlain, J.; Thomas, B.A.; Kirkpatrick, A.

Effect on Breast Cancer Mortality of Annual Mammographic Screening Starting at Age 40
This is a multicentre randomized controlled trial, designed to provide a definitive answer to the open question of the proportion of breast cancer deaths which can be avoided by mammographic screening of women in their 40's. One hundred ninety-five thousand women aged 40 to 41 will be identified from Family Practitioner Committee Registers and randomly allocated to a group of 65,000 in the study group and 130,000 in the control group. Those in the study group will be offered mammographic screening every year for seven years; those in the control group will not be offered screening until they are 50. Both groups will then join the routine 3-yearly breast screening service. During the course of the study all breast cancer cases (no matter how diagnosed) and all deaths among study and control group women will be recorded. Mortality from breast cancer will be compared between the two groups.

UNITED KINGDOM

TYPE: Intervention
TERM: Mammography; Screening
SITE: Breast (F)
TIME: 1990 - 2000

878   Chamberlain, J.   01301
Thames Cancer Registry, Clifton Ave., Belmont, Sutton, Surrey SM2 5PY, United Kingdom (Tel.: (081)6427692)
COLL: Price, J.L.; Joslin, C.A.; Blamey, R.W.; Forrest, A.P.M.

Trial of Early Detection of Breast Cancer
This is an observational trial which aims to determine the effectiveness of alternative early detection programmes in reducing mortality from breast cancer. Two populations of approximately 25,000 women aged 45-64 have been offered screening by clinical examination and mammography. Two others of the same size and age distribution have been offered education in breast self-examination, and four additional similar populations have served as comparison groups, offered no additional detection services above what is normally provided as good current practice. Field-work was completed in 1988 but all the women are flagged so that mortality from breast cancer can be monitored until at least 1992. The main analysis concerns population-based mortality from breast cancer. Other analyses include use of the case-control method regarding screening history in women who have died or have not died from breast cancer. Papers have been published in Br. J. Cancer 44:618-627, 1981, and the Lancet 11:411-416, 1988.

TYPE: Cohort; Mortality
TERM: Mammography; Screening
SITE: Breast (F)
TIME: 1979 - 1993

TAUNTON

* 879   Bowie, C.   04947
Somerset Health Authority, Dept. of Public Health, District Headquarters, Wellsprings Rd, Taunton TA2 7PQ, United Kingdom (Tel.: (0823)333491; Fax: 272710)
COLL: O'Boyle, P.; Ewings, P.; Pocock, R.

Cancer of the Prostate
This study aims to investigate a possible link between prostate cancer and the following: alcohol, tobacco, sexual activity, diet, family history of cancer, occupation (farming in particular), water hardness, use of pesticides, fertilizers and other chemicals, radiation exposure, vitamin supplements, past illness, zinc and cadmium exposure. It is anticipated that approximately 180 cases of prostate cancer will be identified by the Hospitals' Pathology Service during the period May 1989 to June 1991. There will be two control groups, one consisting of patients with benign prostatic conditions and one of hospital patients whose reason for admission was not for urinary or prostatic problems. The controls will be age-matched with the cases. The cases and controls will be interviewed by trained interviewers with a structured questionnaire. Multivariate and univariate methods will be used.

TYPE: Case-Control
TERM: Alcohol; Diet; Fertilizers; Occupation; Pesticides; Radiation, Ionizing; Sexual Activity; Tobacco (Smoking); Vitamins; Water
SITE: Prostate
CHEM: Cadmium; Zinc
TIME: 1989 - 1992

* 880   Bowie, C.   05101
Somerset Health Authority, Dept. of Public Health, District Headquarters, Wellsprings Rd, Taunton TA2 7PQ, United Kingdom (Tel.: (0823)333491; Fax: 272710)
COLL: Bowie, S.H.U.

Radon in Dwellings and Lung Cancer
The hypothesis is that radon found in the dwellings in the U.K. causes lung cancer and that there is a geographical association between districts in England and Wales with high radon levels in dwellings and both local geological features and lung cancer mortality adjusted for smoking and other known causes of lung cancer. The study uses OPCS data on lung cancer mortality for district council areas, radon

UNITED KINGDOM

measurements supplied by NRPB, smoking histories from lifestyle surveys and air pollution data from the Warren Springs Laboratory. This ecological study is designed to correlate lung cancer rates with these population-level data.

TYPE: Correlation
TERM: Air Pollution; Environmental Factors; Radiation, Ionizing; Tobacco (Smoking)
SITE: Lung
CHEM: Radon
TIME: 1990 - 1992

WALLSEND

881   Fletcher, F.J.   00956
       Cookson Minerals Ltd, Willington Quay, Cookson House, Wallsend Tyne & Wear NE28 6UQ, United
       Kingdom (Tel.: (091)2622211; Fax: 2634491 ; Tlx: 537357 zircon g)
COLL: Billington, N.A.

Lung Cancer in the Antimony Industry
A prospective epidemiological survey of workers engaged in the production of antimony metal, antimony oxide, and antimony sulphide is being undertaken to ascertain whether there is an excess of lung cancer by comparison with local death rates and death rates in other occupations. The survey includes all persons working at the plant on 1 January 1961 or joining after that date, and is likely to continue until 1990 or later. Persons leaving employment are traced. The calculations of expected deaths are age-specific and related to man-years of exposure. Other factors studied are smoking, sulphur, arsenic and area of residence.

TYPE: Cohort
TERM: Chemical Exposure; Environmental Factors; Metals; Occupation; Tobacco (Smoking)
SITE: Lung
CHEM: Antimony; Arsenic; Zircon
OCCU: Antimony Process Workers
TIME: 1961 - 1992

WARRINGTON

882   Berry, R.J.   02679
       British Nuclear Fuels Plc, Health and Safety Directorate, Hinton House , Room 525, Risley
       Warrington Cheshire WA3 6AS, United Kingdom (Tel.: (0925)835022; Fax: 822711 ; Tlx: 627581)
COLL: Slovak, A.; Binks, K.; McElvenny, D.

Cancer Risk in Radiation Workers
Mortality and morbidity studies of the British Nuclear Fuels workforce, begun in 1976, have been extended to include both current and, as far as practicable, all former employees. Mortality patterns for the Sellafield workforce have been examined with respect to external ionizing radiation (Smith and Douglas, Br. Med. J. 293:845, 1986). Both of these studies are expected to include cancer morbidity and further follow-up to be completed during 1991. A further extension of the Sellafield study will incorporate organ-specific radiation doses derived from data on plutonium in urine. The total BNFL workforce study concerns approximately 40,000 present and past employees.

TYPE: Cohort
TERM: Late Effects; Occupation; Radiation, Ionizing; Registry
SITE: All Sites
OCCU: Radiation Workers
TIME: 1976 - 1993

UNITED STATES OF AMERICA

ALBUQUERQUE

883 Becker, T.M. 04450
Univ. of New Mexico, School of Medicine, Dept. of Medicine, 900 Camino de Salud, Albuquerque NM 87131, United States of America (Tel.: (505)2775541)
COLL: Helgerson, S.; Wheeler, C.M.; Jordan, S.W.; Rosenfeld, P.; Mertz, G.; Icenogle, J.; Stone, K.M.

Natural History of Human Papillomavirus Infection
This prospective cohort study will evaluate the role of cervical HPV infection as a risk factor for cervical dysplasia. Cohorts totalling 600 Native American, Hispanic, and White women with and without cytologic/ colposcopic evidence of mild cervical dysplasia will be followed for 3-5 years. Presence of type-specific HPV DNA will be determined by nucleic acid hybridization. Multiple logistic regression will be used to explore risk factors for progression of dysplasia, including HPV infection, other sexually transmitted infections, cigarette smoking, and contraceptive use.

TYPE: Cohort
TERM: Contraception; DNA; HPV; Premalignant Lesion; Sexually Transmitted Diseases; Socio-Economic Factors; Tobacco (Smoking)
SITE: Uterus (Cervix)
TIME: 1988 - 1993

884 Samet, J.M. 02430
Univ. of New Mexico, School of Medicine, New Mexico Tumor Registry, 900 Camino de Salud NE, Albuquerque NM 87131, United States of America (Tel.: (505)2775541)
COLL: Morgan, M.V.; Key, C.R.; Pathak, D.R.; Valdivia, A.A.

Lung Cancer Mortality in New Mexico's Uranium Miners
Underground uranium miners are at increased risk of lung cancer because of exposure to radon daughters. In New Mexico, a large population of uranium miners have been exposed to radon daughters at concentrations lower than most previously studied groups. A cohort study of approximately 4,000 male underground miners has been implemented to assess the exposure-response relationship between radon daughter exposure and lung cancer mortality and to descibe the interaction between cigarette smoking and radon daughters. Exposures are documented with multiple sources and information on cigarette smoking is available from records of mining-related clinical examinations. Mortality has been evaluated until 1985 and conventional cohort and multivariate analyses are in progress. A paper has been published in Hlth Phys. 56:415-421, 1989.

TYPE: Cohort
TERM: Metals; Mining; Occupation; Radiation, Ionizing; Tobacco (Smoking)
SITE: Lung
CHEM: Radon; Uranium
OCCU: Miners, Uranium
TIME: 1977 - 1993

ANCHORAGE

885 Lanier, A.K. 03920
Centers for Disease Control, Arctic Investigations Laboratory, 225 Eagle St. , Anchorage AK 99501, United States of America
COLL: AANHS

Cancer in Alaskan Natives
The aim is to maintain an on-going cancer surveillance system to identify all newly diagnosed cancers in Alaskan Natives (Eskimo, Indian, Aleut), resident in Alaska. Baseline demographic and cancer-related data will be obtained on each case and summaries presented to the native health care personnel. The aetiology of cancers which occur in excess among Native people, especially virus-associated cancers (nasopharynx, salivary gland, primary liver, leukaemia-lymphoma, cervix) will be investigated, and viral and other tumour markers will be used in prevention and control programmes. Trends over time in cancer incidence will be followed and correlated with trends in other chronic diseases.

UNITED STATES OF AMERICA

TYPE: Incidence
TERM: Eskimos; Prevention; Race; Virus
SITE: All Sites; Leukaemia; Liver; Lymphoma; Nasopharynx; Salivary Gland; Uterus (Cervix)
TIME: 1975 – 1993

ANN ARBOR

886   **Sheikh, K.**   04203
Univ. of Michigan, School of Public Health, Dept. of Epidemiology, 109 Observatory St., Ann Arbor MI 48109, United States of America (Tel.: (313)7642159)
COLL: Wolfe, R.A.; Higgins, I.T.T.; Carman, W.J.; Harlan, L.

Obesity and Cancer in Women
The objective of the study is to determine whether obesity is associated with common cancers in the female population of Tecumseh, Michigan. The relationship between the type of obesity, body frame, the patterns of change in obesity over time, reproductive factors and risk of cancer of the breast, gastrointestinal tract and reproductive organs will be investigated. Other known or suspected risk factors for cancer to be included in the study are physical inactivity, family medical history, other morbid conditions, use of exogenous hormones, serum cholesterol and uric acid, smoking and alcohol consumption. For a subset of the population, dietary energy and intake of nutrients will also be examined. Since 1959, the Tecumseh population (3,523 women over age 16) have participated in an on-going prospective study. Data have been gathered on demographic and social characteristics, dietary intake, biochemical parameters, exposure to female hormones, morbidity and mortality. This 2-year project will utilize the existing data and newly collected data. Cancer cases have been identified through 1979, a morbidity and mortality follow-up will be carried out to ascertain cancers diagnosed in the female population between 1979 and 1986. 104 cases of breast cancer, 78 cases of uterine cancer and 40 cases of colon cancer are expected to be available for this investigation. Living cases and the relatives of deceased cases will be interviewed and medical records of the cases reviewed to confirm the diagnosis and to determine the approximate time of onset of cancer. In univariate and multivariate analysis, the risk of cancer will be assessed according to risk factor status.

TYPE: Cohort
TERM: Alcohol; Cholesterol; Diet; Familial Factors; Hormones; Obesity; Physical Activity; Physical Factors; Tobacco (Smoking); Uric Acid
SITE: Breast (F); Colon; Female Genital; Ovary; Uterus (Cervix); Uterus (Corpus)
TIME: 1986 – 1992

ARLINGTON

* 887   **Brett, S.M.**   05045
Environ Corporation, 4350 N. Fairfax Drive, Arlington VA 22203, United States of America (Tel.: (703)5162300; Fax: 5162345)
COLL: Rodricks, J.V.; Wilcock, K.; Starr, T.; Chinchilli, V.

Leukaemia Risk Associated with Occupational Benzene Exposure
A re-analysis of the leukaemia risk associated with occupational benzene exposure will be conducted as an update of an earlier study of Pliofilm workers. The risk of leukaemia will be estimated by applying independently developed estimates of cumulative benzene exposure to individuals in this cohort and the use of conditional logistic regression techniques.

UNITED STATES OF AMERICA

TYPE: Cohort
TERM: Occupation; Solvents
SITE: Leukaemia
CHEM: Benzene
TIME: 1990 - 1992

ATLANTA

888　Anderson, M.S.　04518
Center for Chronic Disease Prevention, & Health Promotion, Div. Chronic Dis., Control & Community Intervention, 1600 Clifton Road, NE , Atlanta GA 30333, United States of America (Tel.: (404)4884390)
COLL: Smith, R.A.; Liff, J.; Steiner, C.; Campolucci, S.

Georgia Cervical Cancer Prevention and Control Demonstration Project
The goals of this study are to identify the cultural, psychological, economic, and medical care system barriers to early detection and prevention of invasive cervical cancer; and to design, implement, and evaluate public health interventions to overcome identified barriers. The study includes a population-based, case-control study of black and white women residing in two metropolitan Atlanta counties. Approximately 500 women diagnosed with cervical carcinoma in situ and invasive cervical cancer will be interviewed. Control subjects (n = 1,000) are women in the same age group selected from the same geographical area by random digit dialling techniques. Specific projects include: (1) Interviews eliciting information on demographic characteristics, reproductive and sexual history, cancer screening, and medical care history; (2) Review of prior cytology and histology specimens of diagnosed cases conducted by consultant pathologists and cytotechnologists; (3) Review of medical records of diagnosed cases for information on prior Pap smears, treatment, and follow-up of cervical neoplasia. Intervention will be designed based on information obtained from interviews, and review of slides and medical records. These interventions will include evaluation components so that model cervical cancer prevention programmes can be implemented in other areas.

TYPE: Case-Control
TERM: Cytology; Histology; In Situ Carcinoma; Prevention; Reproductive Factors; Screening; Sexual Activity; Treatment
SITE: Uterus (Cervix)
TIME: 1986 - 1992

889　Anderson, M.S.　04519
Center for Chronic Disease Prevention, & Health Promotion, Div. Chronic Dis., Control & Community Intervention, 1600 Clifton Road, NE , Atlanta GA 30333, United States of America (Tel.: (404)4884390)
COLL: Yerkes, A.M.

Oklahoma Cervical Cancer Surveillance and Intervention Project
This project is designed to improve an existing Pap smear surveillance system, and expand the system to be population-based for the state. These data, along with hospital discharge data and data collected from a mortality review of deaths from cervical cancer during 1987-88 will be used to design interventions. Specific project activities include: (1) expansion of the current Pap smear surveillance system to be state-wide, and linkage of these data to the cancer registry; (2) development of a surveillance system to ensure that at least 90 percent of Oklahoma women with Class III or high Pap smear results receive appropriate medical attention; (3) identification of barriers to adequate medical follow-up and design and implementation of interventions to address identified problem areas; (4) a mortality review of all cervical cancer deaths during 1987-88; (5) a review of hospital discharge data for treatment of cervical lesions; (6) a population-based survey to identify and document barriers to obtaining a Pap smear and adequate medical follow-up.

UNITED STATES OF AMERICA

TYPE: Intervention; Mortality
TERM: Cytology; Screening; Treatment
SITE: Uterus (Cervix)
TIME: 1988 - 1992

890 Blumenthal, D.S. 04735
Morehouse School of Medicine, Dept. of Community Health, and Preventive Medicine, 720 Westview Drive SW, Atlanta GA 30310-1495, United States of America (Tel.: (404)7521620; Fax: 7557318)

Avoidable Mortality from Cancers in Black Populations - Cancer Screening
This project is designed to test an in-home educational intervention delivered by lay health workers. The primary object of the intervention is to improve the regularity with which inner-city black women obtain screening tests for breast and cervical cancer. Lay health workers have interviewed 322 inner-city black women regarding cancer knowledge, attitudes, and practices. The women were randomised into experimental and control groups; women in the experimental group received the educational intervention delivered in three sessions. Both groups of women were re-interviewed to elicit information which can be used to determine changes in knowledge, attitudes and screening practices that result from the intervention. Reports of cancer screening are currently being validated through a review of medical records. Currently the baseline interviews are all completed as well as the three intervention visits. The post or end-point interview is currently being conducted. Analyses will be reported for factors associated with screening compliance in the baseline survey and post intervention.

TYPE: Intervention
TERM: Blacks; Screening
SITE: Breast (F); Uterus (Cervix)
TIME: 1987 - 1992

891 Greenberg, R.S. 01025
Emory University School of Medicine, Dept. Epidemiology & Biostatistics, 1599 Clifton Rd NE, Atlanta GA 30329, United States of America
COLL: Liff, J.; Coates, R.J.; Flanders, W.D.; Eley, J.W.

Metropolitan Atlanta SEER Program
The Atlanta Registry is one of nine funded and sponsored by the NCI of the US. Each represents a unique set of persons with respect to region, race, industrialization, lifestyle, and heritage. Atlanta represents a growing but not particularly industrialized city in south-eastern USA with a large black population and whose longtime resident white population is largely Anglo-Saxon. The study's overall objective is registration and follow-up of every case of cancer in Metropolitan Atlanta and a separate ten county rural area whose population is largely black. About 6,000 patients per year are identified through hospital medical record departments, pathology laboratories, radiotherapy facilities, and death certificates. Demographic information, birthplace, census tract of residence, site and histological type of malignancy, diagnostic tests and procedures employed, extent of disease at diagnosis, therapeutic procedures undertaken including surgery, radiotherapy, chemotherapy, and immunotherapy and the name of the physician following the patient are collected by specially trained abstractors. Patients are followed through re-admissions to hospitals, by information from the follow-up physician, and through death certificates. It is hoped that these data will identify persons at unusually high or low risk of cancer. Because the registry is population-based, it is able to produce incidence data and patterns of diagnostic and therapeutic care and survival information for the area. A number of analytical studies are being conducted with this population, including a seroprevalence study of HTLV and HIV infections and lymphomas, a case-control study of cancers which occur excessively among blacks (i.e. those involving the prostate, pancreas, oesophagus, and multiple myeloma), lymphomas and soft tissue sarcomas in relation to dioxin exposure, passive exposure to cigarette smoke and lung cancer risk and barriers to the early detection of cervical cancer; and a prospective cohort study of racial differences in survival from cancers of the breast, colon, bladder and uterine corpus. Papers have been published in Cancer 60:2476-2483, 1987 and 61:396-401, 1988 and in Am. J. Public Hlth 79:71-73, 1989.

UNITED STATES OF AMERICA

TYPE: Case-Control; Cohort; Registry
TERM: Alcohol; Anatomical Distribution; Blacks; Chemical Exposure; Diet; Environmental Factors; HIV; HTLV; High-Risk Groups; Histology; Infection; Passive Smoking; Registry; Survival; Treatment
SITE: All Sites
CHEM: Dioxins
REGI: Atlanta (USA)
TIME: 1975 - 1992

892 Heath, C.W. 03096
American Cancer Society, 1599 Clifton Rd NE, Atlanta GA 30329, United States of America (Tel.: (404)3297686; Fax: 3251467)
COLL: Thun, M.J.; Calle, E.E.; Garfinkel, L.

Cancer Prevention Study II

Confidential questionnaires were collected in 1982 from 1.2 million persons over age 30 by 77,000 volunteers in a prospective study relating lifestyles and exposures to the development of cancer and other diseases. Mortality follow-up uses linkage to the National Death Index. Death certificates are obtained from state health departments. Mortality analyses are being conducted regarding cancer risk in relation to such factors as diet, drugs and medicines, occupational exposures, history of diseases, low tar/nicotine cigarette smoking, passive smoking, drinking habits, etc. Publications include Cancer Res. 48:6951-5, 1988, and Cancer 62:1844-1850, 1988 and 65:2356-2360, 1990.

TYPE: Cohort
TERM: Alcohol; Diet; Drugs; Lifestyle; Occupation; Passive Smoking; Prevention; Tobacco (Smoking)
SITE: All Sites
TIME: 1982 - 1992

893 Nadel, M. 04560
Centers for Disease Control, Center for Chronic Disease Prevention, & Health Promotion, 1600 Clifton Rd NE MS/F11 , Atlanta GA 30333, United States of America (Tel.: (404)4884390)
COLL: Hernandez, C.; Moser, M.; Friedell, G.

Kentucky Cervical Cancer Prevention and Control Demonstration Project

The Centers for Disease Control established a five year cooperative agreement with the Kentucky Department for Health Services to identify factors that contribute to Kentucky's high cervical cancer mortality rate and to design public health interventions to remedy these deficiencies. The project focuses on the 36-county southeastern Appalachian region of the State where mortality is particularly high. Project activities include: (1) the establishment of a population-based registry of histologically-confirmed clinical dysplasia, carcinoma in situ, and invasive cervical cancer in the target area; (2) personal interviews with a population-based sample of 603 women in the target area and with 500 cases of these three conditions who were diagnosed in 1986 and 1987 to determine knowledge, attitudes, and practices relating to use of gynaecological services; (3) review of cytology and histology specimens of interviewed cases; and (4) the design, implementation, and evaluation of interventions that address factors identified as contributing to cervical cancer morbidity and mortality.

TYPE: Intervention
TERM: Cytology; Histology; Prevention; Screening
SITE: Uterus (Cervix)
TIME: 1985 - 1992

894 Smith, R.A. 04495
CDC, Chronic Disease Prevention, & Health Promot. Center, Div. Chronic, Dis. Control & Community Intervention, 1600 Clifton Rd, NE , Atlanta GA 30333, United States of America (Tel.: (404)4884390)
COLL: DeBuono, B.; Feldman, J.; Buechner, J.; Fulton, J.; DiOrio, D.

Rhode Island Breast Cancer Screening Programme Evaluation Project

In 1986, the Rhode Island Department of Health designed a breast cancer screening programme to address the high mortality rate from breast cancer (second in the nation) in the state. The programme includes the Department of Health's coordination of referrals for breast cancer screening, follow-up, low-cost mammography (50 US-Dollar), mass media campaign, and a programme to monitor and improve the quality of mammography. Specific evaluation activities include: (1) a survey of 300 primary care physicians to determine their practice related to mammography as a screening tool; (2) Pre-and post-intervention cross-sectional surveys of 800 women in the general population aged 40 and over, to

UNITED STATES OF AMERICA

determine their knowledge, attitudes, and practices related to mammography screening; (3) interviews with 300 programme participants to determine effectiveness, and to provide information that might improve screening for all women in the State; (4) training for radiology technicians, radiologists, and peer review of all non-negative mammograms, and a sample of mammograms read as having no significant abnormality, by consultant radiologists; (5) three annual inspections of radiologic facilities and testing of the imaging system, with feedback to the facility to improve the quality of the screening test; (6) establishing standards for positioning and compression, technique settings, image quality, interpretation, and follow-up, and on the basis of regular inspections and peer review, recommend objectives for improving mammography.

TYPE: Intervention
TERM: Mammography; Screening
SITE: Breast (F)
TIME: 1987 - 1992

AUGUSTA

895 Wray, B.B. 04514
Medical College of Georgia, Allergy-Immunology Section, CJ141, Augusta GA 30912-3790, United States of America (Tel.: (404)7213531)
COLL: Cole, R.J.

Serum Levels of Mycotoxins in Human Diseases
The primary objective of the study is to monitor the occurrence of aflatoxin in tissue specimens of patients with malignancies and bone marrow depression of unclear aetiology, and certain congenital malformations. The hypothesis is that environmental exposure to mycotoxins may be related to the development of these problems in particular patients. Serum is systematically being saved from patients seen at the Medical College of Georgia who have any of these disorders. They live in areas with known aflatoxin contamination of foodstuffs, and the serum will also be catalogued for different seasons of the year. A similar number of control patients who have other types of diseases will be analysed.

TYPE: Case Series
TERM: BMB; Congenital Abnormalities; Monitoring; Mycotoxins; Serum
SITE: All Sites; Leukaemia
CHEM: Aflatoxin
TIME: 1983 - 1993

BALTIMORE

896 Giardiello, F.M. 00773
Johns Hopkins Univ., School of Medicine, 600 N. Wolfe St., Blalock Bldg 942, Baltimore MD 21205, United States of America (Tel.: (301)9552635)
COLL: Levin, L.S.; Hamilton, S.R.; Zamcheck, N.; Danes, B.S.; Yardley, J.; Owens, A.H.; Baylin, S.B.; Traboulsi, E.I.; Maumenee, I.H.; Offerhaus, G.J.A.; Vogelstein, B.; Booker, S.V.

Families at High Risk of Colorectal Cancer
Families with hereditary polyposis and hereditary cancer of the large bowel from pedigrees ascertained in six adjacent states of the United States are being studied. Subjects are invited to undergo periodic gastrointestinal, ocular and dental examinations. The areas of interest are: natural history of the disease, detailed formal genetic analysis of age of onset, genetic fitness and evidence of heterogeneity, anomalies such as aneuploidies and polyploidies in tissue culture, histological analysis of biopsy specimen, dermatoglyphics, biochemical and molecular markers, dermatological aspects, longitudinal studies of CEA, development of a regional polyposis registry, supporting experimental viral studies in animals, editing of a quarterly Polyposis Newsletter, and organising a hereditary and colon cancer group, IMPACC (Intestinal Multiple Polyposis and Colorectal Cancer). Papers were published in Am. J. Med. Genet. 29:323-332, 1988; Ophthalmology 95:964-969, 1988; and Semin. Surg. Oncol. 3:137-139, 126-133, 1987.

UNITED STATES OF AMERICA

TYPE: Genetic Epidemiology
TERM: Genetic Factors; Genetic Markers; High-Risk Groups; Histology; Pedigree; Premalignant Lesion; Registry
SITE: Colon; Gastrointestinal; Rectum
TIME: 1973 - 1993

897 Gold, E.B. 01893
Johns Hopkins Univ., Dept. of Epidemiology, 615 N. Wolfe St., Baltimore MD, United States of America
COLL: Leviton, A.; Austin, D.F.; Child, M.A.; Kolonel, L.N.; Aschenbrener, C.; Key, C.R.; Lyon, J.L.; Weiss, N.S.; West, D.W.; Gilles, F.H.; Satariano, W.A.; West, C.; Mueller, N.; Lopez, R.

Aetiology of Childhood Brain Tumours
In this case-control study of childhood brain tumours, data have been collected on 361 cases occurring in children under 16 years of age and 1,083 controls selected by random digit dialling. Data collection ceased in 1981. Most of the analyses to date have been concerned with parental occupational exposures. In one component of the data set, employment in the aerospace industry was not associated with increased risk of brain tumour in an offspring (J. Nat. Cancer Inst. 77:17-19, 1986). Employment of either parent in the medicine and science industries was associated with an increased risk of tumours. Employment in the agriculture industry and white collar employment in general were also associated with childhood brain tumours. More recently, investigations have included analyses of risk of childhood brain tumours associated with familial aggregation of birth defects or malignancies and with risk associated with parental smoking and alcohol and caffeine consumption. A recent publication from this study reported that tumours in grandparents and great-grandparents were associated with a younger age at presentation of brain tumours in children (Cancer Causes and Contr. 1:75-79, 1990).

TYPE: Case-Control
TERM: Alcohol; Childhood; Coffee; Congenital Abnormalities; Parental Occupation; Registry; Tobacco (Smoking)
SITE: Brain
REGI: SEER (USA)
TIME: 1977 - 1992

898 Groopman, J.D. 04605
Johns Hopkins Univ., School of Public Health, 615 N. Wolfe St., Baltimore MD 21117, United States of America (Tel.: (301)9550029; Fax: 9550617)
COLL: Montesano, R.; Hall, A.J.; Wogan, G.N.; Wild, C.; Bosch, F.X.

Monitoring Human Exposure to Aflatoxins in the Gambia
Aflatoxin B1 (AFB1) is one of the most potent liver carcinogens known for experimental animals. This chemical is also present in the environment as a result of mould related spoilage of foods and commodities. Consumption of these foods results in high intake of AFB1 by people living in many regions of Asia and Africa. A number of researchers have used classic epidemiological methods to examine the possible causal relationship between AFB1 in the diet and liver cancer. While strong associations between AFB1 exposure and liver cancer have been found, new technologies have only recently become available to monitor individual exposure and metabolism of aflatoxin B1 in people. The development of biomarkers to assess the exposure to aflatoxin will help in efforts to limit the potential risk in human populations. In this study it is proposed to examine both seasonal and annual variation in aflatoxin intake, metabolism, urinary and albumin disposition in people living in the Gambia, who are at high risk of developing liver cancer. These studies will assess the correspondence between dietary intake of aflatoxins with biological markers, such as excreted AFB-DNA adducts, oxidative metabolites in urine, aflatoxin M1 in human milk and covalently bound aflatoxin to serum albumin, to determine which of these markers have utility for non-invasively assessing the exposure status of people at high risk for liver cancer.

UNITED STATES OF AMERICA

TYPE: Cross-Sectional; Molecular Epidemiology
TERM: Biochemical Markers; Diet; HBV; Metabolism; Monitoring; Mycotoxins
SITE: Liver
CHEM: Aflatoxin
LOCA: Gambia
TIME: 1988 – 1992

* 899   Jacobson, L.P.   04962
Johns Hopkins Univ., School of Hygiene & Public Health, Dept. of Epidemiology, 624 N. Broadway, Baltimore MD 21205, United States of America (Tel.: (301)9554320; Fax: 9557587)
COLL: Detels, R.; Munoz, A.; Phair, J.; Rinaldo, C.; Saah, A.

Aids-Related Kaposi's Sarcoma in the Multi-Centre Aids Cohort Study
The overall objective of this prospective study is to describe the natural history of HIV-I among homosexual/bisexual men in the US. In 1984-1985, 4,954 homosexual men were enrolled in four collaborative centres – Baltimore, Chicago, Los Angeles and Pittsburgh; in 1987 enrollment was re-opened. At semi-annual follow-up visits, these men answer a standardized questionnaire, undergo a physical examination and provide specimens for laboratory analyses. Clinical outcomes, including the development of AIDS, malignancies and mortality, are ascertained continuously. Several investigations are being performed within the study to describe the epidemiology of KS in this population. Initial analyses have demonstrated that incidence is not decreasing. Comparing men with KS to men who first developed a non-KS AIDS diagnosis, this study showed that men with KS had more sexual encounters prior to their diagnosis, thereby supporting the theory that a sexually-transmitted agent may be related to the development of KS. Further investigations to explore the timing of such factors are underway. Other research on KS of the Multi-center AIDS Cohort Study includes examination of the effects of prophylaxis and markers in the progression and prognosis of Kaposi's sarcoma. The study provides a unique opportunity to describe the changes in incidence of KS according to maturity of infection, immunosupression and prophylactic interventions.

TYPE: Cohort
TERM: AIDS; BMB; HIV; Homosexuality; Sexual Activity
SITE: Kaposi's Sarcoma
TIME: 1984 – 1995

* 900   Offerhaus, G.J.A.   05037
Johns Hopkins Medical Institutions, Dept. of Pathology, Carnegie Room 214 , Baltimore MD 21205, United States of America (Tel.: (301)9553511; Fax: 9553438)
COLL: Tersmette, A.C.; Giardiello, F.M.; Goodman, S.N.; Vandenbroucke, J.P.; Tytgat, G.N.; Hoedemaeker, P.J.

Longterm Follow-Up of Gastrectomy Patients for Gastric Stump Cancer
This is a long-term follow-up study of a large post-gastrectomy cohort of patients who had surgery for ulcer between 1931-1960, in order to estimate the risk of gastric stump cancer; to evaluate the impact of smoking; to assess risk of other gastro-intestinal cancer in this cohort; and to evaluate the effect of screening on mortality due to gastric cancer in a subset of patients of this cohort who underwent endoscopy plus biopsy. The cohort has 2,633 patients; follow-up is more than 99% complete; 504 asymptomatic patients participated in the endoscopic screening study.

TYPE: Cohort
TERM: Screening; Surgery; Tobacco (Smoking)
SITE: Bile Duct; Colon; Pancreas; Rectum; Stomach
LOCA: Netherlands
TIME: 1975 – 1992

* 901   Strickland, P.   05099
Johns Hopkins Sch. Hyg. & Public Health, Dept. of Environmental Health Sciences, 615 N. Wolfe St., Baltimore MD 21205, United States of America
COLL: West, S.; Taylor, H.; Nethercott, J.; Whitmore, S.E.; Grossman, L.

Susceptibility to Skin Cancer in Maryland Watermen
The overall goal of this project is to understand the mechanisms involved in determining individual susceptibility to skin cancer. This goal will be addressed by examining a previously identified population of over 800 Maryland watermen who are at high risk for non-melanoma skin cancer (basal cell carcinoma and squamous cell carcinoma). Overall prevalence of non-melanoma skin cancer is 10-15% and

UNITED STATES OF AMERICA

20-25% in older age-groups (70-80 years). In this population, accurate estimates of individual lifetime UV exposures have been calculated from a combination of interview data, working characteristics, and field measurements. Other phenotypic characteristics that influence skin cancer risk such as skin pigmentation, ease of sunburning (skin type), and hair colour, have been recorded for all individuals in this population. This nested case-control study, including approximately 80 cases and 80 controls, will investigate the role of DNA repair in determining individual susceptibility to skin cancer. DNA repair efficiency of cancer cases and age-matched controls will be assessed by two DNA repair assays: a lymphocyte-plasmid reactivation assay and an immunoassay for DNA photoproduct repair in lymphocyte and fibroblasts. The results of this study will (1) contribute to overall understanding of the mechanisms that determine susceptibility to skin (and other) cancers in man, (2) aid in the identification of individuals at elevated risk for skin cancer, and (3) be useful in estimating increased risks to the general population resulting from potential increases in solar UV at the earth's surface.

TYPE: Case-Control; Cohort
TERM: Biochemical Markers; DNA Repair; Occupation; Pigmentation; Radiation, Ultraviolet
SITE: Skin
OCCU: Dockers; Fishermen
TIME: 1990 - 1995

* 902 Strickland, P. 05233
Johns Hopkins Sch. Hyg. & Public Health, Dept. of Environmental Health Sciences, 615 N. Wolfe St., Baltimore MD 21205, United States of America
COLL: Rothman, N.; Poirier, M.C.

Molecular Biomonitoring in Wildland Firefighters

This study was designed to characterize the personal exposure of a representative sample of wildland firefighters to a variety of PAHs, and to examine the usefulness of several molecular biomarkers of PAH exposure (PAH-DNA adducts, PAH-haemoglobin adducts, and PAH urine metabolites). In addition, seasonal changes in pulmonary function and respiratory symptoms were examined. Fieldwork was completed during the 1988 fire season (June to September) in Northern California. Approximately 100 firefighters took part in two study designs. The first study examined a group of 60 firefighters before the fire season began and again after the season was over. The second study examined a group of 20 firefighters with very heavy combustion product exposure at the Yellowstone forest fire on their return to California. Twenty age-, sex-, and race-matched controls were also examined. Blood and urine samples were collected from each individual. Spirometry measurements were taken to assess pulmonary functions. Each volunteer filled out a questionnaire regarding demographics, dietary history, previous exposure to PAH, and recent firefighting activity. Only non-smokers were admitted to the study.

TYPE: Methodology; Molecular Epidemiology
TERM: BMB; Biochemical Markers; DNA Adducts; Occupation; Tissue
SITE: Lung
CHEM: PAH
OCCU: Firemen
TIME: 1988 - 1992

BELMONT

903 Holly, E.A. 04363
Northern California Cancer Center, Dept. of Epidemiology, 1301 Shoreway Rd, Suite 425, Belmont CA 94002, United States of America (Tel.: (415)5914484; Fax: 5923625)
COLL: Whittemore, A.S.; Felton, J.S.; Stegman, D.A.; Smith, S.D.

Mutagenic Mucus in the Uterine Cervix of Smokers

An association has been noted between cancer of the uterine cervix and cigarette smoking after controlling for other known risk factors. In this study, the Ames/Salmonella microsomal test was performed on cervical fluids of 333 smokers and 364 non-smokers to determine whether smokers are more likely to have a positive outcome for mutagenicity. Subjects for this study were accrued by physicians at ten clinics who collected the specimens themselves to assure data-collection uniformity. To collect the specimen, the cervix was washed with 1 ml of sterile nonpyrogenic electrolyte solution, the fluid was drawn into a pipette, the wash was repeated and the fluid was then again drawn into the pipette, placed in a test tube and frozen. Patients were interviewed at the time of fluid collection by a professional interviewer. The structured questionnaire covered smoking habits, passive smoking environment, sexual history, contraceptive use, history of sexually transmitted diseases, and recent diet. Statistical

UNITED STATES OF AMERICA

analyses are being conducted by the Northern California Cancer Center. The relationship between laboratory results and exposure characteristics will be examined using methods developed by Mantel and Haenszel, with a multiple-logistic model to asess possible interaction, to control for potentially confounding variables, and to obtain point estimates and confidence limits for the relative risk for each variable under study.

TYPE: Correlation
TERM: Diet; High-Risk Groups; Mathematical Models; Mutagen; Oral Contraceptives; Passive Smoking; Sexual Activity; Sexually Transmitted Diseases; Tobacco (Smoking)
SITE: Uterus (Cervix)
TIME: 1987 - 1992

904   Holly, E.A.   04456
Northern California Cancer Center, Dept. of Epidemiology, 1301 Shoreway Rd, Suite 425, Belmont CA 94002, United States of America (Tel.: (415)5914484; Fax: 5923625)
COLL: Whittemore, A.S.; McGrath, M.S.; Dorfman, R.R.; Rosenberg, S.A.; Ziegler, J.L.

Non-Hodgkin's Lymphoma and Retroviral Tests
A population-based case-control study of non-Hodgkin's lymphoma is under way to include 900 newly diagnosed patients and 1,400 controls in the San Francisco region. A rapid case-finding system is used to identify cases newly diagnosed in all hospitals in the six-county San Francisco Bay Area region. Controls are identified by random digit-dialling, and are frequency-matched to cases by age, sex, and telephone number (area code and prefix). Structured personal interviews are conducted by trained interviewers in study subjects' homes. Data are collected on occupational factors, common allergies, vaccinations, drugs used for treatment of allergies, other disturbances of the immune system, viral infections, travel to foreign countries, and among males, history of homosexual experiences. A sample of lymphoma patients with AIDS-related diseases, a sample of lymphoma patients without these diseases and a sample of controls will be tested for anitibodies to HIV and other viruses. The laboratory results will be used with the questionnaire response data to assess risk factors for lymphoma by histological type.

TYPE: Case-Control
TERM: AIDS; Allergy; Drugs; HIV; Histology; Homosexuality; Immunology; Occupation; Vaccination
SITE: Non-Hodgkin's Lymphoma
TIME: 1988 - 1993

BERKELEY

* 905   Bates, M.N.   04979
Univ. of California at Berkeley, School of Public Health, Epidemiology Program, Haviland Hall, Berkeley CA 94720, United States of America (Tel.: (415)8431736; Fax: 8435339)
COLL: Smith, A.H.; Cantor, K.C.

Bladder Cancer and Arsenic in US Drinking Water Supplies
Studies of a Taiwanese population using water supplies with high levels of inorganic arsenic have shown high rates of mortality for cancers at a number of sites, including the bladder. The present study is investigating whether the generally much lower levels of arsenic found in water supplies in the United States are also associated with bladder cancer. Data on approximately 3,000 cases and 6,000 controls collected for the National Bladder Cancer Study in 1978, will be merged with arsenic exposure estimations, derived from data on arsenic levels in public water supplies, collected under the National Safe Drinking Water Act. Multiple logistic regression analysis will be the main method of analysis.

UNITED STATES OF AMERICA

TYPE: Case-Control
TERM: Water
SITE: Bladder
CHEM: Arsenic
TIME: 1989 – 1992

906 King, M.C. 02275
University of California, School of Public Health, Dept. Biomed. & Env. Health Sciences, Berkeley CA 94720, United States of America

Genetic Epidemiology of Breast Cancer in Families
The objective of the study is to determine the genetic, cultural, and environmental factors that cause breast cancer to cluster in some families. Large, extended families with high incidence of breast cancer are investigated using tools from genetic analysis and epidemiology. The existence of possible susceptibility alleles, greatly increasing cancer risk in women of the high-risk genotypes, is tested by genetic analysis of DNA polymorphisms that may "track" the inheritance of the susceptibility gene within a specific family. Investigation of other genetic mechanisms of inheritance of susceptibility to breast cancer is in progress. In addition, the interaction of genetic and environmental risk factors influencing breast cancer risk among female relatives in high-risk families is being studied using epidemiological techniques. The prevalence of genetically-influenced breast cancer among American Caucasian women has been estimated. Recent papers were published in Genet. Epidemiol. 3 (Suppl. 1):3-14, 15-36, 87-92, 1987; Am. J. Epidemiol. 125:308-318, 1987 and in Hum. Genet. Mapping 9:157, 1987.

TYPE: Cross-Sectional; Genetic Epidemiology
TERM: Caucasians; Cluster; DNA; Environmental Factors; Familial Factors; Genetic Factors; Genetic Markers; High-Risk Groups; Lifestyle; Pedigree; Prevalence
SITE: Breast (F)
TIME: 1974 – 1992

BETHESDA

907 Alavanja, M.C.R. 04261
NCI,NIH, Epidemiology & Biostatistics Program, Div. of Cancer Etiology, Executive Plaza North, Room 543B, Bethesda MD 20892, United States of America (Tel.: (301)4961611)
COLL: Brownson, R.C.; Boice, J.D.; Lubin, J.; Hrubec, Z.

Lung Cancer Among Non-Smoking Missouri Women with Residential Exposure to Radon
This study will attempt to determine the risk of lung cancer among non-smoking females with domestic radon exposure. A population-based case-control study of all non-smoking female lung cancer cases in the state of Missouri will be conducted. Cases, controls and next-of-kin will be made in their current and previous homes. Approximately 600 cases will be accrued from January 1987 through July 1991. Approximately 1,400 age-matched will be collected during the same time intervals. Controls in age-group 30 to 64 years will be randomly selected from Missouri motor vehicles registration records. Controls who are 65 years old or older will be randomly selected from Health Care Finance Administration files of Missouri residents. Diet, occupations, pregnancy history, and family cancer history will also be assessed.

TYPE: Case-Control
TERM: Abortion; Diet; Familial Factors; Female; Occupation; Parity; Registry; Reproductive Factors
SITE: Lung
CHEM: Radon
REGI: Missouri (USA)
TIME: 1988 – 1992

908 Beebe, G.W. 03961
NCI, Div. of Cancer Etiology, Clinical Epidemiology Branch, Executive Plaza Building North, Room 400E, Bethesda MD 20892-4200, United States of America (Tel.: (301)4965067; Tlx: 248232 nih ur rca)
COLL: Seeff, L.B.; Norman, J.E.; Hoofnagle, J.H.

Primary Hepatocellular Carcinoma in Relation to the 1942 Army Postvaccinal Hepatitis Epidemic
The hypothesis to be examined is that HBV infection in healthy young males does not lead to the carrier state or to excess mortality from cirrhosis or primary liver cancer. The study is composed of (1) a

UNITED STATES OF AMERICA

serological survey of 600 men; (2) a cohort mortality study of 60,000 men; and (3) a case-control study of about 1,400 deaths from primary liver cancer in VA hospitals, 1963-1984, versus 2,800 hospital controls. The serological component is aimed at establishing the identity of the virus responsible for the epidemic of 50,000 cases in 1942. This and the cohort mortality study involve comparisons of three sets of men: Group I, men hospitalized with icteric hepatitis in 1942; Group II, men who received contaminated yellow fever vaccine but did not develop clinical illness; and Group III, men who entered service after the original yellow fever vaccine was replaced with one manufactured without human serum. Serum samples were tested for HBsAG, anti-HBs, anti-HBc, anti-HAV, anti-HDV, and for AFP level, and analysis consisted of inter-group comparisons. The mortality cohort study (1946-1983) is based on death certificates supplemented by available hospital and pathology data. The case-control study involves comparing the two groups as to history of immunization with the original vaccine, especially as to lot number, since only 9 out of 1,414 lots were infective. A paper has been published in New Engl. J. Med. 316:965-970, 1987.

TYPE: Case-Control; Cohort
TERM: Cirrhosis; Infection; Sero-Epidemiology; Virus
SITE: Liver
TIME: 1983 - 1989

909   Biggar, R.J.   04523
NCI, NIH, Environmental Epidemiology Branch, Viral Epidemiology Sect., HNC 3348, Room 434D, Executive Plaza Bldg North, Bethesda MD 20892-4200, United States of America (Tel.: (301)4968115; Fax: 4969146)

HIV Seroconverters Registry
The natural history of HIV infection can be evaluated best when the time since infection with HIV is documented. Because most institutions have relatively few subjects with documented HIV seroconversion, a registry has been prepared of such cases. To be eligible, last negative and first positive serology must be available, along with birth day, sex, risk group and follow-up history of AIDS-related conditions (upon request, questionnaires will be sent to groups interested). Time to AIDS and risk of different manifestations of AIDS by risk group, age and sex, will be evaluated. So far over 1,000 seroconverters are enrolled. Follow-up questions will be sent annually. New participants are welcome.

TYPE: Cohort
TERM: AIDS; HIV; Sero-Epidemiology
SITE: Kaposi's Sarcoma; Lymphoma
TIME: 1987 - 1993

910   Blair, A.E.   04292
NCI, Occupational Studies Sect., Executive Plaza North, Room 418, Bethesda MD 20892, United States of America (Tel.: (301)4969093; Fax: 4969146)
COLL: Alavanja, M.C.R.; Cantor, K.C.; Dosemeci, M.; Fraumeni, J.F.; Gómez, M.; Hayes, R.; Heineman, E.; Hoover, R.N.; Pottern, L.M.; Rothman, N.; Riedel, D.; Stewart, P.A.; Ward, E.M.; Zahm, S.H.

Occupational Studies of Cancer
The Occupational Studies Section conducts a comprehensive programme of occupational studies to evaluate the role workplace exposures play in the origin of cancer. To accomplish this goal the Section conducts: (1) descriptive and hypothesis-generating studies to identify promising new areas of research and to sharpen hypotheses, (2) analytical studies to test hypotheses regarding specific exposures and specific cancers, and (3) methodological studies to develop and improve procedures used to assess exposures and conduct epidemiological studies. Proportionate mortality, cohort mortality, and case-control designs are employed. Study populations are assembled from companies, labour unions, professional associations, government agencies, hospitals, and tumour registries. Studies underway are designed to investigate cancer and a variety of exposures including wood dusts (furniture workers), Chinese workers exposed to benzene, industrial workers exposed to phenol, aircraft maintenance workers, cutting oils (automotive workers), metals and solvents (jewellery manufacturers), pesticides (applicators, employees of a lawn care company, grain millers, noxious weed applicators and Department of Agriculture employees), acrylonitrile (manufacturers and plastic producers), silica (miners, brickmakers, and others), talc (ceramic fixture manufacturers), workers exposed to benzidine and formaldehyde (embalmers, resin producers, and others). Studies of occupations with a variety of exposures include plumbers, shipyard workers, professional chemists, and firefighters. Case-control studies of mesothelioma, leukaemia, non-Hodgkin's lymphoma, multiple myeloma and cancers of the brain, lung, stomach, soft-tissue sarcoma, and bladder have been designed to evaluate a variety of occupational exposures as well as other factors that may be associated with the development of these cancers. Methodological projects to develop and improve techniques used to assess exposures in

UNITED STATES OF AMERICA

occupational studies include reliability of exposure estimation by industrial hygienists, comparison of estimates of exposure to specific substances using varying amounts of industrial hygiene information, experimental modification of workplace parameters to evaluate their effects on airborne concentrations of formaldehyde, comparison of the potential for exposure as estimated by industrial hygienists with that reported by workers, effects of exposure misclassification on risk estimates, and comparison of historical estimates of exposure made by industrial hygienists with actual measured levels.

TYPE: Case-Control; Cohort; Cross-Sectional; Methodology
TERM: Chemical Exposure; Dusts; Metals; Occupation; Pesticides; Petroleum Products; Plastics; Registry; Solvents; Wood
SITE: Bladder; Brain; Haemopoietic; Lung; Lymphoma; Mesothelioma; Sarcoma; Stomach
CHEM: Acrylonitrile; Asbestos; Benzene; Formaldehyde; Silica; Talc
OCCU: Acrylonitrile Workers; Agricultural Workers; Automobile Workers; Cabinet Makers; Chemists; Dry Cleaners; Farmers; Firemen; Heavy Equipment Operators; Herbicide Sprayers; Jewellery Manufacturers; Miners; Morticians
LOCA: China; Denmark; Turkey; United States of America
TIME: 1978 – 1993

911　Blair, A.E.　04820
NCI, Occupational Studies Sect., Executive Plaza North, Room 418, Bethesda MD 20892, United States of America (Tel.: (301)4969093; Fax: 4969146)
COLL: Stewart, P.A.; Zahm, S.H.; Falk, R.T.; Ward, M.; Kross, B.; Popendorf, W.; Burmeister, L.F.

Methodologic Study to Enhance Assessment of Pesticide Exposure in Epidemiological Studies of Cancer

Methods to assess exposure to pesticides in case-control studies of cancer must be improved if further progress in this important area is to be made. This project is designed to obtain information on the detail and quality of information regarding pesticide use that can be obtained from farmers or their surrogates, particularly for exposures many years in the past. Information on pesticide use will be obtained from a detailed questionnaire, from records kept by farmers, from records kept by suppliers, from exposure measured by dermal deposition, and from biochemical monitoring of the blood and urine. Two hundred farmers in Iowa and their next-of-kin (wives, sibs, and children) will participate in the study. Approximately 100 of these farmers will be asked to participate in the biologic monitoring portion of the project. Information from the various sources will be compared to identify combinations of data that appear most promising for use in epidemiologic investigations of pesticides and cancer.

TYPE: Methodology
TERM: Biochemical Markers; Occupation; Pesticides
SITE: Inapplicable
CHEM: 2,4-D; Alachlor; Atrazine; Malathion
OCCU: Farmers
TIME: 1989 – 1993

912　Blot, W.J.　04279
NCI, NIH, Epidemiology & Biostatistics Program, Environmental Epidemiology Branch, 431B, Executive Plaza North, Bethesda MD 20892-4200, United States of America
COLL: Dosemeci, M.; McLaughlin, J.K.; Blot, W.J.; Glen, R.; Reger, R.; Hearl, F.; McCawley, M.; Wu, Z.B.; Chen, J.

Retrospective Morbidity and Mortality Study of Silica Exposed Workers in Wuhan and Other Areas

Workers in mines and factories where silica exposure is high will be enrolled in a retrospective mortality study to test the hypothesis that silica exposure causes lung cancer. The cohort will consist of all persons who worked for at least one year during the period 1 January 1960-31 December 1974. Approximately 80,000 workers are expected to be enrolled. Levels of radon, silica, arsenic, nickel and cadmium will be measured in selected work sites. Since about 1963, the government has required that work sites with silica exposure conduct regular industrial hygiene measures of silica levels at the work site. These historical data will form the basis for exposure assessment. For each member of the cohort, a minimum basic data set of information will be collected for identification and follow-up. For persons included in a silicosis registry, additional information will be collected on medical history and complete work history. For selected silicotics, lung cancer deaths and appropriate controls, a nested case – control study will be conducted which will collect information on cigarette smoking, detailed employment, information and other personal characteristics. Rates from the Chinese national mortality survey of 1973-1975 will be used in generating expected numbers of deaths. Internal comparisons of workers with high vs low silica exposure will also be made. Special attention will be paid to assessing mortality among the sub-cohort

UNITED STATES OF AMERICA

with silicosis. Data from the nested case-control study will be examined using stratified and logistic regression analyses to evaluate risks of lung cancer and silicosis by intensity and duration of silica exposure, controlling for smoking.

TYPE: Case-Control; Cohort
TERM: Chemical Exposure; Dose-Response; Dusts; Metals; Occupation; Tobacco (Smoking)
SITE: Lung
CHEM: Arsenic; Cadmium; Nickel; Radon; Silica
LOCA: China
TIME: 1986 - 1992

913   Boice, J.D.   02363
NCI, NIH, Radiation Epidemiology Branch, 6130 Executive Blvd, Executive Plaza North, R408, Bethesda MD 20892-4200, United States of America (Tel.: (301)4966600)
COLL: Land, C.E.; Beebe, G.W.; Kleinerman, R.A.; Chen, K.W.; Hrubec, Z.; Morin, M.; Curtis, R.E.; Ron, E.; Lubin, J.; Jablon, S.; Inskip, P.D.; Tucker, M.A.; Travis, L.B.; Linet, M.S.; Alavanja, M.C.R.; Friedman, D.; Hatch, E.

Radiation-Induced Cancer

The Radiation Epidemiology Branch (REB) plans and conducts independent and cooperative epidemiological research to identify and quantify the risk of cancer in populations exposed to ionising and non-ionising radiation, especially at low-dose levels (populations with documented therapeutic, diagnostic, occupational, environmental or military exposures are studied). The risk of radiation-induced cancer in terms of tissues at risk, dose-response, radiation quality, fractionation of dose, time since exposure, sex, age at exposure and at observation are characterised, as well as possible modifying influences of other environmental and host factors. The REB develops statistical and epidemiological methodologies to facilitate epidemiological research; explores and formulates models of radiation carcinogenesis that may help define basic mechanisms of cancer induction, including the integration of experimental findings with epidemiological observations; conducts case-control and cohort studies of cancer risk in patient populations given diagnostic or therapeutic radiation alone or in combination with cytotoxic drugs and other forms of treatment; conducts population studies to examine possible analogues of radiation carcinogenesis in man, such as the induction of cytogenetic abnormalities in circulating lymphocytes, and integrates laboratory markers of radiation exposure and tissue response into epidemiological studies designed to clarify the patterns of cancer risk and the mechanisms of action. It also advises and collaborates with other agencies and individuals involved in radiation research and regulatory activities. Recent publications appeared in J. Nat. Cancer Inst. 125:318-325, 1991 (Ron et al); J. Amer. Med. Assoc. 265:1290-1294 (Boice et al), and 1403-1408, 1991 (Jablon et al). et al).

TYPE: Case-Control; Cohort; Methodology
TERM: Age; Chemotherapy; Dose-Response; Mathematical Models; Multiple Primary; Physical Factors; Radiation, Ionizing; Radiation, Non-Ionizing; Radiotherapy
SITE: All Sites; Breast (F); Leukaemia; Lung; Thyroid
LOCA: Canada; China; Czechoslovakia; Denmark; Finland; Germany; Israel; Japan; Norway; Sweden; United Kingdom; United States of America; Yugoslavia
TIME: 1977 - 1993

914   Brinton, L.A.   04936
NCI, Environmental Epidemiology Branch, Executive Plaza North, HNC 3342, Room 443, Bethesda MD 20892-4200, United States of America (Tel.: (301)4961691)
COLL: Potischman, N.A.; Barrett, R.J.; Berman, M.L.; Mortel, R.; Twiggs, L.B.; Wilbanks, G.D.

Case-Control Study of Endometrial Cancer

A multi-disciplinary study of endometrial cancer in five US areas (Chicago, IL; Hersey, PA; Los Angeles, CA; Minneapolis, MN; Winston-Salem, NC) is currently being completed. Approximately 425 cases have been enrolled and will be compared with an equal number of community controls and half as many hospital controls being treated for benign endometrial conditions. Risk associated with consumption of specific nutrients will be assessed from an interview of usual dietary intakes. Fat distribution patterns assessed from anthropometry will also be examined. To explore the relationships between hormones, diet, fat distribution and cancer risk, blood samples will be evaluated for hormone levels and micronutrients. A methodologic study is being conducted in one centre comparing various fat measures, including dietary fat, fatty acids in dietary data, adipose depots, and red blood cell values. Although the major focus of this study is on dietary determinants, the study will also provide the opportunity to examine data relevant to several other unresolved issues for this tumour, including exogenous hormones, alcohol consumption, and smoking.

UNITED STATES OF AMERICA

TYPE: Case-Control
TERM: Alcohol; Blood; Diet; Fat; Hormones; Nutrition; Reproductive Factors; Tobacco (Smoking)
SITE: Uterus (Corpus)
TIME: 1988 - 1992

915 Brinton, L.A. 04937
NCI, Environmental Epidemiology Branch, Executive Plaza North, HNC 3342, Room 443, Bethesda MD 20892-4200, United States of America (Tel.: (301)4961691)
COLL: Hoover, R.N.; Schairer, C.; Sturgeon, S.; Schatzkin, A.; Taylor, P.

Follow-Up Study of Participants in the Breast Cancer Detection Demonstration Project

A five year multi-centre breast cancer screening project, the Breast Cancer Detection Demonstration Project, was completed in 1980. At this time, a cohort of approximately 60,000 of the original 280,000 participants was chosen for further follow-up; included were all of the project-detected breast cancer cases, all those receiving a recommendation for breast biopsy, and a sample of normal screenees. For a five year period, annual telephone interviews were conducted with these subjects to elicit information on vital status, cancer incidence, and changes in breast cancer risk factors. Follow-up is now being performed on a bi-annual basis through a mailed questionnaire; this has enabled an expansion of data collection to include information on dietary factors, anthropometry, and physical activity levels. Analyses are currently underway to examine the effects of menopausal hormones on the risk of breast cancer as well as on all-cause mortality, and mortality due to cardiovascular diseases, other cancers, fractures, accidents, and suicide.

TYPE: Cohort
TERM: Diet; Hormones; Menopause; Physical Activity; Reproductive Factors
SITE: Breast (F); Colon; Lung; Ovary; Uterus (Corpus)
TIME: 1980 - 1992

916 Brinton, L.A. 04938
NCI, Environmental Epidemiology Branch, Executive Plaza North, HNC 3342, Room 443, Bethesda MD 20892-4200, United States of America (Tel.: (301)4961691)
COLL: Schairer, C.; Schiffman, M.H.; Nasca, P.C.; Mallin, K.; Richart, R.M.

Case-Control Study of Cancers of the Vagina and Vulva

A case-control study of cancers of the vagina and vulva has been completed in two US areas (Chicago and Upstate New York). Included were women with newly diagnosed invasive or in situ cancers of these two sites. The study began in February 1986 and included 12 months of retrospectively and 18 months of prospectively ascertained cases. Personal interviews focusing on medical, sexual, reproductive, and hygiene factors were conducted with 209 vulvar cancer patients, 41 vaginal cancer patients, and 445 community controls (identified either through random digit dialling or Health Care Financing lists). Results pertaining to vulvar cancer risk were recently published (Obstet. Gynecol. 75:859-866, 1990) and showed significant relationships of risk with sexual behaviour, smoking, histories of genital warts, and previous abnormal Pap smears. The results for vaginal cancer were published in Gynecol. Oncol. 38:49-54, 1990) and showed that the primary predictors of risk were low socio-economic status, a history of genital warts and previous genial abnormalities. Blood specimens drawn at the time of interview are currently being analysed for antibodies to several sexually transmitted agents, and future analyses will attempt to assess relationships with sexual behaviour in detail.

TYPE: Case-Control
TERM: BMB; Blood; Hygiene; Reproductive Factors; Screening; Sexual Activity; Sexually Transmitted Diseases; Tobacco (Smoking)
SITE: Vagina; Vulva
TIME: 1986 - 1992

917 Brinton, L.A. 04939
NCI, Environmental Epidemiology Branch, Executive Plaza North, HNC 3342, Room 443, Bethesda MD 20892-4200, United States of America (Tel.: (301)4961691)
COLL: Schairer, C.; Storm, H.H.; Persson, I.R.

Follow-up Study of Patients with Gynaecological Operations

The purpose of this project is to evaluate the risk of cancer associated with varying indications for and types of gynaecological operations. Cohorts of approximately 60,000 patients with over 500,000 person years of observation and 15,000 patients with over 150,000 person years will be assembled by the cancer registries of Denmark and Sweden, respectively. Demographic and exposure information, as well as

UNITED STATES OF AMERICA

clinical reasons for the operation, will be abstracted from individual patient records. The rosters of patients will then be linked to cancer registry records so that patients who have developed malignancies can be identified. The observed number of cancers will be compared to the expected number, derived from sex-, age-, and calendar-year specific population based rates.

TYPE: Cohort; Incidence
TERM: Record Linkage; Registry; Surgery
SITE: Breast (F); Colon; Lung; Ovary; Uterus (Corpus)
LOCA: Denmark; Sweden
REGI: Denmark (Den); Sweden (Swe)
TIME: 1990 - 1992

918 Brinton, L.A. 04940
NCI, Environmental Epidemiology Branch, Executive Plaza North, HNC 3342, Room 443, Bethesda MD 20892-4200, United States of America (Tel.: (301)4961691)
COLL: Potischman, N.A.; Daling, J.R.; Liff, J.; Schoenberg, J.

Case-Control Study of Breast Cancer with Emphasis on Women Under the Age of 45 Years

A population-based case-control study has recently been initiated in three areas (Atlanta, GA, Seattle, WA, Trenton, NJ) with the aim of evaluating breast cancer risk in relation to oral contraceptive use, alcohol consumption, dietary patterns, and other factors. Cases will comprise all cases under the age of 45 years newly diagnosed over a two year period. In Atlanta, the age range of cases will be extended to 54. Controls in all areas will be identified through random digit dialling techniques, but in Atlanta an additional group of controls will be identified through a household survey to allow comparative analyses using different control groups. A total of approximately 2,000 cases and 3,000 controls is expected to be recruited for study. Personal interviews will be conducted and a variety of anthropometric measurement obtained. The interview will include a quantitative food frequency instrument, designed to elicit information on usual adult as well as adolescent diet. To validate the information on adolescent exposures, interviews with mothers of study subjects will be attempted. Information on usage of oral contraceptives and previous operations will be validated against review of select medical records. In Seattle, blood samples will also be drawn from a sample of subjects to enable a variety of hormonal and micronutrient assays.

TYPE: Case-Control
TERM: Alcohol; Blood; Diet; Genetic Factors; Hormones; Nutrition; Oral Contraceptives; Reproductive Factors
SITE: Breast (F)
TIME: 1990 - 1992

919 Byrne, J. 02138
NCI, Clinical Epidemiology Branch, EPN 400, Bethesda MD 20892-4200, United States of America (Tel.: (301)4964947; Fax: 4961854 ; Tlx: RCA 248232)
COLL: Nicholson, H.S.; Connelly, R.R.; Fears, T.R.; Gail, M.G.

Late Effects in Long-Term Survivors of Childhood and Adolescent Cancer

The objective of this study is to document the ways in which different treatments for cancer in childhood or adolescence affect the fertility and quality of life of adult surviviors. Current investigations include studies of menarche, fertility, foetal deaths and menopause, psychosocial effects, late mortality and morbidity. A cohort study design is used most often; cancer survivors are matched to sibling controls and interviewed about relevant experiences. Clinical measures such as hormone assays and imaging techniques are incorporated as needed. A paper by Mostow et al. has been published in J. Clin. Oncol. 9:592-599, 1991.

UNITED STATES OF AMERICA

TYPE: Case Series; Case-Control; Cohort
TERM: Adolescence; Chemotherapy; Childhood; Congenital Abnormalities; Drugs; Fertility; Intra-Uterine Exposure; Menarche; Menopause; Psychosocial Factors; Radiotherapy; Sib
SITE: Childhood Neoplasms
CHEM: Cyclophosphamide
TIME: 1978 - 1995

920 **Dosemeci, M.** 04618
National Cancer Inst., NIH, Environmental Epidemiology Branch, Occupational Studies Sect., Executive Plaza North, Room 418H, Bethesda MD 20892-4200, United States of America (Tel.: (301)4969093; Fax: 4961854)
COLL: Hayes, R.B.; Blair, A.E.; Unsal, M.; Vetter, R.; Wacholder, S.

Occupational Exposure and Cancer Risk in Turkey
The objective of this study is to identify the possible associations between various occupational exposures and cancer risk in Turkey. Between 1978 and 1984 about 7,500 cases of cancer (excluding haematopoietic malignancies) were treated among industrial workers in SSK, Okmeydam Hospital, in Istanbul, Turkey. In addition to the general demographic information, full employment histories, including workplace name, job title, duration of employment, tobacco and alcohol use were systematically collected for all admissions to the cancer clinic. Abstraction of data was started in November 1988. Exposure assessment will be carried out semi-quantitatively (none, low, medium an high) using a specially developed job-industry exposure matrix for about 20 different chemicals or chemical groups. Cancer risk analyses for various sites will be carried out using standard epidemiological methods for case-control studies. Selected cases of cancer sites other than the one under consideration will form the control group. In the analyses, standardized odds ratios by industry, occupation, duration of employment, level of exposure, cumulative exposures and probability of exposure will be presented as measures of risks for various sites of cancer.

TYPE: Case-Control
TERM: Chemical Exposure; Clinical Records; Dose-Response; Dusts; Metals; Occupation; Pesticides; Physical Activity; Plastics; Solvents; Wood
SITE: All Sites
CHEM: Acrylonitrile; Arsenic; Asbestos; Benzene; Benzidine; Chromium; Formaldehyde; Hydrocarbons, Chlorinated; Nickel; PAH; Phenol; Silica; Styrene
OCCU: Construction and Maintenance Workers; Engineering Workers; Miners; Petroleum Workers; Potters; Rubber Workers; Shipyard Workers; Textile Workers
LOCA: Turkey
TIME: 1988 - 1994

921 **Herrero, R.** 04830
NCI, NIH, Environmental Epidemiology Branch, Executive Plaza North 443 , 9000 Rockville Pike, Bethesda MD 20892, United States of America (Tel.: (301)4961691; Fax: 4969146)
COLL: Brinton, L.A.; Sierra, R.

Incidence Study of Cervical Intraepithelial Neoplasia in Costa Rica
Complete screening records of 12,000 women who have received at least two cytological examinations at the Costa Rica Ministry of Health in the last three years have been extracted and computerized. The aim of the study is to estimate the incidence of CIN of different grades among women without previous cervical disease. The information will be utilized to estimate the sample size in a future prospective study of human papilloma virus and CIN.

TYPE: Incidence
TERM: Clinical Records; Premalignant Lesion; Screening
SITE: Uterus (Cervix)
LOCA: Costa Rica
TIME: 1990 - 1992

* 922 **Kirsch, I.R.** 04990
National Cancer Institute, NIH, Navy Medical Oncology Branch, Naval Hosp., Bldg 8, Room 5101, 8901 Wisconsin Ave., Bethesda MD 20889-5105, United States of America (Tel.: (301)4960909; Fax: 4960047)
COLL: Garry, V.F.; Blair, A.E.

UNITED STATES OF AMERICA

Hybrid Antigen Receptor Genes, Genomic Instability, and Risk of Leukaemia and Lymphoma
Using PCR, it can be demonstrated that the increased frequency of lymphocyte-specific cytogenetic abnormalities found previously in workers exposed to pesticides and herbicides results from an increased frequency of interlocus V(D)J recombination. The mechanism underlying formation of hybrid antigen receptor genes in lymphocytes appears related to that underlying rearrangements between putative oncogenes and antigen receptor genes in lymphoid malignancies, also associated with exposure to pesticides. In a pilot study some 100 grain workers and other farmers are being examined before and after seasonal exposure to pesticides and herbicides for evidence of genomic instability as manifested by an increased frequency of interlocus V(D)J recombination. The control population will be age- and sex-matched unexposed individuals. If, using the PCR-based assay, significant differences can be detected in the exposed population, the study will be expanded and individuals and populations with an increased incidence of lymphocyte-specific interlocus recombination will be followed to determine comparative risk of development of leukaemia and lymphoma.

TYPE: Cohort; Molecular Epidemiology
TERM: Herbicides; Lymphocytes; PCR; Pesticides
SITE: Leukaemia; Lymphoma
OCCU: Farmers; Grain Millers
TIME: 1990 - 1993

* 923   Osborne III, J.S.   05232
National Inst. of Health, Div. of Research Grants, Epidemiology and Disease Control-1, Bethesda MD 20892-4200, United States of America (Tel.: (301)4967246; Fax: 4021279)
COLL: Kaplan, B.H.; Shy, C.M.

Analysis of Reported Cancer Clusters in Small Populations
The objective of this project is to develop a reliable method for assessing reports of cancer clusters in small stable populations which lack reliable denominator data for rates. The method has been successfully applied in a rural community (N=300) in North Carolina, USA (Am. J. Epidemiol. 132(1):87-95, 1990). It involves a three-step process of determining (1) secular trends in proportionate cancer mortality; (2) standardized proportionate mortality ratios (SPMR) stratified by age, sex, race, calendar period, and any other appropriate variables, and (3) 95% Poisson confidence intervals (CI) for the expected distribution of deaths. Secular trend data serve to address and validate concerns arising in the community, SPMR data assess deviation from expected mortality, and, if SPMR data indicate occurrence has differed from expected, the CI enable the likelihood of this difference arising from random case clustering to be evaluated. If mortality exceeds the upper limit of the CI, then it is not likely to represent random case clustering and an aetiological investigation is appropriate. Evaluation of this approach for morbidity and its application in another community reporting excess cancer mortality are in progress.

TYPE: Methodology
TERM: Cluster
SITE: All Sites
TIME: 1986 - 1995

* 924   Rapoport, S.I.   04969
National Inst. of Aging, NIH, Lab. of Neurosciences, Bldg 10, Room 6C103 , Bethesda MD 20892-4200, United States of America (Tel.: (301)4968970; Fax: 4020074)
COLL: Greig, N.H.; Ries, L.

Primary Brain Tumours in the Elderly
Data from NCI's SEER Program are analysed for annual incidence of primary brain tumours. Data are based on registries representing approximately 10% of the US population. In 1985, incidence rates of primary malignant brain tumours for persons aged 75-79, 80-84, and 85 years of age and over were 187%, 394% and 501% respectively of rates in 1973/1974. Reported incidence in younger persons varied little over the same period of time. In the future, the contribution of new diagnostic procedures introduced or augmented since 1973, including brain imaging techniques (e.g. x-ray computerized tomography), will be assessed and distinguished from a true increase in annual incidence of malignant brain tumours in the elderly.

UNITED STATES OF AMERICA

TYPE: Incidence
TERM: Age; Registry
SITE: Brain
REGI: SEER (USA)
TIME: 1988 – 1994

* 925   **Shaw, G.L.**   04971
NCI, NIH, Environmental Epidemiology Branch, Family Studies Sect., Executive Plaza North, HNC 3347, HNC 3347, Room 439A, Bethesda MD 20892, United States of America (Tel.: (301)4964375; Fax: 4020916)
COLL: Tucker, M.A.; Caporaso, N.E.; Johnson, B.E.; Frame, J.N.; Deslauriers, J.

Biochemical Epidemiology of Lung Cancer: the Debrisoquine Metabolic Phenotype

This study is designed to validate previous findings of increased risk of lung cancer among genetically extensive metabolisers of debrisoquine and to determine whether the tumour itself has a role in increasing debrisoquine metabolism. 390 cases and 350 matched controls are planned to complete a questionnaire and complete debrisoquine phenotyping overnight. Blood for DNA will be obtained on a subset. Analysis will include stratified and logistic regression methods to determine risk of lung cancer by metabolic phenotype.

TYPE: Case-Control; Molecular Epidemiology
TERM: BMB; Biochemical Markers; Blood; DNA; Genetic Factors; Metabolism
SITE: Lung
LOCA: Canada; United States of America
TIME: 1988 – 1992

926   **Sondik, E.J.**   03888
NCI, NIH, Executive Plaza North 211, 9000 Rockville Pike, Bethesda MD 20892, United States of America (Tel.: (301)4968506)
COLL: Janerich, D.T.; Flannery, J.T.; Lerchen, M.L.; Swanson, G.M.; Kolonel, L.N.; Isaacson, P.; Platz, C.; Key, C.R.; Thomas, D.B.; West, D.W.; Greenberg, R.S.; Meinert, L.; Martinez, I.; Zippin, C.

Surveillance, Epidemiology, and End Results (SEER) Program

A cancer registration system is conducted in 12% of the United States population through contracts with non-profit medically oriented institutions. The system is designed to report incidence data within 11 months of the close of the calendar year and survival within 18 months. The goals of the SEER Program are to: (1) Determine periodically incidence of cancer in selected geographical areas of the US with respect to demographic and social characteristics of the population. (2) Estimate cancer incidence for the US on an annual basis (3) Monitor trends in incidence of specific forms of cancer with respect to geographical area and demographic and social characteristics of the population. (4) Determine periodically survival experience for cancer patients diagnosed among residents of selected geographical areas of the US. (5) Monitor trends in cancer patient survival with respect to form of cancer, extent of disease, therapy, and demographic, socio-economic, and other parameters of prognostic importance. (6) Identify cancer aetiological factors by conducting special studies which disclose groups of the population at high or low cancer risks. These groups may be defined by social, occupational, environmental, dietary, or other characteristics, and by drug history. (7) Identify factors related to patient survival through special studies of referral patterns, diagnostic procedures, treatment methods, and other aspects of medical care. (8) Promote specialist training in epidemiology, biostatistics, and tumour registry methodology, operation, and management.

UNITED STATES OF AMERICA

TYPE: Incidence; Registry
TERM: Diet; Drugs; Education; Environmental Factors; Occupation; Prognosis; Socio-Economic Factors; Survival; Treatment
SITE: All Sites
LOCA: Puerto Rico; United States of America
REGI: SEER (USA)
TIME: 1973 - 1993

BIRMINGHAM

927   Austin, H.D.   04418
Univ. of Alabama at Birmingham, School of Public Health, Dept. of Epidemiology, Univ. Station, 202 Tidwell Hall, Birmingham AL 35294, United States of America (Tel.: (205)9347129)
COLL: Delzell, E.; Macaluso, M.; Rotimi, C.

Mortality from Lung and Stomach Cancer at Three Car Plants
The purpose of this study is to determine whether employees of a foundry and two engine plants experienced an elevated mortality rate from lung and stomach cancer. This will be accomplished through the use of a retrospective follow-up study and two nested case-control studies. The follow-up study will include 20,000 hourly-paid workers active at any time between 1 January 1973 and 31 December 1986, and former workers who retired before 1973, but were still active as of 1 January 1970. Cohort members will be followed through 31 December 1986 and their mortality experience compared with that of the general US and Ohio populations. Vital status will be ascertained through plant records, Ohio or Michigan death registries, the Social Security Administration, and the National Death Index. Exposures of potential interest in the nested case-control studies, in addition to smoking, include polycyclic aromatic hydrocarbons, machining oils (mineral and synthetic), silica, and welding fumes. About 150 lung cancer cases and 20 stomach cancer cases are expected. Two controls will be selected for each lung cancer decedent and four for each stomach cancer decedent. Controls will be matched to cases by race and year of birth and must have been under observation at the time of death of the corresponding case. Poisson regression and conditional logistic regression will be used for the analyses of the cohort and case-control studies, respectively.

TYPE: Case-Control; Cohort
TERM: Chemical Exposure; Dusts; Occupation; Tobacco (Smoking); Welding
SITE: All Sites; Lung; Stomach
CHEM: PAH
OCCU: Automobile Workers; Foundry Workers
TIME: 1988 - 1992

BOSTON

* 928   Clapp, R.W.   05213
John Snow Research & Training, JSI/World Education, 210 Lincoln St., Boston MA 02111, United States of America (Tel.: (617)4829485; Fax: 4820617 ; Tlx: 200178)
COLL: Spiegelman, D.; Davis, L.; Martin, T.

Occupational Cancer in African Americans
Mortality due to respiratory cancer and other selected cancers by occupation and race in Massachusetts has been analysed. A larger study, involving a comprehensive analysis of 11 states with large African American populations is proposed. The primary hypothesis is that when an occupational risk is present, African Americans are disproportionately affected. Indirect adjustment for cigarette smoking will be carried out to control for confounding of specific occupation-cancer associations. Public use databases, containing data on cancer mortality, occupation and cigarette smoking prevalence, will be used and results will have relevance to attempts to prevent lung cancer, especially in new occupations and industries. The results will be presented as standardized mortality odds ratios by occupation and industry.

UNITED STATES OF AMERICA

TYPE: Mortality
TERM: Occupation; Race
SITE: Bladder; Leukaemia; Lung
TIME: 1988 – 1993

929   Speizer, F.E.   00821
Harvard Medical School, Brigham & Women's Hosp., Dept. of Medicine, Channing Lab., 180 Longwood Ave., Boston MA 02115, United States of America (Tel.: (617)4322276)
COLL: Rosner, B.R.; Willett, W.C.; Stampfer, M.J.; Colditz, G.A.; Hennekens, C.H.; Peto, R.; Hunter, D.J.; Manson, J.E.

The Nurses' Health Study

The determinants of cancer and cardiovascular disease among women have not been well described. This prospective cohort study aims to evaluate cancer morbidity and mortality for a number of exposures including oral contraceptives, cigarette smoking, post-menopausal oestrogens, hair dyes, and family history of selected diseases. Information has been obtained by mail questionnaire for 120,700 married female registered nurses, aged 30-55 residing in 11 larger US states and listed in the Directory of the American Nurses' Association in 1972. Biennial follow-up mailings have been performed beginning in 1976. In 1980 the questionnaire was expanded to include an extensive dietary component as well as sections on drug usage and exercise. Full dietary assessment was repeated in 1984, 1986 and 1990. Continued emphasis is placed on risk factors associated with breast and other cancers. Related studies include assessment of cardiovascular outcomes and diabetes, both as a risk factor for cardiovascular events and as an independent outcome and the collection of toenail specimens to be used prospectively in a case-control fashion to assess selenium as a potential risk factor in breast and other cancers. In 1989 blood specimens were obtained on approximately 35,000 members of the cohort, and will be used for nested case-control analyses.

TYPE: Cohort
TERM: BMB; Blood; Cosmetics; Diet; Drugs; Dyes; Familial Factors; Female; Hormones; Lipids; Menopause; Metals; Occupation; Oral Contraceptives; Physical Activity; Tobacco (Smoking); Toenails
SITE: All Sites; Breast (F)
CHEM: Oestrogens; Selenium
OCCU: Health Care Workers
TIME: 1977 – 1994

930   Stampfer, M.J.   04570
Brigham & Women's Hosp., Dept. of Medicine, Channing Labs, 180 Longwood Ave., Boston MA 02115, United States of America (Tel.: (617)4322279; Fax: 7311541 ; Tlx: 4974739)
COLL: Hennekens, C.H.; Willett, W.C.; Eberlein, K.A.; Rosner, B.R.; Salvini, S.

Nutritional Biochemical Markers of Cancer and Myocardial Infarction

Pre-diagnostic levels of nutritional biochemical markers of cancer and myocardial infarction are studied in the Physicians' Health Study (PHS) cohort. The PHS is a randomized trial of aspirin and beta carotene in the prevention of cardiovascular diseases and cancer incidence, conducted among 22,071 US male physicians aged 40-84 years. As part of the trial, blood samples were collected prior to randomisation from 14,916 participants, and have been stored at -82°C. All subjects developing cancer or myocardial infarction are matched by age and smoking status to a control from the same population. Objectives are to define if any of the biochemical markers assayed is a predictor of cancer or myocardial infarction. Using a nested case-control study design, the possible effects of plasma levels of retinol, lycopene, alpha and beta carotene, alpha and gamma tocopherol, vitamin B6, selenium and specific fatty acids are being studied. Participants have been followed annually since the beginning of the study and additional information on vitamin supplementation, exercise, alcohol consumption, smoking habits and other risk factors has been collected.

UNITED STATES OF AMERICA

TYPE: Case-Control; Cohort
TERM: Alcohol; BMB; Biochemical Markers; Drugs; Metals; Physical Activity; Tobacco (Smoking); Vitamins
SITE: All Sites
CHEM: Aspirin; Beta Carotene; Retinoids; Selenium
TIME: 1986 - 1992

931 Stern, R.S. 01988
Beth Israel Hosp., Harvard Medical School, Dept. of Dermatology, 330 Brookline Ave., Boston MA 02215, United States of America (Tel.: (617)7354591; Fax: 7354948)
COLL: Abel, E.; Epstein, J.H.; Wintroub, B.; Wolf, J.; Nigra, T.P.; Voorhees, J.; Anderson, T.F.; Harber, L.; Prystowsky, J.; Muller, S.; Taylor, J.R.; Frost, P.; Urbach, F.; Arndt, K.A.; Baughman, R.D.; Braverman, I.M.; Murray, J.; Petrozzi, J.; Fitzpatrick, T.B.; Parrish, J.A.; Gonzalez, E.

Photochemotherapy Follow-Up Study
This 16-centre prospective cohort study will assess the risk and efficacy of oral methoxsalen photochemotherapy, introduced in 1974 for the treatment of psoriasis, and now widely used for this common skin disease. The incidence of adverse events, especially skin cancers and other cancers are being documented. Data are collected on patients prior to entry. At annual intervals, each of the 1,380 patients is interviewed and periodically invited to be examined by a dermatologist. The cohort has been followed for an average of 13 years. Historical controls are being used. Since side effects are expected to be dose-related, the incidence of adverse events in patients receiving varying amounts of life-time exposures to photochemotherapy will be compared. A dose-dependent increase in the risk of squamous cell cancer has been documented. Male genitals are especially susceptible to these carcinogenic effects. Follow-up continues.

TYPE: Cohort
TERM: Drugs; Photochemotherapy; Radiation, Non-Ionizing
SITE: All Sites; Eye; Skin
CHEM: Methoxsalen
TIME: 1976 - 1992

BRONX

* 932 Romney, S.L. 05027
Univ. of Yeshiva, Albert Einstein College of Medicine, Dept. of Gynecology & Obstetrics, 1300 Morris Park Ave., Bronx NY 10461, United States of America (Tel.: (212)4302691)
COLL: Burk, R.; Kadish, A.; Ho, G.; Basu, J.; Palan, P.R.; Klein, S.

Clinical Trial of Beta Carotene in Women with Moderate Cervical Dysplasia
The efficacy of oral supplementation with beta-carotene (BC) in increasing the regression rate of cervical neoplasia is being investigated in a clinical trial of women with moderate cervical dysplasia. Approximately 138 women will be recruited from the colposcopy clinic of a public hospital and randomized into either the treatment group receiving 30 mg of beta-carotene a day or the active placebo group. Subjects will be followed up by colposcopy, repeat Pap smear and HPV DNA hybridization studies at three-month intervals. The outcome, regression from moderate to mild dysplasia or to normal, will be determined by biopsy at the ninth month. Women with endpoint regression will be followed up for another six months for recurrence, and blood and cervico-vaginal lavage samples will be obtained at baseline and at each follow-up visit. These samples will be used to assess levels of BC, ascorbic acid and alpha-tocopherol. Presence of HPV infection will also be determined from the lavage sample by southern blot. Subjects' dietary patterns and exposure to risk factors for cervical dysplasia will be monitored via questionnaires. Regression and recurrence rates will be compared between treatment groups in the whole population as well as in strata defined by HPV positivity or by different levels of vitamin C and E. The question of synergistic interaction of various antioxidant micronutrients will be evaluated.

UNITED STATES OF AMERICA

TYPE: Intervention
TERM: BMB; Biochemical Markers; Blood; Cytology; DNA; Diet; HPV; Nutrition; Premalignant Lesion; Tissue; Vitamins
SITE: Uterus (Cervix)
CHEM: Beta Carotene; Tocopherol
TIME: 1991 - 1995

BROOKLINE

* 933   **Buring, J.E.**   05159
Harvard Medical School, Brigham & Women's Hosp., 55 Pond Ave., Brookline MA 02146, United States of America (Tel.: (617)7324965; Fax: 7324970)
COLL: Hennekens, C.H.; Belanger, C.; Cook, N.; Eberlein, K.A.; Hebert, P.; Satterfield, S.; Rosner, B.R.; Gordon, D.; LaMotte, F.; Manson, J.E.; Peto, R.; Cohen, L.; Prentice, R.; Goodman, D.

Randomized Trial of Beta Carotene, Vitamin E and Aspirin in Cancer and Cardiovascular Disease (Women's Health Study)
This is a randomized, placebo-controlled, double-blind trial of beta carotene and vitamin E in the prevention of cancer and cardiovascular disease, and of low-dose aspirin in the prevention of cardiovascular disease. The trial will be conducted among 40,000 post-menopausal female US nurses, aged 50 years or older, with no previous history of cancer (except non-melanoma skin cancer) or cardiovascular disease. The design of the trial is 2x2x3 factorial, with participants randomly assigned to 15mg beta carotene daily or placebo; to 200 IU vitamin E daily or placebo; and to taking either 300mg aspirin, 100mg aspirin, or a placebo every other day. Primary aims will be to investigate whether beta carotene or vitamin E decreases total cancers of epithelial cell origin, and whether low dose aspirin, beta carotene, or vitamin E reduces the risk of important vascular endpoints. Following a run-in phase and randomization, participants will be followed by mailed questionnaires every six months, and all endpoints will be documented by medical records.

TYPE: Intervention
TERM: Antioxidants; Drugs; Female; Prevention; Vitamins
SITE: All Sites
CHEM: Aspirin; Beta Carotene; Retinoids
TIME: 1991 - 1996

934   **Hennekens, C.H.**   03491
Harvard Medical School, Brigham and Women's Hosp., Channing Lab., Dept. of Medicine, 55 Pond Ave., Brookline MA 02146, United States of America (Tel.: (617)7324965)
COLL: Braunwald, E.; Buring, J.E.; Eberlein, K.A.; Goldhaber, S.; Rosner, B.R.; Speizer, F.E.; Stampfer, M.J.; Willett, W.C.; Doll, R.; Peto, R.; Belanger, C.; Manson, J.E.

Randomized Trial of Aspirin and Beta Carotene among US Physicians
A total of 22,071 US male physicians aged 40 to 84 years have been assigned at random to aspirin and/or beta carotene and/or placebo in a 2x2 factorial design. The objectives are to determine whether low-dose aspirin consumption reduces cardiovascular mortality by 20% and whether beta carotene decreases cancer risk by 30% and whether beta carotene reduces the risk of important vascular events. Prerandomization blood specimens have been obtained from 16,000 participants. The randomized aspirin component was terminated in January 1988, based on a finding of a 44% reduction in the risk of myocardial infarction. Obsrvational data on aspirin consumption and cardiovascular endpoints continues to be collected. The beta carotene component continues unchanged until the end of 1996. Recent papers were published in Am. J. Med. 89:772-776, 1990, Am. J. Prev. Med. 6:290-294, 1990 and Circulation 82:897-902, 1990.

UNITED STATES OF AMERICA

TYPE: Intervention
TERM: Analgesics; BMB; Drugs; Occupation; Prevention; Vitamins
SITE: All Sites
CHEM: Aspirin; Beta Carotene
OCCU: Health Care Workers
TIME: 1981 - 1992

* 935   Rosenberg, L.E.   04949
Boston Univ. Medical School, Slone Epidemiology Unit, 1371 Beacon St., Brookline MA 02146, United States of America (Tel.: (617)7346006; Fax: 7385119)
COLL: Shapiro, S.; Palmer, J.R.; Zauber, A.; Warshauer, M.E.; Stolley, P.D.

Surveillance of Serious Illnesses and Drugs
The Slone Epidemiology Unit obtains comprehensive medication histories and data on demographic factors, reproductive history, medical history, cigarette smoking, and other exposures from hospital patients with any of a wide range of cancers and other illnesses. Special emphasis has been placed on the effect of female hormone preparations on the risk of several cancers, including breast and ovary. Large bowel cancer and other rarer cancers are also studied. The large data file is used for hypothesis generation and for the testing of hypotheses. Papers have been published in Am. J. Epidemiol. 131:483-490, 1990 and 132:1051-1055, 1990 and Int. J. Epidemiol. 19:1045-1049, 1990.

TYPE: Correlation
TERM: Data Resource; Drugs; Hormones
SITE: Breast (F); Colon; Lung; Ovary; Prostate; Uterus (Corpus)
TIME: 1976 - 1993

936   Siguel, E.N.   03685
University Hospital, Clinical Nutrition Unit, Fatty Acid Lab., P.O. Box 5, Brookline MA 02146-0001, United States of America (Tel.: (617)7394887)
COLL: Willett, W.C.; Lerman, R.H.; Jannace, P.

Fatty Acid Patterns in Patients with Cancer
Several projects are either planned or underway to relate the nature and extent of cancer, including the likelihood of having cancer, with the pattern of fatty acids in plasma and adipose tissue. Emphasis is on essential fatty acids and their derivatives, saturated fat and trans-fatty acids. Controls, matched by sex, and age, are drawn from individuals with similar characteristics to the patients with cancer but without any known cancer. An additional purpose is to identify fatty acid patterns or substances that behave chemically like fatty acids which distinguish plasma of subjects with cancer from normal subjects.

TYPE: Case-Control
TERM: Fat
SITE: Breast (F); Colon; Ovary; Prostate
TIME: 1985 - 1992

BUFFALO

937   Graham, S.   04057
State University of New York, Dept. Social & Preventive Medicine, 2211 Main St., Buffalo NY 14214, United States of America (Tel.: (716)8312981)
COLL: Hreschchyshyn, M.; Sufrin, G.; Nemoto, T.; Nolan, J.; Rawls, W.E.; Campbell, C.; Marshall, J.; Vena, J.E.; Zielezny, M.; Freudenheim, J.; Hellmann, R.

Western New York Study of Cancer of Ovary, Endometrium, Breast, Prostate and Pancreas
This case-control study investigates the possible relationships between dietary habits, smoking, alcohol ingestion, occupation and reproductive history with cancers of the male and female reproductive tract and pancreas. For those cases with cancer of the breast and prostate, a blood sample will be obtained and analysed to determine vitamin, lipid and mineral levels, which will be correlated with dietary intake. To avoid hospital-related bias and to obtain detailed information on each case, all new cancer cases in Erie and Niagara Counties in New York State are interviewed, along with 500 controls of each sex, 40 years of age and over, in the counties from which the cases are selected. A two percent random sample of Health Care Finance Administration registrants is being used to obtain controls, age 65 and older, and a one percent sample of Department of Motor Vehicle registrants for those aged 40 to 65. In addition to

UNITED STATES OF AMERICA

interviewing patients, spouses receive a self-administered questionnaire to substantiate the participants' dietary history. The use of spouse interviews is primarily to ascertain validity of dietary history data.

TYPE: Case-Control
TERM: Alcohol; Diet; Infection; Occupation; Reproductive Factors; Tobacco (Smoking); Virus; Vitamins
SITE: Breast (F); Ovary; Pancreas; Prostate; Uterus (Corpus)
TIME: 1986 - 1992

CAMBRIDGE

938   Thilly, W.G.   03938
Massachusetts Inst. of Technology, Center for Environm. Health Sciences, Div. of Toxicology, Room E18-666, 77 Massachusetts Ave., Cambridge MA 02139, United States of America (Tel.: (617)2536220; Fax: 2585424 ; Tlx: 921473 MITCAM)
COLL: Morley, A.A.; Albertini, R.J.; Lambert, B.; Ehrenberg, L.; Hemminki, K.; Wogan, G.N.; Tannenbaum, S.R.

Mutational Spectra as a Means of Diagnosing the Causes of Genetic Change in Man
Based on the observation that individual mutagens produce individual patterns of mutation, it is hypothesized that knowledge of mutational spectra in cell populations from the human body will yield important information about what environmental agents, if any, are causing somatic and/or germinal mutations in man. To date the study has shown (1) that the mutational spectra in human cells varies among mutagens, and (2) that all point mutations within exon 3 of hgprt can be detected in human genomic DNA by the use of a gradient denaturing gel electrophoresis. Means to amplify human DNA sequences with high fidelity have been developed. This provides means to study spectra without growth of cells. It is planned to apply the protocols, when ready, to studies of normal subjects, patients receiving chemotherapy involving known mutagens and cigarette smokers. The first phase of the human studies would involve about 100 patients. At the level of cell culture, the mutational spectra of spontaneous change and, in long term low dose protocols, those of polycyclic aromatic hydrocarbons, active oxygen species, alkylating agents, mycotoxins and halodeoxyuridines are being characterized. Papers appeared in Genome 31:590-593, 1989; Mutat. Res. 231:165-176, 1990, and in Ann. Rev. Pharmacol. Toxicol. 30:369-385, 1990.

TYPE: Methodology; Molecular Epidemiology
TERM: Chemical Mutagenesis; Drugs; Mutagen; Mutation, Germinal; Mutation, Somatic; Mycotoxins; PCR; Tobacco (Smoking)
SITE: Inapplicable
CHEM: Alkylating Agents; Oxygen; PAH
LOCA: Australia; Finland; Sweden; United States of America
TIME: 1980 - 1993

CHAPEL HILL

939   Godley, P.A.   04629
Univ. of North Carolina, Div. of Medical Oncology, 3009 Old Clinic Bldg, CB 7305, Chapel Hill NC 27599-7305, United States of America
COLL: Morris, D.L.; Hulka, B.S.; Mohler, J.; Sandler, R.S.

The Role of Omega-3 Fatty Acid in the Aetiology of Prostate Cancer
This preliminary study will examine the hypothesis that an increased intake of Omega-3 fatty acids, commonly found in fish oils, is a protective factor against prostate cancer. Adipose tissue and erythrocyte membrane levels of Omega-3 fatty acids will be investigated in cases and controls by gas chromatography. Cases and controls will be compared on: (1) tissue levels of erythrocyte and subcutaneous fat Omega-3 fatty acids and (2) dietary history of Omega-3 fatty acid ingestion. New patients with symptoms of bladder neck obstruction, or referred for evaluation of a prostatic module, will be recruited from the outpatient clinics and inpatient service of the North Carolina Memorial Hospital Urology Service. Male patients over the age of 50 will be eligible. The goal is to recruit 50 prostate cancer patients and 100 patients with benign prostatic hyperplasia. Medical records will be abstracted for cases and controls. Based on this review the subjects will be classified as: carcinoma of the prostate (cases), benign prostatic hypertrophy (controls). In addition, food frequency survey will be used to estimate dietary Omega-3 fatty acid consumption.

UNITED STATES OF AMERICA

TYPE: Case-Control
TERM: Diet; Fatty Acids; Lipids
SITE: Prostate
TIME: 1989 - 1992

940 Hulka, B.S. 04631
Univ. of North Carolina, School of Public Health, Dept. of Epidemiology, CB# 7400 Rosenau 207, Chapel Hill NC 27599-7400, United States of America (Tel.: (919)9665731)
COLL: Everson, R.; Griffith, J.; Margolin, B.; Wilcosky, T.; Vine, M.F.

Biological Markers in Epidemiological Research
This three-phase project is designed to study the use of biological markers in epidemiological research. Phase I is the development and publication of a book entitled "Biological Markers in Epidemiology" (OUP, 1990) which will serve as a textbook both for epidemiologists new to this field and for toxicologists with an interest in marker applications. Phase II is the development of pilot studies to assess the use of DNA adducts in epidemiological research. Studies currently in progress are 1) an autopsy study designed to address the issue of tissue specificity in the formation of DNA adducts among persons exposed to environmental agents such as tobacco smoke, 2) an analysis of DNA adduct formation in sperm to determine the potential for heritable mutations upon exposure to tobacco smoke, and 3) a study of DNA adduct formation in different white blood cell populations to assess whether certain populations are better than others for studying markers of environmental exposures. Phase III, now in the planning stage, involves the development of large scale epidemiological field studies integrating the use of laboratory assays of biological markers.

TYPE: Molecular Epidemiology
TERM: Autopsy; Biochemical Markers; DNA Adducts; Mutagen; Passive Smoking
SITE: Inapplicable
TIME: 1987 - 1992

941 Loomis, D.P. 04840
Univ. of North Carolina, School of Public Health, Dept. of Epidemiology, CB 7400, Chapel Hill NC 27599-7400, United States of America (Tel.: (919)9667433; Fax: 9662089)
COLL: Savitz, D.A.

Brain Cancer, Leukaemia and Occupation in the United States
The goals of this study are (1) to discover new associations between brain cancer or leukaemia mortality and occupation, and (2) to test occupation-cancer associations seen in smaller, more geographically-restricted populations, using a large multi-state database. A series of case-control studies is being conducted, using all male 1985-1988 brain cancer (n = 3,300) and leukaemia (n = 5,100) deaths, as well as 88,000 control deaths from other causes reported in 16 states which began coding death-certificate occupation and industry data in 1985. Hypotheses of a priori interest include excess leukaemia mortality among farmers, mechanics, and rubber workers, and excess brain cancer deaths among white collar workers, petroleum refinery employees and health care workers. A paper was published in Am. J. Ind. Med. 19:509-521, 1991.

TYPE: Case-Control
TERM: Occupation
SITE: Brain; Leukaemia
OCCU: Farmers; Health Care Workers; Meat Workers; Mechanics; Petroleum Workers; Rubber Workers; Wood Workers
TIME: 1988 - 1992

942 Rappaport, S.M. 04712
Univ. of South Carolina, School of Public Health, Rosenau Hall, CB 7400 , Chapel Hill NC 27599, United States of America (Tel.: (919)9661023; Fax: 9667141)
COLL: Yager, J.W.; Smith, A.H.

Dose-Response for Occupational Styrene Exposures
This study investigates the linkages between exposure to styrene, absorbed dose, and cytogenetic response resulting from occupational exposures in the reinforced plastics industry. A group of 4-10 workers in a reinforced-plastics facility in Southern California provided air and blood samples for initial testing of measurement methods. A larger group of 30 exposed and 15 non-exposed workers was assembled at a reinforced-plastics facility in Southern Washington. A new method has been developed for measuring styrene in mixed-exhaled air which involves forced expiration through a commercial

UNITED STATES OF AMERICA

charcoal tube. Exhaled-air levels measured with this device correlated well with blood levels; thus, blood styrene concentrations can be estimated by breath concentrations. A toxicokinetic model (two-compartment open model) of styrene disposition associated with realistic exposure scenarios has been developed. Blood and exhaled-air concentrations of styrene which have been measured thus far are consistent with the results of the model. In addition to the measurement of SCEs, other cytogenetic endpoints have been added to the protocol for blood analysis, i.e., micronuclei, DNA adducts, and alkylated hemoglobin. Preliminary work has demonstrated the presence of SCEs, micronuclei, and DNA adducts in the workers' lymphocytes. A method to measure styrene-oxide adducts of haemoglobin is being developed. The major objectives are now to: (1) complete the toxicokinetic modelling (with the 2-compartment open model) and correlate styrene exposure with levels of styrene in blood and breath; (2) correlate the numbers of SCEs and micronuclei in workers' lymphocytes with the intensity and pattern of styrene exposure; (3) isolate DNA from blood samples and measure DNA adducts; (4) continue work on a HPLC method for measuring alkylated hemoglobin in the samples and possibly styrene-oxide hg adducts.

TYPE: Methodology; Molecular Epidemiology
TERM: BMB; DNA Adducts; Dose-Response; Lymphocytes; Micronuclei; Occupation; Plastics; SCE
SITE: Inapplicable
CHEM: Styrene
OCCU: Plastics Workers
TIME: 1986 - 1993

943   Sandler, R.S.   04664
Univ. of North Carolina, CB 7080, 423 Burnett-Womack Bldg, Manning Drive, Chapel Hill NC 27599-7080, United States of America (Tel.: (919)9662511; Fax: 9661757)
COLL: Kupper, L.L.; Anderson, J.J.B.

Risk Factors for Colon Adenomas

The specific aim of this study is to determine possible risk factors for colorectal adenomas. The study will evaluate the role of diet and other factors such as obesity, parity, age at first live birth, physical activity, sedentary occupation, bowel habits and cigarette smoking in adenoma aetiology. The design of the study is a case-control epidemiological study. Cases will be patients with one or more pathologically confirmed adenomas found at colonoscopy. Controls will be colonoscopy patients found to be adenoma-free. Interviews with study subjects focus on diet using a quantitative food frequency questionnaire, lifestyle factors, and medical history. Approximately 350 cases and a comparable number of controls are expected to be recruited over three years.

TYPE: Case-Control
TERM: Diet; Obesity; Occupation; Parity; Physical Activity; Polyps; Tobacco (Smoking)
SITE: Colon; Rectum
TIME: 1988 - 1992

944   Swift, M.R.   04317
Univ. of North Carolina, Div. of Medical Genetics, Biological Sciences Research Center, CB 7250, Chapel Hill NC 27599-7250, United States of America
COLL: Chase, C.; Morrell, D.

Cancer in Families with Ataxia Telangiectasia

Ataxia telangiectasia (A-T) is a rare autosomal recessive syndrome predisposing to cancer. To examine the hypothesis that carriers of a single A-T gene (heterozygotes) may also have an excess risk of cancer, a study looking at the incidence of cancer in over 100 newly identified families of A-T patients in the United States was recently completed. In the principal test of the hypothesis, the cancer rate in the 2,168 adult blood relatives of A-T patients was compared to that in the 1,111 unrelated spouse controls and found to be significantly elevated. Now these relatives are being followed prospectively. In 1989, after about 5 years of follow-up, data will be collected on all cancers that have developed in the A-T relatives since the close of retrospective follow-up in the early 1980s. The prospective cancer rates in blood relatives will be compared with the rates in spouse controls, to produce contemporary and unbiased estimates of the risk of developing cancer in A-T blood relatives, many of whom would be expected to be A-T heterozygotes. Blood is being drawn and lymphoblastoid cell cultures established from A-T blood relatives with and without cancer. When a laboratory test becomes available that can identify A-T heterozygotes and non-heterozygotes within families of A-T patients, the samples will be tested and the relatives classified as carriers or not. Then a case-control analysis will be performed and the proportion of A-T heterozygotes among the A-T blood relatives with cancer will be compared with the proportion of carriers among relatives without cancer. This will permit direct testing for the hypothesis that A-T

UNITED STATES OF AMERICA

heterozygotes may be at increased risk of developing cancer. Recent publications include: Genet. Epidemiol. 3:17-26, 1986; N. Engl. J. Med. 316:1289-1294, 1987 and J. Nat. Cancer Inst. 78:455-458, 1987.

TYPE: Case-Control; Cohort
TERM: Ataxia Telangiectasia; Familial Factors; Genetic Factors; Genetic Markers; High-Risk Groups; Premalignant Lesion
SITE: All Sites; Breast (F); Leukaemia
TIME: 1980 - 1992

CHESTNUT HILL

* 945   Poole, C.   05026
Epidemiology Resource Inc., 826 Boylston St., Chestnut Hill MA 02167, United States of America (Tel.: (617)7349100; Fax: 2770335)
COLL: Rothman, K.J.; Dreyer, N.A.; Loughlin, J.E.

Cohort Mortality Study of Styrene Workers
This is a retrospective cohort mortality study of employees at ARCO Chemical Company's plant at Beaver Valley, Pennsylvania, USA. Since the mid-1940s, the plant has been engaged in the production of styrene, polystyrene and a variety of other chemicals. Cause-specific mortality rates will be compared with rates for the populations of the USA and Pennsylvania. The cohort will contain 6,000-7,000 persons. A much smaller, preliminary study of recently employed males suggested a possible increase in mortality from cancers of the central nervous system (three observed, 0.7 expected).

TYPE: Cohort
TERM: Occupation; Plastics
SITE: All Sites
CHEM: Polystyrene; Styrene
OCCU: Chemical Industry Workers
TIME: 1989 - 1992

946   Rothman, K.J.   04757
Epidemiology Resources, Inc., 826 Boylston St., Chestnut Hill MA 02166, United States of America (Tel.: (617)7349100; Fax: 2770335)
COLL: Wilkinson, G.; Franco, S.; Manzo, N.; Dreyer, N.A.

Brain Tumours among Offspring of Electronics Workers
A follow-up study of 4,700 workers at two plants engaged in the manufacture of color television picture tubes is being conducted to examine the health of workers and their offspring. Workers employed between 1982 and 1987 are being surveyed to gather information about their offspring. Of particular interest are any occurrences of brain or other cancers in childhood. Medical records and death certificates ar being obtained for children with illnesses of interest so that a mortality analysis may be conducted. Death certificates are also being collected for workers in the cohorts. Mortality among the workers will be compared with national and state mortality rates; the mortality rates among workers at each facility will be compared.

TYPE: Cohort
TERM: Occupation
SITE: Brain; Childhood Neoplasms
OCCU: Electronics Workers
TIME: 1988 - 1992

CHICAGO

947   Chmiel, J.S.   02229
Northwestern Univ. Medical School, Cancer Center, Biometry Sect., 680 North Lake Shore Drive, Suite 1104 , Chicago IL 60611, United States of America (Tel.: (312)9088655)
COLL: Berlin, N.I.

Cancer Incidence and Mortality in an Electric Utility Company
In cooperation with the Commonwealth Edison Company, the Cancer Center developed a cancer surveillance programme involving all Commonwealth Edison employees. The primary objective of the

UNITED STATES OF AMERICA

Program was to assess over a period of years certain aspects of disease and related risks in workers in an electric utility company. Primary concerns were: (1) mortaltiy from all causes; and (2) incidence of cancer. Particular attention was devoted to radiation induced cancers, with emphasis on long-term exposure to low dose radiation. From 1983 to 1988, all employees were asked to complete an initial (baseline) questionnaire on cancer history, smoking history, and demographics. Participants were contacted every two years to collect information on events occurring since the time of the last questionnaire. Job history and radiation exposure data were also obtained. Although the collection of questionnaire and medical data has been discontinued, the existing data will continue to be analysed. Employees who will be included in analyses are those who have held full-time jobs at Commonwealth Edison for one year or more. All eligible employees were followed for mortality. 20,754 were being followed for mortality and cancer incidence and an additional 5,800 employees who worked at Commonwealth Edison Company between 1975 and 1983 were being followed for mortality only. Additional follow-up for mortality is anticipated.

TYPE: Cohort
TERM: Monitoring; Occupation; Radiation, Ionizing; Tobacco (Smoking)
SITE: All Sites
OCCU: Radiation Workers
TIME: 1979 - 1992

948 Mallin, K. 04725
Illinois Cancer Council, 200 S. Michigan Ave., Suite 1700, Chicago IL 60604, United States of America (Tel.: (312)9869980; Fax: 9860404)

Mortality of Steelworkers in Northwestern Illinois

Deaths among steelworkers in a plant employing approximately 2,500 workers are being analysed. A previous investigation of bladder cancer incidence in northwestern Illinois detected a number of cases who had worked in this steel mill. Chemical exposures include solvents, degreasers, metals, paints, cutting oils, lubricants, and sealants. A previous inspection by OSHA examined trichloroethylene, methyl ethyl ketone, diisobutyl ketone, and methyl isobutyl ketone. Between 1977 and 1988 there were 583 deaths among workers in this plant. The proportional distribution of cancer deaths will be examined using standard techniques, and appropiate follow-up will be undertaken.

TYPE: Cohort
TERM: Metals; Occupation; Paints; Petroleum Products; Solvents
SITE: All Sites
CHEM: Diisobutyl Ketone; Methyl Ethyl Ketone; Methyl Isobutyl Ketone; Trichloroethylene
OCCU: Steel Workers
REGI: Illinois (USA)
TIME: 1990 - 1992

949 Mallin, K. 04726
Illinois Cancer Council, 200 S. Michigan Ave., Suite 1700, Chicago IL 60604, United States of America (Tel.: (312)9869980; Fax: 9860404)

Investigation of a Bladder Cancer Cluster in Northwestern Illinois

An excess of incident bladder cancer cases was detected in a small community near Rockford, Illinois (Amer. J. Epidemiol.). SIRs for the period 1978-1985 were 1.7 in males, and 2.6 in females, based on 21 and 10 cases, respectively. Two drinking water wells in this community are within half a mile of a landfill site that was closed in 1972. One well was closed due to contamination with trichloroethylene and other solvents. Further investigations of this cluster are under way. Known risk factors for bladder cancer are being compared for the cluster cases and cases in the surrounding towns. Interviews with cases or their next of kin have been completed and include information on smoking, occupational histories, residential histories, fluid intake, and other factors. Predominant occupations among all cases include machinists. Additional analyses using 1976-1978 data from the Illinois State Cancer Registry will also be undertaken. A paper was published in Am. J. Epidemiol. 132:S96-S106, 1990.

UNITED STATES OF AMERICA

TYPE: Cross-Sectional
TERM: Cluster; Occupation; Solvents; Tobacco (Smoking); Water
SITE: Bladder
CHEM: Benzene; Chlorobenzene; Trichloroethylene
REGI: Illinois (USA); Iowa (USA); Wisconsin (USA)
TIME: 1987 - 1992

950   Sharp, D.S.   03999
Amoco Corporation, Medical Dept., 200 Randolph Drive, P.O. Box 87703, Chicago IL 60680-0703, United States of America (Tel.: (312)8563901)
COLL: Hornstra, M.H.

Relationship of Exposure to Refinery Processes and Mortality from Cancer
The objective of this follow-up study is to examine mortality patterns in order to assess any increased risk for a particular cause of death. Vital status for over 26,000 employees from ten Amoco refineries has been updated and collection of death certificates for those the Social Security Administration, National Death Index and Bureau of Motor Vehicles identify as deceased. Standardized mortality ratios relative to the US and to the specific state's population are being calculated.

TYPE: Cohort
TERM: Chemical Exposure; Occupation; Petroleum Products
SITE: All Sites
OCCU: Petroleum Workers
TIME: 1985 - 1992

CINCINNATI

951   Brown, D.P.   03007
NIOSH, Industrywide Studies Branch, Div. Surveillance, Hazard Evaluation, & Field Studies, Biometry Sect., 4676 Columbia Parkway, Cincinnati OH 45226, United States of America (Tel.: (513)6842694)

Update of Completed Cohort Mortality Studies
The Industrywide Studies Branch has completed a number of retrospective cohort mortality studies during the past few years, many of which lacked conclusive results. The primary reasons for inconclusive results are too few deaths (low statistical power) and a short follow-up period (resulting in a short latency period). Thus, many of the original hypotheses tested by these studies remain unresolved questions. Examples of such studies include workers exposed to PCBs, perchloroethylene, chlorinated hydrocarbon pesticides, and styrene butadiene rubber. These cohorts are being followed through the latest possible date. The additional death information will be added to the file and the life-table analysis updated. Additional studies will be selected based on length of time since last follow-up, adequacy of the current data to address the hypothesis, and the current interest or priority of the study. The updates have been completed for the studies of workers exposed to perchloroethylene, PCBs, benzene and cadmium. Other studies currently being updated include workers exposed to attapulgite (a clay fibre), vinyl chloride and mineral wool.

TYPE: Cohort
TERM: Chemical Exposure; Dusts; Metals; Occupation; Pesticides; Plastics; Solvents
SITE: All Sites
CHEM: Attapulgite; Benzene; Butadiene; Cadmium; Hydrocarbons; Hydrocarbons, Halogenated; Mineral Fibres; PCB; Styrene; Vinyl Chloride
TIME: 1982 - 1992

952   Hornung, R.W.   04444
National Inst. of Occupational Safety, and Health, Centers for Disease Control, 4676 Columbia Parkway, Mailstop R-4, Cincinnati OH 45226, United States of America (Tel.: (613)8414211; Fax: 8414500)
COLL: Roscoe, R.J.

Risk Relationship between Radon Daughter Exposure and Cigarette Smoking in Colorado Plateau Uranium Miners (1950-1987)
This study involves an update of lung cancer mortality through 1987 in the Colorado Plateau cohort of 3,346 uranium miners. Previous analyses of this cohort involved the vital status of white male miners through 1982. In addition to five more years of follow-up, the cigarette smoking history of both living and

UNITED STATES OF AMERICA

deceased miners has been updated to include the period 1969 to 1985. One of the prime hypotheses to be tested using the new data will be the existence of a possible trend in lung cancer risk attributed to joint exposure to radon and cigarette smoking from multiplicative towards additive as a function of age of the cohort. Changes in risk coefficients compared to previous analyses will also be examined, with special attention to the decline in relative risk with time since last exposure.

TYPE: Cohort
TERM: Dose-Response; Occupation; Time Factors; Tobacco (Smoking)
SITE: Lung
CHEM: Radon
OCCU: Miners, Uranium
TIME: 1991 - 1992

953 Penn, I. 00777
Univ. of Cincinnati, Medical Center, Dept. of Surgery, 231 Bethesda Ave. (ML 558), Cincinnati OH 45267, United States of America (Tel.: (513)5586867; Fax: 5583580)

Cancer in Organ Transplant Recipients and Other Immunosuppressed Patients
The aim is to study the frequency and types of malignancies and their behaviour in organ transplant recipients receiving immunosuppressive therapy. The patients are subdivided into three groups: (1) those who had tumours before transplantation; (2) those in whom tumours were accidentally transplanted with the allograft; (3) those who developed tumours de novo after transplantation. Data on more than 7,000 patients have been collected up till recently. The study continues. Numerous reports have been published, the most recent appearing in Transplant. Proc. 23:1101-1103, 1991.

TYPE: Cohort
TERM: Chemotherapy; Immunosuppression; Transplantation
SITE: All Sites
TIME: 1968 - 1993

954 Roscoe, R.J. 00031
NIOSH, Div. of Surveillance, Hazard Evaluation and Field Studies, Robert A. Taft Labs, 4676 Columbia Parkway, Cincinnati OH 45226, United States of America (Tel.: (513)8414411)
COLL: Prorok, P.C.

Uranium Miners - Low Dose Study
This project aims: (1) to help evaluate the current standard for radon daughter exposures; and (2) to help assess health implications of energy industries. In addition, indoor, residential exposure to radon has been identified as an important risk factor for lung cancer. Since the early 1950s, NIOSH and its predecessors have been studying a cohort of 4,000 uranium miners (also known as the Colorado Plateau cohort) to determine the adverse health effects of radon daughters and other exposures. Dose-response relationships between exposures to radon daughters and lung cancer have been established and used to help set federal regulations. Data points for the dose-response curve in the area of low lifetime doses (i.e. less than 120 WLM) are, however, still needed for this cohort and for radiation carcinogenesis in general. NIOSH, in collaboration with NCI, has updated the work and smoking histories of this cohort through 1985 (last update was through 1969) and extended the vital status follow-up through 1987 (last follow-up through 1982). This updating and follow-up will allow NIOSH to finalize the occupational exposures for virtually all of the cohort and to extend the median period of observation to about 29 years. A report on lung cancer mortality among the non-smoking members of this cohort was published in J. Am. Med. Ass. 262 (5):pp 629-633, 1989. This project is associated with the project by R. W. Hornung (ID:4444).

UNITED STATES OF AMERICA

TYPE: Cohort
TERM: Dose-Response; Latency; Metals; Mining; Occupation; Radiation, Ionizing; Tobacco (Smoking)
SITE: Lung; Lymphoma
CHEM: Radon; Uranium
OCCU: Miners, Uranium
TIME: 1950 - 1992

955 Schnorr, T.M. 03206
NIOSH, Div. of Surveillance, Hazard Evaluation & Field Studies, 4676 Columbia Parkway, Cincinnati OH 45226, United States of America (Tel.: (513)6843593)

Cohort Mortality Study of Antimony Smelter Workers

Two animal studies and a cohort study in Great Britain suggest that antimony may be a carcinogen, especially for the lung. A cohort study of workers exposed to antimony with minimal arsenic exposure is required in order to evaluate the potential carcinogenicity of antimony. The aim of this study is to determine if workers exposed to antimony have increased cancer mortality. A cohort of 2,000 persons employed in an antimony smelter from 1937 to 1971 has been identified. Industrial hygiene surveys were conducted by NIOSH in 1976. The cohort is followed through the Social Security Administration, Internal Revenue Service and state bureau of motor vehicles for determination of vital status. Death certificates must then be obtained from the appropriate states and standard life table analysis conducted.

TYPE: Cohort
TERM: Chemical Exposure; Metals; Occupation
SITE: All Sites; Lung
CHEM: Antimony; Arsenic
OCCU: Antimony Process Workers
TIME: 1983 - 1989

956 Schnorr, T.M. 03207
NIOSH, Div. of Surveillance, Hazard Evaluation & Field Studies, 4676 Columbia Parkway, Cincinnati OH 45226, United States of America (Tel.: (513)6843593)
COLL: Weisburger, E.

Mortality Study of Workers Exposed to Toluene Diisocyanate

The aim of the study is to determine if workers exposed to toluene diisocyanate (TDI) have increased cancer mortality, particularly of the lung. TDI is carcinogenic in rats and mice causing fibromas, fibrosarcomas, pancreatic adenomas and liver neoplasms. Commercial use of TDI comes from the reaction of TDI with hydroxy compounds to form polyurethane products such as foams, surface coatings and sealants. Widespread commercial use of TDI began in the 1950s and production volume has been steadily increasing. An estimated 50,000-100,000 workers are currently exposed to TDI. In the light of animal data and the large number of workers exposed to TDI, a cohort study is needed to determine its potential for human carcinogenicity. A cohort of 8,000 TDI exposed workers employed in four plants has been identified. However, when persons with insufficient latency and duration of employment are excluded from the cohort, the final cohort size is expected to be about 4,000.

TYPE: Cohort
TERM: Chemical Exposure; Occupation; Plastics
SITE: All Sites
CHEM: Diisocyanates; Hydrocarbons
OCCU: Plastics Workers
TIME: 1983 - 1989

957 Schulte, P.A. 03821
NIOSH, Div. of Surveillance, Hazard Evaluation and Field Studies, Industrywide Studies Branch, 4676 Columbia Parkway, Mail Stop R-13, Cincinnati OH 45226, United States of America (Tel.: (513)8414507/8414207; Fax: 8414550/8414540)
COLL: Hayes, R.; Friedland, J.M; Mazzuckelli, L.F.; Ward, E.M.; Caporaso, N.E.; Aldefer, R.

Biological Markers of Occupational Bladder Cancer

This is a case-control study, involving 150 cases and 150 controls, to identify early markers of bladder cancer to assess whether or not there are metabolic phenotypes that increase the risk of occupational bladder cancer. The markers to be utilized in the study include various oncogenes, tumour suppressor genes, growth factors and their receptors and other cellular markers.

UNITED STATES OF AMERICA

TYPE: Case-Control
TERM: Biochemical Markers; Chemical Exposure; Genetic Factors; Occupation; Oncogenes
SITE: Bladder
CHEM: Amines, Aromatic
TIME: 1990 - 1994

958 Schulte, P.A. 04219
NIOSH, Div. of Surveillance, Hazard Evaluation and Field Studies, Industrywide Studies Branch, 4676 Columbia Parkway, Mail Stop R-13, Cincinnati OH 45226, United States of America (Tel.: (513)8414507/8414207; Fax: 8414550/8414540)
COLL: Boeniger, M.; Walker, J.; Herrick, R.; Halperin, W.E.; Griffith, J.

Biological Markers in Hospital Workers Exposed to Low Levels of Ethylene Oxides

Hospital sterilizer operators using ethylene oxide were studied to determine if there was a relationship between exposure and a panel of biological markers. A total of 74 workers from the United States and Mexico will be assessed for haemoglobin and DNA adducts, SCE, micronuclei, and HPRT mutations.

TYPE: Case Series
TERM: BMB; Biochemical Markers; DNA Adducts; Micronuclei; Occupation; SCE
SITE: Leukaemia
CHEM: Ethylene Oxide
OCCU: Health Care Workers
LOCA: Mexico; United States of America
TIME: 1988 - 1992

959 Steenland, K.N. 03030
NIOSH, Div. of Surveillance,, Hazard Evaluation & Field Studies, Dept. of Epidemiologic Methods Activity, Mail Stop R-13, 4676 Columbia Parkway, Cincinnati OH 45226, United States of America (Tel.: (513)8414207)
COLL: Stayner, L.; Hayes, R.B.

Ethylene Oxide Mortality Study

Ethylene oxide (ETO) is used to sterilize hospital equipment by industrial companies which sell hospital supplies. The mortality experience of individuals who have been exposed while sterilizing will be studied. From two animal and two human studies, it appears that ethylene oxide is a leukaemogen. However, all human studies done so far are flawed by small sample size and mixed exposures. The current study will cover 14 plants where medical supplies are sterilized commercially. Data collection is complete. It is expected to study approximately 20,000 people exposed to ETO for at least three months prior to 1978 over the last 20-30 years Follow-up of the cohort has been completed through 1988. Analyses are underway. These workers have generally not been exposed to other confounding toxins. Papers regarding a feasibility assessment and the exposure model have been published in Am. J. Ind. Med. 12:419-430, 1987 and Scand. J. Work Environ. Health 14 (Suppl. 2):29-30, 1988.

TYPE: Cohort
TERM: Chemical Exposure; Occupation
SITE: All Sites; Leukaemia; Pancreas; Stomach
CHEM: Ethylene Oxide
TIME: 1982 - 1992

960 Ward, E.M. 03034
NIOSH, Div. of Surveillance, Hazard Evaluation & Field Studies, 4676 Columbia Parkway , Alice Hamilton Bldg, Cincinnati OH 45226, United States of America (Tel.: (513)6843593)

Investigation of Workers Exposed to MOCA

MOCA has been shown to be a potent carcinogen in animal studies and is structurally similar to aromatic amines, such as benzidine, known to cause bladder cancer in occupationally exposed workers. No adequate epidemiological studies have examined the carcinogenicity of MOCA in humans. MOCA is used as a curing agent for isocyanate containing polymers. About 33,000 US workers are currently exposed, most in workplaces where 10 or fewer employees work directly with the substance. A registry of workers exposed to MOCA at a US production facility has been established. The registry will be used for prospective surveillance of cancer incidence and mortality. Cancer incidence will be determined by notification of the cohort and telephone interview. In addition, workers from the facility will be offered a screening examination which will utilize conventional techniques to examine urine samples for abnormalities indicative of bladder cancer.

UNITED STATES OF AMERICA

TYPE: Cohort
TERM: Occupation; Screening; Urine
SITE: All Sites; Bladder
CHEM: 4,4'-Methylene-bis(2-chloroaniline)
TIME: 1981 - 1992

961   Ward, E.M.   03212
NIOSH, Div. of Surveillance, Hazard Evaluation & Field Studies, 4676 Columbia Parkway, Alice Hamilton Bldg, Cincinnati OH 45226, United States of America (Tel.: (513)6843593)
COLL: Zahm, S.H.

Mortality Study of Workers Exposed to Chlorinated Naphthalenes
The aim of the study is to determine whether Halowax exposure is associated with an excess risk of soft tissue sarcoma, lymphoma or liver cancer. Chlorinated naphthalenes, sold under the trade name Halowax, are known chloracnegens. Because chlorinated naphthalenes are structurally similar to, and share common toxic effects with other chlorinated hydrocarbons that are chloracnegens (polychlorinated biphenyls (PCBs), chlorinated dibenzodioxins (dioxins, TCDDs) and chlorinated dibenzofurans (furans)), it is hypothesized that they might share common chronic health effects as well. These include an excess risk of soft tissue sarcoma and lymphoma associated with TCDDs in humans and liver tumours associated with PCBs and TCDDs in animal studies. The study population will be 8,000 workers exposed to Halowax in the early 1940's at a plant manufacturing Navy cable. Phase I will include 740 workers identified from medical records as having possible or probable chloracne and should be completed in 1.5 years. Phase II will include all 8,000 workers employed at the plant during the time when Halowax was used, and will be completed in three years. Standard NIOSH methods will be used for coding the masterfile, vital status follow-up, nosology of death certificates and life table analysis for computing standardized mortality ratios. Although the current number of workers exposed to chlorinated naphthalenes is under 5,000, the study will be of value in investigating the ultimate health outcomes of other worker groups exposed to potential chloracnegens.

TYPE: Cohort
TERM: Chemical Exposure; Occupation
SITE: Liver; Lymphoma; Soft Tissue
CHEM: Naphthalenes, Chlorinated; PCB
OCCU: Cable Manufacturers
TIME: 1983 - 1992

COLUMBIA

* 962   Coker, A.   05104
Univ. of South Carolina, Dept. of Epidemiology and Biostatistics, School of Public Health, Sumter and Greene, Columbia SC 29208, United States of America (Tel.: (803)7717353; Fax: 7774783)
COLL: Pirisi, L.

Human Papilloma Viruses and CIN among Low-Income Black Women in South Carolina
This nested case-control study will investigate the relationship between HPV, by DNA subtype, and CIN 1, CIN 2, and CIN 3, in a low-income, largely minority population. The approximately 15,000 women receiving Pap smears each year at SC DHEC clinics in the Trident Health District will form the sampling frame. From December 1990 through November 1991, the spatula used to collect each smear will be saved in individually labeled test tubes for subsequent HPV DNA typing of selected subjects. Smear results will be used to select 450 CIN cases and 200 controls with normal cervical cytology. Adjusted odds ratios and 95% confidence intervals will be used to test the following study hypotheses: (1) women with any CIN will be more likely to have HPV 16/18/33 compared with women having normal cervical cytology when controlling for confounders; (2) cases will be more likely to have HPV 6/11 compared with controls when controlling for confounders; (3) the adjusted odds ratio for the HPV 16/18/33 (or HPV 6/11) and CIN relationship will be largest for the comparison of CIN 3 cases with controls relative to the comparison of CIN 2 or CIN 1 cases with controls.

UNITED STATES OF AMERICA

TYPE: Case-Control
TERM: BMB; Blacks; Cytology; HPV; Socio-Economic Factors
SITE: Uterus (Cervix)
TIME: 1991 - 1992

DES MOINES

963 Young, D.C. 04004
University of Iowa, Dept. Preventive Medicine/Radiology, 974 73rd St., Des Moines IA 50312, United States of America
COLL: Young, C.S.; Wallace, R.B.

Women's Health Trial
Using previously identified risk factors women at high risk for breast cancer will be solicited to volunteer as participants in a study to evaluate what role, if any, dietary fat may play in the induction of breast cancer. A total of 30,000 women will be divided into a control group as well as an intervention group. The intervention group will be motivated to reduce dietary fat intake to a level of 25-30% of total calories. The end point will be breast cancer in any form, with an annual mammography being provided.

TYPE: Cohort; Intervention
TERM: Diet; Fat; Mammography; Prevention
SITE: Breast (F)
TIME: 1985 - 1992

DETROIT

964 Demets, R.Y. 04265
Wayne State Univ., Dept. of Family Medicine, 4201 St. Antoine (NHL-4J), Detroit MI, United States of America (Tel.: (313)5775074)

Colorectal Neoplasia among Pattern and Model Makers
This is a case-control study to test the hypothesis that work with cutting oils leads to an increased incidence of colorectal adenomas. Cases are patients with histologically proven adenomas, matched 2:1 by age with controls in the same trade. Exposure data, smoking habits and trade years are independent variables. The study subjects are participants in a longitudinal cancer screening programme. Cases are reviewed and examined every three years for 15 years. The total number of participants is 2,000, from which approximately 300 cases will be recruited. All polyps discovered are biopsied and reviewed by a single university pathologist.

TYPE: Case-Control
TERM: Chemical Exposure; Metals; Occupation; Plastics; Screening; Tobacco (Smoking)
SITE: Colon; Rectum
CHEM: Formaldehyde; Mineral Oil; Nickel
OCCU: Model and Pattern Makers
TIME: 1982 - 1997

965 Haas, G.P. 04383
Wayne State Univ., School of Medicine, Sect. of Uro-Oncology, 4160 John R. Rd, Suite 1017, Detroit MI 48201, United States of America (Tel.: (313)7457381; Fax: 7450464)
COLL: Miles, B.J.; MacIntyre, R.

Benign Testicular Tumours
Based on previously published data, the incidence of the diagnosis of benign testicular tumours seems to be increasing. Since testicular malignancy is usually treated by radical orchiectomy, and some scrotal masses are misdiagnosed as cancer, some patients may undergoe unnecessary orchiectomy. To make an appropriate diagnosis prior to orchiectomy, the epidemiology of testicular tumours as well as the results of available diagnostic modalities will be investigated. This multicentre study comprises retrospective and prospective analyses of data from patients with benign and malignant testicular masses. Analysis of the results of the retrospective part is now in progress. For the prospective part it is expected to enrol 100 patients over a 3-year period. Factors investigated are age, race, duration of symptoms, testicular ultrasound findings, frozen section diagnosis and final pathological diagnosis. Papers have appeared in J. Urol. 136:1219, 1986 and Urology 28:28, 1986.

UNITED STATES OF AMERICA

TYPE: Case Series
TERM: Age; Histology; Race; Time Factors; Ultrasound
SITE: Benign Tumours; Testis
TIME: 1986 – 1992

* 966 Haas, G.P. 05022
Wayne State Univ., School of Medicine, Sect. of Uro-Oncology, 4160 John R. Rd, Suite 1017, Detroit MI 48201, United States of America (Tel.: (313)7457381; Fax: 7450464)
COLL: Edson Pontes, J.; Sakr, W.; Chrissman, J.; Heilbrun, L.K.; Schwartz, S.

Age- and Race- Related Trends in the Prevalence of Prostate Cancer

Carcinoma of the prostate is the most frequent cancer diagnosed and the third cause of cancer related deaths. While the clinical presentation is most common in elderly males, the disease is known to have a long latency period and very little data are available on the prevalence of prostate cancer in the younger age groups. In addition, there are marked differences between various racial groups. The aim of the study is to identify the prevalence of prostate cancer in black and white males with particular attention paid to the younger age groups (less than 50 years of age). Autopsy specimens will be obtained from men who die from unrelated causes and will be histologically step-sectioned to identify the presence of malignant or pre-malignant changes. The accumulated data will demonstrate the prevalence of prostate cancer in black and white males of various age groups and will serve as a data base for further studies of potential predisposing factors, molecular changes and the targeting of populations to be screened.

TYPE: Cross-Sectional
TERM: Age; Autopsy; BMB; Race; Screening
SITE: Prostate
TIME: 1991 – 1995

967 Jackson, C.E. 00977
Henry Ford Hosp., Clinical Genetics Sect., Dept. of Medicine, 2799 W. Grand Blvd, Detroit MI 48202, United States of America (Tel.: (313)8762834)
COLL: Ghironzi, G.; Schuman, B.M.; Salerno-Mele, P.

Gastric Cancer in San Marino

The purpose of this project is to define the genetic and environmental factors causing the high frequency of gastric cancer observed in San Marino. Initial studies have revealed a higher percentage of patients with first degree relatives affected than controls. Screening of segments of the population by gastroscopy is being undertaken in attempts to detect the neoplasms early. The project is a cooperative one between physicians in San Marino and in Detroit, Michigan, USA which has a large San Marinese population. Salmonella mutagenicity testing of urine of patients, relatives and controls as well as ingested material is planned. Gastroscopy has been performed on a total of about 350 first degree relatives of patients with gastric cancer. A high frequency of gastric mucosal abnormalities has been found along with some early gastric cancer. Blood samples for pepsinogen I and I/II ratios have been obtained from a group of 165 patients having gastroscopy and correlation of values obtained with biopsy results is planned. A survey of Pepsinogen I & II values in the Detroit San Marino population is being undertaken.

TYPE: Case-Control
TERM: Biopsy; Environmental Factors; Familial Factors; Mutagen Tests; Screening
SITE: Stomach
LOCA: San Marino; United States of America
TIME: 1975 – 1992

968 Park, R. 04657
International Union, United Auto Workers, Health and Safety Dept., 8000 E. Jefferson St., Detroit MI 48214, United States of America (Tel.: (313)9265563)
COLL: Mirer, F.E.; Siverstein, M.A.

Occupational Cancer Mortality in the Automobile Industry

Several studies are under way to assess work-related mortality in: (1) seven automotive machining plants, with primary attention to non-Hodgkin's lymphoma in transmission and chassis workers (approximately 4,000 deaths in total); (2) five automotive stamping plants, with emphasis on gastrointestinal cancer (approximately 3,000 deaths in total), and (3) a forge plant, with special focus on lung cancer (approximately 1,000 deaths in total). PMRs will be calculated using US deaths as the reference population.

UNITED STATES OF AMERICA

TYPE: Mortality
TERM: Occupation; Petroleum Products; Welding
SITE: Gastrointestinal; Lung; Non-Hodgkin's Lymphoma
OCCU: Grinding Material Workers; Machinists; Metal Workers; Tool Makers; Welders
TIME: 1985 – 1992

969 Schwartz, A.G. 03470
Michigan Cancer Foundation, Dept. of Epidemiology, 110 E. Warren Ave. , Detroit MI 48201, United States of America (Tel.: (313)8330710; Fax: 8318714)
COLL: Moll, P.P.; Boehnke, M.

Familial Heterogeneity of Cancer in Metropolitan Detroit
Approximately 55,000 individuals, from 3,300 families identified through healthy Detroit area women, are being evaluated for familial heterogeneity of cancer risk. Familial heterogeneity of all sites combined and on a site-specific basis will be analysed using a newly developed family risk index method. This method allows adjustment for age, sex, race, and birth cohorts for all family members, as well as adjustment for family size. Families at high and low risk of cancer will be characterised. Results obtained using this new method will be compared with results obtained using more conventional methods. A paper was published in Am. J. Epidemiol. 128:524-535, 1988.

TYPE: Cohort
TERM: Birth Cohort; Familial Factors; Race
SITE: All Sites
TIME: 1982 – 1992

* 970 Schwartz, A.G. 05088
Michigan Cancer Foundation, Dept. of Epidemiology, 110 E. Warren Ave. , Detroit MI 48201, United States of America (Tel.: (313)8330710; Fax: 8318714)
COLL: Moll, P.P.; Swanson, G.M.

Familial Risk of Lung Cancer
A study of the familial risk of lung cancer and other respiratory disease among families identified through a population-based series of 401 non-smoking lung cancer cases aged 40-84, 135 lung cancer cases less than 40, and age-, sex-, race-matched population-based controls is being conducted in the Detroit Metropolitan area. Telephone interviews are used to collect data on cancer history, respiratory disease history, smoking, passive smoking, occupation, and demographic variables for each of the estimated 7,000 first-degree relatives of the cases and controls. Various measures of familial risk will be used in the determination of disease risk among relatives after adjustment for individual risk factors. The modification of a new statistical method, the family risk index method, appropriate for the analysis of family data, is being undertaken to measure familial heterogeneity of cancer risk. In addition, logistic regression and proportional hazards modelling will be used for data analysis.

TYPE: Case-Control; Methodology
TERM: Familial Factors; Occupation; Passive Smoking; Tobacco (Smoking)
SITE: Lung
TIME: 1990 – 1995

EAST LANSING

971 Swanson, G.M. 03431
Michigan State Univ., Comprehensive Breast Cancer Center, A211 East Fee Hall , East Lansing MI 48824, United States of America (Tel.: (517)3538828; Fax: 3361798)
COLL: Satariano, E.; Brissette, P.; Iwamamoto, K.

Occupational Cancer Surveillance: New Approaches
This project is operating an occupational cancer monitoring system. The system complements the Metropolitan Detroit Cancer Surveillance System (MDCSS), which is a participant in the SEER programme of the NCI. Occupational histories and related information are obtained by telephone interview. A total of approximately 20,000 case interviews are projected during the course of the study. Controls are population controls selected by random digit dialling procedures. Cases selected for the study will include all newly diagnosed cases of the specified sites between the ages of 40 and 84. New leads about occupational risks associated with cancer incidence are expected from the proposed study, particularly for women and blacks. These leads will be disseminated to local public health officials,

UNITED STATES OF AMERICA

NIOSH, and appropriate groups of workers and industries. Analytical investigation suggested by study results will follow. Monitoring occupation in terms of cancer incidence permits earlier detection of potential hazards than can be achieved by mortality studies. Evaluation of death certificate occupational information for study subjects who are deceased is underway. Papers have been published in J. Occup. Med. 27:439-444, 1985; Am. J. Publ. Hlth 77:1532-1534, 1987; and in Am. J. Med. 14:121-136, 1988.

TYPE: Case-Control; Incidence; Methodology
TERM: Blacks; Environmental Factors; Female; Lifestyle; Occupation; Registry
SITE: Bladder; Colon; Eye; Liver; Lung; Melanoma; Mesothelioma; Oesophagus; Rectum; Salivary Gland; Stomach
REGI: Michigan (USA)
TIME: 1984 - 1992

972   Swanson, G.M.   04335
Michigan State Univ., Comprehensive Breast Cancer Center, A211 East Fee Hall, East Lansing MI 48824, United States of America (Tel.: (517)3538828; Fax: 3361798)
COLL: Schwartz, A.G.

Investigation of Tumours that Occur Excessively among Blacks

This collaborative study is being conducted by the Detroit, Atlanta and New Jersey registries in the SEER program and the National Cancer Institute. Its objectives are: (1) to identify race-specific risk factors for prostate, pancreas, oesophagus, and multiple myeloma; (2) to estimate the extent to which the risk factors may explain the black/white differences in the incidence rates of these cancers; and (3) to obtain biochemical measures of certain risk factors, such as hormones. The study design is that of a case referent study. The metropolitan Detroit component will enrol 302 male and female cases of multiple myeloma, 325 cases of cancer of the prostate; 422 male and female cases of cancer of the pancreas, and 250 male cases of cancer of the oesophagus over four years of data collection. Interviews will also be conducted with 967 controls. Cases in the age group 30-79 at diagnosis are included in the study. The interview obtains data regarding dietary factors, sexual development and behaviour, occupational history, family history of cancer, medical history, hormonal status, socio-economic status, and tobacco use. For cancers of the pancreas, extensive questions regarding coffee use are included. Data are obtained through in-person interviews, conducted in the home for most cases and controls. For cancers of the pancreas, many interviews are being conducted in hospital. Data collection is projected to continue through 1989. Analyses will be conducted separately for each type of cancer; multiple logistic regression analysis will be performed to control for potential confounders.

TYPE: Case-Control
TERM: Blacks; Diet; Familial Factors; Hormones; Occupation; Premalignant Lesion; Registry; Socio-Economic Factors
SITE: Multiple Myeloma; Oesophagus; Pancreas; Prostate
REGI: Atlanta (USA); Michigan (USA); New Jersey (USA)
TIME: 1986 - 1992

EAST MILLSTONE

* 973   Gamble, J.F.   05108
Exxon Biomedical Sciences, Inc., Div. of Occupational Health, and Epidemiology, Mettlers Rd, East Millstone NJ 08875-2350, United States of America (Tel.: (908)8736004; Fax: 8736009)
COLL: Thar, W.E.; Schnatter, A.R.; Donaleski, D.

Kidney Cancer in Petrochemical Workers

This nested case-control study of petrochemical workers seeks to examine the association of kidney cancer and exposure to gasoline. About 40 cases of kidney cancer will be identified from death certificates and tumour registries from 1972 to the present. Four controls per case are frequency-matched on date of birth and date of first employment and must have lived as long as the case. Exposure-response will be estimated for substances present in the plants and either suspected or known kidney carcinogens. Quantitative exposure scores will be estimated by the product of frequency, intensity and duration of exposure.

UNITED STATES OF AMERICA

TYPE: Case-Control
TERM: Occupation
SITE: Kidney
CHEM: Gasoline
OCCU: Petrochemical Workers
TIME: 1990 - 1992

* 974   **Lewis, R.J.**   05115
Exxon Biomedical Sciences, Inc., Div. of Occupational Health, and Epidemiology, Mettlers Rd, East Millstone NJ 08875-2350, United States of America (Tel.: (908)8736284; Fax: 8736009)
COLL: Lerman, S.E.

Colorectal Cancer Incidence among a Cohort of Polypropylene Workers
This is a historical cohort study of colorectal cancer incidence among 800 polypropylene workers employed for at least six months between 1960 and 1990. The study will update an earlier investigation (N = 335 workers) which found a significant 5.6-fold excess (95% CI = 2.2-11.5) of this cancer through 1985 (J. Occup. Med. 25:100-109, 1988). The cohort has been updated for the period 1986-1990. Case ascertainment methods used in the first study will be used in this investigation. The analysis will compare incidence rates to local and national (SEER) comparison rates. Person years at risk will be calculated from specific intervals after first employment until the earlier of date of diagnosis, termination, or the end of the study period (31 December, 1990). Standardized incidence ratios will be calculated by time since first employment, duration of employment, period of hire, and job category.

TYPE: Cohort
TERM: Occupation; Plastics
SITE: Colon; Rectum
CHEM: Polypropylene
OCCU: Plastics Workers
TIME: 1990 - 1992

* 975   **Lewis, R.J.**   05116
Exxon Biomedical Sciences, Inc., Div. of Occupational Health, and Epidemiology, Mettlers Rd, East Millstone NJ 08875-2350, United States of America (Tel.: (908)8736284; Fax: 8736009)
COLL: Yarborough III, C.M.; Racioppi, L.M.

Leukaemia and Lymphatic Cancer among Research and Development Workers
To follow up earlier mortality findings, a nested case-control study of leukaemia and lymphatic cancer among white male research and development workers is being conducted. The case group consists of 34 white male cohort members whose underlying or contributory cause of death was identified as leukaemia or lymphatic cancer. Incidence density sampling was used to select controls from cohort members alive at the time of deaths of the cases. Controls were matched to cases on year of birth, race, sex, and date of hire (within 2.5 years) using a 4:1 matching ratio. History of exposure to suspect leukaemogens such as benzene, other solvents and radiation (ionizing and non-ionizing) is being assessed via a team of industrial hygienists using exposure monitoring data and employment records. Smoking and medical history are being abstracted from medical records. Odds ratios will be calculated for substance-specific and workplace-specific indices. Potential confounders will be controlled separately in a stratified analysis, and simultaneously using unconditional multiple logistic regression. Potential effect modifiers will also be investigated using logistic regression modelling.

TYPE: Case-Control
TERM: Occupation; Radiation, Ionizing; Radiation, Non-Ionizing; Solvents
SITE: Leukaemia; Lymphoma
CHEM: Benzene
OCCU: Chemists; Engineering Workers; Physicists
TIME: 1990 - 1992

* 976   **Lewis, R.J.**   05117
Exxon Biomedical Sciences, Inc., Div. of Occupational Health, and Epidemiology, Mettlers Rd, East Millstone NJ 08875-2350, United States of America (Tel.: (908)8736284; Fax: 8736009)
COLL: Lerman, S.E.; Hughes, J.I.

Screening for Colorectal Polyps among a Cohort of Polypropylene Workers
The objective of this study is to conduct a second colorectal polyp screening among a cohort of polypropylene workers in Texas. Results from the first screening indicated an elevated prevalence of

UNITED STATES OF AMERICA

polyps among workers compared with a clinic-based population (J. Occup. Med. 31:785-791, 1989). The cohort has been updated and now includes approximately 800 eligible workers employed at least six months between 1960 and 1990. Workers undergo standardized examinations for colorectal polyps at a clinic in Houston, Texas. All examinations are conducted by one of two physicians to reduce inter-examiner variation. Referent subjects will be similar to those used in the first screening; non-Exxon employees (including employees from the headquarters of another petrochemical company) screened at the same clinic will serve as comparison subjects. Prevalence rate ratios will be calculated using data for persons undergoing screening for their first time. For those subjects undergoing a second screening, incidence rate ratios, adjusted for age and time since first screen, will be calculated. Polyp rates will be examined for the original cohort and for the updated cohort. The results of this study will aid in determining whether the original excess of polyps observed was due to chance, bias, or possibly represented a causal association.

TYPE: Cohort
TERM: Occupation; Plastics; Polyps; Screening
SITE: Colon; Rectum
CHEM: Polypropylene
OCCU: Plastics Workers
TIME: 1990 - 1992

EDINBORO

977   Miller, G.H.   04140
Studies on Smoking, Inc., 125 High St., Edinboro PA 16412, United States of America (Tel.: (814)7345538)
COLL: Schneiderman, M.A.; De Pue, R.H.

Comparison of Cancer Incidence in Women Exposed and Non-Exposed to Passive Smoking
The study will compare the specific types of cancer contracted by non-smoking women with no known exposure and by non-smoking women who have had some lifetime exposure. The differences, if any, in cancer incidence between these two levels of exposure, will be examined. Later studies will compare these two groups with smoking women. The present study is based on data from interviews of close relatives of the deceased in Erie County, Pennsylvania during the years 1975-1976 and 1979-1980. A total of 3,295 interviews out of a total population of 10,131 deceased (including males and females) were completed though 1986. Only interviews containing all necessary data on passive smoking were included in the study, resulting in a data base of 934 non-smoking wives (199 had no known exposure and 733 had some lifetime exposure). Preliminary analysis appears to show significant differences in cancer incidence between exposed and non-exposed women, as well as differences in the types of cancer contracted. In order to increase the sample size, additional data on decedents in 1975-1976 and 1979-1980 will be obtained in 1987-1988. A study of breast cancer and passive smoking has been completed. A paper was published in Cancer Prevention and Detection 14:497-503, 1989.

TYPE: Cohort
TERM: Female; Passive Smoking
SITE: All Sites
TIME: 1974 - 1993

* 978   Miller, G.H.   05198
Studies on Smoking, Inc., 125 High St., Edinboro PA 16412, United States of America (Tel.: (814)7345538)
COLL: Cox, C.E.; Depue, R.H.; Gambrill, R.D.; Hawthorne, V.; Novotny, T.; Pierce, J.P.; Schneiderman, M.A.; Wells, A.J.

Can Active or Passive Smoking Cause Breast Cancer?
The aim of this study is to ascertain if there is a relationship between active or passive smoking and breast cancer in women, using a complete population sample of 3,609 deceased women in Erie County, Pennsylvania, who died in the periods 1972-1976 and 1979-1980. Relatives of the deceased were interviewed to determine three categories of exposure to tobacco smoke: (1) no known or minimal exposure; (2) passive smoking exposure, and (3) active smoking exposure. The sample included 371, 2,115 and 1,123 women respectively in these categories. Excluding traumatic causes of death, the proportions of women with breast cancer in each category were 0.5%, 4.4% and 5.5%. The results of the pilot phase appear to show that if both active and passive smoking were eliminated, the incidence of

UNITED STATES OF AMERICA

breast cancer in women should be reduced significantly. Additional research is in progress to clarify the results and to recruit additional subjects.

TYPE: Cohort
TERM: Passive Smoking; Tobacco (Smoking)
SITE: Breast (F)
TIME: 1986 – 1993

EMERYVILLE

979   Reynolds, P.   04854
California Dept. of Health Services, Environmental Epidemiology Sect., 5900 Hollis St., Suite E, Emeryville CA 94608, United States of America (Tel.: (415)5403657; Fax: 5402673)
COLL: Lemp, G.F.; Saunders, L.D.

AIDS and Cancer: A Record Linkage Study

The overall objective of this project is to provide baseline information on the epidemiology of human immunodeficiency virus (HIV)-associated malignancies. This is being accomplished via record linkage between the population-based cancer registry serving the San Francisco Bay Area since 1969 (with over a quarter of a million cancer records) and the San Francisco AIDS Surveillance Program serving San Francisco since 1980 (with over 6,000 case records). A variety of analytical techniques are being used to examine the occurrence and characteristics of HIV-associated cancers.

TYPE: Incidence
TERM: AIDS; HIV; Record Linkage; Registry
SITE: All Sites; Kaposi's Sarcoma
REGI: Bay Area (USA)
TIME: 1988 – 1992

GAINESVILLE

980   Pollock, B.H.   04934
Univ. of Florida, College of Medicine, Div. of Epidemiology & Biostatistics, Dept. of Pediatrics, Suite 22, 4110 SW 34th St. , Gainesville FL 32608-2516, United States of America (Tel.: (904)3925198; Fax: 3928162)
COLL: Mulhern, R.K.; Armstrong, F.D.; Ryan, B.R.

Development of Intervention Strategies to Reduce Delay in Diagnosis of Childhood Cancer

Early detection of childhood cancer usually optimizes response to therapy and thus reduces subsequent mortality and morbidity. As asymptomatic screening is often impractical, the long-range goal is to improve the early symptomatic detection of childhood cancer. The specific aims are: to describe the constellation of signs and symptoms which occur prior to diagnosis; to evaluate the determinants of lagtime (time from the first onset of symptoms until diagnosis); and to assess the relationship between lagtime and prognosis. This cohort study will utilize a self-administered questionnaire, completed by parents of 1,200 children with newly diagnosed malignancies accrued to a Pediatric Oncology Group therapeutic protocol during the period 1990-1992. To determine which factors are associated with lagtime, information will be collected on: demographic factors, socio-economic status, family health-seeking behaviour and access to health services. This questionnaire data along with other data will make it possible to test the hypothesis that lagtime is independently associated with prognosis. Follow-up for survival and disease-free survival will continue up to 1993. Cox proportional hazards regression will be used to assess the association of lagtime with long-term survival with adjustement for potential confounders such as stage at diagnosis and treatment. It is anticipated that these findings will lead to the development of cancer control interventions.

UNITED STATES OF AMERICA

TYPE: Cohort
TERM: Prognosis; Socio-Economic Factors; Survival; Time Factors
SITE: Childhood Neoplasms
LOCA: Canada; United States of America
TIME: 1990 - 1993

HAGERSTOWN

981   Comstock, G.W.   00283
Johns Hopkins Univ., School of Hygiene and Public Health, Training Center for Public Health Res., Box 2067, Hagerstown MD 21742-2067, United States of America (Tel.: (301)7913230)
COLL: Bush, T.L.

Serological Precursors for Cancer Risk Factors
In 1974, 15 ml of blood were collected from approximately 25,600 persons in Washington County, Maryland. The serum was separated from the blood cells and stored at -70 degrees C. In 1989 20 ml of blood were collected from 32,941 persons in the same county. Plasma and white blood cells were also stored at -70 degrees C. In addition to basic identification and personal information obtained when blood was drawn, participants in the 1989 study were asked to complete a dietary questionnaire at their leisure and to return them along with a toenail clipping for trace metal analysis. Dietary histories and information from plasma, white blood cells, and toenails will be analysed for limited nested case-control studies of specific cancers as these develop among participants and are identified by the cancer registry. Examples of potential cancer risk factors include vitamins (A, D, E, carotenoids), trace elements (selenium, zinc), hormones, lipids, white blood cell components (DNA adducts), and antibodies (herpes, cytomegalovirus, papillomovirus, Epstein-Barr virus). Comparison between estimated intake of various nutrients can be made. The stability of plasma components on prolonged storage or on repeated thawing and freezing will be tested in participants who do not live in the immediate area of Washington County, MD.

TYPE: Case-Control; Cohort
TERM: BMB; DNA Adducts; Diet; EBV; HPV; HSV; Hormones; Lipids; Plasma; Serum; Toenails; Trace Elements; Virus; Vitamins; White Cells
SITE: All Sites
CHEM: Beta Carotene; Retinoids
REGI: Maryland (USA)
TIME: 1974 - 1993

HANOVER

* 982   Baron, J.A.   04978
Dartmouth Medical School, 2 Maynard St., Hanover NH 03756, United States of America (Tel.: (603)6465542; Fax: 6466313)
COLL: Rothstein, R.; Beck, G.; Summers, R.; Haile, R.W.C.; Mandel, J.S.; Sandler, R.S.

Calcium in the Prevention of Neoplastic Polyps
This is a randomized, double-blind, placebo-controlled clinical trial of the efficacy of nutritional supplementation with calcium carbonate in preventing recurrence of neoplastic polyps of the large bowel. It is a collaborative investigation conducted at six clinical centres (Cleveland Clinic Foundation, Dartmouth-Hitchcock Medical Center, Univ. of California at Los Angeles, Univ. of Iowa, Univ. of Minnesota and Univ. of North Carolina) with a central data coordinating centre at Dartmouth. The study will involve 850 patients newly treated for neoplastic polyps, who have had at least one neoplastic polyp removed within three months of recruitment, have no known polyps remaining in the colon, and no contra-indications to calcium supplementation. After completing a three-month placebo run-in period, subjects are allocated at random to receive either placebo or calcium carbonate, three grams per day in divided doses with meals. They are monitored by questionnaire for toxicity and compliance to the drug regimen at six-month intervals throughout the study. One year and four years after the qualifying colonoscopy, each subject will receive a complete colonoscopic examination, with removal and histological examination of all polyps. The primary analysis will compare the polyp recurrence rate in the placebo group with that in the calcium group. The study will have at least 80% power to detect a 35% reduction in polyp recurrence from a 10% per year recurrence rate. It will also permit investigation of the relationship between baseline dietary calcium and polyp recurrence.

UNITED STATES OF AMERICA

TYPE: Intervention
TERM: BMB; Polyps; Prevention; Serum
SITE: Colon
CHEM: Calcium
TIME: 1990 - 1993

HIGHLAND HEIGHTS

983   Richmond, R.E.   04485
Northern Kentucky Univ., Dept. of Biology, Highland Heights KY 41076, United States of America (Tel.: (606)5725306)
COLL: Rickabaugh, J.; Carrer, J.; Carrer, H.

Incidence and Mortality of Colorectal Cancer in the Northern Kentucky Region
This study examines the high incidence and mortality of colorectal cancer in the northern counties of Northern Kentucky. The data have been collected from state agencies, tumour registries, and hospital pathology departments throughout the area. All cases of colorectal cancer are confirmed through pathological records. Also, the residence history is compiled on each case to establish that the resident has been living in the given area of Northern Kentucky for at least 5 years prior to diagnosis. The study has involved over 500 confirmed cases of colorectal cancer for the years 1976 through 1983. The cases have been separated, by residence, into urban or rural home locations as a possible risk factor. Urban residence has been found to be a significant risk factor for colorectal cancer. This urban risk factor is being further examined, currently, on the hypothesis that drinking water is a major environmental influence between the urban and rural populations of Northern Kentucky. The low rural colorectal cancer rate may be related to the use of rain water cisterns for drinking water by this population.

TYPE: Case Series
TERM: Chlorination; Registry; Rural; Urban; Water
SITE: Colon; Rectum
TIME: 1980 - 1993

HONOLULU

984   Goodman, M.T.   04440
Univ. of Hawaii, Cancer Research Center of Hawaii, 1236 Lauhala St., Suite 407, Honolulu HI 96813, United States of America (Tel.: (808)5488451; Fax: 5483411)
COLL: Kolonel, L.N.; Nomura, A.M.Y.; Hankin, J.H.; Yoshizawa, C.N.

Dietary Influences on Endometrial Cancer Risk
The primary objective of this 5-year matched case-control study is to examine the role of dietary fat, vitamin A and its precursors, and vitamin C in the development of cancer of the endometrium. A total of 391 cases, consisting of 158 prevalent cancers and 233 incident cancers, are expected to be interviewed during the study period. Eligible cases will be defined as all endometrial cancer patients admitted to the seven largest Oahu hospitals who have a histologically confirmed diagnosis of primary cancer. Two population-based controls will be matched to each case on the basis of ethnicity and age (within 5 years). In addition to the dietary data, the questionnaire will include the following subjects: demographic information, occupational history, anthropometric measurements, reproductive history, contraceptive and non-contraceptive hormone use, medical history, history of cigarette use, and physical activity. Analysis of data will consist of an assessment of the dose-response relation of total, saturated and unsaturated dietary fat, vitamin A, beta carotene and other carotenes, and vitamin C to endometrial cancer while adjusting for the effects of potential confounders.

UNITED STATES OF AMERICA

TYPE: Case-Control
TERM: Contraception; Diet; Dose-Response; Ethnic Group; Fat; Hormones; Nutrition; Obesity; Occupation; Physical Activity; Reproductive Factors; Tobacco (Smoking); Vitamins
SITE: Uterus (Corpus)
CHEM: Beta Carotene
TIME: 1988 - 1992

985 Hankin, J.H. 04630
Univ. of Hawaii, Cancer Research Center, 1236 Lauhala St., Suite 407, Honolulu HI 96813, United States of America (Tel.: (808)5488452; Fax: 5483411)
COLL: Zhao, L.P.; Kolonel, L.N.; Wilkens, L.

Development of a Telephone Interview Method for Quantitative Diet History Assessment

The aim of this study is to design a method for assessing dietary intakes for population-based cancer case-control studies among persons living in widely dispersed geographical areas. The current diet history method, administered in face-to-face interviews, will be modified for administration by telephone, and the comparability of the two methods will be tested among the five major ethnic groups of Hawaii. The current method consists of obtaining frequencies and amounts of selected food items for estimating the usual monthly intakes of calories, nutrients, and other dietary components. Amounts are selected from coloured photographs showing three portion sizes based on the eating patterns of the population. This method will be modified to provide appropriate visual aids that can be mailed to subjects for subsequent use in telephone interviews. Three hundred persons representative of the major ethnic groups will be randomly selected to receive either telephone or face-to-face interviews initially and the alternate method six months later. Dietary intakes from the two interviews will be analysed to assess intra-individual agreement, based on the estimating equation technique. Differences in agreement by sex, age, and ethnicity will be examined.

TYPE: Methodology
TERM: Diet; Ethnic Group; Nutrition
SITE: Inapplicable
TIME: 1989 - 1992

986 Kolonel, L.N. 00057
Univ. of Hawaii, Cancer Research Center, Epidemiology Program, 1236 Lauhala St., Suite 407A, Honolulu HI 96813, United States of America (Tel.: (808)5488451; Fax: 5483411)
COLL: Yoshizawa, C.N.; Hankin, J.H.; Le Marchand, L.; Goodman, M.T.

Lifestyle and Cancer Risk

This is a long-term follow-up study of more than 40,000 Hawaii residents of both sexes, representing a random sample of households throughout the State. The purpose is to test a variety of hypothesis relating lifestyle (diet, smoking, alcohol consumption, etc.) and other factors to cancer risk in Hawaii's multi-ethnic population. Information on demographic and other risk factors was collected by personal interview during the period 1975-1980. Follow-up is achieved largely by linkage of the cohort to the population-based Hawaii Tumor Registry, since out-migration rates from Hawaii are very low.

TYPE: Cohort
TERM: Alcohol; Diet; Ethnic Group; Lifestyle; Tobacco (Smoking)
SITE: All Sites
TIME: 1975 - 1993

987 Le Marchand, L. 04462
Univ. of Hawaii, Cancer Research Center, Epidemiology Program, 1236 Lauhala St., Suite 407, Honolulu HI 96813, United States of America (Tel.: (808)5488453; Fax: 5483411)
COLL: Kolonel, L.N.; Hankin, J.H.; Yoshizawa, C.N.

Diet and Colorectal Cancer in Hawaii

This large-scale, population-based, case-control study is designed to test the independent and joint effects of a number of dietary variables (including fat, calories, fibre, calcium, omega-3 polyunsaturated fatty acid, etc.), body mass and physical activity on the risk of cancer at various locations within the large bowel. Cases are identified through the Hawaii Tumour Registry and will include approximately 1,100 histologically-confirmed colorectal cancer patients in the final sample. Each case is matched on race, sex and age with one population control. A dietary history which assesses total dietary intake is administered in the subject's home. The risk associated with various exposure variables will be examined in crude, stratified and multiple logistic regression analyses.

UNITED STATES OF AMERICA

TYPE: Case-Control
TERM: Diet; Ethnic Group; Fat; Fibre; Obesity; Physical Activity; Registry; Reproductive Factors
SITE: Colon; Rectum
REGI: Hawaii (USA)
TIME: 1988 - 1992

988 Le Marchand, L. 04463
Univ. of Hawaii, Cancer Research Center, Epidemiology Program, 1236 Lauhala St. , Suite 407, Honolulu HI 96813, United States of America (Tel.: (808)5488453; Fax: 5483411)
COLL: Kolonel, L.N.; Yoshizawa, C.N.; Hankin, J.H.

Diet and Malignant Melanoma
This study is examining the role of diet and other possible risk factors, including cumulative and intermittent solar exposure, in the aetiology of melanoma in the multi-ethnic population of Hawaii. Cases are identified through the Hawaii Tumour Registry and will include approximately 350 histologically-confirmed patients with cutaneous malignant melanoma (CMM). Each case is matched on race, sex and age with one population control. Each subject is administered a detailed questionnaire in the home to obtain information on diet, solar exposure, hormone use, physical characteristics and mole count. In addition, a blood sample is collected to determine plasma level of vitamins and selenium. The relationship of biochemical and dietary measures of nutritional status to the risk of CMM will be examined.

TYPE: Case-Control
TERM: Diet; Hormones; Metals; Naevi; Nutrition; Physical Factors; Premalignant Lesion; Radiation, Ultraviolet; Registry; Vitamins
SITE: Melanoma
CHEM: Selenium
REGI: Hawaii (USA)
TIME: 1988 - 1992

989 Yoshizawa, C.N. 04580
Univ. of Hawaii, Cancer Research Center of Hawaii, Epidemiology Program, 1236 Lauhala St., Honolulu HI 96813, United States of America (Tel.: (808)5483411)
COLL: Kolonel, L.N.; Hankin, J.H.; Le Marchand, L.

Diet and Cancer Incidence among Migrants in Hawaii
This is an ecological study which will examine the relationship between diet and cancer incidence among migrants in Hawaii. Using incident cases of cancer from the Hawaii Tumour Registry and population estimates developed from the Hawaii State Department of Health and the 1980 U.S. Census, average annual (1975-84) age-adjusted cancer incidence rates will be calculated by site, sex ethnicity, and place of birth for Filipinos, Caucasians, and Japanese born either in Hawaii or in their parent country (the Philippines, mainland U.S., or Japan, respectively). Corresponding age-adjusted averages of dietary intake and parameters of smoking and alcohol exposure will be calculated from data on an existing cohort of approximately 42,000 individuals, which is representative of the State population. For each site, measures of correlation between the age-adjusted cancer incidence rates and the age-adjusted mean intakes will be obtained for the 12 ethnic-sex-place-of-birth groups. Secondly, for each ethnic group and sex, site-specific cancer incidence rates in the parent country will be compared with those of the first generation migrants to Hawaii, and the second and subsequent generations born in Hawaii. The objective of this study is to develop new aetiological leads, and help focus future research on the relevant period of life when dietary risk factors are most influential.

TYPE: Correlation; Incidence
TERM: Alcohol; Diet; Ethnic Group; Migrants; Registry; Tobacco (Smoking)
SITE: All Sites
REGI: Hawaii (USA)
TIME: 1988 - 1992

HOUSTON

990 Anderson, D.E. 04715
Univ. of Texas, M.D. Anderson Cancer Center, Sect. of Human Genetics, 1515 Holcombe Blvd, Houston TX 77030, United States of America (Tel.: (713)7922586; Fax: 7928149)
COLL: Morton, N.E.

UNITED STATES OF AMERICA

Genetic Epidemiology of Breast Cancer
The aim is to determine whether familial aggregation in an unselected series of families is best explained by the transmission of a major gene, a transmissible multifactorial component acting alone or in concert with a major gene, or by an independent environmental contribution. Segregation analysis under a mixed model will be applied to three data sets: the families of 199 unselected female breast cancer patients, the families of 124 consecutive breast cancer patients with bilateral disease, and the families of 115 unselected male breast cancer patients. The use of three independent data sets will increase the power to resolve the question of the cause(s) underlying the familial aggregation of breast cancer, and whether the cause(s) are the same or different in the three familial groups.

TYPE: Case Series; Genetic Epidemiology
TERM: Familial Factors; Genetic Factors; Segregation Analysis
SITE: Breast (F); Breast (M)
TIME: 1986 - 1992

991 Anderson, D.E. 04941
Univ. of Texas, M.D. Anderson Cancer Center, Sect. of Human Genetics, 1515 Holcombe Blvd, Houston TX 77030, United States of America (Tel.: (713)7922586; Fax: 7928149)
COLL: Ferrell, R.E.

Genetics of Breast Cancer
Families with inherited forms of breast cancer, namely, Li-Fraumeni syndrome, site-specific breast cancer, breast-ovarian cancer, Cowden's disease, and the Torre-Muir syndrome are being investigated in the study. Each of these is being evaluated for evidence of genetic linkage between the appropriate disease susceptibility allele and an array of polymorphic genetic markers in an attempt to localize the susceptibility allele to a specific chromosomal site. Initial effort has been directed to the breast-ovarian cancer and site-specific families. Twenty families comprise the breast-ovarian cancer series and 30 families the site-specific series. The data are being analysed by the lod score method using the computer program LIPED which incorporates options for dealing with age-dependent expression of clinical phenotype penetrance.

TYPE: Genetic Epidemiology
TERM: Chromosome Effects; Genetic Factors; High-Risk Groups
SITE: Breast (F); Ovary
TIME: 1982 - 1992

* 992 Black, H.S. 05034
Baylor College of Medicine, Dept. of Dermatology, Baylor Plaza, Houston TX 77030, United States of America (Tel.: (713)7954411/7066; Fax: 7954411/7601)
COLL: Wolf, J.; Herd, A.J.; Goldberg, L.; Rosen, T.; Bruce, S.; Thornby, J.I.; Foreyt, J.P.; Tschen, J.

Skin Cancer Prophylaxis by Low Fat Dietary Intervention
The specific aim of this protocol is to determine, in a closely-controlled clinical setting, whether dietary modification of lipid intake can alter the course of skin cancer development. 700 patients will be randomized into two groups of 350 each. One will be a non-intervention (NI) group in which no change in dietary habits will be initiated and from which the control rate of carcinoma occurrence will be derived. The dietary intervention (DI) group will, after an intensive dietary modification training programme, adopt a diet characterized by reduced fat intake (20% of total caloric intake). Both groups are to be examined at four-monthly intervals, over 24 months, for carcinoma incidence. Based upon a 24-month 25% rate of carcinoma occurrence in the NI group, a 35% expected improvement in the DI group, and a 35% loss to follow-up, 70% of 20 computed simulations yielded significant differences between the NI and DI groups. This intervention design will not only address the question of whether dietary intervention may be an effective tool for prevention and/or management of a common malignancy induced by ultraviolet radiation, but will also provide direct and definitive evidence for involvement of dietary lipid in carcinogenic development.

UNITED STATES OF AMERICA

TYPE: Intervention
TERM: Diet; Lipids; Nutrition
SITE: Skin
TIME: 1990 - 1994

* 993 Buffler, P.A. 05102
Univ. of Texas, School of Public Health, Health Science Center at Houston, SW Center for Occup. Health and Safety, Reuel A. Stallones Bldg, P.O. Box 20186, Houston TX 77225, United States of America (Tel.: (713)7924638)
COLL: Cooper, S.P.; Burau, K.; Delclos, G.; Downs, T.D.; Key, M.; Tucker, S.; Whitehead, L.

University of Texas Skin Health Study
The specific aim of this research is to evaluate the contribution of occupational exposure to the prevalence of keratosis, a usually benign skin lesion that can be a precursor to skin cancer, among workers in a Paraquat production plant. The study design is a cross-sectional study which will compare the prevalence of keratoses between the current workers of the plant and an age, race, sex frequency-matched group of their non-family friends who have never worked at the plant. Approximately two friends were selected for each worker yielding a total of 364 study participants. Three major sources of data for this study include: (1) an interview questionnaire to ascertain demographic and medical information and sunlight exposure history; (2) a full body dermatology examination by a board of certified dermatologists to ascertain skin conditions; and (3) company records to ascertain a complete work history at the plant from which an exposure score will be developed.

TYPE: Cross-Sectional
TERM: Herbicides; Occupation; Premalignant Lesion; Radiation, Ultraviolet
SITE: Skin
OCCU: Miners, Copper
TIME: 1990 - 1992

994 Jansson, B. 04135
Univ. of Texas, M.D. Anderson Cancer Center, Dept. of Biomathematics, 1515 Holcombe Blvd, Box 501, Houston TX 77030, United States of America (Tel.: (713)7923392)
COLL: Becker, F.F.

Dietary, Total Body, and Intracellular Potassium/Sodium Ratio and its Influence on Cancer
The hypothesis that the ratio K/Na is negatively associated to cancer rates has been tested in a number of different ways: geopathological, dietary and clinical. It has also been found that certain agents or conditions which may reduce cancer risk increase the intracellular K/Na ratio, while other exposures or conditions which may increase cancer risk decrease the intracellular K/Na ratio. This study uses data from various sources, such as "Cancer Mortality by County in USA", "Third National Cancer Survey", published data, to test the hypothesis in two ways: (1) phenomena known to affect K/Na are studied regarding associations with cancer, and (2) phenomena known to affect cancer rates are studied regarding associations with the K/Na ratios. The hypothesis is now also being tested in mice by measuring the cancer incidence rates, induced either by DMH (intestinal cancer), or by DMBH (mammary cancer), for groups with high, normal or low dietary K/Na ratios. Modeling, Thompson and Brown (eds), Marcel Dekker, 1987, pp. 1-59.

TYPE: Correlation
TERM: Diet; Fibre; Metals; Minerals; Vitamins
SITE: All Sites; Colon
CHEM: Potassium; Sodium Chloride
TIME: 1986 - 1992

* 995 McPherson, S. 05203
Univ. of Texas Health Science Center, School of Public Health, 1200 Herman Pressler St., Houston TX 77030, United States of America (Tel.: (713)7924660; Fax: 7944876)
COLL: Winn, R.J.; Nichaman, M.Z.; Levin, B.

Dietary Calcium and Risk of Colorectal Adenomas
Basic research, clinical investigations, and epidemiological studies have provided data to support the role of calcium in the control of cell proliferation in individuals with colorectal cancer. The purpose of this study is to determine if there is an association between dietary intake of calcium and the presence of colorectal adenomatous polyps. The specific aims are: (1) to estimate the intake of total calories, total fat, calcium and fibre; (2) to determine if there is an independent association between low fat intake of dietary

UNITED STATES OF AMERICA

calcium and the presence of colorectal adenomatous polyps; and (3) to determine if there is an independent association between low intake of dietary sources of calcium and the presence of colorectal adenomatous polyps. To address this hypothesis a case-control study is being conducted of 546 subjects from seven clinics in the area of Houston. Cases are those individuals who have undergone a colonoscopy examination and have incident villous, tubular, or tubulovillous adenomas. The comparison subjects are those individuals who also undergo a colonoscopy but are free of polyps. Interviews will be conducted to collect dietary intake information, family history, and smoking history. Data analyses will include the evaluation of nutrients, foods and food groups as risk factors. Univariate odds ratios and their 95% confidence intervals will be calculated comparing the prevalence of exposure in the diseased and non-diseased individuals. Linear trend will be tested by dividing total intake of dietary calcium and other nutrients into tertiles and using the high intake category as the referent group. Stratified analyses and multiple logistic regression will be done in order to determine interactions and possible confounding by other variables.

TYPE: Case-Control
TERM: Diet; Fat; Fibre; Nutrition; Polyps; Premalignant Lesion
SITE: Colon; Rectum
CHEM: Calcium
TIME: 1991 - 1994

996   Strong, L.C.   03572
Univ. Texas System Cancer Center, M.D. Anderson Hosp. & Tumor Inst., Dept. of Experimental Pediatrics, 6723 Bertner Ave. , Box 501, Houston TX 77030, United States of America (Tel.: (713)7922589)
COLL: Williams, W.R.

Genetic Epidemiology of Childhood Cancer

Childhood cancer probands are systematically ascertained from the M.D. Anderson Hospital Tumor Registry according to specific histological tumour types. For a given patient cohort, the frequency and distribution of cancer in first and second degree relatives as determined by interview and confirmed by medical record or death certificate are determined overall, and for each kindred. Segregation analysis is then applied to the entire data set, to determine the most likely model to account for the distribution of cancer in families, to characterize the gene frequency and penetrance if a major gene is identified, to determine the minimum fraction of childhood cancers attributable to a major gene, to identify the proband characteristics most likely to be associated with a major gene, and to identify the specific kindreds in which the cancer pattern suggests a major gene. A paper has been published in J.Natl. Cancer Inst. 79:1213-1220, 1987.

TYPE: Genetic Epidemiology
TERM: Biochemical Markers; Familial Factors; Genetic Factors; Genetic Markers; Registry; Sero-Epidemiology
SITE: Childhood Neoplasms
TIME: 1984 - 1989

997   Vogel, V.G.   04734
Univ. of Texas, M.D. Anderson Cancer Center, 1515 Holcombe Blvd, Box 501, Houston TX 77030, United States of America (Tel.: (713)7928515; Fax: 7969155)
COLL: Peters, G.N.; Winn, R.J.; Bondy, M.D.

Stage Distribution and Risk Factor Survey in Women Screened for Breast Cancer

The goal of the American Cancer Society 1987 Texas Breast Screening Project was to attract women for low-cost screening mammography. The aim of this study is to ascertain whether screening for breast cancer in community centres can yield the same proportion of early-stage breast cancer as that reported in controlled trials. The project yielded more than 64,000 examinations at 306 participating facilities. Findings were radiographically classified as negative, indeterminate (further evaluation required), or suspicious for malignancy. Among the project mammograms were 2,172 suspicious examinations. A study is being conducted to determine: (1) the stage at diagnosis and the clinico-pathological outcomes of the malignancies detected at screening; and (2) the ratio of benign to malignant biopsies. Follow-up data are available for 75 % of the screened women in whom 1,122 biopsies were done; 214 cancers have been documented to date. In addition, questionnaires were given to all screenees, and 32,000 have been coded and entered into a computer database. Data collected included demographic variables, family history of breast cancer, history of benign breast disease, health beliefs and prior health practices, and exposures such as cigarette smoking and oral contraceptive use. Names and addresses are available, and prospective studies of women with positive family histories are planned to improve rates of early

UNITED STATES OF AMERICA

detection of breast malignancy in this cohort. Papers have been published in Cancer 66:1613-1620, 1990 and Cancer Detect. & Prev. 14:573-576, 1990.

TYPE: Cohort
TERM: Familial Factors; Mammography; Oral Contraceptives; Screening; Tobacco (Smoking)
SITE: Benign Tumours; Breast (F)
TIME: 1986 - 1992

998 Wargovich, M.J. 04866
Univ. of Texas, M.D. Anderson Cancer Center, 1515 Holcombe Blvd, Box 501, Houston TX 77030, United States of America (Tel.: (713)7922828; Fax: 7911536; Tlx:9108811556)
COLL: Lynch, P.; Levin, B.; Winn, R.J.

Calcium Intervention in Subjects with Sporadic Colon Adenoma

The hypothesis under study is that calcium in daily doses of 1500-4000 mg modifies colonic epithelial proliferation in subjects with resected adenoma. Biomarkers of cellular proliferation in this study include 3H-TdR, bromo-deoxyuridine incorporation, and proliferating cell nuclear antigen. 105 subjects are randomized to placebo or one of three graded doses of calcium and maintained at their dosage level for three months. Colon biopsies will be assessed at baseline, on study, and two months post-study to determine the degree of calcium inhibition of epithelial proliferation. These data will be used to implement a secondary phase III chemoprevention trial in the same population with the additional endpoint of adenoma recurrence.

TYPE: Intervention
TERM: Biochemical Markers
SITE: Colon
CHEM: Calcium
TIME: 1990 - 1992

HYATTSVILLE

999 Winn, D. 04116
National Center for Health Statistics, Div. of Health Interview Statistics, 6525 Belcrest Rd, Hyattsville MD 20782, United States of America (Tel.: (301)4367085)

Studies of Cancer Risk Factor Distributions, Cancer Epidemiology, Cancer Morbidity and Mortality, and Cancer Prevention Practices in National Samples

The National Center for Health Statistics has numerous on-going, large, population-based, multiple-purpose surveys which afford many opportunities for studies of cancer risk factor distributions, cancer aetiology, cancer morbidity and mortality, and cancer prevention practices in national samples of the US population. The present activities, usually conducted in collaboration with other US Federal and State agencies, include the on-going collection and analysis of cancer death rates for the US population by sex, age and other factors. In addition, the National Health Interview Survey in 1987 was devoted to the assessment of the distribution of cancer-related behaviours including health habits, usual diet, and screening practices in more than 40,000 persons in the US. Data analysis is now under way. In the NHANES I follow-up study over 14,000 US adults were interviewed and examined in 1971-1975 for a variety of demographic and health characteristics. The cohort has been followed for vital status, hospitalizations and re-interviews during 1982-1984, in 1986 and in 1987. Epidemiological studies involving diet and other factors in the aetiology of selected major cancers are in progress using this data base. In the National Mortality Follow-Back Survey cancer risk factors were collected using mail and telephone interviews with relatives of decedents in a national probability sample of 20,000 deaths occurring in the US in 1986, including deaths due to a number of rare cancers (e.g. male breast cancer). Cancer risk factor data collected in the 1986 NMFS included health behaviours with respect to diet, alcohol use, the smoking history of decedents and spouses, and the use of smokeless (snuff and chewing) tobacco. Epidemiological studies using these data are now underway. The multi-purpose NHANES III survey, which started in 1988, involves medical examinations of a national probability sample of 30,000 Americans. Several projects concerning cancer risk factors are being contemplated including assessments of nutritional markers (e.g., carotenoids and cotinine); longitudinal follow-up is also planned which will permit studies of cancer occurrence in relation to nutritional biochemistries, health habits, diet, physical measures, and other factors. Data from the National Hospital Discharge Survey are used to examine trends and patterns in hospital use for specific cancers, including the use of surgical and non-surgical treatments. Biological marterials are being stored for future use in the NHANES III study.

UNITED STATES OF AMERICA

TYPE: Cohort; Incidence; Mortality
TERM: Biochemical Markers; Diet; Hygiene; Prevalence; Prevention; Screening
SITE: All Sites
TIME: 1957 - 1993

IRVINE

1000 Anton-Culver, H. 04415
Univ. of California at Irvine, Dept. of Community, & Environmental Medicine, Irvine CA 92717, United States of America (Tel.: (714)8567416)

Smoking and other Risk Factors for Lung Cancer by Cell Type

Few studies have examined the importance of smoking and other risk factors for the different histological types of lung cancer. The histological distribution differs dramatically between men and women, with squamous cell carcinoma being the most common cell type in men, and adenocarcinoma predominating in women. The proposed case-control study will examine the importance of smoking, occupation, and family history of cancer for specific histological types of lung cancer in men and women using data from the population-based registry of the Cancer Surveillance Program of Orange County. Cases will include all cases of small cell carcinoma and adenocarcinoma of the lung diagnosed in Orange County during 1984-1986, a total of approximately 2,500 cases. Each histological type will be compared with all other cancers excluding those thought to be associated with smoking, occupation or family history. Data available include demographic information, industry/occupation, smoking history, family history of cancer, pathology and stage of disease, treatment and survival status. The relative odds associated with risk factors under study will be compared between different histological types of lung cancer and between men and women. A multiple logistic model will be used to evaluate the independent contributions to risk of environmental and host factors and to study possible interactions between smoking and other risk factors.

TYPE: Case-Control
TERM: Familial Factors; Histology; Occupation; Registry; Tobacco (Smoking)
SITE: Lung
REGI: Orange County (USA)
TIME: 1988 - 1992

ITHACA

*1001 Turnbull, B. 05188
Cornell University, School of Operational Research, & Industrial Engineering, 334 Upson Hall, Ithaca NY 14853, United States of America

Statistical Methodology for Monitoring Clusters of Disease

The aim is to develop and evaluate statistical methodology to monitor the occurrence of clusters of chronic disease. Geographical incidence patterns are examined and the 'randomness' hyothesis is tested. For statistical tests with a general 'omnibus' alternative, a paper was published in Am. J. Epidemiol. 132:S136-S143, 1990, using leukaemia incidence data from the New York State Cancer Registry and population data from the US Census. Tests now being developed are designed to be sensitive to alternatives in which the incidence rates are elevated in the vicinity of (multiple) putative sources of hazard. Applications will be made with toxic dump site locations and leukaemia in New York State and with nuclear installations and childhood leukaemia in Sweden.

UNITED STATES OF AMERICA

TYPE: Methodology
TERM: Cluster; Radiation, Ionizing; Waste Dumps
SITE: Leukaemia
LOCA: Sweden; United States of America
TIME: 1987 - 1992

KANSAS CITY

1002 Neuberger, J.S. 03986
Univ. of Kansas, Medical School, Dept. of Preventive Medicine, Rainblow Blvd at 39th St., Kansas City KS 66103, United States of America (Tel.: (913)5882775)
COLL: Morantz, R.A.; Chin, T.D.Y.

Greater Kansas City Brain Tumour Epidemiology Study
Although a number of causal hypotheses have been advanced, the aetiology of brain tumours is not well understood. Kansas City is a diversified community and should present a wide range of exposure possibilities. This study investigates primary site brain tumours in five Missouri and four Kansas counties in the Kansas City metropolitan area. The study group will comprise subjects with brain tumours, both benign and malignant, at all ages. All tumours will be histologically verified. In year one, brain tumour patients (or respondents) will be asked to complete a self administered questionnaire. In years two through six, one or more control groups will be added to the study. The questionnaire will obtain information on occupation, residence, illnesses, chemical exposures, medical and dental exposures, as well as such personal variables as age, sex, and race. Controls added to the study in 1987 were matched to cases on both these personal variables and residence in the metropolitan area.

TYPE: Case-Control
TERM: Chemical Exposure; Cluster; Environmental Factors; Occupation
SITE: Brain
TIME: 1986 - 1992

LA JOLLA

1003 Barrett-Connor, E.L. 03913
Univ. of California at San Diego, Dept. Community & Family Medicine, 9500 Gilman Drive, La Jolla CA 92093-8625, United States of America (Tel.: (619)5343720; Fax: 5348625)
COLL: Garland, C.F.

Lipids, Diabetes and Cancer
This study was designed to look at diabetes and heart disease prospectively in subjects attending a lipid research clinic. Extensive behavioural and dietary data have also allowed cancer to be examined. Data on mortality from cancer (from death certificates) for all years and incidence according to clinical records (for the ninth and tenth years only) have been collected. There are over 4,000 adult subjects in the geographically defined population, but not all participants have all data, e.g., 24-hour diet recall is available for approximately 1,000 subjects. Data are now available on sex hormones measured in plasma obtained from approximately 1,100 men and 550 women in 1972-1974.

TYPE: Cohort
TERM: Diabetes; Diet; Lipids; Prevalence
SITE: All Sites; Breast (F); Colon; Lung; Prostate
TIME: 1972 - 1993

1004 Garland, C.F. 03630
Univ. of California at San Diego, Dept. Community & Family Medicine, M-007, La Jolla CA 92093-0607, United States of America (Tel.: (619)4523720; Fax: 5580796)
COLL: Shekelle, R.B.; Barrett-Connor, E.L.; Criqui, M.H.; Rossof, A.H.; Paul, O.

Colo-Rectal Cancer Risk in a Cohort of Men
1,954 men employed by the Western Electric Company and enrolled in the Western Electric Heart Study in 1957-1958 are being assessed for colo-rectal cancer risk. The men were aged 40-55 years at entry. They completed dietary histories (2-8 day) at entry and one year later, and 99.9% of the men have been followed until the twentieth anniversary of entry. Incidence of colo-rectal cancer will be ascertained in the

UNITED STATES OF AMERICA

population, and a prospective design will be used to assess risk according to nutritional characteristics at entry.

TYPE: Cohort
TERM: Diet; Vitamins
SITE: Colon; Rectum
TIME: 1983 - 1992

1005   Johnson, B.C.   03381
       Michigan Dept. of Public Health, Environmental Epidemiology Div., Center for Environmental
       Health Science, 3500 N. Logan St., Lansing MI 48909, United States of America (Tel.:
       (517)3358350)

Case-Control Study of Soft Tissue Sarcoma
The study will assess associations between soft-tissue sarcoma cases and environmental variables. Case identification and confirmation methods will be developed to both prevent mis-classification and yield complete case finding. Although the case-control study only depends upon unbiased case finding and not complete case ascertainment, complete case finding will be accomplished to assess incidence rates within the geographical area. Cases of soft-tissue sarcoma, occurring over the past years within an eight county geographical area of eastern-central Michigan, will be identified by pathology report review. The diagnosis will be confirmed by a pathological panel review of histological slides. Controls will be selected both from hospital discharges and the resident population. Detailed occupational, personal habit and place of residence data will be obtained. Within this geographical area resides a group of workers who synthesized 2,4,5-T. Environmental exposure data will be utilized in the study analysis.

TYPE: Case-Control
TERM: Chemical Exposure; Environmental Factors; Herbicides; Histology; Lifestyle; Occupation; Pesticides
SITE: Soft Tissue
CHEM: 2,4,5-T; Phenoxy Acids
TIME: 1984 - 1992

LIVERMORE

1006   Moore II, D.H.   02630
       Lawrence Livermore National Lab., Biomedical Sciences Div. L-452, 700 E. Ave., P.O. Box 5507,
       Livermore CA 94550, United States of America (Tel.: (415)4225631; Fax: 4222282 ;
       Tlx:9103868339)
COLL: Schneider, J.S.

Melanoma among Employees at the Lawrence Livermore National Laboratory
Causes for a two- to three-fold increase in melanoma incidence among the 10,000 employees of the Lawrence Livermore National Laboratory (LLNL), a high energy research facility located in a sunny valley in northern California, are being investigated. The study includes (1) a study of mortality among the approximately 20,000 former employees to determine whether melanoma mortality is also high; (2) interviews of all cases and a matched control for each case to determine work and sun exposure history; (3) comparison of lesion characteristics (e.g. thickness, type and level) with other cases from the surrounding community to determine whether earlier diagnosis at LLNL is a contributing factor. Papers were published in Arch. Dermatol. 126:767-769, 1990 and in the Lancet, June 23, pp 1523-1524, 1990.

TYPE: Case-Control; Cohort
TERM: Chemical Exposure; Occupation; Radiation, Ionizing; Radiation, Ultraviolet
SITE: Melanoma
OCCU: Chemists; Laboratory Workers; Physicists
REGI: Bay Area (USA)
TIME: 1981 - 1992

LOMA LINDA

1007   Fraser, G.E.   04344
       Loma Linda Univ., Center for Health Research, Nichol Hall, Loma Linda CA 92350, United States of
       America (Tel.: (714)8244753)
COLL: Mills, P.; Beeson, W.L.; Lindsted, K.; Sabaté, J.

UNITED STATES OF AMERICA

Adventist Health Study
The purpose of the study is to test the effect of consumption of foods on risk of malignant tumours. Special emphasis has been put on fruits, vegetables and meat. This cohort study enrolled 34,198 Californian Seventh-Day Adventists in 1974-1976. A follow-up study has been carried out for six years since 1982. Information on exposure has been gathered by mailed questionnaire. All hospitalisations of study subjects have been ascertained by annual mail contact. Study field representatives have visited relevant hospital and screened charts for new cancer diagnoses. The Center has collaborated with two California tumour registries to ascertain cases. Cases have been matched with California State Death Tapes.

TYPE: Cohort
TERM: Diet; Fruit; Registry; Religion; Vegetables
SITE: All Sites
REGI: Bay Area (USA); Los Angeles (USA)
TIME: 1974 - 1993

LOS ALAMOS

1008 **Wiggs, L.D.** 00347
Los Alamos National Lab., Epidemiology Sect., Health, Safety and Environment Div., P.O. Box 1663, Los Alamos NM 87545, United States of America
COLL: Voelz, G.L.; Galke, W.A.

Health Study of Plutonium Workers
This epidemiological programme investigates the risk of adverse human health effects, especially cancer risks, associated with internal desposition of plutonium and associated external radiation doses. Retrospective cohort mortality analyses are being conducted at the Rocky Flats, Los Alamos, and Mound facilities by comparing mortality rates of exposed versus unexposed persons at the same facility and by determining standardised mortality ratios based on US death rates. These cohorts include over 50,000 workers. Nested case-control studies are used to investigate significant mortality excesses. Upon completion of mortality studies for these cohorts, a pooled analysis of data from all facilities will be used for more statistical power. A cohort study of disease (morbidity) in plutonium-exposed workers is also planned in order to measure non-fatal health effects and important covariates, such as smoking and lifetime occupational hazards. Data will be collected initially from 3,000 persons by mail questionnaire and telephone follow-up. Risk coefficients will be derived from these studies, especially for the alpha-particle radiation of plutonium, if health effects are identified.

TYPE: Case-Control; Cohort
TERM: Dose-Response; Metals; Occupation; Radiation, Ionizing; Tobacco (Smoking)
SITE: All Sites
CHEM: Plutonium
OCCU: Plutonium Workers
TIME: 1975 - 1993

LOS ANGELES

1009 **Bernstein, L.** 04563
Univ. of Southern California, School of Medicine, Dept. of Preventive Medicine, 1420 San Pablo St., PMB A-202, Los Angeles CA 90033-9987, United States of America (Tel.: (213)2247650; Fax: 2246417)
COLL: Ross, R.K.; Henderson, B.E.

Breast Cancer in Young Women
Although there is no strong evidence that oral contraceptives (OC) have an effect on breast cancer risk when used during the middle reproductive years, women who use OCs for long periods of time in the post-menarcheal period may be at increased risk of breast cancer. Previous studies that have examined this relationship disagree as to whether there is a significantly increased risk with early long-term OC use or no risk. The specific aims of this study of breast cancer in young women are (1) to examine the hypothesis that early long-term OC use is associated with an increased risk of breast cancer and to determine whether any increased risk is modified by reproductive factors and other breast cancer risk factors, the particular formulation of OC used or temporal factors such as calendar period of use, age at first use, time since first use and duration of use; (2) to establish the reliability of retrospective recall of OC

UNITED STATES OF AMERICA

use by comparing interview histories with medical records; and (3) to investigate the relationship between breast cancer and diagnostic and therapeutic radiation, abortion, regularity of menstrual cycle, physical activity, alcohol and caffeine consumption and smoking habits. A matched case-control design is being used. Cases are white, English-speaking women aged 40 and under, diagnosed with breast cancer between January 1985 and December 1990, and identified by the population-based tumour registry. Controls are individually matched to cases by age, race, parity (ever vs. never had a full-term pregnancy) and neighbourhood of residence. Structured personal interviews are being conducted. To facilitate recall, a photograph album of all OCs sold in the United States and a comprehensive calendar to record life events is used during the interview. 720 case-control pairs will be interviewed.

TYPE: Case-Control
TERM: Abortion; Alcohol; Menstruation; Oral Contraceptives; Physical Activity; Radiation, Ionizing; Radiotherapy; Registry; Reproductive Factors; Time Factors; Tobacco (Smoking)
SITE: Breast (F)
CHEM: Caffeine
REGI: Los Angeles (USA)
TIME: 1987 - 1992

1010 Bernstein, L. 04768
Univ. of Southern California, School of Medicine, Dept. of Preventive Medicine, 1420 San Pablo St., PMB A-202, Los Angeles CA 90033-9987, United States of America (Tel.: (213)2247650; Fax: 2246417)
COLL: Levine, A.M.

Epidemiology of HIV-Related Lymphoma
This study is designed to define the aetiology of high-grade lymphoma in the setting of HIV-infection by looking at the specific hypothesis that EBV-induced chronic B-lymphocyte stimulation, in the setting of altered immune function leads to unregulated B-cell proliferation, specific chromosomal translocation, c-myc activation, and subsequent development of monoclonal, B-cell lymphoma. In this population-based case-control study, cases with HIV-positive high-grade lymphoma (n = 200) will be compared with controls who are HIV-positive individuals without lymphoma and to HIV-negative lymphoma patients. Population-based controls will be matched to this latter group of patients as well. In addition to interviews, all patients will have HIV and EBV antibody testing and HLA typing. A subgroup of patients (cases and controls) will be studied at the tissue level to determine (1) if EBV genome is present and its extent; (2) if HIV provirus is present and expressed within lymphoma tissues (PCR and in situ hybridization; (3) the molecular genotype of HIV+ and HIV-negative lymphomas by cytogenetic and molecular genetic analyses and (4) if monoclonality of lymphoma tissue will differentiate HIV+ lymphoma cases from HIV+ individuals without lymphoma (kappa/lambda immunophenotype and immunoglobulin gene rearrangement). A serum and tissue bank will be established for use in future studies.

TYPE: Case-Control
TERM: BMB; EBV; HIV; HLA; PCR; Registry; Serum; Tissue
SITE: Lymphoma
REGI: Los Angeles (USA)
TIME: 1989 - 1994

1011 Enstrom, J.E. 00527
Univ. of California, School of Public Health, 405 Hilgard Ave., Los Angeles CA 90024, United States of America (Tel.: (213)8252048)

Epidemiology of Cancer among Mormons
The SMR for total cancer in 350,000 California Mormons during 1960-1987 and 750,000 Utah Mormons during 1970 and 1975 compared with US whites is about 50% for "active" males, about 65% for all males, and about 75% for all females. For active males the SMR is about 25% for smoking-related cancer sites and about 65% for all other sites. Mormons are a health-conscious religious group, who appear to be similar to US whites in many respects, such as diet, socio-economic status, and urbanization, but who use about one half as much tobacco, alcohol, coffee, and tea as the general population. "Active" Mormons, defined to be church leaders known as High Priests, abstain almost entirely from tobacco, alcohol, and caffeine. In late 1979 a questionnaire survey and prospective mortality follow-up of 10,000 "active" California Mormon adults, specifically 5,400 High Priests and 4,600 wives was initiated. The survey obtained detailed demographic, lifestyle, dietary, occupational, and medical history data. Church-based and California State mortality data have been collected on the 10,000 cohort members and on all other California Mormons during the years 1980-1987. The data on lifestyle and health-related

UNITED STATES OF AMERICA

characteristics of "active" California Mormons will be compared with their detailed cancer mortality rates. Distinctive differences between these Mormons and other comparison groups will be highlighted. Using variation in key health practices (physical activity and sleep), a subgroup of middle-aged high priests has been identified with an SMR of 22% for all causes and 34% for all cancers. These analyses have provided significant understanding as to why Mormons have low cancer mortality rates. A paper has been published in J. Nat. Cancer Inst. 81:1807-1814, 1989.

TYPE: Cohort
TERM: Alcohol; Diet; Lifestyle; Occupation; Religion; Socio-Economic Factors; Tobacco (Smoking)
SITE: All Sites
TIME: 1973 - 1992

1012 Enstrom, J.E. 03009
Univ. of California, School of Public Health, 405 Hilgard Ave., Los Angeles CA 90024, United States of America (Tel.: (213)8252048)

Cancer among Low-Risk Populations

This is a prospective epidemiological investigation of several populations that are at low risk of cancer relative to the general US population. The populations include: California physicians, whose cancer mortality trends are being examined with respect to their substantial degree of smoking cessation since 1950; Alameda County (California) residents, whose cancer rates are being examined with respect to several good health practices; and various cohorts of health conscious Californians, whose cancer rates are being examined with respect to vitamin supplement usage and other lifestyle variables. Papers appeared in Am. J. Public Health 76:1124-1130, 1986.

TYPE: Cohort
TERM: Lifestyle; Occupation; Vitamins
SITE: All Sites
OCCU: Health Care Workers
TIME: 1976 - 1992

1013 Haile, R.W.C. 03639
UCLA School of Public Health, Div. of Epidemiology, 405 Hilgard Ave. , Los Angeles CA 90024, United States of America (Tel.: (213)8258193)
COLL: Sparkes, R.S.; Paganini-Hill, A.; Thomas, D.C.; Thompson, D.W.; Siemiatycki, J.; Gatti, R.A.; Greenland, S.

Genetic-Epidemiological Study of Bilateral Breast Cancer

The objective is to identify a gene or genes involved in the aetiology of breast cancer and to investigate gene-environment interactions. Data on family history and exposure to risk factors for breast cancer will be available for members of 442 families where the index case has bilateral breast cancer diagnosed before 50 years of age. Blood samples will be available from 68 multiple-case families potentially informative for linkage. The following analyses have been completed: case-control analyses of environmental risk factors, familial risk calculations, complex segregation analyses, and linkage analyses of 28 phenotypic genetic markers. The linkage work will be extended by using selected RFLPs and will focus on segments of chromosomes 1, 11 and 13, based on preliminary suggestions of linkage from the data and results published in the literature. A sequential approach will be taken to decide which RFLPs to investigate with the full data set of 68 families. Candidate RFLPs will be first tested on 10 of the more informative families, determined by a new computer program that estimates power while allowing for reduced penetrance and age of onset corrections. Those RFLP's that yield lod scores suggestive of linkage will be considered for further analyses with more families. Linkage analyses will be conducted using LIPED and an affected-pedigree-member method that analyses all pairs of affected relatives. In addition, a regression method of linkage analysis will be used, as well as further case-control and proportional hazards analyses.

UNITED STATES OF AMERICA

TYPE: Case Series, Genetic Epidemiology
TERM: BMB; Blood; DNA; Environmental Factors; Familial Factors; Genetic Factors; Genetic Markers; Linkage Analysis; RFLP; Segregation Analysis
SITE: Breast (F)
LOCA: Canada; United States of America
TIME: 1984 - 1992

*1014   London, S.J.   05072
Univ. of Southern California, School of Medicine, Parkview Medical Bldg, B-306, 1420 San Pablo St., Los Angeles CA 90033, United States of America (Tel.: (213)3421092; Fax: 3421237)
COLL: Idle, J.; Adams, J.

Debrisoquine Metabolism: A Genetic Marker of Lung Cancer Risk among Blacks and Whites
The specific aims of this study are: (1) to examine the association between the ability to extensively metabolize debrisoquine and lung cancer risk and (2) to examine possible differences in the prevalence of this potential genetic marker of lung cancer risk between whites and blacks. 320 incident lung cancer cases will be enrolled from several Los Angeles hospitals chosen to provide an equal number of black and white cases. Population controls will be chosen. Debrisoquine genotype and phenotype will be determined. Questionnaire data on smoking, diet, family history, and occupation are being obtained.

TYPE: Case-Control
TERM: BMB; Diet; Genetic Factors; Genetic Markers; Metabolism; Occupation; Tobacco (Smoking)
SITE: Lung
TIME: 1990 - 1993

1015   Mack, T.M.   04079
Univ. of Southern California, School of Medicine, Dept. of Preventive Medicine, 2025 Zonal Ave., PMB-B105, Los Angeles CA 90033, United States of America (Tel.: (213)2247255; Tlx:9103212434)
COLL: Thomas, D.B.

Geographical and Chronological Patterns of Cancer Occurrence in Los Angeles
Using cancer cases reported to the Los Angeles County Cancer Surveillance Program (roughly 30,000 annually since 1972), patterns of risk by census tract and by cancer site in the entire county are being assessed and mapped in order that the role of point sources of putative carcinogens (polluting industries, toxic dump sites, etc.) can be collectively and individually evaluated and accurately communicated. Similarly, observed frequencies of geographically, chronologically and familially related same-site cases are being compared with expectations.

TYPE: Correlation
TERM: Environmental Factors; Familial Factors; Geographic Factors; Mapping; Time Factors
SITE: All Sites
TIME: 1985 - 1993

1016   Mack, T.M.   04081
Univ. of Southern California, School of Medicine, Dept. of Preventive Medicine, 2025 Zonal Ave., PMB-B105, Los Angeles CA 90033, United States of America (Tel.: (213)2247255; Tlx:9103212434)
COLL: Deapen, D.

Case-Control Studies of Cancer in Twins
Twins who have had cancer diagnoses are ascertained by newspaper advertising. Diagnoses are validated by review of pathology reports and histological sections. Mail questionnaires are distributed to affected and unaffected twins. Returned questionnaires are edited, coded and analysed using standard univariate and multivariate methods. Hypotheses under test depend upon site. In all, it is expected that completed questionnaires will be received from more than 5,000 twin pairs. Studies in tandem with these are being carried out in twins in Denmark, Finland and Sweden.

UNITED STATES OF AMERICA

TYPE: Case-Control
TERM: Twins
SITE: All Sites
LOCA: Canada; United States of America
TIME: 1982 - 1993

1017 Menck, H.R. 04097
Univ. of Southern California, Cancer Surveillance Program, 1721 Griffin Ave., Los Angeles CA 90031, United States of America (Tel.: (213)2247641)
COLL: Mack, T.M.; McLaughlin, J.K.

Case-Control Study of Gallbladder Cancer

The objective is to perform a case-control study of gallbladder cancer in Los Angeles County Hispanics, to evaluate the role of reproductive history, oestrogen activity, Hispanic diet, Amerindian inheritance, past gastrointestinal and metabolic disease, and environmental exposures. An attempt will be made to distinguish between the causes of gallbladder cancer and the causes of cholelithiasis. The cases will consist of all female cases of gallbladder cancer occurring in Los Angeles County in five years (n = 200); controls will consist of matched persons chosen from the neighbourhood of the case and matched cases undergoing cholecystectomy at the hospital in which the cancer case was diagnosed. A personal interview will elicit information about reproductive history, medical history and drug use, Hispanic diet, and environmental exposures.

TYPE: Case-Control
TERM: Diet; Drugs; Environmental Factors; Ethnic Group; Female; Hormones; Reproductive Factors
SITE: Gallbladder
TIME: 1984 - 1992

1018 Paganini-Hill, A. 04027
Univ. of Southern California, School of Medicine, Dept. of Preventive Medicine, 1441 Eastlake Ave., Los Angeles CA 90033-0800, United States of America (Tel.: (213)2246412; Fax: 2246417)
COLL: Henderson, B.E.; Ross, R.K.

Role of Oestrogen and Vitamin A in Disease Prevention

This cohort study is designed to measure the risks and benefits of menopausal oestrogen replacement therapy in terms of incident disease and mortality (endometrial cancer, breast cancer, hip fracture, and acute myocardial infarction); to study the effect of dietary and supplemental intake of vitamin A on the development of epithelial cancer, especially lung cancer; and to evaluate the risks and benefits associated with other health-related practices in older adults. Detailed health surveys have been collected on 13,986 residents of Leisure World, Laguna Hills, CA. Follow-up of this cohort is maintained by abstraction of hospital discharge diagnoses (from three local hospitals), cancer pathology reports (from five local hospitals) and death certificates. Follow-up questionnaires were also sent to all living cohort members in 1983 and 1985. Statistical analysis will include the use of Poisson regression models and the proportional hazards model of Cox. These multivariate methods will be used to sort out possible confounding and modifying effects. Recent papers appeared in Prev. Med. 19:323-334, 1990, Arch. Intern. Med. 151:75-78, 1991 and in Epidemiology 2:16-25, 1991.

TYPE: Cohort
TERM: Age; Diet; Drugs; Hormones; Vitamins
SITE: All Sites; Breast (F); Lung; Uterus (Corpus)
CHEM: Oestrogens
TIME: 1981 - 1992

1019 Paganini-Hill, A. 04562
Univ. of Southern California, School of Medicine, Dept. of Preventive Medicine, 1441 Eastlake Ave., Los Angeles CA 90033-0800, United States of America (Tel.: (213)2246412; Fax: 2246417)
COLL: Ross, R.K.

Hormone Replacement Therapy and Breast Cancer

This case-control study is designed to determine the effect on breast cancer risk both of cyclic oestrogen-progestogen hormonal replacement therapy and of unopposed oestrogen replacement therapy. Cases are white English-speaking women aged 55-64 years at diagnosis of breast cancer and identified in the population-based tumour registry between 1 March 1987 and 1 March 1990. Controls will be individually matched to cases by age, race and neighbourhood of residence. A structured personal interview will be conducted with validation of hormone therapy by review of physician records.

UNITED STATES OF AMERICA

A total of 1,650 case-control pairs will be analysed, which will allow for the evaluation of the effects of hormone replacement therapy in the presence of possible confounding variables such as age at and type of menopause.

TYPE: Case-Control
TERM: Hormones; Registry
SITE: Breast (F)
CHEM: Oestrogens; Progestogens
REGI: Los Angeles (USA)
TIME: 1987 - 1992

*1020   **Paganini-Hill, A.**   04853
Univ. of Southern California, School of Medicine, Dept. of Preventive Medicine, 1441 Eastlake Ave., Los Angeles CA 90033-0800, United States of America (Tel.: (213)2246412; Fax: 2246417)
COLL: Dworsky, R.

Premenopausal Bilaterial Breast Cancer Family Registry
The aim of this project is to establish a cohort of premenopausal bilateral breast cancer patients and their first degree female relatives. Cases are ascertained through the Los Angeles County Cancer Surveillance Program, a population-based tumour registry established in 1971. 674 cases have been identified to date and approximately 50 new cases are added each year. Each case and her female relatives are asked to complete a postal questionnaire requesting information on known breast cancer risk factors, patterns of health care and family history. DNA and plasma samples are collected on a subset of the study group.

TYPE: Genetic Epidemiology
TERM: BMB; DNA; Familial Factors; Genetic Factors; Plasma; Registry
SITE: Breast (F)
REGI: Los Angeles (USA)
TIME: 1991 - 1993

1021   **Preston-Martin, S.**   02914
Univ. of Southern California, School of Medicine, Dept. of Preventive Medicine, Parkview Medical Bldg, B301, 1420 San Pablo St., Los Angeles CA 90033, United States of America (Tel.: (213)3421310; Fax: 3421237)
COLL: Yu, M.C.

Case-Control Study of Cancers of the Nose, Sinuses and Nasopharynx
The aims of this study are to (1) investigate the association between cancers of the nose, sinuses, and nasopharynx and occupational exposure to dust and fumes and to identify high risk occupations and industries; (2) investigate the relationship of these cancers to tobacco use, to a history of upper respiratory infection or irritation, and to radiation exposure. The responses of cases and controls to a questionnaire will be compared in order to define factors associated with the development of these diseases. Cases will be identified by the Los Angeles County, University of Southern California Cancer Surveillance Program and will include non-orientals aged 25-69, diagnosed with nose, sinus or nasopharyngeal carcinoma between 1979 and 1985. It is estimated that it will be impossible to interview 20% of the eligible cases because of deaths and refusals. A person of the same sex, race and birth year (within five years) will be identified for each case from among those living in the neighbourhood where the case lived at the time of his diagnosis. Telephone interviews with cases and controls will be conducted. The questionnaire elicits a detailed occupational history as well as information on tobacco use and radiation exposure. Additional questions elicit a history of infectious, allergic, and other conditions which caused irritation or anatomical alteration of the nasal passages. Standard matched pair methods will be used in the statistical analysis.

TYPE: Case-Control
TERM: Chemical Exposure; Dusts; Environmental Factors; Infection; Occupation; Radiation, Ionizing; Tobacco (Smoking)
SITE: Nasal Cavity; Nasopharynx
TIME: 1982 - 1992

1022   **Preston-Martin, S.**   03104
Univ. of Southern California, School of Medicine, Dept. of Preventive Medicine, Parkview Medical Bldg, B301, 1420 San Pablo St., Los Angeles CA 90033, United States of America (Tel.: (213)3421310; Fax: 3421237)
COLL: Thomas, D.C.

UNITED STATES OF AMERICA

Lip Cancer in Women
The aims of this case-control study are to investigate: (1) whether use of lip coverings protects against cancer of the lip; (2) the relationship of lip cancer to sun exposure; and (3) the association of lip cancer to other suggested risk factors such as use of tobacco. 80 female cases of lip cancer under age 75 diagnosed in Los Angeles County from 1972-1985 will be included. Controls will be selected by random digit dialling. Cases and controls will be interviewed by telephone, using a standard questionnaire which obtains a lifetime history of outdoor jobs, recreational exposure to sunlight, use of various types of lip coverings, and use of tobacco products. The analysis will use unmatched case-control comparisons.

TYPE: Case-Control
TERM: Cosmetics; Female; Radiation, Ultraviolet; Tobacco (Smoking)
SITE: Lip
TIME: 1986 - 1992

1023   Preston-Martin, S.   04069
Univ. of Southern California, School of Medicine, Dept. of Preventive Medicine, Parkview Medical Bldg, B301, 1420 San Pablo St., Los Angeles CA 90033, United States of America (Tel.: (213)3421310; Fax: 3421237)
COLL: Thomas, D.C.

X-Rays and Drugs in the Aetiology of Acute Myelogenous Leukaemia
This study aims to investigate whether adult onset of acute myelogenous leukaemia is related to (1) diagnostic X-rays in the ten years prior to diagnosis, (2) use of phenylbutazone and other bone marrow depressant drugs, or (3) other suggested leukaemia risk factors such as occupational exposure to chemicals or electromagnetic fields. Information from interviews and from medical charts will be collected on 400 leukaemia cases and 400 matched neighbourhood controls.

TYPE: Case-Control
TERM: Chemical Exposure; Drugs; Electromagnetic Fields; Immunosuppression; Radiation, Ionizing
SITE: Leukaemia (AML)
CHEM: Phenylbutazone
TIME: 1987 - 1996

1024   Preston-Martin, S.   04565
Univ. of Southern California, School of Medicine, Dept. of Preventive Medicine, Parkview Medical Bldg, B301, 1420 San Pablo St., Los Angeles CA 90033, United States of America (Tel.: (213)3421310; Fax: 3421237)
COLL: Thomas, D.C.; Daling, J.R.; Holly, E.A.

Childhood Brain Tumours and N-Nitroso Exposures
Childhood brain tumours (CBT) are the most common solid tumour in children aged 0-19 years. Epidemiological studies to date suggest that suspected risk factors such as ionising radiation, head trauma and known genetic predispositions account for only a small proportion of incident cases. A compelling experimental model suggests that N-nitroso compounds (NOC) may cause brain tumours in humans, particularly when exposure is transplacental. Since the tumours that are induced by transplacental NOC exposure of non-human primates occur in young adults as well as in immature animals it seems reasonable to include all cases diagnosed to age 20. It is proposed to study all cases of primary CBT diagnosed from 1984-1990 among children aged 0-19 living in all areas of the West Coast that are covered by population-based tumour registries; this includes 19 counties in the Los Angeles (LA), San Francisco (SF) and Seattle areas. In both SF and Seattle the biological mothers of 105 children with CBT diagnosed during this 7 year period will be interviewed; in LA 300 case mothers will be interviewed. In each region random-digit-dial controls, stratified by age and sex, will be selected with a control: case ratio of 2:1 for SF and Seattle and 1:1 in LA. It is expected to complete interviews with mothers of 510 CBT cases, 720 controls and 1,000 fathers. The questionnaire asks about sources of NOC exposure such as diet, drinking water, drugs, cosmetics, rubber products and parental occupational exposures. Pathology slides of all cases will be reviewed by a single paediatric neuropathologist. Data for each of the three West Coast regions will be analysed separately and will also be pooled. The combined West Coast regions will have sufficient statistical power to conduct separate analyses for the two most common histological types of CBT - astrocytoma and medulloblastoma. Data will also be comparable with data from a concurrent international study.

UNITED STATES OF AMERICA

TYPE: Case-Control
TERM: Childhood; Cosmetics; Diet; Drugs; Histology; Intra-Uterine Exposure; Occupation; Registry; Rubber; Water
SITE: Brain
CHEM: N-Nitroso Compounds
REGI: Bay Area (USA); Los Angeles (USA); Seattle (USA)
TIME: 1988 – 1994

1025 Preston-Martin, S. 04801
Univ. of Southern California, School of Medicine, Dept. of Preventive Medicine, Parkview Medical Bldg, B301, 1420 San Pablo St., Los Angeles CA 90033, United States of America (Tel.: (213)3421310; Fax: 3421237)
COLL: Bowman, J.D.; Sobel, E.; Peters, J.M.; Kaune, W.

Brain Tumours in Children and Exposure to Magnetic Fields
The objective of this study is to test epidemiologically whether magnetic field exposure is a risk factor for brain tumour in children. Subjects are taken from an on-going study in Los Angeles County. Cases include all children under age 20 with a primary tumour of the brain, cranial nerves or cranial meninges diagnosed from 1984-1990. Controls will be children of similar ages selected by random digit dialling. Subjects will total 300 cases and 300 controls. Wiring configurations and exterior magnetic fields will be measured at all subject residences within Los Angeles County from birth until diagnosis; a total of 1,200 such residences is expected. Interior exposure will be modeled from exterior measurements, and validated from interior measurements taken in a subset of 150 residences.

TYPE: Case-Control
TERM: Childhood; Electromagnetic Fields; Registry
SITE: Brain
REGI: Los Angeles (USA)
TIME: 1990 – 1994

*1026 Preston-Martin, S. 05083
Univ. of Southern California, School of Medicine, Dept. of Preventive Medicine, Parkview Medical Bldg, B301, 1420 San Pablo St., Los Angeles CA 90033, United States of America (Tel.: (213)3421310; Fax: 3421237)
COLL: Blot, W.J.; Sarkar, S.

Follow-Up Study of Oral and Pharyngeal Cancer
The aims of this study are to examine the influence of a variety of possible risk factors such as tobacco and alcohol use and diet on the development of a second cancer in persons with a cancer of the oral cavity or pharynx (ICD-O codes 141, 143-146, 148-149). Of the 595 cases of oral or pharyngeal cancer identified in 1984-1985, 41 cases subsequently developed a second cancer up to June 30, 1989. Three matched controls have been identified for each case. A control is defined as a person who had the initial oral or pharyngeal cancer but had not developed a second cancer by the time the case developed the second cancer. The cases and controls will be interviewed with regard to their use of tobacco and alcohol and diet since the development of their oral or pharyngeal cancer.

TYPE: Case-Control
TERM: Alcohol; Diet; Multiple Primary; Registry; Tobacco (Smoking)
SITE: Oral Cavity; Pharynx
REGI: Los Angeles (USA)
TIME: 1990 – 1992

*1027 Sarkar, S. 05087
Univ. of Southern California, School of Medicine, Dept. of Preventive Medicine, Parkview Medical Bldg, B301, 1420 San Pablo St., Los Angeles CA 90033, United States of America (Tel.: (213)3421663; Fax: 3421237)
COLL: Preston-Martin, S.; Blot, W.J.

Feasibility Study of Human Papilloma Virus and Oral and Pharyngeal Cancer
The objective of the study is to determine whether HPV is associated with oral and pharyngeal cancers. At present, the oral and pharyngeal biopsy tissues of 20 cases of oral and pharyngeal cancer are being studied, using the PCR technique to detect the presence of DNA sequences related to HPV. If a higher prevalence of HPV in the tumour tissue than that reported in the general population is found, it is

UNITED STATES OF AMERICA

proposed to do a large case-control study using 550 cases and 550 controls identified from an earlier study of oral and pharyngeal cancer.

TYPE: Case Series
TERM: BMB; Biopsy; HPV; Registry
SITE: Oral Cavity; Pharynx
REGI: Los Angeles (USA)
TIME: 1991 - 1992

LOUISVILLE

1028   Steiner, R.W.   04669
Univ. of Louisville, Dept. of Family Practice, 530 S Jackson St., Louisville KY 40292, United States of America (Tel.: (502)5881779)
COLL: Morris, D.L.; Greenberg, R.A.

Prospective Dietary Predictors of Prostate Cancer
The purpose of this study is to investigate the association of selected dietary factors, including fat, fibre and total calories, with prostate cancer. Both total calories ingested and total fat calories have been implicated as possible aetiological factors in prostate cancer, but the high correlation among these nutrients has been noted. The main thrust of this study is to overcome the problem of co-linearity of fat and total calories using the calorie-adjusted method for transformation of nutrient data, as proposed by Willet and Stampfer (Am. J. Epidemiol., 124(1):17-26, 1986). Through this analytical technique, the new (transformed) variables are statistically independent of total caloric intake, so that the simultaneous assessment of specific nutrients and total calories in mathematical models is possible. The role of dietary fibre as a protector against prostate cancer will also be examined with this design, with incident cases of prostate cancer and matched controls within a data set derived from the National Health and Nutrition Examination Survey. Conditional logistic regression analysis will be used to provide estimates of the exposure odds ratios for the specific nutrients of interest as independent risk factors for prostate cancer.

TYPE: Case-Control; Cohort
TERM: Diet; Fat; Fibre; Nutrition; Obesity
SITE: Prostate
TIME: 1989 - 1992

1029   Tamburro, C.H.   02475
Univ. of Louisville, Sch. of Medicine, Depts of Medicine, Pharm. & Toxicology, Div. Occup. Toxicol., Liver Res. Center, HSC 119A, 500 S. Jackson St., Louisville KY 40292, United States of America (Tel.: (502)5885252; Fax: 5886867)
COLL: Greenberg, R.S.; Wong, J.

Medical Surveillance of Vinyl Monomer Exposure in Chemical Industry Workers
The programme is the long term continuation of the "medical surveillance of occupational environments" which has become the prototype of prospective comprehensive medical surveillance systems for chemical industries. The original surveillance system was directed at cancer control and involves collection and storage of basic core data on job classification, work history records, chemical exposure rating, and medical illness and disease record data. Subsequently each cancer case and other organ diseases as identified are studied, with special emphasis on the development of liver cancer. Research is directed to the development of molecular biological monitoring. The vinyl monomer programme continues the original data base composed of 5,500 previous employees, 1,200 previously active employees and 600 presently active employees. Annually over 70% of the work force is re-examined and new information added to the data base. All vinyl chloride and related chemicals continue to be monitored. Data collection and storage methods, and subsequent study of disease, have already been shown to be cost effective. Recent analysis of the medical surveillance of vinyl monomers data has identified the human hepatotoxic, carcinogenic, and biological threshold level for vinyl chloride induced hepatic angiosarcoma and hepatic fibrosis. Study continues regarding gastrointestinal lung, brain and haematopoietic cancers. Papers have been published in Hepatol. 4:413-418, 1984; and in Am. J. Med. 78:68-76, 1985 (Liss, G.M. et al.).

UNITED STATES OF AMERICA

TYPE: Cohort; Methodology
TERM: Chemical Exposure; Environmental Factors; Metals; Occupation; Petroleum Products; Plastics; Solvents
SITE: Brain; Gastrointestinal; Haemopoietic; Liver; Lung
CHEM: Acrylamides; Acrylonitrile; Butadiene; Hexane; Hydrocarbons; Hydrocarbons, Halogenated; Mercury; PVC; Phenol; Styrene; Toluene; Vinyl Acetate; Vinyl Chloride; Vinylidene Chloride
OCCU: Chemical Industry Workers
TIME: 1974 - 1993

1030 Tamburro, C.H. 03526
Univ. of Louisville, Sch. of Medicine, Depts of Medicine, Pharm. & Toxicology, Div. Occup. Toxicol., Liver Res. Center, HSC 119A, 500 S. Jackson St., Louisville KY 40292, United States of America (Tel.: (502)5885252; Fax: 5886867)
COLL: Greenberg, R.S.; Wong, I.

Assessment of Health Effects of Acrylonitrile Exposure in Chemical Industry Workers

This program is an off-shoot of the medical surveillance of occupational environments program. It applies the health surveillance system to the assessment of possible health effects of acrylonitrile related compounds in exposed worker populations. The study covers a population of approximately 600 to 800 previous workers and 200 to 300 active employees. The retrospective assessment data include a work exposure history, and a rank ordered chemical exposure rating of 10-14 chemicals. These include acrylonitrile, acrylamides, acrylic acid, butadiene, hydrocarbons, halogenated hydrocarbons, methanol, phenol, toluene and caprylyl chloride. Medical data and cancer occurrence data for the retrospective cohort are available. The prospective programme includes annual data on medical examinations and laboratory screening studies for major organ systems. A sub-cohort population has, in addition, radiological screening evaluations for the lung and liver. Environmentally validated job specific determinations of exposure for acrylonitrile are also available for the years 1978, 1979, and 1980. Evaluation is underway to determine the biological threshold for hepatotoxicity and carcinogenicity of acrylonitrile and related compounds. Hepatol. 2:692, 1982 (Liss, G.M.); Seminars in Liver Disease 4:159-169, 1984 (Tamburro, C.H.); Am. J. M. 78:68-76, 1985.

TYPE: Cohort
TERM: Chemical Exposure; Occupation; Plastics; Rubber; Solvents
SITE: Colon; Liver; Lung
CHEM: Acrylamides; Acrylic Acid; Acrylonitrile; Butadiene; Caprylyl Chloride; Hydrocarbons; Hydrocarbons, Halogenated; Methanol; Phenol; Toluene; Vinyl Chloride
OCCU: Acrylonitrile Workers
TIME: 1974 - 1993

1031 Tamburro, C.H. 04762
Univ. of Louisville, Sch. of Medicine, Depts of Medicine, Pharm. & Toxicology, Div. Occup. Toxicol., Liver Res. Center, HSC 119A, 500 S. Jackson St., Louisville KY 40292, United States of America (Tel.: (502)5885252; Fax: 5886867)
COLL: Wong, J.

Health Effects of Heavy Metal Exposure in Chemical Catalyst Workers.

This is a prospective medical surveillance program of heavy metal catalyst workers utilising the health surveillance system to assess and identify any possible health effects of heavy metal and metallic compounds. These include aluminium, arsenic, antimony, barium, beryllium, cadium, chromium, cobalt, copper, iron, lead, magnesium, manganese, nickel, molybdenum, selenium, silver, uranium and zinc. The programme involves approximately 600 workers from two separate plants dealing with complex processes of different composition. Retrospective work histories and ranked chemical exposure ratings are available on all 83 catalyst processes. Exposure data will include past work history, chemical exposure assessment for each year, job classification, and work exposure rank, based on building and process. Annual and biannual medical examination including laboratory screening for major organ systems and radiological, electrocardiographic and pulmonary function studies are performed regularly. Environmental exposure data is validated by atomic absorption determination of environmental levels of certain heavy metals. Medical disorders with a potential relationship to heavy metal exposure, including cancer, are being studied in relation to these exposures.

UNITED STATES OF AMERICA

TYPE: Cohort
TERM: Chemical Exposure; Metals; Occupation
SITE: Colon; Kidney; Liver; Lung; Skin
OCCU: Metal Workers
TIME: 1983 – 1993

1032 Tamburro, C.H. 04846
Univ. of Louisville, Sch. of Medicine, Depts of Medicine, Pharm. & Toxicology, Div. Occup. Toxicol., Liver Res. Center, HSC 119A, 500 S. Jackson St., Louisville KY 40292, United States of America (Tel.: (502)5885252; Fax: 5886867)
COLL: Fortwengler, P.; Wong, J.

Health Effects of Polychlorinated Hydrocarbon Exposure
Retrospective analysis of various populations of individuals exposed to polychlorinated hydrocarbons through accidental environmental exposure or through occupational contact is being carried out. Populations with documented environmental exposure to various polychlorinated hydrocarbons and a work or environmental exposure history will be collected. Exposures will be rank ordered, and environmental (water, air or soil) measurements of the index chemical(s) taken in addition to either blood and/or fat biopsy level determinations of the toxic materials under study. In most cases, detailed medical examination with comprehensive histories and laboratory screening studies for major organ systems are also included. Methodology is being developed to determine if any causal relationships exist between disease(s) (including cancer), and environmental exposure to polychlorinated hydrocarbons. It is hoped to provide data on actual hazardous levels, duration of body burden and biological threshold for disease development (including cancer).

TYPE: Cohort
TERM: Chemical Exposure; Environmental Factors; Occupation
SITE: All Sites
CHEM: Dioxins; Furans; PCB
TIME: 1980 – 1993

LOWELL

1033 Eisen, E.A. 03769
Univ. of Lowell, Dept. of Work Environment, College of Engineering, One University Ave., Lowell MA 01854-2881, United States of America (Tel.: (508)9343278; Fax: 4525711)
COLL: Monson, R.R.; Tolbert, P.E.; Smith, T.

Cancer Mortality among Autoworkers exposed to Machining Fluids
NIOSH has estimated that over one million workers in the US are currently exposed to machining fluids. Previous studies have suggested an association between this exposure and respiratory, digestive and skin cancers. The present cohort study was initiated to assess whether long/term exposure to machining fluids in the course of machining, grinding, or other cutting operations, is associated with excess cancer mortality. The cohort includes all hourly workers who worked at least three years at one of three US automobile production plants. A cohort of 46,384 workers was defined based on company personnel records, and verified by Social Security employment records. An extensive exposure assessment was carried out at each plant. Lifetime exposures to straight, soluble and synthetic cutting oils, as well as to specific components and additives are currently being estimated for each subject in the cohort. The study population is 79% white, 90% male, and, on average, worked for 15 years. The average date of birth varies across the three plants from 1920 to 1939. The follow-up period extends from 1943 to 1984, by the end of which time 21% (10,142) of the cohort had died according to Social Security and National Death Index records. To date, cause of death information has been obtained for 93% of those known to be dead. Based on these data, SMRs have been estimated for each plant, using both the US and local county populations as reference. The overall SMR for death from all causes and from all cancer was 0.9. Slight excesses of 1.1 to 2.0, have been observed for some of the respiratory and digestive cancers. Dose-response models for specific cancers are being fitted, and the observed excess mortality from cancers of larynx and brain is being investigated with a nested case-control study.

UNITED STATES OF AMERICA

TYPE: Case-Control; Cohort
TERM: Chemical Exposure; Occupation
SITE: All Sites; Brain; Larynx
CHEM: Mineral Oil
OCCU: Automobile Workers
TIME: 1985 - 1993

MADISON

1034 Messing, E. 04208
Univ. of Wisconsin, School of Medicine, Div. of Urology, 600 Highland Ave., G5/347, Madison WI 53792, United States of America (Tel.: (608)2631359)
COLL: Young, T.B.; Hunt, V.B.; Roecker, E.B.

Early Detection of Urological Cancer by Home-Screening for Haematuria

To determine whether haematuria screening with urinary dipsticks can detect early-stage transitional cell carcinoma (TCC) of the bladder and renal cell carcinoma (RCC), a home testing programme was designed. A total of 2,982 healthy men age 50 or older, were recruited from five primary health care clinics. Subjects were solicited through their primary care physicians and asked to test their urine once a day for two weeks during month one and nine. Subjects were also asked to complete a questionnaire to gather information on socio-demographic factors, occupational exposures, smoking, medical histories, and other factors. If any dipstick test is at least "trace" positive, a urological evaluation is conducted, which consists of a physical examination and routine laboratory tests, intravenous urogram, cystoscopy, and bladder wash cytology. A total of 1,737 men have agreed to participate and were eligible to do so. Of these 1,340 men actually completed the testing procedure. After the first testing period, 283 had tested positive for haematuria and 192 received a complete urological work-up. Of those completed worked up, 16 had urological malignancies (9 transitional cell carcinoma, 1 renal cell carcinoma, 6 cancer of the prostate) and 47 had other serious urological diseases.

TYPE: Cohort
TERM: Occupation; Screening; Socio-Economic Factors; Tobacco (Smoking); Urine
SITE: Bladder; Kidney
TIME: 1985 - 1992

1035 Newcomb, P.A. 04469
Univ. of Wisconsin, Clinical Cancer Center, 420 N. Charter St., Room 6730, Madison WI 53706, United States of America (Tel.: (608)2637890)
COLL: MacMahon, B.; Willett, W.C.; Longnecker, M.

Alcohol Consumption, Lactation, and Breast Cancer Risk

A population-based case-control study is underway to evaluate the relationship between age-specific alcohol consumption and lactation and the occurrence of breast cancer. Recent evidence suggests that both factors are determinants of risk with modest, but potentially important, magnitudes of effect. Specific hypotheses to be addressed include: (1) cessation of alcohol intake is associated with a lower risk of breast cancer than is continued consumption, (2) alcohol intake before age 30 is associated more strongly with increased breast cancer risk than is consumption at later ages, and (3) increasing duration of lactation is associated with decreased risk of breast cancer among premenopausal women. To test these and other hypotheses, telephone interviews will be conducted with approximately 6,000 women with newly diagnosed breast cancer and 6,000 randomly selected community members. Cases will be identified from tumour registries in Wisconsin, Massachusetts, Maine, and New Hampshire. Controls will be selected from general population lists (driver's license lists and Medicare beneficiary lists). Cases and controls will be queried about past and current alcohol consumption and lactation history, as well as other medical and demographic factors known to influence breast cancer occurrence. The University of Wisconsin Clinical Cancer Center and the Harvard University School of Public Health both serve as coordinating centers for this multi-state study.

UNITED STATES OF AMERICA

TYPE: Case-Control
TERM: Age; Alcohol; Lactation
SITE: Breast (F)
REGI: Maine (USA); Massachusetts (USA); New Hampshire (USA); Wisconsin (USA)
TIME: 1988 - 1992

1036 Newcomb, P.A. 04850
Univ. of Wisconsin, Clinical Cancer Center, 420 N. Charter St., Room 6730, Madison WI 53706, United States of America (Tel.: (608)2637890)

Alcohol Consumption and Risk of Large Bowel Cancer in Women
The relationship between alcohol consumption and risk of cancers of the colon and rectum is currently being evaluated in a population-based case-control study. Approximately 1,000 women newly diagnosed with colorectal cancer during 1990-1992 will be interviewed by telephone. The interview elicits past and current alcohol consumption history as well as reproductive experiences and other medical and demographic factors known to influence the occurence of colorectal cancer. In addition, subjects will complete a food frequency questionnaire regarding dietary practices prior to diagnosis. For comparison, women without colorectal cancer will be selected from the general population using drivers' license and medicare lists and interviewed concurrently with the case group.

TYPE: Case-Control
TERM: Alcohol; Diet; Female; Registry; Reproductive Factors
SITE: Colon; Rectum
REGI: Wisconsin (USA)
TIME: 1990 - 1992

MARSHFIELD

1037 Banerjee, T.K. 04767
Marshfield Clinic, 1000 N. Oak Ave., Marshfield WI 54449, United States of America (Tel.: (715)3875134)
COLL: Layde, P.; Nordstrom, D.; Weber, J.

Pilot Study of Familial Factors in Cancer in a Defined Population Area
It is uncertain whether the aetiological significance of familial clustering of cancer represents inherited susceptibility or common familial exposure to some cultural, dietary, or other environmental stimulus. A case-control study is being conducted in a defined population area to describe cancer patterns and to compare the prevalence of a history of cancer among relatives of a sample of 100 new cancer cases to that among relatives of a sample of 100 age-and sex-matched controls randomly selected from the same postal code areas as the cases and to analyse the validity of information obtained from cases and controls about cancer in their relatives. Cases will be chosen prospectively from those which are newly diagnosed and reported to the population-based Marshfield Tumor Registry from the time when the study begins. Odds ratios for the association of family history of various cancers and the risk of cancer will be calculated using matched pair and conditional logistic regression techniques.

TYPE: Case-Control
TERM: Familial Factors; Genetic Factors
SITE: All Sites
TIME: 1990 - 1992

MAYWOOD

*1038 Swinnen, L.J. 05004
Loyola Univ. of Chicago, Fac. of Medicine, 2160 S. First Ave., Maywood IL 60153, United States of America (Tel.: (708)2168539; Fax: 2169335)
COLL: Costanzo-Nordin, M.R.; Fisher, S.G.

Lymphoproliferative Disorder Following Cardiac Transplantation
In a prior study, a significant increase in the incidence of post- transplant lymphoproliferative disorder (PTLD) was found to be associated with the use of OKT 3. The current study examines the effect of discontinuing the prophylactic use of OKT 3. All patients transplanted at Loyola University will be considered and compared with existing historical control groups of previously transplanted patients with

or without anti-T-cell therapy. The total patient number is expected to be approximately 250. Multivariate analyses will be performed to assess the influence of other immunosuppressive agents.

TYPE: Cohort
TERM: BMB; Immunosuppression; Transplantation; Treatment
SITE: Lymphoma
TIME: 1990 - 1992

MEMPHIS

1039   Roberson, P.K.   01283
St Jude Children's Research Hosp., 332 N. Lauderdale, P.O. Box 318, Memphis TN 38101, United States of America (Tel.: (901)5220370; Fax: 5272270)
COLL: Pratt, C.P.

Epidemiology of Childhood Cancer

This is a long-term study in which data from families of children with malignancies are being gathered. The subjects are children who are admitted to a large children's cancer research hospital. The overall objectives are to better characterize the features of such children and their families and to search for aetiological clues. Methods used include the following: questionnaire/interview of the families; retrospective review of medical records; follow-up interviews and updates of initial information and progeny of long-term survivors; and case-control studies where feasible. Areas to be covered include: socio-economic/occupational history, exposure to known carcinogens/pathogens (including ionizing radiation), previous diseases, other environmental factors and host factors familial cancer, congenital abnormalities, etc.). An analysis of some of the data collected on 1270 patients has been completed.

TYPE: Case Series; Case-Control
TERM: Chemical Exposure; Congenital Abnormalities; Familial Factors; Occupation; Radiation, Ionizing; Socio-Economic Factors
SITE: Childhood Neoplasms
TIME: 1979 - 1993

*1040   Sharp, G.B.   05003
Univ. of Tennessee, College of Graduate Health Sciences, Dept. of Biostatistics & Epidemiology, 877 Madison Ave., Suite 330, Memphis TN 38163, United States of America (Tel.: (901)5778220; Fax: 5286517)
COLL: Cole, P.

Epidemiology of Vaginal Clear Cell Adenocarcinoma

The major objective of this case-control study is to describe risk factors for this cancer among women who were exposed in utero to DES. Self-administered questionnaires were returned by 108 DES-positive cases and by 456 DES-positive controls. Initial studies have focused on risk factors for endometrial cancer, including height and body mass level during adolescence. Future studies will determine if there are case/control differences in age at menarche and in nutritional intakes during adolescence. Papers have been published in the Am. J. Obstet. Gynecol. 162(4):994-1001, 1990 and Cancer 66(10):2215-2220, 1990.

TYPE: Case-Control
TERM: Diet; Drugs; Intra-Uterine Exposure; Menarche; Nutrition
SITE: Vagina
CHEM: DES
TIME: 1990 - 1992

MIDLAND

1041   Cook, R.R.   04295
Dow Chemical Company, Dept. of Epidemiological Health, & Environmental Science, 1803 Building, Midland MI 48674, United States of America (Tel.: (517)6361383; Fax: 6361875 ; Tlx: 227455)
COLL: Bond, G.G.; Olsen, G.W.; Bodner, K.M.; Bloemen, L.J.

UNITED STATES OF AMERICA

Epidemiological Surveillance of Chemical Industry Workers
Cause-specific mortality surveillance is being carried out on approximately 100,000 present and former employees of The Dow Chemical Company employed since 1940. Nested case-control and chemical-specific cohort mortality studies are conducted. Among the chemicals produced at the major production sites are herbicides and insecticides, epoxy resins, benzene, styrene, magnesium, chlorinated solvents, vinyl chloride, polyethylene and various consumer products.

TYPE: Case-Control; Cohort
TERM: Chemical Exposure; Herbicides; Insecticides; Occupation; Pesticides; Plastics; Solvents
SITE: Brain; Kidney; Liver; Lung
CHEM: Benzene; Epoxy Resins; Ethylene; Hydrochloric Acid; Phenoxy Acids; Styrene
OCCU: Chemical Industry Workers
LOCA: Canada; Netherlands; United States of America
TIME: 1979 - 1993

MINNEAPOLIS

1042 Filipovich, A.H. 00417
Univ. of Minnesota, Dept. of Pediatrics, Div. of Immunology, P.O.Box 610 UMHC, 420 Delaware St. SE, Minneapolis MN 55455, United States of America (Tel.: (612)6246676)
COLL: Mertens, A.C.; Frizzera, G.; Robison, L.L.

Immunodeficiency Cancer Registry
The Immunodeficiency Cancer Registry is a non-population based, worldwide registry which collects selected demographic, laboratory, clinical, and pathological data from individuals with a prior history of immunodeficiency who have developed a cancer. The following types of information are available on these cases: laboratory and clinical features of the underlying immunodeficiency and its clinical course and diagnostic data on the malignancy including its therapeutic outcome. The registry independently reviews the pathology of all tumours including the evaluation of non-Hodgkin's lymphomas using the criteria established by the National Cancer Institute. Two investigations, involving the families of immunodeficient children from the registry, are currently being conducted. The first is a collaborative study, involving the autosomal recessive disorder Ataxia-Telangiectasia (A-T), which is investigating an association between A-T heterozygote individuals, and their predisposition to cancer. The second study involves the X-linked recessive disorder, Wiskott-Aldrich syndrome, and is evaluating the cancer type and frequency in female carriers in comparison to non-carrier family controls. Defining a worldwide population of immunodeficient patients who are at risk of developing malignancy is an on-going goal. The registry is interested in acquiring cases of cancer or pre - malignant lesions diagnosed in patients with one of the naturally occurring immunodeficiencies defined by WHO or any pre-existing immunodeficiency not secondary to iatrogenic immunosupression. Recent publications form part of the mini-symposium on Immundeficiency and Cancer (The American Journal of Pediatric Hematology/Oncology, 9(2):178, 1987).

TYPE: Case Series
TERM: AIDS; Familial Factors; Genetic Factors; Immunodeficiency; Immunosuppression; Premalignant Lesion; Virus
SITE: All Sites
TIME: 1973 - 1993

1043 Folsom, A.R. 03627
Univ. of Minnesota, Div. of Epidemiology, School of Public Health, 515 Delaware St. S.E., 1-210 Moos Tower, Minneapolis MN, United States of America (Tel.: (612)6249950; Fax: 6258950)
COLL: Sellers, T.A.; Potter, J.D.; Munger, R.G.; Wallace, R.B.; Prineas, R.; Kushi, L.

Iowa Women's Health Study
A population-based cohort of 41,837 women aged 55-69 years was recruited to test the hypothesis that increased abdominal fat distribution, independent of body mass, is associated with increased incidence of endometrial cancer, breast cancer, and total mortality. Considerable exposure data on the cohort were obtained in a baseline survey in 1986, and cancer incidence and mortality data are now being obtained by linkage to the Iowa Health (SEER) Registry and National Death Index. Published two-year results indicated that abdominal adiposity was independently associated with increased risk of breast cancer, total mortality and several other nonfatal cancer endpoints ascertained by self report. However, the association of abdominal adiposity with endometrial cancer risk was not independent of body mass. Follow-up for cancer incidence and mortality is being continued through at least eight years. Further

UNITED STATES OF AMERICA

follow-up will allow verification of early results and extend them to rarer neoplasms which also may be related to fat distribution. Extended follow-up will also permit testing of dietary hypotheses for cancer, including possible interactions of diet with fat distribution. Further follow-up will enable comparison of cancer rates among women living in farms versus those not living in farms. A nested case-control study of families will also be conducted to test whether body size or shape explains the familial clustering of breast and endometrial cancer.

TYPE: Cohort
TERM: Cluster; Diet; Fat; Obesity; Physical Factors; Registry
SITE: All Sites; Breast (F); Uterus (Corpus)
REGI: Iowa (USA)
TIME: 1985 - 1995

1044　Gilbertsen, V. 03971
Univ. of Minnesota, Dept. of Surgery, 420 Delaware St. SE, Box 195 UHMC, Minneapolis MN 55455, United States of America (Tel.: (612)6274151)
COLL: Mandel, J.S.; Bond, J.; Snover, D.; Bradley, M.; Ederer, F.; Schuman, L.M.; Williams, S.

Trial of Faecal Occult Blood Screening for Colorectal Cancer Detection
This is a randomized clinical trial to determine whether screening for occult blood using the haemoccult test will result in a reduction in mortality from colorectal cancer. Over 45,000 volunteers who are residents of Minnesota between the ages of 50 and 80 have been randomized to three groups. One group is screened by annual faecal occult blood tests for five years, the second is screened every two years over the same period, while the controls are followed-up by annual questionnaire. Follow-up of almost the entire population has been maintained over several years and information on all deaths is carefully assessed. The study was described in detail in J. Chron. Dis. 33:107-114, 1980.

TYPE: Mortality
TERM: BMB; Screening
SITE: Colon; Rectum
TIME: 1975 - 1995

1045　Mandel, J.S. 03975
Univ. of Minnesota, School of Public Health, Div. Environmental & Occupat. Health, 420 Delaware St. SE, Box 197, UHMC, Minneapolis MN 55455, United States of America (Tel.: (612)6274200)

Cancer Risk in X-Ray Technologists
The objective of this study is to evaluate the health status, primarily with respect to cancer, of all current and former members of the American Registry of Radiologic Technologists (ARRT). About 180,000 individuals whose location is known are being sent a detailed questionnaire to obtain data on exposure to radiation, a history of cancer and potential confounding factors. In addition, dosimetry data will be obtaines. All cancers identified from the questionnaire will be validated by sending a diagnostic form to the person's physician. In addition to evaluating morbidity data, death certificates are being obtained on all deceased ARRT subjects. These data will be coded and analysed to examine relationships between mortality and morbidity patterns and radiation exposure. This study is important for three reasons: first, the large, well defined cohort readily available for study of a variety of cancers associated with radiation exposure, second, the varying levels of radiation exposure and the availability of considerable dosimetry data and third, the large proportion of females (70%) provides an opportunity to assess radiation effects in this group.

TYPE: Cohort
TERM: Chemical Exposure; Dose-Response; Occupation; Radiation, Ionizing
SITE: All Sites; Breast (F); Leukaemia; Lung; Thyroid
OCCU: Health Care Workers
TIME: 1982 - 1992

1046　Robison, L.L. 04756
Univ. of Minnesota, Div. of Pediatric Hematology/Oncology, Box 422 UMHC , Minneapolis MN 55455, United States of America (Tel.: (612)6262778; Fax: 6262815)
COLL: Buckley, J.D.; Kersey, J.; Neglia, J.P.; Reaman, G.H.; Severson, R.K.; Hammond, G.D.; Heerema, N.A.; Miller, D.; Potter, J.D.; Sather, H.N.; Trigg, M.

UNITED STATES OF AMERICA

Acute Lymphoblastic Leukaemia in Children
The Children's Cancer Study Group is conducting a case-control study of the role of risk factors for childhood ALL within biologically-defined subgroups. An estimated total of 2,000 newly diagnosed cases of ALL ascertained from January 1989 to December 1992 will be included. Patient subgroups are defined on the basis of FAB morphology, immunophenotyping, and cytogenetics. One control per case is selected by random digit dialling and is individually matched to the case on age, sex, and race. Telephone interviews are being conducted with the parents of both cases and controls. Postulated risk factors under investigation include maternal reproductive history, exposure during the index pregnancy, exposure to ionizing radiation, exposure to pesticides and solvents, timing and extent of infectious exposures, and familial/ genetic components.

TYPE: Case-Control
TERM: Childhood; Familial Factors; Genetic Factors; Intra-Uterine Exposure; Pesticides; Radiation, Ionizing; Reproductive Factors; Solvents
SITE: Leukaemia
TIME: 1988 - 1993

1047 Robison, L.L. 04782
Univ. of Minnesota, Div. of Pediatric Hematology/Oncology, Box 422 UMHC , Minneapolis MN 55455, United States of America (Tel.: (612)6262778; Fax: 6262815)
COLL: Hammond, G.D.; Buckley, J.D.; Arthur, D.C.; Woods, W.G.; Pentz, M.A.; Wilkins, J.R.; London, S.J.

Acute Nonlymphoblastic Leukaemia in Children
The Children's Cancer Study Group (CCSG) has recently completed an analytic study designed to assess specific environmental exposures as potential risk factors for ANLL in children. The findings from this study provide the direction for the proposed epidemiological investigation of childhood ANLL. The study will include approximately 630 cases of ANLL diagnosed during a 4-year period in persons up to 18 years of age. The primary objectives of the study are (1) to confirm the association of pesticide exposure and ANLL risk and to identify the substances or class of pesticides responsible; (2) to confirm the association of solvents and petroleum products and ANLL risk and determine the class of solvent responsible; (3) to confirm the association of maternal marijuana use and ANLL risk; and (4) to identify subgroups, defined by age at diagnosis, FAB morphology, cytogenetic abnormality, or clinical features of the leukaemia, in which the associations with pesticides, solvent, or marijuana exposure are strongest. Cases will be ascertained through the member institutions of the Children's Cancer Study Group. Cases of ANLL will be compared with a series of age-, sex-, and race-matched regional controls selected using random digit dialling. Parents of cases and controls will be interviewed by telephone. Within most of the morphologic subgroups of ANLL (FAB morphology classification: M0-M7), the study will have a sample size that provides sufficient power for assessing the study hypotheses. Classification of cases according to FAB morphology and cytogenetics wil be made by centralised review.

TYPE: Case-Control
TERM: Childhood; Classification; Marijuana; Pesticides; Petroleum Products; Solvents
SITE: Leukaemia (ANLL)
TIME: 1989 - 1994

*1048 Sellers, T.A. 05143
Univ. of Minnesota, School of Public Health, Div. of Epidemiology, 515 Delaware St. SE , Minneapolis MN 55455, United States of America (Tel.: (612)6245400; Fax: 6258950)
COLL: Potter, J.D.; Rich, S.S.; King, R.A.; Anderson, V.E.; Kuni, C.C.; Kushi, L.; Snover, D.

Epidemiological and Genetic Studies of Breast Cancer
A historical cohort study of families identified by the University of Minnesota over 40 years ago is being conducted to identify and quantify gene-environment interactions in the pathogenesis of breast cancer. The exposed group is defined as the first- and second-degree relatives of breast cancer patients treated at the University of Minnesota between 1944 and 1952; unexposed women will be the spouses of first- and second-degree male relatives. Women in both groups will be administered questionnaires on diet, reproductive history, etc., and will be referred for a mammogram to confirm disease status. Collection of blood samples on selected families will permit linkage analysis to confirm the presence of a breast cancer susceptibility gene(s).

UNITED STATES OF AMERICA

TYPE: Cohort; Genetic Epidemiology
TERM: BMB; Blood; Diet; Genetic Factors; Linkage Analysis; Reproductive Factors
SITE: Breast (F)
TIME: 1990 - 1996

*1049 Severson, R.K. 05002
Univ. of Minnesota, Div. of Hematology/Oncology, Box 422,UMHC, 420 Delaware St. SE, Minneapolis MN 55455, United States of America (Tel.: (612)6274438)
COLL: Robison, L.L.; Pollock, B.H.; Hammond, G.D.; Ross, J.

Paediatric Cancer Distribution in the US: A Geographical Analysis

The primary objective of this ecological study is to determine areas within the US where the observed to expected ratios of paediatric cancer differ notably from 1. It has been estimated that almost 10,000 incident cancers are diagnosed annually in the US in persons under the age of 20. During 1989, over 6,000 new cases were diagnosed by member institutions of the Children's Cancer Study Group (CCSG) and the Paediatric Oncology Group (POG). These observed cases have been coded geographically to specific regional areas within the US. Using census data and SEER program rates, expected numbers for the age groups of 0-4, 5-9, 10-14 and 15-19 have been generated. Regional maps of age-, sex-, and race-specific observed/expected numbers are being created for specific cancer sites including leukaemia, lymphoma and brain tumours. Preliminary analyses have indicated several geographical patterns where the observed number of childhood cancer patients is appreciably less than the expected number. Further data will be collected for 1990 incident cases.

TYPE: Incidence
TERM: Mapping
SITE: Childhood Neoplasms
TIME: 1990 - 1992

*1050 Woods, W.G. 05096
Univ. of Minnesota Hospital & Clinic, Box 454 UMHC, 420 Delaware St. SE, Minneapolis MN 55455, United States of America (Tel.: (612)6262778; Fax: 6262815)
COLL: Lemieux, B.; Tuchman, M.

Screening for Neuroblastoma in Infants

The Quebec Neuroblastoma (NB) Screening Project was initiated to assess the clinical and biological aspects of screening infants for the presence of NB in North America. The experimental cohort, all infants born in the province of Quebec during the five-year period 1989-1994, are eligible for screening at three weeks and six months of age. Parents mail urine-saturated and dried filter papers to the screening laboratories, where the catecholamine metabolites VMA and HVA are assessed by a two-stage screening assay, thin-layer chromatography followed by gas chromatography-mass spectrometry. Infants with elevated levels after repeat testing are offered a thorough non-invasive diagnostic evaluation looking for the presence of NB. Patients diagnosed with NB are then uniformly staged, treated, and followed, similarly to clinically detected patients either never screened or missed by screening. Mortality from NB of all children diagnosed as part of the cohort will be compared to results in several population-based, non-screened control groups. Secondary endpoints include NB incidence, clinical staging, and survival. In the process, specific biological parameters will be studied to determine (1) if known prognostic factors in NB detected clinically are of equal prognostic importance in screen-detected tumours; and, (2) if such studies shed light on the hypothesis that NB is more than one distinct disease. These parameters include tumour N-myc oncogene amplification and expression, DNA content, cytogenetics, gangliosides, and histology; serum ferritin, neuron-specific enolase, and gangliosides; and urinary catecholamine metabolites. Finally, as secondary aims, the effect of preclinical detection of NB on health care costs and the maternal psychosocial consequences of cancer screening in infants will be determined. The results obtained in this study will help determine whether screening of infants for the detection of NB will lead to a reduction in population-based mortality, and to shed light on the biology of this malignancy.

UNITED STATES OF AMERICA

TYPE: Cohort
TERM: BMB; Registry; Screening
SITE: Neuroblastoma
LOCA: Canada; United States of America
REGI: Canada (Can); Delaware (USA); Florida (USA); Minnesota (USA)
TIME: 1989 - 1997

MORGANTOWN

1051   Bouquot, J.E.   03400
West Virginia Univ., Dept. of Oral Pathology, School of Dentistry, Morgantown WV 26506, United States of America (Tel.: (304)2932149)

Prospective Study of Oral Precancerous Lesions
Few premalignant lesions in humans have been as extensively studied as leukoplakia of the mouth. Many of these lesions never become malignant and many others disappear in time, yet no US investigator has followed a large number of such cases in order to determine their exact biological behaviour. Follow-up studies in India and Denmark have indicated conflicting "malignant potential" for oral leukoplakia lesions. It is important to determine the proper clinical management of such lesions because: (1) almost 4% of US adults have them, (2) they are readily visualized by aware clinicians, (3) they can be easily followed by dentists during routine annual oral examinations. This study involves the clinical evaluation (with histology, where necessary) of up to 1,200 patients with oral leukoplakia, and their follow-up at 5-year intervals thereafter (with local dentists performing annual examinations). Those who cannot be examined by the principal investigators will be examined by a local oral surgeon, according to the study protocol. Patients in whom leukoplakic lesions have disappeared will continue to be followed; screening tests, such as topically applied metachromatic dyes, will be used where appropriate. Papers were published in Cancer 61:1691-1698, 1988, Cancer 61:1685-1690, 1988, Oral Surg. 65:199-207, 1988.

TYPE: Case Series
TERM: High-Risk Groups; Premalignant Lesion; Screening
SITE: Oral Cavity
TIME: 1983 - 1992

NASHVILLE

1052   Dupont, W.D.   04015
Vanderbilt Univ., School of Medicine, Dept. of Preventive Medicine, Div. of Biostatistics, A-1124 Medical Center North, Nashville TN 37232-2637, United States of America (Tel.: (615)3222001; Fax: 3438722)
COLL: Page, D.L.; Parl, F.F.; Cole, P.

Breast Cancer Risk after Atrophic or Apocrine Changes
Preliminary studies suggest that the presence of atrophy in benign breast biopsies is an important risk factor for subsequent breast cancer. This risk may be as low as one-third or as high as five times that of the general population, depending on other prognostic factors such as age at biopsy or a family history of breast cancer. This wide variation may result from different types of atrophy, similar in morphological appearance but with different aetiologies and associated with different cancer risks. In this case there may be a positive correlation between the type of atrophy associated with high cancer risk and these other factors. Papillary apocrine change is a distinctive lesion seen in 27% of benign breast biopsies, and has never been rigorously evaluated as a breast cancer risk factor. In a retrospective cohort study of approximately 8,100 consecutive women who underwent benign breast biopsy at one of three Nashville hospitals between 1952-1968 (Vanderbilt or Baptist hospitals) or 1950-1968 (St. Thomas Hospital), all biopsies will be reviewed and reclassified for the presence and degree of atrophic or papillary apocrine changes. Information on subsequent breast cancer outcome and epidemiological risk factors will be obtained from interviews with patients or their next of kin. This information has already been obtained on 60% of the study subjects who have participated in prior studies. 90% follow-up is expected. Variation in risk will be assessed by means of intra-study control groups of biopsied patients, as well as external control groups from the Third National Cancer Survey in Atlanta and the Cancer in Connecticut data base. Data will be analysed using hazard regression methods and other statistical techniques to elucidate cancer risks associated with various constellations of risk factors from a longitudinal data base. This study will substantially increase the precision with which breast cancer risk can be predicted on the basis of histological and non-anatomical findings.

UNITED STATES OF AMERICA

TYPE: Cohort; Methodology
TERM: Biopsy; Premalignant Lesion
SITE: Benign Tumours; Breast (F)
TIME: 1986 – 1992

*1053   Dupont, W.D.   05020
Vanderbilt Univ., School of Medicine, Dept. of Preventive Medicine, Div. of Biostatistics, A-1124 Medical Center North, Nashville TN 37232-2637, United States of America (Tel.: (615)3222001; Fax: 3438722)
COLL: Rogers, W.H.; Page, D.L.; King, M.C.

Epidemiology of Molecular Risk Factors for Breast Cancer
A nested case-control study of women with benign breast disease is being conducted to determine the effect of oncogenic and histological factors on breast cancer risk. The study cohort will consist of all women without a prior history of breast cancer who underwent benign breast biopsies at Vanderbilt Hospital between 1959 and 1988, at St. Thomas Hospital between 1964 and 1988 or at Baptist Hospital between 1982 and 1988. Biopsy slides and paraffin-embedded biopsy tissue are available for all these women. The slides of the entry biopsies will be reviewed blindly and classified according to Page's histological criteria (New Engl. J. Med. 312:146-151, 1985). The study cohort contains approximately 10,913 women. Cases will be cohort members who develop breast cancer during follow-up (approximately 586 women). Two controls matched for age, year, benign histological diagnosis and hospital of their entry biopsy at risk for developing breast cancer will be chosen from the cohort when their matched case patient develops this disease. In situ hybridization techniques will be used to look for altered expression, amplification or transcription of specific oncogenes within different types of breast lesions in the entry biopsies of cases and controls. The oncogenes which will be evaluated include the HER-2/neu, c-myc, c-myb, c-rasHa and rb oncogenes. Other oncogenes will be evaluated as evidence accrues suggesting that they may have a role in the pathogenesis of breast cancer.

TYPE: Case-Control; Cohort; Molecular Epidemiology
TERM: BMB; Biopsy; Histology; Oncogenes
SITE: Breast (F)
TIME: 1990 – 1995

NEW HAVEN

1054   Berwick, M.   04313
Yale Univ., Cancer Prevention Research Unit, 26 High St., New Haven CT 06510, United States of America (Tel.: (203)4327390; Fax: 4327377)
COLL: Berwick, M.; Barnhill, R.; Portlock, C.; Kegeles, S.S.; Rosner, G.

Lethal Melanoma and Skin Examination
Prevention of mortality from malignant melanoma is based on the untested concept that visual inspection of the skin will lead to early detection and effective treatment. This case-control study will test the hypothesis that lethal cutaneous melanoma is inversely associated with skin examination. All incident cases of cutaneous melanoma (CM; expected number of cases 1,150) newly diagnosed in Connecticut residents from September 1986 to May 1989 will be identified by rapid ascertainment. About 1,200 population-based controls, frequency-matched to the age-sex distribution of cases, will be obtained by random digit dialling. Cases and controls will be interviewed in person regarding their skin examination practices and risk factors for CM. From the case group, lethal CM (n = 220 to 250) will be compared with the controls with respect to skin examination practices, particularly self-examination. Adjustment will be made for factors that explain the majority of CM in the general population. Non-lethal, invasive CM and in situ CM will also be compared with the general population after adjustment for aetiological factors. Two sources indicate that the prevalence of self-examination in healthy persons at risk for melanoma from the general population is 32% or greater. With 32% as the prevalence of self-examination the study has a statistical power of 80% to detect a relative risk of 1.7 for lethal CM in those not practising self-examination.

UNITED STATES OF AMERICA

TYPE: Case-Control
TERM: Prevention
SITE: Melanoma
TIME: 1986 – 1992

1055 Janerich, D.T. 04137
Yale University, School of Medicine, Dept. Epidemiology & Public Health, 60 College St., P.O. Box 3333, New Haven CT 06510-8034, United States of America (Tel.: (203)7852102)
COLL: Kasl, S.V.; McCrea Curnen, M.G.; Thompson, W.D.; Roush, G.C.; Hayden, C.L.; Kegeles, S.S.

Connecticut Cancer Control Research Unit
The Connecticut Cancer Control Research Unit brings together a multidisciplinary team of scientists to conduct applied research and intervention trials in cancer prevention and control. Researchers from the behavioural sciences, biostatistics, cancer control science, epidemiology, microbiology, medicine and public health are collaborating on several studies which include: (1) analysis of all cases of invasive cervical cancer in Connecticut for three years to identify factors that prevented early detection and treatment of the disease; (2) identification of social, economic, psychological and medical care factors which may determine why cancers of the breast, prostate, colo-rectum and corpus uteri are often not diagnosed in black patients until they have reached a more advanced stage than in white patients; (3) assessment of the effectiveness of two experimental interventions for improving women's use of breast cancer screening; (4) investigation into whether regular self-examination of the skin reduces mortality from malignant melanoma; (5) research to assess microbial and biochemical predictors of colorectal carcinoma; and a paired community study of colorectal cancer control; and (6) identification of factors which influence adolescents' use of smokeless tobacco, and development of interventions advising them of the effects of this new and potentially important carcinogen.

TYPE: Intervention
TERM: Biochemical Markers; Prevention; Psychological Factors; Race; Screening; Socio-Economic Factors; Stage; Tobacco (Chewing)
SITE: Breast (F); Colon; Prostate; Rectum; Skin; Uterus (Cervix); Uterus (Corpus)
TIME: 1986 – 1992

NEW ORLEANS

1056 Fontham, E.T. 04052
Louisiana State Univ., Medical Center, Dept. of Pathology, 1901 Perdido St., New Orleans LA 70112, United States of America (Tel.: (504)5686094; Fax: 5686037)
COLL: Correa, P.; Chen, V.W.; Greenberg, R.S.; Liff, J.; Coates, R.J.; Buffler, P.A.; Alterman, T.; Wu-Williams, A.; Reynolds, P.; Boyd, P.; Austin, D.F.

Lung Cancer in Non-Smoking Women
This is a population-based case-control study of women who have never used tobacco products and develop lung cancer. The study is designed to test the hypothesis that involuntary exposure to environmental tobacco smoke (ETS) increases the risk of lung cancer in non-smoking women. Cases are ascertained in five major US metropolitan areas: Atlanta, Houston, Los Angeles, New Orleans, and San Francisco Bay Area. Interview data will be compared with two control groups: a population group selected by random digit dialling (ages 20-64) and from the Health Care Finance Administration roles (ages 65 +) and a group of women diagnosed with colon cancer. Both control groups are composed of women who report themselves to be lifetime non-smokers. Detailed information is obtained on involuntary exposure to ETS during childhood and adult life. Information is also sought on other lung cancer risk factors such as occupation, diet, residential factors, family history of cancer, etc. Urine specimens and hair samples are collected from study subjects for analysis of cotinine content to validate subjects' reported status with respect to recent smoking habits. Radon monitors were placed in the current residence of a sample of cases and controls. Histopathological review of diagnostic material is evaluated by one pulmonary pathologist. Stratified analyses and logistic regression techniques will be used to examine risk of lung cancer by histological type associated with exposures to ETS while controlling for confounding variables.

UNITED STATES OF AMERICA

TYPE: Case-Control
TERM: Biochemical Markers; Passive Smoking
SITE: Lung
CHEM: Radon
TIME: 1985 - 1992

NEW YORK

1057 Brandt-Rauf, P.W. 04602
Columbia University, School of Public Health, Division of Environmental Sciences, 60 Haven Ave., New York NY 10032, United States of America (Tel.: (212)3053464; Fax: 3058119)
COLL: Niman, H.L.; Neugut, A.I.; Perera, F.P.

Serum Oncogene Protein Screening
The purpose of this research is to develop a new molecular epidemiological method for the surveillance of occupational cancer. This method is based on the use of monoclonal antibody immuno-blotting assays for the detection of oncogene protein products in serum that can be used to screen for early detection of neoplastic changes in occupational cohorts at risk for malignant disease due to workplace exposures. This approach is currently being applied to several such cohorts (including hazardous waste workers and foundry workers) and matched normal controls (total numbers of approximately 100), and further studies are planned on other occupational cohorts and on cohorts of cancer patients to follow their response to therapy.

TYPE: Methodology; Molecular Epidemiology
TERM: Antibodies; Monitoring; Mutagen Tests; Occupation; Screening
SITE: All Sites
OCCU: Foundry Workers; Waste Incineration Workers
TIME: 1987 - 1992

1058 German III, J.L. 02265
The New York Blood Center, Lab. of Human Genetics, 310 E. 67th St. , New York, United States of America
COLL: Passarge, E.

Bloom's Syndrome Registry
Bloom's syndrome is a rare, recessively transmitted disorder which predisposes to cancer of the wide variety of sites and types that affects the general population. Diagnosed cases throughout the world comprise the Bloom's Syndrome Registry (132 accessions to date), and contact with most affected families is maintained, either directly or through their physician. 47 of the persons included in the registry have developed at least one cancer. 63 cancers have been diagnosed at mean age 24. 15 were acute leukaemia (mean age at diagnosis 17 years), and 33 were carcinoma (mean age at diagnosis 32 years). The others were lymphoma and Hodgkin's disease (11), osteosarcoma, Wilms' tumour, and medulloblastoma. Progress reports are made in the journal Clinical Genetics, particular emphasis being laid on cancer occurrence in homozygous affected persons and their parents (obligate heterozygotes). Recent papers appeared in Clin. Genet. 35:57-69 and 93-110, 1989.

TYPE: Registry
TERM: DNA; Genetic Markers; High-Risk Groups; Premalignant Lesion
SITE: All Sites
TIME: 1963 - 1999

1059 Guillem, J. 04441
Columbia Univ., College of Physicians and Surgeons, Dept. of Surgery, 622 W. 168th St., New York NY 10032, United States of America (Tel.: (212)3052268)
COLL: Forde, K.A.; Treat, M.R.; Neugut, A.I.

Colonoscopy Screening in Asymptomatic First-degree Relatives of Colon Cancer Patients
The aim of this prospective study is to compare the colonoscopic detection rate for colonic neoplasms in asymptomatic first-degree relatives of patients with colon cancer to that in asymptomatic individuals without any family history of colon cancer. Since autopsy studies have demonstrated a 2 to 3 fold increase of colon cancer in first-degree relatives of colon cancer patients, as well as an earlier age of onset and a predominance of proximally (right-sided) located lesions, complete colonoscopic screening of this high-risk population at an 'earlier' age has been suggested. At what age screening

UNITED STATES OF AMERICA

should be started, how often it should be done and what the yield of such a screening programme is, are questions that remain to be answered. It is hoped to obtain data which will establish clear screening guidelines for high-risk and low-risk populations.

TYPE: Case Series
TERM: High-Risk Groups; Screening
SITE: Colon; Rectum
TIME: 1986 - 1992

1060 Kerner, J. 04695
Mem. Sloan-Kettering Cancer Center, Dept. Epidemiology & Biostatistics, Div. of Cancer Control, Box 60, 1275 York Ave., New York NY 10021, United States of America (Tel.: (212)7946998)
COLL: Andrews, H.; Zauber, A.; Streuning, E.

Small Area Analysis of Cancer Incidence and Mortality

The purpose of this study is to evaluate the benefit, for planning, implementing, and evaluating cancer prevention and control programmes, of targeting these programmes in small geographical areas where historical cancer incidence and/or mortality have been highest. The study uses New York City census data, and cancer incidence and mortality data from the New York State Cancer Registry and New York City Department of Health. Published work from this project to date (e.g. J. Clin. Epid. 41:543-553, 1988) has evaluated the stability of small area rates over time, and the incremental improvement in the predictive value of a positive screening examination when one targets a programme at areas of stable, high prevalence. Current efforts are using multiple regression analysis of small area incidence rates, with census and mortality data as predictor variables, and evaluating the utility of targeting cancer screening programmes based on estimated incidence rates. The final evaluation of the benefit of targeting cancer screening programmes into high risk areas in the observed "down-staging" of breast, cervix, or colorectal cancer and the eventual reduction of cancer mortality in the high-risk neighbourhoods.

TYPE: Incidence; Methodology; Mortality
TERM: Screening
SITE: Breast (F); Colon; Rectum; Uterus (Cervix)
TIME: 1983 - 1993

1061 Neugut, A.I. 03164
Columbia Univ., College of Physicians & Surgeons, Dept. of Medicine and Public Health, 630 W 168th St., New York NY 10032, United States of America (Tel.: (212)3053921; Fax: 7951182)
COLL: Garbowski, G.; Forde, K.A.; Treat, M.R.; Waye, J.; Fenoglio-Preiser, C.

Colorectal Polyps

The objective is to identify potential risk factors for adenomatous polyps of the colon. Questionnaires were administered to patients undergoing colonoscopic polypectomy. The control group consisted of patients in whom colonoscopic examination showed no neoplasia. Currently data are being analysed from telephone interviews with 303 adenomatous polyp patients (with no history of colon cancer), 509 normal controls (with no history of colon cancer or polyps), and 113 colon cancer patients.

TYPE: Case-Control
TERM: Premalignant Lesion
SITE: Colon; Rectum
TIME: 1983 - 1992

*1062 Neugut, A.I. 05014
Columbia Univ., College of Physicians & Surgeons, Dept. of Medicine and Public Health, 630 W 168th St., New York NY 10032, United States of America (Tel.: (212)3053921; Fax: 7951182)
COLL: Garbowski, G.; Benson, M.; Buttyan, R.; Romas, N.; Rosner, W.; Brandt-Rauf, P.W.

Prostate Cancer Among Ethnic Groups in New York

The objective is to identify risk factors and biological markers which distinguish prostate cancer among different ethnic groups with varying risk in New York City, specifically whites, blacks and Hispanics. Cancer cases are collected at Columbia-Presbyterian Medical Center and its affiliates and data collected on various risk factors, as well as serum and tissue specimens. Controls are other hospitalised men. To date 100 cases and 180 controls have been collected. Case-control comparisons are done overall and by ethnic group.

UNITED STATES OF AMERICA

TYPE: Case-Control
TERM: Ethnic Group; Race
SITE: Prostate
TIME: 1986 - 1995

1063   Pasternack, B.S.   01874
New York Univ. Medical Center, Dept. of Environmental Medicine, 550 1st Ave., New York NY 10016, United States of America
COLL: Shore, R.E.; Toniolo, P.G.; Levitz, M.; Strax, P.; Bruning, P.F.

Endocrine and Environmental Factors in Breast Cancer

This study comprises a prospective investigation of endocrinological factors and dietary components involved in initiation or promotion of breast cancer. Some 15,000 women, both premenopausal and postmenopausal will have a complete breast examination, including questionnaire, clinical examination and mammography. Annual blood samples taken from these women will be frozen and stored. Pre-menopausal women are given menstrual cycle recording calendars, which will allow each blood sample to be dated relatively to the end of the current menstrual cycle and provide evidence of extra-short or extra-long cycles. Telephone interviews are conducted on women who develop breast cancer and their matched controls to obtain more detailed medical, reproductive and socio-economic information. Subsequently, stored blood samples for the matched sets are retrieved for laboratory analyses. To date, over 14,000 women have been enrolled and more than 180 invasive cases have been identified, 55 of which were diagnosed at the time of entry into the cohort and the remaining at or between subsequent screening.

TYPE: Case-Control; Cohort
TERM: BMB; Diet; Drugs; Hormones; Nutrition; Screening
SITE: Breast (F)
CHEM: Oestrogens; Prolactin
TIME: 1984 - 1993

1064   Perera, F.P.   03588
Columbia Univ., School of Public Health, Div. Environmental Health Sciences, 60 Haven Ave. B-109, New York NY 10032, United States of America (Tel.: (212)3053465)
COLL: Santella, R.M; Jeffrey, A.M.; Warburton, D.; Brandt-Rauf, P.W.; Tsai, W.Y.; Pero, R.W.; Hemminki, K.; Walles, S.A.S.; Mayer, J.

Molecular Epidemiology - Chemical Carcinogenesis: Adducts, Oncogenes and Cytogenetic Markers.

The objective of these studies is to develop, validate and apply a battery of biologic markers of carcinogenesis in human subjects. Several related studies of model populations (lung cancer patients, exposed workers, groups with community exposure, tobacco smoke exposed children, chemotherapy patients, coal-tar treated patients) and appropriate controls are being carried out. A battery of measurements (DNA and protein adducts, sister chromatid exchange chromosomal aberrations, oncogene activation and mutation) is being obtained from the same biological samples along with detailed health and exposure histories. Papers have been published in J. Natl. Cancer Inst. 79:449-456, 1987 and in Cancer Res. 48:2288-2291, 1988 and 49:4446-4451, 1989.

TYPE: Correlation; Methodology; Molecular Epidemiology
TERM: Biochemical Markers; Chemical Exposure; Chromosome Effects; DNA Adducts; DNA Binding; Genetic Markers; Passive Smoking; Protein Binding; SCE; Tobacco (Smoking)
SITE: All Sites; Lung
TIME: 1983 - 1993

1065   Roush, G.C.   04251
Cancer Prevention Research Inst., 24 E. 22nd St., New York NY 10010, United States of America
COLL: Barnhill, R.; Hayden, C.L.; Savage, D.

Colorectal Cancer, Acrochordons, and Other Markers

This study tests whether acrochordons (skin tags) are associated with colorectal neoplasms. The design involves comparisons among three groups: 150 persons with colorectal cancer, 150 persons without cancer but with varied types of colorectal polyps, and 150 or more non-polyp, non-cancer controls. Eligible persons are whites undergoing colonoscopy at the Yale Endoscopy Clinic. Consent is sought first from the physician and then from the potential respondent. Prior to the clinic visit, a half hour telephone interview is obtained for information on referral pattern, demographic features, exercise and dietary history. At the clinic, usually in the 5 minutes just prior to endoscopy, the gowned respondent is

UNITED STATES OF AMERICA

examined and axillary and neck regions photographed for skin tags and other skin markings. The three groups will be classified according to stage of colorectal cancer, type of polyp and size of polyp. The groups will be compared regarding number, location and size of acrochordons after adjustment for other factors, including referral patterns and prevalence factors. For this primary hypothesis, the statistical power will exeed 80% to detect an odds ratio of 2 or greater. Other hypotheses include dietary factors and exercise. The study will ultimately be linked with biochemical studies of bile acids and bacterial flora. A publication appeared in Cancer Risk and Incidence Trends, Hemisphere, Washington, DC, 1983, pp 88-103.

TYPE: Case-Control
TERM: Diet; Physical Activity; Premalignant Lesion
SITE: Colon; Rectum
TIME: 1986 - 1992

1066 Roush, G.C. 04663
Cancer Prevention Research Inst., 24 E. 22nd St., New York NY 10010, United States of America
COLL: Pero, R.W.; Miller, D.G.; Osborne, M.; Cahan, A.

Enzymatic Measures of Oxidative Stress and Breast Cancer

Enzymatic measures reflecting oxidative stress have been associated with colorectal neoplasia and a similar assay of unscheduled DNA synthesis (UDS) was associated with breast cancer. The primary aim of this study is to examine the hypothesised association between breast cancer and inter-related enzymatic measures of oxidative stress in a case-control study of 440 women (160 breast cancers, 160 benign breast disease controls without atypia, 120 non-neoplasia controls). All participants provide an interview for risk factors and venous blood for assays. Overall statistical power is 85-99 % (2-sided $p < 0.05$) to detect odds ratios of 2.5 to 3.0. Sex hormones might distort the relationship of these markers to breast cancer, and the UDS marker may help avoid this problem via its non-chemical induction of UDS by UV light.

TYPE: Case-Control; Methodology
TERM: BMB; Biochemical Markers; DNA Repair; Enzymes; Hormones
SITE: Breast (F)
TIME: 1989 - 1992

1067 Roush, G.C. 04857
Cancer Prevention Research Inst., 24 E. 22nd St., New York NY 10010, United States of America
COLL: Pero, R.W.

Enzymatic Measures of Oxidative Stress and Breast Cancer

Unscheduled DNA synthesis and ADPRT are two biochemical measures in peripheral mononuclear leucocytes which have been strongly associated with risk for large bowel neoplasms and with lung cancer by Pero and colleagues. In this study, an evaluation of the association between breast cancer and these two biochemical measures will be undertaken. Breast cancer cases will be obtained from New York University Medical Center and controls will include benign breast disease and other controls from the same clinic population. The study will undertake a careful methological sampling, standardization of transport of specimens, and interview for variables including reproductive factors, body weight, family history and dietary variables. It is hoped to achieve a total sample size of approximately 450 individuals with approximately 200 breast cancer cases.

TYPE: Case-Control
TERM: Biochemical Markers; DNA Repair; Diet; Familial Factors; Reproductive Factors
SITE: Breast (F)
TIME: 1990 - 1992

*1068 Toniolo, P.G. 05129
New York Univ. Medical Center, Dept. of Environmental Epidemiology, 341 E. 25th St., New York NY 10010, United States of America (Tel.: (212)2636500; Fax: 2638570)
COLL: Pasternack, B.S.; Shore, R.E.

Mammographic Parenchymal Patterns and Breast Cancer Risk

This case-control study nested within a cohort aims to explore the relationship between mammographic parenchymal patterns, the dietary intake of fat, serum levels of endogenous reproductive hormones (oestrogens, androgens and prolactin) and the subsequent risk of breast cancer. The study is prospective in design, in that parenchymal patterns, dietary data, and endocrine function are assessed

UNITED STATES OF AMERICA

prior to breast cancer detection, often years in advance. Information is being collected on approximately 250 histologically confirmed breast cancer cases, diagnosed between 1985 and 1992, and on 650 individually matched controls, selected at random within risk sets defined by matching criteria of age, duration of storage of serum specimens, number of blood donations, and menopausal status from among the members of the same cohort from which the cases arose. The study requires consensus classification of mammograms according to Wolfe's patterns and percent densities by two radiologists. Extensive background information on potential confounders is available.

TYPE: Case-Control; Cohort
TERM: BMB; Diet; Fat; Hormones; Mammography; Serum
SITE: Breast (F)
TIME: 1991 - 1994

1069 Wallach, R.C. 01194
New York Univ. School of Medicine, 530 First Ave.., New York NY 10016, United States of America (Tel.: (212)2635199)

Risk Factors for Ovarian Cancer
Failure to develop techniques for early diagnosis of ovarian cancer leads to the late presentation of most cases at a time when cure is unlikely. No significant major risk factors have been identified in the genesis of ovarian cancer. Patients with ovarian cancer are identified and their records are studied in an attempt to correlate factors in personal, medical, social, occupational and other environmental conditions that may exert a carcinogenic effect related to the development of ovarian cancer.

TYPE: Case Series
TERM: Clinical Records; Occupation; Reproductive Factors; Socio-Economic Factors
SITE: Ovary
TIME: 1975 - 1995

1070 Wynder, E.L. 01900
American Health Foundation, Div. of Epidemiology, 320 E. 43rd St., New York NY 10017, United States of America (Tel.: (212)9531900)
COLL: Kabat, G.C.; Moore, M.; Harris, R.E.; Hebert, J.R.; Spritz, N.; Wood, W.; Colberg, J.; Levin, R.; Stolley, P.D.; Rai, K.R.; Grunwald, H.

Epidemiology of Smoking-Related Diseases
The objective of this case-control study is to identify and examine the effect of risk factors associated with cigarette smoking, plus confounding factors, on lung, bladder, pancreatic and kidney cancer, myocardial infarction and cancers of the upper respiratory and upper alimentary tracts. The effects of smoking cigarettes with varying levels of tar and nicotine on the risk of disease will be measured and the use of the newer "less hazardous" cigarettes monitored. Questionnaires are administered to hospitalized patients with an index disease, and to hospitalized controls in 18 US hospitals. A total of 1,250 cases and an equal number of hospital controls are interviewed annually. The major parameters of tobacco usage investigated include: (1) type, duration, and amount of tobacco used by brand, (2) tar and nicotine yields of cigarette smoked, (3) inhalation practices and (4) length of cigarette smoked. These data are analysed in relation to various demographic characteristics, and risk ratios are calculated. One of the fundamental purposes of the study is to determine the risk of the various tobacco-related diseases for individuals with long histories of smoking filter cigarettes in general and low tar/low nicotine yielding cigarettes in particular. A secondary purpose is to investigate the role of nutritional factors in the aetiology of these diseases. Recent papers appeared in Cancer 62:1223-1230, 1988; Int. J. Cancer 43:190-194 and 42:325-328, 1988 and Cancer Res. 48:4405-4408, 1988.

TYPE: Case-Control
TERM: Drugs; Tobacco (Smoking)
SITE: Bladder; Kidney; Larynx; Lung; Oesophagus; Oral Cavity; Pancreas
CHEM: Carbon Monoxide; Nicotine; PAH; Tars
TIME: 1982 - 1992

1071 Wynder, E.L. 03924
American Health Foundation, Div. of Epidemiology, 320 E. 43rd St., New York NY 10017, United States of America (Tel.: (212)9531900)
COLL: Kabat, G.C.

UNITED STATES OF AMERICA

Environmental Tobacco Smoke in Relation to Lung Cancer in Non-Smokers
The purpose of this case-control study is to determine whether lung cancer in non-smokers is associated with exposure to second hand tobacco smoke. Detailed information will be collected on lifetime exposure to other people's tobacco smoke from 180 newly diagnosed lung cancer patients who have never smoked and from a comparison group of 540 lifetime non-smokers. Major components of environmental tobacco smoke (ETS) exposure addressed in the questionnaire include exposure in childhood, in adulthood, at home, at work, in transportation, and in public places. For each member of the household (both in childhood and adulthood) who smoked, the average number of hours per day of exposure, the number of years of exposure, and the quality (intensity) of exposure are asked. Lifetime non-smoking status of study subjects as well as the smoking habits of the current spouse will be validated by reviewing the hospital chart and contacting the spouse. The effect of ETS exposure will be assessed individually for the major components of childhood and adulthood. In addition, a cumulative exposure index will be devised which takes into account exposure from all sources. Information will be collected on other factors which may affect the risk of lung cancer in non-smokers, i.e., occupation, diet, residence, and past exposure to therapeutic radiation. These factors will be controlled for in the analysis. The effect of ETS exposure will be assessed for all histological types of lung cancer combined, as well as for the major specific types (i.e. squamous cell and adenocarcinoma). An effort will be made to obtain information on the location of the lesion within the lung (i.e. central/peripheral) to assess whether it is associated with ETS exposure.

TYPE: Case-Control
TERM: Diet; Environmental Factors; Histology; Occupation; Passive Smoking; Radiation, Ionizing
SITE: Lung
TIME: 1985 - 1992

NEWARK

1072 Mashberg, A. 02582
Univ. of Medicine and Dentistry, New Jersey Medical School, Div. of Surgical Oncology, 185 S. Orange Ave., Newark NJ 07103, United States of America (Tel.: (201)6761000/1627; Fax: 6767752)
COLL: Boffetta, P.

Alcohol and Oral Squamous Carcinoma
The object of the final segment of this study is to evaluate the carcinogenic effects of alcohol and tobacco on specific anatomical sites of the oral cavity/oropharynx in a population of heavy drinkers and cigarette smokers. Series of 353 oral cavity cancer male patients with 347 invasive cancers and 64 carcinomas in situ, collected at a New Jersey Veteran Administration Medical Center, were interviewed with respect to tobacco smoking and alcohol drinking. Sites of origin of cancers are classified as floor of the mouth (N = 144), oral tongue (N = 51), anterior tonsillar pillar (N = 46), soft palate (N = 41), lingual aspect of retromolar trigone (N = 10), alveolar ridge, buccal mucosa and hard palate (less than 10 cases). The remaining 51 cases have cancers from multiple sites. Odds ratios of tobacco and alcohol consumption will be calculated by comparing each site with floor of the mouth. Preliminary analysis suggests that tobacco smoking appears more strongly associated with cancers of soft palate, retromolar trigone and oral tongue, whereas alcohol seems to exert a stronger carcinogenic effect on the floor of the mouth and oral tongue.

TYPE: Case-Control
TERM: Alcohol; Anatomical Distribution; High-Risk Groups; Tobacco (Smoking)
SITE: Oral Cavity; Oropharynx
TIME: 1981 - 1992

OAKLAND

1073 Friedman, G.D. 00851
Kaiser Permanente Medical Care Program, of Northern California,, Div. of Research, 3451 Piedmont Ave., Oakland CA 94611, United States of America (Tel.: (415)4286715)
COLL: Selby, J.V.

Surveillance for Drugs that May be Carcinogenic
The objective is to evaluate medications for possible carcinogenic effects. To seek hypotheses, 143,000 persons will be classified according to use and non-use of various medications by means of computer-stored records collected from 1969-1973 by the San Francisco Kaiser-Permanente Clinic

UNITED STATES OF AMERICA

(1.3 million prescriptions for 3,400 drugs). The development of various kinds of cancer will be followed up by surveillance of hospital records. Drug-cancer associations found would be subject to further study. More detailed evaluation of hypotheses are by case-control studies using the Kaiser-Permanente records, with attention to potential confounding variables. Papers were published in Int. J. Epidemiol. 15:424-426, 1986; Int. J. Cancer 41:677-682, 1988 and in Cancer Res. 49:5736-5747, 1989.

TYPE: Cohort
TERM: Clinical Records; Drugs; Monitoring; Record Linkage
SITE: All Sites
TIME: 1977 - 1992

1074 Friedman, G.D. 01977
Kaiser Permanente Medical Care Program, of Northern California,, Div. of Research, 3451 Piedmont Ave., Oakland CA 94611, United States of America (Tel.: (415)4286715)
COLL: Sidney, S.

Surveillance of Health Effects of Potentially Less Hazardous Cigarettes

This is a prospective epidemiological study designed to investigate the health effects of smoking, with particular interest in the recently introduced smoking products which have reduced yields of selected toxic constituents. The differences between health effects of the older and newer smoking products are to be evaluated. A self-administered questionnaire was completed by most Kaiser health plan subjects given a multiphasic health checkup (MHC) at the KFRI Oakland-San Francisco facilities, 1979 - 1986. Approximately 115,000 subjects are involved. The questionnaire asks for details of smoking habits, as well as including more general questions on occupation, demographic/socio-economic characteristics, and dietary practices. Questionnaire responses, in conjunction with mortality and morbidity data (hospitalization, physician visits, days of incapacity) and MHC data, are to be analysed for relationship between smoking habits and adverse health effects. Long-term follow-up is in progress. Publications appeared in West. J. Med. 145:651-656, 1986; Am. J. Epidemiol. 129:1305-1309, 1989, and in Am. J. Public Health 79:1415-1416, 1989.

TYPE: Cohort
TERM: Diet; Occupation; Socio-Economic Factors; Tobacco (Smoking)
SITE: All Sites
TIME: 1979 - 1992

1075 Goldhaber, M.K. 04793
Kaiser Permanente Medical Care Program, of Northern California, Div. of Research, 3451 Piedmont Ave., Oakland CA 94611, United States of America (Tel.: (415)9872188)
COLL: Friedman, G.D.; Armstrong, M.A.; Golditch, I.

Retrospective Follow-up Study of the Sequelae of Tubal Sterilization

This is a study to determine if certain forms of female sterilization are associated with increased incidences of serious disease, including cancer, and death. Subjects for the study (n = 42,922) were all women sterilized in the Kaiser Permanente Medical Care Program in Northern California during 1971-1984, who had given birth to their most recent child in California. Data on age, race, interval since last birth, and parity at the time of the sterilization were extracted from the children's birth certificates and used to match sterilized women to controls (n = 48,215) within the Kaiser membership. Study and control cohorts were computer-linked to cancer registry and hospitalization files at Kaiser Permanente. Relative risks will be calculated by the Cox proportional hazards model for blocked data.

TYPE: Cohort
TERM: Tubal Ligation
SITE: Breast (F); Ovary; Uterus (Cervix); Uterus (Corpus)
TIME: 1986 - 1992

1076 Hiatt, R.A. 04748
Kaiser Permanente Medical Care Program, Div. of Research, 3451 Piedmont Ave. , Oakland CA 94611, United States of America (Tel.: (415)9872109)
COLL: Krieger, N.

Feasibility of a Cohort Study of Women Exposed to Diethylstilbestrol (DES) during Pregnancy

In order to study further the aetiological role of DES in transplacental carcinogenesis, additional cohorts of women exposed during pregnancy are needed. The possibility of developing such a cohort from the medical records of pregnant women who were members of a large pre-paid health plan in Northern

UNITED STATES OF AMERICA

California between 1948 and 1960 is being explored. Feasibility will be assessed by determining the total number of pregnancies, by reviewing the records of a random sample of medical charts for DES exposure and by attempts to follow-up these women and their offspring to 1990.

TYPE: Cohort
TERM: Hormones; Intra-Uterine Exposure
SITE: Breast (F); Vagina
CHEM: DES
REGI: Bay Area (USA)
TIME: 1990 - 1992

1077   Hiatt, R.A.   04131
    Kaiser Permanente Medical Care Program, Div. of Research, 3451 Piedmont Ave., Oakland CA 94611, United States of America (Tel.: (415)9872109)
COLL: Krieger, N.

Benign Breast Disease, Exogenous Hormone Use and Breast Cancer

In order to clarify the relationship between benign breast disease (BBD), exogenous hormones and breast cancer (BC), a three part study has been undertaken using a cohort of approximately 2,500 women with biopsy documented benign breast disease. BC risk in this cohort was previously found to be directly related to the degree of atypia (Cancer 39(6): 2603-2607, 1977). This previous study has been updated by determining all additional BCs that have occurred in this group through 1985 and making some corrections in the dataset. In a nested case-control study, BC cases will be compared with women with BBD and without cancer for hormone use after the diagnosis of BBD as documented in their medical records. Third, in a second nested case-control study, hormone use before BBD will be examined among a sub-sample of women with BBD and matched health plan controls. (Am. J. Epidemiol. 126: 735, 1987.)

TYPE: Case-Control; Cohort
TERM: Drugs; Hormones; Premalignant Lesion
SITE: Benign Tumours; Breast (F)
CHEM: Oestrogens
TIME: 1986 - 1992

*1078   Hiatt, R.A.   05234
    Kaiser Permanente Medical Care Program, Div. of Research, 3451 Piedmont Ave., Oakland CA 94611, United States of America (Tel.: (415)9872109)

Renal Cell Cancer and Use of Thiazide

Previous studies have suggested that thiazide use may increase the risk of renal cell carcinoma. These studies need to be confirmed and, in addition, data are needed to assess duration, dose and any effect of the indication for thiazide use. The current case-control study selected all cases of renal cancer among persons who had a multiphasic health check-up while members of a large pre-paid health plan from 1964-1988. Controls also took such an examination and were matched on age, sex, and year of examination. Data on body mass index, smoking and alcohol use, coffee and tea consumption, race/ethnicity and family history of cancer will be taken from the answers to questions recorded at the time of examination. Medical record reviewers will abstract information on thiazide use.

TYPE: Case-Control
TERM: Alcohol; Clinical Records; Coffee; Diuretics; Race; Registry; Tea; Tobacco (Smoking)
SITE: Kidney
REGI: Bay Area (USA); California (USA)
TIME: 1991 - 1992

1079   Selby, J.V.   04784
    Kaiser Permanente Medical Care Program, Div. of Research, 3451 Piedmont Ave., Oakland CA 94611, United States of America (Tel.: (415)9872106; Fax: 9873027)
COLL: Friedman, G.D.; Quesenberry, C.P.

Case-Control Evaluations of Screening Tests for Colorectal Cancer

This three-year study evaluates the efficacy of digital rectal examination, sigmoidoscopy, and faecal occult blood testing for preventing mortality from colorectal cancer. In separate studies, cases of fatal colorectal cancer within reach of digital examination will be compared with controls who did not die by the date of death of the case, matched to the case for age, sex and length of Program membership.

UNITED STATES OF AMERICA

Exposures to the screening test of interest, other screening tests, and a history of prior colorectal neoplasms or a family history of colorectal cancer will be sought by review of medical records. Estimates of efficacy for the most recent screen will be obtained from the odds ratios for exposure to the test, adjusted for the number of prior tests, for other screens, and for other risk factors. Estimates of the most efficacious screening interval will also be calculated. The number of cases in each study is: sigmoidoscopy, 300 cases, two controls per case; digital rectal examination, 250 cases, one control per case; faecal occult blood, 650 cases, one control per case.

TYPE: Case-Control
TERM: Familial Factors; Prevention; Screening; Time Factors
SITE: Colon; Rectum
TIME: 1989 - 1992

1080 Wong, O. 03549
Environmental Health Associates, 520 Third St., Suite 208, Oakland CA 94607, United States of America (Tel.: (415)4511888)

Prospective Mortality Registry of Refinery Employees in the Petroleum Industry
The objective is to develop and implement a mortality registry of refinery employees in the petroleum industry. In its first year of operation, the registry will consist of 43,000 employees from 53 refineries (13 companies). By its fifth year, it will likely have 55,000 participants. This prospective data base will be used to monitor the cause-specific mortality pattern of petroleum refinery employees.

TYPE: Cohort; Mortality; Registry
TERM: Chemical Exposure; Occupation; Petroleum Products; Registry
SITE: All Sites
OCCU: Petroleum Workers
TIME: 1984 - 1992

1081 Wong, O. 03704
Environmental Health Associates, 520 Third St., Suite 208, Oakland CA 94607, United States of America (Tel.: (415)4511888)

Mortality of Petroleum Industry Employees Exposed to Downstream Gasoline
The objective of this study is to determine whether employees exposed to downstream gasoline in the petroleum industry are at higher risk of mortality from malignant and non-malignant diseases, in particular kidney cancer, when compared to a suitable control population. The study will consist of approximately 50,000 employees from five major US petroleum companies, and the study period will cover 1941-1985. Cause-specific SMRs, relative risks and other appropriate statistics will be calculated by exposure, latency, etc.

TYPE: Cohort
TERM: Chemical Exposure; Latency; Occupation; Petroleum Products
SITE: Kidney
CHEM: PAH
OCCU: Petroleum Workers
TIME: 1986 - 1993

OLYMPIA

1082 Milham Jr, S. 04908
Washington State Dept. of Health, ED-13, Olympia WA 98504, United States of America (Tel.: (206)1536408)

Washington State Occupational Surveillance System
The occupational and cause of death information for 600,000 male deaths 1950-1989 and 100,000 female deaths 1974-1989 will be analysed using an age- and year of death- standardized proportionate mortality ratio programme. Detailed cause of death analysis (160 causes) will be published for 220 occupational categories for males and 50 categories for females. Previous analyses of this data set detected a multiple myeloma excess in nuclear workers, a lung cancer excess in copper smelter workers and an excess of leukaemia and non-Hodgkin's lymphoma in electrical workers.

UNITED STATES OF AMERICA

TYPE: Mortality
TERM: Occupation
SITE: All Sites
TIME: 1968 - 1992

OMAHA

1083   Purtilo, D.T.   00963
Univ. of Nebraska, Medical Center, Dept. of Pathology and Microbiology, 600 S. 42nd St., Omaha NE 68198-3135, United States of America (Tel.: (402)5594244; Fax: 5596018)

Registry of X-Linked Lymphoproliferative Disease
This registry is evaluating males suspected of having an X-linked lymphoproliferative disease characterized by immune deficiency to EBV and expressed phenotypically by fatal infectious mononucleosis, virus-associated hemophagocytic syndrome or agammaglobulinaemia or malignant B-cell lymphoma following EBV infection. The males show defective antibody responses to EBV-specific antigens and also cellular immune functions. The registry laboratory is able to detect 95% of female carriers by noting linkage to the XLP locus with DXS42 and DXS37, using restriction fragment polymorphisms of these known genes in the X-chromosome. The patients have immune regulatory disturbance with excessive suppressor activity resulting in agammaglobulinaemia and defective recognition functions resulting in the lymphoproliferative phenotypes. Ig therapy is provided as prophylaxis against EBV infection. Papers have been published in Cancer Genet. and Cytogenet. 47:163-169, 1990 and 51:143-153, 1991 and in Immunol. 83-10-16, 1991.

TYPE: Registry
TERM: EBV; Genetic Markers; High-Risk Groups; Immunodeficiency; Premalignant Lesion; RFLP
SITE: Lymphoma
TIME: 1975 - 1999

PHILADELPHIA

1084   Blumberg, B.S.   00351
Vice President for Population Oncology, Fox Chase Cancer Center, 7701 Burholme Ave. , Philadelphia PA 19111, United States of America (Tel.: (215)7282203)
COLL: London, W.T.; Millman, I.; Lustbader, E.; McGlynn, K.; Murphy, E.D.; Venkateswaran, P.

Prevention of Primary Cancer of the Liver: Amelioration of the Carrier State for Hepatitis B Virus
During the past 12 years, a substantial body of evidence has accumulated to support the hypothesis that persistent infection with HBV is required for the development of primary hepatocellular carcinoma, and extensive local, regional and national programmes are now in progress or being planned for protection against infection. In 1969, a vaccine against HBV was introduced here and this is now in general production and widely available. A programme of testing and vaccination in the Asian population of the Philadelphia area (approximately 100,000 individuals) is now being conducted. Pregnant women who are carriers of HBV are identified, and hepatitis B immune globulin and vaccine offered to their newborn children. Other susceptible members of these populations are also vaccinated. Current work is directed towards elimination of the carrier state or to a decrease in replication of the hepatitis virus, so that infectiousness is decreased or eliminated, and the long term outcome in terms of development of chronic liver disease and primary cancer of the liver improved. To this end, new chemotherapies to decrease viral replication and/or eliminate infection are being investigated. In vitro animal and human studies are in progress or planned. 5,000 Asian chronic carriers in the Philadelphia region are being recruited into a cohort to be followed for five years. Carriers will be monitored for elevations of AFP in order to detect liver cancers at an early treatable stage. Risk factors for liver cancer other than chronic infection with HBV will be evaluated. The natural history of chronic HBV infection will be described.

UNITED STATES OF AMERICA

TYPE: Cohort; Intervention
TERM: Asians; Biochemical Markers; Chemotherapy; Environmental Factors; HBV; High-Risk Groups; Prevention
SITE: Liver
TIME: 1982 - 1992

*1085 Bunin, G.R. 05046
Children's Hosp. of Philadelphia, 34th and Civic Center Blvd, Philadelphia PA 19104, United States of America (Tel.: (215)5901445)
COLL: Buckley, J.D.; Hammond, D.; Ruccione, K.; Sather, H.N.; Woods, W.G.

Parental Occupation and Childhood Cancer
The Children's Cancer Study Group is conducting a hypothesis-generating case-control study of parental occupation and all major types of childhood cancer. The study uses data on parental occupation from a self-administered questionnaire. The cancers to be studied and the approximate number of cases to be included are: acute lymphocytic leukaemia 1500, acute non-lymphocytic leukaemia 200, Hodgkin's disease 200, non-Hodgkin's lymphoma 200, neuroblastoma 300, Wilm's tumour 200, osteosarcoma 100, astrocytoma 100, primitive neuroectodermal tumour 100, rhabdomyosarcoma 100, and germ cell and gonadal tumour 90. Data on 800 controls selected by random-digit dialling are also available. The occupational exposures of interest include exposure to paints, solvents, metals, and pesticides.

TYPE: Case-Control
TERM: Metals; Occupation; Paints; Parental Occupation; Pesticides; Solvents
SITE: Childhood Neoplasms
OCCU: Farmers; Metal Workers; Welders
LOCA: Canada; United States of America
TIME: 1990 - 1992

*1086 Bunin, G.R. 05047
Children's Hosp. of Philadelphia, 34th and Civic Center Blvd, Philadelphia PA 19104, United States of America (Tel.: (215)5901445)
COLL: Yandell, D.; Meadows, A.T.

Correlation of Mutations in the Retinoblastoma Gene with Parents' Exposures
This pilot study will investigate the possible association of specific exposures that may increase the risk of retinoblastoma with specific types of alterations in the retinoblastoma gene. 10-50 cases from an epidemiological study of retinoblastoma will be studied. The cases to be included are those without a family history who had an exposure that was associated with increased risk. DNA from the case, the tumour, and the parents of the case will be analysed by molecular biological techniques to locate and characterise the mutation in the retinoblastoma gene. The type and location of mutations in exposed cases will be compared to those in unexposed cases.

TYPE: Molecular Epidemiology
TERM: DNA; Mutation, Germinal; Parental Occupation
SITE: Retinoblastoma
OCCU: Metal Workers; Welders
LOCA: Canada; United States of America
TIME: 1991 - 1993

1087 London, W.T. 02480
Fox Chase Cancer Center, 7701 Burholme Ave., Philadelphia PA 19111, United States of America (Tel.: (215)7282204; Fax: 7283574)
COLL: Buelow, K.B.; MacMahon, B.; Redeker, A.; Smith, M.

Relationship of HBV and Genetic Changes to Aetiopathogenesis of Hepatocellular Carcinoma
Chronic infection with HBV is associated with 80% of hepatocellular carcinoma (HCC) cases worldwide. The precise role of HBV in pathogenesis of HCC is not understood. The aim of this study is to examine the relationship of HBV DNA integration to genetic changes in HCC cells. Tumours will be identified prospectively in a cohort of 5,000 HBV carriers in the Philadelphia area, in a cohort of 1,500 carriers in Alaska and in referred cases in Los Angeles. Status of HBV DNA in tumour and non-tumour tissues will be compared as will the pattern on southern blots of DNA probes for chromosomal regions suspected of containing genes critical to HCC development (i.e. 4, 11, 13). Specifically, loss of heterozygosity in

UNITED STATES OF AMERICA

tumour tissues compared with normal tissues at a rate significantly above background will be considered as genetic changes in the pathogenetic pathway.

TYPE:	Cohort
TERM:	DNA; HBV; Oncogenes
SITE:	Liver
TIME:	1987 – 1992

1088   Meadows, A.T.   04309
Children's Hosp. of Philadelphia, Div. of Oncology, 34th and Civic Center Blvd , Philadelphia PA 19104, United States of America (Tel.: (215)5902804)
COLL: Bunin, G.R.; Nass, C.

Epidemiology of Childhood Gliomas

This case-control study is an investigation of the role of gestational and preconception exposures, especially those occurring through parental occupation, in the aetiology of childhood gliomas. The specific hypothesized risk factors include solvents, metals, N-nitroso containing substances, alcohol and neurally active drugs. Cases are identified through paediatric medical centres in Pennsylvania, New Jersey, and Delaware and must be less than 15 years old at diagnosis. Controls matched to cases by race and birthdate (within 2 years) are obtained through random digit dialling. Parents of cases and controls are interviewed on the telephone. The expected sample size is 200 case – control pairs.

TYPE:	Case-Control
TERM:	Alcohol; Chemical Exposure; Childhood; Drugs; Metals; Parental Occupation; Solvents
SITE:	Brain
CHEM:	N-Nitroso Compounds
TIME:	1987 – 1992

1089   Meadows, A.T.   00318
Children's Hosp. of Philadelphia, Div. of Oncology, 34th and Civic Center Blvd , Philadelphia PA 19104, United States of America (Tel.: (215)5902804)
COLL: Banfi, A.; Baum, E.; d'Angio, G.J.; Green, D.; LeMerle, J.; Morris-Jones, P.H.; Nesbit, M.; Newton, W.A.; Obringer, A.; Sallan, S.; Siegel, S.; Voûte, P.; Woods, W.G.; Zipursky, A.

Second Malignant Neoplasms after Childhood Cancer

The Late Effects Study Group registers second malignant neoplasms (SMN) occurring in individuals treated for childhood cancer at any of the 13 member institutions and surviving for at least two years after diagnosis. The registry, which now includes over 300 individuals, serves as a resource for studying many aspects of SMN: (1) incidence; (2) histopathology; (3) identification of risk factors such as treatment and genetic factors; (4) unusual tumour associations and (5) congenital anomalies associated with SMN. A recent paper described 91 patients who developed bone sarcomas as SMNs (Cancer 67:193-201, 1991).

TYPE:	Registry
TERM:	Chemotherapy; Congenital Abnormalities; Genetic Factors; Late Effects; Multiple Primary; Mutation, Somatic; Radiotherapy
SITE:	Childhood Neoplasms
LOCA:	Canada; France; Italy; Netherlands; United Kingdom; United States of America
TIME:	1972 – 1995

1090   Meadows, A.T.   04310
Children's Hosp. of Philadelphia, Div. of Oncology, 34th and Civic Center Blvd , Philadelphia PA 19104, United States of America (Tel.: (215)5902804)
COLL: Bunin, G.R.; Albright, L.A.; Allen, J.C.; Boessel, J.; Buckley, J.D.; Neuberg, R.W.; Rorke, L.B.; Shiminski, T.; Smithson, W.A.; Deutsch, M.

Brain Tumours in Young Children

The role of genetic predisposition, preconception exposure to mutagens, and gestational exposure to carcinogens in the aetiology of astrocytoma and primitive neuroectodermal tumour medulloblastoma is being investigated. As these factors are likely to influence the development of brain tumour early in life, the study is restricted to patients diagnosed at ages 0-5. Specifically, family history of seizures and cancer, preconception exposures through parents occupation, gestational exposure to alcohol, medication and nitrosamines, are being examined. Cases are identified through the Childrens Cancer Study Group, a multi-institution cooperative group conducting clinical trials. The group consists of 34

UNITED STATES OF AMERICA

member institutions in the US and Canada. Controls matched to cases on race and birthdate (+ 1 year) are obtained by telphone random digit dialing. Parents of cases and controls are interviewed by telephone. Data on 330 case-control pairs have been collected. Separate analyses for astrocytoma and primitive neuroectodermal tumour/medulloblastoma will be conducted.

TYPE: Case-Control
TERM: Alcohol; Childhood; Familial Factors; Intra-Uterine Exposure; Metals; Occupation; Solvents
SITE: Brain; Childhood Neoplasms
CHEM: N-Nitroso Compounds
LOCA: Canada; United States of America
TIME: 1987 - 1992

1091 Strom, B.L. 03382
Univ. Pennsylvania, School of Medicine, Clinical Epidemiology Unit, Department of Medicine, Room 225L NEB/S2, Philadelphia PA 19104, United States of America (Tel.: (215)8984623)
COLL: Soloway, R.D.; Stolley, P.D.; West, S.; Litvak, J.; Ríos-Dalenz, J.L.; Rodriguez-Martinez, H.

Biochemical Epidemiology of Biliary Tract Cancer

Biliary tract cancer, uncommon in the US but of considerable international importance, provides a unique opportunity to combine biochemistry and epidemiology to shed light on the possible aetiology. The marked differences in incidence between countries have led to many hypotheses about aetiology, including biochemical hypotheses arising out of preliminary work by the investigators: marked differences have been demonstrated in the bile and gallstone constituents of normal, cholelithiasis, and cancer patients in Bolivia and US. In order to further investigate the risk factors and aetiology of biliary tract cancers, a case-control study will examine subjects collected from two collaborating Latin American centres: La Paz, Bolivia and Mexico City, Mexico. Three case groups will be recruited: patients undergoing abdominal surgery who have newly-diagnosed (1) cancer of the gallbladder, (2) cancer of the extrahepatic biliary tract, and (3) cancer of the ampulla of Vater. Each group of cases will be compared to two set of controls: (1) abdominal surgery patients with cholelithiasis and (2) abdominal surgery patients who have no biliary tract disease. Personal interviews will be used to obtain demographic characteristics, prior medical history, family history, and exposure to agents presumed to be risk factors for biliary cancer. Bile, blood, and, where appropriate, gallstone specimens will be obtained from cases and controls during surgery and analysed for various biochemical parameters. 55 cases and 440 controls are expected annually.

TYPE: Case-Control
TERM: Biochemical Markers; Gallstones
SITE: Bile Duct; Gallbladder
LOCA: Bolivia; Mexico; United States of America
TIME: 1984 - 1992

1092 Weiss, W. 00021
Hahnemann Univ., 3912 Netherfield Rd, Philadelphia PA 19129, United States of America (Tel.: (215)8494971)

Respiratory Effects of Chloromethyl Ethers

A prospective study of 125 male workers at a chemical plant by periodic chest X-rays and questionnaires was conducted from 1963 to 1968. In 1965 a spirogram was done in 103 men. Additional follow-up has been continued to the present. 94 men exposed to chloromethyl methyl ether containing bis(chloromethyl)ether as a contaminant were divided into an unexposed and three exposed groups according to the magnitude of individual exposure indexes. There was a dose-response relationship between cumulative chemical exposure and the prevalence of chronic cough and expectoration, and the incidence of bronchogenic carcinoma. There was an inverse relationship between smoking habits and the risk of lung cancer. Of 94 exposed men, 22 have developed lung cancer (15 small cell). Only 1 of 31 unexposed men has developed lung cancer, not small cell. The shape of the epidemic curve has been investigated graphically and by calculating period SMRs. Since 1975-1979 the curve has been descending. Recent analysis shows that the way the cohort was assembled in 1963 resulted in a biased sample due primarily to inadvertent selection of men with higher exposures. Follow-up continues. Papers have been published in J. Occup. Med. 22:527-529, 1980; J. Nat. Cancer Inst. 69:1265-1270, 1982 and in J. Occup. Med. 31:102-105, 1989.

UNITED STATES OF AMERICA

TYPE: Cohort
TERM: Chemical Exposure; Dose-Response; Histology; Occupation; Tobacco (Smoking)
SITE: Lung
CHEM: BCME; CMME
TIME: 1963 - 1993

PITTSBURGH

1093 Grufferman, S. 04136
Univ. of Pittsburgh, School of Medicine, Dept. Clin. Epidemiol. & Prev. Medicine, M-200 Scaife Hall, Pittsburgh PA 15261, United States of America (Tel.: (412)6488933; Fax: 6489114)
COLL: Olshan, A.F.; Gula, M.J.

Hodgkin's Disease in Childhood

This is the first interview case-control study of childhood Hodgkin's disease (HD). Environmental exposures, particularly likelihood of exposure to infectious agents during childhood, appear to play an important aetiological role in young adult HD. The aim is to study environmental exposures as risk factors for childhood HD, by evaluating whether childhood HD is epidemiologically distinct from the young adult and old adult diseases. The risk factors under study are: patterns of infectious diseases, day-care, breast feeding, and other determinants of risk of childhood infectious diseases; socio-economic status; parental occupational exposures; wood and chemical exposures of the child; familial aggregation of HD, and increased risks of other malignancies in case families. This study uses methods developed for a collaborative therapeutic clinical trials group for aetiological research, which greatly facilitates studies of extremely rare cancers, such as childhood HD, by permitting relatively rapid accrual of sufficient numbers of subjects. All cases have their diagnoses confirmed by central expert pathological review. An expected total of 300 new cases aged 0-14 years from all geographical areas of the US will be obtained over three years from the Pediatric Oncology Group and the Children's Cancer Study Group which are national collaborative clinical trials groups. Case and control parents will be interviewed by telephone after consent is obtained. For each case under age 10 years, three controls of the same race, sex and age will be selected from the same communities by random-digit dialling. For cases aged 10-14, one control per case will be selected. An estimated 92 cases will be under the age of 10 and 208 cases will be in the age-group 10-14, so that 484 controls are expected.

TYPE: Case-Control
TERM: Chemical Exposure; Childhood; Environmental Factors; Familial Factors; Infection; Socio-Economic Factors
SITE: Hodgkin's Disease
TIME: 1987 - 1993

1094 Grufferman, S. 04534
Univ. of Pittsburgh, School of Medicine, Dept. Clin. Epidemiol. & Prev. Medicine, M-200 Scaife Hall, Pittsburgh PA 15261, United States of America (Tel.: (412)6488933; Fax: 6489114)
COLL: Bontempo, L.C.; Rinaldo, C.; Ndimbie, O.K.

Epidemiological Studies of HIV-Associated Malignancies

This study will investigate the distribution and determinants of cancer occurrence in a defined population of homosexual men infected with HIV or at high risk of HIV infection. Subjects will be the 4,955 homosexual men enrolled in the Multicenter AIDS cohort Study (MACS) plus a supplemental cohort currently being enrolled. This population is drawn from four different regions of the US and has been intensively followed for six years. In addition to extensive longitudinal questionnaire information regarding lifestyle, drug use, sexual practices, past medical history, AIDS therapy, etc., a bank of biological specimens (white cells, sera, semen, urine, saliva, etc.) collected at frequent intervals from these subjects has been maintained. Of this cohort, nearly 2,000 were HIV-antibody positive at time of entry and approximately 390 of the initially sero-negative subjects have sero-converted to HIV-positive. To date, 359 cancers have occurred in the total cohort. The design of the MACS allows for the performance of cohort study analyses for those variables on which data have been collected on all subjects (i.e., lifestyle factors, T4/T8 ratios, etc.) . Additionally, the banking of biological specimens allows for the use of nested case-control study analyses for virological and other laboratory studies. Autopsies will be performed on subjects dying of AIDS to assess the prevalence of occult malignancies. A tumour bank will be assembled and maintained for use in this and future research. The data generated from this research will be merged with the other MACS data so that comprehensive analyses can be done.

UNITED STATES OF AMERICA

TYPE: Cohort
TERM: AIDS; BMB; Drugs; HIV; High-Risk Groups; Homosexuality; Lifestyle; Saliva; Semen; Serum; Sexual Activity; Tissue; Urine; White Cells
SITE: All Sites
TIME: 1988 - 1992

1095   Marsh, G.M. 03562
Univ. of Pittsburgh, Graduate School of Public Health, Dept. Biostat. & Environm. Epidemiology, 130 De Soto St., Pittsburgh PA 15261, United States of America (Tel.: (412)6243032)
COLL: Enterline, P.E.

Mortality Surveillance in Fibreglass and Mineral Wool Workers

This is an update and extension of a previous historical-prospective mortality study by the same investigators, of 15,016 fibreglass production/maintenance workers fromm 11 plants and 1,874 mineral wool workers from six plants, who were employed a year or more (six months for two plants) between 1 January 1945 and 31 December 1963 (1 January 1940-1963 for one plant). The current study will include male and female workers hired until 1978 and will examine the mortality experience of the cohort ultimately until 1989. The primary objective is to determine whether exposures associated with the production of man-made mineral fibres are associated with excess mortality risks due to malignant or non-malignant respiratory disease. The current study will also include the collection of data on smoking habits and an expanded assessment of historical environment exposures in the plants.

TYPE: Cohort
TERM: Chemical Exposure; Dusts; Environmental Factors; Occupation; Tobacco (Smoking)
SITE: Respiratory
CHEM: Glass Fibres; Mineral Fibres
OCCU: Mineral Fibre Workers
TIME: 1987 - 1992

1096   Marsh, G.M. 04138
Univ. of Pittsburgh, Graduate School of Public Health, Dept. Biostat. & Environm. Epidemiology, 130 De Soto St., Pittsburgh PA 15261, United States of America (Tel.: (412)6243032)
COLL: Leviton, L.C.; Talbott, E.O.

Health Registry Study of Drake (Kilsdonk) Chemical Company

The Drake "Superfund" waste site is located in Lock Haven, Pennsylvania, on the site of the former Drake Chemical Company. Drake and its predecessors produced, used or stored many chemicals including beta-naphthylamine, benzidine and benzene. The EPA has been involved in the clean-up of the waste site since February 1982. Several studies have revealed excess bladder cancer rates among males in the Lock Haven area and among former employees of the Drake and predecessor companies. The purpose of the current research is to register and undertake medical surveillance of a cohort of workers known to be at increased risk of bladder cancer due to occupational exposure to beta-naphthylamine. The registry has been designed to: (1) determine the vital status of all former Drake workers (2) determine the current addresses of all living cohort members, (3) notify and enrol as many cohort members as possible, (4) establish and maintain a programme of medical surveillance for bladder cancer, (5) analyse the results of the surveillance programme, (6) evaluate the bladder risk in the cohort and (7) evaluate the total and other cause-specific mortality risks in the cohort.

TYPE: Cohort
TERM: Analgesics; Chemical Exposure; Occupation; Solvents
SITE: All Sites; Bladder
CHEM: 2-Naphthylamine; Benzene; Benzidine
TIME: 1986 - 1992

1097   Marsh, G.M. 03660
Univ. of Pittsburgh, Graduate School of Public Health, Dept. Biostat. & Environm. Epidemiology, 130 De Soto St., Pittsburgh PA 15261, United States of America (Tel.: (412)6243032)

Update of the Indian Orchard Plant Mortality Study

This is an update of a previous historical prospective mortality study of 2,479 male hourly employees who worked one or more years between 1 January 1949 and 31 December 1966 at a plastics producing plant in Massachusetts. Vital status was determined for the cohort through 1976. In the original study, comparisons with the local county white male mortality experience revealed a slight excess in digestive system cancer and a statistically significant excess ($P < 0.05$) in genito-urinary cancer. A secondary

UNITED STATES OF AMERICA

nested matched case-control study revealed possible associations between rectal cancer and cellulose nitrate production and between prostatic cancer and polystyrene processing. The original cohort also served as a data base for a proportional mortality study designed to examine the long term health effects of formaldehyde exposure. Reports of the original findings have been published (Br. J. Industr. Med. 39:313-322, 1982; J. Occup. Med. 25:219-230, 1983). The updating of this cohort through 1983 will add to the number of deaths and the duration of follow-up and will include complete work history information for the entire study population. These new data will allow an historical prospective analysis of total and cause specific mortality patterns relative to specific factors in the occupational environment of these men. The updated study will also include a matched case-control analysis of all respiratory cancer, buccal/pharyngeal cancer deaths, and a prospective assessment of mortality patterns among workers with previous employment in the vinyl chloride polymerization and processing areas.

TYPE: Case-Control; Cohort
TERM: Chemical Exposure; Environmental Factors; Occupation; Plastics
SITE: All Sites
CHEM: Formaldehyde; Hydrocarbons; Hydrocarbons, Halogenated; Nitrates; Styrene; Vinyl Chloride
OCCU: Chemical Industry Workers; Plastics Workers
TIME: 1985 - 1992

1098 Marsh, G.M. 04139
Univ. of Pittsburgh, Graduate School of Public Health, Dept. Biostat. & Environm. Epidemiology, 130 De Soto St., Pittsburgh PA 15261, United States of America (Tel.: (412)6243032)
COLL: Esmen, N.A.

Mortality Patterns among Chemical Plant Workers Exposed to Formaldehyde and Other Substances
In 1986 the National Cancer Institute and Formaldehyde Institute reported on a cohort mortality study of 26,651 workers employed at some time before 1966 in one or more of 10 facilities using formaldehyde in the United States. This study revealed a statistically significant excess of nasopharyngeal cancer (4 observed, less than 1 expected) in a sub-cohort of 4,389 workers from one plant using formaldehyde. It was not, however, possible to draw definitive conclusions concerning this excess due to the small number of deaths and to difficulties in assessing the contribution of extraneous or confounding factors. The aim of the present study is to perform an enhanced, extended, and independent historial-prospective cause-specific mortality study of the Connecticut plant workforce. The cohort will include about 6000 employees hired between 1941 and 1986. Using both external and internal comparisons the mortality experience of the cohort will be examined from 1 January 1945 through 31 December 1984 with emphasis on mortality from malignant neoplasms and particularly nasopharyngeal cancer. This study will test the hypothesis that workers exposed not to formaldehyde alone but to formaldehyde in conjunction with particulates are at an increased risk of developing cancers in the nasopharynx.

TYPE: Cohort
TERM: Chemical Exposure; Occupation; Plastics
SITE: All Sites; Nasopharynx
CHEM: Formaldehyde
TIME: 1986 - 1992

1099 Olshan, A.F. 03770
Univ. of Pittsburgh, Dept. of Clinical Epidemiology, and Preventive Medicine, M200 Scaife Hall, Pittsburgh PA 15261, United States of America (Tel.: (412)6488933; Fax: 6489114)
COLL: Breslow, N.E.; Falletta, J.M.; Grufferman, S.; Pendergrass, T.W.; Robison, L.L.; Strong, L.C.; Waskerwitz, M.; Woods, W.G.

Risk Factors for Wilms' Tumour
The purpose of this study is to examine the relationship between specific environmental exposure and the risk of Wilms' tumour. These exposures include paternal occupational hydrocarbon and lead exposure and maternal exposure to progestins during pregnancy. Information will be collected on additional exposures which may be used to develop new aetiological hypotheses. The environmental factors will be examined within hereditary and non-hereditary sub-groups of Wilms' tumour patients to better understand the role that genetic-environmental interactions play in the development of Wilms' tumour. The nationwide case-control study will identify its cases from the National Wilms' Tumour Study. Approximately 200 cases and 200 matched controls will be obtained for one year of accrual. Controls will be identified using random digit dialling. Information collected from the questionnaire will include: socio-demographic factors, parental occupational and environmental exposures, mothers' exposures during pregnancy, and family medical history.

UNITED STATES OF AMERICA

TYPE: Case-Control
TERM: Chemical Exposure; Environmental Factors; Familial Factors; Heredity; Intra-Uterine Exposure; Occupation; Socio-Economic Factors
SITE: Wilms' Tumour
LOCA: Canada; United States of America
TIME: 1984 - 1992

*1100 Olshan, A.F. 05173
Univ. of Pittsburgh, Dept. of Clinical Epidemiology, and Preventive Medicine, M200 Scaife Hall, Pittsburgh PA 15261, United States of America (Tel.: (412)6488933; Fax: 6489114)
COLL: Grufferman, S.; Pollock, B.H.; Stram, D.; Neglia, J.P.; Seeger, R.; Shah, N.

Risk Factors for Neuroblastoma

This will be the first large case-control interview study of risk factors for childhood neuroblastoma. About 640 cases (under age 19 years at diagnosis) will be identified through the Childrens Cancer Study Group and the Pediatric Oncology Group. Parents of cases and of 640 controls matched for age, sex, and race will be interviewed by telephone. Controls will be selected by random digit dialling. The study will evaluate the relative importance of risk factors for neuroblastoma reported from previous studies. These include: maternal use of medications during pregnancy (specifically alcohol, diuretics and a group of drugs that includes amphetamines, antidepressants, antipsychotics and tranquillizers); parental employment in the electronics industry and occupational exposures to electromagnetic fields; maternal age at birth and the subject's length of gestation and birth weight. The study will also evaluate drugs used by the mother during pregnancy for their potential to cause transplacental exposure of the foetus to N-nitroso compounds. Information on other potential risk factors such as family medical, gestational and delivery history will also be collected. The study will examine the possibility of neuroblastoma having a similar set of risk factors to that of other childhood cancers. Since most patients will have data collected on clinical, biological and genetic markers, potential environmental risk factors will be analysed separately for subgroups of patients defined by such characteristics as disease stage and N-myc oncogene amplification.

TYPE: Case-Control
TERM: Age; Alcohol; Diuretics; Drugs; Electromagnetic Fields; Genetic Markers; Intra-Uterine Exposure; Oncogenes; Parental Occupation; Stage; Time Factors
SITE: Neuroblastoma
CHEM: N-Nitroso Compounds
LOCA: Canada; United States of America
TIME: 1991 - 1996

1101 Redmond, C.K. 00126
Univ. of Pittsburgh, School of Public Health, Dept. of Biostatistics, 130 De Soto St., Pittsburgh PA 15261, United States of America
COLL: Costantino, J.P.; Rockette, H.E.; Mazumdar, S.; LeGasse, A.A.; Bass, G.I.

Cancer Mortality in Steel Workers

The objective of this study is the identification of work areas, occupations, and processes within the steel industry that are associated with unusual mortality from specific diseases indicative of health hazards in the work environment. Two cohorts of steel workers consisting of (1) all male workers employed at seven US Allegheny County steel plants in 1953 and (2) all oven workers employed at twelve coke plants, plus controls, in the USA and Canada in 1951-1955 have been followed for mortality, 1953-1975. Mortality rates for specific causes of death among men employed in various jobs and work areas are compared with rates for the total steel worker population. Reports have been prepared on coke oven workers, crane operators, and open hearth workers, sheet and tin mill workers, and masons. Excess cancer mortality was observed in several work areas including coke ovens (respiratory), foundry (genito-urinary), electric furnace (urinary), sheet and tin mill (lymphatic and haemopoietic) and blacksmith shop (respiratory) (Env. Hlth Perspect. 52:67-73, 1983). Efforts to update until 1982 work histories and vital status of all coke-oven workers and selected portions of the remaining steelworkers cohort until 1982 have been completed. Currently this updated information is being used to develop modelling methodology for carcinogenic risk assessment of exposure to environmental pollutants. Some preliminary work has been published (Am. J. Epidemiol. 128:860-873, 1988 and Environm. Hlth Perspect. 90:271-277, 1991).

UNITED STATES OF AMERICA

TYPE: Cohort
TERM: Metals; Occupation
SITE: All Sites; Genitourinary; Haemopoietic; Lymphoma; Respiratory
CHEM: Steel
OCCU: Coke-Oven Workers; Steel Workers
LOCA: Canada; United States of America
TIME: 1971 - 1992

1102   Redmond, C.K.   01962
Univ. of Pittsburgh, School of Public Health, Dept. of Biostatistics, 130 De Soto St., Pittsburgh PA 15261, United States of America
COLL: Costantino, J.P.; LeGasse, A.A.; Bass, G.I.

Mortality Study of High Nickel Alloy Workers

There is evidence of increased incidence of respiratory cancers, including cancers of the nasal sinuses, among workers in the nickel refining industry. In order to determine whether any such relationship might exist among workers in the high nickel alloy (more than 20% Ni) industry, cause-specific mortality and especially site-specific cancer mortality among workers in 11 study plants across the USA were investigated. The mortality experience of nickel-exposed workers in the study plants was compared to the population of workers involved in all phases of operation in the study plants, and to the total US mortality experience. The study was a historical prospective follow-up on a cohort of approximately 25,000 workers with at least one year of job experience during the period 1956-1960. A report on cause specific mortality patterns as of 1977 has been prepared (IARC Scient. Publ. No 53, 1984). Overall, no increased risk was observed for cancers of the nasal sinuses, larynx or kidney. Cancer of the liver and large intestine did demonstrate statistically significant standard US mortality ratios. Observed increases were found primarily among longer term workers, but were not concentrated in a particular work area or job category. In addition, excess mortality from lung cancer was found among men employed in maintenance jobs, but it is unclear whether the greater risk is directly associated with nickel exposure since such an excess was not evident in other job categories where nickel exposure was also present. No conclusions regarding a causal association with nickel exposure have been drawn at this time. Continued observation of the cohort is recommended, since the cohort is still relatively young and additional follow-up would result in a substantial increase in expected cancer deaths. Currently, efforts are underway to update the vital status follow-up of the cohort through 1988.

TYPE: Cohort
TERM: Metals; Occupation
SITE: All Sites; Colon; Liver; Nasal Cavity; Rectum; Respiratory
CHEM: Nickel
OCCU: Nickel Workers
TIME: 1978 - 1993

PORTLAND

*1103   Nussbaum, R.H.   04996
Portland State Univ., Depts of Physics and, Environmental Sciences and Research, P.O.Box 751, Portland OR 97207-0751, United States of America (Tel.: (503)2225643; Fax: 7254882)
COLL: Stewart, A.M.; Cummins, H.W.

Late Effects of Low-Level Radiation

A cohort of about 18,000 military personnel who were stationed at Camp Hanford (Washington State) during several years of large releases of radioactive substances has been identified. In cooperation with the US Veterans Administration Environmental Epidemiology Branch, death certificates will be traced. As a first step, cancer mortality among Hanford personnel will be compared with that of a matched cohort of up to 100,000 military personnel, stationed during the same period at Fort Lewis (Washington State) Camp. A ten-year follow-up is planned and other end-points might be added. Statistical analysis will be carried out at Birmingham University (United Kingdom). The hypothesis to be tested is that cancer mortality among a radiation-exposed cohort cannot be distinguished from that of an unexposed cohort.

UNITED STATES OF AMERICA

TYPE: Cohort
TERM: Occupation; Radiation, Ionizing
SITE: All Sites
OCCU: Military Servicemen
TIME: 1987 - 1992

*1104 Weinmann, S.A. 05131
Univ. of Washington, School of Public, Health and Community Medicine, Kaiser Permanente, Northwest Region, 3715 N Interstate Ave. , Portland OR 97227, United States of America (Tel.: (503)2493318; Fax: 2493320)
COLL: Weiss, N.S.; Psaty, B.; White, E.; Siscovick, D.; Ragunathan, T.E.; Glass, A.G.

Risk Factors for Renal Cell Carcinoma

The primary objective of this case-control study is to examine the role of diuretic drugs in the aetiology of renal cell carcinoma. Other prescription medications, including anti-hypertensives, sex hormones, and diet pills will also be investigated, as will personal and family medical history, patterns of adult weight, and tobacco use. The study will involve 212 cases (127 males and 85 females). Controls will be health plan members of the Kaiser Permanente, Northwest Region, individually matched to cases on sex, age, months of health plan membership, date of entry into plan, and presence in plan on case diagnosis date. Exposure information will be obtained from out- and in-patient medical records with a possible interview component, to be determined from the results of a pilot study comparing chart data with interview responses. Results will be analysed using standard statistical methods for case-control studies including logistic regression analysis.

TYPE: Case-Control
TERM: Diuretics; Drugs; Hormones; Tobacco (Smoking)
SITE: Kidney
TIME: 1989 - 1992

PROVIDENCE

1105 Morrison, A.S. 04110
Brown Univ., Dept. of Community Health, 97 Waterman St., Providence RI 02912, United States of America (Tel.: (401)8633172)
COLL: Lev, R.

Rectosigmoid Polyps and Risk of Colorectal Cancer

The long-term goal of this study is to develop knowledge of the relationship of benign rectosigmoid polyps and cancer of the colon (including the rectum). The principal specific aims are evaluation of the colon cancer mortality rate in persons who have had one or more surgically-confirmed polyps and comparison of colorectal cancer incidence and death rates according to histological features of the previously removed polyps. A retrospective follow-up study is proposed. A population-based group of 2,500 men and women with surgically-confirmed large bowel polyps (primarily rectosigmoid) will be identified by use of pathology records in Rhode Island hospitals Subjects will have had their index admission in the period 1959-1975, and will have been 35-79 years old at the time. The cohort will be followed-up by use of medical records, death certificates, city directories, and other existing resources. Slides of the benign tissue will be retrieved for all the cases of colon cancer that are identified, and for a sample of non-cases. The slides will be reviewed according to a standard protocol in order to assess the features of the polyps, especially to distinguish neoplastic from hyperplastic polyps, and to assess the villous component (if any) and the extent and severity of dysplasia. The death rate from colon cancer in the cohort will be compared to the rate experienced by Rhode Island residents with the same age and sex distribution. By use of the case-control within a cohort approach, the incidence and mortality rates from colon cancer will be correlated with histological features of the benign tissue removed. Death rates from other causes also will be evaluated, and rates of recurrence of large bowel polyps will be estimated.

UNITED STATES OF AMERICA

TYPE: Case-Control; Cohort
TERM: Clinical Records; Histology; Premalignant Lesion
SITE: Benign Tumours; Colon; Rectum
TIME: 1986 - 1992

RALEIGH

*1106 Aldrich, T.E. 05190
North Carolina Cancer Registry, North Carolina Div. of Statistics, & Information Serv., P.O. Box 27687, Raleigh NC 27611-7687, United States of America

Monitoring of Rare Cancers as Sentinel Events
The Cancer Surveillance Section routinely evaluates spatial and temporal patterns of selected rare cancers, the hypothesis being that shifts in the occurrence of these events can serve as sentinels of potential public health significance. Currently studies of leukaemia, paediatric cancers and cancers of liver, pancreas, soft tissue and non-Hodgkin's lymphoma are under way. The leukaemia study is a 38-county evaluation from 1980-1989 with emphasis on changes in sub-group rates, cell-type proportions and incidence in relation to recognized point sources of potential leukaemogens (e.g. benzene). The paediatric study is a statewide investigation of rates for 1975-1989, focussing on geographical variation of specific sites. Liver cancer, soft tissue sarcoma and non-Hodgkin's lymphoma are being studied in counties with paper mills. Disease rates for communities around paper mills will be compared with control communities. The pancreas study is a two-county study evaluating potential pesticide exposure and disease incidence between 1985-1990. Sample sizes range from 30 to 300 cases, depending on the time period and geographical area involved. Statistical studies focus on geographical, temporal and personal characteristics (age, race, sex) and disease traits (cell type) to evaluate evidence of increased risk.

TYPE: Case Series; Correlation; Incidence
TERM: Cluster; Geographic Factors; Monitoring; Pesticides; Registry; Solvents; Time Factors; Trends
SITE: Childhood Neoplasms; Leukaemia; Liver; Non-Hodgkin's Lymphoma; Pancreas; Soft Tissue
CHEM: Benzene; Dioxins
REGI: N. Carolina (USA)
TIME: 1990 - 1992

*1107 Aldrich, T.E. 05191
North Carolina Cancer Registry, North Carolina Div. of Statistics, & Information Serv., P.O. Box 27687, Raleigh NC 27611-7687, United States of America
COLL: Morris, P.; Tribble, T.; Lindsey, J.; Carter, D.

Geographical Pathology of Brain Cancer
This is an exploratory case-control study of residential proximity to potential environmental hazards. Case residencies for a six-county region will be evaluated for spatial patterns with respect to many point sources: industry discharges, hazardous waste sites, nuclear power plants, high voltage power lines, etc. Approximately 300 incident cases from 1980-1989 will be identified from hospital records and a case series of residences, identified from death certificates (septicaemia deaths) will be used as controls. The analyses will feature simple contingency comparisons for residential proximity to several point sources. Some of the space-time cluster methods (Knox, Barton-David, nearest neighbour, etc.) may also be used. The main goal is to see how well the cases can be mapped using this linkage with environmental databases.

TYPE: Case-Control
TERM: Cluster; Electromagnetic Fields; Environmental Factors; Geographic Factors; Mapping; Radiation, Ionizing; Registry; Solvents
SITE: Brain
CHEM: Benzene
REGI: N. Carolina (USA)
TIME: 1990 - 1992

*1108 Aldrich, T.E. 05192
North Carolina Cancer Registry, North Carolina Div. of Statistics, & Information Serv., P.O. Box 27687, Raleigh NC 27611-7687, United States of America
COLL: Morris, P.; Lindsey, J.

UNITED STATES OF AMERICA

Evaluation of Cancer Cluster Reports
Reports of increased cancer occurrence from members of the public are systematically evaluated using the computer program CLUSTER. Spatial characteristics of case occurrence, temporal trends and personal traits (age, race, sex, occupation) are evaluated. In all cases, the hypothesis being tested is: is there an increased occurrence of cancer, and if so, does the pattern of increase suggest an environmental risk factor? Most studies are very small (3 to 30 cases). No control groups are involved: increases are evaluated based on national incidence and evidence of environmental patterns is evaluated based on randomness of case occurrence. Surveillance will be carried out, in most cases for a minimum of five years. Special follow-up studies are also being performed. Poisson distribution tests are used for many techniques, extra-Poisson for two of them. Chi-square techniques are also used with likelihood ratio solutions for two of the methods. One approach uses a combinatorial solution (Grimson 1979). These studies all involve local incidence data from North Carolina communities.

TYPE: Incidence; Methodology
TERM: Cluster; Environmental Factors; Registry; Trends
SITE: All Sites
REGI: N. Carolina (USA)
TIME: 1989 - 1993

RESEARCH TRIANGLE PARK

1109 Tennis, P. 04336
Burroughs Wellcome Co., Div. of Epidemiology, Information, & Surveillance, 3030 Cornwallis Rd, Research Triangle Park NC 27709, United States of America (Tel.: (919)2483437)
COLL: Andrews, E.; Tilson, H.H.; Bombardier, C.; Mitchell, D.

Retrospective Cohort Study of Rheumatoid Arthritis and the Risk of Neoplasia
This population-based cohort study is based on automated data sets from the Canadian province of Saskatchewan and assesses cancer incidence in 1,300 rheumatoid arthritis patients compared with two comparably ill age- and sex-matched control populations. Rheumatoid arthritis and control populations are identified via hospital discharge diagnoses selected from the Saskatchewan Hospital Services Plan (SHSP) files for 1978-80. Controls are matched to the rheumatoid arthritis population on age, sex, and timing of hospitalization and consist of 1,900 osteoarthritis controls and 7,000 controls with other diagnoses. Data gathered for each study subject include prescription drugs (from the Prescription Drug Plan), subsequent and past hospital admissions and associated diagnoses (from the Hospital Services Plan), documented neoplasias (from the Cancer Registry), and vital status (from the Health Registration). Validity studies to verify hospital diagnosis recorded in the SHSP data files with a sample of actual hospital records are being performed. Analysis focuses on a relative risk measure based on person-years at risk. Person years at risk are measured using the index admission date and date of death or last date of follow-up. Stratified analyses of cancer risk are conducted by age, sex, and primary site of cancer. Using multivariate modelling, the effects of drug therapy are controlled for in order to assess the effect of rheumatoid arthritis itself on the risk of cancer. To supplement the main analysis, cancer risks are compared with those of the Saskatchewan population for the same time period.

TYPE: Cohort
TERM: Arthritis; Drugs; Mathematical Models; Record Linkage; Registry
SITE: All Sites; Haemopoietic; Lymphoma; Skin
LOCA: Canada
REGI: Saskatchewan (Can)
TIME: 1986 - 1989

RICHLAND

1110 Kathren, R.L. 02842
Hanford Environmental, Health Foundation (HEHF), Dept. of Research, P.O.Box 100, Richland WA 99352, United States of America (Tel.: (509)3768650)
COLL: Omohunndro, E.L.; Mahaffey, J.; Gilbert, E.S.

Department of Energy Hanford Health and Mortality Study
The health effects of occupational low-level ionizing radiation on nuclear industry workers have been of concern since the association of leukaemia, osteosarcoma, and aplastic anaemia with high-level exposure to ionizing radiation was noted early in this century. This prospective cohort study was

UNITED STATES OF AMERICA

designed originally to evaluate the mortality patterns of Hanford Site employees as a function of their radiation exposure. Its scope has been expanded to include reproductive outcomes and assessment of occupational non-radiation exposures and ongoing monitoring of major medical problems among currently active workers. These objectives require: (1) the maintenance and updating of a personnel roster representing 15,000 current employees; (2) active mortality follow-up of translocated and retired workers; (3) periodic medical examination of active employees; (4) collection of individual internal and external ionizing radiation exposure data; and (5) interval data analysis for detection of evolving detrimental health effects. To avoid the underassessment of risk inherent in a "healthy worker" population, controls consist of internal cohort risk sets with comparable demographic and work history characteristics. The potential effects of radiation exposure are assessed with a modified Cox proportional hazards model. Hanford site data also have been submitted to external investigators for corroboration of conclusions regarding the health effects of low-level radiation exposure. A paper has been published in Health Phys. 55:1945-1981, 1988.

TYPE: Cohort
TERM: Occupation; Radiation, Ionizing; Radiation, Non-Ionizing
SITE: All Sites
OCCU: Radiation Workers
TIME: 1944 - 1993

1111 Stevens, R.G. 04571
Battelle Pacific Northeast Labs, P.O.Box 999, Richland WA 99352, United States of America (Tel.: (509)3756941)

Iron and Risk of Cancer
The first National Health and Nutrition Examination Survey (NHANES I) cohort of 14,000 subjects drawn as a sample of the United States population is being followed for mortality and morbidity outcome. During the period 1971-75, the subjects were identifid and given a medical examination, detailed dietary questionnaire, and a series of blood tests. The project is concerned with determining whether and to what extent body iron stores, and dietary intake influence cancer risk. The hypothesis is based on two possible biological mechanisms: (1) iron can catalyse the production of oxygen radicals, and these may be proximate carcinogens or promoters, and (2) iron may be a limiting nutrient to the survival and growth of an existing cancer cell. Of the original 14,000 adult subjects aged 25-79 identified in 1971-1975, 3,355 men survived at least four years after a blood test for transferrin saturation. Of these, 242 developed cancer by 1981-1984. Transferrin saturation was significantly higher in the men who developed cancer than in those who did not. Among women, only those 8% with the highest transferrin saturation showed any suggestion of increased cancer risk. These results were reported in N. Engl. J. Med. 319(16):1047-1052, 1988. Future studies are directed toward the effect of iron in prognosis after cancer diagnosis, effect of iron in radiation sensitivity, and effect of iron on risk of a second cancer after cancer therapy.

TYPE: Cohort
TERM: Diet; Nutrition; Promotion; Trace Elements
SITE: All Sites
CHEM: Iron
TIME: 1985 - 1993

ROCHESTER

*1112 Talley, N.J. 05007
Mayo Clinic and Foundation, 200 First St., Rochester MN 55905, United States of America (Tel.: (507)2556027; Fax: 2556318)
COLL: DiMagno, E.P.; Blaser, M.J.

Association of Helicobacter Pylori with Gastric Adenocarcinoma
The aim is to determine whether H. pylori is a risk factor for gastric adenocarcinoma at different sites. A case-control study is being undertaken including patients with gastric adenocarcinoma, patients with cancers elsewhere (colon, oesophagus, lung) and cancer-free controls (asymptomatic persons, benign non-gastric diseases). Demographic and clinical data have been collected and serum obtained for analysis.

UNITED STATES OF AMERICA

TYPE: Case-Control
TERM: BMB; H. pylori
SITE: Stomach
TIME: 1990 - 1992

ROCKVILLE

1113 Devesa, S.S. 03968
NCI, Div. of Cancer Etiology, Epidemiology & Biostatistics Program, Biostatistics Branch, Executive Plaza Bldg North, Room 415F, Rockville MD 20852, United States of America (Tel.: (301)4964153)

Analysis of Cancer Incidence and Mortality Patterns in the United States
Utilizing data derived from several population-based NCI surveys, the Connecticut Tumor Registry, and the SEER program, incidence patterns are being assessed among residents of several geographical areas. Mortality data provided by the National Center for Health Statistics are also being analysed. Current analysis focuses on the patterns for more specific forms of cancer among various racial/ethnic groups and by histologic category.

TYPE: Incidence; Mortality
TERM: Blacks; Caucasians; Registry; Trends
SITE: All Sites
REGI: Connecticut (USA); SEER (USA)
TIME: 1987 - 1993

1114 Pottern, L.M. 03350
NCI, NIH, Environmental Epidemiology Branch, Occupational Studies Sect., Executive Plaza North 418, Rockville MD 20892, United States of America (Tel.: (301)4969093)
COLL: Miller, B.A.; Blair, A.E.; Stewart, P.A.; McCammon, C.; Zey, J.

Mortality Study of Workers Exposed to Acrylonitrile
A cohort mortality study will be conducted of approximately 7,000 workers occupationally exposed to acrylonitrile (AN) from plants that began producing or using AN between 1952 and 1962. Workers will be identified from historical company records and traced to the present to determine vital status. Job histories, industrial hygiene measurements, limited biological monitoring data, smoking information, and death certificates will be used to evaluate associations between AN exposure and cause-specific mortality. The mortality of the cohort will be compared with that of the general US population, regional populations, and an unexposed worker population, if feasible.

TYPE: Cohort
TERM: Chemical Exposure; Environmental Factors; Occupation; Plastics; Tobacco (Smoking)
SITE: All Sites; Lung
CHEM: Acrylonitrile
OCCU: Acrylonitrile Workers
TIME: 1985 - 1992

1115 Tucker, M.A. 04950
National Cancer Institute, NIH, Environmental Epidemiology Branch, Family Studies Sect., 6130 Executive Blvd, Rockville MD 20892, United States of America (Tel.: (301)4961691; Fax: 4020916)
COLL: Hartge, P.; Holly, E.A.; Sagebiel, R.W.; Halpern, A.; Clark, W.

Case-Control Study of Malignant Melanoma
The purpose is to estimate the effects of host and environmental factors on malignant melanoma aetiology. 600 cases will be identified over a 2-year period at two pigmented lesion clinics; 1,200 controls will be selected from clinics within the same university hospital system. Whole body skin examinations, photography, biopsy data, personal interviews, and self-administered diet questionnaires will be used. The study is a collaborative project of the NCI, the University of California, and the University of Pennsylvania. The separate effects of normal and dysplastic naevi and the interaction of sunlight and moles will be estimated with this large dataset.

UNITED STATES OF AMERICA

TYPE: Case-Control
TERM: Biopsy; Diet; Naevi; Radiation, Ultraviolet
SITE: Melanoma
TIME: 1990 - 1993

*1116 Ward, M. 05032
NCI, Environmental Epidemiology Branch, Occupational Studies Section, 6130 Executive Blvd, Rockville MD 20892, United States of America (Tel.: (301)4969093; Fax: 4969146)
COLL: Cantor, K.P.; Stewart, P.A.; Zahm, S.H.; Heineman, E.; Blair, A.E.

Gastric Cancer in Eastern Nebraska
A case-control study of stomach cancer, using the histological classification of Lauren, will be conducted in the 66 counties of eastern Nebraska. The main objective is to evaluate the association of agricultural exposures with the two types of stomach cancer. Agricultural exposures, which are also hypothesized risk factors for stomach cancer, include fertilizer and pesticide use, dust exposure and ingestion of nitrate from contaminated drinking water. About 300 cases from the period 1985-1991 will be identified from the Nebraska Tumour Registry. Population controls are frequency-matched to the age, sex and year of eath of the cases. Interviews will be conducted with subjects and next-of-kin in 1991.

TYPE: Case-Control
TERM: Dusts; Fertilizers; Histology; Occupation; Pesticides; Registry; Water
SITE: Stomach
CHEM: Nitrates; Nitrites
REGI: Nebraska (USA)
TIME: 1991 - 1993

1117 Zahm, S.H. 03192
NCI, Environmental Epidemiology Branch, Occupational Studies Sect., 6130 Executive Blvd, R418K, HNC 3344, Executive Plaza Bldg North, Rockville MD 20892-4200, United States of America (Tel.: (301)4969093)
COLL: Thomas, T.L.; Roscoe, R.J.; Beaumont, J.J.

Calculation of Mortality Rates for Blue-Collar Workers
Mortality studies conducted by epidemiological research organizations usually use United States or other general population death statistics for comparison. While general population statistics are good because usually based on very large numbers, their validity in worker comparisons is questionable because of differences between workers and the general population. Such differences include the following: (1) workers appear to smoke more (on the average) than the general population, (2) they are economically more homogeneous, and (3) they are generally healthier than the general population because of selection at the time of hiring and the physical activity associated with many jobs. These drawbacks to general population comparisons are well documented in the occupational health literature. Epidemiologists have little alternative to general population comparisons because comparison rates based upon sufficiently large worker populations do not exist. NCI and NIOSH are collaborating to develop a computerized data base from completed studies of worker populations. About 280,000 workers from NCI cohort studies and 230,000 workers from NIOSH studies will be included in the data base. The pooled worker population will be used as a comparison group in future mortality studies. Updating will include vital status follow-up and obtaining death certificates for the deceased. Calculation of death rates will be accomplished using programs written for the purpose. A paper has been published in Am. J. Epidemiol. 123:918-919, 1986.

TYPE: Methodology; Mortality
TERM: Occupation
SITE: All Sites
TIME: 1983 - 1992

1118 Zahm, S.H. 03646
NCI, Environmental Epidemiology Branch, Occupational Studies Sect., 6130 Executive Blvd, R418K, HNC 3344, Executive Plaza Bldg North, Rockville MD 20892-4200, United States of America (Tel.: (301)4969093)
COLL: Walrath, J.; Mead, M.; Li, F.P.; Fraumeni, J.F.

Cancer Mortality among Members of the American Chemical Society
A retrospective cohort study of men who were members of the American Chemical Society will be conducted to provide evidence on whether cancer mortality among chemists differs from that expected.

UNITED STATES OF AMERICA

Approximately 50,000 men were members prior to 1955 and were active members in 1965. They will be followed to December 1978. SMRs will be calculated. Particular attention will be given to malignant lymphoma, leukaemia, and cancer of the pancreas. Length of membership, age at entry into the Society, and type of employment will be considered.

TYPE: Cohort
TERM: Chemical Exposure; Occupation
SITE: Leukaemia; Lymphoma; Pancreas
OCCU: Chemists
TIME: 1980 – 1992

1119 Zahm, S.H. 04288
NCI, Environmental Epidemiology Branch, Occupational Studies Sect., 6130 Executive Blvd, R418K, HNC 3344, Executive Plaza Bldg North, Rockville MD 20892-4200, United States of America (Tel.: (301)4969093)

Mortality Study of Lawn Care Workers: A Retrospective and Prospective Study

Phenoxy acetic acid herbicides have been associated with lymphoma and soft tissue sarcoma. Pesticide-exposed groups may be at higher risk of these tumours as well as leukaemia, lung cancer, and multiple myeloma. To investigate these hypotheses, a study of over 20,000 former and current lawn care service workers in one company is being conducted. The applicators have been exposed to 2,4-D and other pesticides for up to 90 days per year. A retrospective study of workers employed between 1969-1980 will be done to compare their mortality to that of the US general population. A prospective component will allow future study of these workers and newly-hired workers.

TYPE: Cohort
TERM: Chemical Exposure; Herbicides; Occupation; Pesticides
SITE: All Sites; Brain; Leukaemia; Lung; Multiple Myeloma; Non-Hodgkin's Lymphoma; Soft Tissue
CHEM: 2,4-D; Phenoxy Acids
OCCU: Herbicide Manufacturers; Herbicide Sprayers
TIME: 1987 – 1993

1120 Zahm, S.H. 04289
NCI, Environmental Epidemiology Branch, Occupational Studies Sect., 6130 Executive Blvd, R418K, HNC 3344, Executive Plaza Bldg North, Rockville MD 20892-4200, United States of America (Tel.: (301)4969093)
COLL: Garrity, T.J.; Telles, J.L.

Mortality Study of Philadelphia Firefighters

A mortality study of firefighters is being conducted to investigate the hypothesis that the introduction of synthetic building materials and exposure to other products of combustion may result in increased cancer risk. The cohort will consist of approximately 10,000 firefighters hired since 1910. Approximately 5,600 have been actively employed since 1945, when development of new building materials began. Mortality will be examined according to employment characteristics such as calendar time, job title, duration of employment, assignment to engine or ladder companies, number of runs per year for assigned company, area of city, and age and year of first employment. Introduction of diesel equipment and potential for exposure in five houses will be assessed.

TYPE: Cohort
TERM: Age; Chemical Exposure; Occupation; Time Factors
SITE: All Sites; Brain; Colon; Leukaemia; Rectum
OCCU: Firemen
TIME: 1987 – 1992

SALT LAKE CITY

1121 Lyon, J.L. 03896
Univ. of Utah, School of Medicine, Dept. of Family & Preventive Medicine, 50 N Medical Dr., Salt Lake City UT 84112, United States of America (Tel.: (801)5817234)
COLL: Archer, V.E.; Schiager, K.J.

Cancer Risk from Radon in Homes

A case-control study is being done to evaluate the risk of radon in homes among smokers and non-smokers separately. Lung cancer cases will be collected over a three year period: approximately

UNITED STATES OF AMERICA

450 smokers and 350 non-smokers. Controls will be selected by random digit dialling supplemented by records of health financing for the elderly. Controls will be matched on age, sex, race and cigarette usage. A telephone interview will obtain information an all residences the person has lived in, plus supplementary information on lifestyle and diet. Radon is measured in as many of the residences as practical by a year-long track etch technique. Adjustments will be made for changes in the houses over time. For houses not measured, estimates will be made based on geology and house features. Lifetime radon exposures are calculated. Odds ratios will be obtained for different lifetime radon and cigarette exposures. Regression analyses will be used. An attempt will be made to determine at which age the radon exposures were most important.

TYPE: Case-Control
TERM: Age; Radiation, Ionizing; Time Factors; Tobacco (Smoking)
SITE: Lung
CHEM: Radon
REGI: Idaho (USA); Utah (USA)
TIME: 1989 - 1994

*1122 Slattery, M.L. 05089
Univ. of Utah, School of Medicine, Dept. of Family and Preventive Medicine, 50 N Medical Drive, Room 1C26, Salt Lake City UT 84132, United States of America (Tel.: (801)5817234)
COLL: Potter, J.D.; Friedman, G.D.; Caan, B.

Diet, Physical Activity and Reproduction as Risk Factors for Colon Cancer

The purpose of this study is to assess the interaction between dietary intake, physical activity, reproductive history, body size, and genetic factors as they relate to the development of colon cancer. This is a large multi-centre case-control study of 2,400 cases and 2,400 controls. Cases and controls will be population-based in Utah and Minnesota; they will come from the membership records at the Kaiser Health Plan. Detailed dietary intake data will be ascertained using a previously validated diet history questionnaire. Detailed information on physical activity, reproductive history, and family history of cancer will also be ascertained in an interviewer-administered questionnaire. Cases should be interviewed within three months of diagnosis since a rapid-reporting system will be used to identify cases. Data will be analysed to determine how diet, physical activity, reproductive history, age, sex, and tumour site interact in the development of colon cancer.

TYPE: Case-Control
TERM: Diet; Genetic Factors; Physical Activity; Registry; Reproductive Factors
SITE: Colon
REGI: Bay Area (USA); Minnesota (USA); Utah (USA)
TIME: 1991 - 1996

SAN FRANCISCO

1123 Cleaver, J.E. 03930
Univ. of California, Lab. of Radiobiology, and Environmental Health, Box 0750, San Francisco CA 94143-0750, United States of America (Tel.: (415)6664563)

Molecular Studies of DNA, Mutagenicity, and Carcinogenicity

Human DNA repair deficient diseases, xeroderma pigmentosum, and Cockayne syndrome, which involves hypersensitivity to ultraviolet light and other clinical symptoms, are being investigated at molecular and genetic levels. Particular emphasis is placed on rapid diagnostic methods for patient and prenatal diagnosis and studies of the fine structure of regulation and control of DNA repair and cloning of relevant genes. Additional studies are being made into mechanisms of reversion and detailed rates of repair of thymine dimers and (6-4) photoproducts. Recent publications include : Exp. Cell Research 22, 513-520, 1989; Teratogenesis Carcinogenesis and Mutagenesis 9, 147-155, 1989; Carcinogenesis, 10, 1691-1696, 1989.

UNITED STATES OF AMERICA

TYPE: Molecular Epidemiology
TERM: Biopsy; Congenital Abnormalities; DNA Repair; Premalignant Lesion; Radiation, Ultraviolet; Xeroderma Pigmentosum
SITE: Skin
TIME: 1968 - 1993

*1124   Moscicki, A.B. 05036
Univ. of California at San Francisco, 400 Parnassus Ave., San Francisco CA 94143, United States of America (Tel.: (415)4762184; Fax: 4766106)
COLL: Palefsky, J.; Bolan, G.; Darraugh, T.; Brescia, B.; King, E.; Schoolnik, G.; Winkelstein, W.; Schachter, J.; Daniels, T.; Benowitz, N.

Natural History of Human Papillomavirus
The aims are: (1) to observe the natural history of HPV and (2) to determine risk factors related to the development of cervical neoplasia in adolescents. This population was chosen for several reasons: adolescents with HPV and limited years of sexual activity demonstrate more clearly an early infection (short exposure time) than most adult populations who have been sexually active for many years. Since rates of neoplasia are increasing in younger women, co-factors identified in this group may represent important factors for the acceleration of neoplastic development. Finally, adolescence is a time of abundant metaplastic activity, so presenting an ideal population to study the role of cervical immaturity in neoplasia. 1,568 females aged 13-19 years and positive for HPV DNA will be asked to undergo a colposcopic examination and face-to-face interview. Females sexually active for less than two years with a positive HPV DNA test and no evidence of cervical neoplasia (N = 960) will be asked to enroll in a longitudinal study. Patients will be examined at the initial visit and every four months or until CIN develops. Colposcopic examinations, cytology, test for sexually transmitted diseases and face-to-face interviews for information on sexual behaviour, contraceptive use, and cigarette use will be performed at scheduled intervals. 400 HPV negative control subjects will be followed similarly for comparison. This study will describe the natural history of latent HPV infection and calculate risk factors for the accelerated development of CIN in a case-control and cohort population. The results from this study will be related to prevention and education in young women as well as to identifying a clinical model for the role of viruses in abnormal cellular development.

TYPE: Case-Control; Cohort
TERM: Adolescence; Contraception; Cytology; HPV; Sexual Activity; Sexually Transmitted Diseases; Tobacco (Smoking)
SITE: Uterus (Cervix)
TIME: 1990 - 1995

SEATTLE

1125   Beresford, S.A.A. 04818
Univ. of Washington, Sch. Publ. Health & Community Medicine, Dept. of Epidemiology, SC-36, Seattle WA 98195, United States of America (Tel.: (206)5439512; Fax: 5438525)
COLL: Weiss, N.S.; McKnight, B.; Wilbur, D.

Endometrial Cancer Risk and Postmenopausal Hormone Use
This case-control study is designed to increase knowledge on the role of exogenous oestrogens and progestins in altering the risk for endometrial carcinoma. Although the adverse effects of oestrogens alone are well established, the effect of adding progestins has not been adequately quantified. The specific aims are: (1) to estimate the risk of endometrial cancer associated with combined oestrogen and progestogen therapy, relative both to no therapy, and to oestrogen-only therapy in post-menopausal women; (2) to determine whether this risk is further influenced by the number of days in the cycle on which progestogen is given in addition to oestrogen; (3) to determine the length of the oestrogen-free interval after which the risk of endometrial cancer approaches background rates, for different durations of prior oestrogen use. The cases for this study are women aged 45-74 years, residing in King, Pierce or Snohomish counties, Washington, USA, who were diagnosed with endometrial cancer between January 1987 and December 1990. Cases are being identified through the Cancer Surveillance System, a population-based cancer registry that has served the area since 1974. All endometrial cancer cases are subject to independent histological review. It is anticipated that approximately 660 incident endometrial carcinoma cases will be enrolled. About 660 controls are being recruited through random digit dialling. Identical in-person interviews are administered to cases and controls to collect information on hormone use, reproductive history and other variables of interest. Color photographs of both contraceptive and

UNITED STATES OF AMERICA

noncontraceptive hormones are used to facilitate recall. A self-administered dietary questionnaire (developed by NCI) is also collected from each respondent.

TYPE: Case-Control
TERM: Diet; Hormones; Oral Contraceptives; Registry; Reproductive Factors
SITE: Uterus (Corpus)
CHEM: Oestrogens; Progestogens
REGI: Seattle (USA)
TIME: 1991 - 1993

1126 Davis, S. 04377
Fred Hutchinson Cancer Research Center, Program in Epidemiology, 1124 Columbia St., Seattle WA 98104, United States of America (Tel.: (206)4674787)

Identification and Characterization of Case Aggregation Patterns in Hodgkin's Disease
The study aims primarily to investigate the potential role of an infectious process in the aetiology of Hodgkin's disease (HD). Case aggregations in the past will be determined and aetiological relationships characterized and quantified. Data from a population-based case-control study conducted in north-western Washington are used. Cases were Caucasian residents of King, Pierce, and Snohomish counties diagnosed with HD between January 1974 and December 1982, who were 15-74 years of age at diagnosis. Of 313 cases eligible, 281 (90%) were located and personally interviewed. For each case, one control was selected at random from the same population, matched according to sex, approximate age, and general socio-economic status. The research program aims to: (1) determine whether cases aggregate through prior interpersonal contact in school, employment or residential settings more than would be expected to occur by chance in the same population; (2) improve previously developed measures of interpersonal contact and develop additional indicators among study subjects; (3) characterize and quantify any patterns of case aggregation observed in terms of complexity, first-order interdependence and various postulated periods of biological activity (i.e., susceptibility and infectivity) and (4) evaluate the role of host factors related to immunocompetence and antigenic stimulation in any case aggregation patterns observed.

TYPE: Case-Control; Methodology
TERM: Caucasians; Cluster; Infection
SITE: Hodgkin's Disease
TIME: 1988 - 1992

1127 Davis, S. 04451
Fred Hutchinson Cancer Research Center, Program in Epidemiology, 1124 Columbia St., Seattle WA 98104, United States of America (Tel.: (206)4674787)
COLL: Bley, L.D.

Nutritional Factors and Non-Hodgkin's Lymphoma
This study is a population-based case-control study of non-Hodgkin's lymphoma (NHL) in adults. The primary purpose is to investigate the role of animal protein, fats, and Vitamin A in the development of NHL. Cases will consist of all persons aged 20-64 newly diagnosed with NHL in King, Pierce, and Snohomish Counties in western Washington State between 1 June 1988 and 31 May 1989 (140 cases anticipated). An equal number of controls will be randomly selected from the populations of the same three counties, matched to the cases regarding age and sex. Detailed dietary data will be collected using a self-administered questionnaire developed at the National Cancer Institute, and a short telephone interview will be conducted with study subjects to elicit information regarding lifestyle and demographic factors. Results from this investigation are expected to provide the first epidemiological evidence regarding a possible role of specific dietary constituents in the development of NHL.

TYPE: Case-Control
TERM: Diet; Fat; Lifestyle; Nutrition; Protein; Vitamins
SITE: Non-Hodgkin's Lymphoma
TIME: 1988 - 1992

1128 Davis, S. 04824
Fred Hutchinson Cancer Research Center, Program in Epidemiology, 1124 Columbia St., Seattle WA 98104, United States of America (Tel.: (206)4674787)
COLL: Kopecky, K.J.; Hamilton, T.E.; Amundson, B.; Ruttenber, J.; Sage, M.

UNITED STATES OF AMERICA

Hanford Thyroid Disease Study
This cohort study is designed to determine whether thyroid morbidity (including hypothyroidism and benign and malignant neoplasms) is increased among persons exposed to atmospheric releases of I-131 between 1944 and 1957 from the Hanford Nuclear Site in eastern Washington. A pilot phase will be conducted initially, to include 450 persons who represent both those most likely to have been exposed to I-131 releases as well as those presumed to be unexposed to such releases. Attempts will be made to locate all study subjects and to conduct personal interviews with each (or their next-of-kin, if deceased) regarding residential, medical, and personal habit histories. Each subject's mother or other close relative will also be located and interviewed regarding aspects of the subject's childhood that are relevant to I-131 exposure. Based upon interview data and in collaboration with the Hanford Environmental Dose Reconstruction Project, estimates of thyroid radiation dose will be obtained for each individual. Medical and pathology records will be sought to verify self-reported thyroid disease. Each study subject will be examined independently by two endocrinologists, and blood specimens will be obtained for thyroid function tests. For those found to have a thyroid nodule, permission will be sought to conduct a fine needle aspiration. Data obtained in the pilot study will be used to perform power calculations and to design a full study to estimate the risk of benign and malignant forms of thyroid disease among exposed subjects.

TYPE: Cohort
TERM: Childhood; Radiation, Ionizing
SITE: Thyroid
CHEM: Iodine
TIME: 1989 - 1993

*1129 Evanoff, B.A. 05107
Univ. of Washington, Robert Wood Johnson Scholar Programs, HQ-18, Seattle WA 98195, United States of America(Fax: (206)6852473)
COLL: Gustavsson, P.; Hogstedt, C.

Cancer Incidence and Mortality among Swedish Chimney Sweeps: An Extended Follow-Up
This study is an update of a previous analysis of cancer incidence and mortality among a cohort of Swedish chimney sweeps. The aim is to test the hypothesis that exposure to combustion by-products is associated with an excess risk of lung, oesophageal, and other cancers. The study will analyse changes in the sweeps' excess cancer risk over time, in an attempt to identify changes in exposure which affect cancer incidence. The study uses a retrospective cohort design with registry linkage. Members of the chimney sweeps in Sweden, were identified by union records. 5,464 trade union members employed between 1918 and 1980 are eligible for inclusion in the study. Cancer mortality will be ascertained from death certificates for the period 1951-1990; cancer incidence data will be obtained from the Swedish National Cancer Registry for the period 1958-1987. National rates will be used as a comparison; SMRs are calculated using the man-years method.

TYPE: Cohort
TERM: Occupation; Registry
SITE: Haemopoietic; Lung; Lymphoma; Oesophagus; Prostate
CHEM: PAH
OCCU: Chimney Sweeps
LOCA: Sweden
REGI: Sweden (Swe)
TIME: 1990 - 1992

1130 Henderson, M.M. 03725
Fred Hutchinson Cancer Research Center, Cancer Prevention Research Unit, 1124 Columbia St., Seattle WA 98104, United States of America (Tel.: (206)4674678)
COLL: Anderson, G.; Baker, M.; Bowen, D.; Chu, J.; Goodman, G.; Grizzle, J.; Kinne, S.; Kristal, A.; Omenn, G.; Peterson, A.; Sarason, B.; Sarason, I.; Thompson, B.; Thornquist, M.; Urban, N.; White, E.; Beresford, S.A.A.; Bankson, D.; Feng, Z.; Kestin, M.; Koepsell, T.; Taplin, S.; Wagner, E.

Cancer Prevention Research Unit
The Cancer Prevention Research Unit is a federally funded programme with a core of cancer research scientists and multiple interrelated research studies in cancer prevention and control. Studies include the efficacy of vitamin A/beta carotene in preventing lung cancer in high risk groups (smokers/asbestos workers); the efficacy of folic acid in reversing cervical dysplasia; whether a school-based prevention programme can deter children from using smokeless tobacco; whether women trained to adopt a low-fat diet will maintain the change over time and whether their husbands adopt a similar eating pattern;

UNITED STATES OF AMERICA

whether behavioural changes brought about through a community organisation (the Seattle Archdiocese) will motivate large numbers of people to make dietary (low fat/high fibre) changes; whether breast cancer screening rates among women over 50 can be improved , utilising a community organisation intervention strategy; which methods of communication are most effective in encouraging women to take part in mammography screening programmes; whether worksite quit-smoking rates can be improved with a comprehensive multi-level, or "stepped" approach; and, in a multi-institutional statistical/methodological study, development of new strategies for community-based studies. Pilot studies investigate whether precursors of colon cancer can be predicted through new screening tests, and the effect of worksite smoking policies on workers' smoking habits.

TYPE: Cohort; Correlation; Cross-Sectional; Incidence; Intervention; Methodology
TERM: Biochemical Markers; Diet; Dusts; Fat; Fibre; High-Risk Groups; Mammography; Nutrition; Occupation; Prevalence; Prevention; Screening; Tobacco (Smoking); Vitamins
SITE: All Sites; Breast (F); Colon; Lung; Uterus (Cervix)
CHEM: Asbestos; Beta Carotene
OCCU: Asbestos Workers
TIME: 1982 - 1993

*1131   Omenn, G.   05078
        Univ. of Washington, School of Public Health, and Community Medicine, SC-30, Seattle WA 98195,
        United States of America (Tel.: (206)5431144; Fax: 5433813)
COLL: Goodman, G.; Grizzle, J.; Thornquist, M.; Rosenstock, L.; Barnhart, S.; Anderson, G.; Cone, J.;
        Balmes, J.; Valanis, B.; Glass, A.G.; Cullen, M.; Cherniack, M.; Keogh, J.; Meyskens, F.L.

Chemoprevention of Lung Cancer in High-Risk Populations

CARET, the beta-carotene and retinol efficacy trial, is a two-armed, double-blind, randomized chemoprevention trial to test the hypothesis that oral administration of beta-carotene 30 mg/day plus retinol (retinyl palmitate) 25,000 IU/day will decrease the incidence of lung cancer in two high-risk groups: heavy smokers and asbestos-exposed workers who have smoked. The two agents are thought to contribute complementary and potentially beneficial mechanisms of action: anti-oxidant role of beta-carotene and nuclear ligand stimulation of maintenance of differentiated state of epithelial cells by retinol. Smokers are recruited from health insurance rolls; asbestos-exposed workers are recruited from workers' compensation systems, lawyers, physicians, and unions. The study is designed to detect a "maximal potential chemopreventive effect" of 33% reduction in lung cancer incidence, allowing a two-year time-lag to full effect (linear with time), accrual over five years, overall adherence to medication schedule 67% for smokers and 80% for asbestos-exposed workers at three years, placebo groups (due to secular promotion of multivitamins containing active agents of the Trial), loss to competing causes of death 2%/year, and loss to follow-up 2% over full study. Statistical testing procedures involve weighted log-rank test, 80% power, 0.05 level, two-sided. Sample projected: 4277 asbestos-exposed and 13,629 smokers, for a total of 17,906. Pilot studies 1985-1988 yielded a Vanguard population of 1,845.

TYPE: Intervention
TERM: BMB; Dusts; High-Risk Groups; Prevention; Registry; Tissue; Tobacco (Smoking)
SITE: Lung; Mesothelioma
CHEM: Asbestos; Beta Carotene; Retinoids
REGI: SEER (USA)
TIME: 1985 - 1999

*1132   Sherman, K.J.   04972
        Fred Hutchinson Cancer Research Center, Div. of Public Health Sciences MP-381, 1124
        Columbia St., Seattle WA 98104, United States of America (Tel.: (206)6674630; Fax: 6675948)
COLL: Beckmann, A.M.; Daling, J.R.; McDougall, J.K.; Schubert, M.; Ashley, R.L.

Oral Cancer: Epidemiology, Biochemistry, and Immunology

This case-control study will test the hypothesis that infection with HPV and/or HSV type 1 or 2 is related to an increased risk of oral cancer. The study will be conducted in three counties of western Washington. Using the local population-based registry, all men and women aged 18-65, diagnosed with squamous cell carcinoma of the tongue, gum, floor of the mouth, other and unspecified parts of the mouth, and oropharynx, who are residents of three metropolitan counties in western Washington, will be identified and invited to participate. Eligible subjects will be all incident cases (180 men, 120 women) from December 1988 to November 1994. General population controls (360 men, 240 women) will be selected by random digit dialling. Cases and controls will be interviewed regarding history of sexually transmitted diseases, sexual practices, as well as known or suspected risk factors for oral cancer. Tissue specimens and exfoliated oral cavity cells will be collected from each case. A biopsy specimen will be obtained from

UNITED STATES OF AMERICA

a segment of the interviewed controls, and exfoliated oral cavity cells will be obtained from all controls. These tissue samples will be examined using molecular hybridisation techniques for HPV. Blood samples will be collected for all cases and controls and analysed for evidence of prior exposure to HSV types 1 and 2 and several HPVs.

TYPE: Case-Control
TERM: Biochemical Markers; HPV; HSV; Immunology; Registry; Sexual Activity; Sexually Transmitted Diseases
SITE: Oral Cavity; Oropharynx
REGI: Seattle (USA)
TIME: 1990 - 1995

1133 Stanford, J.L. 04315
Fred Hutchinson Cancer Research Center, Univ. of Washington, Div. of Public, Health Sci., Program in Epidemiology, 124 Columbia Parkway, Seattle WA 98104, United States of America
COLL: Weiss, N.S.; Daling, J.R.

Breast Cancer in Middle-Aged Women
The objective is to test the hypothesis that oral progestin use, with or without exogenous oestrogens, reduces the risk of breast cancer in menopausal women. Secondary aims include assessment of prescribing patterns of exogenous hormones in postmenopausal women, and the potential interrelationships between hormone replacement therapy and other hormonally related risk factors for breast cancer, e.g. age at first live birth, parity. The study is a population-based case-control investigation of incident breast cancer in women 50-64 years of age, residing in King County, Washington, USA. Cases will consist of about 500 white women diagnosed with primary breast cancer between 1 January 1988 and 30 June 1990. A population control group of 500 white women, residing in the same county, and who do not have a prior history of breast cancer will be selected by random digit dialling. Controls will be frequency matched to cases by five-year age groups. Cases and controls will be interviewed in person regarding use of exogenous hormones and other recognized risk factors for breast cancer including: detailed aspects of menstrual and reproductive history, use of exogenous hormones (oral contraceptives and hormone replacement therapy), medical history, family history of cancer, and socio-demographic factors. A standard case-control analysis will be performed, including descriptive, stratified, and multivariate analyses. Odds ratios and 95% confidence intervals will be estimated by stratified techniques and logistic regression. The risk of breast cancer will be estimated in relation to "ever use" of progestins and whether the progestin was used with or without oestrogens. Potential confounding variables will be controlled in the analysis. Possible effect modification will also be examined to assess whether specific subgroups of women experience different risk patterns.

TYPE: Case-Control
TERM: Age; Familial Factors; Hormones; Oral Contraceptives; Reproductive Factors; Socio-Economic Factors
SITE: Breast (F)
CHEM: Progesterone
TIME: 1987 - 1992

1134 Thomas, D.B. 02038
Fred Hutchinson Cancer Research Center, 1124 Columbia St., Seattle WA 98104, United States of America (Tel.: (206)4675134)
COLL: Berry, G.; Molina, R.; Modan, B.; Rodriguez Cuevas, H.; de la Cruz, J.; Supawat, C.; Suporn, K.; Suporn, S.; Kenya, P.; Cuadros, A.; Tao, Y.; Ebeling, K.; Stanford, J.L.; Rosenblatt, K.

WHO Collaborative Study of Neoplasia and Steroid Contraceptives
The purposes of this project are to determine whether use of oral or injectable steroid contraceptives alters risks of cancer of the breast, cervix uteri, corpus uteri, ovary, liver and gallbladder. It is multinational hospital-based, case-control study. Data were collected from 1979-1988 in three developed countries (Australia, the GDR, and Israel) and seven developing nations (Chile, China, Colombia, Kenya, Mexico, Philippines, and Thailand). Approximately two controls per case were selected from among new hospital admissions, but they were not matched to individual cases, so all cases of each type could be compared to all controls. All study subjects were interviewed in hospital to ascertain information on contraceptive practices and risk factors for all of the neoplasms under study. All cases were histologically confirmed locally (except some liver cancers), and slides were sent to reference pathologists for standardized review and histological classification. Since 1983, serum samples have been obtained from cervical cancer cases and selected controls from Thailand and Kenya, and interviews of male partners of cervical cancer cases and controls were initiated in Thailand in 1986. All data are processed at the coordinating

UNITED STATES OF AMERICA

center in Seattle, where data analyses and report writing are also performed. Reports have been published or submitted for publication on oral contraceptives and risks of all of the neoplasms under study. The next analyses will primarily focus on depo-medroxyprogesterone acetate and risks of these cancers.

TYPE: Case-Control
TERM: BMB; Contraception; Oral Contraceptives; Serum
SITE: Breast (F); Gallbladder; Liver; Ovary; Uterus (Cervix); Uterus (Corpus)
CHEM: Steroids
LOCA: Australia; Chile; China; Colombia; Germany; Israel; Kenya; Mexico; Philippines; Thailand
TIME: 1979 - 1992

1135 Thomas, D.B. 03433
Fred Hutchinson Cancer Research Center, 1124 Columbia St., Seattle WA 98104, United States of America (Tel.: (206)4675134)
COLL: Vaughan, T.L.; Davis, S.; Menard, T.

Alcohol and Cancers of the Larynx, Oesophagus, Mouth and Stomach

This population-based case-control study is designed to elucidate further the aetiology of cancers of the larynx, oesophagus and oral cavity, in particular to (1) identify the mechanisms by which alcohol enhances the risk of these cancers; (2) determine whether host susceptibility to alcohol is related to risk; (3) determine whether dietary deficiencies of such nutrients as vitamins A and C, zinc, iron and others enhance risk; (4) clarify the role of dental health and care in genesis; (5) identify occupational exposures that increase risk; and (6) compare epidemiological features according to specific subsites. Cases, identified through the population- based SEER tumour registry in Northwestern Washington State consist of all persons diagnosed with one of the three cancers under study in King, Pierce and Snohomish counties between 1 September 1982 and 28 February 1987. An equal number of controls were randomly selected from the same three county populations and have the same age and sex composition as the overall case group. Subjects were interviewed in their homes to ascertain detailed information on prior use of tobacco and alcohol, dietary intake, dental health, and other factors of aetiological interest. In addition, dental information was abstracted from each subject's dental records. Toe nail specimens were also collected from all subjects for measurement of a variety of trace elements. It is anticipated that the results of this investigation will also provide comprehensive information regarding alcohol use and abuse and diet and nutrition in a general population sample of individuals. Analyses of data are underway. Additional cases of adenocarcinomas of the oesophagus, gastro-oesophageal junction and gastric cardia and controls will be accrued for an additional three years.

TYPE: Case-Control
TERM: Alcohol; BMB; Diet; Hygiene; Registry; Toenails; Trace Elements; Vitamins
SITE: Larynx; Oesophagus; Oral Cavity; Stomach
CHEM: Beta Carotene; Iron; Zinc
REGI: Seattle (USA)
TIME: 1983 - 1992

1136 Vaughan, T.L. 04283
Fred Hutchinson Cancer Research Center, Div. of Public Health Sciences, 1124 Columbia St. (W404), Seattle WA 98104, United States of America (Tel.: (206)4675134)
COLL: Berwick, M.; Swanson, G.M.; Isaacson, P.; Lyon, J.L.; Morgan, M.

Epidemiological Study of Nasopharyngeal Cancer

A population-based case-control study of nasopharyngeal cancer is being carried out (a) to determine if occupational and residential exposure to formaldehyde increases the risk of nasopharyngeal cancer and (b) to identify other medical, environmental and lifestyle factors associated with the disease in a low-incidence population. Eligible cases (estimated number 245) will include all persons aged 18-74 years who develop nasopharyngeal cancer between 1 April 1987 and 30 June 1991, and who reside in areas covered by six population-based cancer registries in the United States funded by the SEER Program of the NCI. A random digit dialling technique is being used to select one control per case from among residents of the same areas as the cases. Subjects are interviewed by telephone to determine their occupational and residential histories, along with other factors suspected to be associated with risk of nasopharyngeal cancer, including medical, tobacco, alcohol and dietary histories. Using exposure assessment methods already developed in a preliminary study, indices of formaldehyde exposure will be calculated both from home and workplace sources. Blood samples are being collected and stored for future analysis of HLA type and EBV antibody response. Data processing and analysis are being performed at the coordinating centre in Seattle.

TYPE: Case-Control
TERM: Alcohol; Chemical Exposure; Diet; Environmental Factors; Lifestyle; Tobacco (Smoking); Wood
SITE: Nasopharynx
CHEM: Formaldehyde
TIME: 1988 - 1993

1137 Weiss, N.S. 04763
Univ. of Washington, Sch. of Public Health & Comm. Medicine, Dept. of Epidemiology, 1124 Columbia St., SC-36, Seattle WA 98195, United States of America (Tel.: (206)6674630/6851788; Fax: 5438525)
COLL: Nazar, V.; Eaton, D.; Motulsky, A.; White, E.

Metabolic Susceptibility to Lung Cancer among Smokers
A case-control study is under way to investigate whether individual differences in the metabolism of a carcinogen in cigarette smoke, benzo(a)pyrene, are associated with differences in risk for lung cancer development among cigarette smokers. Cases and controls are identified through pathology services at several hospitals from among patients who undergo lung surgery or autopsy. Cases are long-term users of cigarettes with primary malignant lung cancer. Controls have a cause-of-death or cause-for-surgery unrelated to smoking and are matched broadly to the cases for history of cigarette use and age. Fresh uninvolved lung tissue is collected from each subject and assayed for the presence and activity of enzymes involved in the metabolism of benzo(a)pyrene: glutathione S-transferase and epoxide hydrolase. Existing assay methods have been modified for use with human lung tissue. Data on smoking history and potential confounders are collected through questionnaires mailed to subjects or next-of-kin, and from hospital records. Stratified analysis and logistic regression will be used for data analysis.

TYPE: Case-Control
TERM: Biochemical Markers; Genetic Factors; Tobacco (Smoking)
SITE: Lung
CHEM: Benzo(a)pyrene
REGI: Seattle (USA)
TIME: 1989 - 1992

*1138 Weiss, N.S. 05132
Univ. of Washington, Sch. of Public Health & Comm. Medicine, Dept. of Epidemiology, 1124 Columbia St., SC-36, Seattle WA 98195, United States of America (Tel.: (206)6674630/6851788; Fax: 5438525)
COLL: Daling, J.R.; Rossing, M.A.; Self, S.G.; Moore, D.E.

Risk of Cancer in a Cohort of Infertile Women
Infertile women, particularly women with ovulatory abnormalities, may provide a unique opportunity to assess the role of endogenous hormones in cancer aetiology. In addition, concerns have been raised regarding the potential neoplastic effects of the powerful ovulatory stimulants currently used in the treatment of infertility. This study is a retrospective case-cohort study of cancer incidence among women who sought evaluation and treatment for infertility at any one of several infertility clinics in King County, Washington from 1974-1985. It is anticipated that approximately 4,000 women will be enrolled. Cancer incidence will be ascertained by linking cohort members to the records of the population-based tumour registry of western Washington. Medical records of a subset of women selected for the case-cohort study will be reviewed in order to assess the type of infertility, hormonal treatments received and other factors of interest. Both within-cohort comparisons and comparisons with population rates will be performed in order to assess the risk of various cancers among infertile women according to type of infertility and infertility treatments received.

UNITED STATES OF AMERICA

TYPE: Cohort
TERM: Hormones; Infertility; Registry
SITE: All Sites
REGI: Seattle (USA)
TIME: 1990 - 1992

*1139 Weiss, N.S. 05133
Univ. of Washington, Sch. of Public Health & Comm. Medicine, Dept. of Epidemiology, 1124 Columbia St., SC-36, Seattle WA 98195, United States of America (Tel.: (206)6674630/6851788; Fax: 5438525)
COLL: Stanford, J.L.; Beresford, S.A.A.; Herrinton, L.; Scott, C.R.

Consumption and Metabolism of Lactose and Galactose and the Occurrence of Epithelial Ovarian Cancer

The aims of the study are to test the hypothesis that ovarian cancer risk is increased by (1) consuming relatively high levels of dietary lactose and galactose; (2) having relatively high levels of urinary galactose following ingestion of lactose; and (3) having relatively high erythrocyte levels of metabolic intermediates of galactose following ingestion of lactose, or being heterozygous for a low-activity variant of a specific enzyme of galactose metabolism. Caucasian residents of western Washington who are diagnosed with Stage I epithelial ovarian cancer during the period 1989 to 1991, will be identified through the Cancer Surveillance System, a population-based tumour registry. Each case will be asked to identify a friend control, matched to the case on age. Study subjects will complete a self-administered food frequency questionnaire and an in-person interview, and will undergo an oral lactose challenge followed by collection of blood and urine samples. The samples will be assayed for enzymes and intermediates of lactose and galactose metabolism. 121 cases are expected.

TYPE: Case-Control
TERM: BMB; Blood; Diet; Enzymes; Metabolism; Registry; Urine
SITE: Ovary
REGI: Seattle (USA)
TIME: 1991 - 1993

1140 White, E. 04576
Fred Hutchinson Cancer Research Center, 1124 Columbia St. W202, Seattle WA 98104, United States of America (Tel.: (202)4674678)
COLL: Daling, J.R.; Weiss, N.S.

Physical Activity and Colon Cancer

Studies of job activity and colon cancer suggest a decreased risk associated with high activity jobs. In addition, there has been a decline in colon cancer incidence among men and women under age 50 in western Washington, which may be related to an increase in leisure time activity. The specific aims of this study are: (1) to determine if physical activity (job activity plus leisure activity) is related to reduced risk of colon cancer, and (2) to determine whether leisure time activity affects colon cancer incidence independently of job related physical activity. The first aim addresses an aetiological question, the second will help answer the cancer control issue of whether an exercise programme during leisure time could be an intervention to prevent colon cancer. To study the relationship between physical activity and colon cancer, a population-based case-control study of 400 cases and 400 controls is proposed. The case group will be incident colon cancer cases age 25-59, diagnosed between January 1986 - December 1990 in three counties in Washington State identified through the Seattle-Puget Sound SEER Cancer Registry. Controls will be an age and sex stratified sample identified through random digit dialling. Cases and controls will be interviewed by telephone. Questions will concern leisure time physical activity, job activity, household activity, diet, weight, and demographic characteristics. Data analysis will focus on the effect of physical activity on colon cancer, using quantitative estimates of energy expenditure categorised by type of activity (e.g., leisure activities) and by intensity of activity. Various components of diet including calories, fat, fibre, calcium and vitamin A will also be considered in the analyses.

UNITED STATES OF AMERICA

TYPE: Case-Control
TERM: Diet; Fat; Fibre; Nutrition; Occupation; Physical Activity; Prevention; Registry; Vitamins
SITE: Colon
REGI: Seattle (USA)
TIME: 1987 - 1992

SOUTHFIELD

1141 Reid, R. 04484
The Reid Inst. for Research, and Education, 29355 Northwestern Highway, Suite 215, Southfield MI 48034, United States of America (Tel.: (313)3544338; Fax: 3500851)
COLL: Greenberg, M.D.; Campion, M.; Omoto, K.H.; Rutledge, L.H.; Lovincz, A.T.; Rothrock, R.

Polyploid Papillomaviral Infections and Cervical Cancer
This multi-center cohort study will evaluate the role of cervical HPV infection as a risk factor for progression of dysplasia. Women with abnormal Pap smears will undergo colposcopy and cervicography. A cohort of 300 women with biopsy-confirmed CIN 2 or lower will be followed for three years. Presence of type-specific HPV will be determined by nucleic acid hybridization. Other factors being studied include nicotine/cotinine levels in serum and cervical mucus and DNA ploidy. Differences in progression among women with various types of HPV will be analysed.

TYPE: Cohort
TERM: Cytology; DNA; HPV; Premalignant Lesion; Sexually Transmitted Diseases
SITE: Uterus (Cervix)
TIME: 1987 - 1992

*1142 Tilley, B.C. 05030
Henry Ford Health System, Div. of Biostatistics, & Research Epidemiology, 23725 Northwestern Highway, Southfield MI, United States of America (Tel.: (313)3548060; Fax: 3544812)
COLL: Visintainer, P.F.; Vernon, S.W.; Myers, R.; Sower, M.F.; Glanz, K.

Colorectal Cancer Screening and Nutrition Intervention
This project involves evaluation of a nutrition and screening enhancement intervention programme among pattern and model automotive workers of the General Motors Corporation (GM). Pattern and model makers have been shown to have an elevated risk of colorectal cancer mortality. Consequently a colorectal cancer screening programme has been administered to eligible employees since 1980. For this intervention trial, 26 worksites, with more than 6000 employees, will be randomly selected to receive either both interventions (screening enhancement and nutrition), or the usual GM screening and any health promotion programmes currently available to employees. The nutrition intervention will combine education, environmental change, and regular contact to achieve a reduction in total fat intake, percent calories from fat, and an increase in average dietary fibre intake. The screening intervention will use constructs of the Health Belief model and Social Learning Theory to promote employee participation in the corporate screening programme. To evaluate the interventions, intervention and control worksites will be compared using a generalization of the paired t-test that accounts for cluster design.

TYPE: Cohort; Intervention
TERM: Diet; Fat; Fibre; Nutrition; Occupation; Screening
SITE: Colon; Rectum
OCCU: Model and Pattern Makers
TIME: 1991 - 1996

STANFORD

1143 Paffenbarger Jr, R.S. 04350
Stanford Univ., School of Medicine, Dept. of Health Research and Policy, HRP Bldg, Room 113, Stanford CA 94305-5092, United States of America (Tel.: (415)7236417/4975460; Fax: 7256951; Tlx: 3731148 sumc/348402)
COLL: Whittemore, A.S.; Wing, A.L.

Physical Activity, Body Size and Cancer Incidence
Recent epidemiological evidence has associated sedentary work and leisure-time activities with increased risk of colon and breast cancer; overweight and endocrine patterns have for some time been linked with colon, breast, prostate and other cancers. A data base on 25,000 US college alumni will be

UNITED STATES OF AMERICA

constructed using retrospectively collected data for 1916-1950 (physical examination, social, and athletic records), contemporary alumni data since 1962 on the same study subjects (mail questionnaire responses pertaining to exercise, other lifestyle elements, personal health, and family disease patterns), and mortality certification. Data on 6,000 San Francisco Bay Area longshoremen are available from multiphasic health examinations, annual job classifications, and death certificate assessments. A projected questionnaire will update the alumni follow-up to 23 years, while survey of annual job assignments will extend the longshoremen follow-up to 34 years. Multivariate analyses will examine absolute measurements and changing patterns of physical activity, body size, and other lifestyle characteristics. Subjects will be classified as to physical activity by type, frequency, intensity, duration, kilocalorie expenditure, and postural nature (sitting, standing, and locomotion). Expected numbers of cancer cases offer statistical power sufficient to test hypotheses of physical activity and body size as causation for cancers of the colon, rectum, and prostate, and perhaps others such as pancreas, lung and breast. Within these two diverse populations, differences in physical activity, body size, cigarette habits, blood pressure levels, etc. significantly altered incidence of cardiovascular-respiratory diseases and estimates of longevity. Those findings invite parallel study of cancer risk, especially since much of the information gained thus far has been applicable to public health.

TYPE: Cohort
TERM: Familial Factors; Lifestyle; Physical Activity; Physical Factors; Tobacco (Smoking)
SITE: Breast (F); Colon; Lung; Pancreas; Prostate; Rectum
TIME: 1987 - 1992

*1144   Paffenbarger Jr, R.S.   04997
Stanford Univ., School of Medicine, Dept. of Health Research and Policy, HRP Bldg, Room 113, Stanford CA 94305-5092, United States of America (Tel.: (415)7236417/4975460; Fax: 7256951; Tlx: 3731148 sumc/348402)
COLL: Whittemore, A.S.; Wu, A.; Gallágher, R.P.; Howe, G.R.

Prostate Cancer in High, Medium and Low Risk Populations
This study is a population-based case-control study of modifiable risk factors for prostate cancer among blacks, whites and Asians in Los Angeles, San Francisco and Hawaii. Investigators will use a common protocol and questionnaire to administer personal interviews to approximately 500 black patients, 500 white patients, and 500 Asian patients with histologically confirmed carcinoma of the prostate, and 1,500 black, white and Asian population-based controls. Controls will be matched to cases on age, ethnicity and region of residence. A Canadian component (Vancouver and Toronto) will contribute Asian cases and controls only. The study will examine within each ethnic group and each of the two age groups (less than 70 years, 70+ years) how prostate cancer risk varies with diet, physical activity, and body size. Serum collected from controls will be analysed for hormones and prostate-specific antigen.

TYPE: Case-Control
TERM: BMB; Diet; Physical Activity; Physical Factors; Registry; Serum
SITE: Prostate
LOCA: Canada; United States of America
REGI: Bay Area (USA); Br. Columbia (Can); Hawaii (USA); Los Angeles (USA); Ontario (Can)
TIME: 1989 - 1993

TUCSON

*1145   Clark, L.C.   05136
Univ. of Arizona, College of Medicine, Dept. of Family and Community Medicine, Epidemiology Sect., 2504 E. Elm St., Tucson AZ 85716, United States of America (Tel.: (602)6264890; Fax: 3217774)
COLL: Combs, G.; Turnbull, B.

Colon Cancer Screening of a Defined Population
The aim of this study is to test the hypothesis that low selenium status is related to the presence of colonic neoplasms in a defined population. This is a cross-sectional study of 1,700 patients enrolled in a pair of double-blind, placebo-controlled clinical trials on the prevention of non-melanoma skin cancers with a supplement of selenium. The supplementation levels are 200 mcg and 400 mcg per day in the form of high selenium Brewer's yeasts as the treatment and a Brewer's yeast placebo. The patients will be screened according to the ACS-NCI colon cancer screening guidelines. This study will generate two cohorts of individuals that can be followed in a prospective cohort study to evaluate the relationship of selenium status and the formation of colonic neoplasms. Patients are from seven clinical centres in the

UNITED STATES OF AMERICA

Eastern coastal plain of the US. Data will be collected in the dermatology and collaborating gastroenterology clinics and processed in the study office in Tucson, Arizona.

TYPE: Cohort; Cross-Sectional
TERM: Screening
SITE: Colon
CHEM: Selenium
TIME: 1990 – 1993

*1146   Clark, L.C.   05144
Univ. of Arizona, College of Medicine, Dept. of Family and Community Medicine, Epidemiology Sect., 2504 E. Elm St., Tucson AZ 85716, United States of America (Tel.: (602)6264890; Fax: 3217774)
COLL: Combs, G.; Turnbull, B.

Nutritional Prevention of Non-Melanoma Skin Cancer

The aim of this study is to test the hypothesis that improved selenium status will prevent the formation of non-melanoma skin cancers. Treatment consists of a supplement of 200 mcg of selenium, in the form of high selenium Brewer's yeast; the placebo is an identical tablet of Brewer's yeast. The patients are drawn from seven dermatology practices in the Eastern coastal plain of the US. Patients must be under the age of 80, white, have a 5-year life expectancy, and have a negative history of hepatic or renal disease. 1,300 patients have been randomized into this study; over 3,400 person-years of observation have been collected in the dermatology clinic sites.

TYPE: Intervention
TERM: BMB; Prevention
SITE: Skin
CHEM: Selenium
TIME: 1983 – 1993

*1147   Clark, L.C.   05145
Univ. of Arizona, College of Medicine, Dept. of Family and Community Medicine, Epidemiology Sect., 2504 E. Elm St., Tucson AZ 85716, United States of America (Tel.: (602)6264890; Fax: 3217774)
COLL: Combs, G.; Turnbull, B.

Prevention of Squamous Cell Carcinoma of the Skin with a Nutritional Supplement of Selenium

The aim of this study is to test the hypothesis that improved selenium status will prevent the formation of squamous cell carcinomas of the skin. A supplement of 400 mcg of selenium, in the form of high selenium Brewer's yeast is the treatment; the placebo is an identical tablet of Brewer's yeast. The patients are drawn from a private dermatology practice in Macon, Georgia. 356 of the proposed 400 patients have been randomized into the study. Patients must be under the age of 80, white, have a 5-year life expectancy, and have a negative history of hepatic or renal disease.

TYPE: Intervention
TERM: BMB; Prevention
SITE: Skin
CHEM: Selenium
TIME: 1989 – 1994

*1148   Garcia, F.   05053
Univ. of Arizona, Cancer Center, Sect. of Epidemiology and Biometry, 1515 N Campbell St., Room 2942, Tucson AZ 85724, United States of America (Tel.: (602)6264010)
COLL: Stark, A.; Moon, T.E.; Cartmel, B.; Villar, H.; Giordano, G.; McNamara, D.

Tissue Lipids and Dietary Markers for Breast Carcinogenesis

A case-control study is in progress in south-eastern Arizona to assess the relationship between breast carcinogenesis and nutrient status as reflected by tissue lipids and food frequency questionnaire. Forty-eight Hispanic and non-Hispanic caucasian women with pathologically confirmed primary, non-metastatic breast carcinoma have been enrolled in the study. Using a random-digit dialling procedure, 135 community controls have been selected. Adipose tissue biopsies, and a sample of buccal epithelial cells were collected from each consenting subject. The specimens will be analysed using gas liquid and thin layer chromatography in order to identify the relative proportions of 13 individual fatty acids. A food frequency questionnaire concerning over 250 food items was completed by all

UNITED STATES OF AMERICA

subjects. Additionally a detailed questionnaire on established risk factors, menstrual history, anthropometric data, and exposure to hormones, tobacco and ethanol was completed by all cases and most controls.

TYPE: Case-Control
TERM: BMB; Diet; Lipids; Nutrition; Registry
SITE: Breast (F)
REGI: Arizona (USA)
TIME: 1988 - 1992

1149 Lebowitz, M.D. 03312
Univ. Arizona, College of Medicine, Arizona Medical Center, Division of Respiratory Sciences, Tucson AZ 85724, United States of America
COLL: Burrows, B.; Knudson, R.

Longitudinal Study of Obstructive Airways Disease
A 20-year longitudinal study of obstructive airways diseases in a community population of 4,800 persons is being undertaken. The population is age-stratified, so that many older cohorts are included. It is family-oriented for family concordance studies. It is believed that sufficient cases of lung cancer will be seen to test hypotheses on smoking, the interrelationship between lung cancer and obstructive airways diseases, immunological states and other environmental factors that may be related to the development of lung cancer. Death certificates will be examined. Chest X-rays, medical history, immunological and physiological tests will also be conducted. Questions will be specifically asked about occupational and other environmental exposures. At the end of the first 12 years, 12% had died; they were older, included more smokers, males, and subjects with chronic disease. Lung cancer was the underlying cause in 36/547 (6.6%). Of these, 40% had moderate or severe obstructive airways disease, but only 11% had such reported on the death certificate. There are common risk factors for both causes of death.

TYPE: Cohort
TERM: Environmental Factors; Occupation; Premalignant Lesion; Tobacco (Smoking)
SITE: Lung; Respiratory
TIME: 1972 - 1992

1150 Moon, T.E. 03567
Univ. Arizona Cancer Center, Coll. Med., Epidemiology & Biometry Section, Dept. Family & Community Medicine, 1515 North Campbell St., Tucson AZ 85724, United States of America (Tel.: (602)6264010; Tlx:9109521238)
COLL: Levine, N.; Alberts, D.; Schreiber, M.M.; Cartmel, B.

Chemoprevention of Skin Cancer by Vitamin A
The objective of the first study is to evaluate the role of retinol (Vitamin A) in the prevention of skin cancer. Persons with a clinical history of at least ten actinic keratoses and no prior history of cancer are enrolled in a prospective randomized double blind placebo-controlled intervention trial. A total of 2,800 such subjects identified by all dermatologists in Arizona will be included. The study will evaluate the efficacy of continued pharmacological doses of retinol in the reduction of risk of skin cancer, the effects of medication compliance, motivation methods, etc. The objective of the second study is to evaluate the role of retinol (Vitamin A) or 13-cis retinoic acid (synthetic Vitamin A) in the prevention of new skin cancers. Persons with a diagnosis of 4 or more non-melanoma skin cancers are enrolled in this prospective double blind placebo controlled intervention trial. A total of 711 eligible subjects identified by community dermatologists or use of newspaper advertisements are included.

TYPE: Intervention
TERM: Diet; Metabolism; Prevention; Vitamins
SITE: Skin
CHEM: Retinoids
TIME: 1983 - 1992

1151 Moon, T.E. 04141
Univ. Arizona Cancer Center, Coll. Med., Epidemiology & Biometry Section, Dept. Family & Community Medicine, 1515 North Campbell St., Tucson AZ 85724, United States of America (Tel.: (602)6264010; Tlx:9109521238)
COLL: Alberts, D.; Levine, N.; Schreiber, M.M.; McNamara, N.

UNITED STATES OF AMERICA

Epidemiology of Cutaneous Cancers
The objective of these two studies is to evaluate the relative and attributable risk of skin cancer for demographic, phenotypic, environmental (including diet) and biochemical factors. The first case - control study includes at least 400 subjects with newly diagnosed primary melanoma and 800 age, sex, race, and geographical region matched controls. The second case-control study includes at least 400 subjects with newly diagnosed basal or squamous cell skin cancer with no history of cancer. Controls for this second study include 800 age, sex, race and geographical region matched subjects. Controls will be identified by a random phone digit procedure. Biochemical measurements include blood retinyl palmitate, beta carotene and selenium and tissue levels of ornithine decarboxylase, and fatty acids.

TYPE: Case-Control
TERM: Diet; Environmental Factors; Enzymes; Lipids; Metals; Vitamins
SITE: Melanoma; Skin
CHEM: Beta Carotene; Retinoids; Selenium
TIME: 1983 - 1992

1152   Moon, T.E.   04653
Univ. Arizona Cancer Center, Coll. Med., Epidemiology & Biometry Section, Dept. Family & Community Medicine, 1515 North Campbell St., Tucson AZ 85724, United States of America (Tel.: (602)6264010; Tlx:9109521238)
COLL: Cartmel, B.; Villar, H.; King, D.; Micozzi, M.; McNamara, D.

Case-Control Study of Breast Cancer
This research applies complementary methods to quantitate macro-and micro-nutrient intake and nutrient-mediated factors to evaluate the aetiology and predict breast cancer risk in a defined white Hispanic and white non-Hispanic population in Arizona. Nutrients of primary focus include dietary lipids, retinoids, carotenoids, selenium and total calories. Nutrient-mediated factors include body size, body fat, tissue and blood fatty acids, plasma retinoids, carotenoids and selenium. Exposure estimates are based on interview, biochemical, anthropometric, and bioelectrical impedance of 600 post menopausal cases of breast cancer and 600 controls randomly selected from Arizona residents.

TYPE: Case-Control
TERM: Biochemical Markers; Diet; Fat; Lipids; Metals; Nutrition
SITE: Breast (F)
CHEM: Beta Carotene; Retinoids; Selenium
TIME: 1988 - 1994

1153   Surwit, E.A.   03568
Arizona Cancer Center, Dept. of Biometry, Computing & Epidemiology (BICEPs), 1501 N. Campbell St., Tucson AZ 85724, United States of America (Tel.: (602)6264010; Tlx:9109521238)
COLL: Meyskens, F.L.; Slymen, D.

Chemoprevention of Cervical Dysplasia
The objective of the first study is to determine the relationship between serum micronutrient levels and cervical dysplasia. Measurements of beta carotene, vitamin A and vitamin C dietary intake and serum levels are determined using high pressure liquid chromatography assays. Dietary intake is measured using a food frequency instrument administered by a nutritionist and food intake records completed by the subjects. Thirty patients with moderate or severe cervical dysplasia, as documented by Papanicolaou slide and colposcopy will be entered on study. Cases are identified through out-patient clinics at the Arizona Health Sciences Center. A control group of women without dysplasia consisting of 30 individuals matched for age, age at first intercourse and number of sexual partners will be identified by clinics throughout the Tucson community. The objective of the second study is to evaluate the efficacy of retinoids in the prevention of cervical cancer. A dose-ranging Phase I Study evaluating Beta-All-Trans Retinoic Acid was carried out with 36 patients with moderate to severe cervical dysplasia. A second phase I Study evaluating retinyl acetate involving 27 patients is nearing completion. A Phase II Trial of 18 patients with moderate to severe dysplasia who received 0.372 grams of Beta-All-Trans Retinoic Acid daily for four days during four consecutive weeks nears completion. A randomized prospective clinical trial to evaluate Beta-All-Trans Retinoic Acid has begun. Three hundred patients will be enrolled and randomly allocated to the retinoic acid or placebo.

UNITED STATES OF AMERICA

TYPE: Cross-Sectional; Intervention
TERM: Diet; Premalignant Lesion; Prevention; Sexual Activity; Vitamins
SITE: Uterus (Cervix)
CHEM: Beta Carotene; Retinoids
TIME: 1984 - 1992

TYLER

1154 McLarty, J.W. 03768
Univ. of Texas Health Center, P.O. Box 2003, Tyler TX 75710, United States of America (Tel.: (214)8773451)
COLL: Kummet, T.D.

Beta Carotene, Retinol and Lung Cancer Chemoprevention
A randomized clinical trial of beta carotene and retinol versus placebo as lung cancer preventive agents is being conducted. This five-year study will enrol at least 600 persons occupationally exposed to asbestos. Bronchial epithelial changes, as evaluated by repeated sputum cytology, will be the primary measure of response. Serum levels of beta carotene will be studied with respect to the degree of sputum metaplasia or dysplasia found at initial examination and with respect to bronchial epithelial changes throughout the clinical trial.

TYPE: Intervention
TERM: Cytology; Dusts; Occupation; Premalignant Lesion; Prevention; Sputum; Vitamins
SITE: Lung
CHEM: Asbestos; Beta Carotene; Retinoids
TIME: 1984 - 1992

WASHINGTON

1155 Gibb, H. 04438
US Environmental Protection Agency, Off. Health & Environmental Assessment, 401 Main St. SW, RD-689, Washington DC 20460, United States of America (Tel.: (202)3825720)
COLL: Lees, P.; Stewart, W.

Derivation of Exposure-Response Relationships from Analysis of Retrospective Industrial Hygiene and Epidemiological Data: Chromium Compounds and Lung
Hexavalent chromium compounds are acknowledged to be human carcinogens. This conclusion is based on the results of occupational studies of chromate production workers and trivalent chromium and animal bioassays of hexavalent chromium compounds. Data from occupational studies of chromate production workers to date have not been detailed enough to examine the carcinogenic risk of exposure to trivalent and hexavalent chromium compounds separately. This study will examine the lung cancer mortality risk for a cohort of chromate production workers by exposure to both hexavalent and trivalent chromium compounds. Lung cancer response by solubility of the chromium compounds will also be examined, and a lung cancer dose-response analysis for the various chromium compounds will be attempted. The cohort contains over 2,500 workers. Over 200,000 industrial hygiene measurements, some of which date back to the first date of entry (1950) of workers into the cohort, will be used to estimate exposure. Smoking data on part of the cohort is available.

TYPE: Cohort
TERM: Dose-Response; Metals; Occupation
SITE: All Sites; Lung
CHEM: Chromium
OCCU: Chromate Pigment Workers
TIME: 1988 - 1992

*1156 Kikendall, J.W. 04963
Walter Reed Army Medical Center, Gastroenterology Serv., Washington DC 20307-5001, United States of America (Tel.: (202)5761768; Fax: 5762478)
COLL: Bowen, P.; Burgess, M.; Magnetti, C.; Woodward, J.; Sobin, L.; Langenberg, P.

Risk Factors for Colonic Neoplasia
361 subjects undergoing colonoscopy because of occult bleeding or barium enema suggestive of polyp are being studied. All were in good general health at recruitment. Each subject underwent a dietary

UNITED STATES OF AMERICA

assessment and evaluation of serum nutrients and growth factors, as well as of other cancer risk factors. Subjects were classified as controls, adenomas, or cancer according to the results of colonoscopy. The three groups are compared using univariate and multivariate analysis to define potential risk factors. To date, the data have shown that cigarettes and alcohol are independently associated with colonic adenomas and that serum gastrin is not elevated in subjects with colonic neoplasia. Assessment of the following risk factors is in progress: carotenoids, retinol, tocopherols, zinc, calcium, vitamin D, cruciferous vegetables, insulin-like growth factor 1.

TYPE: Case-Control
TERM: Alcohol; Diet; Nutrition; Polyps; Tobacco (Smoking); Vegetables; Vitamins
SITE: Colon
CHEM: Calcium; Retinoids; Tocopherol; Zinc
TIME: 1984 - 1992

1157   McGowan, L.   01419
George Washington Univ. Medical Center, Div. Gynecologic Oncology, Dept. of Obstetrics & Gynecology, 2150 Pennsylvania Ave. NW, Washington DC 20037, United States of America (Tel.: (202)9944218; Fax: 9940815)
COLL: Hoover, R.N.; Lesher, L.; Hartge, P.; Morris, H.J.

Ovarian Cancer in Greater Washington
Ovarian cancer risk has been estimated according to various reproductive and other suggested risk factors, using interview and medical records data from a case-control study. Identification was attempted of all women aged 20-79 years residing in the   Washington, DC, metropolitan area who were first diagnosed at surgery with microscopically confirmed primary epithelial ovarian cancer from August 1978 to June 1981. The discharge lists of all 33 area hospitals that treated ovarian cancer were regularly checked. Cases included women with tumours of low malignant potential and frankly malignant tumours. For all potential cases, microscopic slides made from the tumour tissue were obtained. Those found not to have definite primary ovarian cancer of the epithelial type by clinical and microscopic evaluation were excluded from the study. 400 cases aged 20-79 years with histologically confirmed epithelial cancer were identified, of whom 296 were interviewed. Controls were identified from hospital discharge lists and matched to cases according to age, race, hospital, and date of discharge. A woman was not eligible to be a control if her discharge diagnosis was potentially related to the exposures under study, and it was ascertained from physician's records that the woman had at least one ovary intact. 439 women aged 20-79 years, eligible to be controls, were identified, of whom 343 were interviewed. Trained, experienced medical interviewers administered a standardized questionnaire in the patient's home. The interviewers obtained a detailed history of all gynaecological surgery and elicited which organs were involved in each procedure. Effects on ovarian cancer risk were measured by the estimated rate ratio, the ratio of ovarian cancer incidence in the exposed group to that in the unexposed. The estimates were adjusted for the effects of confounding variables by stratified contingency table analysis and by logistic regression models. Papers appeared in Oncology 3:51-62, 1989; Gynecol. Oncol. 33:129-132, and in Am. J. Obstet. Gynecol. 161:10-16, 1989.

TYPE: Case-Control
TERM: Drugs; Hormones; Infertility; Virus
SITE: Ovary
CHEM: Oestrogens; Steroids
TIME: 1978 - 1992

WAYNE

1158   Lucas, L.J.   03966
American Cyanamid Company, One Cyanamid Plaza, Wayne NJ 07470, United States of America (Tel.: (201)8312318; Fax: 8312821 ; Tlx: 219136)
COLL: Utidijian, H.M.D.; Caporossi, J.C.; Swaen, G.M.H.

Mortality of Employees Exposed to Acrylamide, Acrylonitrile and Methyl Methacrylate
Retrospective cohort mortality studies have been undertaken to determine if there is an excess cancer risk among employees exposed to the following chemicals: acrylonitrile, acrylamide and methyl methacrylate. The numbers of persons exposed are, respectively: 1,700, 3,000 and 1,300. In each study the mortality experience of exposed workers will be compared with that of an internal plant population, and with that of the general regional and national populations. Reports on cause specific mortality patterns as of 1983 have been prepared. Overall, no increased risk was observed for cancer of all

UNITED STATES OF AMERICA

combined sites. Efforts are currently underway to update work histories and vital status of all cohort members. These data will be integrated with exposure data to evaluate exposure-response relationships. Papers have been published in J. Nat. Cancer Inst. 78(1):192-193, 1987; J. Occup. Med. 31(1):41-46, and 31(7):614-617, 1989 (Collins, J.J. et al.).

TYPE: Cohort
TERM: Chemical Exposure; Occupation; Plastics
SITE: All Sites
CHEM: Acrylamides; Acrylonitrile; Methyl Methacrylate
OCCU: Chemical Industry Workers
LOCA: Netherlands; United States of America
TIME: 1983 - 1992

1159 Lucas, L.J. 04842
American Cyanamid Company, One Cyanamid Plaza, Wayne NJ 07470, United States of America (Tel.: (201)8312318; Fax: 8312821 ; Tlx: 219136)
COLL: Utidijian, H.M.D.; Caporossi, J.C.

Mortality Study of Employees Exposed to Ethylene Oxide

A retrospective cohort mortality study will be conducted to test the hypothesis that workers exposed to ethylene oxide during the industrial production of ethylene oxide sterilized medical products are at increased risk of mortality from malignant neoplasms, in particular leukaemia. The cohort will include about 4,600 employees identified from company records who were hired between 1952 and 1988. Using national and regional external comparisons the mortality experience of the cohort will be examined from 1 January 1957 through 31 December 1990. Exposure-response relationships will be assessed from work history data and industrial hygiene measurements.

TYPE: Cohort
TERM: Occupation
SITE: All Sites; Leukaemia
CHEM: Ethylene Oxide
TIME: 1990 - 1993

WORCESTER

*1160 Hebert, J.R. 05023
Univ. of Massachusetts, Medical Center, 55 Lake Ave. North, Worcester MA 01655, United States of America (Tel.: (508)8564129; Fax: 8563840)
COLL: Reale, F.R.; Jederlinic, P.; Lew, R.; Rogers, E.

Relationship Between Precancerous Conditions of the Lung and Dietary Factors in High Risk Populations

This is a multi-phase study of the relationship between dietary factors, especially dietary lipids and lipid-soluble vitamins, and cytological conditions and DNA ploidy characteristics that precede frank squamous cell carcinoma of the lung. The first phase of establishing the utility of the cell markers is currently under way in a study of 30 non-smokers, 30 current smokers of > /20 cigarettes per day, and 30 patients newly diagnosed with squamous cell carcinoma of the lung. The follow-up studies planned include an antioxidant vitamin intervention trial of current smokers and a cross-sectional study of black and white Americans from a Boston inner city neighbourhood. The latter is motivated by concern at the large and growing divergence in lung cancer rates between balcks and whites that is not attributable to cigarette smoking.

TYPE: Cross-Sectional; Intervention
TERM: Biochemical Markers; DNA; Diet; High-Risk Groups; Lipids; Nutrition; Premalignant Lesion; Vitamins
SITE: Lung
TIME: 1990 - 1996

URUGUAY

MONTEVIDEO

1161 De Stéfani, E. 03622
Hosp. de Clínicas "Dr. Manuel Quintela", Depto de Patologia, Ave. Italia y Las Heras, Montevideo, Uruguay (Tel.: 781045; Tlx: 22348 urexpor uy)
COLL: Pellegrini, H.D.; Carzoglio, J.; Cendan, M.; Olivera, L.

Gastric Cancer in Uruguay

Although rates from stomach cancer in Uruguay have been declining since 1964, counties in the north display high rates and adjusted rates for the whole country are in the order of 23/100,000 in males and 10/100,000 in females. Ingestion of salted meat and exposure to a local infusion known as "mate" have been suspected as aetiological factors. This project attempts to evaluate the possible association of these variables with gastric cancer in a case-control study. Only incident and histologically proven cases will be eligible. Lauren's classification in intestinal and diffuse types will be employed. According to admission rates at the University Hospital, an average of 70 cases per year is expected. In the study period a total of 200 cases will be assembled and interviewed. In order to increase the statistical power of the study, four controls per case will be selected randomly from patients admitted to the same hospital. Patients with oesophageal, oropharyngeal, laryngeal and lung cancer will not be eligible, in order to assess tobacco smoking and alcohol drinking effects.

TYPE: Case-Control
TERM: Alcohol; Cooking Methods; Diet; Mate; Metals; Tobacco (Smoking)
SITE: Stomach
CHEM: Sodium Chloride
TIME: 1985 - 1992

1162 De Stéfani, E. 03624
Hosp. de Clínicas "Dr. Manuel Quintela", Depto de Patologia, Ave. Italia y Las Heras, Montevideo, Uruguay (Tel.: 781045; Tlx: 22348 urexpor uy)
COLL: Olivera, L.; Pellegrini, H.D.; Carzoglio, J.; Ordoqui, G.

Dark Tobacco Exposure and Lung Cancer

It has been suggested that dark tobacco exposure could be associated with a greater risk of developing lung cancer than smoking the Virginia brands. Uruguay is particularly well suited for such assessment due to the rather similar prevalence of both kinds of exposures. This project aims to assess relative risk (RR) estimates for dark and Virginia tobacco. Only incident cases of lung cancer occurring in males aged 30-79 will be selected and classified histologically according to the WHO system. In order to detect an RR of 2.5 with a significance level of 5% and a power of 0.80, at least 80 cases of each major histological type will be assembled. The control group (one control per case) will be selected from the same hospital and will include males of the same age range. Patients with respiratory conditions and neoplasms with well documented associations with smoking will be excluded. Usual residence, occupation, type of tobacco smoked, dose (in cigarettes per day) duration of smoking, years since quitting, use of hand rolled cigarettes and filter use are the variables included in the study. Preliminary analysis of data collected on 804 cases and 680 controls shows a RR 1.6 higher for dark tobacco compared with Virginia tobacco.

TYPE: Case-Control
TERM: Occupation; Tobacco (Smoking)
SITE: Lung
TIME: 1985 - 1992

1163 Oreggia Coppetti, F.V. 04142
Univ. of Montevideo, Hosp. de Clínicas, Dept. of Otolaryngology, French 2013 - Carrasco, Montevideo, Uruguay (Tel.: +598 2 604648)
COLL: De Stéfani, E.; Rivero, S.; Fierro, L.

Risk Factors in Cancer of the Oral Cavity, Pharynx and Larynx

An earlier study has been extended in order to gain further insight into tobacco and alcohol carcinogenesis in the oral cavity, pharynx and larynx. The effect of smoking cessation will be examined according to duration, intensity of habit, and type of tobacco smoked. Data will be collected according to a structured questionnaire at the University Hospital and Oncology Institute. The number of cases expected per year is 200; each case will be matched on age, sex, and place of residence to two controls, selected from the same hospitals, with diseases not related to tobacco and alcohol.

URUGUAY

TYPE: Case-Control
TERM: Alcohol; Time Factors; Tobacco (Smoking)
SITE: Larynx; Oral Cavity; Pharynx
TIME: 1991 – 1993

*1164 **Ronco, A.L.** 05241
Registro Nacional de Cáncer, Eduardo Acevedo 1530, 11200 Montevideo, Uruguay (Tel.: +598 2 486594)
COLL: Vassallo, J.A.

Atlas of Cancer Mortality in Uruguay 1989-1992
The aim of the study is to present the patterns of geographical distribution of cancer mortality in Uruguay for the period 1989-1992. The source is death certificates provided by the Statistics Bureau of the Ministry of Health, collected and coded (ICD) in the National Cancer Registry. The data are being processed to establish crude and age-adjusted rates for the most important types of cancer in the 19 Departments of Uruguay. The patterns of cancer distribution are also compared with those of the main non-cancer causes of death (cardiovascular, respiratory) for the same period.

TYPE: Mortality
TERM: Geographic Factors; Mapping; Registry
SITE: All Sites
REGI: Uruguay (Uru)
TIME: 1990 – 1993

1165 **Vassallo, J.A.** 04510
Registro Nacional de Cáncer, Eduardo Acevedo 1530, 11200 Montevideo, Uruguay (Tel.: +598 2 486594)
COLL: Leibovici, S.; Ronco, A.L.

Cancer Incidence and Mortality in a Department of Uruguay
Uruguay is administratively subdivided into 19 Departments. One of them, Cerro Largo, which has the highest cancer rates in the country, and adequate medical resources, is being used for detailed study of the high cancer mortality rates in Uruguay. This correlation study attempts to identify the pattern of cancer mortality in this chosen geographical region, and compare it with the whole country. The risk factors to which the population of Cerro Largo is exposed (consumption of tobacco, meat, the infusion known as "mate", etc.), will be analysed, with the object of correlating them with the magnitude of the tumour incidence and mortality rates.

TYPE: Correlation; Incidence; Mortality
TERM: Diet; Fat; Mate; Registry; Tobacco (Smoking)
SITE: All Sites
REGI: Uruguay (Uru)
TIME: 1988 – 1992

*1166 **Vassallo, J.A.** 05243
Registro Nacional de Cáncer, Eduardo Acevedo 1530, 11200 Montevideo, Uruguay (Tel.: +598 2 486594)
COLL: Ronco, A.L.

Cancer in Frontier Inhabitants of Uruguay
Uruguay has long (500 Km) frontiers with Argentina and Brazil. It is well known that the lifestyle and habits in frontier regions are different from the rest of the countries involved. The cancer rates for Uruguay will be compared with those for Argentina and Brazil. The study will try to assess if the influence from the neighbouring countries on the habits of the frontier cities modifies the cancer rates. Data will be obtained from the National Cancer Registry.

URUGUAY

TYPE: Incidence
TERM: Environmental Factors; Geographic Factors; Lifestyle; Registry
SITE: All Sites
REGI: Uruguay (Uru)
TIME: 1991 - 1994

*1167   Vassallo, J.A.   05244
Registro Nacional de Cáncer, Eduardo Acevedo 1530, 11200 Montevideo, Uruguay (Tel.: +598 2 486594)
COLL: Ronco, A.L.

Geographic Pathology of Cancer in Uruguay and Its Correlation with Risk Factors
A study of the distribution of cancer in Uruguay is being carried out. The main goals are (1) to establish correlations with risk factors (lifestyle, habits, diet) in different geographical areas; (2) to assess the magnitude of cancer in the different areas; and (3) to obtain the necessary information to develop local and/or regional strategies for cancer prevention.

TYPE: Correlation; Incidence; Mortality
TERM: Diet; Environmental Factors; Geographic Factors; Lifestyle; Prevention; Registry
SITE: All Sites
REGI: Uruguay (Uru)
TIME: 1989 - 1992

USSR

LENINGRAD

1168 Abdulkadirov, K.M. 04766
Leningrad Res. Inst. of Hematology, & Blood Transfusion, Clinic of Hematology, 2nd Sovetskaya St. 16, Leningrad 193024, USSR (Tel.: 2772550)
COLL: Samuskevich, I.G.

Haemopoietic Malignancies in RSFSR Areas Contaminated by Radioactive Fallout from the Chernobyl Accident
The aim of the present study is to estimate the possible effect of environmental radioactive contamination on the development of haemoblastoses and suppressed haematopoiesis in the population of the Briansk region (RSFSR), where the Chernobyl accident took place. Morbidity rates in the population exposed to irradiation in areas of the Briansk region (300,000 men) are compared with those in areas beyond the irradiated zone (1 million men). A retrospective analysis of haemoblastosis and suppressed haematopoiesis for 1979-1985 (background data) has shown that morbidity averaged 10,950 per 100,000 men. Taking into account that the ionizing radiation is a risk factor rates from 1986 onward are being compared with those from 1979-1985. Variations in haemoblastosis and suppressed haematopoiesis morbidity during 1986-1989 are from 9,637 to 10,842 per 100,000 men, i.e. do not exceed the background rates. The study has been carried out using expedition-epidemiological, medico-demographic and cartographic methods.

TYPE: Incidence
TERM: Radiation, Ionizing
SITE: Haemopoietic
TIME: 1989 - 1995

1169 Semiglazov, V.F. 04859
N.N. Petrov Research Inst. of Oncology, USSR Ministry of Health, Leningradskaya St. , Pesochny-2, Leningrad 189646, USSR (Tel.: 2378748; Fax: 2378947)
COLL: Sagaidak, V.N.; Ebeling, K.

USSR/WHO Randomized Controlled Trial of Breast Self Examination in Breast Cancer Mortality
The major objective of the study is to determine the effect of a breast self-examination (BSE) programme on mortality from breast cancer. The assessment will be made in a controlled trial. A population of about 100,000 women aged 40-64 will be randomized in Leningrad. After a one-year feasibility study in 1984, the project will continue for 15 years. It consists of a four-year intensive education progamme (1985-1988) during which and afterwards until 1994 the detected breast cancer cases will be registered and treated, and followed-up for five years until 1999. The key issue of the study is the compliance of the population to the BSE programme. The frequency and competence of BSE practice will be defined in a subsample of 400 randomly selected women by means of a survey made at the sixth month and annually since the start of the project. The study is expected to result in the accrual of more than 400 new breast cancer cases and 200 deaths. It is also intended to evaluate a list of risk factors and take the opportunity to pin-point high-risk groups.

TYPE: Cohort; Intervention
TERM: Screening
SITE: Breast (F)
TIME: 1985 - 1999

MOSCOW

1170 Basieva, T. 04892
Inst. of Carcinogenesis, All-Union Cancer Research Centre, Academy of Medical Sciences of the USSR, Kashirskoye shosse, 24, Moscow 115478, USSR (Tel.: +7 095 3241470; Fax: 2302450)
COLL: Zaridze, D.; Kabulov, M.

Spatial, Temporal and Ethnic Distribution of Oesophageal Cancer in Karakalpakstan
The Soviet republic of Karakalpakstan, an area situated in the central Asian part of the Soviet Union, is thought to be endemic for oesophageal cancer. Age-standardized rates for oesophageal cancer for Karakalpakstan as a whole are 50.2 and 50.9 for males and females respectively. The incidence rates vary substantially in different regions of Karakalpakstan: the highest (120.0 males, 150.6 females) are reported from the Muinak region located in the north. Incidence of oesophageal cancer also varies between

different ethnic groups and is highest among Kazakhs, followed by Karkalpaks; the lowest incidence is reported in Russian ethnics. The aim of the study is to analyse the pattern of oesophageal cancer, including geographical distribution, incidence in different ethnic groups, and time trends. Incidence maps of oesophageal cancer in 14 districts will be drawn. Time trends for the period 1973-1989 will be analysed, using age-period-cohort methods.

TYPE: Incidence
TERM: Ethnic Group; Geographic Factors; Mapping; Trends
SITE: Oesophagus
TIME: 1990 - 1993

1171 Bulbulyan, M. 04902
Inst. of Carcinogenesis, All-Union Cancer Research Center, Academy of Medical Sciences of the USSR, Kashirskoye shosse, 24, Moscow 115478, USSR (Tel.: +7 095 32241470; Fax: 2302450 ; Tlx: 411015)
COLL: Savitskaya, T.Y.

Cancer among Workers Exposed to Aromatic Amines

This is a cohort study of about 4,000 male and female employees at a chemical plant, with initial exposure after 1940, and who were alive on 1 January 1975. The cohort consists of one group exposed to aromatic amines and three control groups. The aim of the study is to determine the risk of bladder cancer and other malignant neoplasms before and after improvement of work conditions. Information about cancer occurrence, including histological type, will be obtained from the Moscow Cancer Registry.

TYPE: Cohort
TERM: Occupation
SITE: All Sites; Bladder; Lung; Stomach
CHEM: Amines, Aromatic
OCCU: Chemical Industry Workers
REGI: Moscow (USSR)
TIME: 1987 - 1992

1172 Garkavtseva, R.F. 04626
All-Union Scientific Cancer Center, Kashirskoye shosse 24, Moscow 115478, USSR (Tel.: +7 095 3244257)
COLL: Kazubskaya, T.P.; Nephedov, M.D.

Assessment of the Contribution of Hereditary Factors to Common Neoplasms

The aim of this investigation is the development of methods to prevent breast and stomach cancer (genetic counselling, follow-up of "risk groups" for early diagnosis etc). Pedigrees of 1,046 patients with breast cancer and 951 patients with stomach cancerr have been collected. The general population of Moscow serves as control group. An investigation of lung cancer genetics in approximately 500 probands and their relatives is planned. Preliminary results obtained by means of a multifactorial model of the heritability have shown that the genetic contribution to breast cancer was 52% (pre-menopausal 62.4%; post-menopausal 39.1%), stomach cancer 22% for men and 41% for women of which 19% have been x-linked. Tables of recurrent risk have been drawn up to assess the probability of a family member developing a tumour.

TYPE: Case Series
TERM: Genetic Factors; Heredity; High-Risk Groups; Prevention
SITE: Breast (F); Lung; Stomach
TIME: 1985 - 1992

*1173 Levshin, V.F. 05167
Academy of Medical Sciences of the USSR, All-Union Cancer Research Centre, Kashirskoye Shosse 6, Moscow 115478, USSR (Tel.: 3229134)

Smoking Cessation and Vitamins A and E for Lung Cancer Prevention

The main aim is to compare the efficacy of smoking cessation and diverse regimens of medication with vitamins A and E in the prevention of lung cancer among heavy smokers. In addition, the study aims to examine the association of lung cancer with life history factors other than smoking. Men aged 50-69 with a lifetime smoking history of more than 100,000 cigarettes will be included. Four comparable groups of such men will be recruited to receive different types of preventive intervention: (1) special help to stop smoking; (2) the combination of retinol palmitate 50,000 IU and alpha-tocopherol acetate 0.1 mg all the

USSR

year round; (3) the same agents, but only for a four-month period per year; (4) placebo. It will take three years to recruit all the participants. The interventions will continue for five years. The follow-up period will last an average 10 years.

TYPE: Intervention
TERM: BMB; Hair; Prevention; Tobacco (Smoking); Vitamins
SITE: Lung
CHEM: Retinoids; Tocopherol
TIME: 1991 - 2001

*1174 **Remennick, L.I.** 05154
All-Union Cancer Research Centre, Dept. of Cancer Epidemiology, Karshiskoye Shosse 6, Moscow 113478, USSR
COLL: Koshkina, V.S.; Okuneva, L.A.

Reproductive Factors and Abortion in Breast and Genital Cancer
This hospital-based case-control study will examine the interaction of various reproductive events and sexual characteristics in causation of breast and gynaecological cancers in a female population with low fertility and a high abortion rate. All new cases of cancers of the breast, cervix and corpus uteri, and ovary (about 300 over three years) arising in the city of Magnitogorsk will be compared to a similar number of non-cancer patients individually matched by age (within two years), residence and socio-economic status. The questionnaire will include items on occupation and living standards, detailed reproductive history (with special emphasis on abortion, including late, out-of-hospital, complicated and other types of terminations), history of gynaecological diseases and screening. All cases and controls will be interviewed by trained clinical nurses. The results will be analysed by univariate and multivariate methods.

TYPE: Case-Control
TERM: Abortion; Lifestyle; Reproductive Factors
SITE: Breast (F); Ovary; Uterus (Cervix); Uterus (Corpus)
TIME: 1991 - 1995

1175 **Smulevich, V.B.** 04761
Academy of Medical Sciences of USSR, All-Union Cancer Research Center, Lab. of Occupational Cancer, Kashirskoye Shosse 24, Moscow 115478, USSR (Tel.: +7 095 3214409; Tlx: 411015 KNIFE)
COLL: Belyakova, S.V.; Remennick, L.I.; Garkavtseva, R.F.; Bukhny, A.E.

Parental Occupation as Cancer Risk Factor for Children: A Population-Based Case-Control Study
Parental exposure to occupational carcinogens and other factors related to cancer in children under 15 is being studied in a population-based case-control study in Moscow. In contrast to most previous studies, a detailed occupational history of both parents throughout the whole period between the first job and cancer diagnosis in a child is being assessed, with special emphasis on simultaneous exposure in both parents. The study will include all new cancer cases in the city of Moscow for the period 1986-90 as registered by the childhood cancer registry of the All-Union Cancer Research Center. Two control children, free of chronic diseases and congenital malformations, are matched to each cancer child by sex, year of birth and residence. Interview with the parents of diseased and control children is performed in a standard way in the medical setting (hospital or out-patient clinics). Besides items on occupational history and hazards, the questionnaire includes information on parental chronic diseases, obstetric history of the mother, smoking and drinking habits and genetic susceptibility. Results will be analysed by cancer site with application of standard statistical techniques.

TYPE: Case-Control
TERM: Alcohol; Childhood; Genetic Factors; Intra-Uterine Exposure; Occupation; Tobacco (Smoking)
SITE: Kidney; Leukaemia; Lymphoma; Nervous System; Neuroblastoma; Retinoblastoma
TIME: 1986 - 1993

*1176 **Smulevich, V.B.** 05128
Academy of Medical Sciences of USSR, All-Union Cancer Research Center, Lab. of Occupational Cancer, Kashirskoye Shosse 24, Moscow 115478, USSR (Tel.: +7 095 3214409; Tlx: 411015 KNIFE)
COLL: Solionova, L.G.; Gorelikova, O.N.

USSR

Cancer Incidence and Mortality among Public Transport Drivers
There is substantial epidemiological evidence of higher cancer risks in drivers, presumably related to their occupational exposures (polluted microclimate, noise, vibration, etc.) and lifestyle (intense smoking, alcohol abuse, irregular eating, stress, etc.). In previous studies, significant increases in chronic morbidity (gastritis, ulcers, bronchitis, etc.) directly related to length of service were demonstrated. In an extension of these studies, the cohort of 3,038 public transport drivers of Moscow city is being followed from 31 December 1969 to 31 December 1989. Of these, 1,442 are bus drivers (control group). Female drivers and retired male drivers will be analysed as separate subgroups. Cancer incidence and mortality measures (SIRs and SMRs) will be calculated, expected figures being based on the rates of Moscow population.

TYPE: Cohort
TERM: Lifestyle; Occupation; Stress
SITE: All Sites
OCCU: Bus Drivers
TIME: 1990 – 1995

1177 Solionova, L.G. 04760
All-Union Cancer Research Center, Academy of Medical Sciences of the USSR, Lab. of Occupational Cancer, Kashirskoye shosse 24, Moscow 115478, USSR (Tel.: (095)3214409)
COLL: Smulevich, V.B.; Remennick, L.I.; Belyakova, S.V.; Boiko, L.P.; Szeszenia-Dabrowska, N.

Cancer Incidence and Mortality among Workers in the Rubber Industry
Production of rubber goods has been shown to be carcinogenic for men, but no epidemiological data for the USSR exist. To assess incidence and mortality from cancer and other causes among workers, the cohort of 3,679 employees (60% women) of one of the oldest rubber footwear plants in Moscow has been assembled, including the workers with ten or more years of employment between 1979 and 1983, both working and retired. This closed cohort has been followed from 1 January 1979 up to 31 December 1988; total number of person-years of observation is about 25,000 and the mean follow-up period is about nine years. Vital status and cause of death are identified through employment records, death certificates, medical histories and post-mortems. The results will be presented as SIRs and SMRs, expected figures being derived from cancer rates in the population of Moscow. Besides cancer, mortality from cerebrovascular diseases and accidents will be assessed. Cancer mortality in the Moscow cohort will be compared with that in a cohort of rubber plant workers in Lodz, Poland.

TYPE: Cohort
TERM: Occupation; Rubber
SITE: All Sites
OCCU: Rubber Workers
LOCA: Poland
TIME: 1986 – 1992

1178 Zaridze, D. 03301
Inst. of Carcinogenesis, All-Union Cancer Research Center, Academy of Medical Sciences of the USSR, 24, Kashirskoye Shosse, Moscow 115478, USSR (Tel.: +7 095 3241470; Fax: 2302450; Tlx: 411015 KNIFE)
COLL: Boyle, P.; Trapeznikov, N.N.; Poddubni, B.K.; Kuvshinov, J.P.; Matiakin, E.P.; Rosin, M.P.; Stich, H.F.; Thurnham, D.; Hoffman, D.

Precancerous Lesions of the Mouth and Oesophagus in Uzbekistan (USSR)
The survey is an expansion of a screening programme in an area of high oral cancer incidence and moderately high oesophageal cancer incidence (Narpai Region of Samarkand oblast). All 2,150 men aged 55-69 years resident were invited to attend medical examination. The 1,569 respondents underwent interview and oral examination. Endoscopy was performed on 1,344 men. The questionnaire contained information on socio-demographic characteristics, use of nass, cigarettes and alcohol, dietary habits, medical history and family history of cancer. Oral and oesophageal lesions were photographed, material was taken for cytology from oral lesions and for biopsy and cytology from oesophageal mucosa. Oral and oesophageal smears were also taken for the micronuclei formation frequency test. Blood was collected from 22% of persons for analysis of levels of riboflavin, retinol and beta carotene. Biochemical analysis revealed that 86%, 79% and 14% of men have low blood levels of riboflavin, carotenoids and retinol, respectively. A randomized trial for the prevention of precancerous lesions of the mouth and oesophagus is presently being carried out. All eligible persons with oral leukoplakia and chronic oesophagitis (800 in all) have been randomly allocated to four treatment groups receiving vitamin A, beta carotene, riboflavin and placebo. One year after the initiation of treatment a

USSR

second examination will be performed, during which the procedures used during the first examination will be followed.

TYPE: Intervention
TERM: Alcohol; Biopsy; Cytology; Diet; Familial Factors; Histology; Nass; Premalignant Lesion; Tobacco (Chewing); Tobacco (Smoking); Vitamins
SITE: Oesophagus; Oral Cavity
CHEM: Beta Carotene
TIME: 1983 - 1992

1179 Zaridze, D. 04916
Inst. of Carcinogenesis, All-Union Cancer Research Center, Academy of Medical Sciences of the USSR, 24, Kashirskoye Shosse, Moscow 115478, USSR (Tel.: +7 095 3241470; Fax: 2302450; Tlx: 411015 KNIFE)
COLL: Zemlianaya, G.M.

Lung Cancer in Non-Smokers in Moscow

This case-control study aims to assess the effect of indoor and outdoor air pollution as a possible risk factor for lung cancer in lifetime non-smokers in Moscow. Cases are lung cancer patients who never used tobacco products, residents in Moscow, and admitted to the two largest oncological hospitals in Moscow. To control groups are collected. Hospital controls are other cancer patients, excluding those with respiratory cancer, and the second control group includes non-smoking residents of Moscow who do not have cancer. Information on demographic characteristics, residential history (to assess lifetime exposure to outdoor air pollution), passive smoking, radon exposure (the direct measurement of radon concentrations in the present residence of the study subjects), occupation, dietary habits (consumption of food rich in vitamins), and alcohol intake is obtained from patients and controls. About 400 cases will be collected in the years 1991-1994.

TYPE: Case-Control
TERM: Air Pollution; Alcohol; Diet; Occupation; Passive Smoking
SITE: Lung
CHEM: Radon
TIME: 1990 - 1994

1180 Zaridze, D. 04917
Inst. of Carcinogenesis, All-Union Cancer Research Center, Academy of Medical Sciences of the USSR, 24, Kashirskoye Shosse, Moscow 115478, USSR (Tel.: +7 095 3241470; Fax: 2302450; Tlx: 411015 KNIFE)
COLL: Zemlianaya, G.M.

Case-Control Study of Lung Cancer with Particular Emphasis on the Role of Air Pollution

The objective of this case-control study is to clarify the relationship between lung cancer and outdoor and indoor air pollution in two industrial towns with developed coal and chemical industries. In each town 200 cases of lung cancer and two population controls per case, matched by sex and age (+/- 5 years) will be recruited. The role of several aetiological factors will be assessed, including: residential history (exposure to outdoor and indoor air pollution), smoking, occupation, consumption of food rich in vitamins and alcohol intake.

TYPE: Case-Control
TERM: Air Pollution; Alcohol; Diet; Occupation; Tobacco (Smoking); Urban; Vitamins
SITE: Lung
TIME: 1990 - 1994

1181 Zaridze, D. 04918
Inst. of Carcinogenesis, All-Union Cancer Research Center, Academy of Medical Sciences of the USSR, 24, Kashirskoye Shosse, Moscow 115478, USSR (Tel.: +7 095 3241470; Fax: 2302450; Tlx: 411015 KNIFE)
COLL: Lifanova, Y.E.; Bukin, Y.V.; Babaeva, R.Y.; Levtchuk, A.A.; Shevchenko, V.Y.; Basalyk, L.; Kushlinski, V.

Diet and Breast Cancer

The aim of this case-control study is to examine the relationships between diet and risk of breast cancer. All consecutive patients with newly diagnosed breast cancer (without distant metastasis), resident in Moscow, and admitted to the breast cancer clinic, will be interviewed; blood and mammary tissue samples will be collected. Age-matched neighbourhood controls will be sampled from among the persons attending, for minor complaints, the same regional out-patient clinic from which the cases are recruited. Statistical analysis of 140 pairs of breast cancer patients and controls has shown that a decreased risk of post-menopausal breast cancer was associated with high intakes of cellulose, mono- and disaccharides, vitamine C, beta-carotene and also polyunsaturated fatty acids. High intakes of total fat resulted in a statistically insignificant decrease in the odds ratio, while saturated fats slightly increased risk of breast cancer. Alcohol use significantly increased risk of breast cancer. The concentrations of total oestradiols (E2) and free E2 were higher in breast cancer patients than in controls. The levels of two principle polyunsaturated fatty acids were lower in cases than in controls. The study continues.

TYPE: Case-Control
TERM: Alcohol; Diet; Hormones; Nutrition; Reproductive Factors
SITE: Breast (F)
CHEM: Beta Carotene
TIME: 1987 - 1993

1182   Zaridze, D.   04919
Inst. of Carcinogenesis, All-Union Cancer Research Center, Academy of Medical Sciences of the USSR, 24, Kashirskoye Shosse, Moscow 115478, USSR (Tel.: + 7 095 3241470; Fax: 2302450 ; Tlx: 411015 KNIFE)
COLL: Filipchenko, V.V.

Diet and Colorectal Cancer

The aim of the study is to investigate the role of diet in the aetiology of colorectal cancer. The study population consists of colorectal cancer patients, resident in Moscow, admitted to the Central Moscow Oncology Clinic, and hospital controls. Cases and controls are interviewed using a food frequency questionnaire. In addition, other lifestyle and environmental factors related to colorectal cancer risk will be assessed. A parallel study has been set up in Chabarovsk. In all, 150 pairs of colorectal cancer patients and controls have been interviewed in Moscow and 150 pairs of cases and controls in Chabarovsk. Preliminary results indicate a protective effect of cellulose, beta-carotene and vitamin C, as well as of vegetable consumption. The increased risk was associated with high meat/vegetable, protein/cellulose, fat/cellulose ratios.

TYPE: Case-Control
TERM: Diet; Environmental Factors; Lifestyle
SITE: Colon; Rectum
TIME: 1988 - 1992

1183   Zaridze, D.   04920
Inst. of Carcinogenesis, All-Union Cancer Research Center, Academy of Medical Sciences of the USSR, 24, Kashirskoye Shosse, Moscow 115478, USSR (Tel.: + 7 095 3241470; Fax: 2302450 ; Tlx: 411015 KNIFE)
COLL: Basieva, T.H.

Atlas of Cancer Mortality in the USSR

Maps of cancer mortality in the USSR will be produced using computer graphic techniques. Mortality data for 22 cancers in 170 areas of the USSR for 1988-1989 will be provided by the Central Departments of Statistics of the USSR. The denominator will be based on the 1989 population census, provided by the same Department.

TYPE: Mortality
TERM: Mapping
SITE: All Sites
TIME: 1990 - 1992

1184   Zaridze, D.   04922
Inst. of Carcinogenesis, All-Union Cancer Research Center, Academy of Medical Sciences of the USSR, 24, Kashirskoye Shosse, Moscow 115478, USSR (Tel.: + 7 095 3241470; Fax: 2302450 ; Tlx: 411015 KNIFE)
COLL: Necrasova, L.I.; Basieva, T.H.

USSR

Bladder Cancer
The study aims to investigate the associations of bladder cancer with occupation, smoking, dietary habits, coffee and alcohol consumption, use of drugs, artificial sweeteners, drinking water, diseases of kidney and bladder and family history of cancer. The study comprises histologically verified bladder cancer cases and population controls, individually matched for age, sex and neighbourhood. 150 pairs of bladder cancer patients and controls have been collected. Preliminary statistical analysis has shown that the most important risk factor for bladder cancer in Moscow is smoking. Elevated risk of bladder cancer was observed in drivers. The risk was also found to be associated with family history of bladder and cancers of other sites. The study continues.

TYPE: Case-Control
TERM: Alcohol; Coffee; Diet; Drugs; Familial Factors; Occupation; Sweeteners; Tobacco (Smoking); Water
SITE: Bladder
TIME: 1989 - 1993

1185 Zaridze, D. 04923
Inst. of Carcinogenesis, All-Union Cancer Research Center, Academy of Medical Sciences of the USSR, 24, Kashirskoye Shosse, Moscow 115478, USSR (Tel.: +7 095 3241470; Fax: 2302450; Tlx: 411015 KNIFE)
COLL: Bulbulyan, M.A.; Maximovich, D.M.; Willett, W.C.; Maclure, M.

Physicians Health Study
This prospective cohort study aims to evaluate the relationship between cancer risk and a number of environmental factors, including diet, alcohol consumption, cigarette smoking, and family history of cancer. Information will be obtained by mail, using self-administered questionnaire from 50,000 physicians, males and females, aged over 40 years and living in Moscow and the Moscow region. Information about cancer occurrence, including histological type, will be obtained from the Moscow Cancer Registry.

TYPE: Cohort
TERM: Alcohol; Diet; Familial Factors; Occupation; Tobacco (Smoking)
SITE: All Sites
OCCU: Health Care Workers
REGI: Moscow (USSR)
TIME: 1989 - 2010

SVERDLOVSK

1186 Kogan, F.M. 04382
Medical Scientific Center, for Prophylaxis and Health Protection, of Industrial Workers, 30 Popov St., Sverdlovsk 620014, USSR (Tel.: 517645)
COLL: Yatsenco, A.S.; Gurvich, E.B.; Kuzina, L.E.

Cancer Mortality in the Brake Lining and Asbestos Textile Industries
A study of some 300 persons in a brake lining plant showed that lung cancer mortality was surprisingly lower than in a similar group of asbestos textile workers living in the same city and having similar lifestyle. A further study was carried out on a cohort of about 3,000 workers in an asbestos friction products plant situated in another city. This cohort was devided into three sub-cohorts, exposed in the past to different kinds of dusts and other hazards: asbestos, vapours of phenol and fromaldehyde or benzene, asbestos bakelite or asbestos rubber dusts. The observed/expected mortality ratio for stomach cancer was more than 1 in the first sub-cohort only. In the second and third sub-cohorts no excess mortality was observed, as well as in the total cohort. In addition an experimental study was carried out on three groups of rats injected twice intraperitoneally with these dusts at one month interval. The asbestos bakelite and asbestos rubber dusts induced a much smaller number of tumours than chrysotile asbestos. In the last sub-group survival was significantly lower than in the two other groups. The limit exposure for asbestos bakelite and asbestos rubber dusts may of course be higher than for asbestos.

USSR

TYPE: Cohort
TERM: Dusts; Occupation; Plastics; Resins; Rubber; Textiles
SITE: Gastrointestinal; Lung
CHEM: Asbestos; Formaldehyde; Mineral Fibres; Phenol
OCCU: Asbestos Textile Workers; Asbestos Workers
TIME: 1988 - 1992

*1187 Kogan, F.M. 05150
Medical Scientific Center, for Prophylaxis and Health Protection, of Industrial Workers, 30 Popov St., Sverdlovsk 620014, USSR (Tel.: 517645)
COLL: Kashansky, S.V.; Plotko, E.G.; Berzin, S.A.; Bogdanov, G.B.

Environmental Asbestos and Lung Cancer
Despite the well known carcinogenicity of asbestos dust, the effect of low exposure is rarely studied. The aim of this study is to compare the lung cancer mortality of the population never having worked in asbestos plants with the rates for the region or other industrial towns without asbestos plants. The results will be used to formulate proposed time-weighted average exposure limits for asbestos in ambient air.

TYPE: Mortality
TERM: Air Pollution; Dusts
SITE: Lung; Respiratory
CHEM: Asbestos
TIME: 1991 - 1993

ULYANOVSK

1188 Storozhuk, M. 04694
Ulyanovsk District Oncology Dispensary, 92, 12th September St., Ulyanovsk 432700, USSR
COLL: Loginov, A.P.; Yaroslavtsev, V.N.

Programme for the Fight against Cancer in Ulyanovsk District
According to the decision of the "Local Soviet" a programme of social and medical cancer prevention was adopted in 1988. The social programme includes air protection and reasonable use of soil and water resources. The medical programme includes early detection of cancer, rapid treatment of cancer patients and identification of high risk groups. The creation of a high-risk group data bank is planned. People are selected according to risk factors. Data on each person are stored until death or until a precancerous lesion becomes a cancer. In December 1988, data on 3,848 persons had been collected. It is anticipated that 10-15% of the total population will be included in the data bank. This will permit follow-up of high cancer risk populations, detection of cancer at initial stages and estimation of frequency of transition of cancerous lesions into cancer. The remaining population will be used as a control group.

TYPE: Cohort; Incidence
TERM: Data Resource; High-Risk Groups; Premalignant Lesion; Prevention; Treatment
SITE: All Sites
TIME: 1988 - 1993

YUGOSLAVIA

BELGRADE

1189 Djordjevic, M. 03078
Inst. of Oncology & Radiology, Pasterova 14, 11000 Belgrade, Yugoslavia (Tel.: +38 11 6855755/423; Tlx: 11077 onkos yu)
COLL: Gec, M.

Adolescent Obesity or Malnutrition as Risk Factors for Cancer

The hypotheses being tested are that adolescent obesity or, alternatively, malnutrition in the adolescent period, are risk factors in cancer aetiology. Cancer of the genital organs and the digestive system are now being studied, while the study of breast cancer has been completed. It is planned to interview 500 patients with a corresponding control group matched by age, sex, occupation, demography, etc.

TYPE: Case-Control
TERM: Adolescence; Diet; Nutrition; Obesity; Physical Factors
SITE: Breast (F); Gastrointestinal; Genitourinary
TIME: 1982 - 1992

1190 Djordjevic, M. 04740
Inst. of Oncology & Radiology, Pasterova 14, 11000 Belgrade, Yugoslavia (Tel.: +38 11 6855755/423; Tlx: 11077 onkos yu)
COLL: Vuletic, L.; Mitrovic, N.; Jovicevic-Bekic, A.; Kanjuh, V.

Follow-up of Risk Groups for Breast Cancer

The aim of this prospective study is to justify the use of selective screening in breast cancer. Through a case-control study of breast cancer completed in 1988, 27 risk factors for breast cancer have been identified, including age, occupation, parity, oral contraception, previous diseases and injuries. Based on these data, a table of graduated risk factors by age has been created. In this prospective study, randomly selected women from the general population will be interviewed with a questionnaire to elicit these risk factors. According to the presence, combination or absence of various risk factors, women will be included in three groups with low, medium and high risk for breast cancer (at least 1000 women each group) and followed for 10 years in order to determine the incidence of breast cancer in each group. If the high-risk group should reveal significantly higher breast cancer incidence, this might open up the possibility of selective screening in a high-risk population.

TYPE: Cohort; Methodology
TERM: High-Risk Groups; Screening
SITE: Breast (F)
TIME: 1990 - 2000

1191 Mitrovic, N. 04754
Inst. of Oncology & Radiology, Pasterova 14, 11000 Belgrade, Yugoslavia (Tel.: +38 11 685755; Tlx: 11077 onkos yu)
COLL: Djordjevic, M.; Vuletic, L.; Kanjuh, V.; Jovicevic-Bekic, A.; Matijasevic, A.

Ovarian Cancer in Serbia

In the first part of the investigation, the aim is to determine standardized mortality and incidence rates for ovarian cancer, trend over a five-year period and demographic differences concerning the ethnic and religious characteristics of the female population. This analysis will be by histological type. In a case-control study with 300 cases and a control group of 900 disease-free women matched for age, occupation and residence the following factors will be investigated: positive family history of cancer (using Chase's life-table method in order to obtain cumulative incidence of family cancer), links with other sites (cancer of the endometrium, breast cancer and cancer of the colon), parity, hormonal impairments and oral contraception.

YUGOSLAVIA

TYPE: Case-Control
TERM: Ethnic Group; Familial Factors; Histology; Oral Contraceptives; Parity; Religion; Trends
SITE: Ovary
TIME: 1990 - 1995

LJUBLJANA

1192 Pompe-Kirn, V. 04888
Inst. of Oncology, Cancer Registry of Slovenia, Zaloska 2, 61000 Ljubljana, Yugoslavia (Tel.: +38 61 316490; Fax: 329177)
COLL: Vlaisavljevic, V.; Kaucic, M.; Jelinncic, V.; Us, J.; Rudolf, Z.; Novak, F.

Pilot Study of Breast Cancer Screening in Six Communes of Slovenia
A pilot randomized trial on the efficiency of an organized breast cancer screening programme has been started in three different regions of Slovenia. About 12,400 randomly selected women aged 50-64 years received a personal invitation letter. Every participant has had a physical examination, breast self examination instruction, and a single mediolateral view mammogram. In the first round the response rate was 55%-75%, better in little towns and rural populations than in urban populations. In an interval of 18 - 24 months mammography and physical examination will be repeated. The first analysis of mortality and incidence data on breast cancer in the screened and in the unscreened group of the study population is planned for the year 2000.

TYPE: Cohort; Intervention
TERM: Screening
SITE: Breast (F)
REGI: Slovenia (Yug)
TIME: 1989 - 2000

1193 Primic-Zakelj, M. 04482
Inst. of Oncology, Unit of Epidemiology, Zaloska 2, 61000 Ljubljana, Yugoslavia (Tel.: +38 61 314344)
COLL: Ravnihar, B.; Pompe-Kirn, V.; Oblak, B.; Kosmelj, K.

Breast Cancer Risk in Relation to Oral Contraceptive Use in Slovenia
In a hospital-based case-control study (1980-83) a positive association between breast cancer and oral contraceptives was established. A population-based study is now added. Women aged 25-54, residents of Slovenia since 1965, with breast cancer newly diagnosed anywhere in the republic, are studied as cases, with one control, matched by age and municipality of residence, selected from the general population for each case. During the three years of data collection (1988-1990) 624 matched pairs were expected to be personally interviewed in hospitals or at home. Additional data regarding the disease were obtained from hospital records and confirmation of oral contraceptive use was sought from prescribing doctors. Relative risk estimates will be calculated by logistic regression. Analysis started in 1991.

TYPE: Case-Control
TERM: Oral Contraceptives; Registry
SITE: Breast (F)
REGI: Slovenia (Yug)
TIME: 1988 - 1992

1194 Zwitter, M. 04587
Inst. of Oncology, Zaloska 2, 61105 Ljubljana, Yugoslavia (Tel.: +38 61 327955)
COLL: Pompe-Kirn, V.; Primic-Zakelj, M.

Possible Role of Pregnancy in the Pathogenesis of Hodgkin's Disease
Hodgkin's disease (HD) may be considered as one of the cancers of the immune system, and pregnancy is certainly a period of marked change in immune reactivity. HD soon after delivery is not a rare experience in oncology; however, a causal relationship between the two events has not been confirmed. This study will include about 150 women with HD aged 17-50 years at the time of diagnosis, who were reported to the Cancer Registry of Slovenia in the years 1966-1988. Medical reports will be reviewed, and a written or personal contact with the patient or her relatives will be made. It is hoped that an analysis of the age at the first and subsequent pregnancies, and the age at first symptoms and at the

YUGOSLAVIA

diagnosis of HD will disclose links between the two events, and thus point to any role of pregnancy in the pathogenesis of HD.

TYPE: Case Series
TERM: Age; Pregnancy; Registry; Time Factors
SITE: Hodgkin's Disease
REGI: Slovenia (Yug)
TIME: 1988 - 1992

MARIBOR

1195 Vlaisavljevic, V. 04512
Hosp. Maribor, Dept. of Gynaecology, Ljubljanska 5, 62000 Maribor, Yugoslavia (Tel.: +38 62 37221)
COLL: Harper, P.

Incidence of Breast Cancer in Patients with Breast Cysts

The aim of the project is to study the incidence of breast cancer in patients followed-up because of clinically manifest breast cysts. The patients are volunteers, symptomatic patients visiting the Center for Breast Diseases. All those in which a cyst is certified are included in the study. All patients are self-selected. Recruitment began in 1980 and will terminate in 1995. The hypothesis is that the incidence of breast cancer is higher in patients with breast cysts than in the normal population. The control group for this natural incidence of breast cancer would be that published by the Cancer Registry for the observed region and the population of the same age distribution and exposure period. All patients once registered will be followed by clinical examination, mammography or ultrasound. All refusers of the study will be interviewed every year about their breast (clinically manifest breast cancer).

TYPE: Cohort
TERM: Breast Cysts; High-Risk Groups; Premalignant Lesion; Registry
SITE: Breast (F)
REGI: Slovenia (Yug)
TIME: 1980 - 1995

ZAGREB

1196 Bauman, A. 03176
Inst. for Medical Research, and Occupational Health, Dept. for Radiation Protection, Ksaver 158, P.O.Box 291, 41000 Zagreb, Yugoslavia (Tel.: (041)434188; Fax: 274572)
COLL: Kovac, J.; Novakovich, M.; Lokobauer, N.; Marovic, G.; Ambrosic, D.

Cancer Risk from Mutagenic Agents at a Coal Fired Power Plant

In a coal-fired power plant in the western part of Yugoslavia, extensive risk assessment has been carried out for many years. Athracite coal with high uranium content (25 ppm) and 7-9% of sulphur, meavy metals and benzo(a)pyrene is again being used. Epidemiological studies and constant monitoring are being done. A special case is a former radioactive slug waste pile site from the highly contaminated coal mine abandoned 90 years ago and on which houses are built and where two generations of the same families live. Both sites are investigated in cooperation with local medical services. Radon is being measured in houses, some of which are built with radioactive slag. At present teratogenic effects are being investigated with special emphasis on the child-bearing population. Many women are descendents of seven generations of coal miners, which were working in the radioactive coal mine. The results will be added to the existing cancer register.

YUGOSLAVIA

TYPE: Case-Control; Incidence; Mortality
TERM: Air Pollution; Coal; Dose-Response; Environmental Factors; Metals; Occupation; Radiation, Ionizing; Record Linkage
SITE: Gastrointestinal; Leukaemia; Lung
CHEM: Benzo(a)pyrene; PAH; Radon; Sulphur Dioxide; Uranium
OCCU: Miners, Coal
TIME: 1990 - 1995

1197　Bauman, A.　03609
Inst. for Medical Research, and Occupational Health, Dept. for Radiation Protection, Ksaver 158, P.O.Box 291, 41000 Zagreb, Yugoslavia (Tel.: (041)434188; Fax: 274572)
COLL: Kovac, J.; Cesar, D.; Marovic, G.; Horvat, D.; Konjevic, R.; Pticar, M.

Cancer Risk in the Fertilizer Industry

The aim of this study is to assess the risk of working in, or living near, a phosphate agricultural chemical plant. A uranium extraction pilot plant installed in Zagreb has now been abandoned. Carcinogens such as uranium, thorium, sulphuric and phosphoric acids, fluorides and nitrates are sources of air and soil pollution in the working environment and around the plant. At present 450 workers have been examined and chromosome aberrations, SCE and internal contamination with lead-210 have been detected. Carcinogenic effects are investigated. The awareness of the population that phosphate fertilizers are radiation sources when applied in agriculture is a psychological threat.

TYPE: Incidence; Mortality
TERM: Chemical Exposure; Environmental Factors; Fertilizers; Metals; Mutagen; Occupation; Radiation, Ionizing
SITE: Gastrointestinal; Lung
CHEM: Fluorides; Nitrates; Phosphates, Inorganic; Sulphuric Acid; Thorium; Uranium
OCCU: Agricultural Workers
TIME: 1990 - 1995

List of Investigators

Index of Terms

Index of Sites

Index of Types of Study

Index of Chemicals

Index of Occupations

Index of Countries

Index of Cancer Registries

INDEX OF INVESTIGATORS

This index identifies projects by investigator. It includes both principal investigators and collaborators.

Each project is identified by its **serial number**, not the page number.

For studies in which the person indexed is the principal investigator, the serial number is shown in **bold type**.

For studies included in the Directory for the first time, the serial number is preceded by an **asterisk (*)**.

INVESTIGATOR

AANHS	885	Anderson, D.E.	**990, 991**
Aaran, R.K.	243	Anderson, G.	1130, *1131
Abdulkadirov, K.M.	**1168**	Anderson, J.J.B.	943
Abdulnour, E.	97	Anderson, L.	**203**
Abe, R.	527	Anderson, M	223
Abel, E.	931	Anderson, M.S.	**888, 889**
Abel, U.R.	**352**, 353	Anderson, T.F.	931
Abid, L.	847	Anderson, V.E.	*1048
Ábrahám, E.	**381**	Andersson, I.	722
Abramson, M.	*33	Andersson, M.A.	220
Achmad Ghozali	431	Andersson, S.O.	727
Adam, H.	*764	Andrade, A.	*794
Adamec, M.	*194	Andrews, E.	1109
Adami, H.O.	42, 206, **725, 726, 727, 728, 729, *730**, 731, 734, 736, 737, 738	Andrews, H.	1060
		Anton-Culver, H.	**1000**
		Anttila, S.	241
		Anzai, S.	533, **566**
Adams, J.	*1014	Aoki, K.	529, **533**, 535, 566
Adenis, L.	**267**, 269	Aoyama, T.	*551
Ades, A.E.	807	Arblaster, L.	*784
Adhvaryu, S.G.	**394**	Arbman, G.	**686**
Adler, Z.	441	Arce, V.	189
Agarwal, S.S.	418	Archer, V.E.	1121
Aggazzotti, G.	***493**	Aristizabal, N.	283
Aghi, M.B.	411	Arlett, C.F.	763
Agius, R.	777, 788	Armstrong, B.G.	807
Agnarsson, B.A.	390	Armstrong, B.K.	27, 29
Ahlbom, A.N.	298, 302, **703, 704, *705**	Armstrong, F.D.	980
Ahmed, M.	644	Armstrong, M.A.	1075
Ahonen, R.	*257	Arndt, K.A.	931
Ahrens, W.	292, 341	Arnorsson, J.V.	390
Airaksinen, O.	*257	Aromaa, A.	243, 244
Akdas, A.	**749, *750**	Arsenault, A.	74
Åkesson, B.	*695	Arslan, A.	277
Alavanja, M.C.R.	**907**, 910, 913	Arthur, D.C.	1047
Alberti Fidanza, A.	498	Arumugam, S.	415
Albertini, R.J.	938	Arveux, P.	263
Alberts, D.	1150, 1151	Aschenbrener, C.	897
Albin, M.P.	**689**, 699	Ascunce, N.	283
Albright, L.A.	1090	Ashby, D.	791, *792
Aldefer, R.	957	Ashley, R.L.	*1132
Aldrich, T.E.	***1106, *1107, *1108**	Ashmore, J.P.	**76**
Alexander, F.E.	*783	Ashraf, S.	643
Alexeyeff, M.	24	Askari, A.	644
Alfthan, G.	243	Asp, S.	249
Alhonen, L.	*257	Assennato, G.	***456, *457**
Alican, Y.	*750	Astrup-Jensen, A.	*280
Allan, R.N.	**752**	Atjak, M.T.J.	614
Allan, S.	62	Atkin, W.S.	**793**
Allen, J.C.	1090	Attam, K.	418, *419
Allouache, A.	1	Attewell, R.	689
Alonso de Ruiz, P.	283	Auburtin, G.	**323**
Alpers, M.P.	**645**	Auerbach, A.D.	366
Alterman, T.	1056	Augustin, J.	287
Alvarez, N.	*284, *285	Austin, D.F.	897, 1056
Amadori, A.	496	Austin, H.D.	**927**
Amadori, D.	*454	Autrup, H.	**204**
Ambrosic, D.	1196	Avanzi, G.C.	510
Amemiya, T.	*530	Avella, A.	675
Amerio, P.L.	501	Avni, A.	444
Amestoy, G.	*359	Axelson, O.	686, **687**, 688, 714, 720
Amichetti, M.	*515	Axerio, M.	481
Amorosi, A.	467	Ayme, S.	50
Amundson, B.	1128	Ayzac, L.	278, *279
Andersen, A.	*280, *281, 295, 636		

493

INVESTIGATOR

Baanders, A.N. 610
Baanders-van Halewijn, E.A. **607, 608**, *611
Babaeva, R.Y. 1181
Bach, F.H. **617**
Bachi, R. 579
Backley, D. 262
Badia, A. 673
Badosa, E. 673
Báez, S. *123
Baghurst, P.A. 300
Bah, E. ***334**, 335
Bahmer, F.A. 336
Bailey, A. 818
Bain, C.J. 7
Baines, C.J. 105, 106, 111
Bairati, I. 90
Baker, G. 39
Baker, M. 1130
Baki, M. 385
Balakrishnan, T.S. 418
Balanda, K.P. *9, *10
Balar, D.B. **395**, 396, 397, *398, *399, 400, 401, *402
Baldasseroni, A. 481
Ball, D. 77
Balmes, J. *1131
Baltrusch, H.J.F. 350, *351
Balzi, D. 286, 471
Balzi, M. 467
Banasik, R. 648, 649
Band, P.R. **113, 114,** *119
Banerjee, T.K. **1037**
Banfi, A. 1089
Bang, L.J. 282
Bankier, A. *21
Bankson, D. 1130
Bánóczy, J.E. **382**
Barker, D.J.P. 862, 875
Barlow, L. 287
Barnhart, S. *1131
Barnhill, R. 1054, 1065
Baron, J.A. 453, *982
Barra, S. 453
Barreto, J.H.S. 65
Barreto, S.M. *794
Barrett, L. 266
Barrett, R.J. 914
Barrett-Connor, E.L. **1003**, 1004
Barry, W. *433
Bartolome, M. 677
Bartsch, H. 303, 512
Basalyk, L. 1181
Basieva, T.H. 1183, 1184
Bass, G.I. 1101, 1102
Bassett, A.A. 111
Bastecky, J. 375
Basu, J. *932
Basu, M.K. 757
Bat, L. 452
Bates, M.N. ***905**
Bathers, S. *759
Batiste, E. 673
Battistutta, D. 4, 7

Baughman, R.D. 931
Baum, E. 1089
Bauman, A. **1196, 1197**
Bausch-Goldbohm, R. 595, 596, 613
Baylin, S.B. 896
Bayo, J. *359
Beard, M. *627
Beaumont, J.J. 1117
Becciolini, A. 467
Becher, H. 294, 370
Bechi, P. **467**
Bechtel, P. *333
Beck, G. *982
Beck, J. 665
Beck, P. 103
Becker, F.F. 994
Becker, N. 298, 302, 352, **353**
Becker, T.M. **883**
Becker, W. 731
Beckers, R. 50
Beckman, M. 715
Beckmann, A.M. *1132
Bedenne, L. 263
Beebe, G.W. **908**, 913
Beeson, W.L. 1007
Beggi, A. *475
Beijerinck, D. 609
Belanger, C. *933, 934
Belfiore, A. **466**
Bell, C.M.J. *876
Bellander, T. *280, *697, 702, 710, 713
Belli, S. 298, 507
Bellini, A. 487
Belon-Leneutre, M. 260
Belyakova, S.V. 1175, 1177
Ben Harush, M. 438
Benatti, P. 494
Benediktsdottir, K. 390
Bengtsson, C.B. **680**
Bengtsson, N.O. 720
Benhamou, E. 324, 326
Benhamou, S. 292, 298, **324, 325, 326**, 741
Benito, E. 273, **674**, 675
Benlatreche, K. 1
Benn, T. *280
Benowitz, N. *1124
Benraadt, J. *590
Benson, M. *1062
Beral, V. **795, 796, 797**, 822, 841
Bercovich, J. *359
Beresford, S.A.A. **1125**, 1130, *1139
Bereza, D. 199
Berg, J. 544
Bergeret, A. 278, *279
Bergkvist, L. 731, 737
Bergström, R. 728, 731
Berkel, J. 60, 62
Berlin, N.I. 947
Berman, M.L. 914
Bernard, E. 307
Bernard, J.L. **307**
Bernstein, L. 288, *291, **1009, 1010**
Bernstein, R. 669

INVESTIGATOR

Berrino, F.	293, 296, **481**, **482**, 485	Bock, J.	210
Berry, G.	1134	Bode, P.	595
Berry, R.J.	**882**	Bodmer, J.	842
Bert, J.L.	114	Bodmer, W.	782
Bertazzi, P.A.	294, 295, **483**, ***484**	Bodner, K.M.	1041
Berwick, M.	***14**, **1054**, **1054**, 1136	Bodrogi, I.	385
Berzin, S.A.	***1187**	Body, J.J.	314
Bethune, A.	802	Boehnke, M.	969
Bethune, G.	111	Boeing, H.	293, **355**, **356**, 361, 365, ***456**, 847
Bethwaite, P.	***627**		
Betta, P.G.	**462**	Boeniger, M.	958
Bhaduri, A.	400, 401	Boessel, J.	1090
Bhambhani, S.	418	Boffetta, P.	***270**, ***280**, ***281**, 292, 294, 295, ***306**, 508, ***794**, 1072
Bharati Arumugam, S.	**415**		
Bhat, B.	69		
Bhatavdekar, J.M.	**396**, **397**, ***398**, ***399**	Bogani, S.	467
Bhatnagar, P.	418	Bogdanov, G.B.	***1187**
Bhattathiri, V.N.	***427**	Boice, J.D.	217, 220, 222, 450, 907, **913**
Biancalani, M.	469		
Bianchi, C.	**495**	Boiko, L.P.	1177
Biasco, G.	262	Bokey, L.	7
Bicevskis, M.	15	Bokros, F.	386
Bidoli, E.	453	Bolan, G.	***1124**
Bierre, A.	618	Bolm-Audorff, U.	341
Biggar, R.J.	213, **909**	Bolund, C.	**706**
Biggeri, A.	286, 468, 471	Bombardier, C.	1109
Bilir, N.	749	Bonaguri, C.	503
Billington, N.A.	881	Bonanomi, A.	467
Binder, V.	***226**	Bond, G.G.	1041
Bingham, S.	293	Bond, J.	1044
Binks, K.	882	Bondi, R.	468
Biocca, M.	***280**	Bondy, M.D.	997
Biran, T.	446, 447	Bontempo, L.C.	1094
Birkeland, S.A.	**228**	Booker, S.V.	896
Bisanti, L.	487	Booth, M.	796, 797
Bishop, D.T.	851	Bopp, M.	743
Bistoletti, P.	**685**	Bordi, C.	503
Bithell, J.F.	839	Boreiko, C.	***270**
Bjarnason, Ó.	390	Borgia, P.	507
Bjerk, J.E.	***280**	Borlee, I.	50
Bjørkheim, A.	640	Borman, B.	***620**
Bjornsson, J.	390	Borroni, G.	501
Bjurstam, N.	722	Bos, R.P.	584, **597**
Black, H.S.	***992**	Bosch, F.X.	**271**, **272**, **273**, 283, 335, 496, 646, 674, 747, ***748**, 873, 898
Black, R.	287		
Blair, A.E.	**910**, **911**, 920, ***922**, 1114, ***1116**		
		Bosnyák, M.	382
Blamey, R.W.	878	Botta, M.	**463**, **464**
Blanc, C.	**313**	Bouchardy, C.	288, **741**
Blankenstein, M.A.	608	Bouquot, J.E.	**1051**
Blanks, R.G.	807, 808	Bourke, G.J.	298, ***433**
Blaser, M.J.	***1112**	Boutron, M.C.	262, 263
Blasi, B.	498	Bowen, D.	1130
Blettner, M.	***224**, **354**, 791, ***792**	Bowen, P.	***1156**
Bley, L.D.	1127	Bowie, C.	***879**, ***880**
Bleyen, L.J.	***52**	Bowie, S.H.U.	***880**
Bloemen, L.J.	1041	Bowman, D.M.	111
Blondal, H.	390	Bowman, J.D.	629, 1025
Blot, W.J.	134, 135, ***165**, **912**, **912**, ***1026**, ***1027**	Boyd, P.	1056
		Boyes, D.A.	111
Blumberg, B.S.	**1084**	Boyle, P.	300, 301, 302, 303, 304, 305, 743, 1178
Blumenthal, D.S.	**890**		
Boal, W.	***281**	Braas, P.A.M.	588
Bobev, D.	287	Brabin, B.	645

INVESTIGATOR

Bradley, M. 1044
Bradley, W.E.C. 74
Braga, M. 487, 507
Brancker, A. 86, *291
Brand, H. *340
Brandi, M.L. 314
Brandt-Rauf, P.W. 1057, *1062, 1064
Brasilino de Carvalho, M. **59**
Braunwald, E. 934
Braverman, I.M. 931
Breibart, E.W. 336
Bremond, A.G. **274**
Brenes, F. 189
Brescia, B. *1124
Breslow, N.E. 1099
Brett, S.M. ***887***
Bridges, B.A. **763**, 875
Briet, H.A. 585
Brignone, G. *497
Brinton, L.A. *225, **914, 915, 916, 917, 918, 921**
Brisanti, L.A. 507
Brissette, P. 971
Brisson, J. **88**
Brochard, P. *333
Brodie, P. *766
Brögger, A. *694
Brogan, J. *433
Brok, K.E. **229**
Broll, D. 337
Brollo, A. 495
Bron, D. 276
Broustet, A. *279
Brown, A. *25
Brown, D.P. **951**
Brownbill, P.A. 838
Brownson, R.C. 907
Bruce, Å. 728, **731**
Bruce, S. *992
Brugère, J. 316, 317
Brullet, E. 673
Bruning, P.F. 586, 1063
Brunnemann, K.D. 118
Bruno, C. 502
Brusa, M. 463, 464
Bruzzi, P.A. 303, *454
Bryson, D. 843
Buchanan, D. *778
Buckley, J.D. 1046, 1047, *1085, 1090
Budiningsih, S. 534
Buechner, J. 894
Buelow, K.B. 1087
Bülow, S. **227**
Bueno de Mesquita, H.B. 294, 300, *591, *592
Buffler, P.A. *993, 1056
Buhr, P. 358
Buiatti, E. 286, 469, 471, 473, 846
Bukhny, A.E. 1175
Bukin, Y.V. 1181
Bulbrook, R.D. 819
Bulbulyan, M.A. 1185
Bunch, K.J. 838, 854

Bundgaard, T. **200**
Bunin, G.R. *1085, *1086, 1088, 1090
Burau, K. *993
Burg, G. 336
Burgess, M. *1156
Buring, J.E. *933, 934
Burk, R. *932
Burmeister, L.F. 911
Burrows, B. 1149
Busellu, G.P. 480
Bush, H. 111
Bush, T.L. 981
Bussey, R. 793
Buttyan, R. *1062
Bylin, G. 709
Byrne, J. **919**
Caan, B. *1122
Cabeza, E. 674, 675
Cade, J.E. *780
Cahan, A. 1066
Cai, S.X. *124, *126
Cai, X.C. *126
Calabro, A. 50
Calabro, L. 496
Calatayud Sarthou, A. *676
Calavrezos, A. 346
Calheiros, J.M. 292, 847
Calle, E.E. 892
Callet, B. 313
Calmettes, C. **314**
Calvo, A. *123
Calzolari, E. 50
Camargo, B. 65
Cambier, L. 267, 269
Cameron, A.H. 755
Campbell, C. 937
Campbell, M. *780
Campbell, T.C. 125
Campion, M. 1141
Campolucci, S. 888
Cannon, L. *9, *10
Cano, E. *284
Cantin, J. 111
Cantor, K.C. *905, 910
Cantor, K.P. *1116
Caperle, M. 503, *505
Capizzi, R. 500
Caporaso, N.E. 512, *925, 957
Caporossi, J.C. 1158, 1159
Caraballoso, M. 190
Carbone, A. 455
Carden, A. *21
Cardis, E. 287
Cardona, T. 673
Carle, F. 479
Carli, P. **468**
Carli, S. 473
Carlin, J. *22
Carman, W.J. 886
Carmentano, P. 507
Carpenter, L.M. 796, 797
Carpenter, M. 86
Carrer, H. 983
Carrer, J. 983

INVESTIGATOR

Carretta, D.	516	Chen, S.Y.	*127
Carstensen, J.M.	688, 717	Chen, V.W.	1056
Carta, P.	**458, *459, *461**	Chen, X.L.	172
Carter, D.	*1107	Chen, X.Q.	*162
Carter, J.	*627	Chen, Y.X.	177
Carter, J.F.	618	Chenevix-Trench, G.	4
Cartmel, B.	*1148, 1150, 1152	Cheng, S.M.	*126
Cartwright, R.A.	305, **781, 782, *783**, 798	Cheng, Z.Y.	*148
Carzoglio, J.	1161, 1162	Cherchi, P.	*459, *461
Casale, V.	503	Cherian, T.	*427
Casella, C.	478	Cherian, V.	*422, *424, *425, *426
Cassiman, J.J.	54	Cherniack, M.	*1131
Castagneto, B.	463, 464	Cherrie, J.	295
Castellaneta, A.	503	Chevalier, A.	313
Castiglione, G.	*484	Chew, L.N.	579
Castilho, E.A.	*57, *58	Chianale, J.	122
Castillo, R.	271	Chichareon, S.	*748
Castrén, O.	256	Chick, J.	11
Castro, D.	*284	Chieco-Bianchi, L.	**496**
Catton, G.	111	Child, M.A.	897
Cecconi, R.	471	Chilvers, C.E.D.	298, 840, 852
Ceci, A.	511	Chin, T.D.Y.	1002
Celleno, L.	501	Chinchilli, V.	*887
Cendan, M.	1161	Chittukadu, G.K.	**416**
Centonze, S.	*456	Chmiel, J.S.	**947**
Ceppi, M.	*475, 477, 478	Choi, B.C.K.	**100**
Cerimele, D.M.	**500, 501**	Choi, N.W.	78, 91, 116, **120**, *121, 301, 302
Cesar, D.	1197	Choi, W.C.	300
Cesarini, J.P.	304	Chongsuvivatwong, V.	***748**
Chacko, P.	*417	Chotard, J.	335
Chadha, B.	418, *419	Chrissman, J.	*966
Chadha, P.	418, *419	Christensen, J.M.	204
Chaffey, C.	*82	Christie, D.G.	***25**
Chagas, J.F.S.	59	Chu, J.	1130
Chaitchik, S.	**446, 447**	Chui, S.X.	128
Cham, K.	335	Ciatto, S.	473
Chamberlain, J.	828, 877, **878**	Ciccone, G.	479, **509, 510**, 513
Chameaud, J.	264	Cicioni, C.	499
Chan, J.	380	Cicollela, A.	*279
Chan, W.C.	380	Cipkowska, B.	*658
Chang, A.R.	**619**	Cislaghi, C.	743
Chang, F.	256	Clapp, R.W.	***928**
Chang, S.E.	805	Clark, L.C.	***1145, *1146, *1147**
Chang, Y.S.	135	Clark, R.E.	**768**
Chang-Claude, J.	**357**, *359, 367	Clark, W.	1115
Chapuis, P.	7	Clarke, E.A.	110
Charlin, B.	95	Claudiani, J.	*359
Charvátová, S.	193	Clausen, N.	642
Chase, C.	944	Clavel, F.	293, **327, *328**
Cheang, J.	380	Cleaver, J.E.	**1123**
Cheirsilpa, A.	747	Clemmesen, I.H.	*225
Chellini, E.	**469**, 471, 474, 507	Clifford, C.	13
Chen, D.	130	Clutton, S.	***9, *10**
Chen, G.C.	167	Coates, M.S.	**35**
Chen, H.Q.	172	Coates, R.J.	891, 1056
Chen, J.	912	Cocco, P.L.	458, *459, **460**, *461
Chen, J.K.	***99**	Cochran, S.	832
Chen, J.S.	125	Cocito, V.	463, 464
Chen, K.	*143, *144, *145, *147, *148, *149	Coebergh, J.W.W.	287, **605, 606**
Chen, K.W.	913	Coggon, D.	*156, *280, *281, 294, **862, 863, *864**, 867
Chen, N.H.	*126	Cohen, D.L.	853
Chen, R.	***156**	Cohen, L.	*933
Chen, S.H	133		

497

INVESTIGATOR

Cohn, R.	669		Cuadros, A.	1134
Coker, A.	*962		Cui, Y.S.	545
Colberg, J.	1070		Cullen, M.	*1131
Colditz, G.A.	929		Cummins, C.W.	758
Coldman, A.J.	112		Cummins, H.W.	*1103
Cole, J.	763		Curado, M.P.	64
Cole, P.	*1040, 1052		Curtis, R.E.	223, *224, 913
Cole, R.J.	895		Cuschieri, A.	50
Coleman, M.P.	1, 277, 725, 847		Cusimano, R.	*497
Colin Jones, D.G.	859		Cuzick, J.M.	793, **798**, **799**, *816, *817
Collalto, A.	*359		Czanik, P.	381
Collet, J.	305		Czeizel, A.	**383**, **384**
Collette, C.	609		D'Albasio, G.	503
Collette, H.J.A.	293, 607, 608, **609**, *611		d'Angio, G.J.	1089
Collins, J.	*22		D'Avanzo, B.	*454, 488
Colonna, M.	*310		Daftary, D.K.	411
Colonna, S.	514		Dahl, C.	210
Colucci, G.	*456		Dahlgren, L.	723
Comba, P.	**502**, 507, 508, 509		Dai, X.D.	**150**, *151
Combs, G.	*1145, *1146, *1147		Dalcin, P.	54
Comstock, G.W.	**981**		Dale, G.	824
Cone, J.	*1131		Daling, J.R.	*119, 301, 918, 1024,
Connelly, R.R.	919			*1132, 1133, *1138, 1140
Connolly, J.	100		Daly, L.	*433
Conso, F.	*328		Damber, L.	714, 721
Consonni, D.	483, *484		Damiecki, P.	277
Conti, E.M.S.	*504, *505		Dan, R.P.	179
Cook, N.	*933		Dananche, B.	278, *279
Cook, R.R.	**1041**		Danes, B.S.	896
Cook-Mozaffari, P.J.	**831**, **832**, **833**, **834**, 839		Daniels, T.	*1124
Cooke, K.R.	624		Daoud, S.	235
Cooper, D.	*38		Darby, S.C.	774, 833, **835**, **836**, **837**
Cooper, S.P.	*993		Dardanoni, L.	*497
Coopmans de Yoldi, G.	*492		Darra, F.	516
Coppock, E.A.	86		Darraugh, T.	*1124
Corbett, S.J.	**36**		Darwis, I.	534
Corbin, A.	312		Das, B.C.	418
Cordier, S.E.	301, 302, *328, ***329**		Das, D.K.	418
Cornain, S.	534		Dauda, G.	388
Cornelissen, M.A.	585		Daures, J.	**311**
Corrah, T.	873		Dave, B.J.	394
Corrao, G.	**479**, **480**		Davey-Smith, G.	810
Correa, P.	*284, 357, 1056		David, G.	54
Cortes Vizcaino, C.	*676		Davies, A.P.	**800**
Costa, A.	*492		Davies, J.	*270
Costa, G.	507		Davies, J.A.	63
Costantino, J.P.	1101, 1102		Davies, J.M.	834
Costanzo-Nordin, M.R.	*1038		Davies, N.F.	*766
Coste, I.	274		Davies, P.	756
Cottoni, F.	501		Davies, T.W.	***767**
Courtial, I.	274		Davis, L.	*928
Cowen, A.	7		Davis, S.	**1126**, **1127**, **1128**, 1135
Cox, C.E.	*978		Day, N.E.	293, 335, 423, *424, *425,
Cox, N.H.	***772**			768, 804
Craft, A.W.	**824**, 825, *826, *827		Dayer, P.	741
Crawford, N.	777		de Backer, G.	*52, 847
Crespi, M.	282, 357, **503**, *504, *505		de Boorder, T.	594
Criqui, M.H.	1004		de Bruin, M.	595, 612
Cristofolini, M.	453		de Contreras, O.	*284
Crognier, E.	*315		De Gaudemaris, R.	266
Crommelin, M.A.	606		de Grandi, P.	745
Cronjé, H.S.	**662**		De Groot-Vlasveld, M.	588
Crosignani, P.	462, 482, **485**, 513		de Klerk, N.H.	**26**, 30, 31, *32
Crouch, P.	3		de la Cruz, J.	1134

INVESTIGATOR

De Lucia, G.	478	Dimitrova, E.	**191**
de Metz, C.	62	Ding, C.Y.	*126
de Nully Brown, P.	218	Ding, J.H.	**158**
De Oliveira, H.	262	Dini, S.	469
De Palo, G.	*492	Dinya, E.	381
De Pue, R.H.	977	DiOrio, D.	894
De Quint, P.	***49**	Djordjevic, M.	**1189, 1190**, 1191
De Rijke, M.E.	588	Dockerty, J.D.	***620, *621**
De Rossi, A.	496	Doerken, H.	**349**
de Sanjosé, S.	283, *284, *285	Dolezel, L.	199
De Stéfani, E.	**1161, 1162**, 1163	Doll, R.	774, 833, 835, 836, 837, 843, 934
de Thé, G.	*315		
de Vathaire, F.	330	Dols, P.W.	*591
De Vet, H.C.W.	***593**	Dombernowsky, P.	220
de Villiers, E.M.	210	Domenici, R.	516
de Waard, F.	300, 607, 608, 609, **610**, *611	Domergue, J.	**312**
		Dominguez, F.V.	**2**
De Wals, P.	50	Donaldson, L.	861
Deacon, M.C.	24	Donaleski, D.	*973
Deapen, D.	1016	Donelli, S.	*484
DeBuono, B.	894	Dong, J.	154, *155
Decarli, A.	*454, **486**, 488, 489, 490, 491, 743	Donna, A.F.	462
		Donnan, G.A.	*34
Decinti, E.	*123	Doornbos, G.	*591
Degerth, R.	246	Dorant, E.	596, 613
Degiovanni, D.	463, 464	Dore, J.F.	304
Dei, R.	467	Dorfman, R.R.	904
Del Mistro, A.	496	Doria, M.	*475
Del Moral, A.	293	Dosemeci, M.	910, 912, **920**
Delaloye, J.F.	745	Dosman, J.	**91**, 92
Delclos, G.	*993	Douglas, A.	796, 815
Delendi, M.	296	Dousset, M.	**265**
Delisle, M.J.	**318**	Downey, R.	*81
Della Foglia, M.	*484	Downs, T.D.	*993
Delzell, E.	927	Drake, J.J.	63
Demets, R.Y.	**964**	Draper, G.J.	831, **838, 839**, 848, 849, 854
Den Engelse, L.	584		
Den Tonkelaar, I.	***611**	Dreyer, N.A.	*945, 946
Depoorter, A.M.	*49	Driscoll, T.R.	**37**, 39
Depue, R.H.	*978	Duan, B.R.	*127
Derazne, E.	449	Dubeau, H.	74
Déri, Z.	387	Duca, G.	507
Desai, P.B.	***403**, *410	Duca, P.	**487**
Deschenes, L.	111	Duchene, Y.	*259
Deslauriers, J.	*925	Duclos, J.C.	**275**
Desmeules, M.	*81	Dueng, W.Z.	186
Dessi, S.	460	Duffy, S.W.	423, *424, *425
DeStavola, B.	798	Dufour, R.	77
Detels, R.	*899	Dupont, W.D.	**1052, *1053**
Deurenberg, J.J.M.	609	Dupras, G.	74
Deutsch, M.	1090	Durand, G.	263
Devesa, S.S.	**1113**	Duruz, G.	744
Devey, P.	*25	Dutreix, A.	330
Dhom, G.	369	Dutta, S.	418, *419
di Orio, F.	479, 480	Dworsky, R.	*1020
Di Placido, R.	480	Dwyer, T.	***14**
Diaz, A.	190	Easton, D.	*270
Dicato, M	303	Eaton, D.	1137
Dich, J.	718	Eaton, R.S.	78
Dietz, A.	*360	Eaves, E.R.	7
Dijkstra, B.K.S.	**666**	Ebbesen, P.	213
Dillner, J.	685	Ebeid, N.I.	**234, 235, 236**
Dillner, L.	685	Ebeling, K.	1134, 1169
DiMagno, E.P.	*1112	Eberlein, K.A.	930, *933, 934

INVESTIGATOR

Ebigbo, P.O.	375	Fan, J.X.	**157**
Eccles, J.	26, 30, *32	Fan, R.L.	133
Echavé, L.V.	97	Fang, F.	*132
Economidou, J.	377, 378	Fanning, D.	*270
Ederer, F.	1044	Fantuzzi, G.	*493
Edling, C.	682, **732, 733**	Farid, N.R.	**98**
Edson Pontes, J.	*966	Farmer, K.C.R.	*814
Ehrenberg, L.	938	Farmer, P.B.	204
Einhorn, N.	712	Fasoli, M.	488, 490
Eisen, E.A.	**1033**	Faure, J.R.	**266**
Eisenberg, E.	*99	Fava, A.S.	59, 64
Ekbom, A.	*730, **734**	Fears, T.R.	919
Eklund, G.	**707, 708**, 718	Feldman, J.	894
El Asser, A.K.	236	Felton, J.S.	903
El Din Shebl, S.	236	Fenaux, P.	***268**, *770
Elbrond, O.	200	Feng, H.M.	135
Elder, J.	847	Feng, Z.	1130
Eley, J.W.	891	Fenoglio-Preiser, C.	1061
Elfgren, K.	685	Fenwick, M.	24
Eliasziw, M.	*80	Ferencz, T.	386
Elliott, P.	***801**	Ferguson, D.A.	39
Ellman, R.	786	Ferguson-Smith, J.	843
Elwood, J.M.	304, 305, *620, *621, **622**, **623**, 624, *626	Fernández, L.M.	**190**
		Ferrario, F.	481
Elwood, P.C.	846	Ferraroni, M.	*454
Emmelin, A.	**719**	Ferrell, R.E.	991
Endang Soetristi	432	Ferri, G.M.	*456, *457
Enderlin, F.	744	Ferro, G.	*281, 294, 295
Engels, H.	**48**	Ferro-Luzzi, A.	*505
Engholm, G.	*212, 223, *224	Feuer, G.M.	***101**
English, D.R.	**27**, 29, 276	Févotte, J.	278, *279
Enstrom, J.E.	**1011, 1012**	Feychting, M.	704
Enterline, P.E.	1095	Fichtinger-Schepman, A.	276
Eomois, M.	237	Fidanza, F.	498
Epstein, J.H.	931	Fière, D.	*279
Epstein, L.M.	437	Fierro, L.	1163
Eriksson, A.B.	686	Filakti, H.H.	**802**
Erny, R.	**308**	Filiberti, R.	*454
Ersev, D.	749	Filipchenko, V.V.	1182
Eschwege, E.	*315	Filipovich, A.H.	**1042**
Eskelinen, M.	*257	Filippini, G.	301
Esmen, N.A.	1098	Fincham, S.F.	**60**, 91, 116
Esteva, M.	273, 674	Fincham, S.M.	**61**
Estève, J.	262, 273, **276, 277**, 512, 659, 747	Finkelstein, M.M.	**102, 103, 104**
		Fiordi, T.	499
Evanoff, B.A.	***1129**	Firth, H.M.	**624**
Evans, G.	24	Fischer, J.	314
Everhart, J.	*56	Fisher, S.G.	*1038
Everson, R.	940	Fissi, R.	482
Ewertz, M.	**205, 206**, 215	Fitzpatrick, T.B.	931
Ewings, P.	*879	Fiumara, A.	466
Eylenbosch, W.J.	48	Flamant, R.	264, 327, **330**
Faber Vestergaard, B.	210	Flander, L.	*22
Fabia, J.	111	Flanders, W.D.	891
Fabry, J.	278	Flandrin, G.	*328
Facchini, L.	*281	Flannery, J.T.	926
Fagerberg, G.	722	Flaten, T.P.	**634**
Fair, M.E.	76, *85, 86	Flatten, G.	**371**
Fairley, C.K.	***33**	Fletcher, C.D.M.	**803**
Fairweather, D.	858	Fletcher, F.J.	**881**
Faivre, J.	**262, 263**	Flodin, U.	682
Falcini, F.	503	Flore, C.	*459, *461
Falk, R.T.	911	Folsom, A.R.	**1043**
Falletta, J.M.	1099		

INVESTIGATOR

Fontana, V.	*475	Gallegaro, L.	496
Fontham, E.T.	**1056**	Gallimore, P.	760
Forastiere, R.	507	Galt, M.	623
Ford, J.M.	41	Galteau, M.M.	741
Forde, K.A.	1059, 1061	Gamble, J.F.	***973***
Foreyt, J.P.	*992	Gambrill, R.D.	*978
Forman, D.	293, *778, 833, **840**, **841**, **842**, **843**, **844**, **845**, **846**, **847**	Gangadharan, P.	*421, *422, 423, *424, *425
		Gange, D.	24
Formelli, F.	*492	Gao, P.	*148
Forrest, A.P.M.	878	Gao, R.N.	**163**, **164**, 170, 180
Forsén, A.	375	Gao, Y.J.	150
Fortvin, M.	335	Gao, Y.T.	163, 164, ***165***, 166, 170, 174, 178, 180
Fortwengler, P.	1032		
Fossati-Bellani, F.	511	Garabrant, D.H.	629
Foster, A.	812	Garau, I.	**675**
Fracheboud, J.	609	Garbe, C.	**336**
Fradet, Y.	90	Garbowski, G.	1061, *1062
Frame, J.N.	*925	Garcia, A.	314
Franceschi, S.	303, **453**, ***454***, 455, 488, 491	Garcia, F.	***1148***
		Gardner, M.J.	295, **865**, **866**, **867**, **868**, **869**, **870**, **871**
Franch, P.	675		
Franco, E.L.F.	59, **64**, **65**	Garfinkel, L.	892
Franco, S.	946	Garkavtseva, R.F.	**1172**, 1175
François, P.	330	Garland, C.F.	1003, **1004**
Frankel, S.	*769	Garland, S.	*33
Franze, A.	503	Garofalo, R.	466
Fraser, G.	*436, **1007**	Garrity, T.J.	1120
Fraser, P.M.	795, 796, 797, **804**	Garry, V.F.	*922
Fraumeni, J.F.	217, 910, 1118	Garvicz, S.	642
Fredriksson, M.	686	Gaspari, R.	471
Freitag, S.	77	Gatta, G.	485
Frentzel-Beyme, R.R.	295, 352, 353, **358**	Gatti, R.A.	1013
Freudenheim, J.	937	Gaudette, L.A.	**77**, *81, 86, *216
Friden-Kill, L.M.	606	Gavino, V.	69
Friedell, G.	893	Gavosto, F.	455
Friedl, W.	*363	Gec, M.	1189
Friedland, J.M	957	Geddes, M.	286, **470**, **471**
Friedman, D.	913	Gehde, E.	375
Friedman, G.D.	**1073**, **1074**, 1075, 1079, *1122	Geirsson, G.B.	390
		Geissler, E.	**339**
Friis, L.	733	Genka, K.	535
Frisell, J.	722	Gennaro, V.	***475***
Frizzera, G.	1042	Genovese, O.	*504
Frost, P.	931	Gentile, A.	488
Fryns, J.P.	54	Genuardi, M.	506
Fuerst, C.J.	711	George, M.	335
Fujiki, H.	566	George, W.O.	303
Fujimoto, I.	544, 550, 576	Georgescu-Tulcea, N.	661
Fujita, M.	518	Gerhardsson, L.	*691, *724
Fukao, A.	537, **563**, ***564***, 847	Gerhardsson, M.R.	**709**
Fukuda, K.	**528**, 533	Gerin, M.	66, 316
Fukuma, S.	517, 535	German III, J.L.	**1058**
Fulton, J.	894	Gewelke, U.	*360
Funatsu, H.	567	Gey, F.	574
Furuyama, J.	523	Ghadirian, P.	**69**, 300, 303
Gabiano, P.	322	Ghironzi, G.	967
Gabrielli, M.	503	Ghosh, M.K.	*420
Gafà, L.	293	Ghys, R.	**70**
Gail, M.G.	919	Giacosa, A.	262, *454, 503
Galanti, C.	50	Giannetti, A.	501
Galke, W.A.	1008	Giannotti, B.	468
Gallágher, R.P.	113, 114, 116, *119, 304, *1144	Giardiello, F.M.	**896**, *900
		Gibb, H.	**1155**

INVESTIGATOR

Gibbs, A.R.	*784, 860	Granberg, S.B.O.	**681**	
Gies, H.P.	*14	Grandjean, P.	**230**	
Gignoux, M.	**261**	Grassi, A.	282, 503	
Gilbert, E.S.	1110	Grassi, L.	375	
Gilbertsen, V.	**1044**	Grattan, H.	7	
Giles, G.G.	12, 13, **17, 18, *19, *20, *21, *22, *34**	Gravelle, I.H.	*771	
		Greaves, J.	791	
Gili, M.	283	Greaves, M.F.	669	
Gilles, F.H.	897	Greco, M.	498	
Gillman, J.C.	618	Green, A.C.	8	
Gilman, E.A.	753	Green, D.	1089	
Gimenez Fernandez, F.J.	*676	Green, L.M.	294	
Ging, L.	297	Green, M.	*22	
Giordano, G.	*1148	Greenberg, M.D.	1141	
Giri, D.D.	*398, ***399,** *402	Greenberg, R.A.	1028	
Gissmann, L.	*359	Greenberg, R.S.	**891**, 926, 1029, 1030, 1056	
Giuffrida, D.	466			
Giwercman, A.	**207**, 215	Greenland, S.	1013	
Glanz, K.	*1142	Greenwood, B.	335	
Glashan, P.	798	Greggi, S.	**506**	
Glass, A.G.	*1104, *1131	Greig, N.H.	*924	
Glattre, E.	634, **635**	Greiser, E.	***340**	
Glen, R.	912	Griciute, L.	302	
Godard, C.	313	Griffith, J.	940, 958	
Godley, P.A.	**939**	Grifols, R.	271	
Gois Filho, J.F.	59	Grignoli, M.	502, 507, 508	
Gold, E.B.	**897**	Grinstein, S.	*359	
Goldberg, L.	*992	Grizzle, J.	1130, *1131	
Goldberg, M.	75, 313	Grogan, D.	76	
Goldblatt, P.O.	785	Groopman, J.D.	**898**	
Goldbourt, U.	443	Grosclaude, P.	***259**, 324	
Goldhaber, M.K.	**1075**	Grossman, L.	*901	
Goldhaber, S.	934	Grufferman, S.	**1093, 1094**, 1099, *1100	
Golding, J.	***764, *765**	Grulich, A.	*811	
Golditch, I.	1075	Grunwald, H.	1070	
Goldsmith, J.R.	**434**	Guarnieri, C.	503	
Goldstein, B.	**667**	Guenel, P.	313	
Gómez, M.	910	Guercilena, S.	483	
Gonzales, M.	*19	Guerrero, E.	283	
González, C.A.	292, 293, **673**	Guglielmotti, M.B.	2	
Gonzalez, E.	931	Guillem, J.	**1059**	
Gonzalez, L.C.	283	Gula, M.J.	1093	
Goodman, D.	*933	Gulati, S.S.	*409	
Goodman, G.	1130, *1131	Gulie, C.	744, 745	
Goodman, K.	670	Gundy, S.	**385**	
Goodman, M.T.	**984**, 986	Guo, B.C.	163	
Goodman, S.N.	*900	Guo, J.	129	
Gordon, D.	*933	Guo, L.P.	128	
Gordon, I.	16	Guo, Y.R.	139	
Górecka, D.	656	Gupta, M.M.	418	
Gorelikova, O.N.	*1176	Gupta, P.C.	299, **404, 405,** 408	
Górski, T.	**656**	Gupta, S.	418, *419	
Gorst, D.W.	***779**	Gurevicius, R	302	
Goss, B.	844	Gurucharri, C.	*359	
Goto, R.	560	Gurvich, E.B.	1186	
Gou, Y.R.	138	Gustavsson, N.	248	
Goujard, J.	50	Gustavsson, P.	702, **710**, *1129	
Goulston, K.	7	Guttmann, R.	**71**	
Gouvernet, J.	307	Gutzwiller, F.	*615	
Govindan, S.	415	Haas, G.P.	**965, *966**	
Grace, J.R.	114	Habbema, J.D.F.	603, *604	
Graham, S.	**937**	Hackman, P.	241	
Graiff, C.	*515	Haelterman, M.	*55	
Gramenzi, A.	491	Hämäläinen, E.	*257	

INVESTIGATOR

Hafez, A.H.	234	Hatch, E.	913
Hagberg, S.	*684	Hatschek, T.	688
Hagenbeek, A.	276	Hatton, F.	298
Hagmar, L.E.	*690, *691, *692, *693, *694, *695, *696, *697, *698	Hattori, M.	518
		Haubrich, T.	847
		Hausken, T.	640
Haguenoer, J.M.Y.	*268, 320	Hawkins, M.M.	848, 849, 850
Haider, M.	46	Hawthorne, V.	*978
Haile, R.W.C.	169, *982, 1013	Hayakawa, K.	539
Haimovici, L.	*359	Hayakawa, N.	520, 533, 570
Hakama, M.K.	243, 255, 258	Hayashi, M.	517
Hakulinen, T.R.	243, 248, 250	Hayashi, Y.	535
Haley, N.J.	292	Hayat, M.	276
Hall, A.J.	335, 496, 872, 873, 898	Hayata, Y.	567
Hall, J.	276	Hayden, C.L.	1055, 1065
Hallgrímsson, J.	390	Hayes, M.V.	63
Halperin, W.E.	958	Hayes, R.	910, 957
Halpern, A.	1115	Hayes, R.B.	*604, 920, 959
Hamada, G.S.	573, *575	Hayward, J.L.	819
Hamdi-Chérif, M.	1	Hayward, R.J.	*766
Hamilton, S.R.	896	He, W.	130
Hamilton, T.E.	1128	He, Y.P.	*124
Hammond, D.	*1085	Head, J.	*811
Hammond, G.D.	1046, 1047, *1049	Hearl, F.	912
Hanai, A.	544, 576	Heath, C.W.	892
Hanai, J.	562	Hebert, J.R.	1070, *1160
Hanke, J.	653	Hebert, P.	*933
Hankey, B.	544	Heederik, D.	*281
Hankin, J.H.	984, 985, 986, 987, 988, 989	Heenan, L.D.B.	*620
		Heenan, P.T.	29
Hankinson, D.	24	Heerema, N.A.	1046
Hanley, J.	71	Heikkilä, L.	241
Hannaford, P.C.	822, 841	Heikkilä, P.	245
Hansen, E.S.	208	Heilbrun, L.K.	*966
Hansen, H.S.	314	Heim, S.	*694
Hansen, J.	30	Heineman, E.	910, *1116
Hansen-Koenig, D.	50	Heinonen, K.	*257
Hansluwka, H.E.	287	Helgerson, S.	883
Hanson, J.	61	Heller, W.D.	*360
Hansson, L.E.	728	Hellers, G.	752
Hansteen, I.L.	*694	Hellmann, R.	937
Hantman, J.M.	*761	Helmick, C.	726, 734
Harber, L.	931	Hemminki, K.	938, 1064
Hardcastle, J.D.	828, *829, *830	Hen, Z.X.	135
Hardell, L.O.	720, 728	Henderson, B.E.	178, 617, 1009, 1018
Harding, S.	802	Henderson, M.M.	1130
Hare, C.	448	Hendriks, J.H.C.L.	598, 600, 602
Harlan, L.	886	Hengels, K.J.	847
Harnden, D.G.	782	Henneberger, P.	*281
Harper, P.	1195	Hennekens, C.H.	929, 930, *933, 934
Harrington, J.M.	*759	Henry-Amar, M.	276, 331, 332
Harris, C.C.	512	Herbison, G.P.	624
Harris, F.	50	Herd, A.J.	*992
Harris, R.E.	1070	Herfarth, C.	*363
Hart Hansen, J.P.	219	Herity, B.	*433
Hart, A.A.M.	586	Hermus, R.J.J.	595, 596, 613, 614, 616
Harte, G.A.	*766	Hernandez, C.	893
Hartge, P.	1115, 1157	Hernandez, J.M.	271
Harvey, I.M.	*769	Hernandez, J.M.	582, 583
Harvey, V.J.	618	Hernberg, S.G.	245, 246, 249
Hasegawa, Y.	521	Herrero, R.	921
Hashimoto, T.	523, 533	Herrick, R.	958
Hassan, T.J.	643	Herrinton, L.	*1139
Hassoun, J.	307	Herva, M.A.	253

INVESTIGATOR

Hesseling, P.B.	671	Hufnagl, A.	45
Heuch, I.	633	Hughes, J.I.	*976
Hiatt, R.A.	**1076, 1077, *1078**	Hulka, B.S.	939, **940**
Hickling, C.	31	Hulshof, K.F.A.M.	*592
Hietanen, E.	241, 303	Hung, K.L.	*126
Higashiiwai, H.	519	Hunt, K.	857
Higgins, I.T.T.	886	Hunt, V.B.	1034
Hill, C.	330	Hunter, D.J.	929
Hill, G.B.	*81, 91, 111	Hurley, S.	***34**
Hill, M.J.	262	Husgafvel-Pursiainen,	
Hillerdal, G.N.	**735**	K.T.	**241**
Hint, E.	**237, *238**	Huttunen, J.K.	*615
Hippeläinen, M.	256, *257	Iannarilli, R.	508
Hirayama, T.	529	Icenogle, J.	883
Hirohata, I.	528	Ichikawa, S.	568
Hirohata, T.	528, 529, 566	Icsó, J.	**198**
Hirsch, I.	197	Ida, Y.	*564
Hirvonen, A.	241	Idle, J.	*1014
Hisamichi, S.	519, 533, 537, 563, *564	Ignatyeva, R.	*764
Ho, G.	*932	Ihamäki, T.	240
Ho, J.H.C.	379	Ikeda, M.	526, 552, 553, 554, 555
Hobbs, M.S.T.	26, 30, 31, *32	Ikeda, N.	565
Hodgson, J.T.	787, 788, *790	Ikeda, T.	**531**, *532
Hoedemaeker, P.J.	*900	Ilker, Y.	*750
Högstedt, B.	*694	Illiger, H.J.	375
Hoff, G.	**639**, 640	Imai, K.	*556, 557, 558, 559
Hoffman, D.	1178	Imanishi, K.	565
Høgaard Nielsen, N.	219	Inaba, Y.	533, 566, **568**
Hogstedt, C.	**701**, *1129	Indrawijaya,	429, 430
Holá, N.	195, 196	Infante-Rivard, C.	***72**
Holdaway, I.M.	618	Inskip, H.M.	*334, **335**, 873
Holländer, H.C.	218	Inskip, P.D.	*730, 913
Holland, R.	598, 600, 602	Ippolito, O.	466
Hollard, D.	*279	Ireland, P.	18
Hollingworth, T.	799	Irianiwati,	432
Holly, E.A.	301, **903, 904**, 1024, 1115	Iriya, K.	*575
Holm, L.E.	**711**, 717, 718	Isaacson, P.	926, 1136
Holm, N.V.	**231**	Isaksson, H.J.	390
Holmberg, L.	206, 731, **736**	Iscovich, J.M.	288, **442**, 444, 445
Holmstock, L.	*55	Iselius, E.L.	874
Hong, C.J.	*171	Ishibashi, H.	533
Hoofnagle, J.H.	908	Ishikawa, A.	517
Hoover, R.N.	737, 910, 915, 1157	Iskovich, J.	*291
Hopper, J.L.	13, 18, ***22**	Ito, H.	*522
Hornstra, M.H.	950	Ivanov, E.	287
Hornung, R.W.	**952**	Iversen, O.H.	298
Horsman, D.E.	**115**	Iwamamoto, K.	971
Horvat, D.	1197	Izarzugaza, I.	283
Horwich, A.	*876	Izquierdo, M.	677
Hoshiyama, Y.	***556**, 558, 559	Jablon, S.	913
Hossfeld, D.K.	349	Jack, A.	*334, 335
Hours, M.	278, *279, *281	Jackson, A.	*780
Howe, G.R.	**105, 106, 106**, 109, 111, 112, 292, 300, 302, *1144	Jackson, C.E.	**967**
		Jacobs, A.	*770
Howel, D.	*784	Jacobs, I.	**28**
Hreschchyshyn, M.	937	Jacobsen, A.	220
Hrubec, Z.	907, 913	Jacobson, L.P.	*899
Hsu, C.	380	Jäger, K.	*363
Hu, J.	**152**, *153	Jägeroos, H.	*257
Hu, M.X.	**140**, *156, 297	Jänne, J.	*257
Hu, X.L.	*126	Jäppinen, P.T.	**252**, *281
Huang, M.Y.	*124, *126	Järvholm, B.	682, 683
Huang, Y.H.	157	Jafarey, N.A.	644
Hubert, A.	*315	Jahn, I.	341

INVESTIGATOR

Jain, M.	106, 300	Jovicevic-Bekic, A.	1190, 1191
Jaiswal, M.S.D.	417	Juel, K.	230
Jakobsson, K.	689	Juhász, L.	**388, 389**
James, W.P.T.	303	Jung, G.	336
Janerich, D.T.	926, **1055**	Jupe, D.M.L.	15
Jannace, P.	936	Jurikovic, M.	**199**
Jansson, B.	**994**	Jussawalla, D.J.	**407**, *411, *412
Jaskiewicz, K.	663, *664	Kaaks, R.	293
Jass, J.	793, 845	Kaatsch, P.	364, *373, *374
Jayant, K.	*406, 407, *409, *411, *412	Kabat, G.C.	1070, 1071
		Kabulov, M.	1170
Jeannel, D.	*315	Kabuto, M.	574
Jebbink, M.	585	Kadam, V.T.	*412
Jederlinic, P.	*1160	Kadish, A.	*932
Jedrychowski, W.	651	Kadlubar, F.	512
Jeffrey, A.M.	1064	Kafka, S.	193
Jelinncic, V.	1192	Kahan, E.	**448, 449**
Jellum, E.	**636**	Kahn, T.M.	*359
Jenkins, D.	*816, *817	Kaipainen, P.	246
Jenkinson, C.	623	Kaiser, D.	337
Jensen, J.	62, 217	Kalandidi, A.	292
Jensen, O.M.	42, 203, 205, 210, 214, 215, 219, 220, 222, 223, 230, 231, 304, 729	Kaldor, J.	*38, 276, 286, 287
		Kallinikos, G.	377, 378
		Kallio, P.	256
Jewell, D.P.	752, 846	Kalyanaraman, S.	415
Ji, B.T.	164, *165, 180	Kamat, M.R.	*409
Jiang, J.S.	150	Kamat, V.	413
Jiao, D.A.	*143, *144, *145, *147, *148	Kamiyama, S.	533
		Kampman, E.	616
Jílek, V.	**193**	Kamps, W.A.	605
Jin, F.	*165, **166**, 170, 174	Kanaka, T.S.	415
Jin, M.L.	135	Kanda, J.L.	59, 64
Jin, T.H.	167	Kane, M.	335
Jin, Z.G.	*159, *160, 176	Kanjuh, V.	1190, 1191
Jindal, S.K.	292	Kanka, J.	197
Jöckel, K.H.	341	Kapadia, A.	*398, 400
Johannesson, B.	*392	Kaplan, B.H.	*923
Johanning, H.	*211	Kaprio, J.	**242**
Johansen, H.	203, **209**, 220	Karácsonyi, L	381
Johansson, H.	725	Karjalainen, A.	241
Johansson, J.E.	727	Karjalainen, S.	206, 252, 287
Johansson, L.G.	689, **699**	Kark, J.D.	**443**
Johansson, R.	*257	Karkut, G.	50
Johnson, B.C.	**1005**	Karner-Hanusch, J.	**45**
Johnson, B.E.	*925	Kashansky, S.V.	*1187
Johnson, E.	294	Kashyap, V.	418, *419
Johnson, N.W.	**805**	Kasl, S.V.	1055
Johnson, P.J.	**806**	Kasno	430
Johnsson, R.R.	65	Kasper, H.	262
Jónasson, J.G.	390	Kataja, V.	256
Jones, D.R.	**785**, 786	Kathren, R.L.	**1110**
Jones, R.D.	866	Kato, H.	*551, 567
Jones, W.R.	276	Kato, I.	536
Jongeneelen, D.J.	597	Katsouyanni, K.	293, 303
Jongeneelen, F.J.	584	Katsouyiannopoulos, V.	*281
Jonmundsson, G.K.	642	Katz, L.	442, **444**, 445
Jonsson, B.	679	Katz, R.	*99
Joossens, J.V.	*53	Kaucic, M.	1192
Jordan, J.	760	Kaufmann, M	352
Jordan, S.W.	883	Kaul, A.	368
Jordan-Simpson, D.	86	Kaune, W.	1025
Jørgensen, K.E.	**232**	Kauppinen, T.P.	245, 246, 294, 298
Joslin, C.A.	878	Kawai, K.	533
Joveniaux, A.	269	Kawajiri, K.	557

INVESTIGATOR

Kay, C.R.	822	Kobayashi, Y.	576
Kay, L.	841	Köck, M.	*44
Kay, R.G.	**618**	Köpf, I.	681
Kaye, A.	*19	Koepsell, T.	1130
Kaye, S.	276, 847	Kogan, F.M.	**1186, *1187**
Kazantzis, G.	*270, **807, 808**	Kogevinas, M.	*270, *280, *281, 294, 295, *306, *794
Kazubskaya, T.P.	1172		
Keefer, L.	135	Kohlmeier, L.	**337, *338**
Kefford, R.F.	43	Kohlmeier-Arab, L.	*615
Kegeles, S.S.	1054, 1055	Kohyama, N.	546, *548
Kekki, M.	239, 240	Kok, F.J.	**614, *615,** 616
Kellerman, R.	450	Kolk, J.	*601
Kellokoski, J.	256	Kolmannskog, S.	641
Kendall, G.M.	773, 774	Kolonel, L.N.	617, 897, 926, 984, 985, **986,** 987, 988, 989
Kenya, P.	1134		
Keogh, J.	*1131	Kolosza, Z.	648, 649
Kerenyi, N.A.	*101	Komada, J.	199
Kerner, J.	1060	Komatsu, S.	519, *564
Kernohan, E.	*761	Kondo, H.	*532
Kersey, J.	1046	Kondo, S.	*532
Kesteloot, H.	*53	Konjevic, R.	1197
Kestin, M.	1130	Kono, S.	552, **565**
Keszler, A.	2	Kono, T.	521
Kettunen, K.	*257	Koo, L.C.	**379**
Kew, M.C.	**668**	Kopecky, K.J.	1128
Key, C.R.	884, 897, 926	Korman, M.	7
Key, M.	*993	Koshkina, V.S.	*1174
Key, T.	293, 841, **851**	Koskenvuo, M.J.	242
Khaw, K.T.	293	Koskinen, H.	**245,** 246
Khlat, M.	286, *290	Kosma, V.M.	*257
Khoo, S.K.	5	Kosmelj, K.	1193
Khuhaprema, T.	747	Koutová, E.	193
Kiemeney, L.A.L.	*601	Koutras, D.	376
Kikendall, J.W.	***1156**	Kovac, J.	1196, 1197
Kikuchi, S.	568	Kowalska, S.	655
Killingback, M.	7	Kowalski, L.P.	59, 64, *575
King, D.	1152	Krajewska, B.	653
King, E.	*1124	Kramárová, E.	191, 192
King, M.C.	**906,** *1053	Krause, B.E.	62
King, R.A.	*1048	Krejcirik, L.	193
Kingston, J.E.	848, 850	Kresbach, H.	336
Kinne, S.	1130	Kressner, U.	737, 738
Kinnier Wilson, L.M.	848, 850	Krewski, D.	76
Kirkpatrick, A.	877	Kreysel, H.W.	336
Kirsch, I.R.	*922	Kriauciunas, R.	287
Kiss, J.	191	Kricker, A.	**29**
Kitinya, J.N.	746	Krieger, N.	1076, 1077
Kjaerheim, K.	*637	Kriek, E.	**584**
Klaassen, D.J.	92	Krishnan Nair, M.	117, *421, *422, 423, *424, *425, *426, *427
Kleevens, J.W.L.	380		
Klein, S.	*932	Kristal, A.	1130
Kleinerman, R.A.	222, 913	Kristinsson, J.R.	642
Kluck, H.M.	606	Kromhout, D.	293
Kneale, G.W.	753	Kroon, B.B.R.	586, 587
Knekt, P.B.	**243, 244**	Kross, B.	911
Kneller, R.	135	Kruger Kjær, S.	**210**
Knight, N.	4	Krugliak, P.	*436
Knight, T.	***778,** 847	Krumhaar,	337
Knipschild, P.G.	*593	Krutchkoff, D.	*99
Knox, E.G.	**753,** 760	Ktenas, D.	18
Knudsen, L.	*694	Kubik, A.	***194**
Knudson, R.	1149	Kubo, N.	534, 535
Kobayashi, N.	**569, 570,** 571	Kujawska, A.	*657
Kobayashi, S.	536	Kulkarni, M.G.	579

INVESTIGATOR

Kumar, A.	414	Launoy, G.	261
Kumar, D.	418, *419	Laurence, K.	50
Kummet, T.D.	1154	Laurent, P.	*333
Kundi, M.	46	Laurenti, R.	573
Kung, I.T.M.	380	LaVecchia, C.	303
Kuni, C.C.	*1048	Lawrence-Jones, C.	821
Kupper, L.L.	943	Lawson, D.H.	859
Kuratsune, M.	553	Layde, P.	1037
Kurihara, M.	520	Laydevant, G.	*310
Kurihara, N.	566	Le Mab, G.	324
Kuroki, T.	566	Le Marchand, L.	617, 986, **987**, **988**, 989
Kurppa, K.	245, *280	Leake, R.E.	303
Kuse, R.	349	Leandro, G.	**465**
Kushi, L.	1043, *1048	Lebesque, J.	*590
Kushlinski, V.	1181	Lebowitz, M.D.	**1149**
Kushwaha, M.S.	**414**	Lechat, M.F.	**50**
Kusuki, Y.	*530	Leclerc, A.	**316, 317**
Kuvshinov, J.P.	1178	Leduc, C.P.	**95**
Kuzina, L.E.	1186	Lee, J.	39
Kvåle, G.	**633**	Leech, S.	843
Kwa, H.G.	819	Lees, A.W.	**62**
Kyrtopoulos, S.	276, 847	Lees, P.	1155
L'Abbé, K.A.	**107**	Lefebvre, J.L.	267, **269**
L'Huillier, M.C.	*319	LeGasse, A.A.	1101, 1102
La Rosa, F.	**498**	Lehtonen, J.	*257
La Rosa, G.L.	466	Leibovici, S.	1165
La Vecchia, C.	453, *454, 486, **488**, **489**, 490, 491, 743, 744, 745	Leigh, J.	37, **39**, **40**
		Lejeune, F.	304
Lach, B.	*81	LeMerle, J.	1089
Lachlan, G.	845	Lemieux, B.	*1050
Laciak, J.	655	Lemp, G.F.	979
Lacroix, A.	69	Len, C.Y.	*151
Länsimies, E.	*257	Lence, J.	190
Lagorio, S.	**507**	Lennox, E.L.	854
Lahaye, D.	51	Lenoir, G.	314
Lahdensuo, A.	245	Leon, D.A.	789
Lahermo, P.	248	Lepine, J.	**96**
Lai, K.D.	*148	Lepore, A.R.	479
Laleman, G.R.	***55**	Lerchen, M.L.	926
Lam, S.Y.	380	Lerman, R.H.	936
Lam, T.H.	**380**, 808	Lerman, S.E.	*974, *976
Lam, W.K.	380	Lesaffre, E.	*53
Lambert, B.	*694, **712**, 938	Lesher, L.	1157
Lambert, J.	7	Lessner, K.	*360
Lammes, F.	585	Lestani, M.	516
LaMotte, F.	*933	Létourneau, E.G.	**78**
Land, C.E.	913	Lev, R.	1105
Landi, M.T.	483	Levi, F.G.	292, 303, *454, 743, 744, **745**
Langenberg, P.	*1156		
Langlois, S.P.	97	Levin, B.	*995, 998
Langman, M.J.S.	859	Levin, L.S.	896
Langmark, F.	287	Levin, R.	1070
Lanier, A.K.	*216, **885**	Levine, A.M.	1010
Lanning, M.	642	Levine, C.	443
Laplanche, A.	327	Levine, N.	1150, 1151
Laporte, J.	298	Leviton, A.	897
Larkins, R.	18	Leviton, L.C.	1096
Larsen, S.	639	Levitz, M.	1063
Larsen, T.E.	**638**	Levshin, V.F.	*1173
Larsson, L.G.	**721**, 722	Levtchuk, A.A.	1181
Larsson, S.	**682**	Lew, R.	*1160
Larusdottir, M.K.	*392	Lewin, F.	*716
Latan, A.	**184**	Lewis, M.E.	*621
Laudico, A.V.	*289	Lewis, R.J.	*974, *975, *976

INVESTIGATOR

Li, C.C.	**141, 142**	Lopes, L.F.	65
Li, F.M.	128	Lopez, A.D.	**742**
Li, F.P.	1118	Lopez, B.H.	646
Li, G.	134	Lopez, R.	897
Li, J.	125	Lotz, I.	*340
Li, J.Y.	135	Loughlin, J.E.	*945
Li, L.	154, *162	Louhivuori, K.	258
Li, S.	182	Louwrens, H.	663
Li, W.	182	Love, E.J.	*155
Lian, X.X.	*126	Lovincz, A.T.	1141
Liang, S.	***127**	Lowenfels, A.B.	296
Liang, X.Z.	*126	Lowenthal, R.M.	15, *770
Liao, C.S.	177	Lu, J.B.	**185, 186**
Liberati, C.	488	Lu, S.H.	**128**
Liddell, F.D.K.	*73	Luande, J.	367
Lifanova, Y.E.	1181	Lubbe, J.T.N.	603
Liff, J.	888, 891, 918, 1056	Lubin, J.	907, 913
Liippo, K.	245	Lucas, L.J.	**1158, 1159**
Lilleorg, A.	237	Luce, D.	316
Lillis, D.F.	50	Luesley, D.	760
Lim, A.	578, 580	Lund, V.J.	*809
Limasset, J.G.	*328	Lund-Larsen, P.	636
Limbert, E.	314	Lundberg, I.S.	*280, **713**
Lin, C.Y.	150	Lundegårdh, G.	726
Lin, M.	*160	Lundell, M.	711
Lin, Y.T.	158	Lundström, N.G.	***724**
Lindblad, P.	729	Luo, F.J.	128
Lindefors, B.M.	706, 708	Luria,	449
Lindsey, J.	*1107, *1108	Lustbader, E.	1084
Lindsted, K.	1007	Luthra, U.K.	**418, *419**
Linet, M.S.	305, 913	Lutz, J.-M.	*281, 287, ***310**
Ling, M.	176	Lutz, W.	653
Linnainmaa, K.	*694	Lynch, P.	998
Linos, D.	376	Lynge, E.	210, ***211, *212**, 214, 230, *280, *281, 294
Little, J.H.	6, 8, 304, *829, *830		
Littorin, L.	294	Lyon, J.L.	897, **1121**, 1136
Litvak, J.	1091	Ma, X.Y.	*144, *145, *147
Liu, C.B.	*151	Maas, M.J.	612
Liu, M.Z.	141	Maase, H.	215
Liu, P.L.	**167, 168**	Maatela, J.	243
Liu, Q.	140	Macaluso, M.	481, 927
Liu, Q.F.	**161**	Macdonald-Davies, I.	802
Liu, W.D.	135	MacDougall, L.G.	**669**
Liu, X.Q.	135	MacGibbon, B.H.	**773, 774**
Liu, Y.Y.	*153	Machinami, M.	541
Lloyd, O.L.L.	**775**, *776	MacIntyre, R.	965
Lluch, A.	271	Mack, T.M.	**1015, 1016**, 1017
Lo Curto, M.	511	MacKenzie, G.	***751**
Lo, K.K.	380	Mackintosh, L.	856
Loach, M.J.	854	MacLaren, W.M.	**777**
Loddenkemper,	337	MacLennan, R.	4, **6, 7**, 8, 11, 35
Lodi, P.	507	Maclure, M.	1185
Lönnberg, G.	723	MacMahon, B.	303, 1035, 1087
Logamuthukrishnan, T.	415	MacOnochie, N.	795, 797
Logan, R.F.	***829, *830**	Macrae, F.A.	7, *21, 24, 820
Loginov, A.P.	1188	Madarnas, P.	95
Lohde, D.	337	Mäntyjärvi, R.	*257
Lokobauer, N.	1196	Maès, P.	318
Lomuto, M.	501	Magnani, C.	463, 464, 508, 511, 516
London, S.J.	***1014**, 1047	Magnetti, C.	*1156
London, W.T.	1084, **1087**	Magnússon, B.	390
Longnecker, M.	1035	Mahaffey, J.	1110
Looman, C.	*604	Mahlamäki, E.	*257
Loomis, D.P.	**941**	Mahon, G	303

INVESTIGATOR

Maiani, G.	*505	Masuyer, E.	287
Maier, H.	*360	Mathew, B.	117, *426
Mainz, D.	337	Mathews, J.D.	12, 13, 294
Maitland, N.	760, *864	Matiakin, E.P.	1178
Makimoto, K.	552	Matijasevic, A.	1191
Malfatto, F.	462	Matos, E.	*291, *306
Malker, H.	*724	Matsui, I.	569, 570, 571
Mallin, K.	916, 948, 949	Matsuo, T.	531
Maltoni, C.	*492	Matsuura, M.	520
Manca, P.	460	Maturana, M.	*123
Mancini, M.	511	Mauad, M.A.	65
Mancuso, S.	506	Maumenee, I.H.	896
Mandel, J.S.	*982, 1044, 1045	Maximilien, R.	298
Manders, T.	*225	Maximovich, D.M.	1185
Mann, J.R.	755, 855	May, S.	*627
Mann, T.I.	853	Mayadevi, S.	424
Manniello, M.A.	476	Mayer, J.	1064
Manousos, O.	847	Mays, C.W.	372
Mansel, R.E.	*771	Mazumdar, S.	1101
Manso, C.S.	*615	Mazzoleni, C.	482, 486
Manson, J.E.	929, *933, 934	Mazzuckelli, L.F.	957
Mant, D.	852	McBride, M.L.	116, 305
Mantel, M.	446, 447	McCaig, R.H.	*790
Manti, A.	*475	McCammon, C.	1114
Mantout, B.	321	McCarthy, W.H.	43
Manzo, N.	946	McCartney, A.C.E.	812
Mao, Y.S.	76, 79, *80, *81, *83, 84, *85, 87	McCawley, M.	912
		McCrea Curnen, M.G.	1055
Marcos, G.	673	McCredie, M.R.E.	41, 42, 301, 729
Marek, K.	*657	McDonald, A.D.	*73
Margolin, B.	940	McDonald, J.C.	*73
Mariani, R.	307	McDougall, J.K.	*1132
Mark, E.	84	McDuffie, H.H.	91, 92, 93, 94
Marks, R.	17, *769	McElvenny, D.	882
Marmot, M.G.	288, 810, *811	McGlynn, K.	1084
Marovic, G.	1196, 1197	McGowan, L.	1157
Marrett, L.D.	63, *80, 108	McGrath, M.S.	904
Marsden, K.	15	McGregor, R.G.	78
Marsh, G.M.	1095, 1096, 1097, 1098	McIntosh, W.A.	666
Marshall, J.	937	McIntyre, O.R.	7
Marten, A.	189	McKinney, P.A.	781
Marth, E.	*44	McKnight, B.	1125
Martin, N.G.	8	McLarty, J.W.	1154
Martin, T.	*928	McLaughlin, J.K.	42, 91, *165, 729, 912, 1017
Martin-Moreno, J.	303, *615		
Martina, G.	514	McLeish, J.	7
Martinez, C.	293	McLeod, G.R.C.	6, 8
Martinez, I.	926	McMichael, A.J.	3, 300, 302
Martinsohn, C.	443	McNamara, D.	*1148, 1152
Martos, M.C.	673	McNamara, N.	1151
Martuzzi, M.	508	McNeil, J.J.	*33, *34
Marubini, E.	487, *492	McPherson, C.K.	813, 855
Marynen, P.	54	McPherson, S.	*995
Masaki, M.	572	Mead, M.	1118
Masala, G.	470	Meadows, A.T.	*1086, 1088, 1089, 1090
Masera, G.	511	Meanwell, C.	760
Mashberg, A.	1072	Meara, J.	858
Masina, A.	507	Mecucci, C.	54
Maskens, A.	262	Medes, A.	292
Mason, B.H.	618	Meenakshi, M.	*428
Masood, M.A.	644	Megha, T.	469
Mastrandrea, V.	498, 499	Mehta, F.S.	404, 405, 408
Mastroiacovo, P.	50	Meinert, L.	926
Masuoka, H.	560	Meirik, O.	708

INVESTIGATOR

Meisels, A.	88	Modan, B.	302, **450**, **451**, 1134
Melbye, M.	**213**	Moe, P.J.	**641**, 642
Mellemgaard, A.	**214**	Möller, T.R.	*690, *695, *697
Menard, T.	1135	Möse, J.R.	*44
Menck, H.R.	**1017**	Möslein, G.	*363
Mendy, M.	335	Mohapatra, S.C.	*428
Ménégoz, F.	302	Mohler, J.	939
Meneses, F.	582, 583	Molina, R.	1134
Menon, R.	418	Moll, P.P.	969, *970
Merabishivili, V.	287	Møller, H.	**215**, 847
Mercier, M.	260	Moller, P.	314
Meriläinen, P.	*257	Mollichella, E.	499
Merler, E.	469, 474, 507, 516	Mornas, I.	311
Merletti, F.	*281, 292	Mommsen, S.	229
Merlo, F.	*454, 477	Monfardini, S.	455
Mertens, A.C.	1042	Monnier, V.	745
Mertz, G.	883	Mononen, I.	*257
Merzenich, H.	**361**	Monson, R.R.	1033
Messing, E.	**1034**	Montella, M.	*454
Messing, K.	**74**	Montesano, R.	128, 898
Meyer, F.	**89, 90**	Montpetit, V.	*81
Meyskens, F.L.	*1131, 1153	Moodie, P.F.	120
Mgaya, H.N.	367	Moody, J.	114
Miazzo, G.	496	Moon, T.E.	*1148, **1150, 1151, 1152**
Michaelis, J.H.	287, 364, *373, *374	Moore II, D.H.	**1006**
Micheli, A.	482	Moore, D.E.	*1138
Micozzi, M.	1152	Moore, M.	1070
Middlecote, B.D.	667	Morantz, R.A.	1002
Miers, M.E.	*771	Moreno, V.	674, 675
Mikkelsen, T.	*81	Moreo, P.	283
Milan, C.	263	Morettini, A.	503
Miles, B.J.	965	Morgan, G.G.	40
Milham Jr, S.	**1082**	Morgan, M.	1136
Miligi, L.R.	474, 513	Morgan, M.V.	884
Miller, A.B.	75, 77, 105, **109, 110, 111,** **112**, 300	Mori, H.	531, *532
		Mori, M.	560
Miller, B.A.	1114	Mori, W.	541
Miller, D.	1046	Morimoto, K.	**545**
Miller, D.G.	1066	Morin, C.	88
Miller, G.H.	**977**, *978	Morin, M.	913
Millman, I.	1084	Morinaga, K.	**546, 547,** *548, 550
Mills, C.	*80	Morio, S.	577
Mills, P.	1007	Morison, D.	*82
Milner, M.	235	Morita, N.	519
Minami, Y.	563	Morley, A.A.	938
Mine, M.	531, *532	Morrell, D.	944
Minelli, L.	498	Morris, D.L.	939, 1028
Minowa, M.	529	Morris, H.J.	1157
Mirer, F.E.	968	Morris, P.	*1107, *1108
Mirra, A.P.	288	Morris-Jones, P.H.	1089
Misciagna, G.	*456	Morrison, A.S.	**1105**
Misdorp, W.	587	Morrison, H.	*82, *83, *85
Misra, N.C.	414	Morson, B.	793
Misra, P.K.	414	Mortel, R.	914
Missale, G.	503	Mortensen, P.B.	**233**
Mitchell, D.	1109	Morton, N.E.	**874**, 990
Mitelman, F.	305, *694, **700**	Moscicki, A.B.	*1124
Mitrovic, N.	1190, **1191**	Moser, M.	893
Mitsuhashi, T.	560	Moss, D.	645
Miyake, H.	*281, **560, 561**, 562	Moss, S.M.	**877**
Miyake, K.	533	Moszczynski, P.	**647**
Mmuni, K.	746	Mott, M.	*765
Mo, C.C.	161	Motulsky, A.	1137
Mo, S.Q.	164	Moulin, J.J.	298, **320, 321**

INVESTIGATOR

Mravunac, M.	598, 600, 602	Nervi, F.	**122**
Müller, C.	*343	Nesbit, M.	1089
Müller, H.J.	**739**	Nethercott, J.	*901
Mueller, N.	897	Neuberg, R.W.	1090
Müller-Hermelink, K.H.	348	Neuberger, J.S.	**1002**
Muirhead, C.R.	773, 774, 839	Neuberger, M.	**46**, 294
Mulet, M.	273	Neugut, A.	441
Mulhern, R.K.	980	Neugut, A.I.	1057, 1059, **1061**, *1062
Muller, J.M.	*9, *10	Neusetzer,	337
Muller, S.	931	Nevin, N.	50
Munaka, M.	520	Newall, D.	847
Mund-Hoym	*338	Newcomb, P.A.	**1035, 1036**
Munger, R.G.	1043	Newell, D.	*778, 846
Munoz, A.	*899	Newell, F.	845
Muñoz, N.	185, 271, 272, 273, **282**, **283**, *284, *285, 357, 646, 674, 747, *748	Newman, P.	618
		Newton, W.A.	1089
		Ngelangel, C.A.	*289
Mur, J.M.	321	Ngendahayo, P.	348
Murata, M.	**517**, 533	Nichaman, M.Z.	*995
Mure, K.	545	Nicholls, R.J.	820
Murphy, E.D.	1084	Nicholson, H.S.	919
Murray, J.	931	Nie, G.H.	157
Murthy, N.S.	418	Nie, V.	*25
Musch, G.	487	Nielsen, N.H.	*216
Musk, A.W.	26, **30, 31**, *32	Niessner, H.	**47**
Musto, R.	120	Nigra, T.P.	931
Muth, H.	368	Nikitin, Y.P.	*216
Muti, P.	296, 482	Nikkari, T.	243
Mwakyusa, D.	**746**	Nilssen, S.	633
Myers, R.	*1142	Nilsson, E.	686
Mylvaganam, A.	3	Nilsson, T.	*684
Nadel, M.	**893**	Niman, H.L.	1057
Naéssen, T.	738	Nishi, M.	561, 562
Nagao, K.	535	Nishijama, S.	*359
Nagashima, Y.	517	Nishikawa, H.	565
Nair, M.	423	Nisse, C.	*268
Nakachi, K.	*556, **557**, 558, 559	Nolan, J.	937
Nakamura, K.	533, **572**	Nomizu, T.	**527**
Namihisa, T.	568	Nomura, A.M.Y.	984
Nanni, O.	*454	Noordhoek, J.	612
Narendran, P.	415	Noorlind Brage, H.	687, 688
Nasca, P.C.	916	Nordberg, G.F.	*270, *724
Nasrallah, S.	441	Nordberg, M.	*724
Nass, C.	1088	Nordenson, I.	*694
Nasser, A.	236	Nordstrom, D.	1037
Natarajan, N.	276	Norell, S.E.	709
Natekar, M.V.	*411	Norhanom, W.	*581
Nath, P.	414	Norman, J.E.	908
Naukkarinen, A.	*257	Norpoth, K.	*343
Navarro, C.	283, 293	Nose, T.	533
Nazar, V.	1137	Notani, P.N.	*409, *412
Ndimbie, O.K.	1094	Notkola, V.J.	**254**
Necrasova, L.I.	1184	Novak, F.	1192
Neglia, J.P.	1046, *1100	Novakovich, M.	1196
Negri, E.	*454, 486, 488, 489, 743, 744, 745	Novotny, T.	*978
		Noyon, R.	589
Negri, G.	469	Nurhayati, Z.A.	*581
Neil, H.A.W.	**853**	Nurminen, M.	245
Nelemans, P.J	**599**	Nuruddin, R.	578
Nelson, N.A.	120	Nussbaum, R.H.	758, *1103
Németh, G.	375	Nygaard, R.	**642**
Nemoto, T.	937	Nyong'o, A.	48
Nephedov, M.D.	1172	Nyrén, O.	728
Neri, G.	506	Nyström, L.	719, **722**

INVESTIGATOR

O'Boyle, P.	*879	Otu, A.A.	**630, 631**
O'Connor, C.M.	839	Overvad, K.	205
O'Conor, G.T.	305	Owens, A.H.	896
O'Dea, K.	18	Owor, R.	347
O'Higgins, N.	303	Paci, E.	469, 470, **472**, 473
O'Neill, B.J.	36	Packer, P.	843
Oberhausen, R.	369	Paddle, G.M.	**821**
Oberlin, O.	330	Padmakumary, G.	*422, 423, *424, *425
Oblak, B.	1193	Padmanabhan, T.K.	*421
Obrador, A.	273, 674	Padmavathy Amma, B.	*426
Obrecht, J.P.	739	Padubidri	*419
Obringer, A.	1089	Paffenbarger Jr, R.S.	**1143**, *1144
Obsiniková, A.	192	Paganini-Hill, A.	1013, **1018, 1019**, *1020
Odes, S.	**435**, *436	Page, D.L.	1052, *1053
Offerhaus, G.J.A.	896, *900	Pageau, R.	95
Ogata, M.	552	Pagliaro, L.	*497
Ogimoto, I.	526	Pahwa, P.	91
Ogura, H.	545	Painter, R.	856
Oh, H.C.	120, *121	Pal, R.	603
Ohashi, K.	*549	Palan, P.R.	*932
Ohba, S.	560	Palazzotto, G.	*497
Ohlander, E.-M.	731	Palefsky, J.	*1124
Ohmine, K.	535	Palli, D.	*454, **473**, 846, 847
Ohno, Y.	533, **534, 535**	Palmer, J.R.	*935
Oho, K.	567	Parnart, B.	*268
Ohtaki, M.	520	Pangalis, G.A.	276
Ojeda, R.	*359	Panizzon, R.	336
Okada, S.	*549	Pannett, B.	*280, 863, 867, 868, 869, 870
Okamoto, N.	**577**		
Okojie, C.G.	**632**	Paoletti, C.	324, 741
Okojie, P.I.	632	Paoletti, L.	469
Okong, P.	347	Paolucci, G.	511
Okubo, T.	525	Papaevangelou, G.J.	**377, 378**
Okumura, Y.	*532	Paramsothy, M	578
Okuneva, L.A.	*1174	Parazzini, F.	*454, 488, 489, **490, 491**
Olafsdottir, G.	*391, *393	Parida, B.B.	118
Oliveira, B.V.	64	Park, R.	**968**
Oliver, R.T.D.	840, 842	Parker, L.	**825**, *826, *827
Oliver, W.	*284, *285	Parkes, S.E.	755
Olivera, L.	1161, 1162	Parkin, D.M.	*194, 277, *285, **286, 287, 288**, *289, *290, *291, 299, 335
Olsen, G.W.	1041		
Olsen, J.	**201**, 202, 295		
Olsen, J.H.	**217, 218**	Parkkinen, S.	256
Olshan, A.F.	1093, **1099**, *1100	Parks, S.	*764
Omar, S.	236	Parl, F.F.	1052
Omenn, G.	1130, *1131	Parrish, J.A.	931
Omohunndro, E.L.	1110	Parsons, W.J.	12
Omoto, K.H.	1141	Partanen, K.	*257
Oosterlinck, A.	54	Partanen, T.J.	245, **246**
Ordoqui, G.	1162	Parukutty Amma, K.	*422
Oreggia Coppetti, F.V.	**1163**	Passarge, E.	1058
Orengo, A.	478	Pasternack, B.S.	**1063**, *1068
Orfanos, C.E.	336	Patel, D.	400, 401, *402
Orjaseter, H.	636	Patel, N.L.	*398, 400, *402
Orlandini, C.	478	Patel, R.D.	395, 400, 401
Oruzinsky, J.	199	Patel, T.B.	395, 396, 397, **400**, 401, *402
Osborne III, J.S.	*923		
Osborne, M.	1066	Paterson, J.C.	821
Oshima, A.	550	Pathak, D.R.	884
Osman, J.	294, 787, 788, 789, *790	Pathmanathan, R.	578
Osmond, C.	868	Patrone, M.G.	476
Østerlind, A.	304	Paul, C.E.	**625**
Osterman-Golkar, S.	251	Paul, E.	**345**
Otto, L.P.	603	Paul, O.	1004

INVESTIGATOR

Pawlega, J.	**650**	Pinzone, F.	***497**
Pawlita, M.	367	Pippard, C.	867, **875**
Pearce, N.E.	*281, 294, *627, **628, 629**	Pirastu, R.	507
Pecorelli, S.	506	Pires Torres, C.	*359
Pedersen, M.	220	Pirisi, L.	*962
Pedroni, M.	494	Pisa, R.	*497
Peer, P.G.M.	602	Pisani, P.	293, 481, 482, 513
Peeters, P.H.M.	600	Piva, C.	516
Pellegrini, H.D.	1161, 1162	Plasencia, J.	271
Peltonen, K.	251	Platz, C.	926
Pendergrass, T.W.	1099	Plesko, I.	191, **192**, 193, 287, 303
Penfold, J.C.B.	7	Plotko, E.G.	*1187
Penn, I.	**953**	Pochart, J.M.	318
Penttilä, I.	*257	Pocock, R.	*879
Pentz, M.A.	1047	Poddubni, B.K.	1178
Perani, B.	*515	Poirier, M.C.	*457, *902
Peraza, S.	*284, *285	Pollen, C.	*279
Percy, C.	544	Pollock, B.H.	**980**, *1049, *1100
Perera, F.P.	1057, **1064**	Pompe-Kirn, V.	287, **1192**, 1193, 1194
Peretz, H.	449	Ponder, B.A.J.	314
Perez Viguera, J.	677	Ponz de Leon, M.	**494**
Perez, J.	438	Ponzio, G.	510
Peris-Bonet, R.	301	Pool, J.E.	669
Pero, R.W.	1064, 1066, 1067	Poole, C.	***945**
Perrimond, H.	307	Pooley, F.D.	860
Perry, M.	*780	Popendorf, W.	911
Pershagen, G.	292, **714**	Popiela, T.	**651**
Persson, B.	*281, *684	Popp, W.	*343
Persson, I.R.	*225, **737, 738**, 917	Porro, A.	*456, *457
Pesatori, A.C.	483, *484	Portefaix, P.	321
Peters, G.N.	997	Portlock, C.	1054
Peters, J.M.	629, 1025	Portman, L.	744
Peterse, J.L.	586, 587, *590	Potischman, N.A.	914, 918
Peterson, A.	1130	Potter, A.	16
Peto, J.	*270, 272, 834	Potter, J.	847
Peto, R.	125, 243, 299, 929, *933, 934	Potter, J.D.	1043, 1046, *1048, *1122
Petrinelli, A.M.	498, 499	Pottern, L.M.	910, **1114**
Petrozzi, J.	931	Pottier, D.	261
Pettersson, B.	725	Powell, C.A.	865, 868, 869, 870
Pexieder, T.	50	Powell, D.E.	**762**
Pfäffli, P.	*280	Powell, J.E.	755
Pfeifer, H.	356	Powles, J.W.	18
Pfeiffer, K.P.	*44	Prasad, U.	**578, 579, 580**
Phair, J.	*899	Pratt, C.P.	1039
Pham, Q.T.	**322**	Predieri, G.	*493
Pharoah, P.	**791, *792**	Prener, A.	**219**
Philippe, J.	*279	Prentice, R.	*933
Phillips, M.	28	Preston-Martin, S.	301, **1021, 1022, 1023, 1024, 1025**, *1026, *1027
Phillips, R.	*814		
Pichler-Semmelrock, F.	*44	Preudhomme, C.	*268
Pierce, J.P.	*978	Price, J.L.	878
Pierotti, M.	482	Prihartono, J.	534
Piettanen, P.	303	Prijono Tirtoprodjo	431
Piffer, S.	*515	Primic-Zakelj, M.	**1193**, 1194
Pignat, J.C.	275	Prineas, R.	1043
Pike, M.C.	840, 852	Proietto, J.	18
Pilacik, B.	653	Prorok, P.C.	954
Pilawska, H.	***658**	Prystowsky, J.	931
Pinczowski, D.	734	Psaty, B.	*1104
Pindborg, J.J.	404, 405, 408	Pticar, M.	1197
Pines, A.	**452**	Pu, L.M.	*155
Pingitore, R.	469	Pujol, H.	312
Pinto, C.B.	***57, *58**	Pukkala, E.I.	242, 246, **247, 248**, 249, 250, 258

INVESTIGATOR

Puntoni, R.	*475, **476, 477**, 478	Reissigova, J.	*194
Puppinck, C.	313	Remennick, L.I.	***1174**, 1175, 1177
Purde, M.	237	Ren, U.F.	177
Purtilo, D.T.	**1083**	Renard, H.	277
Qin, D.	166	Rennert, G.	**437, 438, 439, 440**, 441, 442
Qiu, S.L.	**187**, 188, 357		
Quesenberry, C.P.	1079	Renz, K.	354
Quinn, M.A.	*33	Reuterwall, C.	*694, **702**
Quiquandon, I.	*268	Reyes, M.G.	*289
Quiros Garcia, J.R.	293	Reynolds, P.	**979**, 1056
Raafat, F.	755	Riboli, E.	**292, 293**, 297, 673
Racioppi, L.M.	*975	Ricci, M.	513
Radden, B.G.	23	Ricci, P.	**516**
Radic, A.	50	Rich, A.M.	**23**
Radóczy, M.	387	Rich, S.S.	*1048
Raedsch, R.	355, 357	Richards, W.	666
Ragni, G.	499	Richardson, A.K.	***626**
Ragunathan, T.E.	*1104	Richardson, L.	66
Rahal, D.	1	Richardson, S.	309
Rahu, M.A.	287	Richart, R.M.	916
Rai, K.R.	1070	Richmond, R.E.	**983**
Rajeevkumar, S.	*422	Rickabaugh, J.	983
Ramachandran, T.P.	*421	Ricketts, T.J.	781
Ramael, M.	48	Ridanpää, M.	241
Ramamurthy, B.	415	Rider, P.L.	*761
Ramani, L.	495	Ridolfi, R.	503
Ramani, P.	*427	Riedel, D.	910
Ramazzotti, V.	*504, *505	Riemersma, R.	*615
Ramgren, O.	715	Ries, L.	544, *924
Ramli, M.	534	Rifa, J.	675
Rammeloo, J.A.	605	Righi, E.	*493
Rampal, L.	579, 580	Rigo, O.	382
Rampen, F.H.J.	599	Riihimaki, V.	**249**
Ran, R.Q.	161	Rinaldo, C.	*899, 1094
Randell, P.L.	29	Ring, I.	8, *9, *10, **11**
Randow	*338	Rinsurongkawong, S.	747
Randriamiharisoa, A.	743	Ríos-Dalenz, J.L.	1091
Rao, D.N.	*406	Rivero, S.	1163
Rao, N.D.	***413**	Rivosecchi, P.	498
Rapoport, A.	59	Rizetto, M.	873
Rapoport, S.I.	***924**	Rizzelli, R.	507
Rappaport, S.M.	**942**	Roberson, P.K.	**1039**
Rashad, S.	235	Roberts, R.J.	791
Raue, F.	314, ***362**	Robertson, C.M.	854
Rauh, M.	345	Robin, P.E.	757
Ravnihar, B.	1193	Robinson, E.	**441**
Rawls, W.E.	937	Robinson, K.	16
Raybaud, C.	307	Robinson, R.	442
Raymond, L.	*281, 287, 324	Robison, L.L.	1042, **1046, 1047**, *1049, 1099
Raynor, D.	113		
Reade, P.C.	23	Robra, B.P.	369
Reale, F.R.	*1160	Robson, D.	*81, 91, 116
Reaman, G.H.	1046	Robson, S.C.	*664
Recchia, C.	480	Robutti, F.	462
Redeker, A.	1087	Roca, P.	674, 675
Redmond, C.K.	**1101**, **1102**	Rochon, M.	95
Reed, R.	*20	Rockette, H.E.	1101
Rees, D.	**670**	Rodary, C.	330
Rege Cambrin, G.	510	Rodella, S.	513
Reger, R.	912	Roder, D.M.	3
Reggiardo, G.	478	Rodliffe, R.S.	*766
Reginald, S.	415	Rodricks, J.V.	*887
Rehman, G.	644	Rodrigues, V.	*281
Reid, R.I.	**1141**	Rodriguez Cuevas, H.	1134

INVESTIGATOR

Rodriguez, F.F.	292		Roy, P.	276, 847
Rodriguez, M.C.	271		Rozen, P.	355
Rodriguez-Martinez, H.	1091		Ruccione, K.	*1085
Rodvall, Y.	703		Rudolf, Z.	1192
Roecker, E.B.	1034		Ruiter, D.J.	599
Rödelsperger, K.	346		Runia, S.	*592
Roels, H.	*52		Rusciani, L.	500
Römer, W.	346		Russell, I.	*20, *22
Röntgen-Pieper, E.	*591		Russo, R.	479
Roffino, R.	514		Rutegård, J.N.	**679**
Rogers, A.J.	37, 39, 40		Rutledge, L.H.	1141
Rogers, E.	*1160		Rutqvist, L.E.	708, ***716***, 722
Rogers, W.H.	*1053		Ruttenber, J.	1128
Rojas, R.	582, 583		Ryan, B.R.	980
Rolón, P.A.	*290, **646**		Saah, A.	*899
Romano, F.M.	*497		Saarikoski, S.	256
Romas, N.	*1062		Saba, G.	499
Romazzini, S.	*310		Saba, L.M.B.	65
Romieu, I.	582, **583**		Sabaté, J.	1007
Romney, S.L.	***932***		Sabater Pons, A.	*676
Ron, E.	913		Sabroe, S.	201, **202**
Ronco, A.L.	***1164***, 1165, *1166, *1167		Sacripanti, P.	503
			Sällsten, G.	683
Roncucci, L.	494		Sagaidak, V.N.	1169
Rookus, M.A.	588, *590		Sage, M.	1128
Rooney, C.	795		Sagebiel, R.W.	1115
Roos, G.	679		Saigi, E.	673
Rorke, L.B.	1090		Saito, Y.	567
Rørth, M.	215		Saiz Sanchez, C.	*676
Roscoe, R.J.	952, **954**, 1117		Saji, F.	***549***
Rose, C.	220		Sakai, R.	**541**, **542**, ***543***
Rosef, O.	639		Sakamoto, G.	534
Rosen, T.	*992		Sakato, J.	547
Rosenberg, L.E.	***935***		Sakr, W.	*966
Rosenberg, S.A.	904		Salerno-Mele, P.	967
Rosenblatt, K.	1134		Sali, A.	381
Rosenfeld, P.	883		Sallan, S.	1089
Rosenstock, L.	*1131		Salmonsson, S.	710
Roser, M.	336		Salomaa, S.	*694
Rosin, M.P.	1178		Salonen, J.T.	**255**, 303
Rosner, B.R.	929, 930, *933, 934		Salvini, S.	*454, 486, 930
Rosner, G.	1054		Salzburg, M.	302
Rosner, W.	*1062		Samet, J.M.	**884**
Ross, J.	*1049		Samuskevich, I.G.	1168
Ross, R.K.	1009, 1018, 1019		Sanchez, V.	*284
Rosselli del Turco, M.	473		Sancho-Garnier, H.	*315, 324, 741
Rossi, M.	511		Sanders, B.M.	838, 854
Rossing, M.A.	*1138		Sandler, R.S.	939, **943**, *982
Rosso, P.	511		Sandström, A.	**723**
Rosso, S.	514		Sang, J.Y.	186
Rossof, A.H.	1004		Sankaranarayanan, R.	117, *421, *422, **423**, ***424***, ***425***, ***426***
Roth, E.	45			
Roth, Z.	197		Sankila, R.	**250**
Rothman, K.J.	*945, **946**		Sant, M.	485
Rothman, N.	*902, 910		Santamaria, M.	283
Rothrock, R.	1141		Santella, R.M	1064
Rothstein, R.	*982		Saracci, R.	*270, *280, *281, 292, 293, **294**, **295**, 296, 297, 298, *306, 370
Rotimi, C.	927			
Rotoli, M.	501			
Roumagnac, M.	*259		Saragoni, A.	503
Roumeliotou, A.	377, 378		Sarason, B.	1130
Roush, G.C.	1055, **1065**, **1066**, **1067**		Sarason, I.	1130
Rovenský, E.	199		Sarjadi,	**429**, **430**
Roy, C.R.	*14		Sarkar, S.	*1026, *1027

INVESTIGATOR

Sarrazin, D.	330	Schuman, B.M.	967
Sasaba, T.	*556, **558**, 559	Schuman, L.M.	1044
Sasco, A.J.	**296, 297, 298, 299**, *705	Schunk, W.W.	*342
Sassatelli, R.	494	Schvartz, C.	318
Satariano, E.	971	Schwartz, A.G.	**969**, *970, 972
Satariano, W.A.	897	Schwartz, S.	*966
Sather, H.N.	1046, *1085	Schweiger, M.	*784
Sato, E.	**524**	Scott, C.R.	*1139
Sato, H.	527	SEARCH Programme	**300, 301, 302, 303, 304, 305**
Sato, N.	568		
Satoh, Y.	*548	Sebastian, P.	*427
Satterfield, S.	*933	Secreto, G.	482
Sauar, J.	639, **640**	Seeff, L.B.	908
Saunders, L.D.	979	Seeger, R.	*1100
Savage, D.	1065	Seghal, A.	418
Savelli, D.	502	Seidell, J.C.	*611
Savitskaya, T.Y.	1171	Seip, M	641
Savitz, D.A.	941	Seitz, G.	354
Saw, D.	380	Sekfali, N.	1
Saxén, E.A.	243	Selby, J.V.	1073, **1079**
Scala, S.	503	Self, S.G.	*1138
Scalmati, A.	494	Sell, A.	202
Scarpelli, A.R.	474	Sellers, T.A.	1043, *1048
Scazziga, B.	744	Semenciw, R.	76, 79, *80, *82, *83, **84, *85**
Schachter, J.	*1124		
Schäfer, T.	**344**	Semiglazov, V.F.	**1169**
Schairer, C.	915, 916, 917	Seneviratne, S.	380
Schatzkin, A.	915	Seniori Costantini, A.R.	469, **474**, 507, 513
Schedel, I	350	Seppälä, K.	239
Scheepers, P.T.J.	597	Seppänen, R.	244
Scheiner, C.	307	Serra, I.	*123
Schiager, K.J.	1121	Serraino, D.	453
Schifflers, E.	277	Sessink, P.J.M.	597
Schiffman, M.H.	916	Settimi, L.	**508**
Schlaefer, K.O.H.	*363	Severson, R.K.	1046, *1049
Schlag, P.	352	Shah, K.V.	283
Schlehofer, B.	**364, 365**	Shah, N.	*1100
Schlehofer, J.	364	Shah, N.G.	*398
Schmauz, R.	**347, 348**	Shairer, C.	737
Schmitz, P.I.M.	589	Shang, E.R.	*162
Schnatter, A.R.	*973	Shanta, V.	416, *417
Schneider, J.S.	1006	Shao, Y.W.	*145
Schneiderman, M.A.	977, *978	Shapiro, S.	*935
Schneijder, P.	616	Sharma, B.K.	418
Schnorr, T.M.	**955, 956**	Sharma, J.K.	418, *419
Schnyder, U.W.	336	Sharp, D.S.	**950**
Schoenberg, J.	918	Sharp, G.B.	*1040
Schoolnik, G.	*1124	Sharples, K.J.	*620, *621
Schouten, H.J.A.	*593	Shaw, G.L.	*925
Schrameck, C.	326	Shaw, H.M.	**43**
Schraub, S.	**260**	Sheikh, K.	**886**
Schreiber, M.M.	1150, 1151	Shekelle, R.B.	1004
Schrijver, J.	614	Shemesh, E.	452
Schroeder-Kurth, T.M.	**366**	Shen, F.M.	**169**
Schubert, M.	*1132	Shen, H.	135
Schüler, G.	743	Shenberg, C.	446, 447
Schürfeld, C.	*343	Sherman, K.J.	*1132
Schütz, A.	*691, *696	Shetye, S.B.	413
Schull, W.J.	521	Shevchenko, V.Y.	1181
Schulte, P.A.	**957, 958**	Shi, Y.B.	*151
Schulte-Hermann, R.	47	Shibata, A.	528
Schultz, H.	202	Shigematsu, I.	**521**
Schulz, A.	346	Shilkin, K.B.	39
Schulz, G.	352	Shimamura, K.	517

INVESTIGATOR

Shiminski, T.	1090	Smith, J.	*826, *827
Shimizu, H.	**519**, 537	Smith, M.	1087
Shimizu, T.	539	Smith, P.G.	335, *794, 796, **815**, 835, 836
Shimizu, Y.	*551		
Shimokawa, I.	531	Smith, R.A.	888, **894**
Shimono, M.	552	Smith, S.D.	903
Shinchi, K.	565	Smith, T.	1033
Shirakawa, T.	545	Smithson, W.A.	1090
Shoa, Y.W.	*147	Smulevich, V.B.	**1175**, *1176, 1177
Shore, R.E.	1063, *1068	Snover, D.	1044, *1048
Shu, W.X.	*147	Sobala, G.	*284
Shu, X.O.	**170**, 179, 180, 301	Sobel, E.	1025
Shukla, H.S.	*428	Sobin, L.	*1156
Shy, C.M.	*923	Sobrinho, J.A.	59
Sichieri, R.	***56***	Sobue, T.	547
Siddiqi, A.M.	644	Soda, M.	531
Siddiqui, M.	644	Sodhani, P.	418
Sidney, S.	1074	Sörensen, S.	682
Siegel, S.	1089	Soeripto	**431, 432**
Siemiatycki, J.	**66, 67**, *68, *72, 1013	Sofer, T.	434
Sierra, R.	921	Søgaard, J.	202
Signer, E.	739	Soimakallio, S.	*257
Siguel, E.N.	**936**	Solanilla, P.	673
Sigurdsson, K.	*392	Solionova, L.G.	*1176, **1177**
Siimes, M.A.	642	Soloway, R.D.	1091
Silcocks, P.B.	837, 851	Somers, R.	276, 589
Silins, J.	**86**	Sommelet, D.	*319
Silman, A.	820	Somogyi, J.	191
Simard, A.	111	Sondik, E.J.	**926**
Simonato, L.	292, 295, 305	Song, E.	668
Simor, I.	111	Song, W.Z.	138, 139
Simsek, F.	749, *750	Soo, R.	142
Simson, I.W.	667, 670	Sorahan, T.M.	**756, 757**
Singaram, S.	578	Sorsa, M.I.	204, **251**, *694
Singer, A.	799, *816, *817	Soskolne, C.L.	*281
Singh, K.	418, *419	Sower, M.F.	*1142
Singh, M.	418, *419	Soyer, H.P.	336
Singh, V.	418	Spännare, B.	703
Sinues, B.	**677**	Sparén, P.	736
Sipponen, P.I.	**239**, 240	Sparkes, R.S.	1013
Siracusa, A.	499	Sparling, T.	115
Siscovick, D.	*1104	Spaziani, E.	*505
Sitas, F.	845, 846	Spears, G.F.S.	625
Sittenfeld, A.N.A.	**189**	Speck, B.	739
Siurala, M.	239, **240**	Speizer, F.E.	**929**, 934
Siverstein, M.A.	968	Spickett, J.T.	28
Sjödahl, R.	686	Spiegelman, D.	*928
Skakkebæk, N.E.	207, 215	Spiess, H.	**372**
Skalska-Vorbrodt, J.	648	Spigelman, A.D.	*814
Skegg, D.C.G.	*620, *621, 625	Spinelli, J.J.	91, 113, 114, *119
Skerfving, S.	*691, *694	Spritz, N.	1070
Skinnider, L.	91	Sprovieri, O.	*359
Slattery, M.L.	*1122	Srám, R.J.	**195, 196**
Sleijfer, D	276	Sreedevi Amma, N.	*421, *426
Slimane, B.	*315	Sreelekha, T.T.	*427
Slimani, N.	293	Sri Roostini, E.	534
Slotboom, B.J.	607, 608, 609	Srivatanakul, P.	**747**
Slovak, A.	882	St John, D.J.B.	7, *21, **24**
Slymen, D.	1153	Stähelin, H.B.	**740**
Smagghe, G.	321	Stagnaro, E.	477, 513
Smans, M.	659	Staines, A.	*783
Smeets, F.W.M.	*592	Stalder, G.A.	739
Smith, A.	823	Stampfer, M.J.	929, **930**, 934
Smith, A.H.	*905, 942	Staneczeck, W.	287, 339

517

INVESTIGATOR

Stanford, J.L.	**1133**, 1134, *1139	Sun, S.	**182**
Stangel, W.	*351, 375	Sun, X.W.	*151
Stankiewicz, A.K.	**652, 653,** ***654**	Sunderwa, J.	*419
Stansfeld, S.	810	Sunyer, J.	*281
Staples, M.	17	Supawat, C.	1134
Stark, A.	*1148	Suporn, K.	1134
Starr, T.	*887	Suporn, S.	1134
Starreveld, A.A.	62	Sureshchandra Dutt, G.	*426
Starzynski, Z.	***657**	Surwit, E.A.	**1153**
Stayner, L.	959	Sutrisno, E.	534
Steele, R.H.	39	Suzuki, R.	533
Steenland, K.N.	*270, **959**	Svel, I.	50
Stegman, D.A.	903	Svensson, B.G.	*690
Steiner, C.	888	Svirchev, L.M.	114
Steiner, R.W.	**1028**	Swaen, G.M.H.	**594**, 1158
Steinitz, R.	442, 444, **445**	Swaminathan, R.	*417
Stejskal, J.	*764	Swanson, G.M.	926, *970, **971, 972**, 1136
Stenkop, E.	663	Swerdlow, A.J.	288, *291, *794, *811, 844
Stenling, R.	679	Swift, M.R.	**944**
Stenszky, V.	98	Swinburne, L.	*784
Stern, R.S.	**931**	Swinnen, L.J.	*1038
Stevens, M.C.G.	755	Syrjänen, K.J.	**256,** ***257**
Stevens, R.G.	**1111**	Syrjänen, S.	256, *257
Stewart, A.M.	753, 754, **758**, *1103	Szadkowska-Stanczyk, I.	*281
Stewart, A.W.	618	Szalai, J.P.	*101
Stewart, P.A.	910, 911, 1114, *1116	Szarewski, A.M.	***816,** ***817**
Stewart, W.	1155	Szeszenia-Dabrowska, N.	655, *657, 1177
Stich, H.F.	**117, 118**, 1178	Szollosová, M.	198
Stiggelbout, A.M.	589	Szwarcwald, C.L.	*57, *58
Stiller, C.A.	287, 838, 839, **854**	Szymczak, W.	655, *657
Stocker, H.	*83	Tabar, L.	722
Stoll, C.	50	Tabor, E.	630
Stolley, P.D.	*935, 1070, 1091	Tadera, M.	572
Stone, D.	50	Tafur, L.	283
Stone, K.M.	883	Tagashira, Y.	*556, 558, **559**
Storm, H.H.	203, *216, 219, **220, 221, 222, 223,** *224, *225, 287, 917	Tahan, T.	95
		Tahara, E.	***522**
Storozhuk, M.	**1188**	Taikina-aho, O.	241
Stoter, G.	276	Takács, S.	**386, 387**
Stovall, M.	220, 222, 223	Takahashi, K.	525
Straatman, H.	**600**, 602	Takano, A.	*564
Strain, J.J.	*615	Takano, J.	*530
Stram, D.	*1100	Takashima, Y.	574
Strax, P.	1063	Takasugi, N.	562
Streuning, E.	1060	Takayama, K.	517
Strickland, P.	*457, *901, *902	Takeda, T.	561, **562**
Strom, B.L.	**1091**	Takeshita, T.	545
Strong, L.C.	**996**, 1099	Takeuchi, T.	545
Struyk, A.P.H.B.	585	Takita, K.	527
Stücker, I.	***333**	Takizawa, Y.	533
Sturgeon, S.	915	Tala, E.	245
Sturmans, F.	*593, 594, 595, 596, 613	Talamante Serrulla, S.	*676
Su, C.Y.	*124	Talamini, R.	453
Suchánková, A.	197	Talaska, G.	512
Sufrin, G.	937	Talbot, I.	*814
Sugahara, T.	*551	Talbott, E.O.	1096
Sugawara, N.	519, 563, *564	Talley, N.J.	*1112
Sugondo, T.	429	Tamburro, C.H.	**1029, 1030, 1031, 1032**
Sulkowski, W.J.	**655**	Tanaka, H.	533
Sullivan, K.R.	807, 808	Tang, H.X.	*147
Summers, R.	*982	Taniguchi, H.	544
Sun, Q.	136	Tanimura, M.	569
Sun, Q.R.	*147		

INVESTIGATOR

Tanizawa, O.	*549		Thyss, A.	307
Tannenbaum, S.R.	512, 938		Tilgen, W.	336
Tanprasert, S.	747		Tilley, B.C.	*1142
Tao, B.	*160, 176		Tilson, H.H.	1109
Tao, S.C.	**129**		Tin, D.Y.	*137
Tao, X.G.	*171, *181		Tirelli, U.	**455**
Tao, Y.	1134		Tirmarche, M.	**264**
Tao, Z.	130, 136		Tironi, A.	483
Taplin, S.	1130		Tirtosugondo	430
Tarkowski, W.	*291		Tjahjadi, G.	534
Tarquini, M.	503		Tjahjono	429
Tatár, A.	386		Tjindarbumi, D.	534
Taylor, H.	*901		Tjønneland, A.	205, 847
Taylor, I.	*780		To, T.	91
Taylor, J.R.	931		Tockman, M.S.	*457
Taylor, P.	915		Todde, P.F.	460
Taylor, R.H.	*766		Törnberg, S.A.	**717**
Taylor, S.M.	63		Tognoni, G.	488
Tekkel, M.	*238		Tokudome, S.	533, **552, 553, 554, 555**
Telenius Berg, M.	314		Tokunaga, M.	524
Telles, J.L.	1120		Tola, S.	252
Temmerman	48		Tolbert, P.E.	1033
ten Bokkel Huinink, G.	276		Tolomei, S.	*280
Ten Kate, L.P.	50		Tomatis, L.T.	335
Tenconi, R.	50		Tomenius, L.	707
Tennis, P.	**1109**		Tominaga, S.	**536**, 566
Teppo, L.	242, 243, 244, 246, 250, 295		Toniolo, P.G.	1063, *1068
			Torén, K.O.	**683**, *684
ter Meulen, J.	367		Torgrimsen, T.	639
Ter Schegget, J.	**585**		Torhorst, J.K.H.	739, 740, 744
Terracini, B.	287, 463, 464, 508, **511**, 512, 516		Torloni, H.	*575
			Torrent, M.	293
Tersmette, A.C.	*900		Tortorelli, A.	503
Tessier, C.	298		Tosi, P.	469
Tessy, G.	799		Tossavainen, A.	245
Thamm, M.	*338		Totis, A.	482
Thar, W.E.	*973		Toyoshima, H.	533, ***538**
Theodorsen, L.	636		Traboulsi, E.I.	896
Thériault, G.P.	**75**, 91, 116		Traina, A.	*497
Thiessen, J.W.	521		Trapeznikov, N.N.	1178
Thijssen, J.H.H.	608		Travis, L.B.	913
Thilly, W.G.	**938**		Treat, M.R.	1059, 1061
Thomas, B.A.	786, 877		Tres, S.	677
Thomas, D.B.	926, 1015, **1134, 1135**		Tretli, S.	206
Thomas, D.C.	629, 1013, 1022, 1023, 1024		Tribble, T.	*1107
			Trichopoulos, D.	302, 303
Thomas, H.F.	**861**		Trichopoulou, A.	293, *615
Thomas, J.	*81		Trigg, M.	1046
Thomas, T.L.	1117		Trivedi, A.H.	394
Thompson, B.	1130		Trosko, J.E.	521
Thompson, D.W.	110, 1013		Tryggvadottir, L.	***391, *392, *393**
Thompson, R.	37		Tsai, W.Y.	1064
Thompson, W.D.	1055		Tschen, J.	*992
Thomson, J.	812		Tsuchihashi, Y.	529
Thomson, M.H.	262		Tsuchiya, A.	527
Thony, C.	266		Tsuchiya, K.	**525**
Thoresen, S.Ø.	635		Tsugane, S.	573, 574, *575, 847
Thorhallsson, P.	390		Tsunematsu, Y.	576
Thornby, J.I.	*992		Tu, J.T.	164, **172**
Thornquist, M.	1130, *1131		Tuchman, M.	*1050
Thorogood, M.	853, **855**		Tucker, M.A.	913, *925, **1115**
Threlfall, W.J.	113		Tucker, S.	*993
Thun, M.J.	892		Türkeri, L.	749, *750
Thurnham, D.	282, 357, 1178		Tukuma, H.	547

INVESTIGATOR

Tulinius, H.	390, *391, *392, *393, 847	Vargha, M.	287
Tulli, A.	500	Varis, K.S.	239
Tumino, R.	293, 513	Vasanthi, L.	*417
Tuomisto, J.	*257	Vasen, H.F.A.	314
Turazza, E.	*359	Vassallo, J.A.	*1164, 1165, *1166, *1167
Turk, J.	*816, *817		
Turnbull, B.	*1001, *1145, *1146, *1147	Vatn, M.H.	639
		Vaudrey, C.	318
Twiggs, L.B.	914	Vaughan, T.L.	1135, 1136
Tyczynski, J.	288, *291, 659, 660	Vehviläinen-Julkunen, K.	*257
Tytgat, G.N.	*900		
Ubukata, T.	550	Velema, J.	300
Uchimura, H.	552	Vena, J.E.	937
Ujházy, V.	193	Venkateswaran, P.	1084
Ujszászy, L.	386	Verbeek, A.L.M.	598, 599, 600, *601, 602
Ulrich, J.	740	Vercauteren, P.	54
Unsal, M.	920	Vercelli, M.	478
Urbach, F.	931	Verd, S.	*764
Urban, N.	1130	Verdier, P.	*85
Urquhart, J.	305	Verduijn, P.G.	*604
Urso, C.	468	Verganelli, E.	*475
Us, J.	1192	Verge, J.	673
Utidijian, H.M.D.	1158, 1159	Verhagen-Teulings, M.T.	606
Uys, C.J.	667	Vermorken, J.B.	506
Väyrynen, M.	256	Vernon, S.W.	*1142
Vahrenholz, C.	*343	Veronesi, U.	*492
Vaidya, N.S.	413	Vessey, M.P.	852, 855, 856, 857, 858, 859
Vaidya, R.	*420		
Vaidya, S.G.	413	Vetrugno, T.	502
Vainio, H.	241, *281, *306	Vetter, R.	920
Valacco, A.	*359	Victoria, J.	274
Valanis, B.	*1131	Vida, F.	673
Valdivia, A.A.	884	Viel, J.F.	309
Valentini, A.	*515	Viganò, C.	485
Van Damme, K.	*55	Vijayakumar, T.	*427
van de Ven, W.	54	Viladiu, P.	283
Van den Berghe, H.	51, 54	Villar, H.	*1148, 1152
Van den Brandt, P.A.	595, 596, 613	Villelminot, S.	293
Van den Oever, R.	51	Vincent, T.J.	839
Van den Tweel, J.	585	Vine, M.F.	940
Van der Does-van den Berg, A.	605	Vineis, P.	293, 479, 509, 510, 512, 513, 514
van der Gulden, J.W.J.	*601	Vioque, J.	672
van der Heijden L., H.M.	606	Visintainer, P.F.	*1142
Van der Hoordaa, J.	585	Visona, K.	189
Van der Kuip, A.M.	*591	Vitali, R.	498, 499
van der Maas, P.J.	*604	Vivas, J.	*284, *285
Van Dongen, J.A.	586, *590	Viver, J.	673
Van Holten, V.	544	Vlaisavljevic, V.	1192, 1195
Van Kaick, G.	368	Vlasák, V.	192
Van Leeuwen, F.E.	54, 276, 298, 584, 586, 587, 588, 589, *590	Vlietinck, R.	54
		Vobecky, J.	95, 96, 97
Van Marck, E.	48	Vobecky, J.S.	95, 96, 97
Van Mieghem, E.	*55	Voelz, G.L.	1008
van Noord, P.A.H.	608, *611, 612	Vogel, V.G.	997
van Poppel, G.	614	Vogelstein, B.	896
Van Schooten, F.J.	584	von Kries, R.	*765
van Wering, E.R.	605	Vonka, V.	197
Van't Veer, P.	595, 613, *615, 616	Vooijs, P.G.	603, 614
Vandenbroucke, J.P.	*900	Voorhees, J.	931
Vandevelde, E.	*52	Voûte, P.	1089
Vannucchi, G.	470	Vuletic, L.	1190, 1191
Varcoe, R.	*627	Wabinga, H.	348
Vardi, H.	435	Wacholder, S.	920

INVESTIGATOR

Wagner, E.	1130	Weiss, N.S.	*119, 897, *1104, 1125, 1133, **1137**, *1138, *1139, 1140
Wagner, G.	368		
Wagner, J.C.	**860**, 871		
Wahba, R.	234	Weiss, W.	**1092**
Wahlqvist, M.L.	7	Weissel, M.	314
Wahrendorf, J.	42, 282, 293, 302, 352, 354, 356, 357, *363, 364, 365, **369, 370**, 443, 729	Welinder, H.	689, *692, *693
		Wells, A.J.	*978
		Wennborg, H.	*705
Walczak, A.	*658	Werner, M.	76
Wald, N.J.	**818**	Wesenberg, F.	641
Waldherr, R.	355, 365	Wessels, G.	671
Walker, A.M.	300	West, C.	897
Walker, J.	958	West, D.W.	897, 926
Walker, W.B.	78	West, P.	120
Wall, C.	110, 111	West, R.	91
Wall, S.	719, 722, 723	West, R.R.	*770, *771
Wallace, D.M.A.	*759	West, S.	*901, 1091
Wallace, R.B.	963, 1043	Westberg, H.	*684
Wallach, R.C.	**1069**	Westenbrink, S.	616
Walles, S.A.S.	1064	Westerholm, P.	295
Walrath, J.	1118	Westerlund, B.	701
Walter, S.D.	63, *194, 304	Weyler, J.	48
Wammanda, D.R.	632	Wheeler, C.M.	883
Wan, X.R.	150	White, D.	91
Wang, A.R.	*124	White, E.	*1104, 1130, 1137, **1140**
Wang, D.Y.	**819**	White, M.D	94
Wang, L.H.	157	Whitehead, L.	*993
Wang, Q.L.	*137	Whitehead, S.	622
Wang, Q.S.	**183**	Whitmore, S.E.	*901
Wang, Y.	***173**, 182	Whittemore, A.S.	903, 904, 1143, *1144
Wang, Z.Q.	175	Whittle, H.C.	335, 496, 873
Wang, Z.Y.	*126	Wiebecke, B.	262
Warburton, D.	1064	Wiebelt, H.	369
Ward, E.M.	910, 957, **960, 961**	Wiggs, L.D.	**1008**
Ward, M.	7, 911, *1116	Wigle, D.T.	76, 79, *82, 84, **87**
Wargovich, M.J.	**998**	Wiklund, K.	**718**
Warnakulasuriya, K.A.	**678**, 805	Wilbanks, G.D.	914
Warner-Learmonth, G.	**665**	Wilbur, D.	1125
Warshauer, M.E.	*935	Wilcock, K.	*887
Waskerwitz, M.	1099	Wilcosky, T.	940
Watanabe, F.	527	Wilczynska, U.	655, *657
Watanabe, H.	**529**, 533	Wild, C.	276, 845, 847, 898
Watanabe, S.	529, 534, **540**, 573, 574, *575	Wild, P.	*281
		Wildt, J.	200
Waterhouse, J.A.H.	755	Wilkens, L.	985
Waters, W.E.	*780	Wilkins, J.R.	1047
Watts, C.	*21	Wilkinson, G.	946
Watts, P.	832	Willett, W.C.	929, 930, 934, 936, 1035, 1185
Waye, J.	1061		
Weatherall, J.W.	50	Williams, C.B.	*814, **820**
Webb, P.	846, 847	Williams, F.L.R.	775, ***776**
Weber, J.	1037	Williams, J.	18
Weber, W.	739	Williams, M.	11
Webster, I.	667	Williams, S.	1044
Weckbecker, J.	336	Williams, S.M.	*626
Wei, L.	**130**	Williams, W.R.	996
Weiland, M.	369	Williamson, A.L.	*664
Weiner, J.A.	*724	Wilmer, J.W.G.M.	614
Weininger, J.	446, 447	Wilpart, M.	262
Weinmann, S.A.	*1104	Winck, J.C.	292
Weir, H.K.	108	Wing, A.L.	1143
Weisburger, E.	956	Wingren, G.B.	*684, **688**
Weisgerber, U.	355	Winkelmann, R.	*280, *281, 294
Weiss, J.	336	Winkelstein, W.	*1124

INVESTIGATOR

Winn, D.	**999**	Yang, G.	***144**, *145, *148, *149, 164
Winn, R.J.	*995, 997, 998	Yang, G.R.	187, **188**, 357
Winnett, A.	645	Yang, W.X.	128, *148
Winter, J.	358	Yang, X.Z.	177
Winter, P.D.	861, 862, 863, 865, 866, 869, 870, 871	Yang, Z.T.	135
Wintroub, B.	931	Yao, K.Y.	*144
Wogan, G.N.	512, 898, 938	Yarborough III, C.M.	*975
Woitowitz, H.J.	**346**	Yardley, J.	896
Wolf, J.	931, *992	Yaroslavtsev, V.N.	1188
Wolfe, R.A.	886	Yasui, W.	*522
Wolff, H.H.	336	Yatsenco, A.S.	1186
Wong, C.M.	380	Yeh, F.S.	161
Wong, I.	1030	Yeole, B.B.	407, ***411**, ***412**
Wong, J.	1029, 1031, 1032	Yerkes, A.M.	889
Wong, M.	113	Yin, D.M.	170
Wong, O.	*270, **1080**, **1081**	Yin, S.	**134**
Wong, Y.F.	168	Yliskoski, M.	256
Wood, W.	1070	Yoshida, K.	560
Woodman, C.B.J.	**760**	Yoshimura, T.	**526**, 533
Woods, W.G.	1047, ***1050**, *1085, 1089, 1099	Yoshizawa, C.N.	984, 986, 987, 988, **989**
		You, W.C.	**135**
Woodward, A.	**3**	Young, C.S.	963
Woodward, J.	*1156	Young, D.C.	**963**
Woordes-van Baalen, M.	588	Young, G.P.	24
		Young, J.	544
Wray, B.B.	**895**	Young, L.	760
Wright, H.	863	Young, T.B.	1034
Wu, A.	*1144	Youngson, J.H.	**823**
Wu, H.C.	*126	Yu, E.X.	163
Wu, J.M.	*147	Yu, H.	***144**, ***145**, *147, *148, *149
Wu, J.N.	*162		
Wu, M.	**131**	Yu, M.C.	178, 1021
Wu, X.N.	163	Yu, S.Z.	***171**, **177**
Wu, Z.B.	912	Yu, Z.F.	**154**, ***155**
Wu, Z.R.	133	Yuan, J.M.	**178**
Wu-Williams, A.	1056	Yuan, L.	182
Wyatt, J.	845	Yuan, R.	130
Wynder, E.L.	**1070, 1071**	Yuan, Y.C.	***144**, ***145**, *147
Wynne, H.J.A.	*611	Yunyuan, L.	152
Xia, L.F.	*148	Zack, M.M.	726, 734
Xiang, Y.B.	**174**	Zahm, S.H.	910, 911, 961, *1116, **1117, 1118, 1119, 1120**
Xie, C.L.	*126		
Xie, D.	***132**	Zaidi, S.M.H.	643, **644**
Xie, H.J.	**133**	Zakelj, M.P.	847
Xie, X.L.	*162	Zaman, S.	806
Xu, G.W.	135	Zamcheck, N.	896
Xu, G.X.	164	Zampi, G.C.	503
Xu, J.L.	177	Zanetti, R.	**514**
Xue, S.Z.	*159, *160, **175, 176**	Zanghieri, G.	494
Yadav, M.	***581**	Zappa, M.	469, 470, 472
Yager, J.W.	942	Zaridze, D.	303, 305, 1170, **1178, 1179, 1180, 1181, 1182, 1183, 1184, 1185**
Yamaguchi, M.	574		
Yamaki, Y.	527		
Yamakido, M.	*522	Zatonski, W.A.	287, 288, *291, 300, 303, *615, **659, 660**, 847
Yamamoto, R.	**537**		
Yamamoto, Y.	*551	Zauber, A.	*935, 1060
Yamazaki, S.	**518**	Zaun, H.	336
Yanagawa, H.	533	Zembala, M.	651
Yanai, F.	565	Zemla, B.	**648, 649**
Yandell, D.	*1086	Zemlianaya, G.M.	1179, 1180
Yang, B.Q.	135	Zey, J.	1114
Yang, C.	***119**, *126	Zha, Y.	130
Yang, C.S.	128, 185, 187	Zhang, C.Y.	150

INVESTIGATOR

Zhang, L.	135	Zhou, D.N.	*126
Zhang, M.Q.	*126	Zhou, L.	*144, *145, *147, *148, *149
Zhang, S.Q.	**146**		
Zhang, S.Z.	**136**, ***137**	Zhou, Q.L.	*124
Zhang, X.M.	*126	Zhu, H.G.	***181**
Zhang, Y.	138, 139	Ziegler, J.L.	904
Zhang, Z.	*124	Zielezny, M.	937
Zhang, Z.Q.	*124	Zielonka, I.	648
Zhang, Z.W.	**138, 139**, 375	Zippin, C.	926
Zhao, L.	135, 138, 139	Zipursky, A.	1089
Zhao, L.P.	985	Zocchetti, C.	***484**
Zhao, Q.Y.	*181	Zocchetti, E.	483
Zhao, R.X.	135	Zu, Z.H.	545
Zheng, S.	***144**, ***145**, ***147**, ***148**, ***149**	Zuberi, S.J.	643
		Zucchi, A.	487
Zheng, W.	170, **179, 180**	Zuch, C.	495
Zhong, S.C.	161	Zuchairi Dahlan	431
Zhou, C.T.	157	zur Hausen, H.	*359
Zhou, D.H.	175	Zwitter, M.	**1194**

INDEX OF TERMS

Completely new key-words in this issue are:

Areca Nut
Familial Adenomatous Polyposis
HCV
HDV

Parental Occupation
Tobacco (Snuff)
Waste Dumps

'Fibre' now only applies to dietary fibre. Mineral and glass fibres are key-worded to 'Mineral Fibres' and 'Glass Fibres' in the CHEMICALS index.

The rarely used terms 'Aplastic Anaemia' and 'Gastric Bacteria' were deleted.

'Medical History', although used for many years, was considered to be of little value to identify a project of interest and was also deleted.

The list of index entries below identifies all key-words in use in this Directory. General terms such as 'Reproductive Factors' are only used when more specific terms cannot easily be assigned: such specific terms are shown beneath each general term. Cross-references to the OCCUPATIONS and CHEMICALS indexes are also shown. Thus a study indexed to 'Occupation' here will usually also be indexed to one or more specific occupations, while a study indexed to 'Drugs' or 'Metals' will also be indexed to one or more members of these groups in the CHEMICALS index.

Abortion
Adolescence
Age
AIDS
Air Pollution
Alcohol
Allergy
Analgesics
Anatomical Distribution
Animal
Antibodies
Antigens
Antioxidants
Areca Nut
Arthritis
Asbestosis
Asians
Ataxia Telangiectasia
Atomic Bomb
Atrophic Gastritis
Autopsy

BCG
Betel (Chewing)
Bile Acid
Biochemical Markers
    See also: Vitamins; Lipids; Trace Elements
Biopsy
Birth Cohort
Birth Order
Birthweight
Blacks

Blood
Blood Group
BMB
Breast Cysts

Caucasians
Chemical Exposure
    See also specific entries in CHEMICALS index
Chemical Mutagenesis
Chemotherapy
Childhood
Chinese
Chlorination
Cholesterol
Chromosome Effects
Cirrhosis
Classification
Clinical Effects
Clinical Records
Cluster
CMV
Coal
Coffee
Condyloma
Congenital Abnormalities
Contraception
Cooking Methods
Cosmetics
Cost-Benefit Analysis
Crohn's Disease
Cryptorchidism
Cytology

Dairy Products
Data Resource
Demographic Factors
Diabetes
Diet
Diuretics
DNA
DNA Adducts
DNA Binding
DNA Repair
Dose-Response
Down's Syndrome
Drugs
    See also: Analgesics; Chemotherapy;
    Hormones; Oral Contraceptives; Progestins;
    See also specific entries in CHEMICALS index
Dusts
    See also specific entries in CHEMICALS index
Dyes

EBV
Education
Electromagnetic Fields
Environmental Factors
Enzymes
Epilepsy
Eskimos
Ethnic Group

Faeces
Familial Adenomatous Polyposis
Familial Factors
Family History
Fat
Fatty Acids
Female
Fertility
Fertilizers
Fibre
Fibrosis
Fluorescent Light
Fruit
Fungicides

Gallstones
Gastrectomy
Gastric Ulcer
G6PD
G-Banding
Genetic Factors
Genetic Markers
Geographic Factors
Glass

H. pylori
Haemophilia
Hair
HBV
HCV

Herbicides
Heredity
High-Risk Groups
Histology
HIV
HLA
Homosexuality
Hormones
HPRT-Mutants
HPV
HSV
HTLV
Hygiene
Hyperthyroidism
Hysterectomy

Immunodeficiency
    See also: AIDS
Immunologic Markers
Immunology
Immunosuppression
In Situ Carcinoma
Indians
Infection
Infertility
Insecticides
Intra-Uterine Exposure
Iodine Deficiency

Japanese

Lactation
Late Effects
Latency
Leather
Lifestyle
Linkage Analysis
Lipids
Liver Disease
Lymphocytes

Maghrebians
Male
Mammography
Mapping
Marijuana
Marital Status
Mate
Mathematical Models
Menarche
Menopause
Menstruation
Metabolism
Metals
    See also: trace elements, specific metals and
    metal compounds in CHEMICALS index
Micronuclei
Migrants
Minerals
Mining

Monitoring
Multiple Primary
Mutagen
Mutagen Tests
Mutation Rate
Mutation, Germinal
Mutation, Somatic
Mycotoxins
   See also specific entries in CHEMICALS index

Naevi
Nass
Nutrition

Obesity
Occupation
   See also specific entries in OCCUPATIONS index
Oncogenes
Oophorectomy
Oral Contraceptives

Paints
Parasitic Disease
Parental Occupation
Parity
Passive Smoking
PCR
Pedigree
Pepsinogen
Pesticides
   See also: Fungicides; Herbicides; Insecticides
   See also specific entries in CHEMICALS index
Petroleum Products
Photochemotherapy
Physical Activity
Physical Factors
Pigmentation
Plants
Plasma
Plastics
   See also specific plastics, resins and monomers in CHEMICALS index
Pneumoconiosis
Polyps
Pregnancy
Premalignant Lesion
 See also: Atrophic Gastritis; Polyps
Prevalence
Prevention
Prognosis
Projection
Promotion
Protein
Protein Binding
Psychological Factors
Psychosocial Factors

Race
   See also: Asians; Blacks; Caucasians; Chinese; Eskimos; Ethnic Group; Indians; Japanese; Maghrebians
Radiation, Ionizing
Radiation, Non-Ionizing
Radiation, Ultraviolet
Radiotherapy
Record Linkage
Recurrence
Red Cells
Registry
Religion
Reproductive Factors
   See also: Abortion; Birthweight; Contraception; Contraceptives, Oral; Fertility; Infertility; Lactation; Menarche; Menopause; Menstruation; Parity
Resins
RFLP
Rubber
Rural

Saliva
SCE
Screening
   See also: Mammography
Seasonality
Segregation Analysis
Semen
Sero-Epidemiology
Serum
Sex Ratio
Sexual Activity
Sexually Transmitted Diseases
Sib
Silicosis
Socio-Economic Factors
Soil
Solvents
   See also specific solvents in CHEMICALS index
Sputum
Stage
Stress
Surgery
Survival
Sweeteners
   See also specific entries in CHEMICALS index

Tea
   See also: Mate
Textiles
Time Factors
Tissue
Tobacco (Chewing)
Tobacco (Smoking)
   See also: Passive Smoking
Tobacco (Snuff)
Toenails
Tonsillectomy

Trace Elements
Transplantation
Trauma
Treatment
   See also: Chemotherapy; Drugs; Radiotherapy; Surgery
Trends
Tubal Ligation
Tumour Markers
Twins

Ulcerative Colitis
Ultrasound
Urban
Uric Acid
Urine

Vaccination
Vegetables
Vegetarian
Virus
   See also: CMV; EBV; HBV; HCV; HDV; HIV; HPV; HSV; HTLV
Vitamins

Waste Dumps
Water
Welding
White Cells
Wood

Xeroderma Pigmentosum

TERM

Abortion
Breast (F)	*119, *392, 625, 708, 1009, *1174
Childhood Neoplasms	849
Choriocarcinoma	490
Hydatidiform Mole	490
Lung	907
Ovary	396, *1174
Uterus (Cervix)	*1174
Uterus (Corpus)	*1174

Adolescence
Benign Tumours	27
Breast (F)	89, 1189
Childhood Neoplasms	919
Gastrointestinal	1189
Genitourinary	1189
Melanoma	27
Odontogenic Neoplasms	2
Skin	*772
Stomach	526
Testis	*767
Uterus (Cervix)	665, *1124

Age
All Sites	190, 228, 250, 471, 477, 485, 488, 560, 589, 661, 726, 743, 913, 1018, 1120
Benign Tumours	965
Brain	*924, 1120
Breast (F)	169, 190, 220, 231, *391, 400, 610, *611, 632, 913, 1018, 1035, 1133
Colon	1120
Gastrointestinal	231
Hodgkin's Disease	1194
Leukaemia	231, 589, 605, 913, 1120
Leukaemia (ALL)	605
Liver	668
Lung	190, 589, 913, 1018, 1121
Melanoma	231
Myelodysplastic Syndrome	589
Neuroblastoma	*1100
Oesophagus	357
Prostate	*966
Rectum	1120
Stomach	544
Testis	231, 965
Thyroid	725, 913
Uterus (Cervix)	88, 210, 231, 400
Uterus (Corpus)	231, 1018
Wilms' Tumour	65

AIDS
All Sites	213, 496, 979, 1042, 1094
Hodgkin's Disease	213, 455
Kaposi's Sarcoma	*38, 377, 455, 803, *899, 909, 979
Lymphoma	*38, 909
Non-Hodgkin's Lymphoma	348, 455, 904
Soft Tissue	803

Air Pollution
All Sites	*44, 193, 487, 495, 507, *801
Bladder	208
Brain	208
Gastrointestinal	208, 1196

Numbers indexed are project numbers, asterisks identify new projects

TERM

Haemopoietic	208
Larynx	487, *801
Leukaemia	*801, 1196
Liver	*801
Lung	*171, 292, 487, *776, *801, *880, 1179, 1180, *1187, 1196
Lymphoma	*801
Male Genital	208
Mesothelioma	*784
Respiratory	208, *1187
Sarcoma	208
Skin	208

Alcohol

All Sites	60, *85, 87, 113, 247, 327, 488, 495, 533, 545, 552, 557, 559, 572, 596, 650, *654, *690, 861, 891, 892, 930, 986, 989, 1011, 1185
Benign Tumours	586
Bladder	66, 311, 1184
Bone	170
Brain	170, 897, 1088, 1090
Breast (F)	*22, *56, *85, 255, 297, 303, *409, 482, 534, 583, 586, 588, *590, 613, 625, 650, 886, 918, 937, 1009, 1035, 1181
Breast (M)	296
Childhood Neoplasms	*764, 1090
Choriocarcinoma	490
Colon	*56, 66, *85, *148, 164, *165, 255, 261, 262, 303, *410, 565, 566, 572, 613, *676, 720, 886, 1036, *1156
Female Genital	453, 886
Gallbladder	*676
Gastrointestinal	*73, 453
Head and Neck	95, 666
Hodgkin's Disease	66
Hydatidiform Mole	490
Hypopharynx	269, 317, *420
Kidney	42, 66, *73, 325, 529, 729, *1078, 1175
Larynx	51, 64, *73, 180, 317, *360, *420, *637, 655, 666, *716, 1135, 1163
Leukaemia	170, 1175
Lip	64
Liver	66, 136, 163, 170, 261, 528, 541, 566, 568, 572, *637, 668, *676, 747
Lung	*56, 66, *73, 255, 322, 535, 541, 566, 572, 613, 648, 650, 1179, 1180
Lymphoma	170, 453, 1175
Melanoma	66
Mesothelioma	51
Nasal Cavity	180, 316
Nasopharynx	51, 1136
Nervous System	1175
Neuroblastoma	*1100, 1175
Non-Hodgkin's Lymphoma	66
Oesophagus	*56, 66, *165, 261, 357, *410, *420, *637, 646, *676, *716, 1135, 1178
Oral Cavity	23, 59, 64, 180, 200, 269, *359, *420, *637, *716, *1026, 1072, 1135, 1163, 1178
Oropharynx	59, 269, *359, 397, *420, 1072
Ovary	886, 937
Pancreas	66, *165, 246, *676, 937
Peritoneum	*73
Pharynx	64, 180, *637, *716, *1026, 1163
Pleura	*73
Prostate	66, *879, 937
Rectum	*148, 164, *165, 255, 261, 303, *410, 565, 572, 613, *676, 1036
Respiratory	*359
Retinoblastoma	1175
Salivary Gland	757
Skin	255, 453
Small Intestine	*676
Stomach	*56, 66, 136, 152, 184, 322, *410, 480, 526, 550, *556, 566, 572, 613, 650, *676, 728, 1135, 1161
Testis	66, 215

Thyroid	453
Urinary Tract	453
Uterus (Cervix)	*359, *581, 650, 886
Uterus (Corpus)	566, 886, 914, 937
Wilms' Tumour	65

Allergy
All Sites	353, 709
Non-Hodgkin's Lymphoma	904

Analgesics
All Sites	934, 1096
Bladder	176, 1096
Kidney	42, 325, 729

Anatomical Distribution
All Sites	191, 192, 661, 891
Colon	616
Melanoma	192
Oral Cavity	1072
Oropharynx	1072
Skin	192

Animal
Inapplicable	251
Leukaemia	349
Lung	337

Antibodies
All Sites	1057
Female Genital	256
Kaposi's Sarcoma	631
Leukaemia	364
Liver	630
Male Genital	256
Nasopharynx	141, 142, 150, 578
Stomach	845, 846
Uterus (Cervix)	197, 760

Antigens
Kaposi's Sarcoma	501
Liver	271, 630, 747, 872
Lung	*457

Antioxidants
All Sites	*933
Breast (F)	*338, *615
Stomach	*284

Areca Nut
Oral Cavity	394

Arthritis
All Sites	1109
Haemopoietic	1109

TERM

Lymphoma	1109
Skin	1109

Asbestosis
All Sites	*654
Mesothelioma	502
Peritoneum	502
Pleura	502

Asians
All Sites	*761
Liver	1084

Ataxia Telangiectasia
All Sites	875, 944
Breast (F)	944
Leukaemia	944

Atomic Bomb
All Sites	520, 521, 531, *532
Retinoblastoma	*530

Atrophic Gastritis
Stomach	135, 503, 536, 663, *778

Autopsy
All Sites	495
Appendix	390
Bile Duct	390
Colon	390
Inapplicable	940
Lung	860, *864, 871
Mesothelioma	860, 871
Nervous System	390
Oesophagus	390
Oral Cavity	390
Pancreas	390
Pleura	860
Prostate	390, 448, *966
Small Intestine	390
Soft Tissue	390
Testis	207

BCG
Stomach	651

Betel (Chewing)
Oral Cavity	117, 404, 405, 678
Oropharynx	397

Bile Acid
Colon	*363
Gastrointestinal	*814
Rectum	*363

TERM

Biochemical Markers

All Sites	236, 573, 636, 853, 930, 999, 1064, 1130
Angiosarcoma	677
Bile Duct	1091
Bladder	677, 957
Breast (F)	237, *399, 446, 447, 527, 573, *615, 1055, 1066, 1067, 1130, 1152
Childhood Neoplasms	996
Colon	447, 527, *830, 998, 1055, 1130
Gallbladder	1091
Hodgkin's Disease	414
Inapplicable	911, 940
Leukaemia	414, 958
Liver	236, 573, 677, 898, 1084
Lung	241, 292, 584, 677, *902, *925, 1056, 1064, 1130, 1137
Lymphoma	414
Multiple Myeloma	414
Nasopharynx	578
Neuroblastoma	824
Oesophagus	282
Oral Cavity	805, *1132
Oropharynx	397, *1132
Ovary	506
Prostate	229, 573, 1055
Rectum	447, 527, *830, 1055
Respiratory	236
Skin	*901, 1055
Stomach	527, 573, *575, *778, 846
Uterus (Cervix)	272, 283, *932, 1055, 1130
Uterus (Corpus)	1055

Biopsy

Appendix	390
Benign Tumours	1052
Bile Duct	390
Breast (F)	312, 1052, *1053
Burkitt's Lymphoma	645
Colon	390, 452
Female Genital	256
Kaposi's Sarcoma	803
Liver	*664
Male Genital	256
Melanoma	1115
Mesothelioma	735
Nasopharynx	142
Nervous System	390
Oesophagus	282, 357, 390, 1178
Oral Cavity	23, 390, *1027, 1178
Pancreas	390
Peritoneum	735
Pharynx	*1027
Pleura	735
Prostate	390
Rectum	452
Skin	1123
Small Intestine	390
Soft Tissue	390, 803
Stomach	467, 846, 967
Testis	207
Uterus (Cervix)	168, 272, 585

Birth Cohort

All Sites	190, 743, 969
Bladder	489
Breast (F)	190

TERM

Childhood Neoplasms	753
Kidney	489
Larynx	489
Leukaemia	753
Lung	190, 489
Neuroblastoma	562
Non-Hodgkin's Lymphoma	753
Oesophagus	489
Oral Cavity	489
Pancreas	489
Pharynx	489
Thyroid	725
Uterus (Cervix)	154

Birth Order

Leukaemia	*417
Lymphoma	*417

Birthweight

Testis	202

Blacks

All Sites	891, 1113
Bladder	971
Breast (F)	890
Colon	971
Eye	971
Leukaemia	669
Liver	668, 971
Lung	971
Melanoma	971
Mesothelioma	670, 971
Multiple Myeloma	972
Oesophagus	971, 972
Pancreas	972
Prostate	972
Rectum	971
Salivary Gland	971
Stomach	971
Uterus (Cervix)	662, 890, *962

Blood

All Sites	293, 573, *690, *691, 929
Benign Tumours	*257
Bladder	512
Breast (F)	*257, 573, 841, 855, 918, 929, 1013, *1048
Childhood Neoplasms	*764
Colon	*363, 498, *830, 841
Gastrointestinal	855
Liver	573
Lung	*456, *457, *925
Ovary	*1139
Prostate	573, 841
Rectum	*363, 498, *830
Stomach	*284, 573, 841
Thyroid	314
Uterus (Cervix)	582, *932
Uterus (Corpus)	914
Vagina	916
Vulva	916

TERM

Blood Group
All Sites	228, 633
Breast (F)	*393, 633
Colon	498
Gastrointestinal	633
Ovary	633
Rectum	498
Uterus (Cervix)	633
Uterus (Corpus)	633

BMB
All Sites	205, 243, 293, *505, 533, *538, 557, *564, 573, 636, 680, *690, *691, 895, 929, 930, 934, 981, 1094
Angiosarcoma	677
Benign Tumours	*257
Bladder	512, 677, 843
Breast (F)	18, *22, 62, 237, 243, *257, *428, 482, *492, *505, 573, 583, 595, 609, *615, 819, 841, 855, 929, 1013, *1020, *1048, *1053, 1063, 1066, *1068, 1134, *1148
Childhood Neoplasms	*764
Colon	18, 243, 262, 355, *363, 498, *505, 595, 612, 679, *830, 841, *982, 1044
Gallbladder	1134
Gastrointestinal	*814, 855
Inapplicable	204, 942
Kaposi's Sarcoma	*38, *899
Larynx	*637, 1135
Leukaemia	895, 958
Liver	335, 573, *637, 677, 1134
Lung	*132, 241, 243, *333, *456, *457, 595, 677, 843, *902, *925, *1014, *1131, *1173
Lymphoma	*38, 1010, *1038
Melanoma	*809
Mesothelioma	*1131
Nasal Cavity	*809
Nasopharynx	*315
Neuroblastoma	*1050
Oesophagus	357, *637, 843, 1135
Oral Cavity	*637, 805, *1027, 1135
Ovary	506, 1134, *1139
Pharynx	*637, *1027
Prostate	18, 243, 573, 595, 727, 841, 851, *966, *1144
Rectum	18, 243, 355, *363, 498, 595, 612, 679, *830, 1044
Skin	*1146, *1147
Stomach	135, 243, *284, *285, *505, 573, 595, *778, 841, 843, 847, *1112, 1135
Testis	842
Thyroid	314, *362
Uterus (Cervix)	*33, 197, 243, 272, *581, *748, *816, *817, *932, *962, 1134
Uterus (Corpus)	1134
Vagina	916
Vulva	916

Breast Cysts
Breast (F)	1195

Caucasians
All Sites	1113
Breast (F)	906
Hodgkin's Disease	1126
Leukaemia	669

Numbers indexed are project numbers, asterisks identify new projects

TERM

Chemical Exposure
All Sites	16, 28, 39, 61, *68, *80, 86, 114, 183, 193, 230, 249, 266, *281, 298, *310, 320, 321, 327, *329, 370, 464, 477, 479, 483, 495, 516, 545, 553, 594, 628, *696, *698, 701, 702, 713, *766, 821, 861, 865, 869, 891, 920, 927, 950, 951, 955, 956, 959, 1032, 1033, 1045, 1064, 1080, 1096, 1097, 1098, 1114, 1119, 1120, 1158
Bladder	66, *80, 100, 208, 278, 311, 910, 957, 1096
Bone	170
Brain	*25, *80, 170, 208, 302, 479, 703, 910, 1002, 1029, 1033, 1041, 1088, 1119, 1120
Breast (F)	1045
Childhood Neoplasms	*621, 1039
Colon	66, *80, 686, 720, 791, 964, 1030, 1031, 1120
Gastrointestinal	208, 865, 1029, 1197
Haemopoietic	16, 208, 479, 647, 910, 1029
Hodgkin's Disease	66, 462, 1093
Hypopharynx	317
Inapplicable	50
Kidney	66, *80, 481, 808, 1031, 1041, 1081
Larynx	317, 321, 1033
Leukaemia	28, *80, 116, 170, 201, 483, *627, 713, 781, *783, *790, 959, 1045, 1118, 1119, 1120
Leukaemia (ALL)	*72, *621
Leukaemia (AML)	1023
Liver	66, 170, 961, 1029, 1030, 1031, 1041
Lung	40, 66, 92, 198, 199, 278, 321, 322, 341, 379, 380, 401, 464, 499, 553, 594, 653, 791, 808, 865, 871, 881, 910, 912, 927, 955, 1029, 1030, 1031, 1041, 1045, 1064, 1092, 1114, 1119, 1197
Lymphoma	12, 16, *80, 170, 249, 481, 483, 628, *698, 713, 910, 961, 1118
Male Genital	208
Melanoma	66, 599, 1006
Mesothelioma	39, 346, 667, *784, 865, 871, 910
Multiple Myeloma	*698, 1119
Myelodysplastic Syndrome	15, *268, *770
Nasal Cavity	201, 1021
Nasopharynx	150, 1021, 1098, 1136
Non-Hodgkin's Lymphoma	66, 462, *790, 1119
Oesophagus	66
Oral Cavity	118, 321
Pancreas	66, 246, 959, 1118
Peritoneum	39, 667
Pharynx	321
Pleura	39, 346, 667
Prostate	66, *80, 594, *750, 795, 808
Rectum	*80, 686, 791, 964, 1120
Respiratory	201, 208, 463, 683, 865, 1095
Sarcoma	12, 208, 249, 910
Skin	208, 479, 1031
Soft Tissue	28, 483, 509, 628, 961, 1005, 1119
Stomach	66, *80, 322, 651, 683, 791, 910, 927, 959
Testis	66, 749
Thyroid	1045
Urinary Tract	*759
Wilms' Tumour	65, 1099

Chemical Mutagenesis
Inapplicable	383, 384, 938

Chemotherapy
All Sites	276, 331, 589, 712, 913, 953
Breast (F)	96, *590, 913
Childhood Neoplasms	919, 1089
Choriocarcinoma	5
Gastrointestinal	332

Haemopoietic	768	
Hydatidiform Mole	5	
Inapplicable	196, 597, 656	
Leukaemia	331, 332, 589, 641, 642, 669, 804, 848, 913	
Leukaemia (ALL)	641, 768	
Liver	1084	
Lung	331, 332, 589, 913	
Lymphoma	641	
Myelodysplastic Syndrome	589, 768	
Non-Hodgkin's Lymphoma	331, 332	
Testis	385	
Thyroid	913	

Childhood

All Sites	*55, 432, 450, 451, 496, 576, 711, 825
Benign Tumours	27, 451
Brain	41, *49, 301, 897, 1024, 1025, 1088, 1090
Breast (F)	825
Childhood Neoplasms	569, 753, 831, 854, 919, 1090
Haemopoietic	305
Hodgkin's Disease	511, 1093
Kidney	1175
Leukaemia	*49, *55, 116, 287, 305, 364, *417, 605, 641, 642, 669, 711, 753, *790, 831, 848, 854, 1046, 1175
Leukaemia (ALL)	511, 605, 641
Leukaemia (ANLL)	511, 1047
Lymphoma	*49, *55, *417, 641, 1175
Melanoma	27, 844
Nervous System	1175
Neuroblastoma	*49, 511, 854, 1175
Non-Hodgkin's Lymphoma	511, 753, *790
Odontogenic Neoplasms	2
Retinoblastoma	1175
Sarcoma	825
Soft Tissue	825
Stomach	845, 862
Thyroid	1128
Wilms' Tumour	511

Chinese

Breast (F)	297
Liver	*160
Lung	379, 380
Nasopharynx	579, 580
Oesophagus	*160
Stomach	*160

Chlorination

All Sites	*80
Bladder	*80
Brain	*80
Colon	*80, 634, 639, 983
Kidney	*80
Leukaemia	*80
Lymphoma	*80
Prostate	*80
Rectum	*80, 634, 639, 983
Stomach	*80

Cholesterol

All Sites	*505, 572, 818, 853
Breast (F)	*505, 717, 886

TERM

Colon	498, *505, 572, *830, 886
Female Genital	886
Liver	572
Lung	97, 572
Ovary	717, 886
Rectum	498, 572, *830
Stomach	*505, 572
Uterus (Cervix)	717, 886
Uterus (Corpus)	717, 886

Chromosome Effects

All Sites	54, 521, 545, *694, 700, 712, 1064
Angiosarcoma	677
Bladder	677
Breast (F)	991
Colon	45
Inapplicable	195, 196, 251, 383, 656
Leukaemia	54, 510
Liver	*162, 677
Lung	677, 1064
Lymphoma	54
Myelodysplastic Syndrome	510
Oral Cavity	394
Ovary	*162, 681, 991
Rectum	45
Uterus (Cervix)	*162, 418
Uterus (Corpus)	*162

Cirrhosis

Colon	261
Liver	261, 465, 554, 568, 630, *664, 806, 908
Lung	653
Oesophagus	261
Rectum	261

Classification

All Sites	742
Appendix	390
Bile Duct	390
Brain	415
Colon	390
Leukaemia	641, 669
Leukaemia (ALL)	641
Leukaemia (ANLL)	1047
Lymphoma	641
Nervous System	390
Oesophagus	390
Oral Cavity	390
Ovary	108
Pancreas	390
Prostate	390
Small Intestine	390
Soft Tissue	390
Testis	108

Clinical Effects

Childhood Neoplasms	754
Colon	227
Leukaemia	754
Rectum	227

Clinical Records
All Sites	46, 109, 432, 576, 920, 1073
Benign Tumours	1105
Brain	*81
Childhood Neoplasms	*765
Colon	369, 452, 1105
Kidney	*1078
Leukaemia	*417, 641
Leukaemia (ALL)	641
Lung	46
Lymphoma	*417, 641
Mesothelioma	46
Ovary	1069
Rectum	369, 452, 1105
Thyroid	318
Uterus (Cervix)	665, 921

Cluster
All Sites	*801, 832, *923, 1043, *1108
Bladder	949
Brain	1002, *1107
Breast (F)	906, 1043
Childhood Neoplasms	*373, 434, *620, 753, 839, *1106
Gastrointestinal	*57
Hodgkin's Disease	1126
Larynx	*801
Leukaemia	349, 753, 762, *801, 839, *1001, *1106
Liver	*801, *1106
Lung	*801
Lymphoma	762, *801
Nasopharynx	*402
Non-Hodgkin's Lymphoma	753, *1106
Pancreas	*1106
Soft Tissue	*1106
Uterus (Corpus)	1043

CMV
Uterus (Cervix)	168, *581, *748

Coal
Gastrointestinal	1196
Hodgkin's Disease	36
Inapplicable	597
Leukaemia	1196
Leukaemia (CLL)	36
Lung	40, 133, *151, *171, 1196
Non-Hodgkin's Lymphoma	36

Coffee
All Sites	443
Bladder	311, 1184
Brain	897
Breast (F)	70
Colon	443, *676
Gallbladder	*676
Kidney	42, 325, 729, *1078
Liver	*676
Lung	443
Oesophagus	*676
Pancreas	246, *676
Rectum	*676
Small Intestine	*676

Numbers indexed are project numbers, asterisks identify new projects

TERM

Stomach	*676
Wilms' Tumour	65

Condyloma

Female Genital	256
Male Genital	256
Uterus (Cervix)	88

Congenital Abnormalities

All Sites	130, 895
Brain	897
Childhood Neoplasms	434, 571, 754, *826, *827, 849, 919, 1039, 1089
Inapplicable	50, 383, 384
Leukaemia	116, 571, 754, 895
Skin	1123
Testis	215

Contraception

All Sites	856
Breast (F)	*22, 416, 625, 856, 1134
Choriocarcinoma	490
Gallbladder	1134
Hydatidiform Mole	490
Liver	1134
Ovary	1134
Uterus (Cervix)	619, 856, 883, *1124, 1134
Uterus (Corpus)	984, 1134

Cooking Methods

Lung	133, *151, 380
Stomach	673, 1161

Cosmetics

All Sites	929
Brain	1024
Breast (F)	929
Lip	1022
Skin	514

Cost-Benefit Analysis

Breast (F)	274, 600
Colon	*145, *147
Head and Neck	382
Liver	747
Neuroblastoma	562, 824
Oral Cavity	382
Rectum	*145, *147
Uterus (Cervix)	603

Crohn's Disease

All Sites	752
Colon	*226, 734, 752
Rectum	734, 752
Small Intestine	*226

Cryptorchidism

Testis	215

Numbers indexed are project numbers, asterisks identify new projects

Cytology
All Sites	856
Breast (F)	856
Head and Neck	382, 666
Larynx	666
Lung	199, 567, 1154
Oesophagus	186, 357, 1178
Oral Cavity	59, 382, 1178
Oropharynx	59
Uterus (Cervix)	*9, 48, 272, 367, 418, *419, 603, 662, 665, 760, *817, 856, 888, 889, 893, *932, *962, *1124, 1141

Dairy Products
Colon	616
Testis	*767

Data Resource
All Sites	*334, *497, *801, 1188
Breast (F)	*935
Colon	*935
Larynx	*801
Leukaemia	*801
Liver	*801
Lung	*801, *935
Lymphoma	*801
Ovary	*935
Prostate	*935
Urinary Tract	*759
Uterus (Corpus)	*935

Demographic Factors
Brain	*19
Spinal Cord	*19

Diabetes
All Sites	853, 1003
Bladder	105
Breast (F)	416, 1003
Colon	1003
Kidney	42, 729
Lung	105, 1003
Pancreas	105, 246
Prostate	1003

Diet
All Sites	*44, *53, 84, *85, *143, 205, 247, 293, 356, 439, 443, *454, 476, 485, 488, *505, 533, *538, *543, 557, 559, 560, 572, 573, 650, *690, 740, 891, 892, 926, 929, 981, 986, 989, 994, 999, 1003, 1007, 1011, 1018, 1043, 1074, 1111, 1130, 1165, *1167, 1185
Benign Tumours	491
Bile Duct	300, *592
Bladder	66, 1184
Brain	302, 1024
Breast (F)	18, *56, 69, *85, 89, 106, 125, 129, *144, *155, 244, 255, 297, 303, *399, *409, *424, *428, 482, 491, *505, 534, 573, 583, 588, 595, 613, *615, 623, 650, 731, *780, 841, 855, 886, 915, 918, 929, 937, 963, 1003, 1018, 1043, *1048, 1063, 1067, *1068, 1130, *1148, 1152, 1181, 1189
Breast (M)	296
Childhood Neoplasms	*764
Choriocarcinoma	490

Numbers indexed are project numbers, asterisks identify new projects

TERM

Colon	18, *56, 66, 69, *85, 106, 125, *144, *148, *149, 158, 164, *165, 214, 244, 255, 262, 273, 303, 355, *363, *410, 443, 449, 498, *505, 537, 565, 566, 572, 595, 613, 616, 639, 674, *676, 686, 720, 740, *829, 841, 886, 915, 943, 987, 994, *995, 1003, 1004, 1036, 1065, *1122, 1130, 1140, *1142, *1156, 1182
Female Genital	453, 886
Gallbladder	300, 643, *676, 1017
Gastrointestinal	453, 476, 542, 855, 1189
Genitourinary	1189
Head and Neck	95
Hodgkin's Disease	66
Hydatidiform Mole	490
Hypopharynx	*420
Inapplicable	985
Kidney	66, 106, 365, 529, 729
Larynx	64, 180, *360, *420, 423, 655, *716, 1135
Leukaemia	125, 179
Lip	64
Liver	66, 125, 136, *144, 566, 572, 573, *676, 898
Lung	31, *56, 66, 97, 125, *144, *151, 244, 255, 292, 379, 380, 423, 443, *456, 535, 566, 572, 595, 613, 648, 650, 740, 907, 915, 1003, *1014, 1018, 1071, 1130, 1179, 1180
Lymphoma	453
Melanoma	66, 304, 988, 1115, 1151
Mesothelioma	31, 670
Multiple Myeloma	972
Nasal Cavity	180
Nasopharynx	125, 140, 178, *315, 1136
Non-Hodgkin's Lymphoma	66, 1127
Oesophagus	*56, 66, 125, *165, 282, 357, *410, *420, 423, 435, 537, 646, 672, *676, *716, 972, 1135, 1178
Oral Cavity	64, 180, *420, 423, *716, *1026, 1135, 1178
Oropharynx	*420
Ovary	106, *398, 491, 886, 915, 937, *1139
Pancreas	66, 106, *165, 300, *592, 672, *676, 937, 972
Pharynx	64, 180, *716, *1026
Prostate	18, 66, 69, 90, 229, 244, *259, 324, 573, 595, 727, 841, 851, *879, 937, 939, 972, 1003, 1028, *1144
Rectum	7, 18, 106, 125, *144, *148, *149, 158, 164, *165, 214, 244, 255, 273, 303, 355, *363, *410, 498, 537, 565, 572, 595, 613, 639, 674, *676, 686, *829, 943, 987, *995, 1004, 1036, 1065, *1142, 1182
Skin	255, 453, *992, 1150, 1151
Small Intestine	*676
Stomach	*56, 66, 125, 135, 136, *144, 152, 184, 244, 388, *410, 480, 503, *505, 526, 537, *556, 563, 566, 572, 573, *575, 595, 613, 650, 651, 663, 672, 673, *676, 728, 740, *778, 841, 846, 862, 1135, 1161
Testis	66, *767
Thyroid	376, 453, 635, 688, 744
Urinary Tract	453
Uterus (Cervix)	88, *419, *425, 491, *581, 582, *593, 650, *816, 886, 903, *932, 1130, 1153
Uterus (Corpus)	106, 491, 566, 745, 886, 914, 915, 937, 984, 1018, 1043, 1125
Vagina	*1040
Vulva	491
Wilms' Tumour	65

Diuretics

Kidney	325, 729, *1078, *1104
Neuroblastoma	*1100

DNA

All Sites	521, *538, 557, 1058
Bladder	614
Breast (F)	*22, *428, 527, 906, 1013, *1020
Burkitt's Lymphoma	645

Colon	45, 527, 679	
Female Genital	256	
Gastrointestinal	*814	
Inapplicable	74, 597	
Leukaemia	874	
Liver	1087	
Lung	*132, 567, 614, *925	
Male Genital	256	
Oesophagus	128	
Rectum	45, 527, 679	
Retinoblastoma	*1086	
Stomach	527	
Thyroid	314	
Uterus (Cervix)	168, 760, 852, 883, *932, 1141	

DNA Adducts
All Sites	276, 981, 1064
Bladder	512
Inapplicable	204, 940, 942
Leukaemia	958
Lung	241, *457, 584, *902, 1064
Stomach	*575, 847

DNA Binding
All Sites	1064
Lung	1064

DNA Repair
All Sites	739
Bone	739
Breast (F)	739, 1066, 1067
Childhood Neoplasms	739
Gastrointestinal	739
Inapplicable	204
Leukaemia	739
Lymphoma	739
Oesophagus	131
Ovary	739
Skin	*901, 1123
Testis	739

Dose–Response
All Sites	39, 113, 203, 264, 521, *691, *693, 710, 773, 796, 797, 835, 836, 870, 913, 920, 1008, 1045, 1155
Bone	368
Breast (F)	220, 223, *338, 913, 1045
Childhood Neoplasms	*827
Colon	689, 791
Gastrointestinal	*73, 1196
Head and Neck	*604
Inapplicable	942
Kidney	*73
Larynx	*73
Leukaemia	221, 223, *224, 368, *693, 804, 835, 836, 913, 1045, 1196
Leukaemia (ALL)	221
Leukaemia (CLL)	221, 223
Liver	368
Lung	*73, 78, 175, 264, 368, 689, 699, 710, 714, 789, 791, 870, 912, 913, 952, 954, 1045, 1092, 1155, 1196
Lymphoma	368, *693, 954
Melanoma	622
Mesothelioma	39, 689, 870

TERM

Peritoneum	39, *73
Pleura	39, *73
Rectum	689, 791
Spleen	368
Stomach	689, 791
Thyroid	913, 1045
Uterus (Corpus)	221, 984

Down's Syndrome

All Sites	130
Childhood Neoplasms	*826, *827
Inapplicable	384
Leukaemia	874

Drugs

All Sites	203, 209, 217, 228, 276, 488, 552, 596, *698, 702, 712, 861, 892, 926, 929, 930, 931, *933, 934, 1018, 1073, 1094, 1109
Benign Tumours	586, 1077
Bladder	233, 311, 489, 1070, 1184
Bone	170, 368
Brain	41, *81, 170, 301, 1024, 1088
Breast (F)	233, 297, 312, 586, 588, 858, 929, *935, 1018, 1063, 1077
Breast (M)	296
Childhood Neoplasms	217, 571, *621, 754, *764, *765, 919
Colon	164, 686, 720, *829, *935
Eye	931
Female Genital	453, 858
Gallbladder	1017
Gastrointestinal	453, *814, 859
Haemopoietic	1109
Inapplicable	*343, 383, 597, 938
Kaposi's Sarcoma	*38, 377
Kidney	42, 365, 489, 529, 729, 1070, *1104
Larynx	489, 741, 1070
Leukaemia	116, 170, 179, 217, *340, 368, 571, 641, 754, *783
Leukaemia (ALL)	*621, 641
Leukaemia (AML)	1023
Liver	170, 217, 368
Lung	233, 368, 489, 741, *935, 1018, 1070
Lymphoma	*38, 170, *340, 368, 453, 641, *698, 1109
Multiple Myeloma	*698
Neuroblastoma	*1100
Non-Hodgkin's Lymphoma	904
Odontogenic Neoplasms	2
Oesophagus	489, 1070
Oral Cavity	489, 741, 1070
Ovary	608, *935, 1157
Pancreas	489, 1070
Pharynx	489, 741
Prostate	233, *750, *935
Rectum	164, 686, *829
Skin	453, 514, 931, 1109
Spleen	368
Stomach	480
Thyroid	453
Urinary Tract	453
Uterus (Cervix)	233
Uterus (Corpus)	*935, 1018
Vagina	*1040
Wilms' Tumour	65

Dusts
All Sites	39, 46, 86, 236, *281, *306, 464, 474, *475, 487, 495, 516, *654, *657, *695, *698, 710, 787, 865, 866, 869, 870, 920, 927, 951, 1130
Bladder	66, *306, *461, 843, 910
Brain	910
Breast (F)	1130
Colon	30, 66, 689, 720, 1130
Gastrointestinal	67, *73, 102, 865, 1186
Haemopoietic	647, *684, 910
Hodgkin's Disease	66
Hypopharynx	317
Kidney	66, *73, *461, 481
Larynx	51, *73, 317, 487, 655
Leukaemia	201, *461, *627
Liver	66, 136, 236, *461
Lung	30, 31, *32, 46, 66, 67, *73, 92, *127, *156, 198, 199, *333, *342, 380, *461, 464, 469, 472, 487, 547, 577, *657, 682, *684, 689, 710, 788, 843, 860, 865, 866, 870, 871, 910, 912, 927, 1130, *1131, 1154, 1186, *1187
Lymphoma	481, *684, *698, 910
Melanoma	66
Mesothelioma	30, 31, 37, 39, 46, 51, 67, 102, *342, 469, 502, 546, 667, 689, 735, *784, 860, 865, 870, 871, 910, *1131
Multiple Myeloma	*698
Nasal Cavity	201, 275, *306, 316, 1021
Nasopharynx	51, 178, 1021
Non-Hodgkin's Lymphoma	66
Oesophagus	66, 843
Ovary	506
Pancreas	66
Peritoneum	30, 39, *73, 502, 546, 667, 735
Pleura	39, *73, 469, 502, 546, 667, 735, 860
Prostate	66
Rectum	689
Respiratory	102, 201, 236, 463, 683, 787, 865, 1095, *1187
Sarcoma	910
Stomach	30, 66, 136, 683, *684, 689, 843, 910, 927, *1116
Testis	66, 840
Uterus (Cervix)	1130

Dyes
All Sites	*281, *484, 929
Angiosarcoma	677
Bladder	*484, 677
Breast (F)	929
Liver	677
Lung	677
Nasal Cavity	*484
Pharynx	*484
Respiratory	*484

EBV
All Sites	981
Lymphoma	1010, 1083
Nasopharynx	*315, *402, 578, 579, 580
Uterus (Cervix)	*581, 760

Education
All Sites	113, 247, 926
Uterus (Cervix)	*425

Electromagnetic Fields
All Sites	75, 313

TERM

Brain	*25, *34, 704, 1025, *1107
Childhood Neoplasms	*621, 704, 707
Leukaemia	116, *328, *627, 629, 704
Leukaemia (ALL)	*621
Leukaemia (AML)	1023
Neuroblastoma	*1100

Environmental Factors

All Sites	13, 39, *44, 84, 86, 113, 130, 193, 230, 242, 266, 293, 320, 344, 444, 464, 478, 483, 485, 495, 644, 700, 775, *801, 891, 926, 1015, 1032, 1097, *1108, 1114, *1166, *1167
Bladder	66, 208, 386, 971
Bone	644
Brain	208, 301, 703, 704, 1002, 1029, *1107
Breast (F)	*144, *155, 644, 906, 1013
Childhood Neoplasms	*319, *374, 571, 704, 707, 755, *764
Colon	4, *21, 30, 66, *144, 261, 971, 1182
Eye	971
Gallbladder	1017
Gastrointestinal	67, 208, 344, 386, 1029, 1196, 1197
Haemopoietic	208, 305, 1029
Hodgkin's Disease	66, 414, 462, 1093
Inapplicable	384
Kaposi's Sarcoma	377
Kidney	66, 808
Larynx	*801
Leukaemia	305, *340, *374, 414, *417, 483, 571, 704, 762, *801, 1196
Liver	66, *144, 261, 378, *801, 971, 1029, 1084
Lung	30, 40, 66, 67, 78, *132, *144, 198, *212, 379, 464, 682, *801, 808, 837, *880, 881, 971, 1029, 1071, 1114, 1149, 1196, 1197
Lymphoma	*340, *374, 414, *417, 483, 762, *801
Male Genital	208
Melanoma	35, 66, *809, 844, 971, 1151
Mesothelioma	26, 30, 37, 39, 67, 346, 667, 670, *784, 971
Multiple Myeloma	414
Myelodysplastic Syndrome	15, *268
Nasal Cavity	275, *809, 1021
Nasopharynx	*315, 1021, 1136
Neuroblastoma	*374
Non-Hodgkin's Lymphoma	66, 462
Oesophagus	66, 261, 971
Pancreas	66
Peritoneum	30, 39, 667
Pleura	39, 346, 667
Prostate	66, 808
Rectum	4, *21, *144, 261, 971, 1182
Respiratory	208, 344, 1095, 1149
Salivary Gland	971
Sarcoma	208
Skin	35, 208, 500, 1151
Soft Tissue	483, 509, 1005
Stomach	30, 66, *144, 563, 967, 971
Testis	66
Uterus (Cervix)	644
Wilms' Tumour	*374, 1099

Enzymes

All Sites	460
Breast (F)	1066
Larynx	741
Lung	92, 653, 741
Melanoma	1151
Oral Cavity	741
Ovary	*1139
Pharynx	741

Skin	1151	
Stomach	846	

Epilepsy
All Sites	217
Childhood Neoplasms	217
Leukaemia	217
Liver	217

Eskimos
All Sites	77, *216, 219, 885
Leukaemia	885
Liver	885
Lymphoma	885
Nasopharynx	885
Salivary Gland	885
Uterus (Cervix)	885

Ethnic Group
All Sites	77, 113, *216, 219, 443, 617, 986, 989
Childhood Neoplasms	434, 854
Colon	443, 987
Gallbladder	1017
Inapplicable	985
Leukaemia	854
Lung	437, 443
Nasopharynx	580
Neuroblastoma	854
Oesophagus	1170
Ovary	1191
Penis	347
Prostate	448, *1062
Rectum	987
Retinoblastoma	812
Stomach	544
Uterus (Cervix)	347, 619
Uterus (Corpus)	984
Vulva	347

Faeces
Colon	262, 263, 355, *363
Rectum	263, 355, *363

Familial Adenomatous Polyposis
Colon	4, 227, *363
Gastrointestinal	*814
Rectum	4, 227, *363

Familial Factors
All Sites	366, 439, 517, 557, 596, 739, 825, 929, 944, 969, 1015, 1037, 1042, 1185
Benign Tumours	586, 997
Bladder	1184
Bone	739
Brain	1090
Breast (F)	70, *144, 237, 297, 303, 312, 389, *391, 527, 586, 588, *611, 613, 618, 731, 739, 825, 886, 906, 929, 944, 990, 997, 1013, *1020, 1067, 1133, 1143
Breast (M)	990
Childhood Neoplasms	307, 739, 838, 849, 996, 1039, 1090
Choriocarcinoma	490

TERM

	Colon	4, *21, 24, 45, *144, 158, 164, *165, 227, 303, 449, 452, 494, 498, 527, 613, 640, 886, 1079, 1143
	Female Genital	886
	Gastrointestinal	739
	Hodgkin's Disease	1093
	Hydatidiform Mole	490
	Leukaemia	739, 944, 1046
	Liver	*144, 163, 378
	Lung	92, *144, 380, 613, 907, *970, 1000, 1143
	Lymphoma	739
	Melanoma	8, 35, 638
	Multiple Myeloma	972
	Oesophagus	*165, 972, 1178
	Oral Cavity	1178
	Ovary	506, 739, 800, 886, 1191
	Pancreas	*165, 972, 1143
	Prostate	972, 1143
	Rectum	4, *21, 24, 45, *144, 158, 164, *165, 227, 303, 452, 494, 498, 527, 613, 640, 1079, 1143
	Retinoblastoma	838
	Sarcoma	825
	Skin	35
	Soft Tissue	825
	Stomach	*144, 240, 388, 517, 527, 613, 967
	Testis	739
	Uterus (Cervix)	517, 886
	Uterus (Corpus)	886
	Wilms' Tumour	1099

Family History

	Childhood Neoplasms	*319
	Colon	*148, *149, *829
	Leukaemia	*783
	Rectum	*148, *149, *829

Fat

	All Sites	*53, 443, 1043, 1130, 1165
	Breast (F)	*56, 62, *590, *611, *615, 731, *780, 936, 963, 1043, *1068, 1130, 1152
	Colon	*56, 443, 537, 566, 616, *829, 936, 987, *995, 1130, 1140, *1142
	Liver	566
	Lung	*56, 443, 566, 1130
	Non-Hodgkin's Lymphoma	1127
	Oesophagus	*56, 537, 672
	Ovary	936
	Pancreas	672
	Prostate	90, 324, 851, 936, 1028
	Rectum	7, 537, *829, 987, *995, *1142
	Stomach	*56, 537, 566, 672
	Uterus (Cervix)	1130
	Uterus (Corpus)	566, 914, 984, 1043

Fatty Acids

	Prostate	939

Female

	All Sites	929, *933, 977
	Bladder	971
	Brain	94
	Breast (F)	106, 929
	Colon	106, 971, 1036
	Eye	971

Gallbladder	1017
Gastrointestinal	67
Hypopharynx	*411
Kidney	106
Larynx	*411
Lip	1022
Liver	971
Lung	67, 93, *132, 133, *151, 326, 379, 380, *411, 907, 971
Melanoma	971
Mesothelioma	67, 971
Multiple Myeloma	94
Non-Hodgkin's Lymphoma	94
Oesophagus	971
Oral Cavity	*411
Oropharynx	*411
Ovary	106
Pancreas	106
Rectum	106, 971, 1036
Salivary Gland	971
Sarcoma	94
Soft Tissue	94
Stomach	971
Tongue	*411
Uterus (Corpus)	106

Fertility

All Sites	856
Breast (F)	856
Childhood Neoplasms	919
Choriocarcinoma	5
Hydatidiform Mole	5
Leukaemia	642
Uterus (Cervix)	856

Fertilizers

All Sites	79, *82, 114
Bladder	843
Brain	*82
Gastrointestinal	1197
Leukaemia	*82
Lung	*697, 843, 1197
Multiple Myeloma	*82
Non-Hodgkin's Lymphoma	*82
Oesophagus	843
Prostate	*82, *601, *879
Sarcoma	79
Stomach	*697, 843, *1116

Fibre

All Sites	994, 1130
Breast (F)	613, 731, 1130
Colon	613, 616, *829, 987, 994, *995, 1130, 1140, *1142
Lung	613, 1130
Prostate	1028
Rectum	7, 613, *829, 987, *995, *1142
Stomach	613
Uterus (Cervix)	1130

Fibrosis

Lung	241

TERM

Fluorescent Light
All Sites	*101
Melanoma	6, *101, 622

Fruit
All Sites	443, 1007
Colon	443
Larynx	423
Lung	423, 443
Oesophagus	423, 672
Oral Cavity	423
Pancreas	672
Stomach	672
Wilms' Tumour	65

Fungicides
Hodgkin's Disease	462
Lung	92
Non-Hodgkin's Lymphoma	462

G-Banding
Liver	*162
Ovary	*162
Uterus (Cervix)	*162
Uterus (Corpus)	*162

G6PD
All Sites	460, 652, *654

Gallstones
Bile Duct	1091
Gallbladder	122, 643, 1091

Gastrectomy
All Sites	726
Stomach	467

Gastric Ulcer
All Sites	596

Genetic Factors
All Sites	13, 366, 460, 557, 739, 944, 1037, 1042
Benign Tumours	586
Bladder	176, 798, 957
Bone	739
Breast (F)	*22, *144, *155, 231, 527, 586, *590, 739, 782, 906, 918, 944, 990, 991, 1013, *1020, *1048, 1172
Breast (M)	990
Childhood Neoplasms	307, *319, *374, *403, *621, 739, 849, 996, 1089
Colon	24, *144, *211, 273, *363, 527, 782, 896, *1122
Gastrointestinal	231, 739, 896
Inapplicable	523
Kidney	1175
Leukaemia	231, *374, 739, 782, 944, 1046, 1175
Leukaemia (ALL)	*621
Liver	*144
Lung	92, *144, 653, *925, *1014, 1137, 1172

Lymphoma	*374, 739, 782, 1175
Melanoma	8, 231, 844
Nasopharynx	150
Nervous System	1175
Neuroblastoma	*374, 1175
Odontogenic Neoplasms	2
Oesophagus	131
Oral Cavity	394
Ovary	681, 739, 800, 991
Prostate	*750
Rectum	24, *144, *211, 273, *363, 527, 896
Retinoblastoma	812, 1175
Skin	500
Stomach	*144, 527, 563, 1172
Testis	231, 739, 782, 842
Thyroid	98, 314, *362
Uterus (Cervix)	231
Uterus (Corpus)	231
Wilms' Tumour	*374

Genetic Markers

All Sites	366, 700, 739, 944, 1058, 1064
Bone	739
Breast (F)	*22, *144, 169, 739, 782, 906, 944, 1013
Childhood Neoplasms	739, 996
Colon	*144, 782, 896
Gastrointestinal	739, 896
Leukaemia	739, 782, 944
Liver	*144
Lung	*144, 653, *1014, 1064
Lymphoma	739, 782, 1083
Melanoma	8, 43
Neuroblastoma	*1100
Ovary	739
Rectum	*144, 896
Stomach	*144
Testis	739, 782
Thyroid	98

Geographic Factors

All Sites	1, 63, 79, *80, 86, 190, 192, 193, 277, 286, *291, 344, 444, 471, 478, 617, 659, 660, 700, 742, 743, *801, 833, 1015, *1164, *1166, *1167
Bile Duct	558
Bladder	*80, 288, 386, 833
Brain	*80, *1107
Breast (F)	*144, 190, 288, 649
Childhood Neoplasms	*620, 753, 755, 831, 838, 839, 854, *1106
Colon	*80, *144, 261, 288, *676
Gallbladder	*676
Gastrointestinal	*57, 344, 386
Kidney	*80
Larynx	*801
Leukaemia	*80, 605, 753, *801, 823, 831, 833, 839, 854, *1106
Leukaemia (ALL)	605, *779, 823
Leukaemia (AML)	*779
Leukaemia (CLL)	823
Leukaemia (CML)	*779
Lip	*99
Liver	*144, 261, 558, 668, *676, *801, *1106
Lung	*144, 190, 288, 429, 721, *801
Lymphoma	*80, *801
Melanoma	17, 192, 288, *809
Multiple Myeloma	486
Nasal Cavity	*809
Nasopharynx	429

Numbers indexed are project numbers, asterisks identify new projects

TERM

Neuroblastoma	854
Non-Hodgkin's Lymphoma	753, 823, *1106
Oesophagus	261, 288, *676, 1170
Pancreas	*676, *1106
Prostate	*80, 288
Rectum	*80, *144, 261, 288, *676
Respiratory	344
Retinoblastoma	838
Sarcoma	79
Skin	17, 192
Small Intestine	*676
Soft Tissue	*1106
Stomach	*80, *144, 288, *676
Thyroid	725
Uterus (Cervix)	288, 429, 649
Uterus (Corpus)	288, 649

Glass
All Sites	474

H. pylori
All Sites	818
Stomach	*284, *285, *575, *778, 846, 847, *1112

Haemophilia
Kaposi's Sarcoma	377

Hair
Lung	*1173
Prostate	851

HBV
All Sites	*334, 531
Bile Duct	558
Liver	161, 189, 335, 378, 465, 528, 558, 568, 630, *664, 668, 747, 806, 873, 898, 1084, 1087
Uterus (Cervix)	367

HCV
Liver	*664

Herbicides
All Sites	28, 79, *82, 249, 294, 370, *591, 628, 701, 1119
Brain	*82, 1041, 1119
Childhood Neoplasms	434
Hodgkin's Disease	462, 513
Kidney	365, 1041
Leukaemia	28, *82, *340, 513, *627, *922, 1119
Liver	1041
Lung	92, 1041, 1119
Lymphoma	12, 249, *340, 628, *922
Multiple Myeloma	*82, 513, 1119
Non-Hodgkin's Lymphoma	*82, 462, 513, *591, 1119
Prostate	*82
Sarcoma	12, 79, 249, *591
Skin	*993
Soft Tissue	28, 509, *591, 628, 1005, 1119
Wilms' Tumour	65

TERM

Heredity
All Sites	545, 739
Bone	739
Breast (F)	739, 782, 1172
Childhood Neoplasms	*403, 739
Colon	*211, *363, 494, 686, 782
Gastrointestinal	739
Leukaemia	739, 782
Lung	1172
Lymphoma	739, 782
Melanoma	43
Ovary	681, 739
Rectum	*211, *363, 494, 686
Stomach	1172
Testis	739, 782
Wilms' Tumour	1099

High-Risk Groups
All Sites	84, 247, *657, 825, 875, 891, 944, 1058, 1094, 1130, 1188
Bladder	834
Breast (F)	172, 237, *238, 312, 389, 400, 618, 736, *771, 825, 906, 944, 991, 1130, 1172, 1190, 1195
Childhood Neoplasms	*403
Colon	4, *21, 24, *145, *149, *211, 261, 355, *436, 640, 793, 896, 1059, 1130
Gastrointestinal	896
Hodgkin's Disease	455
Inapplicable	523
Kaposi's Sarcoma	455
Kidney	71
Leukaemia	848, 944
Liver	172, 261, 465, 747, 1084
Lung	172, *194, 381, 567, *657, 1130, *1131, 1172
Lymphoma	1083
Mesothelioma	*1131
Non-Hodgkin's Lymphoma	455
Oesophagus	131, 185, 187, 261
Oral Cavity	59, 1051, 1072
Oropharynx	59, 1072
Ovary	506, 991
Rectum	4, *21, 24, *145, *149, 172, *211, 261, 355, *436, 640, 896, 1059
Sarcoma	825
Soft Tissue	825
Stomach	172, 388, 467, 663, 1172
Uterus (Cervix)	112, 172, 177, 400, 903, 1130
Uterus (Corpus)	308

Histology
All Sites	192, 250, 521, 606, 891
Appendix	390
Benign Tumours	965, 1105
Bile Duct	390
Brain	*81, 415, 1024
Breast (F)	*155, 389, 442, 598, 606, *1053
Colon	164, 390, 452, 793, 896, 1105
Gastrointestinal	*814, 896
Head and Neck	267, 382, 666
Hodgkin's Disease	414
Kaposi's Sarcoma	803
Larynx	51, 666
Leukaemia	179, 414
Lung	157, *212, 241, 380, 469, 699, 860, 1000, 1071, 1092
Lymphoma	414, 524
Melanoma	192, 304
Mesothelioma	51, 469, 667, 860

Numbers indexed are project numbers, asterisks identify new projects

TERM

Multiple Myeloma	414
Nasal Cavity	275
Nasopharynx	51
Nervous System	390
Non-Hodgkin's Lymphoma	904
Oesophagus	390, 1178
Oral Cavity	382, 390, 1178
Ovary	108, 1191
Pancreas	390
Peritoneum	667
Pleura	469, 667, 860
Prostate	390
Rectum	164, 452, 896, 1105
Skin	11, 192, 524, *772
Small Intestine	390
Soft Tissue	390, 803, 1005
Stomach	239, 388, *522, 544, 673, *1116
Testis	108, 965
Thyroid	318, 376, 725
Uterus (Cervix)	88, 603, *817, 852, 888, 893

HIV

All Sites	891, 979, 1094
Hodgkin's Disease	455
Kaposi's Sarcoma	*38, 455, 631, 803, *899, 909, 979
Lymphoma	*38, 909, 1010
Non-Hodgkin's Lymphoma	455, 904
Soft Tissue	803
Uterus (Cervix)	367

HLA

All Sites	228
Kaposi's Sarcoma	501
Lymphoma	1010
Nasopharynx	*315
Prostate	*750
Skin	500
Testis	842
Thyroid	98

Homosexuality

All Sites	213, 1094
Hodgkin's Disease	213
Kaposi's Sarcoma	377, *899
Non-Hodgkin's Lymphoma	904

Hormones

All Sites	488, 573, 574, 737, 929, 981, 1018, *1138
Benign Tumours	491, 586, 587, 1077
Bile Duct	300
Breast (F)	*119, 235, 237, 297, 303, 312, *399, 482, 491, 573, 586, 587, 588, *590, 618, 623, 625, 632, 737, 819, 857, 858, 886, 915, 918, 929, *935, 1018, 1019, 1063, 1066, *1068, 1076, 1077, 1133, 1181
Breast (M)	296
Choriocarcinoma	490
Colon	164, 303, 886, 915, *935
Female Genital	453, 858, 886
Gallbladder	300, 1017
Gastrointestinal	453
Hydatidiform Mole	490
Kidney	42, 729, *1104
Liver	573, 747

Lung	574, 915, *935, 1018	
Lymphoma	453	
Melanoma	304, 336, 988	
Multiple Myeloma	972	
Oesophagus	972	
Ovary	108, 396, *398, 491, 506, 607, 608, 737, 886, 915, *935, 1157	
Pancreas	300, 972	
Prostate	229, 573, 727, *935, 972	
Rectum	164, 303	
Skin	453	
Stomach	563, 573, 574	
Testis	108, 215, 840	
Thyroid	453, 744	
Urinary Tract	453	
Uterus (Cervix)	491, 886	
Uterus (Corpus)	235, 491, 737, 745, 857, 886, 914, 915, *935, 984, 1018, 1125	
Vagina	1076	
Vulva	491	

HPRT-Mutants
Inapplicable	74

HPV
All Sites	981
Female Genital	256
Lung	*864
Male Genital	256
Oral Cavity	*359, *1027, *1132
Oropharynx	*359, *1132
Penis	347
Pharynx	*1027
Respiratory	*359
Uterus (Cervix)	*33, 48, 88, 168, 272, 283, 347, *359, 367, 395, 418, *419, *581, 582, 585, 619, 665, 685, *748, 760, 799, 852, 883, *932, *962, *1124, 1141
Vulva	347

HSV
All Sites	981
Oral Cavity	*1132
Oropharynx	*1132
Uterus (Cervix)	48, 168, 197, 283, 418, *419, *581, 582, 665, *748, 760

HTLV
All Sites	531, 891
Kaposi's Sarcoma	377
Leukaemia	*549, 555
Lymphoma	524, 555
Skin	524

Hygiene
All Sites	999
Hypopharynx	269, *420
Larynx	180, *420, 1135
Lung	379
Nasal Cavity	180
Oesophagus	*420, 1135
Oral Cavity	180, 200, 269, *420, 1135
Oropharynx	269, *420
Pharynx	180
Stomach	1135
Uterus (Cervix)	*406, *425

Numbers indexed are project numbers, asterisks identify new projects

TERM

Vagina	916	
Vulva	916	

Hyperthyroidism
Thyroid	466

Hysterectomy
All Sites	*225
Breast (F)	738
Ovary	738

Immunodeficiency
All Sites	1042
Childhood Neoplasms	570
Haemopoietic	305
Leukaemia	305
Lymphoma	1083
Skin	514

Immunologic Markers
All Sites	636
Burkitt's Lymphoma	645
Lung	469, 584
Mesothelioma	469
Pleura	469

Immunology
All Sites	353, 557
Childhood Neoplasms	440
Kaposi's Sarcoma	501
Liver	378
Non-Hodgkin's Lymphoma	904
Oral Cavity	*1132
Oropharynx	*1132
Skin	500

Immunosuppression
All Sites	228, 953, 1042
Breast (F)	352
Colon	352
Kidney	71
Leukaemia (AML)	1023
Lymphoma	*1038
Melanoma	336
Non-Hodgkin's Lymphoma	348
Ovary	352
Rectum	352
Stomach	352

In Situ Carcinoma
Breast (F)	*20, 736
Testis	207
Uterus (Cervix)	197, 888

Indians
All Sites	120
Oesophagus	435

TERM

Infection
All Sites	213, 496, 891
Breast (F)	352, 937
Burkitt's Lymphoma	645
Childhood Neoplasms	*319, 854
Colon	352, *363
Female Genital	256
Hodgkin's Disease	213, 1093, 1126
Kaposi's Sarcoma	377
Kidney	71
Leukaemia	116, *417, 555, 854
Liver	163, 167, 271, 378, 554, 872, 908
Lymphoma	*417, 555
Male Genital	256
Nasal Cavity	1021
Nasopharynx	150, 578, 1021
Neuroblastoma	854
Ovary	352, 937
Pancreas	937
Prostate	937
Rectum	352, *363
Stomach	352, 845
Testis	215
Uterus (Cervix)	197, 283, 395, 619, *748, 852
Uterus (Corpus)	937
Wilms' Tumour	65

Infertility
All Sites	*1138
Childhood Neoplasms	849
Ovary	1157
Testis	207

Insecticides
All Sites	79, *82, 861
Brain	*82, 1041
Kidney	1041
Leukaemia	*82
Liver	1041
Lung	92, 1041
Multiple Myeloma	*82
Non-Hodgkin's Lymphoma	*82
Prostate	*82
Sarcoma	79

Intra-Uterine Exposure
Bone	170
Brain	41, 170, 301, 1024, 1090
Breast (F)	858, 1076
Childhood Neoplasms	218, 438, 571, 715, 754, *826, 838, 919, 1090
Female Genital	858
Kidney	1175
Leukaemia	116, 170, 571, 754, 1046, 1175
Leukaemia (ALL)	*72
Liver	170
Lymphoma	170, 1175
Nervous System	1175
Neuroblastoma	*1100, 1175
Retinoblastoma	838, 1175
Testis	215, 840
Vagina	*1040, 1076
Wilms' Tumour	1099

Numbers indexed are project numbers, asterisks identify new projects

TERM

Iodine Deficiency
Thyroid 98, 744

Japanese
Stomach *575

Lactation
Breast (F) *52, 70, *119, *392, *590, 632, 1035

Late Effects
All Sites 203, 521, 882
Bone 368
Breast (F) 813
Childhood Neoplasms *373, 850, 1089
Leukaemia 368, 642
Liver 368
Lung 368
Lymphoma 368
Spleen 368

Latency
All Sites 113, 203, 451, 478, 797, 870
Benign Tumours 451
Breast (F) 220, 813
Colon 791
Kidney 1081
Leukaemia 555
Lung *153, 789, 791, 870, 954
Lymphoma 555, 954
Mesothelioma 870
Rectum 791
Stomach 152, 791

Leather
All Sites *306, 474, *696
Bladder *306
Kidney 481
Lymphoma 481
Nasal Cavity *306, 316

Lifestyle
All Sites 84, 86, *216, 219, *412, *538, 539, 545, 559, 560, *564, 572, 573, 574, 617, 740, 802, 892, 986, 1011, 1012, 1094, *1166, *1167, *1176
Benign Tumours 27
Bladder 105, 386, 971
Brain 703
Breast (F) 18, 125, 361, 573, *590, 595, 613, 906, 1143, *1174
Childhood Neoplasms *374, 755
Colon 18, 125, *145, *148, 158, 262, 572, 595, 613, 740, 971, 1143, 1182
Eye 971
Gastrointestinal 386
Kidney 529
Leukaemia 125, 179, 364, *374
Liver 125, 528, 572, 573, 971
Lung 105, 125, 379, 572, 574, 595, 613, 740, 971, 1143
Lymphoma *374
Melanoma 17, 27, 336, 971
Mesothelioma 971
Nasopharynx 125, 1136

Neuroblastoma	*374
Non-Hodgkin's Lymphoma	1127
Oesophagus	125, 357, 435, 971
Ovary	*398, *1174
Pancreas	105, 1143
Prostate	18, 229, 573, 595, 1143
Rectum	18, 125, *145, *148, 158, 572, 595, 613, 971, 1143, 1182
Salivary Gland	971
Skin	17
Soft Tissue	1005
Stomach	125, 184, *556, 563, 572, 573, 574, *575, 595, 613, 740, 971
Uterus (Cervix)	*1174
Uterus (Corpus)	*1174
Wilms' Tumour	*374

Linkage Analysis

All Sites	*811
Breast (F)	169, 1013, *1048
Colon	*363
Oesophagus	131
Rectum	*363

Lipids

All Sites	443, 573, 574, 740, 929, 981, 1003
Breast (F)	18, 573, 929, 1003, *1148, 1152
Colon	18, 443, 498, 740, 1003
Liver	573
Lung	443, 574, 740, 1003
Melanoma	1151
Prostate	18, 573, 939, 1003
Rectum	18, 498
Skin	*992, 1151
Stomach	573, 574, 740

Liver Disease

Liver	747, 806

Lymphocytes

All Sites	545, 557, *694, 712
Angiosarcoma	677
Bladder	614, 677
Inapplicable	74, 195, 196, 204, 383, 523, 763, 942
Leukaemia	*922
Liver	*162, 677
Lung	*132, 584, 614, 677
Lymphoma	*922
Ovary	*162
Testis	385
Thyroid	*362
Uterus (Cervix)	*162
Uterus (Corpus)	*162

Maghrebians

Nasopharynx	*315

Male

All Sites	327
Lung	*171, 535, 721

TERM

Mammography
All Sites	1130
Benign Tumours	997
Breast (F)	*10, *20, 111, 237, 274, 312, 482, 598, 602, 723, 731, *771, 877, 878, 894, 963, 997, *1068, 1130
Colon	1130
Lung	1130
Uterus (Cervix)	1130

Mapping
All Sites	86, 344, 444, 659, 660, *751, 1015, *1164, 1183
Brain	*1107
Childhood Neoplasms	753, *1049
Gastrointestinal	344
Leukaemia	753
Non-Hodgkin's Lymphoma	753
Oesophagus	1170
Respiratory	344

Marijuana
Childhood Neoplasms	*621
Leukaemia (ALL)	*621
Leukaemia (ANLL)	1047

Marital Status
All Sites	572, 802
Breast (F)	400
Colon	214, 572
Liver	572
Lung	572
Rectum	214, 572
Stomach	572
Uterus (Cervix)	400

Mate
All Sites	1165
Oesophagus	646
Stomach	1161

Mathematical Models
All Sites	13, 743, *751, 913, 1109
Bladder	489
Breast (F)	253, 600, 602, 913
Colon	253
Haemopoietic	1109
Kidney	489
Larynx	253, 489
Leukaemia	253, 913
Lip	253
Lung	253, 489, 913
Lymphoma	1109
Oesophagus	253, 489
Oral Cavity	489
Pancreas	489
Pharynx	489
Prostate	253
Rectum	253
Skin	1109
Stomach	253
Thyroid	913

TERM

Uterus (Cervix)	112, 253, 603, 903
Uterus (Corpus)	253

Menarche
All Sites	633
Breast (F)	*52, 89, 169, 297, 361, *392, *399, 416, 610, 633
Childhood Neoplasms	919
Gastrointestinal	633
Ovary	396, *398, 633
Uterus (Cervix)	633
Uterus (Corpus)	633
Vagina	*1040

Menopause
All Sites	633, 737, 929
Breast (F)	*52, 70, 129, 169, 231, *392, 416, 633, 737, 857, 915, 929
Childhood Neoplasms	919
Colon	915
Gastrointestinal	231, 633
Leukaemia	231
Lung	915
Melanoma	231
Ovary	396, 633, 737, 915
Testis	231
Uterus (Cervix)	231, 633
Uterus (Corpus)	231, 633, 737, 857, 915

Menstruation
Breast (F)	*22, 70, *392, *424, *590, 1009
Colon	*165
Oesophagus	*165
Ovary	396
Pancreas	*165
Rectum	*165
Stomach	563

Metabolism
Angiosarcoma	677
Bladder	677
Liver	677, 898
Lung	677, *925, *1014
Ovary	*1139
Skin	1150

Metals
All Sites	86, 114, 243, 264, *270, 320, 321, 540, 553, *691, *794, 807, 920, 929, 930, 948, 951, 955, 994, 1008, 1101, 1102, 1155
Bladder	66, 208, *461, 910
Bone	103
Brain	208, 910, 1029, 1088, 1090
Breast (F)	243, 613, 731, 929, 1152
Childhood Neoplasms	218, *1085, 1090
Colon	66, 243, 613, 964, 994, 1031, 1102
Gastrointestinal	208, 1029, 1196, 1197
Genitourinary	1101
Haemopoietic	208, 387, 910, 1029, 1101
Hodgkin's Disease	66
Inapplicable	74
Kidney	66, *461, 481, *722, 808, 1031
Larynx	51, 321, 655
Leukaemia	387, *461, 1196

Numbers indexed are project numbers, asterisks identify new projects

TERM

Liver	66, 234, *461, 525, 1029, 1031, 1102
Lung	3, 66, 78, 157, 198, 243, 264, *270, 321, 322, 387, *461, 525, 553, 613, *722, 777, 807, 808, 881, 884, 910, 912, 954, 955, 1029, 1031, 1155, 1196, 1197
Lymphoma	481, 910, 954, 1101
Male Genital	208
Melanoma	66, 988, 1151
Mesothelioma	51, 910
Nasal Cavity	316, 1102
Nasopharynx	51
Non-Hodgkin's Lymphoma	66
Oesophagus	66, 282
Oral Cavity	321
Pancreas	66
Pharynx	321
Prostate	66, 243, *601, 807, 808, 851
Rectum	243, 613, 964, 1102
Respiratory	208, 1101, 1102
Sarcoma	208, 910
Skin	208, 1031, 1151
Stomach	66, 243, *270, 322, 613, 910, 1161
Testis	66
Uterus (Cervix)	243

Micronuclei

All Sites	545, *694
Angiosarcoma	677
Bladder	614, 677
Inapplicable	204, 942
Leukaemia	958
Liver	677
Lung	*132, 614, 677
Oesophagus	282, 357
Oral Cavity	118

Migrants

All Sites	286, *291, 471, 478, 485, 573, *811, 989
Bladder	288
Breast (F)	18, 288, 573
Childhood Neoplasms	854
Colon	18, 288
Leukaemia	854
Liver	573
Lung	288, 437, 648
Melanoma	288
Nasopharynx	*315
Neuroblastoma	854
Oesophagus	288
Prostate	18, 288, 573
Rectum	18, 288
Stomach	288, 573
Uterus (Cervix)	288
Uterus (Corpus)	288

Minerals

All Sites	464, 787, 994
Colon	355, 994
Lung	199, 464, 788
Rectum	355
Respiratory	787

TERM

Mining
All Sites	39, 86, 264
Brain	*25
Colon	30
Hodgkin's Disease	36
Inapplicable	195
Leukaemia (CLL)	36
Lung	3, 30, *32, 40, *127, *156, 198, 199, 264, 322, 458, *548, 721, 732, 884, 954
Lymphoma	954
Mesothelioma	30, 39, *548
Non-Hodgkin's Lymphoma	36
Peritoneum	30, 39
Pleura	39
Stomach	30, 322, 458

Monitoring
All Sites	54, 60, *493, 773, 895, 947, 1057, 1073
Breast (F)	172
Childhood Neoplasms	*1106
Inapplicable	251, 597
Leukaemia	54, 895, *1106
Liver	172, 898, *1106
Lung	172
Lymphoma	54
Non-Hodgkin's Lymphoma	*1106
Pancreas	*1106
Rectum	172
Soft Tissue	*1106
Stomach	172
Uterus (Cervix)	172

Multiple Primary
All Sites	192, 222, 250, 276, 330, 331, 445, 576, 589, 739, *876, 913
Bone	739
Breast (F)	*144, 222, 223, 441, 442, 527, *590, 739, 913
Childhood Neoplasms	*373, 739, 754, 1089
Colon	*144, 441, 527
Gastrointestinal	332, 739
Haemopoietic	768
Head and Neck	441
Hodgkin's Disease	511
Larynx	64, 232, *360
Leukaemia	221, 223, 331, 332, 441, 589, 642, 739, 754, 804, 848, 913
Leukaemia (ALL)	221, 511, 768
Leukaemia (ANLL)	511
Leukaemia (CLL)	221, 223
Lip	64
Liver	*144
Lung	*144, 222, 232, 331, 332, 589, 913
Lymphoma	441, 739
Melanoma	192
Myelodysplastic Syndrome	589, 768
Neuroblastoma	511
Non-Hodgkin's Lymphoma	331, 332, 511
Oral Cavity	64, *1026
Ovary	739
Pharynx	64, *1026
Rectum	*144, 527
Skin	192, 500, *772
Stomach	*144, 527
Testis	739, *876
Thyroid	441, 913
Uterus (Cervix)	222

TERM

Uterus (Corpus)	221
Wilms' Tumour	511

Mutagen
All Sites	700
Bladder	100
Childhood Neoplasms	*764
Gastrointestinal	1197
Inapplicable	523, 938, 940
Liver	*181
Lung	1197
Stomach	*181
Uterus (Cervix)	903

Mutagen Tests
All Sites	1057
Inapplicable	597
Stomach	967

Mutation Rate
All Sites	130

Mutation, Germinal
All Sites	130
Childhood Neoplasms	838, 849
Colon	227
Inapplicable	50, 383, 384, 938
Leukaemia	874
Rectum	227
Retinoblastoma	838, *1086

Mutation, Somatic
All Sites	54, 700
Childhood Neoplasms	754, 1089
Inapplicable	50, 383, 938
Leukaemia	54, 754
Lymphoma	54
Oral Cavity	394

Mycotoxins
All Sites	895
Inapplicable	938
Leukaemia	895
Liver	161, 189, *664, 747, 873, 898
Stomach	746, 845

Naevi
Benign Tumours	27
Melanoma	27, *290, 304, 336, 468, 599, 638, 844, 988, 1115

Nass
Oesophagus	1178
Oral Cavity	1178

Nutrition
All Sites	*53, *85, *143, 205, 293, 356, *454, 533, 740, 1111, 1130

Bile Duct	*592
Bladder	311
Breast (F)	*85, 96, 125, 361, *428, 583, *615, 731, *780, 918, 1063, 1130, *1148, 1152, 1181, 1189
Colon	*85, 125, 164, *363, 740, *829, *995, 1130, 1140, *1142, *1156
Gastrointestinal	1189
Genitourinary	1189
Inapplicable	985
Leukaemia	125
Liver	125
Lung	97, 125, *456, 740, 1130
Melanoma	988
Nasopharynx	125
Non-Hodgkin's Lymphoma	1127
Oesophagus	125, 672
Pancreas	*592, 672
Prostate	1028
Rectum	125, 164, *363, *829, *995, *1142
Skin	*992
Stomach	125, 135, 663, 672, 740
Uterus (Cervix)	88, 582, *932, 1130
Uterus (Corpus)	914, 984
Vagina	*1040

Obesity

All Sites	*85, 1043
Breast (F)	*85, 96, *409, *611, 618, 623, 886, 1043, 1189
Colon	*85, 674, 886, 943, 987
Female Genital	886
Gastrointestinal	1189
Genitourinary	1189
Kidney	42, 365, 729
Ovary	886
Prostate	1028
Rectum	674, 943, 987
Uterus (Cervix)	886
Uterus (Corpus)	745, 886, 984, 1043

Occupation

All Sites	16, 28, 39, 46, 54, *55, 60, 61, *68, 75, 76, *82, *83, 84, 86, *101, 107, 113, 114, *124, *126, 182, 183, 230, 236, 247, 249, 264, 266, *270, *280, *281, 294, 295, 298, *306, *310, 313, 320, 321, 327, *329, 344, 354, 370, *433, 439, 444, 474, *475, 477, 479, *484, 485, 487, 495, *504, 507, 508, 516, 533, *538, 540, *551, 553, 557, *591, 594, 624, 628, 650, 652, *654, 687, *690, *691, *692, *693, *695, *696, *698, 701, 702, *705, 710, 713, 724, 733, 756, *766, 773, 775, 787, *794, 796, 797, 807, 810, 815, 821, 861, 865, 866, 867, 868, 869, 870, 882, 892, 920, 926, 927, 929, 934, *945, 947, 948, 950, 951, 955, 956, 959, 960, 1008, 1011, 1012, 1032, 1033, 1045, 1057, 1074, 1080, 1082, 1096, 1097, 1098, 1101, 1102, *1103, 1110, 1114, 1117, 1119, 1120, 1130, 1155, 1158, 1159, 1171, *1176, 1177, 1185
Angiosarcoma	677
Bile Duct	*123
Bladder	66, 100, *159, 176, 208, 278, *306, 309, 311, *459, *461, *484, 677, 834, 843, 910, *928, 949, 957, 960, 971, 1034, 1096, 1171, 1184
Bone	170
Brain	*25, *34, *82, 94, 170, 208, 254, 260, 302, 309, 479, *504, *551, 910, 941, 946, 1002, 1024, 1029, 1033, 1041, 1090, 1119, 1120
Breast (F)	*119, 235, 650, 929, 937, 1045, 1130
Childhood Neoplasms	218, 438, 754, *826, 946, 1039, *1085, 1090
Colon	30, 66, 164, *165, 261, 498, 686, 689, 720, 791, *829, 943, 964, 971, *974, *976, 1030, 1031, 1102, 1120, 1130, 1140, *1142
Eye	971
Female Genital	453
Gallbladder	*123

TERM

Gastrointestinal	*73, 102, 208, 344, 453, 865, 968, 1029, 1186, 1196, 1197
Genitourinary	1101
Haemopoietic	16, 208, 305, 479, 647, *684, 910, 1029, 1101, *1129
Hodgkin's Disease	36, 66, 91, 254, 462, 513
Hypopharynx	269, 317
Inapplicable	74, 195, 196, 204, 251, *343, 597, 656, 911, 942
Kidney	42, 66, *73, 309, 325, 365, *461, 481, *722, 729, 808, *973, 1031, 1034, 1041, 1081, 1175
Larynx	51, *73, 180, 317, 321, *360, 487, *637, 655, *716, 863, 1033
Leukaemia	28, 54, *55, *82, 116, 134, 170, 179, 201, 254, 305, 309, *328, *340, 364, *461, 510, 513, *627, 629, *693, 713, 754, 781, *783, 823, 867, *887, *928, 941, 958, 959, *975, 1045, 1118, 1119, 1120, 1159, 1175, 1196
Leukaemia (ALL)	*279, 823
Leukaemia (CLL)	36, 823
Liver	66, *126, 136, 170, 234, 236, 261, *461, 525, *637, 677, 961, 971, 1029, 1030, 1031, 1041, 1102
Lung	3, 30, 31, *32, 40, 46, 66, *73, 78, *83, 92, 104, *124, *127, *151, *156, 157, 175, 198, 199, *212, 241, 245, 252, 254, 264, *270, 278, 292, 295, 321, 322, 323, 326, *333, 341, *342, 379, 380, 401, *457, 458, *459, *461, 470, 472, 487, 499, 525, 535, 547, *548, 553, 567, 584, 594, 648, 650, 677, *684, 689, *697, 710, 714, 719, 721, *722, 724, 732, 777, 788, 789, 791, *792, 807, 808, 843, 860, *864, 865, 866, 870, 871, 881, 884, *902, 907, 910, 912, 927, *928, 952, 954, 955, 968, *970, 971, 1000, *1014, 1029, 1030, 1031, 1041, 1045, 1071, 1092, 1114, 1119, *1129, 1130, 1149, 1154, 1155, 1162, 1171, 1179, 1180, 1186, 1196, 1197
Lymphoma	12, 16, 54, *55, 170, 249, 309, *340, 453, *459, 481, 628, *684, *693, *698, 713, 867, 910, 954, 961, *975, 1101, 1118, *1129, 1175
Male Genital	208
Melanoma	17, 66, *101, 304, 336, 599, 622, 971, 1006
Mesothelioma	26, 30, 31, 37, 39, 46, 51, 102, *342, 346, 502, *548, 667, 670, 689, 735, *784, 860, 865, 870, 871, 910, 971
Multiple Myeloma	*82, 91, 94, 309, 513, *698, 972, 1119
Myelodysplastic Syndrome	15, *268, 510, *770
Nasal Cavity	180, 201, 275, *306, 316, *484, 1021, 1102
Nasopharynx	51, 178, 868, 1021, 1098
Nervous System	1175
Neuroblastoma	1175
Non-Hodgkin's Lymphoma	36, 66, *82, 91, 94, 462, *504, 513, *591, 823, 904, 968, 1119
Oesophagus	66, *165, 261, *637, *716, 843, 971, 972, *1129
Oral Cavity	23, 180, 269, 321, *637, *716
Oropharynx	269
Ovary	108, 506, 937, 1069
Pancreas	66, *165, 246, 309, *459, 937, 959, 972, 1118
Peritoneum	30, 39, *73, 502, 667, 735
Pharynx	180, 321, *484, *637, *716
Pleura	39, *73, 346, 502, 667, 735, 860
Prostate	66, *82, 229, 254, *259, 309, 594, *601, 727, *750, 795, 807, 808, *879, 937, 972, *1129
Rectum	164, *165, 261, 498, 686, 689, 791, *829, 943, 964, 971, *974, *976, 1102, 1120, *1142
Respiratory	102, 201, 208, 236, 344, 463, *484, 683, 787, 865, 868, 1095, 1101, 1102, 1149
Retinoblastoma	1175
Salivary Gland	757, 971
Sarcoma	12, 91, 94, 208, 249, 309, *591, 910
Skin	17, 208, 453, 479, *901, *993, 1031
Soft Tissue	28, 91, 94, *504, 509, *591, 628, 961, 1005, 1119
Stomach	30, 66, 136, *270, 322, 458, *522, 563, 650, 651, 673, 683, *684, 689, *697, 728, 791, *792, 843, 910, 927, 959, 971, *1116, 1171
Testis	66, 108, 215, 840
Thyroid	358, 453, 688, 1045
Urinary Tract	453, *759
Uterus (Cervix)	650, 1130
Uterus (Corpus)	235, 937, 984
Wilms' Tumour	65, 1099

Oncogenes
All Sites	521
Bladder	957
Breast (F)	62, 527, *590, *1053
Colon	527
Liver	1087
Lung	241
Neuroblastoma	*1100
Rectum	527
Stomach	527

Oophorectomy
All Sites	*225
Breast (F)	738
Ovary	738

Oral Contraceptives
All Sites	488, 822, 856, 929
Benign Tumours	491, 586, 587, 997
Breast (F)	70, 235, *392, 491, 534, 586, 587, 588, *590, 623, 625, 813, 822, 856, 918, 929, 997, 1009, 1133, 1134, 1193
Female Genital	453, 822
Gallbladder	643, 1134
Gastrointestinal	453
Liver	668, 1134
Lymphoma	453
Ovary	491, 506, 1134, 1191
Skin	453
Thyroid	453, 744
Urinary Tract	453
Uterus (Cervix)	88, 110, 168, 210, 491, 582, *748, 799, 852, 856, 903, 1134
Uterus (Corpus)	235, 491, 1125, 1134
Vulva	491

Paints
All Sites	948
Childhood Neoplasms	218, *1085
Leukaemia	*627
Prostate	*601

Parasitic Disease
Bile Duct	558
Burkitt's Lymphoma	645
Colon	*145, 146
Liver	234, 558
Rectum	*145, 146

Parental Occupation
Brain	*49, 897, 1088
Childhood Neoplasms	*374, *621, *1085
Leukaemia	*49, *374, *790
Leukaemia (ALL)	*72, *621
Lymphoma	*49, *374
Neuroblastoma	*49, *374, *1100
Non-Hodgkin's Lymphoma	*790
Retinoblastoma	*1086
Wilms' Tumour	*374

TERM

Parity
All Sites	633, 856
Benign Tumours	491
Breast (F)	*52, 70, 169, 231, *399, 400, 416, *424, 491, 610, *611, 632, 633, 856
Choriocarcinoma	490
Colon	674, 943
Gastrointestinal	231, 633
Hydatidiform Mole	490
Leukaemia	231
Lung	907
Melanoma	231
Ovary	396, *398, 491, 633, 1191
Rectum	674, 943
Testis	231
Thyroid	744
Uterus (Cervix)	231, 400, *425, 491, 519, 633, 856
Uterus (Corpus)	231, 491, 633
Vulva	491

Passive Smoking
All Sites	87, 818, 891, 892, 977, 1064
Bladder	614
Breast (F)	*119, *978
Childhood Neoplasms	*621, *764
Hypopharynx	317
Inapplicable	940
Kidney	42, 729
Larynx	317
Leukaemia (ALL)	*72, *621
Lung	97, *121, *151, 292, 326, 379, 380, 401, 614, 714, *970, 1056, 1064, 1071, 1179
Nasal Cavity	316
Nasopharynx	178
Uterus (Cervix)	903

PCR
Inapplicable	938
Leukaemia	*922
Lung	*864
Lymphoma	*922, 1010
Oral Cavity	805
Uterus (Cervix)	*33, 48, 367, *581, 585, 799

Pedigree
All Sites	739
Bone	739
Breast (F)	*144, 739, 782, 906
Childhood Neoplasms	739
Colon	*144, *363, 782, 896
Gastrointestinal	739, 896
Leukaemia	739, 782, 874
Liver	*144
Lung	*144
Lymphoma	739, 782
Nasopharynx	150
Ovary	506, 739
Rectum	*144, *363, 896
Stomach	*144
Testis	739, 782

Pepsinogen
Stomach	*575, 846, 847

TERM

Pesticides

All Sites	28, *82, 86, 249, *281, 294, 370, 479, *504, 508, *591, 628, *696, 701, 861, 920, 951, 1119
Bladder	*159, 309, 910
Brain	*34, *82, 94, 309, 479, *504, 910, 1041, 1119
Childhood Neoplasms	434, *1085, *1106
Colon	720
Female Genital	453
Gastrointestinal	453
Haemopoietic	479, 910
Hodgkin's Disease	91, 462, 513
Inapplicable	911
Kidney	309, 365, 481, 1041
Larynx	655
Leukaemia	28, *82, 309, *328, *340, 513, *627, *922, 1046, *1106, 1119
Leukaemia (ALL)	*72, *279
Leukaemia (ANLL)	1047
Liver	1041, *1106
Lung	910, 1041, 1119
Lymphoma	12, 249, 309, *340, 453, 481, 628, 910, *922
Mesothelioma	910
Multiple Myeloma	*82, 91, 94, 309, 513, 1119
Non-Hodgkin's Lymphoma	*82, 91, 94, 462, *504, 513, *591, *1106, 1119
Pancreas	309, *1106
Prostate	*82, 309, *601, *879
Sarcoma	12, 91, 94, 249, 309, *591, 910
Skin	453, 479
Soft Tissue	28, 91, 94, *504, 509, *591, 628, 1005, *1106, 1119
Stomach	910, *1116
Thyroid	453
Urinary Tract	453

Petroleum Products

All Sites	16, *82, 86, *801, 948, 950, 1080
Bladder	66, 208, 910
Brain	*82, 208, 910, 1029
Colon	66, 791
Gastrointestinal	208, 968, 1029
Haemopoietic	16, 208, 910, 1029
Hodgkin's Disease	66
Inapplicable	195
Kidney	66, 1081
Larynx	*801
Leukaemia	*82, *801
Leukaemia (ANLL)	1047
Liver	66, *801, 1029
Lung	66, 791, *792, *801, 910, 968, 1029
Lymphoma	16, *801, 910
Male Genital	208
Melanoma	66
Mesothelioma	910
Multiple Myeloma	*82
Non-Hodgkin's Lymphoma	66, *82, 968
Oesophagus	66
Pancreas	66
Prostate	66, *82
Rectum	791
Respiratory	208
Sarcoma	208, 910
Skin	208
Stomach	66, 791, *792, 910
Testis	66

Numbers indexed are project numbers, asterisks identify new projects

TERM

Photochemotherapy
All Sites	931
Eye	931
Skin	931

Physical Activity
All Sites	740, 920, 929, 930
Breast (F)	89, 255, 361, 482, 841, 886, 915, 929, 1009, 1143
Colon	*148, 164, *165, 255, 449, 565, 674, 740, *829, 841, 886, 915, 943, 987, 1065, *1122, 1140, 1143
Female Genital	886
Kidney	365
Lung	255, 740, 915, 1143
Oesophagus	*165
Ovary	886, 915
Pancreas	*165, 1143
Prostate	841, 1143, *1144
Rectum	*148, 164, *165, 255, 565, 674, *829, 943, 987, 1065, 1143
Skin	255
Stomach	740, 841
Uterus (Cervix)	886
Uterus (Corpus)	745, 886, 915, 984

Physical Factors
All Sites	205, 572, 633, 650, 913, 1043
Breast (F)	*22, *52, *56, 70, 89, 96, 231, 361, *399, 482, 534, 610, 623, 633, 650, 717, *780, 886, 913, 1043, 1143, 1189
Breast (M)	296
Colon	*56, 498, 572, 886, 1143
Female Genital	886
Gastrointestinal	231, 633, 1189
Genitourinary	1189
Leukaemia	231, 913
Liver	572
Lung	*56, 322, 572, 650, 913, 1143
Melanoma	231, 988
Oesophagus	*56
Ovary	633, 717, 886
Pancreas	1143
Prostate	1143, *1144
Rectum	498, 572, 1143
Stomach	*56, 322, 572, 650
Testis	215, 231
Thyroid	376, 913
Uterus (Cervix)	231, 633, 650, 717, 886
Uterus (Corpus)	231, 633, 717, 886, 1043

Pigmentation
All Sites	*101
Benign Tumours	27
Melanoma	27, *101, 304, 468, 622
Skin	500, *769, *901

Plants
All Sites	*543
Gastrointestinal	542
Nasopharynx	580

Plasma
All Sites	981

Breast (F)	18, *1020	
Colon	18	
Oesophagus	357	
Prostate	18	
Rectum	18	

Plastics
All Sites	86, *126, *280, *281, *306, 327, 594, 687, *692, *695, *696, *698, 713, 756, *801, 868, 920, *945, 951, 956, 1097, 1098, 1114, 1158
Angiosarcoma	677
Bladder	66, *306, 677, 910
Brain	*34, 260, 910, 1029, 1041
Childhood Neoplasms	218
Colon	66, 964, *974, *976, 1030
Gastrointestinal	1029, 1186
Haemopoietic	910, 1029
Hodgkin's Disease	66
Inapplicable	204, 251, 942
Kidney	66, 481, 1041
Larynx	*801
Leukaemia	713, *801
Liver	66, *126, 677, *801, 1029, 1030, 1041
Lung	66, 594, 677, *801, 910, 1029, 1030, 1041, 1114, 1186
Lymphoma	481, *698, 713, *801, 910
Melanoma	66
Mesothelioma	910
Multiple Myeloma	*698
Nasal Cavity	*306, 316
Nasopharynx	868, 1098
Non-Hodgkin's Lymphoma	66
Oesophagus	66
Pancreas	66
Prostate	66, 594
Rectum	964, *974, *976
Respiratory	868
Sarcoma	910
Stomach	66, 910
Testis	66

Pneumoconiosis
All Sites	*657
Lung	*127, 323, *657

Polyps
Colon	*363, *436, 452, 639, 674, 820, *830, 943, *976, *982, *995, *1156
Gastrointestinal	*814
Rectum	*363, *436, 452, 639, 674, 820, *830, 943, *976, *995

Pregnancy
All Sites	856
Breast (F)	*392, 856
Childhood Neoplasms	715, 754, 849
Choriocarcinoma	5
Hodgkin's Disease	1194
Hydatidiform Mole	5
Leukaemia	754
Testis	215
Uterus (Cervix)	856

Premalignant Lesion
All Sites	488, 944, 1042, 1058, 1188

TERM

Benign Tumours	27, 1052, 1077, 1105
Bladder	798
Breast (F)	312, 944, 1052, 1077, 1195
Choriocarcinoma	5
Colon	4, 146, 227, 261, 262, 355, *363, 452, 565, 640, 793, *830, 896, *995, 1061, 1065, 1105
Female Genital	453
Gastrointestinal	453, *814, 896
Head and Neck	382
Hydatidiform Mole	5
Inapplicable	523
Leukaemia	944
Liver	261, 465
Lung	*132, 381, 1149, 1154
Lymphoma	453, 1083
Melanoma	17, 27, 468, 988
Multiple Myeloma	972
Myelodysplastic Syndrome	15, *268
Nasopharynx	141, 142
Oesophagus	185, 186, 261, 282, 357, 972, 1178
Oral Cavity	117, 118, *359, 382, 405, 408, 413, *426, 678, 805, 1051, 1178
Oropharynx	*359
Pancreas	972
Prostate	90, 972
Rectum	4, 7, 146, 227, 261, 355, *363, 452, 565, 640, *830, 896, *995, 1061, 1065, 1105
Respiratory	*359, 1149
Skin	17, 453, *769, *993, 1123
Stomach	135, 239, *284, 467, 503, 536, 663, *778, 845, 846
Thyroid	453
Urinary Tract	453
Uterus (Cervix)	88, 110, 197, *359, 367, 418, *419, 519, 585, 662, 665, 883, 921, *932, 1141, 1153

Prevalence

All Sites	485, 496, 999, 1003, 1130
Breast (F)	598, 906, 1003, 1130
Colon	565, 640, 1003, 1130
Inapplicable	195
Lung	1003, 1130
Melanoma	17
Odontogenic Neoplasms	2
Oesophagus	357
Prostate	448, 1003
Rectum	565, 640
Skin	17
Stomach	467, 746
Thyroid	635
Uterus (Cervix)	48, 662, 1130

Prevention

All Sites	191, 250, 885, 892, *933, 934, 999, 1130, *1167, 1188
Benign Tumours	*257
Breast (F)	*238, *257, 312, 361, *492, *515, 527, 963, 1055, 1130, 1172
Colon	355, 527, *982, 1055, 1079, 1130, 1140
Female Genital	256
Gastrointestinal	*814
Inapplicable	50, 195, 196
Leukaemia	885
Liver	335, 885, 1084
Lung	31, 175, 1130, *1131, 1154, 1172, *1173
Lymphoma	885
Male Genital	256
Melanoma	1054
Mesothelioma	31, *1131

Nasopharynx	885
Neuroblastoma	824
Oesophagus	128, 185, 282
Oral Cavity	117, 408, 413
Ovary	607
Prostate	1055
Rectum	7, 355, 527, 1055, 1079
Retinoblastoma	812
Salivary Gland	885
Skin	1055, *1146, *1147, 1150
Stomach	*284, 527, *556, 1172
Uterus (Cervix)	885, 888, 893, 1055, 1130, 1153
Uterus (Corpus)	1055

Prognosis

All Sites	191, 926
Benign Tumours	*257
Bone	170
Brain	170
Breast (F)	62, *257
Childhood Neoplasms	980
Choriocarcinoma	5
Hydatidiform Mole	5
Leukaemia	170, 641
Leukaemia (ALL)	641
Liver	170
Lymphoma	170, 641
Melanoma	*809
Nasal Cavity	*809
Nasopharynx	578
Stomach	563
Thyroid	318

Projection

All Sites	191
Breast (F)	253
Colon	253
Gastrointestinal	542
Larynx	253
Leukaemia	253
Lip	253
Lung	253
Oesophagus	253
Prostate	253
Rectum	253
Stomach	253
Uterus (Cervix)	253
Uterus (Corpus)	253

Promotion

All Sites	1111

Protein

Colon	*829
Non-Hodgkin's Lymphoma	1127
Rectum	*829

Protein Binding

All Sites	1064
Lung	1064

Numbers indexed are project numbers, asterisks identify new projects

TERM

Psychological Factors
All Sites	138, 375, *564, 706, 810
Breast (F)	138, 255, 1055
Colon	255, 1055
Lung	255
Prostate	1055
Rectum	255, 1055
Skin	255, 1055
Stomach	138
Uterus (Cervix)	1055
Uterus (Corpus)	1055

Psychosocial Factors
All Sites	139, 785, 810
Breast (F)	785, 786
Childhood Neoplasms	919
Colon	164
Haemopoietic	350, *351
Lung	785
Rectum	164
Stomach	785

Race
All Sites	885, 969
Benign Tumours	965
Bladder	*928
Breast (F)	297, 618, 1055
Colon	1055
Kidney	*1078
Leukaemia	885, *928
Liver	885
Lung	*928
Lymphoma	885
Nasopharynx	579, 580, 885
Ovary	506
Prostate	*966, 1055, *1062
Rectum	1055
Salivary Gland	885
Skin	1055
Testis	965
Uterus (Cervix)	885, 1055
Uterus (Corpus)	1055

Radiation, Ionizing
All Sites	54, *55, 76, *83, 86, 109, 130, 182, 193, 203, 209, 217, 264, 265, 330, 331, 354, *421, 439, 450, 451, 520, 521, 531, *532, *551, 702, 711, 758, *766, 773, 774, 796, 797, 815, 835, 836, 882, 913, 947, 1008, 1045, *1103, 1110
Benign Tumours	451
Bone	170, 368, 372
Brain	*25, 41, 170, 301, *551, *1107
Breast (F)	220, 913, 1009, 1045
Childhood Neoplasms	217, *373, *374, *403, 571, *621, 715, 753, 754, *826, *827, 839, 1039
Colon	164
Gastrointestinal	1196, 1197
Haemopoietic	387, 1168
Inapplicable	74, 384, 763
Kidney	372
Larynx	180
Leukaemia	54, *55, 116, 170, 179, 217, 287, *328, 331, *340, 368, *374, 387, 571, 711, 753, 754, *790, 835, 836, 839, 913, *975, *1001, 1045, 1046, 1196
Leukaemia (ALL)	*621
Leukaemia (AML)	1023

TERM

Liver	136, 170, 217, 368, 372
Lung	3, 78, *83, 93, *137, 157, 198, 264, 331, 368, 387, 714, 777, 837, *880, 884, 913, 954, 1045, 1071, 1121, 1196, 1197
Lymphoma	54, *55, 170, *340, 368, *374, 954, *975
Melanoma	1006
Myelodysplastic Syndrome	*770
Nasal Cavity	180, 1021
Nasopharynx	1021
Neuroblastoma	*374
Non-Hodgkin's Lymphoma	331, 753, *790
Oral Cavity	180
Pharynx	180
Prostate	795, *879
Rectum	164
Retinoblastoma	*530
Salivary Gland	757
Skin	514
Spleen	368
Stomach	136
Testis	749, 840
Thyroid	376, *730, 744, 913, 1045, 1128
Wilms' Tumour	*374

Radiation, Non-Ionizing

All Sites	913, 931, 1110
Brain	*25
Breast (F)	913
Eye	931
Leukaemia	913, *975
Lung	913
Lymphoma	*975
Skin	931
Thyroid	913

Radiation, Ultraviolet

All Sites	439, 739
Benign Tumours	27
Bone	739
Breast (F)	739
Childhood Neoplasms	739
Gastrointestinal	739
Leukaemia	739
Lip	*99, 1022
Lymphoma	739
Melanoma	6, *14, 17, 27, 35, *290, 304, 336, 468, 599, 622, 988, 1006, 1115
Ovary	739
Skin	17, 29, 35, 500, 514, *769, *901, *993, 1123
Testis	739

Radiotherapy

All Sites	222, 330, 589, 711, *876, 913
Breast (F)	222, 223, *590, 913, 1009
Childhood Neoplasms	754, 919, 1089
Gastrointestinal	332
Haemopoietic	768
Head and Neck	*604
Leukaemia	221, 223, *224, 332, 589, 642, 711, 754, 804, 848, 913
Leukaemia (ALL)	221, 768
Leukaemia (CLL)	221, 223
Lung	222, 332, 589, 913
Myelodysplastic Syndrome	589, 768
Non-Hodgkin's Lymphoma	332
Testis	*876

Numbers indexed are project numbers, asterisks identify new projects

TERM

Thyroid	913
Uterus (Cervix)	222
Uterus (Corpus)	221

Record Linkage

All Sites	13, 61, 79, *82, 86, 87, 109, 120, *225, 242, 247, 483, 979, 1073, 1109
Bladder	105, 233
Brain	*82
Breast (F)	106, 233, *393, 917
Childhood Neoplasms	838, 839
Colon	106, 734, 917
Gastrointestinal	1196
Haemopoietic	1109
Kaposi's Sarcoma	979
Kidney	106
Leukaemia	*82, 483, 839, 1196
Lung	105, 233, 917, 1196
Lymphoma	483, 1109
Multiple Myeloma	*82
Non-Hodgkin's Lymphoma	*82
Ovary	106, 917
Pancreas	105, 106
Prostate	*82, 233
Rectum	106, 734
Retinoblastoma	838
Sarcoma	79
Skin	1109
Soft Tissue	483
Stomach	550
Uterus (Cervix)	233
Uterus (Corpus)	106, 917

Recurrence

Breast (F)	*492
Leukaemia	642
Skin	*772

Red Cells

Breast (F)	18
Colon	18
Prostate	18
Rectum	18

Registry

All Sites	1, 13, 28, *55, 60, 61, 63, 76, 77, *85, 86, 113, 120, 166, 174, 183, 192, 203, 205, 209, *216, 217, 219, 222, *225, 228, 242, 243, 247, 248, 249, 250, *270, 277, *280, *281, 286, 294, 295, 298, *310, *334, 344, 366, *412, 430, 431, 432, 443, 445, 451, 471, 478, 485, *504, *505, 531, 533, *564, 576, *591, 594, 596, 606, 617, 624, 628, 636, 650, *658, 661, 675, *690, *691, *692, *693, *695, *698, *705, 706, 709, 711, 718, 739, *761, 797, *801, 818, 825, 832, 833, 866, 869, 875, *876, 882, 891, 979, 989, 1007, 1043, 1080, *1108, 1109, 1113, *1138, *1164, 1165, *1166, *1167
Benign Tumours	*257, 451
Bladder	105, 208, 233, 718, 833, 910, 971
Bone	170, 739
Brain	*19, *25, 41, *49, 170, 208, 254, 260, *504, 704, 897, 910, *924, 1024, 1025, *1107
Breast (F)	*20, *22, *52, *85, *119, 169, 220, 222, 223, 233, 237, *238, 243, 244, 255, *257, 274, 389, *391, *392, *393, 407, 416, 441, 442, 482, *505, *590, 595, 606, *611, 613, *626, 650, 708, 717, 723, 736, 738, 739, 786, 813, 825, 855, 857, 917, 1009, 1019, *1020, 1043, *1148, 1193, 1195

TERM

Breast (M)	206
Childhood Neoplasms	217, 218, 307, *319, *373, *374, 561, 571, *620, *621, 704, 707, 715, 739, 754, *826, *827, 831, 838, 839, 854, 996, *1106
Colon	24, *85, 164, *165, *211, 214, 227, 243, 244, 255, 261, 263, 441, 443, 494, *505, 595, 613, 616, 634, 689, 718, 734, 791, 896, 917, 971, 983, 987, 1036, *1122, 1140
Eye	971
Gastrointestinal	208, 344, 739, 855, 859, 896
Haemopoietic	208, *684, 910, 1109, *1129
Head and Neck	441
Hodgkin's Disease	36, 254, 462, 1194
Hypopharynx	*411
Inapplicable	50
Kaposi's Sarcoma	979
Kidney	42, 481, *722, 729, *1078
Larynx	51, 180, 232, *411, *637, *716, 741, *801, 1135
Leukaemia	28, *49, *55, 170, 179, 201, 217, 221, 223, *224, 254, 287, 364, *374, 441, 571, 605, 629, 641, *693, 704, 711, 739, 754, *790, *801, 804, 831, 833, 839, 854, *1106
Leukaemia (ALL)	221, 605, *621, 641
Leukaemia (CLL)	36, 221, 223
Lip	*99
Liver	170, 217, 261, 335, *637, *801, 873, 971, *1106
Lung	3, *32, 105, *121, *212, 222, 232, 233, 243, 244, 254, 255, *270, 295, *411, 437, 443, 472, 577, 594, 595, 613, 650, *684, 689, *697, 714, 718, *722, 741, 777, 791, *792, *801, 866, 907, 910, 917, 971, 1000, *1129, *1131
Lymphoma	12, *49, *55, 170, 249, *374, 441, 481, 628, 641, *684, *693, *698, 739, *801, 910, 1010, 1109, *1129
Male Genital	208
Melanoma	8, 17, 35, 192, 971, 988
Mesothelioma	26, 37, 51, 689, *784, 910, 971, *1131
Multiple Myeloma	*698, 972
Nasal Cavity	180, 201
Nasopharynx	51, 178
Neuroblastoma	*49, *374, 824, 854, *1050
Non-Hodgkin's Lymphoma	36, 462, *504, *591, *790, *1106
Oesophagus	*165, 261, *637, 672, *716, 971, 972, *1129, 1135
Oral Cavity	180, 200, *411, *637, *716, 741, *1026, *1027, *1132, 1135
Oropharynx	*411, *1132
Ovary	108, 717, 738, 739, 917, *1139
Pancreas	105, *165, 246, 672, 972, *1106
Pharynx	180, *637, *716, 741, *1026, *1027
Prostate	233, 243, 244, 254, *259, 324, 594, 595, *601, 972, *1129, *1144
Rectum	24, 164, *165, *211, 214, 227, 243, 244, 255, 261, 263, 494, 595, 613, 634, 689, 718, 734, 791, 896, 971, 983, 987, 1036
Respiratory	201, 208, 344, 683
Retinoblastoma	838
Salivary Gland	971
Sarcoma	12, 208, 249, *591, 825, 910
Skin	11, 17, 35, 192, 208, 255, 514, *772, 1109
Soft Tissue	28, *504, *591, 628, 825, *1106
Spinal Cord	*19
Stomach	240, 243, 244, *270, 388, *505, 544, 550, 595, 613, 650, 672, 683, *684, 689, *697, 718, 791, *792, 910, 971, *1116, 1135
Testis	108, 215, 739, *767, *876
Thyroid	*362, 441, 635, 725, *730, 744
Tongue	*411
Urinary Tract	*759
Uterus (Cervix)	210, 222, 233, 243, 258, 650, 717
Uterus (Corpus)	221, 717, 745, 857, 917, 1043, 1125
Wilms' Tumour	*374

Religion

All Sites	*412, 443, 1007, 1011
Breast (F)	416

TERM

Colon	443
Lung	443
Ovary	506, 1191

Reproductive Factors

All Sites	86, 205, 439, 517, 596, 802
Brain	94
Breast (F)	*22, *155, 297, 303, 361, 389, *409, *424, 482, 534, 588, 618, 623, 819, 915, 918, 937, 1009, *1048, 1067, 1133, *1174, 1181
Breast (M)	296
Colon	*148, 164, *165, 303, 915, 987, 1036, *1122
Gallbladder	643, 1017
Kidney	42, 729
Leukaemia	1046
Lung	907, 915
Multiple Myeloma	94
Non-Hodgkin's Lymphoma	94
Oesophagus	*165
Oral Cavity	*359
Oropharynx	*359
Ovary	506, 915, 937, 1069, *1174
Pancreas	*165, 937
Prostate	937
Rectum	*148, 164, *165, 303, 987, 1036
Respiratory	*359
Sarcoma	94
Soft Tissue	94
Stomach	517, 526
Thyroid	358
Uterus (Cervix)	154, *359, 517, *581, 619, 888, *1174
Uterus (Corpus)	745, 914, 915, 937, 984, 1125, *1174
Vagina	916
Vulva	916

Resins

Gastrointestinal	1186
Lung	1186

RFLP

All Sites	557
Breast (F)	1013
Colon	45
Lymphoma	1083
Melanoma	43
Rectum	45

Rubber

All Sites	1177
Brain	1024
Childhood Neoplasms	218
Colon	1030
Gastrointestinal	1186
Kidney	481
Liver	1030
Lung	*342, 1030, 1186
Lymphoma	481
Mesothelioma	*342

Rural

Bile Duct	*123
Breast (F)	649

Colon	983
Gallbladder	*123
Hodgkin's Disease	462
Non-Hodgkin's Lymphoma	462
Rectum	983
Soft Tissue	509
Uterus (Cervix)	649
Uterus (Corpus)	649

Saliva
All Sites	1094

SCE
All Sites	54, 545, *694, 1064
Angiosarcoma	677
Bladder	614, 677
Inapplicable	*343, 383, 656, 942
Leukaemia	54, 958
Liver	677
Lung	614, 677, 1064
Lymphoma	54
Oral Cavity	394
Testis	385

Screening
All Sites	517, 633, 739, 960, 999, 1057, 1130
Benign Tumours	997
Bile Duct	*900
Bladder	960, 1034
Bone	739
Breast (F)	*10, *52, 70, 106, 111, 172, 274, *289, 371, 473, 598, 600, 602, 609, *626, 633, 723, 739, 786, 877, 878, 890, 894, 997, 1055, 1060, 1063, 1130, 1169, 1190, 1192
Childhood Neoplasms	739
Colon	24, 106, 146, *147, 263, 369, 371, *436, 640, 679, 828, *900, 964, *976, 1044, 1055, 1059, 1060, 1079, 1130, *1142, *1145
Female Genital	256
Gastrointestinal	633, 739
Head and Neck	382
Inapplicable	50
Kidney	106, 1034
Leukaemia	739
Liver	172, 668, 747
Lung	172, *194, 381, 567, 1130
Lymphoma	739
Male Genital	256
Melanoma	638
Nasopharynx	579
Neuroblastoma	562, 824, *1050
Oesophagus	186
Oral Cavity	382, 1051
Ovary	106, 607, 633, 739, 800
Pancreas	106, *900
Prostate	371, *966, 1055
Rectum	24, 106, 146, *147, 172, 263, 369, 371, *436, 640, 679, 828, *900, 964, *976, 1044, 1055, 1059, 1060, 1079, *1142
Skin	371, 1055
Stomach	172, *285, 517, 518, *900, 967
Testis	739
Thyroid	314
Uterus (Cervix)	*9, 48, 112, 172, 177, 258, 371, 418, 517, 603, 633, *748, 888, 889, 890, 893, 921, 1055, 1060, 1130
Uterus (Corpus)	106, 308, 633, 1055

TERM

Vagina	916
Vulva	916

Seasonality
Testis	202

Segregation Analysis
Breast (F)	*144, 169, 990, 1013
Breast (M)	990
Colon	*144
Leukaemia	874
Liver	*144
Lung	*144
Rectum	*144
Stomach	*144

Semen
All Sites	1094

Sero-Epidemiology
All Sites	213, 243
Breast (F)	243
Childhood Neoplasms	996
Colon	243
Female Genital	256
Hodgkin's Disease	213
Kaposi's Sarcoma	377, 631, 909
Leukaemia	*549
Liver	163, 630, 908
Lung	243
Lymphoma	524, 909
Male Genital	256
Nasopharynx	141, 578, 579
Prostate	243
Rectum	243
Skin	524
Stomach	243
Uterus (Cervix)	168, 197, 243, 283, 418, 685

Serum
All Sites	243, 533, *538, *564, 636, 895, 981, 1094
Breast (F)	237, 243, *428, *1068, 1134
Colon	243, 262, *982
Gallbladder	1134
Inapplicable	204
Leukaemia	895
Liver	1134
Lung	243
Lymphoma	1010
Nasopharynx	*315
Ovary	1134
Prostate	243, 727, *1144
Rectum	243
Stomach	135, 243, *285, *778, 847
Uterus (Cervix)	197, 243, *748, 1134
Uterus (Corpus)	1134

Sex Ratio
All Sites	228, 589
Breast (F)	595

Childhood Neoplasms	*422
Colon	595, 793
Leukaemia	589, 823
Leukaemia (ALL)	823
Leukaemia (CLL)	823
Liver	668
Lung	589, 595, *776
Myelodysplastic Syndrome	589
Non-Hodgkin's Lymphoma	823
Prostate	595
Rectum	595
Stomach	595

Sexual Activity

All Sites	213, 1094
Benign Tumours	491
Breast (F)	400, 491
Female Genital	256
Hodgkin's Disease	213
Kaposi's Sarcoma	*38, *899
Lymphoma	*38
Male Genital	256
Oral Cavity	*1132
Oropharynx	*1132
Ovary	491
Penis	347
Prostate	229, *259, 324, *750, 851, *879
Stomach	563
Testis	749, 840
Uterus (Cervix)	88, 154, 168, 210, 283, 347, 395, 400, *406, *419, *425, 491, 519, 582, *593, *748, 760, 799, 888, 903, *1124, 1153
Uterus (Corpus)	491
Vagina	916
Vulva	347, 491, 916

Sexually Transmitted Diseases

Kaposi's Sarcoma	*38
Lymphoma	*38
Oral Cavity	*1132
Oropharynx	*1132
Prostate	324, *750, 851
Uterus (Cervix)	48, 283, 367, 662, 665, 685, *748, 883, 903, *1124, 1141
Vagina	916
Vulva	916

Sib

Childhood Neoplasms	754, 838, 919
Leukaemia	*417, 754
Lymphoma	*417
Retinoblastoma	838

Silicosis

All Sites	*657
Lung	458, *657
Stomach	458

Socio-Economic Factors

All Sites	*44, 87, 247, *412, 443, 572, 775, 785, 802, *811, 926, 1011, 1074
Bile Duct	*123
Bladder	66, 1034
Breast (F)	125, 255, *399, 400, 416, *424, 482, 731, 785, 1055, 1133

TERM

Childhood Neoplasms	839, 980, 1039
Colon	66, 125, 255, 443, 572, 1055
Gallbladder	*123
Hodgkin's Disease	66, 1093
Kaposi's Sarcoma	377
Kidney	66, 529, 1034
Leukaemia	116, 125, *417, 823, 839
Leukaemia (ALL)	823
Leukaemia (CLL)	823
Liver	66, 125, 136, 378, 572
Lung	66, 125, 255, 379, 443, 572, 785
Lymphoma	*417
Melanoma	17, 66
Multiple Myeloma	972
Nasopharynx	125
Non-Hodgkin's Lymphoma	66, 823
Oesophagus	66, 125, 186, 972
Oral Cavity	*359
Oropharynx	*359
Ovary	396, *398, 1069
Pancreas	66, 972
Penis	347
Prostate	66, 972, 1055
Rectum	125, 255, 572, 1055
Respiratory	*359
Skin	17, 255, 1055
Stomach	66, 125, 136, 572, 651, 728, 785, 862
Testis	66
Thyroid	376
Uterus (Cervix)	154, 347, *359, 400, *425, 582, 619, 883, *962, 1055
Uterus (Corpus)	1055
Vulva	347
Wilms' Tumour	65, 1099

Soil

All Sites	193, *658
Haemopoietic	47
Stomach	746

Solvents

All Sites	266, *306, 474, *475, 702, *801, 920, 948, 951, 1096
Bladder	66, *306, 910, 949, 1096
Brain	*34, 910, 1029, 1041, 1088, 1090, *1107
Childhood Neoplasms	218, *1085, 1090, *1106
Colon	66, 1030
Gastrointestinal	1029
Haemopoietic	647, 910, 1029
Hodgkin's Disease	66, 513
Inapplicable	*343
Kidney	66, 481, 1041
Larynx	655, *801
Leukaemia	134, *328, *340, 510, 513, *627, 781, *801, *887, *975, 1046, *1106
Leukaemia (ALL)	*72, *279
Leukaemia (ANLL)	1047
Liver	66, *801, 1029, 1030, 1041, *1106
Lung	66, *801, 910, 1029, 1030, 1041
Lymphoma	*340, 481, *801, 910, *975
Melanoma	66
Mesothelioma	910
Multiple Myeloma	513
Myelodysplastic Syndrome	510
Nasal Cavity	*306
Non-Hodgkin's Lymphoma	66, 513, *1106
Oesophagus	66
Pancreas	66, *1106

Prostate	66, *601	
Sarcoma	910	
Soft Tissue	*1106	
Stomach	66, 910	
Testis	66, 840	

Sputum

Bladder	614
Lung	199, *457, 567, 614, 1154

Stage

Breast (F)	172, 441, 442, 1055
Colon	441, 1055
Head and Neck	441
Larynx	51, *360
Leukaemia	441
Liver	172
Lung	172
Lymphoma	441
Melanoma	345
Mesothelioma	51
Nasopharynx	51
Neuroblastoma	*1100
Ovary	506, 800
Prostate	90, 1055
Rectum	172, 1055
Skin	1055
Stomach	172
Thyroid	441
Uterus (Cervix)	172, 1055
Uterus (Corpus)	1055

Stress

All Sites	375, 557, 785, 810, *1176
Breast (F)	785, 786
Haemopoietic	350, *351
Lung	785
Stomach	785

Surgery

Bile Duct	*900
Breast (F)	917
Colon	*900, 917
Lung	917
Ovary	917
Pancreas	*900
Rectum	*900
Stomach	*900
Uterus (Corpus)	917

Survival

All Sites	174, 276, 330, 520, *532, 606, *657, 726, 802, 853, 891, 926
Brain	*19
Breast (F)	*20, 62, 172, 407, 441, 598, 606, 609, *611, 717
Breast (M)	206
Childhood Neoplasms	*373, *422, 849, 850, 980
Colon	261, 441
Head and Neck	441
Hodgkin's Disease	511
Leukaemia	441, 641, 669, 823, 848
Leukaemia (ALL)	511, 641, *779, 823

TERM

Leukaemia (AML)	*779
Leukaemia (ANLL)	511
Leukaemia (CLL)	823
Leukaemia (CML)	*779
Liver	172, 261
Lung	172, *657
Lymphoma	441, 641
Neuroblastoma	511, 824
Non-Hodgkin's Lymphoma	511, 823
Oesophagus	261
Oral Cavity	23
Ovary	506, 717
Rectum	172, 261
Spinal Cord	*19
Stomach	172
Thyroid	318, 441, 635, 725
Uterus (Cervix)	172, 717
Uterus (Corpus)	717
Wilms' Tumour	511

Sweeteners

Bladder	100, 105, 311, 1184
Lung	105
Pancreas	105

Tea

Colon	*165
Kidney	*1078
Liver	161, 873
Oesophagus	*165
Pancreas	*165
Rectum	*165
Wilms' Tumour	65

Textiles

All Sites	474
Gastrointestinal	1186
Lung	1186

Time Factors

All Sites	190, 191, 228, *270, 330, 331, 471, 477, 478, 726, 737, 742, 743, 797, 835, 1015, 1120
Benign Tumours	965
Bladder	288, 489
Brain	1120
Breast (F)	190, 288, 442, 600, 737, 1009
Childhood Neoplasms	753, 755, 980, *1106
Colon	288, 639, *676, 1079, 1120
Gallbladder	*676
Gastrointestinal	*57, *73
Hodgkin's Disease	1194
Kidney	*73, 489
Larynx	*73, 489, 1163
Leukaemia	331, 641, 753, 762, 835, *1106, 1120
Leukaemia (ALL)	641
Liver	*676, *1106
Lung	*73, 157, 190, *270, 288, 322, 331, 437, 489, 648, 952, 1121
Lymphoma	641, 762
Melanoma	288
Neuroblastoma	*1100
Non-Hodgkin's Lymphoma	331, 753, *1106
Oesophagus	288, 489, *676

Oral Cavity	489, 1163	
Ovary	737	
Pancreas	489, *676, *1106	
Peritoneum	*73	
Pharynx	489, 1163	
Pleura	*73	
Prostate	288	
Rectum	288, 639, *676, 1079, 1120	
Small Intestine	*676	
Soft Tissue	*1106	
Stomach	*270, 288, 322, *676	
Testis	965	
Thyroid	725	
Uterus (Cervix)	210, 288	
Uterus (Corpus)	288, 737	

Tissue

All Sites	1094
Breast (F)	62
Childhood Neoplasms	*764
Colon	*363, 679
Gastrointestinal	*814
Lung	241, *902, *1131
Lymphoma	1010
Mesothelioma	*1131
Oral Cavity	805
Rectum	*363, 679
Stomach	135
Thyroid	314
Uterus (Cervix)	272, *932

Tobacco (Chewing)

All Sites	299
Breast (F)	1055
Colon	1055
Hypopharynx	*411
Larynx	*411, *716
Lung	*411
Oesophagus	*716, 1178
Oral Cavity	117, 118, 200, 394, 404, 405, 408, *411, 413, 678, *716, 1178
Oropharynx	397, *411
Pharynx	*716
Prostate	1055
Rectum	1055
Skin	1055
Tongue	*411
Uterus (Cervix)	1055
Uterus (Corpus)	1055

Tobacco (Smoking)

All Sites	46, 60, *85, 87, 113, 205, 236, 247, 295, 299, 320, 327, 344, 439, 443, 477, 485, 488, 495, 507, 533, 545, 552, 553, 557, 559, 572, 589, 596, 650, *654, *690, 724, 787, 818, 861, 892, 927, 929, 930, 947, 986, 989, 1008, 1011, 1064, 1074, 1114, 1130, 1165, 1185
Angiosarcoma	677
Benign Tumours	997
Bile Duct	*900
Bladder	66, 105, 311, 489, 512, 614, 677, 798, 949, 1034, 1070, 1184
Bone	170
Brain	94, 170, 897
Breast (F)	70, *85, *144, 255, 297, *409, 534, 650, 886, 929, 937, *978, 997, 1009, 1130, 1143
Breast (M)	296

TERM

Term	Projects
Childhood Neoplasms	*764
Choriocarcinoma	490
Colon	30, 66, *85, *144, *148, 164, *165, 255, *410, 443, 498, 566, 572, 674, *676, 686, 720, 886, *900, 943, 964, 1130, 1143, *1156
Female Genital	256, 453, 886
Gallbladder	643, *676
Gastrointestinal	*73, 344, 453
Head and Neck	95, 666
Hodgkin's Disease	66
Hydatidiform Mole	490
Hypopharynx	269, 317, *411, *420
Inapplicable	656, 938
Kidney	42, 66, *73, 325, 365, 489, 529, 729, 1034, 1070, *1078, *1104, 1175
Larynx	51, 64, *73, 180, 317, *360, *411, *420, 423, 489, 655, 666, *716, 741, 1070, 1163
Leukaemia	115, 170, 510, 589, *627, 1175
Lip	64, *99, 1022
Liver	66, 136, *144, 163, 170, 234, 236, 541, 566, 568, 572, 668, *676, 677, 747, 806
Lung	3, 30, 31, 46, 66, *73, 92, 105, *127, *144, *151, *153, *194, 199, *212, 241, 255, 295, 322, 326, *333, 341, 379, 381, 401, *411, 423, 443, *457, 470, 489, 499, 535, 541, 547, 553, 566, 567, 572, 584, 589, 614, 648, 650, 677, 682, 699, 714, 719, 721, 724, 732, 741, *880, 881, 884, 912, 927, 952, 954, *970, 1000, *1014, 1064, 1070, 1092, 1114, 1121, 1130, *1131, 1137, 1143, 1149, 1162, *1173, 1180
Lymphoma	170, 453, 954, 1175
Male Genital	256
Melanoma	17, 66
Mesothelioma	30, 31, 46, 51, *1131
Multiple Myeloma	94
Myelodysplastic Syndrome	115, 510, 589
Nasal Cavity	180, 316, 1021
Nasopharynx	51, 150, 178, 1021, 1136
Nervous System	1175
Neuroblastoma	1175
Non-Hodgkin's Lymphoma	66, 94
Oesophagus	66, *165, 186, 357, *410, *420, 423, 435, 489, 646, *676, *716, 1070, 1178
Oral Cavity	23, 59, 64, 118, 180, 200, 269, *359, 404, 405, 408, *411, 413, *420, 423, 489, 678, *716, 741, *1026, 1070, 1072, 1163, 1178
Oropharynx	59, 269, *359, 397, *411, *420, 1072
Ovary	886, 937
Pancreas	66, 105, *165, 246, 489, *676, *900, 937, 1070, 1143
Peritoneum	30, *73
Pharynx	64, 180, 489, *716, 741, *1026, 1163
Pleura	*73
Prostate	66, *879, 937, 1143
Rectum	*144, *148, 164, *165, 255, *410, 498, 572, 674, *676, 686, *900, 943, 964, 1143
Respiratory	236, 344, *359, 787, 1095, 1149
Retinoblastoma	1175
Salivary Gland	757
Sarcoma	94
Skin	17, 255, 453, 514
Small Intestine	*676
Soft Tissue	94
Stomach	30, 66, 135, 136, *144, 152, 184, 322, *410, 480, 526, 550, *556, 566, 572, 650, 651, 673, *676, 728, *900, 927, 1161
Testis	66, 215, 749
Thyroid	453, 688
Tongue	*411
Urinary Tract	453
Uterus (Cervix)	88, 154, 210, *359, *581, *593, 619, 650, 799, *817, 883, 886, 903, *1124, 1130
Uterus (Corpus)	566, 886, 914, 937, 984
Vagina	916

Vulva	916
Wilms' Tumour	65

Tobacco (Snuff)
All Sites	299
Hypopharynx	*420
Larynx	*420, *716
Oesophagus	*420, *716
Oral Cavity	*420, *716
Oropharynx	*420
Pharynx	*716

Toenails
All Sites	205, 929, 981
Breast (F)	482, 583, 595, *615, 929
Colon	595, 612
Larynx	1135
Lung	595
Oesophagus	1135
Oral Cavity	1135
Prostate	595, 851
Rectum	595, 612
Stomach	595, 1135

Tonsillectomy
Breast (F)	625

Trace Elements
All Sites	557, 573, *658, 981, 1111
Breast (F)	125, 447, 573, 583
Colon	125, 447, 498, 612
Larynx	1135
Leukaemia	125
Liver	125, *173, 234, 573
Lung	125
Nasopharynx	125
Oesophagus	125, 1135
Oral Cavity	1135
Prostate	573
Rectum	125, 447, 498, 612
Stomach	125, 573, 1135
Thyroid	635

Transplantation
All Sites	228, 953
Kidney	71
Lymphoma	*1038

Trauma
Brain	41, 301
Melanoma	*290
Testis	215, 749

Treatment
All Sites	109, 250, 331, 552, 644, 737, 891, 926, 1188
Benign Tumours	*257
Bone	644
Breast (F)	62, *257, 644, 737

TERM

Childhood Neoplasms	671, 850
Female Genital	256
Head and Neck	382, 666
Hodgkin's Disease	511
Kaposi's Sarcoma	*38
Larynx	51, 232, 666
Leukaemia	331, 641, 669
Leukaemia (ALL)	511, 641
Leukaemia (ANLL)	511
Lung	232, 331
Lymphoma	*38, 641, *1038
Male Genital	256
Mesothelioma	51
Nasopharynx	51
Neuroblastoma	511
Non-Hodgkin's Lymphoma	331, 511
Oral Cavity	382
Ovary	506, 737
Stomach	518
Thyroid	318
Uterus (Cervix)	644, 888, 889
Uterus (Corpus)	737
Wilms' Tumour	511

Trends

All Sites	174, 190, 192, *216, 219, 277, 606, 773, *811, *1108, 1113
Bladder	489
Brain	*19, *81
Breast (F)	*58, 190, 606, 649
Breast (M)	206
Childhood Neoplasms	*422, 854, *1106
Colon	*58, 537
Gastrointestinal	*57
Hypopharynx	*411
Kidney	489
Larynx	*411, 489
Leukaemia	287, 605, 854, *1106
Leukaemia (ALL)	605
Lip	*99
Liver	*58, *1106
Lung	*58, 190, *411, 489
Melanoma	192
Neuroblastoma	854
Non-Hodgkin's Lymphoma	*1106
Oesophagus	*58, 489, 537, 1170
Oral Cavity	*411, 489
Oropharynx	*411
Ovary	1191
Pancreas	489, *1106
Pharynx	489
Prostate	*58
Rectum	*58, 537
Skin	192, *772
Soft Tissue	*1106
Spinal Cord	*19
Stomach	*58, 537, 544
Thyroid	635
Tongue	*411
Uterus (Cervix)	*58, 154, 649
Uterus (Corpus)	649

Tubal Ligation

Breast (F)	1075
Ovary	1075

Numbers indexed are project numbers, asterisks identify new projects

Uterus (Cervix)	1075	
Uterus (Corpus)	1075	

Tumour Markers
All Sites	636, 652, *654
Breast (F)	527
Colon	527
Oral Cavity	*427
Rectum	527
Stomach	527
Uterus (Cervix)	685

Twins
All Sites	13, 242, 539, 1016
Breast (F)	231
Childhood Neoplasms	571, 838
Gastrointestinal	231
Leukaemia	231, 571
Melanoma	8, 35, 231
Retinoblastoma	838
Skin	35
Testis	231
Uterus (Cervix)	231
Uterus (Corpus)	231

Ulcerative Colitis
All Sites	752
Colon	*226, 679, 734, 752
Rectum	679, 734, 752
Small Intestine	*226

Ultrasound
Benign Tumours	965
Childhood Neoplasms	754
Leukaemia	754
Testis	965

Urban
Bile Duct	*123
Breast (F)	649
Colon	983
Gallbladder	*123
Lung	1180
Rectum	983
Uterus (Cervix)	649
Uterus (Corpus)	649

Uric Acid
Breast (F)	886
Colon	886
Female Genital	886
Ovary	886
Uterus (Cervix)	886
Uterus (Corpus)	886

Urine
All Sites	205, 293, 573, 574, 960, 1094
Bladder	512, 614, 843, 960, 1034

TERM

	Breast (F)	482, 573
	Childhood Neoplasms	*764
	Colon	355
	Inapplicable	204
	Kidney	1034
	Liver	573
	Lung	574, 614, 843
	Oesophagus	357, 843
	Ovary	608, *1139
	Prostate	573
	Rectum	355
	Stomach	135, 573, 574, 746, 843, 845

Vaccination
	All Sites	*334
	Brain	339
	Childhood Neoplasms	440
	Liver	335
	Non-Hodgkin's Lymphoma	904

Vegetables
	All Sites	443, 1007
	Colon	443, *1156
	Larynx	423
	Lung	423, 443, 547
	Oesophagus	423, 672
	Oral Cavity	423
	Pancreas	672
	Stomach	672
	Wilms' Tumour	65

Vegetarian
	Breast (F)	855
	Gastrointestinal	855

Virus
	All Sites	213, 228, 496, 885, 981, 1042
	Brain	339
	Breast (F)	937
	Burkitt's Lymphoma	645
	Childhood Neoplasms	*373, 753, 754
	Colon	*363
	Hodgkin's Disease	213
	Kaposi's Sarcoma	377
	Kidney	71
	Leukaemia	364, 555, 753, 754, 762, 885
	Liver	163, 167, 271, 554, 568, 872, 885, 908
	Lymphoma	524, 555, 762, 885
	Nasopharynx	150, 885
	Non-Hodgkin's Lymphoma	753
	Ovary	937, 1157
	Pancreas	937
	Prostate	937
	Rectum	*363
	Salivary Gland	885
	Skin	524
	Uterus (Cervix)	197, 619, 852, 885
	Uterus (Corpus)	937

Vitamins
All Sites	243, *505, 557, 573, 574, 596, 740, 818, 930, *933, 934, 981, 994, 1012, 1018, 1130
Breast (F)	*56, 125, 243, *505, 573, 583, 613, 731, 937, 1018, 1130
Childhood Neoplasms	*765
Colon	*56, 125, 243, 261, 449, 498, *505, 537, 566, 613, 740, *830, 994, 1004, 1130, 1140, *1156
Larynx	1135
Leukaemia	125
Liver	125, 261, 566, 573
Lung	31, *56, 97, 125, 243, *456, 566, 574, 613, 740, 1018, 1130, 1154, *1173, 1180
Melanoma	988, 1151
Mesothelioma	31
Nasopharynx	125
Non-Hodgkin's Lymphoma	1127
Oesophagus	*56, 125, 128, 185, 188, 261, 282, 357, 537, 672, 1135, 1178
Oral Cavity	117, *426, 1135, 1178
Ovary	937
Pancreas	672, 937
Prostate	90, 243, 573, *879, 937
Rectum	7, 125, 243, 261, 498, 537, 613, *830, 1004
Skin	1150, 1151
Stomach	*56, 125, 243, *284, *505, 537, 566, 573, 574, 613, 663, 672, 740, 1135
Thyroid	635, 744
Uterus (Cervix)	88, 243, 283, 519, 582, *816, *932, 1130, 1153
Uterus (Corpus)	566, 937, 984, 1018

Waste Dumps
Leukaemia	*340, *1001
Lymphoma	*340

Water
All Sites	*44, *80, 193, 718
Bladder	*80, 311, 386, 718, *905, 949, 1184
Bone	103
Brain	*80, 1024
Colon	*80, *145, 634, 639, 718, 983
Gastrointestinal	386
Haemopoietic	47
Kidney	*80
Leukaemia	*80
Liver	*181
Lung	718
Lymphoma	*80
Oesophagus	186
Prostate	*80, *879
Rectum	*80, *145, 634, 639, 718, 983
Stomach	*80, *181, 718, *1116

Welding
All Sites	516, 927
Gastrointestinal	968
Hodgkin's Disease	513
Leukaemia	510, 513
Liver	234
Lung	927, 968
Multiple Myeloma	513
Myelodysplastic Syndrome	510
Non-Hodgkin's Lymphoma	513, 968
Prostate	*601
Stomach	927

TERM

White Cells
All Sites	981, 1094
Breast (F)	18
Colon	18
Prostate	18
Rectum	18

Wood
All Sites	*306, 474, 920
Bladder	66, *306, 910
Brain	910
Colon	66
Haemopoietic	*684, 910
Hodgkin's Disease	66
Kidney	66, 481
Larynx	51
Leukaemia	201, *627
Liver	66
Lung	66, *684, 910
Lymphoma	481, *684, 910
Melanoma	66
Mesothelioma	51, 910
Nasal Cavity	201, 275, *306, 316
Nasopharynx	51, 1136
Non-Hodgkin's Lymphoma	66
Oesophagus	66
Pancreas	66
Prostate	66
Respiratory	201
Sarcoma	910
Stomach	66, *684, 910
Testis	66

Xeroderma Pigmentosum
Skin	1123

INDEX OF SITES

Entries in this index refer mainly to sites defined at the three-digit level of the ninth revision of the International Classification of Diseases, such as 'Lip' (ICD-9 140) etc. A number of other useful headings have been retained or added, however, such as 'Head and Neck', 'Female Genital', 'Childhood Neoplasms', etc., because many projects address such groups of malignancies, even though they do not fit a precise topographicalcategory. Certain morphological groups have been retained or introduced into this index for the same reason, such as 'Sarcoma', 'Mesothelioma','Neuroblastoma', etc. Where possible, abstracts coded to 'Leukaemia' have been recoded to the main subtype(s) of leukaemia.

General terms (e.g. 'Female Genital') are used only when specific terms cannot be assigned: the specific terms are shown beneath each general term. The entry 'Inapplicable' is used for some genetic and molecular epidemiology projects which involve subcellular phenomena rather than study of a particular cancer site.

The list of index entries below identifies all the cancer sites, types or groups in use in this Directory.

All Sites
Angiosarcoma
Appendix

Benign Tumours
Bile Duct
Bladder
Bone
Brain
Breast (F)
Breast (M)
Burkitt's Lymphoma

Childhood Neoplasms
Choriocarcinoma
Colon

Eye

Female Genital
　See also: Ovary; Placenta; Uterus (Cervix); Uterus (Corpus); Vagina; Vulva

Gallbladder
Gastrointestinal
　See also: Anus; Appendix; Bile Duct; Colon; Gallbladder; Liver; Oesophagus; Pancreas; Rectum; Small Intestine; Stomach
Genitourinary
　See also: Female Genital; Male Genital; Urinary Tract

Haemopoietic
　See also: Leukaemia; Lymphoma
Head and Neck
　See also specific sites
Hodgkin's Disease
Hydatidiform Mole
Hypopharynx

Inapplicable

Kaposi's Sarcoma
Kidney

Larynx
Leukaemia
Leukaemia (ALL)
Leukaemia (AML)
Leukaemia (ANLL)
Leukaemia (CLL)
Leukaemia (CML)
Lip
Liver
Lung

Lymphoma
　See also: Hodgkin's Disease; Lymphosarcoma; Non-Hodgkin's Lymphoma

Male Genital
　See also: Penis; Prostate; Testis
Melanoma
Mesothelioma
Multiple Myeloma
Myelodysplastic Syndrome

Nasal Cavity
Nasopharynx
Nervous System
　See also: Brain; Cranial Nerve; Spinal Cord
Neuroblastoma
Non-Hodgkin's Lymphoma

Odontogenic Neoplasms
Oesophagus
Oral Cavity
　See also: Lip; Parotid; Salivary Gland; Tongue
Oropharynx
Ovary

Pancreas
Penis
Peritoneum
Pharynx
Pleura
Prostate

Rectum
Respiratory
　See also: Larynx; Lung; Mesothelioma; Pleura
Retinoblastoma

Salivary Gland
Sarcoma
Skin
　See also: Melanoma
Small Intestine
Soft Tissue
Spinal Cord
Spleen
Stomach

Testis
Thyroid
Tongue

Urinary Tract
　See also: Bladder; Kidney; Ureter; Urethra
Uterus (Cervix)
Uterus (Corpus)

Vagina
Vulva

Wilms' Tumour

SITE

All Sites

Age	190, 228, 250, 471, 477, 485, 488, 560, 589, 661, 726, 743, 913, 1018, 1120
AIDS	213, 496, 979, 1042, 1094
Air Pollution	*44, 193, 487, 495, 507, *801
Alcohol	60, *85, 87, 113, 247, 327, 488, 495, 533, 545, 552, 557, 559, 572, 596, 650, *654, *690, 861, 891, 892, 930, 986, 989, 1011, 1185
Allergy	353, 709
Analgesics	934, 1096
Anatomical Distribution	191, 192, 661, 891
Antibodies	1057
Antioxidants	*933
Arthritis	1109
Asbestosis	*654
Asians	*761
Ataxia Telangiectasia	875, 944
Atomic Bomb	520, 521, 531, *532
Autopsy	495
Biochemical Markers	236, 573, 636, 853, 930, 999, 1064, 1130
Birth Cohort	190, 743, 969
Blacks	891, 1113
Blood	293, 573, *690, *691, 929
Blood Group	228, 633
BMB	205, 243, 293, *505, 533, *538, 557, *564, 573, 636, 680, *690, *691, 895, 929, 930, 934, 981, 1094
Caucasians	1113
Chemical Exposure	16, 28, 39, 61, *68, *80, 86, 114, 183, 193, 230, 249, 266, *281, 298, *310, 320, 321, 327, *329, 370, 464, 477, 479, 483, 495, 516, 545, 553, 594, 628, *696, *698, 701, 702, 713, *766, 821, 861, 865, 869, 891, 920, 927, 950, 951, 955, 956, 959, 1032, 1033, 1045, 1064, 1080, 1096, 1097, 1098, 1114, 1119, 1120, 1158
Chemotherapy	276, 331, 589, 712, 913, 953
Childhood	*55, 432, 450, 451, 496, 576, 711, 825
Chlorination	*80
Cholesterol	*505, 572, 818, 853
Chromosome Effects	54, 521, 545, *694, 700, 712, 1064
Classification	742
Clinical Records	46, 109, 432, 576, 920, 1073
Cluster	*801, 832, *923, 1043, *1108
Coffee	443
Congenital Abnormalities	130, 895
Contraception	856
Cosmetics	929
Crohn's Disease	752
Cytology	856
Data Resource	*334, *497, *801, 1188
Diabetes	853, 1003
Diet	*44, *53, 84, *85, *143, 205, 247, 293, 356, 439, 443, *454, 476, 485, 488, *505, 533, *538, *543, 557, 559, 560, 572, 573, 650, *690, 740, 891, 892, 926, 929, 981, 986, 989, 994, 999, 1003, 1007, 1011, 1018, 1043, 1074, 1111, 1130, 1165, *1167, 1185
DNA	521, *538, 557, 1058
DNA Adducts	276, 981, 1064
DNA Binding	1064
DNA Repair	739
Dose-Response	39, 113, 203, 264, 521, *691, *693, 710, 773, 796, 797, 835, 836, 870, 913, 920, 1008, 1045, 1155
Down's Syndrome	130
Drugs	203, 209, 217, 228, 276, 488, 552, 596, *698, 702, 712, 861, 892, 926, 929, 930, 931, *933, 934, 1018, 1073, 1094, 1109
Dusts	39, 46, 86, 236, *281, *306, 464, 474, *475, 487, 495, 516, *654, *657, *695, *698, 710, 787, 865, 866, 869, 870, 920, 927, 951, 1130
Dyes	*281, *484, 929
EBV	981
Education	113, 247, 926
Electromagnetic Fields	75, 313

Numbers indexed are project numbers, asterisks identify new projects

SITE

Environmental Factors	13, 39, *44, 84, 86, 113, 130, 193, 230, 242, 266, 293, 320, 344, 444, 464, 478, 483, 485, 495, 644, 700, 775, *801, 891, 926, 1015, 1032, 1097, *1108, 1114, *1166, *1167
Enzymes	460
Epilepsy	217
Eskimos	77, *216, 219, 885
Ethnic Group	77, 113, *216, 219, 443, 617, 986, 989
Familial Factors	366, 439, 517, 557, 596, 739, 825, 929, 944, 969, 1015, 1037, 1042, 1185
Fat	*53, 443, 1043, 1130, 1165
Female	929, *933, 977
Fertility	856
Fertilizers	79, *82, 114
Fibre	994, 1130
Fluorescent Light	*101
Fruit	443, 1007
G6PD	460, 652, *654
Gastrectomy	726
Gastric Ulcer	596
Genetic Factors	13, 366, 460, 557, 739, 944, 1037, 1042
Genetic Markers	366, 700, 739, 944, 1058, 1064
Geographic Factors	1, 63, 79, *80, 86, 190, 192, 193, 277, 286, *291, 344, 444, 471, 478, 617, 659, 660, 700, 742, 743, *801, 833, 1015, *1164, *1166, *1167
Glass	474
H. pylori	818
HBV	*334, 531
Herbicides	28, 79, *82, 249, 294, 370, *591, 628, 701, 1119
Heredity	545, 739
High-Risk Groups	84, 247, *657, 825, 875, 891, 944, 1058, 1094, 1130, 1188
Histology	192, 250, 521, 606, 891
HIV	891, 979, 1094
HLA	228
Homosexuality	213, 1094
Hormones	488, 573, 574, 737, 929, 981, 1018, *1138
HPV	981
HSV	981
HTLV	531, 891
Hygiene	999
Hysterectomy	*225
Immunodeficiency	1042
Immunologic Markers	636
Immunology	353, 557
Immunosuppression	228, 953, 1042
Indians	120
Infection	213, 496, 891
Infertility	*1138
Insecticides	79, *82, 861
Late Effects	203, 521, 882
Latency	113, 203, 451, 478, 797, 870
Leather	*306, 474, *696
Lifestyle	84, 86, *216, 219, *412, *538, 539, 545, 559, 560, *564, 572, 573, 574, 617, 740, 802, 892, 986, 1011, 1012, 1094, *1166, *1167, *1176
Linkage Analysis	*811
Lipids	443, 573, 574, 740, 929, 981, 1003
Lymphocytes	545, 557, *694, 712
Male	327
Mammography	1130
Mapping	86, 344, 444, 659, 660, *751, 1015, *1164, 1183
Marital Status	572, 802
Mate	1165
Mathematical Models	13, 743, *751, 913, 1109
Menarche	633
Menopause	633, 737, 929
Metals	86, 114, 243, 264, *270, 320, 321, 540, 553, *691, *794, 807, 920, 929, 930, 948, 951, 955, 994, 1008, 1101, 1102, 1155
Micronuclei	545, *694
Migrants	286, *291, 471, 478, 485, 573, *811, 989
Minerals	464, 787, 994

SITE

Mining	39, 86, 264
Monitoring	54, 60, *493, 773, 895, 947, 1057, 1073
Multiple Primary	192, 222, 250, 276, 330, 331, 445, 576, 589, 739, *876, 913
Mutagen	700
Mutagen Tests	1057
Mutation Rate	130
Mutation, Germinal	130
Mutation, Somatic	54, 700
Mycotoxins	895
Nutrition	*53, *85, *143, 205, 293, 356, *454, 533, 740, 1111, 1130
Obesity	*85, 1043
Occupation	16, 28, 39, 46, 54, *55, 60, 61, *68, 75, 76, *82, *83, 84, 86, *101, 107, 113, 114, *124, *126, 182, 183, 230, 236, 247, 249, 264, 266, *270, *280, *281, 294, 295, 298, *306, *310, 313, 320, 321, 327, *329, 344, 354, 370, *433, 439, 444, 474, *475, 477, 479, *484, 485, 487, 495, *504, 507, 508, 516, 533, *538, 540, *551, 553, 557, *591, 594, 624, 628, 650, 652, *654, 687, *690, *691, *692, *693, *695, *696, *698, 701, 702, *705, 710, 713, 724, 733, 756, *766, 773, 775, 787, *794, 796, 797, 807, 810, 815, 821, 861, 865, 866, 867, 868, 869, 870, 882, 892, 920, 926, 927, 929, 934, *945, 947, 948, 950, 951, 955, 956, 959, 960, 1008, 1011, 1012, 1032, 1033, 1045, 1057, 1074, 1080, 1082, 1096, 1097, 1098, 1101, 1102, *1103, 1110, 1114, 1117, 1119, 1120, 1130, 1155, 1158, 1159, 1171, *1176, 1177, 1185
Oncogenes	521
Oophorectomy	*225
Oral Contraceptives	488, 822, 856, 929
Paints	948
Parity	633, 856
Passive Smoking	87, 818, 891, 892, 977, 1064
Pedigree	739
Pesticides	28, *82, 86, 249, *281, 294, 370, 479, *504, 508, *591, 628, *696, 701, 861, 920, 951, 1119
Petroleum Products	16, *82, 86, *801, 948, 950, 1080
Photochemotherapy	931
Physical Activity	740, 920, 929, 930
Physical Factors	205, 572, 633, 650, 913, 1043
Pigmentation	*101
Plants	*543
Plasma	981
Plastics	86, *126, *280, *281, *306, 327, 594, 687, *692, *695, *696, *698, 713, 756, *801, 868, 920, *945, 951, 956, 1097, 1098, 1114, 1158
Pneumoconiosis	*657
Pregnancy	856
Premalignant Lesion	488, 944, 1042, 1058, 1188
Prevalence	485, 496, 999, 1003, 1130
Prevention	191, 250, 885, 892, *933, 934, 999, 1130, *1167, 1188
Prognosis	191, 926
Projection	191
Promotion	1111
Protein Binding	1064
Psychological Factors	138, 375, *564, 706, 810
Psychosocial Factors	139, 785, 810
Race	885, 969
Radiation, Ionizing	54, *55, 76, *83, 86, 109, 130, 182, 193, 203, 209, 217, 264, 265, 330, 331, 354, *421, 439, 450, 451, 520, 521, 531, *532, *551, 702, 711, 758, *766, 773, 774, 796, 797, 815, 835, 836, 882, 913, 947, 1008, 1045, *1103, 1110
Radiation, Non-Ionizing	913, 931, 1110
Radiation, Ultraviolet	439, 739
Radiotherapy	222, 330, 589, 711, *876, 913
Record Linkage	13, 61, 79, *82, 86, 87, 109, 120, *225, 242, 247, 483, 979, 1073, 1109

SITE

Registry	1, 13, 28, *55, 60, 61, 63, 76, 77, *85, 86, 113, 120, 166, 174, 183, 192, 203, 205, 209, *216, 217, 219, 222, *225, 228, 242, 243, 247, 248, 249, 250, *270, 277, *280, *281, 286, 294, 295, 298, *310, *334, 344, 366, *412, 430, 431, 432, 443, 445, 451, 471, 478, 485, *504, *505, 531, 533, *564, 576, *591, 594, 596, 606, 617, 624, 628, 636, 650, *658, 661, 675, *690, *691, *692, *693, *695, *698, *705, 706, 709, 711, 718, 739, *761, 797, *801, 818, 825, 832, 833, 866, 869, 875, *876, 882, 891, 979, 989, 1007, 1043, 1080, *1108, 1109, 1113, *1138, *1164, 1165, *1166, *1167
Religion	*412, 443, 1007, 1011
Reproductive Factors	86, 205, 439, 517, 596, 802
RFLP	557
Rubber	1177
Saliva	1094
SCE	54, 545, *694, 1064
Screening	517, 633, 739, 960, 999, 1057, 1130
Semen	1094
Sero-Epidemiology	213, 243
Serum	243, 533, *538, *564, 636, 895, 981, 1094
Sex Ratio	228, 589
Sexual Activity	213, 1094
Silicosis	*657
Socio-Economic Factors	*44, 87, 247, *412, 443, 572, 775, 785, 802, *811, 926, 1011, 1074
Soil	193, *658
Solvents	266, *306, 474, *475, 702, *801, 920, 948, 951, 1096
Stress	375, 557, 785, 810, *1176
Survival	174, 276, 330, 520, *532, 606, *657, 726, 802, 853, 891, 926
Textiles	474
Time Factors	190, 191, 228, *270, 330, 331, 471, 477, 478, 726, 737, 742, 743, 797, 835, 1015, 1120
Tissue	1094
Tobacco (Chewing)	299
Tobacco (Smoking)	46, 60, *85, 87, 113, 205, 236, 247, 295, 299, 320, 327, 344, 439, 443, 477, 485, 488, 495, 507, 533, 545, 552, 553, 557, 559, 572, 589, 596, 650, *654, *690, 724, 787, 818, 861, 892, 927, 929, 930, 947, 986, 989, 1008, 1011, 1064, 1074, 1114, 1130, 1165, 1185
Tobacco (Snuff)	299
Toenails	205, 929, 981
Trace Elements	557, 573, *658, 981, 1111
Transplantation	228, 953
Treatment	109, 250, 331, 552, 644, 737, 891, 926, 1188
Trends	174, 190, 192, *216, 219, 277, 606, 773, *811, *1108, 1113
Tumour Markers	636, 652, *654
Twins	13, 242, 539, 1016
Ulcerative Colitis	752
Urine	205, 293, 573, 574, 960, 1094
Vaccination	*334
Vegetables	443, 1007
Virus	213, 228, 496, 885, 981, 1042
Vitamins	243, *505, 557, 573, 574, 596, 740, 818, 930, *933, 934, 981, 994, 1012, 1018, 1130
Water	*44, *80, 193, 718
Welding	516, 927
White Cells	981, 1094
Wood	*306, 474, 920

Angiosarcoma

Biochemical Markers	677
BMB	677
Chromosome Effects	677
Dyes	677
Lymphocytes	677
Metabolism	677
Micronuclei	677
Occupation	677
Plastics	677

SITE

SCE	677
Tobacco (Smoking)	677

Appendix

Autopsy	390
Biopsy	390
Classification	390
Histology	390

Benign Tumours

Adolescence	27
Age	965
Alcohol	586
Biopsy	1052
Blood	*257
BMB	*257
Childhood	27, 451
Clinical Records	1105
Diet	491
Drugs	586, 1077
Familial Factors	586, 997
Genetic Factors	586
Histology	965, 1105
Hormones	491, 586, 587, 1077
Latency	451
Lifestyle	27
Mammography	997
Naevi	27
Oral Contraceptives	491, 586, 587, 997
Parity	491
Pigmentation	27
Premalignant Lesion	27, 1052, 1077, 1105
Prevention	*257
Prognosis	*257
Race	965
Radiation, Ionizing	451
Radiation, Ultraviolet	27
Registry	*257, 451
Screening	997
Sexual Activity	491
Time Factors	965
Tobacco (Smoking)	997
Treatment	*257
Ultrasound	965

Bile Duct

Autopsy	390
Biochemical Markers	1091
Biopsy	390
Classification	390
Diet	300, *592
Gallstones	1091
Geographic Factors	558
HBV	558
Histology	390
Hormones	300
Nutrition	*592
Occupation	*123
Parasitic Disease	558
Rural	*123
Screening	*900
Socio-Economic Factors	*123
Surgery	*900

Numbers indexed are project numbers, asterisks identify new projects

SITE

Tobacco (Smoking)	*900
Urban	*123

Bladder

Air Pollution	208
Alcohol	66, 311, 1184
Analgesics	176, 1096
Biochemical Markers	677, 957
Birth Cohort	489
Blacks	971
Blood	512
BMB	512, 677, 843
Chemical Exposure	66, *80, 100, 208, 278, 311, 910, 957, 1096
Chlorination	*80
Chromosome Effects	677
Cluster	949
Coffee	311, 1184
Diabetes	105
Diet	66, 1184
DNA	614
DNA Adducts	512
Drugs	233, 311, 489, 1070, 1184
Dusts	66, *306, *461, 843, 910
Dyes	*484, 677
Environmental Factors	66, 208, 386, 971
Familial Factors	1184
Female	971
Fertilizers	843
Genetic Factors	176, 798, 957
Geographic Factors	*80, 288, 386, 833
High-Risk Groups	834
Leather	*306
Lifestyle	105, 386, 971
Lymphocytes	614, 677
Mathematical Models	489
Metabolism	677
Metals	66, 208, *461, 910
Micronuclei	614, 677
Migrants	288
Mutagen	100
Nutrition	311
Occupation	66, 100, *159, 176, 208, 278, *306, 309, 311, *459, *461, *484, 677, 834, 843, 910, *928, 949, 957, 960, 971, 1034, 1096, 1171, 1184
Oncogenes	957
Passive Smoking	614
Pesticides	*159, 309, 910
Petroleum Products	66, 208, 910
Plastics	66, *306, 677, 910
Premalignant Lesion	798
Race	*928
Record Linkage	105, 233
Registry	105, 208, 233, 718, 833, 910, 971
SCE	614, 677
Screening	960, 1034
Socio-Economic Factors	66, 1034
Solvents	66, *306, 910, 949, 1096
Sputum	614
Sweeteners	100, 105, 311, 1184
Time Factors	288, 489
Tobacco (Smoking)	66, 105, 311, 489, 512, 614, 677, 798, 949, 1034, 1070, 1184
Trends	489
Urine	512, 614, 843, 960, 1034
Water	*80, 311, 386, 718, *905, 949, 1184
Wood	66, *306, 910

SITE

Bone
Alcohol	170
Chemical Exposure	170
DNA Repair	739
Dose-Response	368
Drugs	170, 368
Environmental Factors	644
Familial Factors	739
Genetic Factors	739
Genetic Markers	739
Heredity	739
Intra-Uterine Exposure	170
Late Effects	368
Metals	103
Multiple Primary	739
Occupation	170
Pedigree	739
Prognosis	170
Radiation, Ionizing	170, 368, 372
Radiation, Ultraviolet	739
Registry	170, 739
Screening	739
Tobacco (Smoking)	170
Treatment	644
Water	103

Brain
Age	*924, 1120
Air Pollution	208
Alcohol	170, 897, 1088, 1090
Chemical Exposure	*25, *80, 170, 208, 302, 479, 703, 910, 1002, 1029, 1033, 1041, 1088, 1119, 1120
Childhood	41, *49, 301, 897, 1024, 1025, 1088, 1090
Chlorination	*80
Classification	415
Clinical Records	*81
Cluster	1002, *1107
Coffee	897
Congenital Abnormalities	897
Cosmetics	1024
Demographic Factors	*19
Diet	302, 1024
Drugs	41, *81, 170, 301, 1024, 1088
Dusts	910
Electromagnetic Fields	*25, *34, 704, 1025, *1107
Environmental Factors	208, 301, 703, 704, 1002, 1029, *1107
Familial Factors	1090
Female	94
Fertilizers	*82
Geographic Factors	*80, *1107
Herbicides	*82, 1041, 1119
Histology	*81, 415, 1024
Insecticides	*82, 1041
Intra-Uterine Exposure	41, 170, 301, 1024, 1090
Lifestyle	703
Mapping	*1107
Metals	208, 910, 1029, 1088, 1090
Mining	*25
Occupation	*25, *34, *82, 94, 170, 208, 254, 260, 302, 309, 479, *504, *551, 910, 941, 946, 1002, 1024, 1029, 1033, 1041, 1090, 1119, 1120
Parental Occupation	*49, 897, 1088
Pesticides	*34, *82, 94, 309, 479, *504, 910, 1041, 1119
Petroleum Products	*82, 208, 910, 1029
Plastics	*34, 260, 910, 1029, 1041
Prognosis	170

SITE

Radiation, Ionizing	*25, 41, 170, 301, *551, *1107
Radiation, Non-Ionizing	*25
Record Linkage	*82
Registry	*19, *25, 41, *49, 170, 208, 254, 260, *504, 704, 897, 910, *924, 1024, 1025, *1107
Reproductive Factors	94
Rubber	1024
Solvents	*34, 910, 1029, 1041, 1088, 1090, *1107
Survival	*19
Time Factors	1120
Tobacco (Smoking)	94, 170, 897
Trauma	41, 301
Trends	*19, *81
Vaccination	339
Virus	339
Water	*80, 1024
Wood	910

Breast (F)

Abortion	*119, *392, 625, 708, 1009, *1174
Adolescence	89, 1189
Age	169, 190, 220, 231, *391, 400, 610, *611, 632, 913, 1018, 1035, 1133
Alcohol	*22, *56, *85, 255, 297, 303, *409, 482, 534, 583, 586, 588, *590, 613, 625, 650, 886, 918, 937, 1009, 1035, 1181
Antioxidants	*338, *615
Ataxia Telangiectasia	944
Biochemical Markers	237, *399, 446, 447, 527, 573, *615, 1055, 1066, 1067, 1130, 1152
Biopsy	312, 1052, *1053
Birth Cohort	190
Blacks	890
Blood	*257, 573, 841, 855, 918, 929, 1013, *1048
Blood Group	*393, 633
BMB	18, *22, 62, 237, 243, *257, *428, 482, *492, *505, 573, 583, 595, 609, *615, 819, 841, 855, 929, 1013, *1020, *1048, *1053, 1063, 1066, *1068, 1134, *1148
Breast Cysts	1195
Caucasians	906
Chemical Exposure	1045
Chemotherapy	96, *590, 913
Childhood	825
Chinese	297
Cholesterol	*505, 717, 886
Chromosome Effects	991
Cluster	906, 1043
Coffee	70
Contraception	*22, 416, 625, 856, 1134
Cosmetics	929
Cost-Benefit Analysis	274, 600
Cytology	856
Data Resource	*935
Diabetes	416, 1003
Diet	18, *56, 69, *85, 89, 106, 125, 129, *144, *155, 244, 255, 297, 303, *399, *409, *424, *428, 482, 491, *505, 534, 573, 583, 588, 595, 613, *615, 623, 650, 731, *780, 841, 855, 886, 915, 918, 929, 937, 963, 1003, 1018, 1043, *1048, 1063, 1067, *1068, 1130, *1148, 1152, 1181, 1189
DNA	*22, *428, 527, 906, 1013, *1020
DNA Repair	739, 1066, 1067
Dose-Response	220, 223, *338, 913, 1045
Drugs	233, 297, 312, 586, 588, 858, 929, *935, 1018, 1063, 1077
Dusts	1130
Dyes	929
Environmental Factors	*144, *155, 644, 906, 1013
Enzymes	1066
Familial Factors	70, *144, 237, 297, 303, 312, 389, *391, 527, 586, 588, *611, 613, 618, 731, 739, 825, 886, 906, 929, 944, 990, 997, 1013, *1020, 1067, 1133, 1143

SITE

Fat	*56, 62, *590, *611, *615, 731, *780, 936, 963, 1043, *1068, 1130, 1152
Female	106, 929
Fertility	856
Fibre	613, 731, 1130
Genetic Factors	*22, *144, *155, 231, 527, 586, *590, 739, 782, 906, 918, 944, 990, 991, 1013, *1020, *1048, 1172
Genetic Markers	*22, *144, 169, 739, 782, 906, 944, 1013
Geographic Factors	*144, 190, 288, 649
Heredity	739, 782, 1172
High-Risk Groups	172, 237, *238, 312, 389, 400, 618, 736, *771, 825, 906, 944, 991, 1130, 1172, 1190, 1195
Histology	*155, 389, 442, 598, 606, *1053
Hormones	*119, 235, 237, 297, 303, 312, *399, 482, 491, 573, 586, 587, 588, *590, 618, 623, 625, 632, 737, 819, 857, 858, 886, 915, 918, 929, *935, 1018, 1019, 1063, 1066, *1068, 1076, 1077, 1133, 1181
Hysterectomy	738
Immunosuppression	352
In Situ Carcinoma	*20, 736
Infection	352, 937
Intra-Uterine Exposure	858, 1076
Lactation	*52, 70, *119, *392, *590, 632, 1035
Late Effects	813
Latency	220, 813
Lifestyle	18, 125, 361, 573, *590, 595, 613, 906, 1143, *1174
Linkage Analysis	169, 1013, *1048
Lipids	18, 573, 929, 1003, *1148, 1152
Mammography	*10, *20, 111, 237, 274, 312, 482, 598, 602, 723, 731, *771, 877, 878, 894, 963, 997, *1068, 1130
Marital Status	400
Mathematical Models	253, 600, 602, 913
Menarche	*52, 89, 169, 297, 361, *392, *399, 416, 610, 633
Menopause	*52, 70, 129, 169, 231, *392, 416, 633, 737, 857, 915, 929
Menstruation	*22, 70, *392, *424, *590, 1009
Metals	243, 613, 731, 929, 1152
Migrants	18, 288, 573
Monitoring	172
Multiple Primary	*144, 222, 223, 441, 442, 527, *590, 739, 913
Nutrition	*85, 96, 125, 361, *428, 583, *615, 731, *780, 918, 1063, 1130, *1148, 1152, 1181, 1189
Obesity	*85, 96, *409, *611, 618, 623, 886, 1043, 1189
Occupation	*119, 235, 650, 929, 937, 1045, 1130
Oncogenes	62, 527, *590, *1053
Oophorectomy	738
Oral Contraceptives	70, 235, *392, 491, 534, 586, 587, 588, *590, 623, 625, 813, 822, 856, 918, 929, 997, 1009, 1133, 1134, 1193
Parity	*52, 70, 169, 231, *399, 400, 416, *424, 491, 610, *611, 632, 633, 856
Passive Smoking	*119, *978
Pedigree	*144, 739, 782, 906
Physical Activity	89, 255, 361, 482, 841, 886, 915, 929, 1009, 1143
Physical Factors	*22, *52, *56, 70, 89, 96, 231, 361, *399, 482, 534, 610, 623, 633, 650, 717, *780, 886, 913, 1043, 1143, 1189
Plasma	18, *1020
Pregnancy	*392, 856
Premalignant Lesion	312, 944, 1052, 1077, 1195
Prevalence	598, 906, 1003, 1130
Prevention	*238, *257, 312, 361, *492, *515, 527, 963, 1055, 1130, 1172
Prognosis	62, *257
Projection	253
Psychological Factors	138, 255, 1055
Psychosocial Factors	785, 786
Race	297, 618, 1055
Radiation, Ionizing	220, 913, 1009, 1045
Radiation, Non-Ionizing	913
Radiation, Ultraviolet	739
Radiotherapy	222, 223, *590, 913, 1009
Record Linkage	106, 233, *393, 917

SITE

Recurrence	*492
Red Cells	18
Registry	*20, *22, *52, *85, *119, 169, 220, 222, 223, 233, 237, *238, 243, 244, 255, *257, 274, 389, *391, *392, *393, 407, 416, 441, 442, 482, *505, *590, 595, 606, *611, 613, *626, 650, 708, 717, 723, 736, 738, 739, 786, 813, 825, 855, 857, 917, 1009, 1019, *1020, 1043, *1148, 1193, 1195
Religion	416
Reproductive Factors	*22, *155, 297, 303, 361, 389, *409, *424, 482, 534, 588, 618, 623, 819, 915, 918, 937, 1009, *1048, 1067, 1133, *1174, 1181
RFLP	1013
Rural	649
Screening	*10, *52, 70, 106, 111, 172, 274, *289, 371, 473, 598, 600, 602, 609, *626, 633, 723, 739, 786, 877, 878, 890, 894, 997, 1055, 1060, 1063, 1130, 1169, 1190, 1192
Segregation Analysis	*144, 169, 990, 1013
Sero-Epidemiology	243
Serum	237, 243, *428, *1068, 1134
Sex Ratio	595
Sexual Activity	400, 491
Socio-Economic Factors	125, 255, *399, 400, 416, *424, 482, 731, 785, 1055, 1133
Stage	172, 441, 442, 1055
Stress	785, 786
Surgery	917
Survival	*20, 62, 172, 407, 441, 598, 606, 609, *611, 717
Time Factors	190, 288, 442, 600, 737, 1009
Tissue	62
Tobacco (Chewing)	1055
Tobacco (Smoking)	70, *85, *144, 255, 297, *409, 534, 650, 886, 929, 937, *978, 997, 1009, 1130, 1143
Toenails	482, 583, 595, *615, 929
Tonsillectomy	625
Trace Elements	125, 447, 573, 583
Treatment	62, *257, 644, 737
Trends	*58, 190, 606, 649
Tubal Ligation	1075
Tumour Markers	527
Twins	231
Urban	649
Uric Acid	886
Urine	482, 573
Vegetarian	855
Virus	937
Vitamins	*56, 125, 243, *505, 573, 583, 613, 731, 937, 1018, 1130
White Cells	18

Breast (M)

Alcohol	296
Diet	296
Drugs	296
Familial Factors	990
Genetic Factors	990
Hormones	296
Physical Factors	296
Registry	206
Reproductive Factors	296
Segregation Analysis	990
Survival	206
Tobacco (Smoking)	296
Trends	206

Burkitt's Lymphoma

Biopsy	645
DNA	645
Immunologic Markers	645
Infection	645

SITE

Parasitic Disease 645
Virus 645

Childhood Neoplasms
Abortion 849
Adolescence 919
Alcohol *764, 1090
Biochemical Markers 996
Birth Cohort 753
Blood *764
BMB *764
Chemical Exposure *621, 1039
Chemotherapy 919, 1089
Childhood 569, 753, 831, 854, 919, 1090
Clinical Effects 754
Clinical Records *765
Cluster *373, 434, *620, 753, 839, *1106
Congenital Abnormalities 434, 571, 754, *826, *827, 849, 919, 1039, 1089
Diet *764
DNA Repair 739
Dose-Response *827
Down's Syndrome *826, *827
Drugs 217, 571, *621, 754, *764, *765, 919
Electromagnetic Fields *621, 704, 707
Environmental Factors *319, *374, 571, 704, 707, 755, *764
Epilepsy 217
Ethnic Group 434, 854
Familial Factors 307, 739, 838, 849, 996, 1039, 1090
Family History *319
Fertility 919
Genetic Factors 307, *319, *374, *403, *621, 739, 849, 996, 1089
Genetic Markers 739, 996
Geographic Factors *620, 753, 755, 831, 838, 839, 854, *1106
Herbicides 434
Heredity *403, 739
High-Risk Groups *403
Immunodeficiency 570
Immunology 440
Infection *319, 854
Infertility 849
Intra-Uterine Exposure 218, 438, 571, 715, 754, *826, 838, 919, 1090
Late Effects *373, 850, 1089
Lifestyle *374, 755
Mapping 753, *1049
Marijuana *621
Menarche 919
Menopause 919
Metals 218, *1085, 1090
Migrants 854
Monitoring *1106
Multiple Primary *373, 739, 754, 1089
Mutagen *764
Mutation, Germinal 838, 849
Mutation, Somatic 754, 1089
Occupation 218, 438, 754, *826, 946, 1039, *1085, 1090
Paints 218, *1085
Parental Occupation *374, *621, *1085
Passive Smoking *621, *764
Pedigree 739
Pesticides 434, *1085, *1106
Plastics 218
Pregnancy 715, 754, 849
Prognosis 980
Psychosocial Factors 919
Radiation, Ionizing 217, *373, *374, *403, 571, *621, 715, 753, 754, *826, *827, 839, 1039
Radiation, Ultraviolet 739

Numbers indexed are project numbers, asterisks identify new projects

SITE

Radiotherapy	754, 919, 1089
Record Linkage	838, 839
Registry	217, 218, 307, *319, *373, *374, 561, 571, *620, *621, 704, 707, 715, 739, 754, *826, *827, 831, 838, 839, 854, 996, *1106
Rubber	218
Screening	739
Sero-Epidemiology	996
Sex Ratio	*422
Sib	754, 838, 919
Socio-Economic Factors	839, 980, 1039
Solvents	218, *1085, 1090, *1106
Survival	*373, *422, 849, 850, 980
Time Factors	753, 755, 980, *1106
Tissue	*764
Tobacco (Smoking)	*764
Treatment	671, 850
Trends	*422, 854, *1106
Twins	571, 838
Ultrasound	754
Urine	*764
Vaccination	440
Virus	*373, 753, 754
Vitamins	*765

Choriocarcinoma

Abortion	490
Alcohol	490
Chemotherapy	5
Contraception	490
Diet	490
Familial Factors	490
Fertility	5
Hormones	490
Parity	490
Pregnancy	5
Premalignant Lesion	5
Prognosis	5
Tobacco (Smoking)	490

Colon

Age	1120
Alcohol	*56, 66, *85, *148, 164, *165, 255, 261, 262, 303, *410, 565, 566, 572, 613, *676, 720, 886, 1036, *1156
Anatomical Distribution	616
Autopsy	390
Bile Acid	*363
Biochemical Markers	447, 527, *830, 998, 1055, 1130
Biopsy	390, 452
Blacks	971
Blood	*363, 498, *830, 841
Blood Group	498
BMB	18, 243, 262, 355, *363, 498, *505, 595, 612, 679, *830, 841, *982, 1044
Chemical Exposure	66, *80, 686, 720, 791, 964, 1030, 1031, 1120
Chlorination	*80, 634, 639, 983
Cholesterol	498, *505, 572, *830, 886
Chromosome Effects	45
Cirrhosis	261
Classification	390
Clinical Effects	227
Clinical Records	369, 452, 1105
Coffee	443, *676
Cost-Benefit Analysis	*145, *147
Crohn's Disease	*226, 734, 752
Dairy Products	616
Data Resource	*935

SITE

Diabetes	1003
Diet	18, *56, 66, 69, *85, 106, 125, *144, *148, *149, 158, 164, *165, 214, 244, 255, 262, 273, 303, 355, *363, *410, 443, 449, 498, *505, 537, 565, 566, 572, 595, 613, 616, 639, 674, *676, 686, 720, 740, *829, 841, 886, 915, 943, 987, 994, *995, 1003, 1004, 1036, 1065, *1122, 1130, 1140, *1142, *1156, 1182
DNA	45, 527, 679
Dose-Response	689, 791
Drugs	164, 686, 720, *829, *935
Dusts	30, 66, 689, 720, 1130
Environmental Factors	4, *21, 30, 66, *144, 261, 971, 1182
Ethnic Group	443, 987
Faeces	262, 263, 355, *363
Familial Adenomatous Polyposis	4, 227, *363
Familial Factors	4, *21, 24, 45, *144, 158, 164, *165, 227, 303, 449, 452, 494, 498, 527, 613, 640, 886, 1079, 1143
Family History	*148, *149, *829
Fat	*56, 443, 537, 566, 616, *829, 936, 987, *995, 1130, 1140, *1142
Female	106, 971, 1036
Fibre	613, 616, *829, 987, 994, *995, 1130, 1140, *1142
Fruit	443
Genetic Factors	24, *144, *211, 273, *363, 527, 782, 896, *1122
Genetic Markers	*144, 782, 896
Geographic Factors	*80, *144, 261, 288, *676
Heredity	*211, *363, 494, 686, 782
High-Risk Groups	4, *21, 24, *145, *149, *211, 261, 355, *436, 640, 793, 896, 1059, 1130
Histology	164, 390, 452, 793, 896, 1105
Hormones	164, 303, 886, 915, *935
Immunosuppression	352
Infection	352, *363
Latency	791
Lifestyle	18, 125, *145, *148, 158, 262, 572, 595, 613, 740, 971, 1143, 1182
Linkage Analysis	*363
Lipids	18, 443, 498, 740, 1003
Mammography	1130
Marital Status	214, 572
Mathematical Models	253
Menopause	915
Menstruation	*165
Metals	66, 243, 613, 964, 994, 1031, 1102
Migrants	18, 288
Minerals	355, 994
Mining	30
Multiple Primary	*144, 441, 527
Mutation, Germinal	227
Nutrition	*85, 125, 164, *363, 740, *829, *995, 1130, 1140, *1142, *1156
Obesity	*85, 674, 886, 943, 987
Occupation	30, 66, 164, *165, 261, 498, 686, 689, 720, 791, *829, 943, 964, 971, *974, *976, 1030, 1031, 1102, 1120, 1130, 1140, *1142
Oncogenes	527
Parasitic Disease	*145, 146
Parity	674, 943
Pedigree	*144, *363, 782, 896
Pesticides	720
Petroleum Products	66, 791
Physical Activity	*148, 164, *165, 255, 449, 565, 674, 740, *829, 841, 886, 915, 943, 987, 1065, *1122, 1140, 1143
Physical Factors	*56, 498, 572, 886, 1143
Plasma	18
Plastics	66, 964, *974, *976, 1030
Polyps	*363, *436, 452, 639, 674, 820, *830, 943, *976, *982, *995, *1156
Premalignant Lesion	4, 146, 227, 261, 262, 355, *363, 452, 565, 640, 793, *830, 896, *995, 1061, 1065, 1105
Prevalence	565, 640, 1003, 1130
Prevention	355, 527, *982, 1055, 1079, 1130, 1140
Projection	253

SITE

Protein	*829
Psychological Factors	255, 1055
Psychosocial Factors	164
Race	1055
Radiation, Ionizing	164
Record Linkage	106, 734, 917
Red Cells	18
Registry	24, *85, 164, *165, *211, 214, 227, 243, 244, 255, 261, 263, 441, 443, 494, *505, 595, 613, 616, 634, 689, 718, 734, 791, 896, 917, 971, 983, 987, 1036, *1122, 1140
Religion	443
Reproductive Factors	*148, 164, *165, 303, 915, 987, 1036, *1122
RFLP	45
Rubber	1030
Rural	983
Screening	24, 106, 146, *147, 263, 369, 371, *436, 640, 679, 828, *900, 964, *976, 1044, 1055, 1059, 1060, 1079, 1130, *1142, *1145
Segregation Analysis	*144
Sero-Epidemiology	243
Serum	243, 262, *982
Sex Ratio	595, 793
Socio-Economic Factors	66, 125, 255, 443, 572, 1055
Solvents	66, 1030
Stage	441, 1055
Surgery	*900, 917
Survival	261, 441
Tea	*165
Time Factors	288, 639, *676, 1079, 1120
Tissue	*363, 679
Tobacco (Chewing)	1055
Tobacco (Smoking)	30, 66, *85, *144, *148, 164, *165, 255, *410, 443, 498, 566, 572, 674, *676, 686, 720, 886, *900, 943, 964, 1130, 1143, *1156
Toenails	595, 612
Trace Elements	125, 447, 498, 612
Trends	*58, 537
Tumour Markers	527
Ulcerative Colitis	*226, 679, 734, 752
Urban	983
Uric Acid	886
Urine	355
Vegetables	443, *1156
Virus	*363
Vitamins	*56, 125, 243, 261, 449, 498, *505, 537, 566, 613, 740, *830, 994, 1004, 1130, 1140, *1156
Water	*80, *145, 634, 639, 718, 983
White Cells	18
Wood	66

Eye

Blacks	971
Drugs	931
Environmental Factors	971
Female	971
Lifestyle	971
Occupation	971
Photochemotherapy	931
Radiation, Non-Ionizing	931
Registry	971

Female Genital

Alcohol	453, 886
Antibodies	256
Biopsy	256
Cholesterol	886
Condyloma	256

Diet	453, 886
DNA	256
Drugs	453, 858
Familial Factors	886
Hormones	453, 858, 886
HPV	256
Infection	256
Intra-Uterine Exposure	858
Obesity	886
Occupation	453
Oral Contraceptives	453, 822
Pesticides	453
Physical Activity	886
Physical Factors	886
Premalignant Lesion	453
Prevention	256
Screening	256
Sero-Epidemiology	256
Sexual Activity	256
Tobacco (Smoking)	256, 453, 886
Treatment	256
Uric Acid	886

Gallbladder

Alcohol	*676
Biochemical Markers	1091
BMB	1134
Coffee	*676
Contraception	1134
Diet	300, 643, *676, 1017
Drugs	1017
Environmental Factors	1017
Ethnic Group	1017
Female	1017
Gallstones	122, 643, 1091
Geographic Factors	*676
Hormones	300, 1017
Occupation	*123
Oral Contraceptives	643, 1134
Reproductive Factors	643, 1017
Rural	*123
Serum	1134
Socio-Economic Factors	*123
Time Factors	*676
Tobacco (Smoking)	643, *676
Urban	*123

Gastrointestinal

Adolescence	1189
Age	231
Air Pollution	208, 1196
Alcohol	*73, 453
Bile Acid	*814
Blood	855
Blood Group	633
BMB	*814, 855
Chemical Exposure	208, 865, 1029, 1197
Chemotherapy	332
Cluster	*57
Coal	1196
Diet	453, 476, 542, 855, 1189
DNA	*814
DNA Repair	739
Dose-Response	*73, 1196
Drugs	453, *814, 859

SITE

Dusts	67, *73, 102, 865, 1186
Environmental Factors	67, 208, 344, 386, 1029, 1196, 1197
Familial Adenomatous Polyposis	*814
Familial Factors	739
Female	67
Fertilizers	1197
Genetic Factors	231, 739, 896
Genetic Markers	739, 896
Geographic Factors	*57, 344, 386
Heredity	739
High-Risk Groups	896
Histology	*814, 896
Hormones	453
Lifestyle	386
Mapping	344
Menarche	633
Menopause	231, 633
Metals	208, 1029, 1196, 1197
Multiple Primary	332, 739
Mutagen	1197
Nutrition	1189
Obesity	1189
Occupation	*73, 102, 208, 344, 453, 865, 968, 1029, 1186, 1196, 1197
Oral Contraceptives	453
Parity	231, 633
Pedigree	739, 896
Pesticides	453
Petroleum Products	208, 968, 1029
Physical Factors	231, 633, 1189
Plants	542
Plastics	1029, 1186
Polyps	*814
Premalignant Lesion	453, *814, 896
Prevention	*814
Projection	542
Radiation, Ionizing	1196, 1197
Radiation, Ultraviolet	739
Radiotherapy	332
Record Linkage	1196
Registry	208, 344, 739, 855, 859, 896
Resins	1186
Rubber	1186
Screening	633, 739
Solvents	1029
Textiles	1186
Time Factors	*57, *73
Tissue	*814
Tobacco (Smoking)	*73, 344, 453
Trends	*57
Twins	231
Vegetarian	855
Water	386
Welding	968

Genitourinary

Adolescence	1189
Diet	1189
Metals	1101
Nutrition	1189
Obesity	1189
Occupation	1101
Physical Factors	1189

SITE

Haemopoietic
Air Pollution	208
Arthritis	1109
Chemical Exposure	16, 208, 479, 647, 910, 1029
Chemotherapy	768
Childhood	305
Drugs	1109
Dusts	647, *684, 910
Environmental Factors	208, 305, 1029
Immunodeficiency	305
Mathematical Models	1109
Metals	208, 387, 910, 1029, 1101
Multiple Primary	768
Occupation	16, 208, 305, 479, 647, *684, 910, 1029, 1101, *1129
Pesticides	479, 910
Petroleum Products	16, 208, 910, 1029
Plastics	910, 1029
Psychosocial Factors	350, *351
Radiation, Ionizing	387, 1168
Radiotherapy	768
Record Linkage	1109
Registry	208, *684, 910, 1109, *1129
Soil	47
Solvents	647, 910, 1029
Stress	350, *351
Water	47
Wood	*684, 910

Head and Neck
Alcohol	95, 666
Cost-Benefit Analysis	382
Cytology	382, 666
Diet	95
Dose-Response	*604
Histology	267, 382, 666
Multiple Primary	441
Premalignant Lesion	382
Radiotherapy	*604
Registry	441
Screening	382
Stage	441
Survival	441
Tobacco (Smoking)	95, 666
Treatment	382, 666

Hodgkin's Disease
Age	1194
AIDS	213, 455
Alcohol	66
Biochemical Markers	414
Caucasians	1126
Chemical Exposure	66, 462, 1093
Childhood	511, 1093
Cluster	1126
Coal	36
Diet	66
Dusts	66
Environmental Factors	66, 414, 462, 1093
Familial Factors	1093
Fungicides	462
Herbicides	462, 513
High-Risk Groups	455
Histology	414
HIV	455

Numbers indexed are project numbers, asterisks identify new projects

SITE

Homosexuality	213
Infection	213, 1093, 1126
Metals	66
Mining	36
Multiple Primary	511
Occupation	36, 66, 91, 254, 462, 513
Pesticides	91, 462, 513
Petroleum Products	66
Plastics	66
Pregnancy	1194
Registry	36, 254, 462, 1194
Rural	462
Sero-Epidemiology	213
Sexual Activity	213
Socio-Economic Factors	66, 1093
Solvents	66, 513
Survival	511
Time Factors	1194
Tobacco (Smoking)	66
Treatment	511
Virus	213
Welding	513
Wood	66

Hydatidiform Mole

Abortion	490
Alcohol	490
Chemotherapy	5
Contraception	490
Diet	490
Familial Factors	490
Fertility	5
Hormones	490
Parity	490
Pregnancy	5
Premalignant Lesion	5
Prognosis	5
Tobacco (Smoking)	490

Hypopharynx

Alcohol	269, 317, *420
Chemical Exposure	317
Diet	*420
Dusts	317
Female	*411
Hygiene	269, *420
Occupation	269, 317
Passive Smoking	317
Registry	*411
Tobacco (Chewing)	*411
Tobacco (Smoking)	269, 317, *411, *420
Tobacco (Snuff)	*420
Trends	*411

Inapplicable

Animal	251
Autopsy	940
Biochemical Markers	911, 940
BMB	204, 942
Chemical Exposure	50
Chemical Mutagenesis	383, 384, 938
Chemotherapy	196, 597, 656
Chromosome Effects	195, 196, 251, 383, 656

Coal	597	
Congenital Abnormalities	50, 383, 384	
Diet	985	
DNA	74, 597	
DNA Adducts	204, 940, 942	
DNA Repair	204	
Dose-Response	942	
Down's Syndrome	384	
Drugs	*343, 383, 597, 938	
Environmental Factors	384	
Ethnic Group	985	
Genetic Factors	523	
High-Risk Groups	523	
HPRT-Mutants	74	
Lymphocytes	74, 195, 196, 204, 383, 523, 763, 942	
Metals	74	
Micronuclei	204, 942	
Mining	195	
Monitoring	251, 597	
Mutagen	523, 938, 940	
Mutagen Tests	597	
Mutation, Germinal	50, 383, 384, 938	
Mutation, Somatic	50, 383, 938	
Mycotoxins	938	
Nutrition	985	
Occupation	74, 195, 196, 204, 251, *343, 597, 656, 911, 942	
Passive Smoking	940	
PCR	938	
Pesticides	911	
Petroleum Products	195	
Plastics	204, 251, 942	
Premalignant Lesion	523	
Prevalence	195	
Prevention	50, 195, 196	
Radiation, Ionizing	74, 384, 763	
Registry	50	
SCE	*343, 383, 656, 942	
Screening	50	
Serum	204	
Solvents	*343	
Tobacco (Smoking)	656, 938	
Urine	204	

Kaposi's Sarcoma

AIDS	*38, 377, 455, 803, *899, 909, 979
Antibodies	631
Antigens	501
Biopsy	803
BMB	*38, *899
Drugs	*38, 377
Environmental Factors	377
Haemophilia	377
High-Risk Groups	455
Histology	803
HIV	*38, 455, 631, 803, *899, 909, 979
HLA	501
Homosexuality	377, *899
HTLV	377
Immunology	501
Infection	377
Record Linkage	979
Registry	979
Sero-Epidemiology	377, 631, 909
Sexual Activity	*38, *899
Sexually Transmitted Diseases	*38

SITE

Socio-Economic Factors	377
Treatment	*38
Virus	377

Kidney

Alcohol	42, 66, *73, 325, 529, 729, *1078, 1175
Analgesics	42, 325, 729
Birth Cohort	489
Chemical Exposure	66, *80, 481, 808, 1031, 1041, 1081
Childhood	1175
Chlorination	*80
Clinical Records	*1078
Coffee	42, 325, 729, *1078
Diabetes	42, 729
Diet	66, 106, 365, 529, 729
Diuretics	325, 729, *1078, *1104
Dose-Response	*73
Drugs	42, 365, 489, 529, 729, 1070, *1104
Dusts	66, *73, *461, 481
Environmental Factors	66, 808
Female	106
Genetic Factors	1175
Geographic Factors	*80
Herbicides	365, 1041
High-Risk Groups	71
Hormones	42, 729, *1104
Immunosuppression	71
Infection	71
Insecticides	1041
Intra-Uterine Exposure	1175
Latency	1081
Leather	481
Lifestyle	529
Mathematical Models	489
Metals	66, *461, 481, *722, 808, 1031
Obesity	42, 365, 729
Occupation	42, 66, *73, 309, 325, 365, *461, 481, *722, 729, 808, *973, 1031, 1034, 1041, 1081, 1175
Passive Smoking	42, 729
Pesticides	309, 365, 481, 1041
Petroleum Products	66, 1081
Physical Activity	365
Plastics	66, 481, 1041
Race	*1078
Radiation, Ionizing	372
Record Linkage	106
Registry	42, 481, *722, 729, *1078
Reproductive Factors	42, 729
Rubber	481
Screening	106, 1034
Socio-Economic Factors	66, 529, 1034
Solvents	66, 481, 1041
Tea	*1078
Time Factors	*73, 489
Tobacco (Smoking)	42, 66, *73, 325, 365, 489, 529, 729, 1034, 1070, *1078, *1104, 1175
Transplantation	71
Trends	489
Urine	1034
Virus	71
Water	*80
Wood	66, 481

Larynx

Air Pollution	487, *801
Alcohol	51, 64, *73, 180, 317, *360, *420, *637, 655, 666, *716, 1135, 1163

SITE

Birth Cohort	489
BMB	*637, 1135
Chemical Exposure	317, 321, 1033
Cluster	*801
Cytology	666
Data Resource	*801
Diet	64, 180, *360, *420, 423, 655, *716, 1135
Dose-Response	*73
Drugs	489, 741, 1070
Dusts	51, *73, 317, 487, 655
Environmental Factors	*801
Enzymes	741
Female	*411
Fruit	423
Geographic Factors	*801
Histology	51, 666
Hygiene	180, *420, 1135
Mathematical Models	253, 489
Metals	51, 321, 655
Multiple Primary	64, 232, *360
Occupation	51, *73, 180, 317, 321, *360, 487, *637, 655, *716, 863, 1033
Passive Smoking	317
Pesticides	655
Petroleum Products	*801
Plastics	*801
Projection	253
Radiation, Ionizing	180
Registry	51, 180, 232, *411, *637, *716, 741, *801, 1135
Solvents	655, *801
Stage	51, *360
Time Factors	*73, 489, 1163
Tobacco (Chewing)	*411, *716
Tobacco (Smoking)	51, 64, *73, 180, 317, *360, *411, *420, 423, 489, 655, 666, *716, 741, 1070, 1163
Tobacco (Snuff)	*420, *716
Toenails	1135
Trace Elements	1135
Treatment	51, 232, 666
Trends	*411, 489
Vegetables	423
Vitamins	1135
Wood	51

Leukaemia

Age	231, 589, 605, 913, 1120
Air Pollution	*801, 1196
Alcohol	170, 1175
Animal	349
Antibodies	364
Ataxia Telangiectasia	944
Biochemical Markers	414, 958
Birth Cohort	753
Birth Order	*417
Blacks	669
BMB	895, 958
Caucasians	669
Chemical Exposure	28, *80, 116, 170, 201, 483, *627, 713, 781, *783, *790, 959, 1045, 1118, 1119, 1120
Chemotherapy	331, 332, 589, 641, 642, 669, 804, 848, 913
Childhood	*49, *55, 116, 287, 305, 364, *417, 605, 641, 642, 669, 711, 753, *790, 831, 848, 854, 1046, 1175
Chlorination	*80
Chromosome Effects	54, 510
Classification	641, 669
Clinical Effects	754
Clinical Records	*417, 641

Numbers indexed are project numbers, asterisks identify new projects

SITE

Cluster	349, 753, 762, *801, 839, *1001, *1106
Coal	1196
Congenital Abnormalities	116, 571, 754, 895
Data Resource	*801
Diet	125, 179
DNA	874
DNA Adducts	958
DNA Repair	739
Dose-Response	221, 223, *224, 368, *693, 804, 835, 836, 913, 1045, 1196
Down's Syndrome	874
Drugs	116, 170, 179, 217, *340, 368, 571, 641, 754, *783
Dusts	201, *461, *627
Electromagnetic Fields	116, *328, *627, 629, 704
Environmental Factors	305, *340, *374, 414, *417, 483, 571, 704, 762, *801, 1196
Epilepsy	217
Eskimos	885
Ethnic Group	854
Familial Factors	739, 944, 1046
Family History	*783
Fertility	642
Fertilizers	*82
Genetic Factors	231, *374, 739, 782, 944, 1046, 1175
Genetic Markers	739, 782, 944
Geographic Factors	*80, 605, 753, *801, 823, 831, 833, 839, 854, *1106
Herbicides	28, *82, *340, 513, *627, *922, 1119
Heredity	739, 782
High-Risk Groups	848, 944
Histology	179, 414
HTLV	*549, 555
Immunodeficiency	305
Infection	116, *417, 555, 854
Insecticides	*82
Intra-Uterine Exposure	116, 170, 571, 754, 1046, 1175
Late Effects	368, 642
Latency	555
Lifestyle	125, 179, 364, *374
Lymphocytes	*922
Mapping	753
Mathematical Models	253, 913
Menopause	231
Metals	387, *461, 1196
Micronuclei	958
Migrants	854
Monitoring	54, 895, *1106
Multiple Primary	221, 223, 331, 332, 441, 589, 642, 739, 754, 804, 848, 913
Mutation, Germinal	874
Mutation, Somatic	54, 754
Mycotoxins	895
Nutrition	125
Occupation	28, 54, *55, *82, 116, 134, 170, 179, 201, 254, 305, 309, *328, *340, 364, *461, 510, 513, *627, 629, *693, 713, 754, 781, *783, 823, 867, *887, *928, 941, 958, 959, *975, 1045, 1118, 1119, 1120, 1159, 1175, 1196
Paints	*627
Parental Occupation	*49, *374, *790
Parity	231
PCR	*922
Pedigree	739, 782, 874
Pesticides	28, *82, 309, *328, *340, 513, *627, *922, 1046, *1106, 1119
Petroleum Products	*82, *801
Physical Factors	231, 913
Plastics	713, *801
Pregnancy	754
Premalignant Lesion	944
Prevention	885
Prognosis	170, 641
Projection	253

Race	885, *928	
Radiation, Ionizing	54, *55, 116, 170, 179, 217, 287, *328, 331, *340, 368, *374, 387, 571, 711, 753, 754, *790, 835, 836, 839, 913, *975, *1001, 1045, 1046, 1196	
Radiation, Non-Ionizing	913, *975	
Radiation, Ultraviolet	739	
Radiotherapy	221, 223, *224, 332, 589, 642, 711, 754, 804, 848, 913	
Record Linkage	*82, 483, 839, 1196	
Recurrence	642	
Registry	28, *49, *55, 170, 179, 201, 217, 221, 223, *224, 254, 287, 364, *374, 441, 571, 605, 629, 641, *693, 704, 711, 739, 754, *790, *801, 804, 831, 833, 839, 854, *1106	
Reproductive Factors	1046	
SCE	54, 958	
Screening	739	
Segregation Analysis	874	
Sero-Epidemiology	*549	
Serum	895	
Sex Ratio	589, 823	
Sib	*417, 754	
Socio-Economic Factors	116, 125, *417, 823, 839	
Solvents	134, *328, *340, 510, 513, *627, 781, *801, *887, *975, 1046, *1106	
Stage	441	
Survival	441, 641, 669, 823, 848	
Time Factors	331, 641, 753, 762, 835, *1106, 1120	
Tobacco (Smoking)	115, 170, 510, 589, *627, 1175	
Trace Elements	125	
Treatment	331, 641, 669	
Trends	287, 605, 854, *1106	
Twins	231, 571	
Ultrasound	754	
Virus	364, 555, 753, 754, 762, 885	
Vitamins	125	
Waste Dumps	*340, *1001	
Water	*80	
Welding	510, 513	
Wood	201, *627	

Leukaemia (ALL)

Age	605
Chemical Exposure	*72, *621
Chemotherapy	641, 768
Childhood	511, 605, 641
Classification	641
Clinical Records	641
Dose-Response	221
Drugs	*621, 641
Electromagnetic Fields	*621
Genetic Factors	*621
Geographic Factors	605, *779, 823
Intra-Uterine Exposure	*72
Marijuana	*621
Multiple Primary	221, 511, 768
Occupation	*279, 823
Parental Occupation	*72, *621
Passive Smoking	*72, *621
Pesticides	*72, *279
Prognosis	641
Radiation, Ionizing	*621
Radiotherapy	221, 768
Registry	221, 605, *621, 641
Sex Ratio	823
Socio-Economic Factors	823
Solvents	*72, *279
Survival	511, 641, *779, 823
Time Factors	641

SITE

Treatment	511, 641
Trends	605

Leukaemia (AML)

Chemical Exposure	1023
Drugs	1023
Electromagnetic Fields	1023
Geographic Factors	*779
Immunosuppression	1023
Radiation, Ionizing	1023
Survival	*779

Leukaemia (ANLL)

Childhood	511, 1047
Classification	1047
Marijuana	1047
Multiple Primary	511
Pesticides	1047
Petroleum Products	1047
Solvents	1047
Survival	511
Treatment	511

Leukaemia (CLL)

Coal	36
Dose-Response	221, 223
Geographic Factors	823
Mining	36
Multiple Primary	221, 223
Occupation	36, 823
Radiotherapy	221, 223
Registry	36, 221, 223
Sex Ratio	823
Socio-Economic Factors	823
Survival	823

Leukaemia (CML)

Geographic Factors	*779
Survival	*779

Lip

Alcohol	64
Cosmetics	1022
Diet	64
Female	1022
Geographic Factors	*99
Mathematical Models	253
Multiple Primary	64
Projection	253
Radiation, Ultraviolet	*99, 1022
Registry	*99
Tobacco (Smoking)	64, *99, 1022
Trends	*99

Liver

Age	668
Air Pollution	*801
Alcohol	66, 136, 163, 170, 261, 528, 541, 566, 568, 572, *637, 668, *676, 747
Antibodies	630
Antigens	271, 630, 747, 872

SITE

Asians	1084
Biochemical Markers	236, 573, 677, 898, 1084
Biopsy	*664
Blacks	668, 971
Blood	573
BMB	335, 573, *637, 677, 1134
Chemical Exposure	66, 170, 961, 1029, 1030, 1031, 1041
Chemotherapy	1084
Chinese	*160
Cholesterol	572
Chromosome Effects	*162, 677
Cirrhosis	261, 465, 554, 568, 630, *664, 806, 908
Cluster	*801, *1106
Coffee	*676
Contraception	1134
Cost-Benefit Analysis	747
Data Resource	*801
Diet	66, 125, 136, *144, 566, 572, 573, *676, 898
DNA	1087
Dose-Response	368
Drugs	170, 217, 368
Dusts	66, 136, 236, *461
Dyes	677
Environmental Factors	66, *144, 261, 378, *801, 971, 1029, 1084
Epilepsy	217
Eskimos	885
Familial Factors	*144, 163, 378
Fat	566
Female	971
G-Banding	*162
Genetic Factors	*144
Genetic Markers	*144
Geographic Factors	*144, 261, 558, 668, *676, *801, *1106
HBV	161, 189, 335, 378, 465, 528, 558, 568, 630, *664, 668, 747, 806, 873, 898, 1084, 1087
HCV	*664
Herbicides	1041
High-Risk Groups	172, 261, 465, 747, 1084
Hormones	573, 747
Immunology	378
Infection	163, 167, 271, 378, 554, 872, 908
Insecticides	1041
Intra-Uterine Exposure	170
Late Effects	368
Lifestyle	125, 528, 572, 573, 971
Lipids	573
Liver Disease	747, 806
Lymphocytes	*162, 677
Marital Status	572
Metabolism	677, 898
Metals	66, 234, *461, 525, 1029, 1031, 1102
Micronuclei	677
Migrants	573
Monitoring	172, 898, *1106
Multiple Primary	*144
Mutagen	*181
Mycotoxins	161, 189, *664, 747, 873, 898
Nutrition	125
Occupation	66, *126, 136, 170, 234, 236, 261, *461, 525, *637, 677, 961, 971, 1029, 1030, 1031, 1041, 1102
Oncogenes	1087
Oral Contraceptives	668, 1134
Parasitic Disease	234, 558
Pedigree	*144
Pesticides	1041, *1106
Petroleum Products	66, *801, 1029
Physical Factors	572

Numbers indexed are project numbers, asterisks identify new projects

SITE

Plastics	66, *126, 677, *801, 1029, 1030, 1041
Premalignant Lesion	261, 465
Prevention	335, 885, 1084
Prognosis	170
Race	885
Radiation, Ionizing	136, 170, 217, 368, 372
Registry	170, 217, 261, 335, *637, *801, 873, 971, *1106
Rubber	1030
SCE	677
Screening	172, 668, 747
Segregation Analysis	*144
Sero-Epidemiology	163, 630, 908
Serum	1134
Sex Ratio	668
Socio-Economic Factors	66, 125, 136, 378, 572
Solvents	66, *801, 1029, 1030, 1041, *1106
Stage	172
Survival	172, 261
Tea	161, 873
Time Factors	*676, *1106
Tobacco (Smoking)	66, 136, *144, 163, 170, 234, 236, 541, 566, 568, 572, 668, *676, 677, 747, 806
Trace Elements	125, *173, 234, 573
Trends	*58, *1106
Urine	573
Vaccination	335
Virus	163, 167, 271, 554, 568, 872, 885, 908
Vitamins	125, 261, 566, 573
Water	*181
Welding	234
Wood	66

Lung

Abortion	907
Age	190, 589, 913, 1018, 1121
Air Pollution	*171, 292, 487, *776, *801, *880, 1179, 1180, *1187, 1196
Alcohol	*56, 66, *73, 255, 322, 535, 541, 566, 572, 613, 648, 650, 1179, 1180
Animal	337
Antigens	*457
Autopsy	860, *864, 871
Biochemical Markers	241, 292, 584, 677, *902, *925, 1056, 1064, 1130, 1137
Birth Cohort	190, 489
Blacks	971
Blood	*456, *457, *925
BMB	*132, 241, 243, *333, *456, *457, 595, 677, 843, *902, *925, *1014, *1131, *1173
Chemical Exposure	40, 66, 92, 198, 199, 278, 321, 322, 341, 379, 380, 401, 464, 499, 553, 594, 653, 791, 808, 865, 871, 881, 910, 912, 927, 955, 1029, 1030, 1031, 1041, 1045, 1064, 1092, 1114, 1119, 1197
Chemotherapy	331, 332, 589, 913
Chinese	379, 380
Cholesterol	97, 572
Chromosome Effects	677, 1064
Cirrhosis	653
Clinical Records	46
Cluster	*801
Coal	40, 133, *151, *171, 1196
Coffee	443
Cooking Methods	133, *151, 380
Cytology	199, 567, 1154
Data Resource	*801, *935
Diabetes	105, 1003
Diet	31, *56, 66, 97, 125, *144, *151, 244, 255, 292, 379, 380, 423, 443, *456, 535, 566, 572, 595, 613, 648, 650, 740, 907, 915, 1003, *1014, 1018, 1071, 1130, 1179, 1180
DNA	*132, 567, 614, *925

SITE

DNA Adducts	241, *457, 584, *902, 1064
DNA Binding	1064
Dose-Response	*73, 78, 175, 264, 368, 689, 699, 710, 714, 789, 791, 870, 912, 913, 952, 954, 1045, 1092, 1155, 1196
Drugs	233, 368, 489, 741, *935, 1018, 1070
Dusts	30, 31, *32, 46, 66, 67, *73, 92, *127, *156, 198, 199, *333, *342, 380, *461, 464, 469, 472, 487, 547, 577, *657, 682, *684, 689, 710, 788, 843, 860, 865, 866, 870, 871, 910, 912, 927, 1130, *1131, 1154, 1186, *1187
Dyes	677
Environmental Factors	30, 40, 66, 67, 78, *132, *144, 198, *212, 379, 464, 682, *801, 808, 837, *880, 881, 971, 1029, 1071, 1114, 1149, 1196, 1197
Enzymes	92, 653, 741
Ethnic Group	437, 443
Familial Factors	92, *144, 380, 613, 907, *970, 1000, 1143
Fat	*56, 443, 566, 1130
Female	67, 93, *132, 133, *151, 326, 379, 380, *411, 907, 971
Fertilizers	*697, 843, 1197
Fibre	613, 1130
Fibrosis	241
Fruit	423, 443
Fungicides	92
Genetic Factors	92, *144, 653, *925, *1014, 1137, 1172
Genetic Markers	*144, 653, *1014, 1064
Geographic Factors	*144, 190, 288, 429, 721, *801
Hair	*1173
Herbicides	92, 1041, 1119
Heredity	1172
High-Risk Groups	172, *194, 381, 567, *657, 1130, *1131, 1172
Histology	157, *212, 241, 380, 469, 699, 860, 1000, 1071, 1092
Hormones	574, 915, *935, 1018
HPV	*864
Hygiene	379
Immunologic Markers	469, 584
Insecticides	92, 1041
Late Effects	368
Latency	*153, 789, 791, 870, 954
Lifestyle	105, 125, 379, 572, 574, 595, 613, 740, 971, 1143
Lipids	443, 574, 740, 1003
Lymphocytes	*132, 584, 614, 677
Male	*171, 535, 721
Mammography	1130
Marital Status	572
Mathematical Models	253, 489, 913
Menopause	915
Metabolism	677, *925, *1014
Metals	3, 66, 78, 157, 198, 243, 264, *270, 321, 322, 387, *461, 525, 553, 613, *722, 777, 807, 808, 881, 884, 910, 912, 954, 955, 1029, 1031, 1155, 1196, 1197
Micronuclei	*132, 614, 677
Migrants	288, 437, 648
Minerals	199, 464, 788
Mining	3, 30, *32, 40, *127, *156, 198, 199, 264, 322, 458, *548, 721, 732, 884, 954
Monitoring	172
Multiple Primary	*144, 222, 232, 331, 332, 589, 913
Mutagen	1197
Nutrition	97, 125, *456, 740, 1130
Occupation	3, 30, 31, *32, 40, 46, 66, *73, 78, *83, 92, 104, *124, *127, *151, *156, 157, 175, 198, 199, *212, 241, 245, 252, 254, 264, *270, 278, 292, 295, 321, 322, 323, 326, *333, 341, *342, 379, 380, 401, *457, 458, *459, *461, 470, 472, 487, 499, 525, 535, 547, *548, 553, 567, 584, 594, 648, 650, 677, *684, 689, *697, 710, 714, 719, 721, *722, 724, 732, 777, 788, 789, 791, *792, 807, 808, 843, 860, *864, 865, 866, 870, 871, 881, 884, *902, 907, 910, 912, 927, *928, 952, 954, 955, 968, *970, 971, 1000, *1014, 1029, 1030, 1031, 1041, 1045, 1071, 1092, 1114, 1119, *1129, 1130, 1149, 1154, 1155, 1162, 1171, 1179, 1180, 1186, 1196, 1197
Oncogenes	241

Numbers indexed are project numbers, asterisks identify new projects

SITE

Parity	907
Passive Smoking	97, *121, *151, 292, 326, 379, 380, 401, 614, 714, *970, 1056, 1064, 1071, 1179
PCR	*864
Pedigree	*144
Pesticides	910, 1041, 1119
Petroleum Products	66, 791, *792, *801, 910, 968, 1029
Physical Activity	255, 740, 915, 1143
Physical Factors	*56, 322, 572, 650, 913, 1143
Plastics	66, 594, 677, *801, 910, 1029, 1030, 1041, 1114, 1186
Pneumoconiosis	*127, 323, *657
Premalignant Lesion	*132, 381, 1149, 1154
Prevalence	1003, 1130
Prevention	31, 175, 1130, *1131, 1154, 1172, *1173
Projection	253
Protein Binding	1064
Psychological Factors	255
Psychosocial Factors	785
Race	*928
Radiation, Ionizing	3, 78, *83, 93, *137, 157, 198, 264, 331, 368, 387, 714, 777, 837, *880, 884, 913, 954, 1045, 1071, 1121, 1196, 1197
Radiation, Non-Ionizing	913
Radiotherapy	222, 332, 589, 913
Record Linkage	105, 233, 917, 1196
Registry	3, *32, 105, *121, *212, 222, 232, 233, 243, 244, 254, 255, *270, 295, *411, 437, 443, 472, 577, 594, 595, 613, 650, *684, 689, *697, 714, 718, *722, 741, 777, 791, *792, *801, 866, 907, 910, 917, 971, 1000, *1129, *1131
Religion	443
Reproductive Factors	907, 915
Resins	1186
Rubber	*342, 1030, 1186
SCE	614, 677, 1064
Screening	172, *194, 381, 567, 1130
Segregation Analysis	*144
Sero-Epidemiology	243
Serum	243
Sex Ratio	589, 595, *776
Silicosis	458, *657
Socio-Economic Factors	66, 125, 255, 379, 443, 572, 785
Solvents	66, *801, 910, 1029, 1030, 1041
Sputum	199, *457, 567, 614, 1154
Stage	172
Stress	785
Surgery	917
Survival	172, *657
Sweeteners	105
Textiles	1186
Time Factors	*73, 157, 190, *270, 288, 322, 331, 437, 489, 648, 952, 1121
Tissue	241, *902, *1131
Tobacco (Chewing)	*411
Tobacco (Smoking)	3, 30, 31, 46, 66, *73, 92, 105, *127, *144, *151, *153, *194, 199, *212, 241, 255, 295, 322, 326, *333, 341, 379, 381, 401, *411, 423, 443, *457, 470, 489, 499, 535, 541, 547, 553, 566, 567, 572, 584, 589, 614, 648, 650, 677, 682, 699, 714, 719, 721, 724, 732, 741, *880, 881, 884, 912, 927, 952, 954, *970, 1000, *1014, 1064, 1070, 1092, 1114, 1121, 1130, *1131, 1137, 1143, 1149, 1162, *1173, 1180
Toenails	595
Trace Elements	125
Treatment	232, 331
Trends	*58, 190, *411, 489
Urban	1180
Urine	574, 614, 843
Vegetables	423, 443, 547
Vitamins	31, *56, 97, 125, 243, *456, 566, 574, 613, 740, 1018, 1130, 1154, *1173, 1180
Water	718

SITE

Welding	927, 968
Wood	66, *684, 910

Lymphoma

AIDS	*38, 909
Air Pollution	*801
Alcohol	170, 453, 1175
Arthritis	1109
Biochemical Markers	414
Birth Order	*417
BMB	*38, 1010, *1038
Chemical Exposure	12, 16, *80, 170, 249, 481, 483, 628, *698, 713, 910, 961, 1118
Chemotherapy	641
Childhood	*49, *55, *417, 641, 1175
Chlorination	*80
Chromosome Effects	54
Classification	641
Clinical Records	*417, 641
Cluster	762, *801
Data Resource	*801
Diet	453
DNA Repair	739
Dose-Response	368, *693, 954
Drugs	*38, 170, *340, 368, 453, 641, *698, 1109
Dusts	481, *684, *698, 910
EBV	1010, 1083
Environmental Factors	*340, *374, 414, *417, 483, 762, *801
Eskimos	885
Familial Factors	739
Genetic Factors	*374, 739, 782, 1175
Genetic Markers	739, 782, 1083
Geographic Factors	*80, *801
Herbicides	12, 249, *340, 628, *922
Heredity	739, 782
High-Risk Groups	1083
Histology	414, 524
HIV	*38, 909, 1010
HLA	1010
Hormones	453
HTLV	524, 555
Immunodeficiency	1083
Immunosuppression	*1038
Infection	*417, 555
Intra-Uterine Exposure	170, 1175
Late Effects	368
Latency	555, 954
Leather	481
Lifestyle	*374
Lymphocytes	*922
Mathematical Models	1109
Metals	481, 910, 954, 1101
Mining	954
Monitoring	54
Multiple Primary	441, 739
Mutation, Somatic	54
Occupation	12, 16, 54, *55, 170, 249, 309, *340, 453, *459, 481, 628, *684, *693, *698, 713, 867, 910, 954, 961, *975, 1101, 1118, *1129, 1175
Oral Contraceptives	453
Parental Occupation	*49, *374
PCR	*922, 1010
Pedigree	739, 782
Pesticides	12, 249, 309, *340, 453, 481, 628, 910, *922
Petroleum Products	16, *801, 910
Plastics	481, *698, 713, *801, 910
Premalignant Lesion	453, 1083
Prevention	885

Numbers indexed are project numbers, asterisks identify new projects

SITE

Prognosis	170, 641
Race	885
Radiation, Ionizing	54, *55, 170, *340, 368, *374, 954, *975
Radiation, Non-Ionizing	*975
Radiation, Ultraviolet	739
Record Linkage	483, 1109
Registry	12, *49, *55, 170, 249, *374, 441, 481, 628, 641, *684, *693, *698, 739, *801, 910, 1010, 1109, *1129
RFLP	1083
Rubber	481
SCE	54
Screening	739
Sero-Epidemiology	524, 909
Serum	1010
Sexual Activity	*38
Sexually Transmitted Diseases	*38
Sib	*417
Socio-Economic Factors	*417
Solvents	*340, 481, *801, 910, *975
Stage	441
Survival	441, 641
Time Factors	641, 762
Tissue	1010
Tobacco (Smoking)	170, 453, 954, 1175
Transplantation	*1038
Treatment	*38, 641, *1038
Virus	524, 555, 762, 885
Waste Dumps	*340
Water	*80
Wood	481, *684, 910

Male Genital

Air Pollution	208
Antibodies	256
Biopsy	256
Chemical Exposure	208
Condyloma	256
DNA	256
Environmental Factors	208
HPV	256
Infection	256
Metals	208
Occupation	208
Petroleum Products	208
Prevention	256
Registry	208
Screening	256
Sero-Epidemiology	256
Sexual Activity	256
Tobacco (Smoking)	256
Treatment	256

Melanoma

Adolescence	27
Age	231
Alcohol	66
Anatomical Distribution	192
Biopsy	1115
Blacks	971
BMB	*809
Chemical Exposure	66, 599, 1006
Childhood	27, 844
Diet	66, 304, 988, 1115, 1151
Dose-Response	622

SITE

Dusts	66
Environmental Factors	35, 66, *809, 844, 971, 1151
Enzymes	1151
Familial Factors	8, 35, 638
Female	971
Fluorescent Light	6, *101, 622
Genetic Factors	8, 231, 844
Genetic Markers	8, 43
Geographic Factors	17, 192, 288, *809
Heredity	43
Histology	192, 304
Hormones	304, 336, 988
Immunosuppression	336
Lifestyle	17, 27, 336, 971
Lipids	1151
Menopause	231
Metals	66, 988, 1151
Migrants	288
Multiple Primary	192
Naevi	27, *290, 304, 336, 468, 599, 638, 844, 988, 1115
Nutrition	988
Occupation	17, 66, *101, 304, 336, 599, 622, 971, 1006
Parity	231
Petroleum Products	66
Physical Factors	231, 988
Pigmentation	27, *101, 304, 468, 622
Plastics	66
Premalignant Lesion	17, 27, 468, 988
Prevalence	17
Prevention	1054
Prognosis	*809
Radiation, Ionizing	1006
Radiation, Ultraviolet	6, *14, 17, 27, 35, *290, 304, 336, 468, 599, 622, 988, 1006, 1115
Registry	8, 17, 35, 192, 971, 988
RFLP	43
Screening	638
Socio-Economic Factors	17, 66
Solvents	66
Stage	345
Time Factors	288
Tobacco (Smoking)	17, 66
Trauma	*290
Trends	192
Twins	8, 35, 231
Vitamins	988, 1151
Wood	66

Mesothelioma

Air Pollution	*784
Alcohol	51
Asbestosis	502
Autopsy	860, 871
Biopsy	735
Blacks	670, 971
BMB	*1131
Chemical Exposure	39, 346, 667, *784, 865, 871, 910
Clinical Records	46
Diet	31, 670
Dose-Response	39, 689, 870
Dusts	30, 31, 37, 39, 46, 51, 67, 102, *342, 469, 502, 546, 667, 689, 735, *784, 860, 865, 870, 871, 910, *1131
Environmental Factors	26, 30, 37, 39, 67, 346, 667, 670, *784, 971
Female	67, 971
High-Risk Groups	*1131
Histology	51, 469, 667, 860
Immunologic Markers	469

SITE

Latency	870
Lifestyle	971
Metals	51, 910
Mining	30, 39, *548
Occupation	26, 30, 31, 37, 39, 46, 51, 102, *342, 346, 502, *548, 667, 670, 689, 735, *784, 860, 865, 870, 871, 910, 971
Pesticides	910
Petroleum Products	910
Plastics	910
Prevention	31, *1131
Registry	26, 37, 51, 689, *784, 910, 971, *1131
Rubber	*342
Solvents	910
Stage	51
Tissue	*1131
Tobacco (Smoking)	30, 31, 46, 51, *1131
Treatment	51
Vitamins	31
Wood	51, 910

Multiple Myeloma

Biochemical Markers	414
Blacks	972
Chemical Exposure	*698, 1119
Diet	972
Drugs	*698
Dusts	*698
Environmental Factors	414
Familial Factors	972
Female	94
Fertilizers	*82
Geographic Factors	486
Herbicides	*82, 513, 1119
Histology	414
Hormones	972
Insecticides	*82
Occupation	*82, 91, 94, 309, 513, *698, 972, 1119
Pesticides	*82, 91, 94, 309, 513, 1119
Petroleum Products	*82
Plastics	*698
Premalignant Lesion	972
Record Linkage	*82
Registry	*698, 972
Reproductive Factors	94
Socio-Economic Factors	972
Solvents	513
Tobacco (Smoking)	94
Welding	513

Myelodysplastic Syndrome

Age	589
Chemical Exposure	15, *268, *770
Chemotherapy	589, 768
Chromosome Effects	510
Environmental Factors	15, *268
Multiple Primary	589, 768
Occupation	15, *268, 510, *770
Premalignant Lesion	15, *268
Radiation, Ionizing	*770
Radiotherapy	589, 768
Sex Ratio	589
Solvents	510
Tobacco (Smoking)	115, 510, 589
Welding	510

SITE

Nasal Cavity
Alcohol	180, 316
BMB	*809
Chemical Exposure	201, 1021
Diet	180
Dusts	201, 275, *306, 316, 1021
Dyes	*484
Environmental Factors	275, *809, 1021
Geographic Factors	*809
Histology	275
Hygiene	180
Infection	1021
Leather	*306, 316
Metals	316, 1102
Occupation	180, 201, 275, *306, 316, *484, 1021, 1102
Passive Smoking	316
Plastics	*306, 316
Prognosis	*809
Radiation, Ionizing	180, 1021
Registry	180, 201
Solvents	*306
Tobacco (Smoking)	180, 316, 1021
Wood	201, 275, *306, 316

Nasopharynx
Alcohol	51, 1136
Antibodies	141, 142, 150, 578
Biochemical Markers	578
Biopsy	142
BMB	*315
Chemical Exposure	150, 1021, 1098, 1136
Chinese	579, 580
Cluster	*402
Diet	125, 140, 178, *315, 1136
Dusts	51, 178, 1021
EBV	*315, *402, 578, 579, 580
Environmental Factors	*315, 1021, 1136
Eskimos	885
Ethnic Group	580
Genetic Factors	150
Geographic Factors	429
Histology	51
HLA	*315
Infection	150, 578, 1021
Lifestyle	125, 1136
Maghrebians	*315
Metals	51
Migrants	*315
Nutrition	125
Occupation	51, 178, 868, 1021, 1098
Passive Smoking	178
Pedigree	150
Plants	580
Plastics	868, 1098
Premalignant Lesion	141, 142
Prevention	885
Prognosis	578
Race	579, 580, 885
Radiation, Ionizing	1021
Registry	51, 178
Screening	579
Sero-Epidemiology	141, 578, 579
Serum	*315
Socio-Economic Factors	125
Stage	51

Numbers indexed are project numbers, asterisks identify new projects

SITE

Tobacco (Smoking)	51, 150, 178, 1021, 1136
Trace Elements	125
Treatment	51
Virus	150, 885
Vitamins	125
Wood	51, 1136

Nervous System

Alcohol	1175
Autopsy	390
Biopsy	390
Childhood	1175
Classification	390
Genetic Factors	1175
Histology	390
Intra-Uterine Exposure	1175
Occupation	1175
Tobacco (Smoking)	1175

Neuroblastoma

Age	*1100
Alcohol	*1100, 1175
Biochemical Markers	824
Birth Cohort	562
BMB	*1050
Childhood	*49, 511, 854, 1175
Cost-Benefit Analysis	562, 824
Diuretics	*1100
Drugs	*1100
Electromagnetic Fields	*1100
Environmental Factors	*374
Ethnic Group	854
Genetic Factors	*374, 1175
Genetic Markers	*1100
Geographic Factors	854
Infection	854
Intra-Uterine Exposure	*1100, 1175
Lifestyle	*374
Migrants	854
Multiple Primary	511
Occupation	1175
Oncogenes	*1100
Parental Occupation	*49, *374, *1100
Prevention	824
Radiation, Ionizing	*374
Registry	*49, *374, 824, 854, *1050
Screening	562, 824, *1050
Stage	*1100
Survival	511, 824
Time Factors	*1100
Tobacco (Smoking)	1175
Treatment	511
Trends	854

Non-Hodgkin's Lymphoma

AIDS	348, 455, 904
Alcohol	66
Allergy	904
Birth Cohort	753
Chemical Exposure	66, 462, *790, 1119
Chemotherapy	331, 332
Childhood	511, 753, *790
Cluster	753, *1106

Coal	36	
Diet	66, 1127	
Drugs	904	
Dusts	66	
Environmental Factors	66, 462	
Fat	1127	
Female	94	
Fertilizers	*82	
Fungicides	462	
Geographic Factors	753, 823, *1106	
Herbicides	*82, 462, 513, *591, 1119	
High-Risk Groups	455	
Histology	904	
HIV	455, 904	
Homosexuality	904	
Immunology	904	
Immunosuppression	348	
Insecticides	*82	
Lifestyle	1127	
Mapping	753	
Metals	66	
Mining	36	
Monitoring	*1106	
Multiple Primary	331, 332, 511	
Nutrition	1127	
Occupation	36, 66, *82, 91, 94, 462, *504, 513, *591, 823, 904, 968, 1119	
Parental Occupation	*790	
Pesticides	*82, 91, 94, 462, *504, 513, *591, *1106, 1119	
Petroleum Products	66, *82, 968	
Plastics	66	
Protein	1127	
Radiation, Ionizing	331, 753, *790	
Radiotherapy	332	
Record Linkage	*82	
Registry	36, 462, *504, *591, *790, *1106	
Reproductive Factors	94	
Rural	462	
Sex Ratio	823	
Socio-Economic Factors	66, 823	
Solvents	66, 513, *1106	
Survival	511, 823	
Time Factors	331, 753, *1106	
Tobacco (Smoking)	66, 94	
Treatment	331, 511	
Trends	*1106	
Vaccination	904	
Virus	753	
Vitamins	1127	
Welding	513, 968	
Wood	66	

Odontogenic Neoplasms

Adolescence	2
Childhood	2
Drugs	2
Genetic Factors	2
Prevalence	2

Oesophagus

Age	357
Alcohol	*56, 66, *165, 261, 357, *410, *420, *637, 646, *676, *716, 1135, 1178
Autopsy	390
Biochemical Markers	282
Biopsy	282, 357, 390, 1178
Birth Cohort	489

SITE

Blacks	971, 972
BMB	357, *637, 843, 1135
Chemical Exposure	66
Chinese	*160
Cirrhosis	261
Classification	390
Coffee	*676
Cytology	186, 357, 1178
Diet	*56, 66, 125, *165, 282, 357, *410, *420, 423, 435, 537, 646, 672, *676, *716, 972, 1135, 1178
DNA	128
DNA Repair	131
Drugs	489, 1070
Dusts	66, 843
Environmental Factors	66, 261, 971
Ethnic Group	1170
Familial Factors	*165, 972, 1178
Fat	*56, 537, 672
Female	971
Fertilizers	843
Fruit	423, 672
Genetic Factors	131
Geographic Factors	261, 288, *676, 1170
High-Risk Groups	131, 185, 187, 261
Histology	390, 1178
Hormones	972
Hygiene	*420, 1135
Indians	435
Lifestyle	125, 357, 435, 971
Linkage Analysis	131
Mapping	1170
Mate	646
Mathematical Models	253, 489
Menstruation	*165
Metals	66, 282
Micronuclei	282, 357
Migrants	288
Nass	1178
Nutrition	125, 672
Occupation	66, *165, 261, *637, *716, 843, 971, 972, *1129
Petroleum Products	66
Physical Activity	*165
Physical Factors	*56
Plasma	357
Plastics	66
Premalignant Lesion	185, 186, 261, 282, 357, 972, 1178
Prevalence	357
Prevention	128, 185, 282
Projection	253
Registry	*165, 261, *637, 672, *716, 971, 972, *1129, 1135
Reproductive Factors	*165
Screening	186
Socio-Economic Factors	66, 125, 186, 972
Solvents	66
Survival	261
Tea	*165
Time Factors	288, 489, *676
Tobacco (Chewing)	*716, 1178
Tobacco (Smoking)	66, *165, 186, 357, *410, *420, 423, 435, 489, 646, *676, *716, 1070, 1178
Tobacco (Snuff)	*420, *716
Toenails	1135
Trace Elements	125, 1135
Trends	*58, 489, 537, 1170
Urine	357, 843
Vegetables	423, 672
Vitamins	*56, 125, 128, 185, 188, 261, 282, 357, 537, 672, 1135, 1178

SITE

Water	186	
Wood	66	

Oral Cavity

Alcohol	23, 59, 64, 180, 200, 269, *359, *420, *637, *716, *1026, 1072, 1135, 1163, 1178
Anatomical Distribution	1072
Areca Nut	394
Autopsy	390
Betel (Chewing)	117, 404, 405, 678
Biochemical Markers	805, *1132
Biopsy	23, 390, *1027, 1178
Birth Cohort	489
BMB	*637, 805, *1027, 1135
Chemical Exposure	118, 321
Chromosome Effects	394
Classification	390
Cost-Benefit Analysis	382
Cytology	59, 382, 1178
Diet	64, 180, *420, 423, *716, *1026, 1135, 1178
Drugs	489, 741, 1070
Enzymes	741
Familial Factors	1178
Female	*411
Fruit	423
Genetic Factors	394
High-Risk Groups	59, 1051, 1072
Histology	382, 390, 1178
HPV	*359, *1027, *1132
HSV	*1132
Hygiene	180, 200, 269, *420, 1135
Immunology	*1132
Mathematical Models	489
Metals	321
Micronuclei	118
Multiple Primary	64, *1026
Mutation, Somatic	394
Nass	1178
Occupation	23, 180, 269, 321, *637, *716
PCR	805
Premalignant Lesion	117, 118, *359, 382, 405, 408, 413, *426, 678, 805, 1051, 1178
Prevention	117, 408, 413
Radiation, Ionizing	180
Registry	180, 200, *411, *637, *716, 741, *1026, *1027, *1132, 1135
Reproductive Factors	*359
SCE	394
Screening	382, 1051
Sexual Activity	*1132
Sexually Transmitted Diseases	*1132
Socio-Economic Factors	*359
Survival	23
Time Factors	489, 1163
Tissue	805
Tobacco (Chewing)	117, 118, 200, 394, 404, 405, 408, *411, 413, 678, *716, 1178
Tobacco (Smoking)	23, 59, 64, 118, 180, 200, 269, *359, 404, 405, 408, *411, 413, *420, 423, 489, 678, *716, 741, *1026, 1070, 1072, 1163, 1178
Tobacco (Snuff)	*420, *716
Toenails	1135
Trace Elements	1135
Treatment	382
Trends	*411, 489
Tumour Markers	*427
Vegetables	423
Vitamins	117, *426, 1135, 1178

SITE

Oropharynx

Alcohol	59, 269, *359, 397, *420, 1072
Anatomical Distribution	1072
Betel (Chewing)	397
Biochemical Markers	397, *1132
Cytology	59
Diet	*420
Female	*411
High-Risk Groups	59, 1072
HPV	*359, *1132
HSV	*1132
Hygiene	269, *420
Immunology	*1132
Occupation	269
Premalignant Lesion	*359
Registry	*411, *1132
Reproductive Factors	*359
Sexual Activity	*1132
Sexually Transmitted Diseases	*1132
Socio-Economic Factors	*359
Tobacco (Chewing)	397, *411
Tobacco (Smoking)	59, 269, *359, 397, *411, *420, 1072
Tobacco (Snuff)	*420
Trends	*411

Ovary

Abortion	396, *1174
Alcohol	886, 937
Biochemical Markers	506
Blood	*1139
Blood Group	633
BMB	506, 1134, *1139
Cholesterol	717, 886
Chromosome Effects	*162, 681, 991
Classification	108
Clinical Records	1069
Contraception	1134
Data Resource	*935
Diet	106, *398, 491, 886, 915, 937, *1139
DNA Repair	739
Drugs	608, *935, 1157
Dusts	506
Enzymes	*1139
Ethnic Group	1191
Familial Factors	506, 739, 800, 886, 1191
Fat	936
Female	106
G-Banding	*162
Genetic Factors	681, 739, 800, 991
Genetic Markers	739
Heredity	681, 739
High-Risk Groups	506, 991
Histology	108, 1191
Hormones	108, 396, *398, 491, 506, 607, 608, 737, 886, 915, *935, 1157
Hysterectomy	738
Immunosuppression	352
Infection	352, 937
Infertility	1157
Lifestyle	*398, *1174
Lymphocytes	*162
Menarche	396, *398, 633
Menopause	396, 633, 737, 915
Menstruation	396
Metabolism	*1139

SITE

Multiple Primary	739
Obesity	886
Occupation	108, 506, 937, 1069
Oophorectomy	738
Oral Contraceptives	491, 506, 1134, 1191
Parity	396, *398, 491, 633, 1191
Pedigree	506, 739
Physical Activity	886, 915
Physical Factors	633, 717, 886
Prevention	607
Race	506
Radiation, Ultraviolet	739
Record Linkage	106, 917
Registry	108, 717, 738, 739, 917, *1139
Religion	506, 1191
Reproductive Factors	506, 915, 937, 1069, *1174
Screening	106, 607, 633, 739, 800
Serum	1134
Sexual Activity	491
Socio-Economic Factors	396, *398, 1069
Stage	506, 800
Surgery	917
Survival	506, 717
Time Factors	737
Tobacco (Smoking)	886, 937
Treatment	506, 737
Trends	1191
Tubal Ligation	1075
Uric Acid	886
Urine	608, *1139
Virus	937, 1157
Vitamins	937

Pancreas

Alcohol	66, *165, 246, *676, 937
Autopsy	390
Biopsy	390
Birth Cohort	489
Blacks	972
Chemical Exposure	66, 246, 959, 1118
Classification	390
Cluster	*1106
Coffee	246, *676
Diabetes	105, 246
Diet	66, 106, *165, 300, *592, 672, *676, 937, 972
Drugs	489, 1070
Dusts	66
Environmental Factors	66
Familial Factors	*165, 972, 1143
Fat	672
Female	106
Fruit	672
Geographic Factors	*676, *1106
Histology	390
Hormones	300, 972
Infection	937
Lifestyle	105, 1143
Mathematical Models	489
Menstruation	*165
Metals	66
Monitoring	*1106
Nutrition	*592, 672
Occupation	66, *165, 246, 309, *459, 937, 959, 972, 1118
Pesticides	309, *1106
Petroleum Products	66
Physical Activity	*165, 1143

Numbers indexed are project numbers, asterisks identify new projects

SITE

Physical Factors	1143
Plastics	66
Premalignant Lesion	972
Record Linkage	105, 106
Registry	105, *165, 246, 672, 972, *1106
Reproductive Factors	*165, 937
Screening	106, *900
Socio-Economic Factors	66, 972
Solvents	66, *1106
Surgery	*900
Sweeteners	105
Tea	*165
Time Factors	489, *676, *1106
Tobacco (Smoking)	66, 105, *165, 246, 489, *676, *900, 937, 1070, 1143
Trends	489, *1106
Vegetables	672
Virus	937
Vitamins	672, 937
Wood	66

Penis

Ethnic Group	347
HPV	347
Sexual Activity	347
Socio-Economic Factors	347

Peritoneum

Alcohol	*73
Asbestosis	502
Biopsy	735
Chemical Exposure	39, 667
Dose-Response	39, *73
Dusts	30, 39, *73, 502, 546, 667, 735
Environmental Factors	30, 39, 667
Histology	667
Mining	30, 39
Occupation	30, 39, *73, 502, 667, 735
Time Factors	*73
Tobacco (Smoking)	30, *73

Pharynx

Alcohol	64, 180, *637, *716, *1026, 1163
Biopsy	*1027
Birth Cohort	489
BMB	*637, *1027
Chemical Exposure	321
Diet	64, 180, *716, *1026
Drugs	489, 741
Dyes	*484
Enzymes	741
HPV	*1027
Hygiene	180
Mathematical Models	489
Metals	321
Multiple Primary	64, *1026
Occupation	180, 321, *484, *637, *716
Radiation, Ionizing	180
Registry	180, *637, *716, 741, *1026, *1027
Time Factors	489, 1163
Tobacco (Chewing)	*716
Tobacco (Smoking)	64, 180, 489, *716, 741, *1026, 1163
Tobacco (Snuff)	*716
Trends	489

SITE

Pleura
Alcohol	*73
Asbestosis	502
Autopsy	860
Biopsy	735
Chemical Exposure	39, 346, 667
Dose-Response	39, *73
Dusts	39, *73, 469, 502, 546, 667, 735, 860
Environmental Factors	39, 346, 667
Histology	469, 667, 860
Immunologic Markers	469
Mining	39
Occupation	39, *73, 346, 502, 667, 735, 860
Time Factors	*73
Tobacco (Smoking)	*73

Prostate
Age	*966
Alcohol	66, *879, 937
Autopsy	390, 448, *966
Biochemical Markers	229, 573, 1055
Biopsy	390
Blacks	972
Blood	573, 841
BMB	18, 243, 573, 595, 727, 841, 851, *966, *1144
Chemical Exposure	66, *80, 594, *750, 795, 808
Chlorination	*80
Classification	390
Data Resource	*935
Diabetes	1003
Diet	18, 66, 69, 90, 229, 244, *259, 324, 573, 595, 727, 841, 851, *879, 937, 939, 972, 1003, 1028, *1144
Drugs	233, *750, *935
Dusts	66
Environmental Factors	66, 808
Ethnic Group	448, *1062
Familial Factors	972, 1143
Fat	90, 324, 851, 936, 1028
Fatty Acids	939
Fertilizers	*82, *601, *879
Fibre	1028
Genetic Factors	*750
Geographic Factors	*80, 288
Hair	851
Herbicides	*82
Histology	390
HLA	*750
Hormones	229, 573, 727, *935, 972
Infection	937
Insecticides	*82
Lifestyle	18, 229, 573, 595, 1143
Lipids	18, 573, 939, 1003
Mathematical Models	253
Metals	66, 243, *601, 807, 808, 851
Migrants	18, 288, 573
Nutrition	1028
Obesity	1028
Occupation	66, *82, 229, 254, *259, 309, 594, *601, 727, *750, 795, 807, 808, *879, 937, 972, *1129
Paints	*601
Pesticides	*82, 309, *601, *879
Petroleum Products	66, *82
Physical Activity	841, 1143, *1144
Physical Factors	1143, *1144
Plasma	18

Numbers indexed are project numbers, asterisks identify new projects

SITE

Plastics	66, 594
Premalignant Lesion	90, 972
Prevalence	448, 1003
Prevention	1055
Projection	253
Psychological Factors	1055
Race	*966, 1055, *1062
Radiation, Ionizing	795, *879
Record Linkage	*82, 233
Red Cells	18
Registry	233, 243, 244, 254, *259, 324, 594, 595, *601, 972, *1129, *1144
Reproductive Factors	937
Screening	371, *966, 1055
Sero-Epidemiology	243
Serum	243, 727, *1144
Sex Ratio	595
Sexual Activity	229, *259, 324, *750, 851, *879
Sexually Transmitted Diseases	324, *750, 851
Socio-Economic Factors	66, 972, 1055
Solvents	66, *601
Stage	90, 1055
Time Factors	288
Tobacco (Chewing)	1055
Tobacco (Smoking)	66, *879, 937, 1143
Toenails	595, 851
Trace Elements	573
Trends	*58
Urine	573
Virus	937
Vitamins	90, 243, 573, *879, 937
Water	*80, *879
Welding	*601
White Cells	18
Wood	66

Rectum

Age	1120
Alcohol	*148, 164, *165, 255, 261, 303, *410, 565, 572, 613, *676, 1036
Bile Acid	*363
Biochemical Markers	447, 527, *830, 1055
Biopsy	452
Blacks	971
Blood	*363, 498, *830
Blood Group	498
BMB	18, 243, 355, *363, 498, 595, 612, 679, *830, 1044
Chemical Exposure	*80, 686, 791, 964, 1120
Chlorination	*80, 634, 639, 983
Cholesterol	498, 572, *830
Chromosome Effects	45
Cirrhosis	261
Clinical Effects	227
Clinical Records	369, 452, 1105
Coffee	*676
Cost-Benefit Analysis	*145, *147
Crohn's Disease	734, 752
Diet	7, 18, 106, 125, *144, *148, *149, 158, 164, *165, 214, 244, 255, 273, 303, 355, *363, *410, 498, 537, 565, 572, 595, 613, 639, 674, *676, 686, *829, 943, 987, *995, 1004, 1036, 1065, *1142, 1182
DNA	45, 527, 679
Dose-Response	689, 791
Drugs	164, 686, *829
Dusts	689
Environmental Factors	4, *21, *144, 261, 971, 1182
Ethnic Group	987
Faeces	263, 355, *363

SITE

Familial Adenomatous Polyposis	4, 227, *363
Familial Factors	4, *21, 24, 45, *144, 158, 164, *165, 227, 303, 452, 494, 498, 527, 613, 640, 1079, 1143
Family History	*148, *149, *829
Fat	7, 537, *829, 987, *995, *1142
Female	106, 971, 1036
Fibre	7, 613, *829, 987, *995, *1142
Genetic Factors	24, *144, *211, 273, *363, 527, 896
Genetic Markers	*144, 896
Geographic Factors	*80, *144, 261, 288, *676
Heredity	*211, *363, 494, 686
High-Risk Groups	4, *21, 24, *145, *149, 172, *211, 261, 355, *436, 640, 896, 1059
Histology	164, 452, 896, 1105
Hormones	164, 303
Immunosuppression	352
Infection	352, *363
Latency	791
Lifestyle	18, 125, *145, *148, 158, 572, 595, 613, 971, 1143, 1182
Linkage Analysis	*363
Lipids	18, 498
Marital Status	214, 572
Mathematical Models	253
Menstruation	*165
Metals	243, 613, 964, 1102
Migrants	18, 288
Minerals	355
Monitoring	172
Multiple Primary	*144, 527
Mutation, Germinal	227
Nutrition	125, 164, *363, *829, *995, *1142
Obesity	674, 943, 987
Occupation	164, *165, 261, 498, 686, 689, 791, *829, 943, 964, 971, *974, *976, 1102, 1120, *1142
Oncogenes	527
Parasitic Disease	*145, 146
Parity	674, 943
Pedigree	*144, *363, 896
Petroleum Products	791
Physical Activity	*148, 164, *165, 255, 565, 674, *829, 943, 987, 1065, 1143
Physical Factors	498, 572, 1143
Plasma	18
Plastics	964, *974, *976
Polyps	*363, *436, 452, 639, 674, 820, *830, 943, *976, *995
Premalignant Lesion	4, 7, 146, 227, 261, 355, *363, 452, 565, 640, *830, 896, *995, 1061, 1065, 1105
Prevalence	565, 640
Prevention	7, 355, 527, 1055, 1079
Projection	253
Protein	*829
Psychological Factors	255, 1055
Psychosocial Factors	164
Race	1055
Radiation, Ionizing	164
Record Linkage	106, 734
Red Cells	18
Registry	24, 164, *165, *211, 214, 227, 243, 244, 255, 261, 263, 494, 595, 613, 634, 689, 718, 734, 791, 896, 971, 983, 987, 1036
Reproductive Factors	*148, 164, *165, 303, 987, 1036
RFLP	45
Rural	983
Screening	24, 106, 146, *147, 172, 263, 369, 371, *436, 640, 679, 828, *900, 964, *976, 1044, 1055, 1059, 1060, 1079, *1142
Segregation Analysis	*144
Sero-Epidemiology	243
Serum	243
Sex Ratio	595

Numbers indexed are project numbers, asterisks identify new projects

SITE

Socio-Economic Factors	125, 255, 572, 1055
Stage	172, 1055
Surgery	*900
Survival	172, 261
Tea	*165
Time Factors	288, 639, *676, 1079, 1120
Tissue	*363, 679
Tobacco (Chewing)	1055
Tobacco (Smoking)	*144, *148, 164, *165, 255, *410, 498, 572, 674, *676, 686, *900, 943, 964, 1143
Toenails	595, 612
Trace Elements	125, 447, 498, 612
Trends	*58, 537
Tumour Markers	527
Ulcerative Colitis	679, 734, 752
Urban	983
Urine	355
Virus	*363
Vitamins	7, 125, 243, 261, 498, 537, 613, *830, 1004
Water	*80, *145, 634, 639, 718, 983
White Cells	18

Respiratory

Air Pollution	208, *1187
Alcohol	*359
Biochemical Markers	236
Chemical Exposure	201, 208, 463, 683, 865, 1095
Dusts	102, 201, 236, 463, 683, 787, 865, 1095, *1187
Dyes	*484
Environmental Factors	208, 344, 1095, 1149
Geographic Factors	344
HPV	*359
Mapping	344
Metals	208, 1101, 1102
Minerals	787
Occupation	102, 201, 208, 236, 344, 463, *484, 683, 787, 865, 868, 1095, 1101, 1102, 1149
Petroleum Products	208
Plastics	868
Premalignant Lesion	*359, 1149
Registry	201, 208, 344, 683
Reproductive Factors	*359
Socio-Economic Factors	*359
Tobacco (Smoking)	236, 344, *359, 787, 1095, 1149
Wood	201

Retinoblastoma

Alcohol	1175
Atomic Bomb	*530
Childhood	1175
DNA	*1086
Ethnic Group	812
Familial Factors	838
Genetic Factors	812, 1175
Geographic Factors	838
Intra-Uterine Exposure	838, 1175
Mutation, Germinal	838, *1086
Occupation	1175
Parental Occupation	*1086
Prevention	812
Radiation, Ionizing	*530
Record Linkage	838
Registry	838
Sib	838

Tobacco (Smoking)	1175	
Twins	838	

Salivary Gland

Alcohol	757
Blacks	971
Environmental Factors	971
Eskimos	885
Female	971
Lifestyle	971
Occupation	757, 971
Prevention	885
Race	885
Radiation, Ionizing	757
Registry	971
Tobacco (Smoking)	757
Virus	885

Sarcoma

Air Pollution	208
Chemical Exposure	12, 208, 249, 910
Childhood	825
Dusts	910
Environmental Factors	208
Familial Factors	825
Female	94
Fertilizers	79
Geographic Factors	79
Herbicides	12, 79, 249, *591
High-Risk Groups	825
Insecticides	79
Metals	208, 910
Occupation	12, 91, 94, 208, 249, 309, *591, 910
Pesticides	12, 91, 94, 249, 309, *591, 910
Petroleum Products	208, 910
Plastics	910
Record Linkage	79
Registry	12, 208, 249, *591, 825, 910
Reproductive Factors	94
Solvents	910
Tobacco (Smoking)	94
Wood	910

Skin

Adolescence	*772
Air Pollution	208
Alcohol	255, 453
Anatomical Distribution	192
Arthritis	1109
Biochemical Markers	*901, 1055
Biopsy	1123
BMB	*1146, *1147
Chemical Exposure	208, 479, 1031
Congenital Abnormalities	1123
Cosmetics	514
Diet	255, 453, *992, 1150, 1151
DNA Repair	*901, 1123
Drugs	453, 514, 931, 1109
Environmental Factors	35, 208, 500, 1151
Enzymes	1151
Familial Factors	35
Genetic Factors	500
Geographic Factors	17, 192

SITE

Herbicides	*993
Histology	11, 192, 524, *772
HLA	500
Hormones	453
HTLV	524
Immunodeficiency	514
Immunology	500
Lifestyle	17
Lipids	*992, 1151
Mathematical Models	1109
Metabolism	1150
Metals	208, 1031, 1151
Multiple Primary	192, 500, *772
Nutrition	*992
Occupation	17, 208, 453, 479, *901, *993, 1031
Oral Contraceptives	453
Pesticides	453, 479
Petroleum Products	208
Photochemotherapy	931
Physical Activity	255
Pigmentation	500, *769, *901
Premalignant Lesion	17, 453, *769, *993, 1123
Prevalence	17
Prevention	1055, *1146, *1147, 1150
Psychological Factors	255, 1055
Race	1055
Radiation, Ionizing	514
Radiation, Non-Ionizing	931
Radiation, Ultraviolet	17, 29, 35, 500, 514, *769, *901, *993, 1123
Record Linkage	1109
Recurrence	*772
Registry	11, 17, 35, 192, 208, 255, 514, *772, 1109
Screening	371, 1055
Sero-Epidemiology	524
Socio-Economic Factors	17, 255, 1055
Stage	1055
Tobacco (Chewing)	1055
Tobacco (Smoking)	17, 255, 453, 514
Trends	192, *772
Twins	35
Virus	524
Vitamins	1150, 1151
Xeroderma Pigmentosum	1123

Small Intestine

Alcohol	*676
Autopsy	390
Biopsy	390
Classification	390
Coffee	*676
Crohn's Disease	*226
Diet	*676
Geographic Factors	*676
Histology	390
Time Factors	*676
Tobacco (Smoking)	*676
Ulcerative Colitis	*226

Soft Tissue

AIDS	803
Autopsy	390
Biopsy	390, 803
Chemical Exposure	28, 483, 509, 628, 961, 1005, 1119
Childhood	825
Classification	390

Cluster	*1106
Environmental Factors	483, 509, 1005
Familial Factors	825
Female	94
Geographic Factors	*1106
Herbicides	28, 509, *591, 628, 1005, 1119
High-Risk Groups	825
Histology	390, 803, 1005
HIV	803
Lifestyle	1005
Monitoring	*1106
Occupation	28, 91, 94, *504, 509, *591, 628, 961, 1005, 1119
Pesticides	28, 91, 94, *504, 509, *591, 628, 1005, *1106, 1119
Record Linkage	483
Registry	28, *504, *591, 628, 825, *1106
Reproductive Factors	94
Rural	509
Solvents	*1106
Time Factors	*1106
Tobacco (Smoking)	94
Trends	*1106

Spinal Cord

Demographic Factors	*19
Registry	*19
Survival	*19
Trends	*19

Spleen

Dose-Response	368
Drugs	368
Late Effects	368
Radiation, Ionizing	368

Stomach

Adolescence	526
Age	544
Alcohol	*56, 66, 136, 152, 184, 322, *410, 480, 526, 550, *556, 566, 572, 613, 650, *676, 728, 1135, 1161
Antibodies	845, 846
Antioxidants	*284
Atrophic Gastritis	135, 503, 536, 663, *778
BCG	651
Biochemical Markers	527, 573, *575, *778, 846
Biopsy	467, 846, 967
Blacks	971
Blood	*284, 573, 841
BMB	135, 243, *284, *285, *505, 573, 595, *778, 841, 843, 847, *1112, 1135
Chemical Exposure	66, *80, 322, 651, 683, 791, 910, 927, 959
Childhood	845, 862
Chinese	*160
Chlorination	*80
Cholesterol	*505, 572
Coffee	*676
Cooking Methods	673, 1161
Diet	*56, 66, 125, 135, 136, *144, 152, 184, 244, 388, *410, 480, 503, *505, 526, 537, *556, 563, 566, 572, 573, *575, 595, 613, 650, 651, 663, 672, 673, *676, 728, 740, *778, 841, 846, 862, 1135, 1161
DNA	527
DNA Adducts	*575, 847
Dose-Response	689, 791
Drugs	480
Dusts	30, 66, 136, 683, *684, 689, 843, 910, 927, *1116

Numbers indexed are project numbers, asterisks identify new projects

SITE

Environmental Factors	30, 66, *144, 563, 967, 971
Enzymes	846
Ethnic Group	544
Familial Factors	*144, 240, 388, 517, 527, 613, 967
Fat	*56, 537, 566, 672
Female	971
Fertilizers	*697, 843, *1116
Fibre	613
Fruit	672
Gastrectomy	467
Genetic Factors	*144, 527, 563, 1172
Genetic Markers	*144
Geographic Factors	*80, *144, 288, *676
H. pylori	*284, *285, *575, *778, 846, 847, *1112
Heredity	1172
High-Risk Groups	172, 388, 467, 663, 1172
Histology	239, 388, *522, 544, 673, *1116
Hormones	563, 573, 574
Hygiene	1135
Immunosuppression	352
Infection	352, 845
Japanese	*575
Latency	152, 791
Lifestyle	125, 184, *556, 563, 572, 573, 574, *575, 595, 613, 740, 971
Lipids	573, 574, 740
Marital Status	572
Mate	1161
Mathematical Models	253
Menstruation	563
Metals	66, 243, *270, 322, 613, 910, 1161
Migrants	288, 573
Mining	30, 322, 458
Monitoring	172
Multiple Primary	*144, 527
Mutagen	*181
Mutagen Tests	967
Mycotoxins	746, 845
Nutrition	125, 135, 663, 672, 740
Occupation	30, 66, 136, *270, 322, 458, *522, 563, 650, 651, 673, 683, *684, 689, *697, 728, 791, *792, 843, 910, 927, 959, 971, *1116, 1171
Oncogenes	527
Pedigree	*144
Pepsinogen	*575, 846, 847
Pesticides	910, *1116
Petroleum Products	66, 791, *792, 910
Physical Activity	740, 841
Physical Factors	*56, 322, 572, 650
Plastics	66, 910
Premalignant Lesion	135, 239, *284, 467, 503, 536, 663, *778, 845, 846
Prevalence	467, 746
Prevention	*284, 527, *556, 1172
Prognosis	563
Projection	253
Psychological Factors	138
Psychosocial Factors	785
Radiation, Ionizing	136
Record Linkage	550
Registry	240, 243, 244, *270, 388, *505, 544, 550, 595, 613, 650, 672, 683, *684, 689, *697, 718, 791, *792, 910, 971, *1116, 1135
Reproductive Factors	517, 526
Screening	172, *285, 517, 518, *900, 967
Segregation Analysis	*144
Sero-Epidemiology	243
Serum	135, 243, *285, *778, 847
Sex Ratio	595
Sexual Activity	563
Silicosis	458

SITE

Socio-Economic Factors	66, 125, 136, 572, 651, 728, 785, 862
Soil	746
Solvents	66, 910
Stage	172
Stress	785
Surgery	*900
Survival	172
Time Factors	*270, 288, 322, *676
Tissue	135
Tobacco (Smoking)	30, 66, 135, 136, *144, 152, 184, 322, *410, 480, 526, 550, *556, 566, 572, 650, 651, 673, *676, 728, *900, 927, 1161
Toenails	595, 1135
Trace Elements	125, 573, 1135
Treatment	518
Trends	*58, 537, 544
Tumour Markers	527
Urine	135, 573, 574, 746, 843, 845
Vegetables	672
Vitamins	*56, 125, 243, *284, *505, 537, 566, 573, 574, 613, 663, 672, 740, 1135
Water	*80, *181, 718, *1116
Welding	927
Wood	66, *684, 910

Testis

Adolescence	*767
Age	231, 965
Alcohol	66, 215
Autopsy	207
Biopsy	207
Birthweight	202
BMB	842
Chemical Exposure	66, 749
Chemotherapy	385
Classification	108
Congenital Abnormalities	215
Cryptorchidism	215
Dairy Products	*767
Diet	66, *767
DNA Repair	739
Dusts	66, 840
Environmental Factors	66
Familial Factors	739
Genetic Factors	231, 739, 782, 842
Genetic Markers	739, 782
Heredity	739, 782
Histology	108, 965
HLA	842
Hormones	108, 215, 840
In Situ Carcinoma	207
Infection	215
Infertility	207
Intra-Uterine Exposure	215, 840
Lymphocytes	385
Menopause	231
Metals	66
Multiple Primary	739, *876
Occupation	66, 108, 215, 840
Parity	231
Pedigree	739, 782
Petroleum Products	66
Physical Factors	215, 231
Plastics	66
Pregnancy	215
Race	965
Radiation, Ionizing	749, 840
Radiation, Ultraviolet	739

SITE

Radiotherapy	*876
Registry	108, 215, 739, *767, *876
SCE	385
Screening	739
Seasonality	202
Sexual Activity	749, 840
Socio-Economic Factors	66
Solvents	66, 840
Time Factors	965
Tobacco (Smoking)	66, 215, 749
Trauma	215, 749
Twins	231
Ultrasound	965
Wood	66

Thyroid

Age	725, 913
Alcohol	453
Birth Cohort	725
Blood	314
BMB	314, *362
Chemical Exposure	1045
Chemotherapy	913
Childhood	1128
Clinical Records	318
Diet	376, 453, 635, 688, 744
DNA	314
Dose-Response	913, 1045
Drugs	453
Genetic Factors	98, 314, *362
Genetic Markers	98
Geographic Factors	725
Histology	318, 376, 725
HLA	98
Hormones	453, 744
Hyperthyroidism	466
Iodine Deficiency	98, 744
Lymphocytes	*362
Mathematical Models	913
Multiple Primary	441, 913
Occupation	358, 453, 688, 1045
Oral Contraceptives	453, 744
Parity	744
Pesticides	453
Physical Factors	376, 913
Premalignant Lesion	453
Prevalence	635
Prognosis	318
Radiation, Ionizing	376, *730, 744, 913, 1045, 1128
Radiation, Non-Ionizing	913
Radiotherapy	913
Registry	*362, 441, 635, 725, *730, 744
Reproductive Factors	358
Screening	314
Socio-Economic Factors	376
Stage	441
Survival	318, 441, 635, 725
Time Factors	725
Tissue	314
Tobacco (Smoking)	453, 688
Trace Elements	635
Treatment	318
Trends	635
Vitamins	635, 744

SITE

Tongue
Female	*411
Registry	*411
Tobacco (Chewing)	*411
Tobacco (Smoking)	*411
Trends	*411

Urinary Tract
Alcohol	453
Chemical Exposure	*759
Data Resource	*759
Diet	453
Drugs	453
Hormones	453
Occupation	453, *759
Oral Contraceptives	453
Pesticides	453
Premalignant Lesion	453
Registry	*759
Tobacco (Smoking)	453

Uterus (Cervix)
Abortion	*1174
Adolescence	665, *1124
Age	88, 210, 231, 400
Alcohol	*359, *581, 650, 886
Antibodies	197, 760
Biochemical Markers	272, 283, *932, 1055, 1130
Biopsy	168, 272, 585
Birth Cohort	154
Blacks	662, 890, *962
Blood	582, *932
Blood Group	633
BMB	*33, 197, 243, 272, *581, *748, *816, *817, *932, *962, 1134
Cholesterol	717, 886
Chromosome Effects	*162, 418
Clinical Records	665, 921
CMV	168, *581, *748
Condyloma	88
Contraception	619, 856, 883, *1124, 1134
Cost-Benefit Analysis	603
Cytology	*9, 48, 272, 367, 418, *419, 603, 662, 665, 760, *817, 856, 888, 889, 893, *932, *962, *1124, 1141
Diet	88, *419, *425, 491, *581, 582, *593, 650, *816, 886, 903, *932, 1130, 1153
DNA	168, 760, 852, 883, *932, 1141
Drugs	233
Dusts	1130
EBV	*581, 760
Education	*425
Environmental Factors	644
Eskimos	885
Ethnic Group	347, 619
Familial Factors	517, 886
Fat	1130
Fertility	856
Fibre	1130
G-Banding	*162
Genetic Factors	231
Geographic Factors	288, 429, 649
HBV	367
High-Risk Groups	112, 172, 177, 400, 903, 1130
Histology	88, 603, *817, 852, 888, 893
HIV	367

Numbers indexed are project numbers, asterisks identify new projects

SITE

Hormones	491, 886
HPV	*33, 48, 88, 168, 272, 283, 347, *359, 367, 395, 418, *419, *581, 582, 585, 619, 665, 685, *748, 760, 799, 852, 883, *932, *962, *1124, 1141
HSV	48, 168, 197, 283, 418, *419, *581, 582, 665, *748, 760
Hygiene	*406, *425
In Situ Carcinoma	197, 888
Infection	197, 283, 395, 619, *748, 852
Lifestyle	*1174
Lymphocytes	*162
Mammography	1130
Marital Status	400
Mathematical Models	112, 253, 603, 903
Menarche	633
Menopause	231, 633
Metals	243
Migrants	288
Monitoring	172
Multiple Primary	222
Mutagen	903
Nutrition	88, 582, *932, 1130
Obesity	886
Occupation	650, 1130
Oral Contraceptives	88, 110, 168, 210, 491, 582, *748, 799, 852, 856, 903, 1134
Parity	231, 400, *425, 491, 519, 633, 856
Passive Smoking	903
PCR	*33, 48, 367, *581, 585, 799
Physical Activity	886
Physical Factors	231, 633, 650, 717, 886
Pregnancy	856
Premalignant Lesion	88, 110, 197, *359, 367, 418, *419, 519, 585, 662, 665, 883, 921, *932, 1141, 1153
Prevalence	48, 662, 1130
Prevention	885, 888, 893, 1055, 1130, 1153
Projection	253
Psychological Factors	1055
Race	885, 1055
Radiotherapy	222
Record Linkage	233
Registry	210, 222, 233, 243, 258, 650, 717
Reproductive Factors	154, *359, 517, *581, 619, 888, *1174
Rural	649
Screening	*9, 48, 112, 172, 177, 258, 371, 418, 517, 603, 633, *748, 888, 889, 890, 893, 921, 1055, 1060, 1130
Sero-Epidemiology	168, 197, 243, 283, 418, 685
Serum	197, 243, *748, 1134
Sexual Activity	88, 154, 168, 210, 283, 347, 395, 400, *406, *419, *425, 491, 519, 582, *593, *748, 760, 799, 888, 903, *1124, 1153
Sexually Transmitted Diseases	48, 283, 367, 662, 665, 685, *748, 883, 903, *1124, 1141
Socio-Economic Factors	154, 347, *359, 400, *425, 582, 619, 883, *962, 1055
Stage	172, 1055
Survival	172, 717
Time Factors	210, 288
Tissue	272, *932
Tobacco (Chewing)	1055
Tobacco (Smoking)	88, 154, 210, *359, *581, *593, 619, 650, 799, *817, 883, 886, 903, *1124, 1130
Treatment	644, 888, 889
Trends	*58, 154, 649
Tubal Ligation	1075
Tumour Markers	685
Twins	231
Urban	649
Uric Acid	886
Virus	197, 619, 852, 885
Vitamins	88, 243, 283, 519, 582, *816, *932, 1130, 1153

SITE

Uterus (Corpus)

Abortion	*1174
Age	231, 1018
Alcohol	566, 886, 914, 937
Biochemical Markers	1055
Blood	914
Blood Group	633
BMB	1134
Cholesterol	717, 886
Chromosome Effects	*162
Cluster	1043
Contraception	984, 1134
Data Resource	*935
Diet	106, 491, 566, 745, 886, 914, 915, 937, 984, 1018, 1043, 1125
Dose-Response	221, 984
Drugs	*935, 1018
Ethnic Group	984
Familial Factors	886
Fat	566, 914, 984, 1043
Female	106
G-Banding	*162
Genetic Factors	231
Geographic Factors	288, 649
High-Risk Groups	308
Hormones	235, 491, 737, 745, 857, 886, 914, 915, *935, 984, 1018, 1125
Infection	937
Lifestyle	*1174
Lymphocytes	*162
Mathematical Models	253
Menarche	633
Menopause	231, 633, 737, 857, 915
Migrants	288
Multiple Primary	221
Nutrition	914, 984
Obesity	745, 886, 984, 1043
Occupation	235, 937, 984
Oral Contraceptives	235, 491, 1125, 1134
Parity	231, 491, 633
Physical Activity	745, 886, 915, 984
Physical Factors	231, 633, 717, 886, 1043
Prevention	1055
Projection	253
Psychological Factors	1055
Race	1055
Radiotherapy	221
Record Linkage	106, 917
Registry	221, 717, 745, 857, 917, 1043, 1125
Reproductive Factors	745, 914, 915, 937, 984, 1125, *1174
Rural	649
Screening	106, 308, 633, 1055
Serum	1134
Sexual Activity	491
Socio-Economic Factors	1055
Stage	1055
Surgery	917
Survival	717
Time Factors	288, 737
Tobacco (Chewing)	1055
Tobacco (Smoking)	566, 886, 914, 937, 984
Treatment	737
Trends	649
Tubal Ligation	1075
Twins	231
Urban	649
Uric Acid	886
Virus	937
Vitamins	566, 937, 984, 1018

Numbers indexed are project numbers, asterisks identify new projects

SITE

Vagina
Blood	916
BMB	916
Diet	*1040
Drugs	*1040
Hormones	1076
Hygiene	916
Intra-Uterine Exposure	*1040, 1076
Menarche	*1040
Nutrition	*1040
Reproductive Factors	916
Screening	916
Sexual Activity	916
Sexually Transmitted Diseases	916
Tobacco (Smoking)	916

Vulva
Blood	916
BMB	916
Diet	491
Ethnic Group	347
Hormones	491
HPV	347
Hygiene	916
Oral Contraceptives	491
Parity	491
Reproductive Factors	916
Screening	916
Sexual Activity	347, 491, 916
Sexually Transmitted Diseases	916
Socio-Economic Factors	347
Tobacco (Smoking)	916

Wilms' Tumour
Age	65
Alcohol	65
Chemical Exposure	65, 1099
Childhood	511
Coffee	65
Diet	65
Drugs	65
Environmental Factors	*374, 1099
Familial Factors	1099
Fruit	65
Genetic Factors	*374
Herbicides	65
Heredity	1099
Infection	65
Intra-Uterine Exposure	1099
Lifestyle	*374
Multiple Primary	511
Occupation	65, 1099
Parental Occupation	*374
Radiation, Ionizing	*374
Registry	*374
Socio-Economic Factors	65, 1099
Survival	511
Tea	65
Tobacco (Smoking)	65
Treatment	511
Vegetables	65

INDEX OF STUDY TYPES

'Study type' is a very useful (and widely used) descriptive term for epidemiological projects, and this Index enables rapid identification of, say, case-control studies of bladder cancer. Epidemiological studies can, however, be difficult to classify into conventional designs such as cohort, case-control, etc, and there may therefore be a few errors. The **study type** of each project has been assigned by the editors for indexing purposes, but was only shown beneath each abstract for the first time in the 1988 edition. Investigators who notice errors should submit corrections when updating their entry for the next edition.

Projects which involve more than one study type are classified under each type.

The study type 'Registry' is used only for projects in which a new population-based cancer registry is being created, and its methods described. These abstracts are retained in the Directory for two years at most. Studies involving collaboration with a cancer registry are indexed both to the general keyword 'Registry' in the TERMS index, and to the specific cancer registry in the REGISTRY index.

'Mutation Epidemiology' has been replaced by the more pertinent terms 'Genetic Epidemiology' and 'Molecular Epidemiology'.

The list of index entries below identifies all study types in use in this Directory.

Case-Control

Case Series

Cohort

Correlation

Cross-Sectional

Genetic Epidemiology

Incidence

Intervention

Methodology

Molecular Epidemiology

Mortality

Registry

Relative Frequency

STUDY TYPE

Case Series

All Sites	366, 895, 1042
Benign Tumours	965
Bile Duct	*123
Breast (F)	*22, 62, 96, 400, *515, 527, 618, 782, 990, 1013, 1172
Breast (M)	990
Burkitt's Lymphoma	645
Childhood Neoplasms	*403, 753, 919, 1039, *1106
Colon	*436, 452, 527, 640, 782, 983, 1059
Gallbladder	*123
Gastrointestinal	*814
Haemopoietic	768
Head and Neck	666
Hodgkin's Disease	1194
Hypopharynx	269
Inapplicable	523
Kaposi's Sarcoma	803
Larynx	666
Leukaemia	349, *417, 510, *549, 669, 753, 782, 895, 958, *1106
Leukaemia (ALL)	768
Liver	465, *664, *1106
Lung	241, 401, 469, 1172
Lymphoma	*417, 524, 782
Melanoma	8, *809
Mesothelioma	469, 735
Myelodysplastic Syndrome	510, 768
Nasal Cavity	275, *809
Nasopharynx	*402, 579
Non-Hodgkin's Lymphoma	753, *1106
Oral Cavity	23, 269, *359, 805, *1027, 1051
Oropharynx	269, *359
Ovary	*398, 506, 1069
Pancreas	*1106
Peritoneum	735
Pharynx	*1027
Pleura	469, 735
Rectum	*436, 452, 527, 640, 983, 1059
Respiratory	*359
Retinoblastoma	812
Skin	524
Soft Tissue	803, *1106
Stomach	527, 651, 1172
Testis	782, 842, 965
Thyroid	318
Uterus (Cervix)	*359, 395, 400, *581, 665

Case-Control

All Sites	16, 46, *68, 75, 79, *80, *82, 113, 138, 222, 242, 243, 295, 298, *306, 327, *329, 375, 443, 471, 479, 483, 488, *504, 508, 521, *543, 545, 553, 559, 589, *690, *692, *696, *698, 807, 870, 891, 913, 920, 927, 930, 944, 981, 1008, 1016, 1033, 1037, 1097
Benign Tumours	*257, 491, 587, 1077, 1105
Bile Duct	300, 1091
Bladder	66, *80, 100, *159, 233, 278, *306, 311, 386, 512, 798, 834, *905, 910, 957, 971, 1070, 1184
Bone	103, 170
Brain	*34, 41, *49, *80, *81, *82, 94, 170, 260, 301, 302, 479, *504, 703, 704, 897, 910, 941, 1002, 1024, 1025, 1033, 1041, 1088, 1090, *1107
Breast (F)	69, *119, 129, 138, *144, *155, 220, 222, 223, 231, 233, 243, 255, *257, 297, 303, *338, 352, 389, *392, *399, *409, 416, *424, *428, 446, 447, 473, 491, 534, 583, 587, 588, *590, 595, 598, *615, 623, 625, 731, 736, *780, 813, 913, 918, 936, 937, 944, 1009, 1019, 1035, *1053, 1063, 1066, 1067, *1068, 1077, 1133, 1134, *1148, 1152, *1174, 1181, 1189, 1193
Breast (M)	206, 296

STUDY TYPE

Childhood Neoplasms	218, 307, 434, 438, 440, 571, *621, 704, 707, 754, *765, *826, 919, 1039, *1085, 1090
Choriocarcinoma	490
Colon	66, 69, *80, *144, *148, *149, 158, 164, *165, 243, 255, 261, 262, 273, 303, 352, *363, 369, *410, 443, 447, 449, 595, 616, 674, 686, 689, 720, *829, *830, 936, 943, 964, 971, 987, *995, 1036, 1061, 1065, 1079, 1105, *1122, 1140, *1156, 1182
Eye	971
Female Genital	453
Gallbladder	300, 643, 1017, 1091, 1134
Gastrointestinal	*73, 102, 231, 332, 386, 453, 542, 1189, 1196
Genitourinary	1189
Haemopoietic	16, 47, 305, 350, *351, 387, 479, *684, 910
Head and Neck	95
Hodgkin's Disease	36, 66, 91, 462, 513, 1093, 1126
Hydatidiform Mole	490
Hypopharynx	317, *420
Kaposi's Sarcoma	*38, 631
Kidney	42, 66, *73, *80, 325, 365, 481, 529, 729, *973, 1041, 1070, *1078, *1104, 1175
Larynx	64, *73, 180, 317, *420, 423, *637, 655, *716, 741, 863, 1033, 1070, 1135, 1163
Leukaemia	*49, *80, *82, 115, 116, 170, 179, 201, 221, 223, *224, 231, 305, *328, 332, *340, 364, 387, 483, 513, 571, 589, *627, 629, 704, 754, *783, *790, 804, 848, 913, 941, 944, *975, 1046, 1175, 1196
Leukaemia (ALL)	*72, 221, *279, *621
Leukaemia (AML)	1023
Leukaemia (ANLL)	1047
Leukaemia (CLL)	36, 221, 223
Lip	64, 1022
Liver	66, 136, *144, 163, 170, *173, 261, 528, 541, 630, *637, 747, 873, 908, 971, 1041, 1134
Lung	3, 46, 66, *73, 78, 92, 93, 97, 104, *121, 133, *137, *144, *151, *153, 222, 233, 243, 245, 252, 255, 278, 292, 295, 323, 326, 332, *333, 337, 341, 379, 380, 387, 423, 443, *456, 472, 499, 535, 541, 547, 553, 589, 595, 648, 653, 682, *684, 689, 699, 714, 719, 721, 741, 789, 807, 837, *864, 870, 871, 907, 910, 912, 913, *925, 927, *970, 971, 1000, *1014, 1041, 1056, 1070, 1071, 1121, 1137, 1162, 1179, 1180, 1196
Lymphoma	16, *38, *49, *80, 170, *340, 453, 481, 483, *684, *698, 910, *975, 1010, 1175
Melanoma	6, 66, 231, *290, 304, 336, 468, 599, 622, 971, 988, 1006, 1054, 1115, 1151
Mesothelioma	26, 37, 46, 102, 346, 546, 670, 689, *784, 870, 871, 910, 971
Multiple Myeloma	*82, 91, 94, 486, 513, *698, 972
Myelodysplastic Syndrome	15, 115, *268, 589, *770
Nasal Cavity	180, 201, *306, 316, 1021
Nasopharynx	140, 150, 178, *315, 580, 1021, 1136
Nervous System	1175
Neuroblastoma	*49, *1100, 1175
Non-Hodgkin's Lymphoma	36, 66, *82, 91, 94, 332, 462, *504, 513, *790, 904, 1127
Oesophagus	66, *165, 261, 357, *410, *420, 423, *637, 646, 672, *716, 971, 972, 1070, 1135
Oral Cavity	64, 180, 200, *359, *420, 423, *427, *637, 678, *716, 741, *1026, 1070, 1072, *1132, 1135, 1163
Oropharynx	*359, 397, *420, 1072, *1132
Ovary	108, 352, 396, 491, 936, 937, 1134, *1139, 1157, *1174, 1191
Pancreas	66, *165, 246, 300, 672, 937, 972, 1070
Penis	347
Peritoneum	*73, 546
Pharynx	64, 180, *637, *716, 741, *1026, 1163
Pleura	*73, 346, 546
Prostate	66, 69, *80, *82, 90, 229, 233, 243, *259, 324, 595, *601, 727, *750, 795, 807, 851, *879, 936, 937, 939, 972, 1028, *1062, *1144
Rectum	*80, *144, *148, *149, 158, 164, *165, 243, 255, 261, 273, 303, 352, *363, 369, *410, 447, 595, 674, 686, 689, *829, *830, 943, 964, 971, 987, *995, 1036, 1061, 1065, 1079, 1105, 1182
Respiratory	102, 201, *359

Numbers indexed are project numbers, asterisks identify new projects

STUDY TYPE

Retinoblastoma	1175
Salivary Gland	757, 971
Sarcoma	79, 91, 94, 910
Skin	29, 255, 453, 479, 514, *901, 1151
Soft Tissue	91, 94, 483, *504, 509, 1005
Stomach	66, *80, 136, 138, *144, 152, 184, 239, 243, *285, 352, 388, *410, 480, 503, 526, 550, *556, 563, *575, 595, 663, 672, 673, *684, 689, 728, *778, 846, 910, 927, 967, 971, *1112, *1116, 1135, 1161
Testis	66, 108, 202, 215, 231, 749, *767, 840
Thyroid	98, 358, 376, 453, 635, 688, *730, 744, 913
Urinary Tract	453, *759
Uterus (Cervix)	*33, 48, 88, 110, 154, 168, 197, 210, 222, 231, 233, 243, 283, 347, *359, *406, *425, 491, 519, 582, *593, 685, *748, 760, 799, 852, 888, *962, *1124, 1134, *1174
Uterus (Corpus)	221, 231, 491, 745, 914, 937, 984, 1125, 1134, *1174
Vagina	916, *1040
Vulva	347, 491, 916
Wilms' Tumour	65, 1099

Cohort

All Sites	16, 28, 46, *55, 61, 75, 76, 79, *82, *83, *85, 87, 107, 109, *124, *126, 130, 139, 182, 203, 205, 209, 213, 217, 222, *225, 228, 230, 242, 243, 249, 250, 264, *270, *280, *281, 293, 294, 295, 298, 299, *306, *310, 320, 321, 327, *329, 330, 331, 353, 354, 356, 370, *433, 443, 445, 450, 451, 460, 464, *475, 477, 479, 483, *484, 487, *504, 507, 516, 517, 520, 521, *532, 533, *538, 539, 540, *551, 552, 553, 557, 560, *564, 572, 576, 589, *591, 594, 596, 628, 633, 650, *657, 680, 687, *690, *691, *692, *693, *694, *695, *696, *698, 701, 702, *705, 706, 709, 710, 711, 712, 713, 724, 726, 733, 737, 740, 752, 756, 758, *766, 773, 774, 785, *794, 796, 797, 815, 818, 821, 822, 835, 836, 853, 856, 861, 865, 866, 867, 868, 869, 870, 875, *876, 882, 891, 892, 913, 927, 929, 930, 931, *945, 947, 948, 950, 951, 953, 955, 956, 959, 969, 977, 981, 986, 999, 1003, 1007, 1008, 1011, 1012, 1018, 1032, 1033, 1043, 1045, 1073, 1074, 1080, 1094, 1096, 1097, 1098, 1101, 1102, *1103, 1109, 1110, 1111, 1114, 1119, 1120, 1130, *1138, 1155, 1158, 1159, 1171, *1176, 1177, 1185, 1188
Benign Tumours	27, 451, 586, 587, 997, 1052, 1077, 1105
Bile Duct	558, *900
Bladder	105, 176, 208, 233, *306, *459, *461, *484, 843, 910, 960, 1034, 1096, 1171
Bone	368, 372
Brain	*25, *82, 208, 254, 339, 479, *504, *551, 910, 946, 1029, 1033, 1041, 1119, 1120
Breast (F)	18, *52, *85, 89, 106, 111, 222, 231, 233, 237, *238, 243, 244, 255, 361, *392, 441, 442, 482, 586, 587, 595, 598, 602, *611, 613, *626, 633, 650, 708, 717, 723, 736, 737, 738, *771, 785, 786, 819, 822, 841, 855, 856, 857, 858, 878, 886, 913, 915, 917, 929, 944, 963, *978, 997, 1003, 1018, 1043, 1045, *1048, 1052, *1053, 1063, *1068, 1075, 1076, 1077, 1130, 1143, 1169, 1190, 1192, 1195
Childhood Neoplasms	217, 715, 754, *764, *827, 849, 850, 919, 946, 980
Colon	18, 30, *85, 106, 146, *147, 214, *226, 243, 244, 255, 261, 263, *363, 441, 443, 494, 498, 566, 572, 595, 612, 613, 679, 689, 734, 740, 752, 791, 793, 820, 828, 841, 886, *900, 915, 917, *974, *976, 1003, 1004, 1030, 1031, 1102, 1105, 1120, 1130, *1142, 1143, *1145
Eye	931
Female Genital	256, 822, 858, 886
Gastrointestinal	*73, 102, 208, 231, 633, 855, 859, 865, 1029, 1186
Genitourinary	1101
Haemopoietic	16, 208, 479, *684, 910, 1029, 1101, 1109, *1129
Head and Neck	382, 441, *604
Hodgkin's Disease	213, 254, 511
Inapplicable	195, 196
Kaposi's Sarcoma	*38, 377, *899, 909
Kidney	71, *73, 106, 372, *461, *722, 808, 1031, 1034, 1041, 1081
Larynx	*73, 232, 321, *360, 487, *637, 863, 1033

Numbers indexed are project numbers, asterisks identify new projects

STUDY TYPE

Leukaemia	28, *55, *82, 134, 201, 217, 231, 254, 331, 368, 441, *461, 483, 555, 589, 641, 642, *693, 711, 713, 754, 781, 804, 835, 836, 848, 867, *887, 913, *922, 944, 959, 1045, 1118, 1119, 1120, 1159
Leukaemia (ALL)	511, 641
Leukaemia (ANLL)	511
Liver	*126, 217, 261, 271, 335, 368, 372, 378, *461, 465, 525, 554, 558, 566, 568, 572, *637, 747, 806, 872, 908, 961, 1029, 1030, 1031, 1041, 1084, 1087, 1102
Lung	3, 30, *32, 40, 46, *73, *83, 105, *124, *127, 133, *156, 157, *171, 175, *194, 198, 199, 222, 232, 233, 243, 244, 254, 255, 264, *270, 295, 321, 322, 331, *342, 368, 381, 443, *456, *457, 458, *459, *461, 464, 487, 525, *548, 553, 566, 567, 572, 577, 589, 594, 595, 613, 650, 653, *657, *684, 689, *697, 710, *722, 724, 732, 740, 777, 785, 788, 791, *792, 808, 843, 865, 866, 870, 881, 884, 910, 912, 913, 915, 917, 927, 952, 954, 955, 1003, 1018, 1029, 1030, 1031, 1041, 1045, 1092, 1114, 1119, *1129, 1130, 1143, 1149, 1155, 1171, 1186
Lymphoma	12, 16, *38, *55, 249, 368, 441, *459, 483, 555, 628, 641, *684, *693, *698, 713, 867, 909, 910, *922, 954, 961, *1038, 1101, 1109, 1118, *1129
Male Genital	208, 256
Melanoma	27, 231, 304, 844, 1006
Mesothelioma	30, 46, 102, *342, 502, *548, 689, 865, 870, 910
Multiple Myeloma	*82, *698, 1119
Myelodysplastic Syndrome	589
Nasal Cavity	201, *306, *484, 1102
Nasopharynx	141, 578, 868, 1098
Neuroblastoma	511, *1050
Non-Hodgkin's Lymphoma	*82, 331, *504, 511, *591, 1119
Oesophagus	131, 186, 261, 435, *637, 843, *1129
Oral Cavity	59, 321, 382, 404, 405, *637
Oropharynx	59
Ovary	106, 607, 608, 633, 717, 737, 738, 800, 886, 915, 917, 1075
Pancreas	105, 106, *459, *900, 959, 1118, 1143
Peritoneum	30, *73, 502
Pharynx	321, *484, *637
Pleura	*73, 502
Prostate	18, *82, 233, 243, 244, 254, 594, 595, 808, 841, 1003, 1028, *1129, 1143
Rectum	18, 106, 146, *147, 214, 243, 244, 255, 261, 263, *363, 494, 498, 572, 595, 612, 613, 679, 689, 734, 752, 791, 820, 828, *900, *974, *976, 1004, 1102, 1105, 1120, *1142, 1143
Respiratory	102, 201, 208, 463, *484, 683, 865, 868, 1095, 1101, 1102, 1149
Sarcoma	12, 79, 208, 249, *591, 910
Skin	208, 255, 479, *901, 931, 1031, 1109
Small Intestine	*226
Soft Tissue	28, 483, *504, *591, 628, 961, 1119
Spleen	368
Stomach	30, 135, 240, 243, 244, *270, 322, 458, 467, 517, *522, 536, 566, 572, 595, 613, 650, 683, *684, 689, *697, 740, *778, 785, 791, *792, 841, 843, 862, *900, 910, 927, 959, 1171
Testis	231, *876
Thyroid	441, 466, 635, 913, 1045, 1128
Uterus (Cervix)	*9, 112, 177, 197, 222, 231, 233, 243, 258, 418, *419, 517, 585, 603, 633, 650, 717, 760, 799, *817, 856, 883, 886, 1075, *1124, 1130, 1141
Uterus (Corpus)	106, 231, 566, 633, 717, 737, 857, 886, 915, 917, 1018, 1043, 1075
Vagina	1076
Wilms' Tumour	511

Correlation

All Sites	*44, *53, 63, *80, 130, 193, 248, 265, 344, 476, 485, 495, 531, 574, *658, 706, 718, 989, 994, 1015, 1064, 1130, 1165, *1167
Bladder	*80, 309, 489, 718
Brain	*80, 309
Breast (F)	*56, 125, 610, *935, 1130
Childhood Neoplasms	839, *1106
Colon	*56, *80, 125, 146, 537, 634, *676, 718, *935, 994, 1130
Gallbladder	*676

Numbers indexed are project numbers, asterisks identify new projects

STUDY TYPE

Gastrointestinal	344, 476
Haemopoietic	47
Kidney	*80, 309, 489
Larynx	489
Leukaemia	*80, 125, 309, 762, 839, *1106
Liver	125, 167, *173, *181, 189, *676, *1106
Lung	40, *56, 125, 489, 574, 718, *776, 860, *880, *935, 1064, 1130
Lymphoma	*80, 309, 762
Melanoma	17
Mesothelioma	502, 860
Multiple Myeloma	309
Nasopharynx	125
Non-Hodgkin's Lymphoma	*1106
Oesophagus	*56, 125, 489, 537, *676
Oral Cavity	118, 489
Ovary	*935
Pancreas	309, 489, *676, *1106
Peritoneum	502
Pharynx	489
Pleura	502, 860
Prostate	*80, 309, *935
Rectum	*80, 125, 146, 537, 634, *676, 718
Respiratory	344
Retinoblastoma	*530
Sarcoma	309
Skin	17
Small Intestine	*676
Soft Tissue	*1106
Stomach	*56, *80, 125, *181, 537, 574, *676, 718, 746, 847
Uterus (Cervix)	903, 1130
Uterus (Corpus)	*935

Cross-Sectional

All Sites	84, 236, 439, 496, *505, 573, 652, 1130
Bladder	910, 949
Brain	910
Breast (F)	70, 235, 274, 371, *505, 573, 906, 1130
Childhood Neoplasms	*620
Colon	371, *505, 565, 639, 1130, *1145
Gallbladder	122
Haemopoietic	910
Inapplicable	74, 251, 656
Liver	*160, 234, 236, 573, 668, 898
Lung	469, 910, 1130
Lymphoma	910
Melanoma	35
Mesothelioma	469, 910
Nasal Cavity	275
Nasopharynx	142
Non-Hodgkin's Lymphoma	348
Odontogenic Neoplasms	2
Oesophagus	128, *160
Oral Cavity	408
Ovary	681
Pleura	469
Prostate	371, 448, 573, *966
Rectum	371, 565, 639
Respiratory	236
Sarcoma	910
Skin	29, 35, 371, 500, *769, *993
Stomach	*160, *505, 536, 573, 746, 845, 846, 910
Testis	207
Thyroid	635
Uterus (Cervix)	48, 272, 367, 371, 619, 662, 1130, 1153
Uterus (Corpus)	235

Numbers indexed are project numbers, asterisks identify new projects

STUDY TYPE

Genetic Epidemiology

All Sites	739
Bone	739
Breast (F)	*22, *144, 169, *393, 739, 906, 990, 991, 1013, *1020, *1048
Breast (M)	990
Childhood Neoplasms	739, 838, 996
Colon	45, *144, *211, 227, *363, 896
Gastrointestinal	739, 896
Leukaemia	739, 874
Liver	*144
Lung	*144
Lymphoma	739
Melanoma	43
Oesophagus	131
Ovary	681, 739, 991
Rectum	45, *144, *211, 227, *363, 896
Retinoblastoma	838
Stomach	*144
Testis	739
Thyroid	314, *362

Incidence

All Sites	1, 13, 39, 60, 63, 77, 86, *101, 120, 174, 183, 191, 192, *216, 219, 247, 248, 277, 286, *291, 294, 313, *334, *412, *421, 430, 431, 432, 444, 478, 485, *497, 531, 539, 606, 617, 624, 636, 644, 661, 675, *761, 773, 775, 787, *801, 802, 810, *811, 825, 832, 885, 926, 979, 989, 999, *1108, 1113, 1130, 1165, *1166, *1167, 1188
Appendix	390
Bile Duct	390
Bladder	105, 971
Bone	644
Brain	*19, *81, *924
Breast (F)	*20, 274, *391, 407, 598, 606, 644, 649, 819, 825, 917, 1060, 1130
Breast (M)	206
Childhood Neoplasms	307, *373, *374, 561, 569, *620, 671, 753, 755, 831, 838, 839, 854, *1049, *1106
Choriocarcinoma	5
Colon	4, *21, 261, 390, 537, 917, 971, 1060, 1130
Eye	971
Gastrointestinal	67, 1196, 1197
Haemopoietic	47, 1168
Head and Neck	267
Hodgkin's Disease	455
Hydatidiform Mole	5
Hypopharynx	*411
Kaposi's Sarcoma	455, 501, 979
Larynx	51, *411, *801
Leukaemia	287, *374, 605, 669, 753, *801, 823, 831, 839, 854, 885, *1106, 1196
Leukaemia (ALL)	605, *779, 823
Leukaemia (AML)	*779
Leukaemia (CLL)	823
Leukaemia (CML)	*779
Lip	*99
Liver	261, 335, *801, 885, 971, *1106
Lung	67, 105, *212, *411, 429, 437, *801, 917, 971, 1130, 1196, 1197
Lymphoma	*374, *801, 885
Melanoma	17, *101, 192, 345, 971
Mesothelioma	39, 51, 67, 971
Nasopharynx	51, *315, 429, 885
Nervous System	390
Neuroblastoma	*374, 562, 824, 854
Non-Hodgkin's Lymphoma	455, 753, 823, *1106
Oesophagus	185, 261, 390, 537, 971, 1170
Oral Cavity	59, 390, *411
Oropharynx	59, *411

Numbers indexed are project numbers, asterisks identify new projects

STUDY TYPE

Ovary	917
Pancreas	105, 390, *1106
Peritoneum	39
Pleura	39
Prostate	390
Rectum	4, 7, *21, 261, 537, 971, 1060
Respiratory	787
Retinoblastoma	*530, 838
Salivary Gland	885, 971
Sarcoma	825
Skin	11, 17, 29, 192, *772
Small Intestine	390
Soft Tissue	390, 825, *1106
Spinal Cord	*19
Stomach	467, 537, 544, 971
Thyroid	318, 725
Tongue	*411
Uterus (Cervix)	429, 644, 649, 885, 921, 1060, 1130
Uterus (Corpus)	649, 917
Wilms' Tumour	*374

Intervention

All Sites	*334, *933, 934, 1130
Bladder	614
Breast (F)	172, 312, *492, 723, 877, 890, 894, 963, 1055, 1130, 1169, 1192
Colon	24, *145, 263, 355, 828, *982, 998, 1055, 1130, *1142
Female Genital	256
Gastrointestinal	*814
Liver	161, 172, 335, 1084
Lung	31, 172, 614, 1130, *1131, 1154, *1173
Male Genital	256
Melanoma	638
Mesothelioma	31, *1131
Oesophagus	187, 188, 282, 1178
Oral Cavity	117, 413, *426, 1178
Prostate	1055
Rectum	24, *145, 172, 263, 355, 828, 1055, *1142
Skin	*992, 1055, *1146, *1147, 1150
Stomach	172, *284
Uterus (Cervix)	172, *816, 889, 890, 893, *932, 1055, 1130, 1153
Uterus (Corpus)	308, 1055

Methodology

All Sites	63, 114, *143, 166, 277, 356, *454, 474, *493, 743, 832, 833, 913, *923, 1057, 1064, *1108, 1117, 1130
Appendix	390
Benign Tumours	1052
Bile Duct	390, *592
Bladder	66, 833, 910, 971
Brain	910, 1029
Breast (F)	*10, *52, 253, *289, 600, 913, 1052, 1060, 1066, 1130, 1190
Colon	66, 253, 390, 971, 1060, 1130
Eye	971
Gastrointestinal	1029
Haemopoietic	910, 1029
Hodgkin's Disease	66, 1126
Inapplicable	204, 251, *343, 597, 911, 938, 942, 985
Kidney	66
Larynx	253
Leukaemia	253, 833, 913, *1001
Lip	253
Liver	66, 971, 1029
Lung	66, 253, 470, 584, *902, 910, 913, *970, 971, 1029, 1064, 1130
Lymphoma	910
Melanoma	*14, 66, 971

Numbers indexed are project numbers, asterisks identify new projects

STUDY TYPE

Mesothelioma	910, 971
Nervous System	390
Non-Hodgkin's Lymphoma	66
Oesophagus	66, 253, 390, 971
Oral Cavity	390
Pancreas	66, 390, *592
Prostate	66, 253, 390
Rectum	253, 971, 1060
Salivary Gland	971
Sarcoma	910
Small Intestine	390
Soft Tissue	390
Stomach	66, 253, 910, 971
Testis	66
Thyroid	913
Urinary Tract	*759
Uterus (Cervix)	*9, 177, 253, 1060, 1130
Uterus (Corpus)	253

Molecular Epidemiology

All Sites	54, 276, 521, *538, 545, 557, *654, *694, 700, 1057, 1064
Angiosarcoma	677
Bladder	614, 677
Breast (F)	*1053
Haemopoietic	647
Inapplicable	74, 195, 196, 204, 251, *343, 383, 384, 597, 656, 763, 938, 940, 942
Leukaemia	54, 510, *922
Liver	*160, *162, 677, 898
Lung	*132, 241, *333, *457, 614, 653, 677, *902, *925, 1064
Lymphoma	54, *922
Myelodysplastic Syndrome	510
Oesophagus	*160
Oral Cavity	394
Ovary	*162
Retinoblastoma	*1086
Skin	1123
Stomach	*160
Testis	385
Uterus (Cervix)	*162, *581
Uterus (Corpus)	*162

Mortality

All Sites	*44, *53, 63, 86, 130, 174, 190, 191, 265, 266, 277, 286, *291, 313, 520, *532, 552, 624, 659, 660, 742, 743, *751, 775, 787, *801, 833, 999, 1080, 1082, 1113, 1117, *1164, 1165, *1167, 1183
Bladder	105, 288, 489, 833, *928
Breast (F)	*58, 190, 253, 288, 598, 609, 878, 1060
Childhood Neoplasms	753, 831
Colon	*58, 253, 288, 1044, 1060
Gastrointestinal	*57, 67, 968, 1196, 1197
Kidney	489
Larynx	253, 489, *801
Leukaemia	253, 605, 753, *801, 831, 833, *928, 1196
Leukaemia (ALL)	605
Lip	253
Liver	*58, *801
Lung	*58, 67, 105, 190, 253, 288, 489, *776, *801, *928, 968, *1187, 1196, 1197
Lymphoma	*801
Melanoma	288
Mesothelioma	67
Non-Hodgkin's Lymphoma	753, 968
Oesophagus	*58, 185, 253, 288, 489
Oral Cavity	489
Pancreas	105, 489

STUDY TYPE

Pharynx	489
Prostate	*58, 253, 288
Rectum	*58, 253, 288, 1044, 1060
Respiratory	787, *1187
Stomach	*58, 253, 288
Uterus (Cervix)	*58, 154, 253, 288, 889, 1060
Uterus (Corpus)	253, 288

Registry

All Sites	39, 313, *334, *421, 431, *497, 617, 675, 739, 773, 891, 926, 1058, 1080
Bone	739
Brain	*19, 415
Breast (F)	237, 274, 739
Childhood Neoplasms	307, *319, *373, 569, 570, 671, 739, 755, 1089
Choriocarcinoma	5
Colon	4, *21, 24
Gastrointestinal	739
Head and Neck	267
Hodgkin's Disease	414
Hydatidiform Mole	5
Inapplicable	50, 383, 384
Leukaemia	414, 739
Leukaemia (ALL)	*779
Leukaemia (AML)	*779
Leukaemia (CML)	*779
Lymphoma	414, 739, 1083
Mesothelioma	39, 667
Multiple Myeloma	414
Ovary	739
Peritoneum	39, 667
Pleura	39, 667
Rectum	4, *21, 24
Spinal Cord	*19
Stomach	518
Testis	739
Thyroid	318

Relative Frequency

All Sites	644
Bone	644
Breast (F)	632, 644
Childhood Neoplasms	*422
Uterus (Cervix)	644

INDEX OF CHEMICALS

The CHEMICALS index includes many individual substances which are members of a class; eight such classes have been identified:

Drugs	DR	Pesticides	PE
Dusts	DU	Plastics, resins and	
Metals and metal		monomers	PL
compounds	ME	Solvents	SO
Mycotoxins	MY	Sweeteners	SW

The code for each class is shown next to each chemical: projects indexed to a chemical in any class are cross-indexed to the class name in the TERMS index. Thus a study of exposure to nickel compounds will be indexed to 'Nickel' in the CHEMICALS index and 'Metals' in the TERMS index.

Chemicals for which an abbreviation is given in the 'List of Abbreviations' are now listed in abbreviated form in the index, e.g. PAH ('Polycyclic Aromatic Hydrocarbons').

New chemicals in this issue are:

Androgens		Oil Mist	
Chlordimeform		Pethidine	DR
Chloroprene		Piperazine	DR
Glycol Ethers	SO	Polypropylene	PL
Hydrocarbons, Chlorinated		Polystyrene	PL
Mustard Gas		Urethane	PL

'Toluene Diisocyanate' and '4,4'-Diphenyl-methane Diisocyanate' have been replaced by 'Diisocyanates'.

The list below identifies all entries in current use in this Directory.

Acrylamides	PL	Benzidine	
Acrylic Acid	PL	Benzo(a)pyrene	
Acrylonitrile	PL	Benzoyl Chloride	
Aetiocholanolone		Beryllium	ME
Aflatoxin	MY	Beta Carotene	
Alachlor	PE	Butadiene	PL
Alkylating Agents			
Aluminium	DR	Cadmium	ME
Amines, Aromatic		Caffeine	
Androsterone		Calcium	
Antimony	ME	Caprylyl Chloride	
Arsenic		Carbon Disulphide	SO
Asbestos	DU	Carbon Monoxide	
Asbestos, Amosite	DU	Chloramphenicol	DR
Asbestos, Chrysotile	DU	Chlorine	
Asbestos, Crocidolite	DU	Chlorobenzene	
Aspirin	DR	Chloroform	
Atrazine	PE	Chlorophenols	PE
Attapulgite	DU	Chloroprene	
Azathioprine	DR	Chloropyrenes	
		Chromium	ME
Barbiturates	DR	Cimetidine	DR
BCME		Cis-Platinum	DR
Benzene	SO	CMME	

Copper	ME	Nicotine	DR
Cyclophosphamide	DR	Nitrates	
		Nitrites	
2,4-D		Nitrogen Oxides	
Dehydroepiandrosterone Sulphate		N-Nitroso Compounds	
DES	DR		
Diesel Exhaust		Oestradiol	
Diisobutyl Ketone	SO	Oestriol	
Dimethylsulphate		Oestrogens	
Dioxins		Oestrone	
		Oil Mist	
Epichlorohydrin	PL	Oxygen	
Epoxy Resins	PL		
Ethylene		PAH	
Ethylene Oxide		Paracetamol	
		PCB	
Fluorides		Pethidine	
Folic Acid		Phenacetin	
Formaldehyde	PL	Phenol	PL
Furans		Phenothiazines	DR
		Phenoxy Acids	PE
Gasoline		Phenylbutazone	DR
Glass Fibres	DU	Phenytoin	DR
Glycol Ethers	SO	Phosphates, Inorganic	
Gold	ME	Piperazine	DR
		Plutonium	ME
		Polypropylene	PL
Hexane	SO	Polystyrene	PL
Hydrocarbons		Potassium	ME
Hydrocarbons, Chlorinated		Procarbazine	DR
Hydrocarbons, Halogenated		Progesterone	
Hydrochloric Acid		Progestogens	
		Prolactin	
Iodine		PVC	PL
Iron	ME		
Isopropyl Alcohol	SO	Radium	ME
		Radon	
Lead	ME	Reserpine	DR
Lindane	PE	Retinoids	
Magnesium		Selenium	
Malathion	PE	Silica	DU
Melphalan	DR	Silvex	PE
Mercury	ME	Sodium Chloride	ME
Methanol		Steel	ME
Methotrexate	DR	Steroids	
Methoxsalen	DR	Styrene	PL
Methyl Ethyl Ketone	SO	Sulphur Dioxide	
Methyl Isobutyl Ketone		Sulphuric Acid	
Methyl Methacrylate			
4,4'-Methylene-bis(2-chloroaniline)		2,4,5-T	
Methylxanthines		Talc	DU
Mineral Fibres		Tamoxifen	DR
Mineral Oil		Tars	
Mustard Gas		Technetium	ME
		Terpenes	
Naphthalenes, Chlorinated		Testosterone	
1-Naphthylamine		Tetrachloroethyiene	SO
2-Naphthylamine		Thorium	ME
Nickel	ME	Thoron	ME

Thorotrast	DR	Vinyl Chloride	PL
Tocopherol		Vinylidene Chloride	PL
Toluene	SO		
Trichloroethylene	SO	Warfarin	DR
Trihalomethanes			
Uranium	ME	Xylene	SO
Urethane	PL		
Vehicle Exhaust		Zinc	ME
Vinyl Acetate	PL	Zircon	ME

CHEM

EXPERIMENTAL AND EPIDEMIOLOGICAL STUDIES OF CHEMICALS

A number of chemicals being investigated in epidemiological studies are also being tested for carcinogenicity in animals. A list of these chemicals is reproduced below, by kind permission of the authors of the Directory of Agents Being Tested for Carcinogenicity, also published by IARC (see list of IARC publications). The list includes 49 chemicals which are the subject of about 249 current epidemiological studies.

For further information contact Ms M.-J. Ghess, Unit of Carcinogens Identification and Evaluation, IARC.

1. Aflatoxin	10		26. Mineral Fibres	16
2. Antimony	2		27. Mineral Oil	8
3. Asbestos, Amosite	2		28. 2-Naphthylamine	2
4. Asbestos, Chrysotile	6		29. Nickel	17
5. Asbestos, Crocidolite	4		30. Nicotine	3
6. Atrazine	1		31. Paracetamol	2
7. Attapulgite	1		32. PCB	4
8. Azathioprine	1		33. Phenylbutazone	1
9. Benzene	17		34. Phenytoin	1
10. Benzidine	3		35. Plutonium	1
11. Benzo(a)pyrene	5		36. Potassium	1
12. Beryllium	1		37. PVC	2
13. Butadiene	4		38. Reserpine	1
14. Cadmium	8		39. Silica	12
15. Carbon Monoxide	3		40. Styrene	11
16. Chlorine	1		41. Sulphur Dioxide	3
17. Chromium	19		42. 2,4,5-T	3
18. Dioxins	9		43. Talc	3
19. Formaldehyde	17		44. Tetrachloroethylene	2
20. Furans	2		45. Tocopherol	6
21. Gasoline	2		46. Toluene	4
22. Glass Fibres	3		47. Trichloroethylene	1
23. Hexane	1		48. Vehicle Exhaust	8
24. Isopropyl Alcohol	1		49. Vinyl Chloride	13
25. Methoxsalen	1			

CHEMICAL

Acrylamides
All Sites	1158
Brain	1029
Colon	1030
Gastrointestinal	1029
Haemopoietic	1029
Liver	1029, 1030
Lung	1029, 1030

Acrylic Acid
Colon	1030
Liver	1030
Lung	1030

Acrylonitrile
All Sites	594, 920, 1114, 1158
Bladder	910
Brain	*34, 910, 1029
Colon	1030
Gastrointestinal	1029
Haemopoietic	910, 1029
Liver	1029, 1030
Lung	594, 910, 1029, 1030, 1114
Lymphoma	910
Mesothelioma	910
Prostate	594
Sarcoma	910
Stomach	910

Aetiocholanolone
Ovary	608

Aflatoxin
All Sites	895
Leukaemia	895
Liver	161, 189, 335, *664, 747, 873, 898
Stomach	746, 845

Alachlor
Inapplicable	911

Alkylating Agents
Inapplicable	938

Aluminium
Bladder	*459
Lung	*459
Lymphoma	*459
Pancreas	*459

Amines, Aromatic
All Sites	652, 1171
Bladder	176, 512, 957, 1171
Lung	653, 1171
Stomach	1171

Numbers indexed are project numbers, asterisks identify new projects

CHEMICAL

Androgens
Breast (F)	618

Androsterone
Ovary	608

Antimony
All Sites	955
Lung	881, 955

Arsenic
All Sites	553, 724, 807, 920, 955
Bladder	208, *461, *905
Brain	208
Gastrointestinal	208
Haemopoietic	208
Kidney	*461, 481
Larynx	51
Leukaemia	*461
Liver	*461
Lung	*461, 553, 724, 807, 881, 912, 955
Lymphoma	481
Male Genital	208
Mesothelioma	51
Nasopharynx	51
Prostate	807
Respiratory	208
Sarcoma	208
Skin	208

Asbestos
All Sites	39, 86, 464, 474, *475, 495, 516, *654, *695, 787, 920, 1130
Bladder	910
Brain	910
Breast (F)	1130
Colon	689, 720, 1130
Gastrointestinal	1186
Haemopoietic	910
Hypopharynx	317
Kidney	481
Larynx	51, 317, 655
Lung	241, *333, *342, 464, 469, 472, 547, 577, 682, 689, 860, 910, 1130, *1131, 1154, 1186, *1187
Lymphoma	481, 910
Mesothelioma	37, 39, 51, *342, 469, 502, 546, 667, 689, 735, 860, 910, *1131
Nasopharynx	51
Peritoneum	39, 502, 546, 667, 735
Pleura	39, 469, 502, 546, 667, 735, 860
Rectum	689
Respiratory	463, 787, *1187
Sarcoma	910
Stomach	689, 910
Testis	840
Uterus (Cervix)	1130

Asbestos, Amosite
All Sites	865
Gastrointestinal	865
Lung	865, 871

CHEMICAL

Mesothelioma	865, 871	
Respiratory	865	

Asbestos, Chrysotile
All Sites	46, 869
Gastrointestinal	67, *73, 102
Kidney	*73
Larynx	*73
Lung	46, 67, *73, 199
Mesothelioma	46, 67, 102
Peritoneum	*73
Pleura	*73
Respiratory	102

Asbestos, Crocidolite
All Sites	46
Colon	30
Gastrointestinal	102
Lung	30, 31, 46
Mesothelioma	30, 31, 46, 102
Peritoneum	30
Respiratory	102
Stomach	30

Ascorbic Acid
All Sites	596

Aspirin
All Sites	930, *933, 934

Atrazine
Inapplicable	911

Attapulgite
All Sites	951

Azathioprine
All Sites	228

Barbiturates
All Sites	217
Brain	41, 301
Childhood Neoplasms	217
Leukaemia	217
Liver	217

BCME
Inapplicable	195
Lung	1092

Benzene
All Sites	*306, *475, *801, 920, 951, 1096
Bladder	*306, 910, 949, 1096
Brain	910, 1041, *1107
Childhood Neoplasms	*1106

CHEMICAL

Haemopoietic	647, 910
Inapplicable	*343
Kidney	1041
Larynx	*801
Leukaemia	134, *328, *801, *887, *975, *1106
Liver	*801, 1041, *1106
Lung	*801, 910, 1041
Lymphoma	*801, 910, *975
Mesothelioma	910
Nasal Cavity	*306
Non-Hodgkin's Lymphoma	*1106
Pancreas	*1106
Sarcoma	910
Soft Tissue	*1106
Stomach	910

Benzidine

All Sites	*484, 920, 1096
Bladder	*484, 1096
Nasal Cavity	*484
Pharynx	*484
Respiratory	*484

Benzo(a)pyrene

Gastrointestinal	1196
Leukaemia	1196
Lung	*132, 133, 1137, 1196
Stomach	746

Benzoyl Chloride

Kidney	481
Lymphoma	481

Beryllium

Kidney	481
Lymphoma	481

Beta Carotene

All Sites	243, *505, 533, *564, 596, 818, 930, *933, 934, 981, 1130
Bladder	614
Breast (F)	243, *338, *505, 583, 613, *615, 731, 1130, 1152, 1181
Colon	243, *505, 613, *830, 1130
Larynx	1135
Lung	31, 97, 243, 613, 614, 1130, *1131, 1154
Melanoma	1151
Mesothelioma	31, *1131
Oesophagus	185, 282, 1135, 1178
Oral Cavity	117, *426, 1135, 1178
Prostate	90, 243
Rectum	7, 243, 613, *830
Skin	1151
Stomach	243, *284, *505, 613, 1135
Uterus (Cervix)	88, 243, 283, 582, *932, 1130, 1153
Uterus (Corpus)	984

Butadiene

All Sites	951
Brain	1029
Colon	1030
Gastrointestinal	1029

CHEMICAL

Haemopoietic	1029
Inapplicable	251
Liver	1029, 1030
Lung	1029, 1030

Cadmium
All Sites	807, 951
Bladder	*461
Kidney	*461, 481, 808
Leukaemia	*461
Liver	*173, 234, *461
Lung	*461, 807, 808, 912
Lymphoma	481
Prostate	807, 808, 851, *879

Caffeine
Breast (F)	1009

Calcium
Breast (F)	*56
Colon	*56, 355, 616, *982, *995, 998, *1156
Lung	*56
Oesophagus	*56, 187, 188
Rectum	355, *995
Stomach	*56

Caprylyl Chloride
Colon	1030
Liver	1030
Lung	1030

Carbon Disulphide
Leukaemia	781

Carbon Monoxide
All Sites	818
Bladder	489, 1070
Kidney	489, 1070
Larynx	489, 1070
Lung	489, 1070
Oesophagus	489, 1070
Oral Cavity	489, 1070
Pancreas	489, 1070
Pharynx	489

Chloramphenicol
Childhood Neoplasms	*621
Leukaemia (ALL)	*621

Chlordimeform
Bladder	*159

Chlorine
All Sites	*281

Numbers indexed are project numbers, asterisks identify new projects

CHEMICAL

Chlorobenzene
Bladder   949

Chloroform
All Sites   *493

Chlorophenols
All Sites   *281, 294, 370, *696
Colon   720
Kidney   481
Lymphoma   481

Chloroprene
All Sites   *310

Chloropyrenes
Kidney   481
Lymphoma   481

Chromium
All Sites   *124, 320, 321, *475, *484, 540, *696, 920, 1155
Bladder   208, 386, *484
Brain   208
Gastrointestinal   208, 386
Haemopoietic   208
Hypopharynx   317
Inapplicable   *343
Kidney   481
Larynx   51, 317, 321, 655
Liver   525
Lung   *124, 321, *333, 525, 1155
Lymphoma   481
Male Genital   208
Mesothelioma   51
Nasal Cavity   316, *484
Nasopharynx   51
Oral Cavity   321
Pharynx   321, *484
Respiratory   208, *484
Sarcoma   208
Skin   208

Cimetidine
All Sites   596
Gastrointestinal   859

Cis-Platinum
All Sites   276

CMME
Lung   1092

Copper
All Sites   553
Liver   *173, 234

CHEMICAL

Lung	*156, 553
Stomach	663

Cyclophosphamide
Childhood Neoplasms	919
Inapplicable	597

2,4-D
All Sites	249, 701, 1119
Brain	1119
Inapplicable	911
Leukaemia	1119
Lung	1119
Lymphoma	249
Multiple Myeloma	1119
Non-Hodgkin's Lymphoma	1119
Sarcoma	249
Soft Tissue	509, 1119

Dehydroepiandrosterone Sulphate
All Sites	*654
Ovary	608

DES
Breast (F)	858, 1076
Female Genital	858
Vagina	*1040, 1076

Diesel Exhaust
Bladder	66
Colon	66
Hodgkin's Disease	66
Inapplicable	597
Kidney	66
Liver	66
Lung	40, 66, 719
Melanoma	66
Non-Hodgkin's Lymphoma	66
Oesophagus	66
Pancreas	66
Prostate	66
Stomach	66
Testis	66

Diisobutyl Ketone
All Sites	948

Diisocyanates
All Sites	*692, 756, 956

Dimethylsulphate
Kidney	481
Lymphoma	481

Dioxins
All Sites	*281, 294, 370, 483, *591, *690, 891, 1032

Numbers indexed are project numbers, asterisks identify new projects

CHEMICAL

    Childhood Neoplasms   *1106
    Leukaemia   483, *1106
    Liver   *1106
    Lymphoma   483
    Non-Hodgkin's Lymphoma   *591, *1106
    Pancreas   *1106
    Sarcoma   *591
    Soft Tissue   483, *591, *1106

Epichlorohydrin
    Kidney   481
    Lymphoma   481

Epoxy Resins
    All Sites   687
    Brain   1041
    Kidney   1041
    Liver   1041
    Lung   1041

Ethylene
    Brain   1041
    Kidney   1041
    Liver   1041
    Lung   1041

Ethylene Oxide
    All Sites   *693, *698, 867, 959, 1159
    Inapplicable   *343
    Leukaemia   *693, 867, 958, 959, 1159
    Lymphoma   *693, *698, 867
    Multiple Myeloma   *698
    Pancreas   959
    Stomach   959

Fluorides
    All Sites   230
    Bladder   *459
    Gastrointestinal   1197
    Lung   *459, 1197
    Lymphoma   *459
    Pancreas   *459

Folic Acid
    Uterus (Cervix)   582

Formaldehyde
    All Sites   86, *281, *306, *484, *696, 868, 920, 1097, 1098
    Bladder   *306, *484, 910
    Brain   *34, 910
    Colon   964
    Gastrointestinal   1186
    Haemopoietic   910
    Kidney   481
    Lung   910, 1186
    Lymphoma   481, 910
    Melanoma   *809
    Mesothelioma   910

CHEMICAL

Nasal Cavity	*306, 316, *484, *809
Nasopharynx	868, 1098, 1136
Pharynx	*484
Rectum	964
Respiratory	*484, 868
Sarcoma	910
Stomach	910

Furans
All Sites	*690, 1032

Gasoline
Bladder	66
Colon	66
Hodgkin's Disease	66
Kidney	66, *973
Liver	66
Lung	66
Melanoma	66
Non-Hodgkin's Lymphoma	66
Oesophagus	66
Pancreas	66
Prostate	66
Stomach	66
Testis	66

Glass Fibres
All Sites	86, 870
Lung	870
Mesothelioma	870
Respiratory	1095

Glycol Ethers
Leukaemia (ALL)	*279

Gold
All Sites	86

Hexane
Brain	1029
Gastrointestinal	1029
Haemopoietic	1029
Liver	1029
Lung	1029

Hydrocarbons
All Sites	16, *475, 951, 956, 1097
Brain	1029
Childhood Neoplasms	218
Colon	1030
Gastrointestinal	1029
Haemopoietic	16, 1029
Kidney	481
Liver	1029, 1030
Lung	1029, 1030
Lymphoma	16, 481

CHEMICAL

Hydrocarbons, Chlorinated
All Sites	920

Hydrocarbons, Halogenated
All Sites	266, 327, 951, 1097
Brain	1029
Colon	1030
Gastrointestinal	1029
Haemopoietic	47, 1029
Kidney	481
Liver	1029, 1030
Lung	1029, 1030
Lymphoma	481

Hydrochloric Acid
Brain	1041
Kidney	1041
Liver	1041
Lung	1041

Iodine
Thyroid	635, 1128

Iron
All Sites	1111
Larynx	1135
Liver	*173
Lung	*156, 198, 322, *333
Oesophagus	1135
Oral Cavity	1135
Stomach	322, 663, 1135

Isopropyl Alcohol
Larynx	655

Lead
All Sites	114, *270, *691, 807
Bladder	*461
Kidney	*461, 481, *722
Leukaemia	*461
Liver	*173, 234, *461
Lung	*270, 458, *461, *722, 807
Lymphoma	481
Prostate	807
Stomach	*270, 458

Lindane
All Sites	861

Magnesium
All Sites	*658

Malathion
Inapplicable	911

CHEMICAL

Melphalan
    All Sites   712
    Inapplicable   *343

Mercury
    Brain   1029
    Gastrointestinal   1029
    Haemopoietic   1029
    Liver   1029
    Lung   1029

Methanol
    Colon   1030
    Liver   1030
    Lung   1030

Methotrexate
    Leukaemia   641
    Leukaemia (ALL)   641
    Lymphoma   641

Methoxsalen
    All Sites   931
    Eye   931
    Skin   931

Methyl Ethyl Ketone
    All Sites   948

Methyl Isobutyl Ketone
    All Sites   948

Methyl Methacrylate
    All Sites   1158

4,4'-Methylene-bis(2-chloroaniline)
    All Sites   960
    Bladder   960

Methylxanthines
    All Sites   488
    Choriocarcinoma   490
    Female Genital   453
    Gastrointestinal   453
    Hydatidiform Mole   490
    Lymphoma   453
    Skin   453
    Thyroid   453
    Urinary Tract   453

Mineral Fibres
    All Sites   39, 46, 295, 951
    Gastrointestinal   *73, 1186
    Kidney   *73

Numbers indexed are project numbers, asterisks identify new projects

CHEMICAL

Larynx	*73
Lung	46, *73, 199, 241, 295, 469, *548, 860, 871, 1186
Mesothelioma	39, 46, 346, 469, 546, *548, *784, 860, 871
Peritoneum	39, *73, 546
Pleura	39, *73, 346, 469, 546, 860
Respiratory	1095

Mineral Oil

All Sites	266, 1033
Brain	1033
Colon	791, 964
Inapplicable	195
Larynx	51, 1033
Lung	791
Mesothelioma	51
Nasopharynx	51
Rectum	791, 964
Skin	514
Stomach	791
Testis	840

Mustard Gas

Stomach	*522

N-Nitroso Compounds

Bladder	386, 512
Brain	41, 301, 302, 703, 1024, 1088, 1090
Childhood Neoplasms	1090
Gastrointestinal	386
Liver	747
Neuroblastoma	*1100
Oesophagus	128, 357
Oral Cavity	118
Stomach	135, 467, 847

Naphthalenes, Chlorinated

Liver	961
Lymphoma	961
Soft Tissue	961

1-Naphthylamine

Bladder	176

2-Naphthylamine

All Sites	1096
Bladder	176, 1096

Naphthylthiourea, Alpha

All Sites	861

Nickel

All Sites	86, 320, 321, *475, 920, 1102
Bladder	208, 386
Brain	208
Colon	964, 1102
Gastrointestinal	208, 386
Haemopoietic	208

CHEMICAL

Hypopharynx	317
Inapplicable	*343
Kidney	481
Larynx	51, 317, 321
Liver	1102
Lung	321, *333, 912
Lymphoma	481
Male Genital	208
Mesothelioma	51
Nasal Cavity	316, 1102
Nasopharynx	51
Oral Cavity	321
Pharynx	321
Prostate	851
Rectum	964, 1102
Respiratory	208, 1102
Sarcoma	208
Skin	208

Nicotine

Bladder	489, 1070
Childhood Neoplasms	*764
Kidney	489, 1070
Larynx	489, 1070
Lung	489, 1070
Oesophagus	489, 1070
Oral Cavity	489, 1070
Pancreas	489, 1070
Pharynx	489

Nitrates

All Sites	718, 1097
Bladder	386, 718, 843
Colon	718
Gastrointestinal	386, 1197
Lung	*697, 718, 843, 1197
Oesophagus	843
Rectum	718
Stomach	663, *697, 718, 843, *1116

Nitrites

All Sites	213, 718
Bladder	386, 718
Colon	718
Gastrointestinal	386
Hodgkin's Disease	213
Lung	718
Rectum	718
Stomach	467, 663, 718, *1116

Nitrogen Oxide

Lung	458
Stomach	458

Oestradiol

Breast (F)	235, 297
Ovary	608
Uterus (Corpus)	235

Numbers indexed are project numbers, asterisks identify new projects

CHEMICAL

Oestriol
Breast (F)	235
Ovary	608
Uterus (Corpus)	235

Oestrogens
All Sites	737, 929, 1018
Benign Tumours	1077
Breast (F)	588, *590, 625, 737, 929, 1018, 1019, 1063, 1077
Kidney	42, 729
Lung	1018
Ovary	506, 737, 1157
Uterus (Corpus)	737, 1018, 1125

Oestrone
Breast (F)	235
Ovary	608
Uterus (Corpus)	235

Oil Mist
Lung	*792
Stomach	*792

Oxygen
Inapplicable	938

PAH
All Sites	710, 807, 818, 920, 927
Bladder	208, *459, *461, 1070
Brain	*34, 208
Colon	791
Gastrointestinal	208, 1196
Haemopoietic	208, *1129
Inapplicable	597, 938
Kidney	*461, 481, 1070, 1081
Larynx	51, 655, 1070
Leukaemia	*340, *461, 1196
Liver	*461
Lung	40, *333, *457, *459, *461, 584, 710, 791, 807, *902, 927, 1070, *1129, 1196
Lymphoma	*340, *459, 481, *1129
Male Genital	208
Mesothelioma	51
Nasopharynx	51
Oesophagus	1070, *1129
Oral Cavity	1070
Pancreas	*459, 1070
Prostate	807, *1129
Rectum	791
Respiratory	208
Sarcoma	208
Skin	208, 514
Stomach	791, 927

Paracetamol
Kidney	42, 729

CHEMICAL

PCB
All Sites	*690, 951, 1032
Liver	961
Lymphoma	961
Soft Tissue	961

Pethidine
Childhood Neoplasms	*765

Phenacetin
Bladder	311

Phenol
All Sites	920
Brain	*34, 1029
Colon	1030
Gastrointestinal	1029, 1186
Haemopoietic	1029
Liver	1029, 1030
Lung	1029, 1030, 1186

Phenothiazines
Bladder	233
Breast (F)	233
Lung	233
Prostate	233
Uterus (Cervix)	233

Phenoxy Acids
All Sites	*82, 249, 294, 370, *591, 628, 701, 1119
Brain	*82, 1041, 1119
Colon	720
Hodgkin's Disease	513
Kidney	481, 1041
Leukaemia	*82, 513, 1119
Liver	1041
Lung	1041, 1119
Lymphoma	12, 249, 481, 628
Multiple Myeloma	*82, 513, 1119
Non-Hodgkin's Lymphoma	*82, 513, *591, 1119
Prostate	*82
Sarcoma	12, 249, *591
Soft Tissue	509, *591, 628, 1005, 1119

Phenylbutazone
Leukaemia (AML)	1023

Phenytoin
Childhood Neoplasms	*621
Leukaemia (ALL)	*621

Phosphates, Inorganic
Gastrointestinal	1197
Lung	1197

CHEMICAL

Piperazine
All Sites   *698
Lymphoma   *698
Multiple Myeloma   *698

Plutonium
All Sites   1008

Polypropylene
Colon   *974, *976
Rectum   *974, *976

Polystyrene
All Sites   *945

Potassium
All Sites   994
Colon   994

Procarbazine
All Sites   276

Progesterone
Breast (F)   235, 297, 312, 588, 625, 1133
Uterus (Corpus)   235

Progestogens
All Sites   737
Breast (F)   625, 737, 1019
Ovary   737
Uterus (Corpus)   737, 1125

Prolactin
Breast (F)   297, 819, 1063

PVC
All Sites   *695
Brain   1029
Gastrointestinal   1029
Haemopoietic   1029
Liver   1029
Lung   1029

Radium
Bone   103

Radon
All Sites   *83, 182, 264, *657
Gastrointestinal   1196
Haemopoietic   387
Inapplicable   763
Leukaemia   387, *783, 1196

CHEMICAL

Lung	78, *83, 92, 93, *137, 157, 198, 264, 387, 458, *657, 714, 732, 777, 837, *880, 884, 907, 912, 952, 954, 1056, 1121, 1179, 1196
Lymphoma	954
Stomach	458

Reserpine

Bladder	233
Breast (F)	233
Lung	233
Prostate	233
Uterus (Cervix)	233

Retinoids

All Sites	243, 818, 930, *933, 981
Breast (F)	243, *492, 583, 731, 1152
Colon	243, *830, *1156
Lung	31, 243, *1131, 1154, *1173
Melanoma	1151
Mesothelioma	31, *1131
Oesophagus	282
Oral Cavity	117
Prostate	243
Rectum	243, *830
Skin	1150, 1151
Stomach	243
Uterus (Cervix)	88, 243, 283, 519, 1153

Selenium

All Sites	243, 557, 929, 930
Bladder	386
Breast (F)	243, *338, 446, 447, 595, 613, *615, 731, 929, 1152
Colon	*145, 243, 447, 498, 595, 612, 613, *1145
Gastrointestinal	386
Liver	*173
Lung	243, 595, 613
Melanoma	988, 1151
Prostate	243, 595
Rectum	*145, 243, 447, 498, 595, 612, 613
Skin	*1146, *1147, 1151
Stomach	243, 595, 613, 663
Uterus (Cervix)	243, 582

Silica

All Sites	*657, 866, 920
Bladder	66, *461, 910
Brain	910
Colon	66
Haemopoietic	910
Hodgkin's Disease	66
Hypopharynx	317
Kidney	66, *461
Larynx	317
Leukaemia	*461
Liver	66, *461
Lung	*32, 66, *156, 458, *461, *657, 788, 866, 910, 912
Lymphoma	910
Melanoma	66
Mesothelioma	910
Non-Hodgkin's Lymphoma	66
Oesophagus	66
Pancreas	66
Prostate	66

Numbers indexed are project numbers, asterisks identify new projects

CHEMICAL

    Sarcoma   910
    Stomach   66, 458, 910
    Testis   66

Silvex
    All Sites   701

Sodium Chloride
    All Sites   994
    Colon   537, 994
    Oesophagus   537
    Rectum   537
    Stomach   537, 1161

Steel
    All Sites   *794, 1101
    Genitourinary   1101
    Haemopoietic   1101
    Lung   104
    Lymphoma   1101
    Respiratory   1101

Steroids
    Breast (F)   1134
    Gallbladder   1134
    Liver   668, 747, 1134
    Ovary   1134, 1157
    Uterus (Cervix)   1134
    Uterus (Corpus)   1134

Styrene
    All Sites   *280, 713, 920, *945, 951, 1097
    Brain   1029, 1041
    Gastrointestinal   1029
    Haemopoietic   1029
    Inapplicable   204, 942
    Kidney   481, 1041
    Leukaemia   713
    Liver   1029, 1041
    Lung   1029, 1041
    Lymphoma   481, 713

Sulphur Dioxide
    Gastrointestinal   1196
    Haemopoietic   *684
    Leukaemia   1196
    Lung   *171, *684, 1196
    Lymphoma   *684
    Stomach   *684

Sulphuric Acid
    All Sites   114
    Gastrointestinal   1197
    Hypopharynx   317
    Larynx   317, 863
    Lung   1197

CHEMICAL

2,4,5-T
All Sites	249, 701
Lymphoma	249
Sarcoma	249
Soft Tissue	509, 1005

Talc
Bladder	910
Brain	910
Haemopoietic	910
Lung	*342, 910
Lymphoma	910
Mesothelioma	*342, 910
Ovary	506
Sarcoma	910
Stomach	910

Tamoxifen
Breast (F)	312

Tars
All Sites	540, 818
Bladder	*459, 489, 1070
Kidney	489, 1070
Larynx	51, 489, 1070
Lung	*459, 489, 1070
Lymphoma	*459
Mesothelioma	51
Nasopharynx	51
Oesophagus	489, 1070
Oral Cavity	489, 1070
Pancreas	*459, 489, 1070
Pharynx	489
Skin	514

Technetium
Inapplicable	74

Terpenes
Haemopoietic	*684
Lung	*684
Lymphoma	*684
Stomach	*684

Testosterone
Breast (F)	235, 297
Uterus (Corpus)	235

Tetrachloroethylene
Kidney	481
Lymphoma	481

Thorium
Gastrointestinal	1197
Liver	136
Lung	1197
Stomach	136

Numbers indexed are project numbers, asterisks identify new projects

CHEMICAL

Thoron
Haemopoietic	387
Leukaemia	387
Lung	387, 777

Thorotrast
All Sites	203, 209, 217
Bone	368
Childhood Neoplasms	217
Leukaemia	217, 368
Liver	217, 368
Lung	368
Lymphoma	368
Spleen	368

Tocopherol
All Sites	243
Breast (F)	243, *338, *615
Colon	243, *1156
Lung	243, *1173
Prostate	243
Rectum	243
Stomach	243
Uterus (Cervix)	243, *932

Toluene
Brain	1029
Colon	1030
Gastrointestinal	1029
Haemopoietic	647, 1029
Inapplicable	*343
Liver	1029, 1030
Lung	1029, 1030

Trichloroethylene
All Sites	266, 948
Bladder	949
Kidney	481
Lymphoma	481

Trihalomethanes
All Sites	*493

Uranium
All Sites	86, 182, 264
Gastrointestinal	1196, 1197
Leukaemia	1196
Lung	3, 78, 157, 264, 884, 954, 1196, 1197
Lymphoma	954

Urethane
All Sites	*698
Lymphoma	*698
Multiple Myeloma	*698

Vehicle Exhaust
All Sites	487, 507

Hypopharynx	317	
Larynx	317, 487	
Lung	487	
Prostate	*601	

Vinyl Acetate
Brain	1029
Gastrointestinal	1029
Haemopoietic	1029
Liver	1029
Lung	1029

Vinyl Chloride
All Sites	86, *126, 327, *695, *801, 951, 1097
Angiosarcoma	677
Bladder	677
Brain	*34, 260, 1029
Colon	1030
Gastrointestinal	1029
Haemopoietic	1029
Kidney	481
Larynx	*801
Leukaemia	*801
Liver	*126, 677, *801, 1029, 1030
Lung	677, *801, 1029, 1030
Lymphoma	481, *801

Vinylidene Chloride
Brain	1029
Gastrointestinal	1029
Haemopoietic	1029
Liver	1029
Lung	1029

Warfarin
All Sites	861

Xylenes
Haemopoietic	647

Zinc
All Sites	114
Bladder	*461
Colon	*1156
Kidney	*461
Larynx	1135
Leukaemia	*461
Liver	*173, *461
Lung	458, *461
Oesophagus	282, 1135
Oral Cavity	1135
Prostate	*879
Stomach	458, 1135

Zircon
Lung	881

Numbers indexed are project numbers, asterisks identify new projects

INDEX OF OCCUPATIONS

The use of occupational titles is very variable between countries, but every effort has been made to classify occupational cancer studies to the specific group identified by the principal investigator.

All studies indexed to a specific occupation in this index can also be found under the general heading 'Occupation' in the TERMS index.

There is inevitably a degree of overlap with the CHEMICALS index; thus for example, a study of vinyl chloride workers can be found in both indexes.

New entries for this issue are:

Health Care Workers
Herbicide Manufacturers
Miners, Copper
Waiters

'Herbicide Applicators and Manufacturers' has been replaced by 'Herbicide Manufacturers' and 'Herbicide Sprayers'. 'Sprayers' as such has been deleted.

'Laboratory Technicians' has been replaced by 'Laboratory Workers'.

The new occupation 'Health Care Workers' replaces the previously used key-words 'Hospital Workers', 'Medical Workers', 'Nurses', 'Physicians' and 'Radiology Technicians'.

The list below identifies all entries in use in this Directory.

Acrylonitrile Workers
Administrative Workers
Agricultural Workers
Aluminium Workers
Antimony Process Workers
Asbestos Textile Workers
Asbestos Workers
Automobile Workers

Battery Plant Workers
Bus Drivers

Cabinet Makers
Cable Manufacturers
Carpenters, Joiners
Cellophane Manufacturers
Cement Workers
Chemical Industry Workers
Chemists
Chimney Sweeps
Chromate Pigment Workers
Chromium Plating Workers
Coke-Oven Workers
Construction and Maintenance Workers
Cryolite Workers

Dockers
Drivers
Dry Cleaners
Dyestuff Workers

Electrical Workers
Electrode Manufacturers
Electronics Workers
Engineering Workers

Farmers
Fertilizer Workers
Firemen
Fishermen
Forest Workers
Foundry Workers

Gas Workers
Glass Workers
Grain Millers
Grinding Material Workers

Health Care Workers
Heavy Equipment Operators
Herbicide Manufacturers
Herbicide Sprayers
Horticulturists

Insulation Board Manufacturers

Jewellery Manufacturers

Laboratory Workers
Lacquerers
Laminators
Leather Workers

Machinists
Meat Workers
Mechanics
Metal Workers
Military Servicemen
Mineral Fibre Workers
Miners
Miners, Asbestos
Miners, Coal
Miners, Copper
Miners, Fluorspar
Miners, Gold
Miners, Iron
Miners, Uranium
Miners, Zinc-Lead
Model and Pattern Makers
Morticians

Nickel Workers

Office Workers

Painters
Paper (and Pulp) Workers
Pesticide Workers
Petrochemical Workers
Petrol Station Attendants
Petroleum Workers
Pharmaceutical Industry Workers
Physicists
Plastics Workers
Plutonium Workers
Policemen
Potters

Power Plant Workers
Printers

Quarry Workers

Radiation Workers
Railroad Workers
Rubber Workers

Sawmill Workers
Screw Cutters
Sewage Workers
Shipyard Workers
Shoemakers-Repairers
Smelters, Aluminium
Smelters, Copper
Smelters, Lead
Steel Workers

Taxi Drivers
Textile Dryers
Textile Workers
Tool Makers
Transport Workers

Varnishers
Vinyl Chloride Workers

Waiters
Waste Incineration Workers
Welders
Wood Workers

OCCUPATION

Acrylonitrile Workers
All Sites	594, 1114
Bladder	910
Brain	910
Colon	1030
Haemopoietic	910
Liver	1030
Lung	594, 910, 1030, 1114
Lymphoma	910
Mesothelioma	910
Prostate	594
Sarcoma	910
Stomach	910

Administrative Workers
All Sites	*433

Agricultural Workers
All Sites	294, *504, 508
Bladder	311, 910
Brain	*504, 910
Gastrointestinal	1197
Haemopoietic	910
Lung	92, 910, 1197
Lymphoma	910
Mesothelioma	910
Non-Hodgkin's Lymphoma	*504
Sarcoma	910
Soft Tissue	*504
Stomach	910

Aluminium Workers
Lung	584

Antimony Process Workers
All Sites	955
Lung	881, 955

Asbestos Textile Workers
Gastrointestinal	1186
Lung	1186

Asbestos Workers
All Sites	46, 86, 474, 787, 1130
Breast (F)	1130
Colon	689, 1130
Gastrointestinal	1186
Lung	31, 46, 689, 860, 1130, 1186
Mesothelioma	31, 37, 46, 502, 689, 860
Peritoneum	502
Pleura	502, 860
Rectum	689
Respiratory	787
Stomach	689
Uterus (Cervix)	1130

Automobile Workers
All Sites	927, 1033

OCCUPATION

Bladder	910
Brain	910, 1033
Colon	791
Haemopoietic	910
Larynx	1033
Lung	791, *792, 910, 927
Lymphoma	910
Mesothelioma	910
Rectum	791
Sarcoma	910
Stomach	791, *792, 910, 927

Battery Plant Workers

All Sites	*270
Larynx	863
Lung	*270
Stomach	*270

Bus Drivers

All Sites	507, *1176

Cabinet Makers

All Sites	474
Bladder	910
Brain	910
Haemopoietic	910
Leukaemia	201
Lung	910
Lymphoma	910
Mesothelioma	910
Nasal Cavity	201
Respiratory	201
Sarcoma	910
Stomach	910

Cable Manufacturers

Liver	961
Lymphoma	961
Soft Tissue	961

Carpenters, Joiners

Leukaemia	201
Nasal Cavity	201
Respiratory	201

Cellophane Manufacturers

Leukaemia	781

Cement Workers

All Sites	236, 869
Gastrointestinal	102
Liver	236
Mesothelioma	102
Respiratory	102, 236, 463

Chemical Industry Workers

All Sites	86, 114, 294, 370, *591, *698, 868, *945, 1097, 1158, 1171

OCCUPATION

Bladder	834, 1171
Brain	1029, 1041
Gastrointestinal	1029
Haemopoietic	1029
Inapplicable	195, 597
Kidney	1041
Liver	1029, 1041
Lung	175, 1029, 1041, 1171
Lymphoma	*698
Multiple Myeloma	*698
Nasopharynx	868
Non-Hodgkin's Lymphoma	*591
Respiratory	868
Sarcoma	*591
Soft Tissue	*591
Stomach	1171
Wilms' Tumour	65

Chemists

Bladder	910
Brain	910
Haemopoietic	910
Leukaemia	*975, 1118
Lung	910
Lymphoma	910, *975, 1118
Melanoma	1006
Mesothelioma	910
Pancreas	1118
Sarcoma	910
Stomach	910

Chimney Sweeps

Haemopoietic	*1129
Lung	*1129
Lymphoma	*1129
Oesophagus	*1129
Prostate	*1129

Chromate Pigment Workers

All Sites	1155
Lung	1155

Chromate Producing Workers

All Sites	540

Chromium Plating Workers

Liver	525
Lung	525

Coke-Oven Workers

All Sites	540, 1101
Genitourinary	1101
Haemopoietic	1101
Lung	*457
Lymphoma	1101
Respiratory	1101

OCCUPATION

Construction and Maintenance Workers
    All Sites   920

Cryolite Workers
    All Sites   230

Dockers
    Lung   719
    Skin   *901

Drivers
    All Sites   487
    Inapplicable   597
    Larynx   487
    Lung   487

Dry Cleaners
    Bladder   910
    Brain   910
    Haemopoietic   910
    Lung   910
    Lymphoma   910
    Mesothelioma   910
    Sarcoma   910
    Stomach   910

Dyestuff Workers
    Angiosarcoma   677
    Bladder   311, 677
    Liver   677
    Lung   677
    Wilms' Tumour   65

Electrical Workers
    All Sites   313
    Hodgkin's Disease   513
    Leukaemia   513, *627, 629, *783
    Multiple Myeloma   513
    Non-Hodgkin's Lymphoma   513

Electrode Manufacturers
    All Sites   540, 710
    Lung   710

Electronics Workers
    Brain   946
    Childhood Neoplasms   946

Engineering Workers
    All Sites   920
    Leukaemia   *975
    Lymphoma   *975

OCCUPATION

Farmers
All Sites	28, *82, 86, 479, *504
Bladder	309, 910
Brain	*82, 94, 254, 309, 479, *504, 910, 941
Childhood Neoplasms	*1085
Haemopoietic	479, 910
Hodgkin's Disease	91, 254, 462, 513
Inapplicable	911
Kidney	309
Leukaemia	28, *82, 254, 309, 513, *627, *922, 941
Lung	92, 254, 910
Lymphoma	309, 910, *922
Mesothelioma	910
Multiple Myeloma	*82, 91, 94, 309, 513
Non-Hodgkin's Lymphoma	*82, 91, 94, 462, *504, 513
Pancreas	309
Prostate	*82, 254, 309, *601
Sarcoma	91, 94, 309, 910
Skin	479
Soft Tissue	28, 91, 94, *504
Stomach	910

Fertilizer Workers
All Sites	114
Bladder	843
Lung	*697, 843
Oesophagus	843
Stomach	*697, 843

Firemen
All Sites	86, 107, 1120
Bladder	910
Brain	910, 1120
Colon	1120
Haemopoietic	910
Leukaemia	1120
Lung	*902, 910
Lymphoma	910
Mesothelioma	910
Rectum	1120
Sarcoma	910
Stomach	910

Fishermen
All Sites	*690
Skin	*901

Forest Workers
All Sites	294, 701
Hodgkin's Disease	91
Leukaemia	*627
Multiple Myeloma	91
Non-Hodgkin's Lymphoma	91
Sarcoma	91
Soft Tissue	91

Foundry Workers
All Sites	321, 927, 1057
Larynx	321
Lung	321, 927

OCCUPATION

    Oral Cavity   321
    Pharynx   321
    Stomach   927

Gas Workers
    All Sites   313

Glass Workers
    All Sites   474
    Inapplicable   195

Grain Millers
    Leukaemia   *922
    Lymphoma   *922

Grinding Material Workers
    Gastrointestinal   968
    Lung   968
    Non-Hodgkin's Lymphoma   968

Health Care Workers
    All Sites   *551, 929, 934, 1012, 1045, 1185
    Brain   94, *551, 941
    Breast (F)   929, 1045
    Inapplicable   196, *343, 597, 656
    Leukaemia   941, 958, 1045
    Lung   1045
    Multiple Myeloma   94
    Non-Hodgkin's Lymphoma   94
    Sarcoma   94
    Soft Tissue   94
    Thyroid   1045

Heavy Equipment Operators
    Bladder   910
    Brain   910
    Haemopoietic   910
    Lung   910
    Lymphoma   910
    Mesothelioma   910
    Sarcoma   910
    Stomach   910

Herbicide Manufacturers
    All Sites   249, 294, 628, 1119
    Brain   1119
    Leukaemia   1119
    Lung   1119
    Lymphoma   12, 249, 628
    Multiple Myeloma   1119
    Non-Hodgkin's Lymphoma   1119
    Sarcoma   12, 249
    Soft Tissue   628, 1119

Herbicide Sprayers
    All Sites   249, 294, 628, 1119
    Bladder   910

OCCUPATION

Brain	910, 1119
Haemopoietic	910
Leukaemia	*627, 1119
Lung	910, 1119
Lymphoma	12, 249, 628, 910
Mesothelioma	910
Multiple Myeloma	1119
Non-Hodgkin's Lymphoma	1119
Sarcoma	12, 249, 910
Soft Tissue	628, 1119
Stomach	910

Horticulturists
All Sites	*504
Brain	94, *504
Hodgkin's Disease	91
Multiple Myeloma	91, 94
Non-Hodgkin's Lymphoma	91, 94, *504
Sarcoma	91, 94
Soft Tissue	91, 94, *504

Insulation Board Manufacturers
All Sites	865
Gastrointestinal	865
Lung	865, 871
Mesothelioma	865, 871
Respiratory	865

Jewellery Manufacturers
Bladder	910
Brain	910
Haemopoietic	910
Lung	910
Lymphoma	910
Mesothelioma	910
Sarcoma	910
Stomach	910

Laboratory Workers
All Sites	298, *329, *433, 702, *705
Melanoma	1006

Lacquerers
Haemopoietic	647
Leukaemia	201
Nasal Cavity	201
Respiratory	201

Laminators
All Sites	*280
Inapplicable	251

Leather Workers
All Sites	*306, 474, *696
Bladder	*306
Hodgkin's Disease	513
Leukaemia	513
Multiple Myeloma	513

OCCUPATION

Nasal Cavity *306
Non-Hodgkin's Lymphoma 513
Wilms' Tumour 65

Machinists
Colon 791
Gastrointestinal 968
Lung 789, 791, 968
Non-Hodgkin's Lymphoma 968
Rectum 791
Stomach 791

Meat Workers
Brain 941
Leukaemia *627, 941
Lung *864

Mechanics
Brain 941
Hodgkin's Disease 513
Leukaemia 513, 941
Multiple Myeloma 513
Non-Hodgkin's Lymphoma 513
Prostate *601

Metal Workers
Childhood Neoplasms *1085
Colon 1031
Gastrointestinal 968
Kidney 1031
Liver 1031
Lung 968, 1031
Non-Hodgkin's Lymphoma 968
Prostate *601
Retinoblastoma *1086
Skin 1031

Military Servicemen
All Sites *1103

Mineral Fibre Workers
All Sites 86, 295, 870
Inapplicable 204
Lung 295, 870
Mesothelioma 870
Respiratory 1095

Miners
All Sites 86, 182, 920
Bladder 910
Brain 910
Haemopoietic 910
Lung *127, *548, 732, 910
Lymphoma 910
Mesothelioma *548, 910
Sarcoma 910
Stomach 910

OCCUPATION

Miners, Asbestos
Colon	30
Gastrointestinal	*73
Kidney	*73
Larynx	*73
Lung	30, 31, *73, 199, 860
Mesothelioma	30, 31, 37, 860
Peritoneum	30, *73
Pleura	*73, 860
Stomach	30

Miners, Coal
Brain	*25
Gastrointestinal	1196
Hodgkin's Disease	36
Inapplicable	195
Leukaemia	1196
Leukaemia (CLL)	36
Lung	40, 323, 777, 1196
Non-Hodgkin's Lymphoma	36

Miners, Copper
Lung	*156
Skin	*993

Miners, Fluorspar
All Sites	*83
Lung	*83

Miners, Gold
Lung	*32

Miners, Iron
Lung	198, 322, 721
Stomach	322

Miners, Uranium
All Sites	86, 264
Lung	3, 78, 157, 264, 952, 954
Lymphoma	954

Miners, Zinc-Lead
All Sites	114
Lung	458
Stomach	458

Model and Pattern Makers
Colon	964, *1142
Rectum	964, *1142

Morticians
All Sites	86
Bladder	910
Brain	910
Haemopoietic	910
Lung	910

OCCUPATION

    Lymphoma   910
    Mesothelioma   910
    Sarcoma   910
    Stomach   910

Nickel Workers
    All Sites   86, 1102
    Colon   1102
    Liver   1102
    Nasal Cavity   1102
    Rectum   1102
    Respiratory   1102

Office Workers
    Respiratory   463

Painters
    Leukaemia   *627

Paper (and Pulp) Workers
    All Sites   *281
    Haemopoietic   *684
    Lung   252, *684, 721
    Lymphoma   *684
    Respiratory   683
    Stomach   683, *684

Pesticide Workers
    All Sites   861
    Bladder   *159

Petrochemical Workers
    Inapplicable   195
    Kidney   *973

Petrol Station Attendants
    All Sites   487
    Larynx   487
    Lung   487

Petroleum Workers
    All Sites   16, *475, 920, 950, 1080
    Brain   941
    Haemopoietic   16
    Kidney   42, 729, 1081
    Leukaemia   941
    Lymphoma   16
    Wilms' Tumour   65

Pharmaceutical Industry Workers
    Breast (F)   235
    Uterus (Corpus)   235

Physicists
    Leukaemia   *975

OCCUPATION

Lymphoma *975
Melanoma 1006

Plastics Workers
All Sites *280, 687, *692, *695, 713, 756, 956, 1097
Angiosarcoma 677
Bladder 677
Colon *974, *976
Inapplicable 251, 942
Leukaemia 713
Liver 677
Lung 677
Lymphoma 713
Rectum *974, *976

Plutonium Workers
All Sites 1008

Policemen
All Sites 507

Potters
All Sites 474, 866, 920
Lung 866

Power Plant Workers
All Sites *55, 182
Leukaemia *55, *790
Lymphoma *55
Non-Hodgkin's Lymphoma *790

Printers
Lung 789

Quarry Workers
Lung 788

Radiation Workers
All Sites 76, 86, 354, *766, 773, 796, 797, 882, 947, 1110
Childhood Neoplasms *826
Inapplicable 74

Railroad Workers
All Sites 294, 516

Rubber Workers
All Sites 920, 1177
Bladder 311
Brain 941
Leukaemia 941
Lung *342
Mesothelioma *342

OCCUPATION

Sawmill Workers
Leukaemia　　　　　　　　*627

Screw Cutters
All Sites　　　　　　　　266

Sewage Workers
All Sites　　　　　　　　733

Shipyard Workers
All Sites　　　　　　　　477, 920
Lung　　　　　　　　　　860
Mesothelioma　　　　　　860
Pleura　　　　　　　　　860

Shoemakers-Repairers
Inapplicable　　　　　　*343

Smelters, Aluminium
Bladder　　　　　　　　*459
Lung　　　　　　　　　*459
Lymphoma　　　　　　　*459
Pancreas　　　　　　　*459

Smelters, Copper
All Sites　　　　　　　　553, 724
Lung　　　　　　　　　　553, 724

Smelters, Lead
All Sites　　　　　　　　114, *270, *691
Bladder　　　　　　　　*461
Kidney　　　　　　　　*461, *722
Leukaemia　　　　　　　*461
Liver　　　　　　　　　*461
Lung　　　　　　　　　*270, *461, *722
Stomach　　　　　　　　*270

Steel Workers
All Sites　　　　　　　　*794, 948, 1101
Genitourinary　　　　　1101
Haemopoietic　　　　　1101
Lung　　　　　　　　　104
Lymphoma　　　　　　　1101
Respiratory　　　　　　1101

Taxi Drivers
All Sites　　　　　　　　507

Textile Dyers
All Sites　　　　　　　　*484
Bladder　　　　　　　　*484
Nasal Cavity　　　　　*484
Pharynx　　　　　　　　*484
Respiratory　　　　　　*484

OCCUPATION

Textile Workers
All Sites	474, 920
Angiosarcoma	677
Bladder	677
Liver	677
Lung	472, 677

Tool Makers
Gastrointestinal	968
Lung	968
Non-Hodgkin's Lymphoma	968

Transport Workers
Lung	31
Mesothelioma	31

Varnishers
Haemopoietic	647

Vinyl Chloride Workers
All Sites	*126
Angiosarcoma	677
Bladder	677
Brain	260
Liver	*126, 677
Lung	677

Waiters
Larynx	*637
Liver	*637
Oesophagus	*637
Oral Cavity	*637
Pharynx	*637

Waste Incineration Workers
All Sites	1057

Welders
All Sites	320
Childhood Neoplasms	*1085
Gastrointestinal	968
Inapplicable	*343
Liver	234
Lung	968
Non-Hodgkin's Lymphoma	968
Retinoblastoma	*1086

Wood Workers
All Sites	*306, 474
Bladder	*306
Brain	941
Leukaemia	941
Nasal Cavity	275, *306

INDEX OF COUNTRIES

All studies are indexed, by site, under the country (or countries), territory or region where the data are being collected, so that, say, studies of cancer of the pancreas in Poland or nasopharyngeal cancer in Greenland can be readily identified.

Country names used in the Directory conform as far as possible to the UN list of member states.

When data are being collected in a different country from that where the principal investigator is based, or in more than one country, the countries concerned are listed beneath the study abstract. The study will be indexed under each of the countries where data are being collected.

Large international studies are coded to the country of the principal investigator co-ordinating the project.

'Germany, Federal Republic of' has been replaced by 'Germany'.

New entries this year are:

Estonia
Morocco
Venezuela

The list below identifies all entries in use in this Directory.

Algeria	Hong Kong	Paraguay
Argentina	Hungary	Philippines
Australia		Poland
Austria	Iceland	Portugal
	India	Puerto Rico
Belgium	Indonesia	
Bolivia	Ireland	Romania
Brazil	Israel	Rwanda
Bulgaria	Italy	
		San Marino
Canada	Japan	South Africa
Chile		Spain
China	Kenya	Sri Lanka
Colombia		Sweden
Costa Rica	Luxembourg	Switzerland
Cuba		
Czechoslovakia	Malawi	Tanzania
	Malaysia	Thailand
Denmark	Malta	Turkey
	Mexico	
Egypt	Morocco	Uganda
Estonia		United Kingdom
	Namibia	United States of America
Finland	Netherlands	Uruguay
France	New Caledonia	USSR
	New Zealand	
Gambia	Nigeria	Venezuela
Germany	Norway	
Ghana		Yugoslavia
Greece	Pakistan	
Greenland	Papua-New Guinea	

COUNTRY

Algeria
All Sites 1, *281
Stomach 847

Argentina
All Sites 286, *291
Odontogenic Neoplasms 2
Oral Cavity *359
Oropharynx *359
Respiratory *359
Uterus (Cervix) *359

Australia
All Sites 13, 16, 28, 39, 286, *291, 294
Benign Tumours 27
Bile Duct 300
Bladder 288
Brain *19, *25, *34, 41, 301, 302
Breast (F) *10, 18, *20, *22, 288, 1134
Choriocarcinoma 5
Colon 4, 18, *21, 24, 30, 288
Gallbladder 300, 1134
Haemopoietic 16
Hodgkin's Disease 36
Hydatidiform Mole 5
Inapplicable 938
Kaposi's Sarcoma *38
Kidney 42, 729
Leukaemia 28
Leukaemia (CLL) 36
Liver 1134
Lung 3, 30, 31, *32, 40, 288
Lymphoma 12, 16, *38
Melanoma 6, 8, *14, 17, 27, 35, 43, 288
Mesothelioma 26, 30, 31, 37, 39
Myelodysplastic Syndrome 15
Non-Hodgkin's Lymphoma 36
Oesophagus 288
Oral Cavity 23
Ovary 1134
Pancreas 300
Peritoneum 30, 39
Pleura 39
Prostate 18, 288
Rectum 4, 7, 18, *21, 24, 288
Sarcoma 12
Skin 11, 17, 29, 35
Soft Tissue 28
Spinal Cord *19
Stomach 30, 288
Uterus (Cervix) *9, *33, 288, 1134
Uterus (Corpus) 288, 1134

Austria
All Sites *44, 46, 294, 375, 659
Colon 45
Haemopoietic 47
Leukaemia 287
Lung 46
Melanoma 336
Mesothelioma 46
Ovary 506

COUNTRY

    Rectum   45
    Thyroid   314

Belgium
    All Sites   *53, 54, *55, 276
    Brain   *49
    Breast (F)   *52
    Colon   262
    Gastrointestinal   332
    Head and Neck   *604
    Inapplicable   50
    Larynx   51
    Leukaemia   *49, 54, *55, 332
    Lung   332
    Lymphoma   *49, 54, *55
    Melanoma   304
    Mesothelioma   51
    Nasopharynx   51
    Neuroblastoma   *49
    Non-Hodgkin's Lymphoma 332
    Ovary   506
    Stomach   847
    Thyroid   314
    Uterus (Cervix)   48

Bolivia
    Bile Duct   1091
    Gallbladder   1091

Brazil
    All Sites   *281, 286, *291, 573, *794
    Breast (F)   *56, *58, 573
    Colon   *56, *58
    Gastrointestinal   *57
    Liver   *58, 573
    Lung   *56, *58
    Oesophagus   *56, *58
    Oral Cavity   59
    Oropharynx   59
    Prostate   *58, 573
    Rectum   *58
    Stomach   *56, *58, 573, *575
    Uterus (Cervix)   *58
    Wilms' Tumour   65

Bulgaria
    All Sites   659
    Ovary   506

Canada
    All Sites   60, 61, 63, *68, 75, 76, 77, 79, *80, *82, *83, 84, *85, 86, 87, *101, 107,
                                                        109, 113, 114, 120, *216, *281, 286, *291, 294, *306, 913, 1016, 1101,
                                                        1109
    Bile Duct   300
    Bladder   66, *80, 100, 105, *306
    Bone   103
    Brain   *80, *81, *82, 94, 301, 302, 1041, 1090
    Breast (F)   62, 69, 70, *85, 89, 96, 106, 111, *119, 303, 913, 1013
    Childhood Neoplasms   980, *1085, 1089, 1090
    Colon   66, 69, *80, *85, 106, 303
    Gallbladder   300

COUNTRY

Gastrointestinal	67, *73, 102
Genitourinary	1101
Haemopoietic	305, 1101, 1109
Head and Neck	95
Hodgkin's Disease	66, 91
Inapplicable	74
Kidney	66, *73, *80, 106, 1041
Larynx	64, *73
Leukaemia	*80, *82, 115, 116, 221, 305, 913
Leukaemia (ALL)	*72, 221
Leukaemia (CLL)	221
Lip	64, *99
Liver	66, 1041
Lung	66, 67, *73, 78, *83, 92, 93, 97, 104, 105, *121, 292, 860, 913, *925, 1041
Lymphoma	*80, 1101, 1109
Melanoma	66, *101, 304
Mesothelioma	67, 102, 860
Multiple Myeloma	*82, 91, 94
Myelodysplastic Syndrome	115
Nasal Cavity	*306
Neuroblastoma	*1050, *1100
Non-Hodgkin's Lymphoma	66, *82, 91, 94
Oesophagus	66
Oral Cavity	64, 118
Ovary	106, 108
Pancreas	66, 105, 106, 300
Peritoneum	*73
Pharynx	64
Pleura	*73, 860
Prostate	66, 69, *80, *82, 90, *1144
Rectum	*80, 106, 303
Respiratory	102, 1101
Retinoblastoma	*1086
Sarcoma	79, 91, 94
Skin	1109
Soft Tissue	91, 94
Stomach	66, *80
Testis	66, 108
Thyroid	98, 913
Uterus (Cervix)	88, 110, 112
Uterus (Corpus)	106, 221
Wilms' Tumour	1099

Chile

Bile Duct	*123
Breast (F)	1134
Gallbladder	122, *123, 1134
Liver	1134
Ovary	1134
Uterus (Cervix)	1134
Uterus (Corpus)	1134

China

All Sites	*124, *126, 130, 138, 139, *143, 166, 174, 182, 183, *306, 375, 913
Bladder	*159, 176, *306, 910
Bone	170
Brain	170, 301, 910
Breast (F)	125, 129, 138, *144, *155, 169, 172, 297, 303, 913, 1134
Colon	125, *144, *145, 146, *147, *148, *149, 158, 164, *165, 303
Gallbladder	1134
Haemopoietic	910
Larynx	180
Leukaemia	125, 134, 170, 179, 913

COUNTRY

Liver	125, *126, 136, *144, *160, 161, *162, 163, 167, 170, 172, *173, *181, 1134
Lung	*124, 125, *127, *132, 133, *137, *144, *151, *153, *156, 157, *171, 172, 175, 910, 912, 913
Lymphoma	170, 910
Mesothelioma	910
Nasal Cavity	180, *306
Nasopharynx	125, 140, 141, 142, 150, 178
Oesophagus	125, 128, 131, *160, *165, 185, 186, 187, 188, 282, 357
Oral Cavity	180
Ovary	*162, 1134
Pancreas	*165
Pharynx	180
Rectum	125, *144, *145, 146, *147, *148, *149, 158, 164, *165, 172, 303
Sarcoma	910
Stomach	125, 135, 136, 138, *144, 152, *160, 172, *181, 184, 910
Thyroid	913
Uterus (Cervix)	154, *162, 168, 172, 177, 1134
Uterus (Corpus)	*162, 1134

Colombia

Breast (F)	1134
Gallbladder	1134
Liver	1134
Ovary	1134
Uterus (Cervix)	283, 1134
Uterus (Corpus)	1134

Costa Rica

Liver	189
Uterus (Cervix)	921

Cuba

All Sites	190
Breast (F)	190
Lung	190

Czechoslovakia

All Sites	191, 192, 193, 375, 659, 913
Breast (F)	303, 913
Breast (M)	296
Childhood Neoplasms	*764
Colon	303
Inapplicable	195, 196
Leukaemia	287, 913
Lung	*194, 198, 199, 913
Melanoma	192
Rectum	303
Skin	192
Thyroid	913
Uterus (Cervix)	197

Denmark

All Sites	203, 205, 209, 213, *216, 217, 222, *225, 228, 230, *280, *281, 294, 295, 913
Bladder	208, 233, 910
Brain	208, 910
Breast (F)	220, 222, 223, 231, 233, 913, 917
Breast (M)	206
Childhood Neoplasms	217, 218
Colon	*211, 214, *226, 227, 917

COUNTRY

Gastrointestinal	208, 231
Haemopoietic	208, 910
Hodgkin's Disease	213
Inapplicable	50, 204
Kidney	42, 729
Larynx	232
Leukaemia	201, 217, 221, 223, *224, 231, 287, 642, 913
Leukaemia (ALL)	221
Leukaemia (CLL)	221, 223
Liver	217
Lung	*212, 222, 232, 233, 295, 910, 913, 917
Lymphoma	910
Male Genital	208
Melanoma	231, 304
Mesothelioma	910
Nasal Cavity	201
Oral Cavity	200
Ovary	506, 917
Prostate	229, 233
Rectum	*211, 214, 227
Respiratory	201, 208
Sarcoma	208, 910
Skin	208
Small Intestine	*226
Stomach	847, 910
Testis	202, 207, 215, 231
Thyroid	314, 913
Uterus (Cervix)	210, 222, 231, 233
Uterus (Corpus)	221, 231, 917

Egypt

All Sites	236
Breast (F)	235
Liver	234, 236
Respiratory	236
Uterus (Corpus)	235

Estonia

Breast (F)	237, *238

Finland

All Sites	228, 242, 243, 247, 248, 249, 250, *280, *281, 294, 295, 298, *306, 375, *694, 913
Benign Tumours	*257
Bladder	*306
Brain	254
Breast (F)	243, 244, 253, 255, *257, 303, 913
Breast (M)	206
Colon	243, 244, 253, 255, 303
Female Genital	256
Hodgkin's Disease	254
Inapplicable	251, 938
Larynx	253
Leukaemia	221, 253, 254, 287, 642, 913
Leukaemia (ALL)	221
Leukaemia (CLL)	221
Lip	253
Lung	241, 243, 244, 245, 252, 253, 254, 255, 295, 913
Lymphoma	249
Male Genital	256
Nasal Cavity	*306
Oesophagus	253
Pancreas	246

COUNTRY

Prostate	243, 244, 253, 254
Rectum	243, 244, 253, 255, 303
Sarcoma	249
Skin	255
Stomach	239, 240, 243, 244, 253
Thyroid	913
Uterus (Cervix)	243, 253, 258
Uterus (Corpus)	221, 253

France

All Sites	75, 264, 265, 266, *270, 276, 277, *281, 286, *291, 293, 298, *306, *310, 313, 320, 321, 327, *329, 330, 331
Bladder	278, 288, *306, 309, 311
Brain	260, 301, 302, 309
Breast (F)	274, 288, 303, 312
Breast (M)	296
Childhood Neoplasms	307, *319, 1089
Colon	261, 262, 263, 288, 303
Gastrointestinal	332
Haemopoietic	305
Head and Neck	267
Hypopharynx	269, 317
Inapplicable	50
Kidney	71, 309, 325
Larynx	317, 321, 741
Leukaemia	287, 305, 309, *328, 331, 332
Leukaemia (ALL)	*279
Liver	261
Lung	264, *270, 278, 288, 292, 295, 321, 322, 323, 326, 331, 332, *333, 741
Lymphoma	309
Melanoma	288, 304
Multiple Myeloma	309
Myelodysplastic Syndrome	*268
Nasal Cavity	275, *306, 316
Nasopharynx	*315
Non-Hodgkin's Lymphoma	331, 332
Oesophagus	261, 288
Oral Cavity	269, 321, 741
Oropharynx	269
Ovary	506
Pancreas	300, 309
Pharynx	321, 741
Prostate	*259, 288, 309, 324
Rectum	261, 263, 288, 303
Sarcoma	309
Stomach	*270, 288, 322
Thyroid	314, 318
Uterus (Cervix)	272, 288
Uterus (Corpus)	288, 308

Gambia

All Sites	*334, 496
Liver	335, 898

Germany

All Sites	293, 294, 295, 298, *306, 344, 353, 354, 356, 366, 370, 375, 659, 913
Bladder	*306
Bone	368, 372
Brain	302, 339
Breast (F)	*338, 352, 361, 371, *615, 913, 1134
Childhood Neoplasms	*373, *374
Colon	262, 352, 355, *363, 369, 371
Gallbladder	1134

Gastrointestinal	332, 344
Haemopoietic	350, *351
Head and Neck	*604
Inapplicable	50, *343
Kidney	42, 71, 365, 372, 729
Larynx	*360
Leukaemia	287, 332, *340, 349, 364, 368, *374, 913
Liver	368, 372, 1134
Lung	292, 295, 332, 337, 341, *342, 368, 913
Lymphoma	*340, 368, *374
Melanoma	304, 336, 345
Mesothelioma	*342, 346
Nasal Cavity	*306
Neuroblastoma	*374
Non-Hodgkin's Lymphoma	332
Oral Cavity	118
Ovary	352, 506, 1134
Pleura	346
Prostate	371
Rectum	352, 355, *363, 369, 371
Respiratory	344
Skin	371
Spleen	368
Stomach	352, 847
Thyroid	314, 358, *362, 913
Uterus (Cervix)	371, 1134
Uterus (Corpus)	1134
Wilms' Tumour	*374

Ghana
Retinoblastoma	812

Greece
All Sites	276, *281, 293, 375
Brain	302
Breast (F)	303, *615
Breast (M)	296
Childhood Neoplasms	*764
Colon	303
Kaposi's Sarcoma	377
Liver	378
Ovary	506
Rectum	303
Stomach	847
Thyroid	376

Greenland
All Sites	*216, 219

Hong Kong
Lung	379, 380

Hungary
All Sites	375, 659
Bladder	386
Breast (F)	389
Gastrointestinal	386
Haemopoietic	387
Head and Neck	382
Inapplicable	383, 384
Leukaemia	287, 387

COUNTRY

Lung	381, 387
Oral Cavity	382
Ovary	506
Stomach	388
Testis	385
Thyroid	98

Iceland

All Sites	228
Appendix	390
Bile Duct	390
Breast (F)	*391, *392, *393
Colon	390
Leukaemia	642
Nervous System	390
Oesophagus	390
Oral Cavity	390
Pancreas	390
Prostate	390
Small Intestine	390
Soft Tissue	390
Stomach	847

India

All Sites	299, *412, *421, *761
Brain	415
Breast (F)	*399, 400, 407, *409, 416, *424, *428
Childhood Neoplasms	*403, *422
Colon	*410
Hodgkin's Disease	414
Hypopharynx	*411, *420
Larynx	*411, *420, 423
Leukaemia	414, *417
Lung	292, 401, *411, 423
Lymphoma	414, *417
Multiple Myeloma	414
Nasopharynx	*402
Oesophagus	*410, *420, 423
Oral Cavity	117, 118, 394, 404, 405, 408, *411, 413, *420, 423, *426, *427, 805
Oropharynx	397, *411, *420
Ovary	396, *398
Rectum	*410
Stomach	*410
Tongue	*411
Uterus (Cervix)	395, 400, *406, 418, *419, *425

Indonesia

All Sites	430, 431, 432
Breast (F)	534
Lung	429
Nasopharynx	429
Uterus (Cervix)	429

Ireland

All Sites	298, *433
Breast (F)	303, *615
Colon	303
Inapplicable	50
Rectum	303

COUNTRY

Israel
All Sites	*291, 439, 443, 444, 445, 450, 451, 913
Benign Tumours	451
Bladder	288
Brain	302
Breast (F)	288, 441, 442, 446, 447, *615, 913, 1134
Childhood Neoplasms	434, 438, 440
Colon	288, *436, 441, 443, 447, 449, 452
Gallbladder	1134
Head and Neck	441
Leukaemia	441, 913
Liver	1134
Lung	288, 437, 443, 913
Lymphoma	441
Melanoma	288
Oesophagus	288, 435
Ovary	506, 1134
Prostate	288, 448
Rectum	288, *436, 447, 452
Stomach	288
Thyroid	441, 913
Uterus (Cervix)	288, 1134
Uterus (Corpus)	288, 1134

Italy
All Sites	*280, *281, 293, 294, 295, 298, *306, 375, *454, 460, 464, 471, 474, *475, 476, 477, 478, 479, 483, *484, 485, 487, 488, *493, 495, *497, *504, *505, 507, 508, 516
Benign Tumours	491
Bladder	*306, *459, *461, *484, 489, 512
Brain	301, 479, *504
Breast (F)	303, 473, 482, 491, *492, *505, *515
Breast (M)	296
Childhood Neoplasms	1089
Choriocarcinoma	490
Colon	262, 303, 494, 498, *505
Female Genital	453
Gastrointestinal	332, 453, 476
Haemopoietic	305, 479
Hodgkin's Disease	455, 462, 511, 513
Hydatidiform Mole	490
Inapplicable	50
Kaposi's Sarcoma	455, 501
Kidney	71, *461, 481, 489
Larynx	487, 489
Leukaemia	287, 305, 332, *461, 483, 510, 513
Leukaemia (ALL)	511
Leukaemia (ANLL)	511
Liver	*461, 465
Lung	292, 295, 332, *456, *457, 458, *459, *461, 464, 469, 470, 472, 487, 489, 499
Lymphoma	453, *459, 481, 483
Melanoma	468
Mesothelioma	469, 502
Multiple Myeloma	486, 513
Myelodysplastic Syndrome	510
Nasal Cavity	*306, *484
Neuroblastoma	511
Non-Hodgkin's Lymphoma	332, 455, 462, *504, 511, 513
Oesophagus	489
Oral Cavity	489
Ovary	491, 506
Pancreas	*459, 489
Peritoneum	502
Pharynx	*484, 489

COUNTRY

	Pleura	469, 502
	Rectum	303, 494, 498
	Respiratory	463, *484
	Skin	453, 479, 500, 514
	Soft Tissue	483, *504, 509
	Stomach	458, 467, 480, 503, *505, 846, 847
	Thyroid	314, 453, 466
	Urinary Tract	453
	Uterus (Cervix)	491
	Uterus (Corpus)	491
	Vulva	491
	Wilms' Tumour	511

Japan

	All Sites	*281, *306, 517, 520, 521, 531, *532, 533, *538, 539, 540, *543, 545, *551, 552, 553, 557, 559, 560, *564, 572, 574, 576, 913
	Bile Duct	558
	Bladder	*306
	Brain	*551
	Breast (F)	527, 534, 913
	Childhood Neoplasms	561, 569, 570, 571
	Colon	527, 537, 565, 566, 572
	Gastrointestinal	542
	Kidney	529
	Leukaemia	*549, 555, 571, 913
	Liver	525, 528, 541, 554, 558, 566, 568, 572
	Lung	525, 535, 541, 547, *548, 553, 566, 567, 572, 574, 577, 913
	Lymphoma	524, 555
	Mesothelioma	546, *548
	Nasal Cavity	*306
	Neuroblastoma	562
	Oesophagus	537
	Peritoneum	546
	Pleura	546
	Rectum	527, 537, 565, 572
	Retinoblastoma	*530
	Skin	524
	Stomach	517, 518, *522, 526, 527, 536, 537, 544, 550, *556, 563, 566, 572, 574, 847
	Thyroid	913
	Uterus (Cervix)	517, 519
	Uterus (Corpus)	566

Kenya

	Breast (F)	1134
	Gallbladder	1134
	Liver	1134
	Ovary	1134
	Stomach	845
	Uterus (Cervix)	48, 1134
	Uterus (Corpus)	1134

Luxembourg

	Inapplicable	50

Malawi

	Kaposi's Sarcoma	803
	Soft Tissue	803

COUNTRY

Malaysia
Nasopharynx   578, 579, 580
Uterus (Cervix)   *581

Malta
Lung   653

Mexico
Bile Duct   1091
Breast (F)   583, 1134
Gallbladder   1091, 1134
Leukaemia   958
Liver   1134
Ovary   1134
Uterus (Cervix)   582, 1134
Uterus (Corpus)   1134

Morocco
Nasopharynx   *315

Namibia
Childhood Neoplasms   671

Netherlands
All Sites   276, *281, 293, 294, 298, 589, *591, 594, 596, 606, 1158
Benign Tumours   586, 587
Bile Duct   300, *592, *900
Bladder   614
Brain   1041
Breast (F)   586, 587, 588, *590, 595, 598, 600, 602, 606, 609, 610, *611, 613, *615
Breast (M)   296
Childhood Neoplasms   1089
Colon   595, 612, 613, 616, *900
Gallbladder   300
Gastrointestinal   332
Head and Neck   *604
Inapplicable   50, 597
Kidney   1041
Leukaemia   287, 332, 589, 605
Leukaemia (ALL)   605
Liver   1041
Lung   332, 584, 589, 594, 595, 613, 614, 1041
Melanoma   304, 599
Myelodysplastic Syndrome   589
Non-Hodgkin's Lymphoma   332, *591
Ovary   506, 607, 608
Pancreas   300, *592, *900
Prostate   594, 595, *601
Rectum   595, 612, 613, *900
Sarcoma   *591
Soft Tissue   *591
Stomach   595, 613, *900
Thyroid   314
Uterus (Cervix)   585, *593, 603

New Caledonia
All Sites   617

COUNTRY

New Zealand
All Sites	*281, 294, 624, 628
Breast (F)	618, 625, *626
Childhood Neoplasms	*620, *621
Haemopoietic	305
Leukaemia	305, *627, 629
Leukaemia (ALL)	*621
Lymphoma	628
Melanoma	304
Soft Tissue	628
Uterus (Cervix)	619

Nigeria
All Sites	375
Breast (F)	632
Kaposi's Sarcoma	631
Liver	630

Norway
All Sites	228, *280, *281, 295, 633, 636, *694, 913
Breast (F)	633, 913
Breast (M)	206
Colon	634, 639, 640
Gastrointestinal	633
Larynx	*637
Leukaemia	221, 287, 641, 642, 913
Leukaemia (ALL)	221, 641
Leukaemia (CLL)	221
Liver	*637
Lung	295, 913
Lymphoma	641
Melanoma	638
Oesophagus	*637
Oral Cavity	*637
Ovary	633
Pharynx	*637
Rectum	634, 639, 640
Thyroid	314, 635, 913
Uterus (Cervix)	633
Uterus (Corpus)	221, 633

Pakistan
All Sites	644
Bone	644
Breast (F)	644
Gallbladder	643
Uterus (Cervix)	644

Papua-New Guinea
Burkitt's Lymphoma	645

Paraguay
Melanoma	*290
Oesophagus	646

Philippines
Breast (F)	*289, 1134
Gallbladder	1134
Liver	1134

COUNTRY

Ovary	1134
Uterus (Cervix)	1134
Uterus (Corpus)	1134

Poland

All Sites	*281, 650, 652, *654, *657, *658, 659, 660, 1177
Bile Duct	300
Bladder	288
Breast (F)	288, 303, *615, 649, 650
Colon	288, 303
Gallbladder	300
Haemopoietic	647
Inapplicable	656
Larynx	655
Leukaemia	287
Lung	288, 648, 650, 653, *657
Melanoma	288
Oesophagus	288
Pancreas	300
Prostate	288
Rectum	288, 303
Stomach	288, 650, 651, 847
Uterus (Cervix)	288, 649, 650
Uterus (Corpus)	288, 649

Portugal

All Sites	*281
Breast (F)	*615
Colon	262
Lung	292
Stomach	847
Thyroid	314

Puerto Rico

All Sites	926

Romania

All Sites	659, 661

Rwanda

Non-Hodgkin's Lymphoma	348

San Marino

Stomach	967

South Africa

Childhood Neoplasms	671
Head and Neck	666
Larynx	666
Leukaemia	669
Liver	*664, 668
Mesothelioma	667, 670
Peritoneum	667
Pleura	667
Stomach	663
Uterus (Cervix)	662, 665

COUNTRY

Spain
All Sites	*281, 293, 675
Angiosarcoma	677
Bladder	677
Brain	301
Breast (F)	303, *615
Childhood Neoplasms	*764
Colon	273, 303, 674, *676
Gallbladder	*676
Kidney	71
Liver	271, *676, 677
Lung	292, 677
Oesophagus	672, *676
Pancreas	672, *676
Rectum	273, 303, 674, *676
Small Intestine	*676
Stomach	672, 673, *676
Thyroid	314
Uterus (Cervix)	283

Sri Lanka
Oral Cavity	678, 805

Sweden
All Sites	*225, 228, *280, *281, 293, 294, 295, 298, 680, 687, *690, *691, *692, *693, *694, *695, *696, *698, 700, 701, 702, *705, 706, 709, 710, 711, 712, 713, 718, 724, 726, 733, 737, 752, 913
Bladder	718
Brain	302, 703, 704
Breast (F)	708, 717, 723, 731, 736, 737, 738, 913, 917
Breast (M)	206
Childhood Neoplasms	704, 707, 715
Colon	679, 686, 689, 718, 720, 734, 752, 917
Haemopoietic	*684, *1129
Inapplicable	938
Kidney	42, *722, 729
Larynx	*716
Leukaemia	287, 642, *693, 704, 711, 713, 913, *1001
Lung	292, 295, 682, *684, 689, *697, 699, 710, 714, 718, 719, 721, *722, 724, 732, 913, 917, *1129
Lymphoma	*684, *693, *698, 713, *1129
Mesothelioma	689, 735
Multiple Myeloma	*698
Oesophagus	*716, *1129
Oral Cavity	*716
Ovary	506, 681, 717, 737, 738, 917
Peritoneum	735
Pharynx	*716
Pleura	735
Prostate	727, *1129
Rectum	679, 686, 689, 718, 734, 752
Respiratory	683
Stomach	683, *684, 689, *697, 718, 728
Thyroid	314, 688, 725, *730, 913
Uterus (Cervix)	685, 717
Uterus (Corpus)	717, 737, 917

Switzerland
All Sites	*281, 286, *454, 739, 740, 742, 743
Bone	739
Breast (F)	303, *615, 739
Breast (M)	296
Childhood Neoplasms	739

COUNTRY

Colon	303, 740
Gastrointestinal	739
Inapplicable	50
Kidney	71
Leukaemia	287, 739
Lung	292, 740
Lymphoma	739
Melanoma	336
Ovary	506, 739
Prostate	324
Rectum	303
Stomach	740
Testis	739
Thyroid	314, 744
Uterus (Corpus)	745

Tanzania

Stomach	746
Uterus (Cervix)	367

Thailand

Breast (F)	1134
Gallbladder	1134
Liver	747, 1134
Ovary	1134
Uterus (Cervix)	*748, 1134
Uterus (Corpus)	1134

Turkey

All Sites	920
Bladder	910
Brain	910
Haemopoietic	910
Lung	860, 910
Lymphoma	910
Mesothelioma	860, 910
Ovary	506
Pleura	860
Prostate	*750
Sarcoma	910
Stomach	910
Testis	749

Uganda

Non-Hodgkin's Lymphoma	348
Penis	347
Uterus (Cervix)	347
Vulva	347

United Kingdom

All Sites	276, *280, *281, 286, *291, 293, 294, 295, 298, *306, 450, *751, 752, 756, *761, *766, 773, 774, 775, 785, 787, *796, 797, *801, 802, 807, 810, *811, 815, 818, 821, 822, 825, 832, 833, 835, 836, 853, 856, 861, 865, 866, 867, 868, 869, 870, 875, *876, 882, 913
Bladder	288, *306, 798, 833, 834, 843
Breast (F)	288, 303, *615, 623, *771, *780, 782, 785, 786, 813, 819, 822, 825, 841, 855, 856, 857, 858, 877, 878, 913
Breast (M)	296
Childhood Neoplasms	753, 754, 755, *764, *765, *826, *827, 831, 838, 839, 849, 850, 854, 1089
Colon	262, 288, 303, 752, 782, 791, 793, 820, 828, *829, *830, 841

COUNTRY

Female Genital	822, 858
Gastrointestinal	*814, 855, 859, 865
Haemopoietic	305, 768
Inapplicable	50, 763
Kidney	808
Larynx	*801, 863
Leukaemia	287, 305, 753, 754, 762, 781, 782, *783, *790, *801, 804, 823, 831, 833, 835, 836, 839, 848, 854, 867, 874, 913
Leukaemia (ALL)	768, *779, 823
Leukaemia (AML)	*779
Leukaemia (CLL)	823
Leukaemia (CML)	*779
Liver	*801, 806, 872, 873
Lung	288, 295, *776, 777, 785, 788, 789, 791, *792, *801, 807, 808, 837, 843, 860, *864, 865, 866, 870, 871, *880, 881, 913
Lymphoma	762, 782, *801, 867
Melanoma	288, 622, *809, 844
Mesothelioma	*784, 860, 865, 870, 871
Myelodysplastic Syndrome	768, *770
Nasal Cavity	*306, *809
Nasopharynx	868
Neuroblastoma	824, 854
Non-Hodgkin's Lymphoma	753, *790, 823
Oesophagus	288, 843
Oral Cavity	805
Ovary	506, 800
Pleura	860
Prostate	288, 795, 807, 808, 841, 851, *879
Rectum	288, 303, 752, 791, 820, 828, *829, *830
Respiratory	787, 865, 868
Retinoblastoma	812, 838
Salivary Gland	757
Sarcoma	825
Skin	*769, *772
Soft Tissue	825
Stomach	288, *778, 785, 791, *792, 841, 843, 846, 847, 862
Testis	*767, 782, 840, 842, *876
Thyroid	314, 913
Urinary Tract	*759
Uterus (Cervix)	288, 760, 799, *816, *817, 852, 856
Uterus (Corpus)	288, 857

United States of America

All Sites	86, *216, 222, *281, 286, *291, 294, *306, 450, 758, 885, 891, 892, 895, 913, *923, 926, 927, 929, 930, 931, *933, 934, 944, *945, 947, 948, 950, 951, 953, 955, 956, 959, 960, 969, 977, 979, 981, 986, 989, 994, 999, 1003, 1007, 1008, 1011, 1012, 1015, 1016, 1018, 1032, 1033, 1037, 1042, 1043, 1045, 1057, 1058, 1064, 1073, 1074, 1080, 1082, 1094, 1096, 1097, 1098, 1101, 1102, *1103, *1108, 1109, 1110, 1111, 1113, 1114, 1117, 1119, 1120, 1130, *1138, 1155, 1158, 1159
Benign Tumours	965, 997, 1052, 1077, 1105
Bile Duct	1091
Bladder	288, *306, *905, 910, *928, 949, 957, 960, 971, 1034, 1070, 1096
Brain	897, 910, *924, 941, 946, 1002, 1024, 1025, 1029, 1033, 1041, 1088, 1090, *1107, 1119, 1120
Breast (F)	222, 288, 441, 886, 890, 894, 906, 913, 915, 918, 929, *935, 936, 937, 944, 963, *978, 990, 991, 997, 1003, 1009, 1013, 1018, 1019, *1020, 1035, 1043, 1045, *1048, 1052, *1053, 1055, 1060, 1063, 1066, 1067, *1068, 1075, 1076, 1077, 1130, 1133, 1143, *1148, 1152
Breast (M)	296, 990
Childhood Neoplasms	919, 946, 980, 996, 1039, *1049, *1085, 1089, 1090, *1106
Colon	288, 441, 886, 896, 915, *935, 936, 943, 964, 971, *974, *976, *982, 983, 987, 994, *995, 998, 1003, 1004, 1030, 1031, 1036, 1044, 1055, 1059, 1060, 1061, 1065, 1079, 1102, 1105, 1120, *1122, 1130, 1140, *1142, 1143, *1145, *1156
Eye	931, 971

COUNTRY

Female Genital	886
Gallbladder	1017, 1091
Gastrointestinal	896, 968, 1029
Genitourinary	1101
Haemopoietic	910, 1029, 1101, 1109
Head and Neck	441
Hodgkin's Disease	1093, 1126
Inapplicable	523, 911, 938, 940, 942, 985
Kaposi's Sarcoma	*899, 909, 979
Kidney	42, 71, 729, *973, 1031, 1034, 1041, 1070, *1078, 1081, *1104
Larynx	1033, 1070, 1135
Leukaemia	221, 441, 885, *887, 895, 913, *922, *928, 941, 944, 958, 959, *975, *1001, 1045, 1046, *1106, 1118, 1119, 1120, 1159
Leukaemia (ALL)	221
Leukaemia (AML)	1023
Leukaemia (ANLL)	1047
Leukaemia (CLL)	221
Lip	1022
Liver	630, 885, 908, 961, 971, 1029, 1030, 1031, 1041, 1084, 1087, 1102, *1106
Lung	222, 288, 860, 884, *902, 907, 910, 913, 915, *925, 927, *928, *935, 952, 954, 955, 968, *970, 971, 1000, 1003, *1014, 1018, 1029, 1030, 1031, 1041, 1045, 1056, 1064, 1070, 1071, 1092, 1114, 1119, 1121, 1130, *1131, 1137, 1143, 1149, 1154, 1155
Lymphoma	441, 885, 909, 910, *922, 954, 961, *975, 1010, *1038, 1083, 1101, 1109, 1118
Melanoma	*14, 288, 971, 988, 1006, 1054, 1115, 1151
Mesothelioma	860, 910, 971, *1131
Multiple Myeloma	972, 1119
Nasal Cavity	*306, 1021, 1102
Nasopharynx	885, 1021, 1098, 1136
Neuroblastoma	*1050, *1100
Non-Hodgkin's Lymphoma	904, 968, *1106, 1119, 1127
Oesophagus	288, 971, 972, 1070, 1135
Oral Cavity	*1026, *1027, 1051, 1070, 1072, *1132, 1135
Oropharynx	1072, *1132
Ovary	886, 915, *935, 936, 937, 991, 1069, 1075, *1139, 1157
Pancreas	937, 959, 972, 1070, *1106, 1118, 1143
Pharynx	*1026, *1027
Pleura	860
Prostate	288, *935, 936, 937, 939, *966, 972, 1003, 1028, 1055, *1062, 1143, *1144
Rectum	288, 896, 943, 964, 971, *974, *976, 983, 987, *995, 1004, 1036, 1044, 1055, 1059, 1060, 1061, 1065, 1079, 1102, 1105, 1120, *1142, 1143
Respiratory	1095, 1101, 1102, 1149
Retinoblastoma	*1086
Salivary Gland	885, 971
Sarcoma	910
Skin	*901, 931, *992, *993, 1031, 1055, 1109, 1123, *1146, *1147, 1150, 1151
Soft Tissue	961, 1005, *1106, 1119
Stomach	288, 544, 847, 910, 927, 959, 967, 971, *1112, *1116, 1135
Testis	965
Thyroid	441, 913, 1045, 1128
Uterus (Cervix)	222, 288, 883, 885, 886, 888, 889, 890, 893, 903, *932, *962, 1055, 1060, 1075, *1124, 1130, 1141, 1153
Uterus (Corpus)	221, 288, 886, 914, 915, *935, 937, 984, 1018, 1043, 1055, 1075, 1125
Vagina	916, *1040, 1076
Vulva	916
Wilms' Tumour	1099

Uruguay

All Sites	286, *1164, 1165, *1166, *1167
Larynx	1163
Lung	1162
Oral Cavity	1163

COUNTRY

	Pharynx	1163
	Stomach	1161

USSR
	All Sites	*216, 1171, *1176, 1177, 1183, 1185, 1188
	Bladder	1171, 1184
	Brain	302
	Breast (F)	1169, 1172, *1174, 1181
	Childhood Neoplasms	*764
	Colon	1182
	Gastrointestinal	1186
	Haemopoietic	305, 1168
	Kidney	1175
	Leukaemia	287, 305, 1175
	Lung	1171, 1172, *1173, 1179, 1180, 1186, *1187
	Lymphoma	1175
	Melanoma	304
	Nervous System	1175
	Neuroblastoma	1175
	Oesophagus	1170, 1178
	Oral Cavity	1178
	Ovary	506, *1174
	Rectum	1182
	Respiratory	*1187
	Retinoblastoma	1175
	Stomach	1171, 1172
	Uterus (Cervix)	*1174
	Uterus (Corpus)	*1174

Venezuela
	Stomach	*284, *285

Yugoslavia
	All Sites	375, 659, 913
	Breast (F)	913, 1189, 1190, 1192, 1193, 1195
	Breast (M)	296
	Gastrointestinal	1189, 1196, 1197
	Genitourinary	1189
	Hodgkin's Disease	1194
	Inapplicable	50
	Leukaemia	287, 913, 1196
	Lung	913, 1196, 1197
	Ovary	506, 1191
	Stomach	847
	Thyroid	913

INDEX OF CANCER REGISTRIES

The index of Cancer Registries identifies each project in which a registry is involved. Each registry has been given a short title for indexing. The index is in alphabetical order by country and short name.

The index is followed by an address list. In addition to the short title, the name of the director or contact person and the full address of the registry are given.

The list only contains population-based cancer registries. For information about hospital-based registries, consult the UICC International Directory of Specialized Cancer Research and Treatment Establishments (UICC, Geneva, Switzerland, 1986).

In the body of this Directory, studies with cancer registry involvement are identified beneath each abstract under the heading 'REGI', followed by the short title and the country name in abbreviated form.

New registries are included as project abstracts for two consecutive years, then transferred to the address list.

We would be grateful if readers could notify us of any errors or omissions in the list or the index. Registry assignation to studies may be incomplete or incorrect, or the short title we have chosen may be inappropriate. We need your help to create a complete and correct index which reflects the full range of research in which cancer registries are involved.

CANCER REGISTRY

Australia
Brisbane II (Aus)   4, 24
NSW (Aus)   8, *25, 35, 36, 41, 42
Queensland (Aus)   8, 11
S. Australia (Aus)   3
Sydney (Aus)   26, 37, 39
Tasmania (Aus)   15
Victoria (Aus)   12, 17, 18, *19, *20, *22
W. Australia (Aus)   26, *32

Belgium
Belgium (Bel)   48, *49, *52, *55

Canada
Alberta (Can)   60, 61
Br. Columbia (Can)   113, *119, *1144
Canada (Can)   76, *85, 86, 105, *216, *281, *1050
Manitoba (Can)   120, *121
N.W. Territory (Can)   77
Newfoundland (Can)   77
Ontario (Can)   63, 108, *1144
Quebec (Can)   77
Saskatchewan (Can)   92, 93, 1109

China
Shanghai (Chi)   164, *165, 166, 169, 170, 174, 178, 179, 180
Tianjin (Chi)   183

Czechoslovakia
Slovakia (Cze)   192, 287

Denmark
Denmark (Den)   200, 201, 203, 205, 206, 208, 210, *211, *212, 214, 215, *216, 217, 218, 219, 220, 221, 222, 223, *224, *225, 228, 232, 233, *280, *281, 287, 294, 295, 917

Estonia
Estonia (Est)   237, *238, 287

Finland
Finland (Fin)   206, 221, 228, 240, 242, 243, 244, 246, 247, 248, 249, 250, 252, 255, *257, 258, *280, *281, 287, 294, 295

France
Caen (Fra)   261
Dijon I (Fra)   263
Doubs (Fra)   260
Isère (Fra)   *281, 287, *310
Lorraine (Fra)   *319
Marseille (Fra)   307
Pacific Islands (Fra)   617
Rhône (Fra)   274
Tarn (Fra)   *259, 324

Gambia
Gambia (Gam)   *334, 335, 873

CANCER REGISTRY

Germany
Berlin (FRG)   287, 339
Mainz (FRG)   287, 364, *373, *374
Saarland (FRG)   344

Hungary
Szabolcs-Szatmár (Hun)   388, 389

Iceland
Iceland (Ice)   228, *391, *392, *393

India
Bombay (Ind)   407, *411, *412
Madras (Ind)   416

Israel
Israel (Isr)   434, 435, 437, 440, 441, 442, 443, 444, 445, 451

Italy
Genoa (Ita)   478
Latina (Ita)   *504, *505
Modena (Ita)   494
Piedmont (Ita)   287, 514
Tuscany (Ita)   472
Varese (Ita)   481, 482, *484, 485

Japan
Aichi (Jap)   533
Kanagawa (Jap)   577
Miyagi (Jap)   *564
Nagasaki (Jap)   531
Nagoya (Jap)   550
Osaka (Jap)   544

Lithuania
Lithuania (Lit)   287

Martinique
Martinique (Mar)   *259, 324

Netherlands
Amsterdam (Net)   *590, 595, 596, 597, 613
Eindhoven (Net)   *590, 595, 596, 597, 606
Groningen (Net)   595, 596, 597, 613
Leeuwarden (Net)   595, 596, 597
Leiden (Net)   *590, 595, 596, 597, 613
Leiderdorp (Net)   594, 595, 596, 597
Maastricht (Net)   595, 596, 597, 613
Nijmegen (Net)   *590, 595, 596, 597, 599, *601, 613
Rotterdam (Net)   *590, 595, 596, 597, 613
The Hague (Net)   287, 605
Tilburg (Net)   595, 596, 597, 613
Utrecht (Net)   595, 596, 597, *611, 613, 616

CANCER REGISTRY

New Zealand
New Zealand (NZ)　　　　　*281, 294, *620, 624, *626, 628, 629

Norway
Norway (Nor)　　　　　　　206, 221, 228, *280, *281, 287, 295, 634, 635, 636, *637, 638, 641

Poland
Cracow (Pol)　　　　　　　650
Poland (Pol)　　　　　　　287

Spain
Basque (Spa)　　　　　　　672
Mallorca (Spa)　　　　　　675
Navarra (Spa)　　　　　　 672
Zaragoza (Spa)　　　　　　672

Sweden
Lund (Swe)　　　　　　　　689, *691, *716
Stockholm (Swe)　　　　　 *716
Sweden (Swe)　　　　　　　206, 228, *270, *280, *281, 287, 294, 295, 679, 683, *684, 685, *690, *692, *693, *695, *697, *698, 702, 704, *705, 706, 707, 708, 709, 711, 714, 715, 717, 718, *722, 723, 725, 734, 736, 917, *1129
Uppsala (Swe)　　　　　　 729, *730, 738

Switzerland
Basle (Swi)　　　　　　　 739
Geneva (Swi)　　　　　　　*259, *281, 287, 324, 741
Vaud (Swi)　　　　　　　　744, 745

United Kingdom
Birmingham (UK)　　　　　 755
E. Anglia (UK)　　　　　　*767, 804
Mersey (UK)　　　　　　　 804
Newcastle (UK)　　　　　　824, 825, *826, *827
Northern UK (UK)　　　　　*772
OPCS (UK)　　　　　　　　 777, 786, *790, 791, *792, 797, *801, 810, 818, 833, 855, 857, 859, 866, 869, 875
Oxford I (UK)　　　　　　 287, 804
Oxford II (UK)　　　　　　824, 831, 838, 839, 848, 849, 850, 854
S. Western (UK)　　　　　 832
Scotland (UK)　　　　　　 287
Scotland N. (UK)　　　　　813
Thames (UK)　　　　　　　 804, *876
W. Midlands (UK)　　　　　*759, 804
Yorkshire (UK)　　　　　　*761, *784, 804

United States of America
Arizona (USA)　　　　　　 *1148
Atlanta (USA)　　　　　　 891, 972
Bay Area (USA)　　　　　　979, 1006, 1007, 1024, 1076, *1078, *1122, *1144
California (USA)　　　　　*1078
Connecticut (USA)　　　　 *99, 441, 1113
Delaware (USA)　　　　　　*1050
Florida (USA)　　　　　　 *1050
Hawaii (USA)　　　　　　　987, 988, 989, *1144
Idaho (USA)　　　　　　　 1121
Illinois (USA)　　　　　　948, 949
Iowa (USA)　　　　　　　　949, 1043

CANCER REGISTRY

Los Angeles (USA)	1007, 1009, 1010, 1019, *1020, 1024, 1025, *1026, *1027, *1144
Maine (USA)	1035
Maryland (USA)	981
Massachusetts (USA)	1035
Michigan (USA)	971, 972
Minnesota (USA)	*1050, *1122
Missouri (USA)	907
N. Carolina (USA)	*1106, *1107, *1108
Nebraska (USA)	*1116
New Hampshire (USA)	1035
New Jersey (USA)	972
Orange County (USA)	1000
Seattle (USA)	1024, 1125, *1132, 1135, 1137, *1138, *1139, 1140
SEER (USA)	544, 897, *924, 926, 1113, *1131
Utah (USA)	1121, *1122
Wisconsin (USA)	949, 1035, 1036

Uruguay
Uruguay (Uru)   *1164, 1165, *1166, *1167

USSR
Leningrad (USSR)   287
Moscow (USSR)   1171, 1185

Yugoslavia
Slovenia (Yug)   287, 1192, 1193, 1194, 1195

POPULATION-BASED CANCER REGISTRIES

The list below contains the address and the name of the director or contact person for 266 population-based cancer registries in 52 countries around the world. Most registries are general, collecting data on all malignant tumours in both sexes and at all ages for persons resident in their territory, but some are specialised registries, restricted to an age-group (e.g. childhood registries) or a group of sites (haematological, gynaecological tumours, etc.).

Each registry has been assigned a short title for indexing, and this will be found below the abstract of each project in which the registry is collaborating.

For each country, the registries are listed in alphabetical order of their short title.

Please contact the registries directly for further information about possible research collaboration.

Incidence data from many of these registries have been published in Cancer Incidence in Five Continents, Vol. V, eds: C.S. Muir, J. Waterhouse, T. Mack, J. Powell and S. Whelan, IARC Scientific Publications No. 88, 1987; in Cancer Occurrence in Developing Countries, Ed: D.M. Parkin, IARC Scientific Publications No. 75, 1986 and in International Incidence in Childhood Cancer, Eds: D.M. Parkin, C.A. Stiller, G.J. Draper, C.A. Bieber, B. Terracini and J.L. Young, IARC Scientific Publications No. 87, 1988.

Algeria (Alg)

Algiers
Dr L. Abid
Registre des Cancers Digestifs d'Alger
Service de Chirurgie
Hôpital Bologhine
El-Hamamet 160 160
Alger
Algeria

Sétif
Dr M. Hamdi-Chérif
Service de Médecine Préventive
& Epidémiologie
CHU de Sétif
Hôpital Mère-Enfant
Sétif
Algeria

Argentina (Arg)

La Plata
Dr E. Perez Arias
Direccion de Planeamiento
y Desarollo
Ministerio de Salud de la Placia
de Buenos Aires
1900 La Plata
Argentina

Australia (Aus)

Brisbane I
Dr W.R. McWhirter
Queensland Childhood Malignancy Registry
Department of Child Health
Royal Childrens Hospital
Herston, QLD 4029
Australia

Brisbane II
Dr D. Battistutta
Registry of Colorectal
Familial Adenomatous Polyposis
Queensland Institute of Medical Research
Department of Cancer Epidemiology
Bramston Terrace
Herston
Brisbane, QLD 4006
Australia

Brisbane III
Dr S.K. Khoo
Registry of Gestational Trophoblastic Disease
University of Queensland
Department of Obstetrics & Gynaecology
Clinical Sciences Building
Royal Brisbane Hospital
Brisbane, QLD 4029
Australia

Hobart
Director
Registry of Lymphoproliferative
& Myeloproliferative Disorders
University of Tasmania
Department of Community Health
43 Collins Street
Hobart, TAS 7000
Australia

N.Territory
Dr G. Durling
Northern Territory Cancer Registry
Department of Health & Community Services
P.O. Box 1701
Darwin, NT 5794
Australia

CANCER REGISTRY

NSW
Prof. R. Taylor
NSW Central Cancer Registry
P.O. Box 380
North Ryde, NSW 2113
Australia

Perth I
Dr L.E. Dougan
Cancer Council of Western Australia
Leukaemia & Allied Disorders Registry
184 St. George's Terrace
Perth, WA 6005
Australia

Perth II
Dr V.J. Ojeda
Sir Charles Gairdner Hospital
Adult Brain Tumour Registry
of Western Australia
Dept. of Histopathology
Verdun St.
Perth, WA 6009
Australia

Queensland
Dr I. Ring
Epidemiology & Prevention Unit
(Non-communicable Diseases)
Queensland Department of Health
G.P.O. Box 48
147-163 Charlotte Street
Brisbane, QLD 4001
Australia

S. Australia
Dr A.Z. Bonnet
Central Cancer Registry Unit
South Australian Health Commission
Box 6, Rundle Mall P.O.
Adelaide, SA 5001
Australia

Sydney
Dr S.J. Corbett
Australian Mesothelioma Register
National Institute for Occupational Health & Safety
G.P.O. Box 58
Sydney, NSW 2001
Australia

Tasmania
Ms D. Shugg
Tasmanian Cancer Registry
G.P.O. Box 252C
Hobart, TAS 7001
Australia

Victoria
Dr G.G. Giles
Victorian Cancer Registry
Cancer Epidemiology Center
Keogh House
1 Rathdowne Street
Carlton South, VICT 3053
Australia

W. Australia
Dr C.D.J. Holman
Western Australia Cancer Registry
Epidemiology Branch
Health Dept. of Western Australia
Curtin House
60 Beaufort Street
Perth, WA 6000
Australia

Belgium(Bel)

Belgium
Dr M. Haelterman
Belgian Cancer Registry
Belgian Cancer Foundation
Twee Kerkenstraat 21
1040 Brussels
Belgium

Bolivia (Bol)

La Paz
Dr J.L. Rios Dalenz
Registro de Cancer de La Paz
Calle Bollivian 1266
Casilla Postale 2801
La Paz
Bolivia

Brazil (Bra)

Fortaleza
Dr M. Gurgel Carlos da Silva
Registro de Cáncer do Ceará
Rua Papi Junior 1222
Bairro Rodolfo Teófilo
60430 Fortaleza-Ceará
Brazil

Porto Alegre
Dr A. da Motta Lima Netto
Cancer Registry of Porto Alegre
Av. Washington Luiz, 868
90310 Porto Alegre, RS
Brazil

CANCER REGISTRY

Recife
Dr M.R. da Costa Carvalho
Registro de Cancer de Pernambuco
Departamento de Patologia
Faculdade de Medicina
de Universidade Federal de Pernambuco
Cidade Universitária
Rua Prof. Moraes Rego, s/n
50730 Recife, Pernambuco
Brazil

Sao Paulo
Dr A.P. Mirra
Registro de Cancer de Sao Paulo
Avenida Dr Arnaldo, 715-1°A
01255 Sao Paulo, SP
Brazil

Bulgaria (Bul)

Sofia
Prof. C.G. Tzvetansky
National Oncological Centre
Epidemiology and Cancer Control
Plovdivsko pole 6
1156 Sofia
Bulgaria

Canada (Can)

Alberta
Dr H.J. Berkel
Alberta Cancer Registry
Department of Epidemiology
& Preventive Oncology
Alberta Cancer Board
6th Floor, Capital Place
9707 - 110 Street
Edmonton, Alberta
Canada T5K 2L9

Br. Columbia
Dr A.J. Klaassen
British Columbia Cancer Registry
Cancer Control Agency of British Columbia
600 West 10th Avenue
Vancouver, British Columbia
Canada V5Z 4E6

Canada
Mrs L. Gaudette
National Cancer Incidence
Reporting System of Canada
Health Division
Statistics Canada
R.H. Coats Building
Tunney's Pasture
Ottawa, Ontario
Canada K1A 0T6

Manitoba
Dr H. Schipper
Manitoba Cancer Treatment
& Research Foundation
100 Olivia Street
Winnipeg, Manitoba
Canada R3E 0V9

N.W. Territories
Dr L. Barreto
Northwest Territories Region
Health & Welfare Canada
900 Liberty Building
10506 Jasper Ave.
Edmonton, Alberta
Canada T5J 2W9

New Brunswick
Dr G.D. Smith
New Brunswick Provincial
Tumor Registry
c/o Dept. of Radiation Oncology
Saint John Regional Hospital
P.O. Box 2100
Saint John, New Brunswick
Canada E2L 4L2

Newfoundland
Mrs D.L. Ball
Provincial Tumor Registry
25 Kenmount Rd
Saint John's, Newfoundland
Canada A1B 1W1

Nova Scotia
Mrs T. Croucher
Provincial Cancer Registry
of Nova Scotia
Cancer Treatment
& Research Foundation of Nova Scotia
5820 University Avenue
Halifax, Nova Scotia
Canada 1V7

Ontario
Dr E.A. Clarke
The Ontario Cancer Registry
7 Overlea Boulevard
Toronto, Ontario
Canada M4H

Pr. Edward Is.
Dr D.E. Dryer
Division of Cancer Control
Department of Health & Social Services
P.O. Box 3000
Charlottetown, PEI
Canada C1A 7N8

CANCER REGISTRY

Québec
Mr G.-P. Sanscartier
Fichier des Tumeurs du Québec
Ministère de la Santé
& des Services Sociaux
1075, chemin Ste Foy, 3ème étage
Québec, Québec
Canada G1S 2M1

Saskatchewan
Mrs D. Robson
Saskatchewan Cancer Foundation
400-2631 28th Avenue
Regina, Saskatchewan
Canada S4S 6X3

China (Chi)

Shanghai
Dr Y.T. Gao
Shanghai Cancer Registry
Shanghai Cancer Institute
2200 Xie Tu Road
200032 Shanghai
China

Tianjin
Dr Tian-ze Zhang
Tianjin Cancer Institute
Huan Hu Xi Road
Ti-Yuan-Bei
Tianjin 300060
China

Colombia (Col)

Cali
Dr E. Carrascal Cortes
Cancer Registry
Department of Social Medicine
University of Valle
Sede San Fernando
Apartado Aereo 25360
Cali
Colombia

Cuba (Cub)

Cuba
Dra S. Gonzalez
Registro Nacional de Cancer
Instituto Nacional de Oncologia
29 y F. Vedado
Habana
Cuba

Czechoslovakia (Cze)

Czechoslovakia
Dr J. Augustin
Cancer Epidemiology Dept.
Research Inst. of Clinical
& Experimental Oncology
Zluty Kopec 7
656-53 Brno
Czechoslovakia

Slovakia
Dr I. Plesko
National Cancer Registry
of Slovakia
Oncological Centre
Ul. Csl. Armady 21
812-32 Bratislava
Czechoslovakia

Denmark (Den)

Copenhagen
Dr K.W. Anderson
Danish Breast Cancer
Cooperative Group
Finsen Institutet
Strandblvd 49
2100 Copenhagen
Denmark

Denmark
Danish Cancer Registry
Rosenvaengets Hovedvej, 35
Box 839
2100 Copenhagen
Denmark

Estonia (Est)

Estonia
Dr M. Rahu
Estonian Cancer Registry
Inst. of Experimental & Clinical Medicine
42 Hiiu Street
Tallinn 200107
Estonia

Finland (Fin)

Finland
Prof. L. Teppo
Finnish Cancer Registry
Liisankatu 21 B
00170 Helsinki 17
Finland

CANCER REGISTRY

France (Fra)

Ain
Dr J. Bruhière
Registre des Cancers Digestifs
du Département de l'Ain
28 bis, avenue du Mail
01000 Bourg-en-Bresse
France

Angers
Dr B. Thomas
Registre des Tumeurs Bronchiques
en Maine-et-Loire
79, Avenue Pasteur
49100 Angers
France

Ardèche
Dr F. Olaya
Registre d'Apparition des Cancers
en Ardèche du Nord
1, rue Sadi Carnot
07100 Annonay
France

Bas-Rhin
Prof. P. Schaffer
Registre Bas-Rhinois des Tumeurs
Institut d'Hygiène
Faculté de Médecine
4, rue Kirschleger
67085 Strasbourg Cédex
France

Caen
Dr D. Pottier
Registre des Tumeurs Digestives
du Calvados
CHU Côte de Nacre
Pièce 703
14040 Caen Cédex
France

Calvados
Dr J.M. Robillard
Registre Général des Tumeurs du Calvados
Route de Lion sur Mer
14021 Caen Cédex
France

Côte d'Or
Prof. P.M. Carli
Registre des Hémopathies Malignes
en Côte d'Or
Laboratoire d'Hématologie
Hôpital du Bocage
2, blvd de Lattre de Tassigny
21034 Dijon Cédex
France

Dijon I
Prof. J. Faivre
Registre Bourguignon
des Cancers Digestifs
Faculté de Médecine
7, blvd Jeanne d'Arc
21033 Dijon Cédex
France

Dijon II
Dr D.B. Chaplain
Registre Bourguignon
de Pathologie Gynécologique
Centre Georges-François Leclerc
1, rue du Professeur Marion
21034 Dijon Cédex
France

Doubs
Prof. S. Schraub
Registre des Tumeurs du Doubs
Centre Hospitalier Régional
1, blvd Fleming
25030 Besançon Cédex
France

Essonne
Dr H. Gauthier
Registre des Cancers de l'Essonne
2 bis, rue du Ronneau
91150 Etampes
France

Haute Garonne
Dr P. Pienkowski
Registre des Cancers Digestifs
de la Haute Garonne, CHU Rangueil
Chemin du Vallon
31054 Toulouse Cédex
France

Haut-Rhin
Dr A. Buemi
Registre Haut-Rhinois des Tumeurs
CHR Mulhouse
Service d'Anatomie Pathologique
87, avenue d'Altkirch
B.P. 1070
68051 Mulhouse Cédex
France

Hérault
Dr J.-P. Daurès
Registre des Tumeurs de l'Hérault
B.P. 4111
34091 Montpellier Cédex
France

Isère
Dr F. Ménégoz
Registre du Cancer du
Département de l'Isère
21, chemin des Sources
38240 Meylan
France

CANCER REGISTRY

Lille
Prof. L. Adenis
Registre des Cancers
des Voies Aéro-Digestives Supérieures
des Départements du Nord
et du Pas-de-Calais
Centre Oscar Lambret
B.P. 307
59037 Lille
France

Limousin
Dr J. Grasser
Registre Général des Cancers du Limousin
COPAS
6, rue de l'Amphithéâtre
87000 Limoges
France

Lorraine
Prof. D. Sommelet-Olive
Registre Pédiâtrique de Lorraine
Hôpital d'Enfants
Service de Pédiatrie II
Allée du Morvan
54511 Vandoeuvre Cédex
France

Lyon
Dr A. Brémond
Centre de Dépistage
des Maladies du Sein
INSERM U 265
151 cours Albert-Thomas
69434 Lyon Cédex 03
France

Marseille
Prof. J.-L. Bernard
Registre des Cancers de l'Enfant
Faculté de Médecine
27, blvd Jean Moulin
13385 Marseille Cédex 05
France

Nice
Dr M. Héry
Registre des Tumeurs Cérébrales
Primaires de l'Adulte
Centre A. Lacassagne
36 Voie Romaine
06054 Nice Cédex
France

Nouméa
Dr M. Huerre
Registre du Cancer
de Nouvelle-Calédonie
Institut Pasteur de Nouméa
B.P. 61
Nouméa, New Caledonia
France

Pacific Islands
Dr F.H. Bach
Pacific Islands Cancer Registry
Box D5
Nouméa Cédex, New Caledonia
France

Réunion
Dr H. Chamouillet
Registre des Cancers
Ile de la Réunion
Conseil Général DASSD
Service Informatique
Rue Hippolyte Foucques
B.P. 400
97400 Sainte Clotilde (La Réunion)
France

Somme
Prof. A. Lorriaux
Association pour la Recherche
Epidémiologique par les Registres
en Picardie
Registre du Cancer de la Somme
Médecine 4ème étage
B.P. 3006
80030 Amiens Cédex
France

Tarn
Dr M. Roumagnac
Registre des Cancers du Tarn
Chemin des Trois Tarns
81000 Albi
France

Vaucluse
Dr P. Martin
Association pour la Registration
des Tumeurs du Vaucluse (ARTV)
B.P. 843
84003 Avignon Cédex
France

Gabon (Gab)

Libreville
Université Omar Bongo
Centre Universitaire
des Sciences de la Santé
B.P. 4009
Libreville
Gabon

Gambia (Gam)

Gambia
Mrs M. Mendy
The Gambia Cancer Registry
c/o MRC Laboratories
Fajara, P.O. Box 273
Banjul
Gambia

CANCER REGISTRY

Germany (FRG)

Berlin
Dr M. Möhner
Akademie der Wissenschaften
Zentralinstitut für Krebsforschung
Bereich Nationales Krebsregister
& Krebsstatistik
Sterndamm 13
O 1197 Berlin-Johannisthal
Federal Republic of Germany

Hamburg
Dr W. Thiele
Hamburgisches Krebsregister
Gesundheitsbehörde
Postfach 2524
Tesdorpfstrasse 8
W 2000 Hamburg 13
Federal Republic of Germany

Mainz
Prof. J.H. Michaelis
Cooperative Register
of Childhood Malignancies
Institut für Medizin, Statistik
& Dokumentation (IMSD)
Universität Mainz
Langenbeckstrasse 1
W 6500 Mainz
Federal Republic of Germany

Saarland
Mr H. Ziegler
Statistisches Amt des Saarlandes
Krebsregister
Postfach 409
Hardenbergstrasse 3
W 6600 Saarbrücken
Federal Republic of Germany

Tübingen
Prof R. Schrage
Krebsregister Baden-Württemberg
Universitätsfrauenklinik Tübingen
Postfach
W 7400 Tübingen
Federal Republic of Germany

Hong Kong (HK)

Hong Kong
Dr Y.F. Poon
Hong Kong Cancer Registry
Medical & Health Department
Institute of Radiology & Oncology
Queen Elizabeth Hospital
Wylie Road
Kowloon
Hong Kong

Hungary (Hun)

County Vas
Dr E. Kocsis
County Vas Registry
Hospital "Maarkusovszky"
Semmelweis u. 2
9700 Szombathely
Hungary

Hungary
Dr Z. Péter
National Institute of Oncology
National Cancer Registry of Hungary
Dept. of Epidemiology
Rath György-Str. 7-9
Pf. 21
1525 Budapest XII
Hungary

Szabolcs-Szatmar-Bereg
Dr L. Juhasz
Cancer Registry of the County
Szabolcs-Szatmar-Bereg
Dept. of Oncology
County Hospital
Szent István u. 68
4401 Nyiregyhaza
Hungary

Iceland (Ice)

Iceland
Dr H. Tulinius
Icelandic Cancer Registry
Skogarhlio 8
P.O. Box 5420
125 Reykjavik
Iceland

India (Ind)

Bangalore
Dr M. Krishna Bhargava
Population-Based Cancer Registry
Kidwai Memorial Institute of Oncology
Hosur Road, Karnataka
Bangalore 560 029
India

Bombay
Dr D.J. Jussawalla
Bombay Cancer Registry
Indian Cancer Society
Dr E. Borges Marg.
Parel, Bombay 400 012
India

Madras
Dr V. Shanta
Madras Metropolitan Tumor Registry
Cancer Institute (W.I.A.)
Adyar, Tamil Nadu
Madras 600 020
India

CANCER REGISTRY

Ireland (Ire)

Cork
Prof. J. Corridan
Southern Tumour Registry
Ardpatrick House
College Road
Cork
Ireland

Dublin I
Dr F. Breatnach
Irish Paediatric Tumour Registry
Our Lady's Hospital for Sick Children
Crumlin
Dublin 12
Ireland

Dublin II
Dr J.A. Thornhill
Irish Testicular Tumour Registry
5 Northumberland Road
Dublin 4
Ireland

Israel (Isr)

Israel
Dr J. Iscovich
Israel Cancer Registry
20 King David Street
Jerusalem 91000
Israel

Italy (Ita)

Forli
Dr D. Amadori
Hospitals G.B. Morgagni-P. Pierantoni
Department of Oncology
Romagna Cancer Registry
Via Forlanini
Forli
Italy

Genoa
Prof. L. Santi
Registro Ligure Tumori
Istituto de Oncologia
Viale Benedetto XV, 10
16132 Genoa
Italy

Latina
Dr E.M.S. Conti
Centro Oncologico "G. Porfiri"
Registro Tumori di Popolazione
della Provinzia di Latina
Ospedale S.M. Goretti
04100 Latina
Italy

Modena
Dr M. Ponz de Leon
Registry of Tumours
of the Digestive Organs
Istituto di Patologia Medica
Policlinico
Via del Pozzo, 71
41100 Modena
Italy

Padova
Dr L. Simonato
Registro Tumori Veneto
Via Guistiniani 7
c/o Direzzione Amministrativa
35100 Padova
Italy

Parma
Prof. G. Cocconi
Unità Sanitaria Locale N. 4
Ospedale di Parma
Via Gramsci, 14
43100 Parma
Italy

Piedmont
Dr R. Zanetti
Registro dei Tumori
per il Piemonte e la Valle d'Aosta
Via S. Francesco da Paola, 31
10123 Torino
Italy

Ragusa
Prof. L. Gafà
Registro Tumori di Ragusa
Via Virgilio 10
97100 Ragusa
Italy

Roma
Dr S. Greggi
Familial Ovarian Cancer
European Registry
Dept. of Gynaecology & Obstetrics
8 Largo A. Gemelli
00168 Roma
Italy

Torino
Dr G. Pastore
Childhood Cancer Registry
Cattedra di Epidemiologia dei Tumori
Via Santena 7
10126 Torino
Italy

Tuscany
Dr M. Geddes
Registro Tumori Toscana (R.T.T.)
Unita di Epidemiologia
C.S.P.O. Viale A. Volta, 171
50131 Firenze 1
Italy

CANCER REGISTRY

Varese
Dr F. Berrino
Registro Tumori Lombardia
(Provincia di Varese)
Servizio di Epidemiologia
Istituto Nazionale per lo Studio
e la Cura dei Tumori
Via Venezian, 1
20133 Milan
Italy

Jamaica (Jam)

Jamaica
Dr S.E.H. Brooks
Jamaica Cancer Registry
(Kingston & St. Andrew)
Department of Pathology
University Hospital
University of the West Indies
Mona, Kingston 7
Jamaica

Japan (Jap)

Aichi
Dr A. Ikari
Aichi Cancer Registry
Division of Health & Prevention
Department of Health
Aichi Prefectural Government
3-1-2 Sannomaru, Naka-ku
Nagoya 460
Japan

Fukuoka
Dr T. Nagata
Fukuoka Cancer Registry
Medical Association
2-9-30 Hakata-eki-Minami
Hakata-ku
Fukuoka 812
Japan

Hiroshima
Dr K. Mabuchi
Hiroshima Tumor Registry
Radiation Effects Research Foundation
5-2 Hijiyama Park, Minami ku
Hiroshima 732
Japan

Hyogo
Dr T. Ishida
Hyogo-Kenritsu Kenshin Center
13-70, Kita-Oji-cho
Akashi City
Hyogo-ken 673
Japan

Kanagawa
Dr N. Okamoto
Kanagawa Cancer Registry
Clinical Research Institute
Kanagawa Cancer Center
54-2 Nakao
Asahi-ku
Yokohama 241
Japan

Miyagi
Dr A. Takano
Miyagi Prefectural Cancer Registry
c/o Miyagi Cancer Society
Kamisugi 6-chome, Aoba-ku
Sendai 980
Japan

Nagasaki
Prof. T. Ikeda
Nagasaki Tumor Registry
Radiation Effects Research Foundation
1-8-6 Nakagawa
Nagasaki City
Nagasaki-ken 850
Japan

Nagoya
Dr Y. Tomoda
Trophoblastic Tumour Registry
Nagoya University
School of Medicine
Dept. of Obstetrics & Gynaecology
Tsurumaicho 65
Showa-ku
Nagoya
Japan

Osaka
Dr I. Fujimoto
Osaka Cancer Registry
Center for Adult Diseases of Osaka
Department of Field Research
3 Nakamichi 1-chome
Higashinari-ku
Osaka 537
Japan

Saga
Dr S. Tokudome
Saga Cancer Registry
Saga Medical College
Dept. of Community Health Science
Saga City 840-01
Japan

Sapporo
Dr Y. Kurimura
Hokkaido Cancer Registry
Adult Health Division
Dept. of Health & Environment
Hokkaido Government
North 3, West 6, Chuo-ku
Sapporo, Hokkaido 060
Japan

CANCER REGISTRY

Tokyo
Dr K. Minoda
National Registry of Children
with Retinoblastoma
Teikyo University, Ichihara Hospital
Department of Ophthalmology
3426-3 Anesaki, Ichihara-City
Chiba 299-01
Japan

Yamagata
Dr Y. Satô
Yamagata Cancer Registry
Yamagata Medical Center for Adults
7-17 Sakuracho
Yamagata 990
Japan

Kuwait (Kuw)

Kuwait
Dr Y.T. Omar
Kuwait Cancer Registry
Kuwait Cancer Control Centre
P.O. Box No. 42262
Postal Code No. 70653
Shuwaikh
Kuwait

Lithuania (Lit)

Lithuania
Dr L. Griciute
Lithuanian Cancer Research Institute
Lithuanian Cancer Registry
341-111 Santariskiu St.
Vilnius 232021
Lithuania

Mali (Mal)

Bamako
Dr S. Bayo
Institut National
de Recherche en Santé Publique
Section des Maladies Néoplasiques
Route de Koulikoro
B.P. 1771
Bamako
Mali

Martinique (Mar)

Martinique
Dr P. Escarmant
Registre du Cancer de la Martinique
A.M.R.E.C.
Service de Radiothérapie
Centre Hospitalier de Fort-de-France
B.P. 632
97261 Fort-de-France Cédex,
Martinique

Netherlands (Net)

Amsterdam
Dr J. Benraadt
Comprehensive Cancer Centre
Amsterdam (IKA)
Plesmanlaan 125
1066 CX Amsterdam
The Netherlands

Eindhoven
Dr J.W.W. Coebergh
Integraal Kankercentrum Zuid/SOOZ
P.O. Box 231
5600 AE Eindhoven
The Netherlands

Groningen
Dr R. Otter
Building "La Belle Alliance"
Waterlolaan 1/13
9725 BE Groningen
The Netherlands

Leeuwarden
Dr B.G.S. Szabo
Comprehensive Cancer Centre North
p/o Radiotherapeutical Institute
Friesland
Borniastraat 4a
8934 AD Leeuwarden
The Netherlands

Leiden
Dr H. Kruyff
Cancer Registry
IKW: Comprehensive Cancer Center West
Schipholweg 5A
2316 XB Leiden
The Netherlands

Maastricht
Dr W.G.H. Visschers
Integraal Kankercentrum Limburg (IKL)
Postbus 2208
6201 HA Maastricht
The Netherlands

Middelburg
Dr H.T. Planteydt
Mesothelioma Register
Pathologisch Anatomisch Lab.
Streeklaboratorium "Seeland"
Molenwater 47
Middelburg
The Netherlands

Nijmegen
Dr L.A.L.M. Kiemeney
Regional Cancer Registry
Comprehensive Cancer Centre East (IKO)
P.O. Box 1281
6501 GB Nijmegen
The Netherlands

CANCER REGISTRY

Rotterdam
Mrs M.A. Fijn van Draat-Kuiper
Integraal Kankercentrum Rotterdam
P.O. Box 289
3000 AG Rotterdam
The Netherlands

The Hague
Dr W.A. Kamps
Dutch Childhood Leukaemia
Study Group
Central Bureau
p/a Juliana Kinderziekenhuis
Postbus 60604
2506 The Hague
The Netherlands

Utrecht I
Mr G.A. de Winter
LOK: Coordinating Council of
Comprehensive Cancer Centers
Postbus 19001
3501 DA Utrecht
The Netherlands

Utrecht II
Dr C. Gimbrère
Regional Cancer Registry
Integraal Kankercentrum
Midden-Nederland (IKMN)
Servaasbolwerk 14
3512 NK Utrecht
The Netherlands

New Zealand (NZ)

Auckland
Mr H. Gaudin
Herald Centenary Cancer Registry
Auckland Hospital
P.O. Box 5546
Auckland
New Zealand

New Zealand
Mr F.J. Findlay
The New Zealand Cancer Registry
National Health Statistics Centre
Ballantrae House
192 Willis Street
Private Bag 2
Cumberland House Post Office
Wellington
New Zealand

Waikato
Dr E.H.M. Kerr
Waikato Tumour Registry
Waikato Hospital
Private Bag
Hamilton
New Zealand

Nigeria (Nig)

Ibadan
Prof. U. Aghadiuno
Ibadan Cancer Registry
Department of Pathology
University College Hospital
Ibadan
Nigeria

Zaria
Dr E.A.O. Afolayan
Cancer Registry
Department of Pathology
Ahmadu Bello University
Teaching Hospital
Zaria
Nigeria

Norway (Nor)

Norway
Dr F. Langmark
Cancer Registry of Norway
Inst. for Epidemiological Research
Montebello DNR
0310 Oslo 3
Norway

Panama (Pan)

Panama
Dr G. Campos
Registro Nacional del Cancer
Instituto Oncologico Nacional
Apartado 6-108
El Dorado
Panama

Papua New Guinea (PNG)

Papua New Guinea
Dr J.S. Niblett
Papua New Guinea Tumour Registry
Angau Memorial Hospital
Dept. of Health
Box 457
Lae
Papua New Guinea

Philippines (Phi)

Manila
Dr A.V. Laudico
Rizal Medical Center
Cancer Registry
Department of Health
Shaw Boulevard
Pasig
Metro-Manila 3130
Philippines D. 2801

CANCER REGISTRY

Poland (Pol)

Biakystok
Dr M. Grygoruk
Specjalistyczny Onkologiczny
Zespok Opieki Zdrowotnej
ul. Ogrodowa 12
15062 Biakystok
Poland

Bydgoszcz
Dr J. Sniegocki
Wojewodzka Przychodnia Onkologiczna
Ul. Ujejskiego 75
85168 Bydgoszcz
Poland

Cracow I
Dr J. Pawlega
Cracow City & District Cancer Registry
Institute of Oncology
Garncarska 11
31–115 Cracow
Poland

Cracow II
Prof. J. Skolyszewski
Regional Cancer Registry
Marie Curie Research Institute
of Oncology
Oddziak w Krakowie
35115 Cracow
Poland

Gdynia
Dr E. Jordan
Regional Cancer Registry
Szpital Morski im. PCK
Ul. Powstania Styczniowego 1
81519 Gdynia
Poland

Gliwice
Dr S. Majewski
Centrum Onkologii - Instytut
im. M. Sklodowskiej-Curie
Oddziak w Gliwicach
Ul. Wybrzeze Armii Czerwonej 15
44101 Gliwice
Poland

Katowice
Dr B. Zemla
Katowice District Cancer Registry
ul. Armii Czerwonej 15
44101 Gliwice
Poland

Kielce
Dr S. Gozdz
Wojewodzka Przychodnia Onkologiczna
Ul. Kosciuszki 3
25310 Kielce
Poland

Lodz
Dr T. Naganski
Regionalny Osdarek Onkologiczny
Ul. Gagarina 4
93509 Lodz
Poland

Lublin
Dr R. Patyra
Onkologiczny Specjalistyczny
Zespol Opieki Zdrowotnej
ul. Jaczewskiego 7
20090 Lublin
Poland

Opole
Dr K. Drosik
Wojewodzka Przychodnia Onkologiczna
Ul. Katowicka 64
45060 Opole
Poland

Poland
Dr W. Zatonski
Polish Cancer Registry
Curie-Sklodowska Institute
of Oncology
Wawelska Street 15
00-973 Warsaw
Poland

Poznan
Dr K. Gorny
Specjalistyczny Onkologiczny
Zespol Opieki Zdrowotnej
Ul. Garbary 13/15
61866 Poznan
Poland

Szczecin
Dr A. Niezgoda
Specjalistyczny Onkologiczny
Zespol Opieki Zdrowotnej
Ul. Strzalowska 22
71780 Szczecin
Poland

Warsaw
Dr Z. Wronkowski
Regional Cancer Registry
Centrum Onkologii
Ul. Wawelska 13
00973 Warsaw
Poland

Wroclaw
Prof. M. Wawrzkiewicz
Specjalistyczny Onkologiczny
Zespol Opieki Zdrowotnej
Ul. Hirszfelda 12
53413 Wroclaw
Poland

CANCER REGISTRY

Portugal (Por)

Azores
Dr M. Lima
Registro Oncologico
do Centro de Oncologia dos Açores
Rua Principe de Monaco 38
9700 Angra do Heroismo
Portugal

Lisboa
Dr A. da Costa Miranda
IPORG - Centro de Lisboa
Registro Oncológico Regional Sul
1093 Lisboa Codex
Portugal

Porto
Dr J.M. Amado
Inst. de Ciécias Biomédicas
de Abel Salazar
Sect. de Saùde Communitária
Largo Prof. Abel Salazar 2
4000 Porto
Portugal

Vila Nova
Dr J. Teixeira Gomes
Registro Oncologico de V.N. Gaia
Centro Hospitalar
de Vila Nova de Gaia
Vilar de Andorinho
4400 Vila Nova de Gaia
Portugal

Romania (Rom)

Cluj
Dr G. Bolba
Cluj County Cancer Registry
Oncological Institute
of Cluj-Napoca
Dept. of Epidemiology & Statistics
Str. Republicii 34-36
3400 Cluj-Napoca
Romania

Timis
Dr R. Panaitescu
Service Départemental d'Oncologie
Spitalul Clinic No. 2
Str. Marasesti No. 5
Timisoara
Romania

Singapore (Sin)

Singapore
Prof. K. Shanmugaratnam
Singapore Cancer Registry
Department of Pathology
National University Hospital
Lower Kent Ridge Road
Singapore 0511
Singapore

Spain (Spa)

Basque
Dra I. Izarzugaza
Registro de Cancer de Euskadi
Depto. de Sanidad y Consumo
Gobierno Vasco
Dirección de Información Sanitaria
y Evaluación
c/Duque de Wellington, 2
01011 Vitoria-Gasteiz
Spain

Granada
Dr C.L. Martinez Garcia
Escuela Andaluza
de Salud Publica
Registro de Cancer de Granada
Avenida del Sur 11
Ap. de Correos, 668
18014 Granada
Spain

Guipuzcoa
Dr J. Irigaray
Registro de Tumores Hospitalario
de Guipuzcoa
Instituto de Oncologia
Aldako-enea
20012 San Sebastian
Spain

Mallorca
Dr E. Benito
Registre de Cancer de Mallorca
Paseig Mallorca 42
07012 Palma
Spain

Murcia
Dr C. Navarro Sanchez
Consejeria de Sanidad
Seccion de Epidemiologia
Ronda Levante, 11
30008 Murcia
Spain

CANCER REGISTRY

Navarra
Dr A. del Moral Aldaz
Registro de Tumores de Navarra
Ciudadela, 5-1°
31001 Pamplona
Spain

Oviedo
Dr M. Echeverria Rodriguez
Registro de Tumores Poblacional
Consejeria de Sanidad y Servicios Sociales
General Elorza 32
33001 Oviedo
Spain

Tarragona
Dr J. Borras
Associacion Espanola Contra el Cancer de Tarragona
Delegacio Territorial de Sanitat
Avenida Maria Cristina 54
43002 Tarragona, Catalonia
Spain

Tenerife
Dr D. Nunez
Registro de Tumores Poblacional
Rambla de General Franco, 53
Santa Cruz de Tenerife (Canary Islands)
Spain

Valladolid
Dr A. Cabezon
Registro de Tumores Poblacional
Carretera de Rueda Km. 3
47071 Valladolid
Spain

Zaragoza
Dra A. Vergara
Registro del Cancer de Zaragoza
DGA Servicio Regional de Vigilancia Epidemiologica
Edificio Pignatelli
Paseo Ma. Agustin 36
50004 Zaragoza
Spain

Sudan (Sud)

Khartoum
Dr A. Hidayatalla
RICK - Cancer Registry
Radiation & Isotope Center
P.O. Box 677
Khartoum
Sudan

Sweden (Swe)

Lund
Dr T.R. Möller
Southern Swedish Regional Tumor Registry
University Hospital
221 85 Lund
Sweden

Stockholm
Dr L.E. Rutqvist
Stockholm-Gotland Regional Tumor Registry
Karolinska Hospital
104 01 Stockholm
Sweden

Sweden
Prof. J. Ericsson
The Swedish Cancer Registry
The National Board of Health & Welfare
F-enheten
Socialstyrelsen
106 30 Stockholm
Sweden

Umeå
Prof. L.-G. Larsson
Regional Cancer Registry for Northern Sweden
Oncology Centre
Regional Hospital
901 85 Umeå
Sweden

Uppsala
Dr I.R. Hesselius
Regional Tumor Registry
Oncology Centre
University Hospital
750 14 Uppsala
Sweden

Switzerland (Swi)

Basel
Prof. J. Torhorst
Basel Cancer Registry
Institut für Pathologie
Schonbeinstr. 40
4003 Basel
Switzerland

Geneva
Dr P. Vassilakos
Registre Genevois des Tumeurs
55, blvd de la Cluse
1205 Geneva 4
Switzerland

CANCER REGISTRY

Neuchâtel
Dr A.M. Méan
Registre Neuchâtelois des Tumeurs
Les Cadolles 4
2000 Neuchâtel
Switzerland

St. Gallen
Dr F. Enderlin
Krebsregister St-Gallen-Appenzell
Institut für Pathologie
Kantonsspital
Rorschacherstr. 95
9001 St. Gallen
Switzerland

Vaud
Dr F.G. Levi
Registre Vaudois des Tumeurs
CHUV
Falaises 1
1011 Lausanne
Switzerland

Zürich
Dr G. Schüler
Kantonalzürcherisches
Krebsregister
Institut für Pathologie der
Universität
Universitätsspital
8091 Zürich
Switzerland

Thailand (Tha)

Chiang Mai
Dr N.C. Martin
Tumor Registry Cancer Unit
Maharaj Nakorn Chiang Mai Hospital
Faculty of Medicine
Chiang Mai University
Chiang Mai
Thailand 50002

United Kingdom (UK)

Birmingham
Dr J.R. Mann
West Midlands Regional Children's
Tumour Registry
Children's Hospital
Ladywood Middleway
Birmingham B16 8ET
United Kingdom

E. Anglia
Dr T. Davies
Cancer Registration Bureau
of East Anglian Region
Addenbrookes Hospital
Hills Road
Cambridge CB2 2QQ
United Kingdom

Liverpool
Dr J. Osman
Epidemiology & Medical Statistics Unit
Health & Safety Executive
Room 134
Magdalen House
Stanley Precinct
Bootle, Merseyside L20 3QZ
United Kingdom

Manchester I
Dr J.M. Birch
Children's Tumour Registry
Christie Hospital
Kinnaird Road
Manchester M20 9BX
United Kingdom

Manchester II
Prof. S.A. Langley
Manchester Ovarian Tumour Register
Department of Obstetrics & Gynaecology
Manchester University
St. Mary's Hospital
Whitworth Park
Manchester M13 0JH
United Kingdom

Mersey
Mrs S.M. Gravestock
The Mersey Regional Cancer Registry
Hamilton House
24 Pall Mall
Liverpool L3 6AL
United Kingdom

N. Ireland
Dr A.T. Gavin
North of Ireland Cancer Registry
The Lodge
59 Derryhaw Road, Tunan
Armagh
Northern Ireland
United Kingdom

N.W. Region
Prof. A. Smith
North Western Regional
Cancer Epidemiology Unit
Christie Hospital & Holt Radium Inst.
Wilmslow Road, Withington
Manchester M20 9BX
United Kingdom

Newcastle
Dr A.W. Craft
Northern Region
Young Persons Cancer Registry
Dept. of Child Health, Medical School
Sir James Spence Bldg
Queen Victoria Road
Newcastle-upon-Tyne NE1 4LP
United Kingdom

CANCER REGISTRY

Northern UK
Mr R.A. McNay
Northern RHA Cancer Registry
Benfield Road
Walkergate
Newcastle-upon-Tyne NE6 4PY
United Kingdom

OPCS
Dr A.J. Swerdlow
Office of Population Censuses & Surveys
St. Catherine's House
10 Kingsway
London WC2B 6JP
United Kingdom

Oxford I
Miss C. Hunt
Oxford Cancer Registry
Oxford Regional Health Authority
Old Road
Headington
Oxford OX3 7LF
United Kingdom

Oxford II
Dr C.A. Stiller
National Registry of Childhood Tumours
Childhood Cancer Research Group
57 Woodstock Road
Oxford OX2 6HJ
United Kingdom

S. Western
Dr D. Pheby
South Western Regional
Dept. of Epidemiology and Public Health
Canynge Hall
Whiteladies Road
Bristol BS8 2PR
United Kingdom

Scotland
Dr C.S. Muir
The Scottish Cancer Registry
Scottish Health Service
Information Services Division
Trinity Park House
South Trinity Road
Edinburgh EH5 3SQ
Scotland
United Kingdom

Scotland E.
Dr N.R. Waugh
Scotland-East Cancer Registry
Radiotherapy Department
Ninewells Hospital
Dundee DD2 1UB
Scotland
United Kingdom

Scotland N.
Dr M.H. Elia
Northern Cancer Registry
(Highlands & Western Isles)
Department of Radiotherapy
Raigmore Hospital
Inverness IV2 3UJ
Scotland
United Kingdom

Scotland N.E.
Dr G. Innes
Scotland-North East Cancer Registry
Dept. of Clinical Oncology
Aberdeen Royal Infirmary
Foresterhill
Aberdeen AB9 2ZB
Scotland
United Kingdom

Scotland S.E.
Dr G.A. Venters
Scotland-South-Eastern Cancer Registry
Liberton Hospital
Lasswade Road
Edinburgh EH16 6UB
Scotland
United Kingdom

Scotland W.
Dr C.R. Gillis
West of Scotland Intelligence Unit
Ruchill Hospital
Glasgow G20 9NB
Scotland
United Kingdom

Thames
Dr M.P. Coleman
Thames Cancer Registry
15 Cotswold Road
Sutton, Surrey SM2 5PY
United Kingdom

Trent
Mr D.L. Holmes
Trent Regional Cancer
Registration Bureau
Fulwood House
Old Fulwood Road
Sheffield S10 3TH
United Kingdom

W. Midlands
Dr K. Woodman
Birmingham & West Midlands
Regional Cancer Registry
Queen Elizabeth Medical Centre
Edgbaston
Birmingham B15 2TH
United Kingdom

CANCER REGISTRY

Wales
Dr M. Cotter
Wales Cancer Registry
Health Intelligence Unit
6th Floor, Heron House
35/43 Newport Road
Cardiff CF2 1SB
United Kingdom

Wessex
Dr H.F. Sanderson
Wessex Cancer Registry
Wessex Cancer Intelligence Unit
Royal South Hants Hospital
Brinton's Terrace
Southampton SO2 0AJ
United Kingdom

Yorkshire
Prof. C.A.F. Joslin
Yorkshire Regional Health
Authority Cancer Registry
Cookridge Hospital
Leeds LS16 6QB
United Kingdom

United States of America (USA)

Atlanta
Dr R.S. Greenberg
Georgia Center for Cancer Statistics
Dept. of Epidemiology and Biostatistics
School of Medicine
1599 Clifton Rd NE
Atlanta, GA 30329
USA

Bay Area
Dr D. West
San Francisco Bay Area for Cancer Control
Northern California Cancer Center
140 Harbor Bay Parkway, Suite 260
Alameda, CA 94501-7080
USA

California
Dr J. Young Jr
California Dept. of Health Services
Cancer Surveillance Sect.
714/744 P St.
P.O. Box 942732
Sacramento, CA 94234-7320
USA

Colorado
Mrs R. Bott
Colorado Tumor Registry
State Department of Health
4210 E. 11th Avenue
Denver, CO 80220
USA

Connecticut
Mr J.T. Flannery
Connecticut Tumor Registry
Connecticut State Department
of Health Services
150 Washington Street
Hartford, CT 06106
USA

Delaware
Dr A.T. Meadows
Greater Delaware Valley
Pediatric Tumor Registry
Children's Hospital
Children's Cancer Research Center
Division of Oncology
34th & Civic Center Blvd
Philadelphia, PA 19104
USA

Detroit
Dr G.M. Swanson
Metropolitan Detroit
Cancer Surveillance Section
and Comprehensive Cancer Center
Michigan State University
A-211 East Fee Hall
East Lansing, MI 482824
USA

Hawaii
Dr L.N. Kolonel
Hawaii Tumor Registry
Cancer Center of Hawaii
1236 Lauhala Street, Room 402
Honolulu, HI 96813
USA

Idaho
Mrs M.L. England
Idaho Central Tumor Registry
Idaho Hospital Association
P.O. Box 7482
Boise, ID 83707
USA

Illinois
Dr A. Alexander
Illinois State Cancer Registry
Div. of Epidemiological Studies
Illinois State Dept. of Public Health
525 W. Jefferson Street
Springfield, IL 62761
USA

Indiana
Mr R.C. Barnhill
State Cancer Registry
Indiana Board of Health
P.O. Box 1964
1330 West Michigan Street
Indianapolis, IN 46206-1964
USA

CANCER REGISTRY

Iowa
Mrs K. McKeen
State Health Registry of Iowa
Epidemiology Research Center
University of Iowa
S 100 Westlawn
Iowa City, IA 52242
USA

Kansas
Dr F.F. Holmes
Cancer Data Service
University of Kansas
Medical Center
Kansas City, KS 66103
USA

Los Angeles
Dr L. Bernstein
USC School of Medicine
USC Cancer Surveillance Program
of Los Angeles
1721 Griffin Avenue, Room 200
Los Angeles, CA 90031
USA

Louisiana
Mrs J.F. Craig
Louisiana Tumor Registry
State University Medical Center
1901 Perdido St.
New Orleans, LA 70112-1393
USA

Maine
Dr G.F. Bogdan
State of Maine Cancer Registry
Maine Bureau of Health
State House Station II
157 Capitol Street
Augusta, ME 04333
USA

Maryland
Dr I.I. Kessler
Maryland Cancer Registry
University of Maryland School of Medicine
Department of Epidemiology & Preventive
Medicine
Howard Hall, Room 149D
660 W. Redwood Street
Baltimore, MD 21201
USA

Massachusetts
Mr R. Clapp
Massachusetts Cancer Registry
Department of Public Health
150 Tremont Street - 5th Floor
Boston, MA 02111-1126
USA

Michigan
Mr G. Van Amburg
Michigan Cancer Surveillance Program
P.O. Box 30195
3423 North Logan
Lansing, MI 48909
USA

Minnesota
Dr A.P. Bender
Minnesota Cancer Registry
Minnesota Dept. of Health
Room 408
717 S.E. Delaware Street
P.O. Box 9441
Minneapolis, MN 55440
USA

Missouri
Dr J.C. Chang
Missouri Cancer Registry
Missouri Dept. of Health
Cancer Research Center
201 Business Loop 70 West
Columbia, MO 65205
USA

Montana
Ms D. Hellhake
Montana Central Tumor Registry
Department of Health & Environmental Sciences
Cogswell Building
Helena, MT 59620
USA

N. Carolina
Mr D. Atkinson
North Carolina Central Cancer Registry
North Carolina Div. of Statistics
& Information Services
P.O. Box 27687
Raleigh, NC 27611-7687
USA

Nebraska
Dr F.W. Karrer
Nebraska Cancer Registry
8303 Dodge Street
Omaha, NE 68114
USA

New Hampshire
Ms S.K. Foret
New Hampshire State Cancer Registry
Dartmouth Medical School
Hinman Box 7828
Hanover, NH 03756
USA

CANCER REGISTRY

New Jersey
Dr L. Meinert
New Jersey State Cancer Registry
New Jersey Department of Health
3635 Quaker Bridge Road, CN 369
Trenton, NJ 08625
USA

New Mexico
Dr C.R. Key
New Mexico Tumor Registry
Cancer Research & Treatment Center
900 Camino de Salud, N.E.
Albuquerque, NM 87131
USA

New York
Dr W.S. Burnett
New York State Cancer Registry
Cancer Control Bureau
State of New York Dept. of Health
Corning Tower, Room 565
Empire State Plaza
Albany, NY 12237
USA

Orange County
Dr H. Anton-Culver
Cancer Surveillance Program
of Orange County
Department of Community & Environmental Medicine
University of California at Irvine
Irvine, CA 92717
USA

Puerto Rico
Dr I. Martinez
Central Cancer Registry of Puerto Rico
C-18 Rufine Rodriguez
Villa Clementina
Santurce, PR 00657
USA

Rhode Island
Dr J.P. Fulton
Rhode Island Department of Health
Rhode Island Cancer Registry
75 Davis Street, Room 409
Providence, RI 02908
USA

RIMS-CSP
Dr A.R. King
RIMS Cancer Surveillance Program
Loma Linda Univ.
25455 Barton Road Suite B-208
Loma Linda, CA 92354
USA

Rochester
Ms M. Richardson
Rochester Regional Tumor Registry
Univ. of Rochester Cancer Center
P.O. Box 704
601 Elmwood Avenue
Rochester, NY 14620
USA

San Diego
Dr S.L. Saltzstein
SANDIOCC
2251 San Diego Avenue, Suite A-141
San Diego, CA 92110
USA

San Jose
Dr T.E. Davis
Region 1 Cancer Incidence Registry
1762 Technology Drive, Suite 204
San Jose, CA 95110
USA

Seattle
Dr D.B. Thomas
Cancer Surveillance System
of Western Washington
The Fred Hutchinson Cancer Research Center
1124 Columbia Street
Seattle, WA 98667
USA

SEER
Dr E.J. Sondik
National Cancer Institute
Cancer Statistics Branch
Executive Plaza North
Room 343J
Bethesda, MD 20892
USA

Tennessee
Ms D. Weber
Tennessee Cancer Registry
Tennessee Dept. of Health & Environment
C2-233 Cordell Hull Building
Nashville, TN 37219
USA

Tri-Counties
Dr L. Hart
Tri-Counties Regional
Cancer Registry
Santa Barbara County
Health Care Services
300 San Antonio Rd, Room M-340
Santa Barbara, CA 93110
USA

CANCER REGISTRY

Utah
Mrs R. Dibble
Utah Cancer Registry
Research Park
420 Chipeta Way, Suite 190
Salt Lake City, UT 84108
USA

Virginia
Dr W.J. Munsie
Virginia Tumor Registry
James Madison Bldg, Room 715
109 Governor Street
Richmond, VA 23219
USA

Wisconsin
Ms J.L. Phillips
Wisconsin Cancer Reporting System
Department of Health Statistics
1 W. Wilton Street, Room 172
P.O. Box 309
Madison, WI 53701-0309
USA

Uruguay (Uru)

Uruguay
Dr J.A. Vassallo
National Cancer Registry
Eduardo Acevedo 1530
Montevideo
Uruguay

USSR (USSR)

Leningrad
Dr V.M. Merabishvili
Laboratory of Oncological Statistics
N.N. Petrov Institute of Oncology of USSR
Ministry of Health
68, Leningradskaya Street
Pesochny 2
Leningrad 189646
USSR

Moscow
Dr D. Zaridze
Institute of Carcinogenesis
All-Union Cancer Research Centre
Academy of Medical Sciences
of the USSR
Kashirskoye Shosse 24
Moscow 115478
USSR

Vietnam (Vie)

Hanoi
Dr Phan Thi Hoang Anh
Registre de Cancer
Hopital Benh Vien K.
43 Quan Su
Hanoï
Vietnam

Yugoslavia (Yug)

Slovenia
Dr V. Pompe-Kirn
Cancer Registry of Slovenia
The Institute of Oncology
Zaloska 2
6100 Ljubljana
Yugoslavia

Zagreb
Dr M. Strnad
Cancer Registry
Institute of Public Health
Rockfellerova 7
41000 Zagreb
Yugoslavia

Zambia (Zam)

Lusaka
Prof. D. Shibayunji
Cancer Registry of Zambia
c/o Department of Community Health
University of Zambia
School of Medicine
Box 50110
Lusaka
Zambia

Zimbabwe (Zim)

Harare
Dr L.M. Levy
Zimbabwe Cancer Registry
Box A449
Avondale
Harare
Zimbabwe

BIOLOGICAL MATERIALS BANKS

Biological Materials Banks

Banks of biological materials can provide valuable records for individuals and population groups and are increasingly used in epidemiological research. Some of the investigators listed below have expressed an interest in joint laboratory and epidemiological investigations based on the material in their possession. They should be contacted directly. Addresses are given at the end of the section.

Codes:
- ID — Personal identifying information exists
- CD — Collateral data are available (e.g., physical examination, interview results, laboratory analyses)
- FU — Follow-up information is available

Abbreviations:
- LN — Liquid nitrogen
- RT — Room temperature

Cells in Culture $

Name	Persons *	Year	Code	Temp. °C	Remarks
Becciolini, A.	25	1986 –	ID CD FU	–70	Glioma cell lines
De Tribolet, N.	200	1975 –	ID CD	–70	Human fibroblasts
Ebbesen, P.	7	1978 –			Mammary cell lines
Eppenberger, U.	15	1983 –	ID FU	–70	Neuroblastoma cell lines
Evans, A.E.	22	1969 –	CD		Fibroblast lines
Golan, T.D.	25	1982 –		LN	Samples from virologic surveillance and diagnosis programme
Gottlieb–Stematsky, T.		1968 –			
Greene, M.H.	17	1970 –	ID CD FU	–70	Cell lines reference collection, cells and hybridomas
Haas, G.P.	50	1987 –	ID CD FU	LN	
Hay, R.J.	3,100	1962 –	ID	LN	LCL
Ishida, T.	80–100	1982 –	ID	LN	Fibroblast cell lines from xeroderma pigmentosum cases and controls
Jung, E.G.	70	1979 –	ID	LN	Lung cancer
Kato, H.	10	1971 –	ID CD	LN	Cell lines from children
Kumar, S.	10	1965 –	ID		Lymphoblastoid cell lines from Kenya, Tanzania
Levin, A.	30	1972 –		LN	Fibroblasts
Müller, H.	70	1982 –	ID CD FU	LN	Fetal (amniotic) or fibroblast cell lines with chromosomal aberrations
Mitelman, F.	350	1975 –	ID FU	LN	
Mulivor, R.A.	1,052	1972 –			Fibroblast and lymphoblast cell lines in NIA Cell Repository
Mulivor, R.A.	4,409	1972 –			Fibroblast and lymphoblast cell lines in NIGMS Human Genetic Mutant Cell Repository
Noel, B.	100	1971 –	ID CD		Fibroblasts, lymphocytes with chromosome abnormalities
Prasad, U.		1986 –	ID CD FU	37	P3HR-l, Raji, B95-8
Rüdiger, H.W.	1,004	1971 –	ID CD	LN	Diploid human fibroblasts
Rüdiger, H.W.	200	1977 –		LN	Fibroblast lines from patients with lung cancer

$ It was not always clear whether cells in culture related to an established cell line. Items under this heading could thus pertain to either.

* For a few collections, number of samples are given instead: this will be corrected in future.

Cells in Culture $

Name	Persons *	Year	Code	Temp. °C	Remarks
Sasaki, M.S.	500	1972 –		LN	Cell lines from patients with a wide variety of hereditary disorders associated with cancer propensity
Serotanen, K.	1,000	1981 –	ID CD FU	-70	Genital tract biopsies
Shigematsu, I.	1,710	1985 –	ID CD FU	LN	B-cells
Shope, R.E.	8	1954 –			Cell lines
Soeterboek, A.; Van Noord, P.	8,000	1986 –	ID CD FU	-20	Healthy individuals, 65 years old
Stoner, G.	5	1980 –	ID CD		Fibroblast lines
Stoner, G.	5	1983 –			Epithelial cell cultures (human bronchus)
Tamburro, C.H.	5	1975 –	ID CD FU	-70	Liver
Twentyman, P.	25	1982 –	ID CD FU	LN	Small cell lung cancer cell lines

Faeces

Name	Persons *	Year	Code	Temp. °C	Remarks
Bingham, S.A.	240	1982 –	ID CD		Daily samples collected whilst subjects on normal diet
Cummings, J	300	1974 –	ID		Lyophilized
Gotlieb-Stematsky, T.	1,000	1968 –		-12	Samples from virologic surveillance and diagnosis programme, 2–3 ml suspension
Greene, M.H.	10	1970 –	ID CD FU	-70	Lyophilized
Haines, A.P.	7,000	1974 –	ID FU		From patients with tumours
Rudolf, Z.	50	1978 –	ID CD FU		
Soeripto; Haryono Soeharto	150	1952 –	ID CD		
Zubert, S.J.	300	1972 –	ID		

Red Cells

Name	Persons *	Year	Code	Temp. °C	Remarks
Anton–Culver, H.	75	1978 –	ID	-70	
Assennato, G.	200	1990 – 91	ID CD	-20	
Becciolini, A.	200	1988 –	ID CD FU	-80	
Berrino, F.	6,000	1987 –	ID CD		
Cartwright, R.A.	1,000	1978 –	ID FU		
Chen, J.	8,000	1989 –	ID CD	-20	Haemolysates from cancer patients
Elinder, C.G.	470	1980 –	ID CD		
Forman, D.	3,000	1989 –	ID CD	-20	From Stockholm and Västerås
Greene, M.H.	511	1970 –	ID CD FU	-70	

$ It was not always clear whether cells in culture related to an established cell line. Items under this heading could thus pertain to either.

* For a few collections, number of samples are given instead; this will be corrected in future.

Red Cells

Name	Persons *	Year	Code	Temp. °C	Remarks
Ishida, T.	3,000	1970 –	ID	–20	
Kune, G.A.	160	1980 –			From a case-control study of colorectal cancer
Levin, A.	50	1972 –			From Kenya, Tanzania
London, S.	960	1991 – 93	ID CD	–70	
Marcus, Z.H.	682	1981 –	ID CD	–70	
Marshall, W.H.	1,000	1974 –	ID CD FU	–80	
Neel, J.V.	8,000	1964 –	ID	–70	From Amerindians of Central and South America
Rappaport, S.M.	48	1988 –	ID CD FU	–70	
Soeripto; Soeharjanto	17	1952 –	ID CD		
Soeterboek, A.; Van Noord, P.	10,000	1983 –	ID CD FU	–20	Healthy adults, sequential samples
Toniolo, P.	7,000	1987 –	ID CD FU	–80	
Van Noord, P.	4,500	1983 –	ID CD FU	–20	Women taking part in breast cancer screening
Van Poppel, G.	250	1989 – 90	CD FU	–80	
Willett, W.	30,000	1989 – 90	ID CD	LN	
Yarnell, J.	120	1979 –	ID CD FU		

Saliva

Name	Persons *	Year	Code	Temp. °C	Remarks
Assennato, G.	200	1990 – 91	ID CD	–20	
Cartwright, R.A.	100	1978 –	ID FU		
Greene, M.H.	62	1970 –	ID CD FU	–70	From childhood cancers
Haines, A.P.	1,500	1983 –	ID CD FU		
Marcus, Z.H.	982	1981 –	ID CD	–80	
Marshall, W.H.	200	1974 –	ID CD FU	–70	
Prasad, U.	700	1989 –	ID CD FU	–20	
Sasco, A.J.	150	1986 –	ID CD FU	–70	
Van Noord, P.	80	1982 –	ID CD FU		Women taking part in breast cancer screening
Webster, A.D.B.	30	1980 – 82	ID CD FU	–20	

Serum / Plasma

Name	Persons *	Year	Code	Temp. °C	Remarks
Anton-Culver, H.	2,500	1978 –	ID		
Aoki, K.	40,000	1988 – 90	ID CD FU	–80	1900 sera, 600 plasma
Assennato, G.	200	1990 – 91	ID CD	–70	
Barrett-Connor, E.L.	1,500	1985			

* For a few collections, number of samples are given instead; this will be corrected in future.

Serum / Plasma

Name	Persons *	Year	Code	Temp. °C	Remarks
Barrett-Connor, E.L.	2,000	1983 –	ID FU		Plasma
Barrett-Connor, E.L.	4,000	1972 –	ID FU		Plasma
Beaglehole, R.	1,600	1982 –	ID CD		Sera
Becciolini, A.	400	1985 –	ID CD FU	–20	
Bengtsson, C.	1,500	1968 –	ID FU		Sera from 1,500 women
Bengtsson, C.	1,500	1974,1980 –			Sera from 1,500 women
Berrino, F.	6,000	1987 –	ID CD		
Bingham, S.A.	160	1988 –	ID CD	–80	Daily samples collected whilst subjects on normal diet
Blumberg, B.S.	500,000	1958 –			Serum
Burtin, P	2,600	1976 –	ID CD	–80	Plasma, random urban sample, males and females, age 25–64
Calvert, G.D.	15,000	1980 –	CD		Sera
Camargo, M.E.	3,216	1978 –	ID FU		Sera
Cartwright, R.A.	3,000	1978 –	ID CD	–20	
Chen, J.	8,000	1989 –	ID CD	–70	
Chongsuvivatwong, V.	1,000	1991 – 93	ID CD FU	–50/–80	
Christensen, J.M.	700	1983,1989 – 90	ID CD FU	–70	
Clark, L.C.	1,700	1983 –	ID CD FU	–70	
Comstock, G.W.	58,541	1974 –	ID CD FU		Sera
Dawkins, R.L.	3,500	1969 –	ID CD FU		Sera for genetic studies
De-Thé, G.	70,000	1980 –	ID CD FU	–20	Leukocytes from Botswana
Ebbesen, P.	300	1987 –			Serum/Plasma from Denmark and Greenland
Ebbesen, P.	5,000	1978 –	ID CD		Sets of plasma from Stockholm and Västerås
Elinder, C.G.	470	1980 –	ID CD		Sera
Enarson, D.A.	2,500	1981 – 86	ID CU FU		Sera from healthy persons, esp. young adults in North and South America
Evans, A.S.	50,000	1960 –			
Fordyce, M.K.	3,000	1980 –	ID CD	–20/–70	
Forman, D.	5,000	1980 –	ID CD FU	–70	
Fukao, A.	5,000	1989 – 99		–20	
Garkavtseva, R.	50	1987 – 89	ID	LN	
Golan, T.D.	250	1982 –	CD	–20	Sera
Gotlieb–Stematsky, T.	20,000	1968 –			Sera samples from virologic surveillance and diagnosis programme
Greene, M.H.	57,905	1970 –	ID CD FU	–70	Serum/Plasma
Greggi, S.	48	1989 –		–20	Familial Ovarian Cancer European Registry. From probands and first-degree relatives
Grufferman, S.	300	1990 –	ID CD FU	–70	Serum bank for cases of childhood Hodgkin's Disease
Hadziyannis, S.	700	1970 –			

* For a few collections, number of samples are given instead: this will be corrected in future.

Serum / Plasma

Name	Persons *	Year	Code.	Temp. °C	Remarks
Haines, A.P.	3,500	1974 –	ID CD FU	LN	Plasma
Hansen, B.L.	50	1981 – 81		–20	Screening for antibodies reactive with discrete endocrine cells
Hennekens, C.H.	16,000	1982 –	ID CD FU	–85	Plasma
Hill, A.P.	2,500	1981 – 88	ID CD FU	–20	Population based cohort study of the elderly
Hisamichi, S.	2,000	1982 –	ID FU	–70	Serum/Plasma
Husgafvel-Pursiainen, K.	70	1988 –	ID CD	–70	
Ishida, T.	3,000	1970 –	ID	–20	
Jellum, E.	102,808	1973 –	ID FU	–25/–40	
Juhász, L.	30	1990 – 90	ID CD	–20	Sera
Kapur, M.M.	425	1978 –	ID	–60	
Kark, J.D.	10,000	1963 –	ID CD FU		Sera from patients with goitre and thyroid cancer or male infertility
Kark, J.D.	7,000	1976 –	ID CD FU		Samples from cardiovascular studies
Kennedy, S.M.	7,500	1981 – 86	ID CD FU		Samples from cardiovascular studies
Kew, M.C.	400	1972 –	ID CD FU	–20	From persons employed in mining or smelting
Khoo, S.K.	788	1975 –	ID CD		
Kim, J.S.	100	1982 –		–70	Sera from cancer patients
Knekt, P.	47,000	1968 – 80	ID CD FU	–20	Serum/Plasma from hepatoma patients
Knight, T.M.	43	1991 –	ID CD	–20	Sera
Koskimies, S.	700	1978 –	ID CD	–70	Sera
Kumar, S.	1,000	1965 –	ID		
Kune, G.A.	350	1980 –			Serum/Plasma from a case-control study of colorectal cancer
Levin, A.	3,200	1972 –		LN/–20	Sera from Kenya, Tanzania
Levin, A.	500	1985 –	ID CD FU	LN/–20	From Tanzania, Kenya, Zambia, Singapore, Papua New Guinea
Levine, P.H.	10,000	1969 –		–70	Sera from USA, Tunisia, Malaysia, Ghana, other countries
Liu, Q.F.	6,000	1982 –	ID CD FU	–20	
London, S.	960	1991 – 93	ID CD FU	–70	
London, T.W.	300,000	1965 –	ID CD FU	–20	
Luepker, R.V.	30,000	1980 –	ID CD FU	–20	
Lund, B.	7,100	1985 –	ID CD FU	–18	
Mackay, I.R.	30,000	1960 –	CD	–70	Serum/Plasma
Mandeville, R.	1,578	1980 –	ID		Sera
Mann, J.R.	1,300	1975 –		–20	Sera: 500 breast cancer, 106 endometrial cancer, 972 melanoma
Marcus, Z.H.	4,867	1981 –	ID CD		Sera from children with cancer and with various non malignant disorders
Marshall, W.H.	1,000	1974 –	ID CD FU	–80	Sera
Marubini, E.	360	1982 –	ID CD	–70	Plasma
Melbye, M.; Ebbesen, P.	1,000	1981 –	ID		Sera
Melbye, M.; Ebbesen, P.	450	1983 –	ID		Danish gay men
					Greenland Eskimo children

* For a few collections, number of samples are given instead: this will be corrected in future.

Serum / Plasma

Name	Persons *	Year	Code	Temp. °C	Remarks
Mendy, M.	9,000	1986 -	ID CD FU	-20	Serum/Plasma
Moe, P.J.	42,000	1966 -	ID FU		
Mohapatra, S.C.	50	1989 - 92	ID CD	-20	Serum/Plasma
Mohran, Y.	205	1978 -	ID CD		Sera
Mueller, N.E.	600	1975 -	ID CD	-70	Maternal and umbilical cord sera
Naggan, L.	20,000	1976 -	ID		Plasma from Amerindians of Central and South America
Neel, J.V.	8,000	1964 -	ID	LN	
Noel, B.	50	1980 -	ID CD		Plasma from haemophiliacs
Pacsa, A.	135,000	1976 -	ID		Maternal and infant sera
Paganini-Hill, A.	2,500	1981 - 88	ID CD FU	-20	Plasma
Palli, D.	350	1986 - 89	ID CD FU	-20	Coal Miners
Pham, Q.T.	240	1990 - 90	ID CD FU	-70	Sera
Phoon, W.O.	14,426	1970 -	ID		Plasma
Pizzo, S.V.	1,000	1977 -	ID		
Prasad, U.	18,000	1982 -	ID CD FU	-20	
Rappaport, S.M.	48	1988 -	ID CD FU	-70	
Raue, F.	100	1975 -	ID CD FU	-20	
Rawls, W.E.	7,000	1976 -	ID FU	-20	
Roelcke, D.	140	1975 -	ID CD	-20	Serum/Plasma containing high titer cold agglutinins (more than 5 ml)
Rogan, W.	800	1978 -			Serum/Plasma
Romney, S.L.	1,000	1991 -	ID CD	+4	Serum/Plasma from cancer patients
Rosenberg, S.A.	10,000	1975 -	ID CD	LN	Serum/Plasma from cases of non-Hodgkin lymphoma, Hodgkin's disease, melanoma, breast cancer, NPC, prostate cancer and their controls (ALS patients, hospital employees, healthy elderly).
Ross, R.K.	6,000	1972 -	ID		
Rozen, P.	70	1982 -	CD FU		70 Serum/Plasma from subjects on a high residue, low fat diet. CD data on dietary habits. FU on colon cancer incidence.
Rudolf, Z.	11,500	1987 -	ID CD FU	-20	Sera from patients with tumours
Ryan, E.	150	1979 -	ID CD FU		Sera from patients with elevated CEA levels
Sasco, A.J.	180	1986 -	ID CD	-70	
Sharon, R.	3,200	1979 -	ID CD		2,200 sera, 1,000 plasma
Shen, F.M.	1,000	1989 -	ID	-40	
Shigematsu, I.	10,000	1969 - 79	ID CD FU	-70	Frozen
Shigematsu, I.	10,000	1979 -	ID CD FU	-20	Lyophilized

* For a few collections, number of samples are given instead: this will be corrected in future.

Serum / Plasma

Name	Persons *	Year	Code	Temp. °C	Remarks
Shope, R.E.	4,000	1954 -	CD FU	-40	Sera collected worldwide
Shunzhang, Y.	2,000	1987 -	CD FU	-20	
Sigfusson, N.	23,000	1967 -	ID CD FU	-20	
Soeripto; Pramono Sidhi	185	1952 -	ID CD		Serum/Plasma
Soeterboek, A.; Van Noord, P.	10,000	1983 -	ID CD FU	-20	Healthy adults, sequential samples
Srivatanakul, P.	2,000	1986 -	ID CD FU	-20/-70	
Stampfer, M.J.	14,916	1982 -	ID CD FU	-80	
Stjernfeldt, M.	500	1978 -			250 plasma from leukemic children, 250 specimens from healthy individuals
Syrjänen, K.	1,000	1981 -	ID CD FU	-70	
Tamburro, C.H.	600	1974 -	ID CD FU	-80	
Toniolo, P.	14,700	1985 -	ID CD FU		Serum and plasma from 30,000 individuals
Trell, E.	120,000	1974 -	ID CD FU		Women taking part in breast cancer screening
Van Noord, P.	5,000	1975 -	CD FU	-20	
Van Poppel, G.	250	1989 - 90	ID CD FU	-80	100 nasopharyngeal cancer cases, 100 controls
Vaughan, T.L.	200	1989 - 91	ID CD FU	-20	
Webster, A.D.B.	100	1971 -	ID CD	-20	
Willett, W.	30,000	1989 - 90	ID CD	LN	Plasma
Wood, P.	14,000	1978 - 90	ID CD FU	-70	Continued collection. Serum from patients with all types of cancer
Wray, B.B.	150				Serum/Plasma: 266 cancer cases, 90 controls
Wray, B.B.	356	1975 -	CD	-20	Cervical cancer cases
Yadav, M.S.	3,000	1980 -	ID CD FU	-70	Serum
Yadav, M.S.	100	1990 -	ID CD FU	-50	
Yarnell, J.	2,200	1984 - 88	ID	-20	
Yarnell, J.	2,512	1979 - 82	ID		
Yarnell, J.	2,900	1979 -	ID CD FU		Serum/Plasma
Yu, S.	2,000	1987 -	CD FU	-40	
Zuberi, S.J.	6,000	1972 -	ID		Serum/Plasma

Tissue

Name	Persons *	Year	Code	Temp. C	Remarks
Baba, Y.	1,727	1972 -	ID FU	-20	Liver, kidney, pancreas, muscle specimens
Scully, C.		1982 -	ID CD	-20	
Serotanen, K.	1,000	1981 -	ID CD FU	-70	
Becciolini, A.	1,000	1985 -	ID CD FU	-70	Healthy and neoplastic
Bianchi, C.	1,900	1979 -	ID CD FU	RT	Lung
Bianchi, L.	1,000	1979 -			Lung: Biopsies of hepatic tumours
Blumberg, B.S.	250	1958 -			
Cooksley, W.G.	50	1988 -			Liver–HCC
Davies, D.S.; Boobis, A.R.	150	1978 -	ID CD	-70	Liver samples taken for drug metabolism and toxicology studies

* For a few collections, number of samples are given instead: this will be corrected in future.

Tissue

Name	Persons *	Year	Code	Temp. °C	Remarks
De Tribolet, N.	200	1985 -	ID CD	-70	Normal mammary epithelial and stromal cells
Eppenberger, U.	2	1987 -		-70	Breast, ovary, endometrium samples
Eppenberger, U.	300	1983 -		LN	Breast and thyroid tumours
Fu, X.L.	155	1989 -	ID CD FU	-70	Gallbladder
Gonzalez, S.	67	1988 -	ID CD	-70	
Greene, M.H.	100	1970 -		-70	Familial Ovarian Cancer European Registry.
Greggi, S.	8	1989 -	ID CD FU	-70	Ovarian specimens
Haas, G.P.	50	1987 -	ID CD FU	LN	Frozen liver biopsy specimens
Hadziyannis, S.	15	1982 -	ID CD FU		Fatty tissue specimens
Hardell, O.L.	40	1981 -	ID CD FU	-70	Lung: 1) tumour 2) normal peripheral
Husgafvel-Pursiainen, K.	50	1988 -	ID CD	-70	Oesophageal mucosa
Jaskiewicz, K.	3,000	1984 - 90	ID CD FU		Gastric mucosa
Jaskiewicz, K.	586	1988 - 90	ID CD FU		DOK cell line from oral premalignant lesions
Johnson, N.W.	343	1984 -	ID CD FU	LN	
Juhász, L.	30	1990 - 90	ID CD FU	-20	
Juhász, L.; Dauda, G.	1,431	1973 -	ID CD FU		Breast cancer, histological specimens
Juhász, L.; Dauda, G.	354	1984 -	ID CD FU		Benign breast tumours, histological specimens
Juhász, L.; Dauda, G.	506	1973 -	ID CD FU		Stomach, histological specimens
Kjellström, T.	300	1978 -			Liver, brain, hair from stillborn children
Koss, L.G.	4,000	1979 -	ID		Endometrial smears and cell blocks
Kumar, S.	500	1965 -	ID		Tumours
Leigh, J.	1,150	1980 - 86			Lung: 250 mesothelioma, 350 controls, 550 coalminers. In formalin
Levin, A.	120	1972 -		-70	Tumours from Kenya, Tanzania
Levine, P.H.	4,000	1969 -		-70	Tissue from USA, Tunisia, Malaysia, Ghana and other countries
Lundström, N.G.	157	1976 - 87	ID CD FU	-20	Liver, kidney, lung, bronchus, heart muscle, brain, renal pelvis, skin, fat, psoas muscle, femur, vertebra with spinous process, finger bone, sternum, iliac crest, rib, temporal bone, ventriculus, testis
Mehta, F.S.	3,000	1964 -	ID CD	-20	Oral mucosa biopsies
Mohapatra, S.C.	50	1989 - 92	ID CD	-20	Breast
Moorehead, R.J.	200	1991 -	ID CD FU	-20	
Morgan, P.R.	120	1984 -	ID CD	-70	Oral mucosa; normal, dysplastic, malignant
Pallesen, G.	+1,500	1978 -	ID CD FU	-70	Lymphoma, other tumours, normal tissue
Prasad, U.	240	1987 -	ID CD FU	LN	Biopsy material
Remagen, W.	1,000	1972 -	ID CD FU		Bone tumour specimens in formaldehyde
Remagen, W.	5,800	1972 -	ID CD FU	-70	Bone tumour specimens
Rogan, W.	800	1978 -			Placenta
Ryan, E.	3,000	1979 -	ID CD FU		Breast tissue paraffin blocks
Shigematsu, I.	10,800	1948 - 88	ID CD FU	RT	Autopsied
Shigematsu, I.	73,700	1949 -	ID CD FU	RT	Surgical

* For a few collections, number of samples are given instead; this will be corrected in future

Tissue

Name	Persons*	Year	Code	Temp. °C	Remarks
Solanke, T.F.	283	1982 –			Biopsies of breast lumps from a prevalence survey
Srivatanakul, P.	29	1987 –	ID CD FU	–70	Liver
Stoner, G.	10	1983 –	ID CD		Oesophagus specimens
Stoner, G.	80	1980 –	ID CD		Bronchus and bladder specimens
Syrjänen, K.	1,000	1981 –	ID CD FU	–70	Genital tract biopsies
Takács, S.	39	1988 –	ID CD FU	–18	Brain. In dried form
Tamburro, C.H.	5	1974 –	ID CD FU	–70	Liver
Taylor, N.F.	15	1980 –	ID CD	–20	Portions of adrenal gland containing tumours
Warnakulasuriya, K.A.A.S.	250	1980 –	FU		Oral mucosa histological specimens
Yadav, M.S.	70	1990 –	ID CD FU	–70	Cervical carcinoma
Yarnell, J.	120	1979 –	ID CD FU		Adipose tissue

Urine

Name	Persons*	Year	Code	Temp. °C	Remarks
Bengtsson, C.	1,500	1968 –	ID FU		From 1,500 women
Berrino, F.	6,000	1987 –	ID CD	–80	
Bingham, S.A.	160	1988 – 91	ID CD FU		Complete 24-hour specimens obtained from a random sample of men and women aged 45–65
Chen, J.	4,000	1989 –	ID CD	–20	
Christensen, J.M.	700	1989 – 90	ID CD FU	–80	
Elinder, C.G.	470	1980 –	ID CD		From Stockholm and Västerås
Forman, D.	500	1986 –	ID CD FU	–20	
Greene, M.H.	35	1970 –	ID CD FU	–70	
Groopman, J.D.	350	1985 –	ID CD	–20	
Husgafvel-Pursiainen, K.	70	1988 –	ID CD FU	–70	
Knekt, P.	5,000	1968 – 80	ID CD	–20	
London, S.	960	1991 – 93	ID CD	–70	
Mohran, Y.	205	1978 –	ID		
Pershagen, G.	600	1985 – 88	ID CD	–20	
Pham, Q.T.	240	1990 – 90	ID CD FU	–20	Coal miners
Ross, R.K.	1,000	1972 –	ID		From breast cancer patients and controls
Rudolf, Z.	2,000	1978 –	ID CD FU		From patients with tumours
Schunk, W.	200	1986 – 90	ID CD FU		
Soeripto; Soeharjanto	350	1952 –	ID CD		
Soeterboek, A.; Van Noord, P.	1,500	1984 –	ID CD FU	–20	Healthy women 1920–1950, sequential samples
Srivatanakul, P.	1,500	1987 –	ID CD FU	–40	
Taylor, N.F.	50	1975 –	ID CD	–20	From patients with adrenal adenoma or carcinoma, especially children

* For a few collections, number of samples are given instead: this will be corrected in future.

Urine

Name	Persons *	Year	Code	Temp. °C	Remarks
Van Noord, P.	53,000	1975 –	ID CD FU	-20	Women taking part in breast cancer screening
Van Poppel, G.	250	1989 – 90	CD FU	-20	

White Cells

Name	Persons *	Year	Code	Temp. °C	Remarks
Assennato, G.	200	1990 – 91	ID CD	-70	
Berrino, F.	6,000	1987 –	ID CD	-80	
Birkeland, S.A.	2,500	1971 – 90		-90	
Comstock, G.W.	32,941	1989 –	ID CD FU	-70	
Ebbesen, P.; Melbye, M.; Biggar, R.	200	1981 –	ID		From Denmark and Greenland
Forman, D.	2,000	1989 –	ID CD	-70	DNA
Gazkartseva, R.	160	1987 –	ID	-20	
Greene, M.H.	3,928	1970 –	ID CD FU	-70	Familial Ovarian Cancer European Registry. From probands and first-degree relatives
Greggi, S.	46	1989 –		-70	DNA
Husgafvel-Pursiainen, K.	70	1989 –	ID CD	-20	
Ishida, T.	100	1983 –	ID	LN	
Koskimies, S.	300	1978 –	ID CD	-70	
Kune, G.A.	160	1980 –			From a case-control study of colorectal cancer
Levin, A.	1,600	1972 –		LN	From Kenya, Tanzania
London, S.	960	1991 – 93	ID CD	-70	
Müller, H.	272	1982, 1986	ID CD FU	LN	
Mohan, Y.	205	1978 –	ID		
Pham, Q.T.	240	1990 – 90	ID CD FU	-70	Coal miners
Shigematsu, I.	10,000	1982 –	ID CD FU	LN	Lymphocytes
Soeripto; Haryono Soeharto	380	1952 –	ID CD		
Srivatanakul, P.	2,000	1986 –	ID CD FU	-40/-70	
Van den Berghe, H.	3,000		ID CD FU	-70	
Vaughan, T.L.	200	1989 – 91	ID CD	-20	
Willett, W.	30,000	1989 – 90	ID CD	LN	100 nasopharyngeal cancer cases, 100 controls

Whole Blood

Name	Persons *	Year	Code	Temp. °C	Remarks
Christensen, J.M.	1,700	1978 – 90	ID CD FU	-20/-80	
Hardell, O.L.	35	1988 –	ID FU		
Hennekens, C.H.	16,000	1982 –	ID CD FU	-85	
Hernberg, S.G.	1,300	1971 – 90	ID		
Kjellström, T.	100	1978 –			From stillborn children

* For a few collections, number of samples are given instead: this will be corrected in future.

Whole Blood

Name	Persons*	Year	Code	Temp. °C	Remarks
Levin, A.	50	1985 –	ID CD FU	LN	From Tanzania, Kenya, Zambia, Singapore, Papua-New Guinea
Stampfer, M.J.	14,916	1982 –	ID CD FU	–80	

Other

Name	Persons*	Year	Code	Temp. °C	Remarks
Berrino, F.	6,000	1987 –	ID CD		Toenails
Bingham, S.A.	240	1982 –	ID CD		Diet samples
Chongsuvivatwong, V.	500–600	1991 – 93	ID CD	–70	Exfoliated cervical cells and cells from penis
Cummings, J.	150	1974 –	ID		Diet samples
Cummings, J.	60	1980 –	ID		Ileal effluence
Greene, M.H.	137	1970 –	ID CD FU		Semen from persons at high risk of cancer
Hill, A.P.	162	1988 –	ID CD FU	–80	DNA from breast cancer cases and relatives
Kjellström, T.	12,000	1978 –			Hair of new mothers
Kluijber, V.	41	1987 –	ID CD FU	LN	Semen
Levin, A.	300	1985 –	ID CD FU	LN	From Tanzania, Kenya, Zambia, Singapore, Papua New Guinea
Levine, P.H.	500	1969 –			Cerebrospinal fluid from USA, Tunisia, Malaysia, Ghana and other countries
Lundström, N.G.	157	1976 – 87	ID CD FU	–20	Hair, nails
Müller, H.	5		ID CD FU	LN	Keratinocytes
Marcus, Z.H.	600	1981 –	ID CD	–80	Spermatozoa and ejaculates
Mehta, F.S.	2,000	1964 –	ID CD		Oral cytology smears
Moe, P.J.	1,800	1966 –	ID FU		Spinal fluid
Nakachi, K.	1,500	1989 –	ID CD FU	–80	DNA isolated from lymphocytes
Paganini-Hill, A.	162	1988 –	ID CD FU	–80	DNA from breast cancer cases and relatives
Pallesen, G.	70	1984 –	CD	–70	Lymphnodes from patients with HIV (LAV/HTLV-III) infection
Rios-Dalenz, J.L.		1984 – 87			Bile, blood and gallstones from patients with biliary tract cancer and from controls (a) with cholelithiasis and (b) without gallbladder disease. Medical abstracts are available
Rogan, W.	2,000	1978 –			Maternal milk
Romney, S.L.	1,000	1991 –	ID CD FU	RT	Cervico-vaginal lavage specimens
Sarjadi	4,000	1985 –	ID CD	–80	Paraffin block
Stücker, I.	300	1990 – 92	ID CD FU	–80	DNA
Tonioło, P.	3,500	1989 –	ID CD FU	RT	Blood clots
Van den Brandt	80,000	1986 – 86	ID CD FU		Toenail clippings
Van Noord, P.	19,000	1982 –	ID CD FU		Toenails from women taking part in breast cancer screening
Willet, W.	100,000	1982 – 87	ID CD	RT	Toenails
Zuberi, S.J.	175	1972 –	ID		Gallstones

* For a few collections, number of samples are given instead; this will be corrected in future.

Address List

Dr H. Anton-Culver
Univ. of California at Irvine
Dept. of Community Medicine
& Environmental Medicine
Irvine, CA 92717
USA

Prof. K. Aoki
Nagoya Univ.
School of Medicine
Dept. of Preventive Medicine
Tsurumai-Cho, Showa-Ku
Nagoya 466
Japan

Prof. G. Assennato
Inst. of Occupational Health
Univ. of Bari
Piazza Gulio Cesare
70124 Bari
Italy

Dr Y. Baba
Univ. of Occupational
& Environmental Health
1-1 Iseigaoka
Yahatanishi-ku
Kita-Kyushu 807
Japan

Dr E.L. Barrett-Connor
Univ. of California at San Diego
Dept. of Community & Family Medicine
M-007
La Jolla, CA 92093
USA

Dr R. Beaglehole
Univ. of Auckland
Dept. of Epidemiology
Private Bag
Auckland
Netherland

Prof. A. Becciolini
Radiation Biology Lab.
Dept. of Clinical Physiopathology
Viale Morgagni 85
50134 Florence
Italy

Prof. C.B. Bengtsson
Dept. of Primary Health Care
Redbergsv. 6
416 65 Göteborg
Sweden

Dr F. Berrino
Ist. Nazionale per lo Studio
e la Cura dei Tumori
Via G. Venezian 1
20133 Milan
Italy

Dr C. Bianchi
Hosp. of Monfalcone
Lab. of Pathological Anatomy
Via Galvani
34074 Monfalcone
Italy

Prof. L. Bianchi
Institut für Pathologie
Univ. Basel
Schoenbeinstr. 40
Basel
Switzerland

Dr S.A. Bingham
Dunn Clinical Nutrition Centre
100 Tennis Court Rd
Cambridge CB2 1QL
United Kingdom

Dr S. Birkeland
Odense Univ. Hospital
Dept. of Nephrology
5000 Odense C
Denmark

Dr B.S. Blumberg
Vice President for Population Oncology
Fox Chase Cancer Center
7701 Burholme Ave.
Philadelphia, PA 19111
USA

Dr P. Burtin
CNRS
7 rue Guy Mocquet
P.O. Box 8
94802 Villejuif
France

Dr G.D. Calvert
Flinders Medical Centre
Dept. of Clinical Biochemistry
Bedford Park
Adelaide, SA 5042
Australia

Dr M.E. Camargo
Inst. de Medicina Tropical
de São Paulo
Depto di Immunologia e
Seroepidemiologia
Aven. Dr Eneas de Carvalho
Aguiar, 470
05403 São Paulo SP
Brazil

Dr R.A. Cartwright
Univ. of Leeds
Leukaemia Research Fund
Centre for Clinical Epidemiology
Dept. of Pathology
17 Springfield Mount
Leeds LS2 9NG
United Kingdom

Dr J.S. Chen
Inst. of Nutrition & Food Hygiene
Chinese Academy
of Preventive Medicine
29 Nan Wei Rd
Beijing 100050
People's Republic of China

Dr V. Chongsuvivatwong
Prince of Songkla University
Faculty of Medicine
Epidemiology Unit
90112 Hat Yai
Thailand

Mrs J.M. Christensen
Danish National Inst.
of Occupational Health
Dept. Chemistry & Biochemistry
Lersø Parkallé 105
2100 Copenhagen Ø
Denmark

Dr L.C. Clark
University of Arizona
Dept. of Family & Community Medicine
2504 E. Elm St.
Tucson, AZ 85716
USA

Prof. G.W. Comstock
The Johns Hopkins Univ.
P.O. Box 2067
Hagerstown, MD 21742-2067
USA

Prof. W.G. Cooksley
Royal Brisbane Hospital Foundation
Royal Brisbane Hospital
Brisbane, Qld 4027
Australia

Dr J. Cummings
Dunn Clinical Nutrition Centre
100 Tennis Court Rd
Cambridge CB2 1QL
United Kingdom

Professor D.S. Davies
Royal Postgraduate Medical School
Dept. of Clinical Pharmacology
Du Cane Rd
London W12 0HS
United Kingdom

Prof. R.L. Dawkins
Sir Charles Gairdner Hosp.
Dept. of Clinical Immunology
Queen Elizabeth II Medical Centre
Verdun St.
Nedlands, WA 6009
Australia

Dr G. de Thé
CNRS, Fac. de Médecine Alexis Carrel
Lab. d'Epidémiologie
& Immunovirologie des Tumeurs
Rue G. Paradin
69372 Lyon Cédex 08
France

Prof. N. de Tribolet
CHU Vaudois
Serv. de Neurochirurgie
Rue du Bugon
1011 Lausanne
Switzerland

Dr T.R. Driscoll
National Inst. of Occupational Health
and Safety
G.P.O. Box 58
Sydney, NSW 2001
Australia

Prof. Ebbesen
Inst. of Cancer Research
Radiumstationen
8000 Aarhus C
Denmark

Dr C.G. Elinder
National Board
of Occupational Safety & Health
Sect. of Occupational Medicine
Dept. of Research
Ekelundsvaegen 16
171 84 Solna
Sweden

Dr D.A. Enarson
W.M. Health Sciences Centre
2E331
Edmonton, Alberta
Canada T6G 2B1

Prof. U. Eppenberger
Univ. Women's Clinic Lab.
Schanzenstr. 46
4031 Basel
Switzerland

Dr A.E. Evans
Children's Hospital
34th & Civic Bldg
Philadelphia, PA 19104
USA

Prof. A.S. Evans
Yale Univ.
Dept. of Epidemiology
WHO Serum Ref. Bank
333 Cedar St.
P.O. Box 3333
New Haven, CT 06510
USA

Dr M.K. Fordyce
Univ. of Miami
School of Medicine
Dept. of Epidemiology
1550 NW 10th Ave., Suite 100
Miami, FL 33136
USA

Dr D. Forman
Imperial Cancer Research Fund
Radcliffe Infirmary
Gibson Bldg
Oxford OX2 6HE
United Kingdom

Dr X.L. Fu
Tianjin Cancer Inst.
Huan-Hu-Xi Rd
Ti-Yuan-Bei
Tianjin 300060
People's Republic of China

Dr A. Fukao
Tohoku Univ.
School of Medicine
Dept. of Public Health
2-1 Seiryo-machi
Sendai 980
Japan

Dr R.F. Garkavtseva
All-Union Scientific Cancer Center
Kashirskoye shosse 24
Moscow 115478
USSR

Dr G.T. Golan
Haifa Medical Center (Rothschild)
Faculty of Medicine, Technion
Clinical Immunology Service
47 Golomb St.
Haïfa 31048
Israel

Dr S. Gonzalez
Catholic Univ. of Chile
School of Medicine
Dept. of Pathology
Casilla 114-D
Santiago
Chile

Dr T. Gotlieb-Steimatsky
Tel Aviv Univ., Sackler Medical School
Chaim Sheba Medical Center
Central Virology Lab.
Tel Hashomer 52621
Israel

Dr M.H. Greene
NCI
Environmental Epidemiology Branch
3C19 Landow Bldg
Bethesda, MD 20892-4200
USA

Dr S. Greggi
Catholic Univ. of San Cuore
Dept. of Gynaecology & Obstetrics
Largo A. Gemelli, 8
00168 Roma
Italy

Prof. J.D. Groopman
Johns Hopkins Univ.
School of Public Health
615 North Wolfe St.
Baltimore, MD 21117
USA

Prof. S. Grufferman
Univ. of Pittsburgh
School of Medicine
M-200 Scaife Hall
Pittsburgh, PA 15261
USA

Dr G.P. Haas
Wayne State Univ.
Sect. of Uro-Oncology
4160 John R. Rd
Detroit, MI 48201
USA

Prof. S.J. Hadziyannis
Hippocration General Hospital
Academic Dept. of Medicine
114 Vassilissis Sofias Ave.
Athens 115
Greece

Dr A.P. Haines
Northwick Park Hospital
MRC, Epidemiology & Medical Care Unit
Watford Rd
Harrow, Middlesex HA1 3UJ
United Kingdom

Dr B.L. Hansen
Univ. of Copenhagen
Inst. of Medical Microbiology
22 Juliane Maries Vej
2100 Copenhagen Ø
Netherlands

Dr L.O. Hardell
Univ. Hospital
Dept. of Oncology
901 85 Umea
Sweden

Dr R.J. Hay
Cell Culture Dept.
American Type Culture Collection
12301 Parklawn Drive
Rockville, MD 20852
USA

Dr C.H. Hennekens
Harvard Medical School
Brigham & Women's Hospital
Channing Lab., Dept. of Medicine
55 Pond Ave.
Brookline, MA 02146
USA

Prof. S.G. Hernberg
Inst. of Occupational Health
Topeliuksenkatu 41 a A
00250 Helsinki
Finland

Dr S. Hisamichi
Tohoku Univ.
School of Medicine
Dept. of Public Health
2-1, Seiryo-machi
Sendai 980
Japan

Dr K.T. Husgafvel-Pursiainen
Inst. of Occupational Health
Topeliuksenkatu 41 a A
00250 Helsinki
Finland

Dr T. Ishida
Univ. of Tokyo
Faculty of Sciences
Dept. of Anthropology
Bunkyo-ku
Hongo 7-3-1
Tokyo 113
Japan

Dr K. Jaskewicz
Univ. of Cape Town
Medical School
Dept. of Pathology
Observatory
Cape Town 7925
Republic of South Africa

Prof. E. Jellum
Norwegian Cancer Society
Inst. of Clinical Biochemistry
Pilestredet 32
0027 Oslo 1
Norway

Prof. N.W. Johnson
Royal College of Surgeons of England
Dept. of Dental Sciences
35-43 Lincoln's Inn Fields
London WC2A 3PN
United Kingdom

Dr L. Juhasz
County Hospital Polyclinic Unit
Dept. of Oncology
Cancer Registry
of the County of Szabolcs-Szatmar
Voeroes Hadsereg u. 68
4401 Nyiregyhaza
Hungary

Prof. E.G. Jung
Hautklinik Klinikum Mannheim
Postfach 23
6800 Mannheim
Federal Republic of Germany

Dr M.M. Kapur
All India Inst. of Medical Science
Dept. of Surgery
Ansari Nagar
New Delhi 110029
India

Dr J.D. Kark
Hebrew Univ.
Faculty of Medicine
Hadassah School of Public Health
& Community Medicine
Dept. of Social Medicine
P.O. Box 1172
91000 Jerusalem
Israel

Prof. H. Kato
Tokyo Medical College
6-7-1 Nishishinjuku
Shinjuku-ku
Tokyo 160
Japan

Dr S.M. Kennedy
Univ. of British Columbia
2775 Heather St.
Vancouver, BC
Canada V7L 2K5

Prof. M.C. Kew
Univ. of the Witwatersrand
School of Medicine
7 York Rd
Parktown
Johannesburg 2193
Republic of South Africa

Dr S.K. Khoo
Univ. of Queensland
Royal Brisbane Hospital
Dept. of Obstetrics & Gynecology
Clinical Sciences Bldg
Brisbane, Qld 4029
Australia

Dr J.S. Kim
Kyungpook National Univ.
School of Medicine
Dept. of Clinical Pathology
101 Dong-In Dong
Taegu
Republic of Korea

Dr T. Kjellstroem
Organisation Mondiale de la Santé
EHE/PEP
Bureau no. L 144
1211 Genève 27
Switzerland

Dr V. Klujber
National Inst. of Hygiene
Gyáli ut 2-6
1966 Budapest
Hungary

Dr P.B. Knekt
Social Insurance Institution
P.O. Box 78
00381 Helsinki
Finland

Dr T.M. Knight
Univ. of Keele
School of Postgraduate Medicine
and Biological Sciences
Thornburrow Drive
Hartshill
Stoke-On-Trent ST4 7QB
United Kingdom

Dr S. Koskimies
FRC Blood Transfusion Serv.
Kivihanne 7
00310 Helsinki
Finland

Dr L.G. Koss
Albert Einstein Medical College
Montefiore Hospital
& Medical Center
Dept. of Pathology
111 E 210th St.
Bronx, NY 10467
USA

Dr S. Kumar
Christie Hospital
& Holt Radium Inst.
Dept. of Pediatric Oncology
Wilmslow Rd
Withington
Manchester M20 9BX
United Kingdom

Dr G.A. Kune
61 Erin St.
Richmond, Vict. 3121
Australia

Dr J. Leigh
National Institute of Occupational Health
and Safety
G.P.O. Box 58
Sydney, NSW 2001
Australia

Dr A. Levin
Clinical Research Center
Div. of Clinical Sciences
Harrow, Middlesex HA1 3UJ
United Kingdom

Prof. S. Levin
Kaplan Hospital
Dept. of Pediatrics
Pediatric Research Inst.
Rehovot 76100
Israel

Prof. Q.F. Liu
Cancer Inst. of Guangxi
No. 6, Bin-hu Rd
Naning, Guangxi 530027
People's Republic of China

Dr S.J. London
Univ. of Southern California
School of Medicine
Parkview Medical Bldg B-306
1420 San Pablo St.
Los Angeles, CA 90033
USA

Dr W.T. London
Fox Chase Cancer Center
7701 Burholme Ave.
Philadelphia, PA 19111
USA

Dr V. Luepker
Univ. of Minnesota
School of Public Health
Div. of Epidemiology
Stadium Gate 27
Minneapolis, MN 55455
USA

Dr B. Lund
Fred Hutchinson Cancer Research Center
1124 Columbia
MP 859
Seattle, WA 98003
USA

Dr N.G. Lundström
Umea Univ.
Dept. of Environmental Medicine
901 87 Umea
Sweden

Dr I.R. Mackay
Walter & Eliza Hall
Inst. of Medical Research
Royal Melbourne Hospital
Clinical Research Unit
Post Office
Melbourne, Vict. 3050
Australia

Dr R. Mandeville
Inst. Armand-Frappier
Immunology Research Center
531 Blvd des Prairies
Laval, Québec
Canada H7N 4Z3

Dr J.R. Mann
Birmingham Children's Hospital
Dept. of Oncology
Ladywood Middleway
Birmingham B16 8ET
United Kingdom

Prof. Z.H. Marcus
Blood Bank
Edith Wolfson Hospital
P.O. Box 5
Holon 58100
Israel

Dr W.H. Marshall
The General Hospital
Health Science Center
300 Prince Philip Drive
St. John's Newfoundland
Canada A1B 3V6

Dr E. Marubini
Univ. of Milan
Inst. of Biometry & Medical Statistics
Via Venezian 1
20133 Milano
Italy

Dr F.S. Mehta
Tata Inst. of Fundamental Research
Basic Dental Research Unit
Homi Bhabha Rd
Bombay 400005
India

Ms M. Mendy
Gambia Hepatitis Intervention
MRC Laboratories
Fajara near Banjul
Gambia

Prof. F. Mitelman
Univ. Hospital
Dept. of Clinical Genetics
221 85 Lund
Sweden

Dr P.J. Moe
Regionsykehuset
Barneklinikken
Kyrres gate 17
7000 Trondheim
Norway

Dr S.C. Mohapatra
Banaras Hindu Univ.
Inst. of Medical Sciences
Dept. of Preventive and Social Medicine
Banaras Hindu University Campus
Varanasi 221005
India

Dr Y.Z. Mohran
Ein Shams Univ.
Faculty of Medicine
376 Pyramids Rd
Abbasia, Cairo
Egypt

Mr R.J. Moorehead
Queen's Univ. of Belfast
Inst. of Clinical Science
Grosvenor Rd
Belfast BT12 6BJ
United Kingdom

Dr P.R. Morgan
Royal College of Surgeons of England
Dept. of Dental Sciences
35-43 Lincoln's Inn Fields
London WC2A 3PN
United Kingdom

Dr H.J. Mueller
Basel Cantonal Hospital
Dept. of Research
Lab. of Human Genetics
Hebelstr. 20
4031 Basel
Switzerland

Dr N.E. Mueller
Harvard School of Public Health
Dept. of Epidemiology
677 Huntington Ave.
Boston, MA 02115
USA

Dr L. Naggan
Ben Gurion Univ. of the Negev
Epidemiology Unit
Faculty of Health Sciences
P.O. Box 653
Beer Sheva
Israel

Dr K. Nakachi
Saitama Cancer Center
Research Inst.
Dept. of Epidemiology
Komuro 818, Ina-machi
Saitama 362
Japan

Prof. J.V.G. Neel
Dept. of Human Genetics
Univ. of Michigan
Medical School
1137 E. Catherine St.
Ann Arbor, MI 48109
USA

Dr B. Noel
Centre Départemental
de Transfusion Sanguine de Chambéry
Faubourg Mache
73 Chambéry
France

Dr A. Pacsa
Inst. for Microbiology
Univ. Medical School
Sziarti VT 12
Pecs
Hungary

Prof. A. Paganini-Hill
Univ. of Southern California
School of Medicine
Dept. of Preventive Medicine
1441 Eastlake Ave.
Los Angeles, CA 90033-0800
USA

Dr G. Pallesen
Lab. of Immunohistology
Finsensgade 12
8000 Aarhus C
Denmark

Dr D. Palli
Centre for the Study
& Prevention of Cancer
Epidemiology Sect.
Viale Volta 171
50131 Firenze
Italy

Dr G. Pershagen
Karolinska Inst.
National Inst. of Environmental Medicine
Box 60208
104 01 Stockholm
Sweden

Dr Q.T. Pham
INSERM U 115 - "Santé au Travail
& Santé Publique"
Méthodes & Applications
Dépt. d'Epidémiologie Respiratoire
Faculté de Médecine
Route de la Forêt de Haye
54505 Vandoeuvre-les-Nancy
France

Prof. W.O. Phoon
National Inst. of Occupational
Health & Safety
G.P.O. Box 58
Sydney, NSW 2001
Australia

Dr S.V. Pizzo
Duke Univ. Medical Center
P.O. Box 3712
Durham, NC 27710
USA

Dr U. Prasad
Univ. of Malaya
Faculty of Medicine
Dept. of Otorhinolaryngology
NPC Lab.
Lembah Pantai
Kuala Lumpur 59100
Malaysia

Prof. S.M. Rappaport
Univ. of North Carolina
School of Public Health
Rosenau Hall CB7400
Chapel Hill, NC 27599
USA

Dr F. Raue
Medizinische Universitätsklinik
Endokrinologische Ambulanz
Luisenstrasse 5
W 6900 Heidelberg
Federal Republic of Germany

Dr W.E. Rawls
McMaster Univ.
Dept. of Pathology
Room 4H20, 1200 Main St. W.
Hamilton, Ontario
Canada L8N 3Z5

Dr W. Remagen
Universität Basel
Abteilung für Pathologie
Knochentumor-Register
Shoenbeinstr. 40
4003 Basel
Switzerland

Dr J.L. Rios-Dalenz
Hospital Metodista
Calle 12, Obrajes
P.O. Box 4826
La Paz
Bolivia

Prof. D. Roelcke
Inst. für Immunologie und Serologie
Sekt. für Immunohaematologie
Im Neuenheimer Feld Bau 305
6900 Heidelberg
Federal Republic of Germany

Dr W. Rogan
National Inst. of Health
MD A3-02
N.I.E.H.
P.O. Box 12233
Research Triangle Park, NC 27709
USA

Prof. S.L. Romney
Yeshiva Univ.
Albert Einstein College of Medicine
1300 Morris Park Ave.
Bronx, NY 10461
USA

Dr S.A. Rosenberg
NCI
Chief, Surgery Branch
Div. of Cancer Treatment
Bldg 10, Room 10N 116
Bethesda, MD 20892-4200
USA

Dr R.K. Ross
Univ. of Southern California
Norris Cancer Hospital & Research Inst.
School of Medicine
P.O. Box 33800, Room 803
1441 Eastlake Ave.
Los Angeles, CA 90033-0804
USA

Dr P. Rozen
Tel Aviv Univ.
Sackler School of Medicine
Dept. of Gastroenterology
Ichilov Hospital
6 Weizman St.
Tel Aviv 64239
Israel

Dr Z. Rudolf
Inst. of Oncology
Zaloska 2
Ljubljana 61000
Yugoslavia

Prof. H.W. Ruediger
Inst. of Occupational Health
Unit of Toxicogenetics
Adolph Schönfeldenstr. 5
W 2000 Hamburg 76
Federal Republic of Germany

Dr E. Ryan
McMaster Univ.
Dept. of Pathology
Hamilton, Ontario
Canada L8N 3Z5

Dr Sarjadi
ICS, Central Java Branch
Diponegoro Univ., Kariadi Teaching Hospital
Dept. Pathology
Jalan Dr Sutomo 16
Semarang 50231
Indonesia

Dr M.S. Sasaki
Kyoto Univ.
Radiation Biology Center
Dept. of Mutagenesis
Yoshida-konoecho
Sakyo-ku
Kyoto 606
Japan

Dr A.J. Sasco
IARC
Unit of Analytical Epidemiology
150 cours Albert Thomas
69372 Lyon Cédex 08
France

Prof. W.W. Schunk
Medizinische Akademie Erfurt
Institut für Arbeitsmedizin
Gustav-Freytag-Str. 1
W 5082 Erfurt
Federal Republic of Germany

Prof. C. Scully
Centre for Study of Oral Disease
Bristol Dental Hospital and School
Lower Maudlin St.
Bristol BS1 2LY
United Kingdom

Prof. R. Sharon
Hadassah Medical Centre
Director, Blood Bank
Har-Hatzofin
Jerusalem
Israel

Dr F.M. Shen
Shanghai Medical Univ.
School of Public Health
Dept. of Epidemiology
Yi Xue Yuan Rd 138
Shanghai 200032
People's Republic of China

Dr I. Shigematsu
Radiation Effects Research Foundation
5-2 Hijiyama Park
Minami-ku
Hiroshima 732
Japan

Dr R.E. Shope
Yale Univ.
School of Medicine
Yale Arbovirus Research Unit
60 College St.
Box 3333
New Haven, CT 06510
USA

Dr N. Sigfusson
Heat Preventive Clinic
Iceland Heart Association
Lagmuli 9
108 Reykjavik
Iceland

Dr Soeripto
Gadjah Mada Univ.
Faculty of Medicine
Dept. of Pathology
J1. Kesehatan, Sekip
Yogyakarta
Indonesia

Dr A. Soeterboek
Regional Blood Bank
Ds. Fliednerstr. 1
5631 BM Eindhoven
Netherlands

Dr T.F. Solanke
Univ. of Ibadan
Dept. of Surgery
College of Medicine
Ibadan
Nigeria

Dr P. Srivatanakul
NCI
Research Div.
Rama VI Rd
Bangkok 10400
Thailand

Dr M.J. Stampfer
Brigham & Women's Hospital
Dept. of Medicine
Channing Labs.
180 Longwood Ave.
Boston, MA 02115
USA

Dr M. Stjernfeldt
Univ. Hospital
Dept. of Paediatrics
581 85 Linkoeping
Sweden

Dr G. Stoner
Medical College of Ohio
Dept. of Pathology
Health Education Bldg
Room 202, 3000 Arlington Ave.
Toledo, OH 43614
USA

Dr I. Stücker
INSERM U 170
16 Avenue Paul Vaillant-Couturier
94807 Villejuif Cédex
France

Prof. K.J. Syrjänen
Dept. of Pathology
Univ. of Kuopio
Kuopio Cancer Research Centre
P.O. Box 1627
70211 Kuopio
Finland

Dr S. Takács
Public Health & Epidemiological Inst.
of Borsod County
Environmental Lab. of Community Hygiene
Bacso Béla u. 12
P.O. Box 186
3501 Miskolc
Hungary

Dr C.H. Tamburro
Univ. of Louisville
School of Medicine
Div. of Occupational Toxicology
Liver Research Center, HSC
500 S. Preston St.
Louisville, KY 40292
USA

Dr N.F. Taylor
King's College Hospital
School of Medicine
Dept. of Clinical Biochemistry
Bessemer Rd
London SE5 9PJ
United Kingdom

Dr M. Thorogood
Dept. of Community Medicine & General Practice
Radcliffe Infirmary
Woodstock Rd
Oxford OX2 6HE
United Kingdom

Dr P.G. Toniolo
New York Univ. Medical Center
Dept. of Environmental Medicine
341 East 25th St.
New York, NY 10010
USA

Dr E. Trell
Univ. of Lund
Malmoe General Hospital
Dept. of Preventive Medicine
214 01 Malmoe
Sweden

Dr P. Twentyman
MRC Clinical Oncology Unit
Hills Rd
Cambridge CB2 2QH
United Kingdom

Dr H. Van den Berghe
Univ. of Leuven
Centre for Human Genetics
Gasthuisberg Campus 0 en N
Herestr. 49
3000 Leuven
Belgium

Dr P.A. Van den Brandt
Univ. of Limburg
Dept. of Epidemiology
P.O. Box 616
6200 MD Maastricht
Netherlands

Dr P.A.H. Van Noord
Preventicon
Inst. of Public Health & Epidemiology
Div. of Epidemiology
Radboudkwartier 261-263
3511 CK Utrecht
Netherlands

Dr G. Van Poppel
TNO-CIVO
Toxicology & Nutrition Inst.
P.O. Box 360
3700 AJ Zeist
Netherlands

Dr T.L. Vaughan
Fred Hutchinson Cancer Research Center
Div. of Public Health Sciences
1124 Columbia St.
Seattle, WA 98667
USA

Dr K.A.A.S. Warnakulasuriya
Univ. of Peradeniya
Dept. of Oral Medicine
Augusta Rd
Peradeniya
Sri Lanka

Dr A.D.B. Webster
Clinical Research Centre
Immunodeficiency Research Group
Watford Rd
Harrow, Middlesex HA1 3UJ
United Kingdom

Prof. W.C. Willett
Harvard School of Public Health
Dept. of Epidemiology
677 Huntington Ave.
Boston, MA 02115
USA

Prof. P.D. Wood
Stanford University
Medical Center
Center for Research
in Disease Prevention
730 Welch Rd, Suite B
Stanford, CA 94305
USA

Prof. B.B. Wray
Medical College of Georgia
Allergy-Immunology Sect.
CJ 141
Augusta, GA 30912-3790
USA

Dr M.S. Yadav
Univ. of Malaya
Inst. of Advanced Studies
59100 Kuala Lumpur
Malaysia

Dr J.W.G. Yarnell
Medical Research Council
Epidemiology Unit
4 Richmond Rd
Cardiff CF2 3AS
United Kingdom

Dr S.Z. Yu
Shanghai Medical Univ.
School of Public Health
138 Medical Univ. Rd
Shanghai 200032
People's Republic of China

Dr S.J. Zuberi
Jinnah Postgraduate Medical Centre
Pakistan Medical Research Centre
Karachi 35
Pakistan

PUBLICATIONS OF THE INTERNATIONAL AGENCY FOR RESEARCH ON CANCER
Scientific Publications Series

(Available from Oxford University Press through local bookshops)

No. 1 Liver Cancer
1971; 176 pages *(out of print)*

No. 2 Oncogenesis and Herpesviruses
Edited by P.M. Biggs, G. de-Thé and L.N. Payne
1972; 515 pages *(out of print)*

No. 3 N-Nitroso Compounds: Analysis and Formation
Edited by P. Bogovski, R. Preussman and E.A. Walker
1972; 140 pages *(out of print)*

No. 4 Transplacental Carcinogenesis
Edited by L. Tomatis and U. Mohr
1973; 181 pages *(out of print)*

No. 5/6 Pathology of Tumours in Laboratory Animals, Volume 1, Tumours of the Rat
Edited by V.S. Turusov
1973/1976; 533 pages; £50.00

No. 7 Host Environment Interactions in the Etiology of Cancer in Man
Edited by R. Doll and I. Vodopija
1973; 464 pages; £32.50

No. 8 Biological Effects of Asbestos
Edited by P. Bogovski, J.C. Gilson, V. Timbrell and J.C. Wagner
1973; 346 pages *(out of print)*

No. 9 N-Nitroso Compounds in the Environment
Edited by P. Bogovski and E.A. Walker
1974; 243 pages; £21.00

No. 10 Chemical Carcinogenesis Essays
Edited by R. Montesano and L. Tomatis
1974; 230 pages *(out of print)*

No. 11 Oncogenesis and Herpesviruses II
Edited by G. de-Thé, M.A. Epstein and H. zur Hausen
1975; Part I: 511 pages
Part II: 403 pages; £65.00

No. 12 Screening Tests in Chemical Carcinogenesis
Edited by R. Montesano, H. Bartsch and L. Tomatis
1976; 666 pages; £45.00

No. 13 Environmental Pollution and Carcinogenic Risks
Edited by C. Rosenfeld and W. Davis
1975; 441 pages *(out of print)*

No. 14 Environmental N-Nitroso Compounds. Analysis and Formation
Edited by E.A. Walker, P. Bogovski and L. Griciute
1976; 512 pages; £37.50

No. 15 Cancer Incidence in Five Continents, Volume III
Edited by J.A.H. Waterhouse, C. Muir, P. Correa and J. Powell
1976; 584 pages; *(out of print)*

No. 16 Air Pollution and Cancer in Man
Edited by U. Mohr, D. Schmähl and L. Tomatis
1977; 328 pages *(out of print)*

No. 17 Directory of On-going Research in Cancer Epidemiology 1977
Edited by C.S. Muir and G. Wagner
1977; 599 pages *(out of print)*

No. 18 Environmental Carcinogens. Selected Methods of Analysis. Volume 1: Analysis of Volatile Nitrosamines in Food
Editor-in-Chief: H. Egan
1978; 212 pages *(out of print)*

No. 19 Environmental Aspects of N-Nitroso Compounds
Edited by E.A. Walker, M. Castegnaro, L. Griciute and R.E. Lyle
1978; 561 pages *(out of print)*

No. 20 Nasopharyngeal Carcinoma: Etiology and Control
Edited by G. de-Thé and Y. Ito
1978; 606 pages *(out of print)*

No. 21 Cancer Registration and its Techniques
Edited by R. MacLennan, C. Muir, R. Steinitz and A. Winkler
1978; 235 pages; £35.00

No. 22 Environmental Carcinogens. Selected Methods of Analysis. Volume 2: Methods for the Measurement of Vinyl Chloride in Poly(vinyl chloride), Air, Water and Foodstuffs
Editor-in-Chief: H. Egan
1978; 142 pages *(out of print)*

No. 23 Pathology of Tumours in Laboratory Animals. Volume II: Tumours of the Mouse
Editor-in-Chief: V.S. Turusov
1979; 669 pages *(out of print)*

No. 24 Oncogenesis and Herpesviruses III
Edited by G. de-Thé, W. Henle and F. Rapp
1978; Part I: 580 pages, Part II: 512 pages *(out of print)*

Prices, valid for September 1991, are subject to change without notice

List of IARC Publications

No. 25 Carcinogenic Risk. Strategies for Intervention
Edited by W. Davis and C. Rosenfeld
1979; 280 pages (*out of print*)

No. 26 Directory of On-going Research in Cancer Epidemiology 1978
Edited by C.S. Muir and G. Wagner
1978; 550 pages (*out of print*)

No. 27 Molecular and Cellular Aspects of Carcinogen Screening Tests
Edited by R. Montesano, H. Bartsch and L. Tomatis
1980; 372 pages; £29.00

No. 28 Directory of On-going Research in Cancer Epidemiology 1979
Edited by C.S. Muir and G. Wagner
1979; 672 pages (*out of print*)

No. 29 Environmental Carcinogens. Selected Methods of Analysis. Volume 3: Analysis of Polycyclic Aromatic Hydrocarbons in Environmental Samples
Editor-in-Chief: H. Egan
1979; 240 pages (*out of print*)

No. 30 Biological Effects of Mineral Fibres
Editor-in-Chief: J.C. Wagner
1980; Volume 1: 494 pages; Volume 2: 513 pages; £65.00

No. 31 N-Nitroso Compounds: Analysis, Formation and Occurrence
Edited by E.A. Walker, L. Griciute, M. Castegnaro and M. Börzsönyi
1980; 835 pages (*out of print*)

No. 32 Statistical Methods in Cancer Research. Volume 1. The Analysis of Case-control Studies
By N.E. Breslow and N.E. Day
1980; 338 pages; £20.00

No. 33 Handling Chemical Carcinogens in the Laboratory
Edited by R. Montesano et al.
1979; 32 pages (*out of print*)

No. 34 Pathology of Tumours in Laboratory Animals. Volume III. Tumours of the Hamster
Editor-in-Chief: V.S. Turusov
1982; 461 pages; £39.00

No. 35 Directory of On-going Research in Cancer Epidemiology 1980
Edited by C.S. Muir and G. Wagner
1980; 660 pages (*out of print*)

No. 36 Cancer Mortality by Occupation and Social Class 1851-1971
Edited by W.P.D. Logan
1982; 253 pages; £22.50

No. 37 Laboratory Decontamination and Destruction of Aflatoxins B_1, B_2, G_1, G_2 in Laboratory Wastes
Edited by M. Castegnaro et al.
1980; 56 pages; £6.50

No. 38 Directory of On-going Research in Cancer Epidemiology 1981
Edited by C.S. Muir and G. Wagner
1981; 696 pages (*out of print*)

No. 39 Host Factors in Human Carcinogenesis
Edited by H. Bartsch and B. Armstrong
1982; 583 pages; £46.00

No. 40 Environmental Carcinogens. Selected Methods of Analysis. Volume 4: Some Aromatic Amines and Azo Dyes in the General and Industrial Environment
Edited by L. Fishbein, M. Castegnaro, I.K. O'Neill and H. Bartsch
1981; 347 pages; £29.00

No. 41 N-Nitroso Compounds: Occurrence and Biological Effects
Edited by H. Bartsch, I.K. O'Neill, M. Castegnaro and M. Okada
1982; 755 pages; £48.00

No. 42 Cancer Incidence in Five Continents, Volume IV
Edited by J. Waterhouse, C. Muir, K. Shanmugaratnam and J. Powell
1982; 811 pages (*out of print*)

No. 43 Laboratory Decontamination and Destruction of Carcinogens in Laboratory Wastes: Some N-Nitrosamines
Edited by M. Castegnaro et al.
1982; 73 pages; £7.50

No. 44 Environmental Carcinogens. Selected Methods of Analysis. Volume 5: Some Mycotoxins
Edited by L. Stoloff, M. Castegnaro, P. Scott, I.K. O'Neill and H. Bartsch
1983; 455 pages; £29.00

No. 45 Environmental Carcinogens. Selected Methods of Analysis. Volume 6: N-Nitroso Compounds
Edited by R. Preussmann, I.K. O'Neill, G. Eisenbrand, B. Spiegelhalder and H. Bartsch
1983; 508 pages; £29.00

No. 46 Directory of On-going Research in Cancer Epidemiology 1982
Edited by C.S. Muir and G. Wagner
1982; 722 pages (*out of print*)

No. 47 Cancer Incidence in Singapore 1968–1977
Edited by K. Shanmugaratnam, H.P. Lee and N.E. Day
1983; 171 pages (*out of print*)

No. 48 Cancer Incidence in the USSR (2nd Revised Edition)
Edited by N.P. Napalkov, G.F. Tserkovny, V.M. Merabishvili, D.M. Parkin, M. Smans and C.S. Muir
1983; 75 pages; £12.00

No. 49 Laboratory Decontamination and Destruction of Carcinogens in Laboratory Wastes: Some Polycyclic Aromatic Hydrocarbons
Edited by M. Castegnaro et al.
1983; 87 pages; £9.00

No. 50 Directory of On-going Research in Cancer Epidemiology 1983
Edited by C.S. Muir and G. Wagner
1983; 731 pages (*out of print*)

No. 51 Modulators of Experimental Carcinogenesis
Edited by V. Turusov and R. Montesano
1983; 307 pages; £22.50

List of IARC Publications

No. 52 Second Cancers in Relation to Radiation Treatment for Cervical Cancer: Results of a Cancer Registry Collaboration
Edited by N.E. Day and J.C. Boice, Jr
1984; 207 pages; £20.00

No. 53 Nickel in the Human Environment
Editor-in-Chief: F.W. Sunderman, Jr
1984; 529 pages; £41.00

No. 54 Laboratory Decontamination and Destruction of Carcinogens in Laboratory Wastes: Some Hydrazines
Edited by M. Castegnaro et al.
1983; 87 pages; £9.00

No. 55 Laboratory Decontamination and Destruction of Carcinogens in Laboratory Wastes: Some *N*-Nitrosamides
Edited by M. Castegnaro et al.
1984; 66 pages; £7.50

No. 56 Models, Mechanisms and Etiology of Tumour Promotion
Edited by M. Börzsönyi, N.E. Day, K. Lapis and H. Yamasaki
1984; 532 pages; £42.00

No. 57 *N*-Nitroso Compounds: Occurrence, Biological Effects and Relevance to Human Cancer
Edited by I.K. O'Neill, R.C. von Borstel, C.T. Miller, J. Long and H. Bartsch
1984; 1013 pages; £80.00

No. 58 Age-related Factors in Carcinogenesis
Edited by A. Likhachev, V. Anisimov and R. Montesano
1985; 288 pages; £20.00

No. 59 Monitoring Human Exposure to Carcinogenic and Mutagenic Agents
Edited by A. Berlin, M. Draper, K. Hemminki and H. Vainio
1984; 457 pages; £27.50

No. 60 Burkitt's Lymphoma: A Human Cancer Model
Edited by G. Lenoir, G. O'Conor and C.L.M. Olweny
1985; 484 pages; £29.00

No. 61 Laboratory Decontamination and Destruction of Carcinogens in Laboratory Wastes: Some Haloethers
Edited by M. Castegnaro et al.
1985; 55 pages; £7.50

No. 62 Directory of On-going Research in Cancer Epidemiology 1984
Edited by C.S. Muir and G. Wagner
1984; 717 pages (*out of print*)

No. 63 Virus-associated Cancers in Africa
Edited by A.O. Williams, G.T. O'Conor, G.B. de-Thé and C.A. Johnson
1984; 773 pages; £22.00

No. 64 Laboratory Decontamination and Destruction of Carcinogens in Laboratory Wastes: Some Aromatic Amines and 4-Nitrobiphenyl
Edited by M. Castegnaro et al.
1985; 84 pages; £6.95

No. 65 Interpretation of Negative Epidemiological Evidence for Carcinogenicity
Edited by N.J. Wald and R. Doll
1985; 232 pages; £20.00

No. 66 The Role of the Registry in Cancer Control
Edited by D.M. Parkin, G. Wagner and C.S. Muir
1985; 152 pages; £10.00

No. 67 Transformation Assay of Established Cell Lines: Mechanisms and Application
Edited by T. Kakunaga and H. Yamasaki
1985; 225 pages; £20.00

No. 68 Environmental Carcinogens. Selected Methods of Analysis. Volume 7. Some Volatile Halogenated Hydrocarbons
Edited by L. Fishbein and I.K. O'Neill
1985; 479 pages; £42.00

No. 69 Directory of On-going Research in Cancer Epidemiology 1985
Edited by C.S. Muir and G. Wagner
1985; 745 pages; £22.00

No. 70 The Role of Cyclic Nucleic Acid Adducts in Carcinogenesis and Mutagenesis
Edited by B. Singer and H. Bartsch
1986; 467 pages; £40.00

No. 71 Environmental Carcinogens. Selected Methods of Analysis. Volume 8: Some Metals: As, Be, Cd, Cr, Ni, Pb, Se Zn
Edited by I.K. O'Neill, P. Schuller and L. Fishbein
1986; 485 pages; £42.00

No. 72 Atlas of Cancer in Scotland, 1975–1980. Incidence and Epidemiological Perspective
Edited by I. Kemp, P. Boyle, M. Smans and C.S. Muir
1985; 285 pages; £35.00

No. 73 Laboratory Decontamination and Destruction of Carcinogens in Laboratory Wastes: Some Antineoplastic Agents
Edited by M. Castegnaro et al.
1985; 163 pages; £10.00

No. 74 Tobacco: A Major International Health Hazard
Edited by D. Zaridze and R. Peto
1986; 324 pages; £20.00

No. 75 Cancer Occurrence in Developing Countries
Edited by D.M. Parkin
1986; 339 pages; £20.00

No. 76 Screening for Cancer of the Uterine Cervix
Edited by M. Hakama, A.B. Miller and N.E. Day
1986; 315 pages; £25.00

List of IARC Publications

No. 77 Hexachlorobenzene: Proceedings of an International Symposium
Edited by C.R. Morris and J.R.P. Cabral
1986; 668 pages; £50.00

No. 78 Carcinogenicity of Alkylating Cytostatic Drugs
Edited by D. Schmähl and J.M. Kaldor
1986; 337 pages; £25.00

No. 79 Statistical Methods in Cancer Research. Volume III: The Design and Analysis of Long-term Animal Experiments
By J.J. Gart, D. Krewski, P.N. Lee, R.E. Tarone and J. Wahrendorf
1986; 213 pages; £20.00

No. 80 Directory of On-going Research in Cancer Epidemiology 1986
Edited by C.S. Muir and G. Wagner
1986; 805 pages; £22.00

No. 81 Environmental Carcinogens: Methods of Analysis and Exposure Measurement. Volume 9: Passive Smoking
Edited by I.K. O'Neill, K.D. Brunnemann, B. Dodet and D. Hoffmann
1987; 383 pages; £35.00

No. 82 Statistical Methods in Cancer Research. Volume II: The Design and Analysis of Cohort Studies
By N.E. Breslow and N.E. Day
1987; 404 pages; £30.00

No. 83 Long-term and Short-term Assays for Carcinogens: A Critical Appraisal
Edited by R. Montesano, H. Bartsch, H. Vainio, J. Wilbourn and H. Yamasaki
1986; 575 pages; £48.00

No. 84 The Relevance of N-Nitroso Compounds to Human Cancer: Exposure and Mechanisms
Edited by H. Bartsch, I.K. O'Neill and R. Schulte-Hermann
1987; 671 pages; £50.00

No. 85 Environmental Carcinogens: Methods of Analysis and Exposure Measurement. Volume 10: Benzene and Alkylated Benzenes
Edited by L. Fishbein and I.K. O'Neill
1988; 327 pages; £35.00

No. 86 Directory of On-going Research in Cancer Epidemiology 1987
Edited by D.M. Parkin and J. Wahrendorf
1987; 676 pages; £22.00

No. 87 International Incidence of Childhood Cancer
Edited by D.M. Parkin, C.A. Stiller, C.A. Bieber, G.J. Draper, B. Terracini and J.L. Young
1988; 401 pages; £35.00

No. 88 Cancer Incidence in Five Continents Volume V
Edited by C. Muir, J. Waterhouse, T. Mack, J. Powell and S. Whelan
1987; 1004 pages; £50.00

No. 89 Method for Detecting DNA Damaging Agents in Humans: Applications in Cancer Epidemiology and Prevention
Edited by H. Bartsch, K. Hemminki and I.K. O'Neill
1988; 518 pages; £45.00

No. 90 Non-occupational Exposure to Mineral Fibres
Edited by J. Bignon, J. Peto and R. Saracci
1989; 500 pages; £45.00

No. 91 Trends in Cancer Incidence in Singapore 1968–1982
Edited by H.P. Lee, N.E. Day and K. Shanmugaratnam
1988; 160 pages; £25.00

No. 92 Cell Differentiation, Genes and Cancer
Edited by T. Kakunaga, T. Sugimura, L. Tomatis and H. Yamasaki
1988; 204 pages; £25.00

No. 93 Directory of On-going Research in Cancer Epidemiology 1988
Edited by M. Coleman and J. Wahrendorf
1988; 662 pages (*out of print*)

No. 94 Human Papillomavirus and Cervical Cancer
Edited by N. Muñoz, F.X. Bosch and O.M. Jensen
1989; 154 pages; £19.00

No. 95 Cancer Registration: Principles and Methods
Edited by O.M. Jensen, D.M. Parkin, R. MacLennan, C.S. Muir and R. Skeet
1991; 288 pages; £28.00

No. 96 Perinatal and Multigeneration Carcinogenesis
Edited by N.P. Napalkov, J.M. Rice, L. Tomatis and H. Yamasaki
1989; 436 pages; £48.00

No. 97 Occupational Exposure to Silica and Cancer Risk
Edited by L. Simonato, A.C. Fletcher, R. Saracci and T. Thomas
1990; 124 pages; £19.00

No. 98 Cancer Incidence in Jewish Migrants to Israel, 1961–1981
Edited by R. Steinitz, D.M. Parkin, J.L. Young, C.A. Bieber and L. Katz
1989; 320 pages; £30.00

No. 99 Pathology of Tumours in Laboratory Animals, Second Edition, Volume 1, Tumours of the Rat
Edited by V.S. Turusov and U. Mohr
740 pages; £85.00

No. 100 Cancer: Causes, Occurrence and Control
Editor-in-Chief L. Tomatis
1990; 352 pages; £24.00

List of IARC Publications

No. 101 **Directory of On-going Research in Cancer Epidemiology 1989/90**
Edited by M. Coleman and J. Wahrendorf
1989; 818 pages; £36.00

No. 102 **Patterns of Cancer in Five Continents**
Edited by S.L. Whelan and D.M. Parkin
1990; 162 pages; £25.00

No. 103 **Evaluating Effectiveness of Primary Prevention of Cancer**
Edited by M. Hakama, V. Beral, J.W. Cullen and D.M. Parkin
1990; 250 pages; £32.00

No. 104 **Complex Mixtures and Cancer Risk**
Edited by H. Vainio, M. Sorsa and A.J. McMichael
1990; 442 pages; £38.00

No. 105 **Relevance to Human Cancer of N-Nitroso Compounds, Tobacco Smoke and Mycotoxins**
Edited by I.K. O'Neill, J. Chen and H. Bartsch
1991; 614 pages; £70.00

No. 106 **Atlas of Cancer Incidence in the German Democratic Republic**
Edited by W.H. Mehnert, M. Smans and C.S. Muir
Publ. due 1992; c.328 pages; £42.00

No. 107 **Atlas of Cancer Mortality in the European Economic Community**
Edited by M. Smans, C.S. Muir and P. Boyle
Publ. due 1991; approx. 230 pages; £35.00

No. 108 **Environmental Carcinogens: Methods of Analysis and Exposure Measurement. Volume 11: Polychlorinated Dioxins and Dibenzofurans**
Edited by C. Rappe, H.R. Buser, B. Dodet and I.K. O'Neill
1991; 426 pages; £45.00

No. 109 **Environmental Carcinogens: Methods of Analysis and Exposure Measurement. Volume 12: Indoor Air Contaminants**
Edited by B. Seifert, B. Dodet and I.K. O'Neill
Publ. due 1992; approx. 400 pages

No. 110 **Directory of On-going Research in Cancer Epidemiology 1991**
Edited by M. Coleman and J. Wahrendorf
1991; 753 pages; £38.00

No. 111 **Pathology of Tumours in Laboratory Animals, Second Edition, Volume 2, Tumours of the Mouse**
Edited by V.S. Turusov and U. Mohr
Publ. due 1992; approx. 500 pages

No. 112 **Autopsy in Epidemiology and Medical Research**
Edited by E. Riboli and M. Delendi
1991; 288 pages; £25.00

No. 113 **Laboratory Decontamination and Destruction of Carcinogens in Laboratory Wastes: Some Mycotoxins**
Edited by M. Castegnaro, J. Barek, J.-M. Frémy, M. Lafontaine, M. Miraglia, E.B. Sansone and G.M. Telling
1991; 64 pages; £11.00

No. 114 **Laboratory Decontamination and Destruction of Carcinogens in Laboratory Wastes: Some Polycyclic Heterocyclic Hydrocarbons**
Edited by M. Castegnaro, J. Barek, J. Jacob, U. Kirso, M. Lafontaine, E.B. Sansone, G.M. Telling and T. Vu Duc
1991; 50 pages; £8.00

No. 115 **Mycotoxins, Endemic Nephropathy and Urinary Tract Tumours**
Edited by M. Castegnaro, R. Plestina, G. Dirheimer, I.N. Chernozemsky and H Bartsch
1991; 340 pages; £45.00

No. 117 **Directory of On-going Research in Cancer Epidemiology 1991**
Edited by M. Coleman, J. Wahrendorf & E. Démaret
1992; 773 pages; £42.00

List of IARC Publications

IARC MONOGRAPHS ON THE EVALUATION OF CARCINOGENIC RISKS TO HUMANS

(Available from booksellers through the network of WHO Sales Agents)

Volume 1 Some Inorganic Substances, Chlorinated Hydrocarbons, Aromatic Amines, N-Nitroso Compounds, and Natural Products
1972; 184 pages (*out of print*)

Volume 2 Some Inorganic and Organometallic Compounds
1973; 181 pages (out of print)

Volume 3 Certain Polycyclic Aromatic Hydrocarbons and Heterocyclic Compounds
1973; 271 pages (*out of print*)

Volume 4 Some Aromatic Amines, Hydrazine and Related Substances, N-Nitroso Compounds and Miscellaneous Alkylating Agents
1974; 286 pages;
Sw. fr. 18.-/US $14.40

Volume 5 Some Organochlorine Pesticides
1974; 241 pages (*out of print*)

Volume 6 Sex Hormones
1974; 243 pages (*out of print*)

Volume 7 Some Anti-Thyroid and Related Substances, Nitrofurans and Industrial Chemicals
1974; 326 pages (*out of print*)

Volume 8 Some Aromatic Azo Compounds
1975; 375 pages;
Sw. fr. 36.-/US $28.80

Volume 9 Some Aziridines, N-, S- and O-Mustards and Selenium
1975; 268 pages;
Sw.fr. 27.-/US $21.60

Volume 10 Some Naturally Occurring Substances
1976; 353 pages (*out of print*)

Volume 11 Cadmium, Nickel, Some Epoxides, Miscellaneous Industrial Chemicals and General Considerations on Volatile Anaesthetics
1976; 306 pages (*out of print*)

Volume 12 Some Carbamates, Thiocarbamates and Carbazides
1976; 282 pages;
Sw. fr. 34.-/US $27.20

Volume 13 Some Miscellaneous Pharmaceutical Substances
1977; 255 pages;
Sw. fr. 30.-/US$ 24.00

Volume 14 Asbestos
1977; 106 pages (*out of print*)

Volume 15 Some Fumigants, The Herbicides 2,4-D and 2,4,5-T, Chlorinated Dibenzodioxins and Miscellaneous Industrial Chemicals
1977; 354 pages;
Sw. fr. 50.-/US $40.00

Volume 16 Some Aromatic Amines and Related Nitro Compounds - Hair Dyes, Colouring Agents and Miscellaneous Industrial Chemicals
1978; 400 pages;
Sw. fr. 50.-/US $40.00

Volume 17 Some N-Nitroso Compounds
1987; 365 pages;
Sw. fr. 50.-/US $40.00

Volume 18 Polychlorinated Biphenyls and Polybrominated Biphenyls
1978; 140 pages;
Sw. fr. 20.-/US $16.00

Volume 19 Some Monomers, Plastics and Synthetic Elastomers, and Acrolein
1979; 513 pages;
Sw. fr. 60.-/US $48.00

Volume 20 Some Halogenated Hydrocarbons
1979; 609 pages (*out of print*)

Volume 21 Sex Hormones (II)
1979; 583 pages;
Sw. fr. 60.-/US $48.00

Volume 22 Some Non-Nutritive Sweetening Agents
1980; 208 pages;
Sw. fr. 25.-/US $20.00

Volume 23 Some Metals and Metallic Compounds
1980; 438 pages (*out of print*)

Volume 24 Some Pharmaceutical Drugs
1980; 337 pages;
Sw. fr. 40.-/US $32.00

Volume 25 Wood, Leather and Some Associated Industries
1981; 412 pages;
Sw. fr. 60.-/US $48.00

Volume 26 Some Antineoplastic and Immunosuppressive Agents
1981; 411 pages;
Sw. fr. 62.-/US $49.60

Volume 27 Some Aromatic Amines, Anthraquinones and Nitroso Compounds, and Inorganic Fluorides Used in Drinking Water and Dental Preparations
1982; 341 pages;
Sw. fr. 40.-/US $32.00

Volume 28 The Rubber Industry
1982; 486 pages;
Sw. fr. 70.-/US $56.00

Volume 29 Some Industrial Chemicals and Dyestuffs
1982; 416 pages;
Sw. fr. 60.-/US $48.00

Volume 30 Miscellaneous Pesticides
1983; 424 pages;
Sw. fr. 60.-/US $48.00

Volume 31 Some Food Additives, Feed Additives and Naturally Occurring Substances
1983; 314 pages;
Sw. fr. 60-/US $48.00

List of IARC Publications

Volume 32 Polynuclear Aromatic Compounds, Part 1: Chemical, Environmental and Experimental Data
1984; 477 pages;
Sw. fr. 60.-/US $48.00

Volume 33 Polynuclear Aromatic Compounds, Part 2: Carbon Blacks, Mineral Oils and Some Nitroarenes
1984; 245 pages;
Sw. fr. 50.-/US $40.00

Volume 34 Polynuclear Aromatic Compounds, Part 3: Industrial Exposures in Aluminium Production, Coal Gasification, Coke Production, and Iron and Steel Founding
1984; 219 pages;
Sw. fr. 48.-/US $38.40

Volume 35 Polynuclear Aromatic Compounds, Part 4: Bitumens, Coal-tars and Derived Products, Shale-oils and Soots
1985; 271 pages;
Sw. fr. 70.-/US $56.00

Volume 36 Allyl Compounds, Aldehydes, Epoxides and Peroxides
1985; 369 pages;
Sw. fr. 70.-/US $70.00

Volume 37 Tobacco Habits Other than Smoking: Betel-quid and Areca-nut Chewing; and some Related Nitrosamines
1985; 291 pages;
Sw. fr. 70.-/US $56.00

Volume 38 Tobacco Smoking
1986; 421 pages;
Sw. fr. 75.-/US $60.00

Volume 39 Some Chemicals Used in Plastics and Elastomers
1986; 403 pages;
Sw. fr. 60.-/US $48.00

Volume 40 Some Naturally Occurring and Synthetic Food Components, Furocoumarins and Ultraviolet Radiation
1986; 444 pages;
Sw. fr. 65.-/US $52.00

Volume 41 Some Halogenated Hydrocarbons and Pesticide Exposures
1986; 434 pages;
Sw. fr. 65.-/US $52.00

Volume 42 Silica and Some Silicates
1987; 289 pages;
Sw. fr. 65.-/US $52.00

Volume 43 Man-Made Mineral Fibres and Radon
1988; 300 pages;
Sw. fr. 65.-/US $52.00

Volume 44 Alcohol Drinking
1988; 416 pages;
Sw. fr. 65.-/US $52.00

Volume 45 Occupational Exposures in Petroleum Refining; Crude Oil and Major Petroleum Fuels
1989; 322 pages;
Sw. fr. 65.-/US $52.00

Volume 46 Diesel and Gasoline Engine Exhausts and Some Nitroarenes
1989; 458 pages;
Sw. fr. 65.-/US $52.00

Volume 47 Some Organic Solvents, Resin Monomers and Related Compounds, Pigments and Occupational Exposures in Paint Manufacture and Painting
1990; 536 pages;
Sw. fr. 85.-/US $68.00

Volume 48 Some Flame Retardants and Textile Chemicals, and Exposures in the Textile Manufacturing Industry
1990; 345 pages;
Sw. fr. 65.-/US $52.00

Volume 49 Chromium, Nickel and Welding
1990; 677 pages;
Sw. fr. 95.-/US$76.00

Volume 50 Pharmaceutical Drugs
1990; 415 pages;
Sw. fr. 65.-/US$52.00

Volume 51 Coffee, Tea, Mate, Methylxanthines and Methylglyoxal
1991; 513 pages;
Sw. fr. 80.-/US$64.00

Volume 52 Chlorinated Drinking-water; Chlorination By-products; Some Other Halogenated Compounds; Cobalt and Cobalt Compounds
1991; 544 pages;
Sw. fr. 80.-/US$64.00

Supplement No. 1
Chemicals and Industrial Processes Associated with Cancer in Humans (IARC Monographs, Volumes 1 to 20)
1979; 71 pages; (*out of print*)

Supplement No. 2
Long-term and Short-term Screening Assays for Carcinogens: A Critical Appraisal
1980; 426 pages;
Sw. fr. 40.-/US $32.00

Supplement No. 3
Cross Index of Synonyms and Trade Names in Volumes 1 to 26
1982; 199 pages (*out of print*)

Supplement No. 4
Chemicals, Industrial Processes and Industries Associated with Cancer in Humans (IARC Monographs, Volumes 1 to 29)
1982; 292 pages (*out of print*)

Supplement No. 5
Cross Index of Synonyms and Trade Names in Volumes 1 to 36
1985; 259 pages;
Sw. fr. 46.-/US $36.80

Supplement No. 6
Genetic and Related Effects: An Updating of Selected IARC Monographs from Volumes 1 to 42
1987; 729 pages;
Sw. fr. 80.-/US $64.00

Supplement No. 7
Overall Evaluations of Carcinogenicity: An Updating of IARC Monographs Volumes 1-42
1987; 434 pages;
Sw. fr. 65.-/US $52.00

Supplement No. 8
Cross Index of Synonyms and Trade Names in Volumes 1 to 46 of the IARC Monographs
1990; 260 pages;
Sw. fr. 60.-/US $48.00

List of IARC Publications

IARC TECHNICAL REPORTS*

No. 1 Cancer in Costa Rica
Edited by R. Sierra,
R. Barrantes, G. Muñoz Leiva, D.M. Parkin, C.A. Bieber and
N. Muñoz Calero
1988; 124 pages;
Sw. fr. 30.-/US $24.00

No. 2 SEARCH: A Computer Package to Assist the Statistical Analysis of Case-control Studies
Edited by G.J. Macfarlane,
P. Boyle and P. Maisonneuve (in press)

No. 3 Cancer Registration in the European Economic Community
Edited by M.P. Coleman and
E. Démaret
1988; 188 pages;
Sw. fr. 30.-/US $24.00

No. 4 Diet, Hormones and Cancer: Methodological Issues for Prospective Studies
Edited by E. Riboli and
R. Saracci
1988; 156 pages;
Sw. fr. 30.-/US $24.00

No. 5 Cancer in the Philippines
Edited by A.V. Laudico,
D. Esteban and D.M. Parkin
1989; 186 pages;
Sw. fr. 30.-/US $24.00

No. 6 La genèse du Centre International de Recherche sur le Cancer
Par R. Sohier et A.G.B. Sutherland
1990; 104 pages
Sw. fr. 30.-/US $24.00

No. 7 Epidémiologie du cancer dans les pays de langue latine
1990; 310 pages
Sw. fr. 30.-/US $24.00

No. 8 Comparative Study of Anti-smoking Legislation in Countries of the European Economic Community
Edited by A. Sasco
1990; c. 80 pages
Sw. fr. 30.-/US $24.00
(English and French editions available) (in press)

DIRECTORY OF AGENTS BEING TESTED FOR CARCINOGENICITY (Until Vol. 13 Information Bulletin on the Survey of Chemicals Being Tested for Carcinogenicity)*

No. 8 Edited by M.-J. Ghess,
H. Bartsch and L. Tomatis
1979; 604 pages; Sw. fr. 40.-

No. 9 Edited by M.-J. Ghess,
J.D. Wilbourn, H. Bartsch and
L. Tomatis
1981; 294 pages; Sw. fr. 41.-

No. 10 Edited by M.-J. Ghess,
J.D. Wilbourn and H. Bartsch
1982; 362 pages; Sw. fr. 42.-

No. 11 Edited by M.-J. Ghess,
J.D. Wilbourn, H. Vainio and
H. Bartsch
1984; 362 pages; Sw. fr. 50.-

No. 12 Edited by M.-J. Ghess,
J.D. Wilbourn, A. Tossavainen and
H. Vainio
1986; 385 pages; Sw. fr. 50.-

No. 13 Edited by M.-J. Ghess,
J.D. Wilbourn and A. Aitio 1988;
404 pages; Sw. fr. 43.-

No. 14 Edited by M.-J. Ghess,
J.D. Wilbourn and H. Vainio
1990; 370 pages; Sw. fr. 45.-

NON-SERIAL PUBLICATIONS †

Alcool et Cancer
By A. Tuyns (in French only)
1978; 42 pages; Fr. fr. 35.-

Cancer Morbidity and Causes of Death Among Danish Brewery Workers
By O.M. Jensen
1980; 143 pages; Fr. fr. 75.-

Directory of Computer Systems Used in Cancer Registries
By H.R. Menck and D.M. Parkin
1986; 236 pages; Fr. fr. 50.-

---

* Available from booksellers through the network of WHO sales agents.

†Available directly from IARC